PLANT DISEASES

Plant Diseases

Ninth Edition

RS Singh

Former Professor
Department of Plant Pathology
GB Pant University of Agriculture and Technology
Pantnagar, India

Oxford & IBH Publishing Co. Pvt. Ltd.

New Delhi

(*A Unit of* CBS Publishers & Distributors Pvt Ltd)

CBSPD

CBS Publishers & Distributors Pvt Ltd

New Delhi • Bengaluru • Chennai • Kochi • Kolkata • Lucknow • Mumbai
Hyderabad • Jharkhand • Nagpur • Patna • Pune • Uttarakhand

Plant Diseases

Ninth Edition

ISBN-13: 978-81-204-1746-5
ISBN-10: 81-204-1746-1

Disclaimer

Science and technology are constantly changing fields. New research and experience broaden the scope of information and knowledge. The author has tried his best in giving information available to him while preparing the material for this book. Although all efforts have been made to ensure optimum accuracy of the material, yet it is quite possible some errors might have been left uncorrected. The publisher, the printer and the author will not be held responsible for any inadvertent errors, omissions or inaccuracies.

OXFORD & IBH

New Delhi
(A Unit of CBS Publishers & Distributors Pvt Ltd)

Published by **Satish Kumar Jain** and produced by **Varun Jain** for
CBS Publishers & Distributors Pvt Ltd
4819/XI Prahlad Street, 24 Ansari Road, Daryaganj, New Delhi 110 002, India
Ph: 011-23289259, 23266861

Website: www.cbspd.com
e-mail: delhi@cbspd.com

Corporate Office: 204 FIE, Industrial Area, Patparganj, Delhi 110 092, India
Ph: 011-4934 4934   Fax: 011-4934 4935   e-mail: publishing@cbspd.com;
publicity@cbspd.com

Branches

- **Bengaluru:** Seema House 2975, 17th Cross, KR Road, Banasankari 2nd Stage, Bengaluru 560 070, Karnataka, India
  Ph: +91-80-26771678/79   Fax: +91-80-26771680   e-mail: bangalore@cbspd.com
- **Chennai:** 7, Subbaraya Street, Shenoy Nagar, Chennai 600 030, Tamil Nadu, India
  Ph: +91-44-26680620, 26681266   Fax: +91-44-42032115   e-mail: chennai@cbspd.com
- **Kochi:** 42/1325, 1326, Power House Road, Opp KSEB, Power House, Ernakulum Kochi 682 018, Kerala, India
  Ph: +91-484-4059051-65,67   Fax: +91-484-4059065   e-mail: kochi@cbspd.com
- **Kolkata:** 147, Hind Ceramics Compound, 1st Floor, Nilgunj Road, Belghoria, Kolkata-700056, West Bengal, India
  Ph: +033-25633055, 033-25633056   e-mail: kolkata@cbspd.com
- **Lucknow:** Basement, Khushnuma Complex, 7 Meerabai Marg (Behind Jawahar Bhawan), Lucknow-226001, UP, India
  Ph: +91-522-4000032   e-mail: tiwari.lucknow@cbspd.com
- **Mumbai:** PWD Shed, Gala no 25/26, Ramchandra Bhatt Marg, Next to JJ Hospital Gate no. 2, Opp. Union Bank of India Noorbaug, Mumbai 400009, Maharashtra, India
  Ph: 022-66661880/89   e-mail: mumbai@cbspd.com

Representatives

• Hyderabad	0-9885175004	• Jharkhand	0-9811541605	• Nagpur	0-9421945513
• Patna	0-9334159340	• Pune	0-9923910676	• Uttarakhand	0-9716462459

Printed at Chaman Enterprises, Daryaganj, New Delhi, India

Preface to the Ninth Edition

In the year 2007 the book Plant Diseases had completed 44 years of its existence as a popular book in plant pathology in India. Eight editions of the book with many reprints of each edition during these 44 years are ample evidence. The author has always been trying to keep the book up-to-date in information available in India and other countries. In the eighth edition many diseases of vegetable crops were added although these were part of the book Diseases of Vegetable Crops. The ninth edition has most recent research findings on important plant diseases. The survey of literature covers the period up to August, 2008 ever increasing volume of information has warranted deletion of description of some diseases which are part of other titles *viz.* Diseases of Vegetable Crops and Diseases of Fruit Crops to which the descriptions are restored. The book has given emphasis on certain areas that were not adequately covered in earlier editions. Biological control of plant diseases and use of non-toxic chemicals are such areas. They finds place in description of almost all diseases.

I express my thanks to those who have helped me in preparation of this edition.

Pantnagar   R.S. Singh
August, 2009

Preface to the First Edition

This compilation of class lectures by the author is aimed at making available to under-graduate and post-graduate students all existing information on plant diseases occurring in India. Our country is progressing fast in agricultural education and research. Plant pathology is being taught as a compulsory subject in under-graduate courses and is one of the specialized subjects in post-graduate courses. The need for books on the subject giving most recent information on plant diseases occurring in India is being felt badly.

The cost of standard foreign books on plant diseases is beyond the purchasing capacity of most of our students. These foreign books hardly make any reference to mycological and plant pathological researches done in India and thus a number of diseases of crop plants, important to Indian agriculture remain unknown to the reader. Inadequate funds do not permit our college libraries to keep sufficient number of copies of these books. Excessive load of courses for under-graduate students and variety of topics which the same teacher has to teach make the consultation of periodicals, for latest information, difficult.

The only book giving descriptive account of diseases of Indian plants was written as early as 1918 by late Sir E.J. Butler (*Fungi and Disease in Plants*— Thacker Spink & Co., Calcutta). This book, still a classic, has gone out of print and no edition making it up-to-date has come out. A commendable attempt was made by late Dr. B.B. Mundkur through his low-priced book *Fungi and Plant Diseases* (Macmillan and Company, 1953). However, the descriptions being brief and the information being old, this book also does not satisfy the demands of students for detailed accounts with special reference to recent work done in India.

These limitations compel the students to depend on class notes dictated by the teacher. It has been emphasized that this dictation work is detrimental to students. Mimeographed notes also do not meet the requirements because of lack of sufficient books available to students. These circumstances have prompted me to make an attempt to give to the students this compilation of descriptive account of most of our important plant diseases. No attempt has been made to discuss principles of plant pathology.

Omission and mistakes are bound to be found in this attempt and I am prepared to bear the full responsibility of making any misrepresentation of facts. Students and teachers, especially the latter, are requested to comment and give suggestions for further improvements in the book.

In the end I wish to express my gratitude to our past and present pioneers in the field of phytopathology whose work has been freely used in the preparation of this book.

B.A.C.
Sabour
January 1963

R.S. Singh

Contents

Introduction

Plants are primary producers of food for humans as well as animals beneficial for human welfare. Apart from their role as suppliers of food, plants immensely contribute to availability of fibers for clothing, timber for house building and furniture, sources of medicine, etc. Shortage of food is the most important challenge in the present day civilization. With rapid increase in global population the demand on food sources has increased tremendously. Nearly 1400 million hectare land (12% of the earth's surface) is under cultivation and 80% of the cultivated land area is under some form of food crops. In spite of this there is hunger in substantially large areas.

The shortage of food can be attributed to several causes. There is less production due to poor technologies in many parts of the world, losses caused by enemies of plants, and most important, the rising world population needing food and the static character of the land surface available for cultivation can be listed among the causes. The rise in population is particularly alarming in the Asian and African countries where the population is projected to be around 5713 million in 2020. Although food production has been increasing throughout the world, the production and availability of food has not kept pace with the rate of population increase. The demand for food by the rising population counteracts the efforts to raise production.

The trend in rising population and in decreased production is sharply marked in the developing countries where the pressure of human population and dependence mainly on agriculture has resulted in more than 33% small holdings of less than one hectare and small farmers till more than 65% of the cultivated area. Small farmers are the worst hit by loss of production either due to natural calamities or due to poor crop management. In developed countries such as North America, Europe, Japan, Australia, New Zealand, Israel, etc. about 11.5% of the population is engaged in agriculture while in developing countries of Asia and Africa agriculture engages 57.5% of the population. Crop losses are much less in developed countries where the population is relatively low than in developing countries with high population density where more food is required. On a global basis about 34% of crops is lost annually due to diseases, insect pests and weeds. In cereals the annual loss is estimated to be about 18% in developed countries and 46% in developing countries. Percentages of all produce lost to diseases, insects and weeds are 25 in Europe, 29 in North and Central America, 30 in the Russian Federation of States and China, 33 in South America and 42–43% in Africa and Asia. Of the 35% annual crop loss, 12% is due to diseases caused by fungi, bacteria, and viruses, 11% due to diseases caused by nematodes, 7% due to insect pests and 3% due to weeds. Where plant protection measures are not implemented, annual losses of 30 to 50% are common in major crops. In 1993, the USA lost 40500 ha of crops to diseases (*Encyclopedia Britanica*, 2002).

To meet the demand for food for the rising population a major effort is required which may necessitate an increase of about 75% in the present level of production. In India alone, where food production has shown significant increase, the food grain requirement in the year 2020 will be 280 million tonnes against today's 200 million metric tons, to feed the expected population of 1300 million. There is not much scope in increasing the land area under crop production. The technology also seems to have reached a level where further increase through better plants is not possible in near future. Protection of the crop and its produce against pests, diseases, weeds, rodents, etc. is considered an approach to increase the availability of food. The figures for losses mentioned above suggest that if the loss from the disease can be minimized a major portion of the deficit can be wiped out.

Plant Protection aims at achieving this object. It includes practices integrated in the normal agronomic practices which can prevent recurrence of loss causing situations. Plant protection was not considered a necessity by the traditional farmers in those days when human population was low, land was in plenty, and there were no high yielding crop varieties requiring sophisticated crop culture techniques. The traditional agriculture was meant for sustenance of the family although the traditional farming community did have ways and means based on thousands of years of experience to combat plant diseases. Now the improved varieties backed by modern technology can yield much more than is required only for sustenance. Naturally, the loss of even few quintals is economic loss to the farmer who would like to avoid it. This emphasizes the need for crop protection. Among the crop enemies responsible for the 35% loss, diseases caused by various agencies including nematodes are responsible for the maximum loss (about 23%).

What is Plant Disease?

Plant disease is a normal part of nature and is one of many ecological factors that help keep the hundreds of thousands of living plants and animals in balance with one another. When the plant is suffering, i.e., not developing and functioning in the manner it is expected, we call it diseased. However, this does not define the term "disease". Often, the symptoms produced by a disease, the cause of the disease, and the injuries caused to the plant have been considered synonymous. However, they signify only the condition of the plant due to disease or the cause of the disease.

In 1858, Julius Kuhn, in Germany, had defined plant disease as abnormal changes in physiological processes which disturb the normal activity of plant organs. A similar definition was given by H.M. Ward in 1896 who defined disease as a condition in which the functions of the organism are improperly discharged or, in other words, it is a state which is physiologically abnormal and threatens the life of the being or organ. In 1918, E.J. Butler had defined disease as variation from normal physiological activity which is sufficiently permanent or extensive to check the performance of natural functions by the plant or completion of its development. According to American Phytopathological Society (*Phytopathology* 30: 361–368. 1940) disease is a deviation from normal functioning of physiological processes of sufficient duration or intensity to cause disturbance or cessation of vital activities. The British Mycological Society (*Trans. Brit. Mycol. Soc.* 33:154–160. 1950) defined disease as a harmful deviation from the normal functioning of process. Wheeler (1975) broadly conceived Plant Disease as all the malfunctions which result in unsatisfactory performance of the plant or which reduce ability of the plant to survive and maintain its ecological niche.

An analytical approach to definition of the term "disease" was made by Horsfall and Dimond (1959) who clarified many misconceptions. According to them disease (i) is not a pathogen, it is caused by a pathogen, (ii) is not the symptoms or effects seen on the plant, symptoms result from disease, (iii) is not a condition as the condition results from the disease, (iv) is not any injury which results from disease as well as from any traumatic cause, and (v) cannot be catching or infectious, it is actually the pathogen which is catching or infectious or transmitted. They defined disease as a malfunctioning process in the plant body due to continuous irritation which results in some suffering. Hence, a disease is a pathological process in plants and animals (including man).

Singh *et. al.,* (1989) tried to put up a moderately precise definition of disease based on views expressed by above and other scientists. In their opinion disease is "a sum total of the altered and induced biochemical reactions in a system of the plant or plant part brought about by any biotic or abiotic factor(s) or a virus leading to malfunctioning of its physiological processes and ultimately manifesting gradually at cellular and/or morphological level. All these alterations should be of such a magnitude that they become a threat to the normal growth and reproduction of the plant."

A simplified definition of plant disease is given in Encyclopedia Britanica (2002). A plant is diseased when it is continuously disturbed by some causal agent that results in an abnormal physiological process that disrupts the plant's normal structure, growth, function or other activities. This interference with one or more of plant's essential physiological or biochemical systems elicits characteristic pathological conditions or symptoms.

The malfunctioning processes due to inroads of a foreign factor or due to some other biotic cause should make the plant abnormal in the sense that it is losing its economic value. If there is some malfunctioning or abnormality, even if caused by some biotic or abiotic factor, which does not cause loss of economic value or enhances the beauty or value of the plant it should not be called a disease in general sense. There are many examples such as the "broken tulips" in Holland which were actually due to viral infection but they fetched very high market price because of their beauty. Variegation on crotons (ornamental perennials) is another example.

The Historical Importance of Plant Diseases

The late blight of potato, a disease caused by the fungus,*Phytophthora infestans*, is a famous example of what a plant disease can do to change the course of history. In 1845 this disease destroyed the potato crop of Ireland where potato constituted the staple diet of the majority in rural areas. The disease had started in Ireland, England, and parts of the continental Europe as early as 1830 and was causing some damage every year, resulting in food shortage. When the late blight epidemic destroyed the potato crop in 1845 there was famine in Ireland. The demographic data are highly variable (cf. Hampson, 1992) but it was reported that in 1840 the population of Ireland was 8 million which was reduced to 4 million after the famine. Hundreds of thousands perished from hunger and disease. There was large scale migration of the population to other countries including the north American continent where 6 million are reported to have migrated between 1847 and 1854. There are opinions contradicting the belief that the entire catastrophe in Ireland was due to potato famine. Socio-political conditions including the failure of the government to manage the situation were also equally, if not more, responsible. But the fact remains that this single disease forced man to realize the importance of plant diseases. As a result scientific investigations were taken up, the cause of disease was identified, concept of fungus as cause of a disease was finally established, and extensive use of chemicals for plant idsease control came into existence. The late blight epidemic not only brought the science of plant pathology to limelight, it caused many social and political changes in the affected countries. Free trade in England was permitted and import of food grains and other foodstuff was allowed. In order to protect shipping the country had to strengthen its navy which became world's strongest navy.

Wheat rust has been another disease that has appeared in epidemic form from time to time in many countries. This disease has forced the farmers in many parts of the world to change their cropping pattern. Wheat has been replaced by corn (maize) or rye because it was regularly destroyed by rust. This caused change in the food habit of the population in the affected areas.

In the last years of the Second World War (1943) Bengal had to face a serious famine. One of the reasons to which this famine has been attributed was the loss in yield of the rice crop (major diet of the population) due to attack of Helminthosporium leaf spot which had been affecting the crop for the last several years. Situation was similar to the Irish potato famine but not so catastrophic. In this case also many reasons other than loss of rice crop are listed.

In the middle of the nineteenth century coffee and tea were equally consumed in England because these were available in plenty from such occupied countries as India, Sri Lanka and Malaysia (Malaya). Sri Lanka (Ceylon) used to produce maximum coffee in the world. In 1867 coffee rust attacked the plantations in Sri Lanka and by 1893 the export of coffee from Sri Lanka had declined by 93%. The economic crisis forced the planters to cut down coffee plants and take to tea planting. Export of tea revived the economy to some extent and at the same time consumption of tea increased in England. When coffee rust was spreading in Sri Lanka the science of plant pathology was just developing and control measures for the disease were not known. Tea was also attacked by a blight but by that time chemical control measures were known and the situation did not deteriorate. The system of monoculture in coffee plantations of Sri Lanka was considered a contributory factor in devastations caused by the coffee rust which was not prevalent in coffee growing countries of South America. Coffee rust was first seen in the Western Hemisphere in 1979, in Brazil, where it is now spreading. In Brazil coffee trees are often surrounded by other kinds of trees. The decline of coffee cultivation in Sri Lanka gave a boost to coffee industry in Brazil which became a major coffee exporting country in the world. Nearness of this country to USA could be one reason for popularity of this beverage in USA. These instances of plant disease epidemics are worth mentioning because they left their effect not only in the country concerned but also in other countries.

In India, wheat rusts had been considered to cause a loss of over Rs. 40 million annually. In the years of epidemics there have been losses amounting to Rs. 500 million or more. Although introduction of dwarf high yielding varieties has reduced the losses to a great extent even now the farmers lose 8–10% of the expected yield due to rusts. The loose smut of wheat is estimated to cause an average loss of 3% (about Rs. 50 million) every year. The 'Molya' disease, caused by a nematode (cereal cyst nematode), is another example. This disease of wheat and barley prevalent in most parts of Rajasthan, causes a loss of Rs. 30 million in barley and Rs. 40 million in wheat every year. Different smuts of sorghum are responsible for an annual loss of Rs. 100 million. Five to 75% loss in chickpea due to Ascochyta blight was reported from Rajasthan during 1982. Wilt of pigeonpea causes 5–10% loss every year in U.P. and Bihar. At a time when there is shortage of pulses in the country control of only wilt and sterility mosaic of pigeonpea could increase the production by 15–16%. Of about 10% crop losses in the country due to nematodes, there is a loss of about Rs. 20 million every year in coffee alone due to the attack of *Pratylenchus coffeae*. Other plant diseases, such as red rot and wilt of sugarcane, potato viruses, rice blast and blight, Karnal bunt of wheat, root knot of tomato, eggplant and cucurbits, apple scab, mango malformation, bunchy top of banana, and sandal spike are responsible for huge losses.

In addition to direct loss in yields and monetary returns to the farmer, the plant diseases affect the society in many other ways. When foodgrains are attacked by fungi they may contain toxins (such as aflatoxins) which cause insanity, paralysis, stomach disorders and liver cancer. The money spent on management of plant diseases is also a loss because in absence of diseases this money could be saved. The expenditure on raising the crop before it is attacked by the pathogens is also a waste. When there is less production transport industry may suffer due to lack of goods for transport. Industries that consume raw agricultural materials (cotton, jute, oilseeds, vegetable and fruits for processing) face difficulty in utilizing their installed capacity when there is less production due to plant diseases. In order to make up for the loss of foodgrains and other agricultural products such oilseeds the Governments have to import these commodities which means loss of foreign exchange at the disposal of the Government. Plant disease management requires use of toxic chemicals. Excessive use of such chemicals may lead to environmental pollution affecting human health.

Causes of plant Disease

A pathogen is always associated with a disease. The word "pathogen" can be broadly defined as any agent or factor that incites "pathos" or disease in an organism. Thus, in strict sense the pathogens do not necessarily belong to living or animate groups. They may be non-living (inanimate) or in between the living and nonliving (such as viruses and viroids). The plant pathogens are, thus, grouped under the following categories:

A. Abiotic factors: These include mainly the deficiencies or excesses of nutrients, light, moisture, aeration, abnormalities in soil conditions, atmospheric impurities, etc. Examples of diseases caused by abiotic factors are mango tip rot or fruit necrosis, "khaira" disease of rice, hollow and black heart of potato, nutrient deficiency symptoms in various crops.

B. Mesobiotic causes: These are disease incitants which are neither living nor non-living. They are considered to be on threshold of life.

1. *Viroids* are naked infectious strands of nucleic acid. Spindle tuber of potato, citrus exocortis, and tomato bunchy top are some of the examples.

2. *Viruses* are infectious agents made up of one type of nucleic acid (RNA or DNA) enclosed in a protein coat. Examples of virus diseases of plants are leaf roll of potato, leaf curl of tomato and chilli, mosaic diseases of numerous crops.

C. Biotic causes: This category includes diseases caused by animate or living or cellular organisms.

I) PROKARYOTES:

1. *Mollicutes:* These are wall less prokaryotes that include mycoplasma-like organisms (MLO) or phytoplasmas and spiroplasmas. Examples of diseases are grassy shoot of sugarcane, little leaf of eggplant (brinjal), sandal pike, big bud of tomato, papaya bunchy top (all caused by MLO) and citrus stubborn and corn stunt diseases (caused by *Spiroplasma*).

2. *Rickettsia-like bacteria (RLBs):* These are very small, sometimes submicroscopic, walled bacteria causing such diseases as citrus greening and Pierce's disease of grapevine.

3. *True bacteria:* Examples of diseases caused by true bacteria are brown rot or wilt of potato, soft rot of potato and vegetables, leaf blight and leaf streak of rice, citrus canker, sugarcane ratoon stunting disease, etc.

II) EUKARYOTES

1. *Fungi:* Potato wart, cabbage clubroot, potato late blight, downy mildews and powdery mildews, rusts and smuts, red rot of sugarcane, etc. Nearly 70% of diseases (major and minor) in any plant species of economic importance are incited by fungi.

2. *Protozoa:* Hartrot of coconut palm, phloem necrosis of coffee.

3. *Algae:* Red rust of mango, papaya, etc.

4. *Metazoan animals (nematodes):* Root knot of vegetable crops, molya disease of wheat and barley, ear cockle of wheat, citrus decline, etc.

5. *Flowering plant parasites:* Dodder, *Striga, Loranthus, Orobanche.*

The pathogens, especially the animate ones, will not be able to cause a disease unless environmental conditions and suitability of the host are also favourable for survival, multiplication, and entry of the pathogen into the plant and further development of the disease. This interaction of the host, the pathogen, and the environment constitutes the disease triangle as illustrated below:

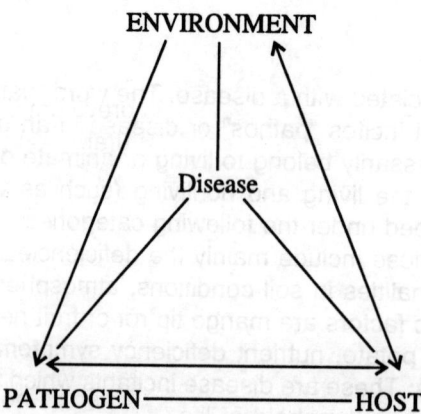

Pythium aphanidermatum or *Phytophthora infestans* may be present in soil but they will not be able to infect a susceptible host unless temperature and moisture conditions are favourable for making the pathogen active and for its dispersal and attack on the plant. Spores of cereal rusts may land on wheat leaves in abundance but infection will occur only when sufficient moisture exists on the leaf surface to facilitate germination of spores, temperature is favourable for germ tube elongation, and stomata are open to allow entry of the germ tube. Thus, pathogens are causal agents and part of a causal complex in which environmental and host conditions play equally important roles. A fourth component - time - can be added to the disease triangle when progress of the disease in the population (epidemiology) is under consideration. Thus, causation of a disease and its progress (amount) in the plant population is determined by:

1. Host: All conditions in the host that favour susceptibility.
2. Pathogen: Total of virulence, abundance, etc.

3. Environment: Total of conditions that favour the pathogen and predispose the plant.

4. Time: Specific point in time at which a particular event in disease development occurs and duration or length of time during which the event takes place. Effective disease control measures aim at breaking this environment-host-pathogen triangle.

Categorization of Plant Diseases

A disease may be localized if it affects only specific organs or parts of the plant. It may be systemic if it affects the entire plant. The diseases are called soil-borne or seed-borne when the causal agent inciting the disease perpetuates through the agency of soil or seed (or any propagating material). However, a "disease" is never soil-borne. It is its causal agent that is soil borne. Similarly, a "disease" is seed-borne only when the seed carrying its causal agent is infected and the pathogen is established in the tissues. Otherwise, in this case also the pathogen is seed-borne, not the disease. The pathogens may be air-borne if they are disseminated by wind. Often, in this case also a disease is called air-borne which is technically incorrect. The symptoms or signs which appear on the affected parts or the entire plant also form a basis for grouping of plant diseases. Thus, we find diseases known as usts, smuts, root rot, wilt, blight, canker, mildew, fruit rot, leaf spots, etc. In all these examples the name of the disease is derived from the most conspicuous symptom of the disease appearing on the host surface. According to the host plants the diseases can be grouped as cereal, forage crop, flax, millet, root crop and plantation crop diseases.

Diseases caused by animate and virus pathogens are often classified in relation to their occurrence under the following groups:

Endemic diseases: The word "endemic" means prevalent in, and confined to, a particular country, district, or location. These diseases are natural to one country or part of the earth. When a disease is more or less constantly present from year to year in a moderate to severe form in a particular geographic region it is classified as endemic to that area. The causal agent is well established in the fields or in the locality by virtue of its ability to survive through soil or other means for long durations and environmental conditions are not adverse to its survival. All those diseases which become persistent through their survival on alternate or wild hosts from one crop season to the next are also included in this group.

Epidemic or **epiphytotic diseases**: The term "epidemic" is derived from a Greek word meaning "among the people" and in true sense was applied to those diseases of humans which appear very violently among a large section of the population. To carry the same sense in the case of plant diseases the term "epiphytotic" was coined. However, in general usage the term epidemic is used for plant diseases also. An epiphytotic disease is one which occurs widely but periodically. It may occur in the locality every year but assumes severe form only on occasions. This may be because the environments or conditions favourable for severe occurrence occur only periodically. It is also possible that the environments are favourable but the pathogen is irregular in its occurrence or its inoculum concentration has not reached the desired level to cause the disease in the plant population.

Sporadic diseases occur at very irregular intervals and locations and in relatively few instances.

A given disease may be endemic in one region and epidemic in another. When epiphytotics become prevalent throughout a country, continent or the world the disease may be called **pandemic**.

Epiphytotics may occur in cycles. When a plant disease appears in a new area, it may grow rapidly to epiphytotic proportions. In time, the disease wanes, and, unless the host species has been completely wiped out, the disease subsides to a low level of incidence and becomes endemic. This balance may change dramatically by conditions that favor a renewed epiphytotic. These conditions are weather (mainly temperature, moisture) very favorable for multiplication, spread and infection by the pathogen, introduction of a new and more susceptible host, development of a very aggressive race of the pathogen, and change in cultural practices favouring disease.

Plant diseases may be infectious or non-infectious. All diseases caused by animate and virus and viroid pathogens under a set of suitable environments are infectious. Association of a definite pathogen is essential with such diseases. In non-infectious diseases no animate, virus or viroid pathogen is associated and, therefore, they remain non-infectious and cannot be transmitted from a diseased plant to a healthy plant. These disorders are due to disturbances in the plant body caused by lack of proper inherent qualities, by improper environmental conditions of soil and air, and by injurious mechanical influences.

General Symptoms of Plant Diseases

Sign or evidence of disease or disorder as shown by the plant or any objective evidence or bodily disorder is called symptom of the disease. Symptoms are seen on the plant either due to character and appearance of the visible pathogen or its structure or organs, or due to some effect upon or change in the host plant due to interaction between the host and the pathogen.

Symptoms due to character and appearance of the visible pathogen: A parasite is present in all parasitic diseases but in most cases the growing vegetative portion of the parasite is within the host tissues and invisible. However, they usually form reproductive or resting structures either outside the plant organs or partly emerging from the host tissues. In either case they become visible provided they are of sufficient size or in sufficient mass. In some diseases, almost the entire body of the parasite including both vegetative and reproductive portions is external to the host and is, then, readily seen, partly on account of its mass. In some diseases large structures of the pathogen constitute the most prominent symptom. Several of these symptoms are described below.

1. *Mildew:* Mildews are plant diseases in which the pathogen is seen as a growth (mildew) on green surfaces of the host. This growth appear as white, gray, brownish or purplish patches of varying size. In downy mildew the superficial growth is a tangled, cottony or downy layer. In powdery mildews enormous numbers of spores are formed on the superficial growth of the fungus giving a dusty or powdery appearance. Black minute fruiting bodies of the fungus may also develop in the powdery mass.

2. *Rust:* The rust diseases appear as relatively small pustules of spores, usually breaking through the host epidermis. The pustules may be either dusty or compact and red, brown, yellow, orange, or black in color.

3. *Smut:* The word "smut" means a sooty or charcoal-like powder. In plant diseases known as smut the affected part of the plant shows a black or purplish black dusty mass composed of the fungus spores. These symptoms appear in floral organs, particularly the ovulary part (ovariculous smuts). The pustules are usually considerably larger than those of the rust. Smut symptoms may also be found on leaves, stems and even roots (culmiculous smuts).

4. *White blisters:* On leaves of crucifers and many other plants there may be found numerous white, blister-like pustules which break open and expose white powdery mass of spores. These pustules resemble rust pustules in texture but are white. Such diseases have, therefore, been often called white rust.

5. *Scab:* The term scab refers to a roughened or crust-like lesion or to a freckled appearance of the diseased organ. In some diseases of this type the parasite appears at a

certain stage, in others it is never seen. Thus, this term may be listed in both major groups of symptoms.

6. *Sclerotia:* A sclerotium is a compact, often hard, mass of dormant fungus mycelium. In some cases, as in ergot of grasses, the sclerotium assumes a characteristic shape, in others the shape may be variable. Sclerotia are most often black, or they may be buff or dark brown or purplish in color.

7. *Blotch:* This symptom consists of a superficial growth giving the fruit a blotched appearance as in sooty blotch and fly-speck disease of apple fruits.

8. *Fruiting bodies:* The wood rotting fungi develop relatively large spore bearing structures (sporophores) which are either fleshy or woody. The parasite can be identified by means of the characteristics of these sporophores.

9. *Exudations:* In bacterial diseases, such as in bacterial blight of rice and fire blight of pome fruits, mass of bacterial cells oozes out to the surface of the affected organ where it may be seen as drops of various size or as a thin smear over the surface.

10. *Tar spots:* These are somewhat raised, black coated fungus bodies with the appearance of a flattened out drop of tar on leaves.

Symptoms due to some effect on, or change in, the host plant: As a result of disease there may be a marked change in the form, size, color, texture, attitude or habit of the plant or some of its organs. Such changes are usually readily observed and often constitute the most prominent symptom of the disease. Two or more of these changes may occur in the same host organ as effects of the same disease. In most diseases, these changes are brought about by the presence and activity or life processes of the pathogen and reaction of the host tissue to such activity. The pathogen may be found within the affected tissue or upon the surface or in some cases it may develop certain structures internally and other structures externally. Fruiting bodies or other structures of the pathogen may thus accompany the more striking changes in the host organs.

1. *Color changes:* Change of color from the normal, mostly green, is one of the most common symptoms due to the effect of a disease in the plant. The green pigment may disappear entirely and its place may be taken by a yellow pigment. When the loss of green color is due to prolonged exposure to darkness the conditions is called **etiolation.** A similar condition may be brought about by the influence of low temperature, lack of iron, excess of lime or alkali, presence of certain virus diseases or from the disturbances caused by fungal or bacterial pathogens. In these cases the yellowing is known as **chlorosis.** When the green pigment is replaced by red, purple, or orange pigment the condition is known as **chromosis.** In some diseases, the leaves are devoid of any pigment. This condition is known as **albinism.**

2. *Overgrowth or hypertrophy:* The most apparent effect in some diseases is the abnormally increased size of one or more organs of the plant or of certain portions of them. This is usually the result of stimulation of the host tissues to excessive growth. It may be brought about by either or both of two processes, hyperplasia and hypertrophy. **Hyperplasia** is the abnormal increase in size of a plant organ due to an increase in number of cells of the organ. In **hypertrophy** the increased size of the organ is due to increase in size of cells. Both these conditions may be simultaneously present. In some diseases the increase in size of the plant organ is due to increase in size of cells and also due to presence of fungus structures. Galls, knots, leaf curl, pockets and bladders, witches' broom, and hairy root are all the result of some form of overgrowth.

3. *Atrophy, hypoplasia* or *dwarfing:* In many diseases one of the results is inhibition of growth resulting in stunting or dwarfing. The whole plant may be dwarfed or only certain organs may be so affected. Sometimes hypertrophy and atrophy both are present in the same organ.

4. *Water-soaking* is a water-soaked, translucent condition of tissues caused by water moving from host cells into intercellular spaces due to damage to cell walls by enzymes and toxins of the pathogen.

5. *Necrosis:* This term is used to indicate the condition in which the death of cells, tissues, or organs has occurred as a result of the parasitic activity. The characteristic appearance of the dead area differs with different hosts and host organs and with different parasites so that there are different types of necrotic symptoms. Spots, streaks, stripes, canker, blight, damping-off, burns, scald or scorch and rot are result of necrosis of tissues.

6. *Anthracnose:* This term is derived from a Greek word meaning ulcer. Ulcer-like lesions, on twigs, stems, pods, and fruits constitute anthracnose diseases caused by a specific group of fungi. Anthracnose is also a type of necrosis.

7. *Die-back:* Die-back is also result of necrosis of terminal tissues of twigs in which the twigs and branches start drying from the tip backwards.

8. *Wilt:* In many diseases the most striking effect of the disease is drying or wilting of the entire plant. The leaves and other green or succulent parts lose turgidity, become flaccid, and droop. This effect is usually seen first in some leaves. Later, the young growing tip or the whole plant suddenly or gradually dries. Wilting may be the result of injury to the root system, to partial plugging of water conducting vessels, or to toxic substances secreted by the pathogen and carried with water to delicate tissues.

9. *Mummification:* Final stage in certain fruit rots in which the dried, shriveled and wrinkled fruit is called a "mummy". The stage is brought about by loss of moisture due to permeation of the flesh by fungus hyphae.

10. *Miscelleneous symptoms:* (i) Alteration in habit and symmetry can occur under the influence of some pathogens. Plants which, under normal conditions, are prostrate or creeping become ascending or even erect. Leaves become lobed from being simple. Inflorescence is changed from a head to a spike. (ii) Premature dropping of leaves, blossoms, fruits or twigs occurs in many diseases of perennial plants. (iii) Complete destruction of organs occurs due to attack of many parasites. In smuts the entire inflorescence or individual flowers are completely destroyed. (iv) Organs are transformed or replaced by new structures. The floral organs can be transformed in to a mass of leafy structures. Ovaries may be transformed in to sclerotia.

Diagnosis of Plant Diseases

Rapid and accurate diagnosis of disease is necessary before proper control measures can be suggested. It is the first step in the study of any disease. Diagnosis of plant disease is a field science and practice is the most sound method of identifying a disease in the field. Many illustrated guides and charts are available which can help in tentative identification. It is easy to identify such diseases as rusts, smuts, downy mildews, and powdery mildews because the structures of the pathogen are prominently visible to the naked eye. However, in many diseases the symptoms are not so useful and they may cut across the area of many different diseases. For example, chlorosis or yellowing of leaves is a symptom which can be caused by a virus, fungus, or a bacterium or even nutritional deficiency without association of a parasite. Similarly, plants may wilt due to fungal infection or bacterial infection. For effective and economical management of any disease accuracy in diagnosis is important because some diseases such as damping off may be caused by many fungi and same treatment may not work against all the fungi. Accurate diagnosis requires systematic field observations (symptoms, structures of the pathogen if any, pattern of occurrence) and some laboratory studies.

In those cases where symptoms are such that may be caused by a variety of living or non-living disease incitants the first step is to determine whether the incitant is infectious or non-infectious. This can be done by observing the pattern of development of the disease in the plant population and possible spread of symptoms on other plants. If the disease is spreading in the plant population it is infectious. An infectious disease may be caused by fungi, bacteria, viruses, or nematodes. These can be determined first by visual observation of the affected parts for presence of fungal structures, bacterial exudates, or nematode cysts or females and then by laboratory studies. If present on the host and examined under the microscope the fungal structures may reveal presence of a particular fungus. If the fungus is not a biotroph and can be cultured, isolation on an artificial medium and tests for Koch's postulates can pinpoint the actual cause of the disease. Bacteria can also be detected in similar manner. Examination of cut pieces of the affected part in water under microscope reveals streaming of bacterial cell masses in water. The bacteria can be isolated on suitable media and Koch's postulates proved. Nematodes, if present, can be seen on the host or in the tissues examined under the microscope. By separating them from the host and multiplying them, under aseptic conditions, on a susceptible host, sufficient number of larvae can be obtained for proving the Koch's postulates.

Viruses can neither be seen with optical microscope nor they can be cultured. Symptoms caused by viruses are mostly similar to those due to nutritional deficiency (such as chlorosis). However, nutritional deficiency is non-infectious while virus disease symptoms spread in the plant population. If in artificial cultures no pathogen is obtained and in tissue examination no fungal structure of a biotroph or structures of nematodes are seen but the disease is infectious, it can be expected that the disease is caused by a virus or mycoplasma-like organism (phytoplasma). Transmission tests by grafting, sap inoculation or use of insect vectors can help in final diagnosis. The differentiation between virus and MLO can be made by electron microscopy and by spraying tetracycline antibiotics which mask symptoms of MLO diseases but not of virus diseases.

Developments in microscopy, serology and immunology, molecular biology, and laboratory instrumentation have resulted in many new and sophisticated laboratory Currently, molecular techniques using DNA array technology with its later improvements are used for multiple diagnosis of pathogens. procedures for the identification of plant pathogens, particularly bacteria, viruses and viroids.

REFERENCES

Agrios, G.N. 1988. *Plant Pathology*, 3rd Ed. Academic Press.

Cramer, H.H. 1967. *Plant Protection and Crop Production*. Transl. From German by J.N. Edwards. *Pflantzenschutz Nachr*. Vol. 20.

Farbenfabriken Bayer A.G, Leverkusen Croxall, H.E. and L.P. Smith. 1984. *The Fight for Food: Factors Limiting Agricultural Production*. Allen and Unwin, London.

Hampson, M.C. 1992. Some thoughts on demography of the Great Potato Famine. *Plant Dis*. 76: 1284–1286

Horsfall, J.G. and A.E. Dimond. 1959. *Plant Pathology-An Advanced Treatise*. Vol. 1:1–17 Academic Press, New York.

Lieven, B. and B.P.H.J. Thomma. 2005. Recent developments in pathogen detection arrayimplications for fungal plant pathogens and use in practice. Phytopathology 95(12): 1374.

Oerke, E.C. and H.W. Dehne. 2004. Sageguarding production. Losses in major crops and the role of crop protection. *Crop Protection* 23(4): 275.

Paddock, W.C. 1967. Phytopathology in a hungry world. *Annu. Rev. Phytopathol*. 5: 375–390.

Schumann, G.L. 1991. *Plant Diseases: Their Biology and Social Impact*. Am. Phytopathol. Soc., St. Paul, Minn., U.S.A.

12

Singh, U.S., R.K. Khetarpal and J. Kumar. 1989. Plant Disease Concept—An overview. In: *Perspectives in Plant Pathology*, pp. 497–503. Eds. V.P. Agnihotri and others. Today and Tomorrow's Printers and Publishers, New Delhi.

Strange, R.N. and P.R. Scott. 2005. Plant Disease: A Threat to Global Food Security. *Annu. Rev. Phytopathol.* 43:83.

Thurston, H.D. 1990. Plant Disease Management Practices of Traditional Farmers. *Plant Dis.* 74: 96–101.

Ward, H.M. 1896. *Disease in Plants.* Society for Promoting Christian Knowledge, London.

Wheeler, H. 1975. *Plant Pathogenesis.* Springer—Verlag, Berlin.

History of Plant Pathology

The enemies of plants have been harming the organized agriculture ever since man started depending on plants for food. Fossil evidence indicates that plants were affected by disease 250 million years ago. Conditions of food shortage and famines have been created by these enemies of plants. Naturally, the man had been obvious of their presence and proposed explanations and methods of combating them according to the knowledge available at that time all over the world. It is not true that attention to plant diseases was first given in the western countries. In terms of mention of plant diseases in the ancient literature the following records are cited:

Probably, development of agriculture started in the orient. There is mention of Chinese practicing crop rotation as early as 3000 BC and in the first century BC they were supposed to keep the field fallow for a year if the field gave poor yield in the second year. Seed health was mentioned in China by Fan Sheng-Chih in the first century BC. These records suggest that agriculture was fairly developed in China centuries before the Christian Era. In the ancient western literature (the Greek and Roman civilization) the central figures of observations on plant life and plant diseases was Theophrastus (c. 372–287 BC), a Greek philosopher and writer on various subjects. He was disciple of Plato (c. 428–348 BC) and colleague of Aristotle (c. 384–322). Although most of the work of Theophrastus is lost, two of his books, *Historia Plantarum* and *De Causis Plantarum*, still find a place as reference. In his writings, Theophrastus elaborately mentioned plant diseases (rusts, mildews, blight, etc.) but expressed the opinion that these diseases were due to bad nutrition and bad air. There was no mention of association of living organisms with plant diseases. Others during that time (Varo, 116–27 BC, Maro, 70–19 BC) did mention specific treatments of seed. In the Indian subcontinent, more than 3 centuries before the Christian era, during the Mauryan Empire (c. 320 BC), agriculture was fairly developed and was the main concern of the government. Kautilya, also known as Chanakya and Vishnugupt, had written (c. 320–296 BC) the classic treatise Arthashastra

(Science of Source of Livelihood) which incorporated many observations and recommendations for healthy crop culture.

Much earlier (about 4700 BC and later) than the time of Theophrastus in the west and Kautilya in India sages had composed the four Vedas (religious verses). These were composed and, in absence of a script, were passed on through word of mouth from generation to generation for centuries. Rigveda (4700 BC) is considered the oldest composition of religious hymns*. It was followed by Yajurveda (hymns and rituals), Samveda (Rigveda recomposed for singing) and Atharvaveda (3000 BC and later) which contained charms and spells for warding off evils and diseases. This composition specifically mentioned blight as a disease and its control. During the Vedic period agriculture in India was fairly developed. Ploughs and other agricultural implements were in use. In addition to such diseases as blight, powdery mildew, rust and tumours on trees, fungi (mushrooms) and algae are also mentioned in the Vedas. In Rigveda a number of verses are devoted as prayer to the Sun God for purifying and protecting everything and for destroying the tiny, invisible or visible creatures that poison the food and cause disease. Similarly, fire (*Agni*) was worshipped for destroying the poisonous beings and purifying articles used in religious activities. Spoiled foodstuff was recommended not to be eaten. Clean water was recommended for men and cattle and vigorous, clean seed was recommended for planting. Obviously, thousands years before the time of Theophrastus the civilization in India was aware of living beings (*Krimi*) that caused disease in man, cattle and plants and also spoiled in the Vedas are (i) visible and invisible creatures entered the body and caused disease and (ii) sun heat and fire kill these poisonous creatures.

More specific references to plant health and disease are found in written literature after the four Vedas when attempts were made to prevent and cure diseases. Susruta, the great Indian pioneer in medicine and surgery who wrote Susruta-Samhita (c. 400 BC) and Charak, also a man of medicine, were aware of diseased conditions in plants and often compared them to diseases in man. Although, Susruta did not deal with plant disease, while advising men for good health, he wrote "just as the proper season, good soil, water and vigorous seed produce a healthy plant". The oldest text on Indian agriculture- Krishi Parashar- was probably written by Parashar (c. 400 BC), before the Arthashastra of Kautilya. Parashar was also a physician and philosopher. His book is considered the first extensive coverage of ancient agriculture of the Aryans. It contains chapters on most aspects of crop culture from sowing to harvest and seed storage. However, plant protection is mentioned directly in one verse that mentions powdery mildew, rust, insects, and larger animals as enemies of crops and invokes the Wind God to move them away from his field. Parashar had declared that origin of plentiful yield is the seed, implying seed health. Kautilya (321–296 BC) in his Arthashastra, listed treatments of seed in addition to recommendations for punishment against sale of spurious seed. Interestingly, Kautilya had recommended that seed should be left in open for 7 or more nights and days. This, probably, is the oldest reference of seed treatment with solar radiation after wetting by dew in the night. Some other recommendations of Kautilya that can be viewed in the light of modern science included (i) sugarcane seed cutting end should be treated with lard and honey and pasted with cow dung and (ii) cucurbits should be cultivated on riverbeds during summer, a practice that has sustained for more than 2000 years and is still followed. These composers and writers belonged to the North and North-West region of the Indian subcontinent. In the south, the Tamil poet Tholkappier (200 BC) considered plants as living beings, mentioned monocot and dicot plants, and wrote about benefits of rice-legume rotation (cf. Jeyrajan, 2000).

---

*The citations about the Vedic literature are mainly based on various publications of Asian Agri-History Foundation, Hyderabad, India (1999–2000).

Varahamihir (505–587 AD) in his Brihat-Samhita included a chapter on science of plant life. Apart for writing about fungi (mushrooms) and algae, he also advocated the importance of good seed and seed treatment for good and healthy seedlings. An interesting recommendation of Varahamihir was that when a piece of land is brought under a crop, sesame should be planted, chopped down before seeds are formed and incorporated in to the soil. The recommendation suggests value of green manuring and organic amendment which are now known to reduce root diseases. Sesame is also a trap crop for *Striga* (Nene, in Sadhale,1996).

More extensive coverage of ailments of plants is found in the text Vriksharyurveda by Surapal (c. 1000 AD). This text mainly deals with cultivation of fruit and flower trees. It is based on previous compositions and writings as well as author's own observations. It covers such topics as importance of trees, their location, soil types, methods of propagation, tree nutrition, diseases and their treatment. Plant protection was already recognized as an important agricultural activity when Surapal wrote his Vrikshayurveda. Surapal was also a physician and had divided plant diseases, like human diseases, into two categories, internal and external. The internal diseases were caused by inroad of foreign organisms in the plant body while the external diseases were attributed to non-parasitic injuries by heat, frost, high winds, soil acidity, water stress, poor quality seed, etc. To avoid both internal and external diseases Surapal prescribed treatment of pits for planting trees, treatment of seed and treatment of standing trees.

The materials used for treatments of seed, field crops and trees during the period of Parashar to Surapal appear to be exclusively organic sources. The list includes cow dung, clarified butter (ghee), decoction of root of five trees, mustard, hog fat, cattle horn, milk, honey, liquid manure, oil of *Madhuca*, ash, cow urine, human urine, beef, extract of *Embelia ribes* (most commonly recommended) and so many other things. These were available in plenty those days and farmers could use them at no cost. They may look strange but modern science is gradually explaining their utility (Singh, 2000) and this text). The application of cow dung to seed, pits, and tree trunks is even now practiced in certain areas and is effective. In the Kumaon Hills (Uttaranchal) the apple tree owners apply a mixture of cow dung and mud to cut ends of branches after annual pruning which protects the wound from infection until self healing. The extracts of decomposed excreta of animals have been found to reduce incidence of powdery and downy mildews of grapevine in Germany and other countries. Plant oils have been used effectively against powdery mildew of apple giving as much as 90% control.

Skimmed milk and cow urine have been demonstrated as disease control materials during the last few decades. Mechanically emulsified rape oil applied as spray is comparable to the use of Karathane against apple powdery mildew (Northover and Schneider, 1993) and grapevine powdery mildew (Azam *et al.,* 1998). Applying a coat of pure mustard oil to ripe mango fruits gives 90% control of Aspergillus rot. Skimmed milk has been found a good seed treatment material against common bunt of wheat. Antimicrobial activity of lactic acid bacteria in milk and of joney is now scientifically proved. Use of bone meal and crab shell is now often mentioned as materials for control of soil-borne root pathogens. Apart from promoting development of biocontrol agents in soil and on plant surfaces, such organic substances are also agents of inducing systemic acquired resistance.

Discovery of the Role of Fungi

In 1675, Leeuwenhoek (Anton van Leeuwenhoek, 1632–1723) developed the first microscope with magnification of 50X–300X and in 1674 he described bacteria seen with this microscope. The Italian botanist Micheli was the first scientist who in 1729 studied fungi and saw their spores under the microscope. He also proved that if these spores are placed on a piece of fruit they grow into a new thallus of the fungus. Although this was a successful experiment it was not universally accepted.

In 1775 the French botanist Tillet published a paper on bunt or stinking smut of wheat. In this paper he described well-planned field experiments and proved that such wheat seeds that contained a black powder on their surface (he did not know that this powder represented spores of the fungus) produced more diseased plants than clean seeds. While emphasizing that the bunt was a contagious disease he observed that its occurrence could be reduced by seed treatment. However, Tillet believed that not the fungus but some toxin produced by the black powder caused the disease.

Although scientists like Persoon (1801) and Fries (1821), who were busy with classification and nomenclature of fungi, believed that microorganisms originated from disease, the French scientist Prevost proved, in 1807, that diseases are caused by microorganisms. Like Tillet he was also working with bunt of wheat. In addition to other details of the disease he studied the germination of spores. By mixing spores with clean seeds he could reproduce the disease. The credit for discovering the life-cycle of the bunt fungus goes to Prevost. In his opinion the bunt spores did not germinate in copper sulphate solution, hence this could be used as a chemical treatment for control of the disease. He also mentioned the fungicidal and fungistatic properties of chemical treatments. These were major discoveries, later confirmed by many scientists. Tulasne brothers (R.L. Tulasne and C. Tulasne) of France, who had produced illustrated description of rust and smut fungi, had also confirmed the findings of Prevost. But because of firm belief in the theory of spontaneous generation among all the contemporary scientists, the acknowledgment came only after 40 years.

During 1830–1845, when late blight of potato was fast spreading in England, Ireland, and the continental Europe, there was no one opinion amongst the scientists about the disease-fungus relationship. While acknowledging the fact that the fungus was invariably associated with late blight the majority believed that the fungus developed from the disease rather than it caused the disease. Berkeley (1846), Morren (1845) and Von Martius (1842) were the few scientists who believed that late blight of potato was caused by the fungus found associated with it. However, they had no experimental evidence to prove it.

The foundation of modern experimental plant pathology was laid by the German scientist Heinrich Anton de Bary (1831–1888). In 1853 he confirmed the findings of Prevost. In 1861 he experimentally proved that the fungus *Phytophthora infestans* was the cause of late blight. The credit for a detailed study of the late blight fungus, its nomenclature, and proof of organisms being plant pathogens goes to the work of de Bary. He studied other diseases also which included rusts, smuts, downy mildews, and rots. In his book *Unter suchingen über* (Research Concerning Fungal Blights), de Bary (1853) had asserted that rust and smut are cause, not effect, of disease. The discovery of heteroecious nature of rust fungi was reported by him in 1885. Probably, the first study of physiology of plant diseases was carried out by de Bary when in 1886 he reported the role of enzymes and toxins in tissue degradation caused by *Sclerotinia sclerotiorum*. De Bary was the first to report that lichen consists of a fungus and an alga and coined the term symbiosis.

Brefeld, a colleague of de Bary, developed methods of artificial culture of microorganisms between 1875 and 1912. With these methods the study of infectious microorganisms became easier. After de Bary suggested the role of enzymes and toxins in pathogenesis many workers successfully attempted to explain such effects of infection on plants as rotting and wilting. In 1905 Jones reported the role of cytolytic enzymes in soft rots caused by bacteria and in 1915 Brown recognized the role of pectic enzymes which was followed by discovery of the role of cellulases. Although de Bary had hinted the possibility of toxins in rots and many others had suspected involvement of toxins in leaf spots and wilt diseases the first experimental proof was obtained by Tanaka in 1933 in black spots of pear caused by *Alternaria*. Subsequently, the role of toxins in vascular wilt disease syndrome and many other blight and similar diseases was established.

With the establishment of role of fungi in plant diseases and observations that there was some degree of variation among different isolates of the same pathogen in their ability to cause disease and among their hosts to suffer different degrees of injuries, attention was diverted to study the genetics of host-parasite interactions and disease resistance. Although Mendel had published his work on genetics of peas in 1866 and by 1898 it was known that resistance to rust in wheat was inherited, the names of Orton (1900–1909) and Biffen (1905–1912) are mentioned as pioneers in this field of resistance breeding. In 1905 Biffen described inheritance of resistance to yellow rust in two varieties of wheat and their progenies on the basis of Mendelian laws of inheritance. In 1909 Orton working with wilt diseases of cotton, watermelon, and cowpea developed varieties resistant to the disease and distinguished disease resistance from disease escape and disease endurance (tolerance). Since then efforts to develop resistant varieties in most crops have been continuing but there are only few crops that have varieties possessing permanent or durable resistance to a disease. One of the causes for this short-lived nature of resistance is the variability among the pathogens.

The phenomenon of variability among fungi was first discovered by the Swedish scientist Erickson in 1894 when he reported the existence of physiologic races in the rust fungi. Almost at the same time Ward (1903) and Salmon (1903–1904) also discovered physiologic specialization in fungi causing rust and powdery mildew of cereals. E.C. Stakman of the University of Minnesota, USA took up this aspect of plant pathology for further investigation in the second decade of the twentieth century. After prolonged studies he came to conclusion that due to continuous evolution of races and biotypes in botanical species of the rust fungi their pathogenic capability goes on changing in their favour and as a result the resistance capability of the host also shows changes.

Resistance to disease in plants was considered to be due to presence of some toxic substances in the host. In 1946 Flor, working with linseed (flax) rust advanced the gene for gene concept of disease resistance and susceptibility. According to him susceptibility to a disease depends on compatibility of genes in the host and the pathogen. For every gene controlling resistance or susceptibility in the host there must be matching genes for avirulence or virulence in the pathogen. This gene for gene relationship is now proved in a large number of host-disease systems. Wherever genetic information is sufficient in both the host and the pathogen it is usual to find a gene-for-gene relationship between the avirulence gene in the pathogen and the resistance gene in the host. In 1963, Van der plank suggested that there are two kinds of resistance: one, controlled by few 'major' genes is strong but race specific (vertical resistance) and the other determined by many 'minor' genes is weaker but effective against all races of a pathogen species (horizontal resistance). The plant cell structures and substances that impart resistance are controlled by genes. The phenomenon of resistance through hypersensitivity was reported by Gaumann in 1946. Muller in 1961 and Cruickshank in 1963 confirmed accumulation of antimicrobial plant metabolites called phytoalexins during pathological processes and their role in resistance.

Resistance breeding by conventional methods (mating of plants of the same species) does not usually allow interspecific crosses and takes years to get a desired resistant variety. During the lapsing period the pathogen may develop new races against which the finally obtained variety may not be resistant. To overcome these shortcomings pathologists and breeders have been trying to utilize the techniques of tissues culture and genetic engineering. More than 20 years ago it was demonstrated that plant cells and protoplasts could be selected in culture for resistance to a pathogen toxin and that plants with an altered response to infection by the pathogen could be regenerated from these cultured cells. Since then the techniques have improved much and a few varieties have also been developed from cell culture. However, the progress has been slow for various reasons, one being that not enough is known about the basic biochemical and genetic events that occur in diseased as well as healthy plants. Of particular interest in these techniques for developing resistant varieties are protoplast fusion methods, ovule and embryo culture and in vitro fertilization, and uptake of organelles, chromosomes and DNA by protoplasts. Transgenic plants in many crops have been developed for resistance to specific diseases.

Discovery of the Role of Bacteria

The discovery that bacteria can act as specific infectious agents of disease was first made in animals through the study of anthrax disease. Rod-shaped bacteria had been seen in the blood stream of diseased animals as early as 1850. Conclusive demonstration and irrefutable proof of the bacterial etiology of anthrax disease was provided by Robert Koch in 1876. He gave the famous Koch's postulates for proving that a particular organism is the cause of a particular disease. He also demonstrated the biological specificity of disease agents. Every bacterium does not cause disease and those that cause disease are specific to certain types of organisms.

In 1882, T.J. Burrill of USA for the first time reported that a plant disease (fire blight of pear) was caused by a bacterium (now known as *Erwinia amylovora*). This report was soon followed by reports of bacterial etiology of yellows disease of hyacinth in 1883 and olive knot disease in 1886. By 1890, over a score of plant diseases had been shown to be caused by bacteria. E.F. Smith of USA was the main contributor to the discovery of most of these bacterial plant diseases since 1895. He is considered father of phytobacteriology for his discoveries and the methodologies he introduced for study of bacterial plant diseases during 1905–1920. Alfred Fischer of Germany, who had studied under de Bary, did not agree with Smith and all others who had claimed to have seen bacteria in plant cells. The heated controversy between Smith and Fischer became one of the best documented cases of scientific disagreement. But methodological studies conducted by Smith won the battle. Smith was also among the first to notice and study the crown gall disease (1893–1894). He considered crown gall similar to cancerous tumors of humans and animals. Later, in 1977, it was demonstrated by Chilton and his team that the crown gall bacterium transforms normal plant cells into tumor cells by introducing into them a part of its plasmid which becomes inserted into the DNA of chromosome of the plant cell. Subsequent pioneers who made significant contributions to plant bacteriology during the first half of twentieth century include C. Elliott (1930–1951), P.A. Ark (1937), H.W. Burkholder (1930–1948), W.J. Dowson (1949–1957) and C. Stapp (1956–1961). Since the 1960s an explosive world-wide development.

Role of Nematodes

Existence of nematodes in nature has been traced to millions of years. A stylet bearing nematode has been found in 26 million years old fossil of the insect *Drosophila* (Poinar in *J. Parasitolo.* 70: 306. 1984). The Guinea worm and roundworms as parasites of man were known to Egyptians as early as 1553–1550 BC. A plant disease with which association of a nematode could be noticed was first reported by Needham in England in 1743 AD. He described the wheat gall nematode now known as *Anguina tritici*. However, for about 100 years after Needham no attention was given to the role of nematodes in plant diseases. In 1857, the life cycle of this nematode was studied by C. Devaine.

Root knot nematode was reported by Berkeley in 1855 although specific mention of the nematode was first made by Cornu in 1879. The stem nematode (*Ditylenchus dipsaci*) was reported by Kuhn in 1857. The cyst nematode of beet (*Heterodera schachtii*) was first noticed in Germany by H. Schacht in 1859. The name was given in 1871 by Schmidt. From 1913 to 1932, N.A. Cobb studied the structure of many nematodes and classified them. He coined the word 'nematology' and developed many techniques for the study of nematodes.

The economic loss due to infestations of nematodes was realized when soil fumigants were developed in the 1940s and with their application high increases in yields were obtained. The association of nematodes with diseases caused by other agents had been noticed as early as 1892 when Atkinson reported that Fusarium wilt of cotton was more severe in the presence of root knot nematodes. In 1901 Hunger showed that bacterial wilt of tomato was facilitated by root knot nematodes. An outstanding discovery on nematodes in relation to plant diseases was the finding of Hewitt and associates in 1958 when they discovered that a virus was transmitted by a nematode. This started studies on nematodes as vectors of viruses. Developments in phytonematology have been more rapid than in other branches of plant pathology and now the science occupies an independent position in many research centres.

Discovery of Role of Viruses

The year 1882 may be considered as the beginning of the era of plant virology when scientific studies were initiated by Adolf Eduard Mayer, a German scientists working in Netherlands. Between 1882 and 1886 he reported that the tobacco mosaic disease was neither due to a microorganism nor due to nutritional imbalance. He demonstrated the contagious nature of the causal agent by artificial inoculation and also showed that boiling of the sap of infected leaves destroyed infectivity of the causal agent. In 1892 Dimitrii Ivanowski, a Russian botanist working in Crimea, reported that he had confirmed the findings of Mayer regarding transmission of tobacco mosaic agent and, in addition, had found that the causal agent could pass through filters with pores small enough to retain bacteria. The filtered sap remained infective for months. Martinus Wilhelm Beijerinck, a Dutch scientist, further confirmed the findings of Mayer and Ivanowski in 1898 and concluded that the causal agent of tobacco mosaic was something other than a microbe. The agent could pass through porcelain filters and could diffuse through agar gel. He was convinced that the agent was not a bacterium but a '*contagium vivum fluidum*' a contagious living fluid. This was a revolutionary idea at a time when all substances could be classified only into two groups, corpuscular (bacteria and blood cells) and dissolved such as salts and other molecules in solution. The idea that the tobacco mosaic agent was fluid, and therefore dissolved, but at the same time living and capable of reproduction and, therefore, infectious in plants seemed extraordinary. Beijerinck is considered founder of virolgy for it was he who firmly established the novel characteristics of tobacco mosaic agent distinct from known agents of infectious diseases. Soon after, many other plant diseases were found to be caused by similar agents.

In 1926, about 40 years after the work of Mayer, the biochemist Maurice Mulvania suggested that the tobacco mosaic virus (TMV) might be a protein of very simple nature having characters of an enzyme. In 1935, W.M. Stanley of U.S.A was able to obtain a crystalline protein by treating juice of TMV infected leaves with ammonium sulphate. This crystalline substance remained infective. He concluded that the virus was an autocatalytic protein that could multiply within living cells. Studies carried out by F.C. Bawden and N.W. Pirie in Britain in 1936 showed that TMV was a nucleoprotein and contained phosphorus. The specific nucleic aid in TMV was identified as ribonucleic acid (RNA). In 1929 H.O. Holmes had provided a method by which the quantity of the virus in tissues could be estimated. He showed that the amount of virus present in a plant sap preparation is proportionate to the number of local lesions produced on an appropriate host plant leaf rubbed with that sap. The leading role of the nucleic acid of plant viruses in the infection process was discovered through the study of bacterial viruses (bacteriophages). In 1956, Gierer and Schramm from Germany and Fraenkel-Conrat and his coworkers from USA showed that TMV nucleic acid free from its protein coat, could alone cause infection provided it was protected from inactivation.

Although the TMV was considered a rod-shaped particle on the basis of various biochemical and biophysical tests its visual observation was possible only when the electron microscope was developed in 1939 and Kausche and colleagues for the first time saw the virus particles with the help of this microscope. The crude pictures obtained confirmed the rod shape of TMV particles. The shadow casting technique developed in early 1940s using heavy metals enabled the scientists to obtain a clearer picture and determine the overall size and shape of particles. Widespread use of ultracentrifuge, the electron microscope, eletrophoresis, and serological techniques during 1940–1950 further helped in the understanding of plant virus structure, chemistry, replication, and genetics. Quick and accurate detection and identification of viruses became easier with the development of agar double diffusion serological tests in 1962 and enzyme-linked immunosorbent assay (ELISA) in 1977 as well as production of monoclonal antibodies in 1975. These techniques are now being used for detection and identification of mycoplasmas and bacteria (*Xanthomonas*) and some fungi also. Subsequently, more sophisticated techniques like immunosorbent electron microscopy (IEM) were also developed. Prior to 1960 the viruses studied were shown to consist of single stranded RNA. With the advances in technique for study some were found to have double stranded RNA (1963), some double stranded DNA (1968) and some single stranded DNA (1977).

Durimg the years that followed purification of TMV, the astonishing stability of the virus was discovered. In infected tobacco sap the virus could remain viable for 3990 days. Purified virions kept at 5°C could remain infective for at least 50 years. The virus could be inactivated only by heating at 80°C for several hours.

The discovery of viroids and virusoids were additions to virology after 1970. In 1971, T.O. Diener reported that potato spindle tuber disease was caused by a small, naked, single stranded circular molecule of infectious RNA which he called viroid. There are now more than a dozen plant diseases known to be caused by viroids. In 1982, a circular, single stranded viroid- like RNA was found encapsidated together with a single stranded linear RNA causing velvet tobacco mottle disease. This RNA molecule was called virusoid which seems to form an obligatory association with the viral RNA in many plant viruses. Viroids are the smallest nucleic acid molecules to infect plants. So far they have not been found in animals although similar proteinaceous infectious particles known as 'prions', were reported in 1982 in scrapie disease of sheep and goats. This disease has been recently mentioned in connection with the "mad cow disease" in U.K.

Phytoplasmas (Mycoplasma- like Organisms, MLOs) and Rickettsia- like Bacteria (RLBs)

Fungi, bacteria, nematodes, and viruses were considered the main incitants of plant diseases up to 1967. The discovery of virus, so small that it could not be seen under the optical microscope and could pass through bacterial filters, induced a feeling that nothing could be smaller than these agents. This, together with the observation that only few types of bacteria were plant pathogens and whatever passed through a bacterial filter must be a virus were responsible for delay in uncovering of a variety of new types of phytopathogenic bacteria (prokaryotes) that were later called the fastidious or hidden vascular pathogens. Fastidious because they were very exacting in their food requirement and did not grow on routine bacteriological media and hidden because they were deep-seated in the plant and could not be seen. Mycoplasmas were known to medical science during the closing years of the nineteenth century. The contagious bovine pleuropneumonia was believed by Luis Pasteur to be caused by a specific microorganism which could neither be seen nor grown in culture. In 1898, E. Nocard and E.R. Roux had succeeded in growing the organism in artificial medium and the organism is now known as *Mycoplasma mycoides*. The organism was placed in the order Mycoplasmatales among bacteria. The organisms in this order differed from true bacteria in having no true cell wall and not responding to antibacterial antibiotics like penicillin which act on cell wall but responding to tetracyclines. In 1967 Doi and his colleagues in Japan observed that mycoplasma-like bodies were constantly present in the phloem of plants suffering from leafhopper transmitted yellows diseases till then considered as virus diseases although virus particles had not been seen. The same year Ishiie and colleagues reported that the mycoplasma-like bodies temporarily disappeared when the plants were treated with tetracycline antibiotics. Since then a large number of plant diseases were conclusively proved to be caused by such organisms. The true mycoplasmas associated with animal diseases were culturable but none of those associated with plants could be grown in cell free media. Thus, they could not be characterized and Koch's postulates could not be proved in most of these diseases. Therefore, they were called mycoplasma-like organisms (MLO). More recently, in 1990s, scientist have studied the sequence analysis of highly conserved 16S RNA in the MLOs amplified by polymerase chain reaction (PCR) technology and have found that these organisms, although Mollicutes, are not members of the order Mycoplasmatales. Instead, they are more close to the order Acholeplasmatales of Mollicutes. They have been placed in the family Acholeplasmataceae with tentative (candidate) genus *Phytoplasma* consisting of groups of phytoplasmas from different plant species. In 1972, Davis and his colleagues observed a motile, helical, wall-less microorganism associated with corn stunt disease. They called it *Spiroplasma* and this organism could be cultured and characterized. Therefore, it has been assigned to a separate family.

Phytoplasma and *Spiroplasma* are phloem-inhabiting bacteria. During closer examinations of phloem for presence of other such organisms another group of fastidious prokaryotes was discovered in 1973 in citrus plants attacked by greening disease. The organism has definite cell wall and is susceptible to both penicillin and tetracycline antibiotics. This suggested that the organism is a true bacterium, not MLO. Gram-negative character of the organism was confirmed in 1974. In late 1990s the organism was identified as *Liberobacter asiaticum* for Asian citrus greening and *Liberobacter africanum* for African citrus greening. Search for similar organisms in xylem enabled the discovery of another Gram-negative bacterium (*Xylella fastidiosa*) in Pierce's disease of grapevine. The causal agent of ratoon stunting disease of sugarcane, long thought to be a virus disease, was identified as a Gram-positive xylem inhabiting bacterium (*Clavibacter xyli*) in 1980 although association of a coryneform bacterium with the disease had been known since 1973. These fastidious bacteria are often referred as Rickettsia-like (RLB) only because of some superficial resemblance to Rickettsia but they are not Rickettsia.

Protozoan Diseases of Plants

In 1909 flagellate protozoa had been found in latex bearing cells of plants in family Euphorbiaceae but were thought to be living in the latex without causing a disease. In 1931 Stahel found flagellates infecting the phloem of coffee trees acausing abnormal phloem formation and wilting of trees. It was further confirmed in 1963 by Vermeulen. In 1976, flagellates were reported to be associated with several diseases of coconut and oil palm trees in South America and Africa.

In the last few decades, with the fast advances in molecular biology, greater. Emphasis has been given to use molecular tools in fungus taxonomy, specific identification of pathogens (including fungi, bacteria and viruses) and their strains, process of infection, interaction between pathogens-plants-environment, and understanding of mechanisms involved in disease resistance.

Developments in Chemical Control of Plant Diseases

By trial and error men had found that chemicals could be useful in disease management much before the discovery of Bordeaux Mixture. In the middle of the 17th century use of brine (salt solution) for removal of bunted grains from seed lot followed the observation that grains salvaged from sea were free of damaged grains Sulphur was known as a pest-averting material in 1000 B.C. and was probably in general use by 1800 AD. It was recommended for the powdery mildew of peach by Robertson in 1824. Admixture of lime to reduce phytotoxicity of sulphur was first recommended by Weighten in 1814. Boiled lime sulphur was recommended by Kenrick in 1834 and Grisson in 1852. In 1902, Lowe and Parrott noticed that apple scab was controlled by lime sulphur. By 1906 it was recommended as a general fungicide. Commercial formulations of lime sulphur continue to be common protectant fungicides even to-day.

Copper sulphate was recommended for wheat seed treatment against bunt by Prèvost in 1807 who for the first time established the fungitoxic value of the compound. It was used as a wood preservative in 1767. Copper sulphate was suggested as a possible control of late blight of potato during the 1844 epidemic and was recommended for roses in 1862. The use of copper sulphate admixed with lime was introduced by Dreisch in 1873 as a treatment for wheat seed infested by bunt. While basic information on copper sulphate with or without lime as a fungitoxicant was known, the modern era of chemical control of plant diseases started with the discovery of Bordeaux mixture during the epidemic of grape downy mildew in France during 1879–1882. The mixture of lime and copper sulphate had been sprinkled on the vines to deter pilferage. Prof. Millardet (Pierre Marie-Alexis Millardet, 1838–1902) of Bordeaux University noticed that vines thus sprinkled had little downy mildew. Millardet was Professor of Botany at Bordeaux University since 1876 and retired in 1899. He developed the mixture and published his observation in *Journal Agriculture Pratique*. The mixture named Bordeaux mixture dominated chemical plant disease control for more than half century since 1885. Burgundy mixture, using sodium carbonate in place of lime, was introduced in 1887. Later, soluble copper fungicides (copper oxychloride) were discovered and gradually replaced Bordeaux mixture. However, Bordeaux mixture continues to be used against many diseases in some countries.

In late 1940s and during the 1950s a large number of synthetic compounds were introduced in the market as fungicides. These included chlorinated hydrocarbons, organophosphates, carbamates, phenoxys and acetamines. By 1990s, there were 113 active ingredients registered as fungicides worldwide (cf Knight *et al.*, 1997). However, all are not equally effective and safe. Salts of toxic metals and organic acids, organic compounds of mercury and sulphur, quinones and heterocyclic nitrogen compounds have been major protectant fungicides in the latter half of the twentieth century.

The introduction of systemic fungicides, that could penetrate tissue and work from within the plant, in 1966, was a major landmark in the history of fungicidal management of plant diseases. It started with the discovery of oxathiins in 1966 by von Schmeling and Kulka. It was soon followed by confirmation of systemic activity of pyrimidines (1968) and benzimidazoles (1968,1969). These fungicides are not effective against Oomycetes. Metalaxyl, effective against oomycetes (Peronosporales), was developed by Ciba-Geigy in 1973 and came in use as a fungicide in 1977. During the same period, an organic phosphate fungicide, fosetyl-Al, was also developed and used against Oomycetes. The efficacy of these systemic fungicides made them highly acceptable to the farmers and for some time greater attention was given to them rather than to protectant fungicides. However, since they are narrow spectrum and site-specific in action against the pathogen, the latter soon developed resistance to them. Nearly 73 species of fungal pathogens have been reported resistant to 62 fungicidal compounds all over the world (cf. Singh, 2001). Thus, their dominance over protectant fungicides was soon over.

Antibiotics also have been used in plant disease control ever since the discovery of streptomycin. The antibiotics used in plant disease control belong to groups known as streptomycin, tetracyclines, polyenes, cycloheximide and griseofulvin. However, because of rigid dosage and danger of phytotoxicity none could become popular except streptomycin. This antibiotic was first used against fire blight of apple and pear in early 1950s. As in the case of systemic fungicides, resistance to streptomycin in the fire blight bacterium was also reported in 1971. Although this happened after the antibiotic had been in use for almost 20 years.

Information about the toxicity of chloropicrin (tear gas) to nematodes was known in 1919 when Matthews successfully controlled root knot nematode population in pots. Field application was demonstrated in pineapples in 1932. Nematicidal properties of dichloropropene-dichloropropane mixture (DD) were for the first time reported by Carter in 1943. It was later found by Newhall and Lear in 1948 that dichloropropene was the actual toxic fraction in the mixture. Almost at the same time, in 1945, Christie reported excellent response of ethylene dibromide (EDB) against root knot nematode. Rapid commercial manufacture of these two fumigants began in 1946. Soon after the success of DD and EDB a large number of other halogenated hydrocarbons were introduced which included chlorobromide, chlorobutene and chloropropane. In 1954, McBeth and in 1955, Raski separately reported successful use of dibromo chloropropane (DBCP). This nematicide could be used in standing crops and was generally accepted but had to be withdrawn because of health hazards. Attempts were also made to bring out combination products to broad base the effective control spectrum.

The fumigants were volatile chemicals and were sold as liquid. There method of application was quite cumbersome. Most of them had to be applied days or weeks ahead of planting. This prompted the concept of chemicals that could be formulated as granules and which could quickly dissolve in soil water and act against the nematode. These were the non-fumigant contact or systemic nematicides such as carbofuran, phorate and aldicarb. Aldicarb was introduced in 1965 by Union Carbide. The advantage with these granular nematicides-insecticides was that they could be applied more conveniently and when systemic they could be absorbed by plant roots and remain effective for several weeks.

Synthetic toxic compounds dominated in the field of plant protection for more than 110 years. They are now considered as pollutants and harmful to human health, In the USA, use of fungicides in 1970s came down to one third of the use during 1940s. Studies in the use of toxicants for disease control were then directed to devise methods for reducing the amount of their use. Many toxic compounds were removed from the market. These included all mercurials and some benzene compounds. Although antagonism among microbes had been known since 1885 and its application thrpugh some cultural practices was made in early 1900s, extensive studies were tasken in mid 1900s. Biological control became a field of unlimited studies. The emphasis has been on antagonism and antagonistic fungi and bacteria. In the last few decades its has been realized that biological control with microbes is not only through antagonism but also through inducing the plant to defend itself (induced resistance). As of now, the global market of biological fungicides is less than 1%. Obviously, as the new crops, new cropping systems and new diseases appear, the fungicides will continue to be in use judiciously, with safeguards, for another century.

Development of Biological Control of Plant Pathogens

The present limitations to use of synthetic fungicides which dominated the plant protection strategies for more than 100 years are rising cost and unfavourable cost: benefit ratio, environmental pollution, health hazards and damage to non-target beneficial organisms. During the last 3–4 decades these aspects of fungicides have prompted search for eco-friendly, cheap and sustainable alternatives. Biological management of pathogens has been found an alternative. By definition, biological control does not mean use of only microbes for suppression of pathogens. Any living entity, except man, including the host plant itself can be biocontrol agents. However, for decades after this approach was taken up for disease control, the major attention had been given to use of fungal and bacterial antagonists.

The present day biological control or management of pathogens or pests with biological entities is actually based on the natural law of maintenance of biological equilibrium. Nature has produced enemies of all forms of life which operate when necessary to keep down over-population of some forms of life. Since ancient times, agriculture has been making use of this phenomenon for better crop production, not exasctly knowing the role of microbes, but through such practices as crop rotation, fallow, mixed cropping, organic amendments (use of animal waste).

The word antagonism was introduced in 1874 when an antibiotic-producing fungus was found to suppress bacteria in liquid cultures. In 1908, it was observed that *Erwinia carotovora* (the soft rot bacteria) in turnip tissues and *Penicillium italicum* of orange fruit rind were killed by the liquid media in which each had grown. More specific indications of biological control of plant pathogens were provided in 1926 and early 1930s. G.B. Sanford suggested that suppression of common scab of potato in green manured soil was due to microbial activity. R. Weindling reported that *Trichoderma lignorum* was a parasite of other soil fungi and culture filtrates of *Trichoderma* contained fungitoxic substances. There was slow progress in the exploitation of these observations in subsequent several decades but the observations founded the base for microbial control of plant diseases. In 1981, J. Katan observed "More than 100 years after the introduction of soil disinfestations and more than 50 years after Sanford's classical publication on biological control, we are frustrated by the large gap between promising results in the greenhouse and failure in the field. We no longer aim to achieve absolute control, but rather an economic reduction in disease level".

Realizing that biological control is not a myth but reality, the pace of studies on biological control has picked up during the last few decades. Hundreds of biocontrol agents (fungi including yeasts and mycorrhizae, free living and rhizobacteria or phylloplane bacteria, bacteriphages, fungivorous soil fauna, nematophagpus fungi) have been identified and demonstrated to suppress pathogens *in vitro* and *in vivo* (mostly greenhouse studies). It has also been realized that biocontrol agents isolated from the same habitat as the pathogen are more effective and show less variability in results. Mixtures of biocontrol agents having different mechanisms have been used to eliminate variability in results. Since late 1960s, organic amendment of soil (addition of decomposable organic matter to the soil) has been extensively studied as a method of inducing natural biological control in the field.

Use of the genes that confer biocontrol ability in microbes have been used to transform plants by genetic engineering. The transgenic plants possess the ability to suppress pathogens by same weapons that the donor microbe uses under natural conditions.

Antagonism is the not only mechanism of suppression of pathogens by microbes. It has been established that quite a large number of microbial agents induce systemic or local resistance by eliciting resistance responses of the host plant.

Plant Pathology in India

The development of science of plant pathology in India, as in other countries followed the development of mycology. Up to 1930, there was more attention to study of fungi than to diseases caused by them. The study of fungi in India was initiated by Europeans in the nineteenth century. They used to collect fungi in India and send the specimen for identification to laboratories in Europe. During 1850–1875, D.D. Cunningham and A. Barclay started identification of fungi in this country. Cunningham made a special study of rusts and smuts. K.R. Kirtikar was the first Indian scientist who collected and identified the fungi in this country.

Organized researches on fungi and plant diseases, based on long-term planning, were started in this country only in the first decade of the twentieth century when the then British government established the Imperial Agricultural Research Institute at Pusa (Bihar). This institute, now known as Indian Agricultural Research Institute (IARI), was shifted to Delhi in 1934. It was in this institute at Pusa in 1901 that E.J. Butler (trained in medical profession) initiated an exhaustive study of fungi and diseases caused by them. During the 20 years Butler stayed and worked in this country, he made a scientific study of most of the fungal plant diseases known in India at that time. In addition, he trained a team of plant pathologists who took over the work from him. The diseases, a detailed account of which was given for the first time by Butler, included wilts of cotton and pigeonpea, different diseases of rice, toddy palm, sugarcane, potato and rusts of cereals. He studied and wrote a monograph on Pythiaceous and allied fungi. His very important contribution to plant pathology in India still exists in the form of the classic text *Fungi and Disease in Plants*, written by him and published from Calcutta in 1918.

J.F. Dastur (1886–1971), a colleague of Butler, was the first Indian plant pathologist who is credited with a detailed study of fungi and plant diseases. His special field of study was the genus *Phytophthora* and diseases caused by it in castor and potato. He is internationally known for the establishment of the species *Phytophthora parasitica* from castor. During the period when Butler was here, plant pathologists trained by him, or his associates had by 1920, made a detailed study of a number of other diseases. G.S. Kulkarni published exhaustive information on downy mildew and smuts of sugarcane and pearl mille and S.L. Ajrekar studied wilt of cotton, smut of sugarcane, and ergot of sorghum. At that time the plant pathologist in India were more inclined towards a descriptive phase of the disease including the pathogen involved and had not paid much attention to the control aspect.

E.J. Butler left India in 1920 to take over as the first Director of the Imperial Mycological Institute (later Commonwealth Mycological Institute, CMI) in England. He had trained a good number of mycologists and plant pathologists who took over the work on plant diseases with more emphasis on control aspects. B.B. Mundkur started work on control of cotton wilt through varietal resistance which ultimately resulted in reduction of losses from this disease in Maharashtra to a great extent. He was also responsible for identification and classification of a large number of Indian smut fungi. The most significant contribution of Mundkur to plant pathology in India will be remembered through the Indian Phytopathological Society which he started almost single handed in 1948 with its journal *Indian Phytopathology*. He also authored a textbook *Fungi and Plant Diseases* which was the second book of its type after the classic work of Butler. Dr. K.C. Mehta of Agra College made outstanding contributions to the knowledge of disease cycle of cereal rusts in India during the first half of the twentieth century. Dr. R. Prasada, trained by Dr. K.C. Mehta, continued the work of rusts and added to the knowledge of linseed rust.

J.C. Luthra and associates developed the solar heat treatment of wheat seed for the control of loose smut. Before the advent of systemic fungicides this was the only sure method of control of this disease. During 1965–1971, Singh and his colleagues revived the ancient practice of using organic matter for plant disease control by demonstrating the efficacy of organic amendment of soil in the control of root knot of tomato and okra and black scurf of potato. They used oil cakes (seed meal after extraction of oil) and sawdust of various trees. This was later confirmed by many workers in India and abroad.

Strong schools of fundamental plant pathology, especially biochemistry of host-parasite relationship, were started at Lucknow and Madras Universities under the leadership of S.N. Dasgupta and T.S. Sadasivan, respectively. Dasgupta carried out extensive studies on the controversial mango black tip disease. Sadsivan's school developed the concept of vivotoxins and worked out the mechanism of wilting in cotton due to *Fusarium oxysporum* f.sp. *vasinfectum*. The production of fusaric acid by the fungus was demonstrated by them. M.K. Patel, V.P. Bhide and G. Rangaswami pioneered the work on bacterial plant pathogen in India. M.J. Thirumalachar conducted exhaustive studies of smuts and rusts. His association with the Hindustan.

Antibiotics Limited at Pimpri (Pune) resulted in discovery several antifungal antibiotics. In more recent period, a notable development was the establishment of the International Crop Research Institute for Semi Arid Tropics (ICRISAT) at Hyderabad. Although confined to work on specific crops, this Institute is responsible for introducing new trends in plant pathological research in the country.

In 1964–65, the role of organic amendment of soil for nematode and fungal disease control was reported by Singh and associates. Since then there has been a tremendous increase in research in this field. This approach now forms a major part of biological control under natural conditions.

Teaching of plant pathology as a major subject in Indian universities started rather late. The first India universities that were established in 1857 at Calcutta, Bombay and Madras emphasized fungal taxonomy. Probably, the University of Madras was the first to take up plant pathology as a university science. University of Allahabad (est. 1887) and University of Lucknow (est. 1921) also took up plant disease aspects of mycology. Organized teaching in mycology and plant pathology as part of agricultural science was being conducted by the Indian Agricultural Research Institute at Delhi which later grew up and started giving post-graduate degree in the subject. However, before this the Agra University had introduced post-graduate degree programme in plant pathology at the Government Agricultural College at Kanpur in 1945. This college was the first agricultural college in the country established in 1906. The first Agricultural University was established in 1961 at Pantnagar where plant pathology research and teaching occupied a prominent place. After the establishment of agricultural universities in the country in 1960 and thereafter, teaching in plant pathology with its supporting courses in mycology, bacteriology, virology and nematology has become an important part of graduate and post-graduate programmes in agriculture. In the field of research publications the country showed phenomenal increase in numbers. After *Indian Phytopathology*, started in 1948, Indian Society of Mycology and Plant Pathology was started in 1971 with *Indian Journal of Mycology and Plant Pathology* (now *J. Mycol. Pl. Pathol.*) as its publication. Same year Nematological Society of India was also established and started *Indian Journal of Nematology* as its publication. Subsequently many other societies and journals were started in the country.

REFERENCES

Creager, A.N.H., K-B. G. Scholthof, V. Citovsky and H.B. Scholthof. 1999. Tobacco Mosaic Virus: Pioneering Research for a Century *Plant Cell* 11: 301.

Dugan, F.M. 2008. *Fungi in the ancient world*. APS Press, St. Pauil, Minn, USA.

Mehta, P.R. 1963. Plant Pathology in India - past, present and prospects. *Indian Phytopath.* 16: 1–7.

Morton, V. and T. Staub. 2008. A short history of fungicides, *APSNet Feature* March Nene, Y.L. 1999. Seed health in ancient and medieval history and its relevance to present day agriculture. *Asian Agri-History* 3(3): 157–184.

Raychaudhuri, S.P. 1972. History of plant pathology in India. *Annu. Rev. Phytopathol.* 10: 21–36.

Raychaudhuri, S.P. 1991. Development of mycological and plant pathological work in India. *Indian J. Mycol. Plant Pathol.* 21: 14–26.

Sadhale, N. (Tr.) 1996. *Surapal's Vrikshaayurveda (The Science of Plant Life by Surapal)*. Agri-History Bull. No.1 Asian Agri-History Foundation, Secunderabad (India).

Sadhale, N. (Tr.) 1999. Krishi-Parashar (Agriculture by Parashar). Agri-History Bulletin No. 2. Asian-Agri-History Foundation, Secunderabad (India).

Scholthof, K.B. G. 2008. Tobacco Mosaic Virus: The beginning of Plant Virology. *APSNet Feature* April 2008.

Starr, M.P. 1984. Landmarks in the development of phytobacteriology. *Annu. Rev. Phytopathol.* 22: 169.

Thurston, H.D. 2001. Tropical Plant Pathology: At Home and Abroad. *Annu. Rev. Phytopathol.* 39:1.

The Prokaryotes
(Plant Pathogenic Bacteria)

*The living organisms are grouped under two categories on the basis of size: the microorganisms and macroorganisms. Early in the history of biology, the biologists recognized only two kinds of the living systems—the division Plantae and the division Animalia. For long the microorganisms (microbes) were considered members of either of these two kingdoms of living organisms. The plants and animals were differentiated on the basis of certain functional and structural characters. This division of the living world was satisfactory so long as highly differentiated multicellular organisms (higher plants and animals) were taken in to consideration. The simple functional or structural criteria such as energy source, carbon source, active movement, presence or absence of cell walls and chloroplasts, and open or closed growth do not apply to the lower organisms (microbes). In 1866, a third kingdom, the Protists was proposed for these organisms which included protozoa, algae, fungi, and bacteria. The protists differ from plants and animals in having unicellular or coenocytic or multicellular body without differentiation of cells and tissues.

The above classification of the living world in to plants, animals, and protists was based on visual observation of structure and function and observations with light (optical) microscopes. With the invention of electron microscopy about the year 1950, detailed examination of cells became easier. It was now possible to study the fine or ultra-structures in the cell such as intranuclear or intranucleolar structures, membranes of the cell and intracellular organelles other than nuclei.

Bacteria seem to be the most primitive living organisms. It is speculated that they originated on the earth when there was no vegetation. Bacterial fossils exist that date back 3.5 billion years. Bacteria were first seen under a crude microscope in 1675.

The association of bacteria with an animal disease (anthrax) had been known since 1850. The first evidence of bacteria being responsible for plant diseases was reported in 1882 when association of a bacterium (now known as *Erwinia amylovora*) was established with the fire blight disease of pear. Since then the number of plant diseases identified as caused by bacteria has risen rapidly.

PROKARYOTES CELL

The prokaryotes are very diverse groups of organisms. At the lower extreme are the smallest known cellular organisms, the wall-less Mollicutes, and at the other extreme, among the true bacteria, are the filamentous forms, the Streptomycetes, that resemble fungi. In between are not only the wall-less,

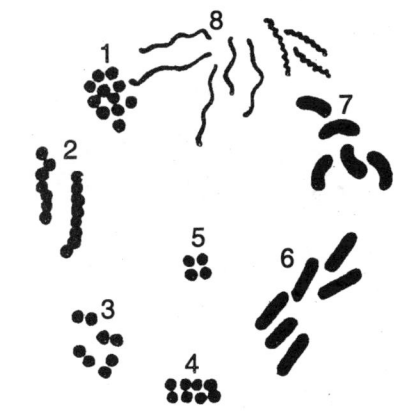

Fig. 1. Morphological types of bacteria (Drawn after Daugerty and Lamberti, 1954).

1–Staphyococci; 2–Streptococcus; 3–Micrococcus; 4–Sarcinae; 5–Tetrads; 6–Bacilli; 7–Spirilla; 8–Spirochetes.

L-phase of some walled bacteria such as *Erwinia* and *Agrobacterium* but also many relatively small but specialized bacteria.

Cell size: Without exception, no individual prokaryote cell can be seen with unaided human eye the resolving power of which is about 2 mm under optimal conditions. The bacterial cells are much smaller, in fact so small that in some, such as in mycoplasmas, the individual cells are not seen in the light or optical microscope which can magnify an object up to 2000 to 3000 times.

Cellular size of bacteria (including mycoplasmas) varies with the stage of life cycle, nutritional status, mode of cell division, and many other factors. The size is variable more in length than in width. The ordinary bacterial cell (not the mycoplasma-like organisms, MLO) is about 1 to 5 microns (μm) long and 0.5 μm wide (1 micron = 1/1000 mm). The size of an individual *Mycoplasma* cell is much smaller than true bacteria. The known range of size extends from 100 to 400 nm or 0.1 to 0.4 microns (1 nm = 1/1000 microns). The size of mycoplasmal cells shows much variation in size, being 60–100 nm for small rounded cells, 150–1100 nm in diameter for large globose cells, to 1–2 to several microns in length for the branched filamentous forms.

Cell shape: The shape of prokaryote cells is governed by the presence or absence of the cell wall and its chemical composition. The bacterial cells are spherical (coccus) or nearly so, elongated cylindrical (rods) or nearly so, or helical (spirals) with rounded ends. The rods and spirals are known as bacillus. All plant pathogenic true bacteria come under this category. Pleomorphism (many shapes) is characteristic of wall-less prokaryotes (mycoplasmas) while irregular shape is found in V-form corynebacteria and L-phase of *Agrobacterium* and *Erwinia*. During the L-phase these bacteria lack a rigid cell-wall.

Cell rigidity and flexibility: Among the walled prokaryotes (true bacteria) the cells are rigid or flexible depending on the chemical nature of the cell wall. Most walled bacterial cells are rigid, i.e. they maintain their shape when they meet an obstruction while swimming in a fluid. This rigidity is imparted by a high peptidoglycan content of the cell wall. The elastically flexible bacterial cells bend and flex when they hit another object while swimming. However, they return to their original form when the stress is removed or passed. The flexibility is attributed to a low amount of peptidoglycan in the cell-wall.

30

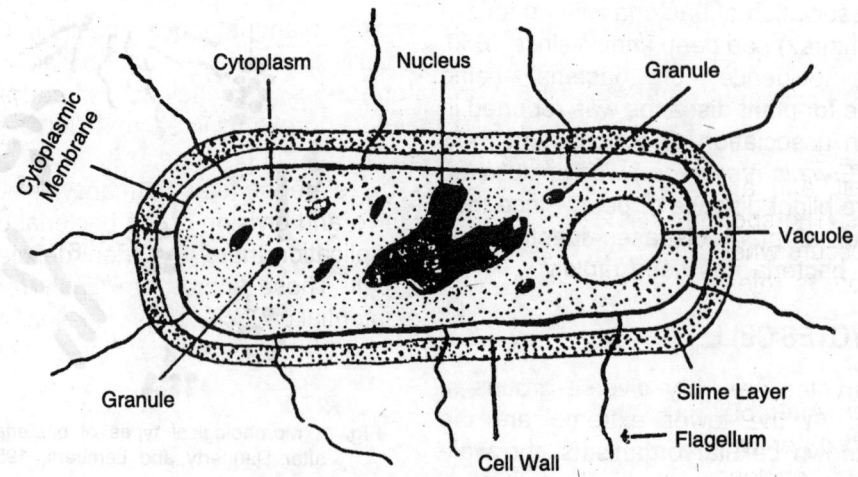

Labels on figure: Cytoplasm, Nucleus, Granule, Cytoplasmic Membrane, Vacuole, Granule, Slime Layer, Flagellum, Cell Wall

Fig. 2. Schematic drawing of hypothetical bacterial cell.

Cell boundary layers (Cell wall and membranes): The cells of prokaryotes vary considerably in the nature of the boundary layers that protect the underlying protoplasm. While the Mollicutes and L-form bacteria have only a membrane around the protoplasm, most prokaryotes have both membranes and the cell wall which is rigid but often flexible and gives the cell its shape. The diversity in boundary layers can be observed by staining properties (Gram's stain reaction), serological characteristics, and ultra-structural appearance as seen in electron microscope.

The term cell envelope refers collectively to one or more structural layers outside the protoplast. The protoplast consists of the plasma membrane and everything bounded by this membrane. The plasma membrane, also known as cytoplasmic membrane is typically a double track (unit) membrane made up of phospholipid and protein. This membrane regulates passage of materials into and out of the cell and is the site of various other metabolic activities. Exterior to the plasma membrane, in all prokaryotes except the mycoplasmas and L-phase of certain walled bacteria, is the layer of a complex polymer known as peptidoglycan or mucin. The peptidoglycan layer and any other boundary layer outside it constitute the cell envelope or the cell-wall. In some bacteria, exterior to or protruding through the cell-wall are various structures such as flagella, pili appendages, glycocalyx, etc.

Multicellularity and filaments: Most ordinary bacterial cells remain as single cells only for the period their division has not occurred, which is for a few minutes or hours, and then the daughter cells are formed. These daughter cells may remain attached to the mother cell giving an appearance of multicellularity. But the cells are independent entities not maintaining any relationship with the mother cell. However, there are some bacterial groups in which there is a multicellular structure with a persistent and characteristic attachment of cells and in which some sort of distinction in structure and function also exists. Examples are actinomycetes and streptomycetes which characteristically form coenocytic filaments. Some species grow a mycelial mat with branched filaments. As a unit this mat often forms reproductive and nutritive branches.

Cell arrangement: The spherical bacteria (cocci) divide by binary fission to form a pair of cells (diplococcus) or chains of cells (streptococcus) or bunch of cells (stephylococcus). These bacteria are not plant pathogens. The rod-shaped bacteria divide at right angle to the long axis. They can remain single (microbacillus), in pairs (diplobacillus) or short chains (streptobacillus). Such cell arrangements are possible in plant pathogenic bacteria. Ordinarily, the bacterial cells in a chain have minimal, essentially tangential, contact between cells.

Spores, cysts, conidiophores and sporangiophores: The bacterial genera *Bacillus* and *Clostridium* which are rarely plant pathogenic, are endospore forming bacteria. MLOs and most plant pathogenic true bacteria are non-spore forming. The endospores are formed singly, one in each cell, and are actually dormant cells. This state of total dormancy is known as cryptobiosis. The spores are not formed during active and division of the bacterial cell. Their formation occurs when after active growth and multiplication there is nutrient deficiency. The mature spore is released by lysis of the mother cell. These spores are highly resistant to ultraviolet light, chemicals and heat and retain the potential capacity of germination after years to form new vegetative cell. Some bacteria form exospores by budding at one end of the cell. These structures are not formed by any plant pathogenic bacteria. Cysts are formed by *Azotobacter*. After binary fission the cell undergoes encystment with synthesis of a thick multilayered outer wall and an inner coat which, in turn, encloses the resting cell.

Conidiospores and sporangiospores are reproductive spores formed by only streptomycetes some of which are plant pathogens. They are formed by multiple cleavage or fragmentation of reproductive branches. When the spores are not contained in any sac they are called conidia or arthrospores. When formed in a sac they are called sporangiospores.

Flagella and other locomotor devices: All prokaryotes are not motile. Self-controlled motility is absent in spherical bacteria while rapid movement of rod-shaped bacteria is very common. Mycoplasmas are also not motile but *Spiroplasma* has characteristic stationary as well as translocational movement without any organ of locomotion. The motile bacteria show several different mechanisms of locomotion depending mainly on the medium in which they are placed. One type of motility—swimming mediated by flagella—depends on a fluid environment. The other modes such as swarming, gliding, twitching, and darting, all depend on a solid surface but have different mechanisms.

Flagella and flagellar motility: Prokaryotic flagellum can be defined as specialized prokaryotic locomotive organelle which consists of a filiform extension, arising from cytoplasm and extending through the

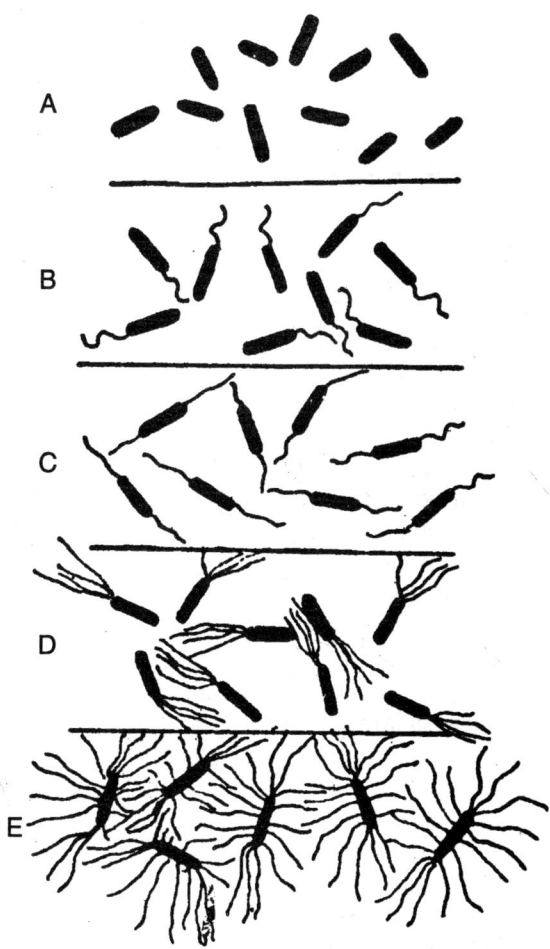

Fig. 3. Bacteria and their flagella.
A–Atrichous; B–Monotrichous; C–Amphitrichous;
D–Lophotrichous; E–Peritrichous.

cell surface. They are rarely bounded by an extension of cytoplasmic membrane, and are composed of a single helically wound filament. All spiral bacteria and nearly half of other rod-shaped bacteria possess flagella. Most plant pathogenic bacteria are flagellated cells. These structures account for the surface translocation of bacterial cells by swimming. In addition to organs of motility, bacterial flagella act also as protein export/assembly apparatus. The position and number of flagella varies with genera and is an aid in their identification. Flagella differ from pili which also are straight hair-like structures originating from protoplast and passing through the cell wall. Both structures have protein as basic chemical material. In properly stained preparations they appear as long, slender, undulating organs with blunt or thickened tip. They originate in the cytoplasm and emerge through the cell-wall. The bacterial flagellum is driven at its base by a rotary motor-like mechanism causing the individual flagellum to describe a helical rotation giving motion to the cell. During forward motion of the bacterial cell flagella are bent backward at an angle of 45°. Reversal of direction occurs by swinging the flagella through an angle of 90°. Turning movements take place by swinging flagella forward only on one side. When the bacterial cell has no flagella it is called atrichous. It is monotrichous when there is a single flagellum at one end of the cell. When two or more flagella are present at one or both ends the bacterial cell is lophotrichous. Large number of flagella all over the surface make the cell amphitrichous.

A constitutive protein, flagellin, in the bacteria flagella functions as elicitor of host responses and decides compatibility and incompatibility. In incompatible host-pathogen combination, hypersensitive reaction is elicited.

Gliding motility: This is a surface translocation in non-flagellated bacteria and occurs only when cells are in contact with a surface not in solution. External slime (microfibrils), probably internal contractile fibrils, and other similar protrusions on cell surface, are supposed to give this type of motility.

Twitching motility: A number of bacterial species including some species of *Pseudomonas*, are known to exhibit a flagellum-independent surface translocation on solid media. This is called twitching motility. This surface bound motion is due to presence and function of thin, usually polar, pili (hair) or fimbriae (fibers) on the cell surface. Such bacteria usually produce spreading colonies which often form channels on the agar medium.

Capsules and other surface structures: Capsule, slime layer, sheath, etc. are present in many groups of bacteria and help in survival, dissemination, and parasitism. These are result of biosynthesis of various organic polymers by the bacterial cell and their deposition as exopolymer around the cell.

The terms capsule and slime layer are often used interchangeably, there being not much difference between their composition and position. Generally relatively thin layers of exopolymers, which are present in almost all bacteria, are called slime layer and when the slime layer is larger and firmly adheres to the cell surface it is called capsule. If it parts freely from the cell it is called free slime or gum. These exopolymers can be observed by special negative staining. Large mucoid colonies on solid media are often indicative of slime or capsule formation. Slime helps in adherence of bacterial cells to surfaces. The free slime in liquid media imparts the characteristic viscosity to the culture. Chemically, the exopolymers are of a wide variety but most commonly the capsules consist of polysaccharides (exopolysaccharides or EPS). The exopolysaccharides of bacteria play many significant roles. They help adhesion to surfaces, protect cells from desiccation, serve as nutrient for bacterial cell, determine the virulence and some other functions. Some bacteria such as *Ralstonia solanacearum* use the slime as sole source of carbon for growth. The EPS is virulence determinant in many phytobacteria such as in *Erwinia amylovora, R. solanacearum* and *Clavibacter michiganensis* subsp. *sepedonicus*. In ring rot of potato (*C. michiganensis* subsp. *sepedonicus*) the EPS intensively bind to protoplast surfaces and microsomal membranes of a susceptible cultivar but not to those of a resistant cultivar (Romanenko *et al.*, 2003). The EPS also protects cells against desiccation and helps in survival of the cell in its air-borne stage.

Bacterial nucleus (nucleoid): Prokaryotes have no membrane bound, organized nucleus as found in eukaryotic cells. Hence, the prokaryote nucleus, more precisely known as nucleoid (genome) is not easily seen when the cells are stained with routine nuclear stains. Basic dyes, which selectively stain the chromatin of he eukaryotic nucleus, stain most bacterial cells densely and evenly suggesting the presence of nucleic acid throughout the cell. This basophilic property is due to an abundance of ribosomes which confer an unusually high ribonucleic acid (RNA) content in the cytoplasm. If the RNA is removed by first treating the fixed cells with ribonuclease or dilute HCL, which hydrolyses RNA, subsequent staining with a basic dye reveals the nucleoid as dense, centrally located region of irregular outline. Two to four such regions may be found in each cell indicating that the bacterial cell is multinucleate. In electron microscope, under optimum conditions of fixation, the nucleoid gives a finely fibrillar appearance which varies little from one prokaryote to another. The single chromosome per nucleus is represented by a double stranded circular DNA molecule (genome) which take up a sizable area of the cell space in the chemically fixed cells but in living cells the DNA is in a less condensed state, probably dispersed throughout the cytoplasm. The total length of the DNA strand is often much longer than the length of the cell (0.25 mm in MLO to about 3 mm in blue green algae or photobacteria).

OTHER INTRACELLULAR ORGANELLES

Structures like Golgi bodies, nucleolus, endoplasmic reticulum, lysosomes, mitochondria, and microtubules found in eukaryotic cells are absent in prokaryote cell except that microtubules have been reported in L-form of certain bacteria. On the other hand, extrachromosomal genetic elements, such as plasmids and episomes, are characteristic organelles in many bacteria.

Plasmid is a broad or general term used to encompass extrachromosomal, stably inherited replicons. A replicon is a genetic element capable of regulating its own replication. The plasmids are circular double stranded DNA molecules capable of autonomous replication in the host cell. An episome is a genetic element that can replicate in either of the two alternative states: (i) integration into, or (ii) independent of, the host chromosome. Certain plasmids can become episomes. They may integrate with the chromosomes of the receptor cells or recombine with other plasmids and may then come under the control of another replicon. Plasmids play an important role in conjugation of bacterial cells. In some bacteria such as in *Agrobacterium* they determine the tumor inducing capacity of the species.

Ribosomes are organelles present in all types of cells. They consist of RNA and proteins. They are the work bench for protein synthesis through ribosomal RNA (rRNA) by translating the message encoded in messenger RNA (mRNA) from the DNA in the nucleoid. Apart from protein synthesis, ribosomes perform many other functions. Their number per cell varies according to rate of multiplication of the cell. A bacterial cell capable of dividing in 5 hours may contain 200 ribosomes while that dividing in 25 min may contain 75,000 ribosomes. This is because more protein synthesis is required in rapidly multiplying cells.

Inclusion bodies such as polyglucoside granules, polyphosphate tubules, striated fibers, etc. are found in many prokaryotes but none of these is enclosed by a membrane (consisting of a lipid bilayer with intrinsic and other proteins).

CELL DIVISION (MULTIPLICATION) OF PROKARYOTES

Three morphologically distinct methods of reproduction are found in prokaryotes binary fission (transverse and usually symmetric), budding and fragmentation. Bacterial cells generally multiply (reproduce) by binary fission. Budding is a rare phenomenon. Fragmentation is common in actinomycetes. One cell gives rise, by fission, to two daughter cells which further multiply in the same manner. Binary fission has two stages: replication of the genome (DNA) and then partition of the cell by a cross wall in such a manner that one replica of the mother genome is placed in each daughter cell. Thus, the daughter cells are true to type. While in binary fission the DNA replication is followed by equal division of the cytoplasm and the synthesis of cell wall, in budding DNA replication is followed by unequal division of cytoplasm.

VARIABILITY IN BACTERIA

The normal cell division in prokaryotes, preceded by replication of the genome and followed by division of the cytoplasm into two equal parts does not leave any chance for variability in the progeny which is necessary for evolution in species. Bacterial reproduction and multiplication is basically production of progenies true to type. Bacterial nucleus is always haploid. But variations do occur in prokaryotes as is evident from the existence of strains in species which differ in pathogenicity, colony characters, and response to chemicals and other stimuli. These races or strains can evolve by mutation or gene recombination. The transfer of genetic material or DNA or characters coded in this material to daughter cells is not possible at the time of binary fission. The variability among bacteria may be caused by genetic changes as a result of the following:

Conjugation: This plasmid-motivated process occurs in two steps, both governed by plasmid genes (i) a series of reactions between the surfaces of the conjugating cells (the donor and the recipient) resulting in the formation of a conjugation bridge, and (ii) passage of a molecules of plasmid DNA through the bridge. Reactions occurring between the donor and the recipient cells are the consequence of specific properties of the surface conferred by the plasmid to the donor cell. Every cell will not conjugate Only cells with transmissible plasmids become donors and conjugate. Prior to conjugation, the plasmid exists in the donor cell as a double stranded circular DNA molecule. The molecules is not transferred as such. Actually, one strand of the plasmid DNA is broken at the replication origin and enters the recipient cell. Complimentary strands are synthesized by DNA polymerase in both donor and recipient ceils, thus both retain replica of the original DNA molecule.

Transformation: In this process, briefly, the bacterial cells absorb the genetic material exuded by a compatible cell or freed by lysis of the cell wall. The compatibility is supposed to be due to the presence of a specific protein on surface of the recipient cell.

Transduction: The bacterial viruses (bacteriophages) play the main role in transfer of genetic material in this phenomenon. A small piece of bacterial chromosome is incorporated into a maturing bacterial virus particle that has infected the donor cell. When, after death of the infected cell, the virus particle is released it carries with it the genetic material to the new bacterial cell which it infects. If this cell does not die, recombination of genetic material takes place.

LIFE-CYCLE OF THE PROKARYOTES

The life-cycle of the prokaryotes is simple. In most bacteria, the vegetative cell multiplies by fission, without much change, every few minutes to hours depending on nutrition and environment. The progeny cells repeat the process. In some groups, however, distinctly different morphologies of cell types occur at different stages of life cycle. In others, there may be a series of biochemical and morphological events, such as endospore formation, which are irreversible once started but are separate from reproductive cycle. They are directed to survival of the cell in adverse environments. In most plant pathogenic bacteria these complexities are absent.

CELL NUTRITION:

Like any other living organism, prokaryote cells also require nutrition from the surrounding environment for synthesis of their cell material and for the generation of energy for running the machinery for biosynthesis (except the Chlamydiae, the simplest cellular organisms). Physiological diversities in prokaryotes are very wide and consequently nutrients are obtained by them in very diverse manner.

The chemical composition of cell, broadly constant throughout the living world, points to the nutritional requirement of bacteria. The major elements required in cell nutrition are hydrogen, oxygen, carbon, nitrogen, phosphorus, and sulphur, in the order of decreasing abundance in the cell. These six elements account for 95% of the cell dry weight. Many other elements are required in traces in the remaining 5% but are essential. These are potassium, magnesium, calcium, iron, manganese, cobalt, copper, zinc and molybdenum.

The prokaryotes have been grouped under different categories on the basis of the sources from which they derive the above nutrients. All pathogenic bacteria are *chemoheterotrophs* which use a chemical energy source and an organic compound as the principal source of carbon. Both these sources can be derived from metabolism of a single compound. There are bacteria that use light as energy source and carbon dioxide in the atmosphere as carbon source (photoautotrophs), those that use light as energy source and an organic compound as the principal source of carbon (photoheterotrophs), those few that use a chemical energy source and carbon dioxide as carbon source (chemoautotrophs).

Often the mode of nutrition in the above groups overlaps. The terms obligate and facultative are used where a group shows presence or absence of nutritional versatility. Plant pathogenic bacteria are basically facultative saprophyte, i.e., in the absence of the living host plant tissue, they can remain active as saprophytes on dead organic matter also. Thus, most of the true bacteria can be easily grown on artificial culture media the composition of which may have to be varied according to some specific preferences of the bacterium. The degree of saprophytic survival ability varies widely depending on the resistance of the bacterium to biological competition and deleterious biochemical agents in soil. The wall-less prokaryotes (Mollicutes) also appear to be facultative saprophytes since the true *Mycoplasma* spp. have been grown on artificial media. However plant pathogenic mycoplasmas (MLOs) have not been grown in artificial media.

CELL RESPONSE TO EXTERNAL ENVIRONMENT

Physical and chemical forces such as temperature, moisture, light, H-ion concentration, pressure, etc. are capable of disturbing the delicate state of balance in the protoplasmic material in the cell. When this balance is disturbed the vital metabolic processes in the cell are also adversely affected. As a result growth and reproduction are affected and the cell may even die.

Temperature: Bacteria can survive at temperatures of less than 0°C to 85°C or even more depending on the species. The thermophilic bacteria isolated from hot springs are capable of growth even at 95°C while the psychrophilic bacteria isolated from cold environments can grow at temperatures as low as minus 10°C provided the medium does not freeze. The following table gives the temperature relations of three groups of bacteria: The temperature at which all bacterial cells in a solution get killed after an exposure for 10 min is known as thermal death point or thermal death rate. Similarly, thermal death time is defined as the time required to kill all cells at a given temperature. Only vegetative cells (applicable to all plant pathogenic bacteria) are considered for these determinations. Spore forming bacteria (very few of plant pathogens) survive even in boiling temperature. To kill them the spores must be first activated to grow into vegetative cells. Thus, such bacteria have two thermal death points, one for vegetative cells and one for spores.

Death of bacterial cells by heat is due to coagulation of proteins in the protoplasm. This coagulation is governed by many factors such as water content of the medium, water content of the cell, pH of the medium, composition of the medium, age of cells, etc. The greater the percentage of water in the medium, the lower will be the temperature required to kill the cells. Moist heat is more effective agent of sterilization than dry heat. In autoclaves, moist heat at comparatively lower temperatures gives perfect sterilization within 15-20 min as compared to dry heat in oven where much higher temperatures and longer time (in hours) are required to sterilize an object. Dry bacterial cells are more tolerant to heat than young actively growing cells with more moisture. Killing of bacterial cells is rapid if the reaction of the medium is alkaline or acidic. Media containing high quantity of protein or albuminous materials require high temperatures to kill bacterial cells suspended in them. The protein forms a protective film around the cells thus hindering penetration of heat. Pathogenicity and metabolic activities of bacteria are influenced by temperature. Some bacteria produce phytotoxins only at low temperatures.

Moisture: Bacteria are more aquatic than terrestrial and thrive in the presence of a high percentage of water. This is not because the bacterial cell itself requires water for its metabolic activities but because water helps in absorption of nutrients and dispersal of cells. However, in water-logged soils where availability of free oxygen is reduced, the activity of strictly aerobic bacteria ceases and anaerobic bacteria thrive. Survival of plant pathogenic bacteria in soil is generally reduced if the soil is wet. If the bacterial cells are dried they may cease activity but they survive and when restored to moist conditions they regain their activity.

Light: Ordinary light visible to human eye does not adversely affect the bacterial activity. The invisible components of the light spectrum, the ultra-violet rays at one end and infra-red rays at the other end, are germicidal. The germicidal action of sunlight is due to these invisible rays, particularly the infra-red rays that produce heat. Ultra-violet rays are used in ultraviolet lamps used for disinfection of the atmosphere in laboratories and surgical rooms.

Pressure: Because of their small size and volume the bacterial cells are not sensitive to ordinary mechanical pressure. But they respond to osmotic pressure which is defined as the diffusion pressure of fluid passing through a semi-permeable membrane (such as the cell-wall). This is an equalizing process in which concentration of fluids of two unequal densities on the two sides of the membrane is equalized. The osmotic pressure of any solution is expressed in terms of osmolarity.

Solutions having equal osmotic pressure are called isotonic while solutions of high density are called hypertonic, in contrast to hypotonic which have a low concentration. The fluid inside the bacterial cell is always hypertonic in relation to the fluid outside in the host cell. Due to the equalizing process fluid from outside enters the cell through the membranes. This is how nutrients enter the bacterial cells. When a cell is placed in hypertonic solution, i.e., where concentration of the fluid outside is greater than the fluid inside, water leaves the cell to equalize the pressure from outside. This continues until the concentration of the protoplasmic fluid and the fluid outside become equal. If the difference in the densities of the two fluids is great, the pressure causes the cell to shrink and ultimately die. This is known as plasmolysis and the cell is plasmolysed. When bacterial cells are placed in hypotonic solution, water enters the cell causing it to swell. Ultimately, the cell wall may burst. This is known as plasmolysis. Use of concentrated sugar or salt solutions (hypertonic) for preserving foodstuff is based on the principles of plasmolysis.

Antibiotics and Antagonism: Antibiotics are organic substances of microbial origin which are either toxic or growth inhibitory for other organisms. Their great value in the therapy of infectious diseases largely depends on their selective toxicity against the infectious agents even at very low concentrations without adversely affective the plant tissue.

Penicillin and some other antibiotics are selectively toxic to prokaryotes but not for eukaryotes because they affect steps in the synthesis of the peptidoglycan of the cell wall which is not present in eukaryote cells. Since the Mollicutes (mycoplasmas) have no cell wall they are also insensitive to penicillins. Many antibiotics interfere with protein synthesis which occurs at ribosomes. Ribosome in prokaryotes has sedimentation coefficient of 70S and eukaryotes of 80S. The antibiotic cycloheximide inhibits the function of 80S ribosome, so, is effective against eukaryotes but not against prokaryotes. Streptomycin, tetracycline, and chloramphenicol inhibit activity of 70S ribosomes, hence, are effective against prokaryotes but not all eukaryotes. In eukaryotes the organelles like chloroplasts and mitochondria have 70S ribosomes and are affected by these antibiotics.

Outlime of Prakaryote Classification

Bacterial taxonomy, the principles and practice of classifying bacteria, consists of classification, nomenclature, and identification, In prokaryotes these three activities have been rapidly changing ever since attempts were started to group the bacteria in a systematic manner. The changes became more frequent in the latter half of the 20th century.

Although all prokaryotes possess the common chemical composition (DNA, RNA, and proteins) many of them exhibit absolutely no relationship with the majority in their structure and function. In absence of knowledge of their ancestry, a hierarchical system of classification (families, orders, etc.) had not been possible and still there are many genera which have not found place in a well-defined family or order. Most bacterial pathogens can be easily assigned to their correct genus but identification of species or subspecies is rather difficult. In eukaryotes (plants, fungi, etc.) the morphological differences are such that it is easy to compare two individuals and distinguish them from each other on the basis of phenotypic differences. In prokaryotes comparison of individuals is difficult because they appear in colonies in which presence of contaminants or variants is possible.

Bergey's Manual of Determinative Bacteriology, started in 1923, had been the major treatise on bacterial taxonomy but it could serve only the purpose of identification. It had been considerably changing with respect to composition of taxa higher than genus (families, orders, tribes) in its successive editions. The eighth edition of this publication had adopted a rational approach but the subsequent changes in bacterial taxonomy based on nature of DNA and chemical composition of the cell and cell wall caused this edition also to fail in providing a durable classification of prokaryotes. The publication of this series was stopped after this edition and replaced by a new series under the title Bergey's Manual of Systematic Bacteriology (Krieg and Holt, 1984; Sneath, 1986).

Young *et al.,* (1992) adopted a classification based on more recent information about the bacterial cell and put up the following system for plant pathogenic bacteria:

Division: Firmicutes (Gram-positive genera).

Families: No family classification in this division.

The genera included are *Arthrobacter, Bacillus, Clavibacter, Curtobacterium, Nocardia* and *Rhodococcus.*

Division Gracilicutes (Gram-negative genera).

Class: Proteobacteria. Alpha subclass has three familes, Acetobacteriaceae (*Acetobacter*), Rhizobiaceae (*Agrobacterium*), and a unclassified family with genus *Rhizomonas.*

Beta subclass has one named family and one unnamed family. The named family Comamonadaceae has the genera *Acidovorax.*

(*Pseudomonas*) with a number of subspecies and *Xylophilus.* The unnamed family has the genus *Pseudomonas.*

Gamma subclass has several families but only Enterobacteriaceae with genera *Enterobacter* and *Erwinia* is well established. The position of Pseudomonadaceae this subclass is doubtful. The genera *Xanthomonas* and *Xylella* have not been assigned a family.

Division Fenericutes.

Class: Mollicutes.

Family: Spiroplasmataceae with genus *Spiroplasma.*

The most recent classification of *Phytoplasma* is superlomgdom Prokaryota, kingdom Monera, division Bacteria, phylum Fermicutes,(low G+C, Gram positive eubacteria), class Mollicutes, *candidatus* (Ca) genus Phytoplasma.

Genera *Rhizobacter* and *Streptomyces* have uncertain affinity.

Many species in genus *Pseudomonas* are considered misclassified.

Since the basic category in biological system is the species, a new bacterial isolate must be identified at this level through certain universally accepted criteria avoiding need for major dependence on morphology alone. The criteria are incorporated in the International Code of Bacterial Nomenclature. A complete and modern description which meets these criteria includes morphology including flagellation, Gram's stain reaction, cultural, biochemical and physiological characters, serology, phage and bacteriocin typing, DNA base composition (the G+C content), DNA-DNA or rRNA-DNA homology, and extensive pathogenicity and host range tests with detailed description of symptoms. Chemical analysis of cell and cell wall is an important tool in identification and classification. Any claim for a new bacterium should always be published in the International Journal of Systematic Bacteriology and type culture must be deposited in internationally recognised type culture collection.

Pathovars in plant pathogenic bacteria

Apprehending that names of many plant pathogenic species, especially of *Pseudomonas* and *Xanthomnas,* may not find place in the Approved Lists, the International Society for Plant Pathology came out with a system of pathovar names (Dye, *et al.,* 1980) in the case of those species names which had been prevalent and had been found otherwise acceptable by phytobacteriologists but could not find place in the Approved Lists for one or the other reason. Thus, when the name *Xanthomonas oryzae* was not approved because it was considered not much different from *X. campestris,* it was named *X. campestris* pathovar (pv) *oryzae.* Same was true for many other names, such as *Pseudomonas syringae* pv. *vesicatoria* for *P. vesicatoria, P. syringae* pv. *phaseolicola* for *P. phaseolicola, X. campestris* pv. *citri* for *X. citri* and *X. campestris* pv. *malvacearum* for *X. malvacearum.* However, later there were other changes and species name *oryzae* was restored while the pathovar name was retained (*viz., oryzae* pv. *oryzae* and pv. *oryzicola*). *X. campestris* pv. *vesicatoria* was restored as *X. vesicatoria*. A number of pathogens in genus *Xanthomonas* were reclassified and transferred to the species *axonopodis* in place of *campestris* but the pathovar system was retained.

Major plant pathogenic genera

Pseudomonas: The genus, as described earlier, contained some of the world's most serious bacterial diseases of plants. The genus has the following characters: Cells single, straight or curved rods but not helical; generally 0.5–1.0 × 1.5–4.0 μm in size; motile by one or more polar flagella, non-spore forming, Gram-negative; strict aerobes; metabolism respiratory, never fermentative. At present, species of *Pseudomonas* are conventionally grouped according to whether or not they produce fluorescent pigment on iron-deficient media and poly-ß-hydroxybutyrate inclusions (cf Young, *et al.,* 1992).

The fluorescens group: The fluorescens group comprises those species which mostly produce fluorescent pigment and which are strictly classified in the Pseudomonadaceae under gamma subclass of the Proteobacteria. Most plant pathogenic members in this group belong to *Pseudomonas syringae* and its pathovars. The recognised number of pathovars was 45 in 1992.

The solanaceous group: A number of non-fluorescent pathogenic species of *Pseudomonas* including *P. solanacearum* were considered part of a solanacearum DNA-rRNA homology group. However, the group always merited a reclassification. *P. solanacearum* was renamed as *Burkholderia solanacearum* and then as *Ralstonia solanacearum*.

The acidovorans group: Species in this group include *P. avenae, P. rubrilineans* and many others but the entire group has been reclassified as the family Comomonadaceae in the beta subclass of the Proteobacteria. All the species are considered to be members of a single species, *P. avenae*, some being synonyms while other being pathovars. In later studies these species were transferred to the genus *Acidovorax*, such as *A. avenae* subsp. *avenae*. Another genus of the group is *Xylophilus*. *X. ampelinus* was originally classified in *Xanthomonas* but was shown to be member of acidovorans group. However, it is recognised as a distinct genus.

Xanthomonas: This genus includes more pathogens than all other genera combined. The members of the genus have following characters: Cells single straight rods measuring 0.2– 0.8 × 0.6–2.0 (usually 04 × 1.0) μm; Gram-negative, motile by a single polar flagellum; non-spore forming; copious extracellular slime produced; strict aerobes; metabolism respiratory never fermentative; oxidase negative and catalase positive; bacterial colonies yellow coloured due to production of xanthomonadin pigment.

In early studies, it was found that most species could not be distinguished from (*Xanthomonas campestris*). Subsequently, 5 distinct species, namely *X. campestris, X. fragerieae, X. albilineans, X. axonopodis* and *X. ampelina* were recognised. All were plant pathogenic. Later, a non-pathogenic species *X. maltophilia* was added. Under the reclassification of *Xanthomonas* (Vauterin, *et al.,* 1995) reinstatement of some species such as *oryzae*, the number of species was raised to 8. The species *X. ampelina* was shown to be a member of the acidovoras group (Comamonadaceae) and is no more a *Xanthomonas*. It belongs to the genus *Xylophilus (Xylophilus ampelinus)*. Phenotypic differences of *X. maltophilia* from other species are so large that it could be reallocated to a separate but related genus (cf. Young, *et al.,* 1992). The reclassification of *Xanthomonas* changed the location of some pathovars of *campestris* also. Thus, *X. campestris* pv. *citri*, instead of being reinstated as *X. citri* was allocated to the species *X. axonopodis* (*X. axonopodis* pv. *citri*) or

X. campestris pv. *malvacearum* instead of being reinstated as *X. malvacearum* was allocated to the species *axonopodis* (*X. axonopodis* pv. *malvacearum*).

Erwinia: The characters of this genus are: cells predominantly single, straight rods measuring 0.5–1.0 × 1.0–3.0 μm; all members are motile by peritrichous flagella except *E. stewarti* which is now named *Pantoea stewarti* subsp. *stewarti* and *E. dissolvens* which is now named *Enterobacter dissolvens*; Gram-negative, aerobic as well as anaerobic. The genus has two major groups of species. The group consisting of bacteria with strong pectolytic capacity includes *E. carotovora* with its subspecies which are the most common soft rot causing bacteria and *E. chrysanthemi* with many pathovars that specialize on specific hosts and cause vascular wilt and parenchymatous necrosis. The group consisting of bacteria that do not cause soft rot but dry necrosis and wilt include *E. amylovora* and *E. tracheiphila*. These species do not produce pectic enzymes or yellow pigment but do cause necrosis and wilt symptoms probably through toxin production. A third group consists of epiphytes that cause neither soft rot nor necrosis (*E. herbicola* = *Enterobacter agglomerans*) and the saprophyte *E. persicinus*.

Agrobacterium: The genus *Agrobacterium* (division Gracilicutes, class Proteobacteria, family Rhizobiaceae) was established to include related soil bacteria and gall forming pathogens. The cells are without endospores, normally rod-shaped and motile by one polar or sub-polar flagellum or 2–6 peritrichous flagella. They are Gram-negative, aerobic and utilize many carbohydrates. Considerable extra-cellular slime is produced on carbohydrate media. All species in the genus except *A. radiobacter* incite cortical hypertrophies on plants, either as root nodules (*Rhizobium*) or as galls, tumours or hairy root (*Agrobacterium* species on diverse hosts). Some strains of *Rhizobium* and *Agrobacterium* show close relatedness in DNA base composition and there have been proposals to merge the two in a single genus. The genus Agribacterium has now been abolished and its species merged with *Rhizobium rediobacter*.

Four species in the genus (*A. radiobacter, A. tumefasciens, A. rubi*, and *A. rhizogenes*) were accepted in the Approved Lists of 1980. The species *radiobacter* and *tumefaciens* and their biovars are identical except that *A. tumefasciens* is pathogenic while *A. radiobacter* is non-pathogenic. Since the species epithet *radiobacter* has priority over *tumefasciens* in nomenclature it was considered logical to merge the two and call them *A. radiobacter* pv. *tumefasciens* and *A. radiobacter* pv. *radiobacter*. However, somehow, *A. tumefaciens* continues to be a recognized species. At present the recognized species are *A. tumefaciens, A. rhizogenes, A. rubi* and *A. vitis* (Young, *et al.,* 1992). *A. vitis* was originally biovar 3 of *A. tumefaciens*. The gall inducing genes are present in a large plasmid in *A. tumefaciens*. Plasmids are present in *A. radiobacter* also but they lack the gall inducing principle. In more recent classification all agrobacteria have been transferred to the genus *Rhizobium*.

Among the Gram-positive plant pathogenic bacteria (division Firmicutes in which there is no family classification) the coryneform bacteria are most important. Prior to 1980, these bacteria were considered species of the genus *Corynebacterium* because of some morphological similarities. Bacterial cells are straight to slightly curved rods, non-motile, non-endospore forming, Gram-positive with irregularly stained segments or granules, not acid fast, often pleomorphic, sometimes ovoid, sometimes club-shaped, rarely thread or hypha-like, and in fresh cultures show rudimentary branching. Multiplication is by simple cell division. Carbohydrate metabolism is fermentative as well as respiratory. They are aerobic or facultatively anaerobic and catalase positive. After 1980, the plant pathogenic corynebacteria were transferred to four different genera, namely, *Arhtrobacter, Curtobacterium, Clavibacter* and *Rhodococcus* on the basis of typical chemistry of cell-wall, G+C content of the DNA and genetic homologies. The genus *Corynebacterium* was restricted to animal and human pathogens.

Clavibacter: The genus was proposed for those phytopathogenic corynebacteria that contain diaminobutyric acid as the major acid in their cell wall and which were formerly included in the *Corynebacterium michiganense* group (*C. michiganense, C. nebraskense, C. insidiosum, C. rathayi, C. iranicus, C. tritici* and *C. sepedonicus*). At the same time a new species *C. xyli* was also created that contained *C. xyli* subsp. *xyli* (sugarcane ratoon stunting disease and *C. xyli* subsp. *cynodontis* for Bermuda grass stunting diseases (Davis, *et al.,* 1984). The cells are Gram-positive, non-acid fast, pleomorphic rods, often arranged at an angle to give V-formation as a result of snapping and bending type of cell division. No coccoid cells are seen. The cells are non-motile. The bacterium is strict aerobe and nutritionally fastidious. The cell wall peptidoglycan contains diaminobutyric acid. G+C content of the DNA is 70.5 mole%. Type species is *Clavibacter michiganensis.* The species *C. tritici* is now reclassified as *Rathayibacter tritici.*

Curtobacterium: This genus has following characters: cells small, short rods, coccoid cells found in old culture, weakly Gram-positive, frequently old cells lose Gram-positivity, generally motile by lateral flagella, cell multiplication by bending type of cell division, pleomorphism only slight, major cell wall amino acid is ornathine, G+C content of the DNA ranges from 68 to 75 mole%. Plant pathogens transferred to this genus were *C. flaccumfaciens* and its pathovars *C.* f. *flaccumfaciens, C.* f. *betae, C.* f. *oortii* and f. *poinsettiae.*

Rhodococcus: *Corynebacterium fascians* was the only species transferred to this genus as *R. fascians.* It is a plant pathogenic bacterium that causes malformation on aerial plant parts (fasciation) or in which leafy galls occur at axillary meristems are formed. Cells in young cultures measure 0.5–0.9 × 1.5–4.0 μm. They are slightly curved rods arranged singly at an angle or as palisade. They are non-motile. The cell wall peptidoglycan contains diaminopimelic acid. The G+C content of the DNA is 62.9–67.3 mole%. The bacteria extensively colonize the host surface where they are surrounded by a slime layer. The epidermal cells of the host collapse and bacteria penetrate intercellularly in to the plant tissue. The onset of symptom development precedes the extensive colonization of the interior.

The genus *Arthrobacter* is suspected to be *Curtobacterium.*

Streptomyces: This genus is of uncertain affinity although earlier it was classified in the family Streptomycetaceae of Actinomycetales. These bacteria are widely distributed in soil. The following characters were given for the genus: Slender, coenocytic filaments, 0.5–2.0 μm in diameter, aerial mycelium at maturity forms chains of 3 to many spores, cell wall contains diaminopimelic acid. It is Gram-positive, aerobic, and in culture produces small colonies of 1–10 mm diameter which are discrete and lichenoid, leathery or butyrous, initially relatively smooth but later develop a weft of aerial mycelium that may appear granular, powdery, velvet or floccose. Although 400 species were listed in the last edition of Bergey's manual only 275 were accepted in the Approved Lists of 1980. Two species, *Streptomyces scabies* (common scab of potato) and *S. ipomea* on sweet potato are recognized as plant pathogens although many others have been isolated from diseased plant specimens.

THE FASTIDIOUS PROKARYOTES

The fastidious prokaryotes are minute bacteria which are very exacting in their nutritional requirements, refusing to grow on routine bacteriological media, and which for long remained hidden in their plant or insect hosts because they could not be cultured or easily seen. The lack of proper use of electron microscopy and failure to grow such organisms in culture media had originally clubbed the diseases caused by them with virus diseases. These organisms are divided in to two groups: the walled Rickettsia-like bacteria (RLB) and the wall-less mycoplasma-like organisms (MLO), now known as *Phytoplasma*. The discovery of RLBs followed the discovery of MLOs. The term rickettsia-like bacteria is misnomer. These fastidious walled bacteria only superficially resemble the Rickettsiae otherwise they are totally distinct. Some of them are species of well known genera listed above.

MYCOPLASMA-LIKE ORGANISMS (PHYTOPLASMAS)

The existence of *Mycoplasma* as disease incitants in animals had been known in the nineteenth century. The contagious bovine pleuroneumonia of cattle was believed by Luis Pasteur to be caused by a specific microorganism that could not be seen or grown in culture. Later, in 1898, E. Nocard and E.R. Roux succeeded in growing the organism in serum enriched broth. The organism is now known as *Mycoplama mycoides* var. *mycoides*. Subsequently, all such organisms associated with hosts other than plants were classified as *Mycoplasma*. The organism was placed in the order Mycoplasmatales among bacteria. Such organisms were later found to have some characters quite distinct from true bacteria. They have no cell-wall and are, therefore, pleomorphic, they do not grow on ordinary bacterial culture media, and they respond to tetracyclines but not to penicillins. The group was, therefore, separated from true bacteria (eubacteria) and raised to the status of a class, the Mollicutes in the division Tenericutes.

The main basis for separation of mycoplasmas from true bacteria (eubacteria) is the absence of true cell wall. They belong to the class Mollicutes (*mollis* = soft or pliable, *cutis*= skin; *mollicute*=with pliable cell boundary, i.e., lacking true cell wall that gives rigidity). The main characteristics that identify an organism as a Mollicute are:

1. Lack of cell wall: cells are bounded by only a single triple-layered membrane.
2. Colony appearance on agar medium: those that have been cultured produce colonies giving 'fried egg' appearance.
3. Filterability through a 450 nm membrane: very small size that permits them to pass through filters with 450 nm diameter pores.
4. Absence of reversion to bacterial form: in some eubacteria as in *Erwinia carotovora* subsp. *atroseptica* bacterial L-forms are produced due to some stress such as effect of penicillin or osmotic pressure. In this stage the bacteria develop characters of Mollicutes. They lack true cell wall and therefore are capable of passing through membrane filters due to pliability. When the stress is removed they revert to the true bacterial form. Mollicutes do not do so. An organism can be considered a Mollicute if it can be passaged five times in a medium without penicillin or increased osmotic pressure and not revert to a walled bacterial form.

The generic characters of *Mycoplasma*, as derived from cultured forms are: Cells highly pleomorphic, varying in shape from spherical to slightly ovoid, 125–250 nm in diameter in early exponential growth phase, to slender, branched filaments of uniform diameter ranging in length from a few to 150 μm. Chains of coccoid bodies develop at a later stage of growth. The modes of reproduction primarily include binary fission of coccoid and filamentous cells but budding and formation of elementary bodies within the filaments are also supposed to be involved. In the latter case, the filaments may also branch forming a pseudomycelium. Dense corpuscles appear in the filaments which are transformed into chains of spherical bodies. These chains fragment liberating free elementary bodies. Cells lack true cell wall and are bounded by a single triple-layered membrane. They are usually non-motile but gliding motility is described in some species. No resting spores are formed. They are Gram-negative. Colonies on culture media are formed with typical fried egg appearance. *Mycoplasma* is resistant to penicillin and ampicillin.

The above characters are for the genus *Mycoplasma* which can be grown *in vitro* on cell-free media. Before 1967 no one had shown that mycoplasmas could be in any way associated with plant diseases. At the same time it had also been realized that many plant diseases showing symptoms of virus infection did not yield any virus-like entity in electron microscopy. Doi *et al.,* (1967) and Ishiie *et al.,* (1967) in Japan were the first to report that close examination, with electron microscope, of sieve elements of many plants showing witches' broom and yellows symptoms revealed the presence of mycoplasmal bodies which were not present in the phloem of healthy plants. Prior to this report these diseases (aster yellows, mulberry dwarf, potato witches' broom and Paulownia witches' broom) were supposed to be virus diseases. These Japanese workers found constant association of the mycoplasmal bodies with diseased plants and also that they were leafhopper transmitted and disappeared with remission of symptoms when tetracycline antibiotics were applied to the plant. During the years that followed this discovery, mycoplasma-like organisms were found associated with disease condition in more than 200 plant species and insects. These organisms, resembling the *Mycoplasma,* found associated with plant diseases lack ability to grow, so far, on artificial media. In absence of isolation and cultivation of the organisms *in vitro* their characterization is incomplete. Hence, they were known as mycoplasma-like rganisms (MLOs) but there is no doubt that they are Mollicutes and once they are cultured it is possible that they prove to be a new taxon in the Mollicutes. During the last 10–12 years, scientists have taken recourse to molecular methods to differentiate, characterize and classify plant pathogenic phytoplasmas on a phylogenetic basis. This is mainly done by sequence analysis of the 16S RNA gene amplified by polymerase chain reaction (OCR) technology. The study of genetic relatedness has shown that phytoplasmas are Mollicutes but have more relatedness to Acholeplasmatales and Acholeplasmataceae than to *Mycoplasma*. They are given this taxonomic position with tentative genus (candidates taxon) *Phytoplasma.* No species epithet is decided. The genus is divided into a number of groups of genetically related phytoplasmas. In this book the terms MLO and *Phytoplasma* both are interchangeably used at some places.

The number of plant diseases of mycoplasmal etiology is large. Some examples are grassy shoot of sugarcane, little leaf of eggplant, and sandal spike. Later, some of the mycoplasmas such as the causal agent of citrus stubborn and corn stunt diseases could be grown in cell-free media. However, they turned out to be different in morphology from *Mycoplasma*. They were helical and showed motility in certain stages of their growth in culture media. The trivial name "spiroplasma" was first suggested in 1973 for mycoplasmas with helical morphology and was adopted for the generic name when the agent of citrus stubborn disease was cultured and characterized. This agent, *Spiroplasma citri*, is the type species of the genus. *S. kunkelii* causes corn stunt disease. General characters of *Spiroplasma* are: Cells are pleomorphic varying in shape from spherical or slightly ovoid to helical and branched non-helical filaments. The spherical or ovoid forms measure 100–2550 nm in diameter. The helical forms are about 129 nm in diameter and 2–4 μm in length in the lagarithmic phase and much longer in post-logarithmic phase in liquid culture medium. In solid media only non-helical forms are seen. Cells lack true cell wall and are bounded by a single triple layered membrane. The helical forms are motile showing two types of motion, a rapid rotary or screw type motion and a slow undulating motion. Colonies on solid media exhibit a typical "fried egg" appearance. The organism is sensitive to tetracyclines but not to penicillin. The helices of *Spiroplasma* commonly possess one tapered end (tip structure) and one blunt or rounded end. The tip structure is morphologically different from the rest of the helix, exhibiting an electron-dense conical or rod-shaped core.

Phytoplasma in Plant Tissue

The morphology of plant MLOs in the host is basically the same as described for the genus *Mycoplasma*. Extensive studies of these organisms have been carried out by thin section electron microscopy of tissues removed from diseased plants (McCoy, 1979) or from their leafhopper vectors or hosts. In the plant tissue the phytoplasmasa are pleomorphic, small round (60–100 nm dia), large globular (190–1100 nm dia) and branched filaments, 1–2 to several microns in length (Davis and Lee, 1982). Average diameter is 0.3–0.8 μm (McCoy, 1981). Small rounded and large globular bodies predominate in the late season or in advanced stage of the disease while branched filamentous bodies are seen in early season or in early stage of the disease (Davis and Lee, 1982). However these observations are based on single ultra-thin section. It is possible that thicker sections or examination of serial sections may, in some cases, reveal that the globular bodies are part of filamentous forms.

Studies have shown that reproduction of mycoplasmas and spiroplasmas is by transverse binary fission, budding, release of inclusion bodies and formation of intracellular elementary bodies. In single ultra-thin section of sieve tubes with phytoplasma some of the bodies appear to be budding and others appear to contain membrane bound inclusions, some are very small, dense elementary bodies. According to McCoy (1981) the examination of serial sections reveals that forms appearing as budding are actually branch points in the filamentous forms; membrane bound inclusions are invagination in one phytoplasma sell into which another cell is closely packed, and the so called elementary bodies are actually constriction points in filaments of overall larger diameter. The reproduction is, therefore, basically by binary fission as is common in all prokaryotes.

The Vascular Fastidious Walled Bacteria

The phytoplasmas and spiroplasmas are phloem inhabiting prokaryotes and are transmitted by insect vectors. During closer examination of phloem for presence of other such organisms another group of fastidious prokaryotes was discovered in 1970–1973 in the citrus greening diseased plants. This phloem inhabiting organism has definite cell wall and is susceptible to both tetracycline and penicillin. This suggested that the organism is a true bacterium rather than wall-less prokaryote. It is Gram-negative but has so far not been brought in culture. It was later identified as *Liberobacter asiaticum* and *L. africanum*.

The search for similar organisms in the plant xylem revealed presence of yet another Gram-negative, xylem inhabiting, fastidious bacterium that causes Pierce's disease of grapevines (*Xylella fastidiosa*). The other two xylem-limited fastidious bacteria are *Pseudomonas syzygii* and *Clavibacter xyli*. *X. fastidiosa* and *P. syzygii* are insect transmitted while no insect vectors are known for *C. xyli*. It is transmitted mechanically. All the xylem inhabiting fastidious bacteria are similar in morphology and all are Gram-negative in stain reaction. In general, they have elongated cells measuring in the range of 0.2–0.5 × 1–4 microns. The cells usually have well defined cell wall and plasma membrane, both triple layered in structure. The walls are ridged or rippled due to periodic infolding of the outer membrane of the wall. These ridges appear to spiral around the long axis of the cell. The cell wall structure of these Gram-negative bacteria consists of an outer membrane, an intermediate electron dense zone and an inner dense peptidoglycan layer (R-layer) which is separated from the plasma membrane by an electron lucent zone. Intracellular organelles are same as in the walled bacteria. The cells multiply by binary fission. The transmission of these bacteria is exclusively by xylem feeding insects or by vegetative propagation. The mechanism of transmission has been studied only in few cases, mostly with *Xylella fastidiosa* in grapevines. *Xylella fastidiosa* in grapevines and in peach is transmitted by its leafhopper vectors in non-circulative but persistent manner. There is no incubation period in the vector body and infectivity of the vector is lost after molting. There is no evidence of transmission of the pathogen to progenies of the vector.

Being restricted to the vascular fluid these bacteria perennate through the living organs of the perennial hosts. Diseases caused by these xylem restricted bacteria produce symptoms very similar to vascular wilts suggesting dysfunction of the xylem. The xylem-restricted fastidious walled bacteria are distinguished from other xylem-invading bacteria such as species of *Pseudomonas, Erwinia*, etc. by their inability to infect tissues other than xylem, their insect but not mechanical transmission, and their inability to grow on conventional bacteriological media. The isolation of xylem-inhabiting fastidious bacteria in pure culture has been more successful than the phloem inhabiting bacteria. However, they have been grown only on highly specialized media.

The phloem-limited fastidious bacteria, erroneously called the RLBs, have so far not been cultured *in vitro*, hence Koch's postulates have not been proved for their association with respective diseases. Two evidences have been cited as proof of pathogenicity of these bacteria and their distinction from MLOs which are also present in the phloem. They are (1) the high correlation between symptom expression and the presence of RLB in plants, and (2) the remission of symptoms accompanied by disappearance of the RLB after treatment with penicillin in contrast to penicillin insensitivity of MLOs. The bacterial cells are found primarily in mature sieve elements, irregularly distributed among the vascular bundles. They are mostly rigid rods, non-motile, and non-pleomorphic. These features distinguish them from *Mycoplasma* and *Spiroplasma* which are also phloem restricted bacteria. The cells measure 0.2–0.5 × 1.0–2.0 (av. 0.3 × 1. 3) μm and are bound by a double membrane, or a cell wall and cytoplasmic membrane. Both membranes are triple layered and are separated by an electron lucent zone. Peptidoglycan or R-layer is not seen.

The symptoms produced by phloem-limited RLBs are characteristic of yellows type which include stunting, yellowing of young leaves, virescence of floral parts, premature flowering and fruit drop, witches' broom, and, often, premature death of the plant. Transmission is by leafhoppers, dodder, and by grafting. In citrus greening disease the agent is transmitted by citrus psylla also. The leafhopper vectors retain infectivity throughout their life and also pass on the inoculum to progenies (transovarial transmission).

Habitats of fastidious prokaryotes

The fastidious eubacteria and the Mollicutes inhabit the vascular tissues, no other tissue of the plants. The mycoplasmas are restricted to phloem while most fastidious eubacteria are restricted to xylem and some to phloem. A common habitat of both groups is the insect body fluid (haemolymph). Most of these xylem or phloem inhabiting prokaryotes are dependent on insect vectors for transmission. No insect host or vector is known for Clavibacter xyli. The special environment of phloem and xylem vessels and insect haemolymph is the key to the fastidiousness of these organisms in regard to nutrition and refusal to grow on ordinary bacteriological media. The fluid in the vessels and insect haemolymph contain complete metabolic systems (substrates and enzymes) of important pathways like glycolysis and tricarboxylic acid cycle.

SURVIVAL AND DISPERSAL OF PLANT PATHOGENIC BACTERIA

In absence of the crop host, the plant pathogenic bacteria survive through seed, perennial plants, insects, plant residue and other non-host materials and as epiphytes. Seed and vegetative propagation materials are an important source of survival as well as dispersal of bacterial pathogens. The bacteria may be present in these plants parts as cells in or on the seed as well as on the plant debris mixed with the seed. During survival on seed, transfer of bacteria from seed to seed may also occur. Most species of *Xanthomonas* and *Clavibacter* are seed-borne having only a limited soil phase, that too only in association with infected crop debris. The phloem and xylem limited fastidious bacteria are located in deeper tissues and are exclusively transmitted by the propagation material such as grafts and sugarcane cutting. Their transmission may also occur during propagation through cutting tools. The length of survival of bacteria in seed may be very long. Access of bacteria to the vascular elements in the seed promotes their survival and subsequent pathogenesis.

Bacteria may persist for years in debris associated. with seed and kept in dry environment. Generally, the diseased crop debris supports the survival for only sometime during its weathering. Plant pathogenic bacteria are weak competitors and have high susceptibility to the activities of microorganisms associated with decomposition of the crop debris. Therefore there is decreased survival in soil. Although species of *Clavibacter, Curtobacterium* and *Rhodococcus* are not considered soil-borne pathogens, some species apparently survive a year or more with plant debris, dry soils permitting longer survival. Species of *Xanthomonas* have the poorest ability to survive outside the host. In crop debris in soil they show a rapid declining phase depending on the rate of decomposition of the debris. In arid regions, where the decomposition of debris is slow, *X. axonopodis* pv. *malvacearum* (cotton blight) can survive for long in cotton plant residue but in moist regions the survival may not be even up to the next crop season. The rice leaf blight bacterium (*X. oryzae* pv. *oryzae*) can survive for some months on rice straw in dry but not in moist conditions. The species of *Pseudomonas* behave in the same manner in their survival with plant debris. There are some bacteria that behave like true soil inhabitants. These include *Ralstonia solanacearum* (potato wilt) and *Agrobacterium tumefaciens* (crown gall bacterium).

Many phytopathogenic bacteria (including the fastidious ones) are known to survive in certain insects which help in their transfer also. The potato black leg pathogen (*Erwinia carotovora* subsp. *carotovora*) can live in all stages of the seed corn maggot (*Hylemya platura*) and persist in the intestinal tract of both adult flies and larvae. It survives pupation of the insect and therefore the eggs laid by the adults are contaminated and carry on the bacterial transmission. Fruit flies (*Drosophila melanogaster*) and beetles (*Diabrotica* sp. and *Chaetocnema* sp.) are also known to harbor plant pathogenic bacteria.

Dispersal of bacteria occurs through movement of soil, use of infected or contaminated vegetative propagating material and through seed and insect vectors. In addition, short range dispersal within the field occurs by raindrop splashes, dew, fog, sprinkler irrigation, air-borne debris, and aerosols. The bacterial slime that covers the cells plays a major role in protecting the cells during transmission. The helical mollicute, *Spiroplasma citri*, causing citrus stubborn disease, is transmitted by *Circulifer tenellus*. All types of salivary glands as well as the adjacent muscle cells are colonized by the bacterium. A surface protein on the spiroplasma is involved in adhesion to cell walls of the vector (Kwon *et al.,* 1999; Yu *et al.,* 2000). Similar adhesion of fastidious xylem-restricted bacteria in their vectors is reported. All xylem-limited fastidious bacteria, except *Leifsonia (Clavibacter) xyli*, are exclusively transmitted by leafhoppers in addition to propagation material. The phloem-limited bacterium of citrus greening is also insect-transmitted.

INFECTION OF PLANTS BY BACTERIA

Adhesion of bacterial cells to the plant surface is key to the subsequent events in the infection process. Adhesion is facilitated by the extracellular slime or layer of polysaccharides and interaction of surface proteins. Bacterial cells have the ability to form biofilm on the contact point. Plant pathogenic bacteria are incapable of direct penetration of cuticularized wall or layers of cork cells of the host. The main points of entry are noncuticularized areas such as root hair, stigma and natural openings like stomata, hydathodes, lenticels. In addition, the majority of infections by bacteria occurs through wounds or punctures caused by cultural operations or by insects and nematodes. Mechanisms involved in the attachment of pathogenic bacteria to plant surfaces are discussed by Romantschuk (1992).

Surface structures on bacterial cells like fimbriae and pili are involved. Pathogenic bacteria that enter tissues of a non-susceptible host plant are normally immobilized by the plant hypersensitive response, which is obviously deleterious for the bacterium.

Mechanism of Action of Bacterial Pathogens

Bacterial pathogens, on entry into the host, target the defence mechanisms of the plant such as programmed cell death (hypersensitive reaction, HR), cell wall alterations and hormonal signaling system.

Rot causing enzymes: The soft rot bacteria damage tissues by virtue of their ability to decompose the pectic substances of the plant cell wall. Various enzymes are extracellularly produced by these bacteria after they have entered the host and different components of the cell wall are progressively dissolved by them before the bacterium moves into the cell cavity. These pectic enzymes (pectinesterases and polygalacturonases) are very important in soft rot diseases caused by *Erwinia* spp.

Hypertrophy inducing factor: The crown gall bacterium (*Agrobacterium tumefaciens*) induces galls or tumours in two steps. The first is the conditioning phase in which a fresh wound is required without the presence of the bacterium. Then the bacterium, enters the host and the second phase starts. The tumor inducing factor or gene is located in a plasmid in the bacterial cell. The bacterium transfers whole or part of the plasmid into the host cell which have been conditioned by the wound to receive it. The extrachromosomal DNA gets integrated with the host DNA. Thus the host cell becomes a tumor cell. There is no further role of the bacterium. The tumor cells multiply and excessive cell division leads to the formation of the gall. The process cannot be checked by killing the bacterium.

Vascular wilt: In vascular wilt diseases caused by eubacteria occlusion of vessels by bacterial cells and their polysaccharides causes hindrance in the water transport and wilt symptoms develop. In addition, the enzymatic breakdown products of the cell walls are carried in the transpiration stream, collect at the vessel ends, form gels and gums that clog the vessel pores. Phenoloxidases secreted by bacteria or released by disrupted cells cause oxidation of phenolics to quinones which then polymerize to form melanins. This imparts the brown colour to the cell walls.

Toxins: Several bacterial diseases are direct effect of toxins either produced by the bacterium or by bacterium-host interaction. Examples are fire blight of pear caused by *Erwinia amylovora* producing amylovorin, leaf spot and blight of fruit trees caused by *Pseudomonas syringae* pv. *syringae* producing syringomycin, wild fire of tobacco caused by *P. syringae* pv. *tabaci* producing tabtoxin or wild fire toxin, halo blight of bean caused by *Pseudomonas savastonoi* pv. *phaseolicola* producing phaseotoxin. Several wilt inducing compounds have been detected in some bacterial diseases. Certain polysaccharides produced by wilt causing bacteria possess wilt-inducing properties. The pectinolytic enzymes also can induce wilting. The role of polysaccharides and enzymes is involved in the effects of *Ralstonia solanacearum* (potato wilt) and *Pantoea stewartii* subsp. *stewartii* (maize wilt).

SYMPTOMS AND TYPES OF BACTERIAL DISEASES

The symptoms produced by phloem-limited fastidious bacteria are characteristic of yellows type which include stunting, yellowing of young leaves, virescence of floral parts, premature flowering and fruit drop, witches' broom, and, often, premature death of the plant. The xylem-limited bacteria produce symptoms of stunting, yellowing or whitening of leaves, blight, etc. *Phytoplasma* and *Spiroplasma* differ in the nature of symptoms they produce on the host. Virescence (reversion of floral organs to vegetative organs, i.e. greening of petals and conversion of petals to leafy structures-phyllody) is found in diseases caused by *Phytoplasma* but not *Spiroplasma*.

Following symptoms are commonly caused by plant pathogenic bacteria:

Blight: The invasion by the bacteria leads to very rapid and extensive necrosis of the affected plant parts resulting in scorched appearance of the surface.

Soft rot: The major effect is softening of the tissues due to dissolution of the middle lamella by enzymes and disintegration of tissues. Very often a dirty liquid oozes out of the affected parts. In many cases, the disintegration of tissue is preceded by change of colour.

Leaf spot: When a specific bacterium invades leaves through stomata the necrosis of tissue around the substomatal space results in the appearance of necrotic areas on the lamina surface. The dead tissues appear water-soaked and the spots generally remain restricted in growth.

Tumors and galls: In some bacterial diseases the effect of invasion is hyperplasia and hypertrophy of invaded tissue. As a result, tumours develop on the affected parts. Crown gall caused by *Agrobacterium tumefaciens* is the best known and extensively studied disease in this group.

Canker: Cankers or warty outgrowths are formed on leaves, twigs, and fruits. They result from necrosis of the tissue and reaction of the undamaged tissues to produce cork cells. Such diseases are mostly localized.

Vascular diseases: In some of the leaf spot or blight diseases the bacteria move into the vascular system of the leaf and become systemic. In others, the invasion from seed or underground parts is concentrated in the vascular tissues causing typical wilt of the plant by plugging of the vessels and also by producing toxins. Common examples of such diseases are bacterial wilt of potato, tomato, eggplant, cucurbits, beans and maize. Such vascular bacteria are different from the xylem or phloem limited bacteria. In the latter the pathogen is exclusively confined to the vessels, no other tissue, while in the former it may be concentrated in the vessels but is present in other tissues also.

REFERENCES

Buchanan, R.F. and N.E. Gibbons. 1974. *Bergey's Manual of Determinative Bacteriology,* 8th Ed, Williams and Wilkins, Baltimore

Carlson, R.A. and A.K. Vidaver. 1982. Taxonomy of *Corynebacterium* plant pathogens including a new pathogen of wheat based on polyacrylamide gel electrophoresis of cellular proteins. *Int. J. Syst. Bacteriol.* 32: 315

Davis, M.J., R.F. Whitcomb and A.G. Gillaspie Jr. 1981. Fastidious bacteria of plant vascular tissue and invertebrates (including so-called rickettsia-like bacteria), pp. 2171–2188. In: *The Prokaryotes*, Vol. II. M.P. Starr *et al.*, (eds). Springer-Verlag, Berlin

Davis, M.J., A.G. Gillaspie Jr., A.K. Vidaver and R.W. Harris. 1984. *Clavibacter*, a new genus containing some phytopathogenic coryneform bacteria including *Clavibacter xyli* subsp. *xyli* and *Clavibacter xyli* subsp.

cynodontis, pathogens that cause ratoon stunting disease of sugarcane and Bermuda grass. *Int. J. Syst. Bacteriol.* 34: 107

Davis, R.E. and J.F. Worle. 1973. *Spiroplasma*: motile, helical microorganisms associated with corn stunt disease. *Phytopathology* 63: 403

Dye, D.W., J.F. Bradbury, M. Goto, A.C. Hayward, R.A. Lelliott and M.N. Schaad. 1980. International standards for naming pathovars of phytopathogenic bacteria and a list of pathovar names and pathovar strains. *Rev. Plant Pathol.* 59: 153

Gardan, L.C. Bollet, M. Abu Ghorrah, F. Grimmont and P.A.D. Grimmont. 1992. DNA relatedness among pathovar strains of *Pseudomonas syringae* subsp. *savastonoi* and proposal of *Pseudomonas savastonoi* sp. nov. *Int. J. Syst. Bacteriol.* 42: 606

Gundersen, D.E., I.M. Lee, S.A. Rehner, R.E. Davis and D.T. Kingsbury. 1994. Phylogeny of mycoplasma-like organisms (Phytoplasmas): a basis for their classification. *Jour. Bacteriol.* 176(17): 5244

Hopkins, D.L. 1983. Gram-negative xylem-limited bacterial pathogens of plants. *Phytopathology* 73: 347

Huang, J.S. 1986. Ultrastructure of bacterial penetration in plants. *Annu. Rev. Phytopathol.* 24: 241

Krieg, N.R. 1988. Bacterial classification. An overview. *Can. J. Microbiol.* 34: 536

Krieg, N.R. and J.G. Holt (eds). 1984. *Bergey's Manual of Systematic Bacteriology.* Vol I. William and Wlkins, Baltimore

Leben, C. 1981. How plant pathogenic bacteria survive? *Plant Dis.* 65: 633

Lee, I-M., D.E. Gundersen, R.E. Davis and M. Bartoszyk. 1998. Revised classification scheme of phytoplasmas based on RFLP analyses of 16S eRNA and ribosomal protein gene sequences. *Intern. J. Syst. Bacteriol* 48: 1153

Lee, I-M., R.E. Davis and D.E. Gundersen-Rindal.2000. Phytoplasma: Phytopathogenic mollicutes. *Annu. Rev. Microbiol.* 54: 221

Macnab, R.M. 2003. How bacteria assemble flagella. *Annu. Rev. Microbiol.* 57: 77

McBride, M.J. 2001. Bacterial gliding motility: Multiple mechanisms for cell movement over surfaces. *Annu. Rev. Microbiol.* 55: 49

McCoy, R.E. 1982. Chronic and insidious diseases. The fastidious vascular pathogens, pp. 475–489. In: *Phytopathogenic Prokaryotes* Vol. II. M.S. Mount and G.S. Lacy (eds). Academic Press, New York

Purcell, A.H. and D.H. Hopkins. 1996. Fastidious xylem-limited bacterial plant pathogens. *Annu. Rev. Phytopathol.* 34: 131

Romantschuk, M. 1992. Attachment of plant pathogenic bacteria to plant surfaces. *Annu. Rev. Phytopathol.* 30: 225

Schaad, N.W., G.H. Lacy, E. Postnikova and E.L. Schuenze. 2006. Taxonomy and phylogeny of phytopathogenic bacteria. *J. Plant Pathol.* 88(3S): S11

Seemuller, E., B. Schbeider, R. Maurer, U. Ahrens, X. Daire *et al.,* 1994. Phylogenetic classification of phytopathogenic mollicutes by sequence analysis of 16S ribosomal DNA. *Intern. J. Syst. Bacteriol.* 44(3): 440

Seemuller, E., C. Marcone, U. Lauet, A. Rogozzino and M. Goschl. 1998. Current status of molecular classification of the phytoplasmas. *J. Plant Pathol.* 80: 3

Vauterin, L., B. Hoste, K. Kersters and J. Swings 1995. Reclassification of *Xanthomonas*. *Int. J. Syst. Bacteriol.* 45: 472

Vauterin, L., J. Rademaker and J. Swings. 2000. Synopsis on the taxonomy of the genus *Xanthomonas*. *Phytopathology* 90: 677

Yabuuchi, E., Y. Kosako, H. Oyaizu, I. Yano, H. Hotta, Y. Hashimoto, T. Ezaki and M. Arakawa. 1992. Proposal of *Burkholderia* gen. nov. and transfer of seven species of the genus *Pseudomonas* homology group II to the new genus, with the type species *Burkholderia cepacia* (Palleroni and Holmes 1981) comb. nov. *Microbiology and Immunology* 36: 1251. *Int. J. Syst. Bacteriol.* 43: 398. 1993

Young. J.M., J.F. Bradbury, R.E. Davis, R.S. Dickey, G.L. Ercolani, A.C. Hayward and A.K. Vidaver. 1991. Nomenclatural revision of plant pathogenic bacteria and list of names 1980-1988. *Rev. Plant Pathol.* 70: 211

Young, J.M., Y. Takikawa, L. Gardan and D.E. Stead. 1992. Changing concepts in the taxonomy of plant pathogenic bacteria. *Annu. Rev. Phytopathol.* 30: 67

Young, J.M., G.S. Saddler, Y. Takikawa, S.H. De Boer, L. Vauterin, L. Gardan, *et al.,* 1996. Names of plant pathogenic bacteria 1864–1995. *Rev. Plant Pathol.* 75: 721

Zgurskaya, H.I., L.I. Evtushenko, V.N. Akino and L.V. Kalakoutskii. 1993. *Rathayibacter* gen. nov., including the species *Rathyaibacter rathayi* comb. nov., *Rathayibacter iranicus* comb. nov. and six strains from annual grasses. *Int. J. Syst. Bacteriol.* 43: 143

CHAPTER

4

Bacterial Diseases of Plants

CITRUS CANKER

Citrus canker or Asiatic citrus canker (also known as true canker or A-form canker) is a widespread disease in citrus growing areas of all the five continents. It is reported to have originated from China but in the herbaria of the Royal Botanic Gardens (Kew, England) canker lesions have been detected in *Citrus medica* specimen collected from India as early as 1827–1833 and in *Citrus aurantifolia* specimen collected from Indonesia in 1842–1844. Thus, the origin of the disease is supposed to be in the tropical areas of Asia such as South China, Indonesia, and India where *Citrus* species are presumed to have originated. The pathogen was distributed through planting material and spread to Europe and to USA (in 1910) and to other citrus growing areas of the world. Citrus canker is a major disease of citrus in India, China, Japan and Java. The disease is economically important because the fruit lesions downgrade the appearance of fruits and when severe, cause premature fruit drop. Heavy foliage infection causes severe defoliation, leaving only bare twigs. Severe infection of newly planted stock may cause delay in growth and can also be fatal.

Although the disease was once reported to have been eradicated from Australia, New Zealand, South Africa and the United States, its reappearance was reported during the 1980s in some parts of Australia, Mexico and Florida. In the state of Florida in USA. where it was first recognized as a new disease in 1913, it became so severe that mass eradication of diseased trees and nursery stock, often the entire orchard, had to be undertaken to eradicate the pathogen from the state in about 10 years. It was claimed that citrus canker had been completely eliminated from USA by 1933. However, it reappeared in Florida in 1986 and the same eradication measures had to be started. The disease was declared eradicated by 1994. It re-emerged in the same area in 1997 (Graham *et al.*, 2004). The form of citrus canker which reappeared in Florida and Mexico seemed to be different from that identified in Asia (the A-form canker).

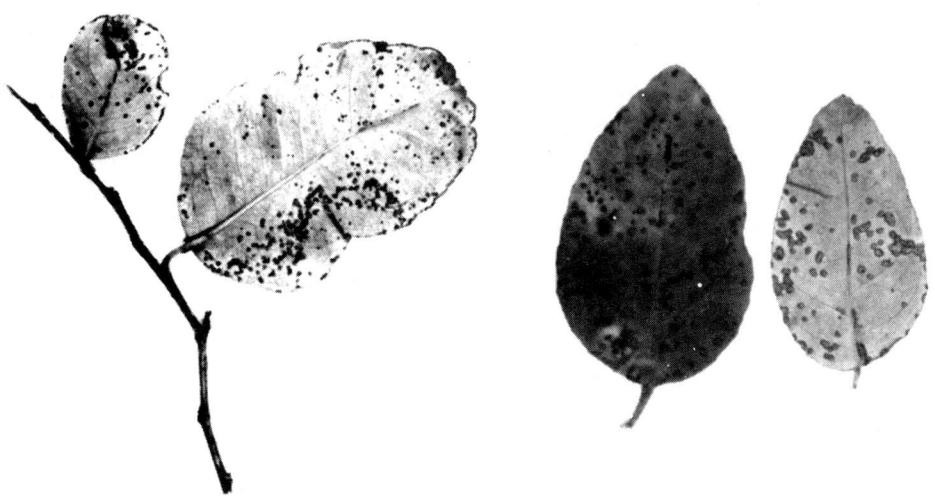

Fig. 4. Cankers on citrus leaves.

Symptoms

The disease occurs on leaves, twigs, thorns, older branches and fruits as necrotic brown spots with rough surface. Canker incidence on exposed roots was known as early as 1936. In the tropical climate of south India, canker lesions were found on roots of seedling plants of 5-year old Kagzi lime (*C. aurantifolia*) up to a depth of 70 cm and in a 20 cm radius. This root infection caused decline of the plants.

Leaf lesions first appear as small, round, watery, translucent spots. They are raised and become yellowish brown. They first develop on the lower surface of the leaf and then on both the surfaces. As the disease advances the surface of the spots becomes white or grayish and finally ruptures in the center giving a rough, corky, or canker-like appearance. The spots increase in size (1 mm to 1 cm in dia) and may coalesce to form elongated lesions on fruits and twigs. The rough lesions are surrounded by a yellowish-brown to green raised margin and watery yellow halo. Spots occurring on petioles and midrib cause premature defoliation. On larger branches the cankers are irregular, more rough and more prominent. Cankers on fruits are similar to those on leaves except that the yellow halo is not visible and a crater-like depression in the center is more prominent. The injury to fruits is only skin deep and no visible effect on pulp or juice is noticed. Cankers on twigs cause them to break. The leaves during their early stage of formation and fruits of about 2–4 cm diameter are most susceptible to infection by the bacterium. As the fruits increase in size they become resistant but water soaking and lesion formation continues to occur as long as the fruit is expanding.

Histopathology: Citrus canker lesions are characterized by over-development of parenchymatous tissues, each consisting of a large number of hypertrophic cells and a limited number of hyperplastic cells. In the early stages of invasion by the bacterium, the spongy cells near the site of infection show increased size as well as increase in the amount of cytoplasm, followed by rapid enlargement. The hypertrophic cells occupy the intercellular spaces. As the cells further increase in size the callus tissue expands, lifting the epidermis, exposing the internal callus tissue. Hyperplasia usually occurs in a few cells adjacent to the healthy tissue. The hyperplastic cells develop into hypertrophic cells without continuous cell division.

The causal organism

The bacterium causing citrus canker was known as *Xanthomonas campestris* pv. *citri* (Hasse) Dye [*X. citri* (Hasse) Dowson]. Although Gabriel, *et al.,* (1989) proposed reinstatement of the name *Xanthomonas citr,.* Vauterin, *et al.,* (1995) finally gave the name *Xanthomonas axonopodis* pv. *citri* (Hasse) Vauterin, *et al.,* causing A form citrus canker. In their classification *X. axonopodis* pv. *aurantifolia* causes B,C,D forms of citrus canker and *X. axonopodis* pv. *citrumelo* causes the E type canker. However, according to Schaad *et al.,* (2006) there is less than 41% DNA/DNA relatedness between citrus xanthomonads and *X. axonopodis* and the former are more closely related to other *Xanthomonas* species such as *X. campestris* pv. *phaseoli* var. *fuscans* and *X. campestris* pv. *alfalfae*. Hence, he has suggested the names *X. citri* subsp. *citri*, *X. fuscans* subsp. *aurantifolii* and *X. alfalfae* subsp. *citrumelonis*.

The cells of the bacterium are rod-shaped and measure 0.5–0.75 × 1.5–2.0 μm. It forms chains and capsules but no spores and the cells are motile by a single polar flagellum (monotrichous). The bacterium is Gram-negative and aerobic. Colonies on beef agar are circular, straw yellow to amber yellow, slightly raised and glistening. The yellow pigment is the characteristic xanthomonadin. Oxidase test is weak or negative. Litmus milk turns blue and milk is peptonized without coagulation. Asparagine is not utilized as a sole source of nitrogen and carbon. The cells are positive for hydrolysis of starch, aesculin, casein, liquefaction of gelatin and pectate gel, and production of tyrosinase and reducing substances from sucrose and hydrogen sulphide. The bacterium is negative for arginine dehydrolase, nitrate reduction, production of 2-ketoglucanate, acetoin, urease, and amino acid dehydrolases and for methyl red test. Growth requires methionine or cysteine and is inhibited by serine. Growth is inhibited by 0.02% triphenyl tetrazolium chloride and by 4% sodium chloride but not by 3% sodium chloride. Xylose, glucose, fructose, sucrose, galactose, mannose, maltose, lactose, trehalose, glycerol, dextrin, starch, malonate citrate, succinate, and malate are utilized as sole source of carbon. L-arabinose, rhamnose, raffinose, malicin, sorbitol, inositol, dulcitol, inulin, glucanate, oxalate, acetate and tartrate are not utilized. Optimal growth temperature is 28°C, minimal 6°–7°C, and maximal 36°–38°C. The doubling time is 79 min.

X. axonopodis pv. *citri* affects Rutaceous plants, primarily *Citrus*, *Fortunella* and *Poncirus*. Important cultivated hosts are lemon, lime, orange, grapefruit and pomelo. Pathotypes within *X. axonopodis* pv. *citri* were identified by host range, geographic origin, bacteriophage sensitivities, plasmid content, and serology. Pathotype A (the A-form canker or Asian citrus canker) has the widest host range and global distribution. Pathotypes B, C, and D were restricted to lemon (*Citrus lemon*) and lime (*Citrus aurantifolia*) in South America and Mexico. The strain B is no more found in nature having been gradually replaced by the A-form which appeared in South America in 1972. Three phages of the bacterium have been isolated and characterized. The phage susceptibility can be used for rapid identification of the citrus canker pathogen. Strains of citrus canker bacterium carry indigenous plasmids. The function of these plasmids is not yet clear. However, plasmids differ in size among different strains (A, B, C). Differentiation based on plasmids is consistent with that observed on the basis of serology and of phage susceptibility.

Disease cycle and environmental relations

Studies have shown that the bacteria of citrus canker have a short life in soil or in fallen leaves. The short longevity in natural soil is attributed to microbial interactions, especially the predatory effect of protozoa. This is generally true when the temperature is warm enough to allow the soil microorganisms to compete with the bacterium. When the source of inoculum is removed (removal of affected leaves and twigs) the survival in soil is considerably reduced. In presence of active host tissue for support such as roots it can survive deep in soil. In tissues of fallen diseased leaves and twigs also the bacterium dies quickly. The survival is not for more than 3 weeks if the lesion-bearing leaves and twigs are wetted on the soil surface or are buried at a depth of 3–6 cm. If the plant debris is maintained under dry conditions the survival is increased to 2–3 months. Bacterial cells in leaf lesions remain viable for as long as the neighboring cells remain viable. Bacterial cells in stem lesions or cankers can remain viable in cankers on 5–7 years old branches. Outside the host, the inoculum associated with exposed symptomless leaves cam survive and remain infective for several months. The bacteria can persist for several weeks on non-host plant material, for longer durations in root zone of grasses under eradicated diseased trees.

The *X. axonopodis* pv. *citri* cells attach to the leaf surface with an adhesion mechanism. They form a complex, structured biofilm (a matrix) which ensures adhesion, infection and ectophytic survival. The extracellular polysaccharides (EPS) are involved. Mutants of the bacterium lacking EPS fail to cause infection and survive. If the EPS dries and is not disturbed the survival of the bacterial cell can be considerably prolonged because EPS forms a protective coat that prevents desiccation of the cells. When the EPS matrix is diluted with water to the level of lower than a million cells/ml the cells are rapidly killed. This *lethal dilution effect* is found in most xanthomonads and some other bacteria. It also implies that during dispersal by raindrop splashes the bacterial cells must carry with them sufficient EPS to remain viable.

Although the bacterium has been detected on certain weeds, this source does not support prolonged survival. Epiphytic populations on citrus leaves away from lesions also quickly decline. However, since the host is a perennial plant the cankers on it can support parasitic survival of the pathogen indefinitely once the plant gets infected. In temperate climate countries where the citrus trees undergo dormancy during winter, survival in holdover cankers on the trees is most important. In the subtropical and tropical countries such as in India attacked twigs bearing old lesions on the standing trees are the main source of perennation of the pathogen.

When water is placed on lesions on detached leaves 10 to 100 thousand bacterial cells/ ml from each lesion are released immediately. The release continues at a high level for 24 hours. Cumulative release is 100 thousand to 1000 thousand cell per lesion (Timmer *et al.,* 1991). Fewer bacterial cells are released and exuded more slowly from old lesions than from young lesions. The concentration of bacterial cells is largely dependent on the age of the lesions. In fresh lesions, bacterial density often reaches 10–100 million/drop. Individual canker lesions ranging 3–9 mm in diameter contain 1–10 million bacterial cells per lesion and lesions formed in spring flush continue to contain this level through summer and rains. In winter the numbers decline.

In short or medium distance dispersal of the bacteria by rain, either the water running on the host surface or splashes of raindrops disperse the bacterial cells. Dispersal by wind-blown rains can be up to a distance of 12 meters although maximum dispersal is within 1 meter. In citrus nurseries with citrus canker, dissemination of the bacteria is primarily by splash dispersal. Rains driven by wind velocity in excess of 8 m/sec aid in effective dispersal. The bacterial numbers per ml of canker washings and their dispersal by rain-driven water splashes are, to some extent, effective as source of inoculum only if sufficient quantity of EPS accompanies the bacterial cells otherwise they may be desiccated before reaching the site for infection. Infection through stomata is aided by wind speed of 17–18 miles per hour. Rains can cause water congestion of tissues. A column of water between the plant surface and the mesophyl is formed through the stomata. Bacteria infect through this column.

The role of strong winds has great importance in the epidemics of citrus canker. Strong winds cause many injuries to leaves and twigs. The nature of wounds varies from easily visible large wounds to small, invisible ones, such as small scratches or removal of the cuticle edges extending over the stomata. However, a single cell of the bacterium is enough to cause infection and the size of wounds does not matter. Storms help in long distance dispersal of raindrop-borne bacterial cells.

Prevailing temperature and duration of leaf wetness influence the extent of infection of leaves by the bacteria. For 100% infection temperature of 25– 35°C and leaf wetness duration of 4 to 24 hours is required. Maximum infection occurs when leaf wetness duration is 24 hours. Minimum temperature for infection is 12°C and maximum 40°C. (Pria *et al.,* 2006).

In addition to driving rains, insects such as the citrus leaf miner (*Phyllocnistis citrella* and *Thosconyrsa citri*) also help in dissemination of the bacterium. Venkataswarlu and Ramapandu (1992) had observed that the percentage of leaves affected by canker in presence of injury by leaf miners was 26 to 48 while in absence of the leaf miner injury it was 3 to 10. Mechanical injuries or injury caused by leaf miner causes reduction in the amount of inoculum required for symptom development. Maximum injury and consequent increased infection is caused by third instar and pupal stages of the leaf miner. The ability of *P. citrella* to transmit *X. axonopodis* pv. *citri* is limited by the rate at which it can acquire inoculum from infected plants. The adults are not efficient vectors and disease is not spread by adults (Belasque *et al.,* 2005). Man himself is the chief agent of dissemination and introduction of the pathogen into new localities through transfer of infected nursery stock.

The result of numerous cells reaching various sites on host surface is the development of numerous secondary foci which later coalesce to form larger patches. The bacterium enters the host through natural openings (stomata) and through wounds such as those caused by insects, movement of thorns, etc. Susceptibility of leaves to infection through wounds is maximum when the ratio of leaf length at a given time to the length of the fully expanded leaf is 0.8: 0.9 or more. Wounds sustained early in spring or late autumn take longer to heal and, therefore, expose the injured tissue to infection for a longer period of time. The greater the number and size of stomata per unit area, the greater the susceptibility of the organ. But the stomatal invasion by the bacterium is governed by developmental stage of the organ. In young organs such as leaves, stems, and fruits, the front cavity of the stomata has a wide opening because the thin cuticular layer of the epidermis is not enough to elongate the edges. As organs approach maturity and the tissues become harder. the cuticular layer of the epidermis becomes thicker so that the edges develop over the stoma, leaving a narrow opening between them. The slit is so narrow that surface tension prevents entry of rainwater carrying the pathogen into the opening of the mature stoma. Thus, availability of young stoma determines the susceptibility of leaves, stems and fruits. In very young leaves, just after emergence, the stoma are immature with no opening and, therefore, only slight infection occurs.

After entry, the bacteria multiply rapidly in the intercellular spaces, dissolve the middle lamella, and establish in the cortical region. The bacterial cells which enter the intercellular spaces adhere to the host cell walls through interaction between EPS and citrus agglutinins. The citrus agglutinins contain 96% proteins and 4% carbohydrates. They are active with EPS of various xanthomonads at pH lower than 6.0. The EPS induces localized water congestion enhancing the growth of the bacterium in the intercellular spaces.

In association with EPS, the bacteria show ethylene biosynthesis for several hours after inoculation. Continuance of ethylene production is followed by leakage of electrolytes and amino acids from the cells indicating damage to cell membrane. A large amount of ethylene is also produced after canker symptoms develop. At this stage ethylene production originates in the hypertrophic host cells within the canker lesions as well as the cells in the peripheral zones, which appear to be under the influence of auxin and sometimes form yellow halo. The high level of ethylene produced at this stage induces the formation of abscission layer at the base of the leaf petiole, resulting in defoliation. Presence of antimicrobial compounds (phytoalexins) in citrus cells has been reported. However, these compounds are present inside the cells and are not leaked into the intercellular spaces to act against the bacterium. Citrusnin-A is one such compound.

Citrus canker is favored by mild temperatures and wet weather. Temperatures between 20° and 30°C with good evenly distributed rains are most suitable. Presence of free moisture on the host surface for at least 20 min is essential for successful infection. The size, density, and age of stomata determine susceptibility and resistance of a citrus cultivar. Under unbalanced conditions of excess nitrogen, citrus trees produce more shoots allowing an increased number of large and tender leaves which bear larger lesions and cankers.

Management of the disease

The only effective method of control of citrus canker is complete destruction of the affected trees by burning. Though drastic and costly, this method has provided its efficacy in USA, Australia, South Africa, New Zealand, and Brazil. The new eradication programme followed in Florida after reappearance of canker involves 1) burning of plants in nursery where an infected plant is found, 2) destroying all trees with canker symptoms within orchards and defoliation of surrounding trees, and 3) using fruits from diseased or exposed trees for only processing. Similar rigid eradication programmes had been implemented in Australia where the first and subsequent outbreaks of the disease had been eliminated by this method. In spite of total destruction of infected and suspected trees and apparent elimination of citrus canker from the country, the reappearance or reintroduction of the pathogen in the same country has pointed out the importance and possible failure of quarantine regulations. Two documented examples are of Florida (USA) and Australia. The introduction of Asiatic citrus canker in to Florida during the 1910s was traced to infected trifoliate orange seedlings imported from Japan for use as rootstock. The reappearance in around 1984 in Florida was again traced to entry of canker affected citrus. Detection of infected material had been made during 1973–1983 at the ports of entry. In Australia, the first outbreak of canker in 1912 was attributed to citrus trees and fruits imported from Japan and China. Mass eradication of trees and rigid quarantine had eliminated the disease but subsequent outbreaks in 1981, 1984, and 1991 are suspected to have originated from illegal importation of citrus into isolated home gardens in one part of the country and subsequent spread to other parts. Both these countries had followed the method of mass destruction of trees and imposition of rigid quarantine regulations not only at the international level but also within the country. In India and in other areas where the disease is well established in most orchards, eradication of trees as a control measure is not feasible.

Since wind driven rains and water soaking of tissues are essential for dissemination and ingress of bacteria and for epidemic development of citrus canker, wind breaks are essential. In absence of this precaution chemical and other methods of disease management remain inadequate. Other recommendations to check the disease are (i) use of disease free nursery stock for planting in new orchards, (ii) spraying the plants before planting in new orchards with a copper fungicide, and (iii) in old orchards pruning of the affected twigs and spraying with copper fungicides at periodical interval, especially during the rainy season. The fallen canker affected twigs and leaves should be collected and burnt. Since inoculum present on fallen leaves is reduced when the leaves are buried deep in soil, periodical ploughing of the orchard floor is helpful. The plant vigor should always be maintained by suitable fertilizers and irrigation. Proper care should be taken to check the attack of leaf miners.

Since the leaf miner, *Phyllocnistis citrella,* is a vector of the bacterium, its control is also warranted. There is natural mortality in this vector by parasites and predators. Latter are more effective. Increasing rates of nitrogen fertilization through urea results in increasing number of live larvae, pupae and parasitized larvae. The number of live larvae and parasitized larvae are greatest at 0.85 kg urea/tree and pupae reach the peak at 1 kg urea/tree. Since natural destruction of the vector is through parasitization of larvae and pupal stage is not affected, application of more than 0.85 kg urea/tree should be avoided. Many parasitoids are known as natural enemies of leaf miner. Biological control of leaf miner by *Semielacher petiolatus* (Hymenoptera: Eulophidae) is reported (Lim and Hoy, 2005).

Streptomycin sulphate or crude agricultural preparations of streptomycin at 100 to 1000 ppm concentration sprayed at 15-days interval had effectively checked the disease on 48-years old lime trees in India while Bordeaux mixture was ineffective. Phytomycin (2500 ppm) was also effective. Four sprays of streptomycin 100 at 500 ppm are most effective in controlling the disease. In general, use of Bordeaux mixture (4 g copper sulphate with 4 g lime in 1 lit water) and Agrimycin or streptomycin is recommended in most countries. In China spraying citrus trees with copper ammonium during the summer and autumn is reported to reduce canker incidence by 86% and 90%, respectively. This control method is environment friendly, easy and cheap. Chemical dip of fruits for shipment was recommended to prevent spread of the disease through infected or contaminated fruits. The treatments include 2-min dip in chlorine (200 μg/ml at pH 7) or 1-min dip in 2.0 % sodium-*o*-phenylphenate. Graham and Leite (2004) tried commercial formulations of defense activators acibenzolar-S-methyl and harpin proteins for control of citrus canker and citrus bacterial spot but found no significant control of the disease incidence Copper oxychloride and copper hydroxide were highly effective.

Although chemical control is claimed effective it is not very successful on all occasions. During rainy conditions, some bacterial cells may achieve direct access to the front cavity of stomata or wounds without being exposed to the chemical left on the leaf surface. This direct ingress of even very low number of bacterial cells may make the chemical sprays less effective. An effective bactericide against the citrus canker bacterium should not only be effective on the host surface but also reach into the substomatal cavity.

In biological control of citrus canker, a strain of *Pseudomonas syringae* is reported to show antagonism to the citrus canker bacterium and also prevent enlargement of canker lesions as well as subsequent defoliation of infected leaves. The antagonist probably stimulates phytoalexin (citrusnins) synthesis in the tissues. *Pseudomonas fluorescens* also is a strong antagonist of *X. axonopodis* pv. *citri.* Kalita *et al.,* (1996) isolated *Bacillus subtilis* and *Aspergillus terreus* from phylloplane of citrus and reported that a strain of *B. subtilis* when sprayed on leaves in high concentration reduced canker incidence by 61.9%. A strain of *A. terreus* also reduced disease incidence by 47.5%. *Pantoea agglomerans* (*Erwinia herbicola*), a common phylloplane microflora, grows more rapidly than the canker bacterium both *in vitro* and *in vivo* and eventually causes quick decline of the pathogen population. However, this bacterium grows only in the area where hypertrophic cells are established, but never in the front boundaries at which the pathogen attacks healthy tissues inducing development of hypertrophic cells. In the state of Andhra Pradesh in India, S. Vaheeduddin had reported in 1959 that spray of neem (margosa) seed cake at the rate of 80 kg/acre is highly effective against citrus canker as well as leaf miner. About 25 kg of the cake is soaked in 100 lit water and allowed to decompose for a week. It is then sprayed without filtration. Some of the cake falls on the ground and becomes manure. Several sprays are required to produce good results. In experiments, two sprays at 3-weeks interval during August-September reduced the disease from 5.8% in unsprayed plots to 2.5% in sprayed plots. Bordeaux mixture was not so effective. This control of canker was probably through enhanced microbial activity and biological control on the leaf surface.

Balogh *et al.,* (2008) have reported control of citrus canker (*X. axonopodis* pv. *citri*) and bacterial spot (*X. a. citrumelo*) with bacteriophage treatment. In a set of greenhouse experiments the phage treatment provided consistent control of canker (50% reduction in disease severity. However the phage treatment was in effective if applied with skimmed milk, a protective formulation that increases phage residua; activity/different species of *Citrus* show different degrees of susceptibility to the disease. These differences result from pathogen strain-host species interaction, ability of the host tissue to release phytoalexins in the intercellular spaces, behaviour of the stomata, level of density and size of stomata, presence or absence of thorns that cause injury to leaves and young twigs, etc. Reddy (1997) had screened 144 varieties of *Citrus* spp. and related genera and had recognized 13 as immune to the disease. Somatic hybridization between *Citrus sinensis* (sweet orange) and *C. deliciosa* has yielded clones that are tolerant or resistant to citrus canker (*X. axonopodis* pv. *citri*) and citrus variegated chlorosis (*Xylella fastidiosa*).

BACTERIAL BLIGHT, ANGULAR LEAF SPOT OR BLACK ARM OF COTTON

Angular leaf spot or black arm of cotton is the most serious bacterial disease of this crop. The disease was first reported from Alabama state of USA in 1891 and is now found in all the major cotton growing regions of the world including the USA, South America, Egypt, Sudan and other African countries, Russia, Sri Lanka, China, Australia, etc. In India the disease was first observed in Tamil Nadu in 1918 and is now known to occur in Maharashtra, Madhya Pradesh, Andhra Pradesh and Uttar Pradesh. Annual losses vary from 5 to 25% (cf. Verma, 1995). Severe epidemics of the disease were reported during 1948–1952 in Tamil Nadu which resulted in rejection of many very promising cotton varieties of all the four species of *Gossypium*.

Symptoms

The bacterium attacks all aerial parts of the plant at different stages of plant growth. The disease has four distinct phases depending on the plant part affected: angular leaf spot (leaf infection), black arm (stem infection), boll rot (boll infection), and seedling blight (seedling infection). The common name of the disease is bacterial blight. The earliest symptom of the disease is seen in the cotyledons of germinating seeds. Minute, water-soaked spots appear on the under surface of the cotyledons. Later, these increase in diameter, turn brown to black and form irregular patches distorting the shape of the cotyledons causing them to dry and wither. The disease spreads to new leaves formed and the seedlings may ultimately collapse and die. On the leaves similar water-soaked spots appear on the under surface first and then on the upper surface. They increase in size, become angular, bound by small veinlets of the leaf, and turn brown to black. Often the disease spreads along the edge

Fig. 5. Black arm and blight of cotton.

of veins, hence called vein blight or black vein. Sometimes, large patches are formed due to coalescing of a number of small spots leading to death and shedding of leaves. The infection may also spread to petioles causing them to collapse. In the affected areas large quantities of bacterial slime are exuded which form a dry film on the brown lesions.

Lesions on stem, petioles, and fruiting branches are dark brown to sooty black. They are elongate and sunken. The affected stems show cracks and gummosis and are easily broken by wind or there may be girdling and death of affected organs. These are the black arm symptoms. On the bolls or fruits the disease is characterized by the appearance of water-soaked lesions on the surface. These lesions turn brown and finally black, and are invariably sunken. Young infected bolls fall down prematurely. If they mature, lint is of not much commercial value. The bacterium within the boll passes along the fibers and infects the seed externally. It may also reach the interior of the seed either through micropyle or through punctures.

The causal organism

Xanthomonas axonopodis pv. *malvacearum* (Smith) Vauterin, *et al.,* The rod-shaped cells of the bacterium measure $0.3–0.6 \times 1.3–2.7$ μm. The bacterium is non-endospore forming, encapsulated, Gram-negative, and motile by a single polar flagellum. It is facultative aerobe. Colonies on beef agar are pale yellow, round, thin, raised, smooth, and shining. Gelatin is liquefied and starch and casein are digested. Milk is peptonized and coagulated. Nitrate is not reduced. Optimum temperature for growth in culture varies from 25°–32°C. Maximum temperature for growth is 42°C and minimum 6°C. Thermal death point is 50°C.

At least 32 races of the bacterium are known which vary in their pathogenicity on different species of *Gossypium.* In India 26 races have been identified. which include races 1, 2, 4, 5, 7–10, 13, 14, 16–25, 27–32. Earlier, race 10 was reported to be the most common in India. Race 20 identified in Africa in 1983 was considered most highly virulent infecting even some immune cultivars. Even less virulent races predominate when susceptible cultivars are available. The race 18 was considered the most predominant and destructive not only in India but in other countries also. It has the ability to overcome 5 major genes for resistance to bacterial blight. Race 32 has now become most predominant and it has also overcome at least 5 resistance genes in cotton.

Disease cycle and environmental relations

The bacteria may remain viable in dry leaves for 17 years and in dry or moist soil for 8 days at 21°–33°C. However, infected stalks buried in moist soil cause death of the bacteria. Infected cotton bolls, leaves and twigs present on soil surface form the important source of carry over of the pathogen. The infected seeds lying dormant in the field and germinating in the crop season prior to the main crop also serve as source of primary inoculum. The main source of primary inoculum is seed. The bacterium may be present as slimy mass on the fuzz or inside the seed. The concentration of the bacterial cells in or on the seed determines the seed quality (seed germination, seedling vigor and dry weight of seedlings). High concentrations reduce seed quality and enhance blight incidence in the crop. On germination of such seeds the bacteria move to cotyledons and then maintain a resident population on the first and second leaves but not on the third leaf. In favorable weather the inoculum from this source is spread to new leaves through wind-splashed rain and dew and further spread continues. Leaves are infected mainly through stomata with abundant infections occurring only when stomata are fully open and the intercellular spaces are filled with water. The pathogen produces pectolytic enzymes cellulase and protease. Once inside the leaf, the bacteria spread from the substomatal cavity, breaking down the walls of the spongy mesophyll cells which are completely destroyed by the time visible lesions become dry and dark in color. Infection may reach the vascular system and spread to leaves, bolls and seed.

Susceptibility of cotton leaves to *X. axonopodis* pv. *malvacearum* is characterized by progressive water-soaking of pin-point size without showing production of anthocyanin. In resistant leaves there is little or no water-soaking but obvious amounts of anthocyanin. However, production of anthocyanin is only a resistance response. It is not essential for resistance (Kangatharalingam *et al.,* 2002).

High humidity and moderate temperature (28°C) favour development of the disease. For initial infection free water is necessary. Soil temperature and moisture at the time of sowing and a few days after are important. Primary infection is favoured by a temperature of 30°C and secondary infection is better at 35°C. Presence of moisture is very important for secondary infection during the first 48 hours. Dry and hot weather retards development of the disease. In Haryana incidence of the disease has been found to be significantly high in late sown crop (end of May) with close spacing of 60 × 15 cm.

Thurbaria thespesoides, Eriodendron afructosum, Jatropha curcas and *Lochnera pusilla* are reported as collateral hosts of the bacterium but their role in disease cycle is not yet established. Cotton cultivars resistant to the disease contain more phenol and diphenoloxidase activity is greater.

Management of the disease

Removal and destruction of diseased plant debris is recommended to reduce the soil-borne inoculum. Deep ploughing after harvest buries the infected stalks and thus reduces survival ability of the bacterium in soil. Pre-sowing irrigation to enable the left over seeds germinate, followed by ploughing and then planting of the main crop is practiced in some countries.

Destruction of possible alternate or collateral hosts is also essential. Crop rotation, late sowing, early thinning, good tillage, early irrigation and addition of potash to soil help in reducing the disease incidence. Absence of cotton for one season in the rotation ensures protection from soil-borne inoculum.

Seed-borne inoculum can be eliminated by seed treatment. External inoculum on the seed is destroyed by delinting of seed with concentrated sulphuric acid. Seeds are immersed in acid for 10-15 min, then rinsed thoroughly by suspending in water to remove acid and finally dried and treated with organo-mercurial compounds such as Agrosan GN, Ceresan, etc. However, use of mercurial fungicides is now discouraged.

A formulation of cuprous oxide is now recommended in some countries. This treatment does not destroy the internally seed-borne inoculum Treating seeds with antibiotics like streptomycin eradicates the internally seed-borne inoculum. Carboxin and oxycarboxin (Vitavax and Plantvax) are as effective as antibiotics in eliminating internal and external seed-borne inoculum. Under Indian conditions they are even better than antibiotics (Verma, 1995). They reduce the chances of resistance development in the bacterium. These fungicides applied as seed dressing at the rate of 2 g/kg seed or as spray at the rate of 1.5–2.0 kg/ha, preferably mixed with 0.25% copper fungicides give excellent control of the disease. They have some curative effect also. The bacterium is known to live in the seed for about a year or so and, therefore, ageing of the seed for two years before sowing has also been recommended. Hot water treatment of seed at 56°C for 10 min destroys the external as well as internal inoculum without affecting seed viability.

The secondary spread can be checked by regular sprays of copper fungicides (0.2–0.3%) alone or with systemic fungicides mentioned above. First spray is given when the crop is

5–6 weeks old. In all, 3–6 sprayings, depending on severity of the disease, are given at 15 days interval. Seed treatment with Agrimycin (3 g/40 kg seed) and its spray (25 ppm) have been reported most promising in controlling the black arm of cotton. In Haryana, bacterial blight was effectively controlled by spraying a mixture of Agrimycin (0.01%) and Blitox (0.2%) at 40–50 day, 70–80 day and 85–95 day stage of plant growth.

Extracts of seeds of neem (margosa, *Azadirachta indica*) are reported to possess extreme potential for control of the disease (Verma, 1995). Hulloli *et al.*, (1998) have reported that certain neem based formulations such as Plantolyte and Agricare were much more effective than the best antibiotics such as aminoglycosides, streptomycin and kanamycin. Combination of half dose of the neem-based formulation with antibiotics not only increased efficiency of the treatment but also excluded the possibility of resistance development in the bacterium against the antibiotics. Satya *et al.*, (2006) has reported control of cotton bacterial blight by foliar spray of 10% aqueous extract of leaves of the medicinal plant zimmu (*Allium cepa* x *Allium sativum*). The number of lesions were reduced by 73%. Treated leaves showed high activity of the enzymes phenylalanine ammonia lyase, peroxidase and polyphenol- oxidase along with rapid accumulation of phenolics. Four days after treatment there was 11-fold increase in chitinase activity. Alkaloids in extracts of *Prosopsis julioflora* (Favaceae) are reported to inhibit *X. axonopodis pv malavacearum*. Cotton seed treatment with such materials improves germination and reduces bacterial blight. Seed treatment and foliar spray of Dravya, a commercial extract of the seaweed *Sargassum wightii* induces resistance in cotton against *X. axonopodis* pv. *malvacearum*. Seed soaking in a 1:500 dilution of the product followed by 3 foliar sprays at intervals of 10 days were evaluated. Foliar sprays, following seed treatment, at 10, 20 and 30 days after sowing resulted in 66%, 70% and 74% reduction in disease incidence, respectively (Raghavendra *et al.*, 2007).

Seed inoculation with certain isolates of *Pseudomonas fluorescens*, *P. putida* and *P. alcaligenes* suppressed the bacterium on seedlings. The suppression of the pathogen by these cotton rhizobacteria was attributed to antibacterial secondary metabolites such as HCN and siderophores (Mondal *et al.*, 2000). Bhowmik *et al.*, (2002) isolated some endophytic strains of *Pseudomonas fluorescens* from cotton seedlings and demonstrated that following seed bacterization with the endophytes their population kept on increasing up to 15 days after seedling emergence, which is considered the active phase when infection of cotyledons occurs. This reduced infection of cotyledons. Saha *et al.*, (2001) identified certain strains of *Pseudomonas* in the phylloplane of cotton that could reduce populations of *X. axonopodis* pv. *malvacearum*, when co-inoculated with the pathogen, by multiplying and heavily colonizing the active multiplication sites on cotton leaves. Strains MMP and Pf1 of *Pseudomonas fluorescens* isolated from cotton rhizosphere were used by Salah Eddin *et al.*, (2007) for seed treatment and foliar spray to control the bacterial blight of cotton. Both isolates showed antagonistic activity against the pathogen in vitro. Seed treatment followed by foliar spray of the antagonist reduced severity of blight with decline of disease index from 43.8 to 14.5. The control was attributed to generation of defense related enzymes. Seed, soil and foliar treatment with talc-based formulations of endohytic strains of *Bacillus* spp. and *Pseudomonas fluorescens* are reported to significantly reduce bacterial blight of cotton. Addition of chitin enhances the biocontrol. (Sarvankumar *et al.*, 2006). The suppression is attributed to induction of defence related enzymes.

Control through development of resistant varieties is possible and provides the best and most effective preventive measure. *Gossypium herbaceum* and *G. arboreum* were considered to be practically immune to the disease whereas *G. barbadense* and *G. herbaceum* var. *typicum* were considered susceptible. The full range of disease expression from fully susceptible to highly resistance is found in *G. hirsutum*. Highest degree of resistance is found in *G. hirsutum* var *punctatum*. Highly resistant strains of cotton have been obtained in Tamil Nadu from arboreums and intra-specific hybridization with *G. hirsutum*. Varieties known to be resistant are HC- 9, BJA-592, P-14-T-12, 101-102B and Reba-B-50. Earlier, in 1974, the cultivars/genotypes 70-IH-480/2, 70-IH-480/3, 70-IH-480/9, K-4005, Badnawar-1, B-1007, Khandwa-2, DHY-286, M-937-CTO-421 of *hirsutum* cotton were reported resistant to bacterial blight.

BACTERIAL LEAF BLIGHT OF RICE

The bacterial leaf blight (BLB) of rice had been known in Japan as an endemic disease since 1881. Later, its incidence in severe form was reported from India also. BLB also occurs in China, Taiwan, Korea, Thailand, Vietnam, Philippines, Indonesia, Malaysia, Bangladesh and Australia. The first Indian record, based on isolation and pathogenicity of the bacterium, was from Maharashtra in 1959 where it was reported to be widespread and destructive since 1951. However, the described symptoms conform to the bacterial leaf streak which is distinct from leaf blight both in respect of the causal organism and the symptoms.

The disease was considered to be localized in Maharashtra until 1963 when in Shahabad district of Bihar a severe blight, previously attributed to nutritional disorder, was identified as bacterial blight. Since then the disease has been reported from other parts of north India also. The rapid spread of the disease to the northern parts of the country was attributed to the introduction of highly susceptible variety Taichung Native (TN) 1. In Punjab, Haryana and Western Uttar Pradesh, major epidemics occurred in 1979 and 1980. Severe Kresek was observed and total crop failure was reported.

The disease is a typical vascular wilt, leaf blight being only the mild phase resulting from secondary infections. Damage is due to the partial or total blighting of leaves or due to complete wilting of the affected tillers leading to unfilled grains. In the wilt or Kresek phase the crop may dry up completely before seed maturation. If the attack is late the loss in yield may be negligible. In Japan the disease is responsible for a loss of 30% or more annually. In India the damage has been estimated to vary from 6 to 60%.

Symptoms

Symptoms of BLB vary considerably with the stage of infection and the prevailing weather conditions. Leaf blight phase is most commonly seen. This phase of the disease is characterized by linear yellow to straw coloured stripes with wavy margin, generally on both edges of the leaf, rarely on one edge. These stripes usually start from the tip and extend downwards. This is followed by drying and twisting of the leaf tip and rapid extension of marginal blight lengthwise and crosswise to cover large areas of the leaf. In occasional cases the linear stripes may develop anywhere on the lamina or along the midrib with or without the marginal stripes. The blighting may extend to the leaf sheaths and culms, killing the tiller or the whole clump. In dry weather opaque and turbid drops of bacterial ooze which dry into yellowish beads can be seen on the leaf surface. These drops are washed down by rains. The glumes of seeds also get infected but the symptoms are not well defined. The blight phase of the disease usually appears 4-6 weeks after transplanting. Since leaf drying can be due to physiological disorders or attack of tungro virus also, the diseased leaves should be subjected to microscopic examination for correct diagnosis. If the affected portions of a leaf is cut and mounted in a drop of water on a glass slide bacterial ooze can be seen as a cloudy mass at the cut ends.

The most destructive form of the disease in the tropics is "Kresek" or wilt phase resulting from early systemic infection either in seedlings grown from infected seeds or in seedlings which come in contact with the bacterium in soil. The leaves roll completely, droop, turn yellow or gray and ultimately the tillers wither away. In severe cases the affected stool may be completely killed. The Kresek affected tillers can be confused with stem borer injury but the latter can be easily pulled out while it is not so with Kresek affected tillers. Another common symptom found in the tropics is the pale yellow leaf phase. Some of the youngest leaves in a clump turn pale yellow or white. These leaves later turn yellowish brown and wither away.

The causal organism

Xanthomonas oryzae pv. *oryzae* (Ishiyama) Swings, *et al.,*1990. *Xanthomonas* is classified under the kingdom Bacteria, phylum Proteobacteria, class Gammaoroteobacteria, order Xanthomonodales and family Xanthomonodaceae.

The rod-shaped cells of the bacterium measure $0.5 - 0.8 \times 1.0 - 2.0$ μm. They are single and form no capsule. Stain reaction is Gram-negative. They are motile by a single polar flagellum. Cultural characters are similar to those of other *Xanthomonas* species. Optimum temperature for growth is $25° - 30°C$. Thermal death point is $53°C$. Colonies are slow-growing, mucoid and straw-colored to yellow. The ability of one strain to inhibit growth of others in the rice plant is reported. However, it seems to have no epidemiological importance (Dardick *et al.,* 2003).

Disease cycle and environmental relations

There can be several modes of perpetuation of the bacterium from season to season. It survives during off-season in seed, weed hosts, rice volunteers and infected rice straw and stubble. In Japan, the weed hosts *Leersia oryzoides* and *L. oryzoides* var. *japonica* constitute the chief source of primary inoculum. The pathogen perpetuates in the rhizosphere of these grass hosts. In India also many of such grass hosts have been identified. These include *Cyperus rotundus* (motha grass), *C. defformis, Leptocorsia acuta* and *Leersia hexandra.* Leaves of *Paspalum scrobiculatum, Leersia hexandra, Panicum repens* and *Cyperus rotundus* harbour the bacteria up to 130–140, 120–130, 100–110, and 50–60 days, respectively, without showing blight symptoms. In double cropped areas, such as in south India, volunteer rice seedlings also perpetuate the pathogen under low-lying field conditions.

Many Japanese and Indian workers have reported perennation of the pathogen in infected rice straw left in the field and also in the stubbles. However, this mode of survival in India was doubted. In areas where two or more crops of rice are taken successively in a year this mode of perpetuation is possible. The bacteria survive in soil only for a short period. Survival in soil is influenced by soil type. In acid sulphate and in saline soils survival is less than 10 days. The pathogen survives better in alluvial than in black calcareous or laterite soils and longer in sterilized than in natural soil. In pond water it has been found to remain viable for 15–18 days. Diseased wild rice growing in ponds may, thus, provide some primary inoculum. Irrigation water contaminated with the bacterium flowing through field to field also provides the primary inoculum. Contact of rice leaves with the contaminated water is essential for initiation of the disease. The pathogen and its phage (virus) survive longer at 15° to 20°C than at 30° to 45°C and more in sterilized than in unsterilized field water. Examination of phage population can forecast onset of disease but not disease severity.

There is evidence that seed from infected crop is one of the major sources of primary inoculum. Viability of the bacterium in naturally infected leaves and artificially inoculated seeds is longest at a very low temperature (4°C) and longer at 24° to 28°C than at 32° to 37°C.

Thus, viability in seed under natural conditions seems to be for short durations at high temperatures. The BLB bacteria may be present in 54% seeds and may survive there for 120–180 days. Such seeds may not produce diseased plants although the bacterium can be isolated from soil around seeds suggesting that the primary inoculum moves from seed through soil. Others, in 1983, had reported that bacterial cells move from the seed into the seedlings. It is also reported that in systemically infected rice tillers, from infected seed, there is a downward pressure of bacterial cells which are maximum in the bottom position of the oldest leaves suggesting invasion through root system (or from seed).

Secondary spread is brought about through wounds and stomata by bacterial cells disseminated by wind- borne raindrop splashes, by irrigation water or rain water coming from infested fields and by contact between diseased and healthy leaves. The leaf hopper (*Nephotettix viriscence*) and the grass hopper (*Hieroglyphus banian*) can transmit the bacterium mechanically on their body. The bacterium does not survive inside the insect body.

The bacteria invade rice plants through hydathodes on leaves, root growth cracks and wounds. Seedling roots are wounded when pulled out from seedbeds and leaf tips are often cut before transplanting. After entry the bacterium multiplies in the intercellular spaces of the host parenchyma and enters the vascular system where it spreads to leaves and finally is exuded through stomata and hydathodes. The bacterium produces *in vitro* and *in vivo* such phytotoxic organic acids as phenyl acetic acid, isovaleric acid, 3-methylthiopropionic acid, fumaric acid, and succinic acid which have been shown to be toxic to cut rice shoots. Xylanase secretion and virulence are closely related. Mutants lacking the system (or genes) producing xylanase are deficient in virulence or virulence deficient isolates are unable to secrete xylanase.

The disease appears in small patches in the field, each patch originating from a single infected plant. It then spreads to neighbouring plants. In the initial stages there is patchy appearance of the disease in the field due to more rapid vertical spread than horizontal spread. Later, due to a large number of such patches, the field appears uniformly affected.

Considerable information is available on topographical and climatic conditions and cultural practices affecting the disease development in Japan. Poorly drained fields along rivers or lakes and mountainous basins, excess of rainfall and humidity, floods and typhoons, higher temperature during tillering stage, low temperature with less sunshine in mid-summer and warm autumn are conducive to disease development. In India cloudy weather with a shower or high humidity is reported to be most conducive for disease development Moderate amounts of rains evenly distributed during the crop season can bring about epidemics.

The development of the disease is favored by temperatures above 25°C. Symptoms never appear at temperatures below 20°C. At mean temperature range of 21.3°C to 32.7°C the size of lesions is longest Combination of rainy weather, strong winds, and temperatures of 20°–26°C favour rapid spread of the disease in the field. Severe Kresek development occurs when there is a combination of maximum temperature of 30°–35°C, minimum temperature of 24°–26°C, uniform high relative humidity of 64–84%, short sunny days and heavy well distributed rainfall. Shade in increasing susceptibility due to physiological weakness of the plant. Water-logging favors the disease as it helps contact of leaves with the floating bacteria.

Disease incidence increases with age of the crop. Hairiness and epicuticular wax of leaves are related to disease resistance Infection of panicles occurs as the crop reaches maturity. Heavy nitrogen application and ill-drained shaded conditions aggravate the disease severity even in moderately resistant varieties Aggravation of disease is not due to increased nitrogen content of leaves. It is possibly due to certain metabolites liberated by the bacterium which help the process of invasion Microbial fertilizers such as *Azotobacter* sp. and *Pseudomonas* sp. aggravate the disease Different concentrations of Mn, Zn, Cu, boron and molybdenum generally increase the length of lesions but doubling the Mn and Cu dose or omitting Mn from the nutrient solution reduce lesion length. Increased supply of calcium reduces susceptibility while magnesium enhances it.

Management of the disease

Many chemicals including antibiotic have been tested and used for the control of bacterial leaf blight of rice but under Indian conditions none has proved satisfactory. Sankel, Phenazine, Cellomate, Streptomycin and Chloramphenicol, etc., which had been recommended in Japan, proved ineffective in India. Partial control of the disease was reported in isolated experiments at different locations in the country.

In 1960s about 95% eradication of seed-borne inoculum by soaking for 12 hours in 0.025% Agrimycin (15% streptomycin and 1.5% oxytetracycline) plus 0.05% wettable Ceresan and then transferring the seed to hot water at 52°– 54°C for 30 min. Subsequenrly, it was shown that simply soaking of seed for 12 hours followed by exposure to hot water at 53°C for 30 min is enough to eradicate seed-borne inoculum. In a screening of 17 chemicals for seed treatment Du-Ter and Captan, dissoived in dichloromethane and Blitox, Terramycin and Benlate dissolved in acetone were reported effective in elimination of the pathogen from seed. Soaking of seed at pH 3.5 for 24 hours and at pH 1.25 for 6 hours also reduced seed infection from 20% to 3.4 and 3.1%, respectively. Bleaching powder (100 μg/ml), copper fungicides, streptomycin (100 μg/ml), triphenyl tin chloride and zinc sulphate (2%) reduce the intensity of bacterial blight of rice when used for seed treatment. In pot experiments. Contamination or infection of rice seedlings could be removed by submerging the seedlings for 24 hours, at the time of transplanting, in 500 ppm solution of Agrimycin-100 or Streptocycline.

Sprays of Agrimycin were recommended in 1964 for preventing secondary spread. Dipping of seed for 8 hours in 0.1% Ceresan wet plus Streptocycline at 0.3 g in 10 lit of water has significant effect in controlling the initial infection. The spread of secondary infection could be checked to a great extent by spraying this antibiotic (3 g/ 100 lit.). Under experimental conditions use of chlorinated water for irrigation or application of bleaching powder to the standing water in the field reduces disease incidence Bleaching powder is toxic to the bacterium at concentrations higher than 100 μg/ml and reduces soil survival of the pathogen when applied to soil. However, in the field the disease is not controlled. Celdion-S (TF-130) was found highly effective against the disease in India but was withdrawn on account of residues in seed at non-permissible level.

The antibiotic Streptocycline had been most commonly recommended chemical for seed treatment and foliar spray against bacterial blight of rice in India. Following numerous reports of unsatisfactory performance of this antibiotic, trials were conducted at several locations These trials finally proved the inefficacy of the chemical which otherwise is very effective against the bacterium *in vitro*. Five sprays of Agrimycin plus copper oxychloride at 12-days intervals can reduce severity of BLB in the field. Fungicides TF-130 and RH-893 were also highly effective. Treatment of rice plants with salicylic acid (1000 μmol/L) triggers synthesis of phenolics, pathogenesis-related proteins and disease resistance response. The treatment causes increase in the level of endogenous SA in the plant. The induced resistance was reported to last for at least 3 days after treatment and before inoculation with the pathogen.

In the absence of specific chemical control measures efforts have been continuing to identify or develop resistant varieties. More than 25 genes conferring resistance to BLB have been identified and incorporated in modern rice varieties. Wild rice species, *Oryza longistaminata* and *O. rufipogon,* are sources of resdistance to the pathogen. However, variability in the pathogen and prersence of specific races restricts the usefulness of such sources. Response of rice to the BLB pathogen varies from highly resistant to highly susceptible depending on host and pathogen genotype. No hypersensitive response is so far known in this disease and a range of lesion size is usually found as expression of resistance. *X. oryzae* pv. *oryzae* as a vascular pathogen elicits a defense response through interaction with metabolically active cells of rice seedlings. In resistant response, secondary walls of the xylem thicken within 48 hours of inoculation and the pit diameter in membrane separating the xylem lumen and the bacteria decreases, thus obstructing the pathogen. In susceptible cultivars these changes do not occur. There is evidence of accumulation of secreted cationic peroxidase in xylem vessel walls and lumen which is associated with secondary thickening (Hilaire *et al.,* 2001).

Ou *et al.,* (1971) tested over 8700 rice varieties from all over the world at the International Rice Research Institute (IRRI) and found most of them susceptible. None was immune. More than half of the 198 wild types of rice were found resistant to 6 races of the bacterium. Varieties resistant at the seedling stage were resistant at the flag leaf stage also. There were indications that a variety that shows some resistance to a few virulent strains of the bacterium could be moderately susceptible to other virulent strains. Cultivars IR-20, IR-22 and Ratna have mild resistance while IR-8, Jaya and IR-24 are mildly susceptible. Somr vultivars are susceptible to only leaf, some to Kresek and others to both. Resistance to BLB, as assessed by lesion length (quantitative resistance) increases considerably with plant age reaching the fastest increase 30–50 days after sowing. Two 2 recessive genes with overlapping effects control resistance in rice to ЬLB. Use of rice cultivars with high levels of partial resistance is considered as a method of bacterial leaf blight control.

Application of acibenzolar-S-methyl induces temporary disease resistance in rice. The resistance lasts for 3 days (Mohan Babu *et al.,* 2003). The pyrazole plant activator CAMPA (3-chloro-1-methyl-1*H*- pyrazole-S-carboxylic acid) induces resistance in rice against bacterial leaf blight by activating a resistance related gene. It is not antagonistic to the pathogen (Nishioka *et al.,* 2005). In biological control of bacterial leaf blight, Vidyasekaran *et al.,* (2001) had reported induction of systemic esistance by *Pseudomonas fluorescens* Pf 1. However, the induced resistance was only transient. Leaves were not protected when challenge inoculated beyond 60 days. Later, Velusamy *et al.,* (2006) reported that 2,4-diacetylphluroglucionil inhibits growth of *X. oryzae* pv. *oryae* in laboratory tests and the isolate of *Pseudomonas fluorescens* producing this antibiotic compound was capable of suppressing BLB in field trials. Rangarajan *et al.,* (2003) have reported that four strains of *Pseudomonas,* tested by them, suppressed BLB as well as sheath blight (*R. solani*). A strain of *Lysobacter antibioticus* isolated from rhizosphere of rice suppresses growth of manyu fungi and bacteria, particularly *X. oryzae* pv. *oryzae*. Whole bacterial culture sprayed on plants has given 69.7% control of BLB. Bacteria suspended in waterm cell-free culture filtrate, and heated culturtes are also highly effective (Ji *et al.,* 2008).

Leaf extracts of *Adhatoda vasica*, *Allium cepa*, neem (margosa) and turmeric are reported to significantly reduce blight incidence and increase yield (Madhiazhagan *et al.,* 2002) Extracts of *Allium sativum* and *Citrus limon* are also reported to suppress the BLB bacteria.

BACTERIAL LEAF STREAK OF RICE

The bacterial leaf streak (BLS) also is a major disease of rice in many Asian countries including Philippines, Thailand, China, Indonesia, Malaysia, Bangladesh and India. This disease was first described as stripe disease by Reinking in 1918 from the Philippines. Identification and description of the causal organism was given by Fang, *et al.,* from China in 1957. In India it was first reported in 1967 in Uttar Pradesh, Madhya Pradesh, Maharashtra and Bihar. Subsequently, it was observed in Karnataka, Haryana, Andhra Pradesh, West Bengal and Orissa. Although Taichung Native-1 was found to be tolerant to a majority of Indian isolates of the bacterium, several other varieties such as IR-8, Jaya, and Padma are highly susceptible. At 100% disease intensity (on the basis of leaf area affected) the loss in yield is about 61%. However, there is an increase in protein and amylose content of the grain.

Symptoms

Bacterial leaf streak is a foliar disease. The first sign of the disease is the appearance of fine, water-soaked to translucent inter-veinal streaks which may be as long as 1 to 10 cm. These streaks are restricted by the veins and soon turn yellow or orange brown. Minute, yellow or amber beads of bacterial exudates are abundant on the streaks. When these beads dry, streaks or rough pustules may be felt on the leaf. These streaks may coalesce to form large patches and cover the entire leaf surface. Eventually, the leaves may be completely blighted. In highly susceptible varieties streaks are surrounded by yellow halo. The infection may reach the leaf sheath and even the seed coat but symptoms are not very clear.

The causal organism

Xanthomonas oryzae pv. *oryzicola* (Fang *et al.,*) Swings *et al.,* 1990. The rod-shaped cells of the bacterium measure 0.5–0.8 × 1.0–2.5 μm and are motile by a single polar flagellum. Stain reaction is Gram-negative. Colonies on beef peptone agar are round, smooth, glistening, and light yellow or golden. It does not reduce nitrate to nitrite. Ammonia is produced from peptone and gelatin is liquefied. Gas is not produced from utilization of glucose, lactose, sucrose, D-fructose, D-mannose and D-galactose.

Disease cycle and environmental relations

The pathogen can survive in infected seed from one season to the next but not in the crop debris in soil or through weeds. The bacteria hibernate under the glumes in mature seed. Seed transmission may not be important for off-season (winter) crops sown with seed produced in the main season (summer) because the disease does not develop in cool, dry conditions. Plumule gets infected from the seed-borne bacteria during germination of the seed. First leaf carries the bacterium to the aerial parts from where secondary spread occurs, as in bacterial leaf blight, through wounds and stomata. Wounds caused by insects are a major source of entry of the pathogen in to the leaves. The pathogen multiplies in the substomatal cavity and intercellular spaces in the parenchyma, Eventually, parenchyma is replaces by bacterial masses. After lesions (streaks) form, the bacterial ooze forms on the leaf surface under moist conditions during night.

Young rice leaves are more susceptible to the disease and become resistant with increasing age. Two to 3 continuous days with high humidity (RH 83–93%) or dew during the morning hours are necessary for infection. If rains stop spread of the disease is also retarded. Lesion enlargement is favoured by moderate temperatures (26°–30.5°C) and retarded at lower temperatures (below 22.4°C) irrespective of relative humidity. Heavy doses of nitrogen, shade, and close planting help in spread of the disease. Although disease development declines with increasing age of the plant, application of nitrogen increases the disease irrespective of age of the plant. Higher the nitrogen dose more is the disease intensity. Lesion size is affected by varietal response while their number is governed by environmental conditions. Hairiness and glabrousness of leaves have some association with disease intensity. Hairy varieties have a higher disease intensity. This is not due to any protoplasmic susceptibility or resistance.

Management of the disease

The seed must be obtained from a reliable source to minimize the danger from seed-borne inoculum. Soaking of seed in 0.025% Streptocycline and hot water treatment at 52°C for 30 min are effective in eradicating the seed infection Spray of Vitavax at 0.15–1.3% is effective in preventing infection and lesion development. Sankel, Captan, and Fytolan are also effective to some extent. Banerjee, *et al.,* (1984) recommended 3 sprays of 100 ppm Streptocycline or Agrimycin-100 at intervals of 10 days starting from the earliest appearance of the disease.

Ou *et al.,* (1970) screened 1118 varieties by artificial inoculation and found that their reaction varied from resistant to very susceptible. None was immune. Only 140 varieties showed few and small lesions indicating resistance. Most varieties were intermediate in reaction. In India, IR-20, Krishna and Jagannath have shown good tolerance to the disease.

Although cereal crops all belong to the grass family (Poaceae), most of their diseases are specific to a particular species. Thus, a given cereal species is typically resistant to diseases of other grasses. This non-host resistance is generally stable. Transgenic rice containing a resistance gene from maize shows resistance to BLS. The same gene confers resistance in maize and sorghum against bacterial streak caused by *Burkholderia andropogonis.*

BACTERIAL BLIGHT AND CANKER OF MANGO

The disease has been known to occur in India for a long time, perhaps as early as 1881, but was first reported in 1948 from Pune (Maharashtra). A similar disease was reported from South Africa in 1915. The disease is present throughout the country but is particularly serious in northern states. It affects leaves, petioles, twigs, tender branches, and fruits and causes leaf spots and cankers. In warm wet weather loss to foliage is considerable. Fruit drop to the extent of 10–15% is also reported. The disease incidence and disease severity have varied from 0.52 to 42.00% and 15 to 90%, respectively in different states. Up to 80% loss may occur in areas with high winds and heavy rains. Different strains cause different predominant symptoms. The leaf spot isolates and fruit or twig canker isolates are different.

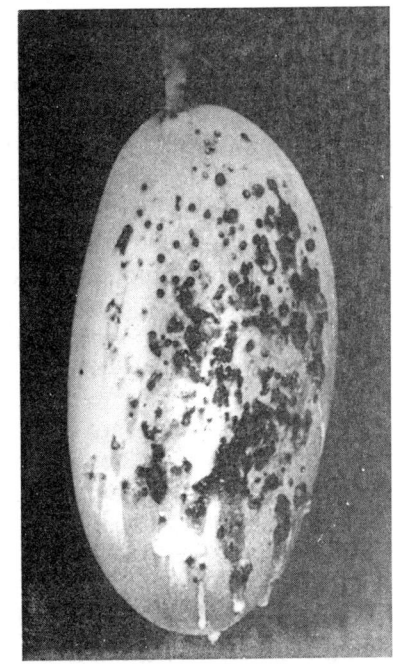

Symptoms

The disease affects all aerial parts of the plant including fruits. Leaf and fruits symptoms are more common. In severe infections twig and branch canker is found. Groups of minute, water-soaked lesions, delimited by veins, appear towards the tip of the leaf. They increase in size to about 1–4 mm, become raised and turn brown to black in color. Sometimes these spots are surrounded by a chlorotic halo. Large necrotic patches may be formed by coalescing of several spots. The patches sometime dry up with decrease of atmospheric humidity. The leaf symptoms are very

Fig. 6. Bacterial canker of mango fruit.

conspicuous on young newly formed leaves. Brown spots are seen on such leaves curl when there is heavy infection. In warm and very humid weather, but when there are no heavy rains, drops of amber colored exudation can be seen on these leaves. The spots are often rough and raised. The young leaves invariably shed and can be seen scattered on the orchard floor. When a major portion of the lamina surface of older leaves is affected they also fall down, otherwise spotted leaves remain on the tree. Petioles, fruits and tender stems are also infected. On young fruits, water-soaked lesions appear and turn dark brown to black. In leaves, the bacteria induce hypertrophy of tissues by enlargement of intercellular spaces of spongy parenchyma early in the infection process. Epidermis is ruptured and bacterial cells ooze out.

Fruit symptoms appear as small water-soaked spots on lenticels. These later become star-shaped, black, erumpent and exude infectious gum. Cracks may appear in the skin of the fruit. Often, a "tear-stain" is observed. Badly affected fruits drop prematurely. Fruits with cracked skin invite other secondary pathogens and post-harvest rot of fruits invariably occurs. Even on the tree, such fruits with cracked skin invite ants and insects and half mature fruits undergo rotting and shed. Xylem plugging and phloem distortion also occurs. Twig cankers are potential source of inoculum and reduce resistance of the branches and twigs to high winds.

The causal organism

Xanthomonas campestris pv. *mangiferae-indicae* (Patel *et al.,*) Robbs, *et al.,* The rod-shaped cells of the bacterium measure 0.36–0.54 × 1.44 μm. The cells are single or in short chains. No spores or capsules are formed. The cells are motile by a single polar flagellum. Stain reaction is Gram–negative. On potato dextrose agar colonies are circular, smooth, glistening, and creamy white. Gelatin is liquefied, casein digested, and litmus reduced by the bacterium. Nitrates are not reduced to nitrite. Optimum temperature for growth of the bacterium is 27°C. Thermal death point is about 55°C.

On the basis of biochemical and pathogenicity studies on 20 isolates of the bacterium from different states of South India, 10 pathotypes or races. have been identified. In this study the cultivar Alphanso was highly susceptible to all the isolates while cultivar Bangalore was resistant to most isolates. Cultural and biochemical variations in the bacterium were reported in 1991. Three or 4 pathotypes are reported from the northern states of Uttar Pradesh and Bihar. In a more specific study, Ab-You *et al.,* (2007) have divided isolates of the bacterium in to 3 groups which were genetically distinct from each other. One group, strains from the old world (India) produced black, raised lesions, consistent with the original description of the pathovar while other groups produced variable symptoms including color of the lesions. These authors have also suggested that *X. campestris* pv. *mangife ae-indicae* be split into three pathovars of *X. axonopodis*, *viz., X. axonopodis* pv. *Mangiferae-indicae*, *X. axonopodis* pv. *anacardii* and *X. axonopodis* pv. *spondiae*.

Disease cycle

In the orchards, the bacteria are phylloplane residents throughout the year. They have been reported as epiphytes on buds where their population is favoured by high relative humidity and temperatures of 15 to 20°C. They also survives in cankers on twigs and smaller branches, stone of the fruit and as phylloplane microflora on weeds. The primary inoculum is brought by contaminated nursery stock, wind driven rains and grove maintenance operations. The disease spreads rapidly during rains. Bacterial cells are rain-splashed from the source of survival to uninfected parts. Long distance spread is caused by infected planting stock. The bacteria enter the fruits through bruises and other types of injuries. Disease development is favoured by high humidity (RH above 90%) and a temperature range of 25°-30°C. Maximum infection of fruits occurs when minimum and maximum (night and day) temperatures are 22°C and 25°C. Rainfall is a major weather factor affecting fruit infection. High wind velocity is also a favourable factor for the disease. Many insects such as ashy weevil (*Myllocerus discolor* var. *variegata*), leaf webber (*Orthega vadrusalia*) and bugs (*Canteconidia furcellata*) mechanically transmit the bacteria on their legs and mouth parts (Kishun, 1986). Some other members of Anacardiaceae such as cashew (*Anacardium occidentale*) are hosts of the bacteria.

Management of the disease

Some workers have claimed control of the disease by streptocycline spray. In north India, Two sprays of 300 ppm streptocycline with 0.3% copper oxychloride in May at 10 days interval. were recommended. In screening of a large number of mango varieties for resistance to the disease. none was found resistant. Langra, Dashehri, Chausa, Bombai Zardalu, Sunder Langra, Gulabkhas, Kesar, and Mankurad were moderately resistant. Bombai Green, Hemasagar, Fazali, Swarnrekha, and Anupam were susceptible. In south Severe natural infection in the varieties Mulgao, Alphanso, Neelam, Rumani, Bangalore and Baneshan were reported in South India but Bombai Green was found resistant. In the Tarai area of Uttaranchal (India), Amrapali and Dashehri have been found highly susceptible to leaf blight but not to fruit cankers. Bombai green was less severely affected while Langra had leaf infection only in traces but sometimes there is heavy fruit infection and loss. Mishra (1995) screened 212 mango varieties and found 95 of them free from the fruit canker.

BLACK ROT OF CRUCIFERS

This serious disease of crucifers is of worldwide occurrence in the temperate and subtropical zones where rainfall and dew are plentiful and average temperatures are between 15° and 22°C during the crop season. Since the bacterium causing the disease is seed-borne it has been distributed to many parts of the world. When the pathogen becomes established in the field early in growing season and favorable environment prevails, it may become extremely destructive, rendering a large percentage of cabbage plants incapable of producing marketable heads and making the remainder of the crop unsafe for storage. The bacterium causing black rot attacks cauliflower and other cruciferous vegetable crops but it seems that there are different races or strains specialized on different hosts.

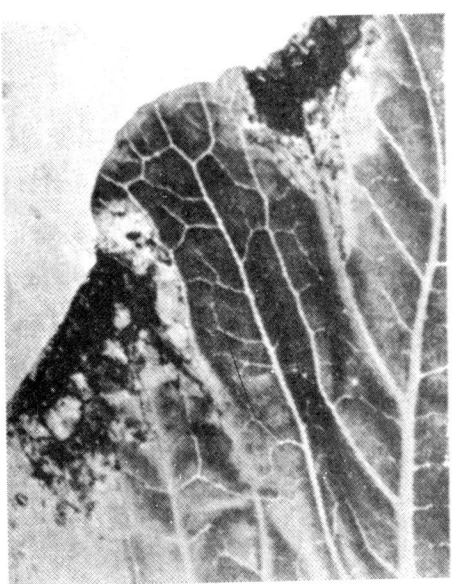

Fig. 6A. Black rot of cabbage.

Symptoms

The plants may be attacked at any time during their growth, from the early seedlings stage until they mature. On young seedlings the cotyledons are infected at the margins which show blackening. Such cotyledons die and drop off. Later, infection of leaves occurs through water pores at the margins or occasionally through wounds caused by chewing insects. The tissue turns yellow and the chlorosis progresses towards the center of the leaf, forming a V-shaped area with base of V towards the midrib. The veins show brown to black discoloration. The vascular discoloration extends to the main stem and progresses upward and downward. The upward advance of the discoloration affects the upper leaves where lesions may appear anywhere on the lamina due to systemic invasion. Leaves so invaded may be stunted unilaterally. There is pre-mature fall of the leaves. Cabbage and cauliflower heads are invaded and discolored. In turnip the fleshy roots are also invaded.

The causal organism

Black rot is caused by *Xanthomonas campestris* pv. *campestris*. The disease is also known as black venation and tracheobacteriosis of *Brassica* spp. including *Brassica campestris*. Some strains of the pathovar produce black rot symptom (V-shaped lesions and black veins on the leaf) and some produce blight symptoms sudden collapse of interveinal tissues following the absence of veinal necrosis at early stages of infection). Blight inducing strains spread to more seedlings than black rot inducing strains by leaf contact and water splash.

Disease cycle and environmental relations

The bacteria survive in diseased plant debris only for some time. The survival in diseased plant debris-infested soil is negatively correlated with soil matric potential and with the abundance of cellulolytic microorganisms. In saturated soil it survives only for 19-28 days (Arias *et al.,* 2000). Kocks and Zadoks (1996) observed that black rot development was more intense with fresh than with old refuse piles placed in the center of the plot. The survival in plant debris in soil depends on temperature and decomposition of the plant debris. Tillage operations distribute the debris-borne inoculum in the field creating multiple foci of infection (Kocks *et al.,* 1998 a,b). The bacteria can also survive as epiphytes on symptomless infected leaves of *Brassica campestris*. Populations of up to 1 million cfu per sq. cm of leaf have been recorded on symptomless leaves. On the phylloplane of cabbage, mustard and lettuce survival may be for 48 days but on rice leaves it survives for only 9 days, suggesting that epiphytic survival depends on plant species. The bacterium produces a stable, low molecular weight diffusible extracellular factor which is essential for xanthomonadin and EPS production. The pigment xanthomonadin, the extracellular polysaccharide (EPS) and the diffusible factor (diffusible signal molecule) are all important for epiphytic survival and/or infection and lesion development The exopolysaccharide (xanthan) induces plant susceptibility to the bacterium by suppressing callose deposition as a resistance response.

The main source of survival of the bacterium is seed where it can survive for 3 years. The bacterium is present in the parenchyma of seed coat and in the endosperm. Symptomless plants may produce infected seed. When germination of infected seed occurs the seed coat that comes out along with cotyledons serves as the source of primary inoculum. The bacteria enter the cotyledons through stomata on the margins. Then from cotyledons the bacteria pass on to true leaves. The true leaves are invaded through the hydathodes (water pores) in the morning hours. Hydathodes are the preferred sites for entry of this pathogen. Even 10 viable bacterial cells applied as droplets can cause infection of leaves. On entry into the host, the bacteria move to the xylem vessels where they multiply rapidly and move up and down the plant. The bacteria secrete different proteases which play a role in black rot pathogenesis. Protease deficient mutants of the bacterium show considerable loss of virulence. The bacterial cells and the xantham gum block the vessels. In the field the secondary inoculum is spread by raindrop splashes, blowing of detached leaves, and occasionally by insects. Lesions on leaves contain up to 100 million cfu per sq. cm. Infected plants usually do not show symptoms until the concentration of the organism reaches a very high level.

Rains and sprinkler irrigation spread the bacteria among the plants. In regions where rainfall is very low during raising of seedlings black rot is not common because of restricted dispersal of inoculum. In the United States, disease development was found to be positively correlated to the number of days with rainfall between 6 and 9 a.m. during May–June (Kocks and Zadoks, 1996). The optimum temperature for growth of the bacterium is about 26.5°– 30°C, minimum being 5°C and maximum 36°C. The lethal temperature is 50°C. At optimum temperature incubation period is 7–14 days. At temperatures below 18–20° symptoms do not develop. Boron status of the soil plays a role in susceptibility of cauliflower to black rot. Susceptibility is higher in boron deficient soil (below 0.4 mg/kg) and boron-excess soil (above 1.6 mg/kg). Significant reduction in black rot incidence is obtained by exogenous application of boron alone (up to 6.4 kg/ha) or with nitrogen. The plants in treated plots have high leaf phenol content.

Management of the disease

The control measures recommended for the black leg disease apply to black rot also. Seed, nursery soil and seedlings should be free from the bacterium. The bacteria live in the field only for one year. Hence, a 2-year rotation is sufficient to check the soil-borne inoculum. For seed treatment a number of antibiotics such as agrimycin (0.01%), streptocycline (0.01%), and aureomycin have been used. Soaking of seed in 0.5% sodium hypochlorite for 30 min provides good control. Physical methods of seed treatment include hot water treatment and aerated steam treatment. Hot water treatment usually involves seed soak at 50°C for 20–30 min. However, the treatment may affect the germinability of the seed, especially the seeds with low vigour. Pre-drying seeds at 40°C for 24 hours followed by air treatment at 75°C for 5–7 days has been found effective without causing seed damage. A 20-min soak of crucifer seed in hot (40°C) acidified (with acetic acid) cupric acetate (0.5%) gives about 95% control of black rot. The treatment is effective against black leg patyhogen also. For cauliflower black rot, hot water seed treatment at 52°C for 30 min followed by 30 min dip in 100 ppm streptocycline was recommended in 1981 in Himachal Pradesh. The seed treatment is followed by 3 sprays of 50 ppm streptocycline at transplanting, curd formation, and at pod formation stages. Application of 10–12 kg/ha stable bleaching powder as soil drench is also effective against black rot and soft rot. Foliar spray of the antibiotic validamycin A is reported to provide effective control of cabbage black rot. The antibiotic suppresses production of extracellular polysaccharides by the bacterium and its multiplication in the plant.

Jalali and Parashar (1995) reported 46% control of the disease in *Brassica juncea* by seed treatment with strains of *Bacillus* isolated from phylloplane of the host. Endophytic *Bacillus* strains also are reported to suppress the disease. Wulff *et al.,* (2002) have reported biological control of black rot in cabbage, cauliflower, rape and broccoli by an endophytic antagonistic strain of *Bacillus subtilis*. The level of control varies with strain of the pathogen, host, season and soil type. During the dry and short rainy season the antagonist suppressed black rot in all the hosts. Biological control was not effective in cabbage and rape during the main rainy season in clay loam soil. Suneja *et al.,* (1994) had claimed that siderophores of *Azotobacter chrococcum* were effective in controlling black rot in mustard. However, the effect is suspected to be due to better plant growth. Other potential antagonistic bacteria are *B. amyloliquefaciens* and *B. pumilus*. In a recent study in Tanzania (Africa), strains of *B. cereus, B. lentimorbus* and *B. pumilus* are reported to suppress black rot symptoms in foliage, stem and heads of a susceptible cabbage cultivar. Root applications of the antagonists were more effective than seed application (Massomo *et al.,* 2004).

In a screening of 30 plant extracts, 8 showed significant antibacterial activity against pathovars of *X. campestris*. These 8 plant species were *Acacia arabica, Achras zapota, Enterolobium saman, Lawsonia inermis, Oxalis corniculata, Prosopsis juliflora, Punica granatum* and *Viscum orientale*. Contrary to general belief, some of the plant species considered as having medicinal value *viz., Azadirachta indica* (neem), *Aegle marmelos* (bel), *Aloe vera, Catharanthus roseus* and many others had no antibacterial activity.

Many commercial varieties such as Sutton Express, Eclipse Drumhead, etc. are resistant to the disease. In Himachal Pradesh the cultivar Symphony (from UK) was found resistant to black rot. In *Brassica carinata* and *B. oleracea* a single dominant major gene controls black rot resistance.

RED STRIPE AND TOP ROT OF SUGARCANE

The disease is common in all the sugarcane growing areas of the world. In India it was first reported in 1933. Since then its occurrence has been recorded in Bihar, Uttar Pradesh, Punjab, Maharashtra, and Tamil Nadu. In certain areas, as in Bihar, the disease causes considerable damage to young plants during the summer.

Symptoms

The disease first appears as water-soaked elongated streaks which soon become chlorotic and carry long, narrow, uniformly dark red stripes which are 0.5–1 mm in width and 5–100 mm or more in length. Sometimes two or more stripes coalesce to form larger bands. The lower half of the leaf is more affected than the tip. The stripes occur mostly on young and middle-aged lleaves rather than on oldest leaves. Whitish flakes of dried bacterial ooze may be seen on lower surface of the leaves.

When young shoots are affected symptoms of top rot appear. In top rot, symptoms include yellowing and wilting of older leaves which may show red stripes. The growing point of the shoot shows many dark red stripes with water-soaked appearance and undergoes rotting. The disease proceeds downwards killing the terminal buds and the leaves. Top rot may also result from stem and bud infections without showing leaf symptoms.

Fig. 7. Red stripes of sugarcane leaf.

The causal organism

Acidovorax avenae subsp. *avenae* (Manns) Willems *et al.,* 1992 {=*Pseudomonas rubrilineans* (Lee, *et al.,*) Stapp}. The rod-shaped cells of the bacterium measure 0.7 × 1.6 μm and occasionally form chains. No endospores or capsules are formed. The cells are motile by 1-3 polar flagella. Stain reaction is Gram-negative. The bacterium is facultative anaerobe. Colonies on beef extract agar are light buff to yellow, opalescent, small, smooth, and glistening. Pellicle is formed on broth. Gelatin is liquefied. Milk is cleared without coagulation. Nitrates are reduced to nitrite. Thermal death point is 51°–52°C.

Disease cycle and environmental relations

The bacterium is not thought to survive well in soil or in plant debris. Perpetual presence of sugarcane fields in the neighbourhood provides good chance of uninterrupted survival of the bacterium. It can infect rice, sorghum, pearl millet, maize, *Sorghum halepense* and *Sorghum sudanense* etc. These hosts may serve as agents for carry over of the bacterium from season to season. Use of seed cuttings from diseased canes also serves as a source of primary inoculum in new fields.

Infection starts through wounds and through stomata. The bacterium is confined to young tissues of the host. It rapidly multiplies in the parenchymatous tissues and moves to vascular bundles. The bacterial slime on the leaves is spread by wind splashed rain drops.

Management of the disease

Once the disease sets in it is very difficult to manage. Systematic cutting down and burning of affected shoots reduces the spread. Use of resistant cultivars is the best method to avoid the disease in areas where its occurrence is common. CO cultivars 6805, 7202, 7321, 7537, 7642 and 8005 and variety Ponda are reported to be resistant.

BROWN ROT AND WILT OF POTATO

The bacterial brown rot and wilt disease of potato is common in tropical and subtropical regions and also in some warm temperate regions of the world where temperature and moisture conditions are favourable for its development. In India the disease is most destructive in mid-hills, the plateau region and in West Bengal. In Karnataka, Madhya Pradesh, Marathwada region of Maharashtra, and West Bengal the disease poses constant threat to cultivation of not only potato but other solanaceous crops also. The disease damages the crop through direct killing of the plants in the field and through rotting of tubers in stores. It is the most serious bacterial disease of potato in India accounting for 10 to 70% loss in yield of tubers.

The losses vary with changes in cropping systems. In Mukteshwar area of Kumaon Hills (Uttarakhand) 37% of potato yield is lost every year. Reported losses are up to 55% in Kumaon Hills, up to 40% in Maharashtra and up to 75% in some localities of Karnataka. Apart from solanaceous crops like chilli, potato, tomato, and eggplant the bacterium is known to attack castor, ginger, groundnut, banana, papaya, cabbage, radish and a large number of other plant species. The very extensive host range of the bacterium includes more than two hundred species belonging to 44 families of plants However, there are anomalies in the distribution of bacterial wilt on certain hosts. Groundnut is extensively attacked only in China and Indonesia but is unimportant in many other major groundnut producing countries. The bacterium often attacks the crop easily when there is root knot nematode infection in the plants.

Symptoms

The characteristic symptoms of the disease are stunting, yellowing of the lower foliage, sudden wilting and finally, collapse of the entire plant. In the hills where bacterial wilt usually appears in July, about a fortnight after the start of the monsoon, as well as in the Ranchi area of Jharkhand many wilted plants show stem rot at the soil level. This symptom may be confused with blackleg disease of potato caused by *Erwinia carotovora*. The brown rot refers to the browning of the xylem in the vascular bundles. This browning is often visible from the surface of the infected stems as dark patches or streaks. The name ring disease is derived from the fact that a brown ring is formed in the tuber due to discolouration of the vascular bundles. This is different from ring rot of potato caused by a different bacterium. The skin of infected tubers is often discoloured. In severely affected tubers the eye buds are blackened. If the infected stems and tubers are cut across and squeezed, grayish-white bacterial ooze comes out of the vascular ring. Microscopic examination of properly stained slides of the ooze confirms the presence of the disease.

Symptoms are variable. In India stunting and yellowing of lower leaves is not commonly seen. Browning of the vascular bundles may or may not be present This browning is due to the production of pigment by the bacterium. Some isolates do not produce the pigment but cause wilt and tuber rot. In such cases the vascular bundles and surrounding parenchyma appear water-soaked.

The causal organism

Ralstonia solanacearum (Smith) Yabuuchi *et al.,* 1996 was earlier known as *Pseudomonas solanacearum* (Smith) Smith, later renamed as *Burkholderia solanacearum* (Smith) Yabuuchi *et al.,* 1993 and finally as *R. solanacearum.* The different forms of the species appear to have originated from a common ancestor, possibly at a single location near the equator. Further evolution of the bacteria then occurred with several wild hosts, possibly in forest ecosystem in geographically isolated areas, creating plenty of diversity within the species.

The rod-shaped cells measure $0.5-0.7 \times 1.5-2.5 \mu m$ and are motile by 1–4 polar flagella. However, virulent strains are usually not flagellated and cells are surrounded by thick polysaccharide material. Stain reaction is Gram-negative. The cells are not readily stained by ordinary methods. Treatment with Ziehl-Nielson's carbol fuchsin stain produces a bipolar effect in which the stain is retained by the ends of the cell, leaving light stained or unstained central portion. The bacterium is aerobic although many strains may grow anaerobically. Two types of colonies are produced on complex media: one type is smooth, fluidous and elevated while the other type is somewhat rough, dry and flat. A large number of organic compounds are utilized by individual strains of the bacterium as carbon source. Most isolates utilize glucose, fructose, and sucrose. The species does not characteristically utilize xylose, arabinose, and arginine. Starch and gelatin are not hydrolyzed. Nitrates are reduced and ammonia and hydrogen sulphide are produced on specific media. On steamed potato cylinders the bacterial slime is at first white, gradually becoming dark brown and black. Optimum temperature for growth varies with races and is between 27º and 35°C, maximum about 41°C and minimum 10°C. Thermal death point lies at about 52°C.

The isolated cultures of the bacterium easily lose their pathogenic virulence when grown on laboratory media. The phenomenon is correlated with colony variation. On tetrazolium chloride medium the virulent wild types produce an irregular to round, white colony with pink centre while the avirulent mutants form round, butyrous, deep red colonies with narrow bluish borders. Chakrabarti *et al.,* (1995) had reported spontaneous reversion of afluidal mutants to fluidal wild type in some strains of the bacterium.

Although as a species *R. solanacearum* has extremely wide host range, its races have relatively limited host range. There are five races on the basis of host affinity and five biovars on the basis of ability to utilize and/or oxidize several hexose alcohols and disaccharides. In general, race 3 (the potato race) and biovar 2 are equivalent. Biovar I is predominant in the Americas and biovar III in Asia. Race 1 infects solanaceous plants and diploid bananas, race 2 infects banana and *Heliconia* spp. and race 3 mainly infects potato and tomato and to a lesser extent eggplant and pepper. Races 1 and 2 have high temperature (35°–37°C) optimum while race 3 has lower temperature (27°C) optimum. Only races 1 and 3 or biovars II, III and IV are present in India. In the plateau area and in West Bengal race 1 is widespread, being particularly severe on eggplant, while in the hills only race 3 has been reported. In the plains and plateau areas of Bihar biovar II is reported to be common. No low temperature race has been reported. In India race 3 (biovar II) is primarily pathogenic to potato and has adapted for persistence to cool humid conditions in Indian hills. Race 1 (biovars III and IV) have wide host range and survive in warm areas. Aggressiveness within strains may differ. Genetic diversity in *R. solanacearum* follows the concept that pests and pathogens are most variable near the centre of origin of their primary hosts (in this case potato). The diversity in the bacterium is less at greater distance from the equator. Strains of race 3 (biovar II) from potato in Australia form a uniform phenotype whereas those from South America show some heterogeneity and can be divided into several distinct phenotypes.

Disease cycle and environmental relations

The pathogen survives between crop seasons in soil, seed tubers and on alternate cultivated or wild host plants. Bacterial populations in soil increase with symptom development in the plant and decline when the plant is dead. The decline continues if there is no cover crop after potato harvest. Populations of the bacterium can survive on roots of non-host plants also. Root systems of even such non-host plants as wheat, sorghum and maize, which are recommended in crop rotations for management of the disease, can harbor low populations of the bacterium. Containers carrying potatoes to market or stores may be contaminated and serve as source of transmission and short term survival. The contamination survives on wooden crates for 4 days (oak wood crates) and 17 days (poplar wood crates. On high density polyethelene survival is zero after two days and on jute bags the contamination drops rapidly after 24 hours.

Soil is considered a potential source of primary inoculum and the disease has been noticed even in the first planting in newly cleared land. The bacteria spread in the soil. The survival ability differs according to soil type and is mainly related to soil pH and accumulation of nitrites. In cultivated soil, survival of the bacterium has been reported for a period of more than two and a half years although some reports indicate that it cannot survive in soil even for one year. Survival of up to 673 days in naturally infested soil stored in plastic bags at 4°C has been reported. Under laboratory conditions the bacterium could survive in soil for at least 16 months and in diseased plant debris for about 9 months. The bacterium can survive in sterile and natural soil for over 250 days. In a temperate climate study of soil survival of biovar 2 (race 3), Van Elsas *et al.*, (2000) have reported that bacterial population in loamy sand and silt loam soil at 20°C declined rapidly in 90–210 days, the greatest decline taking place in the loamy sand soil. A single freezing-thawing cycle caused additional decline. Severe draught also caused significant decline in the bacterial population. Infested soil exposed to 43°C continuously for 4 days or more becomes free of the bacterium. The role of the dormant-like viable-but-nonculturable (VBNC) condition in the survival of *R. solanacearum* is reported. In this condition the bacterial cells can survive for very long periods in soil and maintain their pathogenic ability. However, this condition is reported not to increase soil persistence of fluorescent pseudomonads. Further, prolonged exposure to low temperature in low temperature-induced VBNC, the bacterial cells lose virulence (van Overbeek, *et al.*, 2004). In the mid-hills in north India, it successfully perpetuates in soil. In central Nepal also the bacteria are reported to survive in flooded rice field soil during and after rice harvest. Survival of the bacteria in surface water and transmission to plant roots reaching the water is also reported. Such plants serve as source of inoculum. However, survival of bacterial cells is better in pure water than in drainage water from potato fields in which they survive for a limited period. Temperature effect is most important. Survival is maximum at 12°C, 20°C and 28°C. Microbiota in the drainage water have a strong effect on survival. Vertical and horizontal movement of the bacteria is according to capillary movement of the soil water.

Well drained soil with good water retention capacity, moderate to high soil temperature and low to moderate soil pH promote survival of the bacterium. Soils that allow desiccation of the bacterium or promote antagonistic organisms are detrimental to its survival. In most soils the infestation of the bacterium is highest in the top 30 cm of the profile but low populations have been recovered as deep as 65–75 cm.

Survival of the pathogen in diseased plant debris has been doubted on account of very hot and dry summer that follows the main crop season in the plains of India. In the north Indian hills, however, the potato crop is grown during March-September. At harvest, farmers usually leave the brown rot affected tubers in the field. Since the weather during the off season (September–March) remains cool and humid, desiccation is prevented and also there is little activity of secondary invaders or antagonists. Thus, the pathogen may survive in plant debris under such conditions. In the Nilgiri Hills the bacterium may survive on potato crops grown throughout the year.

Potato tubers carry the bacterium in three ways, i.e. in vascular tissues (active infection), on the tuber surface, and in lenticels. The latent or surface (lenticel) infection has become the most important and significant method of tuber transmission in the last two decades since in this case no symptoms develop in the tubers and the bacterium goes unnoticed. It can express itself only when the tubers are incubated at high temperatures. Tuber transmission of the pathogen is possible if infested or contaminated (or latently infected) tubers are stored for seed at moderate temperatures. Tubers carrying active infection easily rot at high temperatures and are not used for seed. Even if they are planted most of them do not germinate due to rotting in soil. Delayed harvesting increases number of lenticel infected tubers. In infested fields freshly harvested tubers carry about 19.9×10^3 cells per tuber on the surface, 20.8×10^3 cells/g in vascular tissues and 2.3×10^3 cells per lenticel. During storage, bacterial population declines rapidly on tuber surface reaching non-detectable level in 30–60 days at $4°C$, 60–90 days at room temperature ($12.9°–34.4°C$) and in 120–150 days at $10°–15°C$. In the plains of India, where soil survival of the bacterium is doubtful, infected and surface contaminated (latent infection) potato seed tubers appear to be the only source of primary inoculum. In the plateau region there is some native source of primary inoculum. At many places potato is not a popular crop but the disease destroys tomato, brinjal and chilli crops every year. Obviously, in the absence of potato the primary inoculum in these areas has to be from a native source such as soil, perennial weed hosts or plants acting as symptomless carriers or true seed. This contention is further strengthened by the fact that the race flora of brown rot bacterium in the plateau area is different from the hills. However, there is documented evidence that even in this area also apparently healthy looking seed tubers obtained from mid-hills of Garhwal-Kumaon carrying latent infection had been the major source of introduction of the disease in new areas (Sunaina et al., 1989a).

Potato tubers are infected by way of stolons. Root-knot nematodes aid in penetration if they attack the host before it comes in contact with the bacterium. Field resistance in potato is broken down when the plants are infected with M. incognita. In addition, certain gall forming insects, cut worms, white grubs, and root invading parasitic fungi may also provide root injury for entry of the wilt bacterium. During the sowing operation, cutting of tubers with knives, without taking precautions, help in contamination of healthy tubers. After entry, the pathogen moves upwards through the host vessels and colonizes rapidly. The incubation period varies with age of the plant, environment and host variety.

Temperature is the most important factor affecting the host-parasite interaction as well as survival in soil. There is no disease development at 15°C. Most rapid development took place at 37°C. High temperature (30°–35°C) and high soil moisture favour many but not all strains of the bacterium. Plants that are resistant at moderate temperature may become susceptible at a higher temperature. For wilt development caused by race 1 of the bacterium a relatively high temperature range of 28°–30° is required. At lower temperatures infection can occur but symptoms fail to develop. Thus, at lower temperatures chances of latent infection are very high. It is not clear whether temperature sensitivity of resistance is a function of virulence factors which can express fully only at high temperatures or an effect of lack of expression of resistance genes in the host at high temperature. The potato race (race 3) is reported to cause the disease epidemics even at low temperature of 13.1°C In many experiments this race caused more disease at 16°, 20° and 24°C than at 28°C. This feature of the potato race may be because the pathogen has adapted to pathogenesis on potato at lower ambient temperatures such as obtained at the place of origin of potato.

Ralstonia solanacearum cannot withstand desiccation and its populations in soil are considerably reduced in fields given regular ploughing. Many studies have been carried out to establish the close relationship between bacterial wilt epidemics and soil moisture. High soil moisture accumulations resulting from either a high water table or heavy rainfall usually favor development of bacterial wilt. Survival of the pathogen is greatest in wet but well drained soils whereas survival is adversely affected by soil desiccation and by flooding. Not only survival but multiplication in natural soil is also affected by moisture. No infection is obtained below 50% soil moisture level. High soil moisture favours the disease through better survival and spread of the bacterium, better infection, and better disease development after infection. However, temperature and moisture effects are not independent of each other. The ideal temperature (25°–35°C) for disease development in the north-western high hills of India occur during May to 15 June. The crop is planted in March. But the disease starts its appearance from mid-July after rains have started. It is also reported that the bacterial population in soil is highest during July and August when both soil temperature and moisture are high. The incidence of wilt declines when soil moisture drops to 8–10% water holding capacity and maximum/minimum temperature below 20°–15°C (day/night).

The pathogen grows over a wide range of pH. Optimum. is 6.2–6.8, maximum 7.4. However, this may vary with strains. The disease occurs in diverse soil types both acidic and alkaline. It is severe in acidic soils of the Nilgiris (pH 3.6–5.0) as well as in alkaline soils of Madhya Pradesh (pH 7.0–8.0). Amendments of soil affecting pH and nitrite accumulation affect survival of the bacteria. Application of urea at 200 kg N/ha reduces the population immediately and at one week after treatment due to high pH of soil resulting from urea hydrolysis. Growth of the bacterium is decreased if the clay content of soil increases. Heavy infection of eggplant has been reported in years of heavy rainfall. Organic matter promotes growth while inorganic fertilizers decrease it. There is decrease in disease severity with increasing age of the plants. There have been reports of some effect of light intensity and relative humidity on disease incidence in other crops. This is attributed to change in host nutrition and vigor.

Management of the disease

Proper crop rotation is one way of avoiding soil-borne inoculum of the bacterium. Three-year rotation with maize, soybean, and red top grass (*Agrostis alba*) had been found to afford considerable protection to the crop. Maize rhizosphere supports fluorescent *Pseudomonas* species antagonistic to *R. solanacearum*. The survival and multiplication of the bacterium is minimum in rhizosphere of maize and rice. Hence these crops can be included in rotation. In India, rotations consisting of potato-finger millet-potato, potato-wheat- potato, potato-sorghum-potato, potato-maize-potato, and potato-wheat-sunnhemp green manure-potato have been found effective in keeping disease incidence below 3%. Intercropping has also been used as a means of reducing soil populations of the bacterium and its root-to-root transmission. Bean, maize, cowpea, sugarcane have been used as intercrops in different countries.

Seed tubers should always be obtained from a disease free field or locality. The tubers obtained from a healthy crop in an endemic area should be treated with 0.02%. Streptocycline for 30 min after giving a 5 mm deep cut. Rain or irrigation water should not be allowed to flow from infested to healthy fields. At the time of harvest no diseased or rotting tubers and crop debris should be left in the field. After harvest, the field should be ploughed to expose the soil to summer heat of May–June in the plains.

Since the disease is a systemic vascular wilt and the pathogen survives in diverse soil types, its control through chemicals has not been possible. Application of very high doses of urea, toxic chemicals like sulphur, chloropicrin, etc, which are effective, cannot be recommended because of very high cost. Under experimental conditions, foliar application of Agrimycin-100, chloramphenicol or streptomycin sulphate a day prior to inoculation is effective at 1000 ppm but not after inoculation. Application of stable bleaching powder in furrows at planting time at the rate of 12 kg/ha has also been recommended. Organic amendments and crop residue decomposition suppress activity of *R. solanacearum* through biofumigation by release of volatile organic acids.

Differences in susceptibility to brown rot occur in varieties of *Solanum tuberosum* but so far most commercial varieties of potato under cultivation in India have not been found resistant to bacterial wilt and brown rot. Races of the bacterium in India are highly virulent and exotic cultivars resistant to the bacterium lose resistance under Indian conditions. A clone of the wild potato (*Solanum microdontum*) was identified to carry high level of resistance to race 1 and 3 and biotypes II, III, and IV. Culture BRB/A-24 of *S. tuberosum* x *S. microdontum* was found resistant to the disease. Certain clones of *S. phureja* also show resistance to the disease and are being used in hybridization programme. Fock *et al.*, (2000) obtained somatic hybrids between *S. tuberosum* and *S. phureja*. Most hybrids were resistant to race 1 of the bacterium but susceptible to race 3. Amphiploid hybrid clones showed good tolerance to both races. The problem in developing resistant varieties is that resistance found so far is race specific. Hence, resistance to one race/biotype fails to be resistant to other races/biotypes. When the bacterium is inoculated into different varieties it starts multiplying immediately after inoculation in some cultivars while in others there is a lag period of 2–3 days which delays build up of concentration of bacterial cells in the tissues to a level when the disease can develop.

Attempts to control bacterial wilt through biocontrol agents have yielded promising results. (Kumar and Sood, 2001; Singh and Rana, 2000). The agents used are antagonistic rhizobacteria, avirulent strains of *R. solanacearum* and *Pseudomonas fluorescens*. The mechanisms involved are induced systemic resistance and pre-colonization of the rhizosphere by the antagonists. Bacterization of seed tubers with antagonistic bacteria such as *Bacillus cereus* and *Bacillus subtilis* provides better protection against bacterial wilt than cross-protection provided by avirulent strain of the pathogen.

Biological soil disinfestation is effective in suppressing *R. solanacearum*. The infested soil is amended with fresh green plant residue (grass, potato haulms, or any other plant material and covered with airtight plastic sheet. Decomposition increases microbial activity and creates anaerobic condition in absent of re-supply of oxygen. The treatment is reported to reduce soil population of *R. solanacearum* by 92.5% (Messiha *et al.*, 2007a). They have also reported differential responses of R. solanacearum to inorganic and organic fertilizers in Egyptian and Dutch soils. NPK fertilizers reducedbacterial wilt in in conventional Egyptian soils but not in organic Dutch soils. Cow manure amendment significantly reduced disease incidence in organic

Dutch sandy soils but did not affect the bacterial population. However, cow manure amendment reduced bacterial population in the Egyptian soils (Messiha *et al.,* 2007b).

Antibacterial activity of plant extracts against the wilt bacterium is reported by Ooshiro *et al.,* (2004). A 70% aqueous ethanol extract of fresh aerial tissues of *Geranium carolinianum* showed strong antimicrobial activity against *R. solanacearum, Streptomyces scbies* and *Streptomyces ipomoea.* In field test, a treatment combining incorporation of dried aerial tissue into the soil and solarization was highly effective for control of the bacterial wilt of potato.

Genetic engineering techniques are being used for the introduction of lysozyme, cecropins, and other potent antibacterial proteins derived from insects into potato, as a way of augmenting resistance to bacterial wilt and other bacterial diseases. Although the short-term probability of success is low, the importance of the disease, its worldwide distribution and limitations in conventional breeding programmes justify the attempt.

In general, following precautions have been recommended against this disease:

1. Always obtain seed from areas where the disease is not reported.

2. In endemic area the seed tubers should be selected only from healthy crop and should be treated in 0.02% Streptocycline for 30 min after giving a 5 mm deep cut.

3. The plots which show severe incidence of the disease should be put under maize, cereals or soybean for 3 years.

4. Rain or irrigation water should not be allowed to flow from infested to healthy fields.

5. Infected plant residue should always be removed and burnt.

6. After harvest the field should be ploughed to expose soil to summer heat of May-June in the plains.

BACTERIAL WILT OF TOMATO AND OTHER SOLANACEAE

The bacterial wilt of tomato, caused by *Ralstonia solanacearum,* is a major disease of this crop in areas where the bacterium is well established and soil conditions are favorable for it. In Karnataka, Madhya Pradesh, Marathwada region of Maharashtra and in West Bengal this disease poses a constant threat to tomato and eggplant cultivation. The loss in yield may vary between 10.8 and 90.6% depending on the stage of plant growth at which infection occurs and on environmental conditions. Maximum loss occurs during the summer season when the crop is infected within 60 days of planting. The bacterium is same as described under bacterial brown rot and wilt of potato. Races 1 and 3 of the bacterium are mainly associated with tomato and eggplant. The race 1 (high temperature race) is predominantly present in the areas where the disease is destructive on eggplant.

As in potato, the characteristic symptoms of the disease are wilting, stunting, yellowing of the foliage, and finally collapse of the entire plant. Before the plant wilts lower leaves may first show drooping. The vascular system turns brown and if a segment of the lower stem is cut and squeezed it yields bacterial ooze. Development of adventitious roots from the stem is considerably enhanced. In many cases, when nematode infection is also present, the stem at the base becomes dark brown and constricted, leading to collapse of the plant. During a continued wet weather when the temperature is also high the most conspicuous symptom in tomato and eggplans is sudden drooping of the leaves without yellowing, and rotting of the stem from any point. The roots appear healthy and are often well developed. However, the brown discoloration is present inside.

The pathogen is soil-borne, persisting for long periods in some soils. Intact roots are not invaded. Infection always occurs through wounds caused by transplanting, cultural operations or nematode invasion. Root knot infection predisposes the plants to bacterial infection. Rifts in root cortex and breaking of the stem for emergence of adventitious roots also provide openings for entry of the bacteria. The role of weeds in the epidemiology of bacterial wilt of tomato was questioned in Australia. In a field with tomato cropping system, around 35 weeds species were examined and only one species of portulaca was found to harbor the bacterium.

The entry of the bacteria in the taproot is faster in susceptible than in resistant cultivars. However, once inside the root the bacteria colonize, to some extent, almost all regions of the susceptible and resistant plants. But colonization occurs faster in susceptible than in resistant plants. Rate of multiplication is also faster in susceptible plants In resistant tomato, the movement of the bacteria from protxylen or primary xylem to other xylem tissues is hindered (Nakaho *et al.*, 2004). The acidic extracellular polysaccharide (EPS) and plant cell wall degrading polygalacturonases are the major pathogenicity factors in *Ralstonia solanacearum*. Mutants lacking these proteins are less virulent than wild types. They slowly colonize fewer stems and have lower population in the tomato stems. The organism first moves in to the large xylem vessels. The bacterial cells and EPS get distributed throughout the vascular bundles and intercellular spaces of the pith in susceptible cultivars whereas in the resistant cultivars they are restricted to the vascular tissues. The strains of the bacterium lacking EPS do not colonize tomato stems as fast as the strains producing EPS. Resistance to bacterial wilt is measured by the extent of colonization of the stem by the bacterium (Prior *et al.*, 1996).

When a single lateral bundle is invaded drooping of leaves is common but if all bundles are invaded the plant wilts. Tendency of excessive adventitious root formation also depends on number of vascular bundles invaded. Adventitious roots develop just outside the invaded bundle. Nakaho *et al.*, (2000) studied the behavior of the bacteria in tomato tissues. The bacteria colonized both the primary and secondary xylem tissues of a susceptible cultivar. They were abundant in vessels of which the pit membranes were often degenerated. All parenchyma cells adjacent to vessels with bacteria were necrotic and some were colonized by bacteria. In the resistant cultivar the bacteria were observed in the primary xylem tissues but not in the secondary xylem tissues. Necrosis of the parenchyma cells was only occasional. The pit membranes were often thicker with high electron density. The inner electron-dense layer of cell wall of parenchyma cells and vessels was thicker and more conspicuous in resistant than in susceptible plants. The limitation in bacterial movement may be related to the thickening of the pit membrane and/or accumulation of electron dense materials in the vessels and parenchyma cells. In a resistant cultivar, the bacteria are observed in the primary xylem tissues but not in the secondary xylem tissues. Necrosis of parenchyma cells is occasional. The pit membranes are often thicker with high electron density. The limitation in bacterial movement in resistant cultivars may be related to the thickening of the pit membranes and/or accumulation of electron-dense materials in vessels and adjacent parenchyma cells (Nakaho *et al.*, 2000). Rahman *et al.*, (1999) studied the movement of the bacteria in a resistant and a susceptible cultivar of pepper. In resistant cultivar, first, a cell wall coating material developed together with swelling of the primary wall of the xylem vessel, limiting the bacterial spread. Second, formation of various types of vesicles in the vascular parenchyma cells enveloped the bacterial mass, partially restricting bacterial movement. Third, induction of hypersensitive reaction in xylem vessels resulted in the distortion and lysis of the bacteria. Such changes did not occur in the susceptible cultivar. Environmental relations and disease cycle are similar to those described for potato wilt.

Application of superphosphate increases severity of the disease while nitrogen suppresses it. Silicon amendment of soil is known to suppress many fungal diseases of plants by induction of acquired resistance and mechanical tissue barriers. Dannon and Wydra (2005) have, for the first time, reported suppression of a bacterial disease by silicon amendment. They had reported significant reduction of tomato wilt incidence in silicon amended soil planted with susceptible or moderately resistant tomato. The suppression is reported to be due to induced resistance. Silicon induces basal resistance in tomato against *R. solanacearum* by modifying pectic cell wall polysaccharide structure. In plants raised in silicon amended substrate the density of bacterial cells in roots and stems significantly declines (Diogo and Wydra, 2007) Calcium nutrition of tomato affects resistance to wilt. Increased concentration of Ca in a nutrient solution reduced disease severity in a moderately resistance cultivar.

Since the organism has a wide host range (all solanaceous vegetables and a number of weeds have some degree of susceptibility) it is difficult to control the disease by rotation of vegetable crops. Cruciferous vegetables such as cauliflower can be used in the rotation. Changing the fields to cereal cultivation for some years is definitely useful in reducing survival of the pathogen in soil. Soil solarization reduces the population of *Ralstonia solanacearum* to undetectable level. Soil solarization has been found to significantly suppress bacterial wilt in tomato. The suppressive effect is proportionate to duration of solarization. Patricio *et al.*, (2005) observed 100% suppression in a soil solarized for 60 days and 6–22% suppression in soil solarized for 37 days against 43–100% mortality in non-solarized soil Solarization of soil amended with antagonistic bacteria (*Bacillus cereus* and *Pseudomonas fluorescens*) for 8–10 weeks during summer considerably reduces the bacterial wilt (Kumar and Sood, 2001). In addition to *Bacillus* sp. and *Pseudomonas fluorescens, Pseudomonas putida* has also been used as biocontrol agent against bacterial wilt in tomato. The efficacy of these biocontrol agents depends on mobility of the used strains in the rhizosphere. Mobility is defined as ability of a bacterium to follow developing roots in absence of percolating water. Strains having higher mobility give better suppression of wilt than those having none or low mobility. Guo *et al.*, (2004) have studied plant growth promoting rhizobacterial strains of *Serratia, Bacillus* and fluorescent pseudomonads for biological control of bacterial wilt of field grown tomato, *Serratia* sp. strains and *Bacillus* sp strain provided disease control and increased yield. Formulation of the antagonists prepared by them remained effective for one year in storage. Coinoculation of virulent *R. solanacearum* and an avirulent strain of the same gave significant protection to tomato against wilt. Pre-treatment of seedlings with specific bacteriophages before inoculation provides high level of wilt control.

In Taiwan, Chang and Hsu (1988) had reported significant suppression of bacterial wilt of tomato by amendment of soil with a mixture of inorganic and organic waste material which they called S-H mixture. It contained silicon and calcium as some of the many constituents. Amendment of soil with urea + mineral ash also suppressed the disease. The mycelial leachates of shiitake mushroom (*Chentinula edode*) is reported to suppress symptoms of wilt in tomato and lima bean (Pacumbaba *et al.*, 1999). Mycelial leachate or extract of *Agaricus blazei* also suppresses the tomato wilt bacteria. Among indigenous technologies, use of cow urine, tobacco leaf decoction and mixture of asafetida and turmeric are reported to suppress bacterial wilt in tomato, eggplant and potato [Asian-Agri-History 8(4):305]. Cow urine is stored for 2 weeks, diluted with water (1:3) and sprayed on the plants.

Kumar and Sood (2002) studied the effect of host resistance, antagonists and soil amendment on bacterial wilt of tomato. In a moderately resistant genotype of tomato (EC 191536) there was complete control of wilt when seedlings treated with *Pseudomonas fluorescens* or *Bacillus cereus* were planted in soil amended with sodium nitrate or potassium nitrate. In susceptible genotypes also there was significant reduction of wilt incidence. Population of *Ralstonia solanacearum* was reduced to a level where no wilt development took place. In greenhouse experiments with a susceptible tomato cultivar Anith *et al.*, (2004) have reported that the plant growth promoting rhizobacterium *Pseudomonas putida* significantly reduced bacterial wilt incidence when applied to seedlings at the time of transplanting. *Pseudomonas fluorescens* owes its biocontrol efficiency to nitric oxide (NO) production. Strains weak in NO production can be improved by genetic engineering for enhanced NO production (Wang *et al.,* 2005). Presence of antagonistic *Streptomces* of *Aureus* group as endophyte in tomato plants is reported by Tan *et al.,* (2006). Their proportion is high in healthy plants than in wilted plants which had a higher proportion of bacteria with cell wall degrading enzymes.

Pythium oligandrum is a mycoparasite of large number of fungal plant pathogens. Tomato roots treated with the mycoparasite show enhanced resistance to bacterial wilt and reduced disease severity. The disease control is attributed to the fungal cell wall proteins which act as elicitors of defense responses. (Hase *et al.,* 2006, 2008). The mycoparasite is fairly stable in tomato rhizosphere. In a study reported by Takenaka *et al.,* (2008), populations in rhizosphere of tomato grown for 3 weeks in sterilized or unsterilzed field soil increased with the initial application of at least 5×10^5 oospores per plant. The hyphal development was frequent on the root surface and some hyphae penetrated into the root epidermis. However, the rhizosphere population decreased with the time after transplanting in sterile soil. According to these authors the biocontrol is not due to competition for infection sites or for nutrients.

A commercial product containing two *Bacillus* strains and another containing acibenzolar-S-methyl also reduced disease incidence. Acibenzolar-S-methyl alone also reduces tomato bacterial wilt but only when pathogen inoculum load is low. Avirulent mutants of the pathogen provide protection from virulent strains by competition for space in the xylem vessel. The mutants grow faster than the wild type virulent strains (Etcheber *et al.,* 1998) in the xylem when both are co-inoculated. In Himachal Pradesh (India) bacterial wilt of tomato was controlled by application of sodium nitrate or potassium nitrate with *Pseudomonas fluorescens* or *Bacillus cereus.* Amendment of soil with VAM (*Glomus mosseae*) together with antagonistic bacteria (*Pseudomonas fluorescens*) reduces *R. solanacearum* soil population and bacterial wilt of tomato. Localized and systemic increase of phenols in tomato roots induced by the VAM *Glomus versiforme* is reported to inhibit *R. solanacearum* and suppress wilt (Zhu and Yao, 2004). Mixtures of plant growth promoting rhizobactera suppress bacterial wilt of tomato in addition to anthracnose of *Capsicum* and cucumber mosaic virus in cucumber (Jetiyanon and Kloepper, 2002). There is no antagonism among the constituents of the mixture. The suppression is through elicitation of induced systemic resistance. In chilli, *R. solanacearum* is seedborne. Seed treatment with antagonistic *Pseudomonas fluorescens* significantly improves seed quality parameters and drastically reduces incidence of bacteria wilt in the field (Umesha *et al.,* 2007).

Pradhanang *et al.,* (2003) have demonstrated suppression of *R. solanacearum* in soil and reduction of bacterial wilt of tomato by amending the soil with plant essential oils. Populations declined to undetectable level in thymol, palmarosa oil and lemongrass oil treated soil. Tomato seedlings transplanted in infested soil treated with 700 mg/liter of these oils were free from bacterial wilt. Soil amendment with fresh tissues of these oil-yielding plants had no effect. Some thymol producing plants were found to be hosts of the pathogen. Ji *et al.,* (2005) have compared the bacterial wilt control efficacy of thymol and palmrosa oils under field conditions. The oils were applied to naturally infested soil in the field which was sealed with a plastic mulch for up to 6 days. Tomato seedlings were then planted on the seventh day. Thymol treatment reduced incidence of wilt from 65.5% in check to 12% in the treated plot. Palmarosa oil also gave good control but was inferior to thymol. This has demonstrated the potential use of thymol as a biofumigant.

Sharma and Kumar (2004) have studied the effect of crop rotation on population dynamics of *R. solanacearum* in a tomato wilt sick soil. Okra-maize-radish, ragi-French bean-okra, carrot-cucumber-cowpea-maize combinations reduced the soil borne inoculum.

In a screening of 62 exotic and Indian lines/ varieties of tomato 9 lines were immune and many were resistant to tomato wilt but none was a cultivated variety. Most cultivated varieties are highly susceptible. In Himachal Pradesh, genotypes Al 14 and 1794 were found to remain disease free in naturally infested fields and also in artificially inoculated soil for the entire growing season. Cultivar Arka Souran has moderate resistance. The immune lines lose immunity when grown at a different location. Variability among the isolates of the pathogen and weather parameters, especially temperature, determine stability of resistance.

Loss of resistance at high temperatures has been reported. The Mi genes that provide resistance to root knot nematodes protect tomato plants against bacterial wilt also. *Solanum aethiopicum* carries resistance to bacterial wilt. It can be sexually crossed with *S. melongena*. But the fertility of the progeny is very low. In order to transfer the resistance and improve the fertility, somatic hybridization by mesophyl protoplast fusion was tried. This has resulted in many hybrids that are fertile with viable seeds and are resistant to bacterial wilt (Collonmer *et al.,* 2001). Partial but stable resistance in tomato and pepper to bacterial wilt is controlled by oligogenes. Two to five genes with additive effect are estimated to control this type of resistance under tropical conditions (Lafortune *et al.,* 2005).

The chemical elicitor acibenzolar-S-methyl (ASM; Actigard 50 WG) induces systemic resistance in tomato against bacterial wilt (Pradhanang *et al.,* 2005) Used as foliar spray or soil drench enhanced resistance in moderately resistant cultivars. The ASM-mediated resistance was partially due to prevention of internal spread of *R solanacearum* toward upper stem tissues of tomato plants. In field trials, Ji *et al.,* (2007) combined soil fumigation with thymol (73 kg/ha) and foliar application of ASM (25 mg/liter), against wilt and root knot of tomato. The combination proved highly effective against both. The cell wall proteins of the mycoparasitic fungus *Pythium oligandrum* have elicitor activity that induces defence response in plants. Tomato plants treated with the mycoparasite show enhanced resistance to *R. solanacearum* and reduced severity of wilt symptoms. The mechanism of resistance induction involves production of ethylene which acts as a signal molecule.

BACTERIAL ROT OF WHEAT EARS

The yellow ear rot, bacterial spike blight, gumming disease and yellow slime disease of wheat is known in India as Tundu. It was first reported from Punjab in 1917. The disease also occurs in Egypt, China, Australia, Cyprus, and Canada. Similar diseases occur on a number of grasses in many European countries. The disease is usually associated with ear cockle disease (caused by the nematode *Anguina tritici*). In general. *Anguina* species are important, if not essential, vectors of *Clavibacter*, the bacterium that causes yellow ear rot. The bacterium sticks to the cuticle of the nematode. At least in India the disease does not occur on wheat unless the nematode is also present. A species of the bacterium, *Clavibacter toxicus*, with *Anguina* sp., attacks the seedheads of *Agrostis avenacea* and *Polypogon monspeliensis* and fatality among live-stock feeding on these grasses is reported (McKay *et al.,* 1993). *Anguina tritici* is a potential vector of *Clavibacter toxicus* (Riley, 1992). *Anguina australis* is a vector of *Rathayibacter toxicus* on *Ehrharta longiflora* (annual veldt grass).

Symptoms

Wrinkling of the lower and twisting of the central leaves is the first symptom of the disease. This is accompanied by exudation of a bright yellow sticky slime enveloping the entire ear. This slime binds the glumes, stem and leaf sheath thus checking the plant growth and distorting the stem. In wet weather the slime starts trickling down but in dry weather it becomes a deeper yellow, hard and dry. The disease appears only when the crop is reaching maturity.

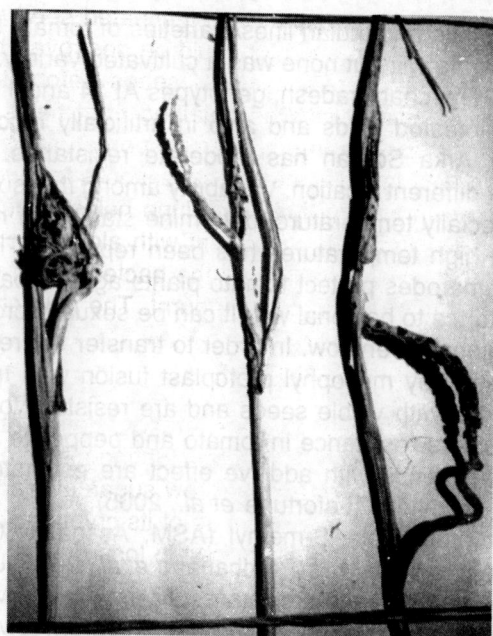

Fig. 8. Bacterial ear rot of wheat.

The causal organism

Rathayibacter tritici (ex Hutchinson) Zgurskaya, *et al.*, 1993. Earlier, it was known as *Clavibacter tritici*. The bacterial cells are Gram-positive, non-motile, pleomorphic rods with maximum temperature for growth at 34 C. Colonies are circular, snall, convex, and yellow pigmented on NBY medium of Vidaver (1980). They produce acid from glucose, fructose, sucrose, maltose but not from lactose. Starch is not hydrolysed. The bacterium can tolerate sodium chloride concentrations of 3–4%. Cell wall amino acid is diaminobutyric acid.

Disease cycle and environmental relations

Field soil is not a source of inoculum. Vasudeva and Hingorani (1952) had suggested an obligate etiological relationship between the bacterium and the nematode *Anguina tritici*. Infection of wheat with the bacterium takes place when seedlings are grown in soil infested with both the bacteria and the nematodes. It was presumed that the nematodes mechanically transmitted the bacterium from the galls to the seedlings. Subsequently, it was confirmed that symptoms of tundu disease appear only when galls as such from rotting ears are used as inoculum. In 1984 it was concluded that the bacterium, like other corynebacteria, does not survive in free state in the soil nor the nematodes separately carry the bacterium on their body. Up to 40 to 55% galls show presence of the bacterium. Only such galls are the main source of survival of the bacterium and primary inoculum. In a similar disease of grasses in Europe, caused by *Rathayibacter rathayi*, the host is free from nematodes.

Management of the disease

Sowing of wheat seeds free from nematode galls is an essential measure for management of this disease. The galls can be removed by floating the seedin brine (20 kg common salt in about 125 lit of water). Hot water treatment of contaminated seed lot is also effective in the removal of nematode galls. Galls in a seed lot could be destroyed by keeping the mixture in water at 54°–56°C for 10-12 min. Pre-soaking of seeds before the host water treatment improves the effect. Seeds should always be obtained from fields which are free from the disease. Since the galls may survive in soil thereby perpetuating the bacterium, attempts should be made to destroy them. In India, the wheat season is followed by hot and dry season. Turning of the soil during the months of May and June helps in destruction of the galls and nematodes by heat and desiccation. Soil treatment with aldicarb or carbofuran destroys the nematode and prevents infection of the plant by the bacterium. In mild attacks, the infected plants should be located, carefully uprooted, and burnt. The field should be well-drained.

RATOON STUNTING DISEASE OF SUGARCANE

This disease is one of the major causes of degeneration, deterioration or "running down" of several commercial varieties of sugarcane. Ratoon stunting disease (RSD) occurs not only in India but also at least in 40 other countries that grow sugarcane and is considered the most important sugarcane disease in the world. Because of its cryptic and chronic nature it is cited as an example of typical endemic disease. Economic loss due to the disease is caused by poor germination, extra cultivation cost in removal of weeds, low yield of canes, and loss of ratooning capacity of the cane. Extent of loss depends on the cultivar and its growing conditions. The loss is increased many fold when RSD occurs along with the mosaic disease of the same crop.

Symptoms

The name 'ratoon stunting' is a misnomer since the disease equally affects plant as well as ratoon crop. Diseased stools usually show stunted growth, reduced tillering, in canes with shortened internodes and yellowish leaves. Such abnormalities in growth are not specific for this disease and could be due to many other factors such as lack of water and nutrients. If such symptoms appear in the crop one has to first exclude the possibility of poor growth due to poor management before concluding the existence of RSD. Germinability of setts from diseased canes is low. Germination is delayed and plant growth is slow. Thus, weeds usually become more predominant and this not only further hampers crop growth but also escalates cultivation cost. During summer diseased plants show signs of wilting.

Typical symptoms of RSD are seen only after splitting open the cane longitudinally. Two types of discolouration are seen in the pith. In mature canes, there is orange red discolouration of vascular bundles at the nodes. In young canes pink colour can be seen near the nodes. However, many varieties are symptomless carriers and the presence of RSD can be detected only by using indicator sugarcane varieties or other artificially inoculated hosts.

The causal organism

For more than two decades after its discovery, RSD was considered a virus disease. In 1973 two independent groups of workers reported that bacteria were associated with the disease. They observed coryneform (pleomorphic) bacteria in the juice of affected canes. The observation was further confirmed by others in 1976. Actinomycete-like or coryneform bacteria were described in 1978. and the bacterium isolated, cultured, characterized and Koch's postulates in 1980. In the reclassification of the genus *Corynebacterium*, in 1984, the RSD pathogen *was* named *Clavibacter xyli* subsp. *xyli* sp. nov. subsp. nov. In 2000, the pathogen was reclassified in the genus *Leifsonia* as *Leifsonia xyli* subsp. *xyli* (Davis *et al.*, 1984) Evtushenko *et al.*

The RSD bacterium differs from other xylem-inhabiting fastidious bacteria in morphology, in being Gram-positive, in having a very narrow host range (sugarcane being the only natural host), and in being only mechanically transmitted during cutting of canes for seed setts.

The bacterium is small, smooth cell-walled, rod-shaped, and filamentous. The cells measure 0.3–0. 5 × 1–3 μm, but length of the filaments may be 10 μm or even more. The cells may be straight, bent, curved and/or swollen at the tip or in the middle. The bacterium is Gram-positive, non-acid fast, non-endospore forming and pleomorphic. Like streptomycetes, the cells divide by formation of a septum and contain mesosomes. They may occur in chains. Branching or pseudobranching has also been reported. Cellular and colony characters resemble those of actinomycetes. Colonies of *L. xyli* subsp. *xyli* on semi-solid media are circular with entire margins, convex, non-pigmented and approximately 0.1–0.3 mm in diameter after 2–3 weeks' growth.

The bacteria can be observed in nodes, internodes, leaves, young tillers and also in tissues just below the growing point. This suggests systemic infection. However, maximum accumulation of the bacterial cells occurs in the basal internode of the cane stalk.

The description of the RSD agent as a virus had included the following biological properties: juice remains infective even in high dilutions; infectivity is retained at room temperature for one day and at 4°–5°C for four days; at 20°C the juice can remain infective for 20 weeks; it is inactivated within 30 min at 50°–51°C.

The RSD bacterium is sap transmitted. No insect vector is known. Soil transmission through bacteria surviving in cane debris is also reported. Field dispersal takes place during harvesting through cutting knives and use of seed setts from diseased canes. Expression of symptoms is related to rate and extent of colonization. The relative population density of the bacterium in extracts of sugarcane is positively correlated with degree of cultivar susceptibility. Spread is faster in susceptible cultivars than in resistant cultivars (Davis *et al.*, 1988; Damann, 1992). Pathogen density in basal internodes is directly correlated with the difference in cultivar yield due to RSD. Sugarcane cultivars resistant to RSD may restrict the interxylar spread of *L. xyli* subsp. *xyli* in resistant plants. Resistant plants have more profuse branching of xylem elements and fewer elements that pass through the node without terminating. Colonization of vascular tissues by the bacterium is most extensive in lower more mature internodes. The number of infected vascular bundles and. the pathogen density estimates within successive internodes both decline distantly from the base of the stalk. The infected vascular tissues also are more extensive in the inner, more mature stalk tissue than in the outer, less mature tissues next to the epidermis. The ease of mechanical transmission of *L. xyli* subsp *xyli* in sugarcane tissues causes rapid secondary (within the crop) spread of RSD in susceptible cultivars. Resistant cultivars have lower density of bacteria in expressed sap than do the susceptible cultivars.

Effective dosage required for infection is lowest in susceptible and highest in resistant cultivars. At a certain level of plant resistance. populations of *L. xyli* subsp. *xyli* could be too low in resistant plants to infect other resistant plants, a condition defined as "**population immunity**". Since there is not much genetic variability in *L. xyli*. subsp *xyli*, the resistance in resistant cultivars is fairly stable. Comstock *et al.*, (1996) studied the relationship between resistance to RSD bacterium and spread of the disease in the field during harvesting by hand cutting and found that spread resulting from hand cutting of canes was very little in resistant cultivars (which have very few or no vascular bundles colonized by the bacterium) and very high in susceptible varieties which have high numbers of colonized vascular bundles. The infection and spread follows the direction of hand harvest within the rows from the infected source. Disease spread and increase is greater in plants grown from stalks collected at the first harvest than in the first ratoon from the harvested field (Hoy *et al.*, 1999).

So far, natural occurrence of the disease has been observed only in sugarcane. In artificial inoculations, the bacterium has been found to infect many plant species such as *Panicum maximum, Echinocloa, Cynodon dactylon* (Bermuda grass), *Brachiaria mutica, B. milliformis, Sorghum verticillatum, Chloris gayana, Sorghum halepense*, and maize. Elephant grass (*Pennisetum purpureum*) is also readily infected.

Management of the disease

If seed setts are disease free the disease can be effectively managed. The following steps are recommended for the management of RSD:

Disease-free nursery: Any good variety can be grown in a separate plot exclusively for seed. The plot should have good drainage and optimum dosages of ertilizers and irrigation should be ensured. The crop should also receive timely hoeing and weeding operations. All weeds growing on the borders should be destroyed. If weak plants are seen in the crop they should be uprooted and burnt. At the same time roguing of stools showing symptoms of wilt, red rot, and smut should also be done. Special attention should be given to even germination of planted seed setts. Any sett that fails to germinate within a reasonable period should be dug out and removed.

Selection of seed from commercial crop: Seed setts should be taken from only those fields or parts of a field where plants are robust. Weak crop should never be used for seed.

Precautions for taking ratoon: Weak or stunted crop should not be ratooned.

Heat therapy of seed: Seed setts may be given hot water, hot air, or aerated steam therapy, wherever feasible. In India, the time-temperature relation recommended for hot water treatment is 50°C for 2 hours or 52°C for half hour. Precautions necessary in this treatment are use of small wire net baskets to facilitate proper and uniform exposure of setts to hot water which should be constant at 50°C or 52°C for the entire duration of the treatment, cooling of setts immediately after treatment, and selection of seed material from a field in which disease incidence is low. The hot air therapy involves dry treatment at 54°C for 8 hours. Bud damage is less in hot air than in hot water treatment. The aerated steam therapy or moist heat therapy is considered better. In spite of various methods of commercial heat treatments the disease level in seed sources may be up to 33%.

Although antibiotics are known against coryneform bacteria, chemotherapy of RSD has not been found successful so far. Application of ammonium sulphate as fertilizer to sugarcane crop is reported to reduce the disease by about 22% and increase the yield by 29%.

The above measures are mostly concerned with phytosanitation. High cost of heat therapy is now being felt even in a country like USA also. The report that the bacterium can be transmitted by soil suggests that precautions like healthy seed and destruction of weeds, etc. may fail to prevent appearance and spread of the disease. Therefore, efforts are being made to develop varieties resistant to the disease which have very few or no vascular bundles colonized by the bacterium. A couple of such varieties have been developed in USA (Comstock *et al.*, 1996). No RSD immune commercial varieties are known anywhere in the world. Application of ammonium sulphate as fertilizer in sugarcane is reported to give 22.9% reduction in RSD incidence with 29% increase in cane yield.

BLACK LEG, WILT AND SOFT ROT OF POTATO

Soft rot of vegetables, including potato, caused by different species of *Erwinia*, is a common storage and transit disease. In potato, these bacteria not only cause soft rot of tubers but also 'black-leg' and wilt of plants in the field. In India, the disease was investigated in some detai' by Hingorani and Addy (1953) with special reference to the properties of the causal organism.

Symptoms

In the field the typical 'black leg' is characterized by a striking brown black or jet black color of the stem at the soil level. This discolouration usually starts from the old seed tuber. The cortical tissues may shrivel and rot. The plants may achieve normal height but usually they remain dwarfed and stunted. Instead of spreading habit. The branches and leaves show a tendency to grow upwards. The foliage turns light green or yellow, with a slight metallic luster, and soon wilts and dies. Curling of leaves similar to that caused by potato leaf curl virus may also be found. Sometimes, young seedlings arising from diseased seed tubers are destroyed before or soon after emergence. When infection occurs late and tubers have developed, they carry the bacterium to the storage godowns.

In soft rot of tubers, which may occur in the field if the soil is moist and temperature is high, or during transit and storage, the tubers are transformed partly or wholly, slowly or rapidly, into a soft decaying pulpy mass. This mass is held together only by the corky epidermis which is not attacked by the bacterium. When a soft rot tuber is cut open the colourless putrid mass turns a pink red on exposure to air, rapidly becoming brownish-red to brown black. The cut tubers have only a slight smell. However, when secondary bacteria have gained entry into the rotting mass a strong repulsive smell is felt.

The causal organisms

The bacterial species involved in black leg and soft rot of potato tubers are *Pectobacterium carotovorum* ssp. *carotovorum* and *P. carotovorum* ssp. *atrosepticum*. Although these names are officially accepted plant pathologist prefer to call them *Erwinia carotovora* subsp. *carotovora* (Jones) Bergey *et al.*, and *Erwinia carotovora* subsp. *atroseptica* (van Hall) Dye. In addition, at high temperatures, *E. chrysanthemi* (25°–35°C) and species of *Bacillus* and *Clostridium* are also encountered in diseased specimen.

E. carotovora subsp. *carotovora* has a very wide host range and is considered the main cause of soft rot of tubers in India while *E. carotovora* subsp. *atroseptica* is reported to be limited to potato in temperate regions and is the main cause of potato black leg (Toth *et al.*, 2003), although at temperatures below 15°C it may affect stems as well as tubers because it produces higher amounts of pectolytic enzymes at low temperatures. *Erwinia aroideae* (Townsend) Bergey, *et al.*, that had been earlier described as a distinct species associated with soft rot is synonym of *Erwinia carotovora* subsp. *carotovora*.

The soft rot causing species of *Erwinia* are straight rods, single cells, 0.5–1.0 × 1.0–3.0 μm, and motile by peritrichous flagella. They are Gram-negative, facultative anaerobes, and possess strong pectolytic activity. They produce acid from fructose, glucose, and sucrose. Gas production is relatively weak or absent. Optimum temperature for growth is 27°–30°C and maximum temperature for growth varies with isolates. While *Erwinia carotovora* subsp. *carotovora* has a maximum temperature for growth at 37°–40°, the subsp. *atroseptica* does not grow at temperatures above 35°C. Both species are tolerant to erythromycin. On crystal violet pectate medium with or without 35 μM/ml erythromycin *Erwinia carotovora* subsp. *carotovora* produces characteristic cavities at 27° and 33.5° but not at 37°C while the subsp. *atroseptica* does so only at 27°C. At these temperatures the species produce more pectolytic enzymes. The G + C content of the DNA of both subspecies ranges from 51 to 53 mole%.

Disease cycle and environmental relations

These bacteria are not seed-borne but in potato they are present in rotting tubers or in tubers which have been contaminated during grading or in the stores. Latent infection of tubers is common in commercial seed tubers. The bacteria are present intercellularly in lenticels, rarely in vascular tissues. The erwinia cells in latent infection appear to remain quiescent or dormant even when relatively large numbers (up to 10^6 cells per g of potato peel are present. Black leg disease is directly related to the level of seed tuber contamination. The higher the level of contamination, the earlier in the season black leg disease appears and the greater is the final level of disease which continues to rise as the season progresses. Finally, the progeny tubers also have the high level of contamination (Toth *et al.,* 2003).

Earlier, it was thought that these bacteria survive in soil as saprophytes for 20 years or more. Studies have now revealed that populations of the bacteria rapidly decline in soil when rotting tubers and plants have been removed. However, low populations continue to exist in soil, especially when protected in precolonized crop debris or as epiphytes on rhizosphere of weed plants, even before potato is planted This also serves as source of primary inoculum. Survival of *E. carotovora* subsp. *atroseptica* is longer in cool wet than in hot dry conditions One of the sources of *Erwinia* inoculum could be ponds or water reservoirs maintained for irrigation and receiving run-off water from infested fields (Norman *et al.,* 2003).

Erwinia carotovora subsp. *atroseptica* can live in the body of all stages of the seed corn maggot, *Hylemya platura* and may persist in the intestinal tract of both adult flies and larvae. The fruit fly (*Drosophila melanogaster*) is another insect host and vector. Such insect hosts and vectors are common on culled potato haulms and rotting tubers around potato fields and help in transmission of the bacteria.

Severity of decay of potato tubers is influence by potato genotype, tuber immaturity, harvest in warm conditions, high water potential, low calcium content, high nitrogen fertilization and interaction with other pathogens. Incidence of soft rot is negligible in cold storages provided proper low temperature is maintained throughout. Although there is no evidence that early varieties are more resistant than late varieties, date of harvesting is related to subsequent soft rot of tubers. According to Somani *et al.,* (1983b) there is progressive increase of soft rot incidence in tubers of cultivar K. Chandramukhi harvested from January to March and kept in country stores. Rapidity of wound cork formation in the tubers is not a criterion of resistance.

Level and source of nitrogen also influence the soft rot incidence in tubers Heavy nitrogen manuring of the crop increases the disease incidence. High calcium in tubers enhances structural integrity of cell wall which impedes growth of the bacterium through tissues thus making tubers resistant to soft rot. Increasing calcium content of soil and its uptake by the plant through use of calcium containing fertilizers is recommended as a method of reducing losses from soft rot. Free water and anaerobiosis weakens tuber resistance to soft rot and provide nutrients for erwinias to multiply. Presence of fungal wilt pathogens (*Fusarium oxysporum, F. solani, Verticillium dahliae*) in the soil enhances the incidence of black leg.

Infection of tubers occurs through bruises, sun scald, insect wounds, and nematode punctures. When the soft rot bacteria enter the wounds, they feed and multiply at first on the liquids released by the broken cells on the wounded surface. Rapid multiplication leads to production of increasing amounts of pectolytic and cellulolytic enzymes which cause maceration of tissues. Pectate lyase enzymes are major virulence factors of *E. carotovora*. These enzymes breakdown the pectic substances in the middle lamella and cell walls in to unsaturated oligogalacturonates, known to elicit plant defence responses. Cellulolytic enzymes partly breakdown the cellulose in the cell walls. Expression of full virulence and production of macerating enzymes is dependent upon uptake of citrate from the host tissue into the bacterium. Isolates lacking the gene for this process cause low level of tissue maceration. In moist fields lenticels of tubers are very much enlarged and, if temperature is high, they provide the maximum opportunity for infection. When cut tubers are planted, high soil moisture combined with lack of oxygen hinders cork formation and so helps in infection by the bacteria which can thrive in low oxygen conditions. When tubers are infected the bacteria enter the vascular bundles and by means of pectolytic enzymes bring about the soft rot condition. Blackleg symptoms are expressed when large numbers of the pathogen present in the rotting mother tubers invade the stems and multiply in xylem vessels under favorable weather conditions. They tend to out-compete other bacteria in tuber rots because of their ability to produce larger quantities of a wider range of cell-wall degrading enzymes (Perombelon, 2002) Bacteria may be present in the plant stem after inoculation but black leg develops only when bacteria from the rotting seed tuber reach the point where stem is attached to the tuber. Motility of the bacterial cells is an important determinant of pathogenesis. Resistance of the stem base is a criterion for identify clones possessing partial resistance to black leg (Allefs *et al.*, 1996). In a growing plant the bacteria become systemic causing browning of vascular bundles along the shoot, down to the stem base and newly formed tubers. Some strains of *E. carotovora* subsp. *carotovora* produce an antibacterial substance which acts like a bacteriocin against other strains (Seo *et al.*, 2004).

Healthy organs (stolons, stems, progeny tubers) from symptomless plants are less frequently contaminated than symptomless organs from diseased plants. The incidence of *E. carotovora* subsp. is generally greater in tissue taken from the stolon ends than from peel samples although there is a positive correlation between the bacterium's presence in the two sample types (De Boer, 2002).

Management of the disease

Only healthy tubers should be used for seed. When cut tubers are used they should be kept, after cutting, at low temperatures (12 – 15°C) in a dark room for about 4 days. This allows cork formation and avoids entry of bacteria from soil after planting. Alternatively, the cut tubers should be dipped in a suitable antibacterial solution. Mercuric chloride solution, organo-mercurial compounds like Agallol, Aretan or Emisan were used in the past. Stable bleaching powder solution (chlorine) is safer than mercurials. Antibiotics have also been recommended for tuber treatment. Stable bleaching powder as soil drench and streptocycline for tuber dip treatment gives effective control of the disease. Stable bleaching powder (chlorine) gives better control than antibiotics. Soil treatment with bleaching powder (12.5 kg/ha) in furrows is more effective than tuber treatment (500–1000 ppm). The bleaching powder treatment inhibits respiratory enzyme activity of the bacteria. Prolonged exposure of seed tubers to chlorine may damage sprouts (Tweddell *et al.*, 2003). In Brazil, seed tuber treatment or plant treatment with the resistance inducer compound, acibenzolar-S-methyl (ASM) at rates below 250 mg a.i./l was found to induce resistance to black leg (*E. carotovora* subsp. *atroseptica*) but the effect was cultivar specific. Some cultivars did not show disease suppression (Benelli *et al.*, 2004). Medina *et al.*, (2004) have tested the antimicrobial effect of 5 naphthoquinones against phytopathogenic bacteria and reported that the comound naphthazarin (NTZ) has the best antibacterial activity. It is bactericidal at 10 μg/ml and can inhibit the soft rot development at a concentration of 2 mg/ml.

Shallow planting, avoiding early planting when soil moisture and temperature are high, avoiding injury to tubers during tillage, and keeping the field well aerated are other precautionary measures. As soon as black leg affected plants are located they should be dug out and burnt. Before storage the tubers should be washed with chlorinated water. The stores should be kept dry and well aerated and maintained constantly at low temperature. Periodical culling of rotten tubers from the store reduces chances of contamination of healthy tubers.

Sharga and Lyon (1998) had reported a strain of *Bacillus subtilis* antagonistic to. soft rot and black leg bacteria. *Pseudomonas fluorescens* strain F113 is also reported to control potato soft rot (*E. carotovora* subsp. *atroseptica*) by production of the antibiotic 2, 4-diacetylephloroglucinol, not through competition, although siderophores produced by the bacteria may play some role. A bacteriocin-like substance is produced by *Bacillus licheniformis* that is inhibitory to growth of *E. carotovora* cells. The substance is effective in preventing spoilage on potato tubers reducing the symptom of soft rot at 240 μg/ml or higher concentrations (Cladera-Olivera *et al.,* 2006).

Bacteriophages are reported as biocontrol agents of *Erwinia*. Two distinct types of bacteriophages that were specifically active against *E. carotovora* subsp. *carotovora* were isolated from calla lily (Ravensdale *et al.,* 2007). The phages pssessed icosahedral heads with long, flexible tails and were placed in the order *caudovirales.*

COMMON SCAB OF POTATO

The disease is of world-wide occurrence. In India, the disease was introduced from Myanmar in 1958 and in the plains of north India it was first seen in Patna in 1958 with as high as 61.7% incidence in some fields. Otherwise it remained restricted to hilly regions. By 1965 common scab was found throughout Bihar. Later, potato scab was reported from U.P. (1979), Punjab (1962), Maharashtra (1976), Gujarat (1979) and M.P. (1984). It is present throughout Himachal Pradesh where it is more severe in Lahaul valley. The percentage of scab bearing tubers had been as high as 80% in some fields in U.P. The greatest loss from the disease is the reduction in market value of the tubers due to rough and blemished skin. When scab is deep seated, there is considerable loss in peeling of tubers. The seed certification laws do not permit more than 3–5% scab in the seed lot, hence the disease is of special significance for seed growers.

Symptoms

There are two types of lesions on the tuber: (i) the shallow and (ii) the deep scab. In shallow scab the affected tubers show superficial roughened areas, sometimes raised above, and often slightly sunken below the plane of the healthy skin. The lesions consist of corky tissues which arise from abnormal proliferation of the cells of the tuber periderm due to attack of the pathogen. The lesions vary widely in size and shape, only sometimes darker than the healthy skin, and often become confluent to give a reticulate appearance.

In deep pitted scab the lesions measure 1–3 mm or more in depth and are darker than the lesions in shallow scab. They also are corky and may join together so that the entire tuber surface becomes affected. The deep-pitted lesions are either extensions of the shallow lesions, combined effect of the scab organism and some chewing insects, or due to some specific strains of the scab organism.

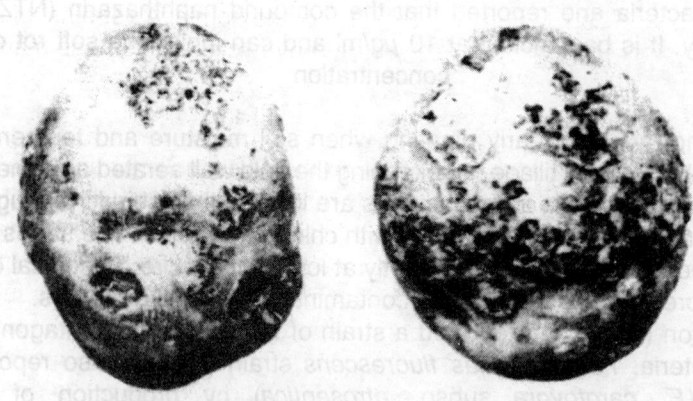

Fig. 9. Common scab affected potato tuber with deep pitted lesions.

The causal organism

Common scab of potato is caused by *Streptomyces scabies* (Thaxter) Waksman and Henrici but many other species such as *S. griseus, S. aureofaciens, S. flaveolus, S. caviscabies, S. acidiscabies, S. turgidiscabies, S. europaeiscabies,* and *S. retuculiscabies* (netted scab) also cause some forms of potato scab. *S. scabies* was first described as *Oospora scabies* by Thaxter in 1891. Later, its name was changed to *Streptomyces scabiei* (ex Thaxter) Lambert and Loria. Description of the species is as follows.

Aerial mycelium is gray. Sporophores are much branched, wavy or slightly curved, occasionally forming spirals. Spores are smooth-walled, cylindrical, 0.8–1.0 × 1.2–1.5 μm in size. On sucrose nitrate agar the growth is abundant, wrinkled, raised, gray to cream colored. Aerial mycelium is cottony and white to grey. On glucose- asparagine agar the growth is restricted, folded, cream coloured and aerial mycelium is scanty, white to grey. On nutrient agar no aerial mycelium is formed and soluble pigment is deep golden brown. Starch is hydrolysed and nitrates are reduced. The bacterium shows strong tyrosinase reaction by forming melanin pigment on tyrosine-casein-nitrate agar medium. However, non-melanin positive strains are also associated with potato scab. Thaxtomin A, a phytotoxic dipeptide produced by many plant pathogenic streptomycetes (*S. scabies, S. acidiscabies, S.turgidiscabies*) is related to pathogenesis. In potato, a positive correlation has been demonstrated between the pathogenicity of different isolates of *S. scabies* and their ability to produce the phytotoxin. The smallest subunits of plant-derived cellulosae (cellobiose and cellotriose) stimulate thaxtomin production by the pathogen. However, it is not the sole cause of symptom development. The disease occurs without evidence of thaxtomin production by some strains. Melanin-negative mutants of *S. scabies* are reported to produce less thaxtomin compared to melanin-positive strains. The phytotoxin targets plasma membrane, various components of the cytoskeleton and the cell wall (Fry and Loria, 2002). Plasmalemma of host parenchyma cells is detached from the cell wall in several places although the plasmalemma is not ruptured. The space between retracted plasmalemma and host cell wall often contains fibrillar material. *In vitro,* the synthesis of thaxtomin A is inhibited by the amino acids tryptophan, tyrosine and phenylalanine. Esterases are involved in thaxtomin biosynthesis. Esterase from *S. scabies* strains allows the release from lipid polymers of plant molecules that may act as inducers for thaxtomin biosynthesis genes or as precursors in the biosynthesis pathway of thaxtomin A. The species causing potato scab may be found occurring simultaneously in the same field, same plant, same tuber and same scab lesion. In addition to thaxtomin A, strains of non-pathogenic *S. turgidiscabies* and *S. scabies* profuce the toxin antimycin A which has mammalian toxicity.

Disease cycle and environmental relations

S. scabies can persist in soil indefinitely, provided the soil conditions are, directly or indirectly, favorable for it. It gets distributed through movement of soil by wind, water, and cultural operations. The infected seed tubers play the main role of its distribution over large areas. Of the total *Streptomyces* spp. isolated from the tubers of potato, 99% are pathogenic from scabby lesions and only 12% are pathogenic from clean surface of the same tuber. The density of pathogenic populations slowly declines from 10^9 to 10^8 CFU per gram of lesion tissue after storage for 18 weeks (Wang and Lazarovits, 2004). The incidence of tuber scab is directly related to the severity of the disease on seed tubers. The bacterium can survive through the digestive tract of cattle and can thus be disseminated by manure.

Infection of tubers occurs through newly formed unsuberized lenticels, stomata, wounds, and directly through the cuticle if it is thin. The young tubers, during formation of internodes and before suberization of lenticels are most susceptible to infection. Once the corky layer develops in the lenticels infection no longer occurs. In absence of natural opening, the phytotoxin thaxtomin aids penetration of developing plant tissues by inhibiting primary cell wall development (Loria *et al.*, 2003). The pathogen normally develops in dead cells on the exterior. Metabolic responses of the underlying cells results in rapid proliferation of tissues which cause the development of lesions. There is little invasion of living tissues.

Severity of the disease is influenced by various soil factors. Like all bacteria, this organism also is susceptible to acidity. The disease is significantly decreased when soil pH is brought down to below 5.2. Within the pH range of 5.2–8.0 the severity of the disease increases with pH. Production of the toxin thaxtomin A is also related to pH. Its production by *S. scabies* declines with decline in pH. However, in *S. acidiscabies* toxin production is not affected by low pH. Biotransformation of thaxtomin A by *Aspergillus niger* into several lesser toxic components is reported (Lazarovits *et al.*, 2004).

Optimum temperature for growth of the organism is 25°–30°C, maximum 40.5°C and minimum 5°C. The disease can develop at a temperature range of 11°–30.5°C with optimum disease development at 20°–22°C. Optimum temperature for infection is slightly below 20°C and for lesion development slightly above 20°C. In general, scab is favored by high temperature and low soil moisture. Early planted crops (September-October) suffer more than the crops planted toward end of November (in India). In a Canadian study. early planting and delayed harvesting enhanced the yield but also increased the number of scab-damaged tubers. The longer harvest was delayed after top-kill the greater was the scab damage. In the presence of potato crop, population of the bacterium increases in soil while in the presence of cereals and legumes the population is decreased.

Unlike other bacteria, *S. scabies* is most active in dry soils. Therefore, the disease is suppressed by watering. While pH has direct effect on the organism, moisture affects its activity indirectly through a system of biological control by increased activity of antagonistic bacteria near the lenticels. The pathogen is highly susceptible to antagonistic effect of the bacterium *Bacillus subtilis*. Certain strains of this bacterium, when present in soil, do not permit establishment of the pathogen in that soil. The population of such bacteria is increased in soil by green manuring with soybean or by cultivation of soybean in the crop rotation. Another method of promoting the activity of such bacteria is continuous cultivation of potato with irrigation. Such soils become pathogen suppressive and scab either declines after some years or at least does not increase.

S. scabies has several strains having different degrees of virulence. Some cause shallow lesions, others different levels of deep lesions. Many avirulent strains of the species are also often isolated from soil infested with the scab organism. Antagonism between virulent and avirulent strains is also reported. *S. scabies* infects carrot and radish also.

Management of the disease

Due to obscure nature of the disease in India not much work on it had been done in the past. Almost in all countries, its control has been difficult because of the soil- and seed-borne nature of the pathogen. No cultivated potato variety in India is resistant to common scab except Patna Red which was reported as highly resistant in 1980. Since soil pH and soil moisture are important in the development of the disease, these have been explored for reducing its incidence. In addition, attempts have been made to induce biological control through such practices as green manuring and soybean cover crop. Several *Streptomyces* strains (including strains of *S. scabies* and *S. diastatochromogenes*) are capable of suppressing potato scab caused by *S. scabies* and have been successful in the biocontrol of the disease in the field. The actinomycetes capable of suppressing *S. scabies* produce inhibitors of aerial mycelium inhibitors/Significant positive correlation between the ability of a strain to inhibit others and the ability of that strain to resist inhibition is reported. The suppression is through production of antibiotics to which resistance may develop in mutants of the pathogen. However, biocontrol success varies significantly among pathogen isolates and is not correlated with *in vitro* sensitivity to antibiotic inhibition. Antagonists also vary in their effectiveness as bio-control agents. Relative effectiveness of different antagonists varies among growing seasons. Further, bio-control varies among potato cultivars in the fields (Ryan *et al.*, 2004). They are less effective when the pathogen is numerically in a strong position. Population densities of a suppressive strain in soil are more highly negatively correlated with scab severity than is the population on roots, suggesting that rhizosphere soil rather than potato roots may be the primary source of inoculum of the suppressive strain (Ryan and Kinkel, 1997). It is not necessary that suppressive strains must show antagonism *in vitro* in cultures. Fertilizer application, especially ammonium, suppresses common scab. The common scab may be managed by minimizing the period the crop is in the ground but this reduces yield. Bacteriophages that infect *S. scabies* have been isolated and found effective in suppression of the pathogen.

Tubers with deep lesions should not be used for seed. As far as possible only spotless tubers should be planted. The tubers can be disinfected by dipping in organo mercurial fungicides. In the past, a 5-min dip in 0.25% suspension of Emisan-6, Agallol-6 or Aretan was recommended. This treatment had been recommended for control of black scurf disease of potato also. Dipping of tubers for 10–30 min in 0.5% Agallol (a mercurial no more recommended) was most effective in eradicating the tuber-borne infection. Streptocycline and Blitox also significantly reduced the disease incidence. Seed tuber treatments recommended for black scurf apply to this disease also. Wilson *et al.*, (1999) found tuber treatment with fluazinon, flusulfamide, fenpiclonil, PCNB and mancozeb effective in suppressing the disease.

Suppression of activity of the pathogen in the soil through modification of soil environments (pH, moisture, and management of soil microbiota) has been extensively studied. Application of elemental sulphur to reduce pH below 5.2 has been a successful method but because of the high cost involved it is not a practical measure. Pavlista (2005) has reported that application of ammonium sulfate or ammonium thiosulfate in furrow at planting time and at tuber initiation stage reduces incidence of common scab. Lapwood and Hering (1970) and Lapwood *et al.*, (1973) had reported control of scab in field plots by maintaining soil moisture near field capacity during tuber formation until all internodes of the tubers were formed. Thus, keeping the potato field soil wet (near field capacity) from the fifth week after planting up to the ninth week by frequent light irrigation is recommended. However, continuously wet soil conditions favour the powdery scab caused by *Spongospora subterranea*. S. scabies is a persistent soil survivor and crop rotation does not control potato scab.

Green manuring and cover crops of legumes promote activity of antagonistic bacteria. In infested fields green manuring before taking a potato crop or cultivation of legumes before potato is a feasible and highly beneficial practice. Green manure treatments may contribute to active management of the pathogen inhibitory activity of the streptomycete community to achieve plant disease control (Wiggins and Kinkel, 2005). A rotation of four years avoiding beet, fleshy rooted crucifers and carrot, which are susceptible to this pathogen, is also beneficial. All Actinomycetes are susceptible to pentachloronitrobenzene. Application of PCNB (Brassicol) at the rate of 20–30 kg/ha reduces the incidence of not only common scab but black scurf or Rhizoctonia disease of potato also. In Canada, Lazarovits *et al.,* (1999) had reported reduction of common scab, Verticillium wilt and nematodes of potato by soil amendments with soymeal, meat and bone meal. Use of sulphate fertilizers enhances biodiversity in the soil and increases the ability of root zone to suppress *S. scabies*. The amendment of soil with lignin-rich biowaste reduces severity of scab by about 55% in the year of application and the suppression lasts ion the next season also.

Jansky and Rouse (2003) have reported that multiple resistance against soft rot, common scab, black scurf, Verticillium wilt and early blight could be found in interspecific *Solanum* hybrids. They have identified one clone (C545) which has exhibited improved resistance to soft rot, scab, pitted scab, early dying disease and early blight. This clone could be an especially valuable source of disease resistance.

CITRUS GREENING DISEASE

Citrus greening disease (Huanglongbing, HLB) is common throughout the world as a destructive disease. The disease is believed to have originated in Africa from where it spread to Europe, India, China and Southeast Asian countries. However, the disease is considered to be present in different areas since ancient times, perhaps even when Africa and Asia formed part of a super continent before the geological divide. In India, it is presumably present since the 18th century under the name die-back and decline. In China it has been known for the last more than 100 years and is known as *huanglongbing* (yellow shoot) which is its internationally accepted official name. In the Philippines, where it was known as mottled leaf disease, it was responsible for loss of 7 million trees in 1962 although it was not a serious problem until 1957. In Indonesia, where the disease was recognized as phloem necrosis and vein phloem degeneration, the disease was responsible for loss of 3 million trees between 1960 and 1970. In India, the disease is particularly serious in the northern states and is considered more dangerous than the tristeza virus because of its widespread occurrence. In most trees greening is found along with tristeza. There is synergistic relationship between the two and the two are jointly responsible for citrus die-back and quick decline. Prior to 2004, citrus greening was known to occur in Asia, from Japan in the east, through southern China, Southeast Asia and the Indian subcontinent to the Arabian peninsula. In Afruca, it can be found throughout eastern, central, and southern Africa. The name "greening" is of South African origin. The disease appeared more recently in Brazil (South America) and Florida (USA). Some regions such as the Mediterranean area, Near and Middle East, Australia are free of the disease but can not hope to remain so for long because of the presence of vectors in contaminated areas in the neighbourhood.

Symptoms

Symptoms of huanglongbing are varied and can resemble other disorders. A tree which becomes infected in the field usually develops one or more yellow shoots, The nama huanglongbib is derived from this characteristic symptom. The name citrus greening is based only on fruit symptoms. Citrus greening, in India, is characterized by yellows type of symptoms. Leaf chlorosis is the main symptom. It resembles the symptoms of zinc deficiency. Since chlorosis can be caused by nutritional disorders many workers had claimed control of greening with micronutrient sprays which only temporarily masked the symptoms of greening. In the yellow tissue of leaf lamina scattered green islands are seen. The leaf veins are also yellow. A characteristic feature of greening is that the yellow areas are surrounded on one side by the midrib and on the other side by lateral veins. The yellowing expands towards the margins. Size of leaves is also reduced. The leaves are thicker than normal and usually remain erect. The internodes of branches are shortened giving a bushy appearance to the branch. Such branches produce excess of buds and later show die-back. The diseased trees look stunted, flower earlier than the healthy trees and produce small fruits. The infected fruits are lopsided and have bitter taste. They often contain aborted seeds. They may fall prematurely. There is considerable reduction in number of roots. Chronically infected trees are sparsely foliated and show extensive twig dieback. The Asian greening symptoms are more severe than African greening symptoms.

The causal organism

The citrus greening was considered a virus disease for a long time. Later, it was considered to be caused by mycoplasma-like organism and some workers even claimed its isolation and culture. A comparative study with citrus stubborn phytoplasma revealed that greening organism has a 25 nm thick envelop while the cytoplasmic membrane of citrus stubborn phytoplasma (*Spiroplasma citri*) is only 10 nm thick. In 1970, D. Lefleche and J.M. Bove' had reported that the disease was caused by a phloem-inhabiting fastidious bacterium with double-membrane cell wall, distinct from MLO cells. In 1981, Naidu and Govindu, in India, had also reported association of a bacterium-like organism with the disease. They had observed the bacteria in sieve elements. The number in sections of invaded tissue varied from a few to a large number. The organism was spherical to ovoid, with rippled cell wall and measured 0.2 to 0.5 μm with an average diameter of 0.35 μm. In transverse and longitudinal sections the organism was surrounded by a 25–30 nm wide envelop. Later, the bacterium was taxonomically identified as *Candidates* (*Ca*) Liberobacter asiaticus for Asian citrus greening and *Ca*. L. africanus for African citrus greening (Jagoueix *et al.,* 1994, 1996). A third species, *Ca*. Liberbacter americanus is reported from Brazil (Teixeira *et al.,* 2005).

The cells of phloem-restricted fastidious bacteria are Gram-negative, non-motile, non-pleomorphic, rigid rods measuring 1.0–2.0 x 0.2–0.5 (1.3 x 0.3) μm. The citrus green bacterial cells, exclusively present in the phloem, are elongated sinuous rods with an uneven diameter of 0.15–0.25 μm. Round forms of larger diameter can also be observed in degenerating host cells. Similar bacterial cells are found in haemolymph and salivary glands of the two insect vectors of the bacterium. These cells are present in mature sieve elements, irregularly distributed among vascular bundles. They are sensitive to different antibiotics including penicillin and tetracyclines. Due to presence of the cell wall they are more sensitive to penicillin which interferes with cell-wall synthesis, than to tetracycline.

Isolates of greening bacterium from Africa, China, Taiwan, India and the Philippines share common serological and protein profile and a high percentage of DNA homology (cf. Da Graca,

1991). However, the isolates of *L. africanus* differ from *L. asiaticus* in their temperature relations. With African greening, severe symptoms appear under cool conditions (22°–24°C) and no symptoms are seen at 27–30°C. In Africa the disrease occurs in cool climates at elevations above 600–1000 meters. The symptoms of Asian greening are pronounced at both low and high temperatures and the organism is not inactivated at extended periods of high temperature (symptoms appearing also above 30°C). Within the species, a number of strains occur. Some cause mild symptoms, others cause severe symptoms. Subandiyah *et al.,* (2000) have observed that some isolates of the greening organism from Japan, the Philippines, Indonesia, Thailand and Nepal constitute one strain which is similar to Indian and Chinese strains but different from African strains. The American *L. americanus* belongs to the heat tolerant group being severe at 22°–24°C and 27–32°C.

Citrus greening is a disease of Rutaceous plants. It severely affects sweet orange, mandarin and tangelo trees. Other species of citrus also show more or less pronounced symptoms. *Citrus aurantifolia* (lime) is less susceptible than sweet orange and mandarin even though it is a preferred host of the vector *Diaphorina citri.*

Alternative hosts of the Asian greening bacterium, outside *Citrus,* are also reported. Chinese box orange (*Severinia buxifolia*) is as good a host for multiplication of the bacterium as the *Citrus* species (Hung *et al.,* 2000; 2001). Wood apple (*Limonia acidissima*) is a transient host in which the bacteria exist temporarily and disappear some months later. Other reported hosts of the greening bacteria and its vector *D. citri* are *Verpris lanceolata, Murraya paniculata* and *M. exotica.* These are citrus related plants. The greening disease could spread from such sources through grafting and the vectors. Ann *et al.,* (2004) have demonstrated that citrus trees suffering from Phytophthora root rot (gummosis) are prone to attack of greening.

The citrus greening bacterium is transmitted through vegetative propagation and by two species of citrus psylla. The transmission by grafting is inconsistent. The pathogen does not readily pass to progeny trees propagated by buds from infected trees possibly because of necrosis of sieve tubes and uneven distribution of the bacteria. No infection is obtained when material from apparently healthy sections of the diseased plant are used for propagation. In a recent study on graft transmission of *Ca* L, americanus on different cultivars and species of *Citrus,* Lopez and Frare (2008) obtained graft transmission by using buds and budsticks from infected plants. Larger the length of budsticks more was the percentage of infection. Both species of *Liberobacter* can be transmitted by dodder (*Cascuta campestris*). Transmission of the bacterium from sweet orange trees to periwinkle through dodder has also been reported. Titeneni et (2008) have recently reported that *Ca*. Linerobacter asiaticus is distributed in bark tissues, leaf midrib, roots and different floral and fruit oarts, but not in endosperm and embryo, of infected trees. Quantitative distyribution is uneven, Maximum concentration of bacterial cells is observed in peduncles.

The psylla species transmitting the citrus greening bacteria are *Diaphorina citri* (Asian psyllid vector) and the African psyllid vector *Trioza erytreae.* Electron microscopic studies have confirmed the role of *T. erytreae* (African psyllid vector) in transmission of the organism. Although other species of psylla are found on citrus there is no evidence that they play any role in transmission. Prevalence of these vectors determines the regional prevalence of greening. Since the bacteria are unevenly distributed in the plant and detection in plants not showing symptoms is uncertain, the detection of the pathogen in the vector is better and more accurate method of following spread of the disease from an infested area to a clean areaLow temperatures (< 10°C) suppress development of *Diaphorina citri.* The spread of greening bacteria by the vector is closely related to its feeding habit and search for suitable sites for laying eggs. A single female can lay 800–1000 eggs during her life. The females searches for young leave which are not yet eaxpanded for oviposition. Fully expanded leaves are not suitable.The eggs hatch to release nymphs. There are five nymphal stages. Within the temperature range of 15°–30°C, longevity of females increases as the temperature decreases. Temperatures above 30°C may shorten the life of adults to less than 30 days, otherwise at 15 to 20°C they can be expected to live 50 to 80 days. Average number of eggs produced per female significantly increases with increasing temperature and reaches the maximum at 28°C. *D. citri* can acquire the pathogen in 15–30 min of feeding with a latent period of 8–12 days while one hour or more is required for 100% transmission. *T. erytreae* acquires the organism after one day of feeding and transmits it 7 days later with an exposure time of one hour (cf. da Graca,1991; Brlansky and Rogers, 2007). After acquiring the bacteria, the vectors remain infective for their whole life. A single individual is enough to transmit the disease. After being acquired by the vector the bacteria have an incubation period of 8–10 days in the vector body before the latter becomes infective. The vector can acquire the bacterium in its larval stage also but cannot transmit it to healthy trees. It can do so only after becoming an adult.

Management of the disease

There are no curative methods to control HLB although antibiotics have been tried with partial success.

Quarantine

If the disease is still absent from a given region very strict quarantine for prevention of infected planting material from contaminated regions is necessary.

Removal of infected (contaminated) trees

If all infected trees were symptomatic, it would be easy to spot them and destroy. Unfortunately, there is a long latency period during which recently infected trees do not show symptoms. Badly affected and uneconomical trees can be spotted and should be cut down and destroyed. New plants raised from indexed stock or from nucellar seedlings should be planted. In areas where greening affected trees and the vectors are present such healthy trees also become infected in a few years through the vectors. Pruming of symptomatic as well as asymptomatic trees does not help. The eradication of vectors is, therefore, the most important control measure.

Thermotherapy to control citrus greening has given variable results. Treatment of budwood at 47°C for 2 hours reduces the disease incidence and longer treatments eliminate the pathogen. Treatment of infected young plants and seedlings budded with infected tissue at 38°–40°C for 6 weeks also kills the pathogen.

Chemical treatment is directed towards suppression of the greening bacterium in the plant tissue and management of the psylla vectors. Appropriate use of insecticides has resulted in good control of vectors in South Africa. For direct control of the bacterium, trunk injection of tetracycline hydrochloride, preferably continuously for 7 days, had controlled the disease in South Africa but this antibiotic is phytotoxic for citrus budwood. In addition, large scale use of the antibiotic has adverse effects on the environment. Kapoor and Cheema (1983) and Cheema *et al.,* (1985) have reported 100% recovery of trees by spraying a mixture of Bavistin (carbendazim) and Ledermycin (500 ppm each) six times at 10 days interval. Ledermycin alone is not effective while Bavistin alone reduces the disease incidence. Streptomycin and chloretetracycline mixture is also reported to suppress the disease.

All methods of control of the disease are insufficient unless there is eradication of the psyllid vectors. Eradication of citrus psylla by regular sprays of 0.02% of such insecticides as diazinon, endrin, or parathion reduces spread of the disease. The efficacy of petroleum oil spray (concentration increasing from 0.25%–1.0%) in the control of psylla (*D. citri*) is reported. Ist and 2nd instars are most susceptible but eggs are highly tolerant (Rae *et al.*, 1997). Repeated sprays make the approach costly.

In biological control, the major parasite of *T. erytreae* is a parasitic wasp (*Tetrastichus* or *Tamarixia radiata* and *T. dryi*). In India, *Tetrastichus* (*Tamarixia*) *radiata* is reported to drastically reduce psylla (*Diaphorina citri*) population. In Africa *T. dryi* is effective against *Trioza erytreae*). However, these parasites of psyllids are themselves hosts of predatory insects. Many predators of *Diaphorina citri* also cause natural mortality of its nymphal stage. Coccinellid beetles are the main predators.

REFERENCES

Ab-You, N. Gagnevin, F. Chinoleu *et al.*, 2007. Pathological variations within *Xanthomonas campestris* pv. mangiferiaeindicae support its separation into three distinct pathovars that can be distinguished by amplified fragment lrmngth polymerphism. *Phytopathology* 97(12): 1568

Anith, K.N., M.T. Momol, J.W. Kloepper, J.J. Marois, S.M. Olson and J.B. Jones. 2004. Efficacy of plant growth promoting rhizobacteria, acibenzolar-S-methyl and soil amendment for integrated management of bacterial wilt on tomato. *Plant Dis.* 88(6): 669

Ann, P.J., W-H. Ko and H-J. Su. 2004. Interaction between Libukin bacterium and *Phytophthora parasitica* in citrus host. *Eur. J. Plant Pathol.* 110(1): 1

Arias, R.S., S.C. Nelson and A.M. Alvares. 2000. Effect of soil matric potential and phylloplanes of rotation crops on the survival of a bioluminscet *Xanthomonas campestris* pv. *campestris. Eur. J. Plant Pathol.* 106(7): 109

Balogh, B., B.I. Canteros Rt. S. 2008. Control of cytud canker and citrud bacterial spot with bacterophages. *Plant Dis.* 92(7): 1048

Belasque Jr., J., A.L. Parra-Pedrazzoli *et al.*, 2005. Adult citrus leafminers (*Phyllocnistis citrella*) are not efficient vectors for *Xanthomonas axonopodis* pv. *citri. Plant Dis.* 89(6): 590

Benelli, A.L.H., N.D. Denardin and C.A. Forcelini. 2004. Action of acibenzolar-S-methyl applied on potato tubers and plants to prevent blackleg caused by *Pectobacterium carotovorum* subsp. *atrosepticum. Fitapat. bras.* 29(3): 263

Bhowmik, B., R.P. Singh, J. Jayaraman and J.P. Verma. 2002. Population dynamics of cotton endophytic *Pseudomonas,* their antagonism and protective action against the major pathogens of cotton. *Indian Phytopath* 55(2): 124

Bove', J.M.1986. Greening in the Arabian Peninsula: Towards new techniques for its detection and control. *FAO Plant Prot. Bull.* 34(1): 7

Bove', J.M. 2006. Huanglongbin: A destructive, newly-emerging, century old disease of citrus. *J. Plant Pathol.* 88:7–37

Brlansky, R.H. and M.E. Rogers. Cutrus Huanglongbing: Understanding the vector-pathogen interaction for disease management. *APSNet Feature Story*, December, 2007

Capoor, S.P., D.G. Rao and S.M. Viswanath. 1974. Greening disease of citrus in the Deccan Trap country and its relationship with the vector *Diaphorina citri*, pp. 43–49. In: *Proc. 6th Conf. Intern. Organiz. Citrus Virologists.* L.G. Weather and M. Cohen (eds), Univ. California, Berkeley.

Chakrabarti, S.K., G.S. Shekhawat and A.V. Gadewar 1995. Phenotypic reversion from afluidal to fluidal colony types in the strains of *Pseudomonas solanacearum. Indian Phytopath.* 48: 353

Cheema, S.S., S.P. Kapur and R.D. Bansal. 1985. Efficacy of various therapeutic agents against greening disease of citrus. *J. Res. PAU.* 22: 479

Cladera-Oliveram F., G.R. Caron, A.S. Motta *et al.*, 2006. Bacteriocin-like substance inhibits potato soft rot caused by *Erwinia carotovora. Can. J. Microbiol.* 52(6): 533

Comstock, J.C., J.M. Shine Jr., M.J. Davis and J.L. Dean. 1996. Relationship between resistance to *Clavibacter xyli* subsp. *xyli* colonization in sugarcane and spread of ratoon stunting disease in the field. *Plant Dis.* 80: 704

Da Graca, J.V. 1991. Citrus greening disease. *Annu. Rev. Phytopathol.* 29: 109

Damann, K.E. Jr. 1992. Effect of sugarcane cultivar susceptibility on spread of ratoon stunting disease by mechanical harvester. *Plant Dis.* 76: 1148

Damann, K.E. Jr. and G.T.A. Benda 1983. Evaluation of commercial heat treatment methods for control of ratoon stunting disease of sugarcane. *Plant Dis.* 67: 966

Dannon, F.A. and K. Wydra. 2004. Interaction between silicon amendment, bacterial wilt development and phenotype of *Ralstonia solanacearum* in tomato genotype. *Physiol. Mol. Plant Pathol.* 64(5): 233

Dardick, C., F.G. de Silva, Y. Shen and P. Ronald 2003. Antagonistic interactions between strains of *Xanthomonas oryzae* pv. *oryzae*. *Phytopathology* 93(6): 705

Davis, M.J., J.L. Dean and N.A. Harrison. 1988a. Distribution of *Clavibacter xyli* subsp. *xyli* in stalks of sugarcane cultivars differing in resistance to ratoon stunting disease. *Plant Dis.* 72: 443

Davis, M.J., J.L. Dean and N.A. Harrison. 1988b. Quantitative variability of *Clavibacter xyli* subsp. *xyli* populations in sugarcane cultivars differing in resistance to ratoon stunting disease. *Phytopathology* 78: 462

De Boer, S.H. 2002. Relative incidence of Erwinia carotovora subsp. atroseptica in stolon ends and peridermal tissue potato tubers in Canada. Plant Dis. 86(9): 960

Diogo, R.V.C. and K. Wydra. 2007. Silicon-induced basal resistance in tomato against *Ralstonia solanacearum* is related to modification of pectin cell wall polysaccharide structure. *Physiol. Mol. Plant Pathol.* 70(4–6): 120

Etchebar, C., D.T. Demery, F. van Gijsegen, J. Vasse and A. Trigalet. 1998. Xylem colonization by an HrcV mutant of *Ralstonia solanacearum* is a key factor for the efficient biological control of tomato bacterial wilt. *Mol. Pant-Microbe Interactions* 11(9): 869

Evtushenko, L.I., L.V. Dorofeeva, S.A. Subbotin, J.R. Cole and J.M. Tiedje. 2000. *Leifsonia poae* gen. Nov. sp. nov., isolated from nematode galls on *Poa annua*, and reclassification of *Corynebacterium aquaticum* Leifson 1962 as *Leifsonia aquatica* (ex Leifson 1962) gen. nov, nom. rev., comb. nov. and *Clavibacter xyli* Davis *et al.*, 1984 with two subspecies as *Leifsonia xyli* (davis *et al.*,1984) gen. nov. comb. nov. *International Jour. Syst. Bacteriology* 50: 371

Fry, B.A. and R. Loria. 2002. Thaxtomin A: Evidence for a plant cell wall target. *Physiol. Mol Plant Path.* 60(1): 1

Gabriel, D.W., M. Kingsley, J.F. Hunter and T.R. Gottwals.1989. Reinstatement of *Xanthomonas citri* and *Xanthomonas phaseoli* to species and reclassification of all *Xanthomonas campestris* pv. *citri* strains. *Int. J. Syst. Bacteriol.* 39: 14

Goto, M. 1992. Citrus canker, pp. 170–208. In: *Plant Diseases of International Importance*, Vol. III. *Diseases of Fruit Crops*. J. Kumar *et al.*, (eds). Prentice Hall, New Jersey

Gottwald, T.R., J.V. da Graca and R.B. Bassanezi. 2007. Citrus Huanglongbin: The pathogen and its impact. *APSnet Feature Story September 2007*

Goyer, C., P.M. Charesi, V. Toussaint and C. Beaulieu. 2000. Ultrastuctural effects of thaxtomin A produced by *Streptomyces scabies* on mature potato tuber tissues. *Can. J. Bot.* 78(3): 374

Graham, J.H. and R.P. Leite, Jr. 2004. Lack of control of citrus canker by induced systemic resistance compounds. *Plant Dis.* 88(7): 745

Guo, J-H., H-Y.Qi, Y-H. Guo *et al.*, 2004. Biocontrol of tomato wilt by plant growth-promoting rhizobacteria. *Biological Control* 29(1); 66

Han, J.S., J.H. Cheng, T.M. Yoon *et al.*, 2005. Biological control agent of common scab disease by antagonistic strain *Bacillus* sp. sunhua. *J. Appl. Microbiol.* 99(1): 213

Hase, S., A. Shimizu *et al.*, 2006. Induction of transient ethylene and reduction in severity of tomato bacterial wilt by *Pythium oligandrum*. *Plant Pathology* 55(4): 537.

Hase, S., S. Takahashi, S. Takenaka *et al.*, 2008. Involvement of jasmonic acid signaling in bacterial wilt disease resistance induced by biocontrol agent *Pythium oligandrum* in tomato. *Plant Pathology* IN PRESS

Hayward, A.C. 1991. Biology and epidemiology of bacterial wilt caused by *Pseudomonas solanacearum*. *Annu. Rev. Phytopathol.* 29: 65

Hilaire, E., S.A. Young, L.H. Willard, J.D. McGee, T. Sweat, J.M. Chittoor, J.A. Guikema and J.E. Leach. 2001. Vascular defence responses in rice: Peroxidase accumulation in xylem parenchyma cells and xylem wall thickening. *Mol. Plant-Microbe Interact.* 14(12): 1411

Hoy, J.W., M.P. Grisham and K.F. Damann. 1999. Spread and increase of ratoon stunting disease of sugarcane and comparison of disease detection methods. *Plant Dis.* 83: 1170

Hulloli, S.S., R.P. Singh and J.P. Verma. 1998. Management of bacterial blight of cotton induced by *Xanthomonas axonopodis* pv. *malvacearum* with the use of neem based formulations. *Indian Phytopath.* 51: 21

Hung, T.H., M.L. Wu and H.J. Su. 2000. Identification of alternative hosts of the fastidious bacterium causing citrus greening disease. *J. Phytopath.* 148(6): 321

Hung, T.H., M.L. Wu and H.J. Su. 2001. Identification of the Chinese box orange (*Severinia buxifolia*) as an alternative host of the bacterium causing citrus Huanglongbin. *Eur. J. Plant Pathol.* 107: 183

Hussain, M.M., S. Shibata, S-I, Airawa and S.Tsuyama. 2005. Motility is an important determinant for pathogenesis of *Erwinia carotovora* subsp. *carotovora*. *Physiol. Mol. Plant Pathol.* 66(4): 134

Jagoueix, S., J.M. Bove' and M. Garnier. 1994. The phloem-limited bacterium of greening disease of citrus is a member of the subdivision of the Proteobacteria.. *Int. J. Syst. Bacteriol.* 44: 379

Jagoueix, S., J.M. Bove and M. Garnier. 1996. PCR detection of the two *Liberobacter* species associated with greening disease of citrus. *Mol. Cellular Probes* 10: 43

Jalali, I. And R.D. Parashar. 1995. Biocontrol of *Xanthomonas campestris* pv. *campestris* with phylloplane antagonists. *Plant Dis. Res.* 10: 145

Jansky, S.H. and D.L. Rouse. 2003. Multiple disease resistance in interspecific hybrids of potato. *Plant Dis.* 87(3): 266

Jetiyanon, K. and J.W. Kloepper. 2002. Mixtures of plant growth promoting rhizobacteria for induction of systemic resistance against multiple plant diseases. *Biological Control* 24(3): 285

Ji, G-H., L-F. Wei, Y-Q. Ho, Y-P. Wu and X-H. Bai. 2008. Biological control of ruce bacterial blight by *Lysobacter antibioticus* strain 13-1. *Biol. Control* 45(3) 388

Ji, P., T. Momol, S.M. Olson, P.M. Pradhanang and J.B. Jones. 2005. Evaluation of thymol as biofumigant for control of bacterial wilt of tomato under field conditions. *Plant Dis.* 89(5): 497

Ji, P., M.T. Monol, J.R. Rich *et al.*, 2007. Development of an integrated approach for managing bacterial wilt and root knot on tomato under field conditions. *Plant Dis.* 91(10): 1321

Kalita, P., L.C. Bora and K.N. Bhagbati. 1996. Phylloplane microorganisms of citrus and their role in management of citrus canker. *Indian Phytopath.* 49: 234

Kangatharalingam, N., M.L. Pierce, M.B. Bayles and M. Essenberg, 2002. Epidermal anthocyanin production as an indicator of bacterial blight resistance in cotton. *Physiol. Mol. Plant Path.* 61(3): 189

Kapur, S.P. and S.S. Cheema. 1983. Chemotherapeutic control of citrus greening disease. *Pesticides* 17: 13.

Kocks, C.G. and J.C. Zadoks. 1996. Cabbage refuse piles as source of inoculum for black rot epidemics. *Plant Dis.* 80: 789

Kocks, C.G., M.A. Ruissen, J.C. Zadoks and M.G. Duijkers, 1998a. Survival and extinction of *Xanthomonas campestris* pv. *campestris* in soil. *Eur. J. Plant Pathol.* 104(9): 911

Kocks, C.G., J.C. Zadoks and T.A. Ruissen. 1998b. Response of black rot in cabbage to spatial distribution of inoculum. *Eur. J. Plant Pathol.* 104(7): 713

Kumar, P. and A.K. Sood. 2001. Integration of antagonistic rhizobacteria and soil solarization for the management of bacterial wilt of tomato caused by *Ralstonia solanacearum*. *Indian Phytopath.* 54: 12

Kumar, P. and A.K. Sood. 2002. Management of bacterial wilt of tomato with VAM and bacterial antagonists. *Indian Phytopath.* 55(4): 513

Lafortune, D., M. Beramis, A.M. Daubeze, N. Borssot and A. Palloix. 2005. Partial resistance of pepper to bacterial wilt is oligogenic and stable under tropical conditions. *Plant Dis.* 89(5): 501

Lapwood, D.H. and T.F. Hering. 1970. Soil moisture and the infection of young potato tubers by *Streptomyces scabies* (common scab). *Potato Res.* 13: 296

Lapwood, D.H., L.W. Wellings and J.H. Hawkins. 1973. Irrigation as a practical means to control potato common scab (*Streptomyces scabies*). *Plant Pathology* 22: 35

Lazarovits, G., K.L. Conn and J. Potter. 1999. Reduction of potato scab, Verticillium wilt and nematodes by soymeal, meat and bone meal in two Ontario fields. *Can. J. Plant Pathol.* 21(4): 345

Lazarovits, G., J. Hill, R.R. King and L.A. Colhoun. 2004. Biotarnsformation of the *Streptomyces scabies* phytotoxin thaxtomin A by the fungus *Aspergillus niger*. *Can. J. Microbiol.* 50(2): 121–126

Lim, U.T. and M.A. Hoy. 2005. Biological assessment in quarantine of *Semielacher petiolatus* (Hymenoptera: Eulophidae) as a potential classical biological control agent of citrus leafminer, *Phyllocnistis citrella* Stainton (Lepidoptera: Gracillariidae), in Florida. *Biological Control* 33(1): 87

Loria, R., R.A. Bukhalid and B.A. Fry. 1997. Plant pathogenicity in the genus *Streptomyces*. *Plant Dis.* 81: 836

Loria, R., J. Coombs, M. Yoshida, J. Kers and R. Bukhalid. 2003. A paucity of bacterial root diseases: *Streptomyces* succeeds where others fail. *Physiol. Mol. Plant Pathol.* 62(2): 65

Madhiazhagan, K., N. Ramadoss and R. Anuradha. 2002. Effect of botanicals on bacterial blight of rice. *J. Mycol. Pl. Pathol.* 32(1): 68

Manjunath, K.L., S.E. Halbert, C.Ramadugu, S. Webbe and R.F. Lee. 2008. Detection of 'Candidatus Liberobacter asiaticus' in *Diaphorina citri* and its importamnce in the management of citrus Huanglongbing in Florida. *Phytopathology* 98(4): 387

Massomo, S.M.S., C.N. Mortensen, R.B. Mabagala, M.A. Newman and J. Hockenhull. 2004. Biological control of black rot (*Xanthomonas campestris* pv. *campestris*) of cabbage in Tanzania with *Bacillus* strains. *J. Phytopath.* 152(2): 98

McKay, A.C., K.M. Ophel, T.B. Reardon and J.M. Gooden. 1993. Live stock deaths associated with *Clavibacter toxicus/ Anguina* sp. infection of seedheads of *Agrostis avenacea* and *Polypogon monspleliensis*. *Plant Dis.* 77: 635

Messiha, N.A.S., A.D. van Diepeningen, M. Wenneker *et al.*, 2007. Biological soil disinfection (BSD), a new control method for potato brown rot caused by *Ralstonia solanacearum* race 3 biovar 2. *Eur. J. Plant Pathol.* 117(4): 403

Messiha, N.A.S., A.D. van Diepeningen, N.S. Farag *et al.*, 2007a. *Stenotrophomonas maltophilia*: a new potential biocontrol agent of *Ralstonia solanacearum*, causal agent of potato brown rot. *Eur. J. Plant Pathol.* 118(3): 211

Messiha, N.A.S., A.H.C. van Buggen, A.D. van Diepper *et al.*, 2007b. Potato brown rot incidence and severity under different management and amendment regimes in different soil types. *Eur. J. Plant Pathol* 119(4): 367

Mishra, A.K. 1995. Control of bacterial canker of mango under suitable weather conditions. *Indian J. Mycol. Pl. Path.* 25: 214

Mohan Babu, R., A. Sajbeena, A.V. Samundeeswar, A. Sreedhar, P. Vidyasekaran and M.S. Reddy. 2003. Induction of bacterial blight (*Xanthomonas oryzae* pv. *oryzae*) resistance in rice by treatment with acibenzolar-S-methyl. *Ann. Appl. Biol.* 143(3): 333

Mondal, K.K., R.P. Singh, P. Dureja and J.P. Verma. 2000. Secondary metabolites of cotton rhizobacteria in the suppression of bacterial blight of cotton. *Indian Phytopath.* 53: 22

Nakaho, K., H. Hibino and H. Miyagawa. 2000. Possible mechanisms limiting movement of *Ralstonia solanacearum* in resistant tomato tissues. *J. Phytopath.* 148: 181

Nakaho, K., H. Inoue, T. Takayama and H. Miyagawa. 2004. Distribution and multiplication of *Ralstonia solanacearum* in tomato plants with resistance derived from different origins. *J. Gen. Plant Pathol.* 70(2): 115

Nishioka, M., H. Nakashita, M. Yoshidam Y. Yoshida and I. Yamaguchi. 2005. Induction of resistance against rice bacterial leaf blight by 3-chloro-1-methyl- 1H-pyrazole-S-carboxylic acid. *J. Pestic. Sci.* 30(1): 47

Ooshiro, A., K. Takaesu, M. Natsume *et al.*, 2004. Identification and use of a wild plant with antimicrobial activity against *Ralstonia solanacearum*, the cause of bacterial wilt of potato. *Weed Biol. Manag.* 4(4): 187

Ou, S.H., P.G. Franck and S.D. Merca. 1970. Varietal resistance to bacterial leaf streak disease in the Philippines. *Philippines Agric.* 54: 8

Ou, S.H., F.L. Nuque and J.P. Silva. 1971. Varietal resistance to bacterial leaf blight of rice. *Plant Dis. Rep.* 55: 17

Pacambaba, R.P., C.A. Beyl and R.O. Pacumbaba Jr. 1999. Shiitake mycelial leachate suppresses growth of some bacterial species and symptoms of bacterial wilt of tomato and lima bean. *Plant Dis.* 83: 20

Pathak, K.N. and R.P. Nath. 1993. Interaction of nematodes with bacterial plant pathogens, pp. 244–253. In: *Handbook of Economic Nematology.* Eds. K. Sitaramaiah and R.S. Singh. Cosmo Publications, New Delhi.

Patricio, F.R.A., L.M.G. Almeida, A.S. Santos *et al.*, 2005. Soil solarizationrvaluation for control of *Ralstonia solanacearum*. *Fitopat. Brasi.* 30(5): 475

Pavlista, A.D. 2005. Early season applications of sulfur fertilizers increase potato yield and reduce tuber defects. *Agron. J.* 97: 599

Perombelon, M.C.M. 2002. Potato diseases caused by soft rot erwinias: an overview of pathogenesis. *Plant Pathology* 51(1): 1

Pradhanang, P.M., M.T. Momol, S.M. Olson and J.B. Jones. 2003. Effects of plant essential oils on *Ralstonia solanacearum* population density and bacterial wilt incidence in tomato. *Plant Dis.* 87(4): 423

Pradhanang, P.M., P. Ji, M.T. Momol, S.M. Olson, J.L. Mayfield and J.B. Jones. 2005. Application of acibenzolar-S-methyl enhances host resistance in tomato against *Ralstonia solanacearum*. *Plant Disease* 89(9): 989

Pria, M.D., R.C.S. Christiano *et al.*, 2006. Effect of temperatures and leaf wetness duration on infection of sweet orange by Asiatic citrys canker. *Plant Pathology* 55(5): 657

Prior, P., S.Bart, S. Leclercq *et al.*, 1996. Resistance to bacterial wilt in tomato as discerned by spread of *Pseudomonas* (*Burkholderia*) *solanacearum* in the stem tissues. *Plant Pathology* 45(4): 720

Rae, D.J., W.G. Liang, D.M. Watson, G.A.C. Beattie and M.D. Huang. 1997. Evaluation of petroleum spray oils for control of the Asian citrus psylla, *Diaphorina citri* (Kuwayama) (Hemiptera: Psyllidae), in China. *International J. Pest Management* 43(1): 71

Raghavendra, V.B., S. Lokesh and H.S. Prakash. 2007. Dravya, a product of extract (*Sargassum wightii*), induces resistance in cotton against *Xanthomonas axonopodis* pv. *malvacearum*. *Phytoparasitica* 35(5): 442

Rahman. M.A., H. Abdullah and M. Vanhaecke. 1999. Histopathology of susceptible and resistant *Capsicum annuum* cultivars infected with *Ralstonia solanacearum*. *J. Phytopath.* 147: 129

Ravensdale, M., T.J. Blom *et al.*, 2007. Bacteriophages and the conrol of Erwinia carotovora aubap. Carotovora. *Can. J. Plant Pathol.* 29(2): 121

Reddy, M.R.S. 1997. Sources of resistance to bacterial canker in citrus. *J. Mycol. Pl. Pathol.* 27: 80

Riley, I.T. 1992. *Anguina tritici* is a potential vector of *Clavibacter toxicus*. *Aust. Plant Pathol.* 21(4): 147

Ryan, A.D. and L.L. Kinkel. 1997. Inoculum density and population dynamics of suppressive and pathogenic *Streptomyces* strains and their relationship in biological control of potato scab. *Biological Control* 10(3): 180

Ryan, A.D., L.L. Kinkel and J.L. Schottel. 2004. Effect of pathogen isolate, potato cultivar, and antagonist strain on potato scab severity and biological control. *Biocontrol Sci. Technol.* 14(3): 301

Saha, S., R.P. Singh, J.P. Verma and J. Jayaraman. 2001. Population dynamics of cotton phylloplane bacteria antagonistic toward *Xanthomonas campestris* pv. *malvacearum*. *Indian Phytopath.* 54(4): 409

Salah Eddin, K., T. Marimuthu, D. Ladhgalakshmi and R. Vrlazhahahn. 2007. Biological control of bacterial blight of cotton caused by *Xanthomonas axonopodis* pv. *malvacearum* with *Pseudomonas fluorescens*. *Arch. Phytopath. Plant Prot.* 40(4): 291

Sarvanakumar, D., L. Rajendra, R. Samiappan and T. Raguchander. 2006. Endophytic bacterial induction of defense enzymes against bacterial blight of cotton. *Phytopath. Mediterr.* 45(3): 203

Satya, V.K., S. Gayathiri, R. Bhaskaran, V. Paranidham and R. Venzhahahn. 2007. Induction of systemicv resistance to bacterial blight caused by *Xanthomonas axonopodis* pv. *malvacearum* in cotton by leaf extyracts from a medicinal plant zimmu (*Asllium sativum* L. x *Allium cepa* L.). *Arch. Phytopath Plant Prot.* 40(5): 309

Schaad, N.W., G.H. Lacy, E. Postnikova and E.L. Schuenzel. 2006. Taxonomy and phylogeny of phytopathogenic bacteria. *J. Plant Pathol.* 88(3): S13

Seo, S-T., N. Furaya, K. Iiyama, M. Takeshita *et al.*, 2004. Characterization of an antibacterial substance produced by *Erwinia carotovora* subsp. *carotovora* Ecc 32. *J. Gen. Plant Pathol.* 70(5): 273

Shargah, B.M. and G.D. Lyon. 1998. *Bacillus subtilis* BS107 as an antagonist of potato blackleg and soft rot bacteria. *Can. J. Microbiol.* 44(8): 777

Sharma, J.P. and S. Kumar. 2004. Effect of crop rotation on population dynamics of *Ralstonia solancearum* in tomato wilt sick soil. *Indian Phytopath* 57)11): 80

Singh, D. and S.K. Rana. 2000. Biocontrol of bacterial wilt/ brown rot (*Ralstonia solanacearum*) of potato. *J. Mycol. Pl. Pathol.* 30: 420

Somani, A.K., V.C. Sharma and G.S. Shekhawat. 1983a. Influence of level and source of nitrogen on the development of bacterial soft rot in potato. *Indian Phytopath.* 41: 238

Somani, A.K., V.C. Sharma and G.S. Shekhawat. 1983b. Influence of date of harvesting on the development of bacterial soft rot and emergence. *Indian Phytopath.* 41: 517

Subandiyah, S., T. Iwanami, S. Tsuyama and H. Icki. 2000. Comparison of 16S rRNA and 16S'23 S. intergenic region sequences among citrus greening organism in Asia. *Plant Dis.* 84: 15

Suneja, S., K. Lakshminarayan and P.P. Gupta. 1994. Role of *Azotobacter chrococcum* siderophores in control of bacterial rot and Sclerotinia rot of mustard. *Indian J. Mycol. Pl. Pathol.* 24: 202

Tan, H.M., L.X. Cao, Z.F. He, G.J. Su, B.Lin and S.N. Zhou. 2006.Isolation of endophytic actinomycetes from different cultivars of tomato and their activities against *Ralstonia solanacearum in vitro*. *Worl. J. Microbiol. Biotechnol.* 22(12): 1275–1280

Tatineni, S., U.S. Sagaram, S. Gowda *et al.*, 2008. *In plants* distribution of *Candidatus* Liberobacter asiaticus as revealed by polymerase chain reaction (PCR) and real-time PCR. *Phytopathology* 98(6): 592.

Teixeira, D.C., C/ Sallard, C. Jagoueix *et al.*, 2005. *Candidatus* Liberobacter americanum associated with citrus huanglongbin (greening disease) in Sao Paulo state, Brazil. *Int. J. Syst. Evol. Microbiol.* 55: 1857

Timmer, L.W., T.R. Gottwald and S.E. Zitko. 1991. Bacterial exudation from lesions of Asiatic citrus canker and citrus bacterial spot. *Plant Dis.* 75: 192

Toth, I.K., K.S.Bell, M.C. Holeva and P.R.J. Birch. 2003a. Soft rot erwinias: from genes to genomes. *Mol. Plant Path.* 4(1): 17–30

Toth, I.K., L. Sullivan, J. Brierley *et al.*, 2003b. Relationship between potato tuber contamination by *Erwinia carotovora* ssp. *atroseptica*, black leg disease development and progeny tuber contamination. *Plant Pathology* 53(2): 119

Trigalet, A. and D. Trigalet-Demery, 1990. Use of avirulent mutants of *Pseudomonasd solanacearum* for the biological control of bacterial wilt of tomatoi plants. *Physiol. Mol. Plant Pathol.* 36(1): 27

Tweddell, R.J., R. Boulanger and J. Arul. 2003. Effect of chlorine atmosphere on sprouting and development of dry rot, soft rot and silver scurf on potato tubers. *Postharvest Biol. Technol.* 28(3): 445

Van Elsas, J.D., P. Kastelein, P.M. de Vries and L.S. vam Overbeck. 2001. Effects of ecological factors on the survival and physiology of *Ralstonia solanacearum* Bv. 2 in irrigation water. *Can. J. Microbiol.* 47(9): 842

Van Overbeck, L.S., J.H.W. Bergervoet, F.H.H. Jacobs and J.D. van Elsas. 2004. The low temperature-induced viable-but–nonculturable state affects the virulence of *Ralstonia solanacearum* Biovar 2. *Phytopathology* 94(5): 463

Velusamy, P., J.E. Imanuel, S.S. Gnnamanickam and L. Thomashow. 2006. Biological control of rice bacterial blight by plant associated bacyeria producing 2,4-diacerylphluroglucinol. *Can. J. Microbiol.* 52(1): 2

Venkataswarlu, Ch. and S.Ramapandu. 1992. Relationship between incidence of canker and leaf miner in acid lime and Sathgudi sweet orange. *Indian Phytopath.* 45: 227

Verma, J.P. 1995. Advances in bacterial blight of cotton. *Indian Phytopath.* 48: 1

Vidyasekaran, P., N. Kamala, A.Ramanathan, K. Rajappan *et al.*, 2001. Induction of systemic resistance by *Pseudomonas fluorescens* PF 1 against *Xanthomonas oryzae* pv. *oryzae* in rice leaves. *Phytoparasitica* 29(2): 1–12

Wang, A. and G. Lazarovits. 2004. Enumeration of plant pathogenic *Streptomyces* on postharvest potato tubers under storage conditions. *Can. J. Plant Pathol.* 26(4): 563

Wang, Y., Q.Yang, Y. Tosa, H. Nakayashiki and S. Mayama. 2005. Nitric oxide-overproducing transformants of *Pseudomonas fluorescens* with enhanced biocontrol of tomato bacterial wilt. *J. Gen. Plant Pathol.* 71(1): 33

Waterer, D. 2002. Management of common scab of potato using planting and harvesting dates. *Can. J. Plant Sci.* 82: 185

Wiggins, B.E. and L.L. Kinkel. 2005. Green manures and crop sequences influence potato diseases and pathogen inhibitory activity of indigenous Streptomycetes. *Phytopathology* 95(2): 178

Wilson, C.R., J.M. Ransom and B.M. Pemberton. 1999. The relative importance of seed-borne inoculum to common scab disease of potato and the efficacy of seed tuber and soil treatment for disease control. *J. Phytopath* 147(1):13

Wulff, E.G., M. Mguni, C.N. Mortensen, C.L. Keswani and J. Hockenhull. 2002. Biological control of black rot (*Xanthomonas campestris* pv. *campestris*) of brassicas with an antagonistic strain of *Bacillus subtilis* in Zimbabwe. *Eur. J. Plant Pathol.* 108(4): 317

Zgurskaya, H.I., L.I. Evtuschenko, V.N. Akimov and L.V. Kalakoutskii. 1993. *Rathayibacter* gen. nov., including the species *Rathayibacter rathayi* comb. nov, *Rathayibacter iranicus* comb. nov. and six strains from annual grasses. *International Jour. Syst. Bacteriol.* 43: 143

Zhu, H.H. and Q. Yao. 2004. Localized and systemic increase of phenols in tomato roots induced by *Glomus versiforme* inhibits *Ralstonia solanacearum*. *J. Phytopath.* 152(10): 537

Phytoplasma Diseases of Plants

Prior to 1967 many plant diseases such as corn stunt, sugarcane grassy shoot and white leaf, little leaf of brinjal (eggplant), sandal spike, and others, broadly termed as 'yellows', often showing witches' broom phenomenon, were considered as viral diseases. This understanding was based on the logic that no fungal, bacterial, nematode or protozoan agents were associated with these diseases which resembled virus diseases, especially in the method of transmission. They were vector or graft transmissible. In some cases dodder transmission was successful. Strain interferences, similar to virus interferences (cross immunity), was also demonstrated in many yellows diseases. They could be filterable also. Experimental evidence was provided for the multiplication of a few of these disease agents in leafhopper vectors and in vector tissues *in vitro*. Certain species of leafhoppers were shown to act as alternate hosts and reservoirs of these disease agents. Although these aspects of yellows type of diseases were paralleling phenomenon in well established plant pathogenic viruses and virus diseases, all attempts to purify the causative agents by virological methods and to characterize them morphologically and chemically as viruses had remained unsuccessful.

In 1967, Japanese workers (Doi *et al.*, 1967; and Ishiie *et al.*, 1967) proposed that the causal agents of yellows diseases might be mycoplasma-like or chlamydia-like organisms. They based their observations on: (a) discovery of such bodies in sieve elements of yellows infected plants, (b) absence of true virus particles in the damaged tissues, and, (c) therapeutic effect of tetracycline antibiotics in such diseases. During the period after this discovery mycoplasma-like bodies were found associated with disease condition in more than 200 plant species and insects. The discovery not only added a new group of plant disease agents, it also induced workers to look for similar organisms in other diseases which resulted in the identification of many other hidden prokaryotes such as the causal agents of ratoon stunting disease of sugarcane and greening disease of citrus.

Symptoms and diseases caused by phytoplasmas

Phytoplasma and *Spiroplasma* differ in morphology in a fluid medium, the latter being helical with one exception, and the former being spherical, ovoid or elongate. There are some differences in the nature of symptoms also. Virescence (reversion of floral organs to vegetative organs, i.e., greening of petals and conversion of petals to leafy structures) is found in diseases caused by *Phytoplasma* but not by *Spiroplasma* sp that have been isolated and characterized (Mc Coy, 1981). Diseases caused by non-helical, non-cultivated phytoplasmas include: Alfalfa witches' broom, Clover, Sesamum and Sunflower phyllody, Coconut lethal yellowing, Eggplant little leaf, Legume little leaf, Elm Yellows or phloem necrosis, Groundnut witches' broom, Little peach, Mulberry dwarf, Papaya bunchy top, Peach decline, Pear decline, X-disease of peach, Potato witches' broom, Potato purple top roll, Potato marginal flavescence, Sandal spike, Sugarcane grassy shoot, Sugarcane white leaf, Tomato big bud, Tomato purple top, Yellow leaf of areca palm, Root (wilt) disease of coconut palms, and Yellow wilt of sugarbeet. These diseases are characterized by yellowing, chlorosis, or bronzing of foliage, stunting (shortening of internodes and reduced size of leaves), proliferation of axillary buds resulting in witches' broom effect, proliferation of secondary roots and abnormal fruits and seeds. Flowers are often sterile. Some diseases show few of these symptoms whereas others may show all or most of these symptoms.

Variation in symptom syndrome may occur according to host species, the time and mode of infection, environmental factors, and mixed infection by more than one type of phloem-limited pathogens such as dual infection of *Phytoplasma* and *Spiroplasma* or *Phytoplasma* and RLB or other bacteria or even virus. Established spiroplasma diseases also show most of these symptoms but virescence is not caused by spiroplasmas. The diseases caused by *Spiroplasma* spp. include Bermuda grass white leaf, Citrus stubborn (*Spiroplasma citri*), corn stunt (*Spiroplasma kunkelii*), Opuntia witches' broom, Rice yellow dwarf, and yellows of aster and lettuce.

Although in India 33 plant diseases with MLO symptoms were listed in 1974 the phytoplasmal etiology was reported only in 13 cases which include arecanut yellow leaf, brinjal little leaf, potato purple top, rice yellow dwarf, sesame phyllody, and sugarcane grassy shoot. Cowpea witches' broom and pigeonpea sterility were listed as diseases with symptoms resembling those caused by *Phytoplasma* but there was no report of MLO etiology. Sterility mosaic of pigeonpea is now confirme as a virus disease. Later, a witches' broom of bitter gourd was shown to be a MLO disease in which presence of *Phytolasma* cells (100–850 nm dia) in phloem sieve tubes were demonstrated. Phyllody of bottle gourd and cucumber had earlier been reported to be possibly caused by a phytoplasma (Singh, S.J., 1992). Little leaf of chilli is another addition to the list of diseases of MLO etiology in India (Singh and Singh, 2000).

GRASSY SHOOT DISEASE OF SUGARCANE

Grassy shoot disease (GSD) was first noticed in India in 1919 in Maharashtra. Subsequently it was found in Andhra Pradesh, Tamil Nadu, Orissa, Bihar, Uttar Pradesh, Punjab, and Rajasthan also. Outside India, grassy shoot is reported from Myanmar, Sri Lanka, and Sudan. In India the affected sugarcane varieties have been Co. 313, 356, 419, 421, 1081, 1148, 1191, 1195, 1238, 975, 1007, 1107, 1286, 1328, 1330, 62303, 62399, 62401, Co J. 46, Bo.10 and 11, Cos 245, 321, 514, 515, 555, and 558. Losses from GSD are in the form of reduced germination of setts from diseased canes, stunted growth and poor juice quality. Infections up to 10% do not cause significant reduction in yield and number of milleable canes. The threshold level for yield reduction is 15% infection.

Symptoms

Symptoms vary at different stages of plant growth and, thus, the disease has been variously described as 'new chlorotic disease', 'albino disease', 'yellowing disease', 'bunchy disease', or 'leaf tuft'. Profuse tillering and grassy appearance of the stools are the main symptoms and, therefore, the standard name for the disease is grassy shoot (GSD). Shoots growing from diseased setts remain dwarfed or stunted. The leaves are narrow and small, like grass leaves; the canes are thin with short internodes, giving a bunchy or grassy appearance to the clump. The leaves appear yellowish and in some cases may be entirely devoid of any pigment (white leaf or albinism). If many tillers are affected in this manner the entire stool dries. In secondary infection, ordinarily the new leaves and internodes show the above symptoms. Leaves exhibit straight, long, white or light green or yellowish steaks. The lower nodes produce large number of grassy shoots. In systemically infected canes the disease appears in May-June. In sprouts raised from top buds the symptoms appear late as compared to sprouts from lower buds. This suggests difference in the concentration of the causal agent in the cane. Anatomical differences in cell characters have also been observed. Enzymes of sucrose synthesis are inhibited due to disturbed inorganic metabolism. This may be more in some varieties than in others. In a study in Cuba, the yellow leaf syndrome of sugarcane was found to modify the composition of juice in polysaccharides, phenols and polyamines (Fontaniella *et al.,* 2003).

Fig. 10. Grassy shoot of sugarcane.

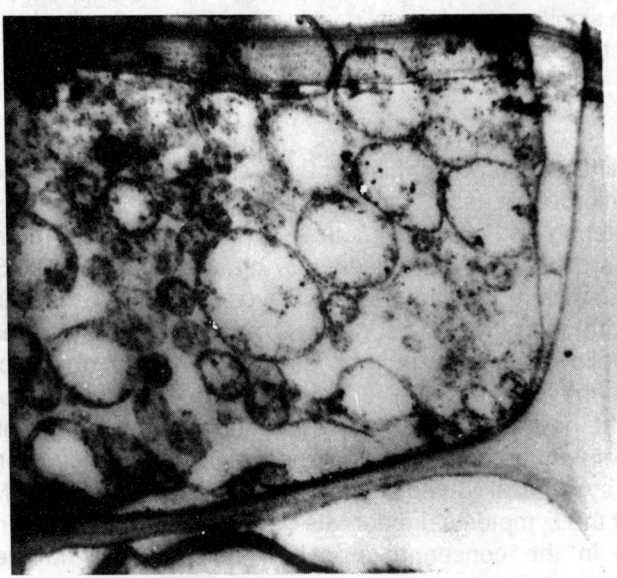

Fig. 11. Phytopiasma bodies seen under electron microscope.

The causal organism

The grassy shoot disease was first considered a virus disease. However, since 1971 there had been increasing evidence to suggest that it is caused by a MLO. Corbett *et al.,* (1971) first suggested that mycoplasma-like bodies were associated with the disease and Rishi *et al.,* (1973) demonstrated with the help of electron microscopy the presence of mycoplasma-like bodies in the phloem of diseased canes. Recently, two Phytoplasmas, different from earlier described have been found in some samples of grassy shoot (Nasare *et al.,* 2007).

The GSD *Phytoplasma* (sugarcane white leaf phytoplasma group) is present in the sieve tubes of the phloem as ovoid, spherical, or irregularly shaped bodies. The sizeof ovoid or spherical bodies is 300–400 nm in diameter and that of protruding filaments 30–50 nm in diameter. These cells are devoid of true cell-wall and are bounded by a single triple layered membrane. They contain ribosomal granules and DNA strand. Mostly the *Phytoplasma* cells are concentrated towards the periphery of the cell, near the cell walls but often the host cells are fully packed with these bodies and die followed by death of the *Phytoplasma* cells due to starvation (Rishi *et al.,* 1973). In some instances the *Phytoplasma* cells are seen lying close to sieve pores of the sieve tubes and turn filamentous while passing through the pores (Rishi *et al.,* 1973). Concentration of the *Phytoplasma* cells is high in canes growing at high temperature (around 30°C). Such canes show severe symptoms. At lower temperatures the symptoms are less severe and the number of MLOs in the cells is also lower. Shukla and Bhansali (1985) had shown that staining of hand sections of diseased canes with Diene's stain gives dark blue color to infected phloem while the healthy phloem remains unstained.

The white leaf disease of sugarcane reported from Taiwan has some difference from grassy shoot disease described in India. While grassy shoot organism is inactivated by hot water at 50°C for 2–3 hours the white leaf disease organism is inactivated by hot water treatment at 55°C for 30 minutes or 54°C for 40 minutes (cf. Rishi *et al.,* 1973). The grassy shoot is transmitted by aphids while white leaf is transmitted by leafhoppers.

The grassy shoot *Phytoplasma* perpetuates through diseased canes used for seed and spreads through diseased setts and cane cutting knives (sap transmission). However, Rishi *et al.,* (1973) did not find proof of sap transmission. Ratooning of a diseased crop is an important means of perennation. Introduction of the pathogen in a new field is mainly through diseased seed cuttings (Shukla, 1985). Many aphids and a leafhopper are reported to spread the pathogen from plant to plant in the field but only to a limited extent but this mode of transmission has not been proved beyond doubt.

The vectors are *Rhopalosiphum maidis* (*Aphis maidis*), and *Melanaphis sacchari* forma *indosacchari*. The minimum acquisition feed period is 34 seconds and inoculation feed period 30 seconds. However, optimum acquisition feed period is 15 minutes. Although a single infected aphid can transmit the pathogen, optimum number is 15. Transmission is in the non-persistent manner. Winged forms are more effective in spreading the diseases in the standing crop. Transmission by dodder (*Cuscuta campestris*) is also reported.

The pathogen has been transmitted from sugarcane to sorghum and from sorghum to sugarcane by using *Melanaphis sacchari*. Sorghum is also a natural host of the GSD phytoplasma. Most sugarcane cultivars have the capacity to throw off the infection after sometime (Shukla, 1985).

Management of the disease

Phytosanitary control measures are important. Movement of seed setts from infected areas to disease-free areas should be prohibited unless setts are certified as being pathogen-free. Growing apical meristems of lateral buds from infected canes in tissue culture yields disease-free plantlets. The number of such plantlets is higher when apices less than 1 mm in length are used. The technique could be used for raising disease-free foundation seed. In farmers' fields, healthy seed setts should be taken from a field where grassy shoot is not present even in traces. During early stage of crop growth insect vectors can be controlled by weekly sprays of 0.16% malathion. Diseased clumps should be dug out and destroyed.

In heat therapy the setts should be kept in hot water at 50°C for 2 hours or at 52°C for one hour or in hot air at 54°C for 8 hours. Moist hot-air treatment (MHAT) at 54°C (90–95% humidity) for one and a half to 2 hours is enough to inactivate the *Phytoplasma* cells (Shukla and Singh, 1990). These treatments are feasible only for seed nurseries. Tetracycline antibiotics (250 ppm) applied to seed setts under negative pressure eliminate symptoms. Single bud setts can be treated with such antibiotics

LITTLE LEAF OF EGGPLANT (BRINJAL)

Little leaf of brinjal (eggplant, *Solanum melongena*) is found throughout India, Sri Lanka, Bangladesh and probably occurs in other neighbouring countries. In India it was first reported from Coimbatore (Tamil Nadu). The disease is a serious threat to profitable cultivation of this vegetable crop in most of the states. When young plants are attacked they do not produce flowers and fruits. Up to 99% loss in fruit yield is

Fig. 12. Littleleaf disease of brinjal (eggplant.)

reported. There is also loss of germinability of seeds from fruits formed on infected plants. Early infection results in reduced root length and fresh and dry root weight more than in late infection.

Symptoms

The main symptom of the disease is production of very short leaves by the affected plant. The petioles are so much reduced in size that the leaves appear sticking to the stem. Such leaves are narrow, soft, smooth, and yellowish in color. Newly formed leaves are further reduced in size. The internodes of the stem are shortened and at the same time a large number of axillary buds are stimulated to grow into short branches with small leaves. This gives the whole plant a bushy appearance. Usually such plants fail to form flowers. Even if flowers are formed they remain green. Fruiting is rare.

The causal organism

Prior to 1968 the causal agent of little leaf of eggplant was considered a virus. Since 1969 evidence had accumulated to show that the disease is caused by a mycoplasma-like organism. It was recognized as a *Phytoplasma* of the clover proliferation group in 1998. The *Phtoplasma* has been described as ovoid or spherical body measuring 40 to 300 nm in diameter and lacking a rigid cell-wall. These structures are present in the phloem sieve tube cells of the stem, petioles, phyllodes, and roots and also in haemolymph and fat bodies of the vector *Hishimonus* (*Cestius*) *phycitis*. Adjacent companion cells of the phloem show wall ingrowths, dilated endoplasmic reticulum, dilated and ruptured cristae, deformed outer mitochondrial membranes and rupture of plastid membranes and tonoplast suggesting involvement of toxin(s) in the etiology. Therapy of diseased plants by etracyclines had earlier given the proof of phytoplasmal nature of the disease.

In nature the little leaf *Phytoplasma* is transmitted by its insect vector. The same *Phytoplasma* occurs on *Datura fastuosa, Vinca rosea* and *Catharanthus roseus*. It has been transmitted by artificial inoculation to tomato, potato and tobacco. Probably, during the off-season for the crop the phytoplasma survives on weed hosts and from there it is transmitted to the main host by its insect vector. Possible relationship between root knot nematode infection and little leaf is also reported.

Management of the disease

Effective control measures against eggplant little leaf disease are not known. Treatment with tetracyclines were claimed to control the disease. In many studies temporary masking of symptoms (mycoplasmastatic effect) was obtained by using achromycin, terramycin, aureomycin and ledermycin. In the absence of definite control measure, eradication of weed hosts and diseased eggplants and control of insect vectors by insecticides are only recommendations against the disease. New crop should be planted only when diseased plants in the field and in the neighborhood have been removed. Metasystox, malathion or a mixture of 0.1% BHC and 0.1% DDT have been recommended for vector control. Cultivars Pusa Purple Cluster, Arka Sheel, Manjari Gota and Banaras Giant, Black Beauty and Brinjal Round show moderate resistance in the field. Cultivars BB-7, BWR-12, Pant Rituraj and H-8 are reported to have some resistance to little leaf in addition to Phomopsis blight and bacterial wilt.

PHYTOPLASMA DISEASES OF POTATO

Three phytoplasma diseases of potato have been described from India. These are Purple top roll (PTR), Marginal flavescence (MF) and Witches' broom (WB). All of them are leafhopper transmitted but not sap transmissible and have been found to be widespread in the hills and the Deccan Plateau area.

Purple Top Roll: This disease, reported for the first time from India, can cause 15–80% loss in yield. It is characterized by rolling and purple or pink coloration of the basal part of the leaflets of top leaves. Some cultivars develop only rolling but no coloration. Stunting, chlorosis, profuse axillary shoots with aerial tubers, and swelling of nodes are other symptoms. There is no wilting of the infected plants and no necrosis of the tubers.

The organism is not sap transmissible. *Orosius albicinctus*, considered as one of the vectors, is not very efficient in transmitting the *Phytoplasma*. Nymphs of *Alebroides nigroscutellatus* (*A. dravidanus*) are better transmitters of the pathogen. The *Phytoplasma* adversely affects its vectors. The urge for mating in the females is delayed and often they become sterile. The hatching time of eggs is enhanced. The longevity of adults is also reduced. This is probably because of invasion of different body organs of the vector by the *Phytoplasma*. The minimum acquisition access period is one hour. Maximum transmission takes place with acquisition access period of 3 days. The *Phytoplasma* needs 7–15 days of incubation in the vector body. One or a few infective vectors are required per plant for maximum transmission. Nymphs are more efficient than adults in transmitting the pathogen. While adult males have better transmitting ability than females, the latter retain the *Phytoplasma* in their body for a longer period. Infected tubers seem to be the most important source of perennation and transmission. The symptoms are pronounced at low temperatures of 15°–20°C and masked at 30°–38°C. The *Phytoplasma* can be transmitted to clover, tomato, *Datura*, and tobacco. Co-existence of two phytoplasmas in the same potato plant, suggesting mixed infection, is reported by Leyva-Lopez *et al.,* (2002).

Marginal Flavescence: This disease is prevalent in the hills and plateau area of the country where cool weather prevails for a longer time during the crop season. The loss due to this disease may vary from 40 to 75% (Nagaich *et al.,* 1974). The symptoms of MF are chlorosis on the margins of upper leaves. In rare cases the severely affected plants, after exhibiting flavescence for quite some time, develop pinkish colour at the base of leaflets and start rolling. Hairy root symptoms on tubers of 60–65% plants affected with MF have been observed. The pathogen is leafhopper and graft transmitted. Five to 35% tubers carry the pathogen and serve as a source of primary inoculum. Hot water treatment of tubers at 50°C for 10–15 min eliminates the pathogen without adversely affecting the tubers.

Witches Broom: This disease results in extreme stunting of the plant and numerous filamentous stems with simple leaves. Very small tubers are formed by the affected plants resulting in total loss of the plant. The pathogen is leafhopper transmitted and perennates through tubers.

Control of these *Phytoplasma* diseases of potato is the same as for virus diseases of the crop.

SANDAL SPIKE DISEASE

Sandal (*Santalum album* L.) is an angiospermic semi-root parasite of many plant species which grows to the size of a tree. It is economically a very important forest tree in India. The species is confined to India and Indonesia. In India the tree is common in the southern parts including states of Karnataka, Tamil Nadu and Kerala. Spike disease is the most destructive of the few disease that attack this plant. Trees of all ages and size are attacked and die within a few years of appearance of symptoms.

Symptoms

Sandal spike is a yellows type of disease with witches broom effect. Two types of symptoms are produced. The more common form, designated as "rosette spike", is characterized by severe reduction in leaf size and shortening of the internodes. This results in crowding of leaves on the leaf bearing branches. The new leaves that develop later are further reduced in size. The reduction is more in width than in length. Such leaves stand out stiffly on the branches like spikes. In advanced stages of the disease just before death of the tree, leaves become yellowish and finally reddish. The flowers show phyllody (virescence). The diseased parts rarely bear any fruit although the healthy parts of the same tree bear normal flowers and fruits. Ends of roots of diseased plants die and haustorial connections with the host plant are damaged. Phloem necrosis is one of the internal symptoms.

The other form of the disease, known as "pendulous spike" is characterized by continuous apical growth of individual shoots without proportionate thickening resulting in a drooping habit of the shoot. The dormant buds do not develop or grow and, therefore, no rosetting effect is seen. In this form of the disease, roots and haustoria are not damaged.

The causal organism

The spike disease of sandal was considered a virus disease. This assumption was based on symptoms and graft and sap transmissibility of the causal agent. In 1969, electron microscopic examinations confirmed that the disease is caused by mycoplasma-like organism (*Phytoplasma*) which is found in the sieve tubes of leaves and twigs showing symptoms. Remission of symptoms and disappearance of MLO after tetracycline treatment further confirmed mycoplasmal etiology of the disease. The sandal spike *Phytoplasma* is a member of the aster yellows phytoplasma group.

The organism shows very high polymorphism. The size of cells of the pathogen in the host varies from 60 to 750 nm, the most commonly occurring forms being ellipsoidal which measure 180–200 × 250–300 nm. Some cells are elongated, measuring up to 750 × 150 nm (Parthasarathi and Venkatesan, 1982). The smooth, unit membrane bounding the cell is about 10–12 nm thick. The pathogen grows in the host best at 30–38°C. Below and above this range the growth declines.

The *Phytoplasma* is reported to infect a large number of plant species such as *Eucalyptus grandis, Vinca rosea, Zizyphus oenoplea, Dodonia viscosa*, etc. The disease can be transferred from *Vinca rosea* to sandal and vice versa through dodder. The spike disease is spread through root contact and insect vectors. A large number of leafhopper species seem to be involved in transmission of the pathogen. These include *Nephotettix virescence* and *Moonia albimaculata*.

Management of the disease

So far no specific measures for control of sandal spike disease have been developed. Attempts had been made to cure the trees by raising temperatures above 38°C by building fires in trenches around the affected trees but it was neither successful nor practical. Eradication of diseased trees and creating disease free strips around affected areas have also failed to arrest its spread. No host resistance has been found. Like other plant diseases of phytoplasma etiology sandal spike also shows remission of symptoms after treatment with tetracycline Ali et al.,. (1987) had used a gravity-flow infusion technique for chemotherapy with tetracyclines. Most tetracyclines, including doxycycline and oxytetracycline were effective when 500 mg per tree was infused.in 500 ml of water. The disease remission lasted for only 4 months. The systemic fungicide Benlate was also reported to bring about temporary remission of symptoms. Application of the antibiotic by the girdling method shows recovery of the trees 25 to 30 days after treatment with terramycin alone or in combination with Benlate and 30 to 35 days after treatment with ledermycin alone or in combination with Benlate. However, these treatments provide only temporary relief for 50–150 days and then the disease reappears.

REFERENCES

Corbett, M.K., S.R. Misra and Kishen Singh. 1971. Grassy shoot disease of sugarcane. IV. Association of mycoplasma-like bodies. *Plant Sci.* 3: 80

Das, A.K. and D.K. Mishra. 1998. Hormonal imbalance in brinjal tissue infected with little leaf phytoplasma. *Indian Phytopath.* 51: 17

Doi, Y., M. Teranaka, K. Yora and H. Asuyama. 1967. Mycoplasma or PLT group-like microorganisms found in the phloem elements of plants infected with mulberry dwarf, potato witches' broom, aster yellows or Paulownia witches' broom. *Ann. Phytopath. Soc. Japan* 33: 259

Fontaniella, B., C. Vicente, M.E. Legaz et al., 2003. Yellow leaf syndrome modifies the composition of sugarcane juices in polysaccharides, phenols and polamines. *Plant Physiol. Biochem.* 41(11–12): 1027

Ishiie, T., Y. Doi, K. Hora and H. Asuyama. 1967. Suppressive effect of antibiotics of tetracycline group on symptom development of mulberry dwarf disease. *Ann. Phytopath. Soc. Japan* 33: 267

Leyva-Lopez, N.E., J.C. Ochoa-Sanchez, D.S. Leal-Klevezas and J.P. Matinez-Soriano. 2002. Multiple phytoplasmas associated with potato diseases in Mexico. *Can. J. Microbiol.* 48(12): 1062

Nasare, K., A. Yadav, A.K. Singh et al., 2007. Molecular and symptom analysis rveal the presence of new Phytoplasmas associated with sugarcane grassyshoot disease in India. *Plant Dis.* 91(11): 1413

Parthasarathi, K. and K.R. Venkatesan. 1982. Sandal spike disease. *Curr. Sci.* 51: 225

Rishi, N., S. Okuda, K. Arai, A. Doi, K. Yora and K.S. Bhargava. 1973. Mycoplasma-like bodies, possibly the cause of grassy shoot disease of sugarcane in India. *Ann. Phytopath. Soc. Japan* 39: 429

Seemiiler, E., C. Marcone, U. Lauer et al., 1998. Currrnt status of molecular classification of the Phytoplasmas. *J. Plant Pathol.* 80(1): 3–26

Shukla, U.S. 1985. Role of secondary transmission in the carry over of grassy shoot disease of sugarcane in north India. *Indian J. Plant Pathol.* 3: 1

Shukla, U.S. and R.R. Bhansali. 1985. Detection of mycoplasma-like bodies in the grassy shoot diseased sugarcane leaves through light microscopy. *Indian J. Virol.* 1: 330

Shukla, U.S. and Vijai Singh. 1990. Inactivation of grassy shoot disease MLOs in diseased setts by heat treatment. *Indian Phytopath.* 43: 561

Siddique, Agarwal, Alam and K. Reddy. 2001. Electron microscopy and inoculum characterization of phytoplasma associated with little leaf disease of brinjal (*Solanum melongena*) and periwinkle (*Catharanthus roseus*) in Bangladesh. *J. Phytopathol.* 149(5): 237

Singh, D. and S.J. Singh. 2000. Chilli little leaf a new phytoplasma disease in India. *Indian Phytopath.* 53: 309

Singh, S., K.S. Bhargava and B.B. Nagaich. 1986. Transmission of the purple top roll pathogen by *Alebroides nigroscutellatus. Indian Phytopath.* 39: 14

Singh, S.J. 1992. Electron microscopic observations of bitter gourd witches' broom. *Indian Phytopath* 45: 354–355.

CHAPTER

6

The Fungi

The lowest forms of cellular organisms associated with plan diseases were discussed in the preceding chapters. Fungi are more highly evolved forms which had been included in the traditional classification in Thallophyta. Thallus is a plant body that is not differentiated into root and shoot. It may be a single cell, a filament of cells, or a complicated multicellular structure. While bacteria, fungi and algae all possess a thallus, the differences are striking enough to place them in separate groups.

Fungi can be described as eukaryotic protists having chlorophyll-less, nucleated, unicellular or multicellular filamentous thallus which reproduces by division of vegetative cells, well defined asexual and/or sexual spores.

VEGETATIVE STRUCTURE OF FUNGI

The thallus of fungi may be plasmodial, unicellular, pseudomycelial or mycelial. In the lowest forms (Myxomycetes, Myxomycota, or slime molds) the somatic or assimilative structure (thallus) is an amoeboid plasmodium lacking a true cell wall and covered by a hyaloplasm. In the true fungi (Eumycota) true cell wall is present whether the thallus is a unicellular or a multicellular branching body. The branches are known as *hyphae* (s. *hypha*) and the entire vegetative body is called *mycelium*. According to their age the hyphae may be uniformly thick or tapering towards the apical end. The thickness of hyphae varies from less than 0.5 μm to over 100 μm. In its size the entire body may be only a few microns (μm) in length (some lower Mastigomycotina and yeast cells) or the mycelium may develop into big sporophores, several inches to feet, as in the wood rotting Basidiomycetes, or into great sheets or strands extending many meters (rhizomorphic fungi). When a mycelial thallus is unicellular but produces small or large buds loosely attached in a chain it is known as *pseudomycelium*. In many fungi the mycelial thallus is present in the host tissue but in artificial cultures it is yeast-like. This type of thallus is called **dimorphic**. In the normal course of development the somatic thallus gives rise to reproductive structures formed asexually or sexually. The most simple types of thalli are wholly converted into reproductive cells and are called **holocarpic**. But in majority of fungi only parts of the thallus become reproductive forming differentiated or undifferentiated reproductive cells. Such thalli are **eucarpic.**

The hyphae are microscopic fungus filaments, usually branched, composed of an outer cell wall and a cavity (lumen) lined or filled with protoplasm. The development of hypha, the aggregation of which constitutes the mycelium, starts from some kind of propagule such as a spore. The spore puts out one or more bud-like processes (germ tubes) which elongate and become hyphae. Further growth and branching of the hyphae forms the mycelium. The hyphae are generally characterized as **septate** or **aseptate** (coenocytic). When the hypha is continuous with nuclei irregularly distributed

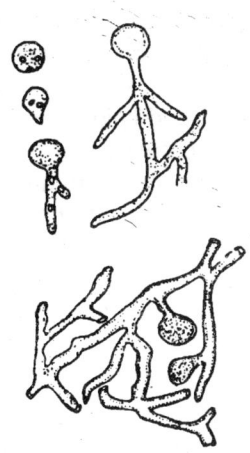

Fig. 13. Development of mycelium from fungal sopre.

throughout the thallus it is aseptate or coenocytic. If it is partitioned into compartments (cells) by means of cross walls (**septa**, S. **septum**) it is known as septate. However, even the hyphae with septa are continuous because of the presence of one or more **septal pores** through which the cytoplasm is in continuity among the adjacent cells. The septa are actually formed in all filamentous fungi. In coenocytic mycelium septa develop to cut off the reproductive units from the hyphae or to prevent death of hyphae due to injury by cutting off the dead portion. Such septa are called **adventitious septa** and have no septal pores. The septa formed to partition the hyphae are called primary septa.

In or on the substrate, such as dead mass of organic matter or living tissues of the host, continuous osmotic absorption of nutrients by hyphae occurs. In soil the hyphae are present in between soil particles in a fluid medium. In plant tissues they are **intercellular** (lying in between the cells) or **intracellular** (lying in the lumen of the cell surrounded by cell protoplasm. When the hyphae are intercellular they send special hyphal structures (**haustoria**, S. **haustorium**) into the cell to absorb nutrients. Haustoria serve not only as organs for nutrient absorption, they also synthesize nutrients for the pathogen and also play a role in expression of avirulence genes of the pathogen.

The hyphae have ability to **anastomose**. Neighbouring hyphae in vegetatively compatible thalli are stimulated to put out short branches which make contact, the intervening wall dissolves, and **H-pieces** are formed. The tips of hyphae also unite in the same manner. Adjacent cells of the same hypha may also be brought into communication with each other by means of loops known as **clamp connections**.

Branching of the hyphae, especially those bearing spores, is of interest since it aids in identification and classification of fungi. Branching is **dichotomous** when the apex of a hyphae ceases elongation and forks into two lateral branches. It may be **lateral** when the main hyphae continues to grow and sub-apical lateral branch is given off. Two lateral branches opposite each other are also produced. In many fungi, a whorl of three or more branches is produced from the same point. This is known as **verticillate** branching. In **sympodial** branching each successive leading apex becomes restricted in growth and is overtaken by a lateral branch from below. In **monopodial** branching the apex of the leading hypha is not suppressed but keeps pace in growth with the most active of the lateral branches from below. The sympodial branching may form a cymose while the monopodial branches form a racemose.

The fungus cell: The cells contain the protoplasm including cytoplasm in the space enclosed by the cell walls. There are a number of membranous and non-membranous bodies in this fluid. These include mitochondria, dictyosomes, vacuoles, ribosomes and nuclei, granules and lipid bodies. The chemical composition of the cell walls is different from that of plants and bacteria and is variable among different groups of fungi and within the same species at different stages of growth. The chief components of the cell wall appear to be various types of proteins and polycarbohydrates or their mixtures such as cellulose, pectose, callose, etc. Cellulose predominates in the cell walls of lower fungi (lower Phycomycetes or Mastigomycotina, now mostly Chromista) while in higher fungi (Ascomycota and Basidiomycota) cellulose is either lacking or its distinguishing reaction is masked by chitin which is the major constituent. Calcium carbonate and other salts may also be deposited upon or within the cell wall. Functions of the cell wall are many. It protects cells from osmotic shock, lytic enzymes of microorganisms, gives the cell its morphology, prevents desiccation and helps in adhesion of the cell to the host surface.

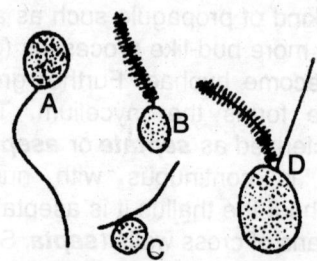

Fig. 14. Types of zoospores in Mastigomycotina. A–Posteriorly uniflagellate; B–Anteriorly uniflagellate; C–Biflagellate zoospore of Plasmodiophorales; D–Biflagellallate zoospore of oomycetes.

Composition of the protoplasm is the same as in the cells of other organisms. It is made up of compounds of carbon, nitrogen, phosphorus, etc. The fluid in the nucleus is called **nucleoplasm** and the fluid outside the nucleus is **cytoplasm**. The cytoplasm is granular to reticulate and includes storage products like globules of amylodextrin, glycogen and fat. The protoplasm possesses the property of secreting several kinds of ferments, enzymes and organic acids which help in the digestion of food and serve to dissolve the walls and protective layers of the host cells.

The cells are usually described according to their nuclear component and nuclear behavior. A **monokaryotic cell** contains a single nucleus (**monokaryon**) with the haploid or basic number (*n*) of chromosomes for the species. It forms part of the **haplophase** in the life cycle of the fungus. The haplophase may be of a very short duration or may be prolonged. A **dikaryotic cell** contains a **dikaryon**, a pair of nuclei which differ in characters but are genetically compatible and which are able to undergo simultaneous or conjugate division to give pairs of daughter nuclei with the same genetic features as the parent nucleus. Each dikaryotic cell thus contains *n+n* chromosomes. The dikaryotic cells form the **dikaryophase** in the life cycle. It is not **diplophase** since the nuclei have not fused. The **synkaryotic cells** contain one nucleus with the diploid number (2 *n*) of the chromosomes after fusion of the two haploid nuclei of the dakaryotic cell. The fusion of nuclei is called **karyogamy.** Sometimes, the dikaryotic condition is extended to **heterokaryotic** condition when the cells are multinucleate (more than two nuclei) and the nuclei are genetically of different types. On the other hand, a multinucleate cell may be a monokaryon if all the nuclei are of the same genetic constitution.

Reproduction of Fungi

The methods of multiplication in fungi may be divided into the following three categories: 1) Growth of the cut off vegetative cells into new thalli. In most fungi any part of the hypha has the capacity to develop into a new thallus if cut off from the parent hypha and placed in suitable environment 2) Reproduction by asexual spores, anamorphic phase of life cycle and 3) Reproduction by sexually produced spores, teleomorphoc phase of life cycle.

Asexual reproduction: In asexual reproduction the hyphae cut off minute spores. The structure and origin of these spores vary greatly and each type has been given a different name. ***Chlamydospores*** are thick-walled, usually round, spores formed by the direct transformation of certain cells of a hypha or cells of a multicellular spore without the production of specialized spore-bearing branches. The term ***conidium*** (plural-***conidia***) is applied to spores which are pinched off or cut off from the ends of special spore-bearing hyphae, the ***conidiophores***. Such spores may be single, in clusters or in chains. Occasionally, there is very little or no difference between the shape of spores and cells of the spore-bearing hypha. Such conidia are called ***conidia***.

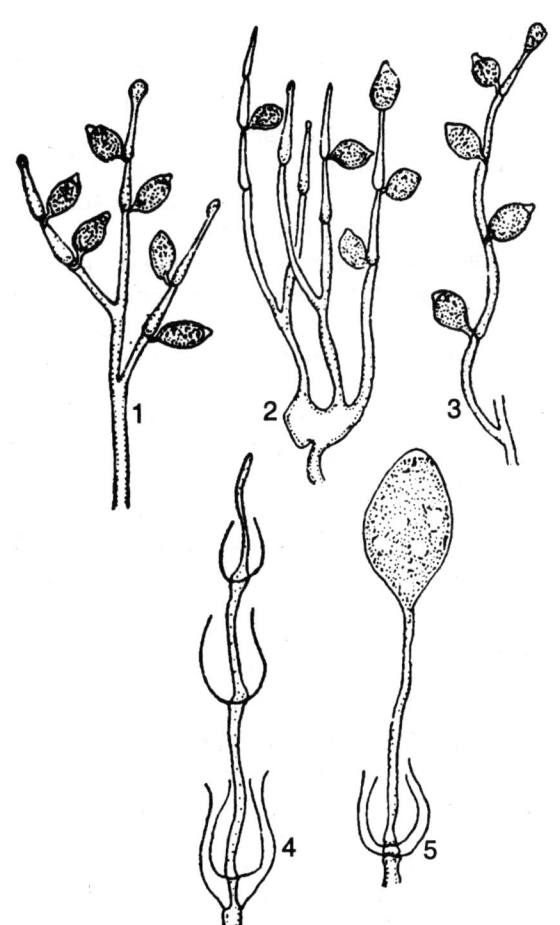

Fig. 15. Types of sporangiophores in Phytophthora.
1–2–Compound sympodial sporangiophore of Phytophthora infestans (1) and of P. phaseoil (2); 3–Simple sympodial sporangiophore of P. cactorum 4–5–Proliferation of sporangiophore of P. megasperma.

Spores such as oidia, conidia or chlamydospores may be structures not enclosed in a fruit or sac. However, in many fungi some types of the asexual spores are produced in spore mother sac, or in different types of fruit bodies or are arranged in different types of stroma. In the lower groups of fungi such as in Phycomycetes (Mastigomycotina) which have aseptate hyphae, spherical, tubular, or ovoid sacs (***sporangium, sporangia***) are formed by swelling of the hyphal tip or by any part of the hypha. In aquatic forms, the sporangia liberate naked protoplasmic bodies, the **swarmspores** or **zoospores**. These spore are provided with delicate, vibratile or undulating filaments or thread-like processes, the cilia or the **flagella** by means of which they are able to swim about in the surrounding film of water. In higher forms of the same group the sporangia liberate walled, motile zoospores or the walled, non-motile **sporangiospores**. The latter are dispersed mostly by air currents. Sometimes, instead of liberating zoospores, the sporangium acts as an independent reproductive unit and functions as a conidium, i.e., it germinates directly by producing a germ tube. This change in the mode of germination of the sporangium depends on environment. Moist conditions favour germination by zoospore and dry conditions induce direct germination. In the Ascomycotina and the Deuteromycotina, the conidia may be produced in special structures variously known as **pycnidia, sporodochia, acervuli**, etc. These are described under the head fruit bodies.

Sexual reproduction: The sexual reproduction involves the act of union of two separate and distinct **gametes** contained in cells called **gametangia** that represent the male and female components. The sexual reproduction by union of two morphologically dissimilar gametes is known as **oogamy**. In most of the higher forms of fungi the sexual process involves the union of two similar gametangia. This is known as **isogamy**.

In oogamous forms the union of gametes results in the development of a simple structure (**oospore**) or a complex fruit (**cleistothecium, perithecium, apothecium**). Compatibility between the gametes is essential for the final production of these fruit bodies. In the isogamous forms the union of two similar motile or non-motile gametes lead to the formation of **zygospore** or **zygote**. Since usually there is no morphological difference in the two gametangia the participating structures are designated as positive (+) or negative (–) gametes. Positive gametes have no affinity for each other. Similarly two negative gametes will not unite. They are sexually incompatible. Union take place only between a positive and a negative gamete. In **heterothallic** species of fungi these gametes are borne on different thalli.

The phenomenon of sexual reproduction in such species is known as **heterothallism**. In those species which possess their sexual gametes on the same thallus the fungus is **homothallic.** Sexual reporoduction in most of the Zygomycetes and all the Basidiomycotina is of isogamous type and heterothallism is common. In most of the higher Ascomycotina also isogamy is common.

Spore fruits: Sporangia are the simplest form of spore fruit. They are tissue-less sacs containing the spores. In higher forms of fungi there is a tendency to organize complex aggregates of spore-bearing hyphae, frequently surrounded by more or less supporting and protective tissues. These structures are the true spore fruits.

In the fungi-imperfecti (Deuteromycotina or anamorphic fungi) which have no sexual stage in their life cycle dense fascicles are formed by aggregation of erect conidiophores. These aggregations are called **coremia** or **synnemata** (S. *coremium* or *synnema*). The spores or spore bearing hyphae are sometime grouped into small or large masses or clusters to which the name **sorus** (pl. **sori**; Gr. *soros*-heap) is applied. The pustules on leaves attacked by white blister disease or pustules of rust diseases are common examples of sori. The term is also applied to spore masses of smuts. A single sorus is an aggregation of spores alone or mixed with spore-bearing hyphae and sometimes sterile filaments. It may be covered with layers of the host cells or may be naked.

The rust fungi produce small, flask-shaped fruiting bodies, the **pycnia** or **spermogonia**. The inner wall of these flask-shaped bodies is lined with numerous hyphal branches which cut off minutes spores known as **pycnospores** or **spermatia**. Soon after the development of pycnospores, the fungus produces a cup-like fruit body which is cytologically related to the pycnia. This fruit body is known as *aecium* or *aecidium* or aecial cup and produces aeciospores.

Sometimes conidiophores and conidia are produced in a flask-shaped flattened or round fruit known as *pycnidium*. The sturcture is the same as that of pycnia of rust fungi. The inner wall is covered with layers of conidiophores which cut off uni- or multicellular spores. The mouth of a pynium or pycnidium is called *ostiole*. The wall of the pycnidium or aecidium is called *peridium*.

The *acervulus* (pl. *acervuli*) is a fruiting body peculiar to order Melanconiales of Fungi-imperfecti. It is a saucer shaped depressed mass of aggregated hyphae and bears conidiophores over its concave surface. Intermingled with conidiophores or at the edges of the acervulus long pointed dark structures may be present. These are called *setae*. On the host the acervulus develops beneath the cuticle but ruptures the cuticle and becomes erumpent on maturity. The *sporodochium* is a fruiting structure similar to acervuli except that there is a well-developed cushion-like aggregation of fungal tissue which breaks through the host tissue and bears conidiophores. These may also be formed in the mass of hyphae lying superficially over a substrate.

In the Ascomycotina, the sexual spores (the *ascospores*) are enclosed in a mother cell, the *ascus*, and one or more of these asci are enclosed in a complex fruit body various known as *cleistothecium, perithecium* or *apothecium*. The cleistothecium is a sexual fruit having a globose shape but no ostiole. It is mostly covered with hyphal outgrowth known as *appendages*. The perithecium resembles a pycnidium in its form, size and character of its wall and possesses a similar opening or ostiole. This flask-shaped fruit body contains asci arising from the basal mass of cells in the inner cavity. Mixed with these asci may be present a large number of sterile filaments known as *paraphyses*. Occasionally similar filaments protrude out through the ostiole. These are known as *periphyses*. The perithecia may be borne singly or in closely associated groups of definite stromata. The apothecium is a disc-like, saucer shaped, or cup-like body seated on or in the substratum or raised on a long or short stalk (*stipe*). The asci are arranged in parallel series on the exposed flat, concave or convex surface. The layer containing the asci is called *hymenium*. Paraphyses may be present.

In the highest forms of fungi (higher Basidiomycotina) the sexual spores, the *basidiospores*, are borne externally on the spore mother sac, the *basidium*. The basidia are borne in groups or layers in complex structures, the *compound sporophores*, which may be enormous in size.

SURVIVAL OF THE FUNGI

During the absence of an active host plant the pathogenic fungi must survive through some source to maintain continuity of the disease cycle and to provide primary inoculum for infection in the next season. The survival in space is covered by dispersal, discussed in the following paragraphs. Survival in time, i.e., survival from one season to the next, is through one or more of the following means:

Soil-borne pathogens: (a) As active saprophytic mycelium in soil or in crop debris, (b) As dormant chlamydospores, conidia, oospores, vegetative cells or sclerotia in soil or in crop debris.

Seed-borne pathogens: (a) As dormant mycelium or spores on the surface of the seed or beneath the pericarp, (b) As dormant mycelium in the embryo or endosperm of the seed.

Perennial hosts, alternative or weed hosts and *off-season crops:* Pathogens of fruit trees generally survive in dormant or active stage on the diseased trees. Infected leaves that defy leaf fall, vegetative buds, dead and dried fruits hanging on the trees, and the infected wood are the main sources of survival. Many pathogens survive in active state on weeds, wild grasses, or off-season crops. The spores of the pathogen are blown by wind from these sources to the main crop in the season.

DISPERSAL OF THE FUNGI

During growth of fungi on a substrate, living or dead, their exit from the substrate and dispersal or dissemination is an important part of the life cycle. In absence of dispersal the development may be checked due to overcrowding, lack of extension to new location and fewer opportunities for genetic variability which is important for survival and adaptation to changing environments.

The two methods of reproduction in fungi, sexual and asexual, are intended to accomplish more or less two distinct acts in the life history of the fungus. The asexual spores are primarily meant for rapid dissemination while the sexual fruits mainly help the fungus to survive through adverse conditions. In addition to sexual or asexual spores the fungi can be dispersed also by unspecialized mycelial fragments, sclerotia, and rhizomorphic strands. The means of dispersal of these other types of propagules may be the same as for spores but the mechanism of their separation from the thallus is mostly passive.

Asexual spores are essentially the main dispersal units. They are the repeating propagules which are produced in enormous quantities to compensate for the loss during dispersal due to nature's adversities. These spores are provided with mechanisms for quick release and transport at the appropriate time, and are capable of immediate germination to establish the individual at the new sites. Such dispersive units have the ability to retain their power of germination during the process of long distance dispersal.

Dispersal of fungal propagules has three distinct stages: liberation or take off (spore release), transport or flight (dispersal) and, finally, deposition or landing on the new site.

Active discharge of spores: In active discharge the spores are provided with some structural mechanisms or physiological responses which enable them to be thrown out with some force.

a) *Liberation due to bursting of turgid cells:* This type of discharge is very common in Ascomycota for the release of ascospores from asci. Asci are turgid cells which burst in regular manner violently liberating the ascospores.

b) *Liberation due to round off of turgid cells:* In this method of separation of the spores from the parent hypha, the violent or sudden discharge of spores is due to rounding off or change in shape of turgid cells in an unstable manner. In *Sclerospora*, there is a flat surface of contact between the spore and the sterigma of the conidiophore. Both spores and conidiophores are turgid cells so that there is a tendency for each to round off at the flat region of contact. Suddenly, the adhesive forces are overcome and the base of the spore and the tip of the stergma round off shooting the spores to a distance of a mm or so. Similar discharge takes place in aecia of the rust fungi.

c) *Ballistospore discharge:* This type of violent discharge is characteristic of the basidiospores of rust and some smut fungi. Very near the point of contact of the spore with the sterigma a gas-filled bubble appears and grows in few seconds to a definite size. Then the spore is shot away carrying the bubble.

d) *Liberation due to hygroscopic movement:* Movement, often involving collapse of sap-filled turgid hyphae and their bursting, is responsible for violent shake off of spores in some fungi such as *Peronospora*. This hygroscopic movement is caused by a sudden fall in humidity resulting in drying of the sporophore bearing the spore. In *Phytophthora infestans* the sporangia are more easily discharged on the leaf surface exposed to the outside atmosphere than on the leaf surface under dense canopies where humidity is retained for much longer periods.

Passive discharge of spores: In passive spore liberation the agency dislodging the spores is mainly the force from some outside source such as strong air currents, raindrops falling on the spore mass, insects visiting the surface containing the spores and workers moving in the field among the diseased foliage. Such spores may be dry or slimy and are found in many groups of fungi.

Dispersal of spores: The agencies which cause spore release or take off may be partly or entirely responsible for their dispersal, transport or flight also. Spore dispersal is essentially a passive process. It is determined by physical agencies such as raindrops, surface water moving in the field, wind, and wind driven raindrops or living organisms such as insects and higher animals including man. Modes of dispersal vary with the habitat of the fungi and their morphological characteristics. These include mycelial growth (mostly in soil fungi), dispersal by soil movement, dispersal by movement of contaminated seed and planting stock, dispersal by air, animals and raindrop splashes.

PHENOMENON OF INFECTION

A parasitic fungus must establish an intimate relationship with the host tissues to absorb the desired nutrients. Successful establishment of this relationship is called infection. The term can be defined as the establishment of a pathogen within the host body. Only attack by a parasite, its arrival on the plant surface, or even its entry into the host is not infection because the true parasitic relationship has not been established.

The phenomenon of infection has three distinct stages: (i) Pre-penetration stages, (ii) Penetration stage, and (iii) Post-penetration stages. In the pre-penetration stage the fungus hyphae or spores come in contact with the host surface. In most fungi, there is close adhesion of the spore surface with the host surface. Chemicals excreted by the spores are involved in the adhesion. The interaction between the spore surface and plant surface can start within a minute of contact. The spores germinate to produce a germ tube or infection thread. The germ tube may undergo some morphological changes to enhance its capacity to adhere and to penetrate the host. This stage may be abortive if environments do not favour growth of the hyphae or germination of spores. During penetration the infection thread enters the host by any of the following methods.

1. Entry through natural openings like stomata, lenticels, hydathodes, etc.
2. Entry through rupture of the host surface due to development of organs like prop roots and through wounds due to mechanical or insect injury.
3. Direct penetration:
    (a) Mechanical pressure by the infection thread through appressoria;
    (b) Chemical action through secretion of enzymes that break down the structural barriers of the host.

Entry through natural openings, rupture of host surface or wounds looks simple because there is no mechanical barrier that could prevent entry into the host. However, it is not essential that the spores must land at a sight where such entry points are present. Two conditions must be fulfilled to make maximum number of spores on the host surface effective. The environment must be favourable for immediate germination of the spores and elongation of the germ tubes. The spores and germ tubes must be supported by some mechanism(s) of adhesion to the host surface so that they are not dislodged. Elongation of germ tubes towards the site of entry is directed by the chemical emanations from such sites. The germ tube may form such structures as appressoria which may be melanised. Infection pegs arise from these appressoria. Many wound parasites first utilize the dead cells for nutrition, increase their mass and produce more cell wall degrading enzymes that help them to grow farther into the wounded tissue.

After entry it is possible that physiological conditions of the host are not found suitable and the infection thread may stop growing or die. If the host tissues are suitable (compatible interaction) the infection cell enters the nearest cell or sends haustorium into it. This results in absorption of nutrients from host protoplasm. Supported by the nutrition the infection thread develops into inter- or intracellular mycelium or supports the ectophytic mycelium. In addition to absorption of food the fungus may interfere with the host activity in various ways. The production of pectinolytic or cellulytic enzymes dissolve the cell walls. Toxins may be produced which may be translocated to other tissues causing various types of damage. Due to the interferences with the normal existence of the host tissue, symptoms of disease develop. The fungus continues growth, produces spores or reproductive units and finally finds exit through the host surface and spreads to repeat the same processes as explained above.

Production of disease by fungal infection is determined by many factors such as the production of host-specific or host-nonspecific toxins, or the extracellular matrix by the infection structures and H_2O_2 generation from penetration pegs. The targets of the toxins are the host plasma membrane, chloroplasts, and mitochondria. There is partial destruction of these targets in a susceptible host. The extracellular matrix produced by the infection structure helps in adhesion, differentiation of the germ tube in to appressoria. The reactive oxygen species are produced by the penetration pegs in the cell walls and plasma membranes which are damaged.

NUTRITION OF THE FUNGI

Like other microflora the fungi take in external materials for nutrition through their membranes, transport them to the active metabolic sites within the mycelium and transform them to provide energy for maintenance and biosynthetic processes. These organisms are heterotrophs for carbon compounds and many are heterotrophic for other materials also. The entire vegetative mycelium may act as an organ of absorption (**holophytic feeding**) or only the haustoria perform this task. They may be obligate or facultative saprophytes, obligate or facultative parasites, or they live with green plants or insects as facultative or obligate symbionts. Obligate parasitism (**biotrophism**) has advanced much above the level of mere feeding on the living tissues. Many fungi have physiologic races which specialize in feeding on specific genera, species, and agricultural varieties of the host family.

Nutritional requirements of fungi are similar to those of higher plants. Carbon and nitrogen are major essential needs. Phosphorus and sulphur are also essential but required in lesser amounts. Under specific situations some fungi are incapable of synthesizing their special vitamin and growth factor requirements and need them from external sources.

Fungi can utilize a very wide variety of carbon sources. Some utilize chemically complex carbon compounds for growth whereas others are much more selective in their requirements. Among carbon compounds shown to be utilized by fungi in synthetic media as sole source of carbon are carbohydrates and related compounds, amino acids and related compounds, organic acids, polycyclic compounds and alkaloids. Within a genus, the species vary greatly in their ability to utilize different carbon sources and within a species similar selectivity is shown by different strains. Among plant pathogens some correlation between preference for a particular carbon source and pathogenesis can be seen.

Some fungi prefer ammonium nitrogen, others nitrate nitrogen or reduced form of nitrogen such as nitrite. Some utilize both with equal efficiency. In mixed sources of nitrogen such as ammonium nitrate, the fungus generally utilizes the ammonium first before reducing nitrate. Amino acids are the main source of organic nitrogen for fungi. Many fungi such as water moulds can utilize only organic nitrogen in the form of amino acids.

As living organisms fungi require minute quantities of vitamins and growth factors. Many fungi can synthesize these substances but some are incapable of doing so. In such cases exogenous supply of the material or its precursor is essential to support the growth and reproduction of the fungus. The vitamins required by fungi are thaimin, biotin and less frequently pyridoxin.

Acquisition of Nutrients by Plant Parasitic Fungi

Nutriuent requirement of pathogenic fungi varies with the stage of infection cycle (spore germination, infection, proliferation, sporulation). These fungi may be **biotrophic** (totally dependent on living host for nutrition and completion of life cycle), **necrotrophic** (obtaining nutrition from dead cells and having no specialized organs for food uptake, or **hemibiotrophic**. In hemibiotrophy, the host remains largely alive while the pathogen is establishing itself in tissues. After a brief period of biotrophy it switches to a necrotrophic lifestyle. The biotrophs and hemibiotrops have specialized organs (haustoria and haustoroid hyphae) for uptake of nutrients whereas necrotrops have no such organs.

CLASSIFICATION OF FUNGI

Classification of fumgi had been based on functional morphology and physiology. This system continued until about two decades back and is still being followed by most plant pathologists. The recent system is based on phylogenetic affinities (DNA/DNA similarity) among various members of Thallophyta. While full taxonomic position of pathogens according to phylogenetic classification is given in chapters of different diseases, the following paragraphs present a comparison of the two systems in respect to pathogens included in this book.

In earlier system, classification had divided fungi (Mycota) into divisions (Myxomycota and Eumycota) with subdivisions in Eumycota (Mastigomycotina, Zygomycotina, Ascomycotina, Basidiomycotina, and Deuteromycotina). The Myxomycota included the plasmodial fungi Plasmodiophora and Spongospora in family Plasmodiophoraceae, order Plasmodiophorales and class Plasmodiophoromycetes. The subdivisions of Eumycota included all the fungi that produce motile spores (zoospores) and have aseptate mycelium (Mastigomycotina), or have aseptate mycelium but nonmotile spores including sexual zygospores (Zygomycotina) or those fungi that have septate mycelium with non motile asexual spores and sexual fruit bodies like perithecia with ascospores (Ascomycotina,) or sexual structures like basidia and basidiospores (Basidiomycotina).

In the phylogenetic system, the classification has division Eumycota in which all members of fungal group are placed in kingdoms, phylum, classes, subclasses, orders and families. The Plasmodiophorales are placed in class Plasmodiophoromycetes, phylum Plasmodiophoromycota and kingdom Protozoa because of more affinity with these organisms. All zoosporic fungi, except Chytridiales (*Synchtrium*) have been placed in the kindom Chromista because of their nearness to algae. Chromista have the phylum Oomycota, class Oomycetes and different orders and families. Chytridiales with family Synchytridiaceae are placed in kingdom Fungi, phylum Chytridiomycota and class Chytridiomycetes.

Apart from these major changes, the remaining taxonomy of fungi has only some minor changes except for new position names and further differentiation of families into subfamilies.

CHAPTER

7

Diseases Caused by Plasmodiophoromycetes, Chytridiomycetes and Oomycetes

CLUB ROOT OF CRUCIFERS

The club-root or finger and toe disease is potentially the most dangerous disease of the mustard family, especially cabbage and closely related plants. The disease is widespread in regions with temperate climate. However, it has no restrictions imposed by temperature and is present as far south as the central part of Sri Lanka. In India, the disease was reported from West Bengal in 1985 but is not widely present. In mustard the loss may be as high as 50% or more. Once the disease has appeared in the area, it becomes difficult to raise a profitable crop of crucifers in that area. Affected plants are usually a total loss. In Nepal, the disease is severe and widespread. Crop losses of 40% are not uncommon in cauliflower. The yield declines from 5–6 metric tons per 1500 m^2 before 2004 to less than 300 kg in 2004.

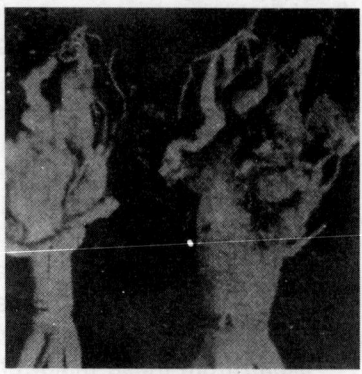

Fig. 16. Club-root of Cabbage.

Symptoms

The disease may progress to a considerable extent without showing any above ground symptoms. The earliest above ground symptoms are unthrifty development of the plants and flagging of the leaves in the hot sunny days, as if the plant is suffering from lack of water. When such plants are uprooted the hypertrophy of the root system can be seen. The infected roots enlarge relatively rapidly to form "clubs". Depending on the host species and nature of infection, the clubbing may be of several types, such as (i) clubbing is complete on main and lateral roots as in cabbage, (ii) clubs are present on only main roots, laterals are free, (iii) clubs are present on laterals but main roots are free, (iv) rootlets are free from clubs, (v) clubs appear as tumors on the roots as in radish, and (vi) dark decomposed spots are present on the root. These variations in the area of root system affected are due to specific roots infected by the pathogen. In cabbage, individual infections on roots progress in both directions along the main axis, and spindle shaped clubs result. The hypertrophy causes malfunctioning of the xylem which results in flagging of leaves. There is no normal development of the cork cambium at the surface and, therefore, the tissues are easily invaded by secondary saprophytic organisms such as soft rot bacteria. This results in conversion of the root system into a black rotting mass of tissues.

The causal organism

Plasmodiophora brassicae Wor. (kingdom Protozoa, phylum Plasmodiophoro-mycota, class Plasmodiophoromycetes, order Plasmodiophorales and family Plasmodiophoraceae. The pathogen was first described by Woronin in 1877. *Spongospora subterranea* f. sp *subterranea* (powdery scab of potato) also belongs to same taxonomic position. The life cycle of the pathogen starts from its resting spores lying in soil or in diseased crop debris. The walls of the spores contain chitin but no cellulose. The spore is spherical and up to 4 mm in diameter. Germination of the spore can occur without any resting period. The spore swells to several times its original size and releases a single anteriorly biflagellate primary zoospore. The larger flagellum points forward and the shorter one is directed almost at the right angle. Both these flagella are truncated whiplash type.

This zoospore is a naked body of uninucleate protoplast and is very active for some time. It moves by irregular jerks with the help of its flagella. Before infection of the host the zoospore becomes amoeboid. In contact with root hair or epidermal cells of the host roots the zoospore discards its flagella and bodily enters the root through a hole dissolved in the cell wall. The hole is then closed by host action. The naked spore then forms a thallus in the lumen of the host cell. Although when there are multiple infections several such thalli may be seen in the cell, such thalli normally do not fuse. The infected host cell is much larger than the non-infected cells. At first a small uninucleate thallus is seen in the host cell. Cytokinins from this plasmodium triggers a local re-initiation of cell division in the root cortex. Consequently, a de-novo meristematic area is established that acts as a sink for host-derived indole-3-acetic acid (IAA), carbohydrates, nitrogen and energy to maintain the pathogen and to trigger gall development (Devois *et al.*, 2006). Later, with enlargement of the amoeboid structure, mitotic divisions of its nucleus occur until a small plasmodium with 30 to 100 nuclei is formed. The plasmodium then cleaves into spherical zoosporangia or gametangia lying packed together in the host cell. This stage can reach within 4 days of infection. The nucleus in the sporangium divides mitotically 2–3 times and 4–8 uninucleate anteriorly flagellate zoospores are formed within the sporangium. These biflagellate zoospores (swarm spores) are secondary zoospores and are smaller than the primary zoospores.

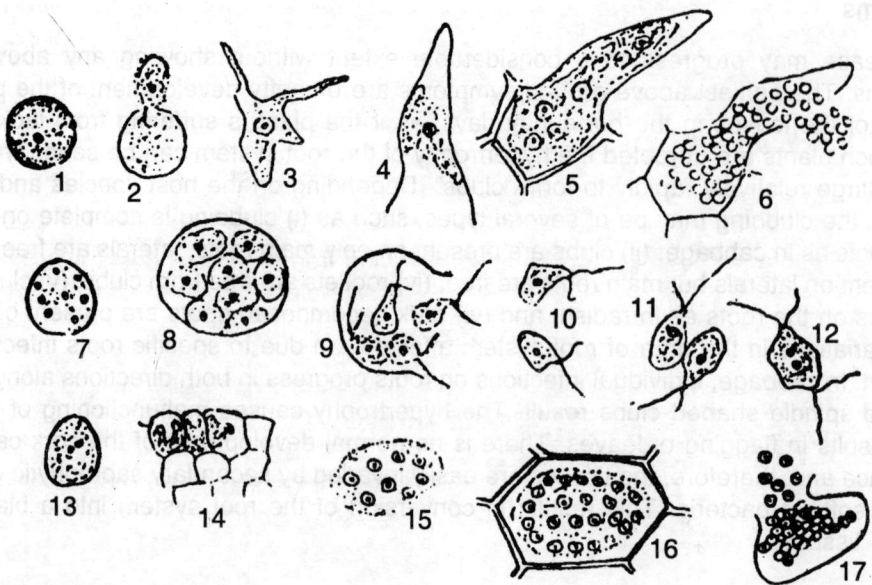

Fig. 17. *Plasmodiophora brassicae*. Stages in life cycle.

Life cycle of *P. brassicae* can be completed on plants infected with dikaryotic resting spore or by two monokaryotic resting spores. It is believed that the swarm spores function as gametes and fuse in pairs. Quadriflagellate or even six-flagellated swarmers have been reported. The fusion of swarmers results in the formation of amoeboid zygote. The zygote is the origin of large plasmodium which gives rise to resting spores. The zygotes and the young plasmodia arising from them may unite to form larger plasmodia. These plasmodia can penetrate the cell walls of the host and thus pass from cell to cell or as the host cell divides the plasmodium is passively distributed to daughter cells. The plasmodium has no specialized feeding structures such as haustoria. It is immersed in the cell cytoplasm surrounded by a thin plasmodial envelop. Following nuclear divisions the plasmodium divides Division of the plasmodium and the host cell is synchronous.

The pathogen can penetrate young roots easily but normally it cannot invade roots with secondary thickening which are infected only through wounds. The secondary biflagellate zoospores are necessary for cortical infection while primary zoospores can infect root hairs within one day of inoculation. Mature sporangia appear after 2 days and secondary zoospores appear the third day when cortical infection is initiated.

The cells containing the plasmodia become hypertrophied. The host nucleus remains functional during this period. Hypertrophy of the host cells is brought about by blocking of the mechanism of cell division and is accompanied by enhanced DNA synthesis. Starch accumulates in infected cells. Eventually, the cell contents are almost completely exhausted and the lumen is almost fully occupied by the plasmodium. The nuclei of this plasmodium undergo two successive divisions, probably meiotic. The protoplasm rounds up into uninucleate spores surrounded by a dark chitinous wall. These are the resting spores which completely fill the cell cavity. With decay of the root tissues these are released into the soil.

Biological and genetic variability among isolates of *Plasmodiophora brassicae* from within a field and between fields is reported. The variability can occur even within a single gall. At least 11 races are known in the species. At least eight races, differing in host range among crucifers, are reported. The race occurring in West Bengal on rapeseed-mustard is not highly pathogenic on crucifer vegetables and many cultivars of cabbage and cauliflower show complete resistance (Chattopadhyay *et al.*, 2001). Although *P. brassicae* is supposed to be a crucifer pathogen in acid soils, there are reports of occurrence of pathotypwa that can cause clubroot in non-acidic soils.

Disease cycle and environmental relations

The pathogen survives through its resting spores lying free or in crop debris. Resting spores of *P. brassicae* have been found in many non-Brassica plants such as *Carica papaya, Beta vulgaris, Reseda alba* and *Tropaeolum majus*. Local dissemination is by drainage water, farm implements, wind-blown soil, moving animals, and, most important, seedlings raised in infested soil. In areas where pond water is used for irrigation and the ponds receive run off water from infested fields, such ponds and irrigation water serve as source of primary inoculum for healthy fields.

Although prolonged resting period is not required for germination of the resting spores, many of them remain viable for years (up to 10 years). The fungus is one of the most persistent soil invaders and crop rotation fails to control the disease. Germinated resting spores have no nucleus. This is used as amarker to count germinated spores (xausing infection) and ungerminated spores Germination of resting spores is stimulated by root exudates of the host (*Brassuca*) and non-host plants. Root exudates of *Lolium perenne* stimulate resting spore germination more than the exudates of other plants including the host (Friberg *et al.*, 2005). The spores germinate poorly or not at all in alkaline media and hence liming of the soil has been considered as one of the methods of control of the disease. Infection can occur at soil temperatures from 9°–30°C. Optimum temperature for infection is 17°–25°C. Root hair and cortical infection decreases with decreasing availability of soil water. Larger water-filled soil pores help in movement of primary zoospores and their fusion to form large secondary zoospores necessary for cortical infection. While infection is limited by low soil moisture (below 45% of water holding capacity) under field conditions even a temporary rise in soil water, such as just after irrigation, can facilitate infection. The disease is reduced by low K and P. There is also a relationship between boron status of the soil and diseases severity. Boron reduces disease severity. Devos *et al.*, (2005) have studied the impacts of infection on cell wall metabolism and hormone balance. Initially, the infection results in an increased total auxin pool which together with action of the xyloglucan endo-transglucosylase hydrolase (XTH) results in wall loosening and consequently in cell expansion.

Differences in physiological response of resistant and susceptible cultivars of *Brassica oleracea* on infection by *P. brassicae* has been studied by Donald *et al.*, (2008). Primary (root hair) and secondary (cortical) stages of the pathogen occurred in both resistant and susceptible cultivars. Symptoms of cortical invasion of cells in both types of host included cell wall breaks, presence of vesicles or inclusion bodies within the cell walls, cell wall thickening in association with plasmodesmata and enlarged and/or disorganized host nuclei. The main difference between the resistant and susceptible host reaction was the absence of degradation of the secondary thickening and cell walls of the xylem in the resistant host. The study has suggested that resistance does not prevent the development of the amoeboid form of the pathogen but reduces its movement as suggested by less cell wall breaks. Although crucifers (*viz., Arabidopsistha; liana*) contain antifungal compounds indole gluosinolates and camalexin (phytoalexin, these do not influence the root gall forming activity of *P. brassicae.*

Management of the disease

Emphasis should be placed on such preventive measures as (i) eradication of crucifer weeds such as wild mustard, (ii) use of well drained, pathogen-free plots, (iii) use of seedlings raised in pathogen-free soil, and (iv) very long crop rotations avoiding any type of crucifer crop. Cultural practices and/or chemical control strategies are generally unsuccessful in giving complete protection to the crop and resistant varieties are the only durable solution for the clubroot problem (Manzanares-Dauleux *et al.*, 2001).

In areas where the pathogen has established itself in soil, seedbeds serve as a source of dissemination of the pathogen. The fields where the crop is already planted and had a diseased crop in the past also help in multiplication and dissemination of the pathogen. Therefore, the first step is to ensure production of disease-free seedlings by treating the seedbeds and then further loss in the main field may be reduced by adopting suitable cultural measures.

The organism can be destroyed by soil fumigation with Vapam, methyl dibromide or its substitutes, etc. The seedbeds should be selected at locations where there was no disease and where the crop refuse had not been dumped. If seedbed site does not strictly meet these conditions, the seedbed soil should be fumigated. Soil solarization to inactivate the inoculum had also been tried. Glasshouse studies have shown a linear relationship between temperature and the time for thermal inactivation of resting spores. However, solarization reduced disease development after a 10-week treatment but not after a 5-week treatment. In laboratory experiments, composting of affected plant roots or application of composted municipal green waste, onion waste or spent mushroom compost are reported to reduce number of resting spores. Incubation of composted affected crop residue at 50°C for 7 days or at 60°C for 1 day with high moisture (60% W/W moisture content) eradicated the inoculum from artificially inoculated Chinese cabbage roots.

Germination of resting spores of *P. brassicae* can be stimulated by certain non-host plants. In absence of the crucifer host this germination leads to self destruction of the pathogen inoculum.

Raising the pH of the field soil to 7 or above by adding lime (calcium carbonate) gives good control of clubroot. A soil with pH 5 may require about 2.5 tons of lime per acre. Lime treatment is done about 6 weeks before planting. Application of 10–20 t/ ha lime increases soil pH from 6.7 to 7.9 and reduces disease incidence from 98% to 0.7%. Sodium carbonate increases the pH to 8. 3 and reduces the disease incidence from 98 to 1.6% while calcium sulphate applied at the rate of 20 tons per hectare only slightly reduces the disease incidence. In highly alkaline sandy loam soil severely infested with the pathogen application of 5–33 t/ha lime is superior to treatment of soil with PCNB or calcium cyanamide in the control of the disease and the effect lasts for 2–3 years. Thorough mixing of lime to soil and use of finely ground lime are essential for effectiveness of lime treatment. Calcium cyanamide reduces disease severity by reducing inoculum density of the pathogen in soil. It has fungicidal activity against the pathogen while benomyl is not fungicidal. The effect of calcium on development of *P. brassicae* in the root hairs is highly dependent on pH. At low pH (5.5), a significant delay in the developmental progress of the pathogen is observed only at the highest level of added calcium. At pH 6.5, all calcium amendments cause a significant delay in the developmental progress of the pathogen. As pH is increased to 8.0 the number of infected roots and the effect of calcium amendment decreases. Calcium nitrate has also been found useful. Uptake of calcium is important for resistance of the plant to the pathogen. This depends on an interaction of pH and extractable Ca-Mg which must be in excess of 14 meq/100 g soil. Calcium uptake by the plant is more at pH 6.8–7.2 than at lower pH values. The soil pH above 7.2 reduces infection and clubbing because thalli abort before producing zoospores. Reduction in disease incidence in mustard by amending the soil with lime to raise the pH to 7.2 is reported from West Bengal. Maximum reduction of the disease was obtained in soil amended with lime (3 mt/ha) plus margosa cake (1.5 mt/ha). Often the application of lime to raise soil pH fails to suppress the clubroot disease. This is explained by the fact that the soil moisture around the rootlets may be acidic because of the carbon dioxide produced by the roots. In moist soil there sufficient movement of the alkaline particles to neutralise this acidity around the rootlets. In dry soil this is not possible. Calcium cyanamide is also effective against the pathogen but is costly. The cost can be reduced by band application along the transplant rows.

In Taiwan, good control of club-root disease was obtained by amending the soil with S-H mixture which consisted of sugarcane bagasse, rice husk, oyster shell powder, urea, potassium nitrate, calcium super phosphate and mineral ash. The last ingredient contained oxide of calcium, magnesium, aluminum, iron and silicon. The mixture was applied at the rate of 0.5 to 1% by weight of soil. Mineral ash alone or mixed with oyster shell powder was equally effective.

PCNB (pentachloronitrobenzene) formulations such as Terraclor, Quintozone, and Brassicol have some effect against the pathogen. About 30 kg of a formulation having 75% active ingredient (PCNB) should be applied to one hectare of the field. This soil application should be done a week before planting. To reduce the cost the fungicides can be used as a water suspension for giving first irrigation to the transplanted seedlings. The suspension should contain 1.5 kg of the fungicide per 200 litres of water. In this method the disadvantage is that while infection of the main roots is checked, the laterals which spread out to the soil zone where the fungicide has not reached may get infected if the field contains a high population of the pathogen. Use of PCNB is being discourage because of its persistence and pollution. The fungicide flusulfamide suppresses clubroot disease by directly affecting the resting spore only and inhibiting their germination through adsorption onto their cell walls. Application of fungicides in bands along the rows of plants has been found more practical and effective. Donald et al., (2001) applied fluazinam into the soil in bands 25 cm wide alongthe transplants row (to a depth of about 15–20 cm) immediately before transplanting for effective control of clubroot.

Bhattacharya and Pramanik (1998) collected antagonistic rhizobacteria (Bacillus subtilis T99, Pseudomonas flourescens H237 and Streptomyces graminofaciens GIH) from different fields of crucifers and tested them and neem (margosa) products (2% neem oil, neem cake and nemin) for suppression of clubroot under laboratory and greenhouse conditions. Pseudomonas and Streptomyces reduced size and growth of clubs as well as rate of root hair infection process while Bacillus stimulated rate of root hair infection and club formation. Neem oil suspension applied through soil was phytotoxic but neem cake and nemin reduced disease severity in field trials.

Suppressiveness to the disease in some soils is reported from Japan. Narisawa et al., (1998) isolated some root colonizing fungi from wheat fields and found them giving complete suppression of clubroot in sterile soil. Heteroconium chaetospira (a fungal root endophyte) was effective in nonsterile soil also. Chinese cabbage seedlings raised from seed treated with the fungus remained healthy in infested soil. Hyphae of the antagonist covered the root surface and extensively colonized the inner cortical tissues. The endophyte has been found associated with atleast 18 different plant species. Chinese cabbage seedlings inoculated with the hyphomycete H. chaetospira and transplanted in P. brassicae-infested soil showed 63–97% reduction in clubroot and 49–67% reduction in Verticillium yellows. The antagonist colonized the cortical cells, especially in the root tip region. The infection process involved formation of appressoria on the cell surface and the subsequent growth of hyphae within cells. No disease symptoms by this fungus were produced. They (Narisawa *et al.,* 2005) studied the effects of soil moisture, soil pH and pathogen resting spore density on the effectiveness of the biological control provided by *H. chaetospira.* In greenhouse tests, the antagonist reduced clubroot by 90 to 100% when the pathogen resting spore density was 10^4 and 10^5 spores/ g soil at pH 5.5, 6.3 and 7.2). When the resting spore density was 10^6 spore/g soil. plants were severely diseased and the antagonist had no effect. The adverse effect of low soil moisture on the disease was accentuated by the antagonist. At high soil moistrure there was no effect of the antagonist on the disease.

Phoma glomerata controls clubroot through its antifungal metabolite epoxydon. Planting of certain decoy crops (oats, spinach) before planting Chinese cabbage reduced inoculum density in the soil. The pathogen infected the root hairs of the decoy plants but could not develop clubbed roots thus reducing the resting spore density.

Inoculation of hairy roots of cabbage, Chinese cabbage and turnip with surface disinfected resting spores of *P. brassicae* show differences among hosts in susceptibility to club-root. Chinese cabbage (*Brassica oleracea* var. *pekinensis*) and turnip (*Brassica rapa* var. *rapifera*) hairy roots support the highest percentage of root hair infection and zoospore-group production (Asano *et al.,* 1999). Gall formation on hairy roots is seen only on Chinese cabbage and turnip but not on cabbage (*Brassica oleracea* var. *capitata*).

Radish (*Raphanus sativus*) has been used as a decoy crop for reducing populayions of resting spores of *P. brassicae* in soil. In infested pots planted with radish the population of spores was reduced by 71% (Murakami *et al.,* 2000). In field trials, 94% reduction was observed when radish was grown in advance of susceptible Chinese cabbage. However, disease severity was not reduced. P. brassicae infected radish hairy roots but no clubs were formed.

All the indigenous varieties belonging to *Brassica juncea, B. campestris* var. *yellow sarson* and *B. campestris* var. *toria* are moderately to highly susceptible to the club-root disease. Exotic cultures belonging to *B. napus, B. nigra* and *B. carinata* are either free from disease or are resistant. Among crucifer vegetable crops, resistance to the rapeseed-mustard isolate of the pathogen in West Bengal has been found in cabbage cultivars KK Cross and Drumhead, cauliflower cultivars Dania and Snowball and knol kohl cultivar White Vienna (Chattopadhyay *et al.,* 2001). The resistant cultivars respond to infection by hypersensitive reaction and the cells die, halting further developments in the pathogen. Calcium is implicated in the cell death as well as alkalizaion of the medium.

WART DISEASE OF POTATO

Black wart, which was first described in 1895 from Hungary, is a serious disease of potato in regions of temperate climate such as Europe, North America as well as South Africa. Even in tropical areas the disease has been found at places located on high altitudes. The pathogen of potato wart co-evolved with the potato in the Andes Mountains of South America from where it was distributed around the world by infected seed tubers and infested soil. It was first introduced to the European countries with potato breeding material from the South America in the aftermath of destruction caused by potato late blight Wart was known to occur in England during 1876–1878. European continent received the wart disease in Czekoslovakia in 1888. It caused great damage to the potato industry in most of the European countries until immune varieties were developed. In India the disease was first reported in 1953 from Rangbul in Darjeeling district of West Bengal and was probably introduced from the Netherlands through seed material. Around 60% of the potato area was affected and the popular variety Darjeeling Red Round suffered most. Periodical surveys are still carried out to monitor spread of the diseases. It co nues to be endemic to that area. Such cases of isolated occurrence of potato wart are consic ered exceptional.

Symptoms

All under ground parts, except roots and basal part of the stem near ground level, exhibit symptoms of the disease. Buds on underground stems, stolons, and the tubers are the centers from where abnormal growth activity starts and leads to development of warts. Lumps of warts are found attached to the affected parts. The warts vary in size from small protruberances to large intricately branched systems. The outgrowth is spherical and usually not a solid structure. These masses vary from less than 1 g to more than 50 g in fresh weigh. Early in the season these outgrowths are green or greenish-white in color if exposed above ground and light and cream coloured when underground. On the tubers the warts are more typical and conspicuous, sometimes covering the whole tuber and larger than the tuber itself. In advanced stages the warts are darker in color, sometimes black, and undergo putrefaction due to attack of secondary saprophytic microorganisms. Morphologically the wart consists of distorted, proliferated branched structures grown together into a mass of hyperplastic tissue.

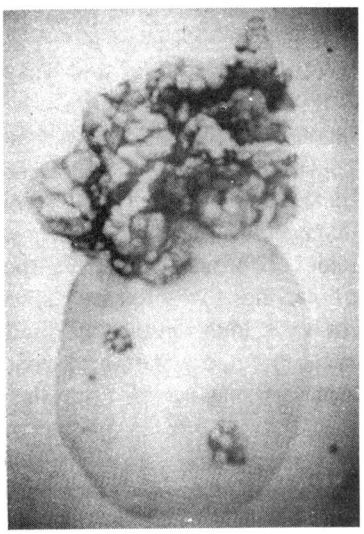

Fig. 18. Potato wart.

The causal organism

Synchytrium endobioticum (Schilb.) Perc is a member of the kingdom Fungi, phylum Chytridiomycota, class Chytridiomycetes, order Chytridiales and family Synchytriaceae. The fungus does not produce hyphae and is biotroph (obligate parasite), holocarpic, endobiotic parasite. In potato tubers the fungus produces warts due to hyperplasia and hypertrophy of host meristem cells. The cells within these warty tissues contain thick-walled resting spores. Each large sized wart has the potential to release 200,000 resting spores which have endogenous dormancy and can remain viable for 40 years. The resting sporangia of the fungus function as prosorus. Infection of the tubers and other underground parts of the plant takes place by posteriorly uniflagellate naked zoospore which are released by sporangia or resting spores present in the wart tissues of the infected tubers, plant debris or in the soil. The outer wall (epispore) of the resting sporangium bursts open by an irregular aperture and the inner wall (endospores) expands into a balloon-like vesicle within which a single sporangium is formed. Zoospores are released from this sporangium. Soil fauna, especially earthworms help in long distance dissemination of the zoospores.

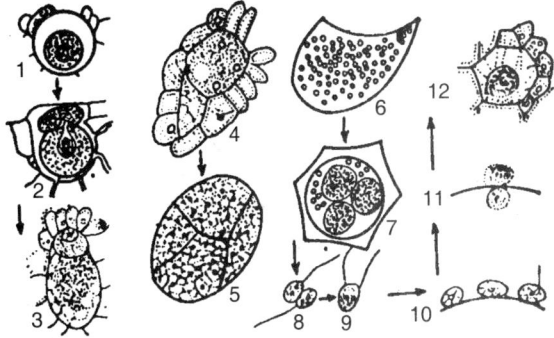

Fig. 19. *Synchytrium endobioticum.*

The zoospores released by the sporangium are capable of swimming for about 20 hours in the soil. They can migrate in soil water to a limited distance of about 50 mm or less. When they come in contact with the host, such as eyes of potato tubers, they come to rest and withdraw their flagellum. The zoospore then bodily enters the host cell and settles down at the bottom of the cell. Here it enlarges in size. Hypertrophy of the host cell occurs. Surrounding cells also expand and form the tumor or the wart. The enlarging thallus fills the lower part of the infected host cell and becomes spherical. A double layered chitinous wall of golden brown colour is secreted around the thallus and it becomes the prosorus or the summer spore. The host cell is by now dead. Until this stage the prosorus has remained uninucleate. On maturity the summer spore germinates within the host cell by protrusion of the inner wall through a pore in the outer wall. The inner wall then expands as a vesicle which enlarges upwards and fills the upper half of the host cell. The contents of the prosorus pass into the vesicle. During its passage into the vesicle the nucleus divides repeatedly so that the vesicle contains about 32 nuclei. The cytoplasmic contents of the vesicle become cleaved into a number of sporangia around the nuclei forming a sorus. The number of sporangia in the sorus varies from 4 to 9. Each nucleus undergoes further divisions until 200 to 300 nuclei have formed in each sporangium or gametangium. The cytoplasm around each nucleus rounds up to form a zoospore or gamete. As the sporangium ripens, it absorbs water and swells causing the host cell to burst open. In the presence of water, sporangia further swell and burst open by means of a small slit through which the zoospores escape. There may be as many as 500–600 zoospores in large sporangia. If suitable host tissues are available they encyst and cause infection. Sometimes a single host cell may be penetrated by several zoospores and, thus, may contain many thalli. Within the host cell the thallus enlarges, forms a prosorus and produces fresh crop of sporangia. This type of repeating cycle continues during the growing season of the host crop (early spring and summer in temperate regions).

In dry conditions or towards the end of the crop season the zoospores function as planogametes. They fuse in pairs (or occasionally in groups of 3 or 4) to form a motile zygote. The flagella are retained and the zygote can swim in water for some time. Since the gametes participating in zygote formation are similar in shape and size the sexual act is isogamous. However, the gametes from same sporangium or gametangium are possibly physiologically similar and do not fuse. Fusion occurs between gametes from different sporangia which may be from the same prosorus. After swimming for some time the biflagellate zygote settles on the host surface and causes infection in the same manner as the zoospore. Nuclear fusion occurs in the zygote before penetration. The results of zygote infection differ from zoospore infection. While zoospore infection is mainly responsible for hypertrophy (increase in size of infected and adjacent cells) the zygote infection mainly causes hyperplasia (repeated cell division).

The endobiotic zygote now settles down at the bottom of the host cell. Repeated cell divisions cover the zygote deep in the warty tissue. The zygote thallus enlarges and becomes surrounded by a 2-layered wall, a thick outer wall which eventually becomes dark brown in colour and forms folds and ridges and a thin hyaline inner layer surrounding the granular cytoplasm. This is the resting spore or resting sporangium. The host cell also dies and some of its contents may deposit on the resting spore. The most common race of the pathogen in India is the biotype 1.

Resting spores, whether in host debris or free in soil, are main survival structures of *S. endobioticum*. As stated above these spores have extreme longevity and enormous numbers of them are contained in the warts and ultimately released in soil. In Canada and USA it has been found that the disease is largely present in home gardens with small plots which optimize conditions for the pathogen through monoculture, infected seed tubers, and crop debris left on soil surface after harvest as well as potato peels on compost heaps. More or less similar small plot situation occurs in the area in India where the disease is endemic.

The dispersal of resting spores is by many methods such as movement of infected seed and infested soil through the agency of man (feet and shoes, tools, and also vehicles moving through the infested area). The resting spores have been detected on vehicles exiting from the infested areas in Canada (Hampson and Wood, 1997; Jennings *et al.*, 1997). The fungus also attacks tomato and many other solanaceous species including *Datura* and *Physalis* but on these hosts it does not produce warts.

Thus, its detection may become difficult if the thick-walled, endogenously dormant resting spores escape from the quarantined area on such transport as a vehicle passing through the area and then cause infection of tomato

Management of the disease

It is difficult to manage the disease once the pathogen has been introduced in a field. The introduction of the pathogen in a field or a locality can be effectively checked by practicing quarantine, i.e., prevention of entry of infected material into healthy areas. Many experiments had been conducted to control the disease by soil treatment. Amongst these were steam sterilization of the soil and application of mercuric chloride, ammonium sulphocyanate, copper sulphate and 5% formalin. However, these treatments were very costly and could not be applied on a large field scale. Further, such treatments were effective in particular types of soil.

The only practical and effective control measure is to cultivate immune or highly resistant varieties. Before 1970, a number of exotic commercial wart immune varieties such as Ackersegen, Ultimus, Pimpernel, and Voran had been introduced in the Darjeeling area but were not generally accepted by farmers. In 1975 Kufri Jyoti was introduced as a wart immune and late blight resistant cultivar. Between 1976 and 1988 many wart immune cultivars were developed and released for cultivation. These are Kufri Muthu, Kufri Sheetman, Kufri Sherpa, Kufri Khasi Garo, Kufri Bahar, and Kufri Kumar. Later, Kufri Kanchan with red skin was found immune to wart. The cultivar Kufri Swarna is also resistant to the disease.

APHANOMYCES ROOT ROT OF PEA

As far as is known this disease does not occur in India although the temperature and moisture conditions favoring its occurrence do exist in this country. It is perhaps the most destructive disease of peas in North America, Europe, Australia, New Zealand and Japan where it is widespread and causes epidemics in most of the seasons. It not only kills the plants but reduces the quality of seeds by making them irregular in shape and also off taste. The pathogen can cause damage to beans and alfalfa also.

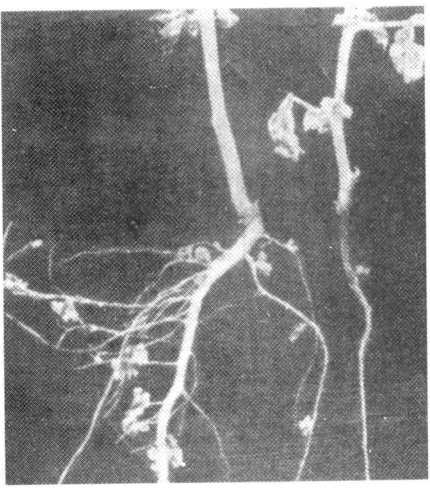

Fig. 20. Aphanomyces root rot of pea.

Symptoms

Severe rotting of the root cortex, hypocotyl and epicotyl is major symptom on peas. The first infection usually takes place on one of the smaller roots. Soon other roots and lower part of the stem are also involved. Tissues of the lower stem and root cortex become water-soaked and soft. Fine roots are decayed and if the plant is pulled out the main root comes out. The water soaked lesions become gray then yellowish or pink, and finally black. These symptoms may be confused with the effects of other root decay causing pathogens. Only microscopic examination of the tissue for the presence of characteristic oospores of the fungus confirms the diagnosis.

The causal organism

Aphanomyces euteiches Drechsler. The genus *Aphanomyces* is placed under the kingdom Chromista and class Oomycetes (Syn. Peronosporomycetes) which is a collection of organisms resembling fungi in morphology and physiology but are phylogenitically related to diatoms, chromophyte algae. The genus is classified under phylum Oomycota, class Oomycetes, order Saprolegniales, family Leptolegniaceae. The oomycete was first recognized as a pathogen of pea in 1925. The species is commonly composed of host specific subpopulations. Of the different isolates from hosts other than pea all have similar morphologic characters as the pea isolate but many are not pathogenic on pea. The subpopulation attacking pea is known as *Aphanomyces euteiches* f. sp. *pisi*. Single zoospore strains of the fungus show wide variation in pathogenicity and genotype. All strains are pathogenic on one or more pea cultivars and some on alfalfa and bean. There is evidence for gene flow between pea and bean pathotypes of *A. euteiches* (Shang *et al.*, 2000). The level of protease enzymes in the isolates of *Aphanomyces euteiches* (and also *A. cochlioides*) determines the host range. Plants resistant to an isolate have inhibitors of the enzymes. In the diseased root tissues the fungus is found in the form of its oospores which are sub-spherical with a 1.5 μm thick colorless wall. The diameter of the oospores is 18–25 μm. This oospore germinates either by producing 1–3 germ tubes which directly cause infection or by a single hyphal filament, 200–350 μm long, which functions as a zoosporangium. The zoosporangium produces 12 or more primary zoospores. The primary zoospores are cylindrical when they come out of the sporangium but soon they round up into spherical masses, secrete a thin peripheral wall and encyst in a clump at the tip of the sporangium. In 1–2 hours the encysted primary zoospores germinate by the formation of a papilla which expands into a vesicle. The protoplasmic contents of the zoospore body passes into this structure which develops two flagella and becomes the secondary zoospore. In *Aphanomyces euteiches*, as in species of *Pythium* and *Phytophthora*, the germination of encysted zoospores (releasing cystospores) is triggered by recognition of root surface components and calcium. The cystospores swims in water for some time and then again encysts. It germinates by 1–3 germ tubes. Production of assimilative hyphae is initiated by pea root exudates. The mycelium which is formed is coenocytic, hyaline, and with frequent short branches. These branches may become zoosporangia. In the host, the mycelium is intracellular. Oogonia develop after vegetative activity has ceased. They are formed terminally on hyphal branches as thin-walled subglobose bodies with densely granular vacuolate contents. Antheridia arise on other branches. One to 4 antheridia are attached to a single oogonium. After fertilization the oogonial wall becomes prominently thick. The oospore nearly fills the oogonium completely.

Other pathogens of pea, *viz.*, *Fusarium oxysporum* f. sp. *pisi*, *F. solani* f. sp. *pisi*, *Mycosphaerella pinodes*, *Pythium* spp., *Rhizoctonia solani* and *Sclerotinia sclerotiorum* are also involved in pea root rot complex.

Disease cycle and environmental relations

A. euteiches can survive in crop debris in soil as oospores for more than 10 years. Oospores in the tissue are spread in the soil during tillage and are mostly found in the ploughed layer. However, oospores have been detected in some soils as deep as 60 cm. Some soils are suppressive for pea root rot. In highly suppressive soils pea monoculture leads to a slow build up of oospores in root tissues and high crop yield is maintained up to the fourth year. In soils with none or less suppressiveness the crop fails in the second year and there is no yield in the fourth year. The pathogen can also be transported with seeds having oospore-containing material as contaminant. The oospore germination is influenced by roots of the host(s). Root exudates are not as effective as the roots (rhizoplane) in stimulating germination (Shang *et al.,* 2000). Oospores of pea pathotypes germinate at a greater frequency on pea, bean and oat roots than on roots of other plants. Primary infection is caused by germ tubes or zoospores produced by these oospores. The fungus produces such a large number of zoospores and oospores that it can easily be disseminated over large areas through running water or in contaminated soil carried from one field to another.

The oomycete can cause infection in relatively dry conditions but it is most destructive in wet soils. However, the damage becomes evident under dry conditions when reduction in water transport due to root rot becomes critical. Zoospore production and their germination is abundant at 14°–20°C with very low germination at 9°C or at 30°C. Infection of roots can occur at any temperature from 16° to 28°C. No infection occurs below 14°C and optimum for infection is from 17°–24°C. Optimum range of pH for growth of the fungus is 4.5–6.5. Under heavy fertilizer dosages, damaged roots are quickly replaced by new roots and plants suffer less from the disease. In inoculations with high concentration of zoospore inoculum, the peak enzymatic activity occurs in 10–14 days when oospore formation starts. The oospore formation is associated with a gradual increase in disease symptoms. Towards the end of the active crop season (last harvest), oospores are found in 90% of the root length but enzymic activity is low suggesting that the pathogen remains active only on living plants and does not grow saprophytically on dead host parts. Synergism between non-pathogenic strains of *Fusarium solani* and *A. euteiches* is reported. Presence of *F. solani* along with the other fungus is common. When co-inoculated, the root rot is significantly increased.

Management of the disease

Keeping the field well drained and following a 3-year rotation avoiding such crops as alfalfa, sweet clover, cowpea, etc. considerably reduces incidence of Aphanomyces root rot. Varietal resistance to the disease is not reported. Application of dinitroaniline herbicides reduces disease severity. Seed treatment with thiram or captan (2.5 g/kg seed) is a commonly recommended practice.

Calcium content of the soil is negatively correlated with disease incidence and severity in a suppressive soil. Zoospore production by oospores of *Aphanomyces euteiches* is inhibited by sub-millimolar concentration of free Ca while mycelial growth is enhanced or is unaffected. Thus, free Ca is a major variable controlling the degree of soil suppressiveness against *A. euteichjes* (Heyman *et al.,* 2007).

In biological control, seed coating with spores and cells of such antagonists as *Trichoderma, Penicillium oxalicum, Gliocladium, Burkholderia cepacia, Pseudomonas fluorescens* and *Bacillus subtilis* gives good control. Many strains of *Serratia* spp. are also antagonists of pea pathogens (Wang *et al.*, 2003). The mechanism of biological control by *B. cepacia* is mainly through antibiosis which retards formation of oogonia of *A. euteiches*. Wakelin *et al.*, (2002) have reported suppression of Aphanomyces root rot in field trials by *Bacillus pumilus, B. subtilis, B. cereus, B. mycoides* and *Paenibacillus polymyxa* although not at significant level. The treatment enhanced root rot by other fungi, particularly pathogenic *Fusarium* spp. Bioprotection of pea roots against *A. euteiches* by the arbuscular mycorrhizal fungus *Glomus mosseae* is reported. It depends on fully established symbiosis in which mycorrhiza-related chitinolytic enzymes are involved. Mucoid strains of the bacterial antagonists or plant growth promoting rhizobacteria form dense and patchy bacterial layer on the host and mycorrhizal fungus structures. This increased adhesive property may lead to more stable interactions between plant root, pathogen and the antagonists. The pea plants preinoculated with the arbuscular mycorrizal fungus *Glomus intraradices* show no symptoms of Aphanomyces root rot even though the pathogen is present and active in these plants. The presence of the mycorrhizal fungus influences enzymatic activities of the root pathogen. The indigenous AM fungi present in the soil do not affect the pathogen in its vegetative stage when it is enzymetically most active, and causes maximum cortical decay of roots. There is no correlation between AMF colonization and disease incidence, severit or pathogen enzymatic activity However, there is a negative correlation with oospore-containing root lengths.

Xue (2003) has reported use of a single biocontrol agent against all pathogens involved in pea root rot complex. When *Clonostachys rosea* (= *Gliocladium roseum*) is grown near the pathogen it is often stimulated to produce lateral branches that grow directly towards the pathogen mycelium typically entwining around the hyphae. When applied to seed, the fungal antagonist multiplies in the rhizosphere and colonizes the seed coat, hypocotyls and roots as the plant develops and grows. It significantly suppresses all the pathogens (including *F. solani*) and increases seed germination, *in vitro*, by 44%, seedling emergence by 22% and reduces root rot by 76%. In soil inoculated with selected pathogens seed treatment increased emergence by 13 to 38%.

Green manuring with a crucifer for two consecutive seasons before pea is reported to reduce the root rot by reducing the number of propagules in soil Glucosinolates in crucifer tissues decompose through action of myrosinase (thioglucoside glucohydrolases) enzymes present in the plant and also released in root exudates. Microorganisms such as *Escherichia coli* and *Pichia pastoris* also produce these enzymes The hydrolytic activity of these enzymes releases highly volatile and highly toxic isothiocyanates in early stages of decomposition and mildly toxic non-glucosinolate S-compounds throughout the decomposition. The latter are quickly metabolized by soil microorganisms. Isothiocyanate have disease suppressive effect. Glucosinolates as such do not suppress the pathogens althouth they contain sulphur. The level of glucosinolates (sinigrin and other S-compounds) varies among crucifer species and within the species in different stages of plant growth. *Brassica juncea* and *B. nigra* are rich in propanyl glucosinolates (sinigrin) in the late flowering stage. Isothiocyanates are also released by the living crucifer plants (*viz.*, canola) in to the rhizosphere. The water-soluble isothiocyanates reduce *A. euteiches* oospore infection and inhibit mycelieal growth thus suppressing disease incidence (Smoolinska *et al.*, 1997, 2003). Earlier, in 1997, suppression of Aphanomyces root rot of pea by green manure with oat, rape and sweet corn had been reported. In a study of combined effect of *Brassica napus* seed meal and *Trichoderma harzianum* on Aphanomyces root rot the disease was significantly reduced by *T. harzianum* alone (100%), by amendment with high glucosinolate seed meal alone (77%) and by *T. harzianum* combination with high glucosinolate seed meal (100%). Amendment with low glucosinolate seed meal did not control root rot and reduced the biocontrol efficacy of *Trichoderma*. Amendment of soil with paper mill residuals suppresses *Aphanomyces euteiches* and *Pythium* spp. on pea in addition to *Colletotrichum lindemuthianum* and *Pseudomonas syringae pv. lachrymans* on cucurbits. Annual application of fresh or composted paper mill residuals to the field are reported to suppress common root rot of snap beans (*Aphanomyces euteiches*). The level of suppression depends on dose of application of the residuals (Leon *et al.,* 2005).

DAMPING OFF OF SEEDLINGS

Damping off is responsible for poor germination and stand of seedlings in nursery beds. The plant species suffering from damping-off are unlimited. The seedlings of ornamental and vegetable plants, field crops, and forest trees in nurseries, all are liable to be affected if and when pre-disposing conditions are there.

Symptoms

In pre-emergence damping off, the young seedlings are killed before they reach the surface of the soil. They may, in fact, be killed even before the hypocotyl has broken the seed coat (seed rot). The radicle and the plumule, when they come out of the seed, undergo complete rotting. Since this happens under the soil surface, the disease is often not detected except for the resulting poor stand.

Fig. 21. Damping-off of seedings.

The post-emergence damping off is very conspicuous. This phase of the disease is characterized by the toppling over of infected seedlings, any time after they emerge from the soil until the stem has hardened sufficiently to resist invasion. Infection usually occurs at or below the ground level and the infected tissues appear water-soaked and soft. As the disease advances the stems become constricted at the base and plants collapse.

Fig. 22. Damping-off of sugarcane seedings.

Seedlings that are apparently healthy one day may have collapsed by the following morning. Generally, the cotyledons and leaves slightly wilt before the seedlings are prostrated, although sometimes they remain green and turgid until collapse of the seedlings occurs. The seedlings from infested soil escaping damping off soon die after transplanting. In nurseries and under field conditions the disease usually radiates from initial infection points, causing large spots or areas in which nearly all the seedlings are killed. In the field the disease is most severe when the soil moisture is medium to high (50% of water holding capacity or more) with comparatively high temperature or temperature unfavourable for growth of seedlings. When conditions are favourable for development of the disease, damping-off is responsible for as much as 90% loss in seedling numbers. In specially susceptible plant species seedling losses of 25–75% occur yearly. Most of the loss is due to pre-emergence damping off.

The causal organisms

A large number of oomycetes and fungi can cause seed and seedling rot but species of *Pythium* are generally known as damping off oomycetes. These include *P. debaryanum, P. aphanidermatum, P. ultimum* and *P. arrhenomanes*. The genus *Pythium* belongs to kingdom Chromista, phylum Oomycota, class Oomycetes, order Saprolegniales (?). In this section only *P. debaryanum* is described.

The mycelium of *Pythium debaryanum* Hesse consists of much branched, hyaline, coenocytic hyphae, the main hyphae being up to 5 μm in diameter. The blunt ends of the hyphae penetrate the cell walls of the hypocotyl and ramify within and between the cells of the cortical parenchyma. For some time there is only hyphal growth but eventually sporangia and oospores of the fungus are produced in the parenchymatous tissues, especially of the cotyledons.

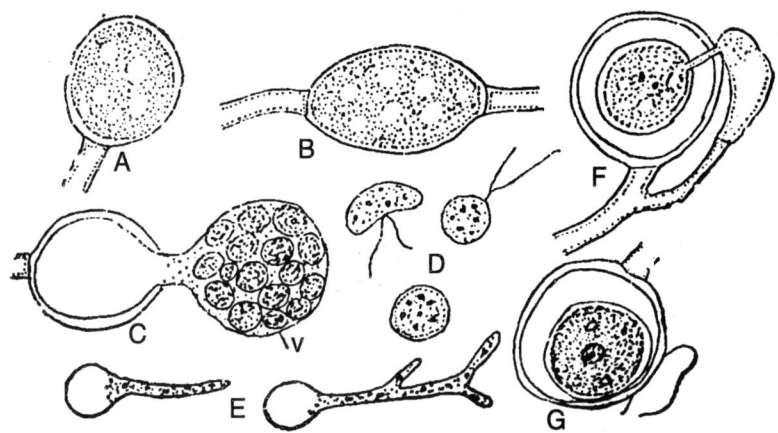

Fig. 23. Pythium debaryanum.
A-B Terminal and intercaary sporangia; C-Vesicle; D-Zoospores;
E-Germinating zoospores; F, G-Sexual reproduction.

Sporangia of the fungus, constituting the asexual reproduction, are globose to oval, unlike the sporangia of *P. aphanidermatum* which are ill-defined in shape. These sporangia may be terminal or intercalary on the somatic hyphae and measure 15 –26 (average 19) μm in diameter. They may remain attached to the parent hypha and germinate *in situ* or they are detached and carried by wind or water to other locations. Germination is either by production of zoospores or direct by a germ tube. Production of zoospores is preceded by formation of a vesicle at the tip of a long tube which is extruded from the sporangium. The sporangial protoplasm flows into the vesicle through the tube and differentiation of zoospores takes place in the vesicle. The vesicle ruptures when zoospores are mature and the latter are released in the film of water present on the host surface or in the soil. The zoospores are kidney-shaped (reniform) with two lateral flagella attached on the concave side. After a period of motility the zoospores lose their flagella and come to rest, encyst, and then germinate by a germ tube. The size of resting zoospores is up to 8 μm in diameter.

Sexual reproduction is oogamous. Oogonia and antheridia are formed in close proximity, often on the same hypha. The oogonium is globose, terminal or intercalary, and with a multinucleate ooplasm surrounded by a layer of periplasm. It measures 15–28 (av. 21) μm in diameter. The antheridia are much smaller and somewhat elongated or club-shaped. One to 6 antheridia may attach to a single oogonium. On gametangial contact a fertilization tube develops from the antheridium and passes through the periplasm into the ooplasm. In the meantime, nuclear changes have taken place in both gametangia and all but one functional nucleus in each are disintegrated. The male functional nucleus passes through the fertilization tube into the ooplasm (oosphere), approaches the female functional nucleus and fuses with it. The oosphere develops into a smooth and thick-walled oospore which is aplerotic (partially filling the oogonial cavity). It measures 12–20 (av. 17) μm. Its wall is 1.5 μm thick. The oospores germinate by producing a germ tube or by functioning as a sporangium.

Disease cycle and environmental relations

Pythium is a weak saprophyte and poor parasite. As a saprophyte it lacks competitive ability against other soil microorganisms for colonization of dead organic matter in soil. As a plant parasite, it can infect only the juvenile or succulent organs of plants, tissues of which are devoid of secondary thickening. Tissues that have matured and developed secondary thickening become resistant to its attack. In soil the species of *Pythium* are susceptible to antagonism of a number of bacteria (*Bacillus, Pseudomonas, Streptomyces*) and fungi (*Trichoderma, Penicillium*) as also some amoebae. However, these weaknesses of the fungus are compensated by its ability to rapidly form oospores. Within the tissues or in the previously colonized dead plant organs the fungus feeds and quickly forms oospores. Very little mycelium of the fungus is seen in the soil within 10 days of its introduction (Peethambaran and Singh, 1978). Most plant parasitic species of *Pythium* enter the soil through pre-colonized host residue carrying oospores and sporangia which are survival structures. Germination of these leads to primary infection of seedlings and the asexual spores later formed carry on the secondary infection and rapid spread of the disease.

Soil type does not much affect the incidence of damping-off caused by *Pythium* spp. But the disease is most severe in ill-aerated, ill-drained soils. Such conditions are common in compact, heavy soils. Loose soils, having a good proportion of sand, exhibit less loss from the disease. *Pythium* spp. are destructive over a wide range of soil temperature but damping-off is severe at temperatures between 24° and 30°C. High soil moisture accelerates development of the disease, the infection being favoured by poor aeration, prolongation of juvenile stage of the seedlings and general weakening of the plants under such conditions. On the whole, the conditions which pre-dispose the seedlings to damping-off are overcrowding, growth under too damp conditions, excess of water in soil, and the presence of too much decaying organic matter.

Management of the disease

The most effective measures against damping-off, irrespective of the causal agent, are use of chemical and/or biological seed protectants to keep away the pre-emergence phase and to adopt sanitary precautions in the nursery to check the appearance of post-emergence damping-off. Seed treatment with fungicides provides good control of pre-emergence damping-off. The chemicals are applied in dry or wet form to the seed and form a protective layer around the seed coat keeping the soil-borne fungi away until the seedlings have emerged. Normally, the fungicides are used in the ratio of one part of the chemical to 500 parts of seeds. The common seed protectants in use these days are thiram, captan and some systemic fungicides such as metalaxyl.

A recent novel approach for suppression of zoosporic fungi is use of surfactants (inert ingredients of fungicides) for seed and seedlings treatment. Surfactants can be produced by bacteria also and many bacterial biocontrol agents agents act against pathogens through it. The surfactants kill zoospores by lysis.

Seed coating with spores of *Trichoderma harzianum, T. hamatum* and *Penicillium oxalicum* and cells of *Burkholderia (Pseudomonas) cepacia* and *Pseudomonas fluorescens* has been found very effective biological control method of damping-off. At low temperature *T. hamatum* is more effective than *T. harzianum* which acts better at high temperatures Jayaraj *et al.,* (2006) evaluated 7 different formulations of *T. harzianum* for biocontrol of tomato damping off (*P. aphanidermatum.* The formulations were talc, lignite, lignite + fly ash based powder, wettable powder, bentonite paste, polyethylene glycol paste, and gelatin-glycerine gel. Seed treatment with these formulations reduced damping off up to 74% and enhanced seedling growth. The shelf life of the formulation was 3 months when stored at 24°C. Application of talc based formulations of *T. viride* and *P. fluorescens* in nursery beds before sowing significantly reduces damping off of tomato and chilli caused by *Pythium aphanidermatum*. Jayaraj *et al.,* (2005) developed and evaluated 6 formulations of *Bacillus subtilis* (strain AUBS-1) for control of tomato damping off caused by *P. aphanidermatum*. The formulations included talc-based powder, lignite-based powder, water-dispersible tablets and bentonite paste. Populations of the antagonistic bacterium in the formulations were stable for up to 2 years storage at room temperature (28°C). One hundred percent viability was found for one year in lignite-based powder, lignite + fly ash, bentonite paste and wettable powder.

In the interaction between *Pythium* and the biocontro agent *Trichoderma hamatum,* growth of the mycoparasite is directed toward its host. In the area of interaction, *T. hamatum* produces appressoria-like structures which attach to the host cell wall. The mycoparasite either grows parallel to and along the host hypha or it coils around its host. It produces hok-like structures in the region of immediate contact and penetrates the host hypha and grows therein. Similar behavious of T. hamatum is seen in *Rhizoctonia solani.*

Trichoderma harzianum extensively colonizes the wounds and broken roots. The glucanase and cellulase enzymes of the antagonist inhibit germination of encysted zoospores and germ tube elongation of *Pythium* at even a very low concentration. *Pythium* stimulates the germination of conidia of *Trichoderma* (Thrane *et al.,* 1997). *Pythium oligandrum* is a mycoparasite of *Pythium aphamidermatum, P. ultimum, P.graminicola* and many other pathogenic *Pythium* spp. in addition to other pathogenic fungi including *Rhizoctonia* and *Aphanomyces.* The mycoparasite has multifaceted processes for antagonism. The effect on structures of the pathogens strats appearing as soon as the mycoparasite contacts them. It produces the antibiotic metabolite oligandrin. In an atypical tomato root colonization by the mycoparasite, Le Floch *et al.,* (2005) observed that one of the most significant effects was the quick colonization of cortical and vascular roots areas. The colonization was similar to a pathogenic *Pythium.* Fungal colonization did not cause host wall disruption and host cell alteration. After 9 hours of inoculation the hyphae of the mycoparasite started showing vacuolation. After 14 hours the host started showing defense responses by producing phytoalexins. *Epicoccum nigrum* is a biocontrol agent of cotton damping-off and root rot caused by *Pythium debaryanum.* It inactivates the cellulase and pectinase activity of *Pythium* Strains of *Pseudomonas fluorescens* produce an antibiotic 2,4-diacetylphloroglucinol which causes disruption of the mycelium of *P. ultimum* and protects tomato seedlings against damping off. De Souza *et al.,* (2003) have reported a strain of *Pseudomonas fluorescens* which has strong zoosporicidal property. Within 30 seconds of exposure to cell suspension or culture filtrate of the bacterium, the zoospores of *Pythium, Albugo candida* and *Phytophthora infestans* become immotile and are subsequently lysed. Ramamoorthy *et al.,* (2002) observed that antagonistic isolates of *Pseudomonas fluorescens* suppressed damping off of tomato and hot pepper not only by antagonism and plant growth promotion but also by inducing resistance against invasion of *Pythium.* Microbial seed treatment with strains of *Pseuodomonas fluorescens, Bacillus cereus, Pantoea agglomerans,* and *Erwinia rhapontici* control *Pythium* damping off of canola, safflower, pea and sugar beet (Bardin *et al.,* 2003). Jayaraj *et al.,* (2007) have isolated a strain of *Pseudomonas fkuorescens* (strain PfT-8) from tomato rhizoplane which is highly antagonistic to *Pythium aphanidermatum.* It produces high levels of chitinase, B-1,3-glucanase, cellulase, fungitoxic metabolites and siderophores. The antagonist is stable for atleast 2 months in formulations based on peat, lignite, lignite+fly ash or bentonite paste and gives high level of damping off control. *Enterobacter cloacae* is an effective antagonist of *P. ultimum* causing damping off in many field crop plants. Its mechanism of action includes inactivation of seed exudation that supports *Pythium*. This type of activity is effective in many, not all, plant species. The protective effect afforded by *Bacillus subtilis* (strain M4) against damping off of bean seedlings (*Pythium ultimum*) is attributed to the role of secreted lipopeptides and induced resistance (Ongena *et al.,* 2005a). In addition to role of antibiosis, Ongena *et al.,* (2005b) have reported that *Bacillus subtilis* strain M4 controls damping-off of tomato (*Pythium aphanidermatum*) through induction of resistance in the plants. Pre-inoculation with strain M4 sensitizes the plants to react more efficiently to subsequent pathogen infection. A similar effect was observed in cucumber against the cucurbit anthracnose fungus *Colletotrichum orbiculare.* Other biocontrol agents that are reported to suppress damping off caused by *Pythium* and *Rhizoctonia solani* are binucleate *Rhizoctonia* strains, *Actinoplanes* spp., *Pseudomonas putida*, *P. aureofaciens*, *P. corrugata*, *Serratia plymuthica* and *Bacillus amyloliquefaciens*. *Pseudomonas marginalis*, *P. syringae* and *P. viridiflava* also suppress damping off caused by *P. ultimum* or *P. aphanidermatum* (Gravel *et al.,* 2005).

Chatterton *et al.,* (2004) have studied the optimal timing of application of *Pseudomonas chlororaphis* for control of *Pythium*. They have observed that the treatment should be timed to maintain about 10^5 CFU of the agent per g fresh roots and more frequent applications are needed to maintain the agent in healthy compared to infected roots. The ideal time for application was 3 days before an attack of *Pythium*. Induction of host resistance is also involved in the control of *P. aphanidermatum* by *Bacillus subtilis* and *Pseudomonas chlororaphis* (*P. aureofaciens*) in damping off of hot pepper (Nakkeeran *et al.,* 2006). Strains of *Serratia marcescens* are highly effective against damping off of cucurbits caused by *P. ultimum*. Live cells or culture filtratethe of the biocontrol agent both provide protection. Several antimicrobial comounds have been isolated from the culture filtrates. In addition to chitinase and protease etanol extracts of culture filtrates yield the antibiotic prodigiosin and the surfactant serrawettin W1. These two compounds are temperature dependent, being propduced by cultures grown at 28°C but not at 37°C. They inhibit sporangia germination and mycelial growth of *P. ultimum*.

Amendment of the growth medium for greenhouse grown watermelon with *Brassica* seed pomace (cake) combined with *Pseudomonas boreopolis*, a bacterium that degrades the glucosinolate sinigrin, gives good control of damping off caused by *Pythium* sp. The control is attributed to the hydrolysis products of glucosinolates in to allyl and butenyl isothicyanates which are toxic to *Pythium aphanidermatum*, *Sclerotium rolfsii*, *Sclerotinia sclerotiorum* and *Phytophthora capsici*. Compost of paper mill sludge, plant waste and manure suppress damping off of cucumber caused by *Pythium ultimum*. Biocontrol agents isolated from compost include *Pseudomonas fluorescence*, *Pseudomonas aeruginosa*, *Graphium putredinis*, *Zygorrhinchinis moelleri* and *Bacillus morinus*. The last two were most effective. Vallad *et al.,* (2003) have reported that composted forms of paper mill residuals induce resistance response in tomato against certain bacterial foliar diseases. Leaf composts may induce soil suppressiveness against *Pythium* and damping off. In cotton, the suppressiveness against *P. ultimum* was studied by McKellar and Nelson (2003). The bacterial community in the suppressive soil rapidly metabolized linoleic acid whereas bacteria from conducive soil did not. Fatty acid-metabolizing bacteria were higher in the suppressive soil than in conducive soil and they rapidly colonized the cotton seed within few hours after sowing. Certain strains of *Pseudomonas corrugata* and *Pseudomonas aureofaciens* applied to roots of cucumber induce systemic resistance against *P. aphanidermatum* causing root rot through accumulation of endogenous salicylic acid and stimulation of peroxidase and polyphenol oxidase activities. Punja and Yip (2003) compared the efficacy of commercial products based on *Streptomyces griseoviride* strain K61, *Trichoderma harzianum* strain T-22, *Trichoderma virens* strain GL21 and

Gliocladium catenulatum strain 11446 for biocontrol of damping off and root rot of greenhouse cucumber caused by *P. aphanidermatum.* The biocontrol agents were applied once at seeding time and again 11 days later following by inoculation with the pathogen. The most effective biocontrol agent was *G. catenulatum* followed by *S. griseoviride. Trichoderma* species were least effective. Kanjanamaneesathian *et al.,* (2003) used *T. harzianum* grown on either sorghum grain, ground mesocarp fibre of oil palm or palm oil shell, both amended with urea fertilizers, for seed treatment of *Brassica alboglabra* (Chinese kale) and obtained good control of damping off of seedlings. *Rhizobium leguminosarum* can also be used for biological control of *Pythium* damping off in addition to its use as a biofertilizer (Bardin *et al.,* 2004).

Often seed-treatment fails to give effective check of damping-off of seedlings. This is because of the fact that many different types of fungi may be involved and a particular chemical is not effective against all of them. A knowledge of the exact cause of the disease is, therefore, essential.

In addition to seed treatment, the treatment of nursery soil, periodically, if not very often, is also important. Such treatments can be done by using formalin (diluted at the rate of one part in 50 parts of water) sprinkled over the loose soil in sufficient amount to soak it to a depth of at least 10 cm. Sowing of seed should be done after several days when it is definite that all traces of formalin have disappeared. Use of formalin dust is also recommended. The dust consists of 85 parts of charcoal ash and 15 parts of formalin. Thirty grams of the dust is mixed thoroughly with about 7 cm depth of soil per square foot. An economical method of disinfection of nursery bed soil is to burn a 30 cm thick stack of farm trash on the nursery beds. Soil solarization has been found very effective against many diseases that start from the nursery soil. The nursery bed is first irrigated and then covered for 40 days under transparent poly sheets. Average maximum temperature of the mulched soil was 49.7°C. *Pythium* and *Fusarium* were killed up to 30 cm depth of soil and seed germination improved by 18-42% in different crops. There is evidence to show that application of biocontrol agenyts to soil after solarization ensures efficiency of biocontrol agents and better damping off suppression.

Disinfestation of the soil and control of post-emergence damping off can also be accomplished by drenching the soil with fungicides effective against specific fungi responsible for the disease. The broad spectrum fungicides such as captan, thiram, and copper oxychloride have proved economical and better than most other fungicides. A 0.2–0.5% suspension of the fungicide can be applied to the soil. Garlic extraxt is fungicidal for species of *Pythium, Phytophthora* and for *Rhizoctonia solani Pythium aphanidermatum* fails to grow in a nutrient solution containing 10% garlic extract. After a single application of a solution containing at least 35% garlic extract or two applications of 25%, *P. aphanidermatum* could not be recovered from infested peat-based root substrate. In sand substrate lower concentrations of the extract were required (Sealy *et al.,* 2007).

Combination of seed- and soil-treatment is best for management of damping-off. Additional precautions recommended are: thin sowing to avoid overcrowding, light sandy soil for nurseries or use of pure fine sand and sawdust mixture for raising seedlings, use of well decomposed manure, light but frequent irrigations and raised nursery beds to drain off excess water.

FRUIT ROT OR COTTONY LEAK OF CUCURBITS

Fruit rot is the most common disease of cucurbits in India. It occurs in most localities during the summer season. The disease is reported on sponge gourd (*Luffa aegyptiaca* and *L. acutangula*), snake gourd (*Trichosanthes anguina*), parwal (*Trichosanthes dioica*), cucumber (kheera, *Cucumis sativus*), bottle gourd (*Lagenaria siceraria*), bitter gourd (*Memordica indica*) and also other cucurbits.

Symptoms

The cucurbit fruit rot is characterized by a luxuriant wooly mycelial weft on the affected fruits which appear as if wrapped in absorbent cotton. The fruits in close contact with the soil suffer most. Before the cottony growth of the fungus appears, the skin of the fruit shows soft, dark green, water-soaked lesions which gradually develop into a watery soft rot. On this rotting portion the cottony growth appears in a wet weather. In watermelons, the decay frequently starts at the blossom end. On the margins of the cottony growth the skin of the fruit looks dark green and water-soaked. This area, which is without any aerial growth of the fungus, indicates the "killing in advance" activity of the pathogen. The tissues in the interior of the fruit become watery and soft and the decaying mass emits a bad odor. Root and crown rot of cucumber is also caused by the same pathogen (*Pythium aphanidermatum*).

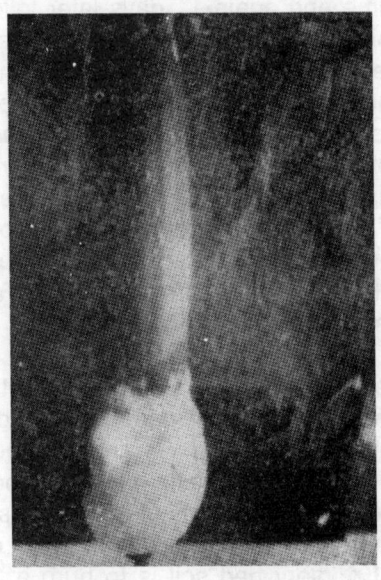

Fig. 24. Fruid rot of *Luffa* with cottony growth of *Pythium*.

The disease is common in the field during and after the rains and most of the fruits lying on the ground or hanging near the soil surface are attacked. The disease spreads among fruits during transit and storage.

The causal organisms

Pythium aphanidermatum (Eds.) Fitz. and *Pythium butleri* Subr. are the common oomycetes associated with the disease, although species of *Fusarium, Rhizoctonia*, and *Phytophthora* are also reported to cause similar rots. Frequently, *Sclerotium rolfsii* is also found growing on the rotting fruits as a saprophyte.

The mycelium of *P. aphanidermatum* consists of intracellular, much branched hyphae. The entire thallus is full of oospores, oogonia and antheridia both within the host tissues and on

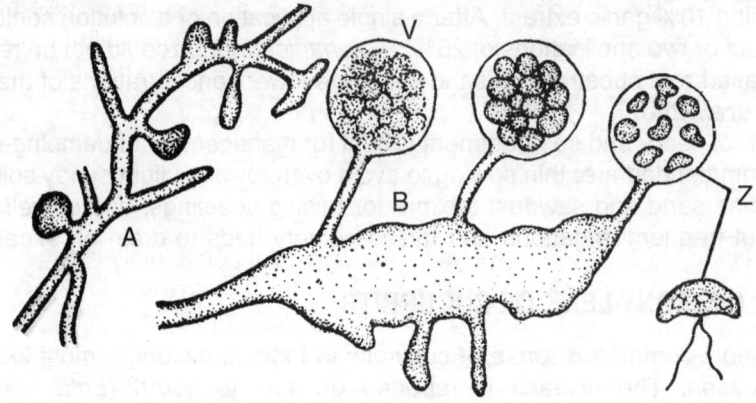

Fig. 25. Pythium aphanidermatum.
A-Mycelium with sporangial sweelings and hyphae; B-Vesicle
(v) formation from sporangial hyphae; Z-Zoospore

the surface. The hyphae are 2.5–8 μm in diameter, hyaline, and coenocytic. However, in old cultures of the fungus and in hyphae present in old tissues irregular septation may be found.

Sporangia are in the form of stout swollen hyphae or lobulate, bud-like outgrowths, up to 500 μm in length and 20 μm in width. These structures are very rich in protoplasm. At the time of germination a bladder-like vesicle is formed at the apex of the outgrowth and the protoplasm flows into it. Later, 30–45 zoospores are formed within the vesicle. The zoospores are similar in shape and mode of germination to those of *Pythium debaryanum*.

The oogonia are smooth, spherical, and usually terminal but may be intercalary, ranging in diameter from 18 to 27 μm. The antheridia are broadly clavate, terminal or intercalary, monoclinous, and one or two may be pressed against a single oogonium. Fertilization occurs as in *P. debaryanum*. The oospores are thick-walled, aplerotic, and 14–25 μm in diameter. They germinate by production of a germ tube.

Pythium butleri, often considered a synonym of *P. aphanidermatum*, produces more copious and robust aerial mycelium, larger, more swollen and more branched sporangia, larger zoospores, oogonia, antheridia and oospores. The hyphae are 9 μm thick. Encysted zoospores are 11 μm in diameter. Oogonia measure 15–33 μm. Average diameter of oospores 22.5 μm.

These oomycetes are soil-inhabitants living in pre-colonized tissues of the host or mostly as oospores. Under suitable conditions these oospores germinate and cause infection. The important pre-disposing factors are abundant moisture and high temperature. A slight mechanical injury to fruit skin due to contact with soil particles facilitates infection. If the fruits can be kept away from soil surface chances of infection are reduced. Soil disinfection is uneconomical in the control of this disease. Solarization of soil during summer reduces inoculum of *Pythium aphanidermatum*.

Growing cucumber plants in the presence of chitosan (100–400 $\mu g/ml$) controls root rot caused by *P. aphanidermatum*. It triggers several host defense responses, including the induction of structural barriers in root tissues and the stimulation of antifungal hydrolases (chitinase, chitosanase, and ß-1,3-glucanase) in roots and leaves. This may induce resistance to fruit rot also. In addition, chitosam treatment causes fungal wall loosening, vacuolation and sometimes protoplasm disintegration. Bacterization of cucumber roots with plant growth-promoting rhizobacteria (*Pseudomonas corrugata, P. aureifaciens*) before inoculation with *P. aphanidermatum* also induces resistance to crown and root rot. In presence of the bacteria and the pathogen, there is increased levels of defense enzymes (phenylalanine ammonia lyase, peroxidase and polyphenol oxidase) which contributes to resistance. Root bacterization of cucumber seedlings with *Pseudomonas corrugata* and *P. fluorescens* alters the attractiveness of root exudates for *P. aphanidermatum*. These bacteria reduce the attraction, encystment, and germination of zoospores.

STEM OR FOOT ROT OF PAPAYA

Also known as collar rot or root rot, this is the most serious fungal disease of papaya (*Carica papaya*). Under conditions favourable for its development it can annihilate the entire plantations within one season making the soil unfit for replanting. It is a widespread disease in India, Sri Lanka, Hawaii, and South Africa. In India it usually appears during the rainy season (July-August) and the severity of the disease depends upon the intensity of rainfall coupled with high temperature.

Symptoms

The earliest symptom of the disease is appearance of water-soaked patches on the stem at the ground level. These patches enlarge and ultimately girdle the entire base of the stem. Due to rotting the diseased tissues become dark brown or black. Simultaneously, the terminal leaves begin to droop and wilt, becoming yellow and falling prematurely. If fruits are formed they also drop down. Due to disintegration of parenchymatous tissues at the base of the stem the entire plant topples down and dies. If the bark is opened the internal tissues appear dry and give a honeycomb appearance. Rotting may spread above and below on the stem down to the roots. Roots deteriorate and may be destroyed.

The typical stem rot is most common in 2-3 years old plants. However, younger plants have also been found dying due to early infection. A damping-off of papaya seedlings in the nurseries is also fairly common and is

Fig. 26. Foot rot of papaya caused by *Pythium*.

caused by the same organisms. Seedlings raised in such nurseries carry the disease when transplanted and under favorable environments develop symptoms of stem rot. Often plants do not die quickly but linger on for some time. In root rot caused by *Pythium aphanidermatum*, roots show dark brown decay, absence of secondary roots and disintegration of internal tissues of the main root.

The causal organism

Several species of *Pythium* are reported to cause stem rot of papaya but in India and Hawaii *Pythium aphanidermatum* is mainly responsible for stem rot as well as damping-off. *Phytophthora parasitical* and *P. palmivora* are also reported normally associated with the disease. In addition, they causes fruit rot also.

P. aphamidermatum is a soil-inhabitant. Papaya residues left in the soil harbor the fungus mostly in its oospore form. Sugars present in the host residues help growth of the fungus on the residue and production of oospores. Optimum temperature for disease development is 36°C. Abundance of moisture around the base of the stem is conducive to disease development and its spread.

Management of the disease

Stem rot of papaya can be avoided if the plants are grown on well-drained land where water-logging does not occur. If any plant is found badly damaged it must be carefully uprooted and destroyed. The same pit should not be used for replanting. Sometimes diseased plants have been cured by removing the affected tissues and applying some fungicidal paste. Thorough drenching of the exposed soil around the stem base and also spraying the stem with 6:6:50 Bordeaux mixture or 0.2% captan considerably reduces the incidence of stem rot. The damping off phase of the disease is effectively checked by soil treatment (drenching) with captan (0.2%)

before and during emergence. Seed treatment with organo mercurial or Thiram or Difolatan (0.25%) improves emergence. Combination of seed and soil treatments is recommended for best results.

For the control of papaya root rot caused by *Phytophthora palmivora* (in Australia), soil application of metalaxyl at transplanting time followed by weekly sprays of potassium phosphonate (1g/L) plus acibenzolar-S-methyl (0.025 g/L) significantly reduced root rot. Application of benzothiadiazole (BTH) induces systemic acquired resistance in papaya against *P. palmivora*. The host response is manifested by increased tolerance to infection, increased ß-1-,3- glucanase and chitinase activities and increased accumulation of PR1 protein (Zhu *et al.*, 2003).

LATE BLIGHT OF POTATO

The late blight attacks and kills the tops of the potato plant and invades the tubers, causing either a dry or a wet rot. It is the most destructive of all the potato diseases when conditions are favourable for its rapid development. The fungus causing this disease, *Phytophthora infestans*, is basically a cold climate pathogen but has tremendous capacity to adopt to environments thus becoming widespread in almost all environments (temperate as well as subtropical) which support potato production.

Potato is a native of Northern Andes in South America. The late blight also occurred on potato in the Andes as an endemic disease. It is believed that the pathogen in the tubers was inactivated while crossing the equator when potatoes were transported to Europe on slow moving boats. With the development of quicker means of transport the pathogen was almost simultaneously taken to Europe and the USA sometime during 1830–840. It is claimed that the potato late blight had reached North America before reaching Europe and that an American botanist (J.E. Teschemacher) of Boston had worked on the disease even before the European scientists (Peterson *et al.*, 1992). The disease was well established in Ireland, England and the continental Europe by 1842. India was not invaded before 1870 and in 1909 Australia, which was long thought to be immune, had the disease recorded in every state.

It is presumed that in India the disease was first introduced into the Nilgiri hills between 1870 and 1880 from Britain. Soon after, it was reported from the Darjeeling district in the Himalayas with the introduction of English potatoes there. During 1899–1900 it was observed for the first time in the plains in Hooghly district of Bengal. The disease spread throughout the state (the then undivided Bengal) in 1901–02 but was again not heard from anywhere in the plains for about a decade. Severe outbreaks of the diseases were reported in 1912–13 from Jorhat (Assam), and in 1913 from Bengal, and Bihar. It was reported from Pusa (Bihar) in February of 1928 and from Patna in February 1933. At Meerut (U.P) and Dehra Dun (Uttarakhand) the disease was observed in the second crop (January to May) in March 1943. It was not present at these places in the first crop (October–January). The appearance and spread of the disease in the plains of Uttar Pradesh coincided with the large-scale introduction of cold storages in the area. Since 1943, the late blight has been making almost regular appearance throughout the plains of northern India. However, its epidemics are restricted to only certain areas.

Although the potato late blight oomycete is reported to infect 89 plant species, only potato and tomato are the main hosts.

Symptoms

At any time during growth of the plant, but mostly in January, the blight appears on the foliage where it causes brown dead spots or extended dead areas, more frequently until the leaves are killed. These blighted areas first appear as faded green patches which soon turn brownish black. These lesions are not delimited in size and enlarge rapidly in a favourable weather. The dead areas appear at the leaf tip or the margins and spread downward and inward, the rate of spread depending on weather. If moist weather prevails, the entire leaf may be killed in 1–4 days. If dry weather follows the appearances of lesions, the invasion of leaf advances slowly and the affected areas curl and shrivel, while under moist conditions they remain limp and soon decay, often giving an offensive smell. Very often, when the weather turns dry after appearance of the spots, the latter remain restricted in size and the dead areas look hard, easily breaking away from rest of the lamina surface. At this stage the symptoms are often confused with early blight disease by the farmers.

Fig. 27. Late blight of potato. Leaf symptoms.

The wet rot in warm muggy weather advances rapidly and involves all the leaves and the stem but it does not develop farther to reach the tubers. The tubers get separate infections. It is generally the lower leaves that first show the disease but all' portions are affected and there are numerous infections in severe cases, primary lesions appearing on petioles and stems as well as on the leaves.

When spots on individual leaves are examined, especially when the leaves are still wet (early morning hours or after humid weather), a zone outside the purplish lesion is found to show a paler than normal green merging into the latter. On the lower side of the leaf against this zone a whitish or grayish mildew growth appears. This whitish haze consists of the aerial fructifications of the pathogen which have grown out through the stomata. In dry weather this mildew disappears. If cool moist weather returns the pathogen becomes active and corresponding symptoms and progress reappear. In this condition there is an abundant production of spores for rapid spread of the disease.

Potato tubers are also infected while in the field and still attached to the plant or they get the infection during harvest and sometimes in storage. Under field conditions the effect is secondary or primary. The early attack of the disease resulting in death of the tops will reduce

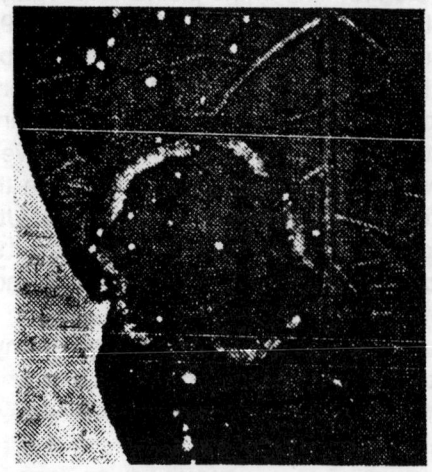

Fig. 28. Late blight of potato. Sporulating lesion.

the size and number of tubers formed but the primary invasions of the tubers by sporangia or zoospores falling from infected leaves and reaching the tubers through soil cause rotting, a dry or wet rot resulting according to moisture and temperature conditions prevailing at the time. The first sign of tuber infection is a brown to purple discoloration of the skin followed by a brownish dry rot which extends to about half an inch below the surface. When the disease proceeds without complications, tuber symptoms are quite distinctive and the entire affected tuber may be rapidly decayed. In storage containers, the wet rot phase of the disease is more common unless ideally cool and dry conditions are maintained. The slimy, soft decay is generally the effect of secondary bacteria which follow the late bight development. In moist atmosphere white tufts of mycelium and sporangiophores of the fungus appear on the surface of the tuber in the stores.

The causal organism

The late blight of potato and tomato is caused by *Phytophthora infestans* (Mont.) de Bary. (Kingdom Chromista, phylum Oomycota, class Oomycetes, order Pythiales, family Pythiaceae) The mycelium of the oomycete consists of hyaline, much branched, coenocytic hyphae which are intercellular with single or double club-shaped haustoria. The sporangiophores arise from the internal mycelium through stomata and through lenticels on the tubers. They are slender, hyaline, branched, and indeterminate. The branching is sparse. The sporangiophores are relatively thick-walled, show cross partitions, and the side branches show bulbous enlargements at intervals. The swellings indicate the position where the sporangia were attached. The sporangium first develops at the tip of a branch. As soon as it is mature, the tip swells slightly, proliferates, and turns the sporangium to a side as elongation of the branch proceeds. The sporangia are multinucleate (7 to 30 nuclei), thin-walled, hyaline, oval or pear-shaped with a definite papilla at the apex. They measure 22–33 × 16–24 μm. These sporangia may germinate directly by means of a germ tube, but most commonly the contents of the sporangium cleave to form a number of zoospores which emerge through the papilla and swim away. The method of germination of sporangia is largely governed by temperature. Low temperature favors zoospore formation while at higher temperatures the sporangium germinates by a germ tube. Sporangia germinating at 24°C or above often form secondary sporangia at the tip of the germ tube. A relative humidity above 90% is necessary for germination of sporangia. The zoospore are biflagellate, one flagellum possesses fine hairs (tinsel type) while the other does not (whip-lash type). After a few minutes of activity the zoospores lose their flagella, come to rest, and then germinate by a germ tube which penetrates the epidermis of the host directly or through stomata.

An isolate of *Phytophthora nicotianae* from pink rot (*Phytophthora erythroseptica*) affected potato tubers is also reported to cause foliage and tuber blight of potato.

Earlier, it was believed that *Phytophthora infestans* is sexually sterile in nature as well as in culture since the perfect stage was found on rare occasions. In the 1950s two mating types (A_1 and A_2) were discovered in Mexico. The sexual reproduction involves the association of these two different mating types of the fungus. Prior to 1984, these two mating types were known to occur in nature only in Mexico. Later, they were detected in Europe, North America and Asia also. In India only one mating type is known. However, selfing in mating type A_1 in presence of other species of *Phytophthora* such as *P. dreschleri* is also reported. In Japan, an A_1 isolate of *P. infestans* is reported to produce self-fertile oospores although it formed antheridia in culture medium (Gotoh *et al.*, 2005). These two facts (natural occurrence of mating

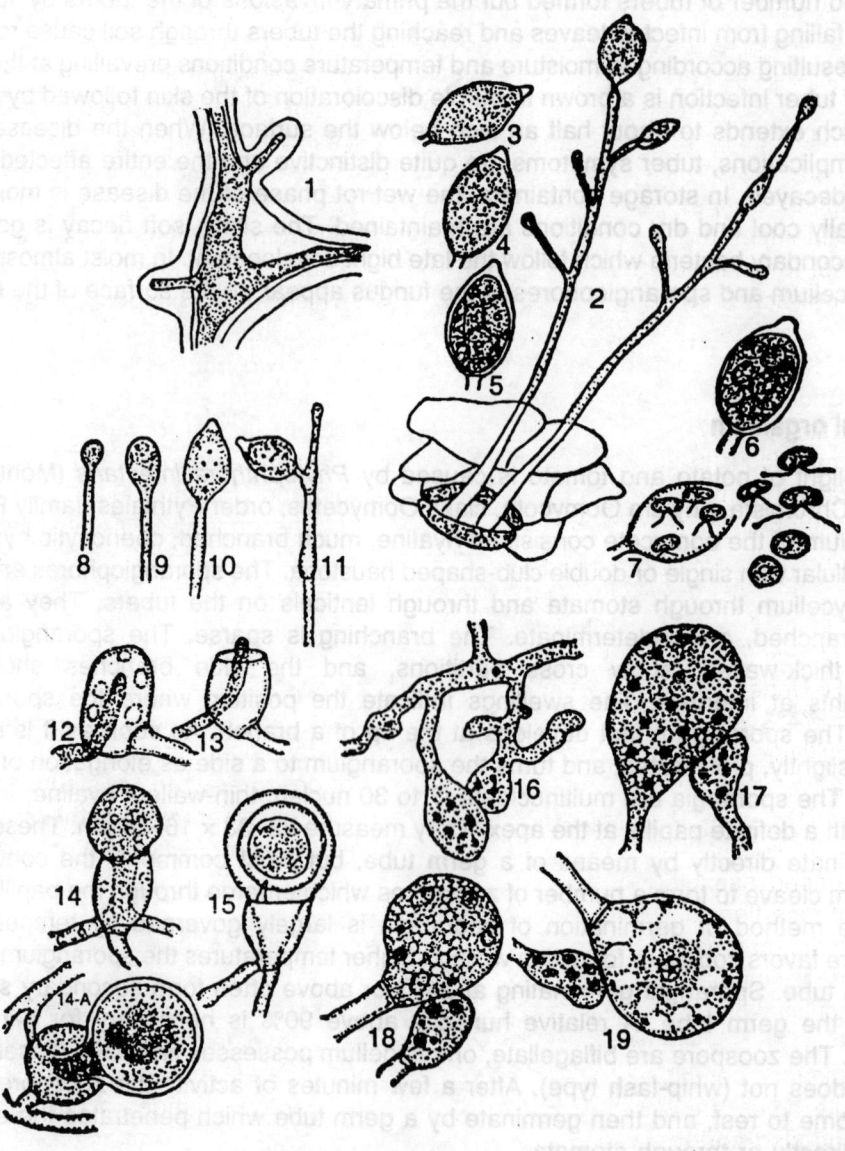

Fig. 29. *Phytophthora* infestans.
1-Intercellular mycelium in potato tuber with finger-like haustorium; 2-Sporangiophore emerging through stomata; 3-4-Sporangia with papilla; 5-6-Differentiation of zoospores; 7-Escape of zoospores from sporangia; 8-11-Development of zoosporangiophore; 12-15-Development of amphigynous antheridium; 16-19-A paragynous species of *Phytophthora (P. cactorum).*

types and selfing in presence of other species of the fungus) present potential means of variability in the species. In self-fertility, which is only to a limited extent, environments and genetic factors play major role. In presence of both mating types in physical proximity, the formation of oospores may be quite common and is promoted by such agricultural activities as sprinkler irrigation (Cohen *et al.,* 2000) and application of fungicides. Maximum production of oospores is obtained in leaves of moderately susceptible cultivars after 10 days of incubation at 15°C and a 16-hour photoperiod (Hammi *et al.*, 2001). The oospores may be produced in tubers also. The fungus produces oospores in fruits and seeds of tomato also (Rubin *et al.*, 2004) which is equally important host of late blight. Some combinations of the two mating types are more reproductive than others and some potato genotypes supported oospore production better than others. In some combinations of A_1 and A_2 multiple oospheres in a single oogonium are reported. Tomato usually supported more oospore formation than potato (Hammi *et al.*, (2001).

Optimal temperature for sexual reproduction ranges from 8–15°C but oospores are formed at 23°C also. In inculation of potato leaves with a mixture of both mating types, oogonia and antheridia develop 6 days after inoculationat 15°C. Thick walled oospores appear 3–4 days later. These oospores are killed when exposed to 40°C but at –80° to 35°F they survive. Continuous supply of moisture to the leaves where the two mating types are interacting is essential. Under optimal conditions of wetness and temperature as many as 100 oospores per sq. mm of tissue may be formed. In Israel, it is reported that in tomato transmission of *P. infestans* may occur by seeds extracted from fruits carrying oospores and less probably by seeds extracted from fruits having no oospores (Rubin and Cohen, 2004).

The antheridia of *P. infestans* are amphigynous. the oogonium is spherical. Oospores measure 24–46 μm in diameter. They germinate by a germ tube. Germination is obtained after 10 days of incubation at 20°C and a 16-hour photoperiod on potato soil extract or potato root extract but only up to 25% oospores are found germinating. Singh *et al.*, (2004) have reported formation of oospores on potato leaves under conditions of high tissue moisture and low ambient RH. Under field conditions oospore formation was detected when the ambient RH had declined to <74% and moisture content of the host tissue had decreased to 83–85%. Response of potato cultivars varies. The time taken for oospore development may be short in relatively susceptible varieties which support higher number of oospores than resistant varieties.

Phytophthora infestans has a large number of races differing in pathogenicity to host genotypes. While presence of the two mating types is one cause of the variability, genetic variation among asexually produced progenies is also common. In India, up to 1965, the physiologic race flora of the pathogen was simple, only races 0, 1 and 4 being present. This was due to the fact that all potato varieties under cultivation at that time were susceptible. But with the intensification of breeding for late blight resistance, introduction of resistant parental lines, and cultivation of certain resistant varieties in different parts of the country, physiologic race flora has rapidly changed and many complex races have developed. At present more than 82 races and race complexes are known in the country. Apart from sexual recombination where the opposite types are present, virulence and other genetic changes occur during asexual reproduction also and these play a major role in changing the race structure of *P. infestans* populations. The multiplicity of races poses new challenges in the use of race-specific resistance.

Pre-disposing factors

Excessive humidity (above 90% RH) coupled with suitable temperature for germination of sporangia and further disease development are the principal pre-disposing factors. In tomato, leaf wetness for minimum 2 hours is necessary. Most infections establish within 6 hours of inoculation (Becktell *et al.*, 2005a). Where the mean atmospheric temperature exceeds 25°C (77°F) the disease is rare or unknown. Cool moist conditions favor dispersal or arrival of viable sporangia. Extended dry periods or rapid dehydration can quickly kill many sporangia but within the temperature range of 15° and 20°C and moisture range of 40 to 88% RH the life time of sporangia is extended. Maximum spread of the disease occurs when conditions are favorable for germination of sporangia by zoospores. Optimum temperature for germination of sporangia by zoospore production is 12°–13°C and by germ tube production 24°C. Optimum for germination of zoospores is also 24°C. The minimum for germination of sporangia is as low as 2°–3°C while the maximum may go up to 24°–30°C .The short lived zoospores, which are motile in water, encyst and germinate to produce the germ tube that penetrates the host. After germination the germ tube elongates most rapidly at 21°C. Penetration can occur at any temperature from 10° to 29°C but successful infection occurs normally when sporangia land on wet host surface and temperature is between 10° and 25°C. *Phytophthora infestans* secretes a host-specific necrosis inducing protease. The process of penetration requires about 0.5–2.5 hours depending on temperature and host plant condition. The incubation period is 3–7 days. It is shortest at 23°C. Although development of sporangia is greatest at 28°C, production is maximum at 18°C. No sporangia are produced at 28°C. Optimum for intercellular growth of the fungus varies from 16°–18°C to 21°C. Sporulation can occur at any temperature from 9° to 26°C. Generally, the conditions for start of an epidemic are low temperature, in a wet atmosphere, for formation of sporangia and their germination by zoospores and then a slight rise in temperature to help optimum growth of the germ tubes and subsequent development of mycelium in the host.

On the basis of the above temperature-moisture-pathogen relations four important conditions for forecasting of late blight occurrence in severe form in temperate climate countries were: (i) night temperature below dew point for at least 4 hour, (ii) minimum temperature of 10°C or slightly above, (iii) clouds on the next day, and (iv) rainfall during the next 24 hours of at least 0.1 mm. In early 1990s, in Punjab, studies had shown that the conditions responsible for onset of late blight are different from those considered necessary under temperate climate. In severe epidemics hours of temperature less than 10°C and above 20°C are important variables. The disease could develop rapidly even in total absence of rains. Dew appears to be the alternative source of free moisture. Survival and availability of inoculum during early crop growth stages seem crucial in deciding late blight behavior in the season.

Increased rate of nitrogen fertilizers have two-fold effect on late blight. In well-fertilized soil the dense foliage of plants provides ideal microclimatic conditions for infection and sporulation of the fungus which may even escape effects of sudden change in outside weather towards dry conditions. In addition, the linear growth of lesions on leaves is also increased. Young plants are most susceptible. In older plants upper leaves show more rapid blighting than lower leaves. Leaf position (apical or basal) is more significant than plant age and leaf age (Visker *et al.,* 2003). Resistance in potato leaves against *P. infestans* always coincides late foliage maturity (Visker *et al.,* 2004).

In tuber infection, greater number of smaller tubers than larger tubers are infected, probably on account of lenticels being resistant in large sized tubers. Late blight development is slow on leaves pre-inoculated with Potato Virus Y. Cross protection is also provided by pre-inoculation with *P. cryptogea* or avirulent strains of *P. infestans*

Disease cycle

The living mycelium in the tubers is killed by exposure to 40°C for 4 hours or to 30°C for 65 hours. The viability of sporangia in soil is also affected by moisture and temperature. Optimum soil moisture is 15 to 20% saturation. Viability and infectivity of sporangia persists for 9–10 weeks at low temperatures and optimum soil moisture conditions. Under favourable conditions 100,000 sporangia per lesion are produced. Majority of sporangia (95%) exposed to solar radiation for one hour lose viability. However, in cloudy weather they remain viable. Germinability and viability of sporangia are strongly affected by external application of calcium or other divalent cations, especially during zoosporogenesis. Both direct and indirect germination of sporangia are suppressed by 1–5 mm concentration of calcium chloride, magnesium chloride or low concentrations of pectin and inorganic phosphate Zoospores in soil lose their viability within 2–3 weeks. Immobilization and lysis of zoospores of *P. infestans* by strains of *Pseudomonas fluorescens* is reported. In the plains of north India, the main potato season is followed by hot summer months. The high temperature and drought obviously kill the fungus in soil. Tubers stored at ordinary temperature during the summer months also become free from living inoculum.

In majority of the potato production areas around the world survival of *P. infestans* from one cropping season to the next is a unique and important phase of the life cycle. In Central Mexico, where oospores are commonly formed, survival of oospores in soil through winter fallow is reported. These oospores cause infection of the next crop (Fernandez-Pavia et al., 2004). In zones of temperate climate, late blight infected tubers left in the field may carry over the living mycelium to the next crop season. Soik survival of active inoculum has significancy in the epidemiology of late blight in such areas. It causes preemergence infection of sprouts which is consistent with the level of disease incidence seen in some field Studies have shown that mycelium of *P. infestans* can tolerate exposure to 0°C for any duration but exposure to temperature below this is fatal depending upon duration of the exposure. Studies in Sweden had shown that when an infected crop of potato was followed by a cereal for one seasonm and then disease-free tubers were planted, late blight appeared and its intensity and distribution was same as in the previous infected potato crop. In the sub-tropical zones the most common method of survival of the pathogen is through the infected tubers, harvested from a diseased crop and stored at low temperatures (in cold storages) for seed. The mycelium in these infected seed tubers serves as the source of primary inoculum. Tubers with even 5% surface area affected can initiate disease in the field. If weather favors this can lead to an epidemic. When both mating types are present the pathogen may become seed tuber-borne through oospores also.

The mycelium from the infected parent tuber (seed) grows upward in the stem and sporulates on the small, dwarfed shoots. Such plants serve as "infector" plants. The fungus can also grow from the tuber into the soil to a limited extent, produce spores which infect lower leaves in contact with the soil and create infector plants. Within 7-8 days, primary infection foci are established around the infector plants. A disease gradient is formed with maximum disease intensity at the source (infector plants) which decreases with increase in distance from the source. Further short distance spread within the field occurs by means of sporangia dispersed by wind, water, and leaf eating insects in a cool weather.

The infection by sporangia or zoospores may take place through any part of the epidermis of leaves and young stems, either through stomata or through unbroken cells. In infection of *Solanum phureja* (Oyarzun et al., 2004), zoospores attached and germinated in random manner on all leaf surface structures (undifferentiated eoidermal cells, veins, trichomes and stomatal complexes). Appressoria and hypha; ramification in epidermal cells occurred more frequently on leaflet tips than on base of leaflets. This occurred both in susceptible and resistant cultivars. Appressorium formation and hyphal ramification in epidermal cells was more frequent when zoospores had landed near stomatal complexes. Germination of zoospores occurred with similar frequency of resistant and susceptible cultivars but the latter supported better appressorium formation and hyphal ramification in epidermal cells.

Tuber infection occurs through eyes, lenticels, or wounds. Susceptibility of eyes and resistance of lenticels increases with storage and maturity of tubers. The general infection of tubers accompanying an attack of late blight is by means of spores which are produced either on the blighted tops during the season or present in the soil from the preceding crop. The spores from the blighted tops are washed down onto the soil where they penetrate to different depths reaching the healthy tubers and infecting them. Loose soil with good proportion of sand allows maximum penetration by the sporangia. More than half of the total blighted tubers are distributed up to a depth of 5 cm from the soil surface and no blight affected tubers are found beneath 15 cm depth. Thus, a soil cover (ridging) of 15 cm over the tubers probably filters out the inoculum coming from the foliage (Arora, 1989). Similar observation regarding soil depth are reported by Porter *et al.,* (2005). Significantly less tuber infection occurs in silt loam than in medium and fine sand. Heavy and frequent rains at a time when about 50% of the foliage is infected causes maximum infection of underground tubers. Contact of healthy tubers with diseased leaves at the time of digging is another source of tuber Infection. There may be some spread from infected tubers to healthy tubers in the stores also. While fungicidal spray on the foliage reduces the amount of inoculum reaching the tubersut does not affect tuber blight development which is correlated with low soil temperature, precipitation and soil moisture, tuber depth in soil, and cultivar resistance. Tuber surface soil microflora influencex infection of tubers by *P. imfestans* spores.

The role of solanaceous weeds in the epidemiology of potato iate blight has been studied in Netherlands. There, *Solanum nigrum* (black nightshade), *S. dulcamara* (woody nightshade) and *S. sisymbriifolium,* used as a trap crop for the cyst nematode, have been found infected. The last named species favors heavy sporangia production.

Host resistance

There are about 200 recognizd wild tuber bearing and cultivated potato species. Most show resistance to one or more of the potato diseases. Most frequently used wild species are *Solanum acaule, S. chacoense* and *S. demissum.* These species carry resistance genes for late blight, wart, viruses, some insects and cyst nematoes.

The central highlands of Mexico are considered to be center of genetic diversity for both the potato late blight pathogen and for the tuber bearing *Solanum* spp. The Mexican *Solanum* species which coevolved with *P. infestans* not only provide sources of resistance (R) genes but also provide a source of quantitative, rate reducing resistance that is highly effective, stable and durable. Early studies on late blight resistance involving *Solanum tuberosum* and *Solanum demissum* had demonstrated that resistance was inherited as a dominant character. These dominant R-genes control the hypersensitive response of the host tissue and are race specific as specified by corresponding pathogen genes for virulence (gene-for-gene hypothesis). Potato breeding efforts had been directed towards development of cultivars with these dominant resistance genes. It is now recognized that breeding for field immunity based on R-genes is rather futile as in practice these cultivars require the same fungicidal coverage as the susceptible cultivars. The pathogen exists as race complexes and is capable of quickly producing new races. Therefore, potato breeding programmes are now emphasizing the establishment of lines with field resistance to late blight as controlled by polygenes. The field or horizontal resistance is a complex of different types of resistance and is found in many cultivated varieties. It is governed by a number of minor, independent genes, and indications are that dominancy is not involved. This resistance is often associated with later plant maturities and involves resistance to infection, inhibition of pathogen growth within the host tissues, and reduced sporulation. This results in a "slow blight" response by cultivars possessing general resistance.

Resistance of the foliage is not necessarily correlated with resistance of tubers. In many cases tuber resistance is due to characteristics of the skin and in others due to characteristics of the flesh. The outer layers of the cortex of the tubers of some potato cultivars are very resistant to colonization by *P. infestans* due to presence of inhibitory factors in a thin layer of cortical tissue, approximately 30 cells deep. If this layer is removed the tissue in the interior becomes susceptible. Tuber resistance is provided by separate genes independent of genes for leaf blight resistance. In a study of infection of tubers in 16 potato varieties possessing different degrees of foliar infection and observed that cultivars with same level of susceptibility to foliar infection (92 to 100%) showed different levels of tuber infection, some showing 20–22% tuber infection while others showing only 9–11% tuber infection. Infection of tubers does not necessarily lead to growth of the fungus in tuber tissues. Tuber resistance varies among cultivars and according to isolates of the pathogen. High level of resistance to both foliage and tuber blight has been observed in some hybrids obtained by protoplasm fusion. Glandular trichomes on leaves of potato, which confer resistance to insects, are also involved in leaf resistance to *P. infestans*. Structure of internal leaf tissue plays a role in post-penetration phase by restricting the growth of the pathogen. If the pathogen is able to invade the epidermal layer, it is encountered by the mesophyll tissue (palisade and spongy tissue). The cells of the spongy tissue are arranged in irregular fashion having intercellular spaces in between them. Loose arrangement of spongy tissue facilitates the rapid and easy spread of the pathogen. Resistant varieties have more compact spongy tissue (Mahajan *et al.*, 2003).

In India, resistance to race O of *P. infestans* had been found in wild species of *Solanum* brought from abroad. These were *S. demissum*, *S. phureja*, *S. spectabile*, *S. stoloniferum*, *S. polyadenium*, *S. andreanum*, *S. brachycarpum*, *S. cardiophyllum* and *S. edinense*. However, only *S. demissum* and *S. phureja* have shown field resistance to the pathogen. The wild diploied potato species *Solanum bulbocastanum* is highly resistant to all races of *P. infestans* and germplasm derived from this species has shown durable and effective resistance in the field (Song *et al.*, 2003). Durable resistance in somatic hybrids of *S. tuberosum* and *S. bulbocastanum* are reported. Resistant hybrids contain an antifungal protein that is similar to patatin, the major storage protein of potato tubers.

Resistant varieties were developed by the Central Potato Research Institute and released for cultivation. However, varieties like Kufri Red and Kufri Neela were soon withdrawn. The variety Kufri Jyoti which was widely accepted by farmers for its resistance to late blight was found susceptible to the disease in Shimla and Shilong hills but has maintained resistance when given low level of fungicidal protection in the plains. In addition this variety is susceptible to bacterial wilt. The varieties recommended against late blight include Kufri Naveen, Kufri Jeevan, Kufri Alankar, Kufri Badshah, Kufri Swarna, Kufri Khasi Garo, Kufri Moti, Kufri Muthu, JH 232, F 5242, AB 286. For the hilly regions of north and north-east varieties Kufri Himalini and Kufri Sherpa were released in 1981. Certain hybrids of wild potato, *S. verrucosum*, *S. phureja*, *S. acaule*, *S. microdontum* and *S. sucrense* possess source of durable resistance to the late blight fungus. A clone (F 87084) of *S. tuberosum* developed from the above species through conventional breeding methodology has been reported to have resistrance to Verticillium wilt, early blight, wart and one race of cyst nematode.

Resistance can also be induced by priming potato cell suspension (during tissue culture) to culture filtrates of *P. infestans* This induction of resistance may not be of much practical value because of presence of races of the pathogen.

Management of the disease

The late blight can be effectively checked by spraying of suitable fungicides but attention should also be given to certain other measures. No control measure is economical and durable unless it is followed on a cooperative basis by the neighbouring growers also.

Selection of seed tubers: Tubers from a diseased crop should never be kept for seed. At the time of planting the tubers should be closely inspected and all suspicious looking tubers rejected. Seed treatment does not check late blight. Use of healthy seed tubes is no insurance against loss but it usually delays onset of the disease and makes fungicidal sprays economical and more effective.

Sanitation: The crop should be kept weed free to check humidity. Weeds, especially solanaceous species, should also not be allowed to grow on borders of the field. At the time of harvest care should be taken to avoid contact between tubers and foliage. It is better to cut off and remove the foliage some days before digging of tubers. The rejected crop debris and tubers should be destroyed.

Harvesting should be delayed: Harvesting of a diseased crop should be delayed until the plants are fully mature. The foliage should be allowed to dry completely before digging of tubers. Killing potato haulms 2–3 weeks prior to harvesting tubers is recommended for the management of potato late blight to eliminate the foliage as source of tuber blight inoculum. This reduces the amount of viable inoculum on the foliage thus avoids infection of harvested tubers. When tubers are harvested within 1–4 days after vine killing, tubers from plants that had been growing longer are more susceptible to late blight tuber rot than tubers from plants that had been growing for shorter period of time.

Sorting potatoes from a blighted field: The digging of tubers should be done in a good dry weather and only sound tubers should be collected first. They should be taken directly to the store and re-sorted before putting in the bins.

Tuber treatment before storage: This saves tubers from secondary rot causing organisms. The tubers may be given a 90-min dip in 1:1000 mercuric chloride solution. Such tubers should be thoroughly washed if used for eating. Postharvest decay of tubers has been prevented by dipping tubers in solution of potassium metabisulfite or sodium hypochlorite. *In vitro*, hydrogen peroxide (1:50) has shown potential for control of tuber rot in storage.

Storage: In a cool, dry, and well aerated store there is less danger of late blight spreading among tubers. The danger is greater in moist, poorly ventilated rooms and when there are too many wounded tubers. The best temperature for storage is 4°–5°C.

High ridging: A 15 cm high ridging at the time of earthing in the crop reduces the chances of tuber infection. Mulching of the plot between rows around stems with black film also reduces disease incidence in tubers. However, barriers to tuber infection are generally effective only when proper control of foliage blight has been achieved . A single spray of 4 kg/ha phosphonic acid (neutralized with potassium hydroxide to pH 6.4) at the mid- or late-season is reported to reduce tuber infection. Weekly foliar spray of phosphonate, phosphate, phosphate of potassium, sodium, ammonium or aluminium suppresses pre-harvest tuber blight infection and the effect is particularly visis after two months of storage of tubers from treated plants (Mayton *et al.,* 2008).

Chemical seed tuber treatments and foliar sprays: When seed tubers are the main source of primary inoculum, control of tuber-borne inoculum with appropriate seed piece treatment results in an increase in plant emergence and improved crop uniformity. Specific tuber treatment chemicals include a thiphanate methyl + mancozeb. Foliar fungicides such as dimethomorph + mancozeb, cymoxanil + mancozeb or propamocarb +chlorothalonil have been used for seed tuber treatment. However, seed treatment gives only initial protection and for protection of the standing crop and foliar sprays are a must. Singh and Pundhir (2004) found that treatment of whole tubers with deep incisions or cut tubers with oxadixyl or Ridomil were less effective because of poor uptake of fungicide. In their opinion sprout treatment was effective and resulted in healthy plants.

In areas where late blight is of common occurrence, fungicide sprays give effective check of the disease. Although unnecessary spraying increases the cost of production and is not recommended, in such areas it is advisable to start spraying operation well ahead of the usual time of appearance of the disease. The first prophylactic spray is recommended when the crop is about 6 weeks old. There are reports suggesting that protection of first leaves by a suitable fungicide spray is important in preventing epidemic development. Further spraying should be decided after watching onset of the disease, its spread, and weather conditions.

During the first half of the last century the chemical control of late blight centered on providing control with Bordeaux mixture. The early sprays were given with 4: 4: 50 strength of the mixture while later sprays used 6: 6: 50 strength. Two to 3 sprays at 15–21 days interval were recommended. Later, other copper fungicides, different formulations of copper oxychloride such as Fytolan, Blitox-50, and Fycol 8E. gradually replaced Bordeaux mixture. These fungicides were used at a concentration of 0.2–0.5%. Subsequently (mid-1950s to mid-1970s), the broad spectrum dithiocarbamate fungicides became the most popular and widely used fungicides for late blight control. In general, the mode of action of these protectant fungicides involved protection of the foliage by halting normal germination and / or penetration by the spores. For zineb (Dithane Z-78) and mancozeb (Dithane M-45) the recommended dose is 2–2.5 kg in 100 litres of water per hectare. Chlorothalonil (Bravo or Daconil) is also a highly effective fungicide against late blight. At 3 sprays (0. 75 kg/ha) it gives as good control as 1.88 kg/ha mancozeb with a cost benefit ratio of 1:13. The efficacy of the fungicide can be increased by increasing the dosage and frequency of application. If used timely and with proper care to wet the entire foliage surface these fungicides give highly effective check of late blight. Some fungicides that were not marketed in the country but are highly effective were Du-Ter (1.8 kg/ha) and Brestan (0.6 kg/ha). Vishwakarma and Singh (1984) had reported that 3-4 sprays of zinc sulphate (5 kg/ha) with lime (2.5 kg/ha) gave as good control of late blight as 4 sprays of mancozeb in the Tarai area of Uttarakhand. This treatment is cheaper than the fungicides and is non-hazardous.

The search for more effective fungicides resulted in the introduction, during 1980s, of metalaxyl, member of a group of acylalanine analogues with systemic activity. It provides control when applied as soil or tuber treatment at planting time and as foliar spray during the growing season. The systemics are superior to protectants since they enter and become distributed in the host tissues and thus are less susceptible to weathering and less frequent applications are required. Metalaxyl was found to provide disease control by reducing sporulation, germination, and intercellular growth of the fungus. It is an effective eradicant halting previously established infections. Ridomil MZ (combination of metalaxyl and mancozeb) reduces both growth and sporulation of the fungus while mancozeb only inhibits spore germination. However, repeated use of this fungicide results in development of resistance in the fungus. The metalaxyl-resistant population of *P. infestans* is resistant to Ofurace (acylalamine fungicide) and Patafol. These two have similar mode of action as metalaxyl. There is no resistance to strobilurins and cymoxanil. This warrants use of combinations of systemic and protectants. The All India Coordinated Potato Improvement Project has given the following recommendation: Two sprays of metalaxyl + mancozeb combination (Ridomil MZ 72 WP containing 8% metalaxyl and 64% mancozeb) at 0.25% concentration should be given at an interval of 15 days starting as soon as first visible symptoms appear in the field. These two sprays should be followed by weekly sprays of 0.2% mancozeb (2 kg / ha) starting 15 days after the second spray of metalaxyl-mancozeb combination. Two to 3 sprays of mancozeb will be sufficient. The cost/benefit ratio of mancozeb treatment is slightly better than Ridomil treatment.

In vitro zoospore encystment and cyst germination are highly sensitiuve to dimethomorph. Direct sporangial germination, hyphal growth and sporulation are less sensitive. These effects are variable with isolates of the pathoghen. However, in later reports (Caldiz *et al.*, 2007) combination of dimethomorph 9% + mancozeb 69% (Acrobat MZ), at a dosage equivalent to 2 kg/ha, has been found a good seed tuber treatment fungicide. It not only prevents tuber to tuber tramsmission of infection during handling but also protects the healthy tubers from infection in soil. At a dosage of 4 kg/ha the fungicides provides protection to the foliage for 28–30 days after crop emergence.

A derivative of phosphoric acid, the systemic fungicide fosetyl-Al (Aliette 80 WP) which is a member of alkyl phosphonate group and contains 80% aluminium-tris-ethyl phosphonate, is also very effective against late blight in the field. This compound reduces sporangial production of *Phytophthora* in soil by 90%. Fosetyl-Al has very little *in vitro* activity against *P. infestans* but provides disease control through inducing host resistance. The fungus is not affected, *in vitro*, even by 1000 ppm of fosetyl-Al although it causes severe morphological changes in *P. cactorum* and *P. capsici* at concentration of 250-750 ppm. The size of lesions and amount of sporulation is significantly reduced by a single application of dimethomorph plus mancozeb or cymoxanil plus mancozeb. Two applications of propiconazole with chlorothalonil are required to restrict lesion expansion and sporulation. The combination is effective even when applied 48 hours after inoculation. Hyphal cells of *P. infestans* are killed within 30 min when exposed to ethaboxam (a thiazole carboximide fungicide) at 0.01 μg/ml. The fungicide is specific for targeting of microtubule disruption in *P. infestans* In Japan spray of a new fungicide cyazofamid (1873 μg a.i./ml at 7-day interval) is reported to give control equal to or superior to standard fungicides. The fungicide is specific for oomycetes and plasmodiophoromycetes. *P. infestans* is sensitive to the new carboxylic acid amide fungicide mandiptopamid. At 2.5 μg/ml it provides good protection. So far no resistance in *P. infestans* to this fungicide has developed even after extensive enforced selection pressure under field conditions (Cohen *et al.*, 2007).

Carboxylic acid amide (CAA) fungicides mandipromide, dimethomorph and iprovalicarb give high level of control of late blight by affecting different developmental stages of the pathogen, especially the germination of infective zoospopres and sporangia, which are most sensitive. Dimethomorph is most effective.

Abiotic suppression of many soil-borne plant pathogenic fungi including *Phytophthora* spp. by aluminium is known. Applications of this Al-mediated suppression are limited by its potential phytotoxicity. Humus or organic amendments of soil reduce phytotoxicity. Silicon also attenuates aluminium toxicity. In *in vitro* tests, addition of aluminium sulphate or chloride inhibits mycelial growth and sporangial production of *P. infestans*. Amendment of peat-based potting media with aluminium sulphate solution (0,0158 gt Al per g of peat) is known to suppress *Phytophthora parasitica*. This aluminium toxicity may be associated with some soils suppressive to growth of the pathogen.

The use of fungicides can be considerably reduced by integrating host resistance with fungicidal control (Naerstad *et al.*, 2007). Greater the resistance to foliage blight lower can be the quantity used for sprays, The usage is reduced by reducing the dose rather than increasing the interval of spray Foliage blight reduction results in lesser tuber blight incidence.

Potato seed tuber treatment with calcium or potassium phosphate (0.1%) is effective against *P. infestans*, *Fusarium solani* and *Rhizoctonia solani*. Phosphites are alkali metal salts of phosphorous acid. The effect is highest against *P. infestans*, intermediate against *F. solani* and low against *R. solani*. Like phosethyl-Al, foliar sprays of these salts provide good control of late blight (Lobato *et al.*. 2008), Beta aminobutyric acid (BABA), a plant activator, gives 52–55% control of late blight when applied to the foliage at 0.4–0.5%. In tomato late blight, spray of BABA provided 82% control of late blight. The alpha and gama amino butyric acid gave only 35% and 6% protection to tomato leaves BABA also reduces the mycelial growth of *P. infestans in vitro*. Thus, its effect in tomato is both fungistatic and induction of resistance. However, on some cultivars of tomato it is toxic to leaves without affecting growth of the plant. Another activator BION has similar effect against *P. infestans* in tomato. Combination of mancozeb with BABA has synergistic effect against *P. infestans* in potato and tomato giving better control than either of them alone (Baider and Cohen, 2003). BABA and mancozeb in 5:1 ratio (w/w, a.i. basis) exhibit a higher synergistic effect than 1 + 1 or 1 + 5 mixture. Andreu *et al.*, (2006) have reported that spray of BABA in early stage of crop growth induces systemic resistance in potato plants against foliage and tuber blight. The treatment enhances the natural resistance in potato against late blight It can help in reducing the use and concentration of fungicides. Spray of acibenzolar-S-methyl as well as dipotassium phosphate (both plant activators) also reduces disease incidence in tomato (Becktell *et al.*, 2005b) Benthiavalicarb isopropyl sprayed at 2.5–7.5 g a.i./ha effectively checks late blight of potato and tomato. The extract of leaves of *Inula viscose* is highly effective against the blight in potato as well ads tomato (Wang *et al.*, 2004). Unger *et al.*, (2006) obtained 90% control of tomato late blight by spraying water extracts of dried biomass of *Penicillium chrysogenum*. The extract had no direct antifungal effect against *P. infestand*. The induction of resistance was accompanied by increase of peroxidase enzymes in the host tissues. Treatment of potato cell suspension (in tissue culture) with culture filtrates of *P. infestans* primes the cells for expression of defense responses.

Priming of defense reactions by an elicitor results in an enhanced ability of the plant to respond to subsequent pathogen challenges. Treatment of potato cell suspension (in tissue culture) with 1 μg/ml of concentrated culture filtrate of *P. infestans* primes the cells for enhanced expression of defense reactions. Compounds like lipopoy saccharide, laminarin, and harpim N needed second elicitation, after some days with the culture filtrate to be effective (Val *et al.*, 2008).

Plant essential oils and formulated products of biocontrol agents have been tried as alternatives to synthetic fungicides. These have given variable results. In laboratory and growth chamber studies, the essential oil oregano, composted tea extract and *Bacillus subtilis* have given 5–40% control of late blight. Application of garlic extract (active principle allicin) to infected tubers or used for fumigation of tubers suppresses *P. infestans* spore germination and disease. In late blight of tomato, pre-inoculation treatment (2-14 days before inoculation) with ethanolic extracts of leaves of *Hedera helix* (Araliaceae) or *Paeonia suffruticosa* (Paeiniaceae) provide protection against infection of *P. infestans* (Rohner *et al.*, 2004).

In biological control pre-inoculation of bottom leaves of tomato with tobacco necrosis virus induces systemic acquired resistance against *P. infestans*. Induction of resistance is accompanied by increased peroxidase activity up to the upper leaves. Soluble organic metabolites produced by the bacterium *Xenorhabdotes bovienii* (Enterobacteriaceae) were reported to suppress late blight. Potato plants were not harmed by the metabolites even at 1000 μg/ ml. Daayf *et al.*, (2003) isolated 43 bacteria from the phylloplane and/or rhizosphere of potato and canola plants and tested their ability to control *P. infestans*. The bacteria with biocontrol activity were from the genera *Bacillus, Pseudomonas, Rahnella*, and *Serratia*. The mechanism of action included antibiosis and induction of resistance in the host. *Bacillus cereus*, applied through seed bacterization, is reported to induce systemic resistance in tomato against *P. infestans, Alternaria solani* and *Septoria lycopersici* (Silva *et al.*, 2004). The treatment can reduce the number of fungicidal sprays by half. Slininger *et al.*, (2007) spray inoculated potato tubers with sporangia of *P. infestans* followed by spray inoculation of *Pseudomonas fluorescens* strains and a strain of *Enterobacter cloacae*. The treatments gave consistent late blight control in tubers along with control Fusarium dry rot.

In tomato, late blight was reduced by treatment with combination of epiphytic antagonists and rhizobacteria (Junior *et al.*, 2006). Bacterization of leaves with *Pseudomonas fluorescens* strain SS101 is effective in preventing infection and reducing the expansion of existing lesions. The active bacterial metabolite involved is massetolide A (Tran *et al.*, 2007). Applicastion of rhizobacteria (*Bacillus cereus*) alone was not effective. The effective epiphytes were *Cellomonas, Candida* and *Cryptococcus*. In some other hosts, the damage to leaves and inflorescence by *P. infestans* is reported to be reduced by inoculation of roots with the vesicular arbuscular mycorrhiza *Glomus etunicatum*. Stephan *et al.*, (2005) have evaluated, by leaf disc assay, a number of commercial biocontrol products and some plant extracts. The preparations Elot-Vis, Serenade and Trichodex and plant extracts of *Rheum rhabarbarum* (rhubarb, Polygonaceae), a plant of temperate and subtropical climates, and *Solidago canadensis* (Compositae), found in North America and Europe, significantly reduced level of infestation by *P. infestans*. However, none of the treatments were better than copper fungicides. In tomato late blight, Soyulu *et al.*, (2006) have reported efficacy of essential oils of many aromatic plants in suppressing growth of *P. infestans*. The volatile phase of the oils was more effective. Son *et al.*, (2008) have isolated two antioomycete compounds from culture filtrates of a strain of Fusarium oxysporum which showed a potent disease control efficacy against late blight of tomayto. These have been identified as bikaverin and fusaric acid. They inhibit the mycelial growth of plant pathogenic oomycetes and fungi. Fusaric acid was effective against plant pathogenic bacteria also. Bikaverin at 300 μg/ml suppressed the dfisease oin tomato by 71%. In addition to late blight fusaric acid suppressed wheat leaf rust by 67%.

Cultivation of resistant varieties in areas where the late blight is common not only ensures best check of the disease it also reduces the cost of chemical sprays. Kirk *et al.*, (2001) emphasized the need for host resistance or low susceptibility to reduce fungicide rate and keep longer intervals between sprays. Diversity in host cultivars within the field reduces spread of late blight from primary and secondary infection. In a study of effect of mixed cropping of susceptible and partially resistant cultivars under natural epidemic conditions, Andrivon *et al.*, (2003) found that late blight severity was significantly lower in a susceptible cultivar growing in rows alternating with partially resistant cultivars than in unmixed plots of the susceptible cultivar alone. The mixture of cultivars reduced disease progress rate and could delay onset of the disease also.

STEM CANKER AND BLIGHT OF PIGEONPEA

During the later part of the last century, the stem blight, stem canker, stem rot or Phytophthora blight joined the list of serious diseases of pigeonpea, others being Fusarium wilt and sterility mosaic. The disease affects the crop at any stage of its growth when environmental conditions are suitable for the pathogen and disease development. The affected plants are usually a total loss.

Symptoms

Infection of seedlings is visible as water-soaked lesions in the primary and trifoliate leaves and within 3 days the lesions become necrotic. The leaflet lesions are circular to irregular and up to 1 cm in diameter. Under conditions of high humidity the foliage gives a blighted appearance.

On stems, brown to dark brown lesions, distinctly marked from the healthy green portion, are formed near the ground level or up the stem. In the latter case they are mostly located on a leaf scar or site of branch initiation and extend in each direction. They enlarge in size and girdle the stem. Girdling of the stem is often seen 1 to 1.5 meters up the stem. Similar lesions develop on the branches. Portion of the plant above the lesions dries but remains attached to the plant. Wind easily breaks the stems at the point of infection. Initially, the stem lesions have a flat surface but later become sunken. Swelling of the stem at the ground level is also common. The stem lesions later develop into cankerous structures at the edges. Sometimes the affected area cracks and shred. With progress of the disease in the field, patches of blighted plants can be seen from a distance. The roots of diseased plants remain healthy although in a similar disease caused by *Phytophthora parasitica* in Australia there is serious root rot.

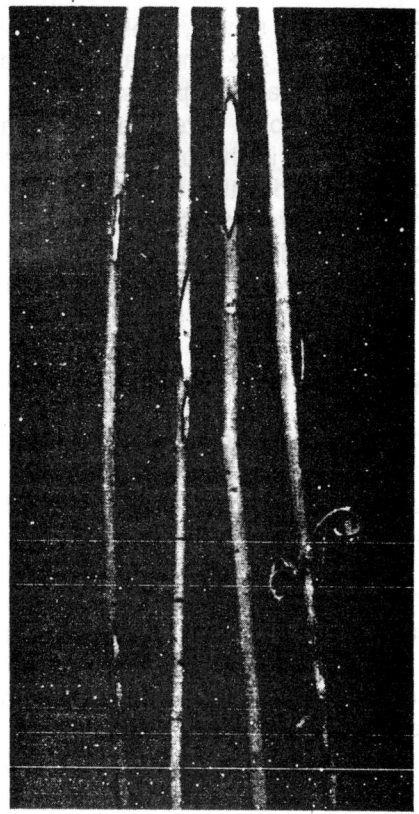

Fig. 30. Stem cankers of pigeonpea caused by *Phytophthora.*

The causal organism

The pigeonpea blight is caused by *Phytophthora drechsleri* f. sp. *cajani.* In culture on corn meal agar medium the fungus produces aerial white mycelium. The hyphae are hyaline, cottony, devoid of granular contents, coenocytic, slender and measure 3–6 μm in diameter. Irregular swellings (12–15 μm dia) with tubular projections are present on the hyphae. These swellings are favoured by low temperature (15°–18°C) but are suppressed by high temperature (35°C).

The sporangiophores are usually hypha-like except for the swollen tip which develops into sporangium. The sporangia are ovate to pyriform, rarely spherical and with a minute papilla on some substrates. In culture they are not formed on solid media but if the culture is transferred to sterilized water they are formed abundantly. The sporangia measure 41–78 × 28–45 μm. Zoospores mature within the sporangia and are released individually after dissolution of the apical portion of the sporangium. Sometimes, the germinated sporangium may produce a secondary sporangium. Zoospores are biflagellate, hyaline, ovoid to reniform, tapering slightly at the anterior end. They swim in water for 2–5 hours, become non-motile and form a spherical cyst. This cyst germinates by one or more germ tubes, sometimes ending in a microsporangium.

Oogonia are hyaline when immature, but become thick-walled and purple yellow to brown after maturity. They are smooth, spherical and 23.5–37.0 (mostly 25.5–28.0) μm in diameter.

Antheridia are simple, hyaline, amphigynous, persistent and measure 12.5–19.0 × 10.0–17.0 μm. Oospores are spherical to globose, 23.5–37.0 (average 30.0) μm. The oospore wall is 1.2 μm thick. Chlamydospores are also formed. These are intercalary as well as terminal and measure 11 to 21 (average 14.4) μm wide.

Variation in the pathogen has been noted on the basis of cultural characters, size of sporangia and pathogenicity. Some isolates are fast growing, some show intermediate growth and others are slow growing. Some host genotypes show susceptibility to all isolates while some show differential reaction. The isolates can also be distinguished on the basis of their temperature response. Some isolates grow at temperatures beyond 35°C while growth of some is totally restricted at temperatures between 30° and 35°C.

Disease cycle and environmental relations

The pigeonpea blight pathogen is capable of surviving in soil (even in absence of a living host) and also in infected debris for at least one year. Oospores and chlamydospores are the main structures of survival. Survival is better in lower soil profile (5–15 cm) than on soil surface. At the onset of rainy season oospores germinate by sporangia and direct germ tube and infection of young seedlings occurs. From the primary infection, a large number of sporangia are produced on the mycelium which serve as secondary inoculum. The secondary inoculum is disseminated by wind, water, movement of soil and raindrop splashes.

High humidity and a temperature range of 28–32°C are conducive for rapid build up of the disease in the field. Optimum temperature for growth, sporangia formation and zoospore germination is around 25° to 30°C. Oospores are formed at a critical temperature of 25°C. The conditions for maximum infection index are 21.1 mm rainfall per day, 100% rainy days, 27.4°C maximum and 21.4°C minimum temperature and 92.4% relative humidity. Light is inhibitory to zoospore germination. Darkness, humidity and temperature being favorable during night, the rate of spread of the disease is high during the night hours.

Addition of nitrogen (26–50 kg/ha) as ammonium sulphate increases the blight whereas addition of potassium (25–50 kg/ha) in the form of potassium sulphate reduces disease incidence regardless of the presence or absence of N or P in the soil. Phosphorus (superphospahte) does not have any effect. In inoculation tests it has been found that incidence of blight was maximum (100%) in seedlings inoculated at 15 days and declines with increasing plant age to a minimum of 23% at 120 days, (Mishra and Shukla, 1986).

Management of the disease

Cultural practices: Pigeonpea should be grown in a field which has no record of blight in the past. Since the inoculum can be brought to the field by running water, low lying fields should be avoided. In disease prone areas, sowing of pigeonpea on ridges has been an effective method of avoiding the initial development of the disease. Crop rotation, wide inter-row spacing and use of potassium fertilizers are other recommended cultural practices. Weed canopy reduces splash dispersal of inoculum from soil to aerial parts of the plant. Interculture of pigeonpea with low height legumes such as urdbean and mungbean has been suggested.

Host resistance: Several thousand germplasm of pigeonpea have been tested against *P. drechsleri* f. sp. *cajani* but no line with a high degree of resistance has been identified. Resistance is unstable when tested against different isolates of the pathogen. Moderate resistance was reported in pigeonpea lines AS-3, 2357 and 4419. Pigeonpea lines ICPL 161 and 366, METH 12, COMP-1, ESR-6, Pusa A-3, Pant A-83-14, GAUT 82-58 and DPPA 85-112 have shown field resistance. Certain genotypes resistant to Fusarium wilt have been found resistant to Phytophthora blight and sterility mosaic.

Biological control: *Trichoderma viride, T. hamatum,* strains of *Pseudomonas fluorescens* and *Bacillus subitilis* are antagonists of *P. drechsleri* f. sp. *cajani*. They are compatible with different levels of fungicides such as metalaxyl, metalaxyl + mancozeb (Ridomil MZ), captan, captafol, thiram and carbendazim. Coating pigeonpea seeds with an antagonist in the presence of a compatible fungicide has no deleterious effect on emergence. A reduction of 28.9% is reported when *P. fluorescens* was used with metalaxyl or metalaxyl + mancozeb. Volatiles in culture filtrates of strains of *Pseudomonas fluorescncs* are known to contain the antibiotics pyoluteorin pyrrolnitrin and also certain levels of HCN. These cause inhibition of zoospore release from sporangia.

Chemical control: Seed treatment with metalaxyl + mancozeb ((Ridomil MZ) provides maximum protection up to 15 days after sowing. Thereafter, the action of the fungicide gradually declines and is almost negligible at 45 days after sowing. High doses of metalaxyl are phytotoxic. But metalaxyl + mancozeb has no phytotoxicity at high doses also. Seed dressing alone is not effective but spray treatment alone or in combination with seed dressing give most effective chemical control of the disease.

The toxicity of metalaxyl or metalaxyl + mancozeb is enhanced by the herbicides fluchloralin and pendimethalin. This has suggested the possibility of using herbicide mixed with metalaxyl formulations for control pigeonpea blight.

GUMMOSIS AND ROOT ROT OF CITRUS

Species of *Phytophthora* cause the most serious soil-borne diseases of citrus throughout the world. Losses occur from damping off of seedlings in the seed beds, root and crown rot in nurseries, and from diseases variously known as gummosis, brown rot gummosis, brown rot, trunk rot, foot rot, collar rot, root rot, fibrous root rot, leaf fall and fruit rot. Serious losses can occur in orchards with trees on susceptible rootstock such as rough lemon (*Citrus jambheri*), or in plantings on resistant rootstock where the graft union is at or below the soil surface exposing scion tissues to the pathogen. If the disease is not checked in time the entire tree may be destroyed. The brown rot of fruits occurs in orchards causing fruit drop and in storage it causes post-harvest decay.

Symptoms

Typical symptoms result when soil- or seed-borne fungus penetrates the stem just above the soil line and causes the seedlings to topple. Seed rot and pre-emergence damping-off may also occur. There is rapid killing when there is abundant soil moisture and temperatures are favorable for the fungus. Once the leaves have emerged the seedlings become resistant. These symptoms are similar to damping-off of any plant caused by *Pythium* or *Rhizoctonia*.

Primary infection by *Phytophthora* normally occurs on the bark at the base of the trunk near the ground level, producing lesions on the trunk and the crown roots. It spreads around the trunk girdling it and killing the tree. The bark and the wood both are affected. The infection can spread upward and down to the roots, often causing fibrous root rot. The root damage is especially serious on susceptible rootstock in nurseries. Fibrous root rot may occur even in bearing trees where the root damage causes tree decline and yield losses. In some tolerant rootstocks the fibrous root rot does not affect the fruit yield.

The main symptom of gummosis is oozing of gum from the affected parts on the trunk. Infected bark remains firm with small longitudinal cracks through which abundant amber colored gum exudation occurs. Citrus gum is water soluble. During ainy season the gum is washed down or gets mixed with soil near the ground level hence this symptom may not be clear. During summer gum deposits dry and stick to the bark making the symptom of gummosis very clear.

The root rot and fibrous root rot symptoms are not seen in the early stages of the disease. However, the root rot destroys a major portion of the root system before well-marked symptoms are seen on the aerial parts. The effect of trunk and root infection is ultimately drying up of the tree. Before death the tree flowers profusely but fruits are small and drop before maturity.

Leaves of the affected trees show symptoms of nutritional deficiency. The veins turn yellow and there is premature leaf fall. In mandarin oranges, infection of leaves by *P. palmivora* is common in heavy rainfall areas. Water-soaked spots appear on the lower leaves and by the time the spots spread over the entire lamina the leaf drops. This usually results in heavy defoliation. The infection reaches the unripe, ripening and ripe fruits and produces water-soaked spots on the skin. This is brown rot of fruits. Direct infection of fruits can occur when water splashed inoculum reaches the fruits near ground level. All the fruits on the tree may gradually become affected under humid conditions. Such fruits become soft and white fungus growth develops on the skin. Ultimately the fruit drops. The fungus continues to grow on fallen fruits.

The fruits which do not show symptoms and are still on the tree carry latent infection. These are harvested and packed. If fruits are untreated, the brown rot spreads from fruit to fruit by contact. In few days of storage these fruits have a characteristic pungent, rancid odor. Brown rot epidemics are usually restricted to areas where heavy rainfall coincides with the early stages of fruit maturity. All cultivars are affected especially the lemons. In Spain, *Phytophthora citrophthora*, one of the causal agents of citrus root rot, is reported to cause branch canker of citrus. The affected trees showed cankers on the scion.

The causal organisms

Three species of *Phytophthora*, viz., *P. palmivora*, *P. parasitica* (*P. nicotianae* var. *parasitica*) and *P. citrophthora* attack citrus. In south India the main cause of gummosis and root rot is recognized as *P. palmivora* although *P. parasitica* is also common as incitant of fruit rot in Karnataka. In other parts of India also *P. parasitica* was reportedly associated with gummosis. In the USA, *P. parasitica* and *P. citrophthora* are the main pathogens of root rot of citrus. In addition, *P. hibernalis* and *P. syringae* attack citrus fruits to a limited extent in areas with cool moist weather. *P. citricola* is reported to attack citrus in some tropical areas.

Phytophthora nicotianae B. de Haan var. *parasitica* (Dastur) Waterhouse: The hyphae are tough, irregularly wide up to 9 μm but without typical hyphal swellings. Sporangiophores are more slender than the mycelial hyphae, irregularly or sympodially branched, the sympodia being close in moist air. Sporangia are papillate, and occasionally have more than one apex. They are broadly ovoid, ellipsoid, obpyriform to spherical, not noticeably narrowed at the apex, occasionally lateral or intercalary. They measure 38–50 × 30–40 μm (av. 40 × 38 μm) and are deciduous with very short (2–5 μm) pedicel. Oogonia are usually produced in single cultures though often very scarcely or not until some weeks, but readily produced in dual cultures with opposite strains (the A_1 and A_2 mating types). Antheridia are amphigynous, spherical or oval, and 12–16 × 18 μm in size. The oogonial diameter is usually 22–29 μm, rarely 31 μm. Oogonia become rough, thick-walled and yellowish brown with age. Oospores are markedly aplerotic and 18–20 (sometimes 20–25) μm in diameter. The oospore wall is about 2 μm thick. Chlamydospores are abundantly produced. They are less than 25 to 60 μm in diameter, with 3-4 μm thick wall. Minimum, optimum and maximum temperatures for growth are 10°C,

30°– 32°C and 47°C, respectively. *P. parasitica* has a broad host range. Virulence of isolates from tomato and other non-citrus hosts towards citrus is low while all isolates of citrus species are pathogenic on tomato.

Phytophthora citrophthora (Smith and Smith) Leonian: Hyphae are fairly coarse and up to 7 *μ*m wide. Sporangiophores are delicate, short, scarcely widening at the base of the sporangium, irregularly branched and with a swelling at the point of branching. Sporangia are rather scanty on some agar media and very variable in shape and size in water; often with more than 1 apices and papilla having 5 *μ*m deep apical thickening. Sporangia are deciduous with a 10–12 *μ*m long pedicel. The average size of sporangia is 40–45 × 27, often 50–55 × 30 or even up to 90 × 60 *μ*m. Chlamydospores may be abundant, few or absent. Most isolates do not form chlamydospores. Their average diameter is 28 *μ*m with 1.5–2 *μ*m thick wall. Sexual reproduction is not seen. Minimum, optimum and maximum temperatures for growth are 5°C, 24°–28°C and 32°C, respectively.

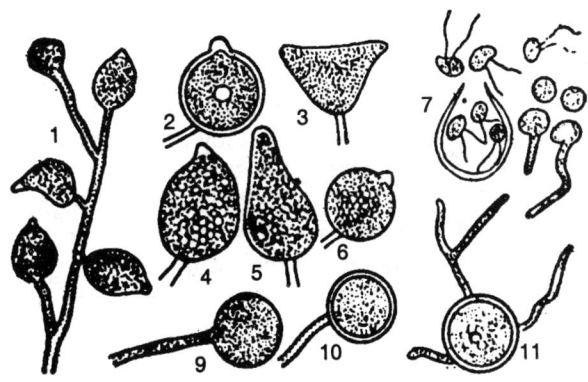

Fig. 31. Phytophthora citrophthora. On PDA at 20–22°C.
1-Sporangiophore; 2-6-Sporangia; 7-8-Release of zoospores; 9-10-Chlamydospores; 11-Germination of chalmydospores.

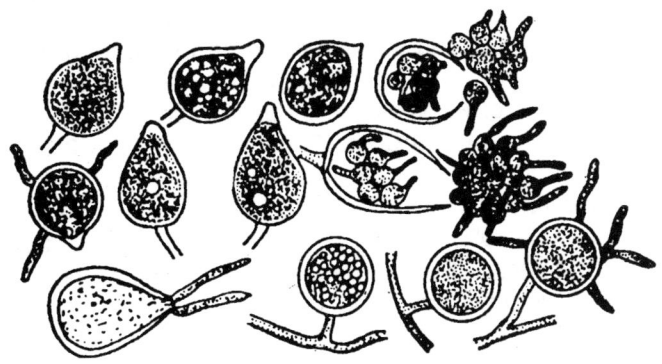

Fig. 32. *Phytophthora palmivora.* Sporangia, their direct and indirect germination; germinating zoospores; and chlamydospores.

Phytophthora palmivora (Butler) Butler: The mycelium is intercellular with haustoria. Hyphae are large and often swollen at regular intervals. They are up to 7 *μ*m wide. Sporangiophores are simple or branched with inverted pear shaped, rarely round and always terminal sporangia which measure 38–72 × 33–42 (av. 50 × 35) *μ*m. Zoospores are large, 8–10 μm in diameter when encysted. Oospores are spherical, 35–45 μm in diameter with 4 μm thick wall. They produce secondary sporangium on germination.

Disease cycle and environmental relations

P. parasitica is a poor saprophyte in soil. It causes noticeable yield losses only when high populations of the fungus are present in the soil. The growth in soil is restricted due to antagonism including intense colonization of hyphae by bacteria leading to rapid lysis and breakdown of cytoplasm. The entire mycelium is soon converted into oospores and chlamydospores which serve as survival structures. Oospores occur in low numbers in soil and probably are resistant to desiccation and cold temperature. The formation of oospore and perpetuation of the fungus through these structures is governed by the presence of mating types in the plant population. Inter-specific crossing in nature is also reported. The oospores mature more slowly than chlamydospores but once matured they germinate in response to nutrients from roots.

Chlamydospores are common and the most important source of soil survival of P. parasitica. These structures are commonly formed when soil moisture is limiting, conditions are cool or where the host roots are not actively growing and producing susceptible tissues. Formation of chlamydospores can also be stimulated by poor aeration and high carbon dioxide concentration in the soil atmosphere. They can survive in soil for several months under unfavorable conditions. Coehlo et al., (2000) studied the effect of high temperature (35°–53°C) on survival of chlamydospores of P. parasitica. Time required to reduce the soil population to residual level (0.2 propagules per g soil) decreased with increasing temperature. Addition of cabbage residue to the soil reduced the time required to inactivate the chlamydospores.

In the presence of host roots, nutrients, aeration, optimum temperature and moisture these resting structures germinate by a germ tube and by sporangia which liberate motile zoospores. Germination of chlamydospores is optimum in well-aerated, moist environments when temperatures are favorable for root growth. Root exudates promote their germination. Chlamydospores cause more uniform infection than zoospores.

Maximum populations of propagules of P. palmivora occur at 0.5–15 cm depth of soil and very low populations are found at 30–45 cm. Optimum soil moisture for survival of propagules was 25–45% WHC at pH 6.5–7.0. The survival could be for up to 18 months in natural soil. Sporangia survive and remain infective after passage through the alimentary canal of two snail species. Ingestion of oospores of P. palmivora by garden snails facilitates their germination.

P. citrophthora grows best at 24°–28°C and P. parasitica at 28°–32°C. Abundant production of sporangia of P. citrophthora occurs at 20° while that of P. parasitica at 30°C. Maximum recovery of P. citrophthora from rootlets in naturally infested soil is obtained at 15°–20°C and that of P. parasitica at 15°–30°C. Thus, in citrus areas where temperatures are low P. citrophthora is more dangerous than P. parasitica. A temperature of 24°C after infection favours development of P. citrophthora in roots.

The number of propagules rapidly increases immediately after irrigation in irrigated orchards or nurseries. In long interval furrow irrigation the initial population of 17 propagules/g soil increased to 77 propagules /g soil two days after irrigation. Very low numbers of oospores of P. parasitica are found in the soil throughout the year. Low levels of dark colored multi-papillate sporangia are also found in the soil. Conditions which allow abundant immature fibrous root development, a highly susceptible rootstock, and favorable soil temperature and moisture promote development of root rot and, consequently, high propagule density of P. parasitica in soil.

In addition to oospores and chlamydospores in the soil as the main survival structures for initiating primary infection. *P. parasitica* also survives on fallen fruits, twigs, leaves, and in cracks on the standing trees. The main source for the spread of these pathogens in Nagpur mandarin orchards in central India is reported to be infested nurseries. More than 20% nursery plants die due to Phytophthora diseases and almost all nurseries are infested.

When chlamydospores and oospores germinate in the presence of nutrients, optimal soil moisture and aeration they form sporangia which liberate zoospores. Zoospore release is optimal in saturated soils. Nutrient depletion stimulates sporangia production. Diurnal temperature changes in soil may serve to synchronize the release of zoospores. Calcium applied as calcium chloride or nitrate to water or calcium-free soluble fertilizer (10–50 meq) suppresses zoospore release from sporangia and their motility. They encyst within 4 hours while in unamended solution they remain motile. The cysts cannot germinate in presence of calcium without help of an organic stimulant. Diplanetism is also suppressed. These effects of calcium reduce infection of seedlings.

The motile zoospores swim to short distances (cm) or are carried by soil water to long distances (meters). Dispersal of sporangia and zoospores is by wind, raindrop splashes, irrigation water, and even insects. Zoospores are attracted to roots by root exudates, especially when roots are damaged but living. The attraction is mediated, in part, by electrotaxis. The plant roots generate electrical currents and associated electrical fields as a consequence of electrogenic ion transport at the root surface (Van West *et al.*, 2002). *Fusarium solani*, a citrus pathogen causing root rot, influences root rot caused by *P. parasitica* and *P. citrophthora*. The influence differs with species of *Phytophthora*. Propagule densities of *P. parasitica* are reduced by 41% when citrus is inoculated 30 days after transplanting in *Fusarium*-inoculated soil or when transplanted in soil inoculated with both fungi. When citrus roots are immersed in zoospore suspension of *P. citrophthora* and transplanted into soil infested with *F. solani* feeder root length is reduced by 68% but feeder root weight and the percentage of living roots are not significantly reduced. The presence of *F. solani* on the roots reduces chemotaxis of *Phytophthora* zoospores.

On the root surface the zoospores encyst, germinate and cause infection by penetrating the cortex. Citrus tissue is more susceptible during the periods of the year when trees are actively growing than during the months when trees are dormant. Colonization of cortical cells is more by *P. palmivora* than by *P. parasitica*. In a susceptible host (sour orange, *Citrus aurantium*) and a tolerant host (trifoliate orange, *Poncirus trifoliata*) penetration by *P. parasitica* and *P. palmivora* occurs within one hour of inoculation. After 24 hours, *P. palmivora* has a higher level of colonization in the susceptible than in the tolerant host but cortical cells remain intact. After 48 hours, the host cells in both hosts are disrupted by either species of the pathogen.

Heavy or fine textured soils where drainage is impeded, high soil moisture, pH of 5.4–7.5 and a temperature of 24°C are conducive to disease development. Soil moisture is the most important plant-soil environmental factor that affects development of Phytophthora root rot. The host is pre-disposed to infection when roots are stressed or damaged in saturated or dry soil. Citrus greening disease (Libukih) affected trees are more prone to root rot caused by *P. parasitica* (Ann *et al.*, 2004). Long periods of soil saturation are required for *P. parasitica* to be an effective root decay agent. The length of saturation time is more important than frequency of saturation. The frequency and duration of irrigation can also influence the activity of the fungus and pre-disposition of roots to rot. In sandy loam soils the greatest destruction of feeder roots occurs in irrigation furrows where saturated conditions favor zoospore production and movement. Roots outside the furrow often remain healthy. If soils are saturated during irrigation, zoospores are released and can infect roots to form more sporangia. When soils do not dry sufficiently between irrigations, sporangia survive until the next irrigation and again release zoospores. Soils with drainage restricted by hard pans or clay layers or those with shallow water table that temporarily rises into the root zone provide ideal conditions for fibrous root rot and build up of *Phytophthora* propagules.

Availability of oxygen in soil atmosphere is closely related to soil moisture because pore space for air is reduced in saturated soil. When roots are subjected to low oxygen conditions they are damaged by reduced forms of minerals and by toxic metabolites of microorganisms on the root surface. Root regeneration is restricted, new roots are not formed and root exudation increases under flooded conditions. Thus, in presence of *Phytophthora*, reduced oxygen level in soil causes greater root decay.

Low grafting and nearness of bud union to soil line increase chances of infection from soil-borne inoculum. Populations of the pathogens around roots in soil are increased where, in addition to the favourable environments, there are abundant immature fibrous roots and a highly susceptible rootstock thus promoting development of root rot. Around resistant rootstock there is low density of the fungi in soil. Most rootstock are at least moderately tolerant to *Phytophthora* but their susceptibility varies. Although mechanisms of resistance are not clearly understood, it is presumed that coumarin phytoalexins in infected roots play some role.

Microbial antagonism in the suppression of *Phytophthora* spp. in soil, rhizosphere or in the infection court is reported. Processes of parasitism, predation and competition may all be operating alone or together in reducing inoculum. The reason for rapid loss of *Phytophthora parasitica* mycelium in soil is attributed to lysis induced by intense bacterial colonization of hyphae. Specific strains of *Streptomyces* and fluorescent pseudomonads (*P. putida*) occur in soil that cause hyphal lysis of *P. citrophthora*. In some areas, *P. citrophthora* infection of citrus roots does not normally occur during the summer although the fungus is abundantly present on alternative hosts in an infested citrus grove. This has been attributed to antagonistic microflora associated with roots and active during summer.

Vesicular-arbuscular mycorrhizae have been implicated in microbial antagonism of *P. parasitica* in citrus but increased tolerance of mycorrhizal plants to root rot is probably due to improved host nutrition, mainly phosphorus. However, induction of resistance by VAM is also known. Mycorrhizal biomass in nurseries can be increased by precropping of nursery soil with mycotrophic crops (Punja and Chaudhuri, 2004).

For brown rot of fruits caused by *P. parasitica* a minimum of 3 hours of fruit wetness at 27–30°C is essential for infection. No brown rot develops at 22°C or below. The optimum temperature for sporulation on fruits is 24°C with sporangium production decreasing above and below this temperature. A few sporangia are formed with 18 hours of fruit wetness and the number increases as duration of fruit wetness increases to 72 hours. In fruit infection, the dispersal of sporangia from lesions is mainly by water droplets, not by wind alone. Propagules are splash dispersed by a single droplet of water horizontally about 350–450 mm and vertically about 30–45 mm (Timmer *et al.*, 2000).

Management of the disease

A number of preventive measures can be recommended to reduce primary infection. The site for citrus orchards should be on well-drained land. Resistant rootstock such as Khatta and trifoliate orange (*Poncirus trifoliata*) may be used in areas where gummosis and virus diseases of citrus are serious problems. The bud union should be about 30–45 cm above the base and at the time of planting care should be taken to keep the bud union well above the soil line. The irrigation system should be planned in such a way that water from below one tree does not flow to the other trees. Every year the trunk should be painted with Bordeaux paste up to a height of about 70 cm. If the soil is known to contain *Phytophthora* the walls of new pits should be dusted with a mixture of zinc sulphate, copper sulphate and lime (5: 1: 4). Cleanliness of the orchard floor is very important. All infected fruits, leaves, twigs, etc. that have fallen should be collected and burnt.

During the summer and rains the orchard should be regularly sprayed with copper fungicides such as Bordeaux mixture (4:4:50 or 5:5: 50) or copper oxychloride formulations such as Blitox-50. Soil drenching with 1000 ppm terrazole is reported to totally inhibit *P. palmivora* up to 2.5 cm depth of soil (Sastry and Hegde, 1992). In more recent chemical treatments, systemic fungicides have been used for soil drench and, sometimes, for trunk spray. Working with root rot and crown rot (*P. citrophthora* and *P. parasitica*) Metalaxyl, fosetyl-Al (phosethyl-Al), and sodium tetrathiocarbonate (Enzone, which releases carbon disulphide in soil) reduce production of sporangia in soil by 90%. There are no lesions when infested soil is treated with 10 μg/ ml metalaxyl. A single application of metalaxyl or fosetyl-Al can provide protection to citrus from colonization by *P. citrophthora* or *P. parasitica* for 2-3 months. In Turkey, fosetyl-Al is reported to protect trees against gummosis for at least one year. Potassium phosphonate is a systemically translocated chemical effective against oomycetes. In *Arabidopsis thalliana* (a crucifer) the treated plants challenged with *Phytophthora palmivora* exhibit rapid cellular responses. There is increase in cytoplasmic activity, development of cytoplasmic aggregates, release of superoxide, localized cell death and enhanced accumulation of phenolic materials around the infected cells (Daniels and Guest, 2006).

Enzone is more effective than metalaxyl in eradicating *Phytophthora* from host debris in soil. Metalaxyl can penetrate and cause inhibition of inoculum up to a depth of 1.25 cm of soil. The density of fungal propagules in soil treated with Ridomil drench or Ridomil spray as well as drench is reduced by 69.1–72.8%. The population of the pathogens in relation to feeder roots density is also decreased. Drench and spray treatment with fosetyl-Al (Aliette) or Ridomil significantly increases the feeder root density.

Matheron and Porchas (2000) studied *in vitro* the impact of azoxystrobin, dimethomorph, fluazinam, fosetyl-Al and metalaxyl on growth, sporulation and zoospore cyst germination of *P. parasitica* and *P. citrophthopra*. Mycelial growth was least affected by azoxystrobin and fluazinam. Reduction of sporangium formation in presence of 1 mg/ml dimethomorph was significantly greater than in same concentration of the other four fungicides. Zoospore motility was most sensitive to fluazinam and least sensitive to fosetyl-Al. Germination of encysted zoospores was most sensitive to dimethomorph, less to fluazinam and metalaxyl and least to azoxystrobin and fosetyl-Al. They subsequently reported (Matheron and Porchas, 2002) their findings of *in vivo* studies of six fungicides in inhibiting development of gummosis of citrus. Dimethomorph, fosetyl-Al and metalaxyl reduced cankers caused by *P. citrophthora* on the trunk better than azoxystrobin or fluazinam. These three fungicides remained more effective than azoxystrobin, fluazinam or zoxamide 5, 30 or 60 days after treatment. Arora *et al.,* (2002) compared many non-systemic and systemic fungicides and found Ridomil MZ soil drench in combination with Aliette as spray best in controlling trunk lesions and reducing propagule density in soil.

For economic control, applications of fungicides should coincide with periods favorable for pathogen activity and disease development. Extremes in temperature (35°C at the max. end) are not favorable for the pathogen and at such periods fungicide application is not required. Application of metalaxyl or fosetyl-Al is beneficial when threshold level of the fungus reaches 10–15 propagules per cubic centimeter of soil. In general, the chemical treatments are required to be done before rains start because the pathogen multiplies rapidly during rains.

Although chemical control of citrus feeder root rot in the field with fosetyl-Al and metalaxyl is effective, it is expensive also. Over the years, emphasis has shifted to the production of nursery trees free of *Phytophthora* spp. by preventive phytosanitary methods such as soil fumigation, treated irrigation water and sound hygiene. Such healthy nursery trees grow consistently better than infected nursery trees. However, nursery plants that are initially certified disease-free and planted in virgin soil eventually become infected by irrigation water sources which are frequently contaminated by *Phytophthora* spp. and nematodes. The contaminated irrigation water supplements existing *Phytophthora* and nematode populations in the soil, making chemical control more difficult. Use of bleaching powder (chlorine) in irrigation water for nursery beds can reduce *Phytophthora* spp. on the planting stock. Chlorine kills the propagules of *Phytophthora parasitica*, *P. citrophthora* and *Fusarium* spp. Nursery bed soil treatment with composted municipal waste has been found beneficial. The beneficial effect depends on the source of the material. Some composts may be harmful because of presence of excess phytotoxic soluble salts and acids. In others, application at 20% (w/w) reduced the incidence of infection from 90%–5%. Storage of the compost reduced its efficacy. *Acremonium* sp. was one mycoparasite found in the compost that parasitized hyphae of *Phytophthora*. Mycorrhization of soil with the arbuscular mycorhizal fungus *Glomus mosseae* reduces the number of locations on roots where infection could occur. Application of sewage sludge can suppress *Phytophthora* in orchard soil and on citrus roots (Leoni and Ghini, 2005) Pre-inoculation of citrus rootstock with avirulent or weaklu pathogenic isolate of *Phytophthora nicotianae* provides protection against virulent *Phytophthora* spp through pre-colonization of the substrate. In irrigated orchards biological control of *P. parasitica* in citrus rhizosphere with an antagonistic strain of *Pseudomonas putida* is reported by Steddom *et al.*, (2002 a, b). However, the treatment is significantly effective only when the bacterium is sprinkler-applied to soil weekly. The antagonist could be distributed up to 75 cm depth of soil. Dipping roots of grafts for 6–10 min in water at 35°C or in 0.02% suspension of captan and soil fumigation with Vapam or Mylone had been routine practices in many citrus growing countries to reduce losses from gummosis and root rot.

Gaur *et al.*, (2002) have suggested a number of integrated approaches for control of the Phytophthora disease in mandarin cv Kinnow. These include (i) Ridomil MZ (1g/L water) spray twice, each during February and August along with stem painting with Ridomil MZ (20 g/L of linseed oil), (ii) Topsin-M (1g/L water) soil drench along with stem painting with Ridomil and (iii) soil application of potash (500 g/tree) or zinc sulphate (150 g/tree) in February and August along with stem painting with Ridomil.

They (Gaur *et al.*, 2004) later reported a modified schedule. Stem painting with metalaxyl (Ridomil MZ) @ 20 g/l linseed oil in asscoaition with soil drenching @ 1 g/l water each during February and August at 15 days interval reduced trunk lesion size and increased fruit production. Soil drenching with thiophanate methyl (0.1%) or carbendazim (0.1%) in association with stem painting were at par and next best. Calcium propionate and calcium lactate applied at 1200 ppm on 4 month old sour orange seedlings have reduced root rot and propagule density of the pathogens in soil (Campanella *et al.*, 2002).

Eradication of infection from standing trees is possible only if the disease is detected in the early stages. If infection is on thin branches they may be cut and burnt. In the early stages of root rot, affected roots may be removed and the soil drenched with a fungicide. On thick branches and trunks the infected portion may be removed by a sharp knife and the wound cleaned with 0.1% mercuric chloride or 1% potassium permanganate solution followed by application of Bordeaux paste on the wound. Compared to Bordeaux paste (1 part monohydrated copper sulphate + 2 parts hydrated lime dust + 3 parts linseed oil), Ridomil paint

(Ridomil MZ 2g/100 ml linseed oil) is more efficient. Jadeja *et al.,* (2000) observed good control of canker and gummosis by using Bordeaux paint and soil drenching with Ridomil MZ as well as fosetyl-Al alone or in combination with streptomycin (100 ppm) + 0.2% copper oxychloride spray.

Generally, the above mentioned chemical treatments reduce the chances of brown rot of fruit. Copper fungicides or captan applied prior to beginning of rains are usually quite effective. Pre-harvest application of systemic fungicides metalaxyl or fosetyl-Al to the canopy provides effective control of brown rot. Post-harvest disinfection of fruits with chlorine or sodium orthophenylphenate, recommended for canker-affected fruits, can also be helpful. Hot water treatment of grapefruit (48°C, 3 min), lemon (52°C, 5–10 min) and orange (53°C, 5 min) has been reported (Barkai-Golan and Phillips, 1991). Hot water treatment of certain citrus fruits has limitations. Although lemons could routinely tolerate immersion in water at 46.1°– 48.9°C for 4 min or longer without injury, release of rind oils leading to oleocellosis could occur if lemons were cold and turgid at the time of treatment. The immersion should, therefore, be delayed by 1–4 days after harvest to allow the rind to lose turgor. Without this conditioning the fruits can be injured even at 37.8°C. Using soap in post-treatment rinse of fruits entraps released oils and terpenoids to further reduce the chances of rind injury.

WHITE BLISTERS OR WHITE RUST OF CRUCIFERS

Species of the genus *Albugo* cause white blisters on aerial parts of a large number of cultivated and wild plant species which include a number of crucifer plants of economic importance such as turnip (*Brassica rapa* subsp. *rapa*), cauliflower (*Brassica oleracea* var. *botrytis*), cabbage (*B. oleracea* var. *capitata*), Chinese cabbage (*B. oleracea* subsp. *pekinensis*), radish (*Raphanus sativus*), mustard or sarson (*B. juncea* var. *juncea*), rape or lahi (*B. napus* var. *napus*), black mustard (*B. nigra*), Brussels sprouts (*B. oleracea* var. *gemmifera*) cress and taramira (*Eruca sativa*). All are hosts of *Albugo candida*. Non-crucifer crops affected by white rust include spinach (*Albugo occidentalis*) and sweet potato (*Albugo ipomoeae panduratae*). In India, white blister disease has been found on turnip, radish, *Eruca sativa*, different species of oil bearing *Brassica* and some weeds such as *Cleome viscosa*. Among the cultivated crops the oilseed crops like *Eruca* and *Brassica* suffer heavy losses due to distortion of floral organs. Combined infection of leaves and inflorescences may cause up to 60% yield loss in *Brassica juncea* (Indian mustard) In late crops the yield loss due to floral infection may be from 23 to 54.5%. Often the disease occurs in association with the downy mildew (*Hyaloperonospora parasitica*) and then the damage is considerable and requires prompt control measures.

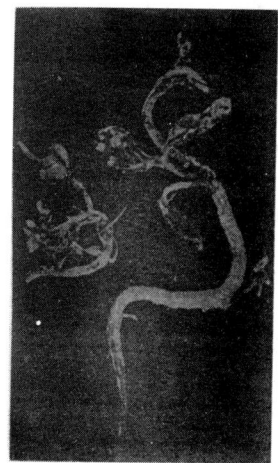

Fig. 33. Hypertrophy and blisters caused by *A. candida* on the infloresence of mustard plant.

Symptoms

The effects on the host plant result from two types of infection: local and systemic. All parts of the plant except roots may show symptoms. In local infections isolated pustules or sori develop on leaves and stems. These pustules or blisters are raised, shiny white areas, measuring 1-2 mm in diameter. They are very variable in shape and size and may arise in close proximity ultimately merging to form larger patches. Often they appear in a circular arrangement around one or two central pustules. The host epidermis ruptures early or sometimes after the pustules are fully formed. The exposed sori reveal a white powdery mass consisting of spores of the fungus. When young stems and inflorescence are infected, the fungus becomes systemic in the tissues and stimulates various types of deformities in which hypertrophy is most pronounced. Blisters may also be formed on the inflorescence and the floral parts. Due to hypertrophy and hyperplasia in the tissues the floral parts show swellings and distortions. The axis of the inflorescence and the flower stalk may be enormously thickened, up to 12–15 times the normal diameter, while the floral organs become wholly or in part swollen, fleshy, green or violet in colour, and persist instead of petals and stamens falling off early. The petals may become sepal-like and stamens become leaf-like or occasionally like carpels. The latter may be open while the ovules are usually atrophied as are also the pollen grains. This results in the sterile condition of the affected ovary. These swollen or hypertrophied parts are full of oospores of the fungus and the cells are rich in starch.

The leaves are not greatly modified on some hosts. However, those found on totally infected stems may be thickened, fleshy, pallid, and distorted or rolled. When the systemic infection occurs early, the entire plant may remain dwarfed and only small leaves may be formed. The swelling on the stems may be restricted in area or they may occupy the entire length. The axis of the inflorescence and the stem often are twisted giving a zigzag appearance. Stimulation of normally dormant buds occurs and they grow into lateral shoots.

The causal organism

Albugo candida (Lev.) Kunz, (Chromista, Oomycota, Oomycetes, Peronosporales, Albuginaceae) is an obligate or biotrophic parasite. The mycelium is intercellular producing knob-shaped haustoria in the host cells. Often a large number of these haustoria are seen in a single cell. They have no nucleus and appear to remain functional for a short time and then generally lose viability. Thus, the large number of haustoria seen in each cell represent different stage of their development.

Hyphae from the endophytic mycelium collect beneath the epidermis into which a large number of haustoria are produced. These dense masses of hyphae form the sporangial beds (sori). The sporangiophores are formed by the vertical growth of broad, short, clavate stalks under the epidermis which is raised up due to pressure of these structures. The raised epidermis is separated from the underlying cells. The sporangiophores are free from each other laterally and are very thick, especially at the base.

The sporangia are formed in basipetal succession in chains, the first formed being thick-walled and at the top closely pressing against the still unruptured epidermis. This spore is incapable of germination. Its only function appears to be to raise the epidermis and finally rupture it thus facilitating the dispersal of the rest of the sporangia. Pads of a gelatinous material are formed between successive sporangia and function as disjunctors. In presence of moisture these pads swell and disintegrate freeing the sporangia from the chains. The sporangia are hyaline, nearly spherical, and 14–16 × 16–20 μm in size. They germinate by means of a germ tube or by formation of zoospores. The conidial method of germination is not common. The germination of sporangia occurs readily when they are completely immersed in water. They can germinate only up to 6 weeks after their formation. The apex of the sporangium is drawn out to form a rounded papilla and a few large vacuoles appear in the protoplasm. After sometime, these become quite spherical. Then the entire mass segments into as many polyhedral cells as there are to be the zoospores. Each of these is provided with a contractile vacuole. The zoospores, 4-8 in number in each sporangium, separate from each other and

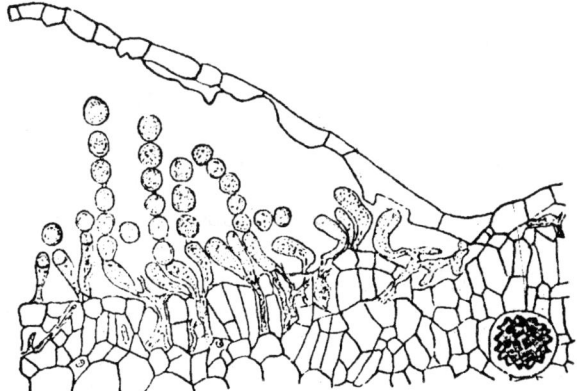

Fig. 34. *Albugo candida.* A sorus showing sporangiophores and chains of sporangia. The epidermis is raised and ruptured. A single oospore is present in the tissue on the right lower corner.

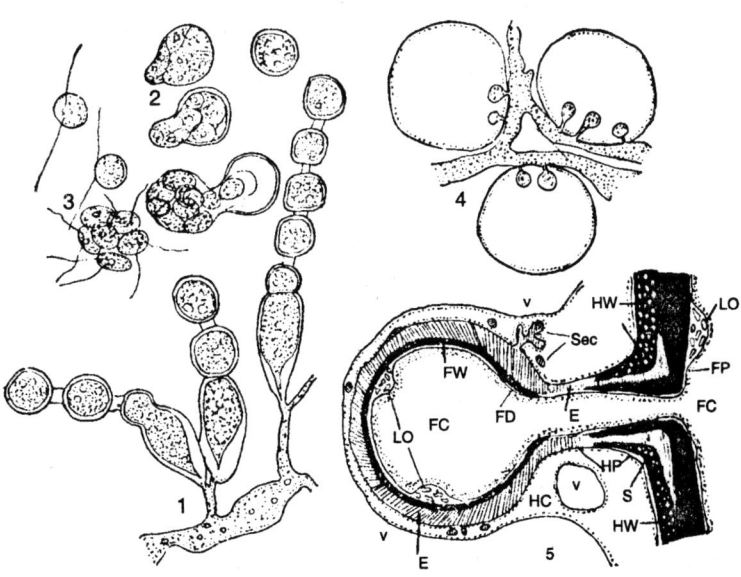

Fig. 35. *Albugo candida.*
1-Sporangiophore and sporangia; 23-Formation and release of biflagellate zoospores;
4-Intercellular hyphae and knob-shaped haustoria; 5-The haustorium as seen under electron microscope. No nucleus is present. FC-fungal cytoplasm; FP-fungal plasma membrane; LO-fungal lomasomes; FW-fungal cell wall; V-host cell vacuoles; HC-host cytoplasm; HP-host plasma membrane HW-host cell-well; S-collar-like sheath; E-encapsulation.

escape to outside, one by one or some times in adherent groups, or the whole mass may be emptied into a sort of bladder formed by the swelling of the papilla. Further developments leading to the germination of the zoospores are the same as in other members of the Peronosporales. The germination of sporangia is closely dependent upon the prevailing temperature, the maximum being 25°C. The optimum temperature for their germination is about 10°C at which the rate of germination and the number of zoospores formed are maximum.

The oogonia and antheridia are formed from the mycelium in the intercellular spaces of the host tissues, particularly in the systemically invaded plants. The oogonium is globose, terminal or intercalary, and consists of about 100 or more nuclei. The contents of the oogonium are clearly demarcated into a periplasm and a single central oosphere. All but one nuclei pass into the periplasm. The central oosphere contains a single female functional nucleus. The antheridium is clavate, paragynous, and contains 6–12 nuclei. At the point of contact with the oogonium, the wall becomes very thin and a papilla of the oogonium protrudes into the antheridium and soon disappears. The fertilization tube from the antheridium passes through the thin spot into the oosphere. The nuclei in the periplasm disintegrate and the nuclei in the antheridium and the oosphere undergo two mitotic divisions. A granular body (coenocentrum) appears in the oosphere and the single female functional nucleus is attached to a point near it. The fertilization tube penetrates the coenocentrum, ruptures and discharges a single male nucleus which fuses with the female nucleus. The fertilization tube collapses and the coenocentrum disappears. The oospore is formed by the development of a wall around the oosphere. This wall consists of a thin endospore and a thick, warty, tuberculate or roughened epispore. Germination of the oospore takes place after a resting period of several months. The epispore bursts and endospore is drawn out as a thin spherical vesicle which may be sessile or formed at the end of a wide cylindrical tube. The fusion nucleus divides rapidly to form about 30 or more nuclei. One of these divisions is reduction division. Within the vesicle 40–60 zoospores are differentiated and are released by rupture of the wall.

The fungus has many specialized races which attack only a particular set of host species. Biologic races 1 and 2 are reported from Kangra Valley in Himachal Pradesh. In North America, 11 pathotypes had been reported on species of *Brassica, Raphanus*, and other hosts. Two additional pathotypes AC 12 and AC 13 were reported from India. Nine of the 11 reported from North America exist in India also (Verma *et al.,* 1999). Gupta and Saharan (2002) have reported four new distinct pathotypes AC-14, AC-15, AC-16 and AC-17) on the basis of their differential interaction with 11 host differentials.

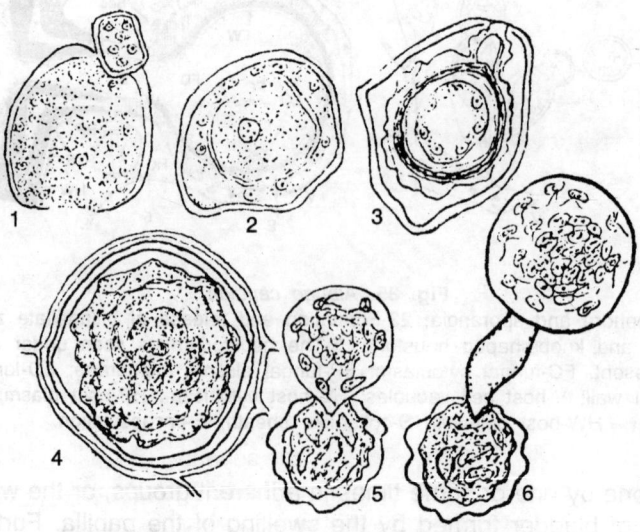

Fig. 36. *Albugo candida.* Sexual reproduction.
1-Oogonium and antheridium with fertilization tube. Female functional nucleus in the centre. Male nucleus at the tip of the tube about of enter; 2-Fusion nucleus in the center of oospore, oospore wall has begun to form; 3-Cross-section of mature oospore, remains of fertilization tube; 4-Mature oospore in host cell; 5-6-Germination of oospores by two methods.

Disease cycle and environmental relations

Albugo candida perpetuates through oospores lying in the soil in diseased plant debris or moving with diseased pieces along with the seed. Perennial weed hosts may also serve as source of primary inoculum. Some cruciferous weeds are known to carry systemic, persistent and nonsymptomatic infection (Jacobson *et al.*, 1998). Oospore production can occur at 10°–27°C, maximum being at 23°C. Earliest development of oospores is observed at 25°C, 6 days after inoculation and incubation and mature oospores are seen at 12 days. Secondary spread is carried out by sporangia and zoospores. Germination of encysted zoospores occurs 2-3 hr after inoculation. Infection is initiated with germ tubes penetrating through stomata. Haustorium formation is seen in the palisade and mesophyll cells adjacent to substomatal cavity 8 hours after inoculation. Up to this stage there is no difference between susceptible and resistant host. Compatible interaction between susceptible host and virulent isolate is characterized by the rapid spread of intercellular hyphae, a high frequency of haustorium formation in mesophyll cells and production of sporangia in the absence of any host cell necrosis. In partially compatible interaction, sporulation occurs but pathogen colonization is restricted. There is striking deposition of ensheathing material containing callose around the haustorium. In highly incompatible interactions, there is no sporulation. Necrotic flecks are seen. Severity of flecking reaction depends on the number of cells affected. There is restriction of fungal growth after penetration. It may grow into mesophyll cells around the penetration site (Soyulu *et al.*, 2004). Soylu (2004) conducted a transmission electron microscopic (TEM) study of the *A. candida-Arabidopsis* pathosystem and observed that the coenocytic hyphae from the substomatal vesicle ramified and spread intercellularly throughout the host tissue, forming haustoria inmesophyl cells. The host does not show any obvious response to this invasion and structural integrity of the host cytoplasm remains unchanged Moist, cool weather favors the disease development. A film of water is essential for sporangial germination. The germination starts at 4.6°C and ceases at 21°C with maximum germination at 12°–14°C. Germination is better in darkness, strong light inhibiting the germination. These conditions of temperature and light are available during night hours. The pustules progress at a faster rate when mean temperature was 11.5°–12.3°C, mean RH above 75%, weather was cloudy with precipitation, and wind velocity was 2.6 km/ h. Progress of leaf blisters is more rapid on susceptible cultivars the leaf blisters progressed at a faster rate on susceptible cultivars from 15th Dec. to 15th Feb (in north India). when the relative humidity is more than 65% and temperature less than 15°C along with little amount of rains. Staghead (floral infection) phase increased at a faster rate between 30 Jan. and 1 March when the relate humidity was more than 70% and temperatures 13°–15°C with 3.9–9.1 mm rains. Rains increased the humidity and lowered the temperature. Single irrigation at pod formation stage results in maximum white rust severity on leaves and on flowers. Reduced sunlight hours (2-6 hours per day) coupled with high total rainfall during flowering period of the crop favors severe incidence of staghead caused by *Albugo candida* as well as downy mildew (*Hyaloperonospora parasitica*). As a result of these environmental conditions for rapid development of the disease, date of sowing may influence incidence and severity of floral malformation. In some years early sowing of rapeseed (first week of October in the Tarai area of Uttarakhand) the crop has escaped the disease. Crops sown on 21st October recorded significantly less disease severity. The disease increased with the delay in sowing time. However. some cultivars may be equally susceptible at all dates of sowing (Gupta *et al.*, 2004). Kumar *et al.*, (1995) have also reported that crops sown late, after 19th October, showed increased incidence of the disease. In comparative study of white rust and downy mildew, Mehta and Saharan (1998) observed that there is more downy mildew infection in early sown crop while white rust infection and development significantly increased in late sown crops. Downy mildew infection started at end of October in early sown crops and progressed up to November. There was no downy mildew infection in crops planted after mid-November. White rust was maximum from last week of December to January. There was no staghead formation in crops sown during September. High incidence and severity of staghead was seen in October-November sown crops. During the season, weather conditions are more favorable for white rust than for downy mildew. According to Kumar et al., (1995) older leaves are more susceptible than younger leaves and symptoms appear earlier in inoculation of lower surface than of upper surface.

Management of the disease

Clean cultivation including use of clean seeds and destruction of weeds in and around the field should be given priority. Crop rotation helps in avoiding the soil-borne primary inoculum. In seed crops of vegetable crucifers and in oilseed crops chemical sprays may be necessary. The fungicides effective against the disease include 0.8% Bordeaux mixture, 0.3 Difolatan, 0.1%, Daconil, 0.2%, Dithane M-45 or 0.1% Ridomil at 8-10 days interval. Ridomil and Aliette (fosetyl-Al) as fungicides for white rust control were recommended. Seed treatment with metalaxyl and three sprays of chlorothalonil (0.1%) or mancozeb at 20, 40, and 60 days after sowing have also been suggested. It may be noted that white rust and downy mildew generally occur together and recommendations of chemical treatments should be combined for both disease.

Sindhan and Hooda (2004) have reported biological control of the staghead phase of the disease. Spray of antagonists (Penicillium citrinum, Aspergillus ochraceus, Bacillus subtilis, Pseudomonas fluorescence and extract of leaves of neem, onoo, garlic, ginger Datura metel provide sime suppression of white rust. Amon microbial antagonists Aspergillus ochraceus gave best result while among plant extracts neem leaf extract was the best. However, none of these treatments were as good as Ridomil.

All cultivars of sarson and toria of Brassica campestris are susceptible to individual or mass isolates of the pathogen. Li et al., (2007) evaluated 22 genotypes of B. juncea from India, 12 from Australia and 10 from Chnia for resistance to Albugo candida by inoculation of cotyledons, leaves and inflorescences/ While 4 genotypes from China and one fro Australia showed consistently high resistance, 3 denotypes from India showed high susceptibility. Dang et al., (2000) and Saharan (2000) have listed a number of species and genotypes in rapeseed-mustard having multiple disease resistance including resistance to white rust and downy mildew. These include Brassica alba, B. carinata (HC-1, PCC-2), Brassica juncea (DIR 1507, DIR 1522) and B. napus (GS-7-27, Midas, Tower). The resistance in some exotic varieties is controlled by a single dominant gene. White rust resistance loci on certain chromosomes may be linked to downy mildew resistance loci. Pre- or co-inoculation with avirulent strain of Albugo candida provides protection against a virulent strain. Pre- and co-inoculation with Peronospora parasitica suppresses the development of Albugo candida on Brassica juncea leaves.

GREEN EAR AND DOWNY MILDEW OF PEARL MILLET

The green ear is a common disease of pearl millet (bajra, Pennisetum typhoides) and occurs in India, Iran, Israel, China, Japan, Fiji, USA and many African countries. It has so far not been reported from South America and Australia. In India, the disease was first described by E.J. Butler in 1907. He considered it to be of sporadic nature, not causing much damage, except in low-lying fields where the loss could be significant. These observations were confirmed in later studies before introduction of hybrid pearl millet. Since the introduction of high yielding hybrid cultivars regular and severe incidence of the disease has shown that it is one of the most serious problems for this crop causing heavy losses. Loss estimates vary from 6–60%

in different countries. Up to 27-30% loss has been estimated in India Up to 60% grain losses have been reported in many African countries.

Symptoms

In systemic infections from soil- or seed-borne inoculum, the plants remain stunted in growth and pale yellow. This can be noticed even when the seedlings are quite young. Sooner or later, the leaves start showing chlorosis in streaks on the upper surface. Just below these streaks on the lower surface a fine downy growth of fungus may appear. Soon, the chlorotic areas turn brown and in advanced stages shredding of leaves along the veins occurs. Very often the nodal buds are stimulated to develop into lateral shoots on the stalk giving it a bunchy appearance. The principal symptoms of this downy mildew are produced in the inflorescence. Abnormalities occur in the ear at the heading stage and it is then that the disease is most readily recognized. The ear deformities are characterized by the transformation of floral organs into twisted leafy structures. This gives the ear an appearance of green leafy mass hence the name 'green ear'. Three types of transformed earheads can be found. The cob length may remain normal but the entire earhead is converted into a leafy mass. The cob length remains normal but only the lower part is converted into a leafy mass and the upper portion bears grains. In the third type, the development of the cob is altogether suppressed and in its place a bunch of small leafy structures is formed.

Fig. 37. Green ear disease of pearl millet.

The bristles of the spikelets become hypertrophied and variously contorted. They may be 2.5 cm long and about 1.5 mm thick and may be flattened, round or angular, the upper part being often twisted or corkscrew-like. Instead of single spikelet two may be formed on the same pedicel. An increase in the number of florets in a spikelet is not uncommon. The stamens become leaf-like or may be elongated to a length of over 1.25 cm or may be atrophied to such an extent that the anther looks sessile. It may remain capillary or become flattened and plumose particularly at the base. The anther may remain unaltered on the modified filament, or it may be suppressed, elongated plumose, sometimes contorted. The whole stamen may be changed to a leafy structure, color-less or brownish, often indicating a division into a sheath-like and blade-like portion. When thus flattened, the middle if often marked by a distinct vein. Sometimes the stamens appear as brown, leathery, pointed bodies with no differentiation into filament and anther, and almost circular in transverse section.

The central portion of the floret is usually elongated into a leafy shoot, consisting of reduced foliage leaves which are much twisted and malformed. The proliferation occurs chiefly in the upper perfect floret of the spikelet, but is accompanied usually by a similar, though less marked, change of lower male or sterile floret. These leafy structures may be up to 7.5 cm in length. Usually a well marked division into sheath and blade occurs, the sheath being more developed than the blade which is imperfectly formed and much in-rolled. The colour may be green or brown. Longitudinal shredding of these leaf-like structures is also common. The pistil rarely develops in severely attacked plants and is usually replaced by small, leafy shoots or horn-like outgrowths.

The leaves of affected plants show considerable changes. In young plants or in early stage of disease development many leaves may be seen with the usual green changed wholly of in part to whitish, and later brown. The whitening of young leaves is seen as streaks often occupying half or more of the lamina surface, extending almost the whole length of the leaf blade. In older plants, the leaves thus affected are chiefly those from whose axil the inflorescence comes out. These are more completely whitened than the leaves of younger plants. The colour rapidly changes to brown and twisting, folding, and shredding towards the tip may occur. At the time when green ears are fully formed the upper leaves are mostly brown and many of them shredded. The cob development if checked is replaced by a small mass of white or brown, twisted leaves enclosed within the contorted outer leaf sheath. The side branches produced on the main stalk bear leafy masses instead of earheads.

The causal organism

Sclerospora graminicola (Sacc) Schroet (Chromista, Oomycota, Oomycetes, Scleroporales, Scleroporaceae). The fungus is a biotroph and genetically highly divergent. Heterothallism was reported in 1982. Compatible mating types are required for sexual reproduction. These are almost equally distributed in dfferent geographic regions regions hence oospore formation occurs everwhere. The mycelium is found in all parts of the systemically affected plants. The hyphae are intercellular with small, bulbous haustoria. They are large, aseptate, and up to 10 μm in diameter. The width of hyphae varies according to space available. These hyphae are found in the ground tissues of the stem and mesophyll of the leaves. Hyphae may penetrate the bundles in the leaf but the xylem and phloem are not penetrated. In between the cells the hyphae branch freely. They have a thick, gelatinous cell-wall and clear protoplasm. In the stem the haustoria are not fully formed and are usually button-shaped. Their full development occurs in the leaf tissues where they are numerous and simple or branched, finger-like, sometimes convoluted, and occupy a major portion of the cell cavity.

In completely infected leaves the hyphae collect mainly between the cells of the mesophyll adjoining the bundles and also in the inner layers between these. When sporangiophores are to be formed, tufts of hyphae reach the air space beneath the stomata which are arranged in parallel rows, one on each side of the vein. Prior to this, the chloroplasts of the leaf cells are wholly or partially destroyed giving a chlorotic appearance to the leaves showing the downy mildew stage of the fungus.

The sporangiophores emerge in clusters through the stomata. Each sporangiophore is a broad, short hypha measuring about 100 microns in length and 12–15 μm in width. It is unbranched in the lower part but usually gives or 2-6 thick, short branches di- or trichotomously at the tip. The sporangiophores are expanded to a diameter of about 30 μm towards the tip where branching occurs. The final branches or the sterigmata are slender, tapering towards the apex, and bear sporangia which are formed by swelling of the tip of the sterigmata and cut off by a septum. These sporangia are hyaline, broadly elliptical, sometimes broadly cylindrical, slightly pointed at the free end with a thin smooth wall. Their size varies on different hosts. On pearl millet they measure 19–31×12–21 μm. On *Setaria viridis* they measure 16–22 x

12–16 μm and on *Setaria magna* 13–36.9 × 11–24.9 μm. Mature sporangia germinate by producing zoospores. Three to 23 zoospores are formed in each sporangium. The zoospores are irregularly reniform, biflagellate, and swim about for some time before coming to rest. The resting zoospores measure 9–12 μm in diameter. They germinate by a germ tube which shows chemotactic response to roots of the host.

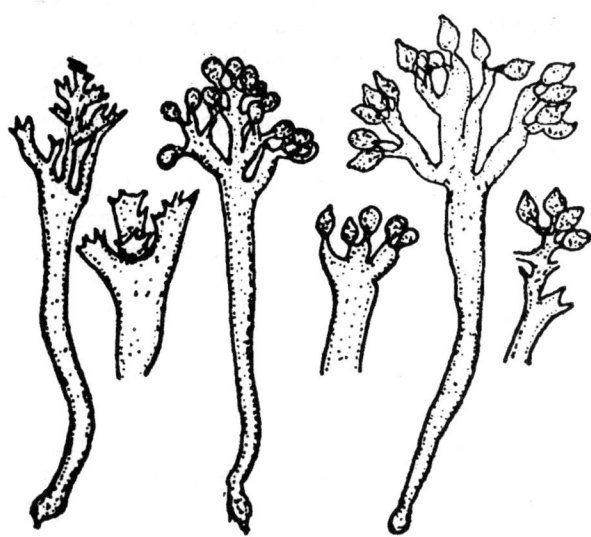

Fig. 38. *Sclerospora graminicola.* Stages in the development of sporangiophore.
(after Weston, 1928).

The sex organs develop within the host tissues, mostly in leaves and malformed floral organs. The oogonia are usually terminal, occasionally intercalary. A mass of protoplasm with about 50 nuclei enters the oogonium from the mycelium. After the oogonium has expanded it is cut off by a septum. The nuclei are at first shriveled in appearance but may enlarge and, as they approach metaphase, become oriented in the region which is to become the boundary between the ooplasm and periplasm. The nuclei undergo a meiotic division. A coenocentrum develops near which a single nucleus is found. The nucleus divides once and one daughter nucleus serves as the female gamete. The antheridium contains 3–4 nuclei and these also undergo meiosis slightly ahead of the division in the oogonium. The fertilization tube penetrates the oogonial wall and passes rapidly to the oosphere where it discharges one nucleus and quickly disintegrates. The coenocentrum also disappears. The male and female nuclei (gametes) remain apart for some time and then unite. Differentiation of the oospore wall follows by secretions from the ooplasm. A thin outer layer make up the exospore and a thicker layer forms the endospore. The original oogonial wall shrinks to touch the new oosporial wall at many points giving the entire oospore an elliptical, angular or irregular shape although the oospore proper is usually spherical. Under the high power of a microscope the mature oospore shows three walls, the exosporium, mesosporium and endosporium. The endosporium is smooth, yellow in colour and of even thickness.

Exosporium is tawny in appearance and deeper coloured. The mature oospores measure 34–52 (average 42) μm in diameter. Singh, S.D. (1995) has given the size as 22–35 (average 32) μm. The oospore proper is 35 μm in diameter. These oospores have prolonged dormancy. Germination occurs by means of 1–4 germ tubes. However, this part of the life cycle is not very clear. Intercontinental variation among isolates of *S. graminicola* are known.

Disease cycle and environmental relations

The inoculum of downy mildew fungus is primarily soil-borne. Oospores, formed abundantly in diseased leaves, fall down on the ground with the debris and perennate during winter. They retain their infectivity better under field conditions than under laboratory conditions. Longevity of oospores is reported to vary from 8 months to 10 years by different workers who studied this aspect under different sets of conditions and with oospores from different sources. Oospores can retain viability in laboratory conditions (15°–35°C) for 14 years. However germination is only 5% in 14 years old oospores. Oospores stored for 4–5 years give up to 60% germination. They require weathering for good germination and infection of the host. One year old oospores cause more infection than fresh or more than one year old oospores. Favorable conditions for germination of oospores in soil are: abundant supply of air heavily charged with oxygen. a low soil moisture content, and a temperature between 20° and 25°C. Time taken for germination is reported to be 24 hours to 6 days by different workers. Wild species of *Panicum* also probably play a role in perennation of the pathogen and its variability. *P. violaceum, P. millissinum, P. purpureum, P. pedicellatum* and *P. polystachyon* are susceptible wild / collateral hosts. *P. schweinfurtii* is resistant to downy mildew and rust and is cross compatible with pearl millet (Singh and Navi, 2000).

On germination the oospores cause infection of seedlings around 9 days after the crop is sown. The pathogen invades the host tissues from the underground parts systemically. The susceptibility of roots and underground portion of the stem decreases with advancing age of the host. Dry soil is more conducive to infection than wet soil. In wet soils seedlings emerge rapidly and thus escape infection. The minimum, optimum and maximum temperature for infection are 11°, 20° and 34°C, respectively. Incidence and severity of the disease is favored by inorganic nitrogen and phosphorus fertilizers.

In addition to soil-borne primary inoculum, seed-borne inoculum is also important. In 1918, E.J. Butler had stated that oospores of *S. graminicola* may be carried with seed during harvesting and threshing. He did not see any mycelium or oospores in the sound seeds collected from partially affected cobs. However, in later studies oospores were seen on the seed and such seeds when planted in sterilized soil produced diseased plants. Oospores carried with seed and present in soil play the main role in establishment and spread of the disease in a field, Surface sterilized seed planted in sterilized soil gave disease free seedlings. Mycelium is also present in the seeds collected from partially affected earheads. Such seeds produced diseased plants. Seed treatment with the systemic fungicide Ridomil at the rate of 0.2% active ingredient for 8 hours resulted in production of disease free plants from such seeds in sterilized soil. The percentage of embryos of seeds of Bajra Hybrid HB-4 showing presence of mycelia was reported to vary from 9.5 to 13.6. The hyphae in the seed were described as inter- and intracellular, aseptate, branched, thick and hyaline. By planting such seeds they could reproduce the disease. About 6-20% of oospores present on stored seeds are viable. Percentage of seeds with internal mycelium varies from 0.1–7.5. Even seeds from healthy plants may be contaminated. Studies conducted in Rajasthan have shown hat maximum incidence of the disease occurs in fields with soil containing oospores and planted with contaminated or infected seed followed by fields in which soil was infested but healthy seeds were planted or when the inoculum was present only on the seed.

Inoculum for secondary spread of the pathogen is believed to consist of sporangia formed on plants that contract infection from soil or seed. Formation of sporangia is favored by a temperature range of 15–25°C. No sporangial stage develops at temperatures above 28°C or 30°C. Although relative humidity above 75% is considered necessary for the stimulation of sporangial production, in general a saturated atmosphere with a film of water on leaf surface is best. Optimum combination of temperature and relative humidity for sporangial production occurs during August and September in north India and the disease is most commonly seen during these months. In arid climate of western Rajasthan (India) shading with netlon net caused drastic reduction in solar radiation and increased relative humidity which favored the disease.

Asexual sporulation of S. graminicola only during the night hours is because of the occurrence of suitable combinations of temperature and moisture for sporulation during night. If suitable combinations can be created artificially sporangial production can occur at any time. Important factors regarding the sporangial stage are (i) an interval of 15–20 hours between successive crops of sporangia to allow organization of new primordia of sporangiophores beneath the stomata, (ii) moisture film on leaf surface, and (iii) temperature around 25°C. By creating these conditions more than 35 thousand sporangia can be produce from one sq. cm leaf area in a single sporangial crop and as many as 11 crops of sporangia can occur on successive days. Although the process of sporulation is complete in 6 hours, high relative humidity (95–100%) is essential only during the last 3 hours. Maximum sporulation occurs when infected leaves are incubated for 6–12 hours at 30°C prior to exposure to high relative humidity. Sporulation occurs at 10°–30°C with optimum at 20°C. The longest sporangiophores are produced at 15°C and largest sporangia at 25°C. Sporangia remain viable for 6 hours at 20–30°C.

Mature sporangia fall off soon. Dispersal of sporangia takes place by means of wind, water (including flowing water in the field), and insects. On the host surface they germinate in the surrounding film of water by producing zoospores. Germination occurs within 30 min at 20–25°C. Adequate moisture, air, and temperatures of 12.5–29°C have been found to permit zoospore liberation from sporangia. Light has no effect on sporangial germination.

The role of sporangia in secondary spread of the disease has been doubted by some workers. Lack of evidence for secondary spread of the disease under field conditions was reported in 1966. Similar observations were reported from India. Sporangia of the fungus germinate early in the morning when there is enough dew on the leaf surface. After 7 a.m. when dew dries, germination stops. Thus, it appears that secondary spread of the disease in the field is limited. However, in a later study of the role of sporangia in epidemiology of the disease, sporangia were found to be abundantly produced on rainy days on systemically infected plants, provide secondary inoculum for spread of the disease. Sporangia falling on the ground liberate zoospores which move against gravity and remain viable and infective for 5 hours. They could serve as source of secondary inoculum under field conditions.

Although in artificial inoculations zoospores are capable of infecting the host through roots, coleoptile, and apical leaf whorl, symptoms develop early in whorl inoculation. Root infection occurs through root hair or epidermal cells in the zone of elongation. Zoospore inoculation of florets shows progress of mycelium in the style, ovary walls, endosperm and embryo. In the style several hyphae fuse to form plasmodium-like structures from which hyphae develop and pass through the style to infect the ovary. Carpels remain susceptible for 4 days after pollination. Such ears show poor seed setting.

Transplant crops suffer less than direct sown crops. Young plants are more susceptible than older plants. The seedlings are very susceptible up to 9–15 days while infection is least in 25 days old seedlings. Pre-treatment of seedlings with a suboptimal concentration of inoculum of the downy mildew pathogen induces resistance that protects the plant (62%) from subsequent infection by the optimum concentration of the pathogen (Kumar et al., 1998). The resistance is associated with increase of beta-1,3-glucanase and peroxidase activity in the plant.

S. graminicola can attack maize, sorghum and sugarcane also. As stated earlier, it has a number of hosts in other tribes of the family Gramineae. Physiologic races of the fungus also exist (Thakur *et al.,* 1999). Some races attack pearl millet while others attack species of *Setaria.* Two pathotypes were identified by Shetty and Ahmed (1981). Variation among single zoospore isolates of the pathogen in respect of latent period, disease incidence and virulence index (incidence × latent period) is reported by Thakur *et al.,* (1998).

Management of the disease

The downy mildew of pearl millet can be kept under check through an integration of several measures, such as choice of a resistant cultivar, crop rotation, chemical treatment of seed, and foliar spray of fungicides. The enzymes polygalacturonase (PG) and B-1,3-glucanase are involved in host biochemical resistance expression. Activities of PG and cellulase increase in leaves of susceptible cultivars while they decline in resistant cultivars (Wadhwa *et al.,* 2002). Lignin and callose deposition are host structural resistance response. In resistant cultivars, accumulation of these polymers is rapid and localized around the pathogen to restrict its entry. In susceptible cultivars the deposition takes long time (Kumudini and Shetty, 2002). According to Geetha and Shetty (2002) host resistance is correlated with oxidative burst in response to pathogenesis. Production of active oxygen species is more intense in incompatible (resistant) interactions. Kumudini *et al.,* (2001) had reported that hypersensitive response, cell death and release of hydrogen peroxide are measure of host defense responses. In susceptible cultivars, treatment with L-methionin causes significant increase in resistance genes and the induced resistance is comparable with a resistant cultivar. (Sarosh *et al.,* 2005).

Although intercontinental variations among isolates of S. graminicola are known, good sources of resistance have been located in many germplasm accessions (mostly from West Africa) and have been used for developing high yielding varieties with total resistance, recovery resistance in which the plant shakes off the pathogen after some time and becomes disease free, and stable resistance to the disease (Singh, 1995). The West African Accessions used were :

Stable resistance: ICML 12 to 16, P 310-17, P 1449.
Recovery resistance: ICMA 1, SD 503, P 1449 and ICMA 841
Complete resistance: IP 18292, IP 18298

In a collaborative international study at 17 locations in India and Africa 11 pearl millet lines were tested for stable resistance to downy mildew. Of these, lines IP 18292, IP 18293, 700651 and P310-17 showed high degree of stable resistance. Resistance in 700651 and P310-17 was highly stable across locations and over years (Thakur *et al.,* 2004).

Open pollinated varieties WC-C 75 and ICTP 8203 have shown durable resistance in India. ICMH 451 and Pusa 23 have also remained free from downy mildew for several years. The open pollinated cultivars Mallikarjun and ICTP are resistant to most pathotypes of the pearl millet isolate (Thakur *et al.,* 1999). Hybrid ICMH 88088 produced by ICRISAT has high level of downy mildew resistance and gives better yield than all other available cultivars. Cultivars HB-5, NHB-10, and NHB-14 are considered resistant to downy mildew. Variety NHB-3 is susceptible although once reported as resistant from Durgapur (Rajasthan). The resistant reaction was due to loss of pathogenicity of the isolate used in the test. The genotype DMRP 202 possesses total resistance through a single dominant gene Rsg (1) and because of the simple mechanism it is not difficult to transfer the resistance to other genotypes (Singh and Talukdar, 1998).

Coleoptile region of pearl millet is the most susceptible site for attack of *S. graminicola.* This region shows presence of some lytic factors that can lyse hyphae of the pathogen.

Significantly higher levels of lytic factors are measured in the coleoptile region of resistant cultivars than in susceptible cultivars. The level of lytic activity correlates with the degree of resistance in field screening (Umesha *et al.*, 2000). Resistant pearl millet seedlings exhibit 2.4 fold increase in lipoxygenase activity as a result of infection. The effect becomes detactable from 4 hours after inoculation and reaches maximum at 18 hours after inoculation (Babitha, 2002; Babitha *et al.*, 2004). Levels of ß-1,3-glucanase in the seedlings determines the level of resistance of a pearl millet genotype to downy mildew. According to Geetha *et al.*, (2005) resistance of pearl millet to downy mildew is correlated with phenylammonia lyase activity. Highest enzyme activity is detected in 4 hours in growing point of resistant seedlings. Higher activity of the enzymes in highly resistant genotype and lower activity in the highly susceptible genotype can be used for screening for resistance (Ramachandra Kini *et al.*, 2000). Accumulation of polyphemol oxidase is implicated in resistance response. The enzyme accumulates more rapidly and in greater amount in seedlings of resistant genotypes than in susceptible genotypes (Niranjan Raj *et al.*, 2006). Shivakumar *et al.*, (2000) have reported ribonucloease activity in homogenates of seedlings of cultivars with different levels of resistance. Inoculation of seedlings with *S. graminicola* reduced Rnase activity by 4-13% in compatible interactions while in incompatible interactions the enzyme activity increased by 10–27%. Somaclonal variants in callus cultures show different degrees of susceptibility and resistant. Some somaclones of a highly resistant variety have shown increased level of resistance. Umesha *et al.*, (1999) examined 201 lines regenerated from seed-derived callus in tissue culture and found only 3 somaclonal variants which consistently proved highly resistant. Seventeen variants were resistant.

Repetitive cultivation of a hybrid in a given area should be avoided. Several hybrids should be grown but withdrawn after 2-3 years. They may be re-introduced after some years. This prevents development of new races and prolongs the resistance in the hybrids.

Part of the infection which is carried with seed can be controlled by seed treatment. In the past, organo-mercurials (now not in use) and thiram had been in use. Seed treatment with 0.1% Agrosan GN + 0.4% thiram used to give about 50% reduction in disease incidence. Among more recent fungicides, metalaxyl compounds (Ridomil and Apron) have been found very effective. Treatment of seed with Ridomil (8 g/kg seed) followed by one spray of 0.1% Ridomil 20 days after planting was recommended as early as 1981. In a study, seed dressing with Apron SD-35 (8 g/kg seed) followed by a foliar spray, 25 days after sowing, with Ridomil-MZ-72 WP (metalaxyl + mancozeb) at 500 ppm concentration or with Ridomil-ZM-280 FW (metalaxyl + ziram) at 1000 ppm concentration was found the best combination. Seed treatment protected the crop up to 20 days. Gupta and Verma (1991) found that seed treatment with Apron SD-35 (2.5 g a.i./kg seed) controlled the disease up to 30 days after sowing. Seed treatment followed by one spray of Ridomil ZM at 23 days after sowing gave the best results with 77.2% disease reduction and 100% increase in yield. Harvest time residues of metalaxyl and ziram were practically nil. Gupta (2001) has reported similar observations. Seed treatment with Apron and two sprays of Ridomil MZ at 20 and 40 days after sowing were effective against the disease. The efficacy of seed treatment followed by an early foliar spray is further confirmed in a recent report by Mani and Heziba (2007). They could reduce the disease incidence from 66%–4.6% by using seed treatment with Apron (2 g/kg seed) followed by one spray of Ridomil (0.6%) 20 days after sowing. Response of pearl millet to metalaxyl can vary with cultivars. The fungicide cyazofamid causes significant reduction in sporangia formation, zoospore release and motility of zoospores. Seed treatment with 0.01–2 mg/ml or foliar spray of 1–10 mg/ml is not phytotoxic under glasshouse conditions. However, seed treatment gives only 19.7% reduction in disease but seed treatment followed by a single foliar spray gives good disease control. Increasing the number of foliar sprays can give more than 90% disease control (Sudisha *et al.*, 2007).

Variable levels of decrease in pearl millet downy mildew by application of dipotassium hydrogen phosphate, 2,3,5-tri-iodo benzoic acid and phosphorous acid (Akomon-40 or Potphos) through seed treatment or as spray is reported by Chaluvaraju *et al.*, (2004). The chemicals acted as growth stimulants by improving seed germination and seedling vigor. Combination of foliar application of phosphorous acid with addition of compost to soil and reduced dosage of metalaxyl was effective against the downy mildew and increased yield in field trials. Sharath Chandra *et al.*, (2004) have reported that combined seed treatment and foliar spray of 7 days old seedlings with elexa, a commercially developed aqueous chitosan formulation, gives 69% protection against downy mildew and reduces severity by 23%. The control is attributed to induction of systemic resistance.

Biological control of pearl millet downy mildew was reported by Umesha *et al.*, (1998). Seed treatment with pure culture or talc-based formulation of a strain of *Pseudomonas fluorescens* improved seedling vigour and suppressed sporulation of the pathogen. Foliar application also suppressed the disease but seed treatment gave better result. Efficacy of the biocontrol was significantly higher when seed treatment was followed by a foliar application. Strains of plant growth promoting rhizobacteria not only promote plant growth but also induce systemic resistance against downy mildew. Niranjan Raj *et al.*, (2003a) have reported that strains of *Bacillus pumilus, Bacillus subtilis* and three other rhizobacteria applied through seed or soil treatment significantly improved seed germination and plant vigour. The incidence of downy mildew was reduced from 43 to 58% although the control was not better than Apron (metalaxyl) treatment. In evaluation of different formulations, Niranjan Raj *et al.*, (2003b) noted that formulations that promoted best growth were the most effective inducers of resistance. Since these bacteria are not affected by metalaxyl their use in seed treatment with lower doses of metalaxyl may be an economical strategy. Both fresh suspension and talc based powder formulations are effective. Seed bio-priming with isolates of *Pseudomonas fluorescens* also enhanced growth of pearl millet plants and induces resistance to downy mildew. The treatment enhances seed germination, seedling vigour, plant height, leaf area, tillering capacity and seed weight per 1000 seeds and total yield. The time required for flowering was advanced by 5 days (Niranjan Raj *et al.*, 2004). They (Niranjan Raj *et al.*, 2005) have reported that a talc based combination product (Trichoshield) containing *Trichoderma harzianum, T. lignorum, Gliocladium virens* and *Bacillus subtilis* provided greater protection than individual antagonist species. The treatment resulted in better seed germination, seedling vigor and earlier flowering. Although Trichoshield provided 52–71% protection against downy mildew, it was not better than metalaxyl. Similar observations are reported by Mani and Hepziba (2007).

They used high level of inoculum of a talc-based formulation of *Pseudomonas fluorescens* for seed treatment. It gave good control, 9.5% incidence compared to 53, 53.3% in check but was inferior to to seed treatment combined with foliar spray of metalaxyl Seed treatment with *Pseudomonas fluoprescens* has reduced disease incidence from 66% to 10.47% and with *Trichoderma viride* to 17.15%). Foliar spray of culture filtrate of *Fusarium longipes* (1000 conidia/ml) 20 days after sowing also reduced disease incidence to 29.76% (Mani and Hepziba, 2008). Strains of *Acetobacter* and *Azospirillum* are al;so reported to provide protection but not as good as *Pseudomonas fluorescens*.

There are reports that seed treatment with crude extract of *Vinca rosea, Oscimum sanctum, Allium sativum, Parthenium hysterophorus, Datura stramonium, Datura metel, Azadirachta indica* and *Thuja sinensis* reduces the incidence of pearl millet down mildew and increases grain yield. *In vitro* antisporulant activity of leaf extracts of many plants is reported by Deepak *et al.,* (2005). Geetha and Shetty (2002) evaluated pear millet seed treatment with 0.75% benzothiadiazole (BTH), 90 mm calcium chloride and 1.0 mm hydrogen peroxide as plant activators for induction of resistance. All the three were efficient in managing the disease giving 70%, 66% and 59% protection, respectively. Hydrogen peroxide treatment enhanced plant growth better than the other two. Seed treatment with the plant activator ß-aminobutyric acid (BABA) improves seedling vigor and protects seedlings from infection by *S. graminicola.* Seeds treated with 50 mm BABA for 6 hours results in 23% disease incidence compared to 98% in untreated check (Shailasree *et al.,* 2001). The BABA-induced resistance is associated with accumulation of defence-related proteins (Shailashree *et al.,* 2007). Unsaturated fatty acids are newly discovered inducers of resistance in plants. Many of these unsaturated fatty acids have been detected in zoospores of *Sclerospora graminicola.* Seed treatment with many of these acids, atleast 4 days before inoculation, have provided resistance (Amruthesh *et al.,* 2005). Pushpalatha *et al.,* (2007) have reported induction of resistance, enhanced plant growth and disease suppression by seed treatment with vitamins. They tested vitamins pyridoxin, folic acid, riboflavin, niacin, D-biotin and menadione sodium bisulphate (MSB). A 6 hour seed-soak treatment with vitamins at 20 mm enhanced germination and seedling vigor and induced downy mildew resistance. MSB treatment offered 73% protection while niacin and riboflavin gave 63% and 62% protection, respectively. No additional benefit was given by combination of the vitamins. In greenhouse studies, Deepal *et al.,* (2007) have reported efficacy of exogenous application of phytohormones in induction of resistance and control of downy mildew. They applied a synthetic jasmonate analog as foliar spray and/or seed treatment. Seed treatment alone provided 50% protection. Combination of seed treatment and foliar spray gave 60% reduction of disease incidence. The resistance induction was attributed to enhanced activities of defense related proteins (enzymes) *viz.,* phenylalanine ammonia lyase, peroxidase and enhamnced level of hydroxiproline-rich glycoproteis. The results could be duplicated under field conditions also. Rogueing of diseased plants within a month of sowing followed by mancozeb spray is also an effective control measure. Crop rotation is also important. Intercropping with mungbean is reported to reduce the incidence of the disease.

DOWNY MILDEW OF SORGHUM

Throughout the peninsular India, downy mildew of sorghum is a common disease. It is rare in north India. The damage depends on the environments and time of infection. The plants may be badly damaged before their full development or, if fully developed, may remain sterile. In a genotype with 100% infection the loss in yield is 74%. Strains of the pathogen of downy mildew of sorghum occur on maize and some grasses also. The sorghum strain causes more downy mildew on sorghum and Johnson grass than on maize.

Symptoms

The symptoms of the disease are almost similar to those of downy mildew of pearl millet. The green ear stage, though reported, is not common. The time of appearance and the effects of the disease depend upon the type of infection. In primary or systemic infection the disease appears in the seedlings soon after their emergence. The affected seedlings have pale yellow, narrow leaves covered with a fine downy growth consisting of the conidial stage of the fungus. The growth may appear on both surfaces of the leaf but is more common on the lower surface. These seedlings continue growth and when about 5-6 weeks old white streaks appear on both surfaces of the upper leaves of the plant. Due to formation of oospores and resulting necrosis of tissues these streaks turn brown. They then tear apart along the veins causing shredding of leaves. This shredding may result in separation of all the tissues from the midrib. In maize leaf shredding is not much. The plants remain stunted and sterile, no ears being]formed. Only individual plants may be affected or the disease may appear on a number of plants in groups or patches scattered in the field which is the result of secondary infection.

The plants escaping systemic infection from the primary inoculum (oospores in soil and infected seeds) may be attacked by secondary or local infection. In this case the first symptoms are noticed when the plants are about 2 months old. The top leaves turn whitish as also do the base of the lower leaves. Irregular brown or yellow streaks then appear and oospores develop in large numbers in the tissues. On the lower leaves, pale yellow patches, bearing the conidial stage may be formed. While considerable shredding of leaves is the rule, in some plants it may not take place and oospores may not be formed. If ears are formed on affected plants they bear only undersized grains.

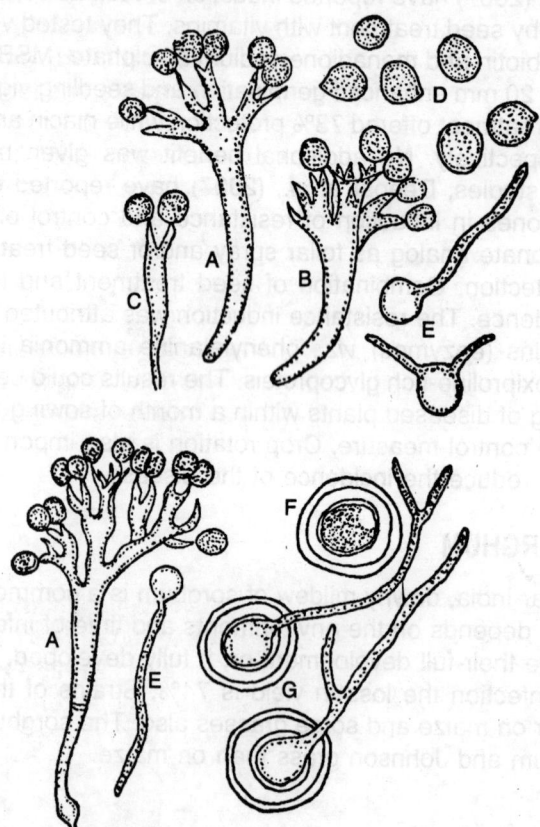

Fig. 41. *Peronosclerospora sorghi.* A-C-Conidiophores with conidia; D-E-Conidia and their germination; F-G-Oospores and their germination.

The causal organism

The sorghum downy mildew is caused by *Peronosclerospora sorghi* (Ito) Shaw (Sclerosporaceae) The mycelial characters of the species do not much differ from those of *Sclerospora graminicola*. The sterigmata are longer than those of *S. graminicola*, being up to

16 μm as against 8 μm in the pearl millet downy mildew fungus. Major difference lies with the sporangia which always germinate by a germ tube (conidia-like), not by producing zoospores. They are more commonly spherical instead of oval (as in *S. graminicola*) and lack the apical papilla. The size of these conidia is also smaller, being 15–29 μm in diameter. The oospore characters are similar to those of *S. graminicola*. The sorghum downy mildew fungus infects maize and Johnson grass and sporulates on these collateral hosts.

Disease cycle and environmental relations

The fungus is soil-borne through its oospores as well as seed-borne. Oospores in plant residue could maintain infectivity for 15 months when exposed to field conditions but when stored in laboratory they become innocuous. Obviously, like in *S. graminicola*, weathering helps in viability and germination of oospores of *Peronosclerospora sorghi*. Since the leaves typically undergo shredding, releasing oospores in the air, wind could play a major role in the dispersal of oospores, perhaps over long distances. Hay stacks and shady areas near sorghum fields serve as source of viable inoculum. The fungus attacks maize and teosinte but not pearl millet and *Setaria*. Some weed species of *Sorghum* commonly found in and around sorghum fields are also infected and the soil-borne inoculum (oospores) can multiply through such weed hosts. The infection from oospores is initiated in the underground parts of the plant from where it systemically spreads upward.

Seed-borne nature of *Peronosclerospora sorghi* in sorghum and maize is well documented. The presence of oospores of the fungus in glumes and pericarp of sorghum seed was reported in 1976. Such seeds were considered responsible for spread of the disease. In maize also the fungus is present as oospores and mycelium in the pericarp, endosperm, and embryo of seeds from systemically infected plants. The fungus becomes established in the sorghum seed either directly from the mother plant or by conidial infection through stigma, style, or ovary during the secondary spread.

Secondary spread of the disease is by conidia dispersed by wind, water and insects. Germ tubes from conidia are attracted equally by roots of host and non host plant as well as by resistant and susceptible cultivars of sorghum and maize. The conidia also cause local nodal infections in which nodal tillers grow in to diseased shoots with symptoms on leaves while the main plant remains free from symptoms. Sorghum plants remain susceptible to downy mildew infection for 38 days after emergence (Narayana *et al.*, 2002).

Optimum temperature range for germination of conidia is 21–25°C. However, isolates from different sources or biotypes vary in their response to temperature. In a study in Zimbabwe (Africa), optimum temperatures for conidial germination, germ tube elongation and infection are 10°–34°C, 20°–33°C and 14°–30°C, respectively. Were reported. Soil temperature of 26.6°C during the period of 16 days after sowing was found to be optimum for infection. Sorghum plants older than 20 days and maize plants older than 15 days are resistant to systemic infection by conidia. It has been observed that the disease is generally (not always) most severe in the wetter parts of the field. It has also been reported that heavy doses of organic manures pre-dispose the plants to heavier infection.

Management of the disease

Deep ploughing (30–35 cm) and regular roguing of diseased plants reduces oospore densities in the top 20 cm soil. Rotation is another cultural practice that reduces loss from the disease.

Potassium azide as a soil fungicide was found useful in controlling the disease. This chemical, applied at 1.12 kg/ha reduces the incidence of sorghum downy mildew by 23%. In general, the chemical control measures recommended for pearl millet downy mildew are applicable for this disease also. Anahosur and Patil (1983) had recommended one spray of metalaxyl (Ridomil 25 WP) at the rate of 2 g active ingredient per litre of water at 10 and 40 or 20 and 50 days after emergence for complete control of sorghum downy mildew. Seedling emergence is not affected by seed treatment with Ridomil 25 WP or Apron 35 SD at 1 or 2 g a.i./kg seed but higher doses are phytotoxic. Treated seeds can be stored for 9 months without loss of viability and efficacy of the fungicide. Seed treatment alone does not control the disease Panicker and Gangadharan (1999) obtained significant control of *P. sorghi* on maize by folair spray of phosphonic acid (20%) and its formulations like Aliette (0.25%) and Akomin (0.25%) 20 days after sowing. Under field conditions the reduction in disease incidence and severity was reflected by an increase of 54–73% in yield. The treatments suppressed sporulation of the fungus on foliage although conidial germination *in vitro* was not affected. This treatments is cheaper than metalaxyl.

Resistance to downy mildew has been located in several genotypes and some have shown stable resistance. Sorghum lines IS 8283 and IS 8607 are resistant to downy mildew as well as rust, anthracnose, zonate leaf spot, grain mold and gray leaf spot (Narayana, *et al.,* 1997). Germplasm accessions IS 1547, IS 4696, IS 5665, IS 5743 and IS 18737 have been found free from downy mildew in multilocational trials in India and abroad. These could be sources of stable resistance in breeding programme.

DOWNY MILDEWS OF MAIZE (CORN)

Doeny mildew fungi reported on maize in India are *Sclerospora philippinensis* Weston (*Peronosclerospora philippinensis*), *Sclerospora maydis* (*Peronosclerospora maydis*), *Sclerospora sorghi* (*Peronosclerospora sorghi*), *Sclerospora sacchari* (*Peronosclerospora sacchari*) and *Sclerophthora rayssiae* var. *zeae.* Another maize downy mildew, the Rajasthan downy mildew caused by a new species, *Peronosclerospora heteropogon*, is reported from Rajasthan of these species the most common and dangerous are *P. sorghi, P. philippinensis, P. sacchari* and *S. rayssiae* var. *zeae.*

Symptoms

The symptoms produced by most of the species are more or less similar on maize leaves. In the sugarcane downy mildew caused by *P. sacchari* the most characteristic symptom on maize is the development of long, rather broad, chlorotic stripes along almost the entire length of the leaf. In some instances, apparently due to secondary infection, the stripes are short and narrow. When a number of such stripes coalesce the margins are lost and irregular elongated patches are formed. The yellowish white colour of the stripes persists for more than a month or at least for three weeks before becoming darker and only in the very late stages browning of the tissues is noticed. Leaf shredding is not common although it has been noticed in very

Fig. 39. Downy mildew of maize caused by *Peronosclerospora sacchari.* Note the long, continuosus yellowish stripes.

severely affected leaves in late stages. Puckering of surface of young leaves bearing the chlorotic stripes is common. The downy growth of the fungus can be seen on both surfaces of the leaf. Young leaves show this white hazy growth more prominently than old leaves. The downy growth also occurs on bracts of green unopened male flowers in the tassel. In a large number of diseased plants of hybrid maize Ganga 3 development of miniature to large leaves on the tassel had been seen. These leaves had well defined white to yellowish stripes with the downy growth. Proliferation of axillary buds on the stalk of tassel as well as on the cob was very common in this variety and leaves thus formed showed the characteristic chlorotic stripes. Multiple ears and elongation of ear shoots was also common. Very often branching of the main stalk and tillering from the base also occurred. The disease becomes apparent before the plants have attained full height. Symptoms are not common on lower mature leaves unless plants are systemically infected but they are very common on upper leaves as a result of rapid secondary infection occurring in a thick stand of the crop. Young unfolding leaves are often uniformly yellow.

Fig. 40. Downy growth of *P. sacchari.* on maize leaves.

Symptoms of downy mildew caused by *P. philippinensis* and *P. sorghi* on leaves are very similar to those described above. The colour of chlorotic stripes caused by *P. philippinensis* is more intense. This fungus produces a characteristic downy growth on leaves followed by yellow discolorations, browning and necrosis of the blade and stunting of the plant.

The lesions of brown stripe downy mildew incited by *Sclerophthora rayssiae* var. *zeae* are shorter than those caused by *Peronosclerospora* species and turn brown rather early. Necrosis of affected tissues is a prominent feature. Their color is reddish to purple from the very beginning compared to almost white to light yellow stripes caused by *Peronosclerospora* species. The two types of downy mildews commonly do not occur on the same plant though occurring in the same field. However, a number of plants have been seen in which all the lower leaves carried the symptoms of *Sclerophthora* (possibly primary infection) and the upper leaves those of *P. sacchari* (possibly secondary infection).

Fig. 41. Brown stripe downy midew of maize caused by *Sclerophthora rayssiae var. zeae*.

Fig. 42. Tassel deformity caused by *Peronosclerospora sacchari*.

The major damage done by these downy mildew diseases is either a complete loss of the plant during an early stage of growth or varying degrees of grain loss if cobs are formed. In majority of plants infected with *P. sacchari*, if plants continue growth, normal cobs are not formed. The deformed cobs bear few grains. When normal looking cobs do develop they rarely reach maturity or bear mature grains. In infections with *Sclerophthora*, if plants continue growth, they produce normal cobs or cobs of different size and normal grains.

The causal organisms

Description of *Peronosclerospora sorghi* is given under downy mildew of sorghum. Conidiophores of *P. sacchari* emerge through stomata singly or in groups of 2, rarely 3. They are massive and wedge-shaped, short, stout structures widening gradually towards the upper portion and dichotomously branching at the tip 2-3 times (sometimes 4). In Taiwan, the length of conidiophores was given as 160–170 μm. In Indian isolates the length of conidiophores including the branched portion varies from 132 to 261 μm with averages of 184.8 to 187.4 μm in different collections. They are thin-walled and 1 or rarely 2-septate. The basal cell of the conidiophore is not swollen and slightly narrows down towards the base. This cell measures 24–36×12–18 μm. The maximum width of the conidiophore is 24–30 μm at the point where branching starts. The branches are stout and sometimes swollen in the middle. The sterigmata are thick in the lower portion, suddenly tapering at the upper end into a long fine tip which bears a single conidium. The length of sterigmata varies from 12 to 27 (average 21.5) μm. The conidia are thin-walled, hyaline, elliptical, cylindrical or ovate oblong with rounded apex. They measure 24–46.6 × 12–20 (mostly 24–36 × 12–18) μm. Oogonia are reddish brown, irregularly elliptical and measure 55–73 × 49–58 μm. Oospores on maize leaves are typically round, thin-walled, and measure 22.5–30 μm (average 24.9 μm) in diameter. The wall of oospores is 3.0-4.5 μm thick. Sun (1970) gave the diameter of oospores on sugarcane as 40–50 μm. He did not find oospores on maize. Conidia and oospores germinate by germ tube. The germ tubes are hyaline and slender, aseptate, and 3.8 μm wide. They develop from one or both ends of the conidium or from its lateral side.

S. philippinensis and *S. sacchari* are morphologically identical but differ from each other in that the former does not produce oospores and does not infect sugarcane. However, infection of sugarcane is reported in the Philippines where its oospores were also observed on leaves that touched or were partially covered by soil. The conidiophores of *P. philippinensis* are reported to be 140–420 μm long. The maximum width is 20–25 μm. The basal cell is 80–100 × 9–11 μm and sterigmata are 6–30, mostly 14–15 μm long. Conidia of this species are mostly barrel-shaped but variable on different hosts. They measure 17–57 × 11–27 (average 34 × 17) μm.

Sporangiophores of *Sclerophthora rayssiae* var. zeae are quite distinct from conidiophores of the above species. They are determinate, unbranched and short. They arise from hyphae congregated in the substomatal space. They produce sporangia sympodially in groups of 2-6. These sporangia are hyaline, ovate, obclavate, elliptical, or cylindrical, smooth, having a truncate or rounded apex. They measure 29–66.5 × 18.5–26 μm. Germination is by production of 4-8 zoospores. Encysted zoospores are spherical, hyaline and 7.5-11 μm in diameter. Sexual organs are abundantly produced in leaf mesophyll or under the stomata. Oogonia are subglobose, thin-walled, hyaline to light straw colored, with 1-2 attached paragynous antheridia. They measure 33-44.5 μm in diameter. Oospores are centrally located in the oogonia. They are spherical or subspherical and have hyaline contents including a prominent oil globule. The wall is uniformly 4 microns thick, smooth and glistening. Diameter of oospores varies from 29.5–37 μm.

Disease cycle and environmental relations

The downy mildew fungi mostly perennate through oospores and alternate hosts. *P. sacchari* was known to produce oospores only on sugarcane and it was considered to alternate between sugarcane and maize. Oospore production on maize was reported in India in 1968 and it can be presumed that the fungus survives in this country also through oospores in maize crop debris. *P. philippinensis* has many alternate hosts such as 'kans' grass (*Saccharum spontaneum*), *Sorghum bicolor* and *S. halepens*. The role of *Digitaria sanguinalis* as a host of *S. rayssiae zeae* in perennation and spread of the fungus is emphasized by many. This species also survives through oospores in crop debris. Once the primary infection is established secondary spread easily occurs through abundant production of conidia or sporangia under conditions of high humidity and moderate temperatures.

Considerable information is now available on seed-borne nature of many of these downy mildew fungi. Transmission of *Sclerophthora macrospora* on maize through seeds, if they are formed on affected plants, was known since 1962. As early as 1916, Rutgers had hinted the possibility of *Peronosclerospora maydis* on maize being internally seed-borne. In 1920, Weston also could trace the mycelium of *P. philippinensis* up to the endosperm but not in the embryo. Purukusumah (1965) reported evidence to prove that *P. maydis* is internally seed borne in maize in Java. In India, in 1967-1968, it was reported that *P. sacchari* and *S. rayssiae zeae* are present as mycelium in the embryo of fresh seed from diseased cobs. The mycelium (hyphae) was mostly in plasmodial form. Presence of mycelium of *P. sacchari* and *P. maydis* in maize seeds was also reported from Taiwan and Indonesia. In Punjab 70% embryos of seeds *P. philippinensis* in Punjab found 70% embryos of seeds from plants systemically infected by *P. philippinensis* contained mycelium of the fungus. About 5% seedlings raised from freshly harvested seeds developed downy mildew. Seeds stored under laboratory conditions for a year produced healthy seedlings. Seed-borne presence of *P. sorghi* was mentioned in downy mildew of sorghum while mycelium is present in all parts of the seed, 80% embryos contained oospores. Presence of oospores in the seed appears to be an important aspect of seed-borne nature of the downy mildew. Generally, when seeds are dried to a moisture content required for storage the mycelium is inactivated but oospores survive. Conidial germ tubes penetrate stigma and style within 6 hours of inoculation and within 48 hours the style, ovary wall and nucellus are invaded. Direct penetration of ovary wall leading to colonization of pericarp, endosperm and embryo also occurs. Infection may take place even after pollination and withering of stigma.

Production and germination of conidia of *P. sacchari* are influenced by temperature, humidity and light. The conidial germination and growth of germ tubes on the host surface in the field in early morning hours when there is heavy dew is so rapid that the surface becomes covered with light weft of hyphae. The optimum temperature for both production and germination of conidia is about 25°C. Conidia are not produced at or below 13°C and at or above 31°C. However, germination can occur even at 10° or 34°C. Leu and Tan (1970) obtained 100% germination of conidia at 8°–32°C. Free water on leaf surface is essential for formation and germination of conidia. Longevity of conidia is affected by humidity. They lose viability at 25°C within one hour at 90% relative humidity. Production of conidia in nature is mostly during the night because favourable temperature and moisture conditions are available during night. Age of the plant also affects conidial production. Fewer conidia are produced on mature leaves. Singh *et al.,* (1970) had reported that zinc and other micronutrient deficiency in plants makes them susceptible to attack of this downy mildew.

In *P. philippinensis* also moisture and night hours are essential for formation of conidia and their germination. Optimum temperature for germination is 19°–20°C. Maximum dispersal of conidia occurs between 4 a.m. and 8 a.m. The optimum temperature for production of sporangia of *S. rayssiae zeae* is 22°–25°C and for zoospore production 20°–22°C. Rather high temperatures (29°–32°C) are conducive to disease development.

Under Tarai conditions of north India, crops planted in May are free from infection and June planted crops have only low incidence of downy mildews while July planted crops have very high incidence. Low temperatures (20°–24°C) and high humidity (90% and above) during germination and early stage of plant growth favour disease incidence. In Punjab, Bains (1982) reported maximum infection in July–August sown crop and the spread was mostly from south-west direction.

Management of the disease

The three approaches to the management of downy mildews of maize are cultural practices, use of fungicides and planting of resistant varieties. Such methods should be, as far as possible, used in combination. In cultural practices prevention is most important. Alternate hosts and diseased crop debris must be destroyed and long rotations should be followed. For valuable seed plots fungicidal sprays are recommended.

In 1970 use of Dithane M-45 at 0.3% (4-6 sprays starting from 10 days after sowing at 7 days interval was suggested against the disease. In Taiwan, Dithane M-22 was found effective. Good control of the Philippine downy mildew of maize was reported with 3 sprays of Dithane Z-78 (0.3%) or 4 sprays of Dithane M-45 (0.3%). Seed treatment with Demosan in combination with rogueing of diseased young plants 20 days after sowing, and one foliar spray · of neem (margosa) oil gives most effective control of the disease.

Hundred per cent control was reported by treating seed with Ridomil MZ 75 WP at the rate of 4 g/kg seed. Metalaxyl (Ridomil MZ 75 WP) was recommended by Figueiredo and Anahosur (1993) as seed dress at 3 g/kg seed to control downy mildew of maize caused by *P. sorghi*. Seed treatment with metalaxyl (Apron 35SD) at a dosage as low as 0.35 g a.i/kg seed gives complete control of sorghum downy mildew in maize. In the event of sudden outbreak of this downy mildew in a valuable crop only one spray of Ridomil MZ at 6 g/lit water may be given 3 weeks after planting. According to dose level required to achieve same level of control the formulation Apron 35 WS is better than Ridomil. Earlier, such fungicides as Brestan, Brestanol and chloroneb (Demosan) had been reported to bring about effective control of the disease through seed treatment.

Some varieties such as Phil. DMR 1 are resistant to downy mildew. No commonly cultivated variety in India is, however, resistant Genotypes of maize resistant to *P. sorghi* are invariably resistant to Rajasthan downy mildew (*Peronosclerospora heteropogoni*) while those resistant to the latter may or may not be resistant to the former.

DOWNY MILDEW OF CRUCIFERS

The downy mildew disease of crucifers especially mustards causes significant damage to young plants, 5-15 cm high, and then to the inflorescences later in the season through sterility in the seed crop. In the oil seed crops, when there is mixed infection of white rust and downy mildew the loss in *Brassica juncea* may be 37–47% in pod yield and 17–32% in seed yield. In addition to oilseed brassicas, the disease occurs on cabbage, cauliflower, turnip and radish among vegetable crops. A downy mildew of cauliflower curd is also reported.

Symptoms

The disease appears as purplish-brown spots on the underside of the leaves. These spots may remain small or may enlarge considerably. The upper surface of the leaf on the lesions is tan to yellow. Downy growth usually appears on the undersurface of these lesions. In the systemically infected plants the symptoms are very conspicuous and almost similar to those produced by *Albugo candida*. In fact, in oil seed crops, the two parasites often attack the same plant at the same time. In downy mildew the greatest deformities appear on the stem while in white rust the inflorescence shows the maximum alterations. The stem swellings may be small or several inches long. The stalks are abruptly bent. The leaves and floral parts do not exhibit swelling. The young ovary may be very much elongated into a twisted body, 5-7 cm in length. More often, the floral buds are atrophied, all parts, *viz.*, sepals, petals, and stamens being shrunken. This leads to sterility of the flowers.

The causal organism

The fungus causing the disease, *Peronospora parasitica* (Pers.) ex Fr., (Chromista, Oomycota, Oomycetes, Peronosporales, Peronosporaceae) is considered a composite species consisting of many smaller morphological and biological species. The species occurring on *Brassica campestris* was often called *Peronospora brassicae*. More recently, the fungus attacking crucifers has been named *Hyaloperonospora parasitica* on the basis of hyaline conidiophores, recurved conidiophore branch tips and rDNA sequence comparisons (Constantinescu and Fatehi, 2002).

The mycelium of *H. parasitica*, a biotroph, is strictly intercellular with large, finger-shaped or clavate, branched haustoria which nearly fill the cell cavity. After a period of vegetative growth numerous erect branched conidiophores emerge through the stomata on the undersurface of leaves. In floral organs they may be absent, only oospores being formed there. However, in some hosts including mustards conidial production on the floral parts is profuse covering the surface with a white downy growth which appears like powdery mildew. The conidiophores arise vertically and their base is flattened, somewhat twisted where it passes through the stomatal opening. These structures are 100–300 μm long and unbranched for a major portion of their length. Dichotomous branching occurs, 6-8 times, at the tip. The sterigmata are long, slender, and pointed. They are at acute angles with each other. A single conidium is borne at the tip of each branch. The conidia are broadly oval, ellipsoidal, hyaline, and measure 24–27 × 15–20 μm. They readily fall off and germinate by a lateral germ tube.

Hyaloperonospora parasitica is heterothallic as well as homothallic. The majority of isolates are heterothallic. Both types occur in the same field. No relationship has been found between geographic origin and mating type. Oospores are produced in the interior of the host tissues at a later period than conidia. The oogonium is spherical and hyaline. The oospores are globose and 26–43 μm in diameter. They are enclosed in crest-like folds and appear pale yellow in colour. Their germination is by a germ tube.

The fungus has physiologic races. The pathotype from cauliflower has failed to infect oil-yielding brassicas. The pathogen from different hosts varies in host range and isolates from same host species or cultivar differ in pathogenicity on different hosts. Isolates similar in morphology may not be so in pathogenicity. The conidial size range from different hosts, as given by these workers, is 18.45–33.97 × 17.77–31.18 μm.

Disease cycle and environmental relations

The pathogen is primarily soil-borne through oospores lying in diseased plant debris. Seeds also carry the oospores in contaminating trash which goes to soil along with seed at the time of sowing. Vishunawat and Kolte (1993) have reported that oospores of *Hyaloperonospora parasitica* occur on the surface and in the hypodermis of seed coat tissue of rapeseed mustard. The seed transmission is to the extent of 0.9% in yellow sarson and 0.4% in toria. Although such seeds produce seedlings with downy mildew symptoms further development of the disease was not noticed and the growing plants were free from symptoms. In a tissue culture study to determine the viability of seed-borne mycelium of *P. parasitica* in cabbage the percentage of seeds showing viable mycelium varied with cultivar. There was direct correlation between embryo infection and seed transmission in callus as well as in symptoms. In a Australian study, no evidence of seed infection by *P. parasitica* in commercial brassica vegetables was found. Absence of oospores due to heterothallism or failure of infected seed to produce symptoms in seedlings were proposed as cause of the failure of the detection of seed-borne inoculum. Wild hosts may also serve as source of primary inoculum. From a study of *Peronospora tabacina*-tobacco pathosystem (blue mold of tobacco), Heist *et al.*, (2002) expressed the possibility that *Peronospora* is capable of moving systemically from shoots to roots and emerge from root tissues as hyphae that could infect neighboring plants and start new systemic infections or could produce oospores.

Conidia are responsible for the secondary spread of the disease. More conidia are produced at 13°C than at 18°C. Conidial germ tubes and appressoria excrete extracellular matrix on their body that helps them in adherence to the host surface. Carzaniga *et al.*, (2001) had studied the differences in structure and composition of these matrices on the three fungal structures. The conidial cell wall comprised an inner electron-lucent layer containing ß-1,3-glucan and an outer electron-opaque layer containing carbohydrates. Germ tubes and appressoria released two types of extracellular matrices, a fibrillar matrix containing ß-1,3-glucan which was confined to the area of contact between germ tube and host surface and a second matrix containing protein which spreads beyond the contact surface as a film. Both types help firm adherence of germ tube and appressorium to the surface. On mechanical removal of these structures, the matrix remains adhering to the surface.

In a study of infection process of *P. parasitica* in *Arabidopsis* (crucifer weed), Soylu, E.M. and Soylu, S. (2003a) have observed that conidia germinated and penetrated through the anticlinal cell walls of two epidermal cells. Rapid hyphal growth and spread with formation of numerous haustoria in mesophyl cells was followed by profuse sporulation without host cell necrosis and loss of cell cytoplasm structural integrity The haustoria were lobed with 67 μm diameter. They were connected with mother hypha without any apparent neck but callose-like deposits were seen at the point of penetration. Race-specific elicitors from the pathogen elicit phenolic accumulation in the host (Soylu and Soylu, 2003b).

In Punjab, 29% mixed infection of *Peronospora parasitica* and *Albugo candida* on *Brassica juncea* inflorescence. Since incubation period of *A. candida* is 6-7 days and that of *P. parasitica* 3-4 days, the latter predominates in mixed infections producing more extensive mycelium and more dense sporulation than *A. candida*. There is also evidence that prior infection of *A. candida* pre-disposes the plants to infection of *P. parasitica*. The incidence of mixed infection on *Brassica juncea* is related with the total amount of rain and relatively low temperature during the flowering period of the crop. Downy mildew infections occur more in early sown crops and no infection occurs in crop sown after mid-November (Mehta and Saharan, 1998). While *Brassica juncea* is susceptible, *B. napus* is resistant. Radish (*Raphanus sativus*) is generally resistant to *Peronospora parasitica*. Many lines in *Brassica oleracea* var. *botrytis* (cauliflower) show differential response to different isolates of the fungus at cotyledon stage. Cotyledon resistance and adult plant resistance of *B. oleracea* are very poorly correlated and the former can not be used for predicting adult plant resistance (Coelho and Mondeiro, 2003). Plants of *B. juncea* infected by mosaic virus are more susceptible to downy mildew. Monot *et al.*, (2002) reported induced systemic resistance in broccoli *(Brassica oleracea* var b*otrytis*) by preinoculation with avirulent isolates of the pathogen. The resistance lasted at least 15 days. Seedlings expressing this resistance accumulated the pathogenesis-related proteins PR-2 and PR-5 but not PR-1, PR-3 and PR-9. Singh, U.S. *et al.*, (2002) studied the biological interaction between *H. parasitica* compatible with *Brassica juncea* and compatible and incompatible isolates of *Albugo candida* (white rust). Pre-inoculation or co-inoculation with incompatible isolate of *A. candida* induced resistance to *P. parasitica.* Pre-inoculation with compatible isolate of *A. candida* increased susceptibility to *P. parasitica*, inoculated subsequently.

Management of the disease

Among cultural practices, crop rotation avoiding crucifers, deep ploughing during summer, and eradication of weed hosts are important. Sprays of fungicides have been found effective. The fungicides recommended against the disease are Dithane Z-78 (0.3%), Dithane M-45 (0.1 to 0.3%), Daconil (0.1%), Difolatan (0.3%), Blitox-50 (0.3%) and Ridomil (0.1%). The spraying should be done at 8-10 days interval Seed treatment with Apron 35 WS (metalaxyl) at the rate of 6 g/ kg seed combined with two sprays of Ridomil MZ at the rate of 0.2% has been found most effective against the disease. Mehta *et al.,* (1996) reported that 3 sprays of metalaxyl + mancozeb at 20 days interval commencing 40 days after sowing gave maximum disease control (82%) followed by seed treatment with metalaxyl combined with two foliar sprays of metalaxyl + mancozeb at 30 days interval. Seed treatment with metalaxyl followed by spray of non-systemic fungicides mancozeb or chlorothalonil gave only 47-49% control.

Brassica alba, *B. carinata* (HC-1), *Brassica juncea* (DIR 1507, DIR 1522) and *B. napus* (GS-7027, Midas and Tower possess multiple disease resistance, including resistance to downy mildew. In susceptible genotypes the fungus rapidly spreads through intercellular hyphae with high frequency of haustorium formation within host mesophyll cells and production of conidia in the absence of host cell necrosis (Soylu, S. *et al.,* 2004). In resistant genotypes the progress is slow, there is hypersensitive response and accumulation of ensheathing material around haustoria. Volatile flavonoids are involved in resistance of wild and cultivated cruvifers to downy mildew.

Ziadi *et al.,* (2001) have reported production of pathogenesis related proteins in cauliflower as a response to treatment with the plant activator (signal molecule) acibenzolar-S-methyl (ASM), a member of benzothiadiazoles (BTH). Induction of resistance to downy mildew in cauliflower by treatment with DL-â-amino-*n*-butyric acid (BABA) is reported by Silue *et al.,* (2002). The protective action of BABA is due to a potentiation of natural defense mechanisms against biotic and abiotic stresses Spray of water solution of the elicitor, benzothiadiazole (BTH), 1-30 days after inoculation induces systemic resistance to downy mildew in cauliflower. Resistance is not noticed if spray is done 1 day before or immediately after inoculation (Godard *et al.,* 1999). When 30 days old cauliflower seedlings are treated with 7.5 ml/L Phytogard (a formulation containing 58% dipotassium phosphate and 42% water) there is complete suppression of sporulation of the downy mildew pathogen. Young plants are completely protected by 10 ml/L of the solution. The induced resistance is not systemic but lasts for at least 15 days. Application through roots protects the foliage. Inoculation of broccoli (*Brassica oleracea* var. *botrytis*) with avirulent strains of *P. parasitica* induces systemic resistance against downy mildew (Monot *et al.,* 2002). Aqueous extract of *Penicillium chrysogenum* mycelium is reported to protect crucifer leaves from *Hyaloperonospora parasitica, Alternaria brassicicola* and *Botrytis cinerea*. The extract by itself is not antimicrobial. It induces insensitiveness in the leaves to these pathogens (Thuerig *et al.,* 2006).

DOWNY MILDEW OF CUCURBITS

Downy mildew of cucurbits was first recorded from Cuba bu Berkeley and Curtis in 1868. It is an important disease of cultivated cucurbits, especially sponge gourd (*Luffa aegyptiaca* =

L. cylindrica), ridge gourd (*Luffa acutangula*), muskmelon (*Cucumis melo*), and cucumber (*Cucumis sativus*) which are more severely affected than bottle gourd (*Lagenaria siceraria*), bitter gourd (*Momordica charantia = M. indica*) and snake gourd (*Trichosanthes anguina*). Pumpkin (*Cucurbita moscata =C. maxima*) and vegetable marrow (*Cucurbita pepo*) are less affected. In addition, the pathogen can infect watermelon (*Citrullus lanatus*), round melon (Tinda, *Citrullus vulgaris* var. *fistulosus*) and some other cucurbits such as Parwal (*Trichosanthes dioca*). The disease is fairly common in north India where it becomes serious during the later part of the rainy season.

Symptoms

The first symptoms on the leaves resemble those of mosaic mottling. The pale green areas are separated by islands of darker green. Soon the spots become well-defined. They are angular, yellow, and often restricted by veins on the upper surface. On the lower side of these spots a purplish downy growth appears in moist weather. However, this growth is not as conspicuous as in other downy mildew diseases. Occasionally, the purplish colour is lacking and the lower side of the spots looks white to almost black. The entire leaf quickly dies. Usually, the central leaves are attacked first and are followed by other leaves until the entire plant is wilted or weakened. On infected vines fruits are few and small with poor taste.

Pathogenesis of *Pseudopereonospora cubensis* on cucurbits results in changes in the metabolic processes within leaves including the rate of transpiration. Due to the negative correlation between transpiration rate and leaf temperature, it has been possible to measure temperature of diseased and healthy leaves even before visible symptoms appear.

The causal organism

Pseudoperonospora cubensis (Berk. and Curt.) Rostowtzew (Chromista, Oomycota, Oomycetes, Peronosporales, Peronosporaceae) is a biotroph. Berkeley and Curtis had named the pathogen as *Peronospora cubensis*. In 1903, the genus *Pseudoperonospora* was erected by Rostowtzew and the cucurbit downy mildew fungus was given its present name. The genus is close to *Peronospora* and *Plasmopara*. The mycelium is coenocytic and intercellular with small, ovate haustoria which sometimes develop finger-like branches. The sporangiophores arise singly or in groups of 2–5 through the stomata. The upper third of the sporangiophore is branched dichotomously or intermediate between dichotomous and trichotomous branching habit. The sterigmata are subacute. The sporangia are greyish to olivaceous purple, ovoid to elliptical, thin-walled, and with a papilla at the distal end. They measure 21–39 × 14–23 μm. The germination of sporangia occurs by production of biflagellate zoospores which are 10–13 μm in diameter when in resting state. Oospore are not common in nature. In India these structures were reported on certain cucurbits in Madhya Pradesh, Punjab and Rajasthan. The oospores are spherical, rarely obovoid to ellipsoid, light yellow, and smooth-walled. They measure 19–22 μm in diameter. The smooth wall is 1.5–3.5 μm thick.

Physiologic specialization exists in *P. cubensis*. Races of the pathogen differ in their ability to infect and sporulate on different hosts as well as in dimensions of sporangiophores and sporangia, number of zoospores per sporangium and in their response to metalaxyl. Five pathotypes were identified in 1987 on the bases of jihjly with specie that could be distinguished on the basis of occurrence of highly compatible reaction with specific cucurbit hosts. The fungus originally isolated from a particular host but grown for several generations on a different cucurbit species loses its affinity for the original host and sporulates better on the host on which it is being currently grown. Shetty *et al.,* (2002) studied the response of cultivars resistant to downy mildew in one geographic area in a different geographic area and observed that resistance response differed in the new area confirming that different local isolates represent different races of the pathogen. Cultivar genotypes resistant in China were resistant in India also but in countries like USA and Poland resistance response was intermediate. Similarly, cultivars resistant in USA and Poland showed intermediate response in India. Cultivars Nongche 4 (China) and M 21 (USA) showed resistance over all locations.

Disease cycle and environmental relations

In describing the parasite and explaining its persistence from season to season, E.J. Butler had, in 1918, stated that the great diversity of climate and season in different parts of India and the abundance of wild cucurbit hosts would probably ensure continuous supply of primary inoculum of the fungus. When winter temperatures are too low for growth of cucurbits, *P. cubensis* perpetuates in the form of active mycelium on self sown or cultivated sponge gourd growing in sheltered places during severe winters and also in the open during mild winters. In areas where oospores are found these may be an important source of survival and primary inoculum for initiation of the disease. The disease spreads in the field through dispersal of sporangia. Young leaves are less susceptible than the older leaves. Infection occurs more readily on the lower surface of the leaf than on the upper surface.

The development of sporangiophores and sporangia is affected by light and temperature. Exposure of the plant to light before exposure to darkness is essential for sporulation. Sporangiophores do not emerge through stomata if moisture saturated atmosphere does not exist on the leaf surface. For the muskmelon isolate of *P. cubensis* 6 hours of wet period after 6 hours of dry period are essential for sporulation. Sporangia do not mature if the sequence is reversed. According to them sporangia production starts before midnight and sporangia mature by 3 a.m. In saturated atmosphere maximum sporulation occurs at 18–28°C. When the plants are exposed to 35°C for more than 9 hours or to 40°C for more than 3 hours before sporulation there is no sporangia production during the subsequent 12 hours irrespective of other favourable environments.

Dispersal of sporangia starts as soon as they are mature and detached. They are carried away mainly by wind or are dispersed by raindrop splashes and then carried by wind. Spotted cucumber beetles also transport the sporangia. Contact between diseased and healthy leaves also causes local spread. Low relative humidity favor dispersal of sporangia. Maximum dispersal occurs between 6 and 10 a.m. The sporangia survive better in dry than in wet conditions. Temperatures of 35–37°C kill the sporangia. Survival of sporangia is greatly reduced if they are wetted for a period too short for germination and exposed to dry conditions.

Free moisture on leaf surface is essential for germination of sporangia. Germination starts within one hour in water drops and reaches the maximum in 2 hours. The optimum temperature for germination and zoospore production lies between 15° and 20°C. Zoospores are motile for a few minutes to several hours depending on the temperature and nature of the substrate. The motility lasts longer at 10° and 15°C than at 30° and 35°C.

During the process of infection the motile zoospores encyst singly on stomata. Germ tubes from the zoospores penetrate the host exclusively through stomata. The relative resistance of the upper leaf surface is due to lesser number of stomata compared to lower surface. This zoospore-stomata association is influenced by light. Success of infection depends on availability of free moisture on the leaf surface. Free moisture is no longer required once penetration has taken place. Short dew period of 2 hours are required for penetration while for haustoria formation 4 hours of dew period are required. Optimum temperature for infection is around 20°C (12–27°C), minimum 1–9°C and maximum 37°C. Incubation period varies from 4 to 12 days depending on moisture, temperature, inoculum density, photoperiod and host. In susceptible and resustant cultivars of muskmelon, the fungus produces intercellular hyphae, rich in β-1,3-glucabs, from which digitate (cauliflower-like) g\haustoria grow into the host cells. There is very little response by a susceptible host upto this stage. In resistant host, significant changes are induce. These include a heavy deposition of callose-like materials along the inner surface of the cerll walls; enrichment of host cell walls with lignin-like materials, and encasement of the haustoria with heavy deposits of callose-like materials.

In muskmelon, low nutritional staus of plants pre-disposes them to infection. The disease development is less on plants grown in high P, low K and high N nutrition solution. Nutrition of the plant affects sporulation and disease intensity. In wild muskmelon chemical composition of leaves is an additional factor for resistance. Leachate of pollen of ridge gourd is reported to enhance zoospore germination and inoculation with mixture of pollen and zoospores increases lesion development.

Management of the disease

Some of the cultural practices that help in reducing losses from the downy mildew are early sowing, destruction of weed hosts, proper nutrition (high N and P and low K), and reducing the amount of moisture among the vines.

Chemical control of the disease can be achieved through protectant and systemic fungicides. Among the protectant fungicides sprays of mancozeb, zineb and Tricop-50 (tribasic copper sulphate) have been recommended. These fungicides do not eradicate the established infection hence prophylactic spray is essential for complete protection. Since infection can occur at any stage of plant growth spraying is to be done at 5-7 days interval. Dithane M-45 and Dithane Z-78 protect the leaves for 9 days after spraying while Tricop-50 gives protection for only 5 days. Other fungicides that have been used are Difolatan and chlorothalonil. Weekly sprays of chlorothalonil with or without azoxystrobin are highly effective against the cucurbit downy mildew. Where proper forecasting method is available the number of sprays can be reduced to one and if the disease appears in the crop then resort to weekly sprays. Metalaxyl is effective in controlling the disease. However, quick emergence of resistant strains of the pathogen has limited its use. Metalaxyl resistance in these strains of the pathogen is fairly stable. Flumorph and dimethomorph are equally good fungicides against P. cubensis but flumorph is less effective than metalaxyl. Chemical control may not be economical in cucurbits. In addition, the nature of the crop (fast growing and difficulty in covering the lower sides of the leaf) also makes chemical sprays uneconomical. Frequent rains during the cucurbit season reduce the fungicidal deposit on leaves. Combined application of beta amino butyric acid (BABA) and mancozeb have synergistic effect against P. cubensis (Baider and Cohen, 2003).

Inula viscose (Compositae) is a perennial weed common in the Mediterranean area as well as in Asia and Africa. Roots of some species of Inula are known to have pharmaceutical value. Wang et al., (2004) obtained leaf extract of I. viscosa in a mixture of acetone and n-hexane (9:1) and prepared a paste by evaporating the solvents. Use of this paste as emulsion in water suppressed downy mildew of cucumber, late blight of potato and tomato, and powdery mildew of wheat. Spray of azoxystrobin and Pseudomonas fluorescens suppresses the downy mildew of cucurbits. The suppression is associated with increased activity of peroxidase, polyphenol oxidase, phenylalanine ammonia lyase and total phenols (Anand et al., 2007) Pre-inoculation treatment of cucumber leaves with ethanol leaf extracts of Paeonia suffruticosa (20% conc) has proved highly effective in suppressing downy mildew.

Breeding for resistance against downy mildew has been successful in many countries such as USA, Japan and France. No specific resistant varieties have been recommended in India. In Punjab, two lines of muskmelon (LC-8 and PPDMR-4)) were identified as resistant to *Peronosclerospora cubensis* showing isolated and slow development of lesions with sparse sporulation.

DOWNY MILDEW OF GRAPEVINES

Downy mildew is the most destructive fungal disease of grapevines and is one of the most widely studied plant diseases. Prior to 1870, this disease was considered endemic to North America where it appears to have co-evolved with wild grapes and then shifted to cultivated grapes. Since 1875 its epidemics were reported in France where it was introduced through grape cultivars imported from the USA as rootstocks resistant to grape phylloxera (*Daktulosphaira vinifoli*). This aphid pest of grapevines was also introduced into Europe from North America before 1863. In western Europe, particularly France, the downy mildew caused heavy losses to the wine industry during 1870s–1880s. It was in the search of chemical control of this disease that Prof. Millardet of the Bordeaux University made the accidental discovery of Bordeaux mixture in 1885. The disease reached England in 1894, Brazil in 1893, South Africa in1907, and Australia in 1916. At present the downy mildew is a major disease of grapes in grapevine-growing areas of the world and has been recorded from 91 countries from temperate zone to the tropics. In India, the disease is known to occur in Maharashtra.

Most of the economic loss is due to cluster destruction and loss of foliage resulting in the loss of photosynthetic area. When conditions of temperature, leaf wetness duration and relative humidity and foliage canopy are favourable, crop losses vary from 50–100%. The damage is greater following early infection. In epidemic years, defoliation of vines prevents maturation of fruits and canes and exposes fruits to sun scald. The downy and powdery mildews of grapevines cause loss of photosynthetic activity of leaves. The reduced assimilation rate is not only due to reduced green leaf area but also through an influence on gas exchange of the remaining green tissues.

Symptoms

Symptoms of the disease appear on all aerial and tender parts of the vines. They are more pronounced on leaves, young shoots, and immature berries which are most susceptible than on mature parts. On the upper surface of leaves irregular light yellow spots are seen. On the opposite surface, below these spots, downy growth of the fungus may be present. Later, due to necrosis, these leaf spots turn dark brown. The growth on the lower surface becomes dirty grey. A number of spots coalesce to form large necrotic patches. The infection of aged leaves results in a mosaic of small, angular, yellow to reddish brown lesions limited by veinlets. The affected leaves fall prematurely. When the leaves with suspected infection are moistened, enclosed in a polyethylene bag and incubated overnight at 20°–25°C in darkness, white cottony growth is produced on active infected tissue. The diseased shoots remain stunted and may be swollen due to hypertrophy. Necrosis may also occur. The shoots are rarely infected and only when about 10–15 cm long. The nodes are more susceptible than internodes. The infected leaves, shoots and tendrils become covered with whitish growth of the fungus. The flowers and young berries are also infected. The flower clusters are highly susceptible. The infected inflorescence first turns oily yellowish-brown and may develop the downy growth during periods of high humidity at night.

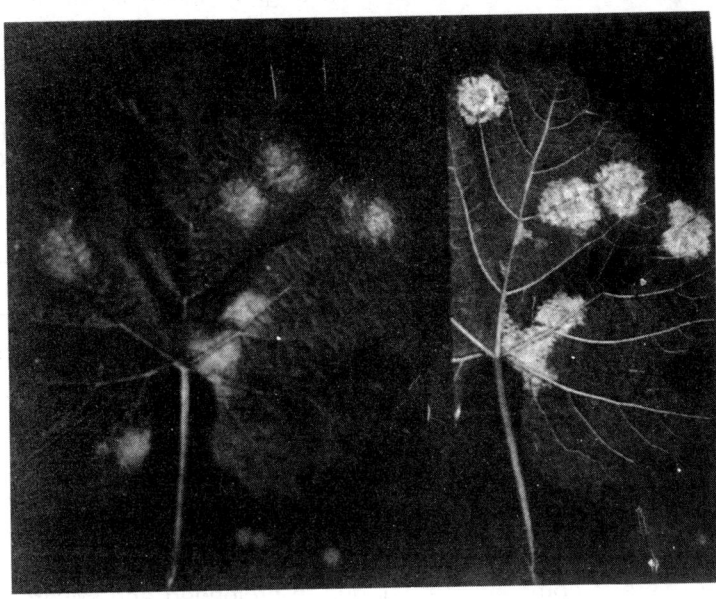

Fig. 43. Downy midew of grapevine leaf.

Infection of peduncles results in death of the entire cluster. The young berries are most susceptible from setting until 5–6 mm in diameter. The berries stop growing, turn hard, bluish green and then brown. They wither and fall down. After the growth of the fungus the entire bunch of berries is destroyed. Sometimes, only a part of the bunch is affected. When the infection of berries occurs in early stages of their development fungal growth is visible on the surface. However, when the fruits are half grown and infection occurs at this stage the fungal growth is confined to the inside of the berries. This results in light green or brown colour of fruits. Normally, the fully grown or maturing berries do not contact fresh infection, probably because the stomata turn non-functional but they are also lost if the supporting pedicels and cluster stem, which remain susceptible, are diseased.

The causal organism

The downy mildew of grapevines (*Vitis vinifera* and other cultivated and wild species) is caused by *Plasmopara viticola* (B. & C.) Berl and de Toni (Peronosporales, Peronosporaceae). The intercellular mycelium of the fungus consists of coenocytic, thin-walled, hyaline, hyphae with granular protoplasm. These hyphae are 7–12 to 60 μm wide according to the size and shape of intercellular spaces through which they pass. Haustoria are spherical or pear-shaped and 4-10 μm in diameter. They are formed predominantly in leaf tissues. Sporangiophores arise from hyphae congregated in the substomatal space. One to twenty sporangiophores emerge from a single stoma. Sometimes they may emerge directly through the cuticle. On young berries they come out through lenticels. These spore-bearing structures are mostly produced during the night when conditions of high humidity are present. They are 300–500 (mostly 140–250) × 7–9 μm in size. The branching of the sporangiphores is almost at right angles to the main axis and at regular intervals. The first branches arise from the apex. From the lower branches secondary branches are also produced. From the apex of each branch 2-3 sterigmata arise and bear sporangia. The sporangia are hyaline, thin-walled, ovoid or lemon-shaped, papillate, and measure 15–30 × 11–18 μm. Their germination is either by production of zoospores formed inside or outside in a vesicle or through germ tubes, depending on humidity. In indirect germination, each sporangium produces 1-6 zoospores. The zoospores are pear-shaped and 7–9 × 4–5 μm in size. They have two apical flagella which may be 30 μm long. On the host, the zoospores encyst and germinate by producing a flexuous aseptate germ-tube, 50–80 μm long. The fungus is heterothallic with two mating types. The oospores are produced mostly in tissues adjacent to the midrib of the leaf. They are thick- and wrinkled-walled, pigmented, 25–36 μm in diameter and contain 14–16 chromosomes. Antheridia are rarely seen. In some species of *Plasmopara* aexual genetic recombination is reported. The oospores germinate by producing a germ-tube of variable length, and 2-3 μm in diameter. The germ-tube bears an apical, pyriform sporangium, 35–40 × 25 μm in size. This sporangium produces 8–20 zoospores. Strains and ecotypes of the pathogen with varying virulence to *Vitis* and related genera are reported.

Disease cycle and environmental relations

Plasmopara viticola is a biotroph. In areas with a short and mild winter where grapevines are evergreen the fungus can survive in its active phase between bud scales and in diseased leaves that remain on the vines until the next bud burst. It can also survive in buds even in those areas where there is complete leaf fall due to prolonged and severe winter. The wild species of *Vitis*, where present, also harbour the fungus. However, the main sources of survival are oospores formed in the leaf and shoot tissue around the end of the growing season when conditions are no longer suitable for asexual reproduction. The fallen leaves contain enormous numbers of oospores. More than 200 oospores/sq. mm are reported.

Some pathologists believe that oospore formation occurs best at 14°–18°C while others feel that temperature is not as important as rainfall. After formation, the oospores undergo a temperature-dependent dormancy period. They can survive in leaf debris for at least 2 years. Generally, the oospores germinate more abundantly and in a shorter time early in the season when they have been subjected to frequent rains and mild temperatures. The risk of downy mildew is highest under these conditions. Rainy season with heavy rains during the oospore formation period hastens their maturity, hence, there is more disease in the next crop season. The date of optimum oospore maturation can be used for predicting disease severity in spring.

Under controlled conditions oospores mature at alternating weekly temperatures of 10° and 5°C. Low temperature reduces the dormancy period. On maturity oospores germinate over a period of 2-3 months in late winter or in spring. Minimum temperatures for germination range from 7°–13°C and optimum 20°–25°C. At 21°C the germination process takes 0.5–2 hours compared to 0.2–6 hours at 10°C. For good and early germination the soil must be saturated. High light intensity may retard or prevent germination.

Primary infection is caused by the soil-borne inoculum and usually the primary infection starts on leaves near the ground level. The sporangia and/or zoospores are splashed or blown to the wet vine surfaces. After sporangia land on the leaves, early stages of the development of the pathogen are guided by host factors. They influence accelerated release of zoospores from mature sporangia, coordinate morphogenesis of the germ tubes and guide them to the stomata (Kiefer *et al.*, 2002). Zoospores are released by sporangia within 30 min at 20°C. The zoospores swim towards the stomata. Width of stomata and possibly exudates from stomata direct the movement. The zoospore swarming may continue for 3–5 hours. Within 20–30 min of reaching the stomata, the zoospores become motionless and encyst in groups of 2–5 by absorbing their flagella and getting enclosed in a membrane. Under favorable conditions they germinate by a germ tube that grows through stomatal aperture into the substomatal cavity. Many germ tubes may penetrate a single stoma. The course of colonization in susceptible and resistance *Vitis vinifera* has been studied by Unger *et al.,* (2007). The pathogen is established within a few hours of inoculation of a susceptible cultivarms primary hyphae and the first haustorium. No further development occurs in the following 10 to 18 hours. After 1.5 days of inoculation the hyphae grow and branch to colonize the intercellular spaces. After 3 days the pathogen has fully colonized the leaf tissues.

In the substomatal cavity, a vesicle is formed. This develops a short hypha at its distal end. When this hypha touches a host cell, a primary haustorium is formed. This occurs within 3.5 hours after the sporangium has come in contact with the susceptible host surface at 23°C. For 12–15 hours there is no development although the size of primary haustorium increases. Within 24 hours of sporangium deposition, hyphae have grown between the cells and produced additional haustoria. After 36 hours, the host cells are noticeably plasmolysed. Hypertrophy of cortical cells may cause a slight swelling of infected stem. Mycelial growth is more abundant in young tissues. The growth ceases at 30°C although the mycelium can remain viable at 42°–43°C for 12 days.

The first crop of sporangia is then produced from these primary infections. Sporulation (sporangia formation) requires continuous RH ranging from 95–100% or saturation. It also requires a minimum of 4 hours of darkness. Moist, dark conditions following a period of light favour maximum sporulation. Under continuous light or darkness and high humidity there is little or no sporulation. Sporulation occurs at temperatures ranging from 9°–34°C. The optimum temperatures have been cited as 18°–24°C and 18°–28°C. The maximum sporangial production is reported to occur between 23° and 28°C. Under favourable conditions sporulation can continue for at least several successive nights (sometimes over a period of 2 months) on the same lesion. Sporulation is also favoured by high water content of the infected tissue.

The grape clusters become resistant to fresh downy mildew infections at stages of development from 1 to 6 weeks postbloom depending on cultivars and climate. The resistance is associated with the loss of infection courts as the stomata are converted to lenticels. This prevents emergens of the pathogen to sporulate. This happens with berries that were infected in early stages of development.

The sporangial dehiscence occurs when the cross wall of the callus between sporangium body and stalk is dissolved in water. Thus, moist air is essential for dissemination of sporangia. The sporangia are detached by movement of leaves and are wind-borne to new sites to cause secondary infection. Water also disperses these sporangia. Leaf hairs are a basic protective barrier against downy mildew of grape. The presence of hairs makes the leaf surface hydrophobic thus not allowing water droplets stay. This excludes the humidity required for spore germination. Cultivars with high hairiness of leaves suffer less from downy mildew. The viability of sporangia decreases with increasing temperature and decreasing RH. The attached sporangia survive longer than detached sporangia. Various reports hyave cited sporangium survival for 4–14 days at temperatures of 10°–20°C. At 30°C sporangia survive only for 6 hours. Sunlight adversely affects sporangial viability. In India, where temperatures reach more than 30°C, sporangia harvested during day hours do not germinate while those harvested during night from 2 a.m. to 6 a.m. germinate within one hour. The optimum temperature for germination of sporangia is 10°–16°C while the minimum is 5°C. Under conditions of optimum temperature and humidity, germination occurs within an hour. Sporangium germination and the liberation of zoospores occurs only in water at temperatures ranging from 2°–9°C to 28°–30°C with optimum between 15° and 23°C.

The germ tube from the zoospore enters the host through the stomata and lenticels. Incubation period varies from 7 or 8 days to 20 days depending on susceptibility of the host variety, air temperature and relative humidity. As temperature increases from 5°C to 26°C, the incubation period decreases. At 100 % RH it is 11–12 days. At 28°C the incubation period is 5–6 days at 100% RH and 8 days at 70 % RH. Gindro et al., (2003) conducted a histological study of downy mildew infection in a resistant and a susceptible cultivar of grape. Seven hours after inoculate of resistant cultivar callose deposition was seen around germinating zoospores on the stomata. Twenty four hours after inoculation, stomata exhibited secretions that enveloped the zoospores and there was necrosis of tissue around the stomata. Stomata in the vicinity of infection site contained callose deposition but there was no necrosis. In susceptible cultivar, neither secretions not callose deposits were seen. According to Gobbin et al., (2005) epidemics of downy mildew the result of the interaction of a multitude of pathogen genotypes, each causing limited (or a few) lesions, and of a dominant genotype able to spread stepwise at plot scale. This contrast with the earlier belief that there is a massive vineyard colonization by one genotype and long distance migration of sporangia.

Disease Management

Sanitation is very important in the management of downy mildew of grapevines. The fallen leaves and twigs should be collected and burnt. The microclimatic conditions under the canopies being closely related to infection, inoculum production, and spread of the disease, vines should be planted with proper spacing and should be trained in such a manner that the leaves are not near the soil surface. These precautions permit free circulation of air and ensure low humidity.

Leaf temperature of grapevines rises atleast 3 days before visible symptoms appear. This change in temperature can be detected by thermal imaging. The chemical control operations can thus be started earlier tham disease appearance The disease has been controlled with well-timed application of copper and dithiocarbamate fungicides. Bordeaux mixture (4: 4: 50), copper oxychloride such as Blitox-50 (0.3%), zineb, maneb and mancozeb (0.2%), and captan (0.2–0.5%) have been successfully used against the disease. Bordeaux mixture in new formulations is still used extensively in French vineyards for management of grape downy mildew. Five sprays are generally recommended at the following stages:

    (i) immediately after pruning,
    (ii) 3-4 weeks after pruning,
    (iii) before the buds open,
    (iv) when berries have formed, and
    (v) during growth of shoots.

Sprays should be stopped 2 weeks before harvest. Some workers have found the antifungal antibiotic Aureofungin effective in controlling the disease. The above schedule of spray is suited to areas where the downy mildew is of known regular occurrence. It may be costly and unproductive in areas where the disease is of irregular occurrence such as in the arid zones. For this, forecasting based on weather parameters and plant growth conditions is recommended so that chemicals are used only when required. Warning systems enable effective control of downy mildew with reduced number of sprays.

Since the mid-1970s the mixture of metalaxyl (Ridomil) with either copper fungicides or mancozeb or folpet has been found to give effective control of the disease when applied before or after the infection. Most mixtures containing metalaxyl persist for 2-3 weeks. Metalaxyl inhibits the formation of secondary haustoria and the growth of mycelium inside host tissue, and stops lesion formation and sporangial formation through action of its volatiles. Metalaxyl can systemically move from roots to leaves but does not enter the berry clusters. But the berries are protected by vapor activity of the fungicide present on nearby leaves (Kennelly *et al.*, 2007). Because of resistance development in the pathogen against metalaxyl, the latter is mostly recommended in combination with protectant fungicides such as mancozeb.

Dimethomorph has better protective and curative activity than metalaxyl-mancozeb combination. It inhibits sporulation when used at 25 μg/L before infection or up to 9 days after infection. Phosphorus (phosphonic) acid or its derivative phosethyl-Al is considered a true systemic downy mildew fungicide and an alternative to metalaxy. Applied at 1.2 g/L up to 12 days after infection, the fungicide reduces the incidence and severity of downy mildew through reduced production of inoculum of *P. viticola*. The fungicide is translocated from leaves via the xylem and to some extent via the phloem to extremities of shoots and to clusters at least until berry set. It provides protection to actively growing foliage and clusters for 14-21 days. In plant tissue, fosethyl-Al rapidly breaks down to phsophorus acid and carbon dioxide. The acid is the active component and acts directly on the fungus. Its indirect effects on host metabolism are secondary.

Azoxystrobin, a strobilurin wide spectrum systemic fungicide, has preventive and curative effect against the grape downy mildew and grape black rot fungi. On young plants, the preventive efficacy of azoxystrobin at 250 g a.i. /ha applied 10 days before inoculation was equal to that of mancozeb at 2800 g a.i./ha. Its efficacy at the same dose was higher than that of mancozeb when applied 7 days before inoculation. A single dose of 0.5 mg/ml azoxystrobun gives 61% control of downy mildew. The same dose of trifloxystrobin and kresoxim methyl gives 41% and 1.1% control, respectively. The mixture of azoxystrobin (187 g a.i.) with cymoxanil (120 g a.i) was also more effective than a mixture of mancozeb (1400 g a.i) and cymoxanil (120 g a.i). When sprayed at 10 days intervals the bunch protection was almost perfect. A comparative study of physical modes of action of azoxystrobin, mancozeb and metalaxyl showed complete control of the disease with 250 mg/ml azoxystrobin, 1790 mg/ml mancozeb and 260 mg/ml metalaxyl applied 1-5 days before inoculation. Post-infection application of azoxystrobin had no effect on disease incidence but reduced sporulation by 96% on the lesions that developed. Post-infection application of mancozeb had little or no effect. Metalaxyl also provided significant control on post-infection (3-5 days after inoculation) application. Trifloxystrobin is highly effective against downy mildew and powdery mildew of grapevines. Two foliar sprays, 2 weeks apart, inhibited downy mildew development and were as effective as the standard metalaxyl-Cu treatment. Prophylactic applications at 100–150 mg/L were superior to sterol inhibitor fungicides (Reuveni, 2001).

The use of systemic fungicides is handicapped by the phenomenon of development of resistance in the pathogen against the fungicide. The resistant races of *P. viticola* have been reported for several phenylamide fungicides including metalaxyl, ofurace, milfuram and cyprofuram. The level of resistance varies. A high level of resistance and a high degree of fitness is a disturbing feature of metalaxyl resistance in fungi. When metalaxyl-mancozeb combinations are used the resistant isolates persist but their frequency decreases if fungicide use in the field is confined to mancozeb or stopped completely. Metalaxyl resistant strains show resistance to other phenylamide systemic fungicides (*viz.,* oxadixyl) also. The aggressiveness of metalaxyl-resistant isolates may be related to heterokaryosis in *P. viticola*. The combination of protectant and systemic fungicides has synergistic effect. In metalaxyl-mancozeb combination, the protectant component may increase the permeability of cell membrane to increase the concentration of the site-specific inhibitor (metalaxyl) reaching the site of action in the fungus. Furthermore, since metalaxyl is usually fungistatic, the activity of sub-lethal concentration of the protectant component may make their combined effect more fungistatic. Control of metalaxyl-resistant isolates has also been achieved by using other types of systemic fungicides such as phosethyl-Al, prothiocarb and cymoxanil. Mixture of oxadixyl and mancozeb is not effective when level of resistance in strains is increased but addition of cymoxanil to the mixture suppresses all levels of resistance. The three-way mixture, oxadixyl/ mancozeb/ cymoxani, is more effective against sensitive and phenylamide-resistant strains of *P. viticola* Benthiavalicarb is a more recent fungicide effective against oomycetes and has been used in the control of grapevine downy mildew (Reuveni, 2003). It does not affect zoospore discharge from sporangia of *Plasmopara viticola* but strongly inhibits zoospore encystment, cyst germination *in vitro* and mycelial growth together witsporangial production *in vivo*. When applied to vines 1,3, and 6 days after inoculation the fungicide protected the plant from downy mildew and inhibited sporulation of the pathogen. The fungicide remains active on plant surface for 28 days. However, it has no translocability and shows only strong prophylactic and local activity. Two foliar sprays of benthiavalicarb at 2 weeks interval is as effective in controlling the disease as metalaxyl + copper fungicide. A combination of benthiavalicarb and folpet is superior to metalaxyl + copper treatment.

The resistance inducing, non-protein amino acid DL-ß-amino-*n*-butyric acid (BABA) is reported to induce local and systemic resistance to downy mildew in grape leaves (Cohen *et al.*, 1999). Foliar application of BABA or a mixture of BABA and different fungicides provide more than 90% control of downy mildew on grape leaves (Reuveni *et al.*, 2001b). The control achieved may be as good as that achieved by metalaxyl-Cu or Acrobat-Plus (dimethomorph + mancozeb). Hamiduzzaman *et al.*, 2005) compared different resistance inducers, *viz.*, BABA, jasmonic acid, salicylic acid and abscisic acid and found BABA most effective, followed by jasmonic acid, in protecting grapevines from the downy mildew.

Callose deposition was one of the defence responses of the host. Jasmonic acid potentiated the action of BABA. Elicitation of defence responses in grapevine against downy mildew and bunch rot (*Botrytis cinerea*) by application of laminarin (beta-1,3-glucan laminarin) a phytochemical derived from brown algae (*Laminaria igitata*) is reported by Aziz *et al.*, (2003). Downy mildew infection was reduced by 75%. Multiple effects including activation of many defence related genes were responsible for reduction of infection. Chitosan (deacetylated derivative of chitin) is reported to be effective in promoting plant defence respons against downy mildew and gray mold (*Botrytis cinerea*). The efficacy of the compound ndepends on nmolecular weight and degree of acetylation. The defence responses include accumulation of phytoalexins. Combination of chitosan with copper sulphate enhances efficacy of the latter (Aziz *et al.*, 2006). Spray of the aqueous extract of dry mycelium of *Penicillium chrysogenum*, an industrial waste product, induces resistance in plants against not only *Plasmopara viticola* (grape downy mildew) and *Uncinula necator* (grape powdery mildew) but also to many diseases/ pathogens such as *Peronospora destructor* (onion downy mildew), *Ventutia inaequalis*, the apple scab fungus and *Phytophthora infestans* on tomato (Thuerig *et al.*, 2006), black spots of crucifers (*Alternaria brassicicola*), cotton wilt (*Fusarium oxysporum* f.sp. *vasinfectum*) and root knot nematodes.

Cohen *et al.*, (2006) have reported control of downy mildew of grapevine by oily paste extracts of leaves of *Inula viscose* (Compositae), made with organic solvents. The effective concentyration required for 90% control of the disease in treated shoots in the field was below 0.125% (paste in water) Under the conditions of natural infection and for whole vine treatment 0.58% concentration gave 90% control. Isoalantolactone in roots of the species *Inula racemosa*, a traditional Chinese medical herb, is known to be highly fungicidal, insecticidal and also phytotoxic if plants are exposed for long hours to its high concentration (Liu *et al.*, 2006),

Rajeswari *et al.,* (2007) tested a large number of plant extracts and some phylloplane microflora for control of grape downy mildew. Neem seed extract at 5% significantly inhibited the sporangial germination (75%). Among microflora *Pseudomonas fluorescens* was highly effective in reducing the sporangia germination. Post-inoculation application of neem seed extract (5%) and *P. fluorescens* (.2%) effectively inhibited the disease development in greenhouse experiments. In field experiment these sprays were effective when applied first as soon as the disease was detected and the second and third spray at 10 days interval.

In Germany, it was reported that leaves sprayed with water extract of composted horse manure-straw-soil mixture develop resistance to infection of *P. viticola*. The extract has no direct effect against the pathogen (Weltzien and Kettere, 1986). Subsequently, it was reported that the disease regulation is the result of biological control through enhanced microbial activity on the leaf surface. Aqueous extracts of fermented products of many types of materials have since been used to control downy mildew, powdery mildew, and other fungal diseases of grapevines. *Fusarium proliferatum* has been identified as a biocontrol agent against *P. viticola*. Post-infection application of microconidial suspension of the fungus on leaf discs reduced sporulation of *P. viticola* by 97% (Falk *et al.,* 1996). In vineyards, weekly application of the conidial suspension reduces the incidence of downy mildew. The hyphae of the mycoparasite coil around and penetrate the sporangiophores of *P. viticola*. A cold tolerant strain of this mycoparasite, isolated through UV mutagenesis of a strain of the mycoparasite, was reported by Bakshi *et al.,* (2001). This strain grows well at the low temperature desirable for the pathogen. *Fusarium proliferatum* is a serious pathogen of garlic and onion and produces potent mycotoxins *Plasmopara vitivola* is an alternative food for generalist predator mites occurring in vineyards (Possebron and Duso, 2008). The mites increase in number as the mildew grows. This relationship with the downy mildew fungus seems to have no comtrol value.

Induction of resistance to downy mildew by treatments with selected strains of *Pseudomonas syriigae* pv. *syringae* or salicylic acid is also reported. Pre-inoculation with *Pseudoperonospora cubensis* (downy mildew of cucucumbe) also induces resistance to *P. viticola* in grapevine leaves. Vecchione *et al.,* (2006) made a large number of isolations from grapebine leaves showing anomalous symptoms of downy mildew. Of about 125 endophytic microorganisms isolated only 5 fungal; isolates showed complete inhibition of sporulation of *P. viticola*. All these were identified as *Alternaria alternata*. Within leaf tissues the endophyte adversely affected *P. viticola* without coming in direct contact indication role of toxin produced by the endophyte. The *P. viticola* mycelium showed severe ultrastructural alteration such as the presence of enlarged vacuoles or vacuoles with electron-dense precipitates. Haustoria appeared nectrotic and irregularly shaped or were enclosed in callose-like substance. Aqueous extract of dry mycelium of *Penicillium chrysogenum* induces resistance in grapevines to downy mildew as well as powdery midew.

The role of pre-formed flavonoids (phenols) in grapevine resistance to downy mildew is domented. The synthesis of flavonoids is assisted by sunlight. Leaves kept exposed to sun light have more flavonoid content than leaves kept in shade. Thus, proper pruning and training of vines to facilitate maximum sun light in the canopy may help in resistance development.

The sources of resistance to downy mildew are present in many species of *Vitis* and these have been used for developing resistant cultivars. Under Indian conditions cultivars Amber Queen, Champion, and Digraset have been found resistant to down mildew while cultivars Athens, Buckland, Sweet water. Essabela, Goethe, James, Malaga, Khalili, Westfield and Solanis × Riperia are moderately susceptible. In another report, the varieties found resistant in South India were Amber Queen, Cardinal, Champa, Champion, Dogridge and Red Sultana. Some promising resistant mutants have also been induced by irradiating seeds.

REFERENCES

Amruthesh K.N., N.P. Geetha, H.J.L.Jorgensen, E. de Neegaard and H.S. Shetty. 2005. Unsaturated fatty acids from zoospores of *Sclerospora graminicola* induce resistance in pearl millet. *Eur. J. Plant Pathol.* 111(2): 125

Anand, T., T. Raghuchander. G. Karthikeyan *et al.*, 2007. Chemically and biolopgically induced systemic resistance in cucumber (*Cucumis sativus*) against *Pseudoperonospora cubensis* and *Erysiphe cicho-racearum*. *Phytopath. Mediterr.*, 46(3): 259

Andreu, A.B., M.G. Guevard *et al.*, 2006. Enhancement of natural disease resistance in potato by chemicals. *Pest Manag. Sci.* 62(2): 162

Andreu, A.R. and D.O. Caldiz. 2005. Early management of late blight (*Phytophthora infestans*) by using systemic fungicides applied to seed potato tubers. *Crop Protection* 25(3): 281

Andrivon, D., J.M. Lucas and D. Ellesseche. 2003. Development of natural late blight epidemics in pure and mixed plots of potato cultivars with different levels of partial resistance. *Plant Pathology* 52(5): 586

Ann, P-J, W-H. Ko and H-J. Su. 2004. Interaction between Libukin bacterium and *Phytophthora parasitica* in citrus host. *Eur. J. Plant Pathol.* 110(1): 1

Arora, R.K. 1989. Distribution of healthy and late blight infected tubers in potato ridges. *Indian J. Mycol. Pl. Pathol.* 19: 232

Asano, T. and K. Kogeyama. 2006. Growth and movement of secondary plasmodia of *Plasmodiophora brassicae* in turnip suspension culture cells. *Plant Pathology* 55(1): 145–153

Aziz, A., B. Poinssot, X. Daire, *et al.*, 2003. Laminarin elicits defense responses in grapevine and induces protection against *Botrytis cinerea* and *Plasmopara viticola*. *Mol. Plant-Microbe Interact.* 16(12): 1118

Aziz, A., P. Trotel-Aziz, L. Dhuicq *et al.*, 2006. Chitopsan oligomers and copper sulphate induce grapevine defence responses and resistance to gray mold and downy mildew. *Phytopathology* 96(11): 1188

Babitha, M.P. 2002. Induction of lipoxygenase in downy mildew resistant seedlings of pearl millet in respsonse to inoculation with *Sclerospora graminicola*. *J. Mycol. Pl. Pathol.* 32(3): 357 (abstract)

Babitha, M.P., H.S. Prakash and H.S. Shetty. 2004. Purification and properties of lipoxygenase induced in downy mildew resistant pearl millet seedlings due to infection with *Sclerospora graminicola*. *Plant Sci.*166(1): 31

Baider, A. and Y. Cohen. 2003. Synergistic interaction between BABA and mancozeb in controlling *Phytophthora infestans* in potato and tomato and *Pseudoperospora cubensis* in cucumber. *Phytoparasitica* 31(4): 399

Bakshi, S., A. Sztejnberg and O. Yarden. 2001. Isolation and characterization of a cold tolerant strain of *Fusarium proliferatum*, a biocontrol agent of grape downy mildew. *Phytopathology* 91: 1062

Bardin, S.D., H.C. Huang, J. Pinto, E.J. Amudsen and R.S. Erockson. 2004. Biological control of Pythium damping off of pea and sugar beet by *Rhizobium leduminosarum* bv. *viceae*. *Can. J. Bot.* 82(3): 291

Barkai-Golan, R. and D.J. Phillips. 1991. Post-harvest heat treatment of fresh fruits and vegetables for decay control. *Plant Dis.* 75: 1085

Becktell, M.C., M.L. Daughtrey and W.F. Fry. 2005a. Temperature and leaf wetness requirements for pathogen establishment, incubation period and sporulation of *Phytophthora infestans* on *Petunia* x *hybrida*. *Plant Disease* 89(9): 975

Becktell, M.C., M.L. Daughtrey and W.F. Fry. 2005b. Epidemiology and management of Petunia and Tomato late blight in the greenhouse. *Plant Dis.* 89(9): 1000.

Bhattacharya, I. and M. Pramanik. 1998. Effect of different antagonistic rhizobacteria and neem products on clubroot of crucifers. *Indian Phytopath.* 51: 87

Caldiz, D.O., D.A. Rolon, J. Di Rico and A.B. Andreu. 2007. Performans of dimethoimorph + mancozeb applied to

Campanella, V., A. Ippolito and F. Nigro. 2002.Activity of calcium salts in controlling Phytophthora root rot of citrus. *Crop Protection* 21(9): 751

Carnegie, S.F. and J.Colhoun. 1983. Effect of plant nutrition on susceptibility of potato leaves to *Phytophthora infestans*. *Phytopath. Z.* 108: 242.

Carzaniga, R., P. Bowyer and R.J. O'Connell. 2001. Production of extracellular matrices during development of infection structures by the downy mildew *Peronospora parasitica*. *New Phytologist* 149(1): 83

Chaluvaraju, G., P. Basavaraju, N.P. Shetty, S.A. Deepak, K.N. Amruathesh and H.S. Shetty. 2004. Effect of phosphorus-based compounds on control of pearl millet downy mildew disease. *Crop Protection* 23(7): 595

Chatterton, S., J.C. Sutton and G.J. Boland. 2004. Timing *Pseudomonas chlororaphis* applications to control *Pythium aphanidermatum*, *Pythium dissotocum*, and root rot in hydroponic peppers. *Biol. Control* 30(2): 360

Chattopadhyay, A.K., A.K. Moitra and C.K. Bhunia. 2001. Evaluation of *Brassica* species for resistance to *Plasmodiophora brassicae* causing club root of rapeseed-mustard. *Indian Phytopath.* 54:131

Coelho, P.S. and A.A. Mondeiro. 2003. Expression of resistance to downy mildew at cotyledon and adult plant stages in *Brassica oleracea* L. *Euphytica* 133(3): 279

Cohen, Y.R. 2002. Beta aminobutyric acid-induced resistance against plant pathogens. *Plant Dis.* 86(5): 448

Cohen, Y., M. Reuveni and A. Baider. 1999. Local and systemic activity of BABA (DL-B-aminobutyric acid) against *Plasmopara viticola* in grapevines. *Eur. J. Plant Pathol.* 105: 351

Cohen, Y., S. Farkash, A. Baider and D.S. Shaw. 2000. Sprinkler irrigation enhances production of oospores of *Phytophthora infestans* in field grown potato. *Phytopathology* 90(10): 1105

Cohen, Y., W. Wang, B.-H. Ben-Daniel and Y. Ben-Daniel. 2006. Extracts of *Inula viscose* control downy mildew of grapes caused by *Plasmopara viticola*. *Phytopathology* 96(4): 417

Cohen, Y., E. Rubin, T. Hadad *et al.*, 2007. Sensitivity of *Phytophthora infestans* to mandipropamid and the effect of enforced selection pressure in the field. *Plant Pathology* 56(5): 836

Constantinescu, O. and J. Fatehi, 2002. Peronospora-like fungi (Chrmista, Peronosporales) parasitic on Brassicaceae and related hosts. *Nova Hedwigia* 74: 291

Daayf, F., L. Adam and W.G.D. Fernando. 2003. Comparative screening of bacteria for biological control of potato late blight (strain US 8), using *in-vitro*, detached leaves, and whole plant testing systems. *Can. J. Plant Pathol.* 25(3): 276

Daniels, R. and D. Guest. 2006. Defence responses induced by potassium phosphonate in *Phytophthora palmivora*-challenged *Arabidopsis thaliana*. *Physiol. Mol. Plant Pathol.* 67(3): 194

Deepak, S.A., G. Oroos, S.G. Sathyanarayana *et al.*, 2005. Antisporulant activity of leaf extracts of Indian plants against *Sclerospora graminicola* causing downy mildew disease of pearl millet. *Arch. Phytopath. Plant Prot.* 38(1): 31

Deepak, S., S. Niranjan-Raj, S. Shailshree, R.K. Kini *et al.*, 2007. Induction of resistance against downy mildew pathogen in pearl millet by a synthetic jasmonate analog. *Physiol. Mol. Plant Pathol.* 71(1-3): 96

De Souza, J.T, M. de Boer, P. de Waard *et al.*, 2003. Biochemical, genetic and zoosporicidal properties of cyclic lipopeptiode surfactant produced by *Pseudomonas fluorescens. Appl. Environ. Microbiol.* 69(12): 7161

Devos, S., K. Visenberg, J-P. Verbelen and E. Prinsen. 2005. Infection of Chinese cabbage by *Plasmodiophora brassicae* leads to a stimulation of plant growth: Impacts on cell wall metabolism and hormone balance. *New Phytologist* 166(1): 241

Devos, S., K. Laukens *et al.*, 2006. A hormone and proteome approach to picturing the initial metabolic eventrs during *Plasmodiophora brassicae* infection on *Arabidipsis. Mol. Plant-Microbe Interact.* 19(12): 1431

Dirac, M.F., J.A. Menge and M.A. Madore. 2003. Comparison of seasonal infection of citrus roots by *Phytophthora citrophthora* and *P. nicotianae* var. *parasitica. Plant Dis.* 87(5): 493

Donald, E.C. I.J. Porter and R.A. Lancaster. 2001. Band incorporation of fluazinan (Shirlan) into soilto control clubroot of vegetable brassica crops. *Aust. J. Exp Agric.* 41(8): 1223.

Donald, E.C., G. Jaudzems, I.J. Porter. 2008. Pathology of cortical invasion by *Plasmodiophora brassicae* in clubroot resistant and susceptible *Brassica oleracea* hosts *Plant Pathology* (Online Early Articles). doi:10.1111/j.1365–3059.2007.01765.x

Drenth, A., E.M. Jensen and F. Gove'rs. 1996. Formation and survival of oospores of *Phytophthora infestans* under natural conditions. *Plant Pathology* 44(1): 86

Falk, S.P., R.C. Pearson, D.M. Gadoury, R.C. Seem and A. Sztejnberg. 1996. *Fusarium proliferatum* as a biocontrol against grape downy mildew. *Phytopathology* 86: 1010

Fernandez-Pavia, S.P., N.J. Grunwald, M. Diaz-Valasia *et al.*, 2004. Soil-borne oospores of *Phytophthora infestans* in Central Mexico survive winter fallow and infect potato plants in the field. *Plant Dis.* 88(1): 29

Figueiredo, N.X. and K.H. Anahosur. 1993. Relative efficacy of two formulations of metalaxyl in the control of downy mildew of maize. *Indian Phytopath.* 46: 180

Friberg, H., J.. Lagerlol and B. Ramert. 2005. Germination of *Plasmodiophora brassicae* resting spores stimulated by a non-host plant. *Eur. J. Plant Pathol.* 113(3): 275

Garrett, K.A., R.J. Nelson, C.C. Mundt, G. Chacon, R.E. Jaramillo and G.A. Forbes. 2001. The effect of host diversity and other management components of epidemics of potato late blight in the humid highland tropics. *Phytopathology* 91(10): 993

Gaur, R.B., M.K. Kaul and R.N. Sharma. 2002. Integrated disease management of Phytophthora rot of kinnow. *J. Mycol. Pl. Pathol.* 32(3): 422 (abstract)

Gaur. R.B., M.K. Kaul and R.N. Sharma. 2004. Integrated disease management of Phytophthora rot in kinnow. *J. Mycol. Pl. Pathol.* 34(2): 465

Geetha, H.M. and H.S. Shetty. 2002. Induction of resistance in pearl millet against downy mildew disease caused by *Sclerospora graminicola* using benzothiadiazole, calcium chloride and hydrogen peroxide. *Crop Protection* 21(8): 601

Geetha, H.M. and H.S. Shetty. 2002. Expression of oxidative burst in cultured cells of pearl millet cultivars against *Sclerospora graminicola* inoculation and elicitor treatment. *Plant Science* 163(3): 653

Geetha, N.P., K.N. Amruthesh, R.G. Sharathcnadra and H.S. Shetty. 2005. Resistance to downy mildew in pearl millet is associated with increased phenylammonia lyase activity. *Functional Plant Biology* 32(3): 267

Gindro, K., R. Pezet and O. Viret. 2003. Histopathological study of the response of two *Vitis vinifera* cultivars (resistant and susceptible) to *Plasmopara viticola* infections. *Plant Physiol. Biochem.* 4(9): 846

Gobbin, D., M. Jermini, D. Loskill, I. Pertot *et al.*, 2005. Importance of secondary inoculum of *Plasmopara viticola* to epidemics of grapevine downy mildew. *Plant Pathology* 54(4): 522

Gotoh, K., S. Akino, T. Kiyoshi and S. Naito. 2005. Sexual mating preferences in vitro and in planta among Japanese isolates of *Phytophthora infestans*. *J. Gen. Plant Pathol.* 71(1): 29

Gravel, V., C. Martinez, H. Antoun and R.J. Tweddell. 2005. Antagonist microorganisms with the ability to control Pythium damping –off of tomato seed in rockwool. *BioControl* 50(5): 771

Gupta, G.K. 2001. Efficacy of metalaxyl formulations in controlling downy mildew of pearl millet and their residue under arid conditions. *Indian Phytopath.* 54: 210

Gupta, K. and G.S. Saharan. 2002. Identification of pathotypes of *Albugo candida* with stable characteristic symptoms on Indian mustard. *J. Mycol. Pl. Pathol.* 32(1): 46

Gupta, R., R.P. Awasthi and S.J. Kolte. 2004. Effect of sowing date and weather on development of white rust (*Albugo candida*) in rapeseed-mustard. *J. Mycol. Pl. Pathol.* 34(2): 441

Hamiduzzaman, M.M., G. Jakab, L. Barnavon *et al.,* 2005. Betya-aminobutyric acid-induced resistance against downy mildew in grapevine acts through the potentiation of callose formation and jasmonic acid dihnalling. *Mol. Plant Microb. Interact* 18(8): 819

Hammi, A., A. Bennani, A. El-Ismaili, Y. Msatef and M.N. Serrhini. 2001. Production and germination of oospores of *Phytophthora infestans* (Mont.) de Bary in Morocco. *Eur. J. Plant Pathol.* 107(5): 553

Hampson, M.C. and S.L. Wood. 1997. Detection of infective resting spores of *Synchytrium endobioticum* in vehicles. *Can J. Plant Pathol.* 19(1): 57

Hanson, K. and R.C. Shattock. 1998a. Effect of metalaxyl on formation and germination of oospores of *Phytophthora infestans*. *Plant Pathology* 47(2): 116

Heist, E.P., W.C. Nesmith and C.L. Schardl. 2002. Interaction of *Peronospora tabacina* with roots of *Nicotiana* spp. in gnotobiotic associations. *Phytopathology* 92: 400

Heyman, F., B. Lindahl, L. Persson *et al.,* 2007. Calcium concentrations of soil affect suppressiveness against Aphanomyces root rot of pea. *Soil Biol. Biochem.* 39(9): 2222

Jadeja, K.B., N.G. Mayani, V.A. Patel and M.T. Ghodasara. 2000. Chemical control of canker and gummosis of citrus in Gujarat. *J. Mycol. Pl. Pathol.* 30: 87

Jakab, G., V. Cottier, V. Toquin, G. Rigoli *et al.,* 2001. â-aminobutyric acid-induced resistance in plants. *Eur. J. Plant Pathol.* 107(1): 29

Jayaraj, J. and N.V. Radhakrishnan. 2008. Enhanced activity of introduced biocontrol agents in solarized soils and implications on the integrated control of tomato damping off caused by Pythium spp. *Plant Soil* 304: 189

Jayaraj, J., N.V. Radhakrishnan, R. Kannan *et al.,* 2005. Development of new formulations of *Bacillus subtilis* for management of tomato damping-off caused by *Pythium aphanidermatum*. *Biocontrol Sci. Technol.* 15(1): 55

Jayaraj, J., N.V. Radhakrishnan, and R. Velazhahan. 2006. Development of formulations of *Trichoderma harzianum* strain MI for control of damping-off of tomato caused by *Pythium aphanidermatum*. *Arch. Phytopath. Plant Prot.* 39: 1–8

Jayaraj, J., T. Parthasarathi and N.V. Radhakrishnan. 2007. Characterization of a *Pseudomonas flkuorescens* strain from tomato rhizoplane and use for integrated management of tomato damping off. *Biocontrol* 52(5): 683

Jhooty, J.S., S.S. Bains and Ved Prakash. 1989. Epidemiology and control of cucurbit downy mildew, pp. 301 – 314. In: V.P. Agnihotri *et al.,* (eds.) *Perspectives in Plant Pathology*. Today and Tomorrow's Printers and Publishers, New Delhi

Junior, V.L., L.A.Mattia *et al.,* 2006. Biocontrol of tomato late blight with the combination of epiphytic antagonists and rhizobacteria. *Biol. Control* 38(3): 331–340 Abs

Kennelly, M.M., M. Gadoury, W.J. Wilcox, P.A. Magarey and R.C. Seem. 2007 Vapour activity and systemic movement of mefanoxam controls grapevine downy mildew. *Plant Dis.* 91(10): 1260

Kiefer, B., M. Riemann, C. Buche *et al.,* 2002. The host guides morphogenesis and stomatal targeting in the grapevine pathogen *Plasmopara viticola*. *Planta* 215(3): 387

Kirk, W.W., K.J. Felcher *et al.,* 2001. Effect of host plant resistance and reduced rates and frequency of fungicide application to control potato late blight. *Plant Dis.* 85: 1113

Krominn, P., A. Taipe, J.L. Andrade-Piedra and G.A. Forbes. 2008. Preemergence infection of potato sprouts by *Phytophthora infestans* in the highland tropics of Ecuador. *Plant Dis.* 92(4): 569

Kumudini, B.S. and H.S. Shetty. 2002, Association of lignification and callose deposition with host cultivar resistance and induced systemic resistance in pearl millet to *Sclerospora graminicola*. *Aust. Plant Pathol.* 31(2): 157

Lai, A., V. Cianciolo, S. Chiavarini and A. Somnino. 2000. Effects of glandular trichomes on the development of *Phytophthora infestans* in potato (*S. tuberosum*). *Euphytica* 114(3): 165

Leon, M.C.C., A. Stone and R.P. Dick. 2005. Organic soil amendments impacts on snap bean common root rot (*Aphanomyces euteiches*) and soil quality. *Appl. Soil Ecol.* 31(3): 199–210

Leoni, C. and R. Ghini. 2006. Sewage sludge effect on management of *Phytophthora nicotianae* in citrus. *Crop Protection* 25(1): 10–22

Li, C.X., K. Sivasithamparam, G. Waltor, P. Salisbury *et al.,* 2007. Expression and relationships of resistance to white rust (*Albugo candida*) at cotyledonary, seedling, ande flowering stages in *Brassica juncea* germplasm from Australia, China and India. *Aust. J. Agric. Res.* 58(3): 259

Liu, C,H,, A.K. Mishra and R.X. Tan. 2006. Repellent, insecticidal and phytotoxic activity of isoalantolactone from *Inula racemosa. Crop Prot.* 25(5): 508

Lobato, C., F.P. Olivier, E.A.G. Altamiranda *et al.,* 2008. Phosphite compounds reduce disease sevcerity in potato seed tubers and foliage. *Eur. J. Plant Pathol.* **IN PRESS** DOI 10.1007/ SI 657

Mahajan, M., T.S. Thind and M. Dhillon. 2003. Structural variability among potato cultivars possessing varying grades of resistance to late blight. *Indian Phytopath.* 56(4): 443

Mani, M.T. and S.J. Hepziba. 2008. Biological management of pearl millet downy mildew caused by *Sclerospora graminicola. Arch Phytopath. Plant Prot.* **IN PRESS**

Mani, M.T. and S.J. Hepziba. 2008. Integrated disease management of pearl millet downy mildew caused by *Sclerospora graminicola. Arch.Phytopath. Plant Prot.* **IN PRESS**

Manzanares-Dauleux, M.J., I. Divaret, F. Baron and G. Tomas. 2001. Assessment of biological and molecular variability between and within field isolates of *Plasmodiophorea brassicae. Plant Pathology* 50(2): 165

Matheron, M.E. and M. Porchas. 2000. Impact of azoxystrobin, dimethomorph, flluazinam, fosetyl-Al and metalaxyl on growth, sporulation and zoospore cyst germination of three *Phytophthora* species. *Plant Dis.* 84: 454

Matheron, M.E. and P. Porchas. 2002. Comparative ability of six fungicides to inhibit development of Phytophthora gummosis on citrus. *Plant Dis.* 86(6): 687

Mayton, H., K. Myers and W.E. Fry. 2008. Potato late blight in tubers.The role of foliar phosphonate applications in suppressing pre-harvest tuber infections. *Crop Prot.* 27(6): 343

McKeller, M.F. and F.B. Nelson. 2003. Compost-induced suppression of *Pythium* damping-off is mediated by the fatty acid-metabolizing seed colonizing microbial communities. *Appl. Environ. Microbiol.* 69(1): 452

Mehta, N. and G.S. Saharan. 1998. Effect of planting date on infection and development of white rust and downy mildew disease complex in mustard. *J. Mycol. Pl. Pathol.* 28: 259

Mohan, C. and T.S. Thind. 1999. Persistence and relative performance of some new fungicides for effective management of potato late blight in Punjab. *J. Mycol. Pl. Pathol.* 29: 32

Monot, C., E. Pajot, D. Le Corre and D. Silue. 2002. Induction of systemic resistance in broccoli (*Brassica oleracea* var. *botrytis*) against downy mildew (*Peronospora parasitica*) by avirulent isolates. *Biological Control* 24(1): 75

Murakami, H., S. Tsushuma, T. Akimoto *et al.,* 2000. Effects of growing leafy daikon (*Raphanus sativus*) on population of *Plasmodiophora brassicae* (clubroot). *Plant Pathology* 49(5): 684

Naerstadm P., A. Hermansen and T. Bjor. 2007. Exploiting host resistance to reduce the use of fungicides to control potato late blight. *Plant Pathol.* 56(1): 156

Nakkeeran, S., K. Kavitha, G. Chandrasekara, P. Renukadevi and W.G.D. Fernandez. 2006. Induction of plant defence compounds by *Pseudomonas chlroraphis* PA 23 and *Bacillus subtilis* BSCBE 4 in controlling damping off of hot pepper caused by *Pythium aphanidermatum. BioControl Sci Technol.* 16(4):403

Narayana, Y.D., R. Bandopadhyay and K.H. Anahosur. 2002. Infection of *Peronosclerospora sorghi* at different growth stages of sorghum.*Indian Phytopath.* 55(2): 203

Narisawa, K., S. Tokumasu and T. Hashiba. 1998. Suppression of clubroot formation in Chinese cabbage by the root endophytic fungus, *Heteroconium chaetospira. Plant Pathology* 47(2): 206

Narisawa, K., M. Shimura, F. Usuki, S. Fukuhara and T. Hashiba. 2005. Effect of pathogen density, soil moisture and soil pH on biological control of clubroot in Chinese cabbage by *Heteroconium chaetospira. Plant Dis.* 89(3): 285

Niranjan Raj, S., G. Chaluvaraju, K.N. Amruthesh, H.S. Shetty, M.S. Reddy and J.W. Kloepper. 2003a. Induction of growth promotion and resistance against downy mildew on pearl millet (*Pennisetum glaucum*) by rhizobacteria *Plant Dis.* 87(4): 380–384

Niranjan Raj, S., S.A. Deepak, P. Basavaraju, H.S. Shetty, M.S. Reddy and J.W. Kloepper. 2003b. Comparative performance of formulations of plant growth promoting rhizobacteria in growth promotion and suppression of downy mildew in pearl millet. *Crop Protection* 22(4): 579

Niranjan Raj, S., N.P. Shetty and H.S. Shetty. 2004. Seed bio-priming with *Pseudomonas fluorescens* isolates enhances growth of pearl millet plants and induces resistance against downy mildew. *International J. Pest Management* 50(1): 41

Niranjan Raj, S., N.P. Shetty and H.S. Shetty. 2005. Synergistic effects of Trichoshield on enhancement if growth and resistance to downy mildew in pearl millet. *BioControl* 50(3): 493

Niranjan Raj, S., B.R. Sarosh and H.S. Shetty. 2006. Induction and accumulation of polyphenol oxidase activities as implicated in development of resistance against pearl millet downy mildew disease. *Functional Plant Biol.* 33(6): 563

Ongena, M., P. Jacques, Y. Toure, J. Destain *et al.,* 2005a. Involvement of fengycin-type lipopeptides in the multifaceted biocontrol potential of *Bacillus subtilis. Appl. Microbiol Biotechnol.* **IN PRESS**

Ongena, M., P. Duby, E. Jourdan, T. Beaudry *et al.,* 2005b. *Bacillus subtilis* M4 decreases plant susceptibility towards fungal pathogens by increasing host resistance associated with differential gene expression. *Appl. Microbiol. Biotechnol.* 67(5): 692

Oyarzun, P.J., J. Yanez and G.A. Forbes. 2004. Evidence for host mediation of preinfection stages of *Phytophthora infestans* on the leaf surface of *Solanum phureja*. *J. Phytopath.* 152(11–12): 651

Panicker, S. and K. Gangadharan. 1999. Controlling downy mildew of maize caused by *Peronosclerospora sorghi* by foliar sprays of phosphonic acid compounds. *Crop Protection* 8(2): 115

Porter, L.D., N. Dasgupta and D.A. Johnson. 2005. Effect of tuber depth and soil moisture on infection of potato tubers in soil by *Phytophthora infestans*. *Plant Dis.* 89(2): 146

Pozzebron, A. and C. Duso. 2008. Grape downy mildew, *Plasmopara viticola*, an alternative food for generalist predatory mites occurring ion vineyards. *Biol. Control* **IN PRESS**

Punja, B.N. and S. Chaudhuri. 2004. Exploitation of soil arbuscular mycorrhizal potential for AM-dependent mandarin orange plants by pre-cropping with mycotrophic crops. *Appl. Soil Ecol.* 26(3): 249

Pushpalatha, H.G., S.R. Mythrashree, R. Shetty, N.P. Geetha *et al.,* 2007. Ability of vitamins to induce downy mildew disease resistance and growth promotion in pearl millet. *Crop Protection* **IN PRESS**

Rajeswari, E., K. Chitra , K. Seetharaman *et al.,* 2008. Exploitating medicinal plants and phyloplane microflora for the management of grapevine downy mildew *Arch. Phytopathol. Plant Prot.* **IN PRESS**

Ramachandra Kini, K., N.S. Vasanthi and H.S. Shetty. 2000. Induction of ß-1,3-glucanase in seedlings of pearl millet in response to infection by *Sclerospora graminicola*. *Eur. J. Plant Pathol.* 106(3): 257

Ramamoorthy, V., T. Raguchander and R. Samiyappan. 2002. Enhancing resistance of tomato and hot pepper to *Pythium* disease by seed treatment with fluorescent pseudomonads. *Eur. J. Plant Pathol* 108(5): 429

Reuveni, M. 2001a. Activity of trifloxystrobin against powdery and downy mildew diseases of grapevine. *Can. J. Plant Pathol.* 23(1): 52

Reuveni, M., T. Zahavi and Y. Cohen. 2001b. Controlling downy mildew (*Plasmopara viticola*) in field-grown grapevine with *B*-aminobutyric acid (BABAB). *Phytoparasitica* 29(2): 1–9

Reuveni, M. 2003. Activity of the new fungicide benthiavalicarb against *Plasmopara viticola* and its efficacy in controlling downy mildew of grapevines. *Eur. J. Plant Pathol.* 109(3): 243

Reuveni, M., T. Zahavi and Y. Cohen. 2001. Controlling downy mildew (*Plasmopara viticola*) in field grown grapevine with beta-aminobutyric acid (BABA). *Phytopathology* 91(2)

Rohner, E., E. Carabet and H. Buchenauer. 2004. Effectiveness of plant extracts of *Paeonia suffruticosa* and *Hedera helix* against diseases caused by *Phytophthora infestans* in tomato and *Pseudoperonospora cubensis* in cucumber. *J. Plant Dis. Prot.* 111(1): 83

Rubin, E. and Y. Cohen. 2004. Oospores associated with tomato seed may lead to seedborne transmission of *Phytophthora infestans*. *Phytoparasitica* 32(3): 237

Saharan, G.S. 2000. Multiple disease resistance in rapeseed-mustard. *Indian Phytopath.* 53: 342

Saharan, G.S. and Indra Hooda. 2004. Control of white rust of mustard with biocontrol agents and plant extracts. *J. Mycol. Plant Pathol.* 34(2): 286

Sarosh, B.R., S. Sivaramakrshnan and H/S/ Shetty. 2005. Elucidation of defence related enzymes and resistance by L-methionine in pearl millet against downy mildew disease caused by *Sclerospora graminicola*. *Plant Physiol. Biochem.* 43(8): 808

Scott. 2005. Seasonal development of ontogenic resistance to downy mildew in grape berries and rachises. *Phytopathology* 95(12): 1452

Sealy, R., M.R Evans and C. Rothrock. 2007. The effect of a garlic extract and root substrate on soilborne fungal pathogens. *Hort Technology* 17: 151

Shailashree, S., B.R. Sarosh, N.S. Vasanthi and H.S. Shetty. 2001. Seed treatment with ß-aminobutyric acid protects *Pennisetum glaucum* systemically from *Sclerospora graminicola*. *Pest Manage. Sci.* 57(8): 721

Shailashree, S., K.K. Ramachandran and S.H. Shetty. 2007. B-aminobutyric acid–induced resistance in pearl millet to downy mildew is associated with accumulation of defence-related proteins. *Aust. Plant Patholo.* 36(2): 204–211

Shang, H., C.R. Grau and R.D. Peters. 2000. Oospore germination of *Aphanomyces euteiches* in root exudates and on the rhizosphere of crop plants. *Plant Dis.* 84: 994

Sharathchandra, R.G., S. Niranjan Raj, N.P. Shetty, K.N. Amruthesh and H.S. Shetty. 2004. A chitosan formulation Elexa induces downy mildew disease resistance and growth promotion in pearl millet. *Crop Protection* 23(10):881

Shetty, N.V., T.C. Wehner, C.E. Thomas, R.W. Doruchowski and K.P.V.Shetty. 2002. Evidence for downy mildew races in cucumber tested in Asia, Europe and North America. *Scientia Horticulturae* 94(3–4: 231

Shivakumar, P.D., H.M.Geetha and H.S. Shetty. 2003. Peroxidase activity and isozyme analysis of pearl millet seedlings and their implications in downy mildew disease resistance. *Plant Science* 164(1): 85

Silue, D., E. Pajot and Y. Cohen. 2002. Induction of resistance to downy mildew (*Peronospora parasitica*) in cauliflower by DL-*B*-N-butanoid acid (BABA). *Plant Pathology* 51(1):97

Silva, H.S.A., R.S. Romeino, R.C. Filho *et al.,* 2004. Induction of systemic resistance by *Bacillus cereus* against tomato foliar diseases under field conditions. *J. Phytopath.* 152(6): 371

Sindhan, G.S. and I Hooda. 2004. Control of white rust of mustard with biocontrol agents and plant extracts. *J. Mycol. Pl. Pathol.* 34(2): 286

Singh, B.P., J. Gupta and D.K. Rana. 2004. Production of *Phytophthora infestans* oospores *in planta* and inoculum potential of *in vitro* produced oospores under temperate highlands and subtropical plains of India. *Ann. Appl. Biol.* 144(3): 363

Singh, R.P. and V.S. Pundhir. 2004. Possible method of eradicating tuber-borne inoculum of *Phytophthora infestans* (Mont) de Bary. *J. Mycol. Plant Pathol.* 34(1):91

Singh, S.D. 1995. Downy mildew of pearl millet. *Plant Dis.* 79: 545

Singh, S.D. and S.S. Navi. 2000. Genetic resistance to pearl millet downy mildew.II. Resistance in wild relatives. *J. Mycol. Pl. Pathol.* 30: 167

Singh, S.D. and B.S. Talukdar. 1998. Complete resistance to pearl millet downy mildew. *Plant Dis.* 82:791

Singh, U.S., K.J. Doughty, N.J. Nashaat, R.M. Bennett and S.J. Kolte. 1999. Induction of systemic resistance to *Albugo candida* in *Brassica juncea* with an incompatible isolate. *Phytopathology* 89: 1236

Singh, U.S., N.I. Nashaat, K.J. Doughty and R.P. Awasthi. 2002. Altered phenotypic response to *Peronospora parasitica* in *Brassica juncea* seedlings following prior inoculation with an avirulent or virulent isolate of *Albugo candida Eur. J. Plant Pathol.* 108(6): 555

Slininger, P.J., D.A. Schisler, L.D. Ericsson *et al.,* 2007. Biological control of post-harvest late blight on potato. *Biocontrol Sci. Technol.* 17(6): 624

Son, S.W., H.Y. Kim, G.J. Choi, H.K. Lim *et al., 2008.* Bikaverin and fusaric acid from *Fusarium oxysporum* show anti-oomycete activity against *Phytophthora infestans. J. Appl. Microbiol.* 104(3): 692

Song, J., J.M. Bradeen and S.K. Naess. 2003. Gene *RB* cloned from *Solanum bulbocastanum* confers broad spectrum resistance to potato late blight. *Proc. Nat. Acad. Sci. USA* 100(16): 9128

Soylu, E.M. and S. Soylu, 2003a. Light and electron microscopy of the compatible interaction between *Arabidopsis* and the downy mildew pathogen *Peronospora parasitica. J. Phytopath.* 151(6): 300

Soylu, E.M., S. Soylu and J.W. Mansfield. 2004. Ultrastructural characterization of pathogen development and host responses during compatible and incompatible interactions between *Arabidopsis thaliana* and *Peronospora parasitica. Physiol. Mol. Plant Pathol.* 65(2): 67

Soyulu, E,M., S. Soyulu and S. Kurt. 2006. Antimicrobial activities of the essential oils of various plants against tomato late blight disease agent *Phytophthora infestrans. Mycopathologia* 151(2): 128

Soylu, S. 2004. Ultrastructural characterization of the host-pathogen interface in white blister-infected *Arabidopsis* leaves. *Mycopathologia* 158(4): 457

Mansfield. 2004. Ultrastructural characterization of interaction between *Arabidopsis thaliana* and *Albugo candida. Physiol. Mol. Plant Pathol.* 63(5): 201

Steddom, K., O. Becker and J.A. Menge. 2002a. Repetitive application of the biocontrol agent *Pseudomonas putida* 06909-rif/mal and effects on populations of *Phytophthora parasitica* in citrus orchards. *Phytopathology* 92(8): 850

Steddom, K., J.A. Menge, D. Crowley and J. Borneman. 2002b. Effect of repetitive application of the biocontrol bacterium *Pseudomonas putida* 06909-rif/nal on citrus soil microbial communities. *Phytopathology* 92(8): 857

Stephan, D., A. Schmitt, S.M. Carvalho *et al.,* 2005. Evaluation of biocontrol preparations and plant extracts for the control of *Phytophthora infestans* on potato leaves. *Eur. J. Plant Pathol.* 113(3): 235

Sudisha, J., M. Shigero, K.N. Amrithesh and H.S. Shetty. 2007. Activity of cyazofamid against *Sclerospora graminicola*, a downy mildew disease of pearl millet. *Pest Manag. Sci.* 63(7): 722

Thakur, R.P., B. Pushpawati and V.P. Rao. 1998. Virulence characteristics of single zoospore isolates of *Sclerospora graminicola. Plant Dis.* 82(7): 747

Thakur, R.P., V.P. Rao, J.G. Sastry, S. Sivaramakrishnan, K.N. Amruthesh and L.D. Brabind. 1999. Evidence for a new virulent pathotype of *Sclerospora graminicola* on pearl millet. *J. Mycol. Pl. Pathol.* 29: 61

Thakur, R.P., V.P. Rao, B.M. Wu, K.V. Subbarao, H.S. Shetty, G.Singh *et al.,* 2004. Host resistance stability to downy mildew in pearl millet and pathogenic variability in *Sclerospora graminicola. Crop Protection* 23(10): 901

Thomas, C.E. and E.L. Jourdain. 1992. Host effect on selection of virulence factors affecting sporulation by *Pseudoperonospora cubensis. Plant Dis.* 76: 905

Thuerig, B., A. Binder, T. Boller *et al.,* 2006. An aqueous extract of the dry mycelium of *Penicillium chrysogenum* induces resistance in crops under controlled and field conditions. *Eur. J. Plant Pathol.* 114(2): 185

Tran, H., A.Ficke, T.Asiimwe *et al.,* 2007. Role of the cyclic lipopeptide massetolide A in biological control of *Phytophthora infestans* and colonization of tomato plants by *Pseudomonas fluorescens. New Phytologist* 175(4): 731

Umesha, S., M.S. Dharmesh, S.A. Shetty, M. Krishnappa and H.S. Shetty. 1998. Biocontrol of downy mildew disease of pearl millet using *Pseudomonas fluorescens. Crop Protection* 17(5): 387

Umesha, S., M.S. Dharmesh and H.S. Shetty. 2000. Lytic activity in pearl millet: its role in downy mildew disease resistance. *Plant Science* 157(1): 33

Ungerm C., L. Wilhelm, R. Jonger *et al.,* 2006. Evidence of induced resistance of tomato plants against *Phytophthora infestans* by a water extract of dried biomass of *Penicillium chrysogenum. J. Plant Dis. Prot.* 113 (5)

214

Unger, S., C. Buche, S. Boso and H. Kassemeyer. 2007.The course of colonization of two different *Vitis* genotypes by *Plasmopara viticola* indicates compatibl and incompatible host-pathogen interactions. *Phytopathology* 97(7): 780

Val, F., S. Desender, K. Bernard *et al.,* 2008. A culture filtrate of *Phytophthora infestans* primes defense reaction in potato cell suspension. *Phytopayhology* 98(6): 653

Vallad, G.E., L. Cooperband and R.M. Goodman. 2003. Plant foliar disease suppression mediated by composted forms of paper mill residuals exhibits molecular features of induced resistance. *Physiol. Mol. Plant Pathol.* 63(2):65

Vishunawat, K. and S.J. Kolte. 1993. Brassica seed infection with *Peronospora parasitica* and its transmission through seed. *Indian J. Mycol. Plant Pathol.* 23: 247

Vishwakarma, S.N. and R.S.Singh. 1984. Possible control of the potato late blight with zinc sulphate-lime mixture. *Indian J. Agric. Sci.* 54: 774.

Visker, M.H.P.W., L.O.P. Keizer, D.J. Budding, *et al.,* 2003. Leaf position prevails over plant age and leaf age in reflecting resistance to late blight in potato. *Phytopathology* 93(6): 666

Visker, M.H.P.W., H.M.G. van Raaijm L.C.P. Keizer, P.C. Struik and L.T. Colon. 2004. Correlation between blight resistance and foliage maturity type in potato. *Euphytica* 137(3): 311

Wadhwa, N., S. Jain, H.K.L. Chawla and M.S. Panwar. 2001. Changes in activities of elicitor-releasing enzymes in pearl millet leaves infected with *Sclerospora graminicola. Indian Phytopath.* 54(1): 414

Wang, H., S.F. Hwabg, K.F.Chang, G.D. Turnbull and R.J. Howard. 2003. Suppression of important pea diseases by bacterial antagonists. *BioControl* 48(4): 447

Wang, W., B.H. Ben-Daniel and Y. Cohen. 2004. Control of plant diseases by extracts of *Inula viscosa. Phytopathology* 94(10): 1042

Weltzien, H.C. and N. Ketterer. 1986. Control of downy mildew, *Plasmopara viticola,* on grapevine leaves through water extract of composted organic wastes. *J. Phytopath.* 116: 186

Zaidi, S., S. Barbedette, J.F. Godard and S. Silue. 2001. Production of pathogenesis related proteins in the cauliflower (*Brassica oleracea* var. *botrytis*)-downy mildew (*Peronospora parasitica*) pathosystem treated with acibenzolar-S-methyl. *Plant Pathology* 50(5): 574

Zhu, Y.J., X. Qiu, P.H. Moore *et al.,* 2003. Systemic acquired resistance induced by BTH in papaya. *Physiol. Mol. Plant Pathol.* 63(5): 237

CHAPTER

8

Diseases Caused by Ascomycota

POWDERY MILDEW OF PEA

This disease of worldwide occurrence is much more serious than the downy mildew of pea because of its more frequent occurrence and coverage of a much larger host surface area. In India it is known since as early as 1910. It usually develops late in the season reaching its maximum intensity when pods are forming. It is worst in dry weather unlike the downy mildew which flourishes in wet weather. Early varieties are less damaged. Varieties maturing in January usually escape the maximum intensity of the disease. The loss is proportionate to the disease intensity and varies considerably depending on the stage of plant growth at which the disease occurs. In a 100% infected crop the reduction in pod numbers is estimated to be about 21–31% and reduction in pod weight about 24–27%.

Symptoms

The powdery mildew first appears on the leaves and then on other green parts of the plant. Its attack is characterized by white floury patches on both surfaces of leaves as well as on tendrils, pods, and stems. These patches originate as minute discolored specks from which a powdery mass radiates on all sides. When the disease has advanced, large areas on the aerial parts of the host may be covered with these white floury patches. The superficial mass consists of mycelium and spores of the fungus causing the disease.

Histological changes in the host include collapse of the necrotic epidermal cells in case of subinfections, collapse of tissue below the penetrated epidermal cells, and movement of the host nuclei towards the haustoria of the fungus. Physiology of the host is also affected. There is increased transpiration, especially during night. Respiration is also increased and photosynthesis decreased.

The causal organism

The disease is caused by *Erysiphe pisi* DC. var. *pisi* (Eumycota, Fungi, Ascomycota, Ascomycetes, subclass Erysiphomycetidae, order Erysiphales, and family Erysiphaceae). Physiologic races of the species were considered to attack common bean, urdbean, lucerne, lentil, groundnut, coriander, turnip, cabbage and many other plant species of different families. Other names for the pathogen listed in literature are *E. pisi* f.sp. *medicaginis-sativae, E. pisi-sativae* and *E. polygoni*.

The mycelium of the pea powdery mildew fungus is generally fine, persistent, rarely thick. It is ectophytic, the entire thallus, except the haustoria in epidermal cells, is present on the host surface. The haustoria develop as outgrowths from lobed swellings (appressoria) on the sides of the hyphae adjacent to the host surface. They first arise as exceptionally narrow tube from the appressorium, penetrating the cell wall and swelling into a rounded sac in the epidermal cell.

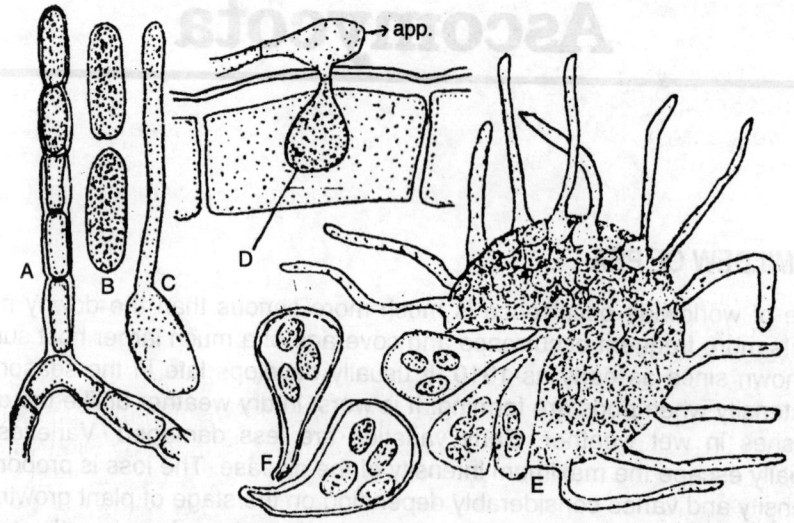

Fig. 44. *Erysiphe pisi.* A-Conidiophore; B-Oidia; C-Germination of oidium; D-Haustorium in host cell; E-Broken cleistothecium with asci; F-Asci with ascospores.

The conidiophores arise vertically from the superficial hyphae on the host surface. They are septate and their cells are not much different from conidia in shape. Conidia are formed single or in short chains. Only the end conidium in the chain is mature at a time as the spores are formed in regular order from the tip backward. The conidia are hyaline, ellipsoid to ovate with vacuolate cytoplasm, and measure 31–38 × 17–21 (25–35 × 13–16) μm. The ripe conidia fall off quickly and are dispersed by wind. Hyphal cells, cells of the conidiophores, conidia and haustoria all are uninucleate.

The sexual fruits (cleistothecia) develop as short, black, minute bodies scattered in the mycelial web. They are round structures measuring about 90 μm in diameter. The peridium of these cleistothecia is provided with a number of myceloid appendages. The appendages may be few or many, free or interwoven and resemble the ordinary hyphae. Usually 2–3 asci are formed in each cleistothecium. These asci originate from a single point in the fruit body and appear in a fan-like manner. They are ovate, nearly sessile, 46–72 × 30–45 μm in size. The asci contain 3–8 ascospores which are hyaline, elliptical, unicellular, and measure 19–25 × 9–14 μm.

Disease cycle and environmental relations

Erysiphe pisi is a biotrophic parasite. The general belief had been that the pathogen survives between crop seasons through cleistothecia in soil. These structures are not common in the crop but were supposed to be formed on fallen leaves conditioned by soil temperature and moisture. They persist in soil until the following crop season when their walls disintegrate and ascospores are released. These spores cause infection of lower leaves near the soil line. The growth that develops produces conidia for secondary spread of the disease. There had also been a belief that the fungus mycelium reaching the seed from infected pods

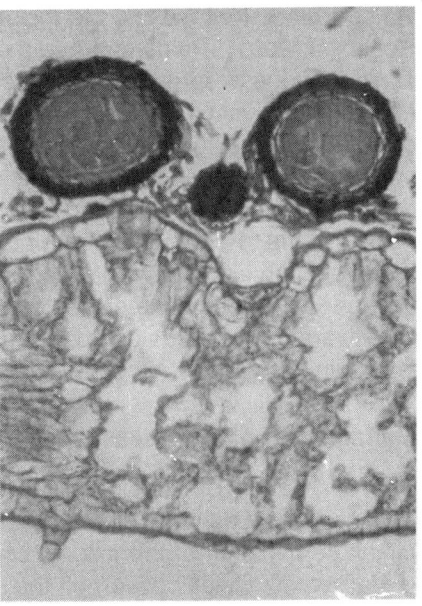

Fig. 45. Erysiphe Cleistothecium on leaf surface cross section.

is carried as dormant mycelium and such seeds initiate the disease in the next crop. Both these views were contradicted in 1990 by workers in Punjab. They did not find any evidence of external or internal seed-borne nature of the pathogen. Cleistothecia were not found by them in the plains of Punjab and rarely in the hills. Cross inoculation with the fungus isolated from many hosts including lentil, *Trigonella polycerata* and *T. foenum-graecum* showed that only the isolates from lentil could produce the disease in pea. Lentil and pea are crops grown in the same season. This excluded the possibility of any role of alternative hosts. In their opinion, since the pea crop is grown throughout the year in one or the other region of Punjab and Himachal Pradesh the pathogen is present in conidial stage throughout the year. It survives in its mycelial and conidial stage on the main host in the plains during the winter months and in cooler regions or on higher altitudes of Himachal Pradesh during the summer months. The primary inoculum in the form of conidia comes from these sources by wind currents to infect the main crop in the plains.

When conidia land on a hydrophobic surface, they quickly secrete extracellular material which helps in adhesion and also in germination. The germ tubes emerge from that side of the conidium which faces direct light *(Fujita et al.,* 2004). Physical structure and chemical compositiopn of the host surface, and especially of epicuticular leaf waxes influence the pre-penetration processes of the pathogen. In general, three closely interrelated principal environmental factors, namely, temperature, humidity and light play significant role in development of the disease in the field. The pea powdery mildew conidia germinate readily in absence of water around them because of their own relatively large water content. Because of their large water content, these conidia are among the shortest lived air disseminated spores. The conidia can germinate at any temperature between 10° and 30°C. Maximum germination occurs at 20°–24°C. Germination of conidia is by a single germ tube on glass slides but by several germ tubes on susceptible host leaves. Each germ tube is effective in infection and forms appressorium. At constant temperature over the range of 10°–25°C the incubation period varies from 4 to 11 days. No visible colonies develop at 30°C even after 19 days (Xu and Robinson, 2001). Rains damage the conidiophores and, therefore, there is less production of conidia and less spread of the disease. Pea powdery mildew flares up when the average of maximum and minimum temperature exceeds 20°C. High humidity is not a pre-requisite for the disease. In Rajasthan, intensity of the pea powdery mildew is maximum in the crop sown in mid-October and minimum in the crop sown in the last week of November (Sharma and Sharma, 2002). Plant spacing has no effect on disease intensity. The early development of mildew is not affected by light but is affected by photosynthetic area of the leaf lamina. It fails to develop on albino mutants of the host where a lesser number of appressoria and haustoria are formed. On highly resistant or resistant genotypes fewer conidiophores with conidia per colony are produced as compared to moderately resistant or susceptible genotypes. Darkness is known to retard development of powdery mildews, possibly due to reduced photosynthesis.

Management of the disease

Soil application of vermicompost is reported to reduce pea powdery mildew severity by enhancing accumulation of phenolics in the leaves. Aqueous extracts of the compost (01–05%) inhibit germination of the mildew conidia. Best protection against powdery mildew of pea is a protective fungicidal cover of the foliage. Sulphur fungicides have been most commonly recommended. Sulphur dust (25–30 kg/ha) was an old recommendation and is even now very effective though costly. Later, the formulations based on lime sulphur became more common. Elasol (0.5%), Morocide (0.1%), Cosan (2 kg/ha), Morestan and, best of all, Sulfex (2.5 kg/ha) were all effective in controlling the disease. Other fungicides are Sultal (0.9 kg a.i./ha), Karathane (0.1%), Calixin (0.1 or 0.05%) and Bavistin (0.1%). Two to three sprays are required. Bitertanol, triarimenol and dichlobutazol are some of the more recent fungicides used against the disease. These are not only protective but curative also in their action. Kapoor and Thakur (1997) suggested economical schedules with better cost: benefit ratio. One spray of Bayleton (triadimefon) followed by 3 sprays of Karathane or one spray of Bavistin (carbendazim) followed by 2 sprays of Karathane are most economical and cost effective schedules. Individually, when used alone, 3 sprays of Karathane or 2 sprays of Bayleton are most economical.

Efficacy of some plant extracts and plant products against the pathogen has been experimentally demonstrated Nemadole (a neem product) and extracts of *Allium cepa* (onion), *A. sativum* (garlic), rhizome of ginger and neem leaves (*Azadirachta indica*) are non-phytotoxic but fungicidal and at par with Karathane in the suppression of pea powdery mildew. The antimicrobial substance in crushed garlic cloves is a volatile sulphur compound allicin (diallylthiosulphimate) which is formed action of the enzyme allin-layse on the precursor sulphur compound alliin. Neemazal, a natural product of neem, induces resistance to pea powdery mildew. It significantly retards several growth parameters of the pathogen, *viz.*, multiple germ tube formation, branches and haustoria and colony size on pea leaves. There is browning of host cells associated with appressoria indicating hypersensitive reaction. Garlic has been shown to inhibit the growth of a variety of microorganisms and some viruses Allicin and ajoen as the major constituent of garlic have been known for the antimicrobial activity. Ajoen, at 100 ppm was found inhibitory to the powdery mildew fungus (Singh, 2000). Rhizome powder of ginger gives as good control as Sulfex and Bavistin.

Several fungi such as *Ampelomyces*, *Cladosporium*, *Tilletiopsis* and *Verticillium* and insects (*Thrips tabaci*) are natural biocontrol agents of the powdery mildew. Seed bacterization with selected strains of *Pseudomonas fluorescens* and *Pseudomonas aeruginosa* alone or in combination with aerial spray of their cell suspension also provides resistance to the disease through induction of host resistance and through antagonism. Singh *et al.,* (2003) have reported that pre-inoculation of pea leaves with powdery mildew fungi (*Oidium, Phyllactinia*), not pathogenic to pea, induces resistance to pea powdery mildew. The resistance could last for 12 days. The effect did not involve hypersensitive cell death or lignin accumulation. Conidial germination, appressorium formation and secondary branch development of *E. pisi* were reduced. Application of the antagonist *Gliocladium roseum* (*Clonostachys rosea*) to roots of tobacco is known to suppress *Erysiphe* in tobacco. This approach does not seem to have been tried in pea against *E. pisi.* Root colonization by the arbuscular mycorrhizal fungus Glomus mosseae enhances resistance to powdery mildew by increasing total phenol accumulation in eaves.

The resistant genotypes of pea include P 185, P 388, P 6563, P 6583, P 6587, P 6588, T 10, T 56, Rachna, DMR 9, DMR 11, Pant P-5, Pant P-8, DPS 3, DPS 6, HUP 2, FC-1, JM-5, JP-4, JP- 83, PM-2. The cultivar P 388 is considered to possess certain degree of stable resistance. In Himachal Pradesh, the pea cultivars identified as resistant to powdery mildew (less than 10% infection) are Sugar Giant, Pant P-8, DPP-54, DPP-+26, PMR-3, JP-71, HPPC-95. In pea, two simgle recessive genes (*er1* and *er2*) have been identified for resistance to powdery mildew. Genotypes containing the genes *er1* show inhibition of penetration of epidermal cells and very few haustoria or colonies are formed. The resistance conferred may be complete or almost complete. There is no visible necrosis at the site of attempted penetration. The genes *er2* operate at post-penetration level. The resistance conferred by these genes increases with temperature and leaf age. Complete resistance is expressed only at high temperature (25°C) (Fondevilla *et al.,* 2006).

POWDERY MILDEW OF CEREALS

The disease is common in temperate climate regions. In India it is found in lower hills of north India and in some parts of Rajasthan. It occurs on wheat, barley, oats, rye as well as many grasses such as species of *Agropyron, Bromus, Dactylis* and *Elymus.* The host is not killed but the general effect is defoliation and weakening of the plant which is more noticeable under glasshouse conditions than in the open. The disease adversely affects ear length and grain weight (Kapoor and Singh, 1993). Losses in cereals grown in rotation and on well drained lands are negligible but if the crop is cultivated on the same land year after year the damage may be very high. Occurrence of powdery mildew along with leaf rust and downy mildew on wheat is also reported.

Symptoms

The symptoms are much alike on all the cereals and generally limited to the upper surface of the leaf which is more severely affected than the lower surface. When conditions are favourable the sheath, stem and glumes are also affected. The fungus develops numerous superficial colonies. The mycelium forms flocculent growth, at first white when conidia are being formed, thereafter changing to gray or reddish brown when cleistothecia are

Fig. 46. Powdery mildew of wheat.

developing. Infected plants remain stunted due to reduction in size and number of leaves. The leaves that are not shed become wrinkled, spirally twisted and deformed. Transpiration and respiration in the affected plants are increased. Photosynthesis is considerably decreased.

The causal organism

Blumeria graminis (DC,) Seer, earlier known as *Erysiphe graminis* DC., is the only species of the genus *Blumeria*. *Blumeria* is differentiated from *Erysiphe* on the basis of certain characteracters such as digitate haustoria, secondary mycelium with bristle-like hyphae, bulbous swellings of the conidiophores and characters of the ascocarps. Taxonomic position of the genus is same as *Erysiphe* (kingdom Fungi, phylum Ascomycotina, class Ascomycetes, subclass Erysiphomycetidae, order Erysiphales, family Erysiphaceae). *Blumeria graminis* is classified into 8 *formae speciales* (f.sp.) The species that attacks wheat is known as *B. graminis* f.sp. *tritici* (*E. graminis* f.sp. *tritici*) and the one that attacks barley is known as *B. graminis* f.sp. *hordei* (*E. graminis* f.sp. *hordei*). *Blumeria graminis* f.sp. *tritici* infects only wheat and *B. graminis* f.sp. *hordei* infects only barley and closely related plants.

The mycelium of *B. graminis* is superficial, sparingly branched with small appressoria and haustoria. Infection is initiated with the germination of conidia (conidiospores) on the plant leaf surface, followed by the formation of appressoria from which develop infection hyphae or penetration pegs. These hyphae breach the host epidermal cell walls. The infection induces dome-shaped extension of the inner surface of the walls probably by physical pressure and enzyme action. The tips of the infection hyphae then expand to form the multifingered haustoria that invaginate but do not penetrate the host plasma membrane. This means all nutrients for fungal growth from the host and potential signal molecules from the pathogen must cross a double membrane interface. The pathogen is an obligate biotroph. To maintain its biotrophic character it induces dynamic changes in the primary metabolism during biotrophic growth in the host.

The superficial septate hyphae with uninucleate cells, 4–5 μm wide, are interlaced to form a web covering the host surface. Conidiophores arise from hemispherical swellings on the mycelial web at right angle to the leaf surface. Each conidiophore forms a chain of 10–20 conidia, the eldest one being at the top and falling off quickly. The life of an individual conidiophore of *B. graminis* *hordei* on barley leaf (from the erection of the conidiophore to the release of the final conidium is 107 hours and during this period each conidiopore produces 33 conidia (Moriura *et al.,* 2006). The conidia are hyaline, elliptical, uninucleate and 25–40 x 8–10 μm in size. These conidia are produced in enormous numbers and are disseminated by wind.

B. graminis appears to be unique among the powdery mildews in that its conidia regularly produce two germ tubes. The first germ tube (primary germ tube or PGT) to emerge remains short (2–10 μm) and does not form appressorium. It Perception of the host surface for suitability of the effective long germ tube was aupposed to be one function of the PGT. It is now reported (Yamaoka *et al.,,* 2007) hat PGT is responsible for the suppression of resistance induction in a host plant cell Shortly after emergence of this PGT, but after the host shows response to its presence, appressorial germ tube (AGT) grows out from the conidium. The growth of the PGT must stop to enable the AGT elongate quickly (Yamaoka *et al.,* 2006) to attain its mature length of about 35 μm while its tip swells to form a hooked appressorium. A septum forms close to the conidial body to separate the appressorium from the conidium.

The initial contact between *B. graminis* f.sp. *hordei* and the host leaf takes place on epicuticular waxes at the surface of the leaf. Here the extent to which chemical composition, crystal structure and hydrophobicity of cuticular waxes affect the pre-penetration processes. In 1990, Carver and Thomas (*Plant Pathology* 39:367) had reported that on intact leaf and on the leaf from which the cuticular wax layer was stripped off, the overall development of B. graminis hordei was very similar except for slightly lower rate of conidial germination and greater length of the appressorial germ tube on the stripped leaf. Of the two major components of waxes on barley leaf, hexacosanol and hexacosanal, the latter significantly stimulates germination and differentiation of the conidia. Removal of total leaf cuticular waxes results in 20% reduction in conidial germination (Zabka *et al.*, 2008).

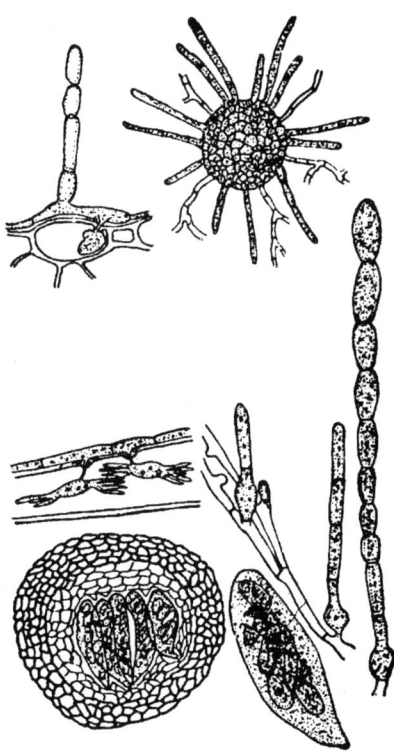

Fig. 47. *Blumeria graminis tritici.*
Haustorium, conidiophores, cleistothecium and asci.

Metabolic activity is required for adhesion of conidia to the host surface. Conidial surface morphology changes due to contact with the host surface. The conidial surface is covered with spine-like wall protrusions. Within 10 min of contact the conidium releases a film that covers the reticulate network. The film thickens. It causes loss of leaf surface integrity and wax crystals disappear. All this happens within 30–60 min before the germination of the conidium. The preparation of infection court is facilitated by release of esterase enzymes in the extracellular film which cause cuticular degradation. Nielson *et al.*, (2000) observed that when conidia land on an appropriate hydrophobic surface (host leaf) the contact between surfaces immediately induces release of an extracellular matrix (ECM) from the body of the conidium These pad-like deposits of extracellular material are the first outcome of the contact between host and conidial surfaces. This may happen within a minute of contact (Wright *et al.*, 2002a). The liquid exudates partially erodes the cuticle of the leaf by means of releasing esterase. In infection of barley coleoptile cells, conidia of *B. graminis hordei*, as also other powdery mildew fungi non-pathogenic on barley, before germination release the ECM, quantity of which depends on the amount of surface contact with the host. More ECM is released if full rather than partial cell contact is made (Fujita *et al.*, 2004). Within 5–6.5 hours of receiving the ECM from non-pathogenic strains, cells show induced inaccessibility to *B. graminis hordei*.

Adhesion of conidia on the leaf surface is weaker than adhesion of the germ tube but sufficient to hold the majority of conidia in place even under very windy field conditions (Wright *et al.*, 2002b). Site of emergence of the primary germ tube from the conidium is the actual point of contact between the two surfaces and is determined within a minute of deposition of the spore (Wright *et al.*, 2002c). An aldehyde component of the cuticular wax is responsible for appressorium differentiation on barley epidermal surfaces. Waxes of non-host plants cause less appressoria formation (Tsuba *et al.*, 2002) although the soluble carbohydrate elicitor from the pathogen is readily recognized not only by leaf surfaces of wheat or barley but also by a broad spectrum of cereals. The recognition activates general defense responses other than hypersensitive response. Cellulase enzymes and mechanical force help penetration by the germ tube. This type of adhesion activity is common in many other fungi. The extracellular matrix material extends to the appressorium and helps its adhesion. Presence of calcium ions at infection site on barley pomotea penetration by *B. graminis hordei* but not by a necrotrophic pathogen *Helminthosporiu*.

The fungus is heterothallic. The cleistothecia are globose, 160–192 × 120–130 µm in size, black and partly immersed in the mycelial weft. Simple or slightly branched pale brown appendages are present on surface of the cleistothecium. Each cleistothecium contains 9–30 cylindrical or ovoid, more or less distinctly pedicellate asci measuring 70–108 × 25–40 µm. Each ascus contains 8, sometimes 4, elliptic, subhyaline to pale brown ascospores that measure 20–23 x 10–13 *µm*.

At least 5 physiological races in *E. graminis tritic* were reported until 1960s. In 1990, 14 races from 37 conidial isolates and 5 races from 12 ascosporic isolates in Himachal Pradesh. Most common races were 3, 12 and 13. Races from conidial isolates were more virulent than those from ascospores. Pathania *et al.*, (1997) studied 84 conidial and ascospore isolates on 11 differentials and grouped them in 40 distinct pathotypes. Pathotype 11 from among conidial isolates was least virulent and pathotype 23 most virulent. Among ascospore isolates the pathotype 27 was least virulent and pathotypes 30, 33, 38 and 40 most virulent.

Disease Cycle and Environmental Relations

In temperate climate regions the pathogen is soil-borne through its cleistothecia. These structures can survive for 13 years at low temperatures. However, survival through this method is considered of secondary importance. Ascospores are formed most rapidly at 20°–27°C if the cleistothecial material is exposed to alternate drying and wetting in soil. Cardinal temperatures for discharge of ascospores of *B. graminis hordei* are 6°, 16°–24° and 24°C. Ascospores germinate best at 16–20°C. In temperate climate regions the infection by ascospores leads to production of conidial inoculum for secondary spread. The mycelium grows best at 20°–21°C. Although conidia can germinate at a temperature range of 3°–31°C, the optimum is about 15°C. Conidia do not remain viable at high temperatures and a temperature of 30°C has deleterious effect on the disease. Low temperatures of 5–9°C are more favorable for starting germination of conidia. The cereal powdery mildew fungus thrives well at low as well as high relative humidity. One hundred percent relative humidity and temperatures of 5–20°C are optimal for conidial germination. Incubation period is about 7 days. Infection is checked when the leaves are kept in a turgid conditions. Early sown crops show more incidence of the disease and heavily infested plants are killed by frost.

Development of the pathogen is bound with the living host during active metabolism. Thus, the infection is favored by all those conditions which favor the vital process of the host. A surplus carbohydrate supply, above the normal requirement, is necessary for disease development. Under normal conditions, light is essential for mildew formation. Over-manuring increases the incidence of the disease due, probably, to the unbalanced application of nitrogen and deficiency of potassic compounds. Application of phosphorus reduces the disease. Silicon status of wheat leaves influence infection by *B. graminis tritici*. Exogenous application of silicon as nutrient solution or calcium silicate slag protects wheat plants from powdery mildew (Belanger *et al.,* 2003). Silicate treated plants responded to the pathogen by defense reaction such as papilla formation, callose production and release of glycosilated phenolics. In a recent study, Guevel *et al.,* (2007) used different formulations of soluble silicon for root and foliar application and observed that root application resulted in accumulation of silicon in the leaves and provided 80% control of powdery mildew. Foliar application with nutrient solution also significantly reduced the disease.

Planting of cultivar mixtures delays and reduces the powdery mildew incidence. Resistance and susceptibility to powdery mildew is determined at the single-cell level at the penetration site. In barley, even in compatible interactions, attacked plant epidermal cells defend themselves against attempted fungal penetration by localized responses leading to papilla deposition and reinforcement of their cell walls. But these responses are not complete and are only temporary. Nonspecific resistance to penetration due to papilla formation and race specific hypersensitive response can both contribute to *Blumeria graminis* resistance in barley. Nitric oxide generation is one of the earliest response of barley epidermal cell defense and may be important in both the initiation and the development of effective papillae and cell death due to HR (Prats *et al.,* 2005) The avirulent strains cause death of the penetrated epiderma cell and make adjoining cells resistant to future attack. The attack of a virulent strain causes invasion of the penetrated and adjoining cells which become susceptible to future attack of avirulent strains.

There have been different opinions about the disease cycle in India. According one opinion, asci and ascospores, even if formed, have practically no chance in nature to germinate and cause infection. According to one view in India, asci remain immature in cleistothecia on fallen leaves but ascospores form in them after 10 months if the leaves are subjected to alternate drying and wetting of the soil. Such conditions exist in the low-lying fields where the disease is also common. However, in 1930, it had been suggested by K.C. Mehta that the disease might be introduced every year through conidial inoculum surviving in cooler environments of high altitudes. Cleistothecia are a source of primary inoculum in temperate regions like Lahaul and Spiti in Himachal Pradesh whereas in other parts of north India conidia introduce the disease. The pathogen present in its conidial stage on suitable hosts throughout the year in cooler areas produces conidia for introducing the disease at places where cleistothecial inoculum is not surviving.

Management of the disease

Powdery mildew of cereals can be checked by adopting the same management strategies as recommended for powdery mildew of pea. In regions where cleistothecia are formed, crop stubble management is recommended. Pyrimidinr systemic fungicides (ethirinol, dimethirimol), Benlate, Karathanand and Calixin had been been recommended for good disease control in early 1970s. The triazole fungicides Baytan, Bayleton and Vigil (0.1–0.2%) give 100% control for 42 days when used as seed dressing and as foliar spray. These compounds have better persistence than ethirimol, Nimrod and Thiovit. Seed treatment with Bayleton (0.1–0.2%) gives protection to plants for at least 7 days and spray of 10ʋ–500 μm/ml suspension gives protection against the disease up to 50 days. Other fungicides used against cereal powdery mildew are triademefon, prochloraz, azoxystrobin and epoxicinazole. These fungicides control the mildew even when sprayed late in the growing season.

Application method influences efficacy of fungicide sprays, particularly when used at reduced rate. Fine or medium spray nozzles give better control even at half dose than coarse nozzles (Barber et al., 2003). The systemic benzophenone fungicide, metrafenone, is a new highly effective compound against cereal powdery mildews. It affects different morphological structures of the pathogen. There is acute malformation and disruption of conidiophores and swelling and distortion of hyphal tips (Opalski et al., 2006; Schmitt et al., 2006). Quinoxyfen, released in late 1990s, is a protectant fungicides that gives excellent control of powdery mildew of cereals and apple. Resistance to this protectant fungicide is also not uncommon. Cyflufenamid is a newly developed fungicide in Japan [J.Pestic. Sci.31(2),95, (2006)] which is highly effective against powdery mildews of most plants at very low concentration (0.8–1.6 ppm). It has excellent preventive, curative, long residual, transliminar and vapor phase activities. Against Monilinia fructicola (brown rot of stone fruits) the fungicide is effective even at 0.01 ppm. The fungicide does not work against infection before the formation of appressoria but significantly inhibits the formation of haustoria, colonies as well as spores.

Spray of chitosan and benzothiadiazole (BTH) 3 days before inoculation of barley with B. graminis hordei induced resistance to powdery mildew. Both compounds induced oxidative burst and accumulation of phenolics which restricted invasion by the pathogen.(Faoro et al., 2007). The greater effect of BTH was due to (i) greater reinforcement papilla (ii) a higher level and the more homogenous diffusion of hydrogen peroxide in the treated leaf tissues and (iii) an induced hypersensitive-like response in many penetrated cells. Foliar spray of a-amino-isobutyric acid one day before or after inoculation was found to reduce the size, number and sporulation of mildew colonies on wheat seedlings and reduced the disease severity. Jasmonic acid (JA) which is known as inducer of defense responses in many host-pathogen combinations, is reported to directly inhibit appressoria differentiation of B. graminis hordei in barley without being involved in any signal transduction mechanism. Although the defense activator acibenzolar-S-methyl (ASM) induces systemic acquired resistance in barley against powdery mildew its effectiveness is impaired by root associated mycorrhizal fungus Glomus sp. and Heterodera avenae. Foliar application of potassium chloride is reported to reduce symptoms of powdery mildew on wheat by reducing spore germination and leaf area affected by the mildew. The mechanism involves osmotic change in the leaf where leaf water potential decreases. Soluble silicon (Si) has been shown to induce resistance in wheat against B. graminis f. sp. tritici. How Si is able to exert the protective activity is yet to be fully elucidated, although roles including providing a physical and/or biochemical defence system have been proposed. One possible mechanism is that deposition of silicon in cell walls hinders penetration and reduces susceptibility to enzymic degradation by the pathogen. An alternative explanation is that Si is a biologically active element and triggers a broad range of natural defences. In silicon treated plants presence of phenolic like material associated with degraded powdery mildew haustoria was noted and suggested production of phytoalexins in response to powdery mildew infection (Belanger et al., 2003; Remus-Borel et al., 2005). Changes in enzyme activities are also reported by growing plants in silicon amended soil. Silicon fertilizers are costly.

The alternative is to incorporate wheat crop residue or straw into the soil. It improves silicon uptake. Extracts of the plant Reynoutria sachalinensis (Polygonaceae) are known to induce resistance to powdery mildew in many plants. The formulated product of the extract, Milsana, inhibits germination of conidia of *B. graminis* f.sp. *tritici* and elicits mild defense responses in wheat (Randoux *et al.,* (2006).

Saccharine (3 mm) applied to first leaves or as soil drench is also reported to induce resistance in barley against *B. graminis hordei* by priming cinnamyl alcohol dehydrogenase (CAD) activity prior to pathogen challenge (Boyle and Walters, 2006) Phytoalexins had never been reported in wheat. Silicon is implicated in resistance development in many fungus-plant combinations Moderate resistance to powdery mildew had been reported in old wheat cultivars NP 710, 716, K-53, E-750, and C-591. Under the cool climate of Shimla only few of the dwarf wheat varieties were found resistant. These included Sharbati Sonora, Sonora 64 and Chhoti Lerma. The cultivar Kalyan Sona was intermediate in reaction in the adult stage. The cultivars HD 19880, HD 2204, HD 2074, HB 208 CPAN 1676, VL 401 are resistant to highly resistant CPAN 1922 was resistant to all races of *B. graminis tritici* known in this country. A major gene for resistance from wild Einkorn wheat (*Triticum monococcum* subsp. a*egilopoides*) has been successfully transferred to hexaploid common wheat (*T. aestivum*) and resistant lines obtained.

Hordeum spontaneum, the progenitor of cultivated barley, is known to be a rich source of disease resistance genes. A large number of accessions of this species have shown high level of resistance or tolerance to *B. graminis hordei* (Fetch *et al.,* 2003; Akhkha *et al.,* 2003). Accessions of *Hordeum chilense* have varying levels of resistance to *B. graminis trici*, *B. graminis hordei* and *B. graminis avenae*.

Certain avirulent strains of *B. graminis hordei* induce systemic resistance against the virulent strains. In susceptible oats and barley, a successful *B. graminis* attack is followed by haustorium formation which renders the attacked cell and to some extent its adjacent cells highly susceptible to later *B. graminis* attacks. In incompatible interactions formation of papilla by the attacked host cells renders the attacked cell and its adjacent cells highly inaccessible to later *B. graminis* attacks. The attack by a highly virulent pathogen strain renders the cells highly accessible to even low-virulence strains (Lyungkjaer *et al.,* 2001). The wheat phylloplane microflora is reported to suppress powdery mildew through production of aromatic substances. High levels of protection against infection of *B. graminis* f. sp. *hordei* in a susceptible barley cultivar were obtained by pre-treatment with mycelial extracts or culture filtrates from seven different fungi (*Bipolaris oryzae, B. sorokiniana, Drechslera teres* f. *maculans, Fusarium culmorum, Trichoderma harzianum, Pythium ultimum* and *Rhizopus stolonifer*). The protection mechanism of the mycelial extracts involves direct antifungal effects and possible induced resistance in case of extract of *P. ultimum* and *B. oryzae*. In general the number of colonies formed was reduced by 70–98%. The size of colonies was smaller than in untreated leaves and there was restricted hyphal growth. Accumulation of defence-related enzymes (peroxidase, chitinase) increases in the leaves treated with mycelial extracts. Nelson (2005) has reported that pre-inoculation of barley seedling roots with *Fusarium oxysporum* f. sp. *radicis-lycopersici* (crown and root rot of tomato) induces resistance to *Bipolaris graminis hordei*. The resistance lasts for 35 days against sporulation and for 22 days against growth of secondary hyphae of the mildew fungus. *Blumeria graminis* is a host of the mycoparasite *Ampelomyces*. In greenhouse studies, but not under field conditions, wheat powdery mildew is reduced by *Pyriformospora indica*.

POWDERY MILDEW OF CUCURBITS

Powdery mildew is sometimes a destructive disease of cucurbits especially pumpkin and bottlegourd. One of the species, *Erysiphe cichoracearum*, is reported to attack potato and tomato seedlings, lettuce, sunflower, mango, castor and many other plant species.

Symptoms

On cultivated cucurbits the first symptoms are tiny, white to dirty grey spots (sometimes with reddish brown tinge) on leaves and stems. Most extensive development of the mildew occurs on the abaxial surface of the leaf. As these spots enlarge the superficial powdery mass may ultimately cover the entire host surface. Black, pin-point bodies, representing the ascigerous stage of the fungus, appear rarely late in the season. They have been found in India only during the winter months. The effect of severe infection may be premature defoliation of the plant. The fruits remain undersized and are often deformed.

Fig. 48. Powdery mildew of cucurbits.

The causal organisms

Two species are reported to cause powdery mildew of cucurbits in India and elsewhere. Their description is given below:

Erysiphe cichoracearum DC ex. Merfat
(Eumycota, Fungi, Ascomycota, Ascomycetes, Erysiphales, Erysiphaceae)

Mycelium is usually well-developed, evanescent but sometimes persistent and effused. Conidia are in long chains, ellipsoidal or barrel-shaped, variable in size, 34.8 × 15.2 μm. Cleistothecia are gregarious or scattered, globose, becoming depressed or irregular, 90–135 μm in diameter. Wall cells of the cleistothecium are usually indistinct, 10–20 μm wide. Appendages are numerous, basally inserted, myceloid, interwoven with the mycelium, and hyaline to dark brown, 1–4 times as long as the diameter of the cleistothecium. They are rarely branched. Ten to 25 asci are present per ascocarp. The asci are ovate to broadly ovate, rarely subglobose, more or less stalked, 60–90 × 25–50 μm in size. Ascospores, two per ascus, rarely three, measure 20–30 × 12–18 μm.

Sphaerotheca fuliginea (Schlecht ex Fr.) Poll
(Podosphaera fuliginea-Podosphaera xanthii) Division Eumycota, Kingdom Fungi, Phyllum Ascomycota, Class Ascomycete, Order Erysiphales, Family Erysiphaceae

Mycelium is hyaline, occasionally brown when old, usually evanescent but sometimes persistent, and forms white circular to irregular patches on the host surface. Conidia are in long chains, often with distinct fibrosin bodies, ellipsoid to barrel-shaped, 27–31 × 15–18 µm in size. In a study of samples collected from 11 states of India, the size of conidia has varied from 28.1 × 15.6 to 34.2 × 18.5 µm. The conidial shape, presence of fibrosin bodies in conidia, conidial dimensions and length/breadth index, morphology of germ tubes and points of their origin, and development of appressoria are consistent characters irrespective of cucurbit host and locality. Cleistothecia are rare, scattered to densely gregarious, 66–98 µm in diameter, usually under 85 mm. The wall (peridium) cells are usually over 25 mm wide. Appendages are variable in number, usually as long as the diameter of the ascocarp, myceloid, tortous, brown, interwoven with the mycelium but sometimes long nearly straight and dark brown. There is a single ascus per ascocarp. It is broadly elliptical to subglobose and 50–80 × 30–60 µm in size. Each ascus contains 8 ascospores which are ellipsoid to nearly spherical and measure 17–22 × 12–20 µm.

Eight race of *S. fuliginea* have been identified in the USA, Africa, Europe and around the Mediterranean Sea. Four new races were reported from greenhouse melons in Japan where prevalence of races varied with season. Race 5 was most common in early season and race 1 in late season. In India, three distinct pathological forms of *S. fuliginea* were reported in 1985 on the basis of differential reactions of cucurbit hosts. In most of the agro-ecological zones of India three races of *S. fuliginea* are present. Race 3 is most widespread and infects most of the cucurbits in different states. Race 2 is restricted to green melon (*Citrulus vulgaris* var. *fistulosus*) and race 1 to spongegourd (*Luffa cylindrical*). In addition, the race 1 infects a few other cucurbits. *Luffa cylindrica*, *Citrullus vulgaris* var. *fistulosus* and *Cucumis sativis* are differential hosts. Race 1 does not infect *C. vulgaris* var. *fistulosus* and *Cucumis sativus*. Race 2 does not infect *L. cylindrica* and *C. vulgaris* var *fistulosus*. The fungus is heterothallic.

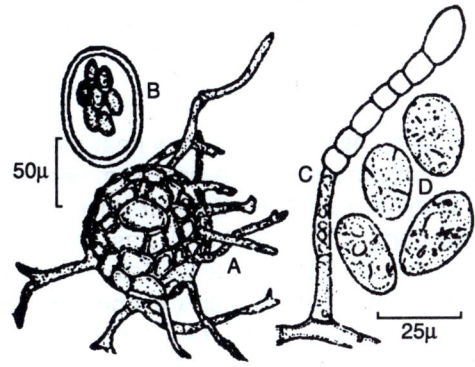

Fig. 49. *Sphaerotheca fuliginea.*
A-Ascocarp; B-Ascus; C-Conidiophore with conidial chain;
D-Conidia with fibrocin bodiies.

Disease cycle and environmental relations

There are many possible means of survival of these fungi between two crop seasons. Where cleistothecia are formed they can explain the mode of perennation from one crop season to the next. In India these sexual fruit bodies develop on leftover cucurbit crops during winter in isolated areas or in the sub-mountainous areas in the north. These may initiate the disease in the local hosts and from there the primary inoculum in the form of conidia might be blown by wind to the main crop in the plains. However, as in the downy mildew, the main source of primary inoculum seems to be the existence of wild and cultivated cucurbits in one or the other locality of the country from where the conidia are blown by wind currents to the new crop.

These powdery mildew fungi are greatly influenced by age of the host plant and air humidity and temperature. Sixteen to 23 days old leaves are highly susceptible while very young leaves are almost immune. The fungi can sporulate and cause infection in a very dry as well as wet atmosphere but infection increases as the atmospheric humidity increases, heavy dew deposits favouring the penetration by germ tubes most. Penetration is direct and confined to the epidermal cells. When conidia land and establish on the leaf surface of a susceptible host one or two germ tubes develop and penetrate one or two epidermal cells The penetration zones are surrounded with callose-like material. In resistant cultivar the fungus develops a single germ tube. Callose-like deposition also occur along with liginification in the epidermal cell. The rapid collapse of the penetrated cell in the resistant cultivar is accompanied by accumulation of callose-like deposits in cell walls and around haustoria, electron-opaque deposits in the plasma membrane and between the cell wall and the plasma membrane. In infection of *Arabidopsis thaliana* by *E. cichoracearum* early in the infection sequence host cell oragenelles move towards penetration site and accumulate near the infection peg.

The minimum and maximum temperature for conidia formation and host penetration are 10° and 32°C, respectively, the optimum being about 26°–28°C. Gupta *et al.,* (2001) have reported 25°C as the optimum for conidial germination at 100% relative humidity. Moderate temperature of 25°C with high relative humidity (<95%) and reduced sunshine hours significantly help in disease development by *Sphaerotheca fuliginea.* Penetration of the host by germ tubes from ascospores is entirely mechanical which results in dislocation of epidermal cells. The cells considerably increase in size. The infection peg does not form haustoria. Instead, it gives rise to several fine branches which extend to neighbouring cells and form primary hyphae. The conidial stage develops from these hyphae.

Management of the disease

The cucurbit weed hosts should not be allowed to grow near the cultivated fields. The diseased crop debris should be burnt. Fungicides are the only commercially available options for the management of powdery mildew of cucurbits. Sulphur dust (15–30 kg/ha) had been an old recommendation for effective control. Elasol (0.5%) was used as a substitute of elemental sulphur. Other effective fungicides are Sulfex (0.2%), Calixin (0.1%), Karathane (0.05–0. 2%), Bavistin (0.1%), Mildex, Ovatram, etc. Use of Sulfex is cheaper. One to two sprays of Calixin or 2–3 sprays of Karathane are required. Since the infection starts on the abaxial surface of leaves which are difficult to reach during the sprays of protectant fungicides, the systemic fungicides such as benomyl, carbendazim (Bavistin) and triazoles were extensively used in many countries. However, soon after these were introduced for use on field scale it was found that *S. fuliginea* quickly develops resistance to all of them. The resistance to triadimefon in the pathogen population spreads rapidly. When less than 50% of the population has developed resistance, application of triadimefon and chlorothalonil is effective but when resistance has developed in 80% of the population triadimefon is ineffective. Similarly, a population with 40% resistance to benomyl shows no effect of this systemic fungicide. Although sensitivity of *S. fuliginea* to triazole fungicides myclobutanil and propiconazole also decreased after they were applied, these fungicides were more effective than triadimefon. To overcome this problem it has been suggested that fungicides like triadimefon should be applied once or twice with a protectant fungicide and then subsequent sprays should be with only protectant fungicides. Thind *et al.*, (2002) have reported a comparative evaluation of three strobilurin fungicides, azoxystrobin Quadris), kresoxim methyl (Flint) and trifloxistrobin (Stroby), for their efficacy against powdery mildew of grapevines and powdery mildew of cucurbits. All the three are highly effective against *Uncinula* (*Erysiphe*) *necator* and *Sphaerotheca fuliginea* on the basis of inhibition of conidial germination. Under field conditions, azoxystrobin proved highly effective in checking powdery mildew on summer squash. It was equal to standard fungicides triademefon and mancozeb. Among more recent fungicides against cucurbit powdery mildew is cyzofamid which has given as good or even better control than fungicides in common use.

Solutions of mono- or dipotassium phosphate and potassium nitrate applied on a 7 or 14 days schedule were highly protective against *S. fuliginea* in cucumber. A single spray of solution of micronutrients (boron, copper, manganese) on upper surface of cucumber leaves provides protection from *S. fuliginea*. The effect is similar to the systemic protection given by potassium phosphate. Mixture of phosphate and micronutrients did not improve control. A single spray of 0.1 M phosphate solution is reported to induce systemic protection against powdery mildew of cucumber (*Sphaerotheca fuliginea*). It suppresses the lesions on diseased leaves. Oxalates and phosphates applied to upper leaf surface of cucumber induce systemic resistance against *S. fuliginea*, *Colletotrichum lagenarium*, *Pseudomonas lachrymans*, *Cladosporium* and *Didymella*. Cucumber plants grown hydroponically with phosphate in the nutrient solution or foliar spray of monopotassium phosphate develop induced systemic resistance against *S. fuliginea*. Mineral oil (1%), potassium bicarbonate (0.5%), sodium bicarbonate (0.5%) and milk powder (10%), whitewash and clay and antitranspirants reduce mildew on leaves as much as a wettable sulphur fungicide.

Wurms, *et al.*, (1999) studies the effect of two activator compound, Milsana and benzothiadiazole (BTH) on haistoria of the powdery mildew fungus in cucumber. Localized application of Milsana caused collapse of the haustoria within 4 days. The haistoria were encapsulated by an amorphous material impregnated by electron-opaque substances. Possible role of phenolics but not of chitimolytic activity was suggested. Application of high doseas of BTH elicited weak resistance response from the host.

In greenhouse hydroponic culture of cucumber, addition of silicon (potassium silicate) to the nutrient solution suppressed powdery mildew (*S. fuliginea*) at 20–25°C but higher temperature inhibited the suppression (Schuerger and Hammer, 2003). Dick *et al.*, (1998) had earlier found 10–16% reduction in diseases incidence by addition of 0.75 mm silicon to a nutrient solution. Application of potassium silicate (25 mg/L) also is reported to suppress powdery mildew in hydroponically grown strawberry (Kanto *et al.*, 2004). Increased yield of cucumbers in presence of infection of *Pythium aphanidermatum* by soluble silicon had been reported in 1994. The mechanisms by which silicon imparts resistance in host plants against fungal pathogens are complex and not entirely explained by the role of silicon as inducer of mechanical resistance. Disruption of pathogen hyphae by silicon is reported for *P. aphanidermatum*as well as *Sphaertotheca*. In healthy cucumber leasves silicon is mainly distributed in cells around the base of trichome hairs. Durimg infection by *S. fuliginea* areas of host cell wall adjacent to the growing germ tube shows altered surface morphology and high concentrations of silicon. Silicon was found in papillae, in the host cell walls, around the haustorial neck and in between the host cell wall and plasma membrane. These morphological alterations in pathogen and host and depositions were thought to be responsible for induced resistance, Silicon treatments significantly reduces the time taken for initiation of production and/or accumulation of phenolic materials in infected host epidermal cells and increases number of infected cells that produce or accumulate phenolics. The number of haustoria produced per colony of *S. fuliginea* is also significantly reduced and conidiophore development is delayed. Silicon-mediated accumulation of flavonoid phytoalexins in cucumber leaves was reported by Fawe *et al.,* (1998). Presence of these antifungal compound induced resistance in leaves to *S. fuligenea.* On the basis of a study of *Arabidopsis-Erysiphe cichoracearm* system, it was reported that treatment with silicon appears to induce the production of an electron-dense fungitoxic substance that deposits in and around the fungus haustoria. Liang *et al.,* (2005) have evaluated effect of foliar and soil application of soluble silicon on resistant and susceptible cultivars of cucumber. Root-applied silicon significantly suppressed the disease, more consistently in resistant than in susceptible cultivar. With root-applied silicon, activities of PR proteins (peroxidase, polyphenyl oxidase and chitinase) were significantly increased in inoculated lower leaves and uninoculated upper leaves. The increase in PR proteins was only in inoculated plants not in uninoculated plants.

Betiol (1999) had reported efficacy of fresh cow milk in suppressing *S. fuliginea* on squash (*Cucumis melo*). Five to 50% concentrations of milk in water were tested for 1–2 sprays on the plants. High concentrations gave better control than fenarimol (0.1 ml/L) or benomyl (0.1 g/L). Ferrandino and Smith (2007) have further confirmed field control of powdery mildew of pumpkin and postharvest rot by milk-based sprays. The control was better than given by sodium bicarbonate spray. Lactic acid bacteria in milk or from other sources like vegetables are known to be antimicrobial. A single spray of 0.5% clay (nonswelling chlorite mica clay) is reported to reduce powdery mildew (*Sphaerotheca fuliginea*) on cucumber without eradicating the fungus. Spray after inoculation was more effective than spray before inoculation.

The mycoparasite *Ampelomyces quisqualis* is a biocontrol agent against powdery mildews caused by *Erysiphe* and *Sphaerotheca*. The parasite is wholly internal within the mycelium, conidiophores, conidia and ascocarps of the mildew fungus. In culture media optimum temperature for spore germination and pycnidium ode of action, host range, biocontrol potential and other aspects of *Ampelomyces* mycoparasites have been reviewed by Kiss *et al.,* (2004). *Verticillium lecanii (Lecanicillium lecani),* as a biocontrol agent, is another valuable alternative to current management strategies for powdery mildew. Its hyphae colonize the structures of *S. fuliginea* by tight binding with the help of mucilage matrix. Within 24 hours of application of the antagonist, increased vacuolation and disorganization of the cytoplasm of the host hyphae occurs. By 36 hours, plasmalemma retraction and local cytoplasmic aggregations are seen. There is no change in cell wall of the host hyphae except at the point of penetration by mechanical pressure. By 72 hours, hyphal cells of *S. fuliginea* collapse, depleted of their protoplasm. The mycoparasite then exits from the dead cells. Species of *Lecanicillium* can provide dual microbial control of powdery mildew and aphids thus reducing virus infection. Dik *et al.,* (1998) compared the above two biocontrol agents with *Sporotrix flocculosa* in biocontrol of *S. fuliginea.* In their experiments, *A. quisqualis* did not control the disease and *V. lecanii* had a small effect. *S. flocculosa* gave the best result. Weekly sprays of this biocontrol agent on a partially resistant cultivar gave as good control as the fungicides bupirimate and imazalil. Romero *et al.,* (2003) have reported that the mycoparasitic fungi *Acremonium alternatum*, *Ampelomyces quisqualis* and *Lecanicillium* (*Verticillium*) *lecanii* suppress powdery mildew (*Sphaerotheca fusca=S. fuliginea*) development on melon leaves when applied in early stages of infection. *L. lecanii* was most effective. Simultaneous suppression of powdery mildew and *Aphis gossypii* (virus vector) by application of 0 is reported by Kim *et al.,* (2008). They used a commercial preparation of the bioagents (Vertalec®) which was highly pathogenic against adult apjids reducing their numbers significantly. It significantly suppressed conidia production by *Sphaerotheca fuliginea*.

Lima *et al.*, (2002) tested several yeasts and synthetic and biofungicides against powdery mildew of cucurbits (*Sphaerotheca fusca*). *Paecilomyces fumosoroseus* is saprophytic and entomopathogenic fungus. Kovkova and Curn (2005) studied the mycoparasitism of this fungus on *Sphaerotheca fuliginia*. Pre-treatment of leaves with the mycoparasite reduced development and spreading of mildew colonies. Spray on mildew colonies also reduced percentage of colonized area. *Trichoderma harzianum* is another fungal biocontrol agent against *Sphaerotheca fusca* in addition to *Pseudoperonospora cubense* (downy milde of cucurbits), *Sclerotinia sclerotiorum* and *Botrytis cinerea*. Involvement of local and systemic resistance has been demonstrated. Cells of the biocontrol agent applied to roots and dead cells applied to the leaves of cucumber plants induced control of powdery mildew. The yeasts *Rhodotorula glutinis*, *Cryptococcus laurentii* and *Aureobasidium pullulans* applied alone or with mineral oil or gum xanthan significantly reduced disease severity on melon leaves. The effect was comparable with a synthetic fungicide Topas and a biofungicide based on *Ampelomyce quisqualis*. The antagonists survived on leaves in the field at high level even in hot dry climate. The yeast-like, blastospore-forming fungi, *Tilletiopsis* spp., also suppress *Sphaerotheca* (*Podosphaera*). These fungi produce exo- and endo-glucanase and chitinase that inhibit the germ tube growth of the powdery mildew fungus and plasmolyze its conidia at 130 μg/ml (Urquhart and Punja, 2002). Spray of the antagonist *Tilletiopsis pallescens* formulated in a natural oil reduce powdery mildew severity. The oil formulation improves biocontrol efficiency of the antagonist, When inoculated on healthy leaves, *Tilletiopsis pallescens* grows better at 90% relative humidity than at 70%, It forms colonies adjacent to leaf veins on healthy leaves. In presence of the powdery mildew the growth of *Tilletiopsis* equally good at both relative humidities and extensive colonies are formed near the base of tichomes, Growth and survival of the fungus are enhamced by high relative humidity. Presence of the powdery mildew and canola oil-lecithin amendments on the leaf surface.

Isolates of several bacterial antagonists from rhizosphere and leaf surface of cucurbits could suppress powdery mildew by 80% (Romero *et al.*, 2004). These bacterial isolates remained stable on the leaf surface and formed microcolonies with extracellular matrix. Some strains of the bacterium *Bacillus subtilis* are reported as very strong antagonists of the cucurbit powdery mildew fungus (*Podosphaera fusca*) giving as good control as the mycoparasites. They inhibit germination of conidia, thus reducing the number of colonies on leaves. The lipopeptides produced by the antagonists cause morphological damage to conidia that includes presence of large depressions in conidial cell wall, loss of turgidity, severe modifications in the plasma membrane and disorganization of the mildew cytoplasma (Romero *et al.*, 2007).

Infection of tobacco necrosis virus induces systemic resistance against *S. fuliginea*. Systemic infections of viruses generally elicit defense responses in plants. Prior inoculation of cucumber leaves with a non-pathogenic isolate of *Alternaria cucumerina* or *Cladosporium fulvum* also induces systemic resistance to powdery mildew. Increased inoculum of the resistance inducer has provided up to 71.6–80% reduction in mildew colonies (Reuveni and Reuveni, 2000).

In a study of effects of plant extracts of *Renautria sachalinensis* on *S. fuliginea* and physiology of cucumber leaves. The extract has no effect on conidial germination but induced rapid accumulation of phenolics phytoalexins) in the leaves which protected the plant against powdery mildew. In greenhouse grown cucumber, even under high disease pressure, *R. sachalensis* extract controlled powdery mildew and increased fruit yield by 49%. Commercial formulation of leaf extract of *R. sachalensis* (Milsana) has been shown to suppress powdery mildew of tomato caused by *Leveeillula tairica*. *Rbinia pseudo-acacia* (Leguminosae) is a plant known for its medicinal and poison value. Zhang *et al.,* (2008) have reported that water-soluble extracts of the plant provide protection to cucumber agaist *Sphaerotheca fuliginea.* Organic solvent-extracts have no effect. The water-soluble extract show presence of at least two bioactive compounds. Ethanolic extracts of pine needles and leaves of *Aloe vera,* well known for medicinal value, reduce the number of mildew colonies on treated cucumber leaves by 30 and 21%, respectively. Extracts of the fruit bodies (basidiocarps) of the higher basidiomycete fungi *Oudemansiella* and *Ganoderma* reduce number of mildew colonies on treated leaves by 79 and 65%, respectively. They also reduce diasmeter of the colonies by 45 and 70%. The extract of *Oudemansiella* reduces conidial germination of S. fuliginea by 71% (Stadnik *et al.,* 2003).

The hydrolytic enzyme â-1,3-glucanase provides defense against powdery mildew. In muskmelon, cultivars Diguria and Haragola are reported to be immune to the disease. In *Cucurbita pepo* (pumpkin, squash) resistance to *Podosphaera xanthii* is conferred by a single incompletely dominant gene (Cohen *et al.,* 2003). In resistant melon, development of the fungus is checked at the primary haustorium stage irrespective of temperature. The resistance genes in the host may be temperature sensitive. When the temperature sensitive resistant cultivar is infected the temperature of incubation of the host has a clear effect on the outcome of infection. At 21°C it takes longer for symptom expression to appear on the resistant cultivar and at 26°C the resistance is complete and no symptoms develop.

Resistance to *E. cichoracearum* and *S. fulginea* occurs in many species of wild cucurbits. African accessions of *Cucurbita ficifolius, C. anguria, C. dinteri* and *C. saggitatus* are reported to possess resistance to both pathogens.

POWDERY MILDEW OF GRAPEVINES

The known history of powdery mildew of grapevine dates back to 1834 when the fungus causing the disease was first described as *Erysiphe necator* in the eastern part of North America. However, it was not considered important until it was reported from England in 1847 and by 1850 it had spread to France and other major grape growing areas of Europe where it caused considerable loss to grapevine growers and the wine industry. Today, powdery mildew can be found in most grape-growing areas of the world including the tropics. It is known to occur in a mild or severe form in North and South America, Europe, parts of Africa, and in Australia. It appears in almost epidemic form in all the vineyards in India when the conditions are favorable for its development. The disease is much more serious than the downy mildew of grapes and is more dangerous than other powdery mildews of different crops. Failure to control the disease results in a chronic reduction in wood maturity (number of nodes) which, by itself, does not reduce yield but losses in the yield of fruits may be up to 40-60%. The most significant effect of powdery mildew is on berry sugar levels and juice color and acidity. The fungus infection makes the grapes unsuitable for wine making. Berries infected by the fungus tend to be higher in acid than healthy berries. This is not considered desirable by the brewers. The fungus itself produces an off-flavour in wine made from infected grapes. The infected berries tend to crack, thus, providing entry to other pathogens and saprophytes.

Symptoms

The disease attacks the vines at any stage of their growth. All the aerial parts of the plant are attacked. As in other powdery mildews, the characteristic symptom of the disease is the appearance of white, powdery patches on affected parts. Cluster and berry infections usually appear first. Cluster infection before or shortly after the bloom results in poor fruit set and considerable crop loss. Young fruits (berries), just after bloom, show whitish mycelial growth on the urface. When the infection of berries occurs before they attain full size, the epidermal cells are killed and the growth of epidermis is prevented. Because the internal pulp continues to grow, the skin cracks. Such berries either dry up or rot. If the attack occurs when fruits are nearing maturity or beginning to ripen, they fail to color properly, become irregular in form and only few of them ripen, remaining undersized with a blotchy surface. Often the infected berries develop a net-like pattern of scar tissues.

Leaf lesions appear late and do not cause much damage. On young leaves, small whitish patches appear on the upper, or sometimes on the lower, surface of the leaf. These patches grow in size and finally coalesce to cover larg areas on the lamina. Similar floury patches are formed on the stem, tendrils, and flowers. The powdery growth gradually turns gray and finally dark colored. Malformation and discoloration of the affected leaves are common symptoms. The stems turn brown in color. The diseased vines have a wilted appearance and remain stunted in growth. Necrosis of the penetrated epidermal cells and even of adjacent cells is a characteristic reaction of resistance in some varieties of the host.

The causal organism

The disease is caused by *Uncinula necator* (Schw.) Bur. (anamorph *Oidium tuckeri* Berk.). The fungus was reclassified as *Podosphaera necator* and then as *Erysiphe necator* which is its original name. The mycelium is entirely superficial on the attacked parts to which it adheres by means of bilobate or multilobate appressoria. The hyphae are hyaline, slender, septate, branched and 4–5 μm in diameter. They turn darker in color when the formation of conidia is over. The conidiophores arising perpendicular to the creeping hyphae on the host surface are simple, multiseptate and erect. They are attached to the mycelial hyphae by a cylindrical foot cell measuring 24–40 μm. Cells of the conidiophores are generally wider than those of the mycelial hyphae, measuring 6.2–7.5 μm. They bear a chain of 3–4 conidia under field conditions. In static humid conditions the chains may contain 8–10 conidia. These hyaline conidia are oval in

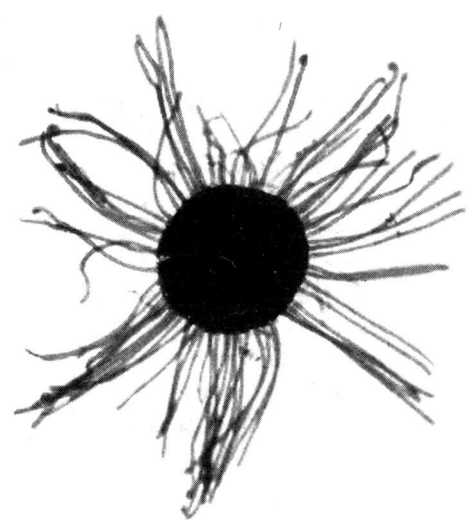

Fig. 50. Uncinula nector Cleistothecia.

shape and measure 25–30 × 15–17 or 27–47 × 14–21 μm. The oldest conidium is at the distal end of the chain. The conidia germinate by a short germ tube terminated by a bilobed or multilobed appressorium.

Cleistothecia of the fungus have been found in North America, Europe, Russia, Peru, and Australia. Under the climatic conditions prevailing in most vine growing areas of India, the perfect stage of the fungus is not found. The fungus is heterothallic and most populations consist of two mutually exclusive mating types. The sexual compatibility of the fungus is controlled by a single mating type gene (MAT1) with two alleles (MAT1-1 and MAT1-2). These may be equally common in natural populations. A small percentage of isolates have the capacity to form cleistothecia in protracted association with isolates that initially appear to be of an incompatible mating type. When the mating types are present, the cleistothecia can form on all infected tissues during the later part of the growing season. They are found embedded in the superficial mycelium on the leaves, or on shoots, chiefly at the nodes or in buds among the scales and hairs. They are hyaline and spherical when young but turn yellow due to the accumulation of a yellow lipid in the ascocarp. When the outer cells of the ascocarp darken the mature cleistothecium turns black, almost round with a flattened top, and measures 75–100 (84–105) μm in diameter. The peridium is covered with 8–25 septate appendages which appear inserted in the equatorial region of the ascocarp. The appendages are coiled (uncinate) at the distal end and are brown at the base. They are 1–6 times as long as the diameter of the cleistothecium. When the cleistothecium is immature it retains functional connections with the hyphal mass. Mature cleistothecium develops a basal concavity and the connections with the hyphal mass die. Each perithecium contains 4–6, sometimes more, ovate to subglobose asci measuring 48–60 × 37–45 μm. In each ascus there are 4–6 ascospores which are oval to ellipsoid in shape and hyaline. The ascospores are low in water content and their measurements are affected by the mounting medium used. Unmounted ascospores measure 23–28 × 14–16 (mean 26 × 15) μm. In lactophenol mount, they measure 18–29 × 10–15 (22 × 12) μm. These ascospore germinate by a short germ tube which terminates into a multilobed appressorium. Multiple germ tubes may also arise from the ascospore.

Disease cycle and environmental relations

The disease cycle of *U. necator* involves both asexual and sexual overwintering stages in temperate regions. Perennation of the pathogen through vegetative mycelium in dormant buds is more significant than perennation through oospores. The conidia can resist desiccation but it is not known how long they can remain viable. The fungus inside dormant vegetative buds does not affect their survival or vigor. The dormant buds carrying the primary inoculum develop into flag shoots early in the spring. These flag shoots provide inoculum for secondary spread of the disease early in the season. The shots commonly have abnormally formed leaves and the borders are often bent upwards. The leaves are covered withy whitish growth of the pathogen consisting of hyphae and conidia. Where cleistothecia are formed they serve as additional, or may be the only, source of perennation. In some areas cleistothecia are the only source of primary inoculum, there being no perennation through mycelium in dormant buds. The ascospore infection occurs late in spring and causes development of isolated colonies while the infection by conidia from flag shoots results in uniform and dense colonies. Miazzi *et al.,* (2003) have hypothesized existence of two genetically separated biotypes in *U. necator* that are related with its two overwintering modes, a biotype overwintering as conidia and mycelium in dormant buds that causes infection of shoots and leaves early in the season and the other biotype overwinters as cleistothecia and infects bunches. The two biotypes are not separated in space and time and sexual crosses between the two occur in the vineyards. In tropical climate the survival is mainly through mycelium and conidia on green tissues remaining on the vines. Environmental factors are very much responsible for the formation of cleistothecia. Rains disperse the cleistothecia to the bark of the vines where they are retained between leaf fall and bud break in the next season. Density of cleistothecia is higher on fallen leaves than on the bark but their viability is more on the bark than on fallen leaves on soil. After the infection from ascospores or from the hibernating mycelium in the host buds enormous numbers of conidia are produced to carry on the conidia-to-conidia life cycle.

The conidia on the leaf surface start germination within 90 min and in about 20 hours about 52% conidia have germinated. Almost all the germ tubes coming in contact with a hard surface form appressoria and in about 5 days after conidial germination on leaves colonies of *U. necator* with profuse conidiophores and conidia are seen. In a study of development and adhesion of infection structures of *U. necator* on grapevine Rumbollz *et al.,* (2000) observed that primary appressoria appeared 3.5 hours after inoculation followed by hyphal growth on the leaf surface in 14 hours, suggesting successful host colonization. Deposits of extracellular material at the contact zone of fungal structures and plant cuticle ensured firm adhesion of the pathogen Esterase activity was associated with conidia and infection structures. Temperature appears to be the major determinant of fungal development. It determines the extent of asexual reproduction of *U. necator* and fluctuating temperatures decide the rate of conidial formation, germination and colony development. In the temperate climate of California (USA) the early warm spring climate usually precedes severe epidemic of powdery mildew.

Rapid germination of conidia, infection, and development of the pathogen occurs at temperatures of 21°– 30°C although the fungus can grow at temperatures from 5° to 30°C. The minimum temperature for germination of conidia is 6°C and for infection and growth 7°C. The optimum for germination is 25°C. Conidial germination ceases at 3°C and above 33°C and at 40°C conidia are killed. At 25°C conidia germinate in about 5 hours and time from inoculation to sporulation is 5 days whereas at 23°C and 30°C it is 6 days. Mildew colonies are killed after exposure to 36°C for 10 hours or to 39°C for 6 hours.

Free water causes poor and abnormal germination of conidia. Rainfall is detrimental because it removes conidia and disrupts the mycelium. Atmospheric relative humidity of 40–100% is sufficient for germination of conidia and infection although germination can occur even at less than 20% RH. Humidity has a greater effect on sporulation than on spore germination. Low diffuse light favors development of powdery mildew. In bright sunshine conidial germination is inhibited.

The fungus perpetuates through hyphae inside dormant vegetative buds (13, 63). The buds are infected in the preceding season. They are most susceptible at the three-to six unfolded leaf stage. Incidence of powdery mildew colonies on the surface of buds is highest at these stages. The colonization of the interior bud tissue from the fungus on the surface is also highest at these stages Hyphae with haustoria, conidiophores and conidia are present on all internal parts of the buds except the meristems (Rumbolz and Gubler, 2005).

In areas where cleistothecia play a major role in survival of the fungus and in primary infection, it is reported that cleistothecia in the bark of the vines discharge ascospores when rainfall occurs between bud burst and bloom. In a 4-year study in Eastern Washington, Grove (2004) did not find any evidence of perennation of the mildew fungus in dormant buds. Cleistothecia were retrieved from bark fissures and senescent leaves and contained viable ascospores at the time of bud burst. Ascospores could be trapped as late as 70 days after bud burst. Ascospore discharge requires free water Temperatures within the range of 10° and 25°C have little effect on ascospore release but a temperature of 4°C or lower can suppress ascospore discharge. In laboratory studies, wetting of leaf discs bearing cleistothecia and incubation in a humid chamber for 48 hours at 20°C were found essential for release of ascospores. Storage of cleistothecia on leaves with periodic wetting at 5°C during 110 days were necessary to induce both ascospore release and their germination ability. Ascospores germinate equally well in free water and in saturated atmosphere. Germination declines rapidly as humidity decreases. Appressoria are not formed by ascospore germ tubes at below 10°C and at above 31°C.

It was reported that climatic conditions during October and November in India are ideal for development of grape powdery mildew. Warm, dry weather with just enough humidity is very favourable. However, a microclimatic study in Maharashtra, in mid 1990s, showed more rapid spore production and disease development during December and January, when the weather is cool and humid, than in November and February. According to this study temperature in the range of 12.2°–30.1°C and relative humidity greater than 57.4% favours sporulation of

U. necator. At temperatures below 8.8°C and above 34°C and RH below 47.4 % the rate of multiplication is zero. The disease development is retarded in sunshine. In south India, the disease incidence is reported to be significantly influenced by relative humidity and maximum temperature whereas influence of minimum temperature, rainfall and total rainy days was not significant. Increase of RH by 1% increased disease incidence by 2.4%. Increase of temperature by 1°C decreased the disease by 4.4%. Maximum temperature range of 27°–31°C along with RH up to 91% favored disease incidence while maximum temperature range of 31°–34°C inhibited the development of powdery mildew. In Punjab, mycelial growth and production of conidia was found to be very rapid at 25°C followed by 20° and 30°C. At 35° and 40°C there was no powdery mildew development. Although infection can occur at zero percent relative humidity extremely low humidity adversely affects growth of the pathogen on the host surface. Humidity plays a significant role in grapevine powdery mildew epidemiology (Carroll and Wilcox, 2003). Incidence and severity of the disease increases with increasing humidity to an optimum near 85% RH and then plateau or marginal decrease at higher values is seen.

Development of ontogenic resistance to powdery mildew, which varies with host genotype and tissue type, is quite rapid in grape berries (Ficke *et al.,* 2003, 2004; Gadoury *et al.,* 2003). The fruit becomes nearly immune to infection by *U. necator* within 4 weeks after fruit set. Only fruits inoculated within 2 weeks of bloom develop severe powdery mildew. Rachises of fruit bunches develop severe mildew when inoculated at bloom and disease increases steadily over the next 60 days. The ontogenic resistance does not affect adhesion of conidia, germination and appressorium formation but checks pathogen ingress at the cuticle before formation of a penetration pore. As the berries age, hyphal elongation and colony growth lowers down and finally there is no development of secondary hyphae. On aged berries, the fungus produces more appressoria in attempt to caus infection. This response of older berries is supposed to be due to synthesis of a specific protein. Ficke *et al.,* (2004) had reported that many factors are responsible for the ontogenic resistance. Cuticle and cell wall thickness, antimicrobial phenolics, pathogenesis related proteins are not the principal causes in halting pathogen ingress on ontogenically resistant berries. The infection is halted by one or more of the following: (i) a preformed physical or biochemical barrier near the cuticle surface, or (ii) the rapid synthesis of an antifungal compound in older berries during the first few hours of the infection process. The protein, thaumatin, is reported to be present in grape berries. In a study of effect of grape proteins on *U. necator, B. cinerea* and *Phomopsis viticola,* Monteiro *et al.,* (2003) have observed that two proteins, osmotin and a thaumatin-like protein, inhibited spore germination and germ tube growth of *U. necator.* Vines infected by *Plasmopara viticola* (downy mildew) show resistance to powdery mildew. While leaves infected with *P. viticola* develop 1.5 powdery mildew colonies per leaf, the uninfected leaves show 23.9 colonies per leaf. The induced resistance is confined to leaf tissue colonized by *P. viticola* and is not transferred to new developing leaves. Conidia of *U. necator* could germinate on downy mildew infected leaves but failed to produce secondary hyphae and colonies. Protection was reversed by application of 0.1% sucrose solution which restored susceptibility of the leaf to powdery mildew.

Management of the disease

Due to the widespread occurrence of powdery mildew in grapevine-growing areas of the world regulatory measures (quarantine) are of no value except where introduction of fungicide-resistant strains is feared. Clean cultivation of vines is an important part of disease management in grapevine orchards. Pruning after shedding of leaves, thinning out and cutting back of laterals and removal and destruction of all diseased parts constitute clean cultivation.

Under no-fungicide treatment conditions, in vertical shoot positioned vines the disease incidence is higher (30% clusters infected) than in free-positioned, topped vines (5% infected clusters). The difference is better marked under low disease pressure. Excessive nitrogen fertilization tends to promote succulent growth which is associated with increased incidence of powdery mildew.

The control of powdery mildew in commercial orchards is generally based on the use of fungicides. Fungicidal control measures should start in the early stages of vine development and repeated at 7–21 days interval depending on the fungicide being used. Dusting of vines with sulphur (300 mesh) had been an effective control method in the past and is still most extensively used chemical control measure. The first dusting should be done when new shoots are 7–15 cm long, second during or just before blossoming. A third application can be made 40–50 days later. Sulphur is also applied as a wettable powder. In dry climates sulphur dust is preferred whereas in areas with plenty of rainfall wettable powder or flowable formulations are recommended for their retention qualities. The optimum temperature for sulphur activity is 25°–30°C. Above 30°C there is the risk of phytotoxicity. The activity of sulphur is reduced in humid air compared to dry air. Because sulphur has poor retention qualities, application schedules of 7–10 days are usually required. Pre-bud swell or dormant stage application of lime sulphur delays the development of epidemics. However, sulphur can leave undesirable residues on table grapes and may taint wine if used within one month before harvest. In addition, sulphur adversely affects beneficial insects. In addition to sulphur dust (15 kg/ha), Sulfex (0.25%), Karathane (0.05%), Calixin (0. 05%), Topsin-M (0.1%), Thiovit (0.25%), all have been found effective against powdery mildew of grapevines.

In most grape-growing areas now, the standard approach is intensive use of sterol demethylation-inhibiting (DMI), quinone outside inhibitor (QoI) and quinoline fungicides. The sterol inhibiting triazole fungicides have proved more effective. Generally, Bayleton (triadimefon) is considered most effective triazole fungicide against grape powdery mildew fungus. Triadimefon at the rate of 40 g/100 lit. or penconazole or cyproconazole at the rate 40 ml/100 liter sprayed in mid-March, last week of April, and first week of May are highly effective against the disease. Fenarimol (Rubigan 12 EC), used in 7 sprays at 0. 05%, at 10 days interval between 15 cm cane length and 60 cm cane length is highly effective against powdery mildew. The fungicide is compatible with wide-spectrum protectants such as mancozeb and its residue persists for 14 days after the last spray but at a very low level. The dissipation rate is high during summer. Under dry temperate conditions of Himachal Pradesh (India) triadimefon (Bayleton), flutriazole (Impact), or hexaconazole (Anival) were more effective than carbendazim (Bavistin), dinocap (Karathane) and sulphur (Sulfex). Single application vapor-action treatment with triadimefon, triadimenol (Bayton), etaconazole (Vangard), flusilazole, mycobutanil and penconazole.was also recommended at 0.2 to 4.0 g a.i. per vine between bloom and two weeks after shatter. The treatment gave effective control of the disease. Cyprodinil and related anilopyrimidine fungicides are more recent compounds being used against powdery mildews. One advantage with the sterol-demethylation inhibiting triazole fungicides is that the interval between sprays can be increased and, thus, the number of sprays decreased. Dinocap (Karathane) is used at 10–14 days interval while the triazoles can be used at 14–21 days interval.

Among the strobilurins, trifloxystrobin is very effective against conidial germination of *U. necator*. It shows strong prophylactic and local activity on leaves. In field control of the disease, it has proved to be as good as sulphut and superior to sterol inhibiting fungicides when applied to heavily mildewed leaves. Different strobilurin fungicides (azoxystrobin, kresoxim-methyl, pyraclostrobin and trifloxystrobin) have different intrinsic activity against *U. necator*.

Although these fungicides were supposed to be free from resistance development in pathogens, there are many reports contradicting the belief. In vineyards where myclobutanil was exclusively used for powdery mildew control, the pathogen developed resistance to it. Strains resistant to myclobutanil were found to be resistant to azoxystrobin also (Wong and Wilcox, 2002).

Isolates of *U. necator* resistant to benomyl, triadimefon, myclobutanil and fenarimol are reported. There is cross resistance between fenarimol and triadimenol. Resistance to benomyl became widespread after 3–4 years of its use in vineyards. Triadimefon resistance in *U. necator* is most likely to manifest itself under high disease pressure which is in part a function of temperature. To overcome the problem of resistance development, integration of sulphur and triazole (DMI) fungicides has been adopted in many countries.

Polar, a polyoxin B compound, is highly effective against powdery mildew of grapevines, apple (*Podosphaera leucotricha*) and nectarines (*Sphaerotheca pannosa*). It gives as good control as standard SI fungicides or sulphur (Reuveni *et al.*, 2000). Alternating Polar spray with SI fungicide sprays reduces the number of sprays of the latter by about 50%. At concentrations of of 1, 10 and 100 mg/L polyoxin B inhibits the germination of *Uncinula necator* conidia by 52.5, 76.6 and 100%, respectively. Foliar spray (100 mg/L) on greenhouse grown grapevine before inoculation provides more than 90% protection while spray on established infection (sporulating colonies) also suppresses the pathogen.

In order to reduce cost of repeated application of fungicides and to avoid resistance development in the pathogen, attempts to encourage the alternative methods of control of powdery mildew, non-conventional chemicals, plant products and many approaches to biological control have yielded encouraging results. Mineral oil (petroleum derived spray oils, PDSO) and plant oils have been used against grape powdery mildew with positive results. Mineral oils are often as effective as myclobutalin. Petroleum oil applied as emulsion (1% v/v) in water has provided moderate protection, excellent pre-lesion and post-lesion curative action and is antisporulant. Plant oils have also showed significant action in pre-lesion treatment and as antisporulants in treatments applied to established lesions. These treatments are not effective against the downy mildew in vineyards. Generally, the recommendation is to use PDSO in early stages of disease incidence as a prophylactic spray to be followed by reduced number of fungicide sprays. Azam *et al.*, (1998) have reported that a rape oil derivative gives good control of grape powdery mildew. Repeated sprays of the rape oil derivative at the rates of 2.0 and 5.0 ml (formulated product) per liter prevented the foliar symptoms as effectively as either wettable sulphur 2 g/L (formulated product) or fenarimol 0.2 ml/L (formulated product). However, in heavily diseased vines the fungicides were better than the oil derivative. Oils of *Allium sativum, Cyperus scariousus, Cuminum cyminum, Cymbopogon citrates* and *Eucalyptus camaldulensis* inhibited conidial germination and germ tube elongation of *U. necator in vitro* and suppressed powdery mildew on potted grapevine plants (Dhaliwal *et al.*, 2002).

Potassium bicarbonate, monopotassium phosphate and sodium bicarbonate are hazard-free alternatives to fungicides. Spray of 1% sodium bicarbonate 3 times during the growing season, commencing after the appearance of first symptoms gives good control of powdery mildew. Sawant and Sawant (2008) have reported good control of grapevine powdery mildew by sprays of potassium bicarbonate (10 g/L) at 10 days intervals, starting 40 days after pruning. *In vitro* the conidial germination is inhibited by 100 pp. When sprayed on infected leaves, potassium, bicarbonate at 10 g/L reduced sporulation of the pathogen A single spray of 0.1 M solution of phosphate is reported to induce systemic protection against powdery mildew and suppress lesions on diseased foliage. Foliar sprays of 0.025 M and 0.04 M solutions of mono- or dipotassium phosphate inhibit the development of the powdery mildew fungus on grape fruit clusters. The control is as good as that given by a systemic fungicide. Alternating phosphate application with a systemic fungicide like penconazole or myclobutanil improves the control efficacy. The efficacy of these inducers of systemic acquired resistance (SAR) is increased by the addition of 0.1% sulphur. In a study in Turkey, efficacy of silicates, phosphate and carbonate for the control of grapevine powdery mildew was tested. While there is low efficiency against leaf infections, bunch infections were inhibited. It was suggested that sodium silicate, potassium silicate, potassium phosphate or sodium carbonate could be included in the spray schedules with sulphur or synthetic fungicides or they could be used in suitable mixtures or alternations with each other.

Spray of milk and whey is also reported to significantly reduce the mildew on leaves. The milk spray causes collapse of hyphae of the fungus and damage conidia within 24 hours. The effect is attributed to production of free radicals and lactoferrin (antimicrobial compound found in bovine milk) The latter ruptures conidia but has no effect on hyphae (Cisp *et al.,* 2006). Spray of soluble silicom (potassium silicate or metasilicate) inhibits formation of mildew colonies on the leaves.

In biological control, suppression of grapevine powdery mildew by the mycoparasite *Ampelomyces quisqualis* Ces. (syn. *Cicinnobolus cesatii* De Bary) is reported. This mycoparasite was described under powdery mildew of cucurbits. However, under natural conditions, populations of *A. quisqualis* lag several weeks behind populations of powdery mildew fungi in their development and its colonies are seen late in the season allowing the mildew to develop to damaging levels. Sprays of concentrated conidial suspensions earlier in the season have given partial control of the powdery mildew of grapevines, particularly in wet seasons. The chitinolytic enzymes of *Trichoderma harzianum* and cells of biocontrol strains of *Enterobacter cloacae* have antifungal activity against *U. necator.* They suppress spore hermination and germ tube elongation. Mixture of the enzymes and bacterial cells has synergistic effect.

English and Norton (2007) have reported 45% reduction in powdery mildew invasion of leaf area by applying the mite *Orthotydeus lambi.* The mite becomes well established in the vines where it is released However some cultivars of grape support high populations of the mite than others. Treatments where this mite is released in the vines are as effective as fungicide. Significantly better disease control is found in treatments with both mites and fungicides. The pathogen could develop cleistothecia only on leaves not inhabited by the mites. Application of myclobutanil and strobilurin fungicide azoxystrobin did not greatly affect abundance of the mites but mancozeb and wettable sulphur greatly reduced their population. Melidossian *et al.,* (2005) have further reported that efficacy of the mite is related to its density per leaf and time of application. Pre-bloom application with high density (30 mites per leaf) gives significant control of foliage mildew and consequently the mildew on berries.

Sprays of aqueous extracts of compost have been used with positive results. Populations of *Bacillus, Pseudomonas, Serraia, Penicillium*, and *Trichoderma*, all are enhanced on the host surface and they provide biological regulation of the plant pathogens. Sendhilvel *et al.,* (2007) have reported effective control of powdery mildew on grape leaves by application of a talc-based formulation of *Pseudomonas fluorescens* strain Pf1 at 2%. *Bacillus subtilis* is an established biocontrol agent against powdery mildew. Fermented cow dung, various neem products, sodium bicarbonate, a formulated rape oil product, and a silica preparation were tried in Germany. Neem products gave reasonable control of powdery mildew but the treatment was very costly. The silica preparation and sodium bicarbonate gave acceptable control. These are inexpensive and environmentally safe. Spray of aqueous extract of dry mycelium of *Penicillium chrysogenum* induces systemic resistance to powder mildew and downy mildew and gives as good control as copper and sulphur fungicides and the plant activators benzothiadiazole or acibenzolar-S-methyl.

The species of the genus *Vitis* and cultivars within the species differ in susceptibility to powdery mildew. Under dry temperate conditions of grape growing areas of Himachal Pradesh, many local varieties resistant compared to other cultivars like Black Prince, Isabella, Sultana Red, Champion, Cardinal, Perlett etc.

POWDERY MILDEW OF APPLE

The powdery mildew is present in apple growing regions of all the countries in the world. Originally, it was considered a disease of nursery stock but now it is recognized as a serious threat to bearing trees also. Economic damage from powdery mildew in bearing trees in apple orchards results from reduction in tree vigor, trunk growth, and blossom bud production, and from aborted blossoms, and fruit russet. Infection can reduce fruit size, weight, and market value. Severe infection can reduce the amount of bloom and almost eliminate the crop in the following season. In nurseries and young plantings, powdery mildew stunts the tree growth and causes poorly formed and misshapen trees. In north India, the disease damages the nursery plants more readily than the older plants during the months of April-June. In certain years it causes considerable damage to leaves and young shoots of bearing trees resulting in reduced yield of fruits. In addition to apple and pear (*Pyrus communis*) the disease occurs on peach (*Prunus persica*), and quince (*Cydonia vulgaris=C. oblonga*) also. Pear cultivars are less susceptible than apple. In pears skin russet of fruits is of concern.

Fig. 51. Powdery mildew of apple.

Symptoms

The disease appears soon after the buds develop into new leaves and shoots. Sometimes, the buds are so heavily infected in the previous season that they are killed before developing into leaves. Even if they grow, the new leaves and shoots are heavily mildewed and weakened.

Early symptoms on such leaves consist of small patches of white or gray powdery mass on the under surface but as the disease progresses both surfaces of leaves and twigs become covered with the powdery mass. Affected leaves grow longer and are narrower than normal leaves. The margin is curled. Later, the leaves turn brown from the tip downward. A severe attack of the disease results in partial defoliation of the tree. In nursery plants the disease prevents formation of wood in the stem. Petals of infected flowers are pale yellow or green and covered with mycelium. The flowers are shriveled and may fail to set fruit. They are more susceptible to frost. Fruit buds suffer more damage than the vegetative buds. They may not bear any fruit at all. When fruits are attacked, they remain small and deformed and tend to develop a rough surface.

Physiological effects of infection include reduced photosynthesis, transpiration, and carbohydrate content of the host, increase in phenolic content, accumulation of sulphur at the site of young mildew colonies, and reduced calcium transport to infected leaves.

The causal organism

Powdery mildew of apple is caused by *Podosphaera leucotricha* (Ellis and Ever.) Salm. The ectophytic mycelium forms saccate haustoria in the epidermal cells. Aerial conidiophores arise from this mycelium on leaves and shoots. Each conidiophore bears a chain of oval, hyaline conidia which measure 20–38 × 12–14 μm. They contain distinct fibrosin bodies.

P. leucotricha is heterothallic. The cleistithecia may form in autumn especially on 1–year-old shoots, on petioles and on midrib, but especially on sucker shoots. Up to 1100 cleistothecia per cm of such shoots have been recorded. The cleistothecia are globose, black, partially embedded in the mycelial web and measure 75–96 μm in diameter. They are densely gregarious. Two types of appendages are formed on the surface of these cleistothecia. Some are long, stiff, diverget and apically formed while others are basal, short, and tortuous and serve to anchor the cleistothecium to the substrate. The apical appendages, (usually 3–5, sometimes 11) are 3–7 times as long as the diameter of the cleistothecium. The top of the appendages is usually unbranched and blunt, rarely, they may be dichotomously branched one or twice. The basal appendages are rudimentary, pale brown, more or less tortuous and simple or irregularly branched. Each cleistothecium contains a single ascus measuring 55–70 × 44–50 μm. Eight ascospores, 20–26 × 12–14 μm in size, are produced in each ascus. Perithecial stage is rare in India. Physiologica races of the pathogen are reported in Europe.

Disease cycle and environmental relations

Probably the cleistothecia of the fungus, produced on heavily infected leaves and shoots, do not play any major role in the disease cycle. The ascospores do not germinate readily. The disease cycle is mostly conidia-mycelium-conidia. At most places the fungus survives in the form of dormant mycelium or encapsulated haustoria in the dormant terminal and lateral shoot buds and in blossom buds produced and infected in the previous growing season. Buds are most susceptible at the three- to six-unfolded leaf stage. Incidence of powdery mildew colonies on surface of buds and colonization of the bud interior via the infected bud surface are also highest at these stages. Hyphae with haustoria, conidiophores and conidia are present on all internal tissues of the bud except the meristems (Rumbolz and Gubler, 2005). Survival is greatly affected by temperature. Extreme cold in winter (–12°C) may kill the mycelium and buds may be freed from infection. When the buds open in the next season the surviving resting structures produce abundance of conidia which are wind-borne and serve as secondary inoculum. Fruit buds are the earliest to emerge and, therefore, provide the earliest source of secondary inoculum. Terminal shoot buds become active slightly later than fruit buds but constitute a more important source of inoculum because of their larger infected area and continued growth, and spore production on new growth well into the season. Lateral buds emerge still later and usually do not grow as long but become dominant if the terminal bud is killed. Usually, the healthy buds have come out slightly earlier than the infected buds. As a result when conidia have been formed and are released there is sufficient fresh young tissue for infection. The maximum number of conidia still attached to conidiophores are found in 7–12 days old colonies. There are three general spore dispersal patterns, *viz.*, diurnal dispersal with peak concentration about mid-day or afternoon, random dispersal in which spore concentrations are not correlated with any measured meteorological parameter, and dispersal associated with the onset of rain. Positive correlation with wind speed and temperature and negative correlation with relative humidity are characteristics of diurnal dispersal.

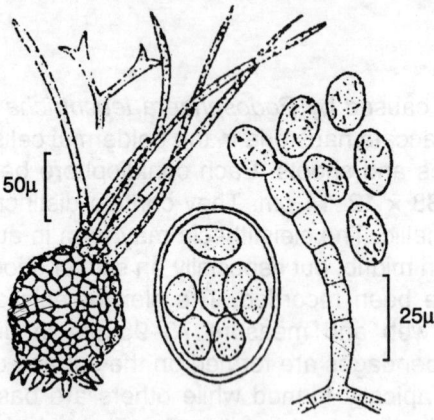

Fig. 52. *Podosphaera leucotricha.* Ascocarp with appendages, ascus and condiophores with conidia.

Leaves are susceptible only for 3–6 days after emergence. Conidia can germinate at temperatures of 5°–30°C but optimum temperature is 22°C. Temperatures of 33°C and above are fatal to conidia. Conidia do not germinate below 88.5% relative humidity. High atmospheric humidity is essential for penetration of the leaf by germ tubes of germinating conidia. Following penetration of the cuticle the hypha becomes thin and peg-like, penetrates the epidermal cell wall and forms a haustorium. The penetration of the cuticle is by action of enzymes. No appressorium formation has been reported. The infection process does not occur when the leaf surface is covered with a water film. Incubation period is usually short and several cycles are completed during the season. The ectophytic mycelium grows well at 20°C and is not responsive to humidity. The rate of development of young colonies depends more on temperature than on moisture stress.

Management of the disease

The old chemical control methods included the use of sulphur dust or spray of lime sulphur. Lime sulphur was recommended for spray according to following schedules:

1. At green tip stage (when buds are green) 1:5 dilution.
2. At the open cluster stage, 1:35 dilution.
3. At blossoming or full pink stage, 1:60 dilution
4. At petal fall stage when about half the petals have fallen, 1:100

More sprays could be given afterwards if necessary. Cupric hydroxide (125 g Cu/100 lit) reduces the disease but causes fruit russet. Addition of slaked lime (2.2 kg/100 lit) reduces fruit russet.

Fungicides in use as substitute for elemental sulphur are dinocap (Karathane), oxythioquinox, benomyl (Benlate), carbendazim (Bavistin), thiophanate methyl (Topsin-M) and the sterol biosynthesis inhibiting fungicides fenarimol (Rubigan), triadimefon (Bayleton), bitertanol (Bacor), triforine (Cella, Funginex or Saprol), bupirimate (Nimrod), myclobutanil (Systhane), etaconazole (Vangard), difenoconazole (Score), penconazole (Topas), flusilazole, triflumazole and fluquinconazole. Tetraconazole (1 lit/ha), hexaconazole (0.28 lit/ha), triadimenol (0.18 lit/ha) and bitertanol (0.4 lit/ha) are reported to give good control of mildew on some apple cultivars. The strobilurin fungicides are also highly effective against the powdery mildew. The compound CGA 279202 of this group at 5–7.5 g/100 L provides high level of protective and curative action. The fungicide inhibits mitochondrial respiration and strongly inhibits conidiophore and conidia formation, spore germination and germ tube elongation. It has high affinity for the waxy layers and stays as a protective reservoir on the leaf surface. It ensures retention against wash-off by rains. Trifloxystrobin, a later formulation of strobilurin fungicides, controls powdery mildews of apple, mango and nectarines trees. In field trials, 0.01–0.015% (v/v) concentration was superior to DMI or sulphur fungicides (Reuveni, 2000).

Combinations of SBI fungicides have also been used to enhance spectrum of disease control. Thus, a mixture of Bacor 25 WP (bitertanol) at 0.05% and Bayleton (triadimefon) at 0.025% gives good control of powdery mildew and scab both. Combination of the SBI fungicide fluquinconazole (Palisade 25 WP) and the anilopyrimidine pyrimethanil (Clarinet 200 SP or Vision) also gives control of both diseases.

In post-symptom activity, Bayleton reduces the number of normal conidia produced 10 days after treatment and etaconazole and sulphur give reductions 20 days after treatment. Four sprays of Bavistin (0.05%) or Morocide (0.1%) starting at bud swell stage and repeated at 15 days interval give control of the disease. Triadimefon (Bayleton) at 0.05% gives the best control by persisting on the host surface for up to 15 days. Baycor, Rubigan and Saprol give a better reduction of conidial production than Karathane. In another study maximum reduction in germination of spores and germ tube growth was obtained by triforine, followed by tridemorph, bitertanol and carbendazim. The best disease control in the field was given by bitertanol followed by carbendazin and triforine. Antisporulant activity of bitertanol and carbendazim was noticed for 21 days. In a comparative study of Bavistin (carbendazim), Bacor (bitertanol) and Karathane (dinocap) four sprays of Karathane or Bacor were found superior to Bavistin. Addition of one dormant or bud swell stage spray of Bavistin to the schedule recommended for apple scab control gives control of powdery mildew also.

The sterol biosynthesis inhibitors are the most advanced mildewcides in use. However, while resistance to benzimidazole fungicides in the fungus is reported, possibility of resistance to SBI fungicides has also been expressed (cf. Singh, 2000). In addition to this problem, the period for application of fungicides is also important for economy and effectiveness of the treatments. Winter application of any fungicide is considered useless since the fungus in the buds is protected by thick scale leaves. In highly susceptible cultivars the most suitable time for starting fungicide sprays is the tight cluster or pre-pink bud stage and continued till mid-summer when terminal shoot growth ceases. Blossoms must be protected as early as pink bud stage to prevent fruit infection. Sometimes, fungicide sprays may be required in the late season to protect unseasonal flush of new growth. Triadimefon can be applied from bloom stage. Phytotoxicity of some fungicides during the growing season affecting pollen germination, or causing fruit russet is reported.

The length of interval between two sprays is more important than the fungicide concentration. The control is enhanced when the interval is shortened rather than increasing the fungicide rate. An interval of 10 days between sprays during the period after bloom until terminal growth stops is recommended. Terminal buds must be protected because they are the source of infection in the next year. Bupirimate and triadimefon applied at consecutive days at two weeks interval are reported to give a better mildew control than when applied at the same rate at one or two weeks intervals. Alternating an appropriate SI fungicide spray with 1% solution of monopotassium phosphate reduces the number of sprays of SBI fungicides while maintaining the same level of control as the scheduled number of sprays of the SI fungicide. Disease forecasting systems should be effective in optimizing timing of sprays. Podem-TM, developed at East Malling (UK) is one such model which simulates epidemics of secondary mildew on vegetative shoots at daily intervals. This model is incorporated in a commercial PC-based software package called Adem TM which also contains models for forecast of scab, fire blight, Nectria fruit rot and bacterial canker on apple and pear.

Plant oils have been more effective against powdery mildew than against scab of apple. Oils of sunflower, olive, maize, soybean, and rapeseed have provided more than 99% control of *P. leucotricha* under controlled conditions when applied to foliage one day before or after inoculation. Mechanically emulsified rape oil is reported comparable to dinocap (Karathane) and gave 99% control when applied 1–7 days after inoculation. Control of powdery mildews by sunflower oil results mainly from the inhibition of conidial germination and suppression of mycelial growth of the pathogen. When acetic acid at 10–12 mg/L is applied to apple shoots known to contain infection of *Podosphaera* powdery mildew fails to appear on the shoots (Sholberg *et al.*, 2005). Acetic acid treatment of root stock and scion shoot eliminated the microflora from the surface and ensures production of healthy nursery stock.

The possibility of biological control of apple powdery mildew exists. *Ampelomyces quisqualis*, described in grapevine and cucurbit powdery mildews, is reported to overwinter in mildewed apple buds and cleistothecia of the fungus and its growth is favoured by wet weather. Szentivanyl and Kiss (2003) have studied the survival of *Ampelomyces* in apple and other hosts. On apple trees the mycoparasite overwintered as resting hyphae in the dried powdery mildew mycelia covering the shoots and in the parasitized ascostromata of *Podosphaera leucotricha* on the bark and the scales of the buds. Overwintered structures collected in spring when placed close to fresh powdery mildew colonies started the life cycle of the mycoparasite. The plant pathogen and the mycoparasite start their life cycle during or soon after bud burst but *Amplemyces* can only slowly follow the spread of its mycohost on infected leaves. Epiphytic yeasts are potential biocontrol agents of *P. leucotricha*. Alaphilippe *et al.*, (2008) have reported an isolate (Y16) that controls not only powdery mildew but also suppresses scab (*Venturia inaequalis*) without inhibiting its conidial germination and leaf penetration. It also significantly reduces the egg-laying of the codling moth (*Cydia pomonella*).

Pruning of dormant shoots infected with mildew in the previous season has been recommended as a means of reducing primary inoculum. Removal of these infected shoots on fungicide treated trees reduces secondary infection by nearly half. Pruning of dormant tips of all shoots longer than 15 cm also reduces the early disease incidence. However, pruning of infected shoots is not economical when there is heavy infection of the tree. In powdery mildew control it is also important to ensure that the pathogen is prevented from establishing in young trees one to three years after planting.

All commercial varieties of apple are moderately to highly susceptible to powdery mildew. The cultivar Jonathan is highly susceptible (Dar and Kaul, 1981) while cultivars Red Delicious and Golden Delicious are rated only slightly susceptible or resistant at some locations. A few apple cultivars with Vf resistance to scab are resistant to powdery mildew also. Cultivars of apple with commercial quality and resistance or immunity to scab, cedar apple rust, and resistance to powdery mildew are now becoming available in USA. Due to existence of races in the pathogen, resistance does not last long. *P12*, a major resistance gene originating from *Malus zumi* is mainly used in apple breeding programmes for resistance to powdery mildew but some virulent strains of the fungus are reported to break the resistance (Caffier *et al.*, 2005)

STEM ROT OF RICE

Stem rot or sclerotial disease of rice was reported from Bengal, Bihar and Tamil Nadu as early as 1911–1912. By 1918, the disease was known to be present in all the rice growing areas of India. In 1944, 50–75% loss in yield was estimated in badly affected crops and 5–15% loss almost every year in Punjab. All yield components are adversely affected by the disease, mainly the number of productive tillers and sound grains per panicle. The disease occurs in many other countries. In the USA it is considered a major disease of rice.

Symptoms

The affected plants have a tendency to produce light ears and to throw out green shoots from the base when rest of the crop is ripening and turning yellow. The base of the stem is slightly discolored at the lowest distinct internode or the next one or two above. Inside the culm, a dark grayish weft of mycelium is found and the inner surface may be dotted with small, round, shining black sclerotia. Lower leaf sheath may also be invaded and within the rotting tissues the sclerotia are present. When one shoot in the clump shows the characteristic late shoots at the base, all other shoots in that clump are, as a rule, similarly attacked and every ear has a large number of light grains.

Fig. 53. Culm rot of rice (*Sclerotium oryzae*).

The causal organism

The disease is caused by *Sclerotium oryzae* Cattaneo. The conidial stage of the fungus is described as *Nakataea sigmoidea* (Cavara) K. Hara (=*Helminthoosporium sigmoideum* Cavara). It has its perfect, ascigerous, stage in *Leptosphaeria salvinii* (Catt.) now renamed as *Magnaporthe salvinii* (Catt.) Krause and Webster (Ascomycetes, Sordariomycetidae, Magnaporthaceae).

The mycelium of the fungus is present in the subepidermal layers of the culm and the sheath. It consists of creeping, much septate, vigorously branched hyphae appearing white inside the host tissue but olivaceous on the surface. They are provided with numerous olivaceous, irregular appressoria. Soon few or numerous sclerotia appear on the white mycelium. They are large, globose, dark black (brown inside) bodies and measure 230–270 μm in diameter. This is the sclerotial stage of the fungus (*Sclerotium oryzae*) and is common on rice and a number of wild grasses.

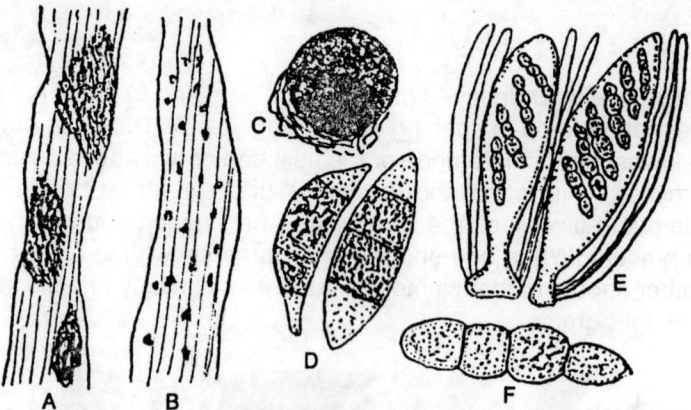

Fig. 54. *Magnaporthe salvinii*. A. Lesion on the culm; B. Sclerotia on old leaves; C. Sclerotium; D. Conidia; E. Asci; F-Ascospore.

Conidia and ascospores are produced about 60–100 days after planting and are not of common occurrence. The conidiophores of *Nakataea* are dark, septate, simple or sparsely branched. The number of septa is 8–10. They measure 100–150 × 3 μm. The conidia are large, falcate-sigmoid, 3-septate and borne singly at the tip of sterigmata on the conidiophores. The middle cells of the conidium are dark brown and apical cells light green or hyaline. The ends are obtuse. The conidia measure 55–65 × 11–14 μm.

The perfect stage, Magnaporthe salvinii, may or may not be always found. The perithecia are black, globose, clustered in the parenchyma of the leaf sheath. They are provided with short but broad beak, flushed with the epidermis of the sheath, and measure 350–400 μm without beak. The asci are clavate, short stipitate, delicate walled, 120 μm long and contain 8 ascospores each. The ascospores are arranged in 2–3 series. They are oblong-fusiform, slightly curved, 3-septate, middle cells constricted at the septum, and measure 60–90 μm.

Pathogenic specialization is reported in S. oryzae. The fungus is heterothallic. In areas where the perfect stage is common, sexual recombinations result in evolution of races that can overcome the resistance in the host.

Disease cycle and environmental relations

Stem rot is a late season disease. It develops rapidly only after plants shift to reproductive stage and most rapidly around the physiological maturity of the host. During absence of the host the fungus survives usually through the sclerotia. These sclerotia are present in enormous numbers in old rice straw and stubbles left in the field and are disseminated by irrigation or rainwater. The sclerotia survive in air-dry soil in the laboratory for 190 days, in moist paddy soil for 133 days, in tap water for 319 days and in corked specimen tubes for 525 days. If exposed to sun on the soil surface they quickly lose their viability. Survival in soil is affected by depth, the sclerotia losing viability quickly at lower depths. Sclerotia survive for 10 months at 10°C, for 16 months at 20°C and for 6 months at room temperature (4–42°C). Viability was very low at the end of each time scale. Sclerotia at 5 cm depth survived better than those at soil surface or at 15 cm depth. The fungus has poor saprophytic survival ability in soil.

On germination the sclerotia initiate the primary infection. The disease is enhanced with increasing levels of nitrogen above 120 kg/ha. Plants receiving split doses of nitrogen exhibit delayed disease development. Thick stand of the crop and high doses of herbicides also favor disease development.

Management of the disease

The straw and stubbles of the infected crop should be burnt in the field. This method of rice residue management is most effective method of stem rot control. In California (USA), spot burning of rice straw is being discouraged. Winter flooding of the field has been found as an alternative for rice residue management. However, the yields are better in plots where the residue was burnt (Cintas and Webster, 2001). Patches of diseased plants should be marked out and should be either uprooted and destroyed or they should not be harvested along with the healthy plants. After harvest these plants can be burnt at the spot. Passage of irrigation or rain-water through or from infested field to healthy fields should be checked. Where rice is an irrigated crop the field should be flooded and then the water should be drained off. The soil should be allowed to dry before the next irrigation is given. Sclerotia survive better in undisturbed soil. Hot weather ploughing helps in decreasing viability of sclerotia although it causes their distribution over the field. In soil solarization experiments, mulching of soil with clear polyethylene sheets for one week is reported to kills 95–100% sclerotia at 5 cm depth.

Although reports of laboratory screening of fungicides against the fungus are many, there are very few studies on fungicidal field control of the disease. Spray of 0.2% Bavistin or Topsin-M effectively reduces the disease incidence. Successful biological control of the pathogen by infesting the field with cultures of Trichoderma has been reported. Trichoderma spp. attack, colonize and kill the sclerotia of the fungus. In Californial rice fields (USA), Ascochyta mycoparasitica was found as a mycoparasite of S. oruyzae.

Most cultivars of rice are susceptible to stem rot. The degree of susceptibility varies with the varieties. Five culture/cultivars, belonging to long duration group, namely, MNP 234, Caloro, IET 5633, ARC 12751 and Tedukar have been found highly resistant to stem rot. Resistant varieties showing less than 20 mm lesion size are Improved Sona, IR 22, Rasi and VL 8. The moderately resistant (lesion size 31–60 mm) were Bala, Basmati 370, BJ 1 (dwarf), IR 20, IR 56, Pant Dhan 4, Pusa 33, Ratna, Suhasino and Tilakchandan.

BLACK LEG OR PHOMA STEM CANKER OF CRUCIFERS

This serious disease of cabbage and oilseed brassicas (canola, *B. napus, B. juncea, B. rapa*) is reported from 49 countries in all the continents. The disease causes seedling death, lodging or early senescence. On rapeseed (*Brassica napus* var. *oleifera*) the disease is common in many temperate climate countries including Canada, United States, Europe and Australia where the damage to the oilseed crop may be as high as 60% through basal cankers and lodging. However, epidemics of the disease are rare in Asia including India although oilseed rape and cabbage are common crops. It could be the most destructive disease of these crops but for the availability of effective control measures.

Symptoms

The plants may be infected in the seedbed or at any stage of their growth. In the seedbed, seedlings are killed by infection of soil-borne pycnidiospores and ascospores. Later, usually the first symptom is an oval or linear, depressed, light brown canker near the base of the stem. The canker enlarges until the stem is girdled. Such dry cankers or dry rot develop at 15°–24°C. At lower temperatures of 7°–12°C there is usually wet rot. Stem lesions at the soil line usually extend to the roots causing dark brown cankers, but only few fungal fruit bodies are formed on the roots.

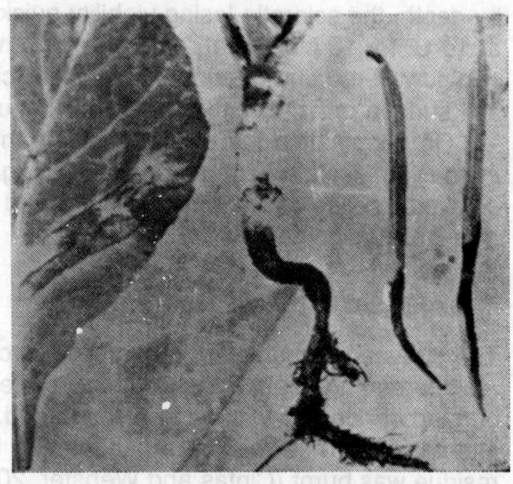

Fig. 55. Black leg of cabbage.

On the leaves the spots are circular, light brown with ashen-grey center. In infection caused by virulent strains the spots enlarge to 1–2 cm in dia. On seed stalks and pods of seed plants of cabbage or oilseed brassicas the lesions are elliptical. Infected seeds are shriveled and discolored. The most important character of the spots or cankers on stems, leaves, etc. is the early appearance of numerous black dots on them. These are fruit bodies of the fungus. In infection caused by less virulent strains the spots are smaller, darker, with few or no pycnidia. Affected plants may die or, if infection occurs late and plant vigor is good, it may survive by producing new roots to compensate for the destroyed roots.

The causal organism

The disease is caused by *Leptosphaeria maculans* (asexual stage *Phoma lingam*). *Leptosphaeria* is placed in family Leptosphaeriaceae, order Pleosporales, subclass Dothideomycetidae, class Ascomycetes. The mycelium is septate, branched, hyaline when young, becoming dark-walled with age. Two types of pycnidia are produced. In the sclerotioid type of pycnidia, the wall has several layers of thick-walled, pseudoparenchymatous cells. These pycnidia measure 200–500 μm in diameter. They are usually found on dead woody tissues and serve as survival structures. The globular type of pycnidia have cell walls that are thickened in the outer cell layers only. They develop in living tissues and in cultures at 20°–24°C. Generally, the pycnidia are flask-shaped and dark colored, sometimes with beaked or papillate ostiole. The conidia in the pycnidia are hyaline, aseptate, and 36 × 12.5 μm in size. These spores are produced in abundance and, in nature, they accumulate within the pycnidium, drying down in a gelatinous matrix. When dew or rain wets the pycnidia, the matrix rapidly absorbs water and the spores expand in long coils consisting of spores and gelatinous matrix. The matrix dissolves in water setting free the spores. Thus, dissemination of spores exclusively depends on wetness of the host surface. In the sexual stage the fungus produces pseudothecia which measure 300–500 μm in diameter. The asci measure 80–125 × 15–22 μm. The ascospores measure 35–70 × 5–8 μm.

Variations in pathogenicity of *Leptosphaeria maculans* have been reported from Australia, Canada, England and the United States. The isolates from different sites differ in pathogenicity. The virulent isolates (group A of the pathogen) are responsible for epidemics. Less virulent isolates (group B) cause only superficial lesions. The two groups are basically different from each other in cultures, genetics, metabolite production, and morphology of pseudothecia. There is no evidence of sexual mating between the two groups. Consequently, the black leg pathogen was considered by many to be a complex of at least two different species. Germ tubes from B-group ascospores are longer than those of A-group ascospores at 5–20°C. The hyphae of A-group isolates are highly branched and grow tortuously where as those of B-group are long and straight. Otherwise, the rate of germination of spores and mode of host penetration do not differ. Presently, the virulent A group is retained in *L. maculans* and the B group in *L. biglobosa* described by Shoemaker and Brun (2002) for the weakly virulent (or non-aggressive) isolates. The virulent group A is further subdivided into many pathogenicity subgroups. Pre-infection by *L. biglobosa* imparts resistance to *L. maculans*. The pathotypes in *L. maculans* are listed as PG-2, PG-3 and PG-4.

The fungus produces a toxin, sirodesmin PL which is decreased when plants are supplied zinc. The toxin is injurious to cotyledons and inhibits seed germination and plant growth. It is non-specific, having similar effect on non-host as well as resistant or susceptible host cultivars. Sirodesmin is not the virulence determinant. Mutants lacking this toxin also can cause infection of cotyledons and their collapse. However, they are less aggressive on stem base. Along with sirodesmin another related phytotoxin, phomalide, is also produced.

Disease cycle and environmental relations

Infected crop stubbles in soil and infected seeds are the main source of survival of *L. maculans*. The fungus survives in diseased crop debris as sclerotioid pycnidia and pseudothecia. After the diseased crop is harvested, the senescent stem tissues are rapidly colonized by the pathogen and pycnidia are produced abundantly, releasing conidia. The conidia colonize stubbles saprophytically, increase the inoculum level and subsequently the numbers of pseudothecia. On stubbles buried in soil in the field along with the pathogen and the indigenous microflora of plant surface, the viability of the inoculum declines over the time up to 10 months while density of the indigenous microflora increases manifold. Raindrop splashes play an important role in dispersal of pycnidiospores from stubbles and infected leaves. The distance of dispersal may be up to 40 cm but about 90% conidia are dispersed to a distance of 14 cm. Pseudothecia of *L. biglobosa* mature more slowly at <10°C. Maturation time for pseudothecia of both species decreases almost linearly with temperature from 5 to 20° under continuous wetness but is longer in natural conditions, especially when periods of dry weather occurred. The mature pseudothecia release ascospores over an extended period of time. These ascospores may remain viable for 6 weeks.

The seed infection level may vary from 0.08 to 3%. A minimum of 0.2% seed infection can lead to an epidemic of the disease. If the infected seed can germinate, the seed coat comes out above the ground along with the fungus. From this point, the fungus causes infection of the adhering cotyledons. These may be killed before the fungus moves to infect the hypocotyl. The seedlings are killed and fungus continues to grow and produce pycnidia and conidia on the dead seedlings. Dispersal of these spores occurs by irrigation water and raindrop splashes. The conidia can spread the disease up to 1 meter from original foci. In vegetable crucifers, the surviving seedlings in the seedbed are infected. These infected seedlings are transplanted and the disease is introduced in the main field. Spores produced on lesions carry on the secondary infection. In oilseed brassicas, ascospores released from stubbles initiate the epidemics. Pseudothecia start to form 1–10 months after harvest on the woody remains of the infected plants and persist as long as crop residue remains. This is an important source of primary inoculum. Ascospore discharge is influenced by rain. Release generally starts within one to a few hours after onset of rains and ceases several hours after rains have stopped. Heavy dew or high humidity are sufficient to initiate and sustain spore release. Ascospores and pycnidiospores can be trapped from the atmosphere between 9 p.m. and 4 a.m. when air-temperatures are 13–18°C and relative humidity is 89% (Guo and Fernando, 2005). Peak ascospore and pycnidiospore dispersal is associated with rain events. Ascospore dispersal occurs several hours after rainfall of >2 mm and continues approximately for 3 days after rains. The ascospores can be disseminated to several kilometers from source of crop debris but level of disease incidence declines as the distance increases. Ascospores land on the leaf surface and start germinating within 2 hours at temperatures of 5°–20°C on oilseed rape (*Brassica napus*) leaves. The peak percentage of germination reaches in 14 hours (Huang *et al.*, 2003). The rate of ascospore germination and length of germ tubes increases with rise in temperature from 5°–20°C. Ascospore germination, penetration and development of symptoms on cotyledons is much earlier than that with pycnidiospores. At 2 hours after inoculation ascospores begin to germinate and by 4 hours about 50% spores have germinated. Pycnidiospores start germination one day after inoculation and reach 50% only by 3 days after inoculation. Incubation period of *L. maculans* is 5 days at 20°C and 2 weeks at 8°C. The germ tubes penetrate the leaf mostly through stomata. Although it is generally believed that infection by air-borne ascospores plays the major role in disease epidemiology and pycnidiospores (conidia) from leaf and stem lesions or stubbles play a minor role, it has been shown that co-inoculation with conidia and ascospores enhances the pathogenicity of conidia (Li *et al.*, 2006). Wound penetration can also occur. This is followed by infection, colonization, systemic biotrophic spread in the lamina leading to latent infection of the stem. The glucosinolate in *Brassica juncea*, which is source of highly fungitoxic compounds after enzymic hydrolysis by myrosinases, has no role in resistance of the oilseed rape or *Brassica juncea* to black leg. Probably, the fungus attacking *B. juncea* detoxifies or evades the toxic effects of isothiocyanates. Glucosinolate levels and myrosinase activity are similar in resistant and susceptible cultivars suggesting that the glucisinalte-myrosinase system does not affect incidence and severity of blackleg in canola. The processes of pycnidiospore attachment, germination, and penetration through the stomata on petioles and stems are similar on susceptible and resistant cultivars. Specific post-penetration defense responses of resistant cultivars have been identified as lignification, suberization, and additional cambium formation (Li *et al.*, 2007). *L. maculans* causes root rot also. Sprague *et al.*, (2007) followed the hyphal growth of the fungus after inoculation of petioles. It grew within stem and hypocotyl tissues during the vegetative stages of plant growth and proliferated into the roots within xylem vessels at the onset of flowering. The hyphae growing in stem tissues were also restricted to the xylem. Direct inoculation of the roots also established the fungus in roots. *L. maculans* also infects intact roots when inoculum is applied directly to them from above and below ground sources. The hyphae enter the root at sites of lateral root emergence without forming any penetration structure.

The fungus grows well at all temperatures at which the host can grow. Optimum temperature for maturation of pseudothecia is 14–15°C. Symptoms readily develop at

15°–28°C. Their development is much delayed at temperatures below 12°C. The conditions that produced the greatest numbers of leaf spot lesions area leaf wetness duration of 48 hours at 20°C. Numbers of lesions decreased with decreasing leaf wetness duration and increasing or decreasing temperature. At 20°C with 48 hours of leaf wetness, one out of 4 spores infected leaves to cause a lesion whereas with 8 hours of wetness only one out of 300 spores caused a lesion. As temperature increased from 8°C–20°C the incubation period decreased from 15–5 days. Sosnowski et al., (2005) have emphasized that in infection of resistant and susceptible cultivars of Brassica napus by pycnidiospores of L. maculans day/night temperatures, wetness duration and spore concentration in the inoculum have significant effects. The greatest number of leaf lesions developed on plants exposed to a day/night temperature of 18/15°C with a 96 hour wetness period. Development of stem infection is greatest at 23/20°C with a wetness period of 48–72 hours. Disease development shows seasonal variation. In glasshouse studies, a resistant cultivar of Brassica napus showed increased diseased severity in summer while a susceptible cultivar showed consistent disease severity throughout the year. In the field, the severity of the disease is in direct proportion to the amount of rainfall in the crop season. In a French study, it is reported that early sowing of oilseed rape (in that region) results in less crown canker. High nitrogen availability during vegetative growth of the crop causes more crown canker.

Management of the disease

Sanitary and other precautions, including a 2 or 3-year rotation, as recommended for other diseases should be followed in this case also. Management of the crop debris through its decomposition and chemical treatment of stubbles reduces inoculum production in soil. Crop isolation may be more important than extended rotation length in the control of blackleg of canola. In south-eastern Australia, the recommendation is to sow canola crops at distances greater than 100 m, preferably 500 m, from the last season's canola stubbles. Hot water seed treatment at 50°C for 30 min destroys the seed-borne inoculum. This treatment is meant for fresh seed. Old, stored seeds lose viability by hot water treatment. Seed treatment with thiram, iprodione or carbathin is reported to be effective. Protection of the crop during the first 4–6 weeks after emergence greatly reduces losses in later stages. Control of the leaf spots by chemical sprays is most effective when started soon after emergence of seedlings. For control of the stem cankers the best timing is when leaves 7–11 are present on most plants and at least 10% of plants are affected by phoma leaf spots (Steed et al., 2007). Fungicides currently being recommended are difenoconazole, difenoconazole + carbendazim and prochloraz + propiconazole. Pre-treatment of the first true leaves of Brassica napus with the plant defense activator menadione sodium bisulphite induces local and systemic resistance. Acibenzolar-S-methyl and benzothiadiazole (BTH) also similarly induce resistance. The treatment leads to accumulation of pathogenesis related genes Paenibacillus polymyxa is a potential biocontrol agent against L. maculans. It inhibits growth of the pathogen through production of antifungal mixture of peptides, fusaricidin group of cyclic depsipeptides (Beatty and Jensen, 2002). Pseudomonas chlroraphis and Pseudomonas aurantiaca also suppress L. maculans through production of phenazine antibiotics (Ramarathnam and Fernandez, 2006). Strains of Bacillus amyloliquefaciens, isolated from among plant-associated Bacillus spp. provides protection against L. maculans and Alternaria brassicae. Pre-treatment of Brassica napus leaves with Leptosphaeria biglobosa or acibenzolar-S-methy induces resistance to L. maculans (Liu et al., 2006). Inoculation of leaves with avirulent or weakly virulent L. biglobosa reduces severity and incidence of the disease. When the pre-inoculated leaves are challenged with the virulent pathoitypes activity of chitinase, beta-1,3-glucanse, peroxidase and phenylalanine ammonia lyase is greatly increased. Rapid necrosis of guard cells of stomata is associated with the arrest of fungal growth in leaves inoculated with an acirulent strain. A hypersensitive response occurs after inoculation ofcotyledon, leaf and stem of *Brassica napus* by an avirulent strain of *L. maculans*. There is condensation of cytoplasma, shrinkage in cell size and nuclear DNA fragmentation in cotyledon and stem tissues.

The most effective way to control Phoma stem canker in oilseed rape could be breeding for resistance. However, major gene or race specific gene resistance is not durable because of wide variability in the pathogen. Strains of *L. maculans* can overcome the single dominant gene-based resistance. Where this happens death os cells is restricted to a few palisade cells even though hyphae are present in the lower tissue layers of the cotyledon. Polyphenolic compounds, associated with resistance response, do not accumulate. Some cultivars of *Brassica napus* have polygenic resistance. In south Australia, the major gene resistance in *Brassica napus* was overcome within three years of commercial cultivation.

ASCOCHYTA BLIGHT OF CHICKPEA

Chickpea, garbanzo bean, gram or Bengal gram (*Cicer arietinum*) is grown over an area of about 6.7 million hectares in India with total production of more than 4.5 million metric tons (FAO data for the year 2002). The world production area was 8.7 million ha with a production of 8.2 million metric tons. About one third of the acreage in India lies in the northern and northwestern states of U.P., Haryana, and Punjab. The three diseases causing significant losses in this crop are Ascochyta blight, Gray mold, and Fusarium wilt. Rust appears late in the season and is not very important as it occurs very late. Ascochyta blight, first described in 1911, is the most damaging disease of chickpea. It is especially important in countries that grow winter-sown chickpea since winter conditions that favour the growth of the crop are also ideal for the development of disease epidemics. Ascochyta blight is reported from 26 countries in the world which include almost all the countries in Europe and North Africa bordering the Mediterranean sea, Iran, Iraq, Pakistan, Portugal, Romania, Spain, USA, former Soviet Union, Mexico, Tanzania, Bangladesh and India. It is extremely important in the areas lying between 30° and 45°N latitude and occasionally important in areas between 26° and 30°N latitude. In many chickpea growing countries such as Nepal, Burma (Myanmar), Argentina, Bolivia, Peru, Chile, Libya, Columbia, Malawi, Zambia, Sudan, Uganda and Yugoslavia there is no report of its occurrence. In the Indian subcontinent it occurs in north west India and in Pakistan. In India chickpea blight is common in Punjab, Haryana, Himachal Pradesh, northwest U.P., and Bihar, only occasionally in Madhya Pradesh but was not reported from Andhra Pradesh.

During the 1930s chickpea blight had caused total loss of the crop in Spain. In the undivided Punjab (now part of Pakistan) it used to appear in severe form and during 1922–1933 the loss was estimated at 25–50% almost every year. In the present day Pakistan the losses amount to about 70%. Five to 75% loss was reported from Rajasthan in 1982. If the environmental conditions favor severe incidence of the disease the yield loss may reach 100%.

Symptoms

The disease appears on all the above ground parts of the plant. Circular spots develop on leaves and pods and elongated spots on petioles and stems. The round or sometimes elongated spots on the leaflets bear depressed brown dots and are surrounded by a brownish

Fig. 56. Ascochyta blight of chickpea. Left-Blighted plant; Right-Lesions on stems and pod.

Fig. 57. Ascochta blight of chickpea. Left-Close up of leaf blight. Right-Lesion and fructification on pod.

red margin. The spots may coalesce and the entire leaf turns brown, presenting a scorched appearance. The usually circular lesions on the green pods are surrounded by a dark margin and bear pycnidia (black dot-like bodies) arranged in concentric circles. Seeds within the pods may also show lesions. On the stems and petioles the lesions are elongated (3–4 cm long), brown, and bear black dots. These lesions may girdle the affected portion. When the spots on the stems completely girdle them the parts above the lesion droop and wilt. If the main stem is girdled at the base (collar region) the whole plant dries. As the disease progresses, patches of drooping and wilting plants becomes prominent in the field, later involving the whole field. In dry weather, the spread of the disease may remain restricted but if there is wet weather (light showers during the crop season) the spread of the disease is very rapid.

The causal organism

The disease is caused by *Ascochyta rabiei* (Pass.) Labrousse, the anamorph of the ascomycetous fungus *Didymella rabiei* (Kov.) von Arx *Didymella* is placed in the order Pleosporales of subclass Dothideomycetidae, class Ascomycetes. Species of *Ascochyta* infect large number of legume species, *viz.*, *A. rabiei* (chickpea, *Cicer* spp.), *A. fabae* (faba bean, *Vicia faba*), *A. pisi* (pea), *A. lentis* (lentil) and *A. viciae-villosae* (hairy vetch, *Vicia villosa*). All these species are host-specific (Hernandez-Bello *et al.*, (2006). In a molecular study of *Ascochyta* spp. in Canada, Chongo *et al.*, (2004) found *A. rabiei* 45% similar to *A. lentis* (lentil blight) and only 14% similar to *A. pinodes* (pea blight).

The mycelium of *A. rabiei* is hyaline to brownish and septate. The pycnidia on stems, leaves and seed pods are immersed becoming erumpent, globose, dark brown, and 140–200 μm in diameter. The wall is composed of 1–2 layers of elongated pseudoparenchymatous cells. The pycnidium has a prominent ostiole which is 30–50 μm wide. Conidia are formed from hyaline, ampulliform phialides from the inner cells of the pycnidium. These conidia are hyaline, oval to oblong, straight or slightly curved, 1-septate, some 0-septate, slightly or not constricted at the septum, rounded at each end, and measure 10–16 × 3.5 μm. If the pycnidia are moistened they absorb water, swell, and a slimy mass of conidia oozes out forming a spore horn. The conidia germinate by long germ tubes. Within the pycnidium the conidia can remain viable for a year or more.

The perfect stage (teleomorph) of *A. rabiei* was observed in Bulgaria by Kovachevski in 1936 and described as *Mycosphaerella rabiei* Kov. It was later renamed as *Didymella rabiei* Both these genera belong to the same taxonomic position. They are characterized by 8-spored asci formed in pseudothecia which are small and immersed in the host tissues (usually dead parts). The perfect stage has been reported from eastern Europe and has not been seen in the Indian subcontinent where hot summer follows the crop season. The amount of variability in the fungus is enhanced by the presence of the teleomorph under field conditions in temperate regions Navas-Cortes (1998) studied the development of *Didymella rabiei* on debris of naturally infected chickpea in Spain. In the area concerned, pseudothecial initials started appearing in November and ascospores were mature by late January to late March. Maximum ascospore discharge occurred 2–4 weeks after ascospore maturation. By the beginning of summer the pseudothecia were exhausted due to ascospore discharge and new asci did not develop in the empty pseudothecia. All activities related to pseudothecial stage were more uniform and occurred later in cooler areas than in warmer areas.

The pseudothecia of *Didymella rabiei* were detected exclusively on plant debris, especially the pods, which had overwintered in the field. They are dark brown or black, globose, with inconspicuous ostiole, and measure 120–250 × 75–152 μm. The asci are cylindrical-clavate, slightly curved, pedicellate, and 48–70 × 9–13.7 μm in size. The ascospores were 1-septate with one cell larger than the other. They were prominently constricted at the septum and measured 12.5–19.0 × 6.7–7.6 μm.

Variability in *A. rabiei* had been indicated by Luthra and co-workers in 1939 Since then many races were identified by different authors on the basis of different criteria in India and Syria. At present 6–7 races are considered common and have been used for studies in disease resistance.

Disease cycle and environmental relations

The pathogen survives as pycnidia on diseased plant debris left in the field and also on or in the seed. However, badly affected seeds do not germinate although they can serve as substrate for growth of the fungus under suitable conditions. Infected seeds always do not give rise to infected seedlings. Often lesions are present on the seed but coleoptile and coleorrhiza remain completely free from infection. Transmission from seed to seedlings is 6.1%. However, about 6% infection from seed is quite enough to start epidemics under favourable weather conditions. Studies with *Didymella rabiei* in temperate climate (Kimber *et al.,* 2006) have shown close positive correlation between the level of seed infection and the incidence of the disease on seedlings. Disease transmission to seedlings is not affected by depth of sowing 1 to 6 cm), resistance or susceptibility of the cultivar and temperatures from 5–19°C).

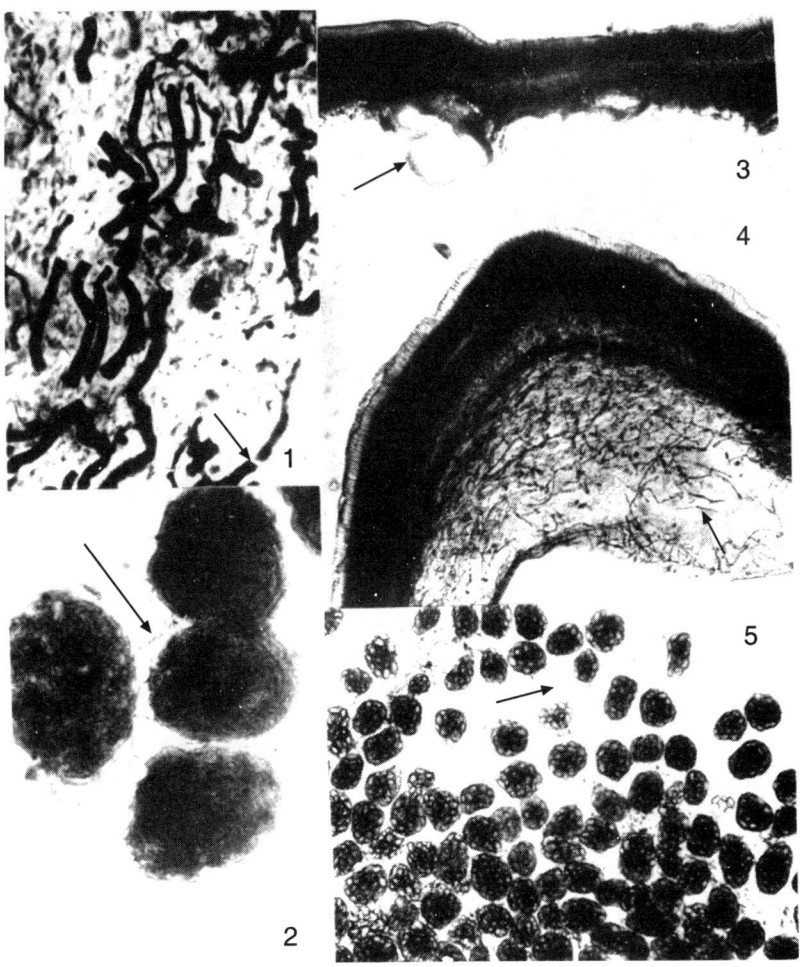

Fig. 58a. Seed-borne inoculum of *A. rabiei.*
1. Mycelium in seed coat, 2. Mycelium in cotyledon, 3. Pycnidium in seed coat, 4. Mycelium in seed coat (vs), 5. Mycelium in cotyledon (vs). Courtsey; Dr. K. Vishunavat.

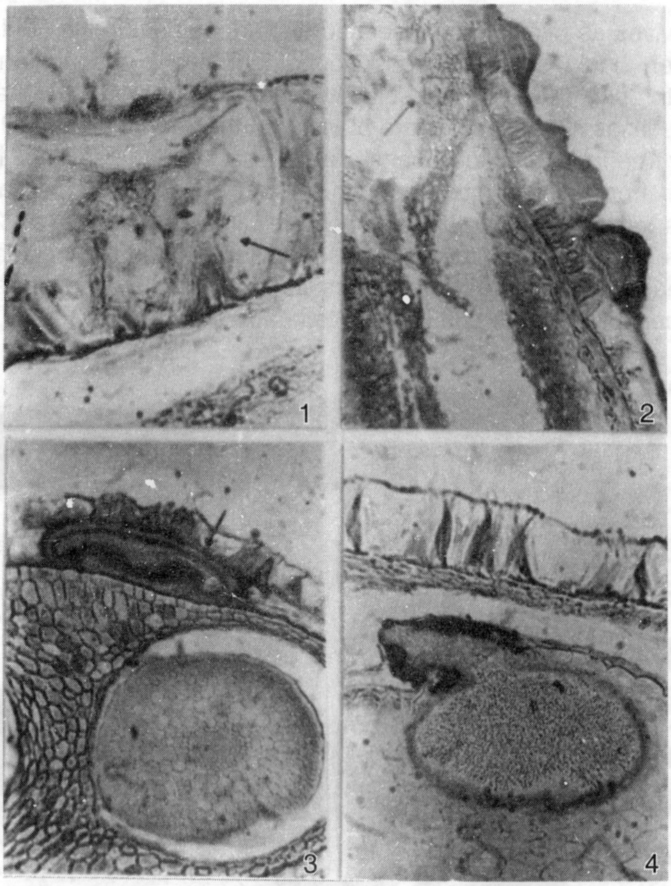

Fig. 58b. Pycnidia of *A. rabiei* in seed tissue. Courtsey. Dr. K Vishunavat.

In temperate climate countries the crop debris left on soil surface and buried under snow with extended periods of freezing reveals pseudothecia with ascospores. Thus, in such regions the perfect stage of the fungus is a source of survival, long distance spread and variability in the pathogen. Pycnidia survival in debris depends on moisture, irrespective of temperature. Pycnidia can survive for over 2 years at 10–35°C if relative humidity on soil surface is 0–3%. They lose viability rapidly if the relative humidity is 65–100% or if the debris is buried deep in soil. Gossen and Miller (2004) have studied survival of the pathogen through buried crop debris in Canadian prairies for up to 4 years. The studies indicated that, under Canadian conditions, the pathogen population increased in the year following an infected crop, and then declined over time. The pathogen was still present at low levels in the field 4 years after the initial crop although the level was very low after 3 years. The pathogen was still present following three intervening non-host crops. This suggested that only a long rotation can be recommended.

When pycnidia are buried at 15 cm depth of soil there is no blight of the aerial parts of the plant. At high temperatures prevailing in the chickpea growing areas of the Indian sub-continent after crop harvest the viability of the fungus in crop debris is completely lost. Under

Indian conditions the main source of gram blight is seed. Mostly the propagules of the fungus are present on or in the seed coat, few on cotyledons and embryo. On contaminated seeds 50% of the conidia survive for 5 months at 25°–30°C but only 5% at 35°C. Mycelium may also be present in the seed coat. If deep lesions are formed on the seed, pycnidia may be present in the seed coat. Presence of sub-epidermal pycnidia in seed coat and mycelium aggregated in the innermost layer of the seed coat and also in cotyledons is reported. The seed-borne infection in seed stored at room temperature declines from 70% in May (after crop harvest) to 4% in October and to 3% in December (the next crop season). The spores present on the seed surface lose viability in October.

In a study under temperate climate regions, where both conidial and perithecial stages are common (Trapero-Casas and Kaiser, 2007), germination of conidia and ascospore began after 2 hours and reached the maximum (> 95%) in 6 hours at 20°C. No germination occurred at 0°C and 35°C. Ascospores germinated more rapidly than conidia at all temperatures. Decline in humidity decreased germination. Optimum temperature for infection and disease development by both types of spores was around 20°C. Disease severity was high when ascospores reached the plant surface directly from the source than when they were used in a suspension. Severity increased as the length of the leaf wetness period increased. Dry periods of 6 to 48 hours after inoculatin decreased severity. Ascospore inoculation caused more disease than conidial inoculation.

After development of the fungus from germinating seed or from infested soil, the primary inoculum multiplies on hypocotyl, epicotyl, and near the stem base and then spreads to aerial parts. In the field, spread of the pathogen occurs through conidia disseminated by raindrop splashes in a windy weather. by insects, and by contact between leaves and movement of animals through the field. Infection by inoculum on dry leaves is significant when the dry leaves are exposed to 98% RH for 48 hours. Disease development in the plant population is influenced by temperature, leaf wetness duration and inoculum concentration (Armstrong et al., 2004). A temperature of 20°C, 48-hour wetness period and 2×10^5 conidia/cubic ml^3 are ideal. Disease severity is substantially reduced by a dry period of 2–6 hours after inoculation. The disease spreads fast if wet weather with strong winds occurs in February and March when the temperatures are around 22°–26°C. Disease severity increases with wetness duration, maximum of 18 hours. The spread in a field is along the direction of the prevailing wind. Years of high rainfall during the crop season are conducive to epidemics of the disease. While in susceptible cultivars the plants are almost equally infected at all stages of growth (seedling to pod-formation stages), in resistant cultivars seedlings are least receptive and plants in pod-formation stage are most receptive (Chongo and Gossen, 2001). Severity of Ascochyta blight is high in plant types having unifoliate-leaves and low in fern-leaved type (Gan et al., 2003).

Although the primary inoculum comes from seed or crop debris in soil, development of severe epidemics of blight in fields where healthy seeds had been planted and the field had no history of chickpea cultivation, is not uncommon. This indicates the possibility of long distance dispersal of inoculum. The fungus has been found on berseem also. Isolates from berseem infect chickpea and vice versa. In Australia, A. rabiei was found to infect 4 cultivars of common bean (Phaseolus vulgaris). However, Frenkel et al., (2007) have reported that A. rabiei can infect chickpea and wild species of Cicer but not other legumes. Phoma pinodella (Ascochyta pinodella) that cause foot rot and blight of pea can infect C. arietinum and other Cicer species.

Ilarslan and Dolar (2002) have studied the pathogenesis on leaves and stems after infection by spores. In susceptible plants the intercellular and subepidermal spread can be seen 3 days after inoculation. By 5 days after inoculation cell plasmolysis, degeneration of organelles and of cellulose, but not of lignified, walls occurs. Tracheidal elements including lignified elements remain intact in both susceptible and resistant plants. There is less cell degeneration and pycnidia formation in resistant plants. According to Jayakumar et al., (2005) during germination and infection the germ tubes secrete a mucilaginous substance that facilitates attachment to the host surface The invading fungus produces several phytotoxins (solanapyrone A, B and C., cytochalasin D and proteinaceous toxin) which seem to play a role in necrosis and cell death. The pathogen can degrade antimicrobial compounds and suppress their production in chickpea.

Management of the disease

Removal and destruction of dead plant debris, rotation, and deep sowing were the earliest recommended control measures. In Ascochyta blight of lentil (Ascochyta lentis) at least two non-host crops between successive lentil crops are required to substantially reduce inoculum following a disease outbreak. One or two years of non-host crops for waem and wet areas and 3–4 year crop rotation for cold and dry areas is recommended (Gan et al., 2006). This reduces the level of stubble-borne inoculum in soil. Field isolation, if possible, should also be followed. It means that chickpea field should be located away from filed planted with chickpea in the previous season. This may prevent airborne inoculum to affect the new field. This is particularly important in cold climate areas where the fungus survives through its perithecial stage and releases ascospores in the chickpea season.

Intercropping of chickpea with cereals to reduce spread of blight was recommended in 1935, In a study of several intercropping systems, Gaur and Singh (1994) found chickpea-barley intercropping as the best combination which reduced the disease incidence and gave maximum produce from the land. Deep ploughing to bury the debris is also recommended. Application of 40–60 kg potash / ha along with 20 kg nitrogen and 40 kg phosphorus is reported to reduce disease severity and increase grain yield. Application of potash alone is not helpful.

Dugan et al., (2005) studied the role of saprophytes colonizing the Ascochyta infected chickpea debris in soil and found many species of Alternaria, Epicoccum and some antagonistic fungi such as Ulocladium spp., Aureobasidium pullulans and Clonostachys rosea, commonly present. Aureobasidium pullulans grew faster than Ascochyta rabiei and excluded it from the substrate. This fungus along with Clonostachys rosea was capable of suppressing A. rabiei and its sexual stage on chickpea debris.

Since the pathogen is externally and internally seed-borne, seed treatment is an important control measure in absence of resistant varieties. Seeds can be treated with organomercurials (Agallol, Agrosan, etc.), copper sulphate, thiram, benomyl or Calixin M. The last named fungicide is reported to completely eradicate the inoculum from the seed. Field control of chickpea Ascochyta blight has been achieved by seed treatment with Bavistin + Thiram in 1:3 ratio at the rate of 2.5 g/kg seed followed by 3 sprays of Bavistin (0.5 kg/ha) at 10 days interval when the disease appears (usually at pod formation stage). Seed treatment with combinations of Bavistin and Thiram (1:1 or 1:3) at the rate of 3 g/kg seed is reported to give 95–100% reduction in disease incidence. Bravo (6 ml/kg), hexacap (4 g/kg) and thiabendazole (5 g/kg) completely eliminated the seed-borne inoculum. Rovral and hexacap (3 g/kg seed) give 98% control.

In mild attack, sprays of zineb, ferbam, maneb, captan, and Daconil are recommended. However, the disease spread is so fast that 4–6 sprays may be required and this will be costly. Foliar spray of Bravo (1500 ml/ha) and Calixin M (900 g/ha) is significantly superior to other fungicides for controlling the secondary infection, followed by Rovral (750 g/ha) and hexacap (900 g/ha) in 300 lit. of water). Post-infection application of tebuconazole or difenoconazole to moderately resistant cultivar can give 95% control if the weather supports mild epidemics but only 65% control when weather favours severe epidemics (Shtienberg *et al.*, 2000). In highly susceptible cultivar, the corresponding figures are 80% and less than 20%. In a Canadian study (Chongo *et al.*, 2003), under high disease pressure a single fungicide application often reduced disease severity but had no effect on yield. Two sprays (early and mid-flowering) of chlorothalonil (Bravo) at 1 kg a.i./ha, two applicarions of azoxystrobin at 125 g a.i./ha or chlorothalonil + azoxystrobin reduced blight and increased yield. The fungicides compliment partial resistance to reduce blight severity and increase seed yield and quality. In the Mediterranean area, single spray of chlorothalonil made before flowering significantly reduced disease severity in susceptible cultivars. Earlier and later spray were not effective. There was no significant difference between untreated control and fungicide application in resistant cultivars. Prevention of early infection from soil-borne inoculum by giving a spray of tebuconazole has resulted in significant reduction in the disease during growing season of the crop.

The use of resistant varieties is the most effective and economical way of management of Ascochyta blight of chickpea. However, currently available cultivars only possess partial resistance to the pathogen, and this level of resistans can breakdown easily because of variability in the fungus. The cultivars become susceptible soon after deployment. Effective control of the disease cannot, therefore, be obtained by total reliance on host resistance. Seed treatment of tolerant cultivars may be much more effective to prevent the increase in pathogen population during conducive weather conditions. The contribution of genotype resistance to disease suppression in a moderately susceptible cultivar varies from < 10% when weather supports severe epidemic to approximately 60% when weather supports mild epidemics. Chemical sprays on a moderately resistant cultivar result in 95% control if weather supports a mild epidemic but only 65% control when weather supports severe epidemic (Shtienberg *et al.*, 2000). A highly resistant cultivar is able to suppress the disease under all weather conditions.

Resistance breeding has been slow due to lack of stable and high level of resistance in cultivated chickpea. Presence of many specialized races of the fungus in the same area is one of the main hindrances in the development of resistant varieties. Erect growth habit, less lateral spread, high hairiness, higher peroxidase activity, higher L-cystine and phenolic contents are attributes of a resistant variety. Different levels of resistance to *A. rabiei* have been identified in some accessions of wild *Cicer* spp. including *Cicer bijugum*, *C. echinospermum*, *C. pinnatifidum* and *C. reticulatum*. Fully fertile hybrids with cultivated chickpea and *C. echinospermum* and *C. reticulatum* are possible.

Stems of resistant plants have more hair than susceptible plants which produce more maleic acid than healthy plants. Kumar *et al.*, (2001) observed that the leaf glandular hair density is more in resistant than in susceptible cultivars. Hair density could create a water-repelling surface thus impeding spore germination. The glandular trichomes secrete a highly acidic fluid which has been shown to promote conidial germination at very low concentrations (0.012–0.06 mg/ml) but inhibit conidial germination at high concentrations of 0.3–1,5 mg/ml (Armstron-Cho and Gossen, 2005). The increased density of glandular trichomes may lead to high concentration of the acid.

In screening work done in India (ICRISAT and many centers for screening trials) sources of resistance have been located in cultivars/germplasm collections F-8, C 325, C 727, I 13, EC 26414, 26435, and 26446. Pal and Singh (1990) had reported Kabuli type lines ILC 3864, 3870, and 4421 resistant to blight. Variety C 215 was at one time considered highly resistant to the disease in Punjab. In general Kabuli type chickpea is more resistant than desi type.

Almost all lines of chickpea which show multiple race resistance are of Kabuli type. This could be because of the fact that the region (Asia Minor) which is supposed to be the original habitat of chickpea almost exclusively grows Kabuli type. In a study with 5 races of the pathogen, it was found that out of 81 chickpea genotypes screened none was resistant to all races. Two genotypes, GG-715 and ICC-76 were resistant to 3 races while three (H-86-8, H-86-100 and HK-86-120) were resistant to 2 races. Resistance is the result of additive and non-additive gene action and is controlled by two dominant complementary genes.

Screening of germ plasm accessions for resistance to 6 races of *A. rabiei* identified in Syria has resulted in identification of many lines possessing resistance to one or more races but not all races. Kabuli type germplasm accessions ILC 202, 3856, and 5928 had resistance to 5 races, ILC 72, 201, 2506, 2956, 3279, and Kabuli type breeding line FLIP 83-48C had resistance to 4 races. The desi type germplasm accession ICC 3996 was resistant to 3 races. Subsequently ILC 3864, 3870 and 4421 were also reported resistant to blight. More recently, the chickpea lines ICC 1467 and 5033 have been reported to be resistant to 10 races, ICC 76 to 7 races and ICC 4370 to 6 races in India. Lines and cultivars resistant to Ascochyta blight: Gaurav, ILC 200, E 110Y, P 1528-1, JM 595, and FLIP 84-92C are listed as resistant in India.

In epiphytotics of chickpea blight, plants are killed within a week and, therefore, only highly resistant varieties could solve the problem. Along with Ascochyta blight, Fusarium wilt is also a major constraint to chickpea production in the Mediterranean region and in India. Losses due to wilt may also be 10–40%. In Kabuli gram, which has more blight resistant genotypes, it is reported that the varieties which are resistant to Ascochyta blight are susceptible to Fusariun wilt and those resistant to wilt are susceptible to blight. Hence, the trend in the resistance breeding programmes is to find sources of resistance to both these diseases. Singh and Reddy (1993) had identified accessions of *Cicer judaicum* and *C. pinnatifolium* as having resistance to *A. rabiei*. These sources of resistance originated from Jordan, Lebanon, Syria and Turkey. Singh *et al.,* (1998) had evaluated accessions of wild *Cicer* species and cultivated chickpea lines for resistance to various biotic and abiotic stresses. *Cicer bijugum* had the maximum number of accessions that showed multiple resistance to biotic stresses (Ascochyta blight and Fusarium wilt). Other wild species showing resistance to at least one stress were *C. pinnatifidum, C. judaicum, C. reticulatum* and *C. chinospermum.*

In absence of cultivars with stable resistance and sincxe the pathogen is seed- and soil borne, only integrated management strategy employing all the methods stated above will keep the disease under check.

ERGOT OF CEREALS

The ergot disease has been recorded on many species of Gramineae which include such grass genera as *Agropyron, Elymus, Agrostis, Poa, Cynodon* and *Paspalum* and the common crops like wheat, rye, oats, rice, pearl millet, sorghum and sugarcane. It is reported that the genus *Claviceps* originated from South America. Evolution of species occurtred with migration of the hosts (grass sub-families) The species involved are *Claviceps purpurea* (cereals), *Claviceps fusiformis* (pearl millet), *Claviceps africana, C. sorghi* and *C. sorghicola* (sorghum) and *Claviceps oryzae-sativae* (rice). The disease is of worldwide occurrence. *C. sorghicola*, reported from Japan in 1999, differs from *C. sorghi* and *C. africana* in colour and texture of sclerotia, colour and morphology of stromata arising from sclerotia, size of ascospores and conidia and also in ergoline alkaloids in sclerotia. Phylogenetically, *C. purpurea* and *C. fusiformis* are in one group while *C. africana, C. sorghi* and *C. sorghicola* are in a different group (Pazoutova and Outova, 2001).

Ergot alkaloids are mycotoxins that affect the nervous and reproductive systems of exposed individuals through interaction with monoamine receptors. They have been studied mainly in the ergot fungi (*Claviceps*) but have been found in some other fungi also (*viz., Aspergillus fumigatus*) where they aid conidiation. The biosynthesis of the toxic ergot alkaloids is genetically controlled in *Claviceps* spp. This determines the level of ergot toxicosis by different species. Although alkaloids are produced by all species, *C. purpurea* seems to produce most toxic alkaloids The presence of such gene(s) in some other fungi on grasses is reported to cause live-stock toxicosis

Symptoms

The cereal ergot (*C. purpurea*) is especially severe on rye where it is grown. Ergot infection is conspicuous from blossoming of the cereals and grasses to maturity of the plants. Infection is first evident in the conidial "honeydew" stage of the pathogen when masses of conidia are exuded in a sugary suspension on the inflorescence. The exudation accumulates in droplets or adheres to the surface of the floral structures. Insects feed on this nectar-like mass and their presence is conspicuous around the infected spikelets. Soon the infected ovaries are transformed into black, horn-like structures, the ergots or sclerotia of the fungus. The sclerotia grow in between the glumes, attain a large size and are quite conspicuous due to their size, shape and color.

Fig. 59. Ergot of cereals.

The sclerotium yields alkaloids such as ergotin. Assimilation of the alkaloid in the blood system of man and cattle causes the disease "ergotism". This disease in animals results in convulsions or nervous breakdown or in gangrene causing shrinkage of tissues of the affected organs, leading to their fall from the body. Due to this effect of the alkaloid pregnant animals grazing on ergoty grasses suffer from abortion. The alkaloid has medicinal value also.

The causal organism

Claviceps purpurea (Fr.) Tul. Taxonomically, *Claviceps* is placed in family Clavicipetaceae, order Hypocreales, subfamily Sordariomycetidae, class Ascomycetes. These fungi display strict organ specificity, attacking exclusively young ovaries of cereals. *C. purpurea* shows three well-marked stages in its life cycle-the sphacelia or honeydew stage, the sclerotium stage and the ascigerous stage. The honeydew stage typically occurs on the host inflorescence. The sclerotial stage develops on the host and carries the fungus to soil. The ascigerous stage occurs outside the host after germination of sclerotia and production of perithecia and ascospores. In temperate climate areas the life cycle of *C. purpurea* and development of the disease on cereal hosts starts from this ascigerous stage.

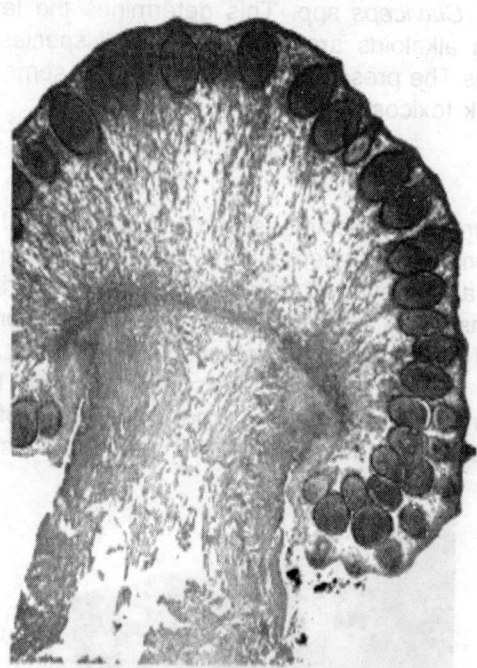

Fig. 60a. Clariceps perithecial head cross section.

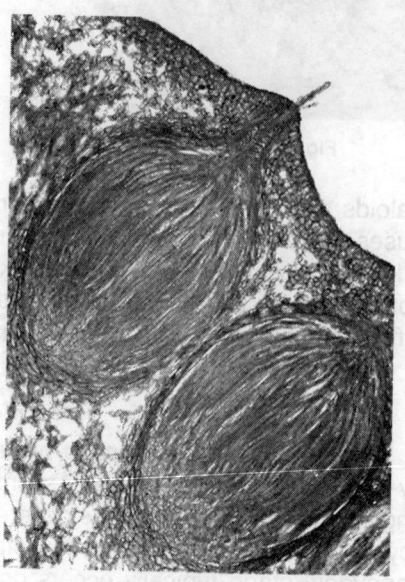

Fig. 60b. Clariceps perithecial

The ascospores are violently ejected out of the perithecia and are carried by wind to healthy flowers at the time of blossoming. On coming to rest on the stigma of the flowers these spores germinate and send out an infection thread which grows down the style and reaches the ovary much in the same fashion as the pollen tubes. The infection is genetically guided toward the base of the ovar. The infection of the ovary gives rise to the honeydew or sphacelia stage. The embryo is attacked and practically destroyed, being replaced by yellowish-white mycelium. The fungus is reported to excrete xylanases in the host tissue as the cell wall degrading enzymes. The secreted extracellular cellulases also interfere with hydrogen peroxide-mediated host defense.

The mycelium consists of septate, branched hyphae which grow rapidly inside the ovary, occasionally creeping out over the surface. Vacuoles in hyphal cells serve as compartments harboring several hydrolytic enzymes. Sclerotial cells have no such ccumulation of hydrolytic enzymes. Sporigenous hyphae arise from the mycelium on the surface of the ovary. These hyphae swell and become conidiophores. Meanwhile, a dense white covering of interwoven hyphae, showing numerous folds and convolutions, develops on the outer surface of the ovary. These folds contain conidiophores bearing small, oval, unicellular conidia produced in succession. The disintegrating tissue of the ovary releases a lot of sweet, viscid fluid and the conidia remain embedded in this fluid. This honeydew attracts insects which act as carriers of conidia causing secondary infection of fresh ovaries. The conidia-infection-conidia cycle goes on for some time during the season.

When the ovary is exhausted or totally replaced by the fungal hyphae, the repeated intertwining of the hyphae produces a compact, hard structure, the sclerotium. At first the lower and then the upper part of the ovary is converted into sclerotium. The sclerotia are soft and light colored in the beginning but soon become hard, black or violet, and slightly curved like a horn. The interior of the sclerotium consists of simple pseudoparenchymatous cells when seen in a cross-section.

The sclerotia (ergots) fall down on the ground or are carried with the seed only to return to the soil at the next sowing time. Thus, the sclerotia serve as the resting structures during the off-season. In the next season when the crop is growing the sclerotia germinate giving rise to the ascigerous stage. A number of swellings develop on the body of the sclerotium. These swellings rupture and the internal loose mass of hyphae from the sclerotium grows out into a vertical column. These are the stromatic stalks or stipes. The fully developed stipe consists of a purplish stalk, 2–3 cm or more in length, and a smooth, later rough, rounded, orange or pink, very small head or the stroma proper. The head bears numerous, spine-like outgrowths representing the neck and ostiole of the perithecia embedded in the stroma. The flask shaped perithecia, as seen in vertical section, are arranged side by side in the peripheral tissue of the stromatic head. Each perithecium contains a number of cylindrical or clavate asci which converge towards the ostiole. The asci are intermingled with numerous paraphyses. Each ascus contains a bundle of 8 ascospores. These spores are slender, filiform and hyaline. The cap of the ascus is burst open and ascospores are ejected out with force above the ground level when they are caught by air currents and disseminated. *C. purpurea* is a low temperature fungus.

Cold (5°–10°C) moist period of about 2 months is found to be the most favourable pre-conditioning treatment to induce sclerotial germination.

The fungus is not highly specialized in its parasitism and spores from a number of infected cereals and grasses can infect each other. Germination of sclerotia is critically influenced by their size, dormancy, humidity and temperature. Alternate freezing and thawing in the presence of moisture accelerates the growth of stipes from sclerotia. The best humidity is near about the saturation point. Prolonged dormancy reduces the viability of sclerotia. Susceptibility of wheat and barley is related to the duration for which the flowers remain open. Since the flowers of male sterile lines of self pollinated cereals remain open for a long time they are more susceptible than others. Sclerotial mass in poorly pollinated male-sterile wheat comprises more than 20% of the threshed grain yield, in contrast to only 0.7% in fertile wheat.

Management of the disease

The possible grass hosts of the parasite should not be allowed to grow near the crop fields. The seed should be clean and no sclerotia should be present as admixture. The seed lots containing sclerotia can be partially cleaned by immersion in 20% solution of common salt. The sclerotia float on the surface and can be manually removed. Rotations avoiding cereals help in reduction of inoculum in the field. Seed treatment with guazatine (Pinoctine 35 LS) or Procloraz + triticonazole (Kinto 80 FS) is reported to reduce germinability of sclerotia to 1–12.8%. Prochloraz, triticonazole, fluquinconazole + prochloraz also suppress sclerotial germination.

Fusarium roseum "Sambucinum" and Fusarium heterosporum are biocontrol agents of cereal ergot. These biocontrol agents parasitize the honeydew and prevent formation of sclerotia.

ERGOT OF PEARL MILLET

The ergot of pear millet was first reported from south India but was not considered a major disease of the crop. The earliest report of its occurrence in epidemic form was in 1956 from the south Satara area of Maharashtra. By 1966, the disease had become a major limitation in the cultivation of improved hybrid pearl millet cultivars. The increased importance of the disease was considered to be the introduction of susceptible hybrids HB-1 and HB-2. Severe epidemics of the disease occurred in Delhi, Uttar Pradesh, Rajasthan, Maharashtra, Karnataka, Tamil Nadu, Andhra Pradesh and Haryana. The average incidence was estimated at about 62% with grain loss of about 58%. In Rajasthan, grain yield loss of 7.64 Q/ha was reported. Hybrid BJ-104 was more severely affected than other varieties. The damage caused by the disease depends on weather at the time of ear formation. Presence of toxic alkaloids in the ergot add to the importance of the disease. The fungus causing the disease has been reported from several African countries.

Symptoms

The disease becomes evident as small droplets of pinkish or light honey-coloured fluid (the honeydew stage) which keep on exuding from the spikelets. Later these droplets become darker, coalesce, and cover larger areas of the cob. In advanced stages, small dark brown sclerotia can be seen projecting from between the glumes. These sclerotia (ergots) contain alkaloids responsible for ergot poisoning in animals.

Fig. 61. Ergot of pearl millet.

The causal organism

The pearl millet ergot fungus was originally described in 1853 as *Kentrosporium microcephalum* Wallr. It was transferred to the genus *Claviceps* by Tulasne who named it *Claviceps microcephala*. Loveless (1967) made detailed study of the fungus from specimen collected in Africa and on the basis of spore size argued that since *Claviceps microcephala* was. considered by Petch (1937) as a synonym of *Claviceps purpurea* and the measurements of pearl millet ergot fungus lie outside the range of *C. purpurea* it should be called *Claviceps fusiformis* instead of *C. microcephala*. The CMI has also listed only *Claviceps fusiformis* Loveless on pearl millet. Confirmation of the identity of the pathogen was further confirmed in 1973 and 1984.

The vegetative characters and honeydew stage of the species are on the same pattern as those of *Claviceps purpurea*. The honeydew produced on the ears is full of conidia. In the initial stages macroconidia are formed and later microconidia are also present. The macroconidia are hyaline, fusiform, and broadly falcate and unicellular. In studies with Indian and African isolates of *C. fusiformis* the average size of macroconidia is given as 16.5 × 3.8 microns. Mostly they are 13–18 × 3–4 microns. Microconidia are globular, hyaline, unicellular and measure 2.4–10.8 × 1.2–4.8 (average 5.9 × 2.5) µm. At 25°C the macroconidia germinate by producing 1–3 polar or lateral germ tubes. These germ tubes produce secondary macro- and microconidia by septation at the tip. Microconidia are in chains. Tertiary and quaternary conidia are formed from secondary conidia. The tertiary and quarternary conidia are infective. Germination of conidia formed on the host is better than those formed in artificial culture.

The honeydew stage is followed by the development of sclerotia which are elongated to round, light to dark brown, hard to brittle in texture, with cavities filled with conidia. The sclerotia measure 3.6–6.1 × 1.3–1.8 µm. Inner pseudoparenchymatous tissue (medulla) of the sclerotium is whitish. Thakur *et al.,* (1984) noted earliest germination of sclerotia about 4 weeks after placement in petri-dishes at 25°C or in pots with soil at 25–30°C. Older sclerotia (12 months old) germinated better than one month old sclerotia. Unlike *C. purpurea*, the pearl millet ergot fungus is warm climate fungus and the sclerotia germinate better when stored in dry conditions and at high temperature (37°C) without chilling.

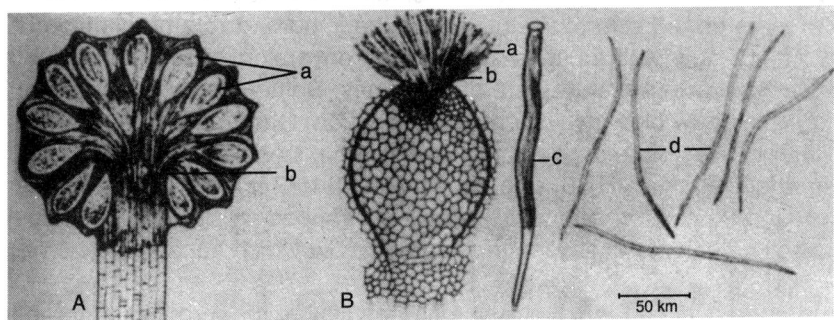

Fig. 62. Ergot of bajra (*C. fusiformis*).
A. Longitudinal section of a stroma showing peripheral layer of perithecia (a) in the stromatic tissues (b). B. A mature perithecium releasing (a) asci and (b) paraphysis. (c) ascus with ascospores, (d) ascospores.

Sclerotia germinate by producing 1–3 or up to 16 fleshy, purplish, 6–26 mm long stipes. These stipes bear perithecial stroma or the capitulum at the tip. The stromata are globular and light to dark brown in colour and show perithecial projections on the surface. Cross section of the capitulum reveals numerous pyriform perithecia embedded in the tissue and arranged in semi-circular manner in fully developed stromata. The protruding neck has an ostiole measuring about 37 (m in diameter and showing presence of periphyses. The asci are numerous, long, cylindrical, slightly tapering towards the base, having short stalk, hyaline, thin-walled and obtuse at the apex where there is a narrow opening. Paraphyses are not seen. Each mature ascus contains 8 ascospores which are long, filiform, hyaline, aseptate and measure 103.2– 176 × 0.4–0.5 μm, average of 100 measurements being 127.7 × 0.5 μm. These uninucleate ascospores germinate to provide primary and secondary conidia. Nucleus from most ascospores migrates to primary conidia without division. But in some ascospores the nucleus divides once so that two primary conidia are formed.

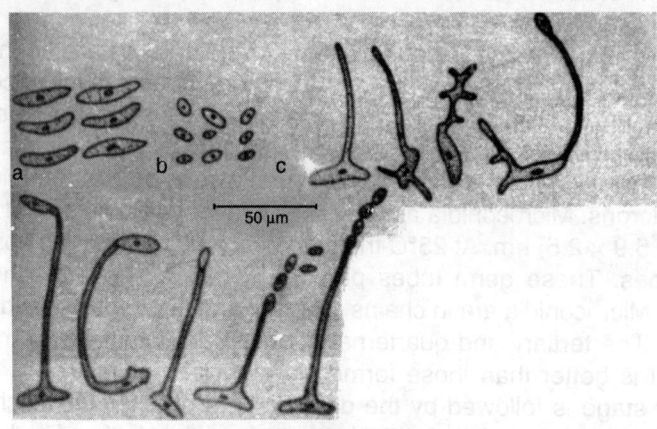

Fig. 63. *Claviceps fusiformis.*
(a) macroconidia, (b) microconidia, (c) germinating macroconidia with 1-3 germ tubes, (d) production of macro- and microconidia on the tips of the germ tubes. (Courtesy: Dr. R.P. Thakur).

Disease cycle and environmental relations

The conidia of C. fusiformis are reported to remain viable for 13 months. These conidia are abundantly present on the sclerotial structures. The important role of this type of inoculum in primary infection is now well established. During conversion of the honeydew stage to the sclerotial stage, sclerotial structures are of two types. Some are aggregations of hyphae and conidia which later show only myceliogenic germination. Others are true sclerotia which show carpogenic germination. Primary infection arises from conidia on sclerotia left on the soil surface. Verma and Pathak (1986) trapped conidia up to 2 meters height in pearl millet fields during August and September. No ascospores were trapped on slides. In their opinion sclerotia reaching the soil at the time of sowing produced conidia which became air-borne and infected the flowers.

Carpogenic germination of true sclerotia results in the production of perithecia and ascospores for primary infection. The sclerotia are easily dispersed as admixture with seed. When introduced into the field with seed they take about 30–45 days after rains to germinate and produce perithecia. The sclerotial germination normally coincides with flowering of the crop. The ascospores are carried by wind currents to fresh flowers. The secondary spread is carried out by conidia in the honeydew mainly by insects, contact and rains. Houseflies are very effective in dissemination of conidia from the honeydew. On the same ear honeydew trickles down to healthy florets.

The infection of florets mainly takes place through the stigma and occasionally by piercing the thin ovary wall before fertilization. It takes about 5–7 days to develop the honeydew stage and, thus, 2–3 conidial generations can be completed within the anthesis period. In a similar ergot disease of sorghum (Claviceps sorghi), secondary conidiation is a factor in epidemiology. After the honeydew dries the hyphae produce conidia that are disseminated by wind. Infection of one ovary affects the grain setting in the adjacent ovaries also, thereby, causing increased reduction in the number of grains. The yield loss is, thus, due to the reduced number and size of grains.

There is definite correlation between the stage of flowering and infection. Rate of infection falls down once fertilization has taken place. Pollination of the inflorescence before or at the time of inoculation has been shown to reduce infection through rapid withering of stigma. There is keen competition between pollen and spores of ergot and smut pathogens to invade the stigmas. When pollen outnumber the spores, disease incidence is minimum. Disease intensity and pollen parent density are inversely proportionate to each other (Gaur et al., 2003) Thus, early increased pollen availability for pearl millet hybrids at the time of flowering by planting early flowering pollen donor lines provide good protection against infection of hybrid varieties that are cross pollinated. In resistant lines protogyny of florets lasts less than 48 hours while in susceptible lines stigmas remain receptive up to 6 days after pollination. The role of longer protogyny and longer stigmas in deciding resistance and susceptibility is affirmed by Thakur et al., (1992). Thus, earlier the withering of stigma through post-pollination stigmatic constriction, the more resistant is the variety. The conidial germination is accelerated by pollen of the main host (pearl millet) as well as of many other allied plant species such as Cenchrus cellaris, Panicum antidotale and Panicum mezianum.

The meteorological factors affecting epidemics of pearl millet ergot disease are high morning relative humidity (85–95%) during flowering and also in the evening (60–90%) as compared to normal evening humidity of 45–50%, cloudy weather (75–100% sky covered with clouds), low sunshine, and daily light showers. The optimum for initiation and spread of the disease as 12 mm mean rainfall, 6 hours per day mean sunshine, 75% mean relative humidity and 20°C mean temperature from protogyny to early flowering stage. Highest ergot severity in susceptible genotypes is found in a flowering season with 12 rainy days and least in a flowering season with no rainy days. Rains washed down the pollen and this increased ergot infection. The disease is less severe in crops planted before July 30 whereas in plantings done on September 15 no grain yield is obtained. In fields where deep ploughing is done after crop harvest disease severity is relatively less. Heavy nitrogen application promotes disease incidence.

The pathogen, C. fusiformis, is reported to occur on a number of other species of Pennisetum such as P. purpureum, P. spicatum, P. alopercuras, P. hohenackeri, P. polystachyon, P. ruppelii, Cenchrus cilaris and Cenchrus setigerus.

Management of the disease

Long crop rotations help in avoiding soil-borne inoculum. The most commonly recommended method of control is use of clean seed. Seeping the seed in 20–32% salt solution floats the sclerotia which can be removed by hand. Sclerotia remain viable for longer time if buried deep in soil. Therefore, repeated deep ploughing, especially during summer, may reduce their viability. Intercropping of pearl millet with mungbean is reported to reduce incidence of the ergot disease. No satisfactory chemical control method has so far been developed. In some studies sprays of ziram, copper oxychloride-zineb and wettable sulphur were recommended. Spray was started just before earhead emergence and repeated 2–3 times at 5–7 days interval. In ergot of sorghum (*Claviceps sorghi*) spray of garlic extract (12% conc.) has provided 90% control of the disease (Singh and Navi, 2000).

In the related species, *C. africana*, in Zimbabwe, a significant reduction in initial disease severity, rate of disease increase, and final disease severity was achieved with one application of Benlate (0.2% a.i.) at heading or stigma exsertion. For the same species, seed treatment with captan and maneb completely inhibited or highly reduced germination of conidia smeared on seed with honey dew. Prom and Isakeit (2003) tried 14 fungicides and have found reduction in spore germination of *C. africana* by all. Pre-inoculation application of propiconazole, tebuconazole, triadimefon, myclobutanil or azoxystrobin (25 μg/L) on full bloom panicles, markedly suppressed ergot severity in a male sterile line of sorghum. Treatment of seed and sphacelia-sclerotia admixture with captan at 94 g a.i./100 kg seed reduces conidial germination and secondary conidiation of *C. Africana* (Fredericksen and Odvody, 2003). In Australia, Ryley *et al.,* (2003) emphasized timing and method of application for effective chemical control of *C. africana*. They found triadimenol spray at 0.125 kg a.i/ha most effective. A combination of activated resistance by acibenzolar-S-methy (0.05 kg a.i./ha) and mancozeb (1.5 kg a.i./ha) was recommended in case the fungus develops resistance to triazoles. The triazoles have no systemic activity on sorghum panicles and repeated applications are required.

The best method of managing the ergot disease in pearl millet is the use of resistant varieties. Several ergot, smut and downy mildew resistant lines with high yield potential were developed at ICRISAT and have been used as source of resistance in breeding programme.

Growth of a *Fusarium* sp. on pearl millet ergot honeydew and immature sclerotia was seen in the wet Tarai area of north India. Tripathi *et al.,* (1981) reported *F. sambucinum* and *Dactylium fusarioides* as mycoparasites of sclerotia. Application of the mycoparasite *F. semitectum* var. *majus* at various stages of ergot development results in 83–93% ovary colonization, 14–52% reduced sclerotia formation, 24–43% sclerotia colonization and 46–48% sclerotia disintegration. In a study of *C. africana* on sorghum. *F. heterosporum* was reported associated with *Claviceps paspali* on *Paspalum dilatatum* (Kodo millet).

Pseudomonas aeruginosa, *Burkholdera cepacia*, isolates of *Penicillium citrinum* and two commercial products of *Trichoderma* spp. are reported to completely inhibit macroconidial germination *in vitro*. However, the bacterial antagonists fail to inhibit infection of sorghum *in vivo* from sorghum ergot in glasshouse tests. All fungal isolates including *Epicoccum nigrum* reduce the percentage of infected spikelets per panicle and in some cases completely suppress ergot formation.

FALSE SMUT OF RICE

False smut occurs in almost all the rice growing areas of the world including India, China, Japan, Southeast Asian countries, North and South America, Myanmar, Sri Lanka,

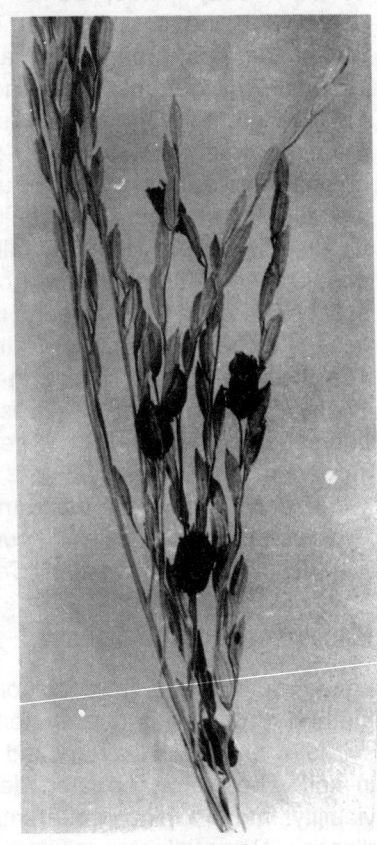

Fig. 64. False smut of rice.

Fiji, and Africa. Epiphytotics of the disease have been reported from India and Philippines. In India, the losses due to false smut have varied between 7% and 75%. Occasionally the disease occurs in destructive form in rice fields of Punjab, Haryana, U.P., and Bihar. Up to 44% loss of grain yield in cultivar Ratna, 17% in IR-8 and 0.6% in cultivar Prasad is reported. In Punjab losses have varied from 1.5 to 16.8% in different varieties. The loss in yield is not only due to conversion of individual grains into smut balls but also due to increased sterility of neighboring florets. There is significant reduction in filled grains and spike weight. When the percentage of smut balls in the ears is 2, 4.5 and 9.6 the percentage of chaffiness is 4.4, 12.1 and 24.2, respectively. Hegde and Anahosur (2000a) noticed significant reduction in seed germination and seedling vigor. Reduction in root length was more pronounced than reduction in shoot length. There was increased chaffiness (40.9%) and decreased 1000 grain weight (32.4%) and panicle weight.

Symptoms

The effects of the pathogen on the host are visible only after flowering when the parasite grows in the ovary of individual kernels and transforms them into large, velvety, green balls (pseudomorphs) which are sometimes more than twice the diameter of normal grains. At first the balls are small and remain confined between the glumes. They gradually enlarge and become 1 cm or more in diameter enclosing the floral parts. Young spore ball is flattened, smooth, light yellow in color and is covered by a membrane. Later, the membrane bursts. The color changes to orange, yellowish green, green, olive green, and finally to greenish black. Inside the ball, the color is orange yellow on the periphery and nearly white in the center. Although an affected glume is encircled by the fungus, in mild infections it remains unaffected and is found closely sticking to the center of the ball which bursts out above and literally from between them.

Usually the disease is sporadic in a field and also on an individual ear only a few grains are affected. However, in severe attack many such balls may aggregate together. Florets beside the balls may remain sterile giving chaffiness to the panicle.

The causal organism

False smut of rice is caused by the fungus *Claviceps oryzae-sativae* Hashioka. Its conidial stage is *Ustilaginoidea virens* (Cooke) Takahashi which occurs on the rice plants. The fungus was first described under the name *Ustilago virens* by Cooke in 1875. Later, in Japan, it was variously described as *Tilletia oryzae* and *Sphacelotheca virens*. Brefeld erected the genus *Ustilaginoidea* and named the rice false smut fungus as *Ustilaginoidea oryzae*. He considered *U. oryzae* and *U. setariae* (from *Setaria*) closely related to *Claviceps* except for the difference in the conidial stage which in this fungus is dry instead of honeydew. Takahashi combined the scientific names and described the fungus as *Ustilaginoidea virens*. The perfect stage with perithecia and asci was described by Sakurai in 1934 as *Claviceps virens* (Cooke) Sakurai. Hashioka further confirmed the observations of Sakurai and renamed the fungus as *Claviceps oryzae-sativae*.

The fructifications replacing the grains represent the conidial, pseudosclerotial and sclerotial stages of the pathogen. They are at first yellow to orange, later turning olive green to black. When young they are fleshy inside becoming hard after sometime. They consist of a central hard mycelial tissue composed of thin, hyaline, septate hyphae, 2–3 μm in width. The central hard core is surrounded by 3 sporiferous layers. The innermost layer is pale yellow, the middle layer orange yellow, and the outermost layer olivaceous greenish black. These layers represent 3 stages of maturity of spores. The spores (conidia) are borne laterally on minute sterigmata on hyaline, septate hyphae. They are spherical, echinulate, olivaceous, and measure 4–6 × 3–5 μm. The younger spores are smaller, smooth, and lighter in color.

Conidia germinate in water by a short germ tube which bears 1–3 secondary conidia after 12–24 hours. Secondary conidia are usually subglobose, often globose to oblong, hyaline, granulate, and 4–8 × 2–5 μm in size. They germinate abundantly in nutrient solutions and bear one to several amerospores which are similar to secondary conidia. Late in the season, the conidial masses in the smut ball harden and become pseudosclerotia.

One to several, mostly two, true sclerotia are formed in each pseudomorph (green ball). At first they are buried inside the conidial mass but after dispersal of conidia they become exposed. These sclerotia are hard, variously shaped, and 5–13 × 2–5 mm in size. On the outside, the sclerotia look black but inside they are white and pseudoparenchymatous. Although the fungus belongs to the genus, *Claviceps* (ergots) it does not contain ergot alkaloids in its sclerotia. Other alkaloids are present. Presence of atropine type alkaloids in the sclerotia is reported. Based on morphological and cultural characters of isolates from different places, Verma and Singh (1988) have categorized the fungus into four groups. A variant (white false smut) producing white pseudomorphs (stroma) has been reported in several countries.

When embedded in moist sand at 24–30°C for 4–5 weeks the true sclerotia germinate to produce stipes which bear perithecial stroma. Singh and Dube (1984) studied the process of germination of sclerotia under controlled conditions. After 27 days of incubation in moist sand at 25°C the sclerotia, that had been earlier exposed to ultra-violet light after soaking in water, produced whitish-yellow fluffy mycelium which changed to yellowish orange. Six to 8 stipes, 9–37 mm long, developed after 10–15 days of mycelial development. These terminated into a cap or stroma (perithecial head) within 16–20 days. The stroma enlarged and became globose, verrucose, greenish yellow to olive and measured 1.0–3.5 mm in diameter. Out of 6–8 stipes only 3–5 produced mature stroma containing asci and ascospores.

The cross section of the mature stroma shows abundant perithecia embedded in the tissue with partly protruding ostioles. Perithecial wall is pseudoparenchymatous. The perithecia are ovate to pyriform, 150.5–430.0 × 86.0 –193. μm. The asci are hyaline, elongated cylindrical, 69–228 × 2–4.7 μm, and contain 8 ascospores. The tip of the ascus is hemispherical with a prominent cap and narrow pore. The upper portion of the ascus has a thickened wall and the lower portion is tapering. Ascospores are hyaline, 1-celled, filiform, and measure 24.33– 86.53 × 0.66–1.33 μm. They germinate in water by one or two germ tubes which bear secondary conidia. The conidia measure 0.88–3.5 μm in diameter. The fungus can be easily isolated and cultured on artificial media by plating conidia or true sclerotia.

Disease cycle and environmental relations

Conidia remain viable for 3 months in a refrigerator and for 2 months at room temperature and under field conditions. In Karnataka, Hegde and Anahosur (2000 b) also observed survival of conidia in paddy straw at 25°–44°C for 4 months. In refrigerator they could survive for 7 months. Earlier reports from Japan had suggested that the conidia could survive for 8 months. Pseudosclerotia, which are dried and hardened conidial masses, remain viable for 9 months at 10°C, for 6 months at room temperature, and for 4 months under field conditions. True sclerotia remain viable for 11 months at room temperature and under field conditions. Thus, true sclerotia are the major source of primary inoculum. In India, these sclerotia were detected in almost all the collections from Kumaon and Garhawal hills of Uttarakhand and occasionally from the plains. Conidia play an important role in the secondary infection. In nature, infection occurs through ascospores produced by overwintered true sclerotia and by conidia. Successful infection can be obtained by inoculating the spores into the leaf sheath cavity just before emergence of the ears. Infection also occurs in seedlings when germ tubes from conidia penetrate the cuticle of coleoptile and penetrate intercellularly causing systemic infection.

Hyphal growth and conidial germination can occur at any temperature from 12°–34°C but the optimum for conidial germination is 28°C. The optimum for *in vitro* growth has been variously reported as 23°, 25°, or 26°C. The conidia do not germinate at a relative humidity less than 92%. Optimum is 98%. The meteorological factors, especially at the time of flowering of the crop, and also the host nutrition have definite influence on the occurrence of false smut, the incidence being favoured by high dosages of fertilizers, especially at the time of flowering. There is direct correlation between the amount of nitrogen applied and disease severity. A lower maximum (31°C) and minimum (23°C) temperature and high relative humidity (96% or above) coupled with less sunshine or cloudy days during the flowering period favor disease development. Since the yield of rice is almost identical when 100, 150 or 200 kg N/ha is applied and since the disease is favored by high nitrogen doses, a minimal dose of 100 kg N/ha should be recommended in the disease prone areas. The possibility of false smut infection helping the plant overcome effects of the rice tungro virus is reported.

A similar fungus is described on maize in the USA, *U. virens* has been recorded on *Oryza officinale*, *Chionachne koenigii*, and on the common weed *Digitaria marginata*.

Management of the disease

Use of sclerotia-free seed and crop rotation are important precautionary measures. At the time of harvesting the diseased plants should be removed first so that sclerotia do not fall on the ground. Some workers have tried seed treatment. Dipping the seed in 1:1000 mercuric chloride solution for 30 min preceded by a soak in plain water for 6 hours and followed by thorough rinsing in water had been recommended as a control measure in the Philippines.

Attempts have been made to control the disease by fungicide sprays. *In vitro* studies have shown that ascospores and conidia are very susceptible to copper and mercurial fungicides while relatively tolerant to organic sulphur compounds. In 1985, copper fungicides (Bordeaux mixture and Blitox) were recommended for spray, thrice at 10-days interval starting when the crop is 60–65 days old. The treatment gave about 90% reduction in disease incidence. Spray of copper oxychloride (0.25%) or propiconazole (0.1%) at 50% panicle emergence stage has been found most effective against the disease. In Japan, application of simeconazole (1.5% granules) to the field with standing crop 3 weeks before heading has given the most effective control of false smut in rice ears.

Majority of cultivated rice varieties have high degree of resistance to the disease. Generally, the susceptible varieties belong to the late maturing group.

APPLE SCAB

Scab is one of the most serious diseases of apple throughout the world, especially in areas with cool and humid weather in spring. The disease was first reported from Sweden as early as 1819. It is now known to occur in different parts of Europe, in North and South America, South Africa, Australia and New Zealand, apart from India and Pakistan. In the Indian subcontinent apple scab was first reported in the Kashmir Valley and in Lahore (now in Pakistan) in 1935. It has been destructive in the major apple growing states of Kashmir and Himachal Pradesh since 1973 when its first epidemic was recorded. The disease remained confined to the Kashmir Valley till 1977. In late 1977 it was seen in a number of apple orchards in Shimla and Kulu areas of Himachal Pradesh. During the next 5 years (1978–1983) the scab affected apple area in Himachal Pradesh increased from 150 to 40,000 hectares. In Kashmir also over 60% of the apple orchard area had become scab affected by 1985. In addition to the Kashmir Valley and Himachal Pradesh, where its destructiveness is now well known, the disease is also reported from some orchards in the Kumaon Hills (Uttarakhand), Sikkim and Arunachal Pradesh in the north-east India, and also in Conoor and the Kodaikanal hills of Tamilnadu in South India.

Losses due to scab result from (i) premature defoliation of the tree, (ii) weakening of the tree, (iii) quantitative loss in fruit yield, (iv) low market price of affected fruits due to malformation and scab spots, and (v) increased cost of cultivation and disposal of fruits. In India, the first epidemic that occurred in 1973 in the Kashmir Valley ruined the apple crop worth Rs. 5.4 million in a single season. In the Shimla hills, each year, 16–34 % fruits are rendered worthless due to scab. In Himachal Pradesh, the loss amounted to Rs. 15 million in 1983.

The apple scab pathogen mainly attacks the genus *Malus*. Most of the cultivated varieties of apple in India including the Delicious group are susceptible. Some varieties such as Cox's Orange Pippin and Jonathan, which are rated as highly susceptible to scab in many European countries, are either free from infection or show only mild infection in India. In addition, the pathogen is reported on *Cotoneaster bacillaris* and *C. aitchinsonii,* both in Rosaceae. Other hosts are crab apple (*Malus buccata*), *Viburnum* (Caprifoliaceae), loquat (*Eriobotrya japonica,* Rosaceae), firethorn (*Pyracantha* spp., Rosaceae), hawthorn (*Crataegus* spp. Rosaceae), and mountain ash (*Sorbus* sp. Rosaceae).

A similar scab of pear caused by *Venturia pirina* Aderhold is also reported from Kashmir. It is an economically important disease of pear throughout the world. Conditions for infection and development of this disease are similar to those for apple scab.

Symptoms

Typical symptoms of scab appear on leaves and fruits and, in a severe attack, on 1–3 years old shoots. Petioles, pedicels, and sepals are also attacked. However, the symptoms on leaves and fruits are very distinct. The first symptoms are seen in spring on young leaves and flower buds. Light brown or olive-green, irregular spots develop on the lower surface of leaves as this surface is exposed to the outside atmosphere at this stage of tree growth. The spots may not be very conspicuous due to hairy underside of the leaf. Later, more pronounced spots are seen on the upper surface of leaves. The spots are darker with velvety, grayish dark surface and are more circular in outline. Soon these lesions become even more circular and metallic black in color. The tissue surrounding these spots is often thickened and sometimes bulged upward. Such spots persist on the leaf which develops resistance to fresh infections with advancing age. There is premature fall of leaves and floral buds. The leaves of susceptible

Fig. 65. Scab spots on apple leaves. Different levels of severity of lesions.
(Courtsy; Dr. G.K. Gupta)

Fig 66. Scab lesions on apple fruit.
(Courtesly; Dr. G.K. Gupta)

apple varieties produce dark colored spots with good sporulation of the fungus while on resistant varieties the spots are of a lighter colour with poor sporulation.

The scab spots on fruits are usually well defined in shape and appearance against the shining and coloured background of the fruit skin. As the fruits mature in the orchard they gain resistance to infection. The size and shape of fruit lesions varies with host variety and stage of development of the fruit at the time of infection. The spots are initially dull in appearance, brownish black and become almost black with passage of time. Early scab infection results in splitting of the skin of the fruit in the area occupied by the spot. Later, this forms corky layers with deep cracks. When infection occurs early near the stem end or on calyx the fruits are often deformed. Late infection of fully grown fruits on the tree does not cause cracking of the skin and formation of corky layer. These spots are usually smaller. McIntosh is more susceptible than Jonathan and Golden Delicious. Scab may also develop on apparently clean fruits during storage. The storage scab results from late season, pre-harvest incipient infections. Apparently healthy Golden Delicious fruits from scab infected trees developed 5–8 lesions after 90 days of storage at 4°C. During storage of scab bearing fruits the scab area increases, new lesions develop, and the entire fruit may rot. In a study in 1997, the storage rot was found to develop after 45 days in a Royal Delicious showed less scab (3–5 lesions). Scab lesions on apple fruits are known to provide entry sites for many fruit rotting pathogens in transit and storage. Also, the storage of fruits results in diminished fruit quality due to shrinking of the fruit. Shrinking during storage is more severe in scabbed fruits.

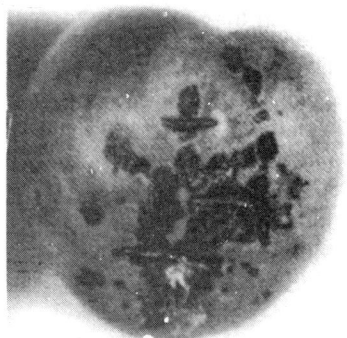

Fig. 67. Scabbed apple fruit showing cracks in skin.
(Courtesy; Dr. G.K. Gupta)

In shoot infection of scab reported from many temperate climate countries as well as from India, the lesions on 1–3 years old twigs are small, raised and cinnamon brown in color. The bark is peeled away at the point of emergence of the lesions and looks silvery gray in appearance. These lesions contain conidia during the active season.

Fig. 68. Twig infection of apple scab.
(Courtesy; Dr. G.K. Gupta)

The causal organism

Apple scab is caused by the fungus *Venturia inaequalis* (Cke.) Wint. (Ascomycota-Ascomycetes-Dothideomycetidae-Pleosporales-Venturiacea). The conidial stage is *Spilocaea pomi* Fr. which was earlier described in India as *Fusicladium dendriticum* (Wallr.) Fckl. A related species, *Venturia asparata* Samuels and Sivanesan, with a *Fusicladium* conidial stage has been reported in New Zealand occurring side by side with *V. inaequalis* on dead fallen leaves of apple. The perfect stage (*Venturia*) is saprophytic and grows on dead leaves while the conidial stage (*Spilocaea*) is parasitic on living host tissues.

In *V. inaequalis* the septate mycelium is at first light in colour but later turns gray in culture and brownish in the host tissues. The cells of the hyphae are all uninucleate. In young leaves the mycelium develops as branched, radiating, parallel strands. In older tissues and in the fruits these strands are compact and thicker. In the living tissues the mycelium is generally located between the epidermal cells and the cuticle. With age it becomes compact and several layers in thickness. In dead leaves, the saprophytic mycelium grows throughout the mesophyll in which it forms a network of hyphae composed of dark brown, irregularly thick-walled cells which ultimately give rise to ascostromata.

The brown, continuous or rarely septate, conidiophores arise from the hyphal strands or from the more compact stroma. This stroma is 10 or more cells in thickness at the center of the scab spot but at the margins it is only one cell thick. The length of conidiophores varies within the same lesion. The continuous conidiophore or each cell of the septate conidiophore are uninucleate. Each conidiophore successively cuts off a number of uninucleate conidia at the tip, the conidiophore elongating after each conidium is produced. The conidia are generally 1-celled but often they become 2-celled through septation. They are ovate to lanceolate, with a truncate base and somewhat pointed apex, and smoky brown in colour at maturity. They measure 12–22 × 6–9 μm on the host and 13–36 × 6–12 μm in culture.

Variability in colony characters and conidial size in culture is also reported. On the host, size of conidia is influenced by cultivar, temperature and relative humidity or rainfall. Size of conidia produced in April is more as compared to those produced later in the growing season. Length of conidia is inversely correlated with temperature and positively correlated with rainfall. Conidial pustules on the fruits show the dark olive conidia surrounded by a fringe of silvery white, torn cuticle.

Late in the season in the dead tissues of fallen leaves, the thick-walled hyphae in the interior begin to form ascostromata (initials of ascocarps or perithecioid pseudothecia). A small coil of hyphae consisting of uninucleate cells initiates the process. As this develops, a coil of multinucleate cells representing the ascogonium is differentiated inside the young stroma and the trichogyne pushes through and protrudes from the stromatal wall. In the meantime an antheridium is formed from a hypha. The fungus is heterothallic and requires two different mating lines for further development of a mature ascocarp. In monoascosporic or monoconidial cultures the above developments will take place. i.e., ascogonium with trichogyne and antheridia will be seen but the antheridium will not fertilize the ascogonium of the same thallus. No further development is possible unless the antheridium is from a hypha of an opposite mating strain. When two mating strains are present contact is soon established between the antheridium and the trichogyne. Through a pore dissolved in the trichogyne wall at the point of contact the antheridial nuclei pass into the ascogonium and pair with the ascogonial nuclei. The nuclear pairs pass into the ascogenous hyphae which now develop from the base of the ascocarp. Asci develop from croziers. In the meantime the stroma continues to grow and forms the pseudothecium (perithecium).

The pseudothecia in dead leaves mostly mature in early spring. Mature ascocarps are spherical, dark brown to black, and possess a short beak (neck) and distinct ostiole around which single-celled dark setae are present. These fruit bodies measure 90–150 μm in diameter. The perithecial wall is composed of brownish cells, 3–5 layers thick. The number of asci in each perithecium may be as high as 242 but the usual number varies between 50 and 100. The ascus is slightly spatulate or saccate in shape and measures $55–75 \times 6–12$ μm. It is thin-walled with a short stalk. The young ascus contains a single, large nucleus formed by the fusion of the two nuclei in the pair. Three successive nuclear divisions in the ascus result in the formation of 8 nuclei around which the 8 ascospores per ascus are formed.

The ascospores are oval to boat-shaped, hyaline, yellowish or olivaceous in color and 2-celled. The upper cell is shorter and somewhat wider than the lower cell. The unequal size of the 2 cells gives the species its name, *inaequalis*. These spores measure $11–15 \times 5–7$ μm (average 13×5 μm). At first the young ascospores are spherical, uninucleate and unicellular. A nuclear division followed by a cell division occurs and each spore is then composed of two equal, uninucleate cells. The spore then elongates rapidly, the two cells growing unequally. On the host the conidia germinate form appressoria. Localized melanization of appressoria is essential for pathogenicity of the fungus.

Variability in monoconidial cultures of *V. inaequalis* had been reported in 1934. At least 7 strains of the fungus are known. These strains differ in their morphologic, physiologic, cultural, and pathogenic reactions. Two strains occur in the Kashmir Valley. The earlier known race which was first detected on the variety Ambri is a mild strain while the strain occurring now on Red Delicious is more virulent with heavy sporulation and resembles Race 1 of the pathogen. It is of worldwide occurrence. A host specific strain occurs on crab apple (*Malus buccata*). The strains occurring on *Cotoneaster* spp. are also other host specific strains. *Spiloceae pyracanthae* that causes scab of *Pyracantha* spp. is also a distinct strain of *Venturia inaequalis*.

Xu *et al.*, (2008) have studied isolates of *V. inaequalis* collected from different cultivars in China, India and U.K. to establish differences in pathogen populations. Within China there was not much diversity among isolates from different regions but the isolates from U.K. and India showed variability in cultivar preference and pathogenicity.

Disease cycle and environmental relations

The apple scab fungus has two distinct stages in its life cycle: the saprophytic pseudothecial or over-wintering stage in dead fallen leaves, and the conidial, summer stage which is parasitic on leaves, flower buds, fruits and shoots on the tree during spring and summer. The saprophytic stage is considered the major source of survival of the pathogen during dormancy of the host in winter and provides the inoculum in the form of ascospores for primary infection of new leaves and flower buds in spring. This initiates the active parasitic stage by formation of lesions on leaves in which millions of conidia are produced. These conidia are responsible for repeated summer cycles of the life history of the pathogen and spread of the disease during the entire summer. Although, susceptibility declines with increasing age of leaves, the conidia can infect the under surface of leaves of any age, ranging from newly emerged leaves to ones that are up to 4 months old (Li and Xu, 2002). On young leaves, the mycelia grow at a faster rate and lead to visible symptoms sooner than on older leaves. Mycelia on very old leaves grow so slowly that they do not lead to visual symptoms at the time of leaf fall, even though some of these symptomless colonies produce conidia which add to the inoculum that finally results in more primary inoculum production in the next spring. Due to the short generation time of 8–10 days under ideal conditions of humidity and temperature several cycles (conidia-infection-conidia) occur within 3–4 months and epidemics develop. With leaf fall in early winter, the pathogen enters its saprophytic stage on dead leaves where ascocarps develop and mature towards the end of the winter just before activity of the tree starts in spring. Good snow coverage and alternate periods of wetness and dryness in winter and early spring favor development of pseudothecia in the leaf tissue. The pseudothecia are formed within 30 days of leaf fall occurring from September to November.

The presence of scab lesions on shoots of standing trees in the orchard was considered another source of survival of the pathogen during winter The role of conidia in perennation and primary infection has been emphasized. Viable conidia were detected in dormant apple buds and on early developing apple tissues. The numbers varied from zero to 142 viable conidia from inner tissues of flower buds. When these conidia were inoculated on apple tissues at early stage of bud growth there was an increase in sepal and fruit infection. The conidia were not detected on surface of buds or lesions on infected shoots. In Netherlands, Holb et al., (2004) found 65% infected shoots collected before bud burst which superficially carried dark mycelium and conidia. There were 581–1033 conidia per 1 cm shoot length. However, viability of these conidia was less than 1.5%. In bud tissues the number of conidia was more than 3000 per 100 buds and conidial viability ranged from 0.7 to 1.9% (for conidia from outer tissues) and 3.7 to 10.5% (for conidia from inner tissues). The shoot lesions as a source of perennation and primary inoculum may be of importance in areas where fallen leaves decompose rapidly and early, eliminating the pseudothecial inoculum. In India, no experimental evidence is available to consider shoot lesions as a source of primary inoculum. Conidia have a short life of about 14 days. It can be presumed that by the time the shoot lesions become active in spring to produce conidia there is already enough primary inoculum in the form of ascospores. Thus, these lesions may be playing only an additive role. The scab bearing shoots are important for long distance dispersal of the pathogen from country to country and, in India, from state to state, when infected material is used for vegetative propagation.

The maturation of ascospores, release of mature ascospores, infection of the host parts, and repetition of conidial cycles are highly influenced by weather conditions such as autumn and spring rains, temperature, and leaf wetness periods. Heavy snow and rains in late winter favour the development of ascocarps and the discharge of ascospores. Ascospore maturity occurs at a wide range of temperatures from 4°C to 24°C with optimum at 16°–20°C while temperatures above 24°C retard ascospore development. Optimum temperature for early growth of the ascocarp is 13°C. Dew and rains wet the leaves on the ground and cause swelling of pseudothecia and favour ejection of ascospores. Dew is not as effective as rains. Intermittent rains at short intervals or even heavy dew influencing the period of wetting and drying of leaves increases the rate of maturity and discharge of ascospores provided the temperature remains below 24°C. Continuous sunshine is lethal for ascospores. In a single scab infected over-wintered leaf up to 2 million ascospores are formed. Each pseudothecium starts producing 1475 ascospores/ml which rises to 69500/ml at petal fall stage of the tree.

Discharge and dissemination of ascospores by *V. inaequalis* during dew is very low or absent. The level of ascospores is 13–20% of season's total spore numbers (Stensvand *et al.,* 1998). Similarly, ascospore release during darkness constitutes a small proportion of total available supply of primary inoculum (Gadoury *et al.,* 1998). At temperatures below 6°C, the discharge of ascospores is very low and increases with rise in temperature. Ascospore release increases with increasing light intensity and with duration of rain. The combined effect of the two in the orchard is a delay in reaching the peak spore release at several hours after sunrise. The infection period, thus, also start after sunrise. The conidia survive radiation doses equivalent to 12 hours in full sunlight. The spores are not effective unless following conditions are met: (i) fair weather days, (ii) cool nights with abundant dew formation, (iii) significant release and dispersal of air-borne spores and (iv) poor drying conditions or additional hours of leaf wetness due to fog or rains. Similar observations were reported by Rossi *et al.,* (2001, 2003). Ascospore release into the atmosphere is influenced by temperature and rains are the most important event helping ascospore discharge and their becoming air-borne in the orchard. Rain event is a period with measurable rainfall (> 0.2 mm/h) lasting for one to several uninterrupted hours or interrupted by dry spell of not more than two hours. From a single source of inoculum, aerial ascospore concentration decreases with increasing distance but could spread up to 18–33 meters depending on cultivar susceptibility in the orchard (Holb *et al.,* 2004). This is not of much epidemiological significance in orchards where leaf litter bearing pseudothecia is randomly distributed on the floor.

The discharge of ascospores starts at a time when the buds on the tree are about to burst open in spring. The total period of ascospore discharge from fallen leaves varies from 4 to 10 weeks depending on weather and then the perithecia are exhausted. It may continue till May, 2 to 3 weeks after petal fall (Agarwala, 1985) depending on the prevalence of cool and mild rainy weather. Mature ascospores are available in the overwintered leaves as early as silver tip stage of apple in March-April with peak ascospore maturity and production occurring at petal fall stage in April-May. The average ascospore emission period is 74 days. On an average nine apple scab infection periods occur in the spring and early summer in Himachal Pradesh (India).

While the ascospore discharge is taking place the lesions on shoots also produce conidia adding to the inoculum load. In cloudy, damp, and windy weather ascospores may be disseminated up to a distance of several kilometers from the source of production. Leaves that fall in September mature pseudothecia at a much faster rate and shall discharge ascospores a month earlier than those falling in November and December. Thus, pseudothecia formed in leaves falling early are not as important as those in leaves falling late for source of primary inoculum.

During formation of appressoria produced from conidia and ascospors, a dark brown ring structure is seenat the base of the appressorium. This melanized appressorial ring structure is attacjed to the leaf surface like a sealing ring and forms the fungus-plant interface. It is believed to be required for pathogen penetration of the cuticle (Steiner and Oerke, 2007).

The efficiency of ascospores, landing on leaves, to form lesions is reported to be 5–14%. The efficiency on cluster leaves of apple flower buds is 6–16%, 3–9%, and 0.4–0.6% at tight cluster, first pink, and full pink to bloom, respectively. Thus, susceptibility of leaves and fruits to scab generally decreases with the increasing age of the tissue. Ascospores begin to germinate on a wet leaf surface in 2–4 hours and penetrate the cuticle by means of penetration peg to form hyphal pads (stroma) between the cuticle and the epidermal wall. Radiating hyphae arise from these pads. Germination of ascospores and subsequent infection of young leaves and flower buds are optimum at 10°–22°C.

At and above 28°C these processes are inhibited. Germination of ascospores and conidia on the host surface depends on the period for which free water or 90% or more relative humidity is present. This determines the period for which the leaf is susceptible on a given day. The leaf wetness period necessary for infection varies with temperature but minimum is 9 hours. It is longer at low temperatures and shorter at relatively higher temperatures. In pear scab (*V. pirina*) minimum wetness period for foliar infection varies from 10 hours at 23.9°C to 25 hours at 7.2°C. It is similar for apple scab infection requirements. These wetness durations determine the infection periods. Sunshine and rise in temperature dries the leaf surface preventing germination of spores and infection. Wetness for a longer period is necessary for the germination of spores at temperatures above or below the optimum. Although, in 1944, Mills had indicated that one third less wetting period is needed for infection with conidia than with ascospores, there are reports that about the same wetting period is required for infection by both types of spores. At 6,11,16 and 22°C disease incidence (scab lesions per plant) increased with increasing dose up to 8100 conidia per ml. The disease was less at 20°C than at other temperatures and incidence increased with increasing duration of leaf wetness.

The primary infection results in the development of visible lesions within 7–8 days at 19°C, within 17 days at 9°C, and in 21 days at 5°C. The vegetative growth of the pathogen in the host can occur at a temperature range of 0°–30°C with the optimum at 19°–20°C. At higher temperatures in this range there is more mycelial development and less sporulation. Usually, more abundant conidial production occurs at 8°–16°C than at 20°C or above. However, some strains of the fungus have the optimum at 16°C and others at 20°C.

Thus, under the ideal temperature and humidity conditions a fresh crop of conidia is produced in 7–8 days. A build up of conidial inoculum for secondary spread is favored by the presence of susceptible varieties, low temperatures, and number of wet days. In the late summer months when temperatures may rise to 30°C or more the exposed conidia are killed. The conidia produced by primary infection are washed down by rains and carried to other green leaves and fruits to start secondary infection. Conidia also germinate best at 10°–20°C, the optimum being 20°C. The conidial germ tubes produce disc-like appressoria from which infection pegs arise. On susceptible leaves several appressoria, and a large number of infection pegs, are formed by a single germ tube.

The spread of the disease continues throughout the summer and early autumn. During spring the primary infection by ascospores and secondary spread by conidia are overlapping. Fertilizer application increases the susceptibility of leaves to the scab fungus. Nitrogen contributes the maximum and potassium the least. However, combination of nitrogen and potash reduces disease incidence. In pear scab (*V. pirina*) susceptibility is enhanced by excess nitrogen which increases total N and Mn and decreases calcium content of the leaf. Some studies in Germany have indicated that manuring and pruning do not significantly affect incidence of apple scab.

Forecasting: To prevent epidemics of apple scab by timely and effective fungicidal sprays forecasting of the disease appearance is considered essential. Control measures are often ineffective if spraying is started after primary infection has taken place on a large scale and primary lesions with conidia have developed. The precise mode of survival on easily located sources (fallen leaves in the orchard) for prediction of primary inoculum and relationship between weather and infection have provided the basis for developing forecasting systems for specific areas. Forewarning of the incidence of apple scab can be done on the basis of an approximate quantity of primary inoculum (ascospores) present at the time when host leaves are susceptible and the frequency of infection periods in the season.

The quantity of primary inoculum and the likely time when ascospore showers will start can be roughly estimated by a periodical examination of the leaf material collected from apple orchard floor. The examination of this material in the laboratory can reveal the stage of ascospore maturity and approximate quantity of ascospores likely to be discharged and at what time. Such examinations are particularly important in early spring when the host parts are just coming out to receive the inoculum. The most critical period, as regards the host, is the time the buds start swelling until about 2–3 weeks after petal fall. A second, though short, critical period occurs in autumn when cool moist weather prevails permitting severe late infections of fruits and leaves. For better precision in making the predictions such additional information as the number of effective ascospores (spores that actually land on and infect a given area of leaf surface) and the relationship between morphological maturity of ascospores and physiological maturity of asci are also required. Lesion forming efficiency of ascospores varies with source of production, host variety and environmental conditions.

Infection period is most important in the prediction of initiation of apple scab. It is based on leaf wetness periods and accompanying temperature prevalent in spring. For instance, if the mean temperature is 17.2°–24°C the incubation period is 9 days provided the leaves are continuously wet. Nine hours of leaf wetness will result in light infection, 12 hours of leaf wetness will result in moderate infection and 18 hours of leaf wetness at the above temperatures will result in heavy infection. Such conditions frequently occur during spring and autumn. For measuring wetness, special recorders are used in the orchards. Hours of wetness period interrupted by more than 8 hours of dry period are not included in determining the infection periods. A similar relationship of fruit scab to fruit age and wetness period is also reported (Xu and Robinson, 2005). On a relatively susceptible variety, a wet period of 9 hours resulted in 90% infection of 4-weeks old fruits but only 9% infection of 9-weeks old fruits. A longer duration of wetness was required for a similar level of fruit scab infection on old fruits and young fruits. Wetness periods of 9 and 32 hours were required for 90% incidence of fruit scab on 4 and 7 weeks old fruits, respectively.

Thakur and Khosla (1999) have investigated the relevance of Mills' scab infection criteria for improving efficiency in the monitoring of apple scab control programme in Himachal Pradesh (India). Observations on potted and orchard plants showed that the light infection period required 3 days more for development of scab symptoms at a leaf wetness duration of 9–13 hours and average temperature range of 8.5°–21.4°C whereas moderate and severe infection periods (> 15 hours of leaf wetness) required 1 day more to exhibit symptoms. Rescheduling of fungicide applications on the occurrence of infection periods could save at least 3 sprays. They concluded that forecasting and monitored spray programme may be devised in accordance with prevailing weather conditions of the region.

Strict adherence to Mills's criteria has not been possible in many countries. In England, the leaf wetness duration has been substituted with the hours of 90% or higher relative humidity after rains. It gives an equally good prediction. In Germany, a mean day temperature aggregate of 105°F (over 15 days) after March 1 favourable for pseudothecial maturity is also taken into consideration along with leaf wetness period. In the Netherlands, warning of the possible disease development can be issued on the basis of (a) ascospores are ready to mature, (b) ascospore release is expected, (c) ascospore release has taken place, and (d) infection periods have occurred.

Gadoury et al., (1992) had qualified the above observations with a report that there is disparity between morphological maturity of ascospores and physiologic maturity of asci which can affect the timing of first fungicidal spray to control primary infection. According to them, at present the criteria used to judge the stage of development of asci include the delimitation of ascospores, ascospore shape and colour, and extrusion of the endoascus in discharged asci. This morphological rating of ascospore maturity may not be related with the physiological maturity of asci. The ability of asci to release ascospores (physiological maturity) lags behind morphological maturity of the ascospores by several days. The disparity is significant for the effects of spraying in early stages of primary infection season. Physiologically, the asci become ready for discharge of mature ascospores several days after bud break and spraying could be postponed beyond the date estimated by spore showers. Predictive models and procedures have been developed in United States and Europe. These PC-based forecast models avoid the cumbersome process of microscopic examination of leaves for ascospore discharge.

Although spray schedules based on weather reduce the number of sprays per season, some studies in New Zealand have suggested that reduced fungicide use by timing sprays with weather information show higher incidence of scab compared to standard calender-based spray schedules. A chemical indicator, described in Russia, reveals scab spots 3–4 days before they become optically visible. This forecasting method has been successfully applied on a field scale. The technique is very useful in adjusting the timing of the number of applications needed every year (*Entomophaga* 41:461, 1996).

Management of the disease

Control of primary infection by ascospores (also conidia at some places) is the most important step in the management of apple scab. The following steps have been suggested to achieve this:

Elimination or reduction of perithecial production: Fallen leaves, bearing stroma for perithecia development, are the main source of survival of the pathogen and supply of the primary inoculum in the spring. Attempts have been made to destroy this source in the orchards. In 1961, it had been demonstrated in Canada that perithecial formation in culture media was inhibited by excess nitrogen. Practical application of this information was made in England where 5% urea was sprayed on infected trees after fruit harvest at the pre-leaf fall stage. This treatment caused about 97% reduction in ascospore discharge from fallen leaves in the next spring Since infected leaves falling late are more dangerous source of viable perithecial inoculum, hastening of leaf fall after fruit harvest assists reduction of inoculum. Application of vegetable oil emulsion to the trees after fruit harvest reduces respiration and stimulates ethylene production in the shoots causing early leaf fall. Curry *et al.,* (2005) have reported that vegetable oil emulsion- induced defoliation (mid-Oct to mid-Nov.) with manual leaf removal from the orchard floor and spray of lime sulfur in early spring reduced leaf and fruit scab to less than 5%. Vegetable oil emulsion at 4% plus 2% lime sulfur and/or 2% urea also effectively suppressed scab.

A pre-bud spray of 2% urea in spring on over-wintered leaves not only reduced ascospore numbers but also prevented release of ascospores from perithecia in the remaining leaves. Urea not only suppresses ascocarp development it also hastens decomposition of leaves thus destroying the perithecia. The populations of such fungal antagonists as *Trichoderma* and *Cladosporium* spp. are increased. Beneficial effects of urea spray are reported in India also. Spraying of trees with 5% urea in autumn prior to leaf fall and again 2% urea just before bud burst was recommended before starting fungicidal sprays. Complete suppression of ascospore discharge from over-wintered leaves occurs when 3–5% urea is sprayed on the trees just before leaf fall (last week of October). However, application of urea to the leaf litter in April before bud burst is reported to reduce ascospore discharge more than the application in November. Post-harvest spray of fungicides, such as 0.4% Benlate or 0.005% organic mercurials also reduces perithecia formation in fallen leaves. Topas-C (penconazole plus captan) and carbendazim spray also is as effective as urea spray. One per cent urea, dodine (Syllit) and fenarimol (Rubigan) were 90% effective. An autumn application of dolomite lime to infected leaves on the orchard floor results in a decrease in the percentage of leaves with pseudothecia, in the number of pseudothecia per leaf, and the number of asci per pseudothecium of *V. inaequalis* and *V. pirina*. Application of 5.08 metric tons per hectare reduced the ascospores dose in the following spring up to 88% for pear and 92% for apple. In an earlier study in 1968 autumn application of lime caused reduction in ascospore discharge in the following spring but urea was more effective than lime. The strategy of accelerating leaf litter decomposition to destroy pseudothecia by using urea, calcium carbonate or dolomite has been found effective in citrus orchards also against greasy leaf spot caused by *Mycospherella citri* (Mondal and Timmer, 2003). In the northwest USA shredding of leaf litter by mowers in November or in April reduced the risk of scab by 80–90% provided the entire litter was shredded. Vincent *et al.,* (2004) have reported that most efficient treatment is application of urea followed by leaf shredding and application of antagonistic fungi (*Athelia bombacina, Microsphaeropsis ochracea*).

Collection and burning of fallen leaves in winter, where feasible, is also recommended for elimination of source of primary inoculum. In Canada, thermal treatment of orchard floor by specially designed propane flamers (temperatures raised to 150°–200°C) has been demonstrated to reduce the ascospore maturation inside fallen leaves and release of ascospore by half. Flaming of the orchard floor is done before the ground becomes covered with green vegetation Fall application of fungal antagonists reduces spring production of ascospores (Carisse and Rolland, 2004). Antagonists such as *Microsphaeropsis ochracea, M. arundinis, Diplodia* sp., *Trichoderma* sp. and *Athelia bombacina* applied to fallen leaves can reduce the amount of ascospores in the spring by 70–79%.

Fungicidal sprays to prevent primary and secondary infection: A large number of fungicides have been found effective against the apple scab pathogen *in vitro* and *in vivo*. The list includes the sulphur and dithiocarbamate fungicides (flowable sulphur, zineb, mancozeb, cuman, polyram, etc.), the benzene fungicides chlorothalonil (Daconil) and dinocap (Karathane), the heterocyclic nitrogen fungicides captafol (Difolatan) and captan, and many other protectant fungicides such as dodine (Cypress or Syllit), Dikar (a combination of mancozeb and dinocap), Glyodex (dodine plus glyodin), the mercurials such as phenyl mercury chloride, the systemic benzimidazole fungicides such as benomyl (Benlate), carbendazim (Bavistin, MBC), thiophanates such as Topsin-M and Cercobin-M, and the sterol biosynthesis inhibiting fungicides such as triforin (Saprol), myclobutanil (Systhane, Eagle, Rally), difenoconazole (Score), fenarimol (Rubigan), and many others.

Among the protectant fungicides, captafol (Difolatan), mancozeb and captan had been extensively used for apple scab control in many countries. Difolatan has high resistance to weathering, persists for a much longer period than other protectant fungicides on the leaf surface and has very low phytotoxicity, thus can be used in high concentrations. In Canada and the USA a single spray of 0.3% Difolatan at the late dormant stage of the trees was found to have a persistent effect against apple scab. Since the massive dose is applied before bud burst, it is exclusively deposited on the bark of woody tissue. This deposit serves as a reservoir during subsequent several weeks and the fungicide is redistributed by rains to emerging foliage which is protected against early infection by the apple scab fungus. The persistence and redistribution of Difolatan is increased throughout the season when it is applied with foliar nutrients (urea and magnesium sulphate) at the pink stage of fruit bud development. In the United States a single application of Difolatan at the green tip stage followed by 0. 2% captan at petal fall stage and for cover sprays was commonly recommended. Mancozeb

(0.25%) is included for two sprays after petal fall in a schedule recommended in India. Six sprays of 0.3% mancozeb were found as effective as carbendazim. Spring and summer sprays of Daconil, Cuman L, Euparen-M, dodine, carbendazim, Aureofungin, flowable sulphur and mancozeb provide almost complete control of both leaf and fruit infection. Difolatan (0.15%) gave 81% and 91% control of leaf and fruit scab, respectively. Residue of dodine applied for scab control at the rate of 3 or 6 kg a.i/ha could be detected for 20 days on fruits and for 30 days on leaves. The safe waiting period was determined as 14 days with a half life of 2.5–4.6 days on fruits and 4–4.9 days on leaves. Although more specific and effective sterol inhibiting fungicides have been introduced, the protectant fungicides are still included in calender-based spray schedules against apple scab.

In on-season sprays the protectant fungicides dodine (Syllit 0.15%) and Difolatan (0.3%) provide better control of primary scab on leaves and fruits than the systemic fungicides Delan, benzimidazoles, Daconil, mancozeb, zineb, and Cuman-L are good for control of secondary infection. Six sprays of 0.05% Bavistin, Dithane (Indofil) M-45 (first spray at 0.4% and subsequent sprays at 0.15%), or Cuman-L (first spray at 0.4% and subsequent sprays at 0.3%) gave 93–100% control of leaf and fruit scab. A schedule of 5 sprays starting with 0.3% Difolatan and followed by 0.05% Bavistin, 0.25%, Dithane M-45, 0.3% Cuman-L, and finally 0.3% Difolatan was highly effective. Effect of pre-symptom sprays on conidial production and secondary spread. Mancozeb retains efficacy (98–100% disease protection in South Africa) in all wetting regimes but some fungicides such as triforine (Saprol) are effective when exposed to light mist only. Increased precipitation reduces their efficacy.

The systemic fungicide Benlate (benomyl) was found an excellent fungicide for apple scab control as early as 1968. In India, Bavistin (carbendazim) was used in place of Benlate. These fungicides have preventive as well as curative action. They are equally effective as post-harvest and pre-bloom sprays. The spray of Benlate or Topsin-M during the growing season prevents the formation of perithecia and ascospores in over-wintered leaves thus reducing the primary inoculum load.

Use of sterol biosynthesis inhibiting fungicides: Attempts have been made to reduce the number of fungicidal sprays (thus reducing the cost of chemical control) by deciding the most suitable timing for the first spray and by chooing sterol biosynthesis inhibiting fungicides which have long duration protective, curative, and eradicative properties. The triazole fungicide difenoconazole (Score) is a highly effective fungicide meeting the above requirements and gives 90–100% control of scab. Fenarimol (Rubigan), flusilazol, or myclobutanil (Enzone) are reported to control the disease by only 4 sprays given at tight cluster, pink bud, petal fall and 10 days after petal fall stages. Curative activity of myclobutanil, fenarimol, hexaconazole, bitertanol and flusilazole was demonstrated in 1992 Curative effect was best when the fungicides were applied 72 hours after inoculation. Beyond this interval the activity decreased or was not reliable. The use of bitertanol (Baycor) at 3.0 kg/ha for commercial sprays was recommended in Himachal Pradesh. Others have also reported the antisporulant activity of myclobutanil (0.075%) and other SBI fungicides (Thakur and Khosla, 1999). Ninety to 100% control of scab by application of the anilopyrimidine fungicide cyprodinil (100 mg a.i./lit) and pyrimethanil is also reported. Cyprodinil significantly reduces fungal infection stages that are formed after penetration of leaf. In apple scab it reduces the growth of subcuticular stroma. The choice of SBI fungicides and anilopyrimidines (cyprodinil) should be based on the prevailing temperature. The SBI fungicides are more effective at higher temperatures and anilopyrimidines at lower temperatures. Resistance development against cyprodinil was not reported even when used 43 times over 4 years (Kunz *et al.,* 1998). Resistance to difenoconazole, tebuconazole, pyrifenox, myclobutanil, fenarimol and fllusilazole in

V. inaequalis is common after these have been frequently used in the orchard. There is high degree of cross-resistance between them. If resistance develops against any one of them it is for all of them. One of the multiple genes that confers resistance to DMI fungicides also lowers sensitivity to cyprodinil (Koller *et al.*, 2005).

Bitertanol (0.075%), hexaconazole (0.03%), myclobutanil (0.04%) and penconazole (0.05%) were tested by Sharma and Verma (1996) for curative effect (after infection), and eradicant (post symptom) activity. Applied within 72 hours of start of the infection period, all the fungicides completely suppressed appearance of scab lesions on leaves. Pre-symptom activity was noticed when the fungicides were applied 168 hours after the infection period. Penconazole was most effective. When applied to lesion bearing leaves, conidia formation was significantly reduced which was more pronounced after 7 days. Myclobutanil was equally effective. Two consecutive sprays of these two fungicides resulted in 89–93% reduction in conidia production by the 14th day. Difenoconazole (Score) and penconazole (Topas) are used on a large scale against apple scab in Russia. In a comparative study of protectant spray programme with curative after-infection sprays, based on predicted infection periods for control of apple scab and grape black rot both provided equally good control but after-infection sprays were more economical. The new DMI fungicide fluquinconazole is reported to be more active than a pyrimidine fungicide in *in vitro* tests. Vision is a specialist polyvalent fungicide comprising two active ingredients in a concentrated suspension. It contains fluquinconazole (50 g/lit) and pyrimethanil (200 g/ lit). It is particularly effective against scab and powdery mildew of apple.

The sterol biosynthesis inhibitor fungicides have special importance for areas where incessant rainfall may continue for days because they can be applied with full efficacy even several days after infection periods. In a Brazilian study (Boneti and Katsurayama, 1999) post-infection applications of fenarimol increased the level of control by increasing the dose (2.4, 3.6, 4.8, and 6.0 g/100 L) when applied 96, 120 and 144 hours after the infection period. Tebuconazole was better than fenarimol when applied 96 hours after the infection period. Application of fenarimol (6 g/100 L) 120 to 144 hours after the infection period followed by another spray 7 days later improved scab control. There was good control when a mixture of fenarimol and dodine (6 g + 39 g/100 L) was applied followed by one application of fenarimol 7 days later.

The strobilurin fungicides were launched in 1996 and within 6 years became most sought after fungicide. The fungicide is derived from the natural antibiotic strobilurin A, produced by the symbiotic fungus *Strobilurus tenacellus*. At least 6 strobilurin active ingredients are commercially available. Azoxystrobin is now the world's biggest selling fungicide. These fungicides have a very broad and balanced spectrum of activity as a foliar fungicide. They dissipate rapidly from soil and surface water and are unlikely to cause undue hazard to non-target organisms. They have preventive and curative action against apple scab and apple powdery mildew. Their mode of action is through inhibition of cytochrome enzyme complex at the Qo-site in the respiration chain of fungal mitochondria (QoI inhibitors). Resistance to the fungicide may develop through development of alternative respiration mechanism. Several fungicides have been developed in this group, *viz.*, kresoxym methyl (Sovran, 50 WG), beta methoxyacrylates, azoxystrobin, trifloxystrobin (Flint, 50 WG), etc. These can work in combination with sterol biosynthesis inhibitor fungicides. Fifty four treatments over 6 years suggested that no resistance to these fungicides had developed at the sites sampled (Kunz *et al.*, 1998). Olaya and Koller (1999) had reported 99% control of apple scab by application of 4 mg/ml of kresoxim methyl. However, in later studies insensitivity of *V. inaequalis* to strobilurin fungicides, kresoxim-methyl and trifloxystrobin, was reported by Koller *et al.*, (2004). The apple scab fungus is generally sensitive to these fungicides although sensitivity varies with the formulation. Resistance to trifloxistrobin was detected in 1997-1999 but no obvious performance deficiencies were reported (Farber *et al.,* 2002). The resistance to these fungicides can be managed by using the maximum recommended level of the fungicide. It gives better check on resistance development than use of low concentration of the fungicide mixed with a protectant fungicide (Turechek and Koller, 2004)

Use of mixtures of protectant and systemic fungicides: The apple scab fungus has the tendency to develop resistance to the single site action systemic fungicide. This has necessitated the use of broad spectrum, multi-cite acting protectant fungicides in combination with the systemic fungicides. Mixtures of protectant and systemic fungicides are preferred over alternate application because they have better chances of preventing resistance development. Combinations of SBI fungicides and protectant fungicides have also been suggested. A combination of tebuconazole and captan has been found to provide excellent control of scab and with no residue problem combination of penconazole with captan (Topas-C) also is used as an effective, anti-resistance combination.

Other approaches to chemical control: Plant oils have also been used against *Ventiria inaequalis* but they are not as effective against scab as against powdery mildew. Six to 10 sprays of 2% hydrated lime are reported to reduce fruit scab by 41% and leaf scab by 83% in pear scab caused by *Venturia pirina. In vitro*, at 0.5%, 1.0% and 2.0% concentration, sodium bicarbonate reduce germination of spores of *V. inaequalis* by 59%, 96.4% and 100%, respectively. The control is comparable with that given by tebuconazole. Application of 1% sodium bicarbonate to trees at 10-day intervals significantly reduced disease incidence and severity on leaves and fruits (Ilhan *et al.,* 2006). Ammonium or potassium carbonate or potassium phosphate at 1% (w/v) are reported to reduce colony diameter of *V. inaequalis* by more than 90% *in vitro*. Spray of 0.5 or 1% aqueous solution of sodium or potassium bicarbonate on younf apple seedlings significantly reduce leaf scab (Jamar *et al.,* 2007). They (Jamar *et al.,* 2008) have reported control of leaf and fruit primary scab by application of a formulated potassium bicarbonate product, Armicab. The control is as good as given by wettable sulphur. This has potential for use in organic productiom of apples.

In Japanese pear scab, caused by *Venturia nashicola*, treatment of leaves with acibenzolar-S-methy (ASM) induces resistance. ASM does not affect spore germination, appressoria formation or penetration. Its effect is visible in the epidermal pectin layers and the middle lamellae of the treated leaves where subcuticular hyphae are at lower frequency than in untreated leaves indicating suppression of the fungal growth. Enzymes induced by the pathogen cause collapse and fragmentation of hyphal cell walls. This is suggested as mechanism of resistance induction (Jiang *et al.,* 2008).

Prohexadione, a plant growth retardant used in the management of severity of fire blight (*Erwinia amylovora*), is reported to induce resistance in apple leaves against the scab pathogen. It activates almost all resistance proteions (PR).

Calender-based spray schedules in India: In Himachal Pradesh (India) the chemical control of apple scab was suggested on the following lines. These spray schedules employ both protectant and systemic fungicides (altetnating) and are used at the critical stages in disease development. One spray of 5% urea on the trees before general leaf fall in October-November is recommended aspost-harvest spray.

Different spray schedule all are effective in controlling leaf scab by 86–96% and fruit scab by 88–99%. These schedules consisted of the recommended protectant and systemic fungicides and a last spray of 5% urea at leaf fall stage. Better protection was given by schedules that started with a protectant fungicide at silver tip stage than those in which the first spray was of a systemic fungicide. On the basis of these observations, modifications have been made in the spray schedules mentioned above. Systemic fungicides are avoided in the sprays during silver tip to green tip stage. Captan is preferred over difolatan because of residue problem in the latter and sterol biosynthesis inhibiting fungicide is introduced in the spray at the fruit set stage. Addition of one spray of a mildewcide at dormant or bud swell stage keeps powdery mildew of apple under check. Two pre-harvest sprays of 0.2% captan, 30 and 15 days before harvest, effectively check development and spread of storage scab up to 90 days in the store. Bitertanol (0.075%) is also effective but carbendazim is not so effective in the control of storage scab. When dodine is used in a pre-harvest spray the incidence of storage scab is reduced. Percival and Boyle (2005) have tested the efficacy of fungicides application *via* microcapsule trunk injection of penconazole, pyrofurox or carbendazim significantly reduced scab incidence for two successive growing seasons and were better than thiabendazole, triadumefon or propiconazole. Post-harvest dip of fruits in 0.075% bitertanol also checks storage scab up to 90 days in stores. Pre-harvest sprays of curative fungicides generally control storage scab under the conditions of low inoculum potential but in moderate and high potential the use of protectant fungicides, during the growing season is necessary. If a high possibility of late infections of fruits exists, the fruits should not be put under long-term storage.

Biological control: In *in vitro* isolates of *Ophiostoma, Chaetophoma, Aureobasidium* and *Phoa* have been found to inhibit the vegetative growth of *V. inaequalis. Athelia bombacina, Chaetomium globosum* Hayes and Andrews, 1983) and *Microsphaeropsis ochracea* (Carisse and Rolland, 2004) are promising biocontrol agents against *Venturia inaequalis*. Used as foliar spray, preferably with calcium nitrate as foliar fertilizer, they reduce the scab intensity on leaves. Timing of application of the antagonists influences their efficacy. Application soon after harvest or just before leaf fall gives best result in reducing perithecial formation and ascospore ejection in spring. Some isolates of epiphytic yeasts on apple leaves are antagonists of *V. inaequalis* and suppress leaf scab. Significant control of leaf scab is achieved by weekly sprays of water extract of anaerobically fermented spent mushroom compost during green tip to petal fall stages. The major inhibitory principle of the spent mushroom slurry is a low molecular weight, heat stable, non-protein metabolite produced by anaerobic microorganisms in the compost. The role of antagonists developing under such conditions in decreasing the ascospore number was mentioned above. Spray of an aqueous extract of dry mycelium of *Penicillium chrysogenum* induces resistance to the scab fungus in apple leaves. The suppression of scab is as good as copper and sulphur fungicides and plant activators like acibenzolar-S-methyl.

Resistant cultivars: Level of phenolics in the host tissues is related to resistance of apple to scab. Resistant cultivars have more phenols (flavonoids) than susceptible or tolerant cultivars. Sources of resistance to the apple scab fungus exist in the East Asiatic wild species *Malus floribunda* (crab apple, *Vf* resistance gene). This was the source of *Vf* resistance used in a number of breeding programmes to develop scab resistant cultivars. The gene confers resistance to 5 races of *V. inaequalis*. Earlier, four receptor-like genes (*Vfa1, Vfa2, Vfa3* and *Vfa4*) were identified within the *Vf* locus but now it is recognized that only *Vfa1* and *Vfa2* are actually functional (Malnoy *et al.*, 2008). *M. micromalus* (Vm resistance gene) and *M. pumila* (Vr resistance) are other sources of resistance. During the last two decades about 30 resistant varieties of apple have been released, mostly in the United States and Canada, a few in England and Europe. These varieties have a high degree of resistance which is mostly monogenic. But these varieties are not resistant to other diseases of apple. The race 6 of *V. inaequalis* is capable of overcoming the monogenic resistance conferred by *Vf* gene although *M. floribunda* is not infected by this race. This suggests incomplete inheritance of resistance gene. A new race (race 7) can infect *M. floribunda* also. Thus, genetic resistance can be used as a valid compliment to the use of fungicides rather than as an alternative to chemical control. Durable resistance, thus, can be obtained only by combining several defense mechanisms.

An endochitinase-encoding gene from *Trichoderma harzianum* has conferred complete resistance to several fungi in many transgenic plants (Lorito *et al.,* 2000). In an attempt to improve scab resistance of apple by transformation with genes encoding chitinolytic enzymes (endochitinase and exochitinase) from *Trichoderma harzianum* success was achieved in reducing disease incidence but the endochitinase had negative effect on the growth of the plant. Faize *et al.,* (2003) have reported similar findings in transgenic lines with *Trichoderma atroviride* endochitinase genes. Along with disease suppression a negative correlation between growth of the transgenic lines and endochitinase activity was observed, suggesting disturbed plant metabolism. Reduction in vigor of plants is associated with high lignin content and high peroxidase and glucanse activity. In similar attempts, the gene encoding for puroindoline B in wheat has been introduced in apple cultivars carrying or not carrying resistance genes. The gene has been found to express in both types of apple cultivars. A significant negative correlation between puroindoline B content and susceptibility to race 6 of *V. inaequalis* has been found (Faize *et al.,* 2004). A polygalacturonase-inhibiting protein in Japanese pear has been found associated with resistance to pear scab (*Venturia nashicola*) in Japan (Faize *et al.,* 2004).

The apple scab resistant cultivars introduced in India from the United States and Canada are Prima, Priscilla, Sir. Prize, Macfree, Coop-12, Red Free, Nova Easygro, Libert, and Freedom. Coop-12 and Red free are being used in a hybridization programme in Himachal Pradesh. Four hybrids, evolved at Mashobra (HP) have shown high field resistance to scab. In comparison to susceptible Red Delicious they exhibit few lesions with poor sporulation under epiphytotic conditions. So far none of the resistant varieties have been accepted in commercial planting mainly because of their uncertain market compared to established varieties which are susceptible. In apple orchards with mixed plantation of genotypes there is less severe scab.

SCLEROTINIA ROT AND WILT

During the cool weather with heavy dew in the winters of north India many field and vegetable crops suffer from rot and wilt diseases induced by *Sclerotinia sclerotiorum* and *S. minor*. In temperate climate countries also these species are involved in similar diseases. The diseases caused by *Sclerotinia* have been described under about 60 different names. The damage may be very high depending on susceptibility of the crop, weather conditions, and nature of infection. Host specificity is not present in either species. An index published in 1994 (Boland and Hall, *Can. J. Plant Pathol.* Vol. 16) has listed susceptible host in 40 subspecies or varieties, in 408 species, 278 genera and 75 plant families spread over gymnosperms and angiosperms, in dicots and monocots. *S. minor* occurs on 94 plant species in 66 genera of 21 plant families. The maximum number of plant species attacked by *S. sclerotiorum* are in the family Compositae, followed by Leguminosae and Cruciferae. In some crops such as sunflower the fungus causes head rot, stem rot, and wilt. In others there may be fruit and blossom rot and general blighting of foliage such as in chickpea, pea, tomato, etc. In Himachal Pradesh, mortality due to curd rot of cauliflower seed crop is reported to occur to the extent of 0.5 to 70% resulting in seed loss of 2 to 250 kg/ha. In oil yielding crucifers, where the infection generally occurs on the stem at the ground level resulting in collapse of the plant, the losses in seed yield vary from trace to 50%. In addition to loss in yields, when the plants survive the produce is adversely affected. Tomatoes may rot during transit. Oilseeds have poor quality and quantity of oil. Infection of mustard seed drastically reduces seed germination.

Symptoms

The rot symptoms appear as white mold (as in beans) and fruit, blossom, seedling or leaf rot. When fruits are attacked, such as in eggplant and tomato, there is rotting of the flesh and in the rotting tissues large number of sclerotia of the fungus can be seen. On pea pods, the symptoms mainly consist of white mycelium sticking to pod surface, mostly at the basal and distal ends, and the tissues show necrosis. Ultimately, the fungus enters the pods and rots the developing seeds. Similar rotting of heads in sunflower and safflower also occurs. The heads show a large number of sclerotia mixed with the seeds. In sunflower infected by *S. sclerotiorum* and wilted at any stage from flowering to near maturity stage the seed yield is significantly reduced. If plants wilt within 6 weeks of flowering the seed quality including oil and protein content also deteriorates. The wilt incidence is generally low during the vegetative stage of sunflower growth.

Fig. 69. *Sclerotinia* wilt of eggplant. Infection at stem base.

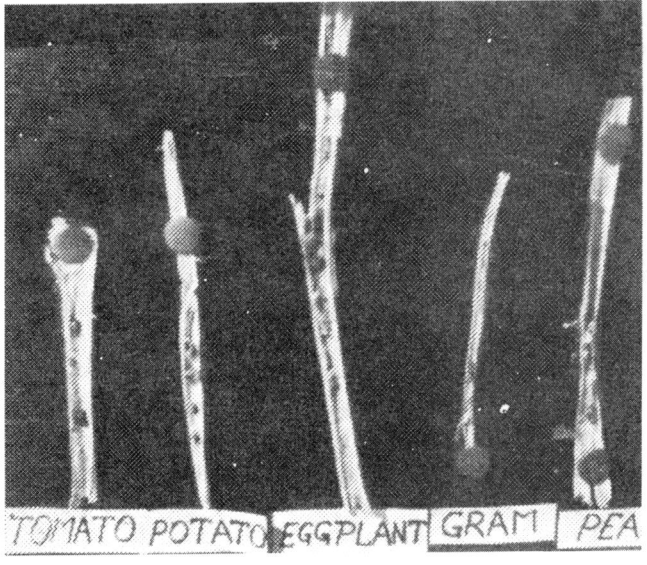

Fig. 70. *Sclerotinia* in stem pith.

Partial or total wilt is the characteristic symptom of infection of stems and branches. Such symptoms are seen in tomato, eggplant, potato, pea, chickpea, and *Hibiscus*. The infection may occur at any part of the foliage, mainly the stems and branches. At the point of infection a dry, discolored spot develops. It gradually girdles the entire stem or the branch and progresses up and down. As a result of tissue necrosis, the portion of the plant beyond the point of infection wilts. If the infection is at the base of the main stem, the entire plant wilts.

If it occurs on branches partial wilting occurs. On opening the dry portion of the stem, the pith can be seen full of fungal sclerotia which may be small or large, elongated or cylindrical, and often attached to each other end to end. Sclerotia also develop on the surface of the affected plant parts in a white mycelial web sticking to the host surface. These sclerotia are generally globose, not elongated or cylindrical. The colour of mature sclerotia is dark brown or back.

In many brassicas, such as in cauliflower, the fungus causes stalk rot. The earliest symptoms are loss of turgidity of leaves during daytime but the leaves recover during night. The bright green sheen of the affected plants is replaced by dull white green colour, ultimately becoming pale yellow. Such leaves shed prematurely. Midrib and petioles of leaves in contact with soil show dark brown to black soft rot and fluffy growth of the fungus during cool humid weather. Rotting from the petioles advances to the stalk where dark brown to black lesions are formed. These enlarge and girdle the stem at the ground level. Inside, the pith of the stem up to the attachment of curd (in cauliflower) undergoes rotting giving way to large cavities lined inside by fluffy mycelium and numerous sclerotia of the fungus. Under cool humid weather the mycelium emerges out on the surface of the affected parts and can be seen sticking to the host surface. In late stages curds are also affected showing brown to dark brown rotting. In seed crops, the fungus attacks the inflorescence on which mycelium and sclerotia can be seen. The affected branches dry and bear only shriveled seeds. In white rot or white mold of pea, bean, etc. the fungus attacks the pods which are covered with white fluffy growth of the fungus and undergo a soft rot.

The causal organism

Sclerotinia sclerotiorum (Lib.) de Bary (Ascomycota, Discomycetes, Helotiales, Sclerotinieaceae), is the recognized name of the fungus causing these diseases. Generic name *Whetzelinia* was also proposed in 1975. The species is highly polyphagous. Its host are from 408 species in 278 genera of plants in 175 families. Except for one host in the division Pteridophyta all hosts are from the Spermatophyta.

Mycelium of the fungus is hyaline, much branched, consisting of closely septate, large hyphae which are both intra- and intercellular and invade all the tissues of the affected portions. The hyphae are 9–18 μm broad and filled with dense protoplasm. *S. minor* differs from *S. sclerotiorum* in always producing small sclerotia (0.5–2 μm diameter) compared to 2–7 μm in *S. sclerotiorum*. In addition, production of apothecia on germination of sclerotia is rare in *S. minor*. Sclerotia germinate by producing mycelium (myceliogenic or eruptive germination).

No true conidia are produced. As the available food supply declines, microconidia (spermatia) are produced on sclerotia, on the discs of over-mature apothecia, and in culture. These spermatia are formed in chains at the tips of short lateral branches of the vegetative mycelium. Sometimes these sporiferous tips bear spermatia endogenously within an old hyphal cell. The spermatia germinate very sparsely in water or in culture media and apparently do not serve as a source of infection or dissemination of the fungus. It is speculated that they only spermatize the ascogonial cells formed beneath the rind of the sclerotium and thus initiate the apothecial development.

When the vegetative growth has ceased the hyphae with thick granular protoplasm and short cells collect in small dense masses which gradually become sclerotia. At first these are pink but later turn dark brown to black and become smooth. They vary in size and shape according to environment and location. Sclerotia formed on the host surface are usually loaf-shaped or globose while those formed in the pith of the stem are elongated according to space available for their growth. On host surface the sclerotia measure 2–12 (average 2.5–6) mm.

Section of sclerotia shows that they are composed of thin-walled, rectangular cells in the centre and thick-walled cells at the periphery. The walls of peripheral cells are impregnated with a gelatinous material. The pigmented cortex cells and 2–3 peripheral cell layers forming the rind give these sclerotia their black appearance.

The sclerotia germinate myceliogenically directly producing mycelium or carpogenically initiating the development of apothecia. In carpogenic germination the sclerotia give rise to 2–5 columnar structures (stipes). Sometimes these stipes are dichotomously branched, each branch being fertile. The length of stipes is 25–88 mm and the width is maximum at the base of the apical apothecium, being about one mm. The exposed portion of the stipe is fawn colored but in the soil the color is brown or dark brown. The apothecial fundaments are borne at the tips of the stipes as minute, brownish or fawn coloured, funnel-shaped cups (apothecia). The apothecial fundaments are usually one mm in diameter and expand into shallow, flat or convex apothecia. The mature apothecium measures 6–9 mm across, generally borne 6–10 mm above the soil surface and becomes darker in colour with age. As the size of sclerotia increases the number of germinating sclerotia and apothecia produced per sclerotium also increases.

Not all surviving sclerotia germinate to produce apothecia. Many show myceliogenic germination. There are reports that Sclerotinia wilt of sunflower is caused by infection of roots by mycelia from myceliogenic germination of sclerotia. The myceliogenic germination of sclerotia is related to the degree of melanization of the sclerotial rind. Only brown sclerotia (incompletely melanized) or injured black sclerotia show myceliogenic germination. An injured sclerotium is capable of healing its wound and start remelanization and undergo dormancy. The remelanization process is prevented by the presence of an oxidase inhibitor. The sclerotia harbour bacteria (*Bacillus subtilis*) that possess antifungal activity through production of a low molecular weight peptide. This protective facility helps sclerotia tolerate antagonism and heal after injury.

The asci are cylindrical and measure 108–153 × 4.5–10 μm, the average being 122.9 × 5.9 μm. The ascospores are violently discharged in abundance from these apothecial cups. Each ascus in the apothecium contains 8 ascospores. The ascus has an apical pore through which ascospores are ejected violently. The ascospores are hyaline, 1-celled, and ovate. Their size falls within the range of 7–16 × 3.6–10 μm. Only these spores are supposed to be the means of spread of the disease during the crop season.

Disease cycle and environmental relations

Between crop seasons, the species of *Sclerotinia* mainly survive through sclerotia which may be present on soil surface in unploughed fields or in crop debris or as admixture with the seed. Twelve month after crop harvest and falling on soil, maximum sclerotia (57.5%) survived on soil surface and 12.5% at 5 cm depth. Sclerotia buried at 10 cm depth showed only 2.5% viability (Duncan *et al.*, 2005). Tan sclerotia produced by some aberrant strains of the pathogen undergo myceliogenic germination and are more pathogenic than normal black sclerotia. However, tan sclerotia have very low survival ability due to easy attack of hyperparasitic fungi.

The new crop is infected by the ascospores produced from germination of these sclerotia. The pathogen is spread from field to field and from one geographic area to another by several means including wind-borne ascospores and soil adhering to seedlings, farm equipments, animals and humans. Where diseased plant tissue is used for bedding and feed for cattle and sheep, the pathogen can be introduced to new uncontaminated fields through manure. About 2% of the sclerotia of *S. sclerotiorum* found in feed eaten by sheep survive the passage through digestive tract and remain viable. Over long distances, the greatest potential disseminator of *Sclerotinia* spp. is probably seed infected with mycelium or contaminated with sclerotia. Seed infection or infestation has been reported in cabbage, cauliflower, clover and groundnut. However, role of seed-borne mycelium in disease cycle has been doubted by some workers who did not get infected seedlings from such seeds. Even if seed-borne inoculum does not play an important role in initiation of the disease, it helps in multiplication of inoculum. The mycelium grows on the rotting seed and produces sclerotia and apothecia (Mueller *et al.,* 1999).

Mycelium and ascospores are short lived and the mycelium present on or in the crop debris does not grow in soil. Survival of sclerotia is governed by many factors such as source and location where formed (cultures, host surface, or host tissues), soil type, moisture and temperature conditions during absence of the host, soil treatments such as flooding and fumigation, depth of burial in soil, and parasites of sclerotia present in soil or induced to grow by special soil treatments. Survival is better when sclerotia are protected by host straw. Deep ploughing (15–20 cm) burries the straw and sclerotia deep in soil, thus making them ineffective. When sclerotia are dried for short periods and then remoistened in wet soil they leak nutrients which attract soil microorganisms. Colonization of sclerotia by soil microorganisms results in rotting of sclerotia. Recovery and viability of sclerotia formed in or on the host is less than sclerotia obtained from cultures and fungal parasites of sclerotia formed on the host are also numerous. Such sclerotia from the host show much more perforations in the rind caused by parasites than the sclerotia from cultures when left in the soil. In sandy clay loam soil the recovery of sclerotia is brought down to zero in 36 weeks but in sandy soils there is only 50% reduction in recovery. Only half of these sclerotia are viable. In a more recent study, Hao *et al.,* (2003) have reported that soil type, temperature and moisture have no effect on viability of sclerotia. In the plains of northern India, the sclerotia from the crop of previous year were found viable and freely germinating. However, in the cool and wet Tarai area of Uttarakhand (India) the free sclerotia present in or on the soil surface decompose and only the sclerotia present in stem pith survive and freely germinate in the next winter after the plant tissues have decayed or disintegrated. Moisture is the limiting factor.

Carpogenic germination of sclerotia is influenced by many factors of which soil organic matter, temperature, moisture and depth of placement of sclerotia in soil are important. Grass mulching of the soil (6–9 cm) reduces aopothecia formation (Ferraz *et al.,* 1999). The optimum temperature for carpogenic germination is 15°C (Hao *et al.,* 2003). Mila and Yang (2008) have studied the effect of fluctuating temperature on sclerotial germination and apothecia formation Taking 20°C as the median and as constant temperature, fluctuation of 8°C resulted in best germination and apothecia production. Exposure to low temperature (near freeze-thaw condition) and strong sunlight are necessary for the germination of sclerotia. Apothecial formation occurs only in saturated or near saturated soil. The moisture content of top 2–3 cm soil layer determines apothecial formation. Conditioning of sclerotia in constant rinsing environment promotes carpogenic germination. Optimum temperature for conditioning is 8°–16°C. In peas the disease develops most frequently in later stages of plant growth, from flowering to maturity. The dense crop canopy usually ensures adequate moisture on soil surface. In soil with high proportion of sand a moisture content of 30% is highly favorable for germination and apothecia formation. The stipes can develop only in the surface soil. Sclerotia buried deep in soil fail to produce apothecia. The number of stipes produced by sclerotia is highly correlated with size of the latter. There is a threshold of sclerotial size below which apothecia are not produced (Hao *et al.,* 2003). Optimal temperature and temperature range for germination of sclerotia are affected by light intensity and moisture level of the sand substrate (Sun and Yang, 2000). Under high light intensity, only few days are required for the initials to develop into full apothecia while under low light intensity it may take several weeks.

Epidemics are initiated by ascospores from sclerotia which germinate during mid-winter (in north India). Once started, the ascospore discharge is continuous, with increasing rate, for 72–84 hours, irrespective of light conditions (Clarkson *et al.*, 2003). The release can be continuous in darkness as well as in light and at high as well as low humidity. These air-borne ascospores are the only primary source of inoculum for *Sclerotinia* infection of pea, bean, tomato, sunflower, *Brassica*, etc. Under field conditions no infection from mycelium growing from sclerotia occurs due to lack of energy source. Ascospore production and release is influenced by relative humidity and temperature. However, Clarkson *et al.*, (2003) have reported that ascospore release occurs in both saturated air (90–95% RH) and at 65–75% RH. Singh and Tripathi (1998) trapped maximum number of ascospores between 4 and 8 in the morning. The ascospore concentration in the atmosphere was directly correlated with relative humidity. After 8 a.m. the rising temperature adversely affected the ascospore release.

Wu *et al.*, (2007) have studied the effect of incubation of excised apothecia on ascospore maturity and discharge. Fresh apothecia collected from germinating sclerotia contain primarily immature or discharged asci. Temperature is an important factor affecting ascus maturation during incubation of excised apothecia. Optimum temperature is 21°C. After incubation at this temperature for 30 hours, the percentage of undischarged mature asci in excised apothecia increased up to 70–80%. This increase is accompanied by a significant increase in ascospore production per apothecium.

Most suitable temperature for ascospore germination lies between 15° and 30°C. At 5°C there is no germination while at 20°–25°C there is 100% germination within 6 hours. Ascospores do not germinate at or below 80% relative humidity. Optimum temperature for infection is in the range of 15.5°–21°C while the optimum growth of the fungus occurs at slightly higher temperature. There is no growth at 33°C. High humidity due to spring rains and /or irrigation and temperatures of 20°–25°C are most conducive for development of Sclerotinia blight of pea. In pea stem blight caused by *S. sclerotiorum* appearance of the disease coincided with post-blossoming stage of the crop, moderate temperatures and high relative himidity favouring the disease. In sunflower, accumulation of soluble phenolic compounds in flower head correlate with resistance. High constitutive and induced phenolic contents as well as phenylalanine ammonia lyase (PAL) activity are present in the most resistant lines (Prats *et al.*, 2003).

Huang *et al.*, (2008) have reported details of ultrastructural and cytochemical study of infection process of *S. sclerotiorum* on leaves and stems of oilseed rape. One day after inoculation with fungal grown agar disc dense mycelial network was usually formed on the host surface. Then, infection cushions of different size were developed. Mucilage produced by the pathogen covered the mycelium and infection cushions. The hyphae forming infection cushions were flattened and of increased diameter. After penetration, fungus hyphae extended between the cuticle and epidermal cell walls as well as inside the epidermal cell walls. During colonization, 5 days after inoculation, marked alterations in the host tissues were detected including disorganization of cytoplasm, cell organelles, disintegration of cell walls and collapse of host cells. Histopathology of infection of flower parts in sunflower was studied by Rodriguez *et al.*, (2004). In a resistant or tolerant cultivar there was cell collapse, changes in cell wall composition and an increase in phenolic compounds in the tissues of corolla and style, which prevented the pathogen from advancing. Such response was weaker in a susceptible cultivar and occurred only in the style. The pathogen was, therefore, not stopped from developing and growing in to the ovary. It colonized ovary, style and base of filaments and produced noticeable colonization of the corolla.

In north India, suitable temperature, moisture and sunlight conditions for germination of sclerotia by apothecia are generally obtained only in winter, about the month of January. This is why the disease appears not earlier on host crops which have been harvested by November-December. This timing of sclerotial germination coincides with two host related factors. By January the host rabi crops are generally in advanced stage of their growth hence the disease is seen mostly at advanced stages of the crop like pea, chickpea, tomato, etc. which ensures high humidity on soil surface for germination of sclerotia and severe outbreaks of pod and fruit rot.

A suitable nutrient base such as injured leaves or flowers and flower parts (senescent stamens and style) remaining firmly attached to the pods during seed development stage is required for ascospore germination and subsequent penetration of the host tissues. In epidemiological studies of the stalk rot of cabbage caused by this fungus, it was found that ascospores disseminated by air currents initiate infection of stems of cabbage seed plants. The leaf scars on the stumps due to detachment of senile leaves, and blossoms are preferred site of infection. However, maximum infection occurs during flowering. No soil level infection was noticed. Freeze and bruise injuries, especially under conditions of continuous leaf wetness, are important factors associated with infection of cabbage by *S. sclerotiorum* (Hudyncia *et al.,* 2000). In pea pod rot and bean white mold as well as in many other hosts infection of tissues occurs by formation of an infection cushion and development of dichotomously branched penetration hyphae and appressoria. There are many reports that anthers and pollen grains stimulate spore germination of fungal pathogens (refer to pearl millet ergot). In the case of *Sclerotinia* also it has been demonstrated that in presence of host pollen, germination of ascospores is much better than in absence of pollen. White mold of bean can be influenced by the microflora of petals (anthoplane). In some crops, as in alfalfa, infected pollen initiates the blossom blight. After infection rapid necrosis of tissues follows. In infection of sunflower head, fungal population of the anthosphere influences suscepotibility or tolerance to infection and it varies with cultivars. Hyphal contact of anthosphere fungi with *Sclerotinia* as well as production of antibiotics play the role of suppressing the pathogenic infection.

S. sclerotiorum secretes extracellular proteases *in vitro* and *in planta* (Billon-Grand *et al.,* 2002). Proteolytic and cellulolytci enzyme-degradation of host tissues was thought to be the mechanism of action of the pathogen. However, pathogenicity and necrotrophic colonization of host tissues by *S. sclerotiorum* is specifically associated with synthesis and secretion of the non-host specific toxin oxalic acid, not pectolytic or cellulolytic enzymes. The toxin is the pathogenicity determinant in species of *Sclerotinia* as well as *Botrytis* (both are of same family). It suppresses the generation of reactive oxygen species to which it is tolerant. Accumulation of oxalates in the fungus is regulated by its carbon nutrition. Recently, Kim *et al.,* (2008) have reported that oxalic acid secretions of *S. sclerotiorum* is not directly toxic to plant cells. The toxin acts as elicitor of programmed cell death (PCD) during disease development. The PCD is essential for fungal pathogenesis. Oxalic acid induces foliar wilting by manipulating guard cells of stomata, inducing the stomata to remain open during night. This increases transpiration. Loss of water causing foliar wilting and also permits the fungus to easily escape from affected tissues and cause secondary infection (Guimaraes and Stotz, 2004). Oxalic acid has affinity for calcium and targets calcium in cell walls forming calcium chelates. The fungus also produces ethylene which helps it in its interaction with the host. Application of an inhibitor of ethylene production (amino ethoxy vinileglycine) suppresses the development of the fungus and reduces disease incidence. On the other hand, application of a ethylene releaser (ethephom) enhances the disease (AlMasri *et al.,* 2005). Plant growth regulators influence veghetative growth of *S. sclerotiorum* (AlMasri *et al.,* 2002). Naphthalene acetic acid (200–400 μg/ml) methyl jasmonate and abscisic acid suppress mycelial growth and disease development while gibberellic acid (50–250 μg/ml) promotes mycelial growth.

Management of the disease

The main strategy for control of *Sclerotinia* is to reduce or eliminate the sclerotial inoculum. Seed must be free from sclerotia and seed infection. Often, sclerotia are carried with seed lot. Removal of sclerotia from seed lot can be done by floatation. Seed-borne mycelium can be inactivated by seed treatment with fungicides such as captan, PCNB, thiabendazole or fludioxonil. Seed treatments with captan + PCNB + thiabendazole completely inhibits mycelial growth from seed. Other seed treatment chemicals are thiram, benomyl, iprodione + thiram, vinclozolin + folpet and thiophanate methyl. Pathak *et al.,* (2001) obtained increased mustard seed germination, plant height and weight by treating seeds with a combination of *Trichoderma viride* or *Trichoderma harzianum* and carbendazim or mancozeb.

Soil-borne inoculum in the form of sclerotia is most important source of initial infection in the crop. Soil treatment with chemicals such as cyanamides is effective but impractical. Stubble burning is one way of destroying the sclerotia on soil surface. An intense fire in a uniformly dense stubbles leads to destruction of sclerotia. Sclerotia that are not burnt have less germinability. Removal of sclerotia bearing plant parts and their destruction by burning is essential. Burying the sclerotia deep in soil by ploughing at least for 30 weeks ensures destruction of most of them. Deep buried sclerotia fail to produce apothecia. Moldboard ploughing buries sclerotia more than 10 cm deep and delayed emergence of apothecia However, in *Sclerotinia minor* (lettuce dro) deep burying of of sclerotias is reporte to enhance their viability. Flooding of soil or cultivation of rice in the field also reduces sclerotial population. Complete destruction of sclerotia during cultural practices used for low land rice cultivation. was reported in 1949–50. Matheran and Porchas (2005) have recommended flooding of soil for 2–3 weeks for reducing population of sclerotia in lettuce fields. Soil solarization with polyethylene sheets inhibits sclerotial germination, disturbs apothecia formation and traps ascospores. In soybean, crop rotation with maize or wheat was found to have significant effect on the production of apothecia of *S. sclerotiorum.* Crop rotation and no-tillage of soybean was most useful combination of treatments to reduce primary inoculum of apothecia. Crop rotation does not affect the number and distribution of sclerotia or incidence of stem rot but id does affect the number of apothecia and the yield. Organic amendment of soil is variously reported to adversely affect survival of sclerotia. Soil solarization kills sclerotia at 5, 10, 30 cm depth in about 90 days. However, solarization added to a system of crop mulch reduces this period to 60 days (Ferraz *et al.,* 2003).

Destruction of sclerotia by antagonistic fungi, bacteria and soil amoebae is very frequently reported and in some cases field control with the antagonist *Trichoderma* has been demonstrated. Appropriate antagonists introduced in standing crop can multiply, even on sclerotia, and spread from plant to plant. Honeybees help in dispersal of *Trichoderma* from flower heads in sunflower (Esconde *et al.,* 2002). Application through seed treatment with the biocontrol agents is most effective in suppressing the disease in the field. Fungal parasites (mycoparasites) of sclerotia include *Gliocladium virens, Teratosperma oligocladum, Sporodesmium sclerotivorum, Coniothyrium minitans, Pythium oligandrum, Trichoderma* spp. and many yeast fungi. Wheat bran-based cultures of Trichoderma koningii applied to soil reduce viability of sclerotia. The hyperparasite penetrates the sclerotia up to the medullar tissues where ots hyphae, copnidia and chlamydospore can be found. The best characterized biocontrol agents are *Sporodesmium sclerotivorum* on *S. minor* and *Coniothyrium minitans* on *S. sclerotiorum.* All these biocontrol agents are anamorphic fungi.

In *S. sclerotivorum, T. oligocladum* and *C. minitans* (all anamorphic fungi) approximately 5 macroconidia per g soil are needed to successfully infect sclerotia and bring about their decay when the soil is mixed every two weeks. Each infected host sclerotium supports production of 15000 new macroconidia of the mycoparasite. Macroconidia of *Sporidesmium sclerotivorum* and *Teratosperma oligocladum* germinate on sclerotia, penetrate the rind through in between cells and proliferate beneath the surface. In the meduallary region of the sclerotia, hyphae of the mycoiparasites branch and grow intercellularly. Sclerotial cells are not penetrated. The mycoparasites grow out on the sclerotium surface and sporulate. *S. sclerotivorum* grows optimally at pH 4.5–5.5. In areas where soil pH is raised to 7 for prevention of clubroot (*Plasmodiophora brassicae*) the biocontrol does not work efficiently.

The hyperparasitism of *Coniothrium minitans* on structures of *Sclerotinia sclerotiorum* is through production of exo-β-1, 3-glucanase. Expression of the gene encoding this enzyme in the mycoparasite increases during parasitization of sclerotia. The mycoparasite is stimulated by exudates from sclerotia and by metabolites of *Sclerotinia* mycelium. In *Sclerotinia minor*, optimum temperature for colonization of sclerotia by the mycoparasite is 14 to 22°C, but declines at temperatures above 28°C. Infected sclerotia provide a unique reservoir for the survival of *C. mitans*. Within 7 days of infection of sclerotia pycnidia of the mycoparasite develop and within 14 days the pycnidia exude conidia on the surface. The pycnidia and dried intact droplets of conidia serve as survival structures. Conidia remain viable for at least 10 months (Bennett *et al.,* 2005) In sunflower where *S. sclerotiorum* causes infection of the head and forms sclerotia, spay of conidia of *C. minitans* parasitizes the sclerotia. Usually such rotting heads fall on the ground and add heavy dose of the mycoparasite to the soil. Seed application of *C. minitans* or application to sclerotia can completely suppress apothecia production and destruction of sclerotia in the soil. The hyphae of *S. sclerotiorum* parasitized by *C. minitans* shrink and collapse. The cells of sclerotia are destroyed. The mycoparasite has the potential to keep contamination of soil with sclerotia low in crop rotation with susceptible crops when applied as spray to crop infected with the pathogen. Infection of sclerotia can be achieved by a single conidium. Sclerotia that have been exposed to temperatures above 80°C and when the soil containing sclerotia is pasteurized or solarized, infection and colonization of sclerotia by *C. minitans* is early and rapid provided the mycoparasite is applied to soil immediately after treatment before recolonization by other microorganisms occurs. (Bennett *et al.,* 2005). Temperature range for conidial germination and pycnidial production of *C. minitans* is between 10–25°C with the optimum at 20°C. This temperature and relative humidity of >95% are optimum for germination, growth and infection of sclerotia by *C. minitans.* Hyphal extension occurs over a greater temperature range between 4 and 25°C with maximu growth occurring at 20–25°C. Under optimum condition two conidia cause infection of 63% sclerotia compared to 90% with 1000 conidia. The best time of spray of the mycoparasite in the field is when infection of the host plant is first visible. The ascospores of *Sclerotinia* and conidia of *C. minitans* both germinate better in presence of senescent petals and pollen than in plain water (< 0.5%). The epidemics being dependent on availability and prior colonization of these two substrates, recolonization by the mycoparasite and its establisment before the colonization by the pathogen is essential for success of biocontrol (Li *et al.,* 2003). Although *C. minitans* can approach healthy sclerotia from colonized sclerortia in close proximity through its conidia and hyphae, soil mesofauna (mites and collembolans) are more important sources of its transmission. Mycoparasitism of *C. minitans* on *S. sclerotiorum* is not the only mechanism of suppression of the pathogen. The mycoparasite degrades oxalic acid, pathogenicity factor of *S. sclerotiorum*, thus reducing its pathogenic capability. Degradation of oxalic acid changes the ambient pH which enhances production of β-1, 3-glucanase by the mycoparasite (Ren *et al.,* 2007). Fro, coponized sclerotia spores of C, minitans can be splash-dispersed by rain drops or water drops of sprinkler irrigation.

Temperature and moisture conditions that influence apothothecial formation influence the mycoparasitic activity also. Inoculum type and concentration and timing of application of the mycoparasite, time of the year and soil temperature are important factors in reducing sclerotial viability and apotyhecial formation by *C. minitans* (Jones *et al.*, 2003, 2004a, b). Maize meal-perlite culture of the mycoparasite is more effective than conidial suspension. Huang and Erickson (2007) have used a liquid medium containing ground sclerotia of *S. sclerotiorum* for mass production of conidia of *C. minitans*. High temperatures delay sclerotial germination or inhibit it and also inhibit infection of sclerotia by *C. minitans*. The control of *Sclerotinia* by *Coniothyrium* is mainly through reduction in the number of sclerotia. It is not effective in preventing secondary spread of the disease by root-to-root contact.

C. *minitans* produces antifungal substances in culture media where its host, *S. sclerotiorum*, is not present. Results of a study by Yang *et al.*, (2007) have shown that mycelial growth of *Sclerotinia sclerotiorum* is reduced by 41.6% and 84.5% on 3 day-old cultures grown on potato dextrose agar amended with 10% (v/v) of culture filtrate of *C. minitans* grown on modified Czapek-Dox agar after incubation for 6 days and 15 days, respectively. In addition to retardation of growth, morphological abnormalities such as hyphal swellings and cytoplasm granulation also occur. Sclerotia of *S. sclerotiorum* soaked in the filtrate for 24 hours remained viable, but their ability to undergo myceliogenic germination on potato dextrose agar is delayed. Germination of ascospores is unaffected but the germ tubes are shortened and deformed.

Whipps and Budge (1993) had reported the role of the collembolans and *Bradysia* sp. in transmission of the mycoparsite in the soil. The mite (*Acarus siro*) and the collembolan (*Folsomia candida*) can transmit the mycoparasite to a distance of 55 mm from colonized sclerotia in wet or dry soils. In soils lacking these orthropods there is negligible spread of the mycoparasite. The conidia ingested by these orthropods are excreted in faeces and intact conidia are viable. Application of the mycoparasite through aerial spray of conidia (100^6 conidia/ml) alone or with reduced dosage of benomyl was found effective against the pathogen. The frequency of colonization of sclerotia by the mycoparasite was 23 to 56% (Li *et al.*, 2006).

The antagonistic yeast *Ulocladium atrum* is an effective biocontrol agent of *S. sclerotiorum* and its efficacy is similar to that of *C. minitans* (Li *et al.*, 2003). Combination of *U. atrum* and benomyl or vinclozolin has been suggested for control of *S. sclerotiorum* Control of stem rot of chickpea caused by *S. sclerotiorum* by seed treatment with mycelial preparation of *Trichoderma harzianum* and field application of the mycelial preparation at the rate of 200 g per sq meter has been suggested. Soil application of and seed treatment with *Trichoderma harzianum* and *T. viride* have given encouraging results in managing white rot of pea. Wheat or maize bran in combination with FYM were used by Kapoor *et al.*, (2002) for mass production of the antagonists. Good control of Sclerotinia rot mustard by seed treatment and foliarb spray of *T. viride* and onion bulb extract is reported. *T. harzianum* causes degradation of cell walls of *S. sclerotiorum* through chitinase and glucanase enzyme activity. Soil application reduces number of germinated sclerotia and improves survival of soybean plants (Menedez and Godeas, 1998). The interaction between *Trichoderma harzianum* and *S. sclerotiorum* in dual cultures in sterilized soil has shown hyphae of the antagonist growing towards and coiled around the hyphae of *S. sclerotiorum* causing degradation of cell walls of the latter. In sterile soil, conidia of *T. harzianum* germinated and the developing mycelium made contact with the mycelium of *S. sclerotiuorum*, forming short branches and appressorium-like bodies which aided in holding and penetrating the host cell wall. Physically damaged sclerotia are more easily and effectively colonized by antagonists (*T. hamatum*) than intact sclerotia. Escande *et al.*, (2002) had demonstrated field application of *Trichoderma* using honeybees as vectors for control head rot of sunflower.

Talaromyces flavus is a destructive hyperparasite of S. sclerotioruim. In dual cultures, hyphae of T. flavus grow toward and coil around the hyphae of S. sclerotiorum. The coiling intensifies as the hyperparasitic hyphae branch repeatedly on the host surface. There may be direct penetration of host hyphal cells also. The result of parasitism is granulation of cytoplasm and collapse of Sclerotinia cells. Huang et al., (2000) conducted field trials to evaluate effectiveness of five biocontrol agents for the control of white mold of dry beans. The agents were Epicoccum purpurascens, Coniothyrium minitans, Talaromyces flavus, Trichothecium roseum and. Trichoderma virens. The disease was significantly reduced by spray of conidial suspension of all the biocontrol agents. C. minitans was the only biocontrol agents recovered consistently from sclerotia and diseased seed present in the harvested samples. It was found to be the most promising biocontrol agent. Sarrocco et al., (2006) have reported a histopathological study of sclerotia of Sclerotinia sclerotiorum and Sclerotium rolfsii colonized by Trichoderma virens. In colonization of sclerotia of S. sclerotiorum by the mycoparasite, its hyphae grew intercellularly in the medulla. Uniform distribution of mycelium of T. virnes just beneath the rind suggested that sclerotia became infected at numerous randomly distributed locations without any preferential point of entry. Penicillium oxalicum is a strong antagonist of many plant pathogenic fungi includimg Sclerotinia sclerotiorum Culture filtrate containing antifungal substances produced by it can suppress mycelial growth and infection even at 10-fold dilution (Yang et al., 2008).

Pythium oligandrum is another potential biocontrol agent of S. sclerotiorum. It colonizes the sclerotia and reduces their germinability through production of enzymes glucosaminidase, enbdochitinase, protease, glucanase, glucosidase and cellobiohydrolase. This mycoparasite grows bigorously in loose type of sclerotia such as Botrytis cinerea permeating throughout the sclerotial structure. Its growth appears to be restricted in tuberoid sclerotia (viz. Sclerotinia minor) where sclerotial cells constitute a harsh environment for the mycoparasite (Rey et al., 2005). Cell walls of P. oligandrum contain elicitor-like proteins. In sugarbeet, these proteins are reported to elicit defense responses. In addition to being a mycoparasite feeding on sclerotial cells, P. oligandrum is reported to have elicitin-like proteins in its cell wall which elicit defense responses from the plant host cells. A strain of Fusarium oxysporum has been found to suppress S. sclerotiorum through production of cyclosporine A which causes growth inhbition and suppression of sclerotia formation in the pathogen (Rodriguez et al., 2006).

Since Sclerotinia and Botrytis depend on oxalic acid as the pathogenicity factor, bacteria capable of degrading oxalic acid can be effective biocontrol agents. Such bacteria have been isolated from cultivated soil. Such bacteria degrade and assimilate oxalic acid. Transgenic tomato containing genes for oxalate oxidase (OxO) is reported. Such lines have reduce symptoms of Botrytis cinerea. Pseudomonas fluoprescens effectively control Sclerotinia wilt of sunflower. Strains of the soil bacterium Serratia plymuthica are potential biocontrol agents of Sclerotinia sclerotiorum and many other filamentous fungi. Serratia plymuthica suppresses apothecial formation through a chlorinated compound that is capable of inhibiting ascospore germination also (Thaning et al., 2001). Strain A153 is reported to produce mycotoxic metabolites that cause complete suppression of apothecium formation and ascospore germination. Bacillus subtilis strain Tu-100 is reported to significantly reduce the incidence of Sclerotinia stem rot of oilseed rape and promote plant growth (Hu et al., 2005). Bacteria isolated from canola and soybean plants produce organic volatile compounds (bezothiazole, cyclohexanol, n-decanal, dimethyl trisulfide, ethyl hexanol and nonanal) which inhibit sclerotia and ascospore germination and mycelial growth of Sclerotinia sclerotiorum (Fernando et al., 2005). Spray of Pseudomonas chlororaphis (PA-23) or Bacillus amyloliquifaciens (BS 6) was found to give control of stem rot of canola comparable with iprodione (Rovral) in a study by

Fernando *et al.*, (2006). The treatment triggered increased levels of hydrolytic enzymes including chitinase and beta-1,3-glucanase and expression of the pathogenesis-related proteins.

Monacrosporium janus is a nematode-trapping hyphomycete fungus recently identified in China. It is also a parasite of *Sclerotinia sclerotiorum sclerotia* with colonization frequencies of 10% and 33%. In addition, it coils around the hyphae of *R. solani* and grows along appressed to hyphae of *Fusarium solani* f.sp. *pisi, S. sclerotiorum* and *Phytophthora cactorum in vitro*. Larvae of *Bradysia coprophila* (fungus gnats) feed on sclerotia of *S. sclerotiorum*. Damage to sclerotia increases as the organic matter content of soil increases. The sclerotia grazed by the larvae become susceptible to attack of *Trichoderma viride*. The salivary deposit of larvae contains chitinase which decreases germinability of sclerotia. One larva produces approximately 10 μg of chitinase before pupation. Myceliogenic eruptive germination of sclerotia is increased by 1.5 μg/ml of the protein but 50, 100 or 150 μg/ml of the protein decrease the myceliogenic germination. When *Trichoderma hamatum* is combined with the larvae, there is more inactivation of sclerotia than either of them alone.

Transferable hypovirulence with double stranded RNA is known to occur in many species of *Sclerotinia* (Boland, 2004) and may be exploited for biological control. Hypovirulence in fungal plant pathogens refers to the reduced ability of the isolate within a population of the pathogen, to infect, colonize, kill, and/or reproduce on susceptible host tissues. It is associated with fungal viruses and other double-stranded RNA elements and is common in many fungi especially species of *Sclerotinia*. At sub-cellular level, sclerotia of hypovirulent strains of *S. sclerotiorum* contain fewer protein bodies. Both sclerotia and hyphae display more granular appearance throughout the cytoplasm and a number of small double membrane bodies are present in the granular cytoplasm. Such structural changes are not seen in virulent strains. In some fungi, the RNA elements have been shown to affect conidia production and protease activity. It was believed that hypovirulence is not due to a typical mycovirus but due to these double membrane bodies which may be unencapsulated mycovirus or virus-like agent. Hypovirulence associated dsRNA with identical nucleotide sequence naturally occur in two taxonomically distinct fungi and indication are that horizontal transmission of the dsRNA virus may have occurred between these fungi. Deng and Boland (2006) have reported that *Ophiostoma mitovirus* 3A in *Sclerotinia homeocarpa* causes attenuation of virulence of the pathogen. Some of the mitoviruses are disruptive and cause mitochondrial disruption. The loss of ability to secrete oxalic acid is also implicated in hypovirulence in virus contaminated isolates.

Two to 4-year rotation with such crops as beet, onion, spinach or maize, which are not attacked by this species, is recommended in many areas against blight. In soybean, crop rotation and no-tillage of soybean was the most useful combination of treatments to reduce primary inoculum of apothecia. In a study of lettuce drop, caused by *Sclerotinia minor,* Hao *et al.*, (2003) found that the density of sclerotia in soil was lowest in lettuce-fallow-lettuce rotaion and highest in successive crops of lettuce. Lettuce rotated with broccoli also decreased sclerotia density in soil. The number of broccoli crops rather than the sequence of lettuce rotation with broccoli was critical for reducing the number of sclerotia in soil. In the same disease soil surface application of finely powdered calcium hydroxide inhibits sclerotial germination and infection at the collar region of lettuce plants. Complete disease suppression is reported with 10 t calcium hydroxide/ha (Wilson *et al.*, 2005). In areas where infection of roots occurs by mycelium from sclerotia, plant spacing of at least 15 cm reduces infection. White mold of bean is less severe in wider rows (60–80 cm) than narrow rows (25–30 cm).

Fungicides have been used against *S. sclerotiorum* in various crops through foliar sprays, soil drench, seed treatment, chemigation and fumigation. However, chemical control is generally not effective. Spray of ziram, ferbam, or systemic fungicides of benzimidazole group (such as Bavistin) have proved effective in checking spread of the disease. *In vitro* growth of *S. sclerotiorum* is completely suppressed by 20 mg/ml carbendazim or metalaxyl + mancozeb. Carbendazim inhibits germination of sclerotia also. In gray mold and white mold diseases of bean benomyl and vinclozolin had been recommended as effective treatments. A combination of Rovral and Topsin M is an effective alternative of these two fungicides, Tetrasodium thiocarbamate and the dicarboximide fungicides such as procymidone, vinclozolin and iprodione are highly effective against *Sclerotinia*. PCNB had been highly effective as soil treatment. Foliar application of fungicides is mainly to protect the plant tissue from infection by ascospores. It is effective only when the senescent tissues such as flower petals are covered by the chemical. However, the chemical treatments alone are not adequate. The best method is to reduce the number of sclerotia in the soil. For this the cultural practices mentioned above should be given priority. Induction of resistance to *S. sclerotiorum* in melons by seed treatment with plant resistance activators acibenzolar-S-methyl and methyl jasmonate is reported by Buzi *et al.,* (2004). Treatment of the lowermost leaf of oilseed rape with 20 mm oxalic acid induces systemic disease resistance in the three next youngest leaves. Resistance is not caused by translocation of fungitoxic concentrations of oxalic acid. Resistance is also expressed in the stem where vertical spread of lesions is halted. Foliar spray of oxalic acid, zinc sulphate or sodium malonate is reported to reduce mortality of chickpea plants caused by *S. sclerotiorum*. Spray of zinc sulphate at 1000 mm reduced mortality from 100% to 13.6% (Sharma *et al.,* 2007).

REFERENCE

Akhkha, A., D.D. Clarke and P.J. Dominy. 2003. Relative tolerance of wild and cultivated barley to infection by *Blumeria graminis* f.sp. *hordei* (Syn. *Erysiphe graminis* f.sp. *hordei*). II. The effects of infection on photosynthesis and respiration. *Physiol. Mol. Plant Pathol.* 62(6): 347

Alaphilippe, A., Y. Elad, D.R. David, S. Derridj and C. Gessler. 2008. Effects of a biocontrol agent of apple powdery mildew (*Podosphaera leucotricha*) on the host plant and on non-target organisms: an insect (*Cydia pomonella*) and a pathogen (*Venturia inaequalis*) *Biocontrol Sci. Technol.* 18(2): 121

Al-Masri, M.L., M.S. Ali-Shtayeh, Y. Elad *et al.,* 2002. Effect of plant growth regulators on white mould (*Sclerotinia sclerotiorum*) on bean and cucumber, *J. Phytopathol.* 150(8-9): 481

Al-Masri, M.J.,Y. Elad, A. Sharon and R. Barakat. 2006. Ethylene production by *Sclerotinia sclerotiorum* and influence of exogenously applied hormone and its inhibitor aminoethoxyvinylglycine on white mold. *Crop Protection* 25(4): 356

Armstrong-Cho, C., B.D. Gossen and G. Chongo. 2004. Impact of continuous and interrupted leaf wetness on infection of chickpea by *Ascochyta rabiei*. *Can. J. Plant Path.* 26(2): 134

Armstrong-Cho, C. and B.D. Gossen. 2005. Impact of glandular hair exudates on infection of chickpea by *Ascochyta rabiei*. *Can. J. Bot.* 83(1): 22

Azam, M.G.N., G.M. Gurr and P.A. Magarey. 1998. Efficacy of a compound based on canola oil as a fungicide for control of grapevine powdery mildew caused by *Uncinula necator*. *Aust. Plant Pathol.* 27: 116

Bahti, P. and R.N. Strange. 2004. Chemical and biochemical reactions of solanapyrone A, a toxin from the chickpea pathogen, *Ascochyta rabiei* (Pass.) Labr. *Physiol. Mol. Plant Pathol.* 64(1):9

Belanger, R.R., N. Benhamou and J.G. Menzies. 2003. Cytological evidence of an active role of silicon in wheat resistance to powdery mildew (*Bluemria graminis* f.sp. *tritici*). *Phytopathology* 93(4): 402

Bennett, A.J., C. Leifert and J.M. Whipps. 2005. Effect of combined treatment of pausetrization and *Coniothyrium minitans* on sclerotia of *Sclerotinia sclerotiorum*. *Eur. J. Plant Pathol.* 113(2): 197

Bennett, A.J., C. Leifert and J. Whipps. 2006. Survival of *Coniothyrium minitans* associated with sclerotia of *Sclerotinia sclerotiorum*. *Soil Biol. Biochem.* 38(1): 161

Betiol, W. 1999. Effectiveness of cow's milk against zucchini squash powdery mildew (*Sphaerotheca fuliginea*) in greenhouse conditions. *Crop Protection* 18(8): 489

Billon-Grand, G., N. Pousserean and M. Fevre. 2002. The extracellular protease secreted *in vitro* and in planta by the phytopathogenic fungus *Sclerotinia sclerotiorum*. *J. Phytopath.* 150(7–8): 507

Boland, G.J. 2004. Fungal viruses, hypovirulence, and biological control of *Sclerotinia* species. *Can. J. Plant Path.* 26(1): 6

Bolar, J.P., J.L. Norelli, K.-W, Wong *et al.,* 2000. Expression of endochinase from *Trichoderma harzianum* in transgenic apple increases resistance to apple scan and reduces vigor. *Phytopathology* 90: 72

Bolton, M.V.D., B.H.J. Thomma and B.D. Nelson. 2006. *Sclerotinia sclerotiorum* (Lib.) de Bary: Epidemiology and molecular traits of a cosmopolitan pathogen. *Nol. Plant Pathol.* 7(1):1

Boneti, J.L.S. and Y. Katsurayama. 1999. Chemical control of apple scab under conditions of prolonged leaf wetness. *Fitopat. Bras.* 24: 31

Boyle, C. and D.R. Walters. 2006. Saccharin-induced protection against powdery mildew in barley: effects on growth and phenylpropanoid metabolism. *Plant Pathology* 55(1):84

Buzi, A., G. Chilosi, D. De Silva and P. Magro. 2004. Induction of resistance in melon to *Didymella bryoniae* and *Sclerotinia sclerotiorum* by seed treatments with acibenzolar-S-methyl and methyl jasmonate but not with salicylic acid. *J. Phytopath.* 152(1): 34

Carisse, O. and D. Rolland. 2004. Effect of timing of application of the biological control agent *Microsphaeropsis ochracea* on the production and ejection pattern of ascospores by *Venturia inaequalis. Phytopathology* 94(12): 1305

Carroll, J.B. and W.F. Wilcox. 2003. Effects of humidity on the development of grapevine powdery mildew. *Phytopathology* 93(9): 1127

Chen, Y., W.G.D. Fernando. 2006. Induced resistance to blackleg (*Leptosphaeria maculans*) disease of canola (*Brassica napus*) caused by a weakly virulent isolate of *Leptosphaeria biglobosa. Plant Dis.* 90(8): 1059

Chongo, G., L. Buchwaldt, B.D. Gossen, G.P. Lafond, W.F. May, E.N. Johnson and T. Hogg. 2003. Foliar fungicides to manage Ascochyta blight (*Ascochyta rabiei*) of chickpea in Canada. *Can. J. Plant Pathol.* 25(2): 135

Cisp, P., T.J. Wicks, G. Troiup and E.W. Scott. 2006. Mode of action of milk and whey in the control of grapevine powdery mildew. *Aust. Plant Pathol.* 35(55): 487

Clarkson, J.P., J. Staveley, E. Phelps, C.S. Young and J.M. Whipps. 2003. Ascospore release and survival in *Sclerotinia sclerotiorum. Mycol. Res.* 107(2): 213

Cohen, R., A. Hanan and H.S. Paris. 2003. Single-gene resistance to powdery mildew in zucchini squash (*Cucurbita pepo*). *Euphytica* 130(3): 433

Curry, E., Z. Ju and Y. Duan. 2005. Apple scab management assistaed by postharvest tree defoliation with vegetable oil emulsion. *Hort Technology* 15: 736

Deng, F. and G.J. Boland. 2006. Attenuation of virulence in *Sclerotinia homeocarpa* during storage is associated with latent infection by *Ophiostoma mitovirus 3a. Eur. J. Plant Pathol.* 114(2): 127

Dhaliwal, H.S., T.S. Thind, C. Mohan and B.R. Chhabra. 2002. Activity of some essential oils against *Uncinula necator* causing powdery mildew of grapevine. *Indian Phytopath.* 55(4): 329

Dik, A.J., M.A. Verhaar and R.R. Belanger. 1998. Comparison of three biological control agents against cucumber powdery mildew (*Sphaerotheca fuliginea*) in semi-commercial-scale glasshouse trials. *Eur. J. Plant Pathol.* 104: 413

Dugan, F.M., S.L. Lupien, M. Hernandez-Bello *et al.,* 2005. Fungi resident in chickpea debris and their suppression of growth and reproduction of *Didymella rabiei* under laboratory conditions. *J. Phytopath.* 153(7-8): 431–439

Duncan, R.W., W.G.D. Fernando and K.Y. Rashid. 2006. Time and burial depth influencing the viability and bacterial colonization of sclerotia of *Sclerotinia sclerotioru Soil Biol. Biochem* 38(2): 275

English-Loeb, G. and A.P. Norton. 2007. Biological control of grape powdery mildew using mycophagous mites. *Plant Dis.* 91(4): 421

Escande, A.R., F.S. Laich and M.V. Pedra. 2002. Field testing of honeybee dispersed *Trichoderma* spp. to manage sunflower head rot (*Sclerotinia sclerotiorum*). *Plant Pathology* 51(3): 346

Faize, M., M. Malnoy, F. Dupuis *et al.,* 2003. Chitinase of *Trichoderma atroviride* induce scab resistance and some metabolic changes in two cultivars of apple. *Phytopathology* 93(12): 1496

Faize, M., S. Sourice, F. Dupuis, L.Parisi, M.F. Gautier and E. Chevreau. 2004. Expression of wheat puroindoline-B reduces scab susceptibility in transgenic apple (*Malus x domestica* Borkh.). *Plant Science* 157(2): 347

Faize, M., T. Sugiyama, L. Faize and H. Ishii. 2004. Polygalacturonase-inhibiting protein (PGIP) from Japanese pear: possible involvbement in resistance against scab. *Physiol. Mol. Plant Pathol.* 63(3): 319

Faoro, F., D. Maffi, D. Cantu and M. Iriti. 2008. Chemical-induced resistance against powdery mildew in barley: the effects of chitosan and benzothiadazole. *BioControl* 56(2): 381

Farber, R.B.K., K.M. Chin and N. Leadbitter. 2002. Sensitivity of *Venturia inaequalis* to trifloxystrobin. *Pest Manage. Sci.* 58(3): 261

Fawe, A., M. Abou-Zaid, J.G. Menzies and R.R. Belanger. 1998. Silicon-mediated accumulation of flavonoid phytoalexins in cucumber. *Phytopathology* 88: 396

Ferrandino, F.J. and V.L. Smith. 2007. The effect of milk-based foliar sprays on yield components of field pumpkins with powdery mildew. *Crop Prot.* 26(4): 657

Ferraz, L.C.L., B. Filho *et al.,* 2003. Viability of *Sclerotinia sclerotiorum* after solarization in the presence of crop mulch. *Fitopat. Bras.* 28(1): 17

Fetch Jr., T.G., B.J. Steffenson and E. Nevo. 2003. Diversity and sources of multiple disease resistance in *Hordeum spontaneum. Plant Dis.* 87(12): 1439

Ficke, A., D.M. Gadoury, R.C. Seem and I.B. Dry. 2003. Effects of ontogenic resistance upon establisment and growth of *Uncinula necator* on grape berries. *Phytopathology* 93(5): 556

Ficke, A., D.M. Gadoury, R.C. Seem, D. Godfrey and I.B. Dry. 2004. Host barriers and response to *Uncinula necator* in developing grape berries. *Phytopathology* 94(5): 438

Frederickson, D.E. and G.N. Odvody. 2003. Inhibition of germination of sphacelial conidia of *Claviceps africana* following treatment of seed-sphacelia admixtures with captan. *Crop Protection* 22(1): 95

Frenkel, O., D. Shtienberg, S. Abbo and A. Sherman. 2007. Sympatric ascochyta complex of wild *Cicer judaicum* and domesticated chickpea. *Plant Pathology* 56(3): 464

Fujita, K., A.J. Wright, A. Meguro, H. Kunoh and T.L. W. Carver. 2004. Rapid pre-germination and germination responses of *Erysiphe pisi* conidia to contact and light. *J. Gen. Plant Pathol.* 70(2): 75

Gadoury, D.M.. R.C. Seem, D.A. Rosenberger, W.F. Wilcvox, W.E. MacHardy and L.P. Berkett. 1992. Disparity between morphological maturity of ascospores and physiological maturity of asci in *Venturia inaequalis. Plant Dis.* 76: 277

Gadoury, D.M., R.C. Seem, A. Ficke and W.F. Wilcox. 2003. Ontogenic resistance to powdery mildew in grape berries. *Phytopathology* 93(5): 547

Gan, Y., T., K.H.M. Suddiqui, W.J. MacLeod and P. Jayakumar. 2006. Management options for minimizing the damage by Ascochyta blight (Ascochyta rabiei) in chickpea (Cicer arietinum L.). *Field Crop Res.* 97(2-3): 121

Gaur, R.B. and R.D. Singh. 1994. Cropping system in relation to chickpea blight (*Ascochyta rabiei*). I. Effect of intercropping on Ascochyta blight spread. *Indian J. Mycol. Pl. Pathol.* 24: 33

Georgioun, C.D., N. Tairis and A. Polycratis. 2002. Production of *b*-carotene by *Sclerotinia sclerotiorum* and its role in sclerotium differentiation. *Mycol. Res.* 105((9): 1110

Gracia-Garza, J.A., S. Neumann, T.V. Vyn and G.J. Boland. 2002. Influence of crop rotation and tillage on production of apothecia by *Sclerotinia sclerotiorum. Can. J. Plant Path.* 24(2): 137

Grove, G.G. 2004. Perennation of *Uncinula necator* in vineyards of Eastern Washington. *Plant Dis.* 88(3): 242

Guevel, H.-H., J.G. Menzies and R.R. Belanger. 2007. Effect of root and foliar applications of soluble silicon on powdery mildew control and growth of wheat plants. *Eur. J. Plasnt Pathol.* 119(4): 429

Guimaraes, R.L. and H.U. Stotz. 2004. Oxalate production by *Sclerotinia sclerotiorum* deregulates guard cells during infection. *Plant Physiology* 136(5): 3703

Guo, X.W. and W.G.D. Fernando. 2005. Seasonal and diurnal patterns of spore dispersal by *Leptsphaeria maculans* from canola stubble in relation to environmental conditions. *Plant Dis.* 89(1): 97

Gupta, S.K., Amita Gupta, K.R. Shyam and R. Bharadwaj. 2001. Morphological characterization and effect of meteorological factors on development of cucumber powdery mildew. *Indian Phytopath.* 54: 311

Hao, J., K.V. Subbarao and J.M. Duniway. 2003. Germination of *Sclerotionia minor* and *S. sclerotiorum* sclerotia under various soil moisture and temperature combinations. *Phytopathology* 93(4): 443

Hegde, Y.R. and K.H. Anahosur. 2000a. Effect of farlse smut of rice on yield components and growth parameters. *Indian Phytopath.* 53: 181

Hegde, Y. and K.H. Anahosur. 2000b. Survival, perpetuation and life cycle of *Claviceps oryzae-sativae*, causal agent of false smut of rice in Karnataka. *Indian Phytopath.* 53: 61

Hernandez-Bello, M.A., M.I. Chilvers, H. Akanatsu and T.L. Peever. 2006. Host specificity of *Ascochyta* spp. infecting legumes of Viciae and Ciceriae tribes and pathogenicity of interspecific hybrids. *Phytopathology* 96(10): 1148

Holb, I. J., B. Heijne and M.J. Jeger, 2004a. Overwintering of conidia of *Venturia inaequalis* and the contribution to early epidemic of apple scab. *Plant Dis.* 88(7): 751

Holb, I.J., B. Heijne, J.C.M. Withagen and M.J. Jeger. 2004b. Dispersal of *Venturia inaequalis* ascospores and disease gradients from a defined inoculum source. *J. Phytopath.* 152(11–12): 639

Holb, I.J., B. Heijne and M.J. Jeger. 2005. The widespread occurrence of overwintered conidial inoculum of *Venturia inaequalis* on shoots and buds in organic and integrated apple orchards across the Netherlands, *Eur. J. Plant Pathol.* 111(2): 157

Hu, X., D.P. Roberts, M. Jiang and Y. Zhang. 2005. Decreased incidence of disease caused by *Sclerotinia sclerotiorum* and improved plant vigor of oilseed rape with *Bacillus subtilis* Tu-100. *Appl. Microbiol. Biotechnol.* 68(6): 802

Huang, H.C. and R.S. Erickson. 2007. Use of sclerotia of *Sclerotinia sclerotiorum* for efficient production of conidia of *Coniothyrium minitans* in liquid culture. *Phytoparasitica* 35(2): 140

Huang, L., H.-Buchenauer, Q. Han, X. Zhamg and Z. Kang. 2008. Ultrastructural and cytochemical studies on the infection process of *Sclerotinia sclerotiorum* in oilseed rape. *J. Plant Dis. Prot.* 115(1)

Ilhan, K., U. Arshan and O.A. Karabulut. 2006. The effect of sodium bicarbonate alone or in combination with areduced dose of tebuconazole on the control of apple scab. *Crop Prot.* 25(9) 963

Jamar, L., B. Lefrancq and M. Lateur. 2007. Control of apple scab (*Venturia inaequalis*) with bicarbonate salts under controlled environment. *J. Plant Dis. Prot.* 114(5)

Jamar, L., B. Lefrancq, C. Fassotto and M. Lateur. 2008. A durimg-infection spray strategy using sulphurcompounds, copper, si;icon, and a new formulation of potassium bicarbonate for primary scab control; in organic apple production. *Eur. J. Plant Pathol/* IN PRESS

Jayakumar, P., B.D. Ggossen, T.D. Warkenttin and S. Banniza. 2005. Ascochyta blight of chickpea: infection and resistance mechanisms. *Can. J. Plant Pathol.* 27(4): 499

Kaiser, W.J. 1997. Inter- and intranational spread of ascochyta pathogens of chickpea, faba bean and lentil. *Can. J. Plant Path.* 19(2): 215

Kaiser, W.J. and L. Kusmenoglu. 1997. Distribution of the mating types and the teleomorph of *Ascochyta rabiei* on chickpea in Turkey. *Plant Dis.* 81: 1284

Kanto, T., A. Miyoshi, T. Ogawa *et al.,* 2004. Suppressive effect of potassium silicate on powdery mildew of strawberry in hydroponics. *J. Gen Plant Pathol.* 70(4):207

Kapoor, A.S., P. Kumar and R. Sharma. 2002. Mass multiplication and viability of *Trichoderma harzianum* vis-à-vis management of white rot of pea. *Indian Phytopath.* 55(3): 374

Kavkova, M. and V. Curn. 2005. *Paecilomyces fumosoroseus* (Deuteromycotina: Hyphomycets) as a potattial mycoparasite on *Sphaerotheca fuliginia* (Ascomycotina: Erysiphales). *Mycopathologia* 159(1): 53

Kim, J.J., M.S. Goettel and D.R. Gillespie. 2008.Evaluation of *Lecanicillium longisporum,* Vertalec for simultaneous suppression of cotton aphid, *Aphis gossypii,* and cucumber powdery mildew, *Sphaerotheca fuligineam* on potted cucumbers. *Biological Control* 45(3): 404

Kim, K.S., J-Y. Min and M.B. Dickman. 2008. Oxalic acid is an elicitor of plant programmed cell death during *Sclerotinia sclerotiorum* disease development. *Mol. Plant Microbe Interact.* 21(5): 605

Kimber, R,B.E., E.S. Scott and M.D. Ramsey. 2006. Factors influencing transmission of *Didymella rabiei* (Ascochyta blight) from inoculated seed chickpea under controlled conditions. *Eur. J. Plant Pathol.* 114(2): 175

Kiss, L., J.C. Russell, O. Szentivanyi, X.Xu and P. Jeffries. 2004. Biology and biocontrol potential of *Ampelomyces* mycoparasites, natural antagonists of powdery mildew fungi. *Biocontrol Sci. Technol.* 14(7): 635

Koller, W., D.M. Parker, W.W. Tureehek, C. Avila-Adame and K. Cronshaw. 2004. A two phase resistance response of *Venturia inaequalis* populations to the QoI fungicides kresoxim-methyl and trifloxystrobin. *Plant Dis.* 88(5): 537

Koller, W., W.F. Wilson and D.M. Parker. 2005. Sensitivity of *Venturia inaequalis* populations to anilopyrimidine fungicides and their constribution to scab management in New York. *Plant Dis.* 89(4): 357

Li, B. and X. Xu. 2002. Infection and development of apple scab (*Venturia inaequalis*) on old leaves. *J. Phytopath.* 150(11–12): 687

Li, G.Q., H.C. Huang and S.N. Acharya. 2002. Sensitivity of *Ulocladium atrum, Coniothyrium minitans* and *Sclerotinia sclerotiorum* to benomyl and vinclozolin. *Can. J. Bot.* 80(8): 892

Li, G.Q., H.C. Huang and S.N. Acharya. 2003. Antagonism and biocontrol potential of *Ulocladium atrum* on *Sclerotinia sclerotiorum. Biological Control* 28(1):11

Li, G.Q.,H.C.Huang, H.J. Miao, R.S. Erockson, D.R. Jiang and Y.N. Xiao. 2006. Biological control of Sclerotinia disease of rapeseed by aerial application of the mycoparasitic *Coniothyrium minitans. Eur. J. Plant Pathol.* 114(4): 345

Li, H., N. Tapper, N. Dean, M. Barbetti and K. Sivasithamparam. 2006. Enhanced pathogenicity of *Leptosphaeria maculans* pycnidiospores from paired co-inoculation of *Brassica napus* cotyledons with ascospores. *Ann. Bot.* 97(6): 1151

Li, H., J. Kuo, M.J. Barbetti and K. Sivasithamparam. 2007. Differences in the responses of stem tissues of spring-type *Brassica napus* cultivars with polygenic resistance and sigle dominant gene-based resistance to inoculation with *Leptosphaeria maculans. Can. J. Bot.* 85(2): 191

Li, H., K. Sivasithamparam, M.J. Barbetti, S.J. Wylie and J. Kuo. 2008. Cytological responses in the hypersensitive reaction in cotyledon and stem tissues of Brassica napus after infection by *Leptosphaeria maculans. J. Gen. Plant Pathol.* 74(2): 120

Liang, Y.C., W.C. Sun, J. Si and V. Rumbold. 2005. Effect of foliar- and root-applied silicon on the enhancement of induced resistance to powdery mildew in *Cucumis melo. Plant Pathology* 54(5): 879

Madsen, A.M. and E. de Neergaard. 1999. Interactions between the mycoparasite *Pythium oligandrum* and sclerotia of the plant pathogen *Sclerotinia sclerotiorum. Eur. J. Plant Pathol.* 105: 761

Malnoy, M., M. Xu, *et al.,* 2008. Two receptor-like genes, Vfa1 and Vfa2, confer resistance to the fungal pathohen *Venturia inaequalis* inciting apple scab disease. *Mol. Plant Microbe Interact.* 21 (4): 448

Matheran, M.E. and M. Porchas. 2005. Influence of soil temperature and moisture on eruptive germination and viability of sclerotia of *Sclerotinia minor* and *S. sclerotiorum. Plant Dis.* 89(1): 50

McGrath, M.T. and N. Shishkoff. 1999. Evaluation of biocompatible products for managing cucurbit powdery mildew. *Crop Protection* 18(7): 471

McGrath, M.J. 2001. Fungicide resistance in cucurbit powdery mildew: Experiences and Challenges. *Plant Dis.* 85: 236

Melidossian. H.S., R.C. Seem. W.F. Wilcox and D.M. Gadoury. 2005. Suppression of grapevine powdery mildew by a mycophagous mite. Plant Dis. 89(12): 1331

Mila, A.L. and X.B. Yang. 2008. Effect of fluctuating soil temperature and water potential on sclerotia germination and apothecia production of Sclerotinia sclerotiorum/ Plant Disease 92(1): 78

Mondal, S.N. and L.W. Timmer, 2003. Effect of urea, CaCO(3), and dolomite on pseudothecial development and ascospore production of *Mycospherella citri. Plant Dis.* 87(5): 478

Monteiro, S., M. Barakat, M.A. Picarra-Pereira *et al.,* 2003. Osmotin and thaumatin from grape: A putative general defence mechanism against pathogenic fungi. *Phytopathology* 93(12): 1505

Mueller, D.S., G.L. Hartman and W.L. Pederson. 1999. Development of sclerotia and apothecia of *Sclerotinia sclerotiorum* from infected soybean seed and its control by fungicide seed treatment. *Plant Dis.* 83: 1113

Nelson, H.E. 2005. *Fusarium oxysporum* f.sp. *radicis-lycopersici* can induce systemic resistance in barley against powdery mildew. *J. Phytopath.* 153(5): 360

Nielson, K.A., R.L. Nicholson, T.L.W. carver, H. Kunoh and R.P. Oliver. 2000. First Touch: An immediate response to surface recognition in conidia of *Bluemeria graminis. Physiol. Mol. Plant Pathol.* 56(2): 63

Olaya, G. and W. Koller. 1999. Baseline sensitivities of *Venturia inaequalis* populations to the strobilurin fungicide kresoxim methyl. *Plant Dis.* 83: 278

Opalski, K.S., S. Trisech, K.-H. Kogel *et al.,* 2006. Metrafenone: studies on mode of action of a novel cereal powdery mildew fungicide. Pest Manag. Sci. 62(5): 303.

Pandey, K., U.S. Singh and H.S. Chaube. 1997. Mode of infection of Ascochyta blight of chickpea caused by *Ascochyta rabiei. J. Phytopath.* 119(1): 88

Pathania, N., P.D. Tyagi and A.K. Basandrai. 1997. Pathogenic specialization in *Erysiphe graminis tritici* in wheat. *J. Mycol. Pl. Pathol.* 27: 35

Pazoutova, S., R. Bandopadhyay, D.E. Frederickson. P.G. Mantle and R.A. Frederickson. 2000. Relations among sorghum ergot isolates from the Americas, Africa. India and Australia. *Plant Dis.* 94: 437

Peever, T.L., M.P. Barve, L.J. Stone and W.J. Kaiser. 2007. Evolutionary relationships among *Ascochyta* species infecting wild and cultivated hosts in the legume tribes Cicereae. *Mycologia* 99(1): 59

Prats, E., L.A.J. Mur, R. Sanderson and T.L.W. Carver. 2005. Nitric oxide contributes both to papilla-based resistance and the hypersensitive response in barley attacked by *Blumeria graminis* f.sp. *hordei. Mol. Plant Pathol.* 6(1): 65

Prom, L.K. and T. Isakeit. 2003. Laboratory, greenhouse and field assessment of fourteen fungicides for activity against *Claviceps africana,* causal agent of sorghum ergot. *Plant Dis.* 87(3): 252

Ramarathnam, R. and W.G.D. Fernandez. 2006. Preliminary phenotypic and molecular screening for potential bacterial biocontroil agents of *Leptosphaeria maculans. Biocontrol Sci. Technol.* 16(6): 567

Randoux, B., D. Renard, E. Nowak *et al.,* 2006. Inhibition of *Blumeria graminis* f.sp. *tritici* germination and partial enhancement of wheat defenses by Milansa. *Phytopathology* 86(11): 127

Remus-Borel, W., J.G. Menzies and R.R. Belanger. 2005. Silicon induces antifungal compounds in powdery mildew-infected wheat. *Physiol. Mol. Plant Pathol.* 66(3):108

Ren, L., G.Li, Y.C. Han *et al.,* 2007. Degradation of oxalic acid by *Coniothyrium minitans* and its effect on production and activity of B-1,3-glucanase of the mycoparasite. *Biological Control* 43(1): 1

Reuveni, M. 2000. Efficacy of trifloxystrobin (Flint), a new strobilurin fungicides, in controlling powdery mildews on apple, mango and nectarines, and rust on prune trees. *Crop Protection* 19(3): 336

Reuveni, M. and R. Reuveni. 1998. Foliar applications of monopotassium phosphate fertilizers inhibit powdery mildew development in nectarines. *Can. J. Plant Pathol.* 20(3): 253

Reuveni, M, and R. Reuveni. 2000. Prior inoculation with non-pathogenic fungi induces systemic resistance to powdery mildew on cucumber. *Eur. J. Plant Pathol.* 106: 633

Reuveni, M., H. Cohen, T. Zahavi and A. Venezian. 2000. Polar- a potent Polyoxin B compound for controlling powdery mildews in apple and nectarine trees and grapevines. *Crop Protection* 19(6): 393

Rey, P., U.L. Flotch, N. Benhamou, M.L. Salerno *et al.,* 2005. Interactions between the mycoparasite *Pythium oligandrum* and twotypes of sclerotia of plant pathogenic fungi. *Mycol. Res.* 109(7): 779

Rodriguez, M.A., G. Cabrera and A. Godeas. 2006. Cyclosporine A from a nonpathogenic *Fusarium oxysporum* suppressing *Sclerotinia sclerotiorum. J. Appl. Microbiol.* 100(3): 578

Romero, D., A. Perez-Garcia, M.E. Rivera *et al.,* 2004. Isolation and evaluation of antagonistic bacteria towards the cucurbits powdery mildew fungus *Podosphaera fusca. Appl. Microbil. Biotechnol.* 64(2): 203

Rossi, V., M. Bolognesi, L. Languasco and S. Giosue. 2006. Influence of envurinmental conditions on infection of peach shoots by *Taphrina deformans. Phytopathology* 96(2): 155

Rossi, V., M. Bolognesi and S. Giosue. 2007a. Seasonal dynamics of *Taphrina deformans* inoculum in peach orchards. *Phytopathology* 97(3): 352

Rossi, V. and L. Languasco. 2007b. Influence of environmental conditions on spore production and budding in *Taphrina deformans*, the causal agent of peach leaf curl. *Phytopathology* 97(3): 359

Rumbolz, J. and W.D. Gubler. 2005. Susceptinility of grapevine buds to infection by powdery mildew *Erysiphe necator* Plant Pathology 54(4): 535

Ryley, M., S. Bhuiyan, D. Herde and B. Gordan. 2003. Efficacy, timing and method of application of fungicides for management of sorghum ergot caused by *Claviceps africana*. *Aust. Plant Pathol.* 32(3): 329

Sawant, S.D. and I.S. Sawant. 2008. Use of potassium bicarbonates for the control of powdery mildew in table grapes. *ISHS Acta Hortic.* 785. International Symp. on grape production and processing, Pune (India) May 2008

Schuenger, A.C. and W. Hammer. 2003. Suppression of powdery mildew in greenhouse-grown cucumber by addition of silicon to hydroponic nutrient solution is inhibited by high temperature. *Plant Dis.* 87(2): 177

Sendhilvel, V., T. Marimuttu and R. Samiappan. 2007. Talc-based formulation of *Pseudomonas fluorescens*-induced defense genes against powdery mildew of grapevine. *Srch. Phytopath. Plant Prot.* 40(2): 81–85

Sharma, B.K., S.A. Basha, D.P. Singh and U.P. Singh. 2007. Use of non-conventional chemicals as an alternative approach to protect chickpea (*Cicer arietinum*) from *Sclerotinia* stem rot. *Crop Prot* 26(7): 1042

Sharma, S.K. and S.K. Sharma. 2002. Influence of time of sowing and spacing levels on powdery mildew and seed yield in pea cv. Arkel. *J. Mycol. Pl. Pathol.* 32(1): 117

Shoemaker, R.A. and H. Brun. 2001. The teleomorph of the weakly aggressive segregate of *Leptosphaeria maculans*. *Can. J. Bot.* 79(4): 412

Sholberg, P.L., P. Randall and C.R. Hampson. 2005. Acetic acid fumigation of apple rootstock and tree fruit scionwood to remove external microflora and potential plant pathogens. *Hort Technology* 15: 422

Sholberg, P., C. Harlton, J. Boule and P. Haag. 2006. Fungicide and clay treatments for control of powdery mildew ingluence wine grape microflora. *Hort Sci.* 41(1): 176

Shteinberg, D., H. Vintal, S. Brener and B. Retig. 2000. Rational management of *Didymella rabiei* in chickpea by integration of genotype resistance and post-infection application of fungicides. *Phytopathology* 90: 834

Singh, R. and N. N. Tripathi. 1998. Weather factors associated with production and release of ascospores of *Sclerotinia sclerotiorum*. *J. Mycol. Pl. Pathol.* 28: 227

Singh, S.D. and S.S. Navi. 2000. Garlic as a biocontrol agent for sorghum ergot. *J. Mycol. Pl. Pathol.* 30: 350

Singh, U.P., B. Prithviraj and B. Sharma. 2000. Development of *Erysiphe pisi* (powdery mildew) on normal and albino mutants of pea (*Pisum sativum*). *J. Phytopath.* 148: 591

Singh, U.P., A. Bahadur, D.P. Singh and B.K. Sharma. 2003. Non-pathogenic powdery mildews induce resistance in pea (*Pisum sativum*) against *Erysiphe pisi*. *J. Phytopath.* 151(7–8): 419

Sosnowski, M.R., E.S. Scott and M.D. Ramsey. 2005. Temperature. Wetness period and inoculum concentration influence infection of canola (*Brassica napus*) by pycnidiospores of *Leptosphaeria maculans*. *Aust. Plant Pathol.* 34(3): 339

Stadnik, M.J., W. Bettiol and M.L. Saito. 2003. Bioprospecting for plant and fungal extracts with systemic effect to control the cucumber powdery mildew. *J. Plant Dis. Prot.* 110(4): 383

Steed, J.M., A. Baierl and B.D.L. Fitt. 2007. Relating plant and pathogen development to optimize fungicide control of phoma stem canker (*Leptosphaeria maculans*) on winter oilseed rape (*Brassica napus*). *Eur. J. Plant Pathol.* 118(4): 359

Steiner, U. and E-C. Yerke. 2007. Localized melanization of appressoria is required for pathogenicity of *Venturia inaequalis*. *Phytopathology* 97(10): 1222

Sun, P. and X.B. Yang. 2000. Light, temperature and moisture effects on apothecial production of *Sclerotinia sclerotiorum*. *Plant Dis.* 84: 1287

Sprague, Susan J., M. Watt, J.A. Kirkegaard *et al.*, 2007. Pathways of infection of *Brassica napus* roots by *Leptosphaeria maculans*. *New Phytologist* 176(1): 211

Thakur, V. S. and K. Khosla. 1999. Relevance of Mills' infection periods to applr scab (*Venturia inaequalis*) prediction and rescheduling fungicide application in Himachal Pradesh. *Indian J. Agric. Sci.* 69: 152

Toal, E.S. and P.W. Jones. 1999. Induction of systemic resistance to *Sclerotinia sclerotiorum* by oxalic acid in oilseed rape. *Plant Pathology* 48(6): 759

Trapero-Casas, A. and W.J. Kaiser. 2007. Differences between ascospore and conidia of *Didymella rabiei* in spore germination and infection of chickpea. *Phytopathology* 97(12): 1600

Tsuba, M., C. Katagirti, Y. Takeuchi, Y. Takada and N. Yamaoka. 2002. Chemical factors of the leaf surface involved in the morphogenesis of *Blumeria graminis* Physiol. *Mol. Plant Pathol.* 60(2): 51

Turechek, W.W. and W. Koller. 2004. Managing resistance of *Venturia inaequalis* to the strobilurin fungicides. *Online Plant Health Progress* doi:10.1094/PHP-2000–0908–01-RS

Urquhart, E.J. and Z.K. Punja. 2002. Hydrolytic enzymes and antifungal compounds produced by *Tilletiopsis* species, phyllosphere yeasts that are antagonists of powdery mildew fungi. *Can. J. Microbiol.* 48(3): 219

Vincent, C., B. Rancourt and O. Carisse. 2004. Apple leaf shredding as a non-chemical tool to manage apple scab and spotted tentiform leafminer. *Agric. Ecosyst. Environ.* 104(3): 595

Whipps, J.M. and S.P. Budge. 1993. Transmission of the mycoparasite *Coniothyrium minitans* by collembolan *Folsomia candida* (Collembola: Entomobrydae) and glassgouse sciarid *Bradysia* sp. (Diptera: Sciaridae). *Ann. Appl. Biol.* 123: 165

Willocquet, L., F. Berud, L. Raoux and M. Clerjean. 1998. Effect of wind, relative humidity, leaf movement and colony age on dispersal of conidia of *Uncinula necator*, causal agent of grape powdery mildew. *Plant Pathology* 47(3): 234

Wilson, C.R., J.A. de Little, J.A.L. Wong *et al.*, 2005. Adjustment of soil-surface pH and comparison with conventional fungicide treatments for control of lettuce drop (*Sclerotinia minor*). *Plant Pathology* 54(3): 393

Wong, F.P. and W.F. Wilcox. 2002. Sensitivity to azoxystrobin among isolates of *Uncinula necator*: Baseline distribution and relationship to myclobutanil sensitivity. *Plant Dis.* 86(4): 394

Wright, A.J., B.J. Thomasa, H. Kunoh, R.L. Nicholson and T.L.W. Carver. 2002a. Influence of substrata and interface geometry on the release of extracellular material by *Blumeria graminis* conidia. *Physiol. Mol. Plant Pathol.* 61(3): 163

Wright, A.J., B.J. Thomasd and T.L.W. Carver. 2002b. Early adhesion of *Blumeria graminis* to plant and artificial surfaces demonstrated by centrifugation. *Physiol. Mol. Plant Pathol.* 61(4): 217

Wu.,B.M., Y.-L. Peng, Q.-M. Qin and K.V. Subbarao. 2007. Incubation of excised apothecia enhance ascus maturation of *Sclerotinia sclerotiorum*. *Mycologia* 99(1): 33

Wurms, K., C. Labbe, N. Benhomou and R.R. Belanger. 1999. Effects of Milsana and benzothiadiazole on the ultrastructure of powdery mildew haustoria on cucumber. *Phytopathology* 89(9): 728

Xu, X-M. and J. Robinson. 2005 Modelling the effects of wetness duration and fruit maturity on infection of apple fruits of Cox's Orange Pippin and two clones of Gola by *Venturia inaequalis*. *Plant Pathology* 54(3): 347

Yamaoka, N., I. Matsumoto and M. Nishiguchi. 2006. The role of primary germ tubes (PGT) in the life cycle of *Blumeria graminis*: The stopping of PGT elongation is necessary for the triggering of appressorial germ tube (AGT) emergence. *Physiol. Mol. Plant Pathol.* 69(4–6): 153

Yamaoka, N., T. Ohta, N. Danno *et al.*, 2007 The role of primary germ tube in the life cycle of *Blumeria graminis*: The primary germ tube is responsible for the suppression of resistance induction of a host plant cell. *Physiol. Mol. Plant Pathol.* 71(4–6): 186

Yang, L, J, Xie, D. Jiang *et al.*, 2008. Antifungal substances produced by *Penicillium oxalicum* strain PY-1, potential antibiotics against plant pa6hogenic fungi. *World J. Microbiol. Bioitechnol.* 24(7):

Yang, R., Y.C. Han, G.Q. Li, D.H. Jiang and H.C. Huang. 2007. Suppression of *Sclerotinia sclerotiorum* by antifungal substances produced by the mycoparasite *Coniothyrium minitans*. *Eur. J. Plant Pathol.* 119(4): 911

Yang, X.B., F. Workneb and P. Lundeen. 1998. First report of sclerotium production by *Sclerotinia sclerotiorum* in soil on infected soybean seeds. *Plant Dis.* 82: 264

Zabka, V., M. Stangl. G. Bringmann *et al.*, 2008. Host surface properties affect pre-penetration processes in the barley powdery mildew fungus. *New Phytologist* 177(1): 251

Zhang, Z.Y., G.H. Dai, Y.Y. Zhengu and Y.B. Li. 2008. Protective activity of *Robinia pseudoacacia* Linn. Extracts against cucumber powdery mildew fungus *Sphaerotheca fuliginea*. *Crop Protection* 27(6): 920

Diseases Caused by
Basidiomycotina: Uredinales

THE CEREAL RUSTS

In India the wheat crop suffers from three rusts, stem or black rust, stripe or yellow rust, and leaf or brown (orange) rust, caused, respectively, by *Puccinia graminis* var. *tritici, Puccinia striiformis* and *Puccinia recondita*. The stem and yellow rusts occur on barley also. Stem rust of oats, *Puccinia graminis* var. *avenae*, is not very common while crown rust, *Puccinia coronata*, is confined to the northwestern part of the Indian subcontinent. A leaf rust of barley caused by *P. hordei* had also been reported from Delhi and its neighborhoods. The damage to wheat is mainly due to brown rust and yellow rust. Stem rust appears quite late in the main wheat belt and usually the crop escapes much damage.

The annual loss from rusts on wheat and barley has been estimated from 40 to 60 million rupees. Nearly one million tons of wheat were lost during 1958–59 due to rusts. Yellow rust alone caused 14% loss of grain weight in barley. Heavy losses due to rusts were reported from Russia, Canada and other countries.

The damage to plants is chiefly due to excessive loss of moisture from the plants at a time when they require moisture to ripen the grains of normal shape and size. Grains from rusted plants are much shriveled and under-sized. In severe and early attack the loss is also due to premature defoliation and breaking of stems.

Currently, *Puccinia* is placed in the family Pucciniaceae, order Uredinales, class Urediniomycetes and phylum Basidiomycotina.

BLACK OR STEM RUST OF WHEAT

Puccinia graminis tritici Eriks. and Henn. usually appears on the wheat crop late in the season when the temperature starts rising. It is often not seen until March or even later, at a time when the wheat is in ear or the crop is maturing.

Symptoms

The onset of the disease is marked by eruption of elongated, brown pustules on the stalk, leaf sheath, and leaves, the stalk (stem) being often most severely affected. These pustules (uredia) may be about 6 mm or more in length and frequently run into one another. They burst early exposing a brown powder (consisting of the urediospores) and are surrounded by prominent epidermal fringes. Telia develop later in the same sorus as uredia or independently. They are darker in colour than the uredia and they burst through the epidermis in the same manner as the uredia, exposing a black bed of spores. Apart from these changes there is no other marked abnormality in the appearance of the plant. Of course, in severe attack plants look unhealthy and fail to form normal ears. The grains are shriveled and lighter in weight.

The rust fungus profoundly affects the physiology of the host plant. Transpiration is increased. Translocation of carbohydrates is retarded in infected tissues of susceptible wheat varieties. A striking increase in respiratory activity in rusted plants, especially at the infection site, has been reported. This increase begins with the appearance of symptoms, rises to a maximum at the time of sporulation and then declines again. The increase is the combined effect of the respiration of the parasite itself and that of the host tissues. An increase of total nitrogen, protein-nitrogen, soluble nitrogen and the ratio of soluble to insoluble nitrogen occurs as the rust develops. It is assumed that amino acids and proteins are synthesized by the rust fungus itself as it develops on its host.

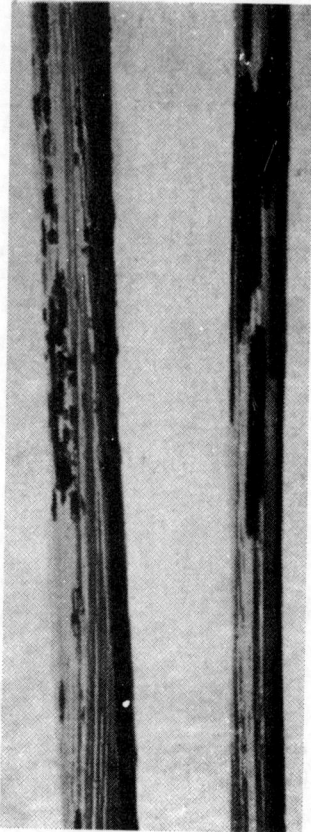

Fig. 71. Black rust of wheat.

The causal organism

Each sorus or rust pustule of *Puccinia graminis tritici* is formed on a limited mycelium resulting from an independent local infection. The hyphae are about 3.5 μm in diameter. They are intercellular and send haustoria into the host cells. A mass of hyphae collects beneath the epidermis and, if early in season, develops into an urediosorus (uredium). From the base of the sorus numerous short, erect stalks arise and a single urediospore is formed at the end of each. The growth in size of these spores ruptures the epidermis and they are set free into the air. Each urediospore is an oval, brown body, measuring 25–30 x 17–20 μm, and consists of a single cell with a thick wall provided with spines. The spore has usually 4 germ pores arranged in an equatorial band. These spores are the repeating spores causing secondary infection of the crop in the same season. Contact between host and pathogen is established when the urediospores are lodged on the surface of the host. Irrespective of whether the host is congenial (compatible) or not (incompatible), the urediospores germinate provided the

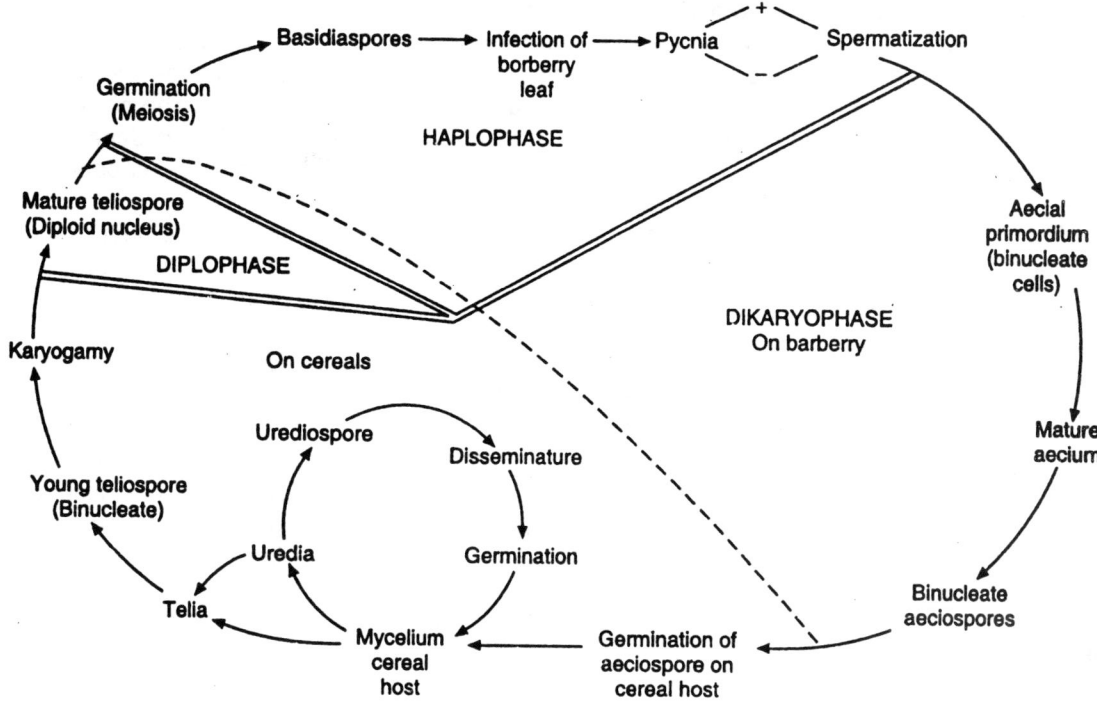

Fig. 72. Life cycle of *Puccinia graminis tritici*.

environmental conditions are favourable. Most important of these are temperature and moisture. If the host is not suitable (non-host or resistant variety) further developments during and after penetration up to haustorium formation stage are checked or slowed down. The germination of urediospores and the development of germ tubes is independent of the substrate, the food material for the process being stored in the spore itself.

Since the germ tube cannot utilize any exogenous source of nutrition it grows only so long as the food reserves in the spore can sustain it and dies if organic relationship with the host is not established soon. It enters the host through stomata. Sometimes 2 germ tubes are produced from different pores of the spores but only the longer one continues its growth. The germ tubes are guided to stomata by topography of leaf surface and emanations from stomata.

On reaching the stomata, growth of the germ tube stops. Its tip swells into an elongated appressorium which lies with its long axis covering the slit-like mouth of a stoma. Thigmodifferentiation (topographical signals) and chemodifferentiation (leaf hexenols) both synergistically induce appressorium formation. When this appressorium has been formed, spore contents, e.g., nuclei, fat particles, pigments, carbohydrates, proteins, organic acids, etc. pass into it leaving the spore as an empty shell. There may be 2 or more appressoria lying over the same stoma. Two appressoria may fuse into one. A small, narrow infection hypha emerges from the appressorium and passing through the stomatal opening, enters the sub-stomatal cavity where a sub-stomatal vesicle is formed. Two such vesicles in a single sub-stomatal cavity may fuse to form one. The cytoplasm from the appressorium passes into the vesicle and the former collapses. Hyphae developing from the sub-stomatal vesicle produce intercellular branches which form pre-haustoria. Probaby, chemical and physical signals are involved in the differentiation of haustorial mother cells of *Puccinia graminis tritici* (Wietholter *et al.,* 2003). Penetration pegs from pre-haustoria reach into the neighboring host cells and form haustoria. These haustoria invade the host protoplasm resulting in the establishment of infection or organic relationship between the host and the parasite. Electron microscopy has shown that haustoria of *P. graminis tritici* are surrounded by an encapsulation. The mycelium continues its growth leading to the development of rust pustules within 10–15 days depending upon the temperature. A single pustule can contain thousands of spores, each capable of initiating a new sorus in another 15 days or less.

The fungus is a biotroph or obligate parasite. Its survival and growth in the host tissue depends on the living condition of the latter. In hypersensitive varieties the death of cells around the infection peg leads to restricted growth of mycelium resulting in minute pustules and less number of urediospores.

Late in the season, the telia arise from the same mycelium. Teliospores differ from urediospores in being firmly fixed to their stalk and are composed of two cells, with a thick and smooth wall. The apex is rounded or pointed. They are chestnut brown in colour and measure 40–46 x 15–20 μm. Each cell has a germ pore, that of the upper being at the apex and that of the lower being at the side just below the septum. The wall of the apex is much thickened. Unlike the urediospore, the teliospore is not capable of immediate germination and must undergo a period of rest for several months. In India, these spores are killed by the hot summer that follows the harvest of the crop.

Puccinia graminis tritici is a polymorphic species, producing a succession of different types of spores. It is heteroecious. Only the uredia and telia are formed on the cereal hosts. The basidiospores are produced on some inactive substrate on which teliospores are present. The pycnia and aecia develop on the alternate hosts, probably species of *Berberis* and *Mahonia* which are not found in the plains of India. In the hills also the aecial stage found on these hosts is not that of *Puccinia graminis tritici*. However, in cold climate countries like the USA and Canada barberry helps in completion of the nuclear cycle of the rust fungus by functioning as a platform on which sexual recombinations occur.

On germination the teliospore produces a 4-celled promycelium from each of its 2 cells. A short branch (sterigma) develops from each cell of the promycelium. It swells at the tip to form a globular basidiospore. These spores are unicellular and uninucleate. Of the 4 basidiospores formed on each promycelium two are of one mating type and the two of the opposite mating type. These spores are violently ejected and are blown about by wind. They are capable of immediate germination in moist air or in water but they can not infect the cereal hosts. They must get the alternate host to develop a mycelium.

When the monokaryotic basidiospore falls on the surface of young barberry leaf it sends out a germ tube which directly penetrates through the epidermis. Different species of *Berberis* show differential response to invasion. The resistance to invasion is correlated with thickness of the cuticle. *Berberis vulgaris* is most susceptible because of the thinness of the cuticle. Inside the barberry leaf tissue, the infection hypha develops into a mycelial mat of uninucleate (monokaryotic) cells. This mat produces pycnia (spermogonia) on the upper surface of the leaf. These flask-shaped bodies have an ostiole protruding out through the cuticle. A number of periphyses are present around the ostiole. Inside these pycnia a large number of spermatia (pycnospores) are produced. These come out through the ostiole in a honey-like fluid. The colour of pycnia and the smell and sweetness of the fluid attract insects which help in dissemination of the unicellular, uninucleate spermatia. These spores and the mycelium from which they develop are all monokaryotic (haploid), belonging to the mating strain to which the basidiospore belonged. Cragie in 1927 had for the first time discovered the role of spermatia in the diploidization of the monokaryotic mycelium of *Puccinia*. Without them the perfect stage (teleomorph) of the fungus could not develop. Later, Fertilization of pycnia by urediospores and aeciospores is also possible.

The dikaryotic mycelium produces aecial cups on the lower surface of the barberry leaves. The aecia contain chains of aeciospores produced on short stalks arranged in a palisade layer. The aecia are yellow in colour. The spores are roundish or angular, measuring 14–26 μm in diameter and have a spiny wall with about 6 germ pores. They can carry the fungus to its cereal host, being incapable of infecting barberry. Infection of cereal host by aeciospores is through stomata.

When the function of the pycnia was discovered it was presumed that the fungus survived through barberry in the absence of its cereal hosts. Later researches in India and in other countries revealed that the nuclear life-cycle may have nothing to do with the actual disease-cycle which is generally completed with the urediospores alone. An account of this phenomenon is given later. The rust fungus is known to attack a number of wild grasses on the hills including *Bromus japonicus* which has been found infected in the plains also.

YELLOW OR STRIPE RUST OF WHEAT

The yellow rust, caused by *Puccinia striiformis* West is generally more destructive than the black or stem rust. In bad years it accounts for very serious losses in the field due to destruction of the foliage, followed in some cases by sterility of spikelets, or in the production of badly shriveled grains. Though there is no substantial evidence to show that the yellow rust is seed-borne, the germinability of the grain is reduced if sori develop on the seed coat.

Symptoms

The disease usually appears earlier than the black rust, coming out as a rule before the grains are formed. In mild attack, the uredia are formed chiefly on the leaves, but in severe attack they appear on the leaf sheath, stalks and glumes also. The green colour of the leaves fades in long streaks on which rows of small urediosori appear. Each row consists of a series of oval, lemon-yellow pustules, arranged end to end, and each distinct from that below and above. In severe attack this serial arrangement is lost and large patches become covered with crowded pustules. The urediospores do not break through the epidermis as quickly as in other rusts but do so eventually and a yellow spore mass is exposed for wind dispersal.

The telia appear late as dull black patches or spots chiefly on the under surface of the leaf. They may form on other parts of the plant also. Like the uredia, they are often arranged in rows. They do not break through the epidermis and remain covered by the epidermis as a flat black crust.

Plants attacked by yellow rust generally show a poorly developed root system. This seems to be the result of heavy leaf infection which hinders translocation of carbohydrates from the leaves to the roots which are starved.

The causal organism

The urediospores of *Puccinia striiformis* are nearly round, binucleate, and unicellular. Their size is very variable, being 23–35 x 20–25 μm. The spore wall is colorless, minutely echinulate, and may possess 6–16 germ pores. On germination of the spore the germ tube forms a small, fragile appressorium over a stoma of the leaf. The infection peg from the appressorium enters through the stomatal opening and forms a large, thick-walled cylindrical sub-stomatal vesicle which is placed just below the stomatal slit. From this an infection hypha arises. The mycelium expands rapidly by means of longitudinal runners which have no haustoria and may remain unbranched for a considerable distance. Other short, branching hyphae form club-shaped haustoria which obtain food from the adjoining cells. These hyphae collect beneath the epidermis to form the uredia. In resistance induced by high temperature in wheat plant hyphal growth of *P. striiformis tritici* is inhibited and organelles develop vacuoles and disintegrate or collapse. Development of haustorial mother cell and haustoria is retarded. These structures are malformed and necrotized. Host cells produce defense structures and material related to infection as well as hypersensitive response. There is formation of cell wall appositions, collars or papillae and encasement of haustoria (Ma and Shang. 2004). Mouldenhauser *et al.,* (2006) studied the infection process of P. striiformis f. sp. tritici in a resistant and a susceptible cultivar of wheat. Their observation regarding appressorium formation are somewhat different. Initially the fungalpenetration of the flag leaf is identical in resistant and susceptible cultivars. The germ tube is directed toward stomata which are penetrated without formation of an appressorium. This is followed by differentiation of substomatal vesicle, infection hyphae, haustorial mother cell and haustoria. For 4 days further development of the pathogen is faster in resistant than in susceptible cultivar. By 7 days after inoculation rapid growth continues only in the susceptible cultivar. Cellular lignification as defense response is rapid in resistant cultivar and within 9 days after inoculation lignified tissues completely surround the fungal colonies. In susceptible cultivar, only isolated lignified cells occur at 6 days after inoculation but long, unbranched fungal hyphae outgrow this defense barrier.

The teliospores are dark brown, often flattened at the tip, and have two cells. They measure 35–63 x 12–20 µm. They may occupy the entire sorus in one group or many groups separated by rows of sterile paraphyses. These spores are capable of immediate germination when mature. However, further development of the parasite is not known. No intervening hosts for pycnial and aecial stages of the fungus have been discovered. The survival of the fungus in India is through its uredial stage present on grass hosts at high altitudes on the hills. The grass hosts include *Bromus japonicus* found in the plains also.

LEAF OR BROWN RUST OF WHEAT

The leaf rust caused by *Puccinia recondita* Rob. ex Desm. is restricted to wheat and certain grasses and is the earliest rust to appear on wheat in India. It has been seen even on 5–6 weeks old crop in the last week of November.

As a rule, the uredia develop on leaves, being rare on the sheath and the stalk. They burst on the upper surface as points of bright orange colour. They are never in rows but may be gathered in small clusters or may be irregularly scattered all over the lamina surface. They are bigger in size than the uredia of yellow rust fungus. When old the uredia of leaf rust cannot be generally distinguished by color from those of the yellow rust except for their irregular arrangement. The sori burst early and shed the spores.

Sometimes teleutopustules (telia) may not develop. When formed they are similar to those of

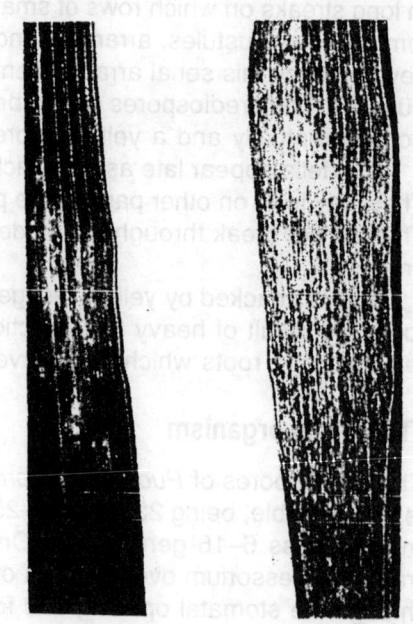

Fig. 73. Brown rust of wheat Uredia on leaf.

yellow rust being on the under surface of the leaf. They are small, oval or linear, dull black, and covered by the epidermis. The sori are divided into compartments by abundantly present paraphyses.

The excessive transpiration in leaf rust affected plants has been studied by many workers and is attributed to increased permeability of the cells in the affected area. Marked interference with leaf functions and the tendency to increased transpiration causes the plants to take a much longer time to produce mature ears. Heavy rusting of the foliage results in poorly developed root system, poor quality and quantity of grains and reduced yield of straw.

The causal organism

The urediospores of *Puccinia recondita* are brown and spherical, 16–28 μm in diameter and with a minutely echinulate wall furnished with 7–10 germ pores. Infection by germ tubes from urediospores occurs through stomata on either side of the leaf. Over a stoma the germ tube forms an appressorium which, in contact with the guard cells, so affects them that the penetrating hypha appears to wait for the stoma to open, penetration being effected at one end of the opening. The function of the overlying appressorium is to exert pressure between the closed guard cells forcing them apart to admit the infection tube. Within the sub-stomatal cavity the invading hypha expands to form a vesicle from which branching hyphae develop to invade the leaf tissues. Hu and Rijkenberg (1998) had reported a scanning electron microscopic study of early events in the infection process of *P. recondita tritici* on wheat. The germ tube extends over the leaf surface and elongates perpendicular to the long axis of the leaf. When the germ tube encounters the stomatal lip, an appressorium forms over the stoma and the pore is entered by an infection peg produced on the surface of the appressorium in contact with the host leaf. At 6 hours after inoculation infection pegs develop from their tip the substomatal vesicles in the substamatal cavity. A septum separates each substomatal vesicle from the interconnective tube. A primarty infection hypha forms terminally from the elongated substomatal vesicle either parallel to the long axis of the stomatal slit or perpendicular to the leaf surface. When a primatry infection hypha attaches to a host cell, a septum forms cutting off the tip of the hypha, delimiting a terminal haustorium mother cell by 12 hours after inoculation. Secondary infection hyphae arise from the haustorium mother cell septum. Additional haustorium mother cells are formed when a secondary hypha or a tertiary hypha adheres to a plant cell. This happens in 24 hours after inoculation. Two or more germ tubes may enter the same stoma, appressoria and vesicle fusing together. The intercellular mycelium with haustoria is soon formed.

The teliospores are similar to those of *Puccinia striiformis*. They are 2-3 celled, smooth and brown. The number of chambers in the sorus is more than in the yellow rust.

For long it was believed that the aecial stage of *Puccinia recondita* occurs on species of *Thalictrum* found in the hills. *Thalictrum polygamum* and about 11 other species harbour an aecial stage of a *Puccinia*. Studies carried out by Prasada (1946) had shown that at least the aecial stage on one species of *Thalictrum* belonged to some other *Puccinia*. Now, this host is not supposed to be connected with the leaf rust of wheat in India.

SURVIVAL OF THE RUST FUNGI

In the plains of India wheat and barley are sown from late October to end of November or sometimes later. The crop is harvested from late March to May depending on regional climate. In cooler areas the crop takes longer to mature than in the warmer areas. Weather conditions and age of the plants during late October till December are quite favourable for infection by the rust fungi. However, in north India initial outbreak of rusts is delayed by 2–3 months in case of leaf and stripe rusts and nearly 4 months or more in the case of stem rust. The latter may not appear at all even if it was present in the crop during the preceding wheat season. This suggests that there is no local source of primary inoculum of any of the wheat and barley rusts. Studies carried out by Dr. K.C. Mehta had established that in the plains of north India, due to prevailing high temperature after harvest of wheat crop, during the summer months, the urediospores and teliospores of the fungus are killed.

The stem rust, *P. graminis tritici*, has its aecial stage on species of *Berberis* which are perennial bushes. In the USA, Canada and other temperate climate countries these bushes and cultivated wheat can be found at the same location. In India barberry is found only in the hills. An aecial stage of a *Puccinia* occurs on several species of barberry. Mehta had stated that the aecial stage on barberry found in the hills of north India has no relationship with *P. graminis tritici* occurring on wheat in the plains or even in the hills. Subsequent studies and reports had shown that these aecia are of *Puccinia* spp. occurring on some grasses, but not on wheat and barley. Similarly, studies with *Thalictrum* spp. have shown that the aecial stage found on these plants is not related to *P. recondita*. Since no alternate hosts of the cereal rust fungi for their aecial stage have been found in India, the teliospores have no role to play in the actual disease cycle. This incomplete lifecycle, thus, leaves only the urediospores as the disseminable propagules for spreading the inoculum of the rust fungi.

Puccinia spp. are common on grasses both in the hills and the plains. The possible role of grass collateral hosts also was investigated by Mehta who could not establish any positive relationship between cereal rusts and the rust found on grasses. However, in later studies the cereal rust fungi have been found to infect a large number of grass species under greenhouse and field conditions. In some cases natural occurrence of the cereal rusts on grasses has also been reported. These hosts may have an impact on epidemiology of the rusts. However, if any of the rusts could have survived on these grass hosts during the summer season, rust should have appeared in the main crop much earlier than its present appearances. Obviously the role of these wild hosts is confined to cooler regions in the hills where they can harbour the fungus in its uredial stage during the summer months.

Having eliminated the possibility of survival of the cereal rust fungi in the plains of north as well as central and south India and having proved the absence of alternate hosts and ineffective role of collateral hosts in the plains, Mehta had considered the possibility of active survival of these fungi during summer in the cooler conditions of the hills such as the Himalayas in the north, particularly central Nepal, and Nilgiri and Pulney Hills in the south. He also pointed out that the Sivalik ranges and Hindukush mountains in the northwest were also the possible places of survival of the rust fungi. From these sources rusts could spread to the plains of Punjab (including the part that is now in Pakistan). He was of the view that for the northern plains central Nepal was the most important place for dissemination of primary inoculum. These presumptions were based on the temperature relations of the three rusts.

The yellow rust (*P. striiformis*) is susceptible to warm weather and does not oversummer below about 1850 meters. It dies in the open as well as in shade at 1723 meters in May–June. Warm weather may induce resistance to yellow rust in wheat. The leaf rust (*P. recondita*) tolerates warm weather better than the yellow rust fungus and survives during summer at 1687 to 2339 meters but during winter it is adversely affected at these heights. It was found in its uredial stage even at an altitude of 985 meters. The stem rust (*P. graminis tritici*) can withstand warm weather even better than the brown rust and could oversummer at 923 meters in the Kumaon Hills of north India. Viability of urediospores of this species is much reduced during winter at 1539 meters.

While most of these observations of Mehta have not been refuted, the studies reported since 1967 have shown that survival of *P. graminis* at low altitude hills in the north has no significance for annual recurrence of this rust in India. Instead, the Nilgiri and Pulney Hills in the south are more important for survival of stem rust and brown (leaf) rust fungi. However, all the studies agree that the three rust fungi survive in their uredial stage at places in the hills where temperature conditions are congenial for their development during summer. These places are in the sub-Himalayan area in the north, mainly for leaf rust and yellow rust, and Nilgiri and Pulney Hills in the south mainly for the stem rust and leaf rust. The annual recurrence takes place through urediospores brought by wind currents from these places to the plains.

DISPERSAL OF UREDIOSPORES

Dissemination of urediospores is mainly through wind. The wind is so effective an agent of dissemination that a little initial inoculum, at least relative to the amount that develops subsequently, increases in various infection centres and is continuously dispersed by wind, until it becomes so large as to enable even a new physiologic race of the rust fungus to become the most prevalent and most widely distributed in a particular area. Due to the possibility of long distance dispersal it becomes difficult to know whether the primary inoculum has been brought from nearby places or from some remote places. This increases the chances of occurrence of more than one physiologic race in the same locality.

If the urediospores of *Puccinia graminis tritici* can reach an altitude of about 1538 meters their dispersal distance in a 38 km per hour wind will be about 1760 km. In the USA it had been presumed that under certain conditions spores of *Puccinia graminis* are blown from the far south (Mexico), where the fungus overwinters, into the Dakotas and Minnesota in the far north, traveling about 1600 km in 2 days and reach Canada a short time thereafter. The spores retain their viability during this long flight. They can travel 50 to 250 km over the sea without losing their viability. Waterhouse had stated that stem rust of wheat might have been introduced into Australia by air-borne spores from India or elsewhere. The similarity in physiologic races of this rust in Australia and New Zealand suggests that the rust was introduced in the latter country from Australia by urediospores crossing over the Tasman sea. Russian scientists believe that wheat rust is probably brought to the Amur region of Siberia by wind-borne spores from North Manchuria where they overwinter in the uredial stage. Similarly, it is claimed that stem rust is introduced every year into North Africa from Sicily and Sardinia across the Mediterranean sea.

Rust pustules of *P. graminis* contain 50,000 to 400,000 spores, enough to infect almost an acre of wheat crop. Each spore is capable of infecting a healthy plant and producing successive generations of spores in about 10–15 days. This means several cycles of urediospores-infection-urediospores during the crop season provided weather conditions are favourable.

ENVIRONMENT AND INCIDENCE OF CEREAL RUSTS

Temperature plays a crucial role in the incidence of rusts. Urediospores of *Puccinia graminis* are not favoured by a very low temperature that usually prevails during winter in the plains of north India. The minimum, optimum, and maximum temperatures for germination of these spores are 20°C, 24°C, and 30°C, respectively. Optimum for growth of the germ tubes and appressorium formation are 20°C and 16°–27°C, respectively. Optimum temperature for penetration and formation of sub-stomatal vesicle is 29°C, minimum is 15°C. However, the infection of cereal hosts by urediospores is governed by temperature only so far as the incubation period is concerned. Infection at low temperatures tends to increase the incubation period, with fewer cycles in the season and very slow development of the stem rust. At 10°–12°C the incubation period is 12–15 days while at 21°–23°C it is only 5–7 days.

The leaf rust (*P. recondita*) has a very wide temperature range for its activity. The minimum temperature for urediospore germination is 2°C and maximum 35°C. The optimum is 20°C. The growth of germ tubes is best at 15°–20°C. No growth occurs below 5°C and above 31°C. Formation of appressoria, penetration, and formation of sub-stomatal vesicle are best at 20°C. Obviously, this rust gets favorable temperatures for a longer time than other rusts during the wheat season in the northern plains of India and is the most common rust on wheat. The yellow or stripe rust (*P. striiformis*) is highly susceptible to high temperatures. Optimum for urediospore germination is 9°–13°C, minimum 0°–2°C and maximum 23°C. Germ tube growth is best at 10°–15°C while the infection process requires the optimum temperature of 8°–13°C. No infection occurs above 23°C and below 2°C. This rust, therefore, has limited period for its activity during the wheat season in the north.

Leaf wetness is another factor that determines the rust infection. The period of exposure to leaf wetness varies in duration and progresses with or without one or more interruptions of variable duration. A study reported by Stuckey and Zadiks (1989), at near optimal temperatures leaf wetness period of 6 hours resulted in 60–65% of the pustules produced with 12–24 hours wetness periods. Interruption of a 6-hour leaf wetness period by a 1hour dry period was most damaging to rust germ tubes. This causes reduction in number of resulting pustules and a prolongation of the median latency period.

For germination of uredospores free water or dew deposit on the leaf surface for nearly two hours is essential in all the 3 rusts. Similarly, high humidity favours elongation of germ tubes and the infection process. During winter such conditions are usually available in the morning hours. Moisture on leaves dries quickly in strong sunshine, and ungerminated spores fail to cause infection.

Taking the factors of temperature and humidity together, the rusts may be severe in a season of abundant moisture when temperatures for major part of the day are around 15°–20°C. The winter rains followed by moderate sunshine usually precede the appearance of rusts in the plains of north India. Neither a continuous cool weather nor dry cool or dry hot periods are favorable for rust epidemics. Patil (1967) has reported that with progressive delay in sowing of wheat in Maharashtra there is progressive increase in stem rust infection. This is attributed to rise in temperature during growth period of the crop and production of more urediospores.

Infection of the host and the subsequent disease development in the plant population is influenced also by such factors as light intensity, host variety, host nutrition, and age of the host. At the time of passage of infection hypha through stomata light intensity plays an important role. Since invasion of the host takes place through stomata, condition of the latter should be such as to facilitate entry of infection hypha. Chief stimulus for the opening of the stomata is direct sunlight. Thus, under reduced light conditions the infectivity of *P. graminis tritici* is hampered. This is not true for *P. recondita* which my cause penetration even through closed stomata. Incubation period may be lengthened a week or more during cloudy weather and the rust does not develop so abundantly as during bright weather. Temperature and light intensity also modify the type of reaction produced by different races of the fungus on the host.

Host nutrition may indirectly influence rust development in the crop. The degree of physiologic susceptibility in susceptible and resistant varieties is not changed by different fertilizers although the morphological resistance could be modified. The increased disease incidence in crops with heavy doses of nitrogen fertilizers is due to the modification in the growth habit of the host thus increasing density of stand and the consequent moisture conditions. Nitrogen favors rust development especially when in excess of P and K, while potassium, in general, has an opposite effect. The variety Thatcher grown at 17°–24°C is resistant to race 56 of *P. graminis tritici* when supplied with ammonium nitrogen but partly susceptible when nitrate nitrogen is used.

Seedling and adult plant resistance is well known in rust diseases. Varieties showing resistance in their seedling stage may be susceptible when mature but more commonly plants approaching maturity are found to be resistant as compared to their behaviour in the seedling stage. This is partly due to the development of thick-walled tissues in mature plants. In seedling stage the wheat plants offer virtually no morphological resistance to the development of rust. Most of the tissues are thin-walled collenchyma with numerous intercellular spaces. The epidermis is thin, delicate and easily ruptured.

PRESENT VIEWS ON ANNUAL RECURRENCE OF CEREAL REUSTS IN INDIA

Stem Rust of Wheat: P. graminis tritici cannot survive the extreme summer temperatures in the Indo-Gangetic plains of the north and also in the plateau area of the south. It can oversummer and overwinter in the hilly regions. While Mehta was of the view that primary inoculum of stem rust comes from central Nepal, the extensive observations made since 1967 have shown that this rust has a unidirectional dissemination from the south to the north (Joshi, 1986; Bahadur *et al.*, 1994). If the Himalayas were the active foci of infection, stem rust should have appeared normally in the foothills and adjoining plains as early as in the first week of February, but it appears much later at the end of March or in April. On the other hand this rust appears as early as December–January in many places in the peninsular India. The reasons for the spread of inoculum in the southern region are prevalence of favourable temperature, wind direction, and other factors in the foci of infection (Nilgiri and Pulney Hills) as well as in the southern plains.

The above is the most important, and perhaps the only, deviation from what Mehta had claimed about annual recurrence of stem rust. Although it is still believed that *Puccinia graminis tritici* can survive during summer months in the northern hills, this source of inoculum is of no significance because the environments in the plains are not favourable for it. The average minimum temperature in the plains during the months of December, January, and February are between 7° and 13°C and in the hills it is much less. Such low temperatures are not congenial for infection, multiplication of inoculum, and spread of the stem rust even if it is present. The fungus is not active below 14°C. Thus, very little inoculum is available in north India before March. On the other hand, there is plenty of inoculum of stem rust (20–25 times more than in the north) and also of leaf rust oversummering and overwintering in the Nilgiri and Pulney hills in the south where climatic conditions are quite favourable. It is, thus, well established that the main source of primary inoculum of stem rust lies in the southern hills. The hills in the north contribute very little, if at all, to the epidemics of this rust. The cyclonic disturbances originating in the Bay of Bengal aid in the dissemination of urediospores through air currents and winter rains. These agencies are important not only for stem rust dissemination but also for leaf rust.

Leaf Rust of Wheat: The early studies reported in 1940 and later observations since 1967 have shown that the first build-up of inoculum of *P. recondita* takes place in the foot hills of Bihar and Uttarakhand in the north and in the plains of Karnataka in the south. The inoculum moves from the south to the north with tropical cyclonic winds (Bahadur, *et al.*, 1994) and then from the north to the south and finally both currents merge. All along the Himalayan foothills in Bihar and in east and central Uttar Pradesh a number of infection foci get established towards the end of December to mid-January. Few isolated pockets of infection also develop in the Tarai area of Uttaranchal and in north-west Punjab. With intensification of the disease in north-east India and with development of western atmospheric disturbances the inoculum moves towards the north-west causing widespread occurrence of the disease in western U.P. and in parts of Haryana. By this time temperature become quite favourable for the rust and the local infection foci also get active, adding to the inoculum load.

Stripe Rust of Wheat: This rust is almost totally absent from south India except in the hills and is confined to cooler parts of the northern wheat belt. It appears earliest in north-west Punjab and the Tarai area of Uttaranchal, by the end of December or early January. From these foci of infection the rust moves southward spreading throughout the Indo-Gangetic plain. The temperature and moisture conditions are ideal for this rust during December–January in north-west India and, therefore, its spread is fast. By the end of February or in early March the temperature starts rising and the uredial stage of the rust is generally terminated and replaced by telia. The disease is essentially a major problem in the north and central northern regions. The temperatures are not very congenial for its development in the eastern and southern parts of the country. It does survive during summer in the southern hills but fails to spread even to the foothills. The exact location of the major source of inoculum of this rust is not clearly known. The possible sources could be high altitude hills in north India and in the Sivalik and Hindukush mountains in the north west.

Physiologic Specialization in Rust Fungi

Physiologic specialization has been defined as "presence of entities within morphologic species, not readily distinguished by structure, but differing from each other physiologically" including pathogenicity, biochemical properties, cultural variability, spore germination, and ecological relationship. These entities have been variously named as formae speciales (f. sp.), biologic species, biologic forms, physiologic races, specialized races, parasitic strains (pathotypes), racial strains, or physiologic forms. The three rusts of wheat can be easily distinguished from each other on the basis of their symptoms and structure, but each rust fungus has distinct entities or groups of forms which cannot be distinguished on morphological basis but still have different parasitic capabilities infecting only certain genera or species or agricultural varieties each of which is known as physiologic race and is given a distinguishing number. The presence of these races poses an important problem in breeding crop varieties resistant to the disease. High yielding varieties of wheat have been evolved and found resistant to one or more of the rusts in a particular locality. The same variety when grown under similar climatic conditions in a different locality is found susceptible to the same rust. Apparently, there were races in this new locality which were missed while the variety was being tested against known races of the fungus.

Two categories of parasitic strains are recognized in *Puccinia graminis:* varieties and physiologic forms. The varieties differ from each other somewhat in size and shape of spores but the principal difference between them is their parasitic capability on host plants. Each variety can parasitize several species of one or more genera of the grass family, but it cannot parasitize members of other genera which may be quite susceptible to other varieties of the rust fungus. A variety, in turn, may contain several physiologic forms which differ from each other principally in their ability to invade varieties within one or more species of a genus. Varieties are given Latin names and physiologic forms are designated by Arabic numerals. Thus, *Puccinia graminis tritici* race 15 means race 15 of the variety *tritici* of the species *P. graminis.* The parasitic specialization may advance further and within a race one or more biotypes may be present. The biotypes differ from the main race only in slight variation in the infection types produced on the host.

In addition to variety *tritici*, other varieties of *P. graminis* are *secalis. avenae, phleipratensis, agrostis* and *poae*. Wheat, in general, is susceptible to variety *tritici*, oats to variety *avenae* and slightly to variety *tritici*, and barley is completely susceptible to varieties *tritici* and *secalis* and slightly to *phleipratensis*. This explains why one kind of cereal may be heavily rusted in the field while other kinds of cereals nearby may be relatively free. It also explains why barberry bushes, the alternate hosts in temperate regions, may be heavily rusted while cereal crop very close to them may remain unaffected.

Pathogenicity for certain selected wheat varieties (the differential hosts) is the only practical and sure method of recognizing physiologic races of cereal rusts. Twelve varieties from 5 species of *Triticum* are used as differential hosts for races of *P. graminis tritici*. When these differential hosts are inoculated with urediospores of the rust collection, there are different resultant types of infection, specific to different physiologic races.

The origin of the physiologic races is probably not much different from that of morphologic species. Adaptation or so-called education of races to become capable of infecting new varieties of the host after being grown for some time on an intermediate host (bridging species) is not considered a mechanism of evolution of new races. Hybridization is the most common method of origin of new races of fungi, followed by mutation. Different varieties or different physiologic races of *P. graminis* mate on barberry leaves. As a result new races unlike their parents may be formed. The expression of dominance and recessiveness and segregation in F2 generation in crosses of rust races follows the well-known laws of heredity in the higher plants. F2 segregates comprise of a variety of types, often including the parent types, together with new, hitherto unknown races. In nature, the rust lines are heterozygous and their selfing on the barberry leaves may result in the production of new races. Since barberry serves as a common platform for the pycnial stage of the varieties and races of *P. graminis*, there are sure chances that new races may originate almost continuously by hybridization and selfing. This makes the number of physiologic races very high in countries where barberry is common and the races occurring on wheat, barley and oats parasitize it.

Mutation in the diploid mycelium, on the cereal hosts, such as is known to occur in *P. graminis* and *P. striiformis*, constitutes the second possible source of origin of physiologic races. In *P. striiformis* mutants at the rate of 1.6 per 100,000 to 200,000 urediospores are reported. Production of new races of *P. graminis tritici* by hyphal fusion on wheat and by diploidization of pycnia by aeciospores and uredospores are also reported. In *P. recondita* f. sp. *tritici* new races are reported arise by somatic hybridization also between pathotypes (Park *et al.,* 1999).

At least the following physiologic races and biotypes of the cereal rust fungi have been reported in India:

Puccinia graminis tritici: Races 11, 14, 15, 17, 21, 24, 34, 40, 41, 42, 72, 75, 117, 122, 126, 184, 194, 222, 295, X, Y, 15-C, 21-A, 21-A-1, 21-A-2, 34-A, 40-A, 42-A, 42-B, 117-A-1, 117-1, 117-2.

Puccinia striiformis: Races 13, 14, 19, 20, 24, 31, 38, 57, A, D, E, F, G, and H.

Puccinia recondita: Races 10, 11, 12, 16, 17, 20, 26, 61, 63, 70, 77, 104, 106, 107, 108, 131, 162, D, 12-B, 77-A, 77-B, 77-A-1, 104-A-1, 104-B, 162-A, 162-B, 107-A, 108-1, 100-1.

The introduction of resistant varieties exerts selection pressure and impels the pathogen to counteract resistance. Being more dominant than others, races 117 and 21 of *P. graminis tritici* have acquired more forms than others. Race 117 was detected from Madhya Pradesh in 1954, its biotype 117-A in 1960, 117-A-1 in 1977, 117-1 in 1987 and 117-2 in 1991.

318

Varietal Resistance to Cereal Rusts

The resistance in the plants may be of the following general types:

1. *Protoplasmic resistance:* This type of resistance is the most common. When a resistant variety is exposed to uredospores, the spores germinate and form appressoria in same fashion as on a susceptible variety. In the resistant variety only a few stomata may permit entry of the infection hypha while in the susceptible variety many infection hyphae will succeed. After penetration, a susceptible variety will permit the infection hypha to form sub-stomatal vesicle, subsequent mycelium, and haustoria which will enable the fungus to develop a pustule. In the resistant variety the cell and protoplasmic contact will inhibit the fungus at any of these stages. A small haustorium may be formed but further development of the pathogen will cease due to death of the host cells and also the haustorium mother cell. This type of resistance is based on gene to gene interaction between the host and the pathogen. Resistance is expressed when there are complimentary genes for avirulence in the pathogen and resistance genes in the host. The gene for gene hypothesis, developed by H.H. Flor after conducting series of experiments with linseed rust, states that during their evolution the host and the parasite have developed complimentary genetic system so that "for each gene conditioning rusts reaction (susceptibility or resistance) in the host there is a complimentary and specific gene in the parasite that determines the virulence and avirulence of the parasite". The genes control the protoplasmic nature of the host as well as its other morphological characters.

2. *Functional resistance:* This type of resistance has to do with the entry of the infection hypha of the rust fungus into the host. Since most rust fungi enter the host through the stomata it is imperative that the number and opening and closing of stomata will have some effect on the success of infection. In the case of *P. graminis tritici* daylight hours are most favourable for infection because the stomata open only when there is direct sunlight. Since the germination of urediospores requires sufficient moisture and humidity is highest in the early morning hours, the critical time for infection is the period when dew is still present on the leaf surface and stomata are open. Opening of stomata is sometimes governed by varietal characteristics also. Varieties having the character of delayed opening of stomata, i.e., stomata open in late morning hours when dew has dried, such as in the variety Hope, possess some resistance to stem rust. Functional resistance is not as important as protoplasmic resistance.

3. *Morphologic or mature plant resistance:* A given variety may be susceptible to a given set of races of the rust fungus in the seedling stage but highly resistant to some races when the plants have matured. The resistance increases with the age of plants. This resistance is based on the relative amount of thick-walled collenchyma developed in different varieties at maturity of the plants. The collenchyma is not invaded by the rust fungus. Stomata in some accessions of *Hordeum chilense* are excessively covered by cuticular waxes that prevent rust fungal germ tubes from perceiving the stomata, resulting in failure of penetration of the pathogen into the leaf (Nika and Rubiales, 2003).

The resistance of a plant to the disease is influenced by environments including the host nutrition and by the environment-physiologic race complex.

Disease escaping varieties: The varieties whose maturity precedes the most destructive phase of disease severity usually escape any substantial loss. It is not actually resistance to the disease but only a method of avoiding contact between the host and the parasite at the critical time.

MANAGEMENT OF THE CEREAL RUSTS

The foregoing discussion of the environmental relations of the rusts, their rapid dissemination, and the frequency of evolution of new physiologic races makes the management of rusts a very intricate problem. All these factors are interrelated. The combined effect of suitable weather, sufficient inoculum, cultivation of susceptible varieties over large areas, and the presence of suitable virulent races result in full-fledged rust epidemics. Attempts have been made to minimize or control the rusts through clean up of alternate and collateral hosts, alteration in manuring practices, and the use of chemicals including antibiotics. However, all these approaches are of secondary importance and breeding for disease resistance remains the most effective and practical method of rust management.

Eradication and clean up. Where the rust fungi depend, even if partly, on the alternate hosts for completion of disease cycle, eradication of the latter is often advised. In India none of the cereal rusts is dependent on any alternate host for survival. In this country the problem is of oversummering on stray or self sown wheat plants and collateral grass hosts on the hills. If these sources could be destroyed during summer months the outbreaks of rusts in the plains in the following winter could be greatly minimized or completely eliminated. This is physically impossible task. These collateral hosts may be present in very inaccessible places where search for them may not be possible. Further, the clean up is to be completed between the time of harvest of the main crop and beginning of the next crop season. A major portion of this period coincides with the monsoons when the clean up work in the hills is not feasible. There has also been a suggestion that if immune or resistant varieties for the hills could be developed and the area is saturated with such varieties, the amount of inoculum of different rusts reaching the plains could be drastically reduced.

Mixed cropping. In the old cropping systems in India, mixed cropping (sowing of seeds of two different crops mixed together) had a popular place for the small farmers. This was an economic necessity. No data are available regarding the effect of mixed cropping on rust incidence in cereals but mixed cropping of cereals with non-cereal crops such as pea and chickpea gives a good crop insurance even if the main cereal crop fails completely due to rusts. Since the susceptible surface area is reduced in mixed crops, numbers of pustules formed are much less. This minimizes the inoculum for secondary spread of the disease. Experiments have shown that in a pure crop of barley, sparse stand increases stem rust incidence (Dill-Mack and Roelfs, 2000). There are also reports claiming that mixed cropping does not help. Under high disease pressure it fails to give any significant relief. Multiline cultivars or cultivar mixtures with similar maturity and other characteristics but having different genetic composition if sown also serve the same purpose. Manthley and Fehrmann (1993) had reported that cultivar mixtures of wheat delay and reduce development of leaf rust (*P. recondite*).

Effect of fertilizers. Since nitrogen fertilizers tend to increase susceptibility while potash has the opposite effect, reduction in the proportion of nitrogen in the NPK ratio can help in reducing rust incidence in a susceptible variety.

Effect of film forming or antitranspirant compounds: Application of film-forming compounds disguises the leaf topography. This interferes with the ability of germ tubes to adhere to the surface and identify the sites for infection. Commercial preparation of such compounds like Bio Film, Foli Cote and Vapot Gard are available. The efficacy of these compounds varies. The films are effective mainly if applied before infection. When such compounds were applied to wheat leaf before inoculation with the leaf rust fungus (*Puccinia recondite*) the number of pustules per square centimeter was reduced. Very good suppression was given by 0.5% Bio Film, 2%.

Golicote or Vapor gard. Germination of urediospores was not affected but the number of appressoria formed was reduced by 20% with Folicote and more than 70% by Bio Film and Vapor gar (Zekaria-Oren *et al.*, 1990; Walters, 2006).

Chemical control of rusts. The use of chemicals for rust control dates back to year 1900. Sulphur had been a widely recommended chemical. In experimental plots efficacy of sulphur dust in controlling rusts was demonstrated in India also long ago. The recommended dosage was 30 kg of fine sulphur dust per hectare at intervals of 4 days during the period when rusts usually appear in the crop. The method was very expensive. The cost of the treatment was prohibitive for average farmers, hence, the treatment was not widely accepted

Effective field control of wheat stem and leaf rusts in field trials by 4–5 sprays of nabam and zinc sulphate, provided the treatment was begun when only a trace of rust was present and the weather was clear, was reported as early as 1956. The treatment was cheaper than sulphur dusting. Four sprays of Parzate liquid (nabam or disodium ethylene bisdithiocarbamate or Dithane D-14) plus zinc sulphate at fortnightly interval starting in the first week of February reduced the rust infection from heavy to traces with corresponding increase in yield of grains. Many other combinations of diathiocarbamates were later recommended. The schedule consisted of 4 foliar sprays at 10–14 days interval, starting when the first pustules were seen. Dithane M-45 and zineb were found to yield better profits when sprayed at 2 kg/ha than at 3 kg/ha. Six sprays of urea (1.2%) plus zineb (0.2%) at 10 days interval, starting in the first week of February, have also been found highly effective. There are many other reports of efficacy of dithiocarbamates in cereal rust control.

Post-infection application of nickel salts were found effective in rust control in 1958 and in 1963 commercial control of *Puccinia striiformis* with nickel fungicides was reported. Value of antibiotics for rust contro was also demonstrated in 1958. Two spraye of cycloheximide (actidione) at 50 ppm. were needed, one at heading and another two weeks later for rust control. The antibiotic persisted for at least 2 weeks after the second spray.

Systemic fungicides gained importance in rust control since 1966 when many studies demonstrated the effectiveness of two oxathiin fungicides, Vitavax and Plantvax, against *Puccinia striiformis, P. recondita*, and *Uromyces phaseoli typica* and the smuts caused by *Ustilago nuda* and *Urocyctsis agropyri*. These fungicides are effective as soil treatment, seed treatment, and as foliar spray. Plantvax was found more effective than Vitavax as foliar spray and seed treatment. Their effectiveness as seed treatment was reported by many in India and abroad. The commercial control of wheat rusts with chemicals had many limitations. Success of protectant, non-systemic fungicides such as dithiocarbamates depends upon the number and timing of spray operations which have to coincide with the time of introduction of inoculum in the field. Spraying done much earlier than the arrival of spores on the host surface may not leave effective concentration of the chemical on the host to protect it. On the other hand long intervals between sprays or sprays applied much later than infection will not be able to eradicate infection. The principal limitation of systemic fungicides is the difficulty of attaining effective quantities of a persistent and non-phytotoxic toxicant within the plant without leaving undesirable residues in the harvested grain. The short duration of activity of these systemic and protectant fungicides compared to the length of the rust epidemic limits their effectiveness. The need for fungicidal activity to persist through the last months of wheat development conflicts with the requirement of minimum residues in the harvested grain. Therefore, effective and economic control of rusts through chemicals depends on very reliable forecasts of the progress of rust epidemics so that foliar sprays can be applied at the most favourable stage of development of the rust and the host.

Some of the above limitations in the use of systemic fungicides appeared to have been removed by the introduction of the triazole fungicide RH-124 (4-n-butyl-1,2,4-triazole). The compound had been used in many countries for successful control of the stripe rust. It was to be introduced in India under the trade name of Indar or Dithane R-24. Greenhouse and field trials had conclusively proved that a single spray of RH-124 at the rate of 1.0 litre a.i./ha gave complete control of leaf rust. The 0.5 litre a.i./ha dose also gave a very high degree of control and was economical. Under north Indian conditions the wheat crop could be sprayed any time between 15 January and 20 February but the best time to give the treatment was when traces of the rust pustules were observed in the field. In similar attempts, a single spray of Benodanil (0.75 litre a.i./ha) or Bayleton (triadimefon, a triazole) at 250 g a.i./ha gave good control of stripe rust of wheat. Single spray of Bayleton 25 WP (0.2%), when the first pustules are seen, gives better control of the leaf rust than 4 sprays of 0.2% mancozeb or Dithane M-45. Triadimefon and fenapanil were reported by Rowell (1981) as good fungicides for control of stem rust. Brahma and Asir (1988) found propiconazole (Tilt) at 0.1% effective in controlling the wheat rusts. The efficacy of the fungicide varies with cultivars (Brahma et al., 1991). Aujla et al., (1993) have reported that a single spray of triadimefon, triadimenol or propiconazole proved very effective against leaf rust and two sprays gave 100% control of the disease. Propiconazole (Tilt) is effective against the other two cereal rusts in addition to powdery mildew and Karnal bunt. It persists for about 12 days and completely inhibits urediospore germination of stem rust and leaf rust fungi. Triadimefon and dichlorobutrazol are next to propiconazole in efficacy. To avoid spray application, triazole-coated superphosphate, applied at sowing time, has been tried in wheat in regions where stripe rust (*Puccinia striiformis* f.sp. *tritici*) appears in the crop early (Brown et al., 1990). The triazole fngicide tebuconazole not only inhibits development of *P. striformis* but also enhances structural and biochemical defence responses of wheat.

Arslan et al., (2006) evaluated food additives and low toxicity compounds to control bean rust (*Uromyces appendiculatus*) and wheat leaf rust (*Puccinia triticina*). Ammonium bicarbonate (0.12 M), potassium bicarbonate (0,03 M), sodium bicarbonate (0.12 M) and sodium citrate (0.03 M) were most effective in reducing rust severity on wheat leaves. The herbicide glyphosate applied to wheat tolerant to the herbicide is reported to control leaf rust and stripe rust. Sodium bicarbonate is effective in inhiting urediospoore germination and germ tube elongation of *P. triticina* Application of sodium bicarbonate at 0.12 or 0.24 mol/ L, sprayed twice at 2-week intervals significantly reduced the disease in the field. The control was comparable with that goiven by tebuconazole (Karabulut et al., 2006).

Resistant varieties. The need for rust resistant varieties of wheat in India was realized in the first decade of this century. Work was started in Maharashtra and at Kanpur (U.P.) and Pusa (Bihar). The earlier work, however, was done in the absence of knowledge of physiologic races present in the country and also of their annual recurrence in the plains. The result was that strains selected as resistant proved susceptible when grown in other areas and in other years. Even then, varieties like NP 12, NP 52, NP 4, NP 165, Pb C 591 showed tolerance to rusts as compared to unimproved local wheats. Systematic breeding of rust resistant varieties was started in 1935. Foreign rust resistant wheats and best Indian varieties, otherwise susceptible to rusts, were used as parents. As a result many varieties of wheat resistant to one or more rusts with desirable commercial qualities were evolved. However, not much success could be achieved in developing varieties resistant to all the three rusts and possessing durable resistance. Rust tolerance had been shown by varieties NP 4, NP 52, NP 125, Pb C 591, Pb 409, C 13, C 46, and BR 319. Varieties NP 710, 720, 733, 770, 775, 797, 783, 784, 785, 789, 790, 798, 799, 809, 822, 823, 825, 827 and Ridley had shown some resistance to the rusts in many areas. However, with the increase in the physiologic race flora they failed at many places. In later years, performance of varieties NP 770, 809, 829, 792, 846, and 830 was found satisfactory in U.P. The rust resistant varieties developed in Madhya Pradesh include Hybrids 65-4, 278, 277-1, 172-2, 11-6, 11-1, 11-8, and 12. In Maharashtra, Kenphad 21, 25, 28, and 32 had shown resistance to rusts. In 1967 it was reported that NI 146, 284 S, 345, and 917 were resistant to stem rust. NI 146, 917, and Hy 65 were resistant to leaf rust and Alternaria blight also.

Although the Mexican dwarf wheats, when originally introduced in India, had shown a high degree of resistance to stem rust, they could not solve the problem of the rusts. Varieties like S 227, S 307, S 308, S 331, Lerma Rojo and Sonora 63 and 64 showed resistance to most races of the three rusts but were susceptible to other races. Other rust resistant varieties are HD 2278, HW 741, WL 614 for stem rust and WL 359, C 306 for leaf and stripe rust. In the North Western Plains of India where leaf and stripe rusts are a problem the wheat cultivars recommended for cultivation are PBW 343, 373 and 435, HD 2687, WH 542, UP 2425 and Raj 3765. HD 2189, HS 424, HUW 543, HW 2044 and HW 2045 are resistant to all pathotypes of black and brown rusts. With intensive cultivation of any given resistant variety over large areas there is breakdown of resistance due to new races of the fungi. Thus, variety Kalyan Sona (S 227), at one time the best and most widely cultivated, had to be discarded because it became susceptible to leaf rust. The variety Sonalika which replaced Kalyan Sona also became susceptible to the leaf rust in most of the wheat tracts of Uttar Pradesh. Thus, recommended resistant varieties can be cultivated for some time but have got to be changed frequently with new varieties.

Triticum dicoccoides has genes for resistance to leaf and stripe rusts. Common wheat transformed with such genes shows resistance to many pathotypes of the rusts. A wild relative of wheat, *Triticum timopheevii* subsp. *armeniacum* (tetraploid) is an important source of resistance to leaf rust. In barley, cultivars BH-11, DL-435, RD-1824, P-147 and Karan-435 are highly tolerant to yellow rust. Lines DWR 44, DWR 45, RD 2627, RD 2640, BH 581, K 707 and K 729 are resistant to all the rusts at seedling and adult stages. Many barley lines have shown multiple disease resistance (rusts and cereal cyst nematode).

Biological control: Biological control of rust pustules on wheat caused by stripe rust, leaf rust and brown rust is reported. Sheroze *et al.,* (2002) evaluated rust antagonistic *Verticillium lecanii* (*Lecanicillium lecanii*) *Paecilomyces fumosoroseus, Beauvaria bassiana, Cladosporium cladosporioides* and *Metarrhizium anisopliae* against leaf rust (*P. recondite*) on wheat. Although all were effective, *B. bassiana* proved the best. The antagonists were sprayed as soon as rust pustules appeared. The treatments prevented expansion of pustules and further spread of the rust. Formulations based on *Trichoderma harzianum, Bacillus subtilis* and the yeast *Saccharomyces cerevisiae* are also reported to give reasonable control of the leaf rust.

RUST OF CHICKPEA (GRAM)

The rust of gram (*Cicer arietinum*) is reported from more than 15 countries in all the continents. The disease is widespread in several parts of India including Maharashtra and Tamil Nadu. In northern India it is common in Bihar, U.P. and Punjab. In the warmer districts of U.P. and Bihar where the crop matures early the damage caused by rust is not heavy but in cooler northern districts of U.P. and in Punjab, the appearance of rust often coincides with the maturation of pods and badly affected plants often show premature death of leaves which results in poor yield of the crop. The disease is common in the countries bordering the Mediterranean where Ascochyta blight is also a problem.

Symptoms

The rust appears as a rule at the end of February when the plants are about 4 months old. In rare cases it has been observed as early as the first week of February. The symptoms of the disease are simple. The leaves become covered with small, round or oval, light brown to dark brown pustules which tend to coalesce and form bigger pustules. They may develop on both sides of the leaf but are more common on the lower surface. In severe attack similar pustules may appear on petioles and stems.

Fig. 74. Rust pustules on chickpea leaves.

The causal organism

The rust of chickpea is caused by *Uromyces ciceris-arietini* (Gregnon) Jacs. Taxonomic position of *Uromyces* is same as of *Puccinia*. The pathogen was first detected and described in France in 1863. The pycnial and aecial stages of the species are so far unknown. The uredia are hypophyllous, scattered, minute, round, powdery when mature, and light brown in colour. The urediospores are globose, loosely echinulate, 20–28 μm in diameter, yellowish brown in colour, with a thick epispore and 4–8 germ pores. The telia appear late (April) and resemble the uredia except for the darker brown colour. The teliospores are variable in shape, round, ovate or angular, with a roundish unthickened apex. The wall is brown and warty. They measure 18–30 x 18–24 μm and have a short, hyaline pedicel. There is one germ pore. Possibility of existence of physiologic races in the species was reported by Bahadur and Sinha (1970a).

Disease cycle and environmental relations

Irrigated crops suffer more from the rust than crops grown under rainfed conditions. The urediospores lose viability in open within 2–3 weeks but can survive for longer periods if kept in sealed vials at 6°C. They germinate at any temperature between 5°C and 30°C but the optimum is 11°–20°C. No germination occurs at 35°C and above. The incubation period depends on prevailing temperature, being 15 days at 11°–12°C, 11 days at 20°–25°C and 9 days at 25°–30° C. The spores are killed when exposed to 45°C for 12 hours, to 40°C for 96 hours and to 35°C for 8 days. The spores of the fungus are, thus, not likely to survive the summer heat in the plains of north India. Attempts to germinate teliospores in laboratory have been unsuccessful. Their formation also does not take place at temperatures below 20°–25°C. It appears that the development of the telial stage depends on a complex of climatic factors that prevail, in the plains of north India, about the end of March and early April. These spores are also destroyed by the summer heat.

During a study of urediospore germination of *Uromyces ciceris-arietini* in leaf exudates of susceptible and resistant varieties of chickpea, Bahadur and Sinha (1970b) observed that percent germination was low in leaf exudates of resistant varieties than in susceptible varieties. Leaves of a resistant variety, Nandriyal 49, contained more malic acid and sucrose than the leaves of a susceptible variety, Agra Local.

Since urediospores and teliospores are inactivated during summer in the plains but the environments are favourable for infection of gram plants from October onward, it is definite that there is no local source of primary inoculum. A legume weed, *Trigonella polycerata*, has been found to harbour this fungus in its uredial stage in the hills up to 1850 meters altitude. Possibly, the rust is introduced in the main crop through urediospores blown by wind from such sources. *Lathyrus* spp. (*L. aphaca* and *L. odoratus*) are known as collateral hosts of *U. ciceris-arietini*. In addition, the rust has been observed on *Vicia biennis*, *V. ervilia* and *V. faba*.

Fig. 75. Telia of *Uromyces ciceris arietini* on chickpea leaf.

Management of the disease

No specific control measures have been suggested. Control with foliar sprays of fungicides has been unsuccessful in many countries. Early sowing may be a method to control the rust as early sown crops escape infection. Many antagonistic fungi in the phyllosphere microflora suppress spore germination of *U. ciceris arietini*. These include *Chaetomium globosum*, *Trichoderma koningii*, *Malustella aeria* and *Fusarium* spp. Only resistant varieties are sure method of avoiding the disease. Chickpea lines NRC 34, NEC 249, JM 583 and 2649, HPC 63, HPC 136 and HPC 147 have been found resistant to rust in Lahaul Valley (H.P.) under natural conditions. Resistance to rust occurs in wild species of *Cicer* and some of them are cross-compatible with *Cicer arietinum*.

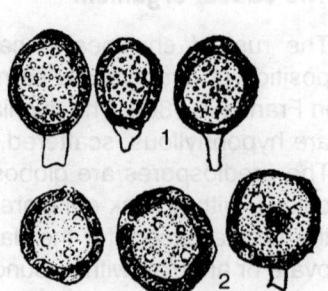

Fig. 76. *Uromyces ciceris-arietini*. Telio- and urediospores.

RUST OF PEA AND LENTIL

Two species of *Uromyces* occur on cultivated peas (*Pisum sativum*). *Uromyces pisi* (Pers.) Wint is a heteroecious species having its aecial stage on *Euphorbia cyparissias* and is not common in India. The other species, *Uromyces fabae* (Pers.) de Bary, is autoecious and is the most common rust of pea and lentil (*Lens esculenta* Moench) in north India. It is known to occur on sweet pea (*Lathyrus* sp.) and broad bean also in the Indo-Gangetic plains. In the north-west Uttar Pradesh the rust causes severe deformity of stems and necrosis, and often death of the plants.

Symptoms

The earliest symptom, is the development of aecia in round or elongated clusters, in February or even later. Pycnia are infrequent or rather inconspicuous. All the spore stages develop on every green part of the host including pods. The formation of aecial stage is preceded by a slight yellowing which gradually turns brown. The uredial pustules develop on both surfaces of the leaves as well as on other parts. They present a powdery, light brown appearance. The telia occur in the same sorus as the uredia and develop from the same mycelium. They are formed on leaves but most commonly on stems and petioles. They are dark brown or almost black in colour.

Fig. 77. Rust of pea.

The causal organism

Uromyces fabae (*U. viciae-fabae*).   Isolates of *U. viciae-fabae* are specialized with respect to host. Each isolate infects only cultivars of the species from which it was collected, although pea isdolates easily infect lentil and vice versa. Host-specialized isolate are also morphologically distinct, differing in both spore dimensions and infection structures, particularly the substomatal vesicle (Emeran *et al.,* 2005).

Pycnia are inconspicuous. Aecia are amphigenous or hypophyllous, usually in groups surrounding the pycnia. They are cupulate, 1–5 mm diameter, short and whitish. The aeciospores are round to angular or elliptical, yellow in colour and possess fine warts. They measure 18–26 μm in diameter. Wall is hyaline, verrucose, 1 μm thick. The urediospores are round to ovate, light brown, spiny, with 3–4 germ pores, and measure 20–30 x 18–26 μm. On lentil the uredia appear very late. The teliospores are subglobose, ovate, or elliptical, with rounded or flattened apex which is considerably thickened and appears papillate. These spores are smooth, brown, and measure 25–38 x 18–27 μm. The stalk is persistent on the detached spores and is pale yellowish-brown, thick, and up to 90 μm long.

The pathogen has a number of races. At least 11 races have been identified using differential hosts from species of *Vicia, Lathyrus* and *Pisum.*

Disease cycle and environmental relations

Major role in the dissemination of lentil rust during the active season is played by aeciospores which form secondary aecia after infection of leaves. Infection at relatively low temperatures of 17°–22°C results in formation of secondary aecia while at 25°C the infection causes development of uredia. No infection by aeciospores occurs at 30°C. These spores remain viable for 8, 6, 4, 3, and 2 weeks at temperatures of 3 -8°, 10–12°, 17–18°, and 30°C, respectively. No viability is retained after 6 weeks showing that the aeciospores do not survive during the off season for the crop.

The optimum temperature for germination of urediospores is 16–22.5°C. No germination occurs at 28–29°C. In the plains of north India where warm season sets in towards the end of March, the urediospores fail to infect lentil after the third week of March. These spores from lentil were found to remain viable for 16–17 weeks when stored at 3–8°C and only for 2 weeks at 36–37°C. Thus, these spores do not survive in the hot summer intervening two successive crops of lentil. The teliospores of the lentil rust fungus have been found to have no dormancy and can germinate at 12–22°C soon after their formation. They can withstand the summer heat in the plains of India thereby showing that the lentil rust fungus perennates in its telial stage in the left over diseased plant trash or mixed with the seed as external contaminant and infects the new crop in the next season. The percentage of quick germination increases with the onset of winter. The number of haustoria per infection unit and their developmental stages are reduced in a resistant line of the host irrespective of whether the resistance is associated with hypersensitivity. The haustoria that enter the host cells are involved not only in nutrient uptake but also in biosynthesis of metabolites such as thiamin (vitamin B1). Kemen *et al.,* have reported that a protein is synthesized by the haustorium and is passed on to the host cytoplasm. It might be playing an important role in the maintenance of the biotrophic interaction. Compared to a susceptible cultivar, in resistant cultivars rust pustules are fewer and smaller with reduced quantity of spores and the incubation period is longer. The molecular basis of the biotrophic uptake of nutrients by rust fungi is mediated by the gene encoding the plasma membrane.

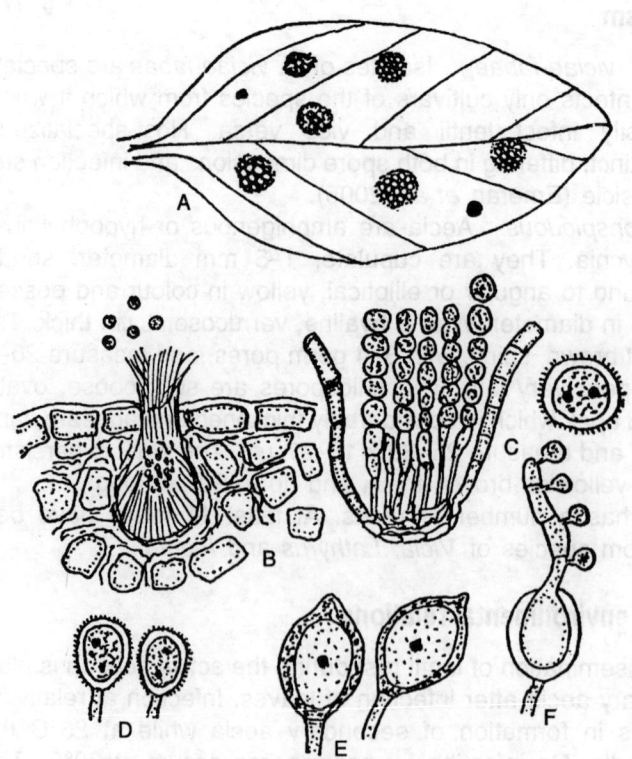

Fig. 78. *Uromyces fabae.* A-Pustules on leaves; B-Pycnium; C-Aecium; D-Urediospores; E-Teliospores; F-Germination of teliospore.

The above mentioned findings about *U. fabae* on lentil are applicable to the fungus on pea also. Aeciospores from lentil infect pea and *Vicia faba* and those from pea can infect lentil. The fungus attacks many other species of *Vicia*. *Pisum* is a more favourable host than *Lathyrus*. Kushwaha *et al.*, (2006) have reported a study of isolates from pea in the North Eastern Plain zone of India. While confirming the major role of aeciospores in repeated cycle of the fungus on pea, they observed that inoculations with aeciospores as well as urediospores resulted in formation of aecia on the leaves. Aecospores were produced at temperatures ranging from 10–25°C, maximum being at around 25°C. At this temperature range about 744 aevcia were found per leaf. Maximum germination of uredispores occurred at 15°C. A relative humidity of 100% favoured aeciospore germination while 98% RH favoured uredispore germination.

Management of the disease

Field sanitation to destroy the crop debris is very important for reducing losses from pea rust. Long crop rotation, avoiding broad bean, *Vicia, Lathyrus*, etc. should be followed in fields where the disease had been present. Use of clean, contaminant-free seed are is also recommended. Seed treatment can also be done with thiram. Fungicide sprays are not very effective due to nature of the crop. Fungicides found effective against the pathogen are mancozeb (0.2%), Bayleton (0.05% a.i.), Calixin (0.2% a.i.) and tebuconazole. First spray is given as soon as the disease is detected in the field and three more sprays are given at 10 days interval. When applied 72 hours after inoculation all the triazole fungicides are reported to provide excellent control (Gupta and Shyam, 2000). Three sprays of tridemorph, chlorothalonil, zineb and mancozeb have provided good control of pea rust. Preinoculation of pea leaves with an avirulent strain of Pseudomonas syringae pv. pisi or treatment with benzothiadiazole induces resistance to the rust fungus. Foliar application of essential oil of basil (*Ocimum basilicum*) is reported to suppress infection by *Uromyces fabae* (Oxenham *et al.*, 2005). Boyle and Walters (2005) have reported that application saccharin to leaves or through soil induces resistance in faba bean (*Vicia faba*) to the rust. Application through soil drench produces effect earlier than application through leaves. Most field pea cultivars are susceptible to this rust. KPMR 551, KPMR 615 and JPF 99025 are reported to possess moderate resistance. Lentil varieties Pant L-639 and Pant L-406 are resistant to rust.

BEAN RUST

In some parts of India this rust disease is of common occurrence on different types of beans (*Phaseolus* species) and on cowpea. It may prove destructive through defoliation of the plant. With 3% infection the yield loss of dry beans may be 2.7% and with 79% infection it may be 36.7%. Occurrence of the disease depends to a great extent on environmental conditions and amount of inoculum present in the locality.

Symptoms

The rust appears mostly on leaves, rarely on stems and petioles. Occasionally, pods also

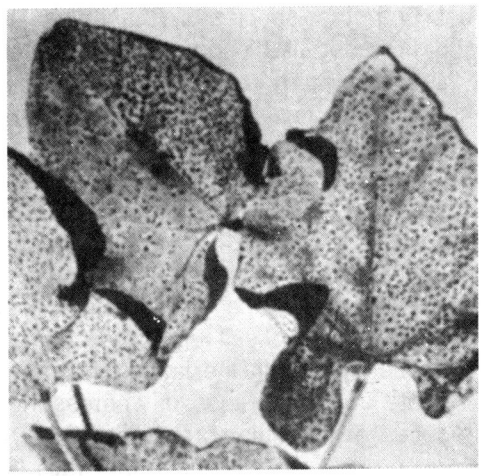

Fig. 79. Rust of bean.

show infection. On leaves, the disease may appear on both surfaces depending on the host species but is more common on the lower surface. The rust pustules appear as minute, almost white, slightly raised spots. On susceptible varieties of the host these spots enlarge to form reddish brown sori, up to 2 mm in diameter, containing the rust spores. Often the sori are surrounded by a ring of secondary sori. With formation of teliospores, the sori turn dark brown or black. When too many of these powdery pustules develop on the leaf they give a rusty appearance to the leaf. Leaves may turn yellow and dry or they may fall off.

The causal organism

Uromyces phaseoli typica Arth. on beans
Uromyces phaseoli vignae (Barcl.) Arth. on cowpea.

Earlier, the fungus was known as *Uromyces appendiculatus*. It is an autoecious long cycle rust fungus and produces all the spore stages on the bean plants. The uredia are brownish and powdery. The urediospores are globoid or ellipsoid, 1-celled, echinulate and 20–33 x 16–23 µm in size. Walls are golden brown with 2 equatorial or sub-equatorial pores and 1–1.5 µm thick. Telia develop in the uredia and appear dark brown or black. Teliospores are globoid or broadly ellipsoid, pedicellate, 1-celled. 24–32 x 20–26 µm in size, smooth with few verrucose marks. Wall of the teliospores is chestnut brown, 3–3.5 µm thick and with a hyaline papilla over the apical pore. Pycnia appear in yellowish spots on the upper surface of the leaf. Orange coloured aecia are formed on the lower surface of the leaf around the pycnia on the opposite side. The aecia are cupulate and 0.25 –0.3 mm in diameter. Aeciospores are ellipsoid, 1-celled, 20–26 x 10–20 µm in size. Their wall is hyaline and 1–1.5 µm thick. It is minutely verrucose. Existence of physiologic races in the species is known. Twenty races have been described in the USA using 19 differential cultivars of bean.

Disease cycle and environmental relations

The bean rust fungus is not seed-borne. In cold climate regions it can survive through teliospores and urediospores in crop debris (Gross and Venette, 2001) as well as on seed left in the field after harvest. Aecia formed on these sources infect volunteer bean plants forming pycnia and aecia (McMillan *et al.,* 2003). In other areas it survives through continuously grown bean and other species of *Phaseolus*. During the crop season the disease spreads through wind dispersed urediospores. Spider mites can also disperse these spores. They are attracted to uredinia (Batra and Stavely, 1994). Intensive cultivation of beans in areas with high atmospheric humidity is one of the conditions favourable for disease development. In these areas if the crop is attacked early there may be total loss. The repeating spores (urediospores) germinate best at 15°–24°C. About 93% germination of urediospores has been observed at 17.5°–22.5°C. The teliospores in crop debris germinate best at 10°–15°C. Spores produced at temperatures above 24°C show reduced germination. Spores in pustules on old leaves also show 30% less germination. Adhesion of the urediospores and germ tubes to the host surface is governed by surface hydrophobicity (Terhune and Hoch, 1993). Mechanical forces imposed by a combination of cell turgor pressure and adhesion of appressoriium to the stomatal surface cause deformities in the latter. The normally erect stomatal guard cell lips usually become prostrate at most stages of appressorium development (Terhune *et al.,* 1993). *Uromyces* produces extracellular proteases which breach the host cell wall by acting on the fibrous hydroxiproline-rich proteins in the walls (Rauscher *et al.,* 1995). The proteins are important in plants for cell wall stability and play a role in defence. Under favourable conditions the spores complete the infection cycle within 5 days and a new crop of urediospores is produced within next 5–10 days. Heavy rains wash down the spores hence there are less chances of spread. Frequency and duration of leaf wetness period are more important than temperature in spread of the disease. Cloudy and humid days, permitting night dew to last on the leaves for sometime in the morning hours favour and a temperature of 20–25°C favour germination of spores and infection. If daily mean temperature reaches 34°C there is no disease development. Long day hours favour spread of the disease.

Severity of rust varies among host species and cultivars. Cultivars having moderate or low susceptibility show only minute pustules. Resistance to rust in common bean (*Phaseolus vulgaris*) is related to leaf pubescence. On leaf surface having large number of trichomes (leaf hair) the infection is prevented by not allowing the germ tubes from urediospores contact the leaf surface (Mmbaga and Steadman, 1992). Cultivars that have resistance to rust through hypersensitive reaction show accumulation of jasmonic acid. This acid, in synergism with ethylene, seems to play key role in activating multiple resistance in various host-parasite combinations (Cavello and Raggi, 2002). In a similar rust of Faba bean caused by *Uromyces viciae-fabae* there is significant increase in aborted stomatal penetration by the fungus in resistant lines. Differences in resistance level are more pronounced in adult plants than in seedlings. Resistance is mainly due to restriction of haustorium formation, reduced colonization of the host and reduced number of haustoria (Sillero and Rubiales, 2002). These reduce the size of uredinia.

Management of the disease

The cultural practices that help in avoiding the disease include (1) removal and burning of crop debris, (2) long crop rotations, and (3) suitable plant spacing and removal of weeds to lower the humidity in the crop. If the disease occurs early and weather is favourable the crop can be sprayed with mancozeb, maneb, zineb, or Daconil at the rate of 2 kg/ha at 10 days interval. Three sprays of 0.1% triadimefon (Bayleton) or tridemorph fungicides, 45, 60, and 75 days after sowing, are reported to give best control of the disease and highest yields of beans. One spray of diclobutrazol is reported to have reduced rust in cowpea. Application of INA (2,6-dichloro-isonicotinic acid) to 16–20 days old bean seedlings has provided protection against rust for at least 5 weeks through induced resistance. The treatment reduced the number of uredenia and also the spread of the rust to upper leaves (Dann and Deverall, 1996). In an attempt to find food additive and low-toxic chemicalm as alternaive to synthetic fungicides, Arslanm *et al.,* (2006) have reported that acetic acid (0.03M), ammonium bicarbonate (0.09 M), potassium acetate (0.03M), potassium benzoate (0.006M), potassium bicarbonate (0.012 M) and many other compounds significantly reduced the disease severity without any injury the plant. The chemicals were toxic to the urediniospores of the rust fungus.

Species of *Bacillus* and *Arthrobacter* are antagonists of the bean rust fungus. Spray application of thick suspension of the antagonists gives about 90% control. Yuen *et al.,* (2001) have reported that strains of *Pontoea agglomerans* (*Erwinia herbicola*) and *Stenotrophomonas multophilia* (*Pseudomonas/Xanthomonas multophilia*) are effective in reducing bean rust severity. Spray of cell suspension of a chitinilytic strain of *S. multophilia* in chitin broth culture gives as good control of bean rust as multiple sprays of thiophanate methyl or thiophanate methyl + maneb. *Erwinia uredovora* (*Pantoea ananas* pv. *uredovora*), and the fungi *Cladosporium tenuissimum* and *Verticillium lecanii* (*Lecanicillium lecannii*) are also listed as biocontrol agents of bean rust. Germination of urediniosopores decreases when in contact with ungerminated conidia of *C. tenuissimum*. The mycoparasite grows towards the bean rust spores and coils around their germ tubes. Penetration of the urediniospores occurs either enzymatically and/or mechanically, through appressorium or infection cushion structures, from which a thin penetration hypha is generated (Assante *et al.*, 2004).

LEAF RUST OF COFFEE

The historical importance of this disease in causing destruction of coffee plantations in Sri Lanka during 1867–1893 was mentioned in the first chapter of this book. The fungus causing coffee leaf rust is a "classic" among plant diseases being one of the first to be fully elucidated after it caused havoc in Sri Lanka and from where it was first authenticated. Although the disease does occur in Sri Lanka even now, it is no more of much economic importance, first, because coffee is now not a major export crop of that country and second, because control measures are now known. In the Western Hemisphere coffee rust was first reported in Brazil in 1970 and has since been spreading there also. In India, the disease was first reported in the year 1870 and occurs every year and threatens coffee cultivation unless strong control measures are applied. It occurs in Karnataka, Kerala and even in Madhya Pradesh where coffee is grown in a small tract.

Fig. 80. Leaf rust of coffee.

Symptoms

The disease is restricted to leaves although sometimes it can be seen on tender shoots and berries. The rust pustules appear as small spots, 1–2 mm in diameter. In early stages these spots are yellowish becoming orange in colour with increase in size. On the upper surface, opposite the spots on the lower surface, the colour is often brownish. In severe attack leaves may dry and wither. The berries remain small and fail to ripen. Extensive defoliation weakens the trees and results in poor yield, severe die-back of twigs and death of trees.

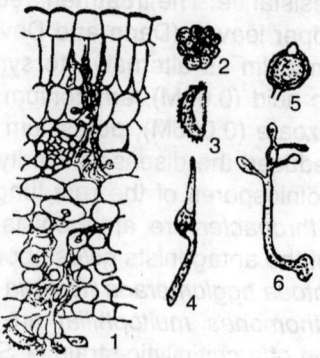

Fig. 81. *Hemileia vastatrix.*
1-Section of infected leaf showing mycelium, hasutoria and spore cluster;
2-Spore cluster, 3-Uredio-spore; 4-Germination of uredio-spore;
5-Teliospore; 6-Germination of teliospore.

The causal organism

Hemileia vastatrix Berk. and Br. is a member of Pucciniaceae. The fungus is co-evolved pathogen of *Coffea* spp. in Africa. It spread rapidly throughout Africa and Asia as the coffee industry of these countries developed. Another species, *Hemileia coffeicola* Maubel and Roger, causes a different type of rust and is restricted to warmer and wetter parts of Africa.

Pycnial and aecial stages of *H. vastatrix* have not been seen and it is not known whether the fungus is autoecious or heteroecious. In the leaf tissues of coffee the mycelium is intercellular with globose haustoria. The sori are hypophyllous, densely scattered and give a powdery appearance on yellowish orange rounded blotches on the leaf. These sori consist of numerous narrow, interwoven "feeder" hyphae and rounded cells below the stomata, bearing clavate filaments emerging through the stomata. The tips of these filaments bear numerous pedicels on which the urediospores are borne. The urediospores are reniform and 28–36 x 18–28 (26–40 x 20–30) μm in size. These spores have one flat and the other convex side. There are 4 germ pores on the urediospores. The humped side is echinulate while the concave or flat side is smooth. Teliospores are formed in the same sori. They are 1-celled and pedicellate, round or turnip-shaped, and measure 18–28 x 14–22 μm. They have no dormancy. Basidiospores do not infect coffee and no alternate hosts are known.

This species is distinguished from *Hemileia coffeicola* in being restricted to yellow orange coloured spots on the leaf, in the production of numerous narrow feeder hyphae forming an interwoven mass in the sub-stomatal cavity and in the urediospores having smaller and more numerous spines.

Disease cycle and environmental relations

The coffee rust fungus exists primarily as mycelium, uredia, and urediospores on infected leaves. The urediospores are mainly disseminated by wind which may be responsible for intercontinental spread of the rust. Rain, contaminated planting material and movement of workers and probably insects also help in dissemination of urediospores. These spores germinate only in presence of free water. Two or three germ tubes simultaneously produced. Urediospores can germainate at temperatures of 13–30°C but germination is quick at 22–28°C. Growth of germ tubes is optimum at 19–22°C. Appressoria form most rapidly at 13–16°C. The broad optimum for germination and total appressoria formation is 16–28°C. The shape of appressoria also varies according to temperature. Strong light inhibits spore germination. The germ tubes, either directly or after branching, produce appressoria on stomata. This happens within 10 hours of germination. On the dorsal surface, the appressorium produces a vesicle which in turn penetrates the stomata by means of an infection thread. The infection hypha ramifies intercellularly in the substomatal cavity and nearby tissues and penetrates cells by means of haustoria. This leads to appearance of lesions on the leaf surface. In absence of free water, exposure to high relative humidity levels is insufficient to induce germination of urediospores. Loss of moisture following germination inhibits the whole process and it does not recover even if water is reintroduced, unless the dry period has been less than a couple of hours. While spore germination and penetration of the host leaf are greatly influenced by leaf wetness the colonization process is independent of leaf wetness but is greatly influenced by temperature and this determines the incubation period which may vary from 10–25 days. Thus, the disease is favored by high moisture and high temperature. Under favorable condition, 4–6 crops of urediopsores are produced in 3–5 months. The urediospores start losing viability soon after they are dispersed.

Young incompletely expanded leaves are resistant to infection while mature leaves are most susceptible. Susceptibility of leaves to support appressoria formation over the stomata increases with increasing age of leaf pairs. Appressoria are formed relatively more readily over stomata of leaf pairs towards the bottom (leaf pair 1, 2, 3, 4, and 6) than on younger leaves at the top. Once uredia develop, premature fall of leaves starts. Even one uredium may cause the leaf to fall. In susceptible coffee leaves, the rust fungus continues growth without apparent inhibition while in resistant leaves it ceases growth after formation of at least one haustorium. The first signs of incompatibility are hypersensitive cell death (HR), host cell wall autofluorescens and haustoria encasement with callose and B-1,4-glucanase. These steps are delayed in the susceptible leaves (Silva *et al.,* 2002). The disease is favoured by high moisture and high temperature. Shelter from strong winds, intermittent rains, dew or mist, ample light, very light overhead shade, and moderately high temperature favour severe incidence of the disease. Plantations at higher and cooler altitudes (1800–2100 m) have little disease incidence because the fungus fails to sporulate. If the dry weather preceding the monsoons is short, if there are rains in February and there is much mist and dew during the dry months one can expect a severe outbreak of the disease. *H. vastatrix* has many races which can infect different *Coffea* spp. *Coffea arabica*, grown in most coffee plantations in south India due to its better quality, is more susceptible than *Coffea robusta*. Varieties of *Coffea canephora* are generally quite resistant.

Management of the disease

Sanitation of the plantation is essential. Fallen leaves should be composted or destroyed. Fungicidal sprays have given effective control of the disease. Copper fungicides are the oldest recommendation. For efficient use of copper fungicides at least 60 mg of copper per square meter of canopy area are necessary. This would mean at least 4.5 kg of metallic copper per hectare per spray. In south India, 2: 2: 40 Bordeaux mixture has been in use. Two sprays are given, one during April–June and the other during September–November. Carboxin 20 EC and Oxycarboxin 20 EC have also been tested in south India. Oxycarboxin (Plantvax) gives effective control by eradicating the rust fungus (Muthappa and Nirmala Kumari, 1976, 1979). The fungicide was sprayed at 0.1% a.i. Carboxin (Vitavax 75 WP) used at 0.03% a.i. gives significant control up to 50 days (Muthappa and Nirmala Kumari, 1981). The triazol fungicide, triadimenol (Bayton), applied to coffee seedlings disrupts rust hyphae in the host tissue 24–48 hours after inoculation. Epoxiconazole, especially in combination with piraclostrobin, is highly effective against coffee leaf rust. Water extracts of coffee leaves and methanol extract of neem seeds are effective but not better than epoxicopnazole + piraclostrobin. Spray of the resistance actuvator acibenzolar-S-methyl alone or in combination with these fungicides is reported to induce resistance to coffee rust.

Possibilities of biological control of the coffee rust are reported. *Verticillium lecanii, V. leptobactrum, V. psalliotae, Cladosporium hemileiae, Paranectris hemiliae* and *Darluca filum* are hyperparasites of the rust fungus Spores contaminated with *V. lecanii* (*Lecanicillium lecanii*) show reduced germination. The hyperparasite penetrates the spores and eventually kills it. *Bacillus megaterium* and *B. subtilis* are pathogens of the rust fungus. Extracts of coffe leaves, acibenzolar-S-methy, *Bacillus subtilis* and *Pseudomonas putida* induce resistance and reduce the disease by more than 77% (Costa *et al.,* 2007).

RUST OF LINSEED AND FLAX

The cultivation of linseed and flax (*Linum usitatissimum*) is greatly handicapped by the incidence of rust in most of the linseed (flax) growing areas of the world. In India the disease is of significance since linseed is a major oilseed crop of commercial importance. The crop is sown from late October to November and harvested in late March to April. The rust disease usually appears in February or later although its appearance as early as November has been reported from some parts of central India. Within a few days of its appearance most of the linseed fields in the locality get rapidly affected. Towards the harvest time plants get a fired appearance due to formation of telia on the foliage. From 16 to 100% loss in seed yield is reported. The injury to the plants results from a reduced amount of foliage and utilization of plant food by the obligate parasite. The disease causes reduction in seed yield in linseed and injury to fiber in flax. The oil content of the seed is also affected.

The rust fungus has quite a wide host range in the genus *Linum*, being reported to occur, on numerous European and North American species of *Linum*. In Australia it is reported on *Linum marginale* and in New Zealand on *Linum monogynum*. In addition to Linum, the fungus is reported on 12 species of *Hesperolinon* (Lawerence *et al.,* 2007). Wild linseed (*Linum mysorens*) harbors the fungus in India although it is not definite if the rust race on this host is the same that attacks cultivated linseed.

Symptoms

The diseased plants are very conspicuous in the field because of the bright orange colour of the affected parts. This is due to the reddish brown uredia which are large and occur on both surfaces of the leaves as well as on other aerial parts of the plant. The small pustules may be surrounded by a chlorotic zone. In early stages the leaves show a little necrosis but later the necrosis becomes more general and the leaves die prematurely. The pustules are nearly round and small on leaves but on the stem they are large, elongated and irregular. Formation of reddish brown teleuto-pustules is not common on leaves as the leaves shed early. However, they are most common and pronounced on the stem. Often, surrounded by the reddish brown telia are the orange yellow uredia on the stem. The telia on the stem do not rupture the epidermis and remain covered appearing glossy.

Fig. 82. Rust pustules on linseed plant caused by *M. lini*.

The causal organism

Melampsora lini (Pers.) Lev. was first described in 1847. It is currently placed in the family Melmpsoraceae, order Uredinales, class Urediniomycetes of the phylum Basidiomycota. It has chromosome number of $n = 18$. It is an autoecious, long cycle or macrocyclic rust, all spore stages occurring on the host plant.

The dikaryotic mycelium growing in susceptible linseed plants produces dikaryotic urediniospores. Uredia occur on both surfaces of the leaves, scattered or in groups, and are usually circular in shape on the leaf but elongated on the stem. The urediospores are ovate, 15–25 x 13–18 μm in size, and are provided with spines on the surface. Capitate paraphyses are abundant in the uredia and the urediospores are intermingled with them. Subsequently, the dikaryotic mycelium produces telia, usually on the stems. The telia are irregularly elongate, sub-epidermal, and form solid crusts which spread along the stem. The teliospores are sessile, cylindrical. 1-celled, 46–80 x 8–20 μm in size. They are reddish brown and have no dormancy period. The long delay in germination is said to be due to the epidermal covering of the host. These spores can germinate readily if the covering is removed (Prasada, 1948a). During teliospore development the two nuclei in the spore fuse to give a diploid nucleus. Soon the nucleus enters into meiosis. However, meiosis remains incomplete during dormancy of the teliospore. It is complete when teliospore is released from dormancy and is germinating.

On completion of meiosis the germinating teliospore has four haploid nuclei which migrate to the metabasidium formed by the germinating teliospore. Septa divide the metabasidium into four cells each of which develops a conical sterigma that bears the single celled haploid

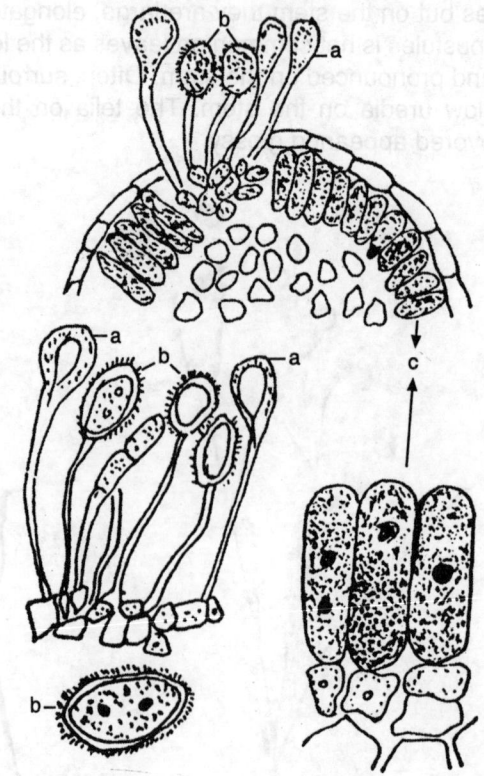

Fig. 83. Sorus and spores of *M. lini*.

basidiosopore. The freshly discharged basidiospore is uninucleate but it undergoes mitosis almost immediately to give mature basidiospore two identical nuclei.

Basidiospores falling on leaves germinate and infect the leaves. They develop monokaryotic mycelium in the leaf tissue. This mycelium develops the pycnium. Pycnial morphology and mating types have been described by Lawrence (1988). Where formed, the pycnia are pale yellow, flask-shaped, and sub-epidermal on the leaf and stem. Sometimes they are diffused having no definite shape. The pycniospores are minute and ovate to globose. And are embedded in a liquid exudates (nectar). *M. lini* is heterothallic. Transfer of the nectar containing pycniospores from one pycnium to another of different matiing type results in dikarytization and development of dikaryotic aecia. The aecia are orange yellow, scattered on the under surface of the leaf, and also on the stem. They are without a peridium and have no paraphyses. The aecisopores are polygonal, 17–27 μm in diameter, and have a thin verrucose outer wall. The aeciospores infect leaves to initiate a dikaryotic mycelium.

Since the rust fungus is autoecious and mating types (Lawrence, 1988) may be common in nature in suitable localities, it can be presumed that continuous hybridization takes place resulting in the evolution of new physiologic races. Nearly 400 races have been identified in USA and Canada. In India, 18 races designated as I-1 to I-17 and 43 have been identified (Saharan, 1991).

Disease cycle and environmental relations

The urediospores are sensitive to heat. At 5–7°C they remain viable for 18–20 weeks but 24 hours exposure to 34°–43°C or 9 hours exposure to 43°–50°C temperatures kills these spores. Thus, they are unable to survive the summer heat of the plains of India. In a film of water the germination of urediospores starts within 80 min. They germinate at 8°–25°C with optimum at 18°C. These spores germinate freely at 3°–30°C with optimum at 15°–16°C. On infection of leaves and young plants, uredial development also takes place at 7°–30°C with optimum at 16–28°C. Infection is slight at 7°–14°C and 26°–30°C. Temperatures between 15° and 21°C are most conducive for infection and disease development.

In susceptible plants urediospores are produced in 9–19 days after inoculation. The incubation period varies with temperature, being 7 days at 13°–21°C, 9 days at 12°–17°C, and 18 days at 2°–10°C. When inoculated plants are kept at 23.8°C and then placed at 2.7°C the incubation period is reduced. Similarly, inoculated plants kept at 2.7°C and then transferred to 20°C develop uredia in 1–3 days. Leaf wetness for 4 hours and temperature of 15°–25°C are optimum for infection. At 10°C minimum duration of leaf wetness required is 8 hours. He also observed that a combination of mean temperature, days with mean temperature below 5°C, cloudy days per week, and some microclimatic and/or biological parameters are important in influencing the progress of rust after infection by urediospores or aeciospores. Gold and Statler (1983) have reported that teliospores are formed 10–15 days after inoculation of stems with urediospores. In seedlings more uredia than telia are formed and on older stems more telia than uredia are formed. Teliospores can be induced to germinate by subjecting them to fluctuating temperature and moisture treatments (Lawrence, 1988).

Telia of the fungus collected in March and stored at 5–7°C during summer gave nearly 60% germination of spores after 8 months. They are killed in the field during the summer in the plains of India but in the hills (1200–1800 meters altitude) viable teliospores have been recovered from plant debris collected with the seed and evidence has been obtained of the active role of such teliospores in initiating fresh outbreaks of the disease. The teliospores in telia covered under the epidermis require exposure to minus 20 to minus 30°F in the open for several months before they can germinate. The weathering of the host debris under these conditions removes the epidermal coating and then the spores are free to germinate. According to Gold and Statler (1983) germination of teliospores is favored by either a period of (i) freeze-thaw and wet dry, or (ii) wet dry. Germination can occur in 25 and 15 days, respectively, under these conditions. Epidermis is ruptured during this period.

In the temperate climate countries flax is sown in spring and harvested in autumn. The plant debris bearing the telial crust lies on the soil in open throughout the following winter and the spores are thus exposed to the freezing temperatures. This enables them to germinate in the following spring and the basidiospores attack the young plants initiating the disease. In India, where linseed is a winter crop and the diseased crop debris is exposed to summer heat instead of freezing, the telia do not play any role in perennation of the pathogen from one crop season to the next in the plains. Urediospores are also destroyed during the summer. It is possible that the rust fungus survives on linseed and other suitable hosts in its uredial and telial stages at high altitudes. From these sources the urediospores are blown by wind to the plains to initiate the disease.

Management of the disease

Some varieties of flax have the ability to retard disease development in the field suggesting tolerance (Rashid, 1997). Fungicide sprays have been found useful against the disease but are not economical. Foliar spray of zineb + copper, mancozeb, ziram, benomyl, tridemorph, triadimenol are reported to suppress the disease. Certain herbicides, such as bentazone, are also reported to suppress the disease. The flax rust could be completely controlled by application of borax at the rate of 60 kg/ha but the method has not been recommended on a field scale. In the hills seed treatment to inactivate the teliospores was also suggested. Sanitary precautions such as destruction of debris of affected crop and weed hosts has also been recommended.

Resistant varieties are the only sure method of management of the rust. In the breeding programmes, specifically for rust resistance, breeders have applied the gene-for-gene concept to produce many resistant cultivars. About 30 resistance genes have been identified in *Linum*. Similarly about 27 virulence genes have been identified in the pathogen (Lawrence, 1988). Breeding for early ripening is suggested as a useful approach for reducing damage. In India resistance has been found in many varieties and lines. The cultivars NP (RR) 9, 10, 56, 95, 218, 297 B, 279 K3, 368, 381, 389, 415, and 501 had been cited as resistant to the disease (Misra and Prasada, 1966). In Madhya Pradesh cultivars Jawahar-7, Jawahar-17, Jawahar -552 and JLS (J)-1 are grown as resistant varieties. In Punjab and Himachal Pradesh, LC-54, LC-115, K-2 and Himaini are grown as resistant varieties. Cultivars LC-216, LC-255, LC-256 are resistant to all races of the pathogen prevalent in the hills (Saharan, 1991). Since the source of primary inoculum lies in the hills seed treatment and resistant varieties are specially recommended for such areas to reduce the inoculum load for the plains. Resistance is associated with increased chitinase activity in the leaves. Pre-inoculation with an avirulent strain induces chitinase activity at a much higher level than inoculation with a virulent strain (McFadden *et al.*, 2001).

REFERENCES

Arslan, U., K. Ilhan and O.A. Karabulut. 2006. Evaluation of food additives and low-toxicoty compounds for the control of bean rust and wheat leaf rust. *J. Phytopath.* 154(8): 534

Assante, G., D. Maffi, M. Saracchi *et al.*, 2004. Histological studies on the mycoparasitism of *Cladosporium tenuissimum* on urediniospores of *Uromyces appendiculatus*. *Mycol. Res.* 108(2): 170

Aujla, S.S., A.K. Basandraj and G.S. Rattan. 1993. Chemical control of leaf rust (*Puccinia recondita*) of wheat. *Indian J. Mycol. Pl. Pathol.* 23: 209

Bahadur, P., D.V. Singh and K.D. Srivastava. 1994. Management of wheat rusts-A revised strategy for gene deployment. *Indian Phytopath.* 47: 41

Baker, C.J., J.R. Stavely and N. Moock. 1985. Biocontrol of bean rust by *Bacillus subtilis* under field conditions. *Plant Dis.* 69: 770

Batra, L.R. and J.R. Stavely. 1994. Attraction of two spotted spider mite to bean rust uredinia. *Plant Dis.* 78: 282

Boyle, C. and D. Walters. 2005. Induction of systemic protection against rust infection in broad bean by saccharin: effect on plant growth and development. *New Phytologist* 167(2): 607

Brahma, R.N. and A. Asir.1988. Chemical control of wheat rusts with Tilt (propiconazole). *Indian Phytopath.* 41: 482

Brahma, R.N., A. Asir and Aloka Saikia. 1991. Efficacy of Tilt (propiconazole) on different wheat cultivars. *Indian Phytopath.* 44: 116

Brown, J.S., M. Hannah and D.J. Ballinger. 1990. The effect of triazole-coated superphosphate, applied at sowing, on stripe rust and yield of wheat. *Aust. Plant Pathol.* 19(3): 79

Cavello, V. and V. Raggi. 2002. Jasmonic acid accumulation in bean hypersensitively resistant to *Uromyces phaseoli*. *J. Plant Pathol.* 84(2)

Chakrabarti, B.P. 1991. Infection processes in wheat stem rust caused by *Puccinia graminis tritici*. *Indian J. Mycol. Pl. Pathol.* 21: 151

Collins, T.J. and N.D. Read. 1997. Appressorium induction by topographical signals in six cereal rusts. *Physiol. Mol. Plant Pathol.* 51(3): 169

Collins, T.J., B.M. Moerschbacher and N.D. Read. 2001. Synergistic induction of wheat stem rust appressoria by chemical and topographical signals. *Physiol. Mol. Plant Pathol.* 58(6): 250-266

Costa, M.J.N., L. Zzambolim and F.A. Rodriguez. 2007. Evalutaion of alternative products and fungicides for the control of coffee leaf rust. *Fitapat. bras.* 32(2): 150

Dann, B.K. and B.J. Deverall. 199C. 2, 6-dichloro-iso-nicotinic acid (INA) induces resistance in beans to the rust pathogen *Uromyces appendiculatus* under field conditions. *Aust. Plant Pathol.* 25(3) 199

Dill-Macky, R. and A.P. Roelfs. 2000. The effect of stand density on the development of *Puccinia graminis* f. sp. *tritici* in barley. *Plant Dis.* 84: 29

Emeran, A.A., J.C. Sillero, R.E. Niks and D. Rubiales. 2005. Infection structures of host-specialized isolates of *Uromyces viciae-fabae* and of other species of *Uromyces* infecting leguminous crops. *Plant Disease* 89(1): 17

Flor, H.H. 1971. Current status of the gene for gene concept. *Annu. Rev. Phytopathol.* 9: 275

Gold, R.E. and G.D. Statler. 1983. Telium formation and teliospore germination in *Melampsora lini*. *Can. J. Bot.* 61: 308

Gupta, S.K. and K.R. Shyam. 2000. Post-infection activity of ergosterol biosynthesis inhibiting fungicides against pea rust. *J. Mycol. Pl. Pathol.* 30: 414

Harding, M.W., J.C. Stutz and R.W. Roberson. 1999. Host-parasite relationship in bean cultivars of varying susceptibility to bean rust. *Can. J. Bot.* 77(11): 1551

Hoch, H.C. and R.C. Staples. 1985. Structural and chemical changes among the rust fungi during appressorium development. *Annu. Rev. Phytopathol.* 25: 231

Hu, G. and F.H.J. Rijkenberg. 1998. Scanning electron microscopy of early structure formation by *Puccinia recondita* f. sp. *tritici* on and in susceptible and resistant wheat lines. *Mycol. Res.* 102: 381

Joshi, L.M. 1986. Perpetuation and dissemination of wheat rusts in India, pp. 41-68. In: L.M. Joshi and K.D. Srivastava (eds) *Problems and Progress of Wheat Pathology in South Asia*. Malhotra Publishing House, New Delhi.

Joshi, L. M., D.V. Singh and K.D. Srivastava. 1985. Status of rusts and smuts in India. *Rachis* 4: 10

Karabulut, O.A., U. Arslam *et al.,* 2006. The effect of sodium bicarbonate alone or in combination with raduced rate of mancozeb on control of leaf rust (*Puccinia triticina*) in wheat. *Can. J. Plant Pathol.* 28(3): 482

Kemen, L., A.C. Kemen, M. Rafiqui *et al.,* 2005. Identification of a protein from rust fungus transgerred from haustoriium into infected plant cells. *Mol. Plant-Microb. Interact.* 18(11): 1130

Kushalappa, A.C. i989. Advances in coffee rust research. *Annu Rev Phytopathol.* 27:503

Kushalappa, A.C. 1990. Development of forecasts: Timing fungicide application to manage coffee rust and carrot blight. *Can. J. Plant Pathol.* 12(1): 92

Kushwaha, C., R. Chand and C.P. Srivastava. 2006. Role of aeciospores in the outbreaks of pea (*Pisum sativum*) rist (*Uromyces fabae*). *Eur. J. Plant Pathol.* 115(3): 323

Lawrence, G.J. 1988. Melampsora lini. Rust of flax and linseed. *Adv. Plant Pathol.* 6: 313

Lawrence G.J., P.N. Dodds and J.G. Ellis. 2007. Rust of flax and linseed caused by *Melampsora lini*. *Mol. Plant Pathol.* 8(4): 349.

Ma, Q. and H.S. Shang. 2004. Ultrastructural analysis of the interaction between *Puccinia striiformis* f. sp. *tritici* and wheat after thermal induction of resistance. *J. Plant Pathol.* 86(1):

338

Manthley, R. and H. Fehrmann. 1993. Effect of cultivar mixtures in wheat on fungal diseases, yield and profitability. *Crop Protection* 12(1): 63

McFadden, H.G., G.J. Lawrence and E.S. Dennis. 2001. Differential induction of chitinase activity in flax (*Linum usitassimum*) in response to inoculation with virulent or avirulent strains of *Melampsora lini*, the cause of flax rust. *Aust. Plant Pathol.* 30(1): 27

McMillan, M.S., H.F. Schwartz and K.L. Otto. 2003. Sexual stage development of *Uromyces appendiculatus* and its potential use for disease resistance screening of *Phaseolus vulgaris*. *Plant Dis.* 87(9): 1133

Mmbaga, M.T. and J.R. Steadman. 1992. Adult plant rust resistance with leaf pubescence in common bean. *Plant Dis.* 76: 1230

Mouldenhauser, L., B.M. Moerschbacher and A.J. van der Westhuiszen. 2006. Histological investigation of stripe rust (*Puccinia striiformis* f. sp. *tritici*) development in resistant and susceptible wheat cultivars. *Plant Pathology* 55(4): 469

Muthappa, B.N. and Nirmala Kumari. 1979. Persistence of curative and prophylactic fungicides in coffee plants for control of leaf rust. *Pesticides* 13(8): 35

Muthappa, B.N. and Nirmala Kumari. 1981. Efficacy of Vitavax 75 WP for control of coffee rust. *J. Coffee Res.* 11(2): 47

Nagarajan, S. and L.M. Joshi. 1975. A historical account of wheat rust epidemics in India and their significance. *Cereal Rusts Bull* 3: 29

Nagarajan, S. and L.M. Joshi. 1978. Further investigations on predicting wheat rust appearance in central and Peninsular India. *Phytopath. Z.* 98: 84

Nagarajan, S., H. Singh and L.M.Joshi. 1975. Climatic factors in relation to stem rust epidemiology. *Plant Dis. Rep.* 59: 670

Oxenham, S.K., K.P. Svoboda and D.R. Walters. 2005. Antifungal activity of the essential oil of basil (*Ocimum basilicum*). *J. Phytopath.* 153(3): 174

Park, R.R., J.J. Burdon and A. Jahoor. 1999. Evidence for somatic hybridization in *Puccinia recondite* f. sp. *tritici*, the leaf rust pathogen of wheat. *Mycol. Res.* 103(9): 715

Prakasam, V. and S. Temburaj. 1991. Efficacy of fungicides in the control of rust of frenchbean caused by *Uromyces phaseoli*. *Indian J. Mycol. Pl. Pathol.* 21: 158

Prasada, R., M. Rai and S.A. Kershi. 1988. Resistance of linseed (*Linum usitatissimum*) germplasm to rust (*Melampsora lini*) and powdery mildew (*Oidium lini*). *Indian J. Agric. Sci.* 58: 549.

Rashid, K.Y. 1997. Slow rusting in flax cultivars. *Can J. Plant Path.* 19: 19

Rauscher, M., K. Mendgen and H. Deising. 1995. Extracellular proteases of the rust fungus *Uromyces viciae-fabae*. *Experimental Mycology* 19: 26

Rowell, J.B. 1981. Control of stem rust of spring wheat by triadimefon and fenapanil. *Plant Dis.* 68: 89

Rubiales, D. and J.C. Sillero. 2003. *Uromyces viciae-fabae* haustorium formation in susceptible and resistant faba bean lines. *Eur. J. Plant Pathol.* 109(1): 71

Saharan, G.S. 1991. Linseed and flax rust. *Indian J. Mycol. Pl. Pathol.* 21119

Schieber, E. and G.A. Zentmeyer. 1984. Coffee rust in the Western Hemisphere. *Plant Dis.* 68: 89

Shaw, D.E. 1999. *Verticillium lecanthi*, a hyperparasite on the coffee rust pathogen in Papua New Guinea. *Aust. Plant Pathol.* 17(1):2

Sheroze, A., A. Rashid, M.A. Nasir and A.S. Shakir. 2002. Evaluation of some biocontrol agents antagonistic microbes against pustule development of leaf rust of wheat caused by *Puccinia recondita* f. sp. *tritici*. *Pak. J. Plant Pathol.* 1(2-4): 51

Sillero, J.C. and D. Rubiales. 2002. Histochemical characterization of resistance to *Uromyces viciae-fabae* in faba bean. *Phytopathology* 92(3): 294

Silva, M.C., M. Nicole, L. Guerra-Guimaraes and C.J. Rodriguez Jr. 2002. Hypersensitive cell death and post haustorial defense responses arrest the orange rust (*Hemileia vastatrix*) growth in resistant coffee leaves. *Physiol. Mol. Plant Pathol.* 60(4): 169-183

Sohn, J., R.T. Voegele, K. Mendgen and Hahn. 2000. High level activation of vitamin B1 biosynthesis genes in haustoria of the rust fungus *Uromyces fabae*. *Mol. Plant-Microbe Interactions* 13(12): 629

Stuckey, R.E. and J.C. Zadoks. 1989. Effect of interrupted leaf wetness periods on pustule development of *Puccinia recondite* f. sp. *tritici* on wheat. *Eur. J. Plant Pathol.* 95(1): 175

Terhune, B.T. and H.C. Hoch. 1993. Substrate hydrophobicity and adhesion of *Uromyces* urediospores and germlings. *Exp. Mycol* 17(4): 241

Terhune, B.T., R.J. Bojko and H.C. Hoch. 1993. Deformation of stomatal guard cell lips and microfabricated artificial topographies during appressorium formation by *Uromyces*. *Experimental Mycology* 17(1): 70

Waller, J.M. 1982. Coffee rust-epidemiology and control. *Crop Protection* 1(4):385

Walters, D.R. 2006. Disguising the leaf surface: The use of leaf coatings for plant disease control. *Eur. J. Plant Pathol.* 114(3): 255

Wietholter, N., S. Horn, K. Reisige, U. Beike and B.M. Moerschbacher. 2003. *In vitro* differentiation of haustorial mother cells of the wheat stem rust fungus, *Puccinia graminis* f.sp. *tritici*, triggered by the synergistic action of chemical and physical signals. *Fungal Genet. Biol.* 38(3): 320

Yuen, G.Y., J.R. Steadman, D.T. Lindgren, D. Schaff and C. Jochum. 2001. Bean rust biological control using bacterial agents. *Crop Protection* 20(5): 395

Zakeria-Oren, J., Z. Eyal and O. Ziv. 1990. Effect of film-forming compounds on the development of leaf rust on wheat seedlings. *Plant Dis.* 75: 231

Rafiee, N. R. Hom, R. Rafaque,   Sikka and B.M. on side effect   2008. Influence on mother cells of the wheat stem cut tissue   Economic genetics   to the   the productivity   wheat of chemical and physical signals. Fungal Genet. Biol. 30(3) 521.

Mewer B.Y.   T.R. Sleathman, D.T. Chopper, P. Setali and C. German   plot field-grown crop in the   pesticide agents. Crop Protect. in 20(3) 385.

Zakaria Omar, L.Z. Eyal and O. Ziv. 1990. Effect of film-forming compounds on the   of chemical sprays on wheat seedlings. Plant Dis. 75. 231.

Diseases Caused by Basidiomycotna (Ustilaginomycetes)

LOOSE SMUT OF WHEAT

Loose smut is a major disease of wheat and is responsible for heavy losses when susceptible varieties with infected seed are grown year after year without proper seed treatment. It occurs in all the wheat growing areas of the world but is more severe in moist than in dry regions. Generally, the loss is proportionate to the percentage of infected heads but other yield contributing components of the host are also affected. Smutted plants produce 22.9% less tillers, 18.7%, 27.5% shorter peduncles and show 15.7% reduction in height of plants. In India, the disease occurs in all the states but the incidence is more in the relatively cooler and moist northern parts. Based on an average of only 3% incidence the disease could cause a loss of more than 50 million. Individual fields may show 23 to 30% smutted heads if untreated seed is used. During 1951–52 loose smut had caused 3–30% damage to wheat crop in several districts of Punjab and the total loss was estimated to be about Rs. 30 million. Extensive cultivation of Kalyan Sona had brought down the average incidence to less than 0.5%. However, later the incidence had gone up to 2–5% (in isolated fields up to 7%). Rice-wheat rotation has increased the incidence probably because of late sowing of wheat (Sinha *et al.*, 2000). In 1988–89 the loss was estimated at Rs 1.05 billion annually. Due to its internally seed-borne nature the disease is dreaded by seed growing agencies who have to adopt costly chemical seed treatment to ensure quality of their product.

Symptoms

The destructive nature of the disease lies in the fact that every head of the affected plant may be converted into a black mass of spores and no grains are formed. In most of the varieties under cultivation the growth of the plant and its general appearance is not affected in any way until ears have appeared. However, in some varieties abnormalities are seen in the flag leaves. In the variety Sonalika (RR-21) characteristic yellowing and chlorotic streaks, later turning necrotic, occurs on flag leaves before emergence of the ear. There may be development of smut sori on these leaves. Significant reduction in height and number of tillers also occurs in this variety. During a comparison of several varieties, Beniwal *et al.,* (1990) observed that there were no visible symptoms on flag leaves of smutted plants of cultivar C-306 while the smutted tillers of Kalyan Sona, HD-2285, WH 283 and WH 291 showed yellowing of tips of flag leaves. Yellow streaks and, later, shredding of flag leaves were observed in Sonalika, WH 147 and DL 153-2 (Kundan). In most varieties heading in the diseased plants is reached at almost the same time, or slightly earlier than, in the normal plants. With the appearance of ears it is found that in place of normal ears there is a mass of blackish powder, the smut spores. These spores, which develop in young spikelets, are initially covered by a delicate silvery membrane. The membrane usually ruptures before the ears are completely out of the sheath exposing the very dark, olive brown, powdery mass in place of the normal spikelets. Only the ends of awns in awned varieties escape transformation. The spores easily separate from the host and are almost all blown off by wind in dry weather leaving a bare rachis (stalk) behind. Partially smutted ears also occur. In many varieties only few tillers show smutted ears while others are free from the disease. The amount of spores formed in these sori depends upon the temperature at which the plant has been growing before flowering.

Fig. 84. Smutted wheat ears.

The causal organism

Loose smut of wheat is caused by *Ustilago segetum* (Pers) Roussel var. *tritici* Jensen. (=*Ustilago tritici* or *Ustilago nuda tritici* Schaf.) *Ustilago* is now classified in class Ustilaginomycetes, subclass Ustilaginomycetidae, order Ustilaginales and family Ustilaginaceae. The spores are olivaceous brown, lighter on one side, spherical or occasionally oval, and measure 5–9 µm in diameter. The epispore (spore wall) has fine spines, especially on the lighter side. Germination occurs in moisture by a germ tube (promycelium) which soon dies unless the spores have fallen on the feathery stigma of healthy wheat flowers. At the time flowers are appearing there is a cloud of these spores over the field. In a crowded wheat field the chances are fairly good that large number of spores will reach the flowers. In dry weather the glumes of sound flowers are widely open and the stigma is sufficiently exposed to the spore dust. In wet, cold weather, however, both spore-diffusion and the opening of the glumes are diminished and few spores reach their destination.

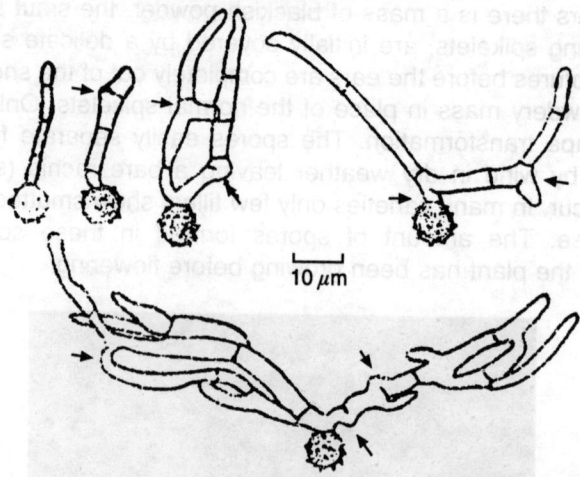

10 µm

Fig. 85. *U. segetum tritici.* Chlamydospore germination.

On germination over the stigma the spores send their simple or branched germ tubes into the style when the ephemeral cells have begun to collapse and dry. No sporidia are produced. Diploidization has already taken place between the cells of the germ tubes. Many spores may lodge on the same stigma and many germ tubes enter in and between the cells and pass down the lower, still alive, part of the style. Here some are checked but others continue growth through intercellular spaces and the channels left by the pollen tubes until, in a week or so, the cavity of the ovary is reached. The ovules are next penetrated through their integuments, entry taking place through small intercellular spaces. The integuments ordinarily become cutinized and impregnable about 10 days after the normal time of fertilization. Thus, successful penetration of the ovule usually occurs between the seventh and the tenth day. The infection hyphae then pass into the space between the endosperm and nucellus. Here they branch freely for the first time. In about 3 weeks branches have reached the lower end of the raphe and then pass round the bottom of the endosperm to reach the scutellum and penetrate the embryo. Some hyphae grow along the rudimentary vascular bundles to occupy the part between the apex and the root. This occurs about 4 weeks after infection. A month later all parts of the embryo except the root primordia contain hyphae and there is a very copious mycelium in the scutellum. The hyphae are mostly 2.5–3.0 μm in diameter, except in the scutellum where they are somewhat swollen. Growth is exclusively intercellular but there are no haustoria and the host cells are not the least affected by the presence of the pathogen. In the ripe grain the hyphae of the fungus are dormant, thick-walled, oily and irregularly swollen.

While the above developments in the host are generally accepted, a study of the disease in England, could find no evidence of infection down the style, but the kernel was infected. There was indication that the infection hyphae caused direct penetration of the ovary.

The dormant mycelium in the seed becomes active when the infected seeds are sown in the next season and their germination starts. It grows just behind the growing point of the host, keeping pace with the apex of the shoot. As the ear formation starts there is a great accumulation of hyphae in the floral parts, which are completely destroyed. The hyphal segments then become swollen, round off, and separate. A thick wall is formed and the cells become chlamydospores (teliospores or smut spores). The hyphal cells and, thus, the young spores, contain 2 nuclei. On maturity of the spore, these nuclei fuse to form a single diploid nucleus. When the spore germinates the diploid nucleus undergoes reduction division followed by mitotic divisions forming haploid nuclei, one for each cell of the germ tube before diploidization.

Simple embryo testing methods are now known for the presence of smut mycelium and for forecasting the disease incidence. (Verma *et al.*, 1985). Seedling crown test can assess the effect of seed treatment on the viability of the smut mycelium in the seed.

The species *Ustilago segetum tritici* has many physiologic races. More than 30 races are now known. Two races, L 1 and L 2 were reported in 1945. Subsequently 8 races were reported from Rajasthan in 1977. Of these 3 were similar to 3 races in the world list. Rewal and Jhooty (1986) reported the occurrence of Canadian races T 1, T 10 and T 11 in India. These 3 races occur to the extent of 8.6, 30 and 61.3%, respectively, depending on predominant cultivars of the host in the region.

Disease cycle and environmental relations

Loose smut is a monocyclic disease. The smut spores can retain viability for 5–6 months. Air dried spores can remain viable for 12 months at 0°–10°C. However, this longevity of smut spores is not important for disease cycle.

There is direct correlation between embryo infection, colonization of seedling tissues and final expression of smut in the crop. Seedlings with less than 50% tissues invaded by the growing loose smut mycelium gradually shake off the pathogen and become free from infection. Although the internally seed-borne pathogen is systemic in the plant, its most active phase is visible only in the floral parts. Environmental conditions in which the plant has grown from the time of sowing infected seed until the heading stage of the plant are important in determining expression of the disease in the field. Deep sowing, about 7.5 cm, favours disease development in the plant. Cardinal temperatures for growth of the fungus are 5°, 20°–25°, and 35°C and optimum pH 3.0. Maximum disease development occurs at 23°C. Atmospheric temperatures during the seedling to spike emergence stages control the number of smutted ears. If plants are grown from infected seed at 29°C, especially between the beginning of internode elongation and boot emergence, spore formation in the smut infected plants is completely prevented and there is no disease. At 24°C only moderate incidence is seen and if the temperature for plant growth has been kept at 19°C there is very heavy incidence of smut.

Optimum time for infection of flowers is when the flowers are in full bloom and the ovary is just commencing to develop after fertilization (early mid-anthesis). A high humidity of 60–85% is essential. The smut spores germinate best at 18°–20°C. The optimum temperature range for seed infection is 22°–5°C. If at the time of flowering these meteorological conditions prevail and inoculum is present in the locality heavy infection of healthy flowers can be expected. Occurrence of light and frequent rains (about 7–21 mm) for 4–5 days accompanied by a temperature range of 6.4°–23.3°C during flowering period favour high seed infection. This study had also found

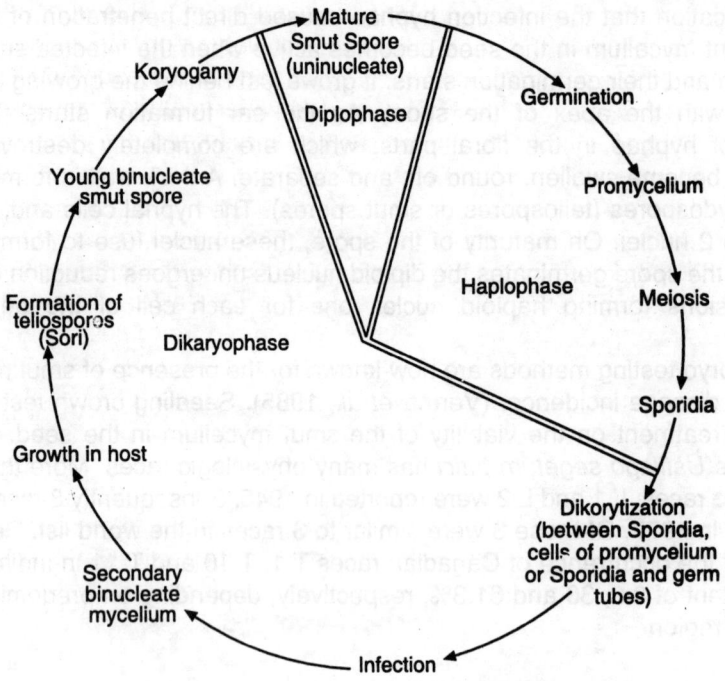

Fig. 86. Life cycle pattern of Ustilaginales.

that crops sown in the second to fourth week of November (in north India) and receiving 120–130 kg N/ha were more susceptible to loose smut. Flowers of cultivars with loose spikelets are more susceptible than those with compact spikelets.

There is some controversy about the effective range of dispersal of spores from smutted ears. In dry air the spores can rise to an altitude of 2–3 thousand feet and hence, theoretically, can be carried to long distances. But due to the above mentioned meteorological restrictions for infection the effective range of dispersal of the smut spores is very short. Various reports suggest a distance of 20 meters to about quarter of a mile.

Management of the disease

Loose smut of wheat, being an internally seed-borne disease, can be effectively managed by eradication of the inoculum from the seed. In the past, before the advent of systemic fungicides, hot water or solar energy methods of seed treatment were the only recommendations. After systemic fungicides were introduced the seed treatment has become easier and, along with resistant varieties, is now the most prevalent method of loose smut control. Another recommendation in the past was to rogue out the smutted plants with the precaution that spores are not dislodged and dispersed among the neighboring healthy flowers. This precaution is still taken in seed plots.

The hot water treatment, originally evolved in 1889 for control of late blight of potato, was first used against loose smut in 1892. In the early stages the method involved use of artificially heated water and maintaining a fixed temperature during soaking of the seed. The method has since undergone many changes. In India, Luthra and Sattar evolved the solar energy treatment against loose smut in 1954. In this method seed was soaked in ordinary water for 4 hours (8 a.m.–12 noon) on a bright summer day (May–June). After this pre-soak, seed was spread in a thin layer and dried in the sun for 4 hours (12 noon–4 p.m.). This work was done in Punjab where summer temperatures during the day go very high. It is not necessary that similar high temperatures may be obtained in other parts of the country. However, Mitra and Taslim had found the solar energy and sun heated water methods suitable for conditions in north Bihar also. Use of a galvanized iron sheet to spread and dry the seed in the sun was recommended for cooler regions. Subsequently, in Punjab, it was demonstrated that a pre-soaking period of 4 hours followed by one hour exposure to sun in summer is enough to devitalize the intraseminal mycelium of the fungus. The rest of the period is required only for drying the seed. Even a 5 min exposure of soaked seed at 12 noon is quite effective in reducing smut infection. Differential lethal effect of hot water treatment on the loose smut mycelium in different varieties of wheat has also been reported.

A number of other non-chemical methods were found effective for loose smut control during the 1950s and before introduction of systemic fungicides. These methods included water soak with or without fungicides or antibiotics at different temperatures. Soaking of naturally infected seed in plain water at 20°C for 41 hours reduced infection from 9.1–3.3% without damage to seed. Soaking in water at 25°C and 30°C for 41 hours and 28 hours, respectively, eliminated the infection completely. Soaking of seeds in 3 and 4% extract of onion or garlic at 30°C for 29 hours brought down infection from 12.5–1.4 and 0.0%, respectively. Soaked and dried seeds could not be stored for more than 10 months without loss of viability.

The advent of systemic fungicides made the control of loose smut of wheat, as well as of most other cereal smuts, easy and highly reliable. In India, Vitavax (carboxin) and Bavistin (carbendazim) have been extensively used for the last 30 years. A dosage of 2.5–3.0 g fungicide per kg seed gives almost complete control of the disease. The dosage can be reduced according to load of infection in the seed lot. If the percentage of seeds found infected in an embryo test is low, lower dosages of the fungicide can be used. Bavistin SD (25% active ingredient) has been found as effective as Bavistin WP (50% a.i.) and is, therefore, cheaper. Resistance in *U. segetum tritici* to carboxin was reported in European field isolates in 2000. The combination of Vitavax with thiram, maneb, or copper quinolate has been found equally effective and reduces the cost of treatment and chances of resistance development. Tebuconazole (Raxil) used at 1.25% is as effective as carbendazim and carboxin but more economical for loose smut control. There is almost 100% suppression of smut spore germination in 200 ppm of the fungicide in *in vitro* tests. It can be used at the rate of 1–2 g/kg seed for loose smut control (Goel *et al.*, 2001) and is cheaper than carboxin treatment. Control of loose smut of wheat and barley by foliar spray of triadimefon was reported by Jones (1997). The strobilurin fungicides (azoxystrobin, picoxystrobin) applied as spray before ear emergence or flag leaf emergence reduce disease levels, delay senescence of the flag leaf and increase grain yield (Ruske *et al.*, 2003).

Johnsson *et al.*, (1998) had reported that a bacterial biocontrol agent *Pseudomonas chlororaphis (Pseudomonas aureofaciens)* strain MA 342 was active against many seed-borne pathogens of cereals. Seeds of barley, oats, wheat and rye were treated with bacterial liquid culture, dried and then sown. The bacterization controlled seed-borne diseases caused by *Dreschlera* spp, *Ustilago avenae*, *U. hordei* and *Tilletia caries* as effectively as fungicides. Diseases caused by *U. nuda*, soil-borne *T. caries* and *T. controversa* were not controlled. In late 1990s, there were reports in India about biocontrol of loose smut by using fungal and bacterial antagonists. Mondal *et al.,* (1996) explained the mechanism of control by *Trichoderma* on the basis of hyphal interactions between the smut fungus and the antagonist. But this study was carried out in petri-dishes using agar medium. Hyphae of the pathogen were raised from chlamydospores sown on the medium. In later studies, certain mutants of *Trichoderma viride* were claimed to be antagonists of the loose smut fungus and they are resistant to carboxin fungicides (Selvakumar *et al.,* 2000). The pathogens listed in the study of Johnsson *et al.,* (1998) were all externally seed borne. They had first to germinate outside the seed and then penetrate the host. Loose smut fungus is entirely internal. Not only this, the amount of hyphae in the host is scanty until the pathogen reaches the inflorescence. Unless *Trichoderma* penetrates the plant in the apical tip of growing radicle there is no contact between the pathogen and the antagonist and no direct antagonism. But possibility of induction of resistance by the antagonist through its enzymes exists. Singh and Maheshwari (2001) have reported that loose smut was almost completely controlled through seed treatment with any of the biocontrol agents like *Trichoderma viride, T. harzianum, Pseudomonas fluorescens* and *Gliocladium virens* in combination with half dose of Vitavax (0.125%).The control was better than full dose of Vitavax (0.25%). Beniwal *et al.,* (2002) had reported that although biocontrol agents like *Trichoderma harzianum, T. virens* and *Bacillus subtilis* control loose smut of wheat, the control is not better than that given by Vitavax. Singh, D.P. (2004) has reported a detailed study of comparative effect of systemic fungicides and *Trichoderma viride* on loose smut of wheat. Carbendazim was not as effective as carboxin (Vitavax) and tebuconazole (Raxil). Low dosage (50 and 100 g/Q) of tebuconazole brought down disease incidence from around 44% (check) to almost nil but at 25 and 12.5 g/Q there was 0.1 to 5.9% incidence. Similarly low dosage (100 g/Q) of Vitavax brought down incidence to 1.0% but at 50 g/Q and 25 g/Q the incidence was 7.3 and 10.0% respectively. Seed soak in *T. viride* spores (10000/ ml) gave highly variable control varying from 3.2% to 37%. Combination of the antagonist with tebuconazole (12.5 g/ Q) reduced smut incidence to 0.0 to 0.4%.

Varieties of wheat that are resistant to loose smut are generally susceptible to rusts and *vice versa.* In Uttar Pradesh, the old improved wheat varieties NP 710, 718, 761, and 770, Bansipali 808, Bansi 224 and P 9D were resistant to the disease while varieties C 13, Pb C 591, NP 4, and NP 12 were susceptible. Cultivars NP 791, 729, and 823 are resistant and Pb C 281, NP 723, 775, 830, Ridley, Sonora 64 and Lerma Rojo were highly susceptible. In Punjab, varieties S 227, PV 18, WG 307, and C 302 were found immune to loose smut. In Madhya Pradesh, the varieties MP 108, 144, 176 and MPO 117, 125, 127A, 128A, 137, 142, 144, 158, 160, 168, 169 were found free from smut in artificial inoculation tests.

A large number of wheat lines resistant to loose smut were listed in 1990 and many of these lines remained smut free for 10 and 8 years in succession. Some were resistant also to yellow or brown rust or both. Sinha *et al.,* (2000) tested a large number of varieties and found that PBW 65, HDR 70, VL 421 and WL 410 were either free or had only trace infection for 8 years. Many of the high yielding rust resistant varieties such as HD 2009, 2285, Lok 1, UP 262, HD 2329, Raj 3077, and WH 147 are susceptible to loose smut. Where susceptible varieties are to be grown it is necessary that only treated seed should be planted.

COVERED SMUT OF BARLEY

Covered smut of barley is common in northern parts of India where it may cause considerable damage in susceptible varieties. It is more common than the loose smut of the same crop. The fungus has been reported on species of *Agropyron* and *Elymus* also.

Symptoms

The smut does not become evident in the field until the ears are formed. The affected ears may emerge about the same time as the healthy ears but remain shorter and are usually retained within the sheath for a longer time before appearing, or may sometimes fail to emerge at all. Every ear in a diseased stool and every grain in a diseased ear is affected. The black spore mass of this smut remains covered by more or less firmly adhering membranes of the grain and the basal part of the glumes. The awns remain intact though they may be sometimes found partly withered. The spore masses are held together due to deposition of a fatty substance, a feature which renders seed treatment difficult unless the fat is removed. As the crop approaches maturity the smutted ears become very conspicuous. The smut sori are mostly broken when the grains are threshed and then the spores get mixed with and stick to healthy seeds.

The causal organism

The disease is caused by *Ustilago hordei* (Pers.) Lagerheim. Many physiologic races are known in the species (Srivastava and Srivastava, 1974). The smut spores are round to elliptical, 6–9 μm or sometimes up to 11 μm in diameter and brown in colour. They appear black in mass. The colour on one side of the spore is lighter than on the other side. The wall is smooth. On germination, which can be easily achieved by placing the spores in moisture, the spores give out a typical 4-celled promycelium. There is no germination of teliospores at 10°C and 40°C. At 15°C there is 13.6% germination and at 20°C 24% germination. At 25°C and 30°C there is 42% and 36% germination, respectively.

Fig. 87. Covered smut of barley.

Sporidia are borne on the promycelium near the septa and at the apex. These sporidia are uninucleate, ovate to oblong, and may form fresh secondary sporidia by budding, especially in presence of nutrition. Inter-sporidial anastomoses are common, nucleus from one sporidium passing into the other. The sporidial as well as the hyphal anastomoses are the means of bringing about the dikaryotic phase of the mycelium. The dikaryotic sporidia germinate to form infection thread which develops into the extensively branched mycelium in the host tissues.

Disease cycle and environmental relations

The fungus perennates through the externally seed-borne smut spores which have contaminated the healthy seed during threshing. Primary infection occurs on the very young seedlings by sporidia when the latter have germinated on the seed coat of the germinating seed. Within 2 hours of contact between sporidia of opposite mating types conjugation tubes develop and fuse. Although on germinating barley shoots several conjugation tubes are produced, only one is involved in mating (Hu *et al.*, 2002). Infection hyphae emerge from either the conjugation tube or conjugated cell body. Hyphae grow along the shoot surface until characteristic crook and appressorium-like structures are formed. From beneath this structure an invading hypha develops and directly penetrates the host epidermal cell. (Hu *et al.*, 2002). In the host tissues the hyphae grow inter and intracellulary without much branching before becoming established in the shoot meristematic region within two days. During this process host plasma membrane remains intact. This process of penetration into the epidermal cells of the coleoptile is similar in susceptible as well as resistant cultivars. In susceptible genotypes, electron-opaque interfacial matrix is formed around the inter and intracellular hyphae. The hyphae grow and extend into the host bundle sheath and invade the parenchymatous cells. About 12 days later, cell wall appositions are formed surrounding the hyphae. In resistant genotypes, cell wall appositions occur as soon as hyphae penetrate the host epidermal cells (Hu *et al.*, 2003). No infection occurs when the primary shoot of the host has grown out above the soil surface. Further development of the parasite follows the growth of the host.

Mycelium rapidly branches in the crown buds and floral structures. Spores are produced by the cells of the hyphae and these replace the kernel. The seedling infection is influenced by soil moisture and soil temperature. Optimum temperature for spore germination is 20°C and the maximum 35°C. Since the period of infection is limited to the period between germination and emergence of seedlings, deep sowing lengthens the period of susceptibility. Shallow sowing avoids too much moisture, a very low temperature, and also helps shoots to come up early thus reducing the chances of infection.

Management of the disease

Since the fungus is externally seed-borne effective and economic control can be achieved by seed treatment with protectant or systemic fungicides. Agrosan GN (an organo mercurial) applied at the rate of 2.5 g/kg seed, had been one of the most commonly recommended fungicide which could control covered smut of barley. These fungicides used to give almost complete control of the disease. Sulphur dust (300 mesh) could also be used for seed treatment. The systemic fungicides named for the control of loose smut of wheat are used against this disease also. Vitavax gives best control of the disease now but is costly for the crop. Treatment with zineb is effective as well as economical. Leaf extract of margosa (*Azadirachta indica*) significantly reduce spore germination of *U. hordei*. Leaf extracts of *Melia azadirachta* and *Cannabis sativa* also suppress germination but at high concentrations (Malik and Singh, 2002).

U. hordei produces mating pheromones which breakdown to smaller peptide compounds that act as potent inhibitors of mating and germination in many fungi (Kostad *et al.*, 2002). The use of pheromone-related antagonists to mating is a promising novel strategy for control of smut and bunt diseases.

Cultivar C 163 had been found immune to covered smut in Punjab Varieties BJ 13, 14, 24, 26, 28 29 are resistant (less than 1% infection).

LOOSE SMUT OF BARLEY

This smut resembles the loose smut of wheat in all important features. However, it is not as common as loose smut of wheat and covered smut of barley.

Symptoms

The sori are formed in the spikelets. In early stages they are covered by a thin, silvery membrane which ruptures while the ear is emerging out of the sheath. The loose spore mass is shed or blown away by wind leaving behind the bare rachis. Early symptoms of the disease, before flowering can be detected sometimes by the discoloration of leaves.

The causal organism

Loose smut of barley is caused by *Ustilago segetum hordei* (*Ustilago nuda* (Jens.) Rostr.). The fungus attacks different species of *Hordeum* and a number of grasses. Thirteen races of the barley loose smut fungus had been reported before 1965.

The spores of *U. segetum hordei* are nearly round, olive brown, lighter on one side than the other, finely echinulate and measure 5–9 μm in diameter. Their germination and mode of infection are similar to those of *U. segetum tritici*. As in loose smut of wheat, in this disease also. The hyphae penetrate the pericarp and pass through the parenchyma along the testa on the ventral side of the grain and cross the basal endosperm to enter the scutellum. From there they permeate most parts of the embryo, especially the hypocotyl and the growing point. Mycelium is mainly intracellular in the pericarp and testa and mainly intercellular in the aleurone, endosperm, scutellum and embryo.

The disease is internally seed-borne and the mycelium can survive in its dormant state for considerably long periods in the embryo. Fezer (1959) had reported that the mycelium could survive for as long as 11 years in barley seed. The blossom or intra-seminal infection takes place through wind-borne spores in the preceding season. The optimum temperature for spore germination is between 20° and 22°C. During the periods of low atmospheric humidity the infection is considerably reduced. Low soil moisture encourages development of the disease within the plant.

Management of the disease

Like the loose smut of wheat, this disease can also be controlled only through the use of resistant varieties, hot water or solar energy treatment or systemic fungicide treatment of the seed. In solar heat treatment the seed is soaked in cold water from 6 a.m.–10 a.m. on a bright sunny day in June and then exposed to sun on a brick floor from 10 a.m.–5 p.m., when the temperature of the floor varies from 40°–50° C. Germinability of the seed is slightly reduced but this can be compensated by increasing the seed rate. Use of cloth sheet instead of brick floor was recommended to avoid seed injury. Chemical control with systemic fungicides, as for the loose smut of wheat, are applicable to this disease also.

Large size seeds are relatively free from infection. Varieties C 44, C 50, CN 292, CN 294 and NP 13 are tolerant.

SMUT OF SUGARCANE

Smut is a well known disease of sugarcane and is reported from India, Java, Taiwan, Philippines, Australia, South Africa, Mauritius, Italy and British Guiana. It is primarily a disease of wild canes and those sugarcane varieties which nearly approach the wild canes are most susceptible to the disease as compared to improved thick canes. However, thick canes in the tropics are by no means immune. According to Padmanabhan et al., (1988) when the smut disease appears during early stages (40–60 days of crop growth) there is total loss. When it appears in a 80–120 days old crop, there is drastic reduction in yield and quality parameters. No marked reduction in yield or quality is observed in plants which show the disease at 210–270 days of crop growth. A negative correlation exists between percent smut and yield. The organism is known to occur on grasses such as "kans", Saccharum spontaneum, which serve as collateral hosts.

Symptoms

Affected plants produce a whip-like black shoot, often very long and much curved on itself. It comes out from the central spindle at the apex. At first the mass of smut powder on this outgrowth is covered by a thin silvery membrane made up of host epidermis. It soon ruptures exposing dense black powdery mass consisting of the smut spores. The spores can be blown about by even a gentle breeze. From the lower part of the systemically infected canes the eyes sprout into lateral shoots which usually also produce the smut whip. Production of these lateral shoots is stimulated if the whip at the top of the main shoot is cut off. In the case of local infection the main shoot may not produce the black outgrowth at the apex but one or two infected lateral buds produce secondary shoots which bear the whip-like structure. Again, if the cane is infected but its growing apex has been damaged as a result of insect or mechanical injury before the whip is produced, several eyes on the cane sprout and invariably produce smut whips. Affected plants usually have slender and thin canes and can be identified before the production of smutted whip. These canes are taller and stand distinctly higher than the rest of the crop. Apart from the whip-like structure and convoluted sori from the lateral buds, the canes may show stem galls and multiple buds.

The spore masses are confined to the few outer cortical layers of the whip. The internal portion is of the normal thin parenchyma and vascular bundles similar to those of central portion of the cane stalk in the region of internodes. There are two main flushes of the production of the whip-like outgrowth, the first during May–June and the second during October–November. All the cases of primary infection, the infection present in the seed pieces or the infection due to contact with spores immediately after sowing, develop the whips early in the season while the secondary infection during crop growth results in appearance of whips late in the season.

The causal organism

The disease is caused by Ustilago scitaminea Syd. The fungus was first described by Rabenhorst in 1870 as Ustilago sacchari. In India it was recorded by Sydow and Butler in 1906. Later, in 1924, Sydow showed that the fungus causing smut of sugarcane in India, Java, and Philippines was different from U. sacchari in the size of spores and named it Ustilago scitaminea. In 1939, Mundkur had further subdivided U. scitaminea into U. scitaminea var.

Fig. 88. Smut of sugarcane. The whip-like structure.

sacchari-barberi and *U. scitaminea* var. *sacchari-officinarum* on the basis of the size, colour, and markings on the spore wall. In more recent molecular studies, the sugarcane smut fungus has been named as *Sporisoreum scitamineum*.

The smut mycelium is intercellular and sends haustoria into the host cells. It gets disorganized quickly and can be located only in the whip-like structure and in the meristem. The spores of *U. scitaminea* are spherical, light brown, with punctate wall and about 5–10 μm in diameter. They are loose, very light, and can be easily disseminated by wind. They germinate quickly under moist conditions forming a short, 3-or 4-celled promycelium. From each cell a sporidium is produced on short stalks and is easily detached. The elongate, 1-celled sporidia germinate to produce the infection thread. The smut spores can germinate and produce sporidia or germ tubes in soil also. Under favourable soil conditions (good humus and moisture) the sporidia may bud off more sporidia in short chains. Sometimes, instead of bearing sporidia the promycelium grows into a branched hypha and functions as infection thread.

Disease cycle and environmental relations

The optimum temperature for spore germination is 25–30°C, the maximum 36–40°C and minimum 5–9°C. The spores are instantaneously killed at 62°C but in ice they survive for more than 3 days. One hundred per cent relative humidity is essential for spore germination. No germination occurs at 90% relative humidity. The spore germination is complete in about 24 hours in water. Under dry conditions the spores remain viable for more than 7 months but under moist conditions they lose their viability within 3 weeks.

The sugarcane crop has no dead season and so the pathogen gets ready hosts. The smut spores from the whip-like structures are blown about by wind and some of them fall on the nearby canes. They are found deposited at the junction of the leaf and the leaf sheath from where they travel down the sheath and reach the tender nodal region and the young eyes. There is plenty of moisture at the base of the leaf sheath facilitating germination of spores. Stimulatory effect of diffusates from buds on the germination of spores is also reported. Copious mycelium can be seen in the growing point of the nodal buds. The infection can also occur through the young germinating shoot. Any injury on the surface of eyes and any part of the cane facilitates infection if the spores fall on the injured surface. Infection is most severe at optimum temperature and humidity for spore germination.

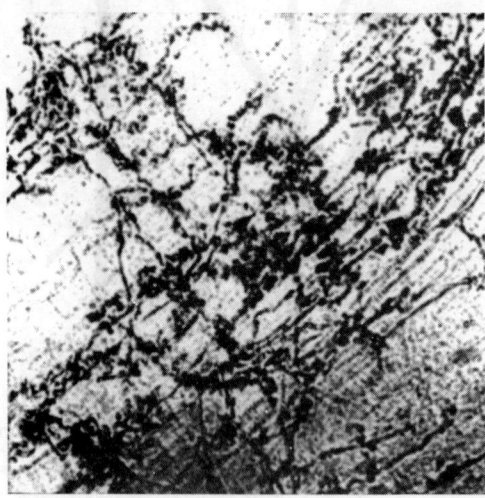

Fig. 89. Hyphae of *Ustilago scitaminea* in the growing point of nodal bud of sugarcane.

The pathogen is perpetuated and spread in any of the following ways:

1. By planting setts of smutted canes. The diseased setts contain living mycelium and on planting may give rise to smutted plants. The germination capacity of the diseased setts is considerably reduced. The sprouts from buds in soil may be killed before coming out.

2. By spores borne on buds. If setts with buds carrying spores are planted the disease may be produced in the canes developing from such buds.

3. By infection of buds on standing canes. It has been shown that during the growing season of the cane, spores from smutted whips may fall on the exposed buds. They germinate on these buds and cause infection. Some of these buds produce smutted shoots during the same season, while others carry the dormant mycelium until setts bearing them are planted in the next season.

4. By ratooning the smutted canes. The ratooned stumps of diseased canes harbour the mycelium of the fungus in the buds and fresh shoots arising from these stumps are invariably smutted.

Management of the disease

Removal of smutted whips from the field. If the attack is sporadic the few smutted whips can be spotted out and carefully removed with precautions to avoid falling of too many spores. After this the entire clump should be dug out and burnt. This reduces secondary infection of buds.

Discouraging the practice of ratooning: In susceptible varieties the practice of ratooning should be entirely given up. A crop which has shown high incidence of the disease should not be ratooned. This prevents perpetuation of the pathogen which may get access to other varieties also.

Avoiding planting of setts from smutted canes. The seed setts should be selected from healthy fields. It is better to grow cane for seed in special seed plots where proper care can be taken to rogue out clumps showing any disease.

Disinfection of setts before planting. Canes in the vicinity of smutted canes may be only apparently healthy but may carry spores on their buds. If sufficient care is not taken at the time of harvest smut whips may also contaminate healthy buds. To guard against these possibilities many chemical and physical treatments of seed setts have been recommended from time to time.

Among chemical treatments a 5-min dip of setts in 0. 25% water suspension of Agallol or Aretan (methoxy ethyl mercury chloride, MEMC), having 6% mercury, had been a common recommendation. Mercuric chloride (0.1%) and formalin (1.0%) for a 5-min dip followed by 2-hour covering under moist cloth, and a 5-min dip in 4:4:50 Bordeaux mixture were also recommended. These treatments ensure only inactivation of spores present on surface of setts or buds or protection from spores present in soil. To eradicate the internally present dormant mycelium attempts have been made to use systemic fungicides like Vitavax, Benlate, Bavistin, etc. for sett treatment. Duration for the dip is usually long. Benlate is highly effective against many pathogens of sugarcane such as *Colletotrichum, Ustilago, Fusarium* and *Cephalosporium* which are transmitted through the seed material. Waraitch (1986) had reported that dipping of setts in 0. 5% Bavistin, Vitavax or Agallol or 0.3% Dithane M-45 controls both internal and external inoculum. Olufolaji (1993) found pre-plant soaking of seed setts in captafol (2 g a.i./L) promising reducing smut incidence and increasing yield. In glasshouse and field trials in Nigeria, Wala (2003) obtained control of smut by giving dip treatment to seed setts for 10 min in solutions or suspensions of fungicides. The best control was obtained with pyroquilon at 4 g a.i./L and carbendazim + maneb at 4.5 g a.i./L.

A hot water treatment was recommended long back as a method of sugarcane smut control. It consists of a 10-min dip in water at 55° to 60°C before planting. Dry or moist hot air or aerated steam treatment of setts have been found effective in the control of not only smut but other diseases like red rot and ratoon stunting also. MEMC hot dip (250 ppm, 40°C/10 min.) is highly effective and at par with hot air (54°C/8 hour) or hot water (52°C/1 hour) gives good disinfection of seed setts. Vijaya (2000) has claimed that fungicidal dip of setts in 0.1% tridemorph solution for 4 hours provides complete control of the smut in the crop.

Use of resistant varieties. Varieties Co 300, 301, 311, 312, 313, 331, 362, 244, 475, 508, 975, 349, 385, 419, 740, 421, 453, 1158, 1287, 7221, 7501, 7801, 7810, 7905, 7908, 7915, 8201, BO 21 are susceptible and in areas where the pathogen is known to be well established in the fields such varieties should not be planted. The list of resistant varieties includes Co 449, Co 527, Co 658, Co 974, Co 1148, Co 6806, Co 7108, Co 7319, BO 11, BO 22, BO 24 (Agnihotri, 1983), Co 62101, Co 6910, Co 7304, 7313, 7321 , Co 7228, 7415, 7536, 7607, 7624, 7805. In an evaluation of sugarcane germplasm against smut, Singh, A.K. (1989) had categorized them as below:

Highly resistant (0–3% incidence): Co 6806, Co Lk 7807, 8001, 8004, 8102, 8402, CoH 29, Co S 8214, S 3-7, S 4 -7, S 4-12, S 15-7, S 15-9, S 20-2 and S 22-9.

Resistant (4–6% incidence): Co Lk 8002, 8003, 8201.

(7–9% incidence): Co Lk 7810, 8005, COH 7801, 7802, Co Pant 84212, S 7918, 8315, S4-18.

(10–12% incidence):   CoH 15, CoH 24, CoH 34, Co J 64, Co Pant 8202, Co Pant 8421 and Co Pant 84213.

In Madhya Pradesh, cultivars Co Jn 86-141, Co 7403, 7321 and 62101 were found resistant to smut.

There is frequent failure of resistance in sugarcane varieties. It has been recommended that in a particular locality even a resistant variety should not be grown continuously for many years. Resistant sugarcane cultivars have high accumulation of free phenols, and high phenyl ammonia lyase and peroxidase activities in leaves. This can be used in screening for resistant varieties.

SMUT OF MAIZE

Maize smut is not a common disease in India and is confined to Kashmir, less common in the Punjab, and rarely seen in the north western parts of Uttar Pradesh or Uttarakhan.

Symptoms

The smut is interesting because of its effects on the host which are not common among smut fungi. The fungus induces gall formation on the infected tissues. The extent of damage depends upon the site of these galls. When formed on the cob they cause extensive damage.

These galls may also appear on the stem, leaves, axillary buds, and parts of the male flower. On the whole, they appear more commonly wherever embryonic tissues are present. The mycelium developing between the thin-walled cells of the embryonic tissues induces hyperplasia and hypertrophy and excessive development of the phloem elements of the bundles. As the galls are enlarging, they appear light coloured to almost white. With the darkening of inner tissues due to spore formation the white outer membrane (epidermis) ruptures and exposes the black spore mass. Each individual tumor usually starts from an independent infection at the site where it is formed. The galls produced by *U. maydis* are edible.

Stem galls result in loss of yield and bending of the stalk. Infection of female flowers gives rise to galls instead of grains. If seedlings are affected they remain stunted and weak. Late development of galls may sometimes kill the entire plant or plant parts. Deep seated alterations may be caused in the inflorescence as a result of infection.

Fig. 90. Smut of maize.

The causal organism

The disease is caused by *Ustilago maydis* (DC) Cda. {syn. *Ustilago zeae* (Schwein) Hunger} The spores are spherical to ellipsoidal, 8–11 μm in diameter, black, and heavily echinulate. They germinate to produce a typical, septate promycelium which bears ovate, unicellular, hyaline sporidia. The germ tube develops through a crack in the spore wall and elongates rapidly reaching 20–30 μm in 24 hours. The sporidia profusely multiply by budding. Ustilaginales in general, and *U. maydis* in particular, are considered model fungal systems in which mating behaviour, morphogenesis, pathogenicity, DNA recombinations and genomics have been extensively studied. In *U. maydis,* during yeast-like budding cell cycle, bud morphogenesis entails a series of shape changes, initially a tubular or conical structure, culminating in a cigar-shaped cell connected to the mother cell by a narrow neck. The growth occurs at the bud tip. The type of germination varies among strains of the fungus. Sporidia may be directly formed from the spore, there may be two opposite strain promycelia from the spore, or the spore may produce a branched promycelium without producing sporidia. Fusion of sporidia of opposite mating types is very common and is essential for infection.

 U. maydis comprises of an indefinite number of lines and biotypes. The smut was known to affect only maize and teosinte (*Euchlaena mexicana*). In a recent study, Leon-Ramirez *et al.,* (2004) have reported that the fungus is able to infect a variety of phylogenetically unrelated plants (monocot, dicot, gymnosperms) grown under axenic conditions. It penetrates, grows into the tissue in the form of pleomorphic mycelium but fails to form teliospores. It even induced formation of lateral buds and tumors in papaya. Mendez-Moran *et al.,* (2005) have studied the interaction of *U. maydis* with a non-natural host *Arabidopsis thaliana* (a crucifer). When plantlets were inoculated with a mixture of compatible haploids, the fungus grew on the plant surface as white mycelium and invaded the tissues but did not form teliospores. Symptoms of invasion were increased anthocyanin formation, development of chlorosis, increased formation of secondary roots, malformed leaves and petioles, tissue necrosis and sometimes stunting.

Disease cycle and environmental relations

Under laboratory conditions, smut spores are known to survive for 5–8 years. Optimum temperature for spore germination is in the range of 20°–25°C. The pathogen is carried over on the seed coat but this method of perennation and spread is not responsible for most of the infections. Manure heaps and crop refuse are the chief sources of harbouring the primary inoculum. Having a strong adaptability to saprophytic life the fungus can survive on these sources. However, germinability of spores is reduced at 37°C and lost at 40°C.

 U. maydis is a dimorphic fungus with a yeast-like, non-pathogenic form and a filamentous (hyphal) pathogenic form. Few smut fungi are known which grow so vigorously as a saprophyte as this species. The saprophytic life can be prolonged indefinitely by transferring the fungus at intervals to fresh nutrient solution. Under field conditions this type of growth may easily occur on manure heaps and in soil. It is because of this saprophytic growth as yeast-like cells and high chances of mating between different mating types that the number of lines and biotypes is very high.

 The non-pathogenic form originates from sporidia on germination of smut spores (teliospores) in soil and is haploid. It cannot cause infection. Fusion of haploid cells and formation of dikaryotic hypha is a pre-requisite for infection. The process is regulated by a pheromone-receptor system. This system coordinates germination, cell elongation, hyphal growth, orientation toward compatible mating type and, thus, the pathogenicity. Pheromones produced by sporidia are perceived by sporidia of opposite mating type which grow into a filament that grows in a zig-zag manner towards the source of the pheromone. The hyphal tip fuses with the sporidia or its hypha if produced and a dikaryote is formed. Sporidia may lack the gene encoding for the pheromone (peptide sex hormone). Such sporidia do not attract hyphae of other mating types but they respond to pheromones from other compatible sporidia and send hyphae towards the source of pheromones. Same is true for sporidia lacking pheromone receptors.

Infection of the host occurs during the period of vegetative development. Maize ears are susceptible to infection by *U. maydis* from silk emergence until 8 to 14 days after emergence. During this period incidence of ears with galls decreases as the silks age. Pollination protects maize ovaries from infection (Snetselaar *et al.,* 2001). Ears pollinated 4 days before inoculation developed only 20% smutted kernels all of which were at the tip of the ear where pollination might not have been effective. In inoculations done 4 days before pollination the infection was 73%. Microscopic examination of silks, channels through which infection occurs, after pollination and inoculation treatments indicated that an abscission zone formed at the bases of pollinated silks and may have prevented fungal infection hyphae from growing in to the ovaries. Similar observations are reported for smut of pearl millet.

Systemic infection in the late seedling stage has also been reported but local infections of exposed embryonic tissues are more common. Once the tissues have ceased growth and have reached maturity the infection cannot occur. Mature stems may be infected if there is mechanical injury to the tissues. On infection by sporidia the host tissues are induced to grow into galls. Hyperplasia in cells surrounding the fungus hypha may appear in advance of the actual invasion of the tissues by the fungus and galls may begin to form even before the fungus actually gets there. The mycelium in the galls is intercellular during most stages of gall formation. Before sporulation, the enlarged host cells are invaded. They collapse and die. The mycelium utilizes the cell contents, grows, and soon the galls entirely consist of dikaryotic mycelium and remains of host cells. Most of the dikaryotic mycelium is converted into teliospores. During the process, they utilize the contents of the other mycelial cells also which look empty. Interaction of maize tissue with *U. maydis* induces production of lytic enzymes (pectate lyase, polygalacturonase, cellulase and xylanase) the activities of which are correlated with chlorosis, gall formation and teliospore formation. Fertilizer rates have been shown to influence maize smut development. Heavy nitrogenous manuring increases disease severity. Application of nitrogen in split doses at planting and then at 16-leaf growth stage reduces smut infection. Herbicides increase disease incidence by reducing weeds and facilitating passage of the fungus from soil to the seedling leaves.

Management of the disease

Crop rotation, field sanitation, and seed treatment may be of considerable help to reduce the incidence of the disease but the best method of management is to grow resistant varieties. Rogueing of diseased plants before the smut galls rupture is a useful cultural measure against the smut. Fungicide sprays give only limited success as control agents. However, carboxin + thiram and benomyl have given systemic protection into the growing season. An amoeba and a myxobacterium have controlled *U. maydis* in inoculated soil.

GRAIN SMUT OF SORGHUM

The most common disease of destructive nature which affects "jowar" (*Sorghum vulgare-S. bicolor*) in India is grain smut, also known as covered smut, kernel smut, or short smut. Severe losses are caused by this disease in regions where the crop is extensively grown and proper control measures are not adopted. Up to 25% of the plants have been found affected in certain areas. It is reported to cause huge losses in the USA, Manchuria, Burma, Tanganyika, South Africa, Italy, Venezuela and many other countries.

Symptoms

Only individual grains are attacked but usually the majority of them, sometimes all, in the diseased ear are involved depending on the susceptibility of the host variety. Each grain is transformed into a spore sac (sorus) which varies in shape and size according to the variety of the host crop. Generally, the grain is replaced by an oval or cylindrical, dirty grey sac, sometimes conical at the tip, and measuring 4–12 mm in length. The sac is surrounded by the unaltered glumes at the base. Sometimes, the stamens develop normally, but more commonly they are absent or are involved in the sorus, being represented by 3 conical protrusions from sides of the sorus. The stigma is often not transformed. Spores are not formed in other parts of the host.

In some varieties of sorghum, the shape and size of smutted grains is not affected and elongated sacs are not formed but the grain is full of smut powder. Such sori easily escape notice and are a dangerous source of contamination of healthy grains during threshing. In such cases the covering of the sorus is generally reddish.

Fig. 91. Grain smut of sorghum.

Fig. 92. Sori of *Sphacelotheca sorghi.*

The wall of the sorus varies according to nature of the attack. In the long protruding sacs, the wall is almost entirely composed of fungus tissues consisting of small-celled pseudoparenchyma. Only at the base the wall is composed partly of host tissues. These sacs rupture and shed their spores much more easily than in the unelongated sacs. In the hidden forms, the wall is, in great part, the ordinary wall of the ripe ovary. It is rough and rigid and usually remains unbroken until after harvest.

The interior of the sorus is completely filled with the spore powder, except a slender, sometimes curved, central column of hard tissues, the columella, which is hollowed into depressions at the surface. These depressions are filled with black spore mass. The columella is composed of the host tissues and consists of parenchyma traversed by fibrovascular bundles. Sometimes the columella is branched at the tip.

The causal organism

Grain smut of sorghum is caused by *Sporisorium sorghi* Ehrenb. Ex Link earlier known as *Sphacelotheca sorghi* (Link.) Clinton. It is placed in the same order and family as *Ustilago*. The spores of the species are round to shortly oval, dark brown in mass but olive brown singly, smooth walled, and 5–9 (generally 6) µm in diameter. They are often united in loose balls which break up into individual spores when placed in water.

Germination of spores may take place immediately or after they have been kept dry for up to six and a half years. In herbarium specimens, the spores have been found viable for as long as 13 years. In India, their germination has been seen after 2 years of storage. In water there are two types of germination. In one case a promycelium of 3 cells is formed and sporidia are budded off laterally and at the apex. The apical sporidium appears as the fourth cell of the promycelium. These sporidia are spindle-shaped, 10–12.5 x 2–3 µm in size and do not bud off secondary sporidia. In the other case, the promycelium directly develops into a branched or unbranched infection hypha. Eight physiologic races of the species are reported. The races are distinguished by their pathogenic ability on certain differential hosts. Races 7 and 8 are reported in India.

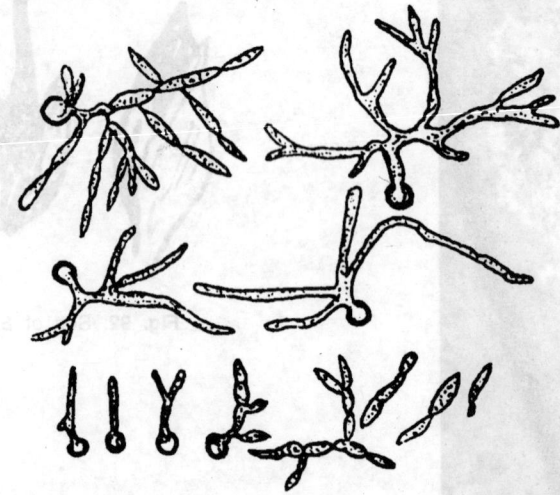

Fig. 93. Forms ot teliospore germination and budding of spordia in *Sphacelotheca sorghi*.

Disease cycle and environmental relations

The pathogen is externally seed-borne and seedling infection occurs at the time of germination and emergence of seedlings. *Cynodon dactylon* (the Doob grass) was reported as an alternative host of *S. sorghi* in 1995. Inoculation of sorghum seed with spores from the grass resulted in infection of the sorghum plant. During threshing the sori are broken and the spores get lodged on the surface of healthy seeds. They remain dormant until the next season when they germinate with the germination of the seed. Seed subjected to a teliospore suspension coupled with partial vacuum and seed directly infested with dried teliospores result in maximum diseased panicles. Inoculations with sporidia in boot stage or 10–12 leaf stage of the plant fail or give minimal disease incidence (Nzioki *et al.,* 2000). The infection can take place only during the period between germination and emergence of seedlings above the soil surface. This period varies according to host variety, soil temperature and soil moisture and depth of planting.

Sorghum germinates best at 36°–40°C, the rate decreasing progressively down to 10°C. The primary shoot grows more rapidly in high soil moisture than in drier conditions. Infection occurs best on slow germinating seeds checked by cold since the optimum temperature for spores germination is only 20–30°C. The germ tube enters mainly through the mesocotyl. Further development of the parasite within the host is like any other smut fungus. High temperatures following sowing result in less smut than do the moderate temperatures regardless of soil moisture conditions existing before or after emergence.

Management of the disease

Since the disease is externally seed-borne, seed treatment with suitable fungicides is quite effective. The use of clean seed from cobs free from smut sori is an additional precaution. Immersion of seeds in 0.5% formalin for 2 hours followed by quick drying, in 0.5–3.0% copper sulphate solution for 10–15 min followed by drying and sowing had been old and very effective recommendation. Dry seed dressing with Agrosan GN (1 : 500) is simplest and a common practice followed. Other mercurial and non-mercurial fungicides such as Arasan, N.I. Ceresan, Tillex, Captan, Ceresan M have also been used with success. The systemic fungicides carboxin and Bavistin, etc. are better than the protectant fungicides.

In Ethiopia (Africa), methanol extracts of roots of a species of *Dolichos* and *Maerua sucordata* (Capparidaceae) and powder of dry berries of *Phytolacca dodecandra* (Phytolaccaceae) are reported to disinfest sorghum seeds when used for seed dressing. The dry berry powder gives as good control of *Sporisorium sorghi* (grain smut) and *Sporisorium cruentum* (loose smut) as the synthetic fungicides (Tegegne and Pretorius, 2007). Crude extracts of *Agapanthus africanus* (Alliaceae), a wild South African plant, also have similar effects against the two smuts.

A solar energy treatment of seed was recommended in 1947. The seeds were soaked in water at ordinary temperature during summer for 4 hours in the morning and then spread out in the sun or shade to dry. This method has been found effective in Uttar Pradesh also.

The sorghum lines and varieties T 29/1, PJ 7K, PJ 23K, Nandyal, Bilichigan, CSH-9, SPV-104, SPV-102, SPV-115, SPV-297, SPV-138, SPV-245, RSV-1-R, SDM-9, CSH-7-R and CSH-5 are reported resistant.

LOOSE SMUT OF SORGHUM

The loose smut of sorghum is not as common as the grain smut in most of the sorghum growing areas. It closely resembles the grain smut in its appearance and sometimes mistaken for it. The disease has been reported from China, Iran, Southern Europe, Africa and the USA. In India, it occurs in Andhra Pradesh, Maharashtra, Karnataka and Tamil Nadu. The effects of this smut are not only on the grain but also on the plant growth. Thus, grain as well as fodder yield may be reduced.

Symptoms

The affected plants are shorter than the healthy plants. The stalk is thinner and produces tillers. Ears come out earlier than in normal plants. Generally, all the spikelets of the head are affected. The floral bracts tend to elongate and proliferate. Frequently, the lemma and palea as well as the ovary contain smut sori. The covering membrane of the sori ruptures early releasing the powdery mass of dark coloured spores. The columella persists after the spores have been discharged.

The causal organism

The disease is caused by *Sporisorium cruentum* (Kuhn) Vanky [syn: *Sphacelotheca cruenta* (Kuhn.) Potter]. The spores are formed in the ovaries and floral bracts. Loose smut sori vary considerably in size (2–4 x 3–18 mm) and are often long and pointed. The covering of the sorus is made up of loosely joined rounded and grey fungal cells which are about twice the diameter of the spores. At first there is a grey enclosing membrane. In some host varieties this wall ruptures early while in others it persists for some time thus showing characters of both loose and covered smuts. In the center of the sorus there is a solid, narrowly conical, curved, pointed columella, extending almost the length of the sorus. The columella consists of host tissues.

The spores are round or shortly elliptical and darker brown than in *S. sorghi*. They have echinulate wall and measure 5–10 μm in diameter. They germinate to form a 4-celled promycelium with laterally borne sporidia. The sporidia are unicellular, hyaline, fusiform and 12–13 μm long. The promycelium may also develop into branched or unbranched hyphae without forming sporidia. At lower temperatures sporidia are formed while at higher temperatures direct germination is common. At least 3 races of the species are known. The fungus can be easily cultured on agar media such as potato dextrose agar and Czapek agar. It hybridizes with *S. reiliana*. The resulting sori are more like those of *S. cruenta*.

Disease cycle and environmental relations

The pathogen is externally seed-borne but may be soil-borne also in dry regions. Seeds are contaminated during threshing. The spores retain viability for 4 years when kept dry. Germination of spores can occur at 8–38°C but the optimum lies between 18° and 32°C. Optimum environmental conditions for maximum infection include temperature of 20–25°C, a pH of 7.2 and 30% soil moisture. However, seedling infection can occur at 15°–30°C with soil water content from 10% to 90%. Slightly acidic soils favor infection. Deep sowing favor infection. The disease is favored by potassium and phosphorus deficiency in fertilizers. Infection of seedlings occurs at the time of germination of seeds and before emergence of seedlings. Entry into the seedlings takes place through the radicle, mesocotyl or hypocotyl. Further development of the pathogen in the host is on the same pattern as in other systemic smuts.

Management of the disease

The seed treatments recommended for the control of grain smut are effective in the control of loose smut also. Where soil survival of spores is possible, crop rotation and field sanitation are recommended.

HEAD SMUT OF SORGHUM AND MAIZE

Head smut is not very common in India and is economically not important except in Kashmir where it causes significant damage to the maize crop. It is reported from Tamil Nadu, Andhra Pradesh, Maharashtra, Karnataka, Madhya Pradesh, U.P., and Bihar also.

Symptoms

In sorghum, the smut fungus reduces plant height. The major reduction occurs in the internodes nearest the panicle and is more severe in naturally infected than in artificially infected plants. Less affected plants develop sterile panicles, and eventually. The inflorescence is invariably totally destroyed in the infected plants. Eventually, the smutted panicles develop phyllodied growth which progresses into leafy shoots. These symptoms suggest that fungal infection interferes with biosynthesis of Gibberellic acids by the host plant.

In maize large smut sori replace the tassel and the ear. The tassel is partly or wholly converted into smut spores. In such cases the floral bracts grow out into leafy structures, sometimes, into small shoots. In sorghum the entire inflorescence is converted into a big sorus about 10 cm in length and 5 cm in breadth. A thin white membrane encloses the sorus in its early stages but during emergence through the boot the membrane is ruptured and spores are exposed. A network of dark fibers traverses the spore mass and remains adhering even after the spores have been blown away.

The causal organism

Head smut is caused by *Sporisorium reilianum* (Kuhn) Langdon and Fullerton [syn. *Sphacelotheca reiliana* (Kuhn) Clinton and *Sorosporium reilianum* (Kuhn) McAlpine]. Phase contrast microscopy of the development of the sorus has shown that instead of a single central column there are several branched columns emerging from the sorus base. As the sorus grows and spores are differentiated, the number of parenchyma cells in the columns is reduced. In fully grown sorus, the columns consist of only vascular bundles. On the basis of these characters, many pathologists still consider the fungus as *Sphacelotheca*.

The spores are reddish brown to black, finely echinulate, irregular to spherical, and 9–12 μm in diameter. Germination results in the development of promycelium and sporidia. However, in contact with the host the germination results in formation of branched infectious mycelium. In *S. reilianum* two pathogenically distinct groups of physiologic race are recognized, one infecting sorghums and the other infecting maize. Each of these groups comprises one or more pathogenic races. At least 4 races in the sorghum group are reported. Only one race is reported on maize.

Disease cycle and environmental relations

The smut spores retain viability for several years (at least two years). Although the pathogen may be externally seed-borne, the major source of infection is soil-borne inoculum. The survival of spores in soil depends on soil temperature and soil moisture and, thus, soil type. Population of spores declines to 42% of the original in 6 months and to 4% after 3 years. Dry and cool soils favour survival while moist and warm soils reduce spore survival. Infections occur from seed germination up to seven-leaf stage (most at the three-leaf stage). The plumule is more susceptible than the roots. The hypocotyls is more severely affected than the coleoptile. The infection site on all organs is the meristematic region. Plasmogamy occurs between adjacent cells and infection occurs directly through the host tissue. In maize head smut (*Sporisorium reilianum* f. sp. *zeae*) the pathogen invades the vegetative shoot apex. It is mostly intracellular and passes through host cell wall by lysis and mechanical pressure. The infected cells appear normal (Martinez *et al.*, 1999b). They (Martinez *et al.*, 2002) have described the biological cycle of *S. reilianum* f. sp. *zeae* during its saprophytic and parasitic phase. In contact with a compatible host, the yeast-like cells of the fungus fused to form dikaryoic hyphae which caused infection through the host roots. The fungus acted as a biotrophic endophyte until sporogenesis in the floral meristem of the maize plant. It is embedded in an amorphous matrix and thus appears isolated from the host cells. Penetration of the host roots is never accompanied by drastic damage to host cells and there is no thickening or apposition of plant material to reinforce the host wall structure.

Optimum infection of seedlings from soil-borne spores occurs under dry conditions at 21–28°C. Application of low C : N organic amendments reduces inoculum density in soil. There is more disease at relatively high soil temperature (25°C) and low soil moisture than in wetter and cooler soils. Thus, crops raised in clay loam soils which have higher soil moisture than silt or sandy loam soils show less disease incidence. Frequent irrigation (total 15–20 cm water) for 18–21 days after sowing has been found to reduce disease incidence. Large size seeds usually produce lower number of diseased plants and infection increases with sowing depth. However, some other workers did not find any effect of sowing depth. Application of urea, ammonium sulphate and triple superphosphate reduces disease incidence.

Management of the disease

The disease can be controlled by a combination of practices such as deep ploughing, sanitation, crop rotation, and seed treatment. Since only few plants are affected in a field it is possible to locate and destroy the infected ears before they shed the spores. Disinfection of seed has been found ineffective in south India while in the USA it has given only partial control of the disease. In Russia, growing barley for 2 years after continuous maize has greatly reduced infection of maize grown subsequently. Some species of Coleoptera (*Phalacrus, Brachytansus, Lystronynchus*) feed on spores of *S. reiliana* and have biocontrol potential. *Bacillus megaterium* is effective against the smut through its antibiotics.

LONG SMUT OF SORGHUM

This smut occurs in Tamil Nadu, Maharashtra, Andhra Pradesh, Karnataka, Madhya Pradesh and Uttar Pradesh. However, it causes little damage to the crop. It is also reported from Egypt, West Africa, Iraq and Pakistan. Its occurrence is sporadic and confined to a few grains in an ear.

Symptoms

Individual and only a few grains are transformed into smut sori. Each sorus is surrounded by healthy grains. The sori are very prominent, long, cyclindrical, slightly curved, and rupture at the apex to release the brownish green spore balls and expose a bundle of 8–10 dark brown filaments.

The causal organism

Tolyposporium ehrenbergii (Kuhn) Pat belongs to family Cintractiaceae of the order Ustilaginales. Spore balls of the fungus remain united in solid balls. The exposed surface of the spores is covered by flattened echinulations. The spores do not have a dormancy period. They germinate *in situ* by the formation of an elongated promycelium which is frequently branching. Sporidia are numerous, single or in chains.

The spore balls are soil-borne. Germination of spores can occur at any temperature from 15°–36°C, the optimum being 28°C. Infection does not occur by either blossom or seed inoculation although there is a report that wind-borne sporidia cause infection through flowers. The sporidia produced by soil-borne spore balls are wind-borne to buds and initiate a systemic mycelium which later expresses itself in the heads. The ovary is converted into smut sorus in the same season. Continuous cultivation of sorghum in the same field is said to increase the incidence of this smut. There is possibility of the primary inoculum being introduced from some alternate host.

Since the inoculum is air-borne seed treatment is of no use. Early sowing of the crop reduces disease incidence. Crop rotation and field sanitation to keep down the build up of inoculum should be practiced. In the south, variety Irungu, in which glumes cover the grains, is usually free from this smut.

SMUT OF PEARL MILLET

Smut of pearl millet (*Pennisetum typhoides*) is very common in many parts of India where it has become a serious problem in the cultivation of the crop. It is reported from Pakistan, several parts of Africa, and the USA. In India, a comparison of yield in untreated plots with yield in plots given 4 sprays of oxycarboxin (Plantvax) showed that yield was 20.6% more in the treated plots.

Symptoms

The infection is visible on scattered grains in the ear in which majority may escape damage. Sometimes, the affected grains are single, sometimes in groups, and very often confined to one side of the ear or towards the base of the ear. Smut sori are pear or oval shaped. They project clearly beyond the glumes and may be from half to twice the diameter of normal grains, being often 3–4 mm long and 2–3 mm broad at the top. The top of the sorus is bluntly rounded to conical in shape. The colour of the affected grains is bright green to chocolate brown in the early stages and becomes dark black on maturity. The colour is due to the membrane covering the sorus, the enclosed spore mass being always black. The covering membrane of the sorus is tough and is made up of host tissues.

50 μm

Fig. 94. Spore balls of *Tolyposporium penicillariae*
(Courtesy: Rao and Thakur).

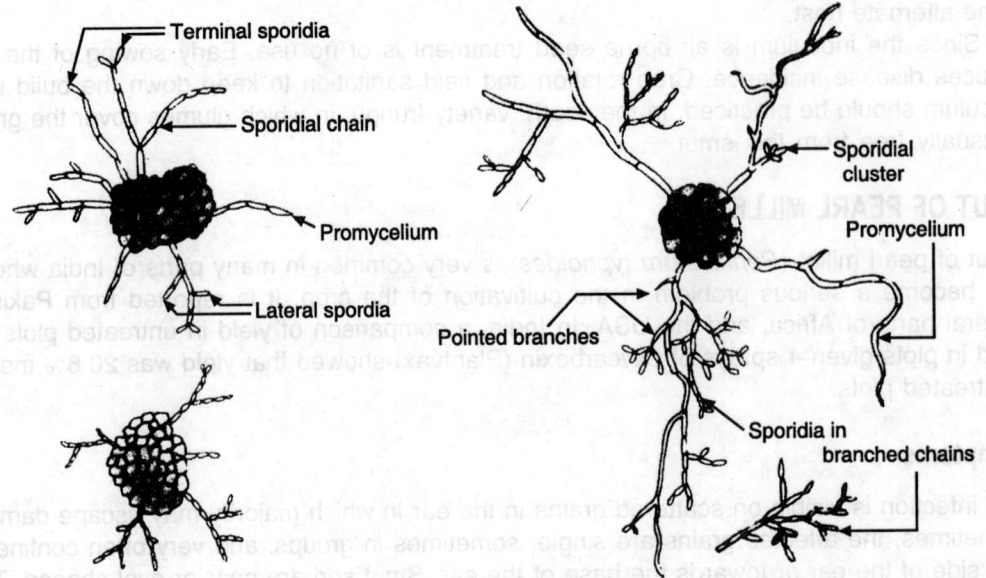

Fig. 95. Germination of teliospore of *T. penicillariae* from the spore balls.
(Courtesy: Rao and Thakur)

The causal organism

Pearl millet smut is caused by *Moesziomyces bullatus* (J. Schrot) Vanky, earlier known as *Tolyposporium penicillariae* Bref. It is classified in same order and family as *Ustilago*. The spores are held together in compact balls which remain persistent even when placed in water. These balls measure 42–325 x 50–175 μm and are not wind-borne. Each ball may contain 200–1400 teliospores packed together. Individual teliospores are round to irregular or angular, 7–12.5 μm in diameter, light brown in colour, and with a slightly roughened wall. They germinate while still held in balls. On germination a 4-celled promycelium is formed. The sporidia may be formed on the promycelium apically and laterally or the cells of the promycelium may separate, fall off, and bud out sporidia. These sporidia are hyaline, 1-celled, spindle shaped and 8–25 μm in length.

Disease cycle and environmental relations

The pathogen is soil-borne. The primary inoculum consists of spore balls that have fallen on the ground. They germinate at the time of ear formation in the next crop season. On germination they give rise to sporidia which are carried by wind currents towards the floral axis where they settle down on the florets and immediately cause infection. The result of this shoot infection is apparent in the same season when smutted grains appear. No resting period for the smut spores is necessary. Secondary infection of late sown crops or of ears appearing late also occurs by spores formed in the same season. The incubation period is only 2 weeks. Pollination with viable pollen 5–8 days after inoculation reduces infection.

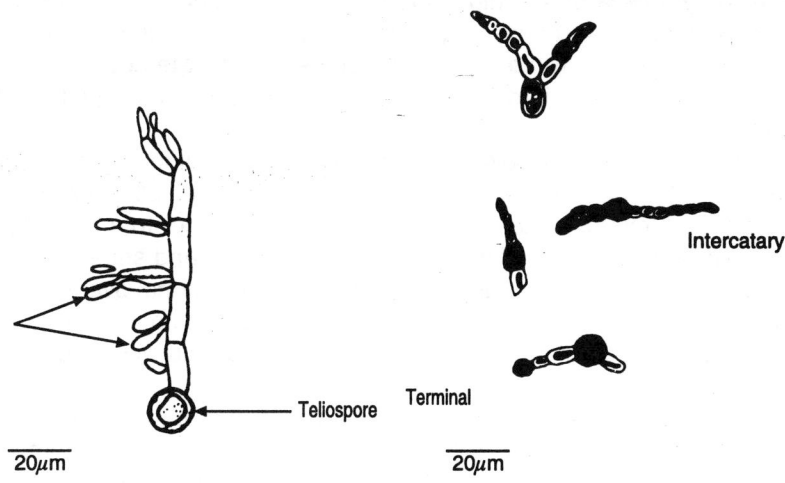

Fig. 96. Diagram showing germination of a single teliospore of *T. penicillariae* with promycelium and spordia (left) and formation of teliospore in culture (right). (Courtesy; Rao and Thakur).

Management of the disease

As the pathogen survives in soil and the inoculum from soil is air-borne, control of the disease is difficult. The control measures suggested for this disease are the removal of smutted ears, clean seed, hot weather deep ploughing, field sanitation, rotation, and use of resistant varieties. Certain genotypes of the host (SSC PS-252-S-4, and ICI 7517-S-1) have consistently shown high level of resistance. There is a report that intercropping of mungbean with pearl millet reduces smut incidence. This may be due to interception of the fungus sporidia passing from soil to the ears.

COMMON BUNT OR STINKING SMUT OF WHEAT

The common bunt, known in India as hill bunt, was one of the most serious diseases of wheat in Europe in the middle ages and was subject of observations and study by many writers since the time of Theophrastus. It has been closely associated with the history of developments in plant pathological research and is responsible for many factual discoveries in the subject. The disease, also known as covered smut of wheat due to nature of its sori, is of worldwide occurrence and is more serious than the loose smut of wheat. However, due to efficacy of seed treatments and the development of resistant varieties its seriousness has been considerably brought down. In India, the disease is confined to cooler regions such as Kashmir, Himachal Pradesh and parts of Punjab and western Uttaranchal. The disease also affects rye and a number of grasses such as *Agropyron* spp. It causes loss in yield, difficulties in threshing, and reduces grain quality. Concentrations of teliospores as low as 0.5% bunt balls (w/w) in the harvested grain reduce the grain quality.

Symptoms

The stinking smut is of two types, one or usually both are present in the same grain. One is caused by *Tilletia tritici* (low smut) in which the affected culms are shorter than normal, and the other is caused by *Tilletia laevis* (high smut) in which the culms are more often as tall as the normal ones. A dwarf bunt caused by a related pathogen (*Tilletia controversa*) is also reported. This bunt is found on winter wheats in temperate climate countries and 85 days of snow cover after sowing are required for infection. Infected plants are stunted, about half the height of normal plants.

The stinking smut is a systemic disease resulting from seedling infection. The symptoms of the disease are not usually evident until the heading stage. Only inflorescences show deviation from normal. A microscopic examination, a few days prior to emergence of the head from the boot, shows some striking differences between normal and smutted florets. The pistil from smutted heads is larger with an ovary double the length of normal ovary and green instead of almost white. The stamens from smutted head are reduced in length and breadth, the anthers are pale yellow instead of green, much reduced in size, and without perfectly organized pollen cells.

After the ears have emerged the presence of the disease can be easily detected. It is much easier in the *compactum* types since in such types the normally compact or square heads are generally changed to a more or less slender type and in many cases exceed the length of healthy ears. In *vulgare* types the change is less striking and does not become evident until the smutted grains begin to expand causing a divergence of the glumes thus giving the head a more loose and open appearance. The smutted ears are darker green than normal and remain green longer. When the wheat is in milk stage, a verification of the presence of smut can be made by pinching the grain with the thumb and the forefinger. The smutted grains yield a soft black pasty mass. In mature grains the black pasty interior changes to an oily powder, the characteristic feature of the spore mass. In some varieties the infected heads stand erect while the normal ones are beginning to droop as a result of the increasing weight of ripening grains. In awned varieties, smut frequently causes shedding of awns with the approach of maturity. The presence of the smut in the field can be detected by a foul smell caused by the presence of a volatile compound, trimethylamine, in the spore mass. About 800 μg of the compound are present per g of spores. It is this smell of rotten fish that gives the name stinking smut to the disease.

Often the badly smutted plants remain undersized and produce smaller heads than normal. Increase in tillering is not uncommon but opposite effect has also been observed. There is also reduction in root development. The affected plants are more susceptible to attack of yellow or stripe rust (*Puccinia striiformis*).

The causal organism

The common bunt is caused individually or jointly by *Tilletia tritici* (Bjerk.) D. Wint in Rabenh. [=*Tilletia caries* (DC.) Tul. & Tul.] and *Tilletia laevis* Kuhn in Rabenh. [=*Tilletia foetida* (Wallr.) Liro]. *Tilletia* is most important genus of order Tilletiales (Basidiomycota, Ustilaginomycetes, Ustilaginomycetidae,). The life history of the two species is similar. One of the chief morphological distinctions is in the markings on the spore wall. In *Tilletia tritici* the wall has reticulations ranging from minute shallow meshes to deep indentations (0.5–1.5 μm deep) while in *T. laevis* the wall is smooth. In *T. tritici* the spores are nearly round and measure 15–23 μm in diameter while in *T. laevis* they are irregular in shape and measure 17–22 μm. In both species the spores are various shades of brown and appear black in mass. They are mixed with sterile cells which are thin-walled, hyaline, and slightly smaller than the spores. The spores germinate similarly in both species. A stout germ tube (promycelium) comes out after rupturing the spore wall and bears at its tip a cluster of filiform, hyaline primary sporidia. The nuclei from the promycelium pass into these sporidia.

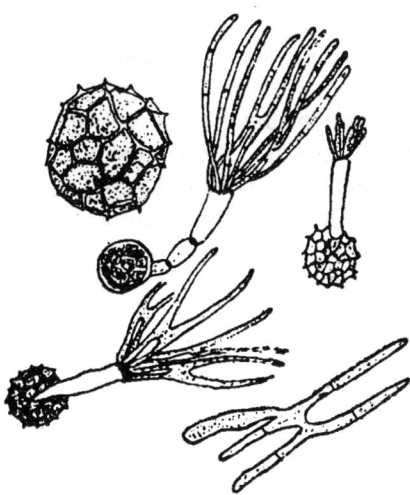

Fig. 97. *Tilletia tritici.* Teliospore, its germination, sporidial whorl at the apex of promycelium, fusion of sporidia and production of secondary sporidia.

The uninucleate primary sporidia fuse in pairs while still attached to the promycelium and form the characteristic H-shaped structures. After fusion the primary sporidia become binucleate. They germinate by giving out a hypha which bears secondary sporidia. These are sickle-shaped, hyaline, and germinate to produce the infection thread. In *T. controversa* the spores are so similar to those of *T. tritici* that it is difficult to distinguish them even with electron microscopy. The promycelium in this species gives out branches on which sporidia are formed, some of which fuse in pairs.

Physiologic races are common in the two species. These races differ in host specialization, nature of symptoms, and presence or absence of trimethlyamine smell in spores. Twenty races (T-1 to T-20) in *T. tritici* and ten in *T. laevis* (L-1 to L-10) were listed in 1965.

Disease cycle and environmental relations

Primary inoculum consists of spores carried on the seed or present in the soil. The latter is not common in India. Species in some genera of collembolans (arthropods) ingest bunt spores in soil. The intact spores recovered from their faeces show only less than 3% germination compared to 75.5% in uningested spores. Addition and incubation of collembolans in soil 10 days before crop sowing reduce bunt incidence from 30% to 3.5% (Dromph and Borgen, 2001). The spores are sticky and easily adhere to the seed during threshing and to soil particles if shed in the field during harvesting. Contamination of the soil also occurs by air-borne spores. These spores can remain dormant in soil for considerably long periods. Spores of *T. tritici* have been found to infect wheat after 10 years of storage in soil and after passage through digestive tract of cattle and sheep but not pigs and poultry.

When the contaminated seed germinates, the spores also germinate. Diploidization occurs and hyphae produced by secondary sporidia form appressoria in contact with cuticle of host seedling. A penetration tube invades the host tissue intercellularly. The infection hyphae grow towards the apical meristem and must reach it before internodes of the plant begin to elongate. When the fungus reaches the culm primordium it grows with the host tissues until heads are produced. It then forms chlamydospores (bunt spores) which begin to mature in the internal tissues of the kernel.

Factors like temperature, moisture, soil type, soil fertility, host variety, physiologic race, spore load, depth and rate of sowing, and day length have marked influence on the germination of spores and subsequent infection of the host seedlings. Minimum temperature for spore germination is 0°–4°C, optimum 18°–20°C and maximum 36°C. Spores take 4–5 days to germinate at 15°C and 10–14 days at 5°C. Different physiologic races differ in their temperature relations. The optimum soil temperature for infection is 9°–12° C. Infection is optimum when soil moisture is midway between field capacity and permanent wilting point with soil temperature of 5°–10°C. Planting during periods of high temperature effectively reduces disease incidence. Shallow sowing also helps in checking the infection.

Management of the disease

In regions where environmental conditions are favourable for perennation of the pathogen and spread of the disease it is always advisable to grow resistant varieties. In order to avoid infection from soil-borne inoculum sanitary precautions (destruction of crop debris) and rotations are recommended. Modifications in the soil temperature and moisture conditions can be brought about by altering the date and depth of sowing. Shallow seeding in warm wet soil drastically reduces the disease incidence.

Seed treatment with appropriate fungicides not only destroys the seed-borne inoculum but also checks infection of seedlings from soil-borne spores. Copper sulphate (2%), formalin (half kg in 200 lit of water), copper carbonate dust (2–3 oz. per 30 kg seed) had been used in the past. Subsequently, organo mercurials such as Agrosan GN, NI Ceresan and Ceresan M were in common use. Grewal *et al.,* (1965), after screening a large number of fungicides, had reported that Panogen at the rate of 2 ml/kg seed and 2% Ceresan give complete control of seed-borne inoculum and corresponding increase in yield. Seed treatment with Benlate (0.3%) and Vitavax (0.2%) also gives complete control of the disease. The efficacy of Vitavax depends on its proper concentration for the level of seed infestation which is determined by the level of disease intensity in the parent crop. The efficacy decreases when lower rates are applied. The triazole, bitertanol, has been recommended against *T. tritici* at the rate of 19 μg/kg seed for seed-borne inoculum and 56 μg/kg seed for soil borne inoculum. Difenoconazole seed treatment is recommended against dwarf bunt, *Tilletia controversa*, and may be used against common bunt also. Tilt (0.01%) is highly effective as seed treatment against the common bunt. Mancozeb and, maneb were also used for seed treatment but were discarded due to phytotoxicity on seed after storage for long periods.

Bacillus licheniformis, B. megaterium, B. pumilus and *B. subtilis* are reported as antagonists of the bunt pathogen in Australia. A strain of *Pseudomonas fluorescens* was reported to provide biological control of common bunt (*T. laevis*) by inhibition of teliospore germination. Hokeberg *et al.,* (1997) had reported that a strain of *Pseudomonas* isolated from roots and applied as seed treatment strongly and reliably suppressed common bunt (*T. caries*). The antagonist could be stored as suspension or on seeds in refrigerator for a month. Johnson *et al.,* (1998) also reported suppression of seed-borne *T. caries* by a strain of *Pseudomonas chlororaphis* but soil borne inoculum was not controlled.

Organic seed treatment as alternative to chemical seed treatment has been investigated. Field trials have shown that seed treatment with skimmed milk powder, skimmed milk and wheat flour used at 160 g per kg of seed reduced common bunt infection levels by 96%, 93% and 62%, respectively. In most cases skimmed milk gave as good control as chemical seed treatment. Seed treatment with yellow mustard meal (60 g/kg seed) is also reported to give good control of seed-borne *T. tritici*.

Wheat cultivars S 227, PV 18, HD 2012, HD 4513, HD 4519, Kalyan Sona, UP 2002, HB 383, WL 410, WL 885, WL 1581, IWP 72, IWP 87, IWP 127 and IWP 129 are listed as resistant, some of them immune, to common bunt.

KARNAL BUNT OF WHEAT

Karnal bunt, known also as New Bunt or Partial Bunt was a minor disease of wheat which became a major problem since 1966 with the introduction of new wheat varieties and concurrent new technology for wheat cultivation. It is known to occur in India, Pakistan, Afghanistan, Iraq, Mexico, and Nepal and has been detected in samples from Syria and Lebanon. The disease was first reported in India from Karnal (Haryana) by Mitra in 1931. It was suspected that the disease was present in the then undivided Punjab (which included the present Haryana and a part of present Pakistan) even as early as 1908. This bunt is more prevalent than the stinking smut or hill bunt which is restricted to cooler hilly areas of north India. Earlier the Karnal bunt also was restricted to certain mountainous districts of Punjab and a part of western Uttar Pradesh adjoining Punjab and the present Haryana state. Surprisingly, during the 1940s and 1950s when Punjab supplied the bulk of wheat seed to most of the northern India the disease did not spread to other areas. Now, it occurs in Delhi, Punjab, Haryana, Jammu and Kashmir, Himachal Pradesh, Rajasthan, Uttar Pradesh, north Bihar, West Bengal, Madhya Pradesh, and Gujarat. It is well established in the major wheat producing areas of the country. Although sporadic occurrences of the disease had been reported from different places it was not considered an important disease of wheat until 1969–70 when most of the dwarf wheat varieties were reported affected by it. In the "tarai" area of north western Uttar Pradesh the disease had once appeared in severe form in some fields in 1966 but was not observed for the next 3–4 years. It again appeared in 1969–70. For the first time its recurrence in two successive seasons was observed during 1974–75 and 1975–76.

In certain varieties such as HD 2009, the percentage of affected grains was as high as 30–40. Loss in yield of about 40000 metric tons of grain per year had been reported in 1975. Normally the disease causes a total loss of 0.3–0.5%. Loss of 42.4% in cv.WL 711 and 57.5% in cv HD 2009 was estimated during the epidemic of the disease in Punjab in 1978–79. Both quantity and quality of wheat grains and seed are adversely affected. Trace infection does not affect the seed germination and seedling vigor. According to Warham (1990) Karnal bunt infection has very little effect on seed viability irrespective of the age of the seed while germination of infected seed appears to depend upon the wheat cultivar and age of the seed In contrast, there is significant reduction in vigor of infected seeds. Such seeds are likely to have lower survival rate in storage compared to healthy seeds of the same seed lot. During the last 6–7 years the incidence of Karnal bunt has come down to a level where it is of not serious concern in India. The fungus causing Karnal bunt of wheat is reported on several wild species of *Triticum* and also on rye and *Triticale*.

Symptoms

The symptoms of the disease are totally different from the hill bunt. The Karnal bunt becomes evident when the grains have developed. It is then found that some grains have been partially, rarely wholly, converted into black powdery mass enclosed by the pericarp. Not all the ears in a stool carry the disease and even in the same ear only few grains are smutted. Since embryo is not always damaged such partially affected grains can germinate and even produce a healthy plant. Due to irregular distribution of infected grains in the ear it was presumed that they are the result of air-borne local infection. As the grains reach maturity, the outer glumes spread out slightly and inner glumes of the spike expand. The spore mass remains covered by the pericarp for some time but later ruptures exposing the black powder to the atmosphere. The presence of foul smell due to trimethylamine is prominent in this disease also and, therefore, the disease had been mentioned by scme early workers as stinking smut.

Fig. 98. Karnal bunt affected wheat ears.

Fig. 99. Karnal bunt affected wheat grains.

The causal organism

The causal organism of Karnal bunt was first reported as *Tilletia indica* by Mitra. The name was later changed to *Neovossia indica* (Mitra) Mundkur in 1940 but was again transferred it to *Tilletia indica* Mitra in 1953. Spore morphology differs from other *Tilletia* species on cereals. Genomic finger printing has also placed the Karnal bunt pathogen separate from *Tilletia* spp. associated with common bunt (McDonald *et al.*, 2000). However, the recent trend is to follow the nomenclature of Fischer and the pathogen is mentioned as *Tilletia indica*. Mitra.

The teliospores of *T. indica* are darker than those of *Tilletia tritici* and *T. laevis* that cause common (hill) bunt. They are spherical to oval, with reticulation on the epispore which appear as curved spines. These spores measure 22–49 μm (average 35 μm) in diameter. There are significant differences in teliospore size from samples collected on the same wheat variety from different locations in India. The range and average size on cultivar Sonalika was 29.4 to 39.2 (36.41) μm in Punjab, 34.8 to 46 (40.6) μm in Haryana and 34.3 to 44.1 (40.64) in Uttar Pradesh. The differences were attributed to environmental conditions. A thin, hyaline membrane, the perisporium or sheath, surrounds the spines on the epispore and persists when the spores are mature. The young spores bear an apiculus which arises from the episporium and not the perisporium. The reticulate projections are composed of two double strands which are cemented together by two cross bands at or near the apex. The endosporium is thick and lamillate. Numerous large sterile cells are present mixed with the spores.

The spore germinates to form a short, stout promycelium at the apex of which a whirl of 60–185 sporidia is formed. These sporidia are long, sickle-shaped and do not fuse in pairs on the promycelium. The fungus is heterothallic and incompatibility and pathogenicity are

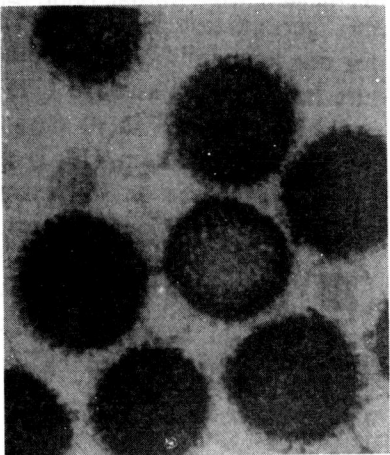

Fig. 100. Teliospores of *Tilletia mdica with strile cells. (Courtesy : Dr. Amerika Singh)*

controlled by multiple alleles. In fungi where proper dikaryotization must occur for successful infection, such as in smuts, the heterothallism has special significance in epidemiology. *Tilletia indica* depends on encounters on wheat spikes between airborne secondary sporidia of different mating types for successful infection and reproduction. As a result there is reduced

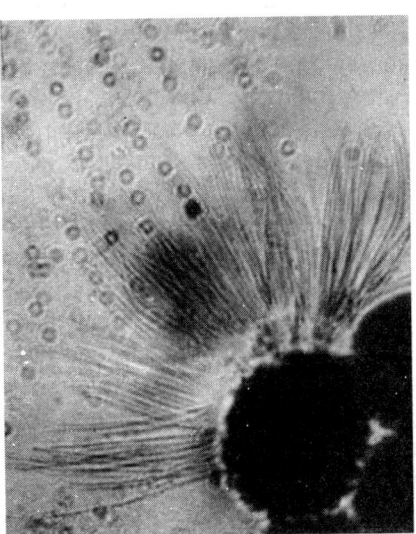

Fig. 101. *Tilletia indica.* Sporidial whorl (Courtesy : Dr. Amerika Singh)

reproductive success in low population densities of sporidia (Garrett and Bowden, 2002). With the help of simple population model it has been possible to demonstrate a theoretical threshold population size below which populations of *T. indica* decline rather than increase. This is known as **Allee effect**. Deployment of partial resistance, use of fungicides and a very limited period for maximum infection of spikelets together with environmental parameters against the pathogen may push its population levels below the threshold (Garrett and Bowden, 2002; Jones, 2007). This may explain the fall in incidence of Karnal bunt throughout the country since late 1990s. It may also partly explain why Karnal bunt remained a minor disease until mid-1960s.

The mature teliospores contain a diploid nucleus which divides meiotically either in the spore or in the promycelium. The unicellular primary sporidia on the promycelium detach *en masse* and develop either into monokaryotic mycelium or produce secondary sporidia. The compatible primary sporidia and secondary sporidia fuse and develop into the dikaryotic mycelium and/or sporidia which indicates the heterothallic nature of the fungus. Infection is caused by the dikaryotic sporidia or hyphae. In light and electron microscopy Cashion and Lutterall (1988) had observed that the fungus is restricted to the pericarp portion of the caryopsis where it is entirely intercellular. Hyphae proliferate in the space formed by disintegration of middle layers of the pericarp during normal development of the grain. The embryo is free from infection even in severely infested grains (Aggarwal *et al.*, 1994; Cashion and Lutterall, 1988). Aujla *et al.*, (1992) studied the sporogenesis and observed that in the developing grains rounding up of the intercalary and terminal cells of sporiferous hyphae was seen 21 days after inoculation. After 28 days apparent symptoms were not seen on the spikes. Possibly by the time the grains matured fully the teliospores formed intercalarly and in chains separated and formed the powdery mass.

The germination of teliospores of *T. indica (Neovossia indica)* for experimental work has always been a problem. Mitra in 1935 and Mundkur in 1943 had emphasized the long dormancy of spores in this species. The germination is unsteady and the exact conditions for germination are unknown. Fresh spores show none or very low germination. Percentage of germination improves with ageing of spores in sori but even 8–9 months old spores give poor germination. While 1–2 years old spores give maximum germination, older spores show decreasing rates of germination. Holton in 1949 induced teliospore germination by soaking in tap water at room temperature for one hour followed by incubation on 2% water agar at room temperature for 7 days. Vasudeva in 1957 had claimed that up to 80% germination could be obtained in 10 days at 15–22°C in extract of farmyard manure. Similar results are reported for spores dusted on soil extract agar and incubating at 10–25°C. Good germination of spores for comparative studies could be obtained by pre-soaking in solutions of aldehydes and fatty acids for 24 hours followed by incubation in moist chamber at 20°–25°C. Many other chemicals and plant extracts have been successfully used for enhancing germination of teliospores of *T. indica.* In a study with spores from India and Mexico teliospores from newly matured wheat spikes showed less than 10% germination which increased to 40–60% in 4 months. Highest germination occurred after 3 weeks of incubation at 15°–20°C in continuous light at pH 6–9.5. Percent germination was similar on soil extract agar and water agar. When spores germinated 2 mm under soil or water agar, the promycelium fails to reach the surface. In a study in Punjab, in 1986, 69% germination in 1:100 dilution of a 1:1 mixture of soil extract and FYM extract in 8 months old spores was obtained. Free floating spores mainly produce sporidia either on a short, stout promycelium or the sporidia arise directly from the teliospores. The promycelium from submerged spores is long and branched. Exposure of teliospores to sunlight at 40°–43°C for 14 days terminates their dormancy. The optimum temperature range for teliospore germination is 20°–25°C while in China optimum has been reported as 15°–22°C with extremely low germination at 2°C and no germination after prolonged exposure to 35°C in darkness. The teliospore contains water soluble self inhibitors which prevent germination. Frequent or excessive irrigation or rains may remove these inhibitors from the spores.

In a study of influence of media on pathogenicity and morphology of secondary sporidia of *T. indica*, Warham and Burnett (1990) had observed that when the pathogen is grown on potato dextrose broth with or without supplemental source of sucrose on a shaker for one week the secondary sporidia produced in the liquid culture are characteristically filiform while on solid potato dextrose agar medium allantoid secondary sporidia are produced. The filiform sporidia are less infectious than the allantoid sporidia. They recommended that for screening varieties for resistance the allantoid sporidia should be used as inoculum. Contamination-free viable teliospores from seed and soil can be obtained by treatment with sodium hypochlorite or acidic electrolysed water (Bonde *et al.*, 2003). The general opinion is that the spores fall on the ground and survive until flowering time of the next crop or they are introduced into the soil through infected or contaminated seeds and remain inactive until flowering time of the crop.

Physiologic specialization in the species is common and the races are fairly stable. In 1987, Aujla *et al.*, had detected 4 distinct pathotypes (K 1 to K 4) on the basis of pathogenicity on genotypes of *Triticum aestivum*, *T. durum*, *Triticale* and *Secale cerealis*. The pathotypes K-1 and K-2 were most common in Punjab. Pannu and Chahal (2000) divided isolates from different locations of northern India into 6 groups based on their pathogenic behaviour on wheat and triticale. Isolates that produced more secondary sporidia were more virulent. On the basis of pathogenicity on 10 differential hosts and biochemical characteristics, Sharma *et al.*, (2002) distinguished three pathotypes (1,2 and 3). The highly virulent isolate had higher lipid, nitrogen, protein, sugars and reducing sugars than less virulent isolates. Subsequently, Sharma *et al.*, (2004) collected T. indica isolates from various location in plains of north India and Zone 1 of Himachal Pradesh and grouped them in to 5 pathotypes (I-V) on the basis of their reaction on a set of 20 genotypes of wheat and triticale.

Disease cycle and environmental relation

The pathogen perennates through teliospores in soil or on the seed. In a recent report, Moosawi and Nejad (2007) have claimed that bunted kernels contain fully developed teliospores as well as some teliosporogenous hyphae which survive with the spores. Under suitable conditions including suitable substrate these hyphae can form teliospores, thus, increasing the amount of inoculum. In cold climate conditions the spores buried in soil survive, as dormant spores, for several years High humidity, frequent light rains, cloudy weather, and low temperature at the time of flowering generally promote disease development. In Punjab the disease is severe when temperatures from February onward are 18°–22°C and relative humidity is above 70%. The meteorological conditions favorable for the disease in the crop usually coincide with flowering time of wheat in north India. Crops given excess of irrigation and nitrogenous fertilizers show heavy incidence of bunt. Fields with high microbial populations, resulting from green manuring and fertilizers, show lower incidence of bunt (cf. Singh, A., 1994). Strains of *Pseudomonas fluorescens* are known to inhibit teliospore germination of the common bunt fungus (McManus *et al.*, 1993) and it is possible that in case of suppression of Karnal bunt in soils with high microbial activity similar bacteria might be playing some role. Practices that stimulate teliospore germination during the off season when wheat crop is not in the field may lead to self destruction of the pathogen. Soil incorporation of sugarcane refuse or wheat straw or growing of soybean, pigeonpea, pearl millet, sorghum or sugarcane before wheat are some of these practices (Singh, A., 1994).

The observations recorded by Singh, A. (1994) and the fact that Karnal bunt remained a minor disease for decades before advent of new varieties and new production technology for the crop suggest that the changes in production technology for wheat are responsible in some way for the increased incidence of the disease. As Bedi (1994) has remarked the popularization of wheat-rice rotation (no place for legumes or a green manure crop as practiced in the past), excessive use of inorganic fertilizers and irrigation (even at the time of flowering which may lead to germination of teliospores and discharge of sporidia), cultivation of high yielding susceptible varieties in disease prone areas, popularization of combine harvesting, etc. are some of the reasons to which the spread of Karnal bunt can be attributed. Sharma *et al.*, (2000) have reported that the mean percent incidence of Karnal bunt in different varieties is much higher in rice-wheat cropping sequence than in the maize-wheat sequence.

The teliospores remain viable in soil for 4–5 years and act as primary source of inoculum. Experiments conducted in Europe (U.K., Norway, Italy) have shown that teliospores can survive in soil at 5–20 cm depth for at least one year (Valvassori *et al.*, 2002). However, prolonged survival occurs only in dry soil. Severe incidence of Karnal bunt in some areas may be due to high density of teliospores in soil. The maximum period for which the teliospores could survive at soil depths of zero, 7.5 and 15 cm is 45, 39, and 27 months, respectively. In storage these spores are known to survive for 7 years. The disintegration of teliospores begins after breakdown of the sheath covering them (Babadoost, *et al.,* 2004). Thus conditions favourable for breakdown of the sheath reduce survival in soil. They can also survive in wheat straw (*bhoosa*) for 2 years and in farmyard manure for 1 year. Ingestion of bunt spores by chickens and grasshoppers or passage through digestive tract of a cow reduced the germinability of spores but does not completely prevent it. Germination of the teliospores placed in the rhizosphere of *Pennisetum* and *Sorghum* is decreased while rhizosphere of rice, cotton and cucurbits tends to improve their germination.

Awn emergence stage is most susceptible for artificial inoculation and related studies. According to Bains (1994) sporidia landing on spikes immediately after their emergence from boot leaf cause more infection than sporidia landing at other stages of the spikes. These findings were further elaborated recently by the work of Goates and Jackson (2006) in the USA. Spikes of a resistant and a susceptible cultivar were inoculated at eight growth stages from awns emerging to soft dough. Spikes became susceptible only after complete emerging from the boot and continued to be susceptible up to soft dough stage. Stages long after anthesis are not susceptible and only low level of disease occurs. Disease severity in both cultivars peaked when spikes were inoculated after complete emergence, but before the onset of anthesis. Disease levels tapered off gradually in spikes inoculated after anthesis. The results broaden the known susceptibility period of wheat to *T. indica* to include stages long after anthesis, and indicate that infection from airborne inoculum is not possible during boot or awns emerging stages, which are commonly referred to as the most susceptible stages. It implies that the spores must remain ungerminated until the spike emergence stage of the host is reached. Soil conditions may induce the spores to germinate early and reduce the risk of establishment of the fungus.

In the field the spores germinate towards the middle of February or early March. Spores from intact spore balls germinate better than loose masses of spores. Mid-day and afternoon conditions exert maximum stress on production of secondary allantoid sporidia. The effect of extremely low (–5° to 10°C) and extremely high (35°–40°C) on germination of bunt spores was studied by Kumar *et al.*, (2003). There is a exposure time-based periodicity in the level of germination. The high and low temperatures induced lethality in the teliospores when exposure time was 10, 15, 30, 15 and 10 days at –5°, 15°, 35° and 40°C, respectively. The extremely low temperature under field conditions is not uncommon. The complete life cycle of the pathogen as proposed by Dhaliwal and Singh (1988) involves germination of sporidia while still attached to the promycelium, in culture or on soil surface, producing allantoid (banana-shaped) and filiform secondary sporidia. The former are violently released and cause infection in nature. The latter serve as reproductive entities to produce more allantoid sporidia. The allantoid sporidia can germinate on leaf surfaces, forming small colonies, and producing more allantoid sporidia. The air-borne sporidia germinate on the glumes and the fungus becomes partially systemic in rachis and rachilla. The disease spreads to adjacent florets and spikelets around the infection site. The germination of teliospores leading to formation of allantoid and filiform sporidia, infection and, finally, teliospore mass formation in the maturing grain are all affected by temperature. The spore mass formation sets in when grains are hardening. It requires a temperature range of 25.3–34.9°C at this stage (Kumar et al., 2003). If temperature range is below this, there is no spore mass formation and grains do not show visible symptoms.

Aujla et al., (1990b) report that under field conditions in Punjab if soil remains moist (15% or above) through rainfall, irrigation, or cloudy and foggy weather even for 7–10 days during December to March and the minimum average temperature is 3.7°C and maximum 27.5°C there is good teliospore germination. The sporidia which are produced are wind-borne and cause infection of individual florets. There are usually 1–4 primary infection sites per spike and then there is spread to other florets under favorable conditions. During extremely favorable weather the simultaneous infection of adjacent spikelets on alternate sides of the rachis and of closely lying ovaries in the same spikelet from the same infection site is fairly common (Bedi and Dhiman, 1984). These workers believed that secondary and tertiary spread of the pathogen within and between spikelets occurs through secondary sporidia or mycelium produced on the initially infected spikelets. When minimum temperature is between 7° and 9.2°C (average 7.9°C) and maximum between 15.5° and 19.8°C (average 16.4°C) with relative humidity above 75% and intermittent rains, there is maximum (up to 81.3%) incidence and spread of the infection in a ear. Resistant wheat cultivars have higher hair density on glumes and rachis than susceptible cultivars (Grewal et al., 1999). Soil conditions may induce the spores to germinate early and reduce the risk of establishment of the fungus.

Management of the disease

Due to lack of definite knowledge of exact disease cycle control measure against Karnal bunt had been controversial. Late sown crops suffer more hence early sowing is better. Nitrogen doses above 80 kg/ha together with irrigation often increase bunt incidence. Crop rotation, to avoid soil-borne inoculum, may not be always of much use due to seed-borne nature of the inoculum and wind dispersal of sporidia. Cases have been reported where the disease appeared in a field which had no history of wheat cultivation and in which supposedly clean seed was sown. At the same time, since the pathogen is soil-borne also and sporidia may be brought into the field from outside by wind, seed treatments alone do not ensure control.

The most effective means of managing Karnal bunt is to develop resistant varieties. Durum and aestivum wheats show promising resistance while Triticum vulgare is susceptible. In screening done in Mexico, no bread wheat were found resistant but 41 of the 448 durum wheat and 151 of 710 triticale lines were bunt free. In varietal trials at the Indian Agricultural Research Institute, varieties HD 1907, HI 358, HP-743, L 176, L 191, M-137A, MW 59-4 x 6A had shown resistance to N. indica. At Pantnagar, under natural conditions of infection, UP 270, UP 368, HD 2222, 2227, and 2235 remained free from bunt in epidemic years. Varieties Arjun, WG 357, WL 711, WL 1562, WH 283, HD 2329 and 2285 and UP 262 and 2003 are susceptible to Karnal bunt. In WL 1562 and DWL 5023 small sori are formed hence there is less inoculum for soil infestation. There was a suggestion that increased cultivation of these varieties had increased the incidence of the disease. Some reports have suggested that there is field resistance in varieties Sonalika, WL 1562, PBW 120, 154, 34, and 65, TL 1210, HD 2281, HD 29, HD 30. Biosynthesis or presence of phenols, especially hydroquinone, is reported to be associated with resistance and testing for these compounds in seeds and seedlings can indicate resistance to the disease (Gogoi *et al.,* 2001). Sharma *et al.,* (2004) have listed gentypes HPW 56, HPW 74, HPW 93, H 567 71/3 PAR, HPW 147, K 8962, PBW 316 and HUW 385 (*T. aestivum*), PBW 34, WH 912, WH 913 Raj 1555, Raj 6513, DWR 202 (*T. durum*) as having combined resistance to Karnal bunt, leaf rust (*Puccinia recondita*), yellow rust (*Puccinia striiformis*) and powdery mildew (*Blumeria graminis tritici*). Resistant wheat cultivars have higher hair density on glumes and rachis than susceptible cultivars (Grewal *et al.,* 1999).

Chemical control of the disease has been tried by many workers. In 1935, Mitra had reported that seed treatment with Agrosan GN, Hortisan A or sulphur dust, even at very low dosages, reduced infection from seed-borne inoculum. Although seed treatment with fungicides does not provide complete control of the disease, the seed-borne inoculum is considerably reduced. In this regard Ceresan and NI Ceresan were found best by him. Different degrees of control (50–100%) of seed-borne inoculum with Agrosan GN, thiram, zineb, benlate, Bavistin, Vitavax and Du-Ter have been achieved under experimental conditions. Singh, M. and A. Singh (1991) conducted laboratory experiments to determine the sensitivity of sporidia to different fungicides and observed that Bavistin was best in reducing sporidial germination and germ tube elongation, followed by Thiram, carboxin and mancozeb. Although most fungicides are capable of inactivating the teliospores, when used for seed treatment they are not as effective because in point infections the spores are well protected by pericarp of the seed. Smilanick *et al.,* (1997) had reported that immersion of seed in water alone at 80°C or with sodium hypochlorite (1.6%) killed the free as well as internal teliospores within 1 min. For general treatment, removal of infected grains before treatment is useful.

The problem of soil-borne inoculum and wind dispersal of sporidia does not permit seed treatment to be effective. Attempts have been made to solve this problem also. As early as 1985, it was reported that infection of wheat flowers can be prevented by a spray of either mancozeb (0.2% a.i. basis) or carbendazim (0.1% a.i. basis) at the early heading stage before flowering of the crop. Fentin hydroxide (0.25% a.i.) is also effective. Bavistin, Baycor, Baytan Bayleton, Benlate, Blitox, Ceresan Dry, Dithane M-45, Topsin M and Vitavax used as spray either at booting or awn emergence or at both stages alone or accompanied by seed treatment lower down disease incidence. Although all the fungicides are effective, Baytan and Bayleton are most effective. Spray applications after pollination have no effect on the disease and seed treatment alone is also not effective. Use of propiconazole (Tilt), a triazole fungicide, has also been found very effective material for spray to control Karnal bunt. A single spray of Tilt reduces bunt incidence by 71.4% and three sprays give 100% control. Goel *et al.,* (2000) obtained 78–87% control in multilocational trials with a single spray of Tilt (250 or 500 ml/ha) at the boot stage. Apart from, Tilt, other triazole fungicides that have been successfully used are tebuconazole (Folicur) and cyproconazole at 500 or 200 ml per ha Sharma and Basandrai (2000) compared Tilt (200 ml/ha), Bacor (250 g/ha) and Bavistin (1000g/ha) and found better control with Tilt. They also demonstrated significant control of Karnal bunt by seed treatment with or foliar spray of conidial suspension of *T. harzianum* and *T. viride*. Leaf extract of *Azadirachta indica* (neem) and *Cassia fistula* (amaltas), used as spray, is reported to give control of bunt not much inferior than propiconazole (Sharma and Basandrai, 2004).

In general, the management practices that should be followed in areas where the Karnal bunt is of regular occurrence include the following. Cultivation of susceptible varieties and continuous cropping of wheat in the same field should be avoided. Practices like deep ploughing during summer to bury the spores deep in soil and inclusion of a green manure crop in the rotation to encourage spore germination and their self destruction should be followed. Some modifications in the date of sowing, amount of fertilizers and number of irrigations may be necessary for some years until the inoculum load in the area has been minimized. Where possible the date of sowing should be such that favorable weather for the disease and flowering of the host do not coincide. Irrigation of the crop just before or during flowering should be avoided to reduce humidity and disease incidence. These steps may result in slightly lower yields but the ultimate gain more than compensates the loss. Only clean seed, free from smut or its spores, should be used. Seed must be treated with Vitavax, Bavistin, Thiram or any organomercurial fungicides. This ensures inactivation of visible or invisible spores on the seed.

To sum up the control strategies for Karnal bunt the following facts may be considered. Every since its discovery in 1930s it remained a micro disease inspite of the fact that the area where it was discovered used to supply wheat seed throughout India. Suddenly, it became a major disease in the latter half of 1960s. The period when remained a minor disease was dominated by a set of cultural practices for wheat. These included rotation of wheat with either fallow or a legume, periodical green manuring of the field and use of mercurials for seed treatment. The practices changed after the 1950s. Rotation was dominated by wheat-rice combination, green manures became rare. However, seed treatment continued. Now research has shown that legume-wheat rotation and green manure to increase humus content of soil reduce the population teliospores in soil while seed treatment reduced the density of teliospores in seed. This, as stated above, with the disease cycle and provides natural control of the disease.

BUNT OF RICE

The bunt of rice, also known as black smut, kernel bunt or kernel smut of rice occurs in Japan, China, Indochina and Cambodia, Philippines, Java, Taiwan, Myanmar (Burma), India, Trinidad, British Guiana. In the USA it is a major disease in Texas. The first report of its occurrence and description was from Japan. In India the disease occurs in Assam, West Bengal, Bihar, Uttar Pradesh, Karnataka, Orissa, Punjab, Rajasthan, Tamil Nadu and Andhra Pradesh. In eastern U.P. the loss has been estimated at about 3.2%. Early maturing varieties are more susceptible than late maturing varieties In the recent past the disease has become important and its incidence is consistently increasing. Up to 87% panicle infection is reported in Pakistan (cf. Chahal *et al.,* 1999)

Symptoms

Only a few grains in the ear, not more than 3–4, are affected and not all the ears in a stool show the presence of bunt balls. Very often the entire grain is not transformed into a sorus, a portion remaining free from spores. Mostly the sori are hidden by the glumes making the smutted kernels inconspicuous in the field or in the threshed grains unless the sorus is broken. However, sometimes the sorus forces the glumes apart slightly when the disease can be detected as minute black pustules or streaks bursting through the glumes at the time of ripening. Awns of partially affected grains remain green for a longer time than on healthy grains. A frequent characteristic is the twisting over and outward of a portion of the unaffected kernel due to partial smutting of the grain. Whole grains may also be destroyed. When crushed with fingers affected grains yield the black powdery spore mass. This powdery mass may shed before harvest and can be seen adhering to awns and on adjacent leaves. If seeds are not too badly damaged they may germinate but the seedlings are weak and stunted.

The causal organism

The disease is caused by *Tilletia barclayana*, earlier known as *Neovossia horrida*, *Tilletia horrida* or *Neovossia barclayana*. There were suggestions that the pecies should be retained under *Neovossia* not in *Tilletia* on the basis of its morphological characters such as the presence of an appendage or apiculus arising from the exosporium and numerous sporidia which do not fuse in pairs. However, according to differentiations based on genomic fingerprinting, the pathogen is now accepted as *Tilletia barclayana* (McDonald *et al.,* 2000).

The spore masses are pulverulent, black, and produced in the ovary of the host. The spores are spherical, black when mature, and measure 22.5–28.7 x 20–25 μm. The exosporium (episporium) is thickly covered with conspicuous spines which are hyaline or slightly coloured and curved. A thin hyaline membrane (perisporium or sheath) surrounds the exosporium. This sheath remains persistent after maturity of the spores because of this sheath the spores are sticky. A hyaline appendage, as observed in *N. indica*, is present in this species also, particularly when the spores are immature. Along with these spores, sterile cells are also found in the powdery mass. Morphological, cultural and pathogenic variations occur among different isolates of the pathogen (Pannu *et al.,* 2002).

Teliospores stored at low temperature germinate readily. They seem to have no dormancy. Good germination can be obtained by soaking them in water for 10 days and then exposing to moist air. Light and good supply of oxygen have a marked influence on their germination which can be hastened by exposure to ultra-violet rays. The spores become lighter in color, absorb moisture, and swell before germination. The mature spore contains a single diploid nucleus which also enlarges and undergoes successive divisions in the spore itself. The exospore cracks open and a short, thick, hyaline, and cylindrical promycelium emerges out. The promycelium isstraight when the spore germinates on water surface. When the teliospore is deeply immersed in water, the length of the promycelium increases to reach the water surface. The four nuclei in the spore migrate into this promycelium. During the process of further elongation of the promycelium and migration of nuclei another mitotic division occurs and 8 nuclei are formed. Subsequently, with full development of the promycelium, further mitotic divisions increase the number of nuclei to 40 or more. These nuclei move towards the tip of the promycelium where small projections have already appeared. These projections ultimately elongate to form sporidia. One nucleus with a small amount of cytoplasm moves into each sporidium. There are 32–76 sporidia in a whorl at the tip of the promycelium. They are filiform or needle-shaped, with a flat base and a pointed tip, curved in various ways, and measure 38–53 x 1.3–1.5 μm. They do not fuse in pair or form H-shaped structures. In addition, allantoid type of sporidia are also produced and these are more pathogenic than filiform sporidia (Chahal *et al.,* 1999).

Fully mature uninucleate sporidia become detached in mass from the tip of the promycelium, become undulate in shape, and elongate a little more. The single nucleus undergoes a mitotic division making the sporidium binucleate. Sometimes a septum is formed. On these sporidia sterigmata develop and bear hyaline, curved, secondary sporidia which measure 7.5–13.7 x 1.2–1.8 μm. The two nuclei from the primary sporidium pass into the secondary sporidium. These nuclei undergo a conjugate division to form nuclei for subsequent secondary sporidia. On germination the secondary sporidia produces an infection hypha which is dikaryotic.

The sporidia of *T. barclayana* are able to germinate on host panicles (rice) as well as on panicles of non-hosts such as *Echinochloa crusgalli*, *Euchlena mexicana*, *Sorghum vulgare* and *Zea mays* (Chahal *et al.,* 1999). Maximum tertiary sporidia are produced on rice followed by *E. crusgalli*.

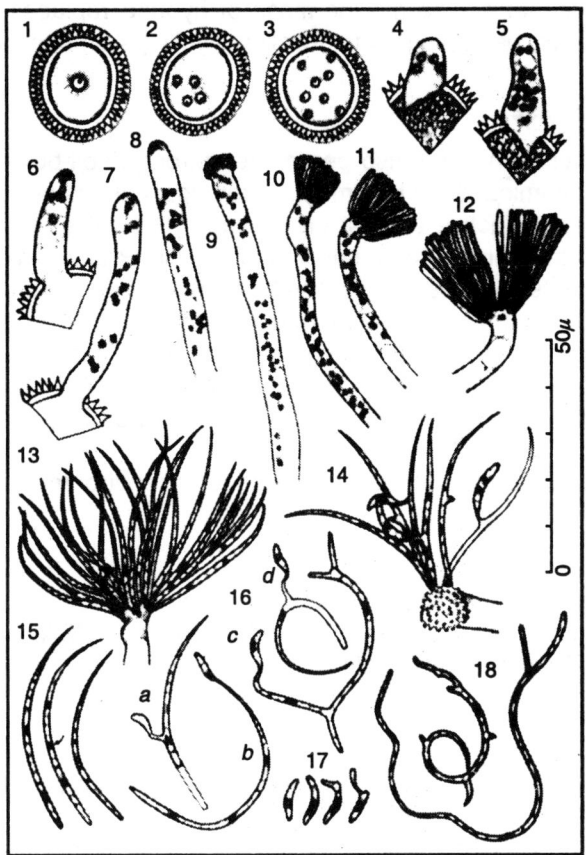

Fig. 102. *Tilletia barclayana.* Cytology of spore germination. (Courtsey : Dr. R.A. Singh).

Disease cycle and environmental relation

The teliospores of the fungus may fall on the ground and survive in soil or they may go with the seed on awns and partially infected grains. They can remain viable for two or more years. The spores require a dormancy period of 5 months. The disease is favoured by light, sandy loam soils. Heavy dosages of nitrogen fertilizers make the plant susceptible to attack. The effect of nitrogen on disease incidence and disease severity depends on timing of fertilizer application. Early varieties of rice suffer more than late maturing varieties. Maximum germination of spores occurs at 29°–30°C. There is no germination at temperatures below 10°C and above 37°C. The acidic medium favors germination. A temperature range of 25°–30°C and high humidity (85% or more) and intermittent light showers at the time of ear emergence are favorable for infection.

Infection of the host takes place when the ears have just emerged out of the leaf sheath. Sporidia from germinating spores in soil are wind blown to fresh flowers. There they germinate and send the infection hypha into the developing ovary. It is a local infection. Production of infected grains in the panicles is much higher with inoculations by allantoid sporidia than with filiform sporidia. Incubation period and production of teliospores differ in different varieties of rice. Sori are formed in the maturing grains where an extensive cavity develops between the seed coat and the endosperm. The hyaline mass of hyphae replaces the starch cells. The spores develop soon after, beginning as thin-walled swellings on short branches.

Management of the disease

Since the inoculum is air-borne and also perennates through soil-borne spores, there is not much advantage from chemical seed treatment. Spray applications of triazole fungicides (*viz.,* Tilt, Folicurm, Bacor) are effective against the disease (Grewal *et al.,* 1996). Early and full flowering stages are most appropriate for spray applications. Propiconazole applied during the booting stage of rice plants reduces the disease by 88.6%. Field sanitation, crop rotation, and use of resistant varieties are practical and more reliable management practices. High doses of nitrogen should be avoided. In U.P., following varieties were reported as resistant:

Early : T 6, T 8, T 27, NCH 20, NCH 43, Russ. 1331, 2877, Hung. 2115.
Medium: HR 22, BJ 1, N 12, N 10B, NCH 7, T 19, N 28, T 108.
Late : T 10, T 22, T 23, T 26, T 33, T 36, T 38.
However, this resistance may be due to the effect of environment.

LEAF SMUT OF RICE

Originally found in India and Myanmar (Burma) the disease is now known to occur in Japan, China, Taiwan, Philippines, Afghanistan, Venezuela and the USA. The fungus has been reported on wild rice, *Zizania aquatica,* also.

Symptoms

The smut is characterized by the appearance of distinct, not confluent, leaden-black spots on the leaves. These spots are linear, rectangular, or angular elliptical and measure 0.5–2.0 x 0.5–1.5 mm. They remain covered by the epidermis which is ruptured when the leaves are soaked in water for some time. Then the black smut spores are exposed.

The causal organism

The smut is caused by *Entyloma oryzae* H. & P. Sydow. The fungus is classified under order Entylomatales and family Entylomataceae (class Ustilaginomycetes, subclass Exobasidiomycetidae) The sorus under the epidermis appears as a black mass resembling sclerotial crust. This sorus consists of a mass of closely packed spores. The spores are angular to globose, smooth-walled, light brown, and measure 6.9 x 7.5–11.5 μm. The epispore is 1–1.5 μm thick. The spores germinate while still held in the crust to form promycelium with sporidia.

In all probability the pathogen perpetuates through hypophyllous sori lying in the field in diseased leaf trash. Infection is caused by sporidia reaching the leaves near the soil level. High nitrogen enhances the disease incidence (Singh and Krishna, 1979).

Management of the disease

Not much work has been done on this smut which is not economically important. Specific control measures have not been worked out. Clean cultivation (destruction of crop debris) and use of resistant varieties may prove helpful. In Japan, application of the triazole fungicide simeconazole (1.5% granules) to the field inn standing crop 1–2 weeks before heading has given highly effective control of rice bunt. A large number of varieties are reported to be resistant. Cultivars Bala, Ch. 45, IR-8, Pusa 2-21, Saket-4 and TN 1 are highly resistant and IR 30 and Ratna are moderately resistant. Other resistant varieties are T 6, T 8, T 10, T 19, T 36, NCH 2, NCH 20, NCH 43, HR 22, BJ 1, N 12, AD 15426, AD 16674, AD 14758, Pankaj, Jagriti, Garima, Patel 5, and Safri 12.

FLAG SMUT OF WHEAT

Flag smut, also known as stem smut or stripes smut, of wheat was first observed in South Australia in 1868 although a similar disease of *Agropyron* was observed earlier in 1848 in Europe. Since then it has been reported from almost all the wheat growing areas of the world, including Japan, South East Asian countries, India, Pakistan, Middle East countries, Europe, South Africa, and the USA. In 1918, in the Indian subcontinent, the disease was known to occur in the area now in Pakistan (the undivided Punjab including Haryana, Baluchistan, North West Frontier Province). The disease has been now observed in Punjab, Haryana, Madhya Pradesh, Delhi, Rajasthan and Himachal Pradesh. During the late 1970s the disease caused 39–78% loss in Rajasthan.

Although its occurrence in the country is not widespread due to absence of favorable environmental conditions its importance cannot be underestimated since, unlike most smut, it is long persisting and if sowing methods and environments are favourable and susceptible varieties are grown the disease may assume serious form. Reduced yield due to complete loss of productivity of infected plants is the most significant effect of the disease.

Symptoms

Flag smut is a disease of stem, culm, and leaf of wheat plants. Leaf and leaf sheaths are most commonly affected. As a rule every shoot of the plant is infected. The infection is evident from the late seedling stage until maturity of the crop. The leaves become twisted and assume a drooping habit (flagging) which is followed by withering. Soon these leaves are shed away and often the whole plant is dead. Frequently, the culm remains sterile, bearing no grains. If grains are formed, they are much shriveled. The outward appearance of the parasite on the host is marked by formation of gray to grayish-black, slightly swollen bands running parallel to the veins of older leaves and leaf sheath. These bands indicate the development of sori under the epidermis in the mesophyll tissues. Later, the epidermis ruptures exposing a black powdery mass of spores. The leaf may shred along the linear bands occupied by the sori. The flag smut infection can be detected in very early seedling stage of plant development by a characteristic twisting and bending of the coleoptile associated with subsequent formation of bleached spots on the coleoptile.

The causal organism

Genus *Urocystis* is placed under family Urocystaceae, order, Urocystales in class Ustilaginomycetes. From 1877 until 1943, *Urocystis tritici* Koern. was accepted as the valid name of the wheat flag smut fungus. Fischer (1942) called attention to the morphological similarity of *U. tritici* to *Urocystis agropyri*, a species that causes flag smut of certain grasses, and a name that takes priority over *U. tritici*. However, the name *U. tritici* is still used by many pathologists even when agreeing that the two species are synonyms. Isolates of *U. agropyri* show marked host specialization. Most wheat isolates do not infect other grass species. However, *Agropyron* is a secondary host of the wheat pathogen.

The spores are borne in tiny balls consisting of 1–6, mostly 3, fertile, bright brown spores surrounded by a layer of flattened sterile peripheral cells. Individual spores are globose to subglobose and reddish to olivaceous, smooth-walled, and measure 8–18 μm in diameter. The spore balls measure 18–52 μm in diameter. The sterile peripheral cells are smaller and more elliptical. The spores germinate *in situ* to form a short promycelium without or with 1–2 septa. At the distal end of the promycelium, 1–4, mostly 3, primary sporidia are formed. They are cylindrical, aseptate or 1–2 septate, and measure 12–15 x 3 μm. They usually remain attached to the promycelium and germinate by a slender germ tube which may be of considerable length and represents the infection hypha.

U. agropyri from wheat is heterothallic. Monosporidial cultures do not form clamp connections. Fusion of sporidia occurs only between two different mating types. Physiologic races of the fungus have been reported from several countries including 12 from China and 2 from the USA. Race 4 is reported to be common in India.

Fresh spores of the fungus are difficult to germinate. A preliminary drying favors germination. In laboratory experiments it has been demonstrated that spores dried for 4 hours over concentrated sulphuric acid germinate readily. Pre-soaking the spores in water for 3 days and then adding small pieces of wheat tissue also gives good germination. The temperatures of 18°–24°C and pH 5.1–5.7 are reported to be favourable for spore germination. Minimum temperature for germination is 5°C and maximum 32°C. Pre-soaking of spore balls in water at 18°–20°C for 7 days followed by incubation at the same temperature in presence of fresh host tissue for 24 hours or at 14°–16°C for 36 hours, gives up to 76% germination of spores. At 18°–20°C promycelia are produced in 12 hours, sporidia in 18 hours and infection threads in 24 hours. At 14°–16°C these periods are 18, 24, and 36 hours, respectively.

Disease cycle and environmental relations

The pathogen is soil- as well as seed-borne. It gains access to the seed and straw during harvest and is distributed as a seed contaminant or on straw or any other material which has become contaminated. It can persist in the soil as spore balls for many years. Seed infestation is more important than soil inoculum alone. The point of entry during infection is restricted to the epidermal cells of the young coleoptile, less than 4 mm long. The penetration is direct. In highly resistant varieties penetration is checked by callus formation at the point of entry. After infection, the fungus grows both inter- and intracellularly until it begins to form spores, generally first in the leaf blades and then in the leaf sheath and other plant parts.

Infection in the seedling stage is influenced by environmental factors such as soil moisture, soil temperature, soil pH as well as by such cultural practices as planting date, sowing depth, host variety and growth stage of the host. Deep sowing increases infection level. Sowing in relatively dry soil favours infection whereas sowing in moist soil is detrimental to disease development. In Egypt, sowing by *herati* method, in which seeds are broadcast on moist land, results in more flag smut than in *afir* method in which seeds are broadcast on dry land, harrowed, and then irrigated. It has been postulated that in the *afir* method of planting since the spores in soil are dry they do not become infective when the seedlings are in susceptible stage of development. On the other hand, in the *herati* method spores receive the necessary pre-soaking and are infective at the proper time, resulting in high percentage of infection.

The minimum, optimum, and maximum temperatures for infection are 5°, 20° and 28°C, respectively. In Australia the disease is less common in calcium deficient soils than in soils having excess of calcium. Deep sowing increases chances of infection due to the fact that the germinating seedlings remain beneath the soil surface for longer period and are thus exposed to infection for longer time than in shallow sowing.

Management of the disease

The effective method of flag smut control is use of resistant varieties along with seed treatment. Where such varieties are available for a particular area, flag smut has almost disappeared. In contrast, where resistant varieties are not grown and the area is known to be infested with flag smut pathogen, it usually persists perennially in spite of careful seed treatment.

Seed treatment with copper sulphate or copper carbonate dust (at the rate of approximately 50 g/50 kg seed) is an effective method of seed disinfection. Steeping the seed in formaldehyde solution (half kg in 200 lit water) had also been recommended in the past. Tetrachloro-nitroanisole (TCNA) is an effective compound for seed treatment. Use of systemic fungicides, as discussed for loose smut of wheat, is effective against flag smut also. Seed treatment with Benlate, Bavistin, Vitavax, fenfuran, triadimefon, and triadimenol gives good control of seed-borne infection and protects the seedlings against soil-borne inoculum. PCNB, Thiram and Captan are also recommended for seed treatment. Raxil (tebuconazole) at 1.5 g/kg seed is reported to give better control than Bavistin and Vitavax or the antagonist *Trichoderma*.

Rotation, early sowing of the crop, burning of stubbles, and hot weather cultivation are some of the cultural practices that reduce incidence of flag smut. A combination of all the practices, as far as possible, is recommended for successful management of flag smut. In Punjab, experiments were conducted to see the combined effect of early sowing, seed treatment, and resistant variety on the incidence of the disease. Wheat was sown towards the end of October instead of the usual time of sowing in November–December. A smut resistant variety NP 165 was sown and the seed was dressed with copper sulphate before sowing. It was found that by this combination treatment the crop gave uniform increase in grain yield during the 3 years of the study.

In India, the variety WG 189 was found resistant with less than 1% infection. Varieties VL 426, HD 2117, HW 161, HB 121 and HB 113 are also reported as resistant to flag smut. In Punjab cv G 377 was found resistant to flag smut and immune to loose smut. Under conditions of natural soil infestation and seed inoculum, 14 varieties and lines of *Triticum durum* and *T. aestivum* were found immune to flag smut. These included CI 13596, 13751, 13755, 14355, 14378, 14394, HP 1102, and HS 86.

REFERENCES

Aggarwal, Rashmi, D.V. Singh and K.D. Srivastava. 1994. Host pathogen interaction in Karnal bunt of wheat. *Indian Phytopath.* 47: 381

Aujla, S.S., A.S. Grewal, G.S. Nanda and Indu Sharma. 1990a. Identification of stable resistance in wheat to loose smut. *Indian Phytopath.* 43: 90

Aujla, S.S., Indu Sharma and K.S. Gill. 1990b. Effect of soil moisture and temperature on teliospore germination of *Neovossia indica*. *Indian Phytopath.* 43: 223

Aujla, S.S., Indu Sharma and Gurdip Singh. 1992. Sporulation of Karnal bunt fungus *in vitro* and *in vivo*. *Indian Phytopath.* 45: 382.

Babadoost, M. 2000. Comments on the zero-tolerance quarantine of Karnal bunt of wheat. *Plant Dis.* 84: 71

Babadoost, M., D.E. Mathre, R.H. Johnston and M.R. Bonde. 2004. Survival of teliospores of *Tilletia indica* in soil. *Plant Dis.* 88(1): 56–62

Bains, S.S. 1994. Influence of wheat spike maturity on susceptibility to infection and growth of sporidia of *Neovossia indica*. *Indian J. Mycol. Pl. Pathol* 24: 111

Bedi, P.S. 1994. Advances in research on Karnal bunt of wheat. *Indian J. Mycol. Pl. Pathol.* 24: 1

Beniwal, M.S., M.L. Chhabra and S.S. Karwasra. 2002. Biological control of smuts of wheat. *Indian Phytopath.* 55(3): 396 (abstr)

Bolker, M. 2001. *Ustilago maydis*-a valuable model system for the study of fungal dimorphism and virulence. *Microbiology* 147: 1395

Bonde, M.R., S.E. Nester, N.W. Schaad *et al.*, 2003. Improved detection of *Tilletia indica* teliospores in seed and soil by elimination of contaminating microorganisms with acidic electrolyzed water. *Plant Dis.* 87(6): 712

Carrisse, L.M., L.A. Castlebury and B.J. Goates. 2006. Nonsystemic bunt fungi-*Tilletia indica* and *T. horrida*: A review of history, systematics and biology. *Ann. Rev Phytopathol.* 44

Cashion, N.L. and E.S. Lutterall. 1988. Host parasite relationship in Karnal bunt of wheat. *Phytopathology* 78: 75

Chahal, S.S. 2001: Epidemiology and management of two cereal bunts. *Indian Phytopath.* 54: 145

Chahal, S.S., G. Singh and P.P.S. Pannu. 1999. Multiplication and infectivity of sporidia of *Tilletia barclayana*. The causal organism of kernel smut of rice. *Indian Phytopath.* 52: 35

Chahal, S.S., S. Kaur, Anita and P.P.S. Pannu. 2003. Compatibility in sporidia of *Tilletia barclayana* and *T. indica*. *Indian Phytopath* 56(1): 78

Dhaliwal, H.S. and D.V. Singh. 1988. Up-to-date life cycle of *Neovossia indica*. *Curr. Sci.* 57: 675

Dromph, K.M. and A. Borgen. 2001. Reduction of viability of soil-borne inoculum of common bunt (*Tilletia tritici*) by collembolans. *Soil. Biol. Biochem.* 33(12–13): 1791

Frederick, R.D., K.E. Snyder, P.W. Tooley *et al.*, 2000. Identification and differentiation of *Tilletia indica* and *T. walkeri* using the polymerase chain reaction. *Phytopathology* 90: 951

Garrett, K.A. and R.L. Bowden. 2002. An allee effect reduces the invasive potential of *Tilletia indica*. *Phytopathology* 92(11): 1152

Goates, B.J. and E.W. Jackson. 2006. Susceptibility of wheat to *Tilletia indica* during stages of spike development. *Phytopathology* 96(9): 962

Goel, L.B., D.P. Singh, V.C. Shukla, D.V. Singh, K.D. Srivastava, *et al.*,. 2000. Evaluation of Tilt against Karnal bunt of wheat. *Indian Phytopath.* 53: 301

Goel, L.B., D.P. Singh, V.C. Sinha, A. Singh, K.P. Singh, A.N. Tiwari *et al.*, (2001). Efficacy of Raxil (tebuconazole) for controlling the loose smut of wheat caused by *Ustilago segatum* var. *tritici*. *Indian Phytopath.* 54: 270

Goel, R.K. 1992. Flag smut of wheat. *Indian J. Mycol. Pl. Pathol.* 22: 113

Gogoi, R., D.V. Singh and K.D. Srivastava. 2001. Phenols as a biochemical basis of resistance in wheat to Karnal bunt. *Plant Pathology.* 50(4): 470–476

Grewal, R.K., S.S. Chahal and K.S. Aulakh. 1996. Effectiveness of triazole fungicides against kernel smut of rice. *Indian Phytopath.* 49: 404

Grewal, T.S., Indu Sharma and S.S. Aujla. 1999. Role of stomata and hairs in resistance/susceptibility of wheat to Karnal bunt. *J. Mycol. Pl. Pathol.* 29: 217

Hokenberg, M., B. Gerhardson and L. Johnsson. 1997. Biological control of cereal seed-borne diseases by seed treatment with greenhouse-selected bacteria. *Eur. J. Plant Path.* 103(1): 25

Hu, G.G., R. Linning and G. Bakkeren. 2002. Sporidial mating and infection process of the smut fungus, *Ustilago hordei*, in susceptible barley. *Can. J. Bot.* 80(10): 1103

Hu, G.G., R. Linning and G. Bakkere, 2003. Ultrastructural comparison of a compatible and incompatible interaction triggered by the presence of an avirulence gene during early infection of the smut fungus, *Ustilago hordei*, in barley. *Physiol. Mol. Plant Path.* 62(3):155

Johnsson, L., M. Hokeberg and B. Gerliardson. 1998. Performance of the *Pseudomonas chlororaphis* biocontrol agent MA 342 against cereal seed-borne diseases in field experiments. *Eur. J. Plant Pathol.* 104(7): 701

Jones, D.R. 2007. Arguments for a low risk of establishment of Karnal bunt disease of whjeat in Europe. *Eur. J. Plant Pathol.* 118(2) 93

Jones, P. 1997. Control of loose smut (*Ustilago nuda* and *U. tritici*) infections in barley and wheat plants by foliar application of triadimefon. *Plant Pathology* 46(6): 946

Kahmann, R. and J. Kamper. 2004.*Ustilago maydis*: how its biology relates to pathogenic development. *New Phytologist* 164(1): 31

Kosted, P.J., S.A. Gerhardt and J.E. Sherwood. 2002. Pheromone-related inhibitors of *Ustilago hordei* mating and *Tilletia tritici* teliospore germination. *Phytopathology* 92: 210

Kumar, J., M.S. Saharan, A.K. Sharma and S. Nagarajan. 2003. Effect of temperature on teliospore germination in *Tilletia indica* under simulated conditions and its relevance in pest risk analysis in wheat. *Indian Phytopath.* 56(1): 14

Kumar, J., M.S. Saharan, A.K. Sharma and S. Nagarajan. 2003. Temperature requirement for teliosporogenesis of *Tilletia indica*, the causal agent of Karnal bunt of wheat. *Indian Phytopath.* 56(4): 439

Leon-Ramirez, C.G., J.L. Cabera-Ponce *et al.*,. 2004. Infection of alternative host plant species by *Ustilago maydis*. *New Phytologist* 164(2): 337

Malik, V.K. and S. Singh. 2002. Effect of leaf extract of different botanicals on spore germination of *Ustilago hordei*. *Indian Phytopath.* 55(3): 395 (abstr.)

Martinez, C., C. Roux and R. Dargent. 1999. Biotrophic development of *Sporisorium reilianum* f. sp. zeae in vegetative shoot apex of maize. *Phytopathology* 89 (3): 247

Martinez, C., C. Roux, A. Jauneau and R. Dargent. 2002. The biological cycle of *Sporisorium reilianum* f. sp. *zeae*: an overview using microscopy. *Mycologia* 94(3): 505

McDonald, J.G., E. Wong and G.P. White. 2000. Differentiation of *Tilletia* species by rep-PCR genomic fingerprinting. *Plant Dis.* 84: 1121

McManus, P.S., A.V. Ravenscroft and D.W. Fulbright. 1993. Inhibition of *Tilletia laevis* teliospore germination and suppression of common bunt of wheat by *Pseudomonas fluorescens* 2–79. *Plant Dis.* 77: 1012

Mendez-Moean, L., C.G. Reynaga-Pena, P.S. Springer and J. Ruiz-Herrera. 2005. *Ustilago maydis* infection of the nonnatural host *Arabidopsis thaliana*. *Phytopathology* 95(5): 480–488

Mondal, G., R. Aggarwal and K.D. Srivastava. 1996. Hyphal interaction between *Trichoderma koningii* and *Ustilago segetum tritici* through scanning electron microscopy. *Curr. Sci.* 70: 425

Moosawi-Jorf, S.A. and R. Farrokhi-Nejad. 2007. Survival of the teliosporogenous mycelia of *Neovassia indica* in infected wheat grains. *Plant Pathol. J.* 6(1): 30

Nagarajan, S., S.S. Aujlam G.S. Nanda, I. Sharma, L.B. Goel, J. Kumar and D.V. Singh. 1997. Karnal bunt (*Tilletia indica*) of wheat-a review. *Rev. Plant Pathol.* 76(12): 1207

Nzioki, H.S., L.E. Claflin and B.A. Ramundo. 2000. Evaluation of screening protocols to determine genetic variability of grain sorghum germplasm to *Sporisorium sorghi* under field and greenhouse conditions. *Intern. J. Pest Management* 46(2): 91

Pannu, P.P.S. and S.S. Chahal. 2000. Variability in *Tilletia indica*, the incitant of Karnal bunt of wheat. *Indian Phytopath.* 53: 279

Pannu, P.P.S., G. Kaur and S.S. Chahal. 2002. Variability in *Tilletia barclayana*, the causal organism of kernel smut of rice. *J. Mycol. Pl. Pathol.* 32(1): 6

Ruske, R.E., M.J. Gooding and S.A. Jones. 2003. The effects of adding picoxystrobin, azoxystrobin and nitrogen to a triazole programme on disease control, flag leaf senescence, yield and grain quality of winter wheat. *Crop Protection* 22(7): 975

Selvakumar, R., K.D. Srivastava, Rashmi Agarwal, D.V. Singh and Prem Dureja. 2000. Studies on development of *Trichoderma viride* mutant and their effect on *Ustilago segatum tritici*. *Indian Phytopath.* 53: 185

Sharma, B.K. and A.K. Basandrai. 2000. Effectiveness of some fungicides and biocontrol agents for the management of Karnal bunt of wheat. *J. Mycol. Pl. Pathol.* 30: 76

Sharma, B.K. and A.K. Basandrai. 2004. Efficacy of fungicides and plant leaf extracts for the control of Karnal bunt of wheat (*Neovassia indica*). *J. Mycol. Plant. Pathol.* 34(1): 102

Sharma, B.K., A.K. Basandrai and K.D. Sharma. 2004. Variability in *Neovossia indica* and sources of multiple disease resistance in *Triticum* spp, and related genera. *J. Mycol. Plant Pathol.* 34(1): 33

Sharma, Indu and G.S. Nanda. 2000. Effect of plant extracts on teliospore germination of *Neovossia indica*. *Indian Phytopath.* 53: 323

Sharma, S.K., P.S. Bagga and Vineet Kumar. 2000. Influence of rice-wheat and maize-wheat cropping sequence on the development of Karnal bunt of wheat. *J. Mycol. Pl. Pathol.* 30: 420

Sharma, S.K., D.V. Singh, R. Aggarwal and K.D. Srivastava. 2002. Pathogenic and biochemical variation in *Neovossia indica*. *Indian Phytopath.* 55(2): 133

Singh, A. 1994. Epidemiology and Management of Karnal Bunt Disease of Wheat. *Res. Bull. No.* 127. 167 p. Directorate of Exp. Sta., G.B. Pant University, Pantnagar, India.

Singh, A. 1997. Karnal bunt of wheat: Current status and future challenges. *J. Mycol. Pl. Pathol.* 27: 117

Singh, D. and V.K. Maheshwari. 2001. Biological seed treatment for the control of loose smut of wheat. *Indian Phytopath.* 54(4): 457

Singh, D.P. 1999. Chemical and biological control of covered smut of barley. *J. Mycol. Pl. Pathol.* 29: 256

Singh, D.P. 2004. Use of reduced dose of fungicides and seed treatment with *Trichoderma viride* to control wheat loose smut. *J. Mycol. Pl. Pathol.* 34(2): 396

Singh, P.J., H.S. Dhaliwal and K.S. Gill. 1989. Chemical control of Karnal bunt of wheat by single spray of fungicides at heading. *Indian J. Agric. Sci.* 59: 131

Singh, S.N. 1999. Evaluation of sugarcane varieties for resistance to smut in Madhya Pradesh. *J. Mycol. Pl. Pathol.* 29: 141

Sinha, V.C., M.S. Beniwal, S. Nagarajan, L.B. Goel, A.S. Grewal, S.S. Karawasra and J. Kumar. 2000. Sources of resistance in wheat and triticale against loose smut caused by *Ustilago segetum tritici*. *Indian Phytopath.* 53: 76

Slaton, N.A., E.E. Gbur Jr, R.D. Cartwright, R.E. DeLong, R.J. Norman and K.R. Brye. 2004. Grain yield and kernel smut of rice as affected by preflood and midseason nitrogen fertilization in Arkansas. *Agron. J.* 96: 91

Smilanick, J.L., J.M. Prescott, J.A. Hoffman, L.R. Secrest and K. Wiese. 1989. Environmental effects on survival and growth of secondary sporidia and teliospores of *Tilletia indica*. *Crop Protect.* 8: 86

Smilanick, J.L., B.J. Goetes, R. Deni-Arrue, G.F. Simmons, G.L. Petrson, D.J. Henson and R.E. Rij. 1994. Germinability of *Tilletia* spp. teliospores after hydrogen peroxide treatment. *Plant Dis.* 78: 861

386

Smilanick, J.L., W. Hershberger, M.R. Bonde and S.E. Nester. 1997. Germinability of teliospores of *Tilletia indica* after hot water and sodium hypochlorite treatment. *Plant Dis.* 81: 932

Snetselaar, K.M., M.A. Carfioli and K.M. Cordisco. 2001. Pollination can protect maize ovaries from infection by *Ustilago maydis*, corn smut fungus. *Can. J. Bot.* 79(12): 1390

Srivastava, K.D., D.V. Singh, R. Aggarwal. A.K. Dixit and P. Bahadur. 1997. Bioefficacy and persistence of tebuconazole against loose smut of wheat. *Indian Phytopath.* 50: 434

Tegegne, G, and J.C. Pretorius. 2007. *In vcitro* and *in vivo* antifungal activity of crude extracts and powdered dry material from Ethioian wild plants against economically important plant pathogens. *BioControl* 52(6): 877

Tegegne, G., J.C. Pretorius and W.J. Swart. 2008. Antifungal properties of *Agapanthus africanus* L. extracts against plant pathogens. *Crop Protection* 27(7): 1054

Valvassori, M., L. Riccioni, A. Inman *et al.*, 2002. Survival of *Tilletia indica* teliospores in soil in Italy. *J. Plant Pathol.* 84(3)

Verma, H.S., A. Singh and V.K. Agarwal. 1984. Application of seedling crown test for screening of systemic fungicides against loose smut of wheat. *Seed Res.* 12 (2): 56

Verma, H.S., A. Singh, V.K. Agarwal and R.K. Khetarpal. 1985a. Comparative morphology of *Ustilago tritici* in wheat seeds, seedlings, spikes and in culture. *Indian Phytopath.* 38: 187

Verma, H.S., A. Singh and V.K. Agarwal. 1985b. An improved embryo count test for detection of loose smut infection in wheat seeds. *Indian Phytopath.* 38: 540

Vijaya, M. 2000. Chemical control of sett-borne infection of smut of sugarcane. *J. Mycol. Pl. Pathol.* 30: 128

Wala, A.C. 2003. Control of sugarcane smut disease in Nigeria with fungicides. *Crop Protection* 22(1): 45

Warham, E.J. 1990. Effect of *Tilletia indica* infection on viability, germination and vigor of wheat seed. *Plant Dis.* 74: 130

Warham, E.J. and P.A. Burnett. 1990. Influence of media on pathogenicity and morphology of secondary sporidia of *Tilletia indica. Plant Dis.*74: 525

Diseases Caused by Anamorphic Fungi or Fungi whose Teleomorph is Uncommon in Nature

EARLY BLIGHT OF POTATO

The early blight is of common occurrence wherever potato is grown in the world. In India it is more common than the late blight and may cause up to 40% loss in yield of tubers. The damage to the crop is considerable as the peak period of attack coincides with the period of tuber formation. The disease can appear when the crop is quite young, much earlier than the late blight. While late blight epidemics are common in cooler and wet areas, early blight is free from inhibitions caused by weather conditions and occurs in cool as well as warm areas. Thus, in India the disease is equally serious in the hills as well as in the plains. The disease occurs on tomato also where the losses may go up to 78%. In chilli (*Capsicum*) fruit rot occurs during transit and storage.

Symptoms

The disease first becomes visible as small, isolated, scattered, pale brown spots on the leaflets. These spots become covered with a deep greenish-blue growth of the fungus. Leaves near the soil line are attacked first and the disease progresses upward. In the necrotic area of the spots concentric ridges develop to produce a target board effect. This is the most characteristic symptom of early blight. This target board effect is due to interruption of fungal growth caused by unfavourable conditions. There is usually a narrow chlorotic zone around the spots which fades into the normal green, and increases with increase in size of the spots. There may be only a few spots or a large number of them may appear occupying a major portion of the leaf surface. When these spots involve larger veins chlorosis commonly extends well beyond the necrotic area. The yellow zone around lesions and beyond them is reported to be due to the diffusion of fungal metabolites like the toxin alternaric acid, produced by the fungus and translocated through the veins. In dry weather the spots turn hard and the leaves curl. In humid weather the spots coalesce and big rotting patches may appear. In severe attack leaves shrivel and fall off. Potato stems are also affected and show brown to black necrotic lesions on the skin. These lesions may cause collapse of the branches or the entire above ground portion of the plant. Infection and rotting of tubers due to early blight is also reported. In tomato, fruit spots occur at the stem-end and are dark, leathery, sunken lesions with target board appearance.

The causal organism

The early blight of potato and tomato is caused by the fungus *Alternaria solani* (Ell. and Martin) Jones and Grout. Two major features of *Alternaria* species are the production of melanin, especially in the spores, and the production of host-specific toxins in the case of pathogenic species. These are imperfect (mitosporic) fungi with no sexual stage known so far.

The inter- and intra-cellular mycelium of *A. solani* consists of septate and branched, light brown hyphae which become darker with age. The intercalary cells contain 0–14 nuclei while the tip cells have 14–36 nuclei. The hyphae in the host are at first intercellular later penetrating the cells. Conidiophores emerge through the stomata from the dead centers of the spots. They are relatively short, 50–90 x 9 μm, and dark coloured. Conidia are 120–296 x 12–20 μm in size, beaked, muriform and dark colored (heavily melanized). Melanins are dark, brown to black, high molecular weight pigments that are produced by microorganisms as well as animals and plants. Structure of melanins varies with source. Apart from their role in conidial development, melanins are virulence factors in many fungi including *Pyricularia grisea* and *Colletotrichum*. However, in *Alternaria* they do not determine virulence.

The conidia of *A. solani* are borne singly on the conidiophores. However, in cultures they form short chains. They develop from a bud formed by the apical cell of the conidiophore. Five to 10 transverse and a few longitudinal septa are present in each conidium. In moist weather these conidia germinate readily and 5–10 germ tubes arise from a single conidium. Germination of conidia occurs within 35–45 min at the optimum temperature of 28°–30°C. The fungus produces chlamydospores also. These are formed by differential swelling of hyphal cells in the curly mycelium. They are single, in chains or in clusters, 1-celled, thick-walled, dark brown, and measure 8–15 (average 11) μm in diameter. Other species of *Alternaria* can also cause early blight of potato. These include *A. tomato* and *A. alternata*. The latter causes stem canker of tomato. It is capable of detoxifying the alpha-tomatine. This steroidal, glucoalkaloidal saponin is toxic to fungi and is present as a biochemical defense barrier.

Races and strains differing in morphology and physiology occur in *A. solani*. In India at least 3 strains have been distinguished on the basis of their cultural characters. Toxins like alternaric acid and alternarine are produced by the fungus. Hosts of *A. solani* include potato, tomato, chilli (*Capsicum*), eggplant (*Solanum melongena*), *Solanum nigrum*, *Nicotiana alata*, *Atropha belladonna*, *Cymphomandra betacea*, *Hyoscymus albus*, and *H. niger*.

Fig. 103. Spods of early blight of potaro.

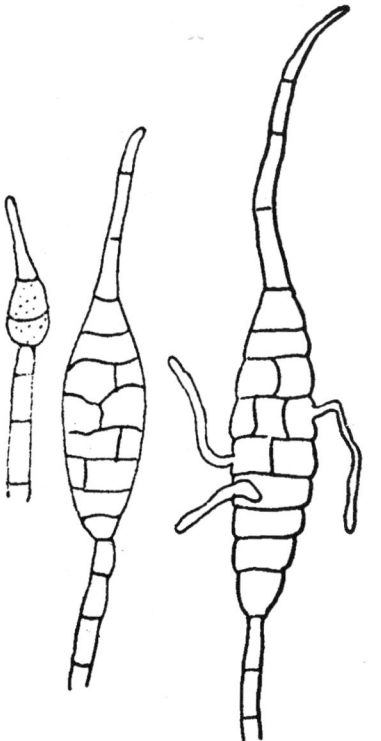

Fig. 104. Conidiophore and conidia of *Alternaria solani*.

Disease cycle and environmental relations

The fungus is extremely resistant to hot and dry conditions. The mycelium remains viable in dry infected leaves for a year or more. Conidia have been found to remain viable for 17 months at room temperature. Mycelium and conidia, thus, survive in the soil on diseased plant debris to cause primary infection in the next crop. Contamination of tubers with conidia or mycelium during harvesting is another source of primary inoculum. In other hosts, seeds are also the source of primary inoculum. In tomato, the fungus produces chlamydospores from conidia and hyphal cells. Conidia require desiccation for production of chlamydospores while desiccation prevents chlamydospore production from hyphae.

Infection of lower leaves first takes place, during periods of warm, rainy, humid weather, through conidia formed on soil. Penetration may be through stomata, direct through cuticle or through wounds. Secondary spread of the disease occurs through conidia developed on primary spots. These conidia are disseminated by wind, rains, and insects. Incubation period varies from 48 to 72 hours. In tomato, the fruit infection occurs while the fruits are still green. Irrespective of level of resistance in different cultivars of potato, the lowest number of lesion and lowest disease severity is seen on the upper leaves, suggesting that younger tissues are less susceptible.

Araujo and Matsuoka (2004) have used light and scanning electron microscopy to study the infection process of A. solani in a resistant and a susceptible cultivar of tomato. Germ tube growth was similar on leaf surface of both cultivars. However, in resistant cultivars the quantity of appressoria, tissue penetration and lesion development were significantly less. The events after penetration were similar in both genotypes and included the initial development of primary and secondary hyphae, colonization processes and lesion development, frequency of formation of papilla and hypersensitive reactions. They concluded that resistance is expressed in the pre-penetration stage. Dita et al., (2007) have studied the histopathology of infection process in potato early blight. Penetration occurs most frequently through the junction of epidermal cells. Penetration through stomata is rarely observed. There is no association between these events and the resistance levels of the host, except for the number of penetration sites showing hypersensitive reaction which is higher in increasing levels of resistance. Hypersensitive response is associated with leaf age. The number of penetration sites with HR is higher in the upper part of the plant.

Shtienberg and Fry (1990) had reported that in fields in which the previous crop was potato or tomato the early blight lesions in the current potato crop appear early in both resistant and susceptible cultivars. Periods of continued drought check the spread of the disease. In general, the disease becomes serious when the season begins with abundant moisture or frequent rains followed by warm and dry weather. These conditions are unfavorable for the host and favor disease development. Tomato plants are susceptible to early blight at all growth stages Young plants of tomato show high degree of resistance. Susceptibility increases as plants mature. Weak plants of potato and tomato are more susceptible than plants with good vigor. Plants grown with balanced fertilizers (100 kg N + 125 kg phosphorus and 125 kg potash per hectare) suffer least damage and give best yield. In tomato the disease is less severe under high nitrogen and high phosphorus conditions. Foliar application of nitrogen through urea or potassium nitrate does not affect host response to A. solani.

Increase of inoculum concentration from 6.2 to 11.5 conidia/ml increases the percentage of leaf area affected and defoliation linearly. As leaf wetness duration increases up to 24 hours, there is an increase in the percentage leaf area affected and level of defoliation but thereafter there is no significant increase in either parameters. In spring (January-April) grown crops higher mean temperatures (19.2°–31.1°C), frequent rains but shorter duration of relative humidity above 80%, the absence of dew during most part of the season, longer photoperiods, and prolonged senescence of the plants are related to low sporulation, restricted size of lesions, and moderate intensity of the disease. In autumn (September–December) grown crop, moderate mean temperatures (13.6°–23.6°C), adequate moisture in the form of dew, infrequent rains but longer periods of relative humidity more than 80%, and shorter photoperiods favored faster development of the disease. The level of total sugars in leaves was generally higher during spring than during autumn and had a negative correlation with disease intensity.

Extracts, leachates and washings of potato leaves infected with potato virus X (PVX) suppress the germination of conidia of *A. solani* and also result in fewer and smaller lesions on the leaves. The increased resistance of leaves appears to be related to fungitoxic compounds produced by interaction of the host with the virus (Kalra *et al.*, 2000). In a more recent study of systemic infection of PVX it has been shown that the plant responds to infection by activating defense-related enzymes.

Management of the disease

Since the pathogen is soil-borne, crop rotation and field sanitation are essential for effective check of the disease. Dead haulms should be raked together and burnt immediately after crop harvest. Every care should be taken to maintain good plant vigor by suitable cultural practices. Timely spray of fungicides is, at present, the best method of protection against the disease. The spraying should be started early, about a month after planting, and continued throughout the period of plant growth at intervals of 10–21 days depending on disease intensity in the locality. The fungicides used against early blight are generally the same as those for control of late blight. Spray of zineb (Dithane Z-78) at 10 days interval was recommended in 1961. Other fungicides that have been recommended against early blight include Dithane M-45 (0.2%), Blitox–50 (0.25%), Difolatan, Daconil, Brestan, Antracol and Captan. Good control of early blight was achieved by 5 sprays of Dithane M-45 (0.2%), Brestan (0.05%), Antracol (0.2%), zineb (0.2%) and Captan (0.2%). Dithane M-45 was most effective and captan least effective. In a comparative study of zineb and ziram 5 sprays of 0.2% zineb, were more effective in controlling the disease. The spray interval was 7–14 days. The protectant, curative, and eradicant fungicide, difenoconazole (Score), has been found highly effective against early blight. Because of its lasting protective activity (up to 3 weeks) it provides great flexibility in number and timing of sprays. The strobilurin fungicides {Q (o) fungicides} azoxystrobin, trifloxystrobin and pyraclostrobin have also been effectively used but strains resistant to these fungicides have also been traced (Pasche *et al.*, 2004). Spray of azoxystrobin or other QoI fungicides at 1, 3, 5 weeks of plant growth alternated with spray of chlorothalonil was recommended to reduce development of resistance to these fungicides. The protectant fungicides are so far the safest and consistently effective. In early blight of tomato, mancozeb persists on the foliage for 14 days. Thus, the interval between two sprays can be two weeks. Oxidate (hydrogen peroxide, 1:50) has been used to prevent tuber rot in storage due to early blight infection. The process is laborious. In a fogging system tubers were exposed to the treatment for 8 hours daily for 2 weeks and then for 1 week for the time of storage.

In tomato almost weekly sprays of fungicide are required for control of early blight. The number of sprays can be reduced by about half if spray schedule is based on a forecasting system. Growing resistant varieties also reduces the number of sprays but in potato there is not much information on potato varieties resistant to early blight. Moderate resistance to early blight is reported in cultivars K. Naveen, K. Sindhuri, and K. Jeevan. Hybrid 66–528/8 (*Solanum tuberosum* x *S. andigena*) is a source of high resistance to early blight in the breeding programme. Combination of such varieties and systemic fungicides like difenoconazole will be much cheaper and more effective management of early blight than repeated sprays of protectant fungicides.

BTH (benzothiadiazole) treatment of the potato foliage at 50 mg a.i./liter induces resistance to early blight and powdery mildew. It may help in reducing the number of sprays with synthetic fungicides (Bokshi *et al.*, 2003). The treatment increases β-1,3-glucanse activity in the following order, leaves > stems > tubers > stolons but not in roots. The increased enzyme activity perists for up to 45 days. Infection of tomato fruits by *Alternaria alternata* the activity of enzymes chitinase and glucanase is related to developmental stage of the fruit and cultivar resistance. Mature but green fruits show more enzyme activity that red ripe fruits.and there is more enzyme activity in resistant than in susceptible cultivars (Cota *et al.*, 2007). In a Brazilian study on early blight of tomato, foliar spray of rhizome extract (1–10%) of turmeric (*Curcuma longa*) or curcumin (active compound of turmeric) provided as good cointrol as copper oxychloride but was inferior to azoxystrobin. Propolis is a resinous substance co0llected by bees with positive effects on human health and inhibitory activity against *Alternaria alternate*. Caffeic acid phenethyl ester is a component of the propolis. In tomato ripe fruit rot caused by *A. alternate* application of caffeic acid ester at 50–100 μg has been found to reduce severity of fruit rot better than captan.

Aureobasidium pullulans, Epicoccum nigrum, Penicillium aurantiogriseum, Rhizopus stolonifer, Trichoderma koningii and *Trichothecium roseum* are reported as natural enemies of *A. solani*. Application of cell suspension of a strain of *Pseudomonas* to tomato leaves reduced leaf spots. Seed coating with strains of *Pseudomonas fluorescens* or *Trichoderma harzianum* is reported to protect tomato plants from early blight. Gomez-Rodriguez *et al.*, (2003) reported allelopathic and microclimatic effect of intercropping of tomato with marigold on early blight. The intercropping induced significant reduction in early blight through (i) allelopathic effect on conidia, (ii) alteration of the microclimatic conditions around the canopy (reduction in number of hours per day with relative humidity > 92%) and (iii) providing physical barrier against spread of conidia. Mycorrhizal tomato plants are reported to show less severity of early blight than non-mycorrhizal plants

BLACK SPOTS OF CRUCIFERS

Several species of *Alternaria* attack cultivated crucifers including oilseed bearing brassicas and vegetables such as cabbage, cauliflower, knol-kohl and radish. Two species, *Alternaria brassicae* and *A. brassicicola* attack all these plant species except radish which is attacked by *A. raphani*. When too many spots are formed on the leaves, the latter die prematurely thus affecting the yield. In seed crops, pods also develop spots and infection of seed results in loss of seed yield and germinability. Depending on host species and variety, loss in seed yield may vary from 9 to 28% and loss of germinability of seed may be 23–27%. Loss of seed yield in rapeseed and mustard has been estimated up to 46 and 35% respectively. In rapeseed (yellow and brown sarson) up to 70% yield reduction is reported. The yield is negatively correlated with number and size of spots, sporulation of the fungus, amount of defoliation and resistance components of the cultivar.

Symptoms

Leaf spots incited by *A. brassicicola* appear as small, dark coloured areas which expand rapidly to form circular lesions up to 1 cm in diameter. In humid weather the fungus may appear as a bluish growth in the center of the spots. Concentric rings may also form in the lesions. The spots caused by *A. brassicae* are similar to those described above but tend to remain smaller and lighter in color. The radish leaf spots, caused by *A. raphani*, appear on plants kept for seed. These spots are yellow, raised, spherical to elliptical, and up to 1 cm in diameter. Black sporulation of the fungus may be seen on the spots. The centre soon dries and may drop out. Seed-borne inoculum could cause pre- and post-emergence damping off.

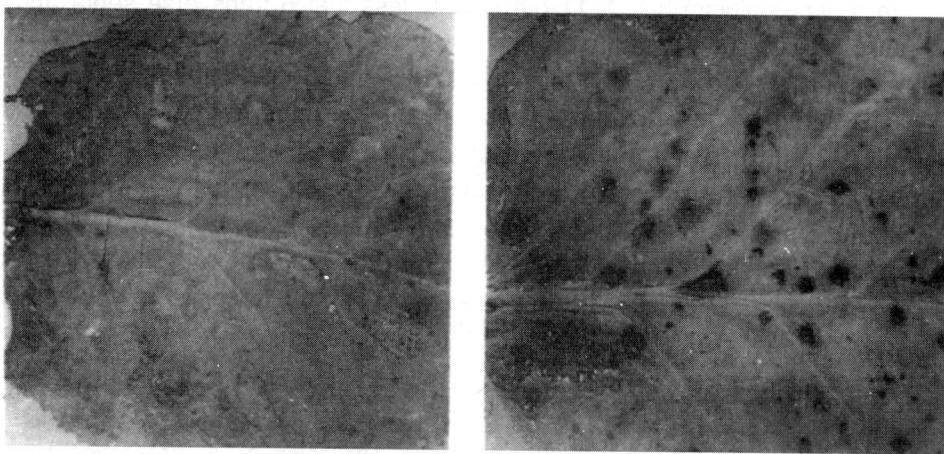

Fig. 105. Alternaria leaf spots of cabbage.

The causal organisms

In the host the hyphae of *A. brassicicola* (Schw.) Wiltshire are branched, septate, hyaline at first, later becoming brown or olivaceous brown, inter- and intracellular, and 1.5–7.5 *µ*m thick.

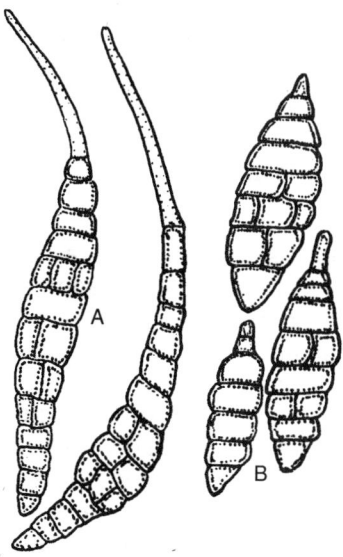

Fig. 106. A. Conidia of *A. brassicae*. B. Conidia of *A. brasicicola*.

Conidiophores arise through stomata singly or in groups of 2–12 or more. They are usually simple, sometimes branched, septate, erect, straight or curved, occasionally geniculate, more or less cylindrical but often slightly swollen at the base, pale or mid-olivaceous brown, smooth, and measure. The conidiophores of *A. brassicicola* (Schw.) Wiltshire are olivaceous, septate, branched and measure 35–45 x 5–7.5 μm. Conidia are linear to obclavate, borne in chains of 8–10 or more spores, septate, muriform, pale or dark brown, and on maturity measure 50–75 x 11–17 μm. Each conidium has 1–11 but usually less that 6 transverse septa and a few, usually up to 6, longitudinal or oblique septa. The beak is about one eighth the length of the conidium. The species is more dangerous than *A. brassicae.*.

In *A. brassicae* (Berk.) Sacc. the conidiophores arise in fascicles and the conidia are dark, obclavate, muriform, borne singly or in short chains of up to 4 spores. The conidia measure 125–225 x 16–28 μm. There are usually 16–19 (usually 11–15) transverse septa. The beak is about one third the length of the conidium and 5–9 μm long. The species produces microsclerotia and chlamydospores but their epidemiological significance is unknown.

Alternaria raphani Groves and Skolko is similar to *A. brassicae* in its morphology but the conidia possess a shorter beak and measure 70–115 x 14–18 μm. It also produces chlamydospores which are olive brown, round and many-celled.

In 1997, phtotoxins were detedted in the spore germination fluid of *A, brassicae* and *A. brassicicola* on leaves of *Brassica napus.* (*Ann,* Appl. Biol. 131, 3, 413–426). The toxins showed selectivity and were effective only in Brassicas, no non-Brassica plants. Within Brassica specxies also there was some selectivity. The toxin of *A. brassicicola* was later named AB-1 toxin. There is differential infection response among the pathogen isolates on different host genotypes. Same is true for toxin production by the isolates. Gupta *et al.,* (2004) used 11 genotypes of *Brassica juncea* as host differentials and identified 4 pathotypes. Pathotypes Bj-4 was highly virulent with minimum incubation period and affected all the 11 genotypes. Kumar *et al.,* (2004) have differentiated 8 pathotypes on the basis of their sensitivity to fungicides and neem-products. The phytotoxin destruxin B is produced by *A. brassicae.* Although considered non-host specific, its toxicity decreases as the genetic distance of the host species from the most sensitive host increases. While confirming non-host specificity of dextruxin *B. Paroda et al.,* (2008) report that the germnation fluid of conidia containes a high molecular weight host-selective toxin that plays a role in the host-parasite interaction.

Disease cycle and environmental relations

These fungi are seed-borne and can cause shriveling of the seed and low germination. Up to 90% incidence of seed infection is reported. In rapeseed mustard, 22–28% bold and symptomless seeds are reported to carry heavy aggregations of mycelium in seed coat epidermis, palisade, endosperm and embryo. *A. brassicae* is recovered more frequently from sections of the seed coat that contain the hilum than from other sections of the seed coat or from the embryo. In the tropics, the seed-borne inoculum is not so important. The fungus starts losing viability in seeds within 2 months of harvest during summer. According to Mehta *et al.,* (2002), *A. brassicae* survives neither on diseased seed nor on diseased crop debris in the subtropical regions. It can survive on crucifer weeds or, in temperate climate regions, in seed and debris from where the inoculum can move to the sub-tropical regions.

Spores and mycelium in diseased plant debris also serve as means of perennation. *A. brassicae* lesions on infected cabbage leaves placed on outdoor soil produce viable conidia for as long as leaf tissues remain intact. *A. raphani* can survive in soil through its chlamydospores for 5 years. Flea beetles (*Phyllotreta cruciferae*) transmit *Alternaria brassicicola* in cabbage. The conidia remain viable after passing through digestive tract of the beetles.

Conidia are abundantly formed in moist atmosphere and are disseminated by wind and rains. The optimum temperature for conidial germination of *A. brassicae* is 21°–28°C. The germ tubes penetrate stomata, cuticle or wounds with or without the formation of small appressoria. Direct penetration through cuticle is mostly by virulent strains. Maximum germination of conidia of *A. brassicicola* occurs at 28°–31°C and arget-board lesions greater than 5 mm diameter are produced at 23–25° after 72 hours. This species requires much longer time and higher temperature to establish disease than *A. brassicae*. This limits its distribution in temperate climate regions. When conidia of *A. brassicicola* land on host leaves, they produce the AB-toxin. The toxin production is aided by a host oligosaccharide. It also produces cutinase during the initial contact of conidia with the cuticle. Lesions larger than 5 mm are produced by exposure of the inoculated plants to 19–23°C for 72 hours. A leaf wetness period of minimum 3 hours at 20°–25°C is required for infection by *A. brassicae*. Disease incidence increases as the leaf wetness period increases from 4 to 24 hours and temperature increases to 20°C. The length of the incubation period of *A. brassicae* decreases as leaf wetness duration is increased from 2–12 hours and temperature from 6–20°C. At mean temperature of 13°C or more spores are produced abundantly after 20 hours of leaf wetness. Fall in relative humidity promotes release of conidia. Chen *et al.*, (2003) have studied the conidial dispersal of *A. brassicicola* on Chinese cabbage in the field and under simulated wind and rain conditions. Conidia were trapped in the field throughout the growing season of the crop. Peaks of high spore concentration were usually associated with dry days, shortly after rain, high temperature or high wind speed. Maximum number of conidia were trapped at around 10 a.m. More conidia were trapped at 25 cm above ground than at 50.75 or 100 cm. Dispersal from leaves with dry surface was more than from wet leaves. On infection the pathogen becomes subcuticular in leaves. After reaching the subcuticular layer, different cutinases that are active during saprophytic growth are induced by cutin monomers. This is followed by colonization of epidermal and mesophyll cells. In resistant plants the phytotoxin destruxin is rapidly detoxified by hydroxylation and glycosilation. In susceptible plants the rate of detoxification is very slow.

Temperature and moisture are major factors affecting disease development. Relative humidity below 67% is not conducive for development of the disease. Optimum conditions for sporulation of *A. brassicae* are RH > 91.5% and a temperature of 18°–24°C. Mustard plants become susceptible with increase in their age and maximum severity of the disease occurs 60–90 days after sowing. Oilseed crops form thick canopy at the flowering stage that is not easily penetrated by wind currents. At this stage dispersal is of short range and largely through splashing rains. During harvesting, conidia are dispersed over long distances by wind. Among oilseed Brassicas nitrogen increases disease severity while potash has suppressive effect on the disease and reduces the effects of toxins produced by *Alternaria*. The combination of N and K is more suppressive than combination of P and K or N, P, K.

Progress of the disease is faster on yellow and brown sarson than in the mustard cultivars. The disease intensity is high in years with high relative humidity during the crop season. Many studies since 1984 are in general agreement that the disease progresses fast when the maximum temperature ranges between around 16° and 26°C, minimum between 1.6° and 9.7°C, maximum relative humidity is between 79 and 96% and wind velocity is between 2.5 and 6 km/hour. Gupta *et al.*, (2003) reported that Alternaria blight on rape seed-mustard was significantly less severe in crops sown on 21st October. Severity increased as the date of sowing was delayed due to prevailing maximum/min temperature and relative humidity.

Management of the disease

Chemical control through seed treatment is neither effective nor needed in the tropics. Earlier, hot water treatment of seed at 40° (30 min) or 50°C for 20 min had been recommended against seed-borne inoculum. In radish, seed treatment with Arasan or Spergon had been suggested. Now iprodione and fenpropimorph are used commercially to treat brassica seed. Foliar infection can be effectively checked by regular sprays of fungicides. Two to 4 sprays of 0.2% Dithane M-45, Dithane Z-78, Duter, Difolatan or Blitox-50 at 15 days interval starting 45 days after sowing, give good control of the disease. Difolatan is best. It is highly persistent on tne foliage. Under humid Tarai conditions, 3–4 sprays of 0.2% Duter, Difolatan or Dithane M-45 have been found effective. First spray in the toria crop is recommended 60 days after sowing and in sarson 60–75 days after sowing. Optimum interval is 15 days. According to Meena et al., (2004) 75 days after sowing is critical age of the host for sprays to control A. brassicae on mustard. Next best is 45 days after sowing. Iprodione (Rovral) has been found highly effective against A. brassicicola in seed and foliar applications but the pathogen can develop resistance if the fungicide is used for many years. The efficiency of fungicidal sprays can be improved if the schedule is based on knowledge of the intensification of disease in time (progress) and movement of the disease in space (spread). Cost/benefit ratio of Bavistin + zineb (Indofil Z-78) is better than other chemical treatments.

In some trials the mustard cultivars RW 351, Pusa Kranti and Seeta have been found resistant to Alternaria blight. In a screening of 350 germplasm lines in different oilseed Brassica species 22 lines showed high degree of resistance to A. brassicae. In Brassica juncea lines PHR-1, PHR-2 and Divya and in B. rapa the line CBYS-7B were resistant. In general, B. carinata, B. napus, B. alba show better resistant reaction than B. campestris and B. juncea. Multiple disease resistance against Alternaria spots, white rust and powdery mildew is reported in lines HC-1 and PCC-2 of B. carinata and GSL-1501 of B. napus (Saharan, 2000). Dang et al., (2000) have also reported multiple disease resistance againt above disease as well as downy mildew in B. alba, B. carinata (HC-1), B. juncea (DIR-157, DIR-1522) and B. napus (GS-7027, Midas, Tower). Coenospecies of Brassica are a rich source of resistance to A. brassicae. Sharma et al., (2002) evaluated 9 genera including cultivated and wild allies of the genus Brassica for resistance to the disease. Eight species (Brassica desnotesii, Camelina sativa, Coincys pseuderucastrum, Diplotaxis berthautii, D. catholica, D. cretacea, D. erucoides and Erucastrum gallicum) were found completely resistant and 12 were moderately resistant. The phytoalexin camalexin, isolated in interaction of Camalina sativa and A. brassicae contributes to resistance against A. brassicicola and A. brassicae in Brassica spp. It inhibits production of the host specific toxin destruxin B by A. brassicae.

Induction of resistance to black spot in a susceptible cultivar against a highly virulent and a moderately virulent isolate of Alternaria brassicae by inoculation with an avirulent strain of A. brassicae was reported by Vishwanath et al., (1999) but inoculation with A. alternata did not induce resistance. On the contrary, it induced susceptibility. Transgenic broccoli plants having endochitinase gene of Trichoderma harzianum through Agrobacterium-mediated transformation show different levels of resistance to A. brassicicola relative to the level of endochitinase. Treatment of Brassica juncea plants with beta-aminobutyric acid (BABA) induces resistance to A. braasicae through an enhanced expression of pathogenesis related protein genes, independent of any role of salicylic acid or jasmonic acid (Kamble and Bhargava, 2007). An aqueous extract of Penicillium chrysogenum mycelium is reported to induce resistance to A. brassicicola through elicitaion of early defense responses in the leaves.

Streptomyces arabicus is a natural enemy of *A. brassicae*. Fistupyrone, a metabolite of *Streptomyces* sp. strain TP-A0569 (Aremu *et al.,* 2003) inhibits spore germination and infection of *A. brassicicola* in Chinese cabbage. The effect is specific for *A. brassicicola*. Other *Alternaria* spp. are not sensitive. Sprays of bulb extract of *Allium sativum* or a strain of *Trichoderma viride* are reported to give as good control of *A. brassicae* as mancozeb. However, efficiency of *T. viride* differs with host cultivars and is more effective in control pod spots than leaf spots.

Mutto *et al.,* (2006) have reported that extracts of dried roots of the weed nightshade (*Solanum nigrum*) contain antifungal compounds against *A. brassicola*. The fraction of the extract effective against the pathogen at 25 ppm suppressed conidial germination and lesion development. The active fraction contains mixture of saponins which have the antigungal activity. Crude methanol leaf extract of *Agave americana* (Agavaceae, century plant or American aloe) has inhibitory effect on conidial germination of *A. brassicae* and on lesion development on *Brassica juncea* Ethanol or methanol extracts of the medicinal plant *Polygomum perfoliatum* inhibit conidial germination of *A. brassicicola* and reduce severity of black spots on cabbage.

LEAF BLIGHT OF WHEAT

This disease was noticed in India as early as 1924 but was considered a minor disease till the late 1950s when it became widespread. Many of the rust resistant varieties introduced during the 1960s were found highly susceptible to this disease. Seedlings are not attacked. The disease makes its first appearance in the field when the crop is 7–8 weeks old. Durum wheat and related varieties are most susceptible. The extent of damage depends on the wheat cultivar and the stage at which infection occurs. In addition to India, the Alternaria blight of wheat is reported from many African countries, South America, Bangladesh, Italy, France, Turkey, Portugal etc.

Symptoms

The disease first appears as small, oval, discoloured lesions which are irregularly scattered on the leaves. These spots become irregular in shape with increase in size and appear brown to gray in colour. A bright yellow halo surrounds the spots. Several lesions coalesce to cover large areas and cause death of the entire leaf. In severe cases the leaf starts drying from the tip. Black powdery conidia of the fungus may cover the lesions under moist conditions. The lowermost leaves, especially on plants along irrigation channels, get the infection first. Similar symptoms appear on the leaf sheath, ears, awns and glumes.

The causal organism

Leaf blight of wheat caused by *Alternaria tenuis* and *Alternaria triticola* had earlier been reported in India. The disease described here is caused by *Alternaria triticina* Prasada and Prabhu which is now recognized all over the world. Species like *A. alternata, A. tenuissima* and *A. arborescens* found associated with the blight have been shown to be non-pathogenic (Vergnes *et al.,* 2006).

The mycelium of *A. triticina* is initially hyaline, later becoming olive buff to deep olive buff. The hyphae are branched, septate and measure 2–7 μm in width. Conidiophores are similar to the mycelium in color. They are septate, usually unbranched, but sometimes branched, erect, single or in fascicles and emerge through the stomata. They measure 17–28 x 3–6 μm. Conidia

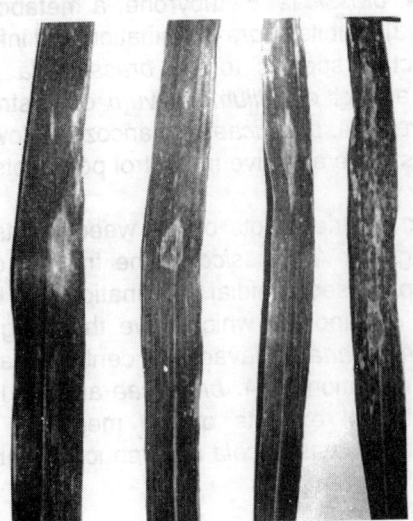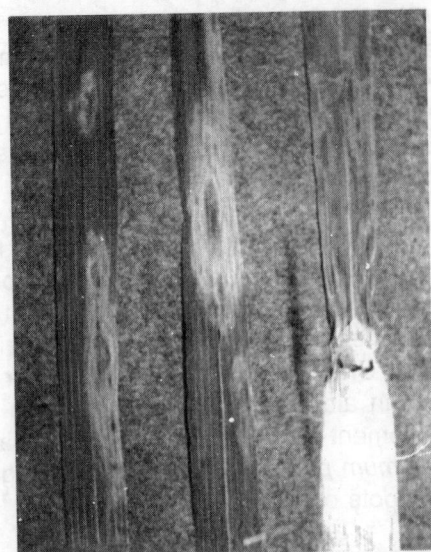

Fig. 107. Alternaria leaf spots of wheat.

are borne singly or in chains of 2–4. They are smooth, irregularly oval, with both ends rounded, or ellipsoid, gradually tapering into a beak. The beak is light brown to dark olive buff, becoming darker with age. The conidia measure 15–89 x 7–30 μm including the beak. The beak is of the same colour as the main body of the conidium. The body of the conidium is light brown to dark olive buff, becoming darker with age. The conidia measure 15–89 x 7–30 μm including the beak. They possess 1–10 transverse septa and 0–5 longitudinal septa. At least 6 physiologic races in the species are suspected.

The conidia germinate at 5°–35°C, the optimum being 15°–27°C. One hundred percent relative humidity is required for conidial germination. Optimum temperature for growth of the fungus in culture is around 25°C. Host range of the species is confined to wheat varieties only.

Disease cycle and environmental relations

The pathogen is soil and seed-borne. The soil-borne inoculum is susceptible to extremes of temperature and at some places it may not survive the summer temperatures on diseased crop debris. In infected wheat debris place on soil surface during summer the fungus could not survive for more than 2 months. They confirmed the seed-borne nature of the pathogen which is present as conidia on the seed surface and as mycelium inside the seed. Parveen and Kumar (2001) have reported that the pathogen invades the epidermal layers of the seed coat and occasionally the endosperm and the embryo. It sporulates in these locations. The extent of seed infection varies among wheat cultivars and is on an average as high as 12.2%. The inoculum carried with seed multiples and establishes in the soil. The leaves in contact with soil get infected first or the spores are splashed by rain to the lowermost leaves. The germ tube from the conidia seldom penetrates the host through stomata or wounds. There is mostly direct penetration on both surfaces of the leaf. Seven to 15 days old seedlings are not susceptible to infection. The susceptibility increases with plant age. The optimum temperature for disease development is around 25°C. When the average maximum temperature coupled with high atmospheric humidity does not exceed 25°C and the plant has reached the boot leaf stage (about 10 weeks old) the development of the disease is maximum. A maximum period of 48 hours of saturated atmosphere is pre-requisite for successful infection.

Management of the disease

The mycelium inside the seed is not inactivated by ordinary seed treatments. Pre-soaking the seed in water for 4 hours followed by 10 min dip in hot water at 52°C gives satisfactory result without affecting seed viability. Later, it was reported that seed treatment with 0.2% Rovral, thiram or Ceresan wet controls the seed-borne infection. Chemical control of the disease with foliar sprays is possible. Sprays of ziram (Cuman), zineb and mancozeb have been recommended. Four to six sprays of zineb or mancozeb at 10–15 days interval give effective control of the disease. Since in some localities the pathogen may subsist on diseased crop debris it is essential to follow suitable sanitary precautions.

Parveen and Kumar (2004) conducted an in vitro study to show antagonistic suppression of *A. triticina* by *Trichoderma viride*. The antagonist coiled around the hyphae of the pathogen forming a rope-like structure and finally disintegrated the hyphae.

The old Indian wheat cultivars like NP 4, NP 52, NP 200, NP 809 and NP 824 were resistant to leaf blight while the varieties NP 710, 718, 761, 788 and 790 were tolerant. Varieties/lines Arnautka, E 6160 and K 7340 show immune reaction while resistance is shown by E 8682, HB 384, HD 2157, HS \74, HW 2449, K 899, K 7333 and VL 417. In Bihar cultivars HP 1163, HD 1941, Janak, M 134 and UP 172 were found resistant under field conditions. In the tarai zone of north West Bengal varieties K 8904, K 8908 and UP 262 have been found resistant. Polyphenol oxidase activity is higher in resistant cultivars than in susceptible cultivars. Active role of peroxidase and polyphenol oxidase in defense mechanism of wheat to *Alternaria* infection was suggested.

LEAF SPOT OR TIKKA DISEASE OF GROUNDNUT

Leaf spot or Tikka disease of groundnut (peanut, *Arachis hypogea*) is the most important fungal disease of this crop. Without foliar application of fungicides the disease can cause up to 100% defoliation prior to harvest and losses in excess of 50% of potential yield. When Tikka disease is associated with rust (*Puccinia arachidis*) the loss may be up to 70%. The spots appear when the plants are 1–2 months old. Due to excessive spotting of the leaves there is defoliation and general weakening and hastening of maturity of the plants which results in fewer and smaller sized nuts and loss in yield.

The groundnut (peanut) leaf spots are caused by two fungi, *Cercosporidium personatum* and *Cercospora arachidicola*, both often occurring on the same leaves. The former causes late leaf spots appearing about 30 days later than the early leaf spots caused by *C. arachidicola*. Late leaf spots are more dangerous than early leaf spots because they are much more numerous and their rate of build up is much more rapid.

Symptoms

Apart from the time of appearance, the symptoms produced by the two organisms differ in size, shape, and color of the lesions. Color of the spots on the leaf is a more reliable criterion than shape for distinguishing the spots produced by the two species. All aerial parts of the plant are attacked by both fungi. In the beginning only leaf spots are common but later stem lesions also develop. In initial stages the two spots are not distinct. The first signs of the disease are slightly pale areas on the upper surface of leaves. The lower surface of the leaf shows the collapse of epidermal cells which lose contact with the underlying mesophyll. Later, the two species are relatively easy to distinguish. The early leaf spots caused by *C. arachidicola* are circular to irregular in outline, 1–10 mm in diameter, and tend to coalesce later. They are surrounded by a yellow halo from the very beginning. On the upper surface the necrotic area is reddish brown to black while on the lower surface it is light brown. The yellow halo is indistinct or not present on the lower leaf surface. *C. personatum* (late leaf spot) produces smaller, more circular spots which are 1–6 mm in diameter and dark brown to black in color.

Fig. 108. Cercospora leaf spots of groundnut.

These brown spots are not surrounded by yellow halo in early stages. The lower surface of the spots is carbon black in colour.

The identification is certain at the time of conidial production. In *C. arachidicola* the conidia are mostly confined to the upper surface of the leaf but are occasionally found on the lower surface also. They are sparse and not formed in concentric rings. In *C. personatum* conidia are restricted to the lower surface and cushions of conidiophores can be seen in concentric rings.

The causal organisms

Cercospora arachidicola Hori (teleomorph *Mycosphaerella arachidicola* W.A. Jenkins) and *Cercosporidium personatum* (Berk. and Curt.) Deighton, earlier known as *Cercospora personata* (Berk. and Curt.) Ell. & Eve. (perfect stage *Mycosphaerella berkeleyii* W.A. Jenkins). In 1983, J.A, von Arx proposed the new name *Phaeoisariopsis personata* (Berk. and Curt.) V. Arx on the basis of the formation of small synnemata or long conidiophores and less thickened, darkened but bulging scars.

The mycelium of *C. personatum* in the host is septate and intercellular sending haustoria into the palisade and mesophyll cells. Conidiophores develop on a dense, globular, brown to black stroma measuring 20–30 μm in diameter. They emerge in dense fascicles by rupturing the epidermis. These conidiophores are uniformly olivaceous brown, continuous, sometimes 1–2 septate, unbranched, and geniculate (having knee-joints) and measure 24–54 x 5–8 μm. Conidia are obclavate to cylindrical, light coloured, 1–7 septate, with bluntly rounded ends and measure 18–60 x 6–11 μm. The perithecial stage, *Mycosphaerella berkeleyii,* belongs to class

Loculoascomycetes of the Ascomycotina in which the perithecia are actually ascostromata. These ascostromata are partly embedded in leaf tissues and are broadly ovate to globose with papillate ostiole. Asci are cylindrical to clavate and contain 8 ascospores. These spores are 2-celled and constricted at the septum. The apical cell is larger than the lower cell. The ascospores measure 35 x 15 μm.

The mycelium of *C. arachidicola* is initially intercellular and then intracellular when the host cells die. No haustoria are produced. The stromata are slight or 25–100 μm in diameter and dark

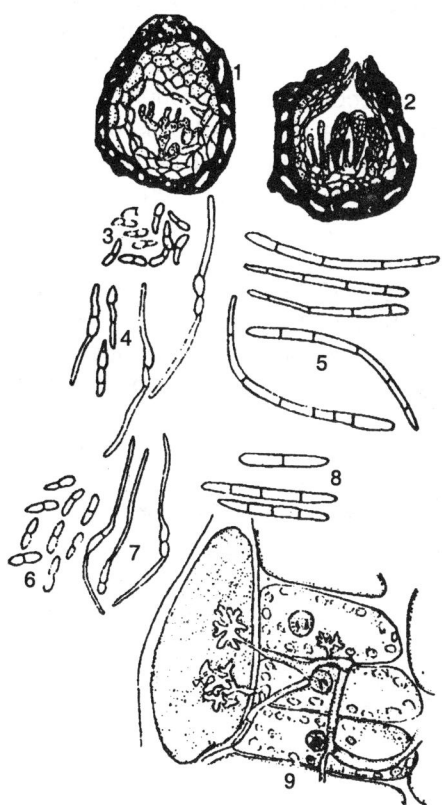

Fig. 109. *Cercospora arachidicola* (1–5) and *Cercosporidium personatum* (6–9).

brown. Conidiophores are olivaceous brown, continuous or 1–2 septate, unbranched, geniculate usually only once and measure 15–45 x 3–5 μm. Conidia are hyaline or pale yellow, obclavate, with rounded to distinctly truncate base and subacute tip. They measure 38–108 x 3–6 μm. The ascostromata are similar to those of *M. berkeleyii*. The major difference between the two species is the regular 1–2 septation and predominance of obclavate conidia in *C. arachidicola* and rare septation and subcylindric conidia in *C. personatum*. Cultural characters of both species are mostly similar.

Races of *C. personatum* are known to occur in nature. Two races have been identified in India. Species of *Cercospora*, including *C. personatum*, produce a phytotoxin, cercosporin, a red-pigmented perylenequinone toxin which lacks toxicity in dark. In the light, these compounds absorb light energy and are converted into triplet molecules that react with oxygen to generate light activated oxygen species. Plant cell membrane is the target of the toxin. Peroxidation of the cell membrane lipids leads to membrane breakdown and cell death. As low as 1 μM cercosporin can kill plant cells.

Disease cycle and environmental relations

The groundnut leaf spot fungi perpetuate through conidia lying in the soil on diseased plant debris and through conidia being carried on the shell of the nuts. The disease onset is earliest and attack most severe when groundnut follows groundnut in the rotation. Although perithecial stage of both species is known, the ascospores are not generally regarded as important source of primary inoculum. Conidia are produced directly from the mycelium in crop debris in the soil following the early rains. Rain splashes deposit these conidia on lower leaves to initiate the disease. Infection of the host is accomplished through both leaf surfaces with penetration directly through the lateral faces of the epidermal cells or by way of open stomata. A great majority of infections originate through the upper epidermis. In artificial inoculations, maximum infection is found to occurs when the plants are sprayed with the inoculum 30 days after sowing. Inoculations after 75 days give less infection. Incubation period varies from 8–10–15–16 days. Both fungi grow at an optimum temperature range of 24°–28°C.

Thermal death point is 51°–52°C. *C. personatum* causes maximum infection at 20°C when the relative humidity is more than 93%. Post-infection development of this pathogen is completely inhibited at 32°C in all genotypes of the host and at 28°C in highly resistant cultivars. *C. arachidicola* thrives best in relative humidity near saturation. Germination of conidia of this species and germ tube elongation is best at 19°–25°C while at 28°–32°C conidial germination is low. At constant high relative humidity (98–99%) conidial production increases with rise in temperature from 10° to 28°C.

Relative humidity is the most important factor for infection. A period of 3 days of high humidity, not leaf wetness, is essential for maximum infection of both species. In a study of the effect of intermittent (wet and dry) and continuous leaf wetness on the pre-infection activities of *C. personatum*, it was observed in intermittent wetness, there were more germ tubes per conidium and more branching of the germ tubes. The germ tubes had tropic growth towards the stomata and subsequent penetration. In continuous wetness of leaves, the germ tube growth did not appear to be directional and the germ tubes commonly passed over the stomatal guard cells without penetration. Sprinkler irrigation promotes the disease. Conidia are detached from the lesions at any time but peak release periods occur when the leaf surface dry in the morning and at the onset of rainfall. Prolonged low temperatures and dew also favor severe infection. In central India the climatic factors in the month of July are not favorable for the disease. The minimum favorable conditions are obtained in the month of August and maximum in the month of September.

Application of nitrogen and phosphatic fertilizers increases disease incidence while potash decreases it only slightly. With gypsum as source of calcium and sulphur and with full NPK fertilizers the development of the disease is least. Susceptibility is directly correlated to increase in ascorbic acid content of leaves while increased riboflavin content is correlated with resistance. Infected leaves have less reducing sugars than comparable uninfected leaves but non-reducing and total sugars are increased presumably as a result of hydrolysis of starch. Resistant varieties have thick epidermis-cum-cuticle and palisade layers, high frequency of trichomes and calcium oxalate crystals while susceptible varieties have high frequency and larger size of stomata. In a study of structural defence mechanisms to late leaf spot, Mayee and Suryawanshi (1995) have also reported that resistant genotypes are characterized by fewer and smaller stomata, compact palisade layer, thicker epidermis-cum-cuticle and presence of trichomes on abaxial surfaces of leaves. Because of reduced infection rate in resistant lines, the resistance to late leaf spot is considered of partial nature, similar to slow rusting phenomenon. Longer incubation period, reduced sporulation (or percentage of lesions that sporulate), stomatal exclusion and absence of directed growth of germ tubes towards stomata are some of the components of host resistance. Reduced lesion numbers and their size and reduced percentage of necrotic area are also components of resistance. Pectic deposits around infection site inducing resistance to colonization by the fungus are reported. Infected leaves produce some gaseous substances which induce leaf abscission.

Management of the disease

Cultural practices: Since there is evidence that the disease is favored by mineral deficiencies such as deficiency of magnesium. Such deficiencies in the crop should be corrected by soil or spray application of nutrients. Plant debris from the previous crop should be burnt to avoid soil-borne primary inoculum. Rotation and deep burying of the debris may also help to destroy the soil-borne inoculum. Weeds should be kept under control because their heavy growth may encourage disease development through modification of the crop microclimate. Seed treatment is essential to eliminate seed-borne inoculum. Disinfection of seed with shell can be accomplished by using sulphuric acid. Seeds without shell are disinfected by a half hour dip in 0.5% copper sulphate solution.

Alteration in the date of sowing can help in avoiding the damage by the disease. In areas where maximum intensity of the disease reaches towards the end of August or in September the sowing time can be advanced to June or even earlier if irrigation facilities are available. Hazarika *et al.,* (2000) have reported that least incidence of leaf spots and rust of groundnut occurs in the crop planted on May 5. Highest pod yield was also recorded in this crop. Maximum incidence of both diseases occurs in the crop planted on June 24 and July 4. There was significant correlation between disease incidence and weather factors (rainfall, relative humidity and temperature).

Resistant varieties: Early maturing varieties (60–100 days) usually escape the loss caused by the disease. Varieties having bushy foliage also are less liable to damage than erect and less foliaged varieties. Resistance to late leaf spot and rust is found in many wild species of *Arachis*. Under greenhouse conditions, some accessions of *A. durasnensis* and *A. hohnei* are resistant or immune to *C. personatum* (Pande and Narayan Rao, 2001). Earlier, *A. glabrata, A. paraguariensis* and *A. burkati* had been reported as highly resistant to *C. personatum*. One accession of *Arachis kuhlmanni* and one of *A. duranensis* are asymptomatic to late leaf spot while 26 accessions were resistant and 10 were moderately resistant. Progenies of the inter-specific crosses show significantly longer incubation period, lower lesion numbers, smaller lesion diameter, lesser sporulation and less leaf area damaged than the susceptible *A. hypogea* (Pande *et al.,* 2002). Late leaf spot resistant genotypes of groundnut have been developed by ICRISAT. Genotypes ICGV 891104 and ICGV 91114 are preferred by farmers in Andhra Pradesh. Some of the leaf spot resistant genotypes are resistant to rust also.

Chemical control: Crop rotation and deep ploughing to bury the crop residues may reduce levels of initial inoculum in soil but these approaches have limited value with a polycyclic disease on a long season crop like groundnut. Foliar application of fungicides are the single most effective method of leaf spot control. Foliar sprays with protectant and systemic fungicides have been recommended and have proved highly effective in checking the secondary spread of the disease occurring through conidia disseminated by wind and insects. Due to nature of the foliage, special care should be taken to ensure that the spray reaches the under surface of all leaves and also the stem. The spraying should be started by the end of July and repeated at 10–15 days interval.

Among effective protectant fungicides are Bordeaux mixture (4:4:50 or 5:5:50), Dithane Z-78 (0. 2%), Dithane M-45 (0. 2%), Fycol 8E, Cosan, and copper sulphate mixture (15–25 kg/ha). Dithane gives better control than Bordeaux mixture. In field trials Dithane M-22, Cosan, Fycol 8E and Dithane Z-78 were effective in controlling the disease in the order listed. Dithane M-22 and Cosan were most profitable. Five to 6 sprays are required. The chemical control is profitable only when the crop is grown under proper conditions of soil fertility. Red oxide of copper, copper dust (300 mesh) with sulphur dust in 1:1 ratio had also been recommended. Three sprays of 0.1% Brestan or 0.2% Cupramar were recommended as economical and effective control measures in early 1970s. First spray was done 40 days after sowing and the rest at 20 days interval. Drenching of soil with these fungicides in combination with 3 sprays proved most effective. The systemic fungicide benomyl (Benlate) at the rate of 0.2 kg/ha, in 200 lit of water, sprayed at 7–14 days interval, is reported to give better control than Dithane M-45 (mancozeb) or copper sulphate dust. Bavistin, Brestanol and Cercobin are also highly effective against Cercospora leaf spots of groundnut. Dubey *et al.,* (1995) considered carbendazim (Bavistin) better than copper fungicides for control of leaf spots as well as rust of groundnut. Similar results were reported in 1981. For combined control of leaf spots and rust of groundnut, a mixture of carbendazim (0.075%) and mancozeb (0.15%) was recommended. One spray was given at the most susceptible stage (around August 15 or 45 days after sowing). It is followed by 1–2 sprays at 15 days interval depending on disease severity. Chandra *et al.,* (1998) used mixture of carbendazim (0.05%) and mancozeb (0.2%). At 10 days interval, between 30 and 80 days after sowing to obtain significant control of the disease.

Chlorothalonil (Daconil or Bravo) spray at 0.2% has been found better than mancozeb, Bavistin, or Blitox. Chlorothalonil at the rate of 1.6 kg a.i./ha has given best results. It has a half life of 10 days on the foliage and where facilities of relative humidity and ambient temperature records on regular basis are available it can prove to be an economical fungicides. Kishore and Pande (2005) have reported that a single spray of chlorothalonil at 45 days after sowing with three sprays of leaf extract of *Datura metel* at 60, 75 and 90 days after sowing effectively reduced late leaf spot and rust of groundnut. Mixture of chlorothalonil copper oxychloride or copper hydroxide has the potential to provide control of groundnut leaf spots comparable to, in some cases superior to, that achieved with standard rate of chlorothalonil alone.

Excellent protective and curative action of difenoconazole (Score) at 75 μg/ml, against *C. arachidicola* leaf spots was reported in 1992. The protective action lasts for 21 days and thus allows longer interval between two sprays. Among other triazole fungicides, Tilt 25 EC (propiconazole) applied as three sprays at 14 days interval, cyproconazole (SAN 619) at the rate of 1.2 g/lit and hexaconazole (0.2%) are reported to be effective. They control not only leaf spots but groundnut rust also. Mixture of cyproconazole (0.062 kg a.i./ha) with chlorothalonil (0.63 kg a.i./ha) gives better control of leaf spots than chlorothalonil alone. Three to 4 sprays of tebuconazole (250 g a.i./ha) are claimed to give best control of Tikka and best increase in yield. Jadeja *et al.,* (1999) compared four triazoles (difenoconazole, propiconazole, hexaconazole and epoxiconazole) with Bavistin and mancozeb for the control of leaf spot and rust of groundnut. All the triazoles reduced both diseases. Use of hexaconazole (0.0025%) gave best cash return. Bavistin (0.025%) was better for leaf spot while mancozeb (0.2%) was better for rust. However, *C. arachidicola* and *C. personatum* both have developed resistance to benomyl and are now threatening to develop resistance to ergosterol biosynthesis inhibiting fungicides such as propiconazole and tebuconazole. To prevent failure of benomyl as a long lasting treatment, mixture of benomyl and chlorothalonil and their alternate use is recommended (Culbreath *et al.,* 2002). Azoxystrobil (strobilurin fungicide) is also effective. Pyraclostrobin is a more recent strobilurin fungicide and has been found effective against leaf spots. Compared to other recommended fungicides (benomyl, chlrothalonil, tebuconazole) it gives comparable or better control at lower dose of 168 g/ha and longer spray intervals (28 days) thus fewer spray application. The number of sprays can be considerably reduced without any adverse effect on disease control and yield.

With proper choice of a fungicides and its time and frequency of application, chemical control is highly effective against early and late leaf spots of groundnut. However, farmers are often reluctant to invest in fungicidal control because of low basic yield of the crop. Even if the chemical control can double the yield, the cost/benefit ratio is not encouraging. Therefore cultural practices and host resistance have better acceptability. Pande *et al.,* (2001) have reported that in many areas of Andhra Pradesh, farmers prefer to use resistant genotypes like ICGV 89104 and ICGV 91114 and chemical seed treatment for yields better than local cultivars. In a study in USA, Monfort *et al.,* (2004) have reported that the number of fungicide applications could be reduced without compromising control of early leaf spot when reduced tillage is used, especially if combined with moderately resistant cultivars.

Biological control: Meena *et al.,* (2000, 2002) have reported biological control of late leaf spot by induction of systemic resistance through the use of *Pseudomonas fluorescens* as biocontrol agent. Seed treatment followed by foliar application of a talc based powder formulation of the bacterium effectively controlled the disease under field conditions and increased pod yield. When the treated seeds are planted the antagonist moves to the rhizosphere and multiplies well there. Maximum protection from the disease by the seed treatment was observed 30 days after sowing. The treatment increased the activity of phenylalanine ammonia-lyase (PAL), phenolic content and lytic enzymes, especially when the treated plants were exposed to pathogen attack. Foliar application of *Serratia marcescens* (strain GPS5) is reported to give significant control of the late leaf spot. Supplementation of the bacterium with chitin gives better control (Kishore *et al.,* 2005a,b). Chitin with *Pseudomonas fluorescens* gave only as much control as the biocontrol agent alone. The disease suppression was by activating the defense-related enzymes. Strains of *Pseudomonas aeruginosa*, tolerant to chlrothalonil, are reported to reduce frequency of lesions on leaves by 60–70%. In the field, a combination of these isolates with 500 μg/ ml chlorothalonil gave as good control as 2000 μg/ml chlorothalonil alone (Kishore *et al.,* 2005a). Chitin-supplemented application of antifungal and chitinolytic *Bacillus circulans* and *Serratia marcescens* have resulted in improved biological control of late leaf spot and yield increase of 62–75% (Kishore *et al.,* 2005b). The mycoparasites *Verticillium lecanii* and *Dicyma pulvinata* parasitize the early and late leaf spots pathogens. These could be explored for biological control Kishore and Pande (2005a,b) have reported that leaf extract of *Datura metel,,* and *Lowsonia ineristhe* provides significant reduction in the intensity of leaf spots. Leaf extract of the legume *Prosopis juliflora* (2% W/V) applied four times at 15 days intervals effectively controls leaf spot and rust diseases of groundnut up to 95 days after sowing. Application of extra. At 45, 75 and 90 days after sowing and chlorothalonil at 60 days after sowing reduced the foliar diseases and increased yield of pods by 81–98%. Spore germination of the groundnut pathogens *Cercospora arachidicola, C. personatum* and *Puccinia arachidis* is inhibited to the extent of >90% by cinnamon citral and clove oils (0.01% v/v). Clove oil (1% v/v) applied as foliasr spray reduces severity of late leaf spot (Kishore *et al.,* 2007). Sprays of extracts of mahogany tree bark and leaves of margosa (*Azadirachta indica*) are reported to suppress severity of leaf spot caused by *C. personata* Extracts of neem seed, garlic cloves, onion bulb, ginger rhizome and papaya leaves also provide provide protection against cercospora leaf spots of groundnut.

STRIPE DISEASE OF BARLEY

This disease of worldwide occurrence occasionally causes considerable damage to the barley crop from seedling stage to maturity. One of the most conspicuous effects of the disease is the failure of the ears to set grains properly. In badly affected cases the entire plant is blighted. Loss of germinability and pre- and post emergence blighting of seedlings may also occur. Under conditions of epidemics, the loss may be as high as 70–72%. There are 22% less grains in diseased ears than in healthy ears and germination is reduced. When disease intensity is 4.4% the yield loss is 4.6% and when intensity is 14.2% the yield loss is 12.2%. In a study of the disease, Kumar *et al.,* (1998) estimated that in the cv.138 with 3.7% disease incidence there were 45.5% less ears, 47.5% fewer grains and the yield loss was 105.7 kg/ha.

Fig. 110. Stripe disease of barley (*D. graminea*).

Symptoms

The symptoms of the disease are conspicuous from the late tillering stage until the crop is mature. The disease starts as small yellow spots on old leaf blades and the sheath. These spots elongate into stripes. Once the leaves start showing symptoms all the successive leaves of the tiller, sometimes the entire stool, show the stripes. The elongation of the culm of the diseased plants varies from rosette-like development to fully developed tillers. The yellow stripes soon turn brown as tissue necrosis progresses and, finally, the tissues dry up and the leaf becomes shredded. A gray to olive gray mass of conidiophores and conidia is visible on the stripes. This usually coincides with maturity of the host. In the case of early and severe incidence spikes fail to emerge. Badly infected seeds of susceptible varieties usually result in some seedling blight.

The causal organism

Stripe disease of barley is caused by *Drechslera graminea* (Rabenh. ex Schlect) Shoemaker. The perfect stage of the fungus is *Pyrenophora graminea* (Rabenh.) Ito and Kuribay in Ascomycetes (family Pleosporaceae, order Pleosporales). Other name of the anamorph is *Helminthosporium, gramineum*. Primary hosts of *D. graminea* are barley and other species of *Hordeum*.

The mycelium in the host tissue is abundant and sub-hyaline to light yellow in color. The hyphae are intercellular in the discolored tissue of the host. In culture media the color of the mycelium varies from gray to olive or black. Sporulation is difficult to obtain in cultures and occurs if cultures are exposed to cyclic changes of light and darkness. The vegetative stage (mycelia and conidia) is haploid. The internal hyphae in the host tissue produce clusters of thick, erect conidiophores (2–6 in each cluster) which emerge through the stomata and bear conidia. The basal segment of the conidiophore is enlarged and distal segment is slender. The color is gray to olivaceous. The conidia are subhyaline to yellowish brown, thin-walled, straight, cylindrical to slightly tapering, having rounded ends, 1–7 septate, and without constriction at the septa. Germ tubes are produced by the end cells (bipolar germination) and only sometimes by the middle cells. The germ tubes and the unpigmented appressoria formed by them are surrounded by a matrix of extracellular, mucilaginous material which provides protection and helps in adhesion to the host surface. The fungus is heterothallic and in areas where perithecia are formed chances of evolution of new races are high.

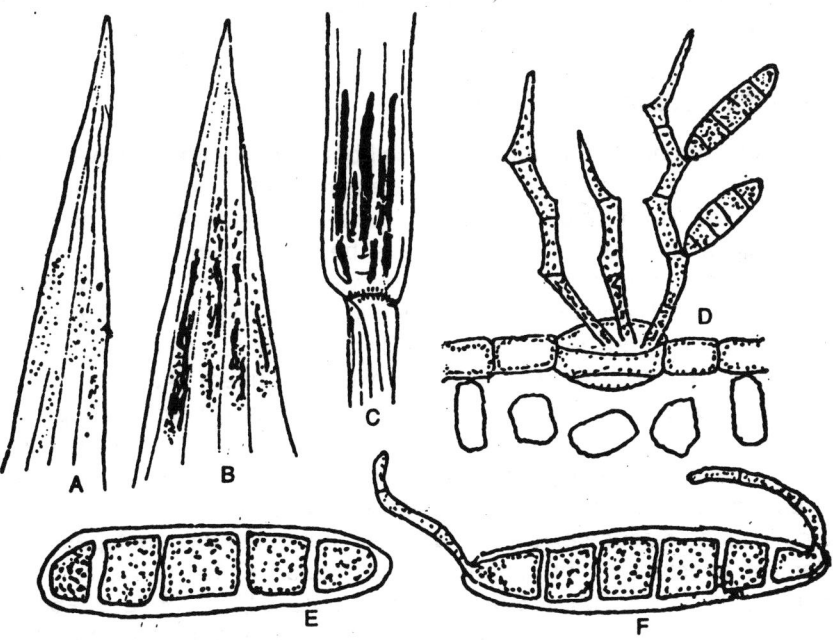

Fig. 111. Stripe disease of barley. A to C-Developing stripes on leaves; D-Conidiophores of *D. graminea* emerging through stomata; E-Conidium; F-Germination of conidium.

Disease cycle and environmental relations

D. graminea is strictly seed-borne and causes a systemic and monocyclic disease. When floral infection occurs the mycelium gets established in the hull, pericarp and seed coat before maturity of the grain. It remains viable in dry seeds for an indefinite period. Although the seed-borne inoculum is the major source of perennation of the pathogen and initiation of the disease in the next crop, some infection may also occur from the mycelium perpetuating in soil on stubbles of the diseased crop of the preceding season. When infected seeds sprout, the hyphae penetrate through the coleoptile, coleorrhiza or root and grow upward through the seedling, permeating the plant extensively, hyphae being present in meristematic tissues and penetrating intercellularly into newly developed leaves and shoots. Soil moisture and soil temperature are critical in assisting the systemic infection during germination of barley seed, especially during the period when the coleoptile reaches apex of the seed and emerges from the soil. Relatively low soil temperatures (around 12°C) and moderate soil moisture are more favorable for systemic invasion. Infection is greatly reduced at soil temperatures above 15°C. For experimental purposes good infection can be obtained by placing the actively growing mycelium of the fungus in contact with the germinating barley seed. This is the only known disease caused by *Helminthosporium* (*Drechslera*) in which extra-embryonic systemic infection occurs. Conidial production usually coincides with maturity of the host.

Abundant conidia are produced on infected leaves during periods of high humidity. The conidia are windblown to neighboring plants where they lodge on flowers and infect seeds at any stage of kernel development up to the milk stage. Sufficient moisture to wet the conidia is necessary for floral infection. Thus, the disease is more common in areas where heavy dew or rainfall occurs during flowering. Irrigation also favors the disease. Infection can occur at temperatures between 10° and 33°C. Infected seed appear healthy. A cool, moist, and fertile soil is favorable for stripe development in seedlings and developing plants. The disease is favored by deep placement of seeds in the soil. Maximum disease occurs when seeds are planted 4–5 cm deep and minimum when planting is done at 2–3 cm depth.

Management of the disease

Field sanitation is recommended to reduce the soil-borne inoculum. However, the pathogen is strictly seed-borne and seed production in dry areas without irrigation is one way of avoiding seed-borne inoculum. Seed treatment with organo-mercurials eradicates the seed-borne inoculum. Carboxin and carboxin + thiram have been recommended in place of organo-mercurials. Imazalil and triadimenol are also effective. Soaking of seeds in solutions of 0.3% copper sulphate or 0.5% ferrous sulphate or zinc sulphate has been found effective against seed-borne inoculum. Diluted neem or marigold leaf extracts used for plant treatment are reported to suppress the disease. The aqueous extract of neem (Azadirachta indica) leaves does not significantly affect conidial germination of *Dreschlera graminea* but provide control of stripe disease through induction of high activity of enzymes phenyl ammonia lyase (PAL) and tyrosine ammonia lyase (TAL) along with rapid accumulation of fungitoxic phenolic compounds. The disease control is as effective as that given by carbendazim.

Resistance to stripe disease exists in a large number of barley genotypes. Barley cultivars highly resistant or resistant to the stripe disease include K 12, K 19, K 24, K 30, K 40, K 69, K 125, K129, K 131, K 133, K 572–64/6, K 572–22–64/29, K 572–28–64/35, K 572/11, K 464/1, K 427–13, C 50, CN 292. Pant and Bisht (1984) reported Morocco, Algerian, VL 10, VL 11, HBL 107 and 108 as highly resistant. In more recent assessment, Kumar *et al.,* (1999) have listed the following as highly resistant: K 492, 1225, 1237, 1239, H 2257, RD 2453, 2461,

2471, 2474, 2478, BH 278, 289, 290, 308, BG 105, C 164, 251, 259. Resistance of barley varieties may not be effective in all areas due to high level of variability in pathogenicity of isolates.

BROWN LEAF SPOT OF RICE

Helminthosporium leaf spot (helminthosporiose) of rice is a major disease occurring in almost all the rice growing areas of the world. In India, it occurs more or less every year in mild or severe form, occasionally as an epidemic. The damage to the crop results from poor germination of seed, leaf spots causing general weakening of the plant and poor grain setting, and infection of grain making them unsuitable for seed. There are indications that damage to seedlings is of no consequence in the plains of India when the crop is raised in summer. It is only the leaf spot phase of the disease which causes the greatest damage.

Symptoms

The symptoms of the disease appear on the coleoptile, the leaves, the leaf sheath, and also the glumes. On the coleoptile the spots are brown, small, and circular or oval. They rarely take the form of long streaks and there is no foot rot of seedlings. On the leaves the spots vary in size and shape from minute dots to circular, eye-shaped or oval lesions measuring 1–14 x 0.5–3 mm. They are distinct and isolated, usually fairly scattered over the leaf surface. The smaller spots are dark brown or purplish-brown. The larger spots are dark brown at the edge but towards the centre they may be pale yellow, dirty white, brown, or grey. Sometimes the spots are surrounded by a yellowish halo and may coalesce and become irregular in shape. Badly affected leaves turn brown and dry out. The symptoms on the leaf sheath are similar to those on the leaf.

In early and severe attack the head may fail to emerge from the sheath and perish without developing fully. When heads emerge they are distorted in various ways. Lesions on the heads first appear on or near the lowest joints of the rachis. Black spots appear on the glumes. These gradually spread over the entire surface as dark brown or olivaceous sporophores of the fungus form on the lesioned areas. The seeds are sometimes shriveled and discolored. The presence of mycelium in infected rice seed was demonstrated in 1966. However, it is seldom that a high proportion of the seeds are affected in this manner.

Fig. 112. *Helminthosporium* leaf spots of rice.

The causal organism

Brown spot of rice is caused by *Drechslera oryzae* (B. de Haan) Subram. & Jain. The fungus was earlier known as *Helminthosporium oryzae* Breda de Haan. *Bipolaris oryzae* (B. de Haan) Shoemaker is another name which was proposed in 1959. The perfect stage was first obtained by Ito and Kuribayashi in 1927 and described as *Ophiobolus miyabeanus*. It was later described and named as *Cochliobolus miyabeanus* (Ito and Kuribay.) Dickson in class Ascomycetes, subclass Dothideomycetidae, order Pleosporales and family Pleosporaceae.

410

The haploid mycelium consists of inter- and intracellular prostrate hyphae and more or less erect conidiophores. The mycelium develops as grayish-brown to dark brown mat on the host parts and in culture. On the host the hyphae are short segmented. The conidiophores arise as lateral branches from these hyphae. They are stout, erect, unbranched except at the base where sometimes they form branches, dark olivaceous near the base identifying with the color of hyphae and lighter in color towards the tip. These conidiophores emerge in tufts through the stomata when developing from the internal mycelium. They are characteristically bent and possess knee-joints, the points where conidia were attached. The lowest conidium is the oldest. The conidiophores may be up to 680 µm long but in India they have not been found longer than 175 µm and 5–7 µm in width. Conidia are 5–10 septate. Their size varies according to environments and physiologic race of the fungus. The smallest mean length is 56 µm and largest 104 µm. Mean width varies from 15 to 20 µm. These conidia germinate characteristically from the two end cells. In the perfect stage, *Cochliobolus miyabeanus*, perithecia in cultures appear in clusters and are globose, pseudoparenchymatous, black and with an ostiolar beak. Asci are cylindrical to long, fusiform, slightly curved, and contain mostly 4– 6 ascospores. The ascospores are filiform and develop in a close helix in the ascus.

Natural occurrence of *D. oryzae* on *Leersia hexandra* was reported in India in 1953. The fungus has been successfully inoculated on *Setaria glauca, S. italica, Zea mays, Echinocloa frumentacea, Sorghum vulgare, Panicum miliaceus, P. milare,* and *Euchlaena mexicana.*

Fig. 113. Rice leaf spots on leaves and grains and conidia of *D. oryzae.*

Wild rice species are also alternate hosts of the fungus. Infection by *H. oryzae* causes a decrease in phenolic content, and in the activities of peroxidase and phenylalanine-ammonia lyase in rice leaves at advanced stages.

The fungus produces several toxins which accumulate in infected tissues at later stages. One of the toxins is cochliobolin. It is highly toxic to rice seedlings inhibiting the growth of the roots and affecting respiration of leaves by destroying the chemical and physical equilibrium of the protoplasm. Vidyasekaran *et al.,* (1986) have reported a toxin of this fungus which is distinct from 3 previously reported toxins. It is host specific and produces brown spot symptoms. Isolates lacking this toxin lose pathogenicity. The toxin suppresses phenol metabolism in rice plants, thus aiding the pathogen to colonize host tissues (Vidyasekaran *et al.,* 1992). Abundant amounts of proteolytic enzymes are also produced by the fungus in cultures as well as in the host tissues. These enzymes cause breakdown of protein fragments of the cell walls resulting in partial disintegration of the integrity of the cell.

Disease cycle and environmental relation

The pathogen is soil- and seed-borne. The mycelium and conidia are carried over from season to season on the seed and in diseased crop debris left in the field. However, the pathogen does not survive for long in natural soil. Earlier, the possible survival of *D. oryzae* in the plains of India in seeds, leaves, nodes and internodes, in stubbles left in the field after harvest, and in the soil till the time of next sowing (December to July) had been studied in 1950s. It was found that the pathogen remains viable only in the seed during this period. The seed-borne inoculum, however, fails to cause infection of the seedlings at the time of seed germination at temperatures of 28°C and above. These findings excluded the possibility of the pathogen being carried over from season to season through soil or seed at least in the plains of India where mostly the crop is raised during summer. Aeroscope studies by these workers revealed that conidia of the fungus are present over the rice fields in the months of April to July, i.e., during the sowing season and earlier. This air-borne inoculum from some external source appears to initiate the primary infection in a locality. These external sources may be perennial grass hosts and early sown or winter rice crops. Another source of these spores may be the paddy straw heaps.

The optimum temperature for formation of conidia is 21°–26°C although they can germinate at temperatures from 5°–36°C. A relative humidity of 92.5% or higher favours conidial production. Aerial dispersal of conidia in relation to weather conditions (minimum and maximum temperature, relative humidity, and wind velocity) was studied in India by many workers. Maximum flight of conidia takes place at a wind velocity of 4.0–8.8 km/hour. Minimum temperature and relative humidity are also important but their effect is masked by the effect of wind velocity. Minimum temperature of 27°–28.5°C, relative humidity of 90–99%, wind velocity of 4.0–8.8 km/hour and low to moderate rainfall of 0.4–14.4 mm were found conducive factors for dispersal of conidia in atmosphere which may be responsible for spread of primary and secondary inoculum.

The optimum temperature for germination of conidia is between 25° and 30°C, the minimum at 2°C and the maximum at 41°C. At 25°C a relative humidity of 92% or more is best for spore germination. The germ tube from the conidium preferentially attaches to thin-walled bulliform cells of the rice leaf. Within a few hours, the tip of the germ tube swells to form a lobed or branched appressorium from which peg-like infection hyphae develop and penetrate through the epidermis. The germ tubes and the appressoria are surrounded by a matrix of extracellular, thick mucilaginous material which enable it to adhere to a hard surface. The adhesion resists washing from the leaf surface. The appressoria are non-pigmented (weakly melanized). The mutants that form melanized appressoria are also equally pathogenic. The direct penetration of the leaf is mainly by enzyme action aided by mechanical pressure. Infection through stomata without formation of appressoria may also occur. The hyphae then spread to the intercellular spaces of the mesophyll. Lateral spread is checked by the vascular bundles which have to be penetrated if the fungus has to pass beyond them. This does not occur in highly resistant varieties. Symptoms may appear within 24 hours of inoculation. From a single leaf spot, clusters of hyphae may arise which, under humid conditions, produce a succession of conidia, each one being produced at the tip of the conidiophore which continues its growth onward at a slight angle to produce a fresh fertile tip and new conidia. Treatments with conidial germination fluid (CGF) of *D. oryzae* induces rapid cell death in leaf tissues suggesting presence of phytotoxion in the CGF. This induces resistance not only to brown leaf spot but blast also in susceptible cultivars.

Free water for about 6 hours after infection is essential for disease development. The best development of spots occurs in shade and high humidity or in a cloudy weather with high humidity and high temperature. Sunlight retards the growth of spots. Seedling blight is more prevalent in cooler soils. At 16°–24°C emergence of seedlings is slow and maximum seedling infection can occur. The disease is common in tracts where rice is grown under irrigation.

Deep sowing resulting in slow emergence of seedlings favours good seedling infection and high pre-emergence loss where the seed-borne inoculum plays an important role. Intensity of infection in any variety is more when grown under transplanted method of cultivation. Youngest top most leaves are most resistant and second leaves from the top are most susceptible to infection. Susceptibility increases with age of the rice plants. Inoculations during September are most effective.

Nutritional imbalances, primarily those of nitrogen and potash, predispose the plants to infection of *D. oryzae*. Severity of brown spot decreases with rise in the potassium level of the soil during growth period of the plants. Conidia from plants grown in potassium deficient soil are more pathogenic than those from plants in soil having optimum potassium. About 90% of conidia of *D. oryzae* penetrate rice leaves via the motor cells, the rest through the stomatal pores. Following invasion, the motor cells develop granular deposits which are abundant in plants grown with excess nitrogen and potash and sparse when there is deficiency of these nutrients. The extent of granular deposits indicates the resistance of the tissue. The infection hyphae appear slightly narrower in cells with granular deposits. Although excess nitrogen supplied through conventional N-fertilizers predisposes rice plants to infection, application of 87–110 kg N/ha through sulphur coated urea or urea super granules (1 g size) reduced disease incidence.

Under conditions of manganese deficiency even relatively resistant varieties become susceptible. However, the concentration of Mn must be critical and appropriate and properly manipulated to bring down the incidence of the disease. There is significant positive correlation between phosphorus content of the host and disease intensity. Higher intensity is positively correlated with N.P. and Mn content and negatively correlated with Ca, Mg, K, Fe, and K/N and Fe/Mn ratios of the plant under phosphorus treatments. Phosphorus content is positively correlated with number of lesions and negatively correlated with lesion types. A positive significant correlation between Fe status and disease expression and negative significant correlation between K, K/N and Mg status and disease expression was observed by them. There was also positive correlation between pH and disease index. Thirty ppm calcium reduces disease while 50 ppm increases it. There is also a negative correlation between Ca and disease severity in clay and loam soils. The disease severity is also positively correlated with P and

Fe/Mn status. It has been concluded that brown spot symptoms are the product of complex nutritional stress. Manganese inside the host tissue and its relationship with other host nutrients determines susceptibility to the disease. Nitrogen helps in establishment of spots and magnesium increases their size while manganese interacting with other nutrients suppresses the disease. In water culture the addition of 5 ppm manganese reduced the disease in a resistant variety. In the absence of manganese even the resistant variety showed susceptibility.

Thick epidermal cells, thick cuticular layer and large number of silicated epidermal cells are also reported to be involved in resistance. Some workers had not found any correlation between silicon content of leaves and resistance to infection in 1950s although they had observed that number of silicated epidermal cells was less in susceptible than in resistant varieties.

Relative preponderance of reducing sugars, non-reducing sugars and acid hydrolyzable polysaccharides in the leaves determines the susceptibility to brown spot. A correlation between the leaf angle, microclimate and susceptibility to leaf spots in rice varieties has been suggested.

Management of the disease

In fields where the disease incidence has been very severe the stubbles should be burnt after crop harvest. Possible alternate hosts like *Setaria, Leersia, Echinocloa,* etc. should be removed or at least their population should be kept at the minimum. In irrigated crops the passage of water from badly infested fields to healthy fields should be checked.

In many areas heavy rainfall. results in leaching down of the more mobile elements like K, Fe, and Mn which are related with resistance. Crops grown in soil subjected to leaching are relatively more susceptible to brown spot. Soil amendments or foliar sprays which restore the balance and make the deficient ions available to the crop are possible approaches to avoid serious losses caused by the disease.

Seed treatment is an effective control measure for areas where soil temperature at the time of sowing or raising seedlings is mild. In Japan dressing of seed with mercuric chloride, silver nitrate, copper sulphate, calcium hypochlorite, formaldehyde, and phenol had been tried and found effective in giving partial or complete control of seed-borne inoculum. Organo mercuria (no more recommended) and other fungicides gradually replaced these inorganic compounds. Agrosan GN or similar fungicides at 2 g/kg seed had been most widely recommended seed treatments in India. However, the mercurias have persistent and polluting effect in soil water and are discouraged in almost all countries.

Since the fungus is sometimes present deep inside the seed, hot water treatment at 55°C for 10 min gives better control than chemical seed treatment. A mixture of the antibiotic Aureofungin (20 ppm) and copper sulphate (20 ppm) was also suggested for destroying the deep seated inoculum. Sulpha drugs like sulphanilamide and antibiotics such as Nystatin and Griseofulvin have been used for seed treatment. Seed treatment with 0.3% Dithane M-45 was also recommended. When seeds of susceptible varieties are germinated in cold water culture extract of *D. oryzae* some degree of resistance is conferred on the seedlings. This effect persists up to the adult stage. The yield of early maturing susceptible varieties is also increased by this treatment. Seeds germinated in solutions of the antibiotics Nystatin and Griseofulvin also confer resistance to seedlings but the treatment may be phytotoxic. Flooding the field to a depth of 10 cm at a time when soil temperature is about 20–28°C enables the seedlings to emerge and survive but prevent conidia from causing infection. Dry conditions of the soil during plant growth should be avoided.

Secondary air-borne spread of the disease can be best managed by fungicidal sprays. Two to 3 sprays with 5:5:50 Bordeaux mixture or 0.2% zineb (Dithane Z-78) at regular intervals had been found very useful in the past. Organo mercurial dusts and antifungal antibiotic Aureofungin had been recommended for spray. In Punjab, 0.1% Hinosan, 25 ppm Aureofungin, 1% Bla-S, 15 kg/ha Ceresan lime dust, 0.3% Dithane Z-78, 0.3% Blitox, 0.2% Benlate or 0.2% Blue Copper controlled the disease when used as foliar sprays at regular intervals. Possible control of leaf spot with brestanol, brestan and Duter was also suggested. Among more recent systemic fungicides two sprays of tricyclazole (0.08%), at mid-tillering and late boot stages, alone or in combination with mancozeb (0.25%) have been found to give good control of the disease. The herbicide Stam F-34 has anti-pathogen activity against *D. oryzae*. Its application provides resistance to rice seedlings against the disease.

Amadioha (2002) has reported that spray of neem extracts suppresses the disease. Spray of leaf extracts of *Nerium oleander* and *Pithecolobium dulce* also suppress brown spots on rice leaves. esses fungicidal activity. It inhibits, *in vitro*, the conidial germination of *D. oryzae* at a minimum concentration of 0.05%. Above this concentration lysis of conidia and inhibition of mycelial growth occurs. Spray of conidial suspension of *Trichoderma harzianum*, *T. reesei* or *T. viride* also provides relief. Two sprays of extracts of neem cake, leaves of *Narium oleander* or *Trichoderma viride*, starting at initial appearance of the disease and given at 15 days interval are reported to give 70, 53 and 48% disease reduction under field conditions (Harish *et al.*, 2008).

Under Indian conditions where seed and soil-borne inoculum may not be very important, the source of primary inoculum seems to be external, and there is no system of disease forecasting that could ensure effectiveness of foliar sprays, resistant varieties are the best method of disease control. Before the introduction of dwarf and semi-dwarf rice varieties over large areas, testing of large number of varieties had shown that CH 13, CH 45, T-141, T- 298–2A, Co 20, BAM 10, T– 998, T- 2118 and T-960 were resistant to b + rown spot. Cultivars Padma and IR-24 are relatively resistant and should be grown in epidemic areas. Among indigenous varieties none has high degree of resistance. Moderate resistance exists in Bajarbang, Gajgaur, Gajraj, Ghagharia, Juhi Bengal, Rambhog, Shyamjira, Sugapankhi and Tulasi Prasad. Hansaraj and Kalanamak varieties are moderately susceptible. The indigenous variety Sokhan was very highly susceptible.

BLAST OF RICE

Rice blast or "rotten neck" is not only one of the earliest known plant disease but also one of the most widely distributed, occurring in every region of the world where rice is grown. It is known to occur in at least 80 countries, being especially serious in humid areas. Since conidia of the fungus causing the disease are not produced at below 88% relative humidity the disease is not severe in areas of low rainfall and humidity.

Although chiefly a foliage disease, the fungus attacks the leaf sheath, rachis, the joints of the culm, and even the glumes. Infection of the culms causes greater damage than leaf infections. Severe leaf infection occurring in the post-transplant stage may, however, lead to total destruction of the foliage. As a result of neck infection, half filled or totally chaffed earheads are formed. There is a tendency in the ears to break and fall off. In India, the damage to rice crop due to this disease is estimated as high as 75%. Many countries have reported devastating epidemics. Even in non-epidemic years the loss is considerable. Often a 5% panicle blast goes unnoticed.

Symptoms

On the leaves symptoms first appear as small bluish flecks, about 1–3 mm in diameter. In older leaves they remain circular but on young leaves they enlarge up to several centimeters long and 1 cm broad. By this time the central portion of the lesion becomes pale green or dull grayish-green and water soaked in appearance. The outer rim is dark. In older leaves the centre becomes gray or almost straw colored. Similar spots are formed on the leaf sheath.

Fig. 114. Blast of rice, Leaf lesions, neck infection and conidia of *Pyricularia grisea*.

Brown to black spots or rings are formed on the rachis of the maturing inflorescence. Ears may also show similar spots. The most characteristic symptoms appear on the culm. The neck becomes shriveled and covered with a gray fluffy mycelium. The affected plants can be very easily identified by examining the bluish patch on the neck or the stem. If the infection has occurred much before the grain formation the latter are not filled and the panicle remains erect. If the attack takes place after some grains have formed the panicle hangs down. However, due to necrosis of neck tissues the ear tends to break and fall off. This stage of the disease causes maximum damage.

The causal organism

The blast disease of rice is caused by *Pyricularia grisea* (Cooke) Sacc. (=*Pyricularia oryzae* Cavara). Earlier, *Pyricularia* occurring on grasses and cereals other than rice was considered the species *P. grisea* which could attack rice also and the rice blast pathogen was called *P. oryzae*. The two species are morphologically identical and the perfect stage of the two is also same. *P. grisea* is the type species of the genus. Rossman *et al.,* in 1990, suggested that the correct name of the rice blast fungus should be *P. grisea*. The fungus belongs to class Hyphomycetes of Deuteromycotina and has been placed under family Moniliaceae of the order Moniliales according to the conventional system of classification.

The perfect stage of *Pyricularia* from cultivated cereals and wild grasses was first reported in 1971. A perfect stage of the fungus was found in culture by mating isolates from distant countries and *Mycosphaerella malinvernians* was suspected as the perithecial stage of *P. oryzae*. This perfect stage was reported to be similar to *Magnaporthe grisea* (T.T. Herbert) Yaegashi and Uddagawa, the perithecial state of *P. grisea*. Genetic relationship between the two was also established (Valent *et al.,* 1986). Finally, with the recognition of *P. grisea* as the correct name for the rice blast fungus the perfect state is recognized as *Magnaporthe grisea* (Herbert) Barra, a haploid fungus. *Magnaporthe* is an ascomycetous fungus in family Magnaporthaceae. The perithecial state was considered rare in nature although fertile strains have now been found common in many countries of Asia. Molecular genetic studies indicate that *P. grisea* exists as a number of genetically distinct clonal (asexually reproducing) populations occurring as a result of strong selection for maintaining host specificity and also due to geographic isolation.

The mycelium of *Pyricularia grisea* consists of septate, mostly uninucleate, branched hyphae. Conidiophores are single or in fascicles and simple or rarely branched. They show sympodial growth and are septate, slender, denticulate and grayish in color. Conidia are produced in succession, one at a time, at the tips of these conidiophores. The conidia are narrowly pyriform (pear-shaped) to obclavate, with rounded base, and narrowed towards the tip which is pointed or blunt depending upon the race of the fungus. The conidium is 2-septate. The cells of the conidium are uninucleate. The nucleus has two large and two smaller chromosomes. Rarely the conidia are 1- or 3-septate. They are hyaline to pale olive and measure 14–40 x 6–13 (mostly 19–23 x 7–9) μm. There is a protruding hilum at the base. The conidiophores produce one conidium every 30–40 min and may bear as many as 20 or more conidia. A typical lesion on leaf produces 4000 to 6000 conidia each night for 2 weeks or more. The conidia are released from the conidiophores by dew or rain and disseminated by wind currents. The spore release behaviour is influenced by light. Even very dim light is sufficient to suppress spore release. Most conidia travel only 1–2 meters from their source of origin before they land on other leaves or other rice plants. When temperature is favorable and in presence of free water these conidia germinate by several germ tubes within 3–4 hours. The conidia of *M. grisea* carry lipophilic self-inhibitors which prevent germination at the site of sporulation. The leaf cuticular wax can relieve this self-inhibition when conidia land on the plant surface. The germ tubes form dome shaped densely melanized appressoria. The environmental factors that induce appressorium formation are hydrophobicity and hardness thigmoytopism) of the contact surface and some chemicals produced by the host. Melanization of the appressoria is related with virulence. Melanin-deficient mutants are defective in penetration. Infection pegs arising from the appressoria penetrate the host tissues. Germination and penetration may be accomplished in 7–8 hours. Lesions appear about 4 days after spore germination and a new crop of conidia is produced in 6–7 days. In culture the fungus produces chlamydospores which are thick-walled and 5–12 μm in diameter. Sclerotia are also produced by the fungus.

The fungus is heterothallic. Two mating types MAT 1–1 and MAT 1–2 are known. Both mating types may be present in the same field at the same time, although Priyadarshini *et al.,* (1999) report that both are not commonly present at a given location. Fertility in *M. grisea* is restricted to non-rice x non-rice or non-rice x rice crosses. Either mating types exist as self-sterile hermaphrodites (Kotasthane and Singh, 2000b). Crosses between isolates from different hosts result mostly in infertile perithecia. Hermophrodites are also found among isolates from ragi. In crosses between a large number of isolates collected from cultivated and wild hosts in India the perfect state of the fungus was not found. In few, only sex organs developed but there was no perithecial formation. Kotasthane and Singh (2000a) tried to trace the development of antheridia and archigonia leading to development of perithecia. They could detect melanized hyphal tips at junction of mating colonies and considered them the aged antheridia. Dayakar et al., (2000) collected 227 isolates of *Magnaporthe grisea* from different states of India and found 39.6% of them fertile. Of the fertile isolates 44.4% produced perithecia, asci and ascospores. This suggests that fertile isolates are not rare. However when monoconidial isolates were mated among themselves, isolates from the same field produced only barren perithecia. Both mating types were found only in Meghalaya and Himachal Pradesh. In Andaman Islands, Andhra Pradesh, Karanataka, Haryana and Punjab only mating type MAT 1–1 was found.

Variability in the pathogen is immense. The sexual cycle being uncommon, it is not the cause of variability. Parasexuality has been considered as the possible mechanism through heterokaryon formation between two strains and subsequent mitotic recombinations. Thus, physiologic races are fairly common in *P. grisea*. There have been controversies about the actual number of races of *P. grisea* in India. Some report say 30 or 33, others 15 and some say only one race caliming that other races are unstable. A 30-year study conducted by Latterell and Rossi (1986) had strongly supported the race concept in *P. grisea* (*P. oryzae*). Each pathotype is characterized by its capacity to attack certain cultivars. The races are fairly stable.

Pyricularia grisea produces the toxin pyricularin which is stimulatory to plant growth in high dilutions but phytotoxic in high concentrations. Like *D. oryzae*, this fungus also produces proteolytic enzymes which help in breakdown of the cell walls. The spore germination fluids (SGF) of *P. grisea* contain some factor(s) which exhibit host species-specific activity and may be the determinants of basic compatibility (Rathour et al., 2002). SGF of rice isolates is highly disruptive for protoplasts of rice but has little effect on protoplast of ragi (*Eleusine coracana*) while ragi isolates have little effect on rice protoplasts. A small icosahedral virus designated Magnaporthe oryzae virus 1 (family *Totiviridae*) is reported in the blast fungus.

Disease cycle and environmental relations

The blast disease is common where rice is grown between 9° and 45° N latitude. The survival of the fungus through hot dry months in the tropics and through cold winters in the sub-tropics and in temperate rice areas is achieved by very different means. One common method that had been suggested is survival through infection of collateral hosts such as sugarcane, *Digitaria marginata*, *Dinebra retroflexa*, *Panicum repens*, *P. proliferum*, *Brachiaria mutica*, *Arundo donax*, *Leersia hexandra*, *Echinochloa crus-gali*, *Digitaria sanguinalis*, *Stenotaphrum secondatum*, and *Eremochloa ophiuroides*. Some workers believe that possibly these grass hosts harbor different pathogenic races of the fungus which may or may not parasitize the local rice varieties but produce enough conidia to be disseminated by air currents to infect crops in other localities. In the tropical climate of south India and Sri Lanka, where several crops of rice are grown in a year survival through the main host is easy, the pathogen maintaining a continuous disease cycle on the rice crop itself. The left over diseased seedlings in nursery beds are another source of primary inoculum.

If chlamydospores and sclerotia observed in artificial cultures are commonly formed in nature they also can serve as an important source of perennation. The fungus has been found to survive the winter snow in Kashmir in the infected rice straw left in the field and produce abundant conidia in April.

The seed-borne inoculum has little, if any, role in disease development. In the plains of north India, the seed-borne inoculum fails to initiate the disease due to high soil temperature in May to end of June when rice is sown. Although there is linear relationship between severity of panicle infection and level of seed infection, the transmission of the fungus from the seed to the seedling is always low. A seed sample with 21% infection results in only 4% seedlings with lesions in a relatively drier soil. However, under milder temperature conditions if infected rice grains are applied to the soil surface at the time of plant emergence the disease can develop among the seedlings. In a study in the USA it is reported that planting naturally infected seed may result in disease development (i) from seedlings grown from infected seed placed beneath the soil surface, (ii) from seedlings grown from germinating seed left on the soil surface, (iii) from seed coats or from ungerminated seed left on the soil surface.

Survival of the pathogen through soil in the tropical climate is restricted. The fungus survives in unsterilized soil for only 19 days due to intense competition with and antagonism of soil microflora including *Streptomyces griseus* and *S. flaveolus*. Thus, for such areas where there is no soil or seed-borne inoculum, the onset of the disease in the main crop must be from some external source (early sown rice on the hills or in the neighboring areas, and weeds). The conidia from these sources are disseminated by air currents to cause primary infection in the new crop or field and spread of the disease. Most conidia that land on leaf surfaces are located on microscopic wart-like protuberances, 2–4 μm high. They become attached to the leaf surface when germ tubes touch the surface. Adhesion is by means of mucilaginous substances excreted by the tip of the germ tube ands is restricted to the area of contact between leaf surface and tip of germ tip (Koga and Nakayachi, 2004).

Reaction of rice plants to blast varies according to the variety. Within the same variety the severity varies according to age of the plant, host nutrition, and meteorological factors such as temperature and moisture. There are three distinct stages of plant growth in which rice is highly susceptible to infection of *P. grisea*, viz., seedling stage, and rapid tillering stage (15–30 days after transplanting) for foliar infection, and ear or neck emergence stage for nodal and neck infection. The susceptibility of the plant at these stages is due to anatomical and physiological changes which take place in the host. Resistance to leaf blast in seedling stage is correlated with shorter leaf angle, high uniform pubescence and more epicuticular wax. Leaf area has no relation to resistance or susceptibility. However, thickness of the cuticle of the leaf *per se* is not the critical factor for immunity to infection. Resistance of panicle to neck blast has been correlated with the number of chlorenchyma strands per cm perimeter of the panicle and epidermal thickness of the neck region. The upper leaf surface is more susceptible than the lower surface.

Conidia of virulent and less virulent strains of the fungus do not differ in germinability and appressorium formation in both resistant and susceptible cultivars. Penetration of the host tissue is also similar in resistant and susceptible cultivars. Thus, resistance to penetration is not of much importance. It is the response of the tissues after penetration has taken place (cellular resistance) that decides response of the particular variety to the disease. In resistant cultivars, such as Tadukan, the number of cells invaded by hyphae from an appressorium is least, in moderately resistant cultivars such as BJ 1, slightly more, and in susceptible cultivars like Co 13 much higher number of cells are invaded by hyphae from a single appressorium. The resistance to penetration is reported to be primarily controlled by polygenes or genes with minor effects whereas the establishment phase or spread among the cells is controlled by major genes with modifiers.

Host nutrition markedly affects reaction of rice to blast, especially in susceptible varieties. Nitrogen utilization efficiency of blast affected rice plants is reduced. In any given cultivar, application of nitrogen above the requirement of the crop increases leaf blast, particularly in susceptible cultivars. Greater the accumulation of nitrogen in the leaf tissue the more susceptible the variety is zero nitrogen level results in significantly less leaf and neck blast but yield are also low (Kapoor and Sood, 2000). Long *et al.,* (2000) have studied the effect of nitrogen fertilization on blast in susceptible and resistant cultivars. They found significant increase in leaf blast (total lesions per plant) except in very highly resistant cultivars. Leaf blast is more severe in susceptible and very susceptible varieties when nitrogen is applied as a single dose than when it is applied in split doses. Nitrogen mainly affects leaf blast, there being no effect on collar rot or neck blast. In upland rice, late application of nitrogen, 30–60 days after sowing, reduces leaf and panicle blast. The increased leaf blast in high nitrogen applications is attributable to increased tissue susceptibility and increased canopy density. Maximum infection of host tissues is at 100 kg N/ha beyond which there is no effect of nitrogen on development of the fungus in host tissues. Leaf blast is suppressed when nitrogen is applied late (30–60 days after sowing). Neither P nor K has any effect on blast incidence in susceptible varieties. However, when rice is grown under P deficiency, correction of the deficiency by application of P tends to reduce disease incidence. This effect is not apparent at high temperatures as P becomes more available from soil even without any addition at that temperature. Enhanced susceptibility to blast in a susceptible variety under conditions of high soil nitrogen and low soil moisture is due to reduced concentration of two cell bound phenols in leaf blade of the susceptible variety. These phenols are B-coumarate and ferulate.

The beneficial effects of silicon fertilization on rice culture are well documented. Although not considered a major nutrient element, silicon is present as 1 to 10% of dry weight of plants. The benefits of silicon include improved fitness in nature, increased productivity and resistance to blast, brown spot and sheath blight through mechanical barriers, accumulation of phenolics and phytoalexins and activation of PR gene (Rodrigues *et al.,* 2004; Rodrigues and Lawrence, 2005). The use of silicon fertilizers is hindered by its high cost but the additional benefits from its application may outweigh its cost. The silicon content of rice leaves has been implicated in resistance to blast disease and to many other diseases of different plants. Silicon content of rice plant and its susceptibility to blast are inversely related. Older leaves are resistant because if silicon is available to roots and is translocated to leaves it combines with one or more constituents of the cell wall to form complex substances resistant to digestion by Rodriguez *et al.,* (2004) have proposed that the mechanism involves biochemical defense through accumulation of momilactone phytoalexins, Silicon stimulates the diterpenoid pathway leading to synthesis of the phytoalexins. When plants are growing in a nitrogen deficient medium there is greater absorption of silicon, which, not the nitrogen deficiency, imparts resistance in the plant to the fungus. They (Rodriguez *et al.,* 2005) have studied the effect of silicon on cytological and molecular events in *P. grisea* compatible and incompatible interactions. They have concluded that while the role of mechanical barriers can not be ignored, silicon plays active potential role in biochemical defense (accumulation of peroxidase, glucanase and PR1). Application of calcium silicate slag as a source of plant available silicon in silicon deficient soils reduces blast as well as brown spot and increases yield of rice. According to Seebold *et al.,* (2001) silicon fertilization affects different components of resistance in the rice plant. The incubation period is lengthened. The number of sporulating lesions, lesion size, rate of lesion expansion, diseased leaf area and number of spores per lesion are reduced. The amendment of soil with silicon is expensive but it can reduce the number and amount of chemical sprays. Kim *et al.,* (2002) suggested that silicon induces cell wall fortification and provides a cellular mechanism of enhanced host resistance. Rodriguez *et al.,* (2003) have provided cytological evidence to correlate the silicon-mediated resistance with specific leaf cell reaction. The typical feature of the cell reaction is accumulation of an amorphous material that restricts the extent of fungal colonization. The occurrence of empty fungal hyphae surrounded or trapped in the amorphous material suggested that phenolic-like compounds or phytoalexins played a primary role. In a later report (Rodriguez *et al.,* (2005) they have further confirmed that while role of physical barriers in silicon amendments, there is possibility of role of enhanced enzymes and PR proteins. There are also reports that resistance in rice to blast is increased by high concentrations of iron, manganese, zinc and copper in the leaf. There is correlation between anaerobic soil conditions and increased resistance to rice blast. It is hypothesized that varying anaerobic conditions mediate production of phytohormones, particularly ethylene, which modify expression of inherent partial blast resistance in some cultivars. Ueno *et al.,* (2004) have reported that treatment of barley leaves with phenol-related compounds, IAA, tryptamine and tryptophan at 50 μg/ml, induces resistance against *M. grisea.*

Meteorological factors are the most important elements in the epidemiology of rice blast. Plant age and nutrition may be of minor importance if temperature and moisture conditions are favorable for infection. These factors help incidence and spread of the disease in many ways, *viz.,* helping production and germination of conidia, their dissemination, and altering the host resistance and nutrition. While temperature during the rice season is normally favorable for the fungus, moisture is dependent on the method of planting, cultivar, regional variations and other factors. *P. grisea* is primarily a night organism. All the vital processes of the disease cycle (spore release, spore germination, infection, and spore production) require free water and the night time dew provides it. Temperature in the tropics is also optimum for the fungus during the night time. Conidia are not produced below 88% relative humidity and at least 90% relative humidity is essential for their abundant production. Diurnal periodicity of spore release in this fungus has been reported since 1960s. In south India peak concentration of conidia in the atmosphere occurs at 4 a.m. in the main crop and at 6 a.m. in the second crop season. Dispersal is highest during the night hours when the temperature is in the range of 25–27°C, relative humidity is between 86 and 98%, and the wind is calm.

The role of night temperature and humidity in blast incidence was stressed by many workers who have suggested that resistance to blast is governed not only by genetic factors but also to a large extent by a set of very critical environmental factors including nyctotemperature (20°C) which influences the metabolic pattern of the host. At low temperatures the absorbed nitrogen tends to accumulate as soluble nitrogen in the leaf. Further, at low temperatures, under high level of nitrogen, very little silicon is absorbed by the plant. The shift in favour of nitrogen metabolism resulting in better nitrate reduction and accumulation of soluble nitrogen such as glutamine causes exhaustion of carbohydrates which could otherwise help in the production of secondary metabolic products imparting resistance. Conversely, high night temperatures favor secondary metabolism leading to increased mechanical resistance and formation of phenolics associated with active resistance.

The susceptibility of rice plants to the rice blast fungus is increased by cold stress. A night temperature of 20°C alternating with day temperature of 30°C with 14 hours of light and 10 hours of darkness is most favorable for infection by *P. grisea.* Infection seldom occurs on resistant varieties under these conditions. At a night temperature of 30°C the number and size of lesions is small in susceptible varieties while resistant varieties remain free from infection. At 20°C the susceptible varieties show a 3-fold increase in the number of lesions and considerable increase in their size while in resistant varieties the lesions remain few and small. At 15°C both susceptible and resistant varieties show very high infection and large sized lesions. Thus, at very low temperature the resistance to blast in resistant varieties may be broken down. According to Koga *et al.,* (2004) resistance expressed by intact leaf sheaths of susceptible rice plants was suppressed by low temperature and by abscisic acid treatment. An inhibitor of abscisic acid biosynthesis suppressed the loss of resistance due to low temperature. Abscisic acid synthesized de novo under low temperature condition is responsible for adverse effect of the low temperature.

The information on the incidence of blast in relation to meteorological conditions has been employed in attempts to develop a method of forecasting outbreaks of the disease so that necessary fungicidal sprays may be started. Forecasting of the disease can be attempted on the basis of a minimum temperature range of 20°–26°C in association with a high relative humidity range of 90% and above lasting for a period of a week during any of the susceptible stage of crop growth, *viz.,* seedling stage, post-transplant tillering stage, and neck emergence stage. Use of trap nurseries, planted with a susceptible variety in advance of the main crop is also recommended as a good warning system. These trap nurseries give a warning 10–15 days ahead of the incidence of the disease in the main crop of the farmer. In the Kagra Valley of Himachal Pradesh, Kapoor *et al.,* (2000) used trap nurseries planted at 15-days interval starting from 1st June and continued high humidity and low night temperatures for prediction of the disease in the main crop.

Management of the disease

The three primary approaches to the management of blast disease are cultural practices, use of fungicides, and resistant varieties. Field sanitation, destruction of weed hosts and seed treatment are precautionary measures that should be followed as far as possible. Early planted crops usually show less disease than late planted crops. Since nitrogen has close relationship with blast incidence, excessive levels of N should be avoided. Vijaya and Balasubramanian (2002) have suggested supply of 50% of the required nitrogen through FYM and 50% through urea to reduce disease severity and get maximum yield. Seed treatment with organo mercurials such as Agrosan GN is effective in eliminating externally seed-borne inoculum. A mixture of 20 ppm Aureofungin and 20 ppm copper sulphate or copper oxychloride can be used as an effective seed treatment. In 1986, the systemic fungicide Fungorene (8 g/kg seed) was found the best treatment which gives protection to plants up to 45 days after sowing. In cultivars showing slow blasting, seed treatment is especially effective in suppressing blast for 62 days.

Direct chemical control by fungicide sprays is highly effective and economical (cost of control about 1.5–4.0% of the value of the produce) if an effective disease forecasting can be implemented. Among fungicides recommended in the past, copper fungicides such as Bordeaux mixture, Blitox, Coppesan, Copper Sandoz, Cupravit, Shell Copper, etc. and organic mercurials such as Ceresan, Agrosan, Verdasan, etc. were popular and effective. However, these two groups of compounds are no more recommended for rice blast control. Copper fails to check the disease in severe epidemics. It is also often phytotoxic to rice and reduces yield. Organo mercurials are highly effective but because of their environment polluting effect they are not recommended.

Among later fungicides Hinosan (1.5 ml/litre water) and Brestanol (0.44 g/litre water) were found highly effective against rice blast. One spray in the seedbed, two at the tillering stage and two at the neck emergence stage were recommended. The number of sprays could be reduced by regulating the volume of liquid required per unit area by using low volume sprayers or mist blowers. In comparative study of fungicides Brestanol, Du-ter, Hinosan 50 EC, and Blitox-50 and the antibiotic Kasumin, Hinosan (1.5 ml/lit water) was found most, followed by Kasumin 2% WP (1 g/lit water), Du-ter (2.5 g/lit) and Brestanol (0.44 g/lit). Among more recent systemic fungicides against rice blast are pyroquilon (Coratop 2.5 G), hexaconazole (Contaf 5EC) at 2 ml/lit and triflumizole (Rilan 70 WP) at 1 g/lit. The fungicide carpropamid, a cyclopropane carboximide (Bayer) is specific against leaf blast. It does not affect spore germination or appressorium formation but germ tubes are longer with smaller, deformed and less melanized or hyaline appressoria. Frequency of penetration into leaves is drastically reduced. The compound inhibits melanin biosynthesis in appressorial cells of *P. grisea* making them hyaline. Penetration by infection hyphae from the hyaline appressoria into rice epidermal cells is substantially hindered. In addition, the spread of the blast spores from primary lesions to other parts of the plant leading to secondary infection is largely prevented when the plants are treated with carpropamid by spray application (Kurahashi *et al.,* 1999). It can be applied through seed and is translocated to leaves where it diffuses on the leaf surface (Rohilla and Singh, 1999, 2001). This exuded fraction probably acts as the anti-penetration compound. Blast suppressive fungicides such as tricyclozolem, pyroquilon and chlobenthiazone inhibit enzyme reactions involed in the biosynthesis of melanin. Carpropamid (1 ml/L) has been found highly effective against neck blast. Tricyclozole and prochloraz + carbendazim were also effective. *P. grisea* can develop resistance to caroripamid by a single step mutation through substitution of amino acid (valine) to methionine. Spray of azoxystrobin at 125 g a.i/ ha is highly effective in controlling rice blast.

The antibiotics were introduced to replace the organo mercurials which were found toxic to animals. The early antibiotics developed and used in Japan were Cephalothecin, Antiblastin, Antimycin-A, Blasmycin and Blasticidin-A but none could become popular due to chemical instability and toxicity to fish. In 1955, Blasticidin-S (Bla-S) was developed and found superior to other antibiotics and many copper and organo mercurial fungicides. This stimulated research on systemic fungicides which resulted in the development of many highly effective antibiotics and organo metallic fungicides such as Kitazin (EBP: O.O-diethyl- S-phosphorothiolate), Hinosan (EDDP: O-ethyl-S-S-diphenyl phosphorodithiolate), Inazin (ESBS: O-ethyl-S-benzylphenyl phosphoro- thiolate), Blastin (PCBA: pentachlorobenzyl alcohol), Oryzon (CPA: pentachlormendel nitrite), Rabcon (CPA: pentachlorophenyl acetate) and Kasugamycin (Kasumin).

In 1969, the International Rice Research Institute reported that the fungicide Benlate (benomyl) has strong systemic action against the rice blast disease. Since then a number of formulations have been reported effective against the disease. Benomyl seed treatment (1: 400 w/w) gives protection to seedlings in nursery for 24–25 days. It inhibits spore germination and appressorium formation.

Benlate, Bavistin, MBC, Hinosan (all 0.1%) and 0.25% Dithane M-45 are effective against the disease, in the order listed, and significantly better than other fungicides. Application of carbendazim (Derosal, Bavistin or MBC) through mud balls, soil drench, and foliar spray at the rate of 0.5 kg a.i./ha give effective control of the disease. Three sprays were given at the tillering stage at 10-day interval and 2 sprays at the neck emergence stage at 5-days interval. Bavistin, used to provide protection against blast, could be detected in shoot tissues 2 days after treated seeds were sown and in adult plants 35 days after spraying the foliage (Parida, T. *et al.,* 1990. *Pesticides Sci.* 30: 303–308). Orysastrobin is a new Qol fungicide with excellent fungicidal efficacy against leaf and panicle blast and against sheath blight.

Spray of calcium chloride, ferric chloride, lithium sulphate, magnesium sulphate or potassium chloride on seedlings is reported to suppress blast in the nurseries probably by triggering resistance. Manandhar *et al.,* (1998) reported that two sprays of ferric chloride, dipotassium hydrogen phosphate or salicylic acid from the seedling to the heading stage suppressed blast by inducing resistance. Best was ferric chloride. A novel, non-toxic chemical control is reported by Stanley *et al.,* (2002). Zosetric acid {*p*-(sulfo-oxy) cinnamic acid] is a naturally occurring phenolic acid in eelgrass (*Zosetra marina*). It is non-toxic to the fungus. When this is applied to leaves, the spores of blast fungus fail to adhere to the surface and produce appressoria thus failing to cause infection. Lesion development is reduced. In greenhouse studies, Amadioha (2000) had reported that spray of water extract of neem (*Azadirachta indica*) leaves or extracts of its seed kernel significantly reduces the development of rice blast. *In vitro*, the radial growth of *Pyricularia grisea* is suppressed by the neem products.

In susceptible cultivars, application of the plant activator benzothiadozole (BTH) increases resistance to blast but not to brown spot. A more recent plant activator, 3-chloro-1-methyl-1*H* l-pyrazole –5-carboxic acid (CMP), is reported to give 80% disease control when applied to soil in pots at very low concentration. This activator is effective against bacterial leaf blight also. The compound has no effect on hyphal growth, conidial germination and haustorium formation of *P. grisea*. The fungicide tiadinil, recently developed in Japan for rice disease control, works as a plant activator and induces resistance response in rice against the blast fungus.

Resistant Varieties: In Tamil Nadu breeding for blast resistance was started in early 1920s and varieties Co 4 and TKM-1 were found resistant to the disease. Hybridization work using Co 4 and a popular variety ADT-10 resulted in two new resistant varieties, Co 25 and Co 26 which replaced ADT-10. The crosses Co 4 x Co 13 and Co 4 x GEB-24 produced the resistant varieties Co 29 and Co 30. Due to existence of physiologic races of the fungus these varieties started showing breakdown of resistance. Varieties T-603 and T-141 (in Orissa) and A-67, A-90, -200 and A-249 (in Maharashtra) have been found resistant. In Bihar varieties Alkulu, Kukulu, GS-397, GS-480, ADT -20, CH 20, CH-13, Suchi, Bj-1, Gennibera, and Kamala were reported to be resistant. A large number of varieties and cultures of rice were screened in Chhotanagpur region of Bihar (Jharkhand). None could be rated as resistant at all stages of plant growth. The best performance was given by the variety Baigan Bichi with moderately resistant reaction in seedling and tillering stages and resistant reaction to neck blast. On the other hand, varieties BK-36, 35/2352 and 37/2342 showed susceptibility during seedling and tillering stages but were resistant at the neck infection stage. In Uttar Pradesh also, 29 varieties and cultures were tested and none was found resistant. Moderate resistance was observed in SM-8, T-100, N-1, T-22-A, T 36, H-755, BJ-1, T-3 and CR-9. Response of varieties depends on physiologic race of the fungus. IR-262A-43-8-11 was found resistant to 8 races while IET-723 (Jaya), IR-8, Co 25, and CR-19–17 were resistant to 5 races. Cultivar Jaya has shown resistance to the disease at Raipur (Chhattisgarh) for 7 consecutive seasons, Pankaj for 4 seasons and Jayanti for 3 seasons. The constancy in resistance suggests different degrees of stable resistance in the variety. Durable resistance in IR- 579 and Bala was reported in Himachal Pradesh. Reddy and Satyanarayana (1984) had screened 1673 germplasm accessions at Hyderabad (A.P.) and found 76% of them having resistant to tolerant reaction to blast. The genotypes Basmati Gola, Basmati Mehtrah, Chandan, G-65, H.KR 97–416, Haryana Basmati 1, Pusa 743 and many others have consistently shown resistance to blast for 3 years in Haryana.

In international cooperative studies to detect varieties having a broad spectrum of resistance (resistance to most races in most countries) the following facts have been established: 1) some varieties have a broad spectrum of resistance, although none seems to be resistant in all tests, 2) varietal reactions usually show regional pattern, i.e., any given variety is resistant or susceptible at all the stations throughout a region or neighboring regions, 3) most exotic varieties show a resistant reaction, i.e., many *japonica* varieties are resistant in South Asia (India and Pakistan) region while numerous *indica* varieties are resistant in temperate Asia (Japan and Korea). Resistance in *indica* varieties has been found to be physiological and independent of environmental factors. The physiological resistance has been attributed to shortening of the moribund stage of infected leaves, causing death of mycelium by release of fungitoxic compounds including phytoalexins. Attempts had been made to induce resistance by mutation using irradiation and chemical mutagens. Certain progenies had shown promising resistance to blast as well as bacterial blight. *Oryza rufipogon*, believed to be progenitor of Asian cultivated rice (*Oryza sativa*), is a good source of lasting resistance against blast.

Through molecular breeding methods transgenic *indica* rice (cv IR50 and cv CO39) resistant to blast and bacterial leaf blight have been developed (Gnanamanickam *et al.*, 2002). Antimicrobial peptides play a role in the immune systems of animals and plants by limiting pathogen infection and growth. The puroindolines are small proteins that give seed hardness in wheat and are antimicrobial. Transgenic rice plants that express puroindoline genes throughout the plant have been produced. The leaf extracts of transgenic rice inhibit growth of *Pyricularia grisea* (blast) and *Rhizoctonia solani* (sheath blight). The transgenic plants having these genes show significantly increased tolerance to these two diseases.

Biological control: Antagonists for biological control of rice blast are also reported. Hypersensitive response inducing bacterium Pseudomonas syringae pv. syringae inoculated on first leaf induces systemic resistance in seedlings. Gnanamanikkam and Mew (1992) identified a strain of *Pseudomonas fluorescens* (strain Pf7-14) which persisted on rice phylloplane for 40 days after application through seed treatment in a Philippine rice field. The biocontrol agent produces antibiotic(s) that suppress *Pyricularia oryzae, Rhizoctonia solani, Pythium ultimum* and *Gaeumannomyces graminis tritici*. It is as effective as the fungicide tricyclozole in controlling rice diseases Significant control of *Magnaporthe grisea* on finger millet by 4 strains of *P. fluorescens* and 2 strains of *P. putida* is also reported. In the same crop, the rhizobacteria *Pseudomonas fluorescens* strain Pf1 induced defense proteins (chitinase, β-1,3-glucanse, peroxidase and polyphenol oxidase for protection against *M. grisea* (Radjacommare *et al.*, 2005) Vidyasekeran *et al.*, (1997) have developed a powder formulation of *Pseudomonas fluorescens* for rice blast control. Krishnamurthy and Gnanamanickam (1998b) also demonstrated the biocontrol of rice blast with *P. fluorescens* Pf 7–14. When applied as a seed treatment followed by three foliar applications, the biocontrol agent provided 59.6% suppression of rice blast in field trials. The bacteria could move to leaves and could be detected up to 110 days. Migration from seed to seedlings could occur only until the seedlings were 16 days old. *Pseudomonas* spp. suppress rice blast through salicylic acid accumulation and induction of systemic resistance (Krishnamurthy, 1998). Strains of *Pseudomonas fluorescens* and their mixtures cause up to 87% reduction in *Magnaporthe grisea* blast of *Setaria italica* (Karthikeyan and Gnanamanmanickam, 2008). They significasntly increase root and shooy length of the host. Viji *et al.*, (2003) isolated strains of *Pseudomonas aeruginosa* from spent mushroom compost that could suppress blast of perennial ryegrass (*Lolium perenne*) caused by *Pyricularia grisea* when used as foliar spray. These strains suppressed *Rhizoctonia solani, R. cerealis, Sclerotinia homeocarpa* and *Fusarium culmorum* also. However, Muralidharan *et al.*, (2004) have reported that field application of *Pseudomonas* formulations have no effect on neck and leaf blast. Strains of *Bacillus* have also been used as biocontrol agents. Bacillus licheniformis has fungicidal effect on *M. grisea*. It produces a peptide, identified as asurfactin, which is the active principle in suppression of growth of the pathogen. Chemotaxis is an important competitive colonization trait. The organic acids and amino acids but not sugars in the root exudates attract the biocontrol bacteria which move towards root tip by flagellar movement (de Weertt *et al.,* 2002). The plant growth promoting bacteria enter the host roots via cracks at the points of lateral root emergence which differs among cultivars. They are present in intercellular spaces, parenchyma and cortical cells and even in the vascular tissues (James *et al.,* 2002). Strain B2 of *Serratia marcescens* also provide biocontrol of blast through chitinolytic enzyme activity, its red pigment antibiotic prodigiosin and induction of resistance. This biocontrol agent is high effective against rice sheath blight caused by *Rhizoctonia solani* AG1 and also *Botrytis cinerea* (Someya *et al.,* 2002, 2003a,b). Prodigiosin is colour-sensitive and loses activity in white and blue light (Someya *et al.,* 2004a). Bacterial isolates from rice also inactivate the antibiotic without suppressing development of *S. marcescens. Erwinia ananas,* transformed with chitinolyctic enzyme gene from *S. marcescens* is reported to be equally effective and undisrurbed by epiphytic rice bacteria (Someya *et al.,* 2002: 2003, 2004b). Kim *et al.,* (1999) had isolated an antibiotic, oligomycin A, from *Streptomyces libani,* which was effective against *Magnaporthe grisea, Colletotrichum lagenarium, Phytophthora capsici* and *Botrytis cinerea* at as low as 3 to 5 μg/ml of minimum inhibitory concentration.

In Japan, Ohyaka *et al.,* (2008) have reported control of the blast on seedlings by treating with freeze-killed (in liquid nitrogen) mycelium of an unidentified fungus (*candidate* MKP5111B). Autoclaved mycelium had no effect. The mechanism is suggested through some elicitor-like compound which was destroyed by heat. Manandhar *et al.,* (1997) have reported suppression of rice blast by pre-inoculation of the host with avirulent *Pyricularia grisea* and a non-rice pathogen *Bipolaris sorokiniana.* Similar suppression of leaf leasions by pre-inoculation with an avirulent strain is reported by Ashizawa *et al.,* (2005). The level of suppression varies with host cultivars and lines. A pheromone from the yeast *Saccharomyces cervisiae* is reported to control rice blast by blocking appressorium formation in *Magnaporthe grisea* (*Science* 276: 1116, 1997). Rice phylloplane fungi could also suppress *Pyricularia grisea* on leaves (Kawamata *et al.,* 2004). Purified preparation of a particular protein from the fungus *Aspergillus giganteus* has strong antifungal activity against *Magnaporthe grisea.* The level at which it causes total inhibition of the pathogen is not phytotoxic to rice protoplast (Vila *et al.,* 2001).

RED ROT OF SUGARCANE

Red rot was first reported in 1893 from Java (Indonesia). It is a destructive disease of sugarcane in the tropics and sub-tropical areas. In India the disease attacks the standing canes often in epiphytotic form and causes huge losses to the cultivators and the sugar industry. Serious epidemics of this disease have occurred in the past in northern India (Uttar Pradesh and Bihar). In these epidemics some of the very promising cane varieties became highly susceptible and had to be withdrawn. Significant local losses due to the disease occur almost every year in Haryana, U.P. and Bihar. In the USA the disease is not serious on standing canes but causes appreciable damage through seed rot and inhibition of germination.

Symptoms

In early stages it is difficult to recognize the diseased plants in the field. First symptoms are seen after the rainy season when the normal plant growth stops and sucrose formation begins. Loss of colour and drooping of third or fourth leaves from the top are the earliest symptoms. Then the entire top withers. In late stages the canes themselves show the effects. They become shriveled, the rind shrinks, and becomes Ingitudinally wrinkled. Such canes are lighter in weight and easily broken. If the diseased canes are split open longitudinally, especially when withering of leaves starts, the pith looks red colored. Characteristic bands of clear white areas

426

Fig. 115. Red rot of sugarcane caused by *C. falcatum*.

are seen running transversely across the full breadth of the reddened pith. In very advanced stages of the disease, the red color may be replaced by dirty brown and the white bands may not be very conspicuous. Cavities filled with grayish or white mycelium are found in the pith. The juice often gives a bad smell and does not set well on boiling due to conversion of sucrose into glucose and alcohol as a result of enzymic action of the pathogen. Late in the season, minute, velvety, dark dots (the acervuli of the fungus) are formed near about the nodes of the diseased canes and also in the shrunken areas. They are also present in the mycelial growth in the pith.

Mid-rib of leaves is also affected by the fungus. Red patches with ash colored centre develop on the mid-rib. Abundant acervuli are present on these patches.

The causal organism

The red rot disease of sugarcane is caused by *Colletotrichum falcatum* Went. Its teleomorph is *Glomerella tucumanensis* (Speg.) von Arx and Muller, a fungus belonging to the family Glomerellaceaeof of Ascomycota. The species is distinguished from other species forming falcate conidia on the basis of morphology of appressoria. Isolates of the species from *Sorghum vulgare* and *S. halepense* can produce red rot symptoms in sugarcane on inoculation.

Once the fungus gains entry into the host, it grows rapidly producing inter- and intracellular septate mycelium. The early development of the fungus inside the host does not show much branching of the hyphae and invaded cells ordinarily do not become filled with the mycelium soon after infection. The hyphae in the ducts of the fibro-vascular bundles or in any centre of

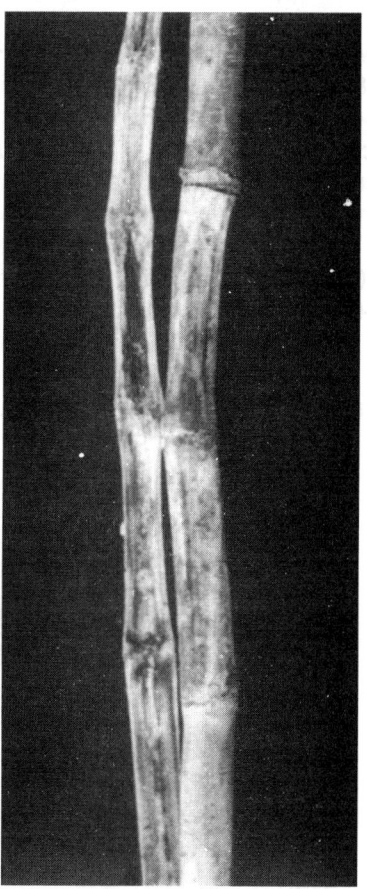

Fig. 116. Dry, shriveled canes due to attack of *C. falcatum*.

infection grow out very rapidly from cell to cell crossing a number of layers without branching much. Eventually, the sugarcane tissues react vigorously to the presence of the fungus and some kind of reaction or change sets in the host cells in advance of the hyphal invasion. The protoplasm changes in color and a gummy dark red material oozes out of the cells filling the intercellular spaces. The soluble pigment present in this ooze is absorbed by the cell walls producing the characteristic red rot appearance. After growing for a period within the host tissues the hyphae produce a large number of chlamydospores in the pith. These chlamydospores can persist in soil for a long time.

Hyphae collect beneath the epidermis and form a stroma of densely packed cells. Hair-like setae, 100–200 µm long and with 4 septa, arise around and in the stroma (acervulus) and push their way through the epidermis exposing the conidia which are borne on small conidiophores packed in the acervulus. They look salmon-pink in mass. These conidia are falcate (sickle-shaped), fusiform, tapered gradually to the base and more abruptly at the apex 20–80 (15.5–26.5) x 5–7 (4–5) µm in size, and possess a large oil globule in the centre. They are easily disseminated by wind, rain or irrigation water, raindrop splashes, and also by insects. The conidia are short lived and readily germinate in presence of moisture by producing a germ tube. This germ tube, on coming in contact with any hard surface such as soil particles, develops an appressorium. This appressorium becomes thick-walled and functions like a chlamydospore.

The appressoria are brown, oval, round, or irregular in shape and measure 6.67–26 or 28 x 6.67–20 (average 14.88 x 12. 36) μm. In some reports the size of appressoria is mentioned as 12.5–14.5 x 9.5–12 μm. Melanization of the appressorium is a pre-requisite for infection process by this structure. Many strains of the fungus produce round, double walled structures (chlamydospores) in old cultures and diseased stalks.

In the perfect stage, perithecia of *G. tucumanensis* are globose, ostiolate, superficial with the bottom embedded in the host tissue, and measure 150–300 μm (average 250 μm) in diameter or 100–240 x 85–250 μm. Asci are numerous, hyaline, clavate, slightly thickened at the apex, and measure 49–66 x 7.0–10.5 (average 56.2 x 9.0) μm. Numerous hyaline paraphyses are present along with these asci. These paraphyses are extremely delicate and disintegrate as oily drops on maturity of asci. Each ascus contains 8 ascospores arranged bi-serially. They are single celled, hyaline, elliptical, and measure 17.5–21.0 x 5.3–7.0 (average 19.6 x 6.3) μm or 18–22 x 7–8 μm.

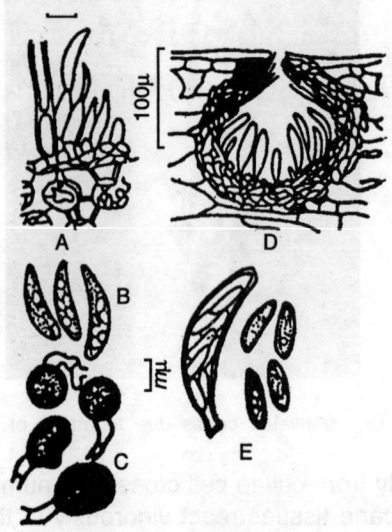

Fig. 117. *Glomerella tucumanensis (C. facatum)*
A—Conidiophores, condia and lower part of setae; B—Conidia; C—Appressoria; D—Perithecium; E—Ascus and ascospores.

Physiologic races are quite common in the fungus and are responsible for incidence of epiphytotics of the disease. Prior to 1932 the race present in India produced dark growth with poor sporulation and, therefore, was not able to cause epidemics of the disease. A new race with light colored growth and heavy sporulation appeared during 1935–36 and was considered as the cause of epidemics during the 1940s. Within this race several strains have been identified. These strains differ from one another not only in their cultural characters such as colour of mycelium, texture of colony, and amount of sporulation but also in their virulence on different varieties of sugarcane. Virulent strains of the light race were reported from West Bengal in 1957. Strains R-117, R-135 and S-244 are the virulent strains common in Uttar Pradesh. In Haryana, 3 distinct strains (strains 1, 2 and 3) were identified on the basis of pathogenic behaviour of the isolates on some differential varieties of sugarcane. Presence of races in different regions may erode the stability of resistance in sugarcane varieties.

The pathogen produces a toxic metabolites also. The phytotoxin is a major pathogenicity determinant. Malathi *et al.,* (2002) reported the detoxification of the phytotoxin by fungal and bacterial antagonists of the pathogen which included strains of *Pseudomonas fluorescens* and *Trichoderma harzianum.* Mohanraj *et al.,* (2003) partially purified a phytotoxin from the red rot fungus. The toxin increased electrolyte leakage in susceptible cultivars and induced higher levels of phytoalexins in resistant cultivars.

Disease cycle and environmental relations

C. falcatum can grow in soil and produce acervuli. The infective propagules in the soil may be conidia, setae, chlamydospores and thick-walled hyphae However, under natural conditions in the field or in manured soil the fungus does not survive in active state beyond 3–4 months due to competitive and antagonistic effects of associated soil microflora. Survival for 63 days during winter and for 34 days during summer in affected stalks and midrib in soil is reported. A density of 50 propagules per gram of soil is sufficient to cause red rot infection in young sugarcane settlings.

Conidia of *C. falcatum* incubated in soil germinate in unipolar or bipolar manner and the germ tubes form appressoria. Thick-walled mycelium is also formed.

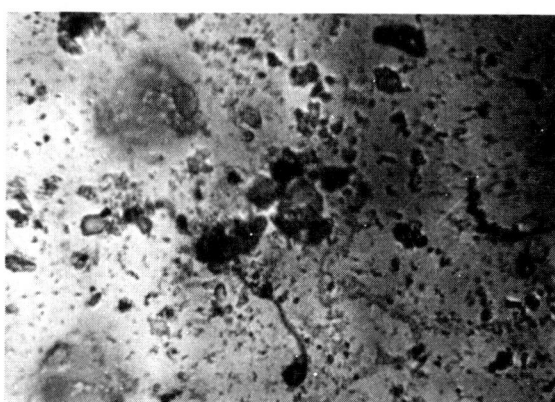

Fig. 118. *C. falcaium.* Conidial germination and appressoria in soil.

These structures may help in soil survival of the fungus. Moisture and nutritional stress enhances the formation of such structures. Since the sugarcane crop has no dead season, even a limited survival of the fungus in soil may be sufficient to carry over the pathogen from the crop being harvested to the newly planted crop. The conidia are short lived but the appressoria and chlamydospores seem to persist in soil. The survival structures are easily disseminated by movement of irrigation water and cultural operations in the field. The red rot fungus can also survive in the rhizosphere of moderately resistant cultivars without showing symptoms.

There is no conclusive evidence to show that the perithecial stage of *C. falcatum* is common in India. If this stage occurs it may be a major source of survival on decaying leaves and may also be the reason for sudden and rapid development of new races of the fungus.

Generally, the seed setts from diseased canes are the main source of primary inoculum and survival of *C. falcatum.* When such setts are planted they invariably produce infected shoots. If such shoots survive and remain in the field, the secondary infection is caused by conidia produced in acervuli on such shoots and transmitted by insects, wind and water. When these conidia fall on wounds caused by insects or on young unfolded leaves, where they travel down to the nodal buds, infection is caused. Presence of septate mycelium, conidia and appressoria in infected buds has been demonstrated. The appressoria in nodal buds can perpetuate the fungus for long periods. Ratoon crops also serve as a source of perennation and multiplication of inoculum. Conidia formed on lesions on the midrib also cause infection of canes. There is possibility that acervuli and chlamydospores ingested by cattle through sugarcane leaf fodder pass through the digestive tract undamaged and are disseminated through manure.

High humidity, water logged conditions, lack of proper cultural operations resulting in growth of weeds, continuous cultivation of the same variety in a particular locality, and the presence of susceptible varieties in the vicinity are some of the factors that lead to the appearance of the disease in a healthy crop and build up of inoculum for epiphytotics. Temperature at different stages of crop growth influences development of red rot. Workers at different places have reported temperatures of 25–30°C, 28–30°C, 24–33°C and 29–31°C during August and September as favorable for rapid development of red rot. Under low temperature conditions (as in January) reaction of cultivars shifts towards resistance while the opposite is true for higher temperatures

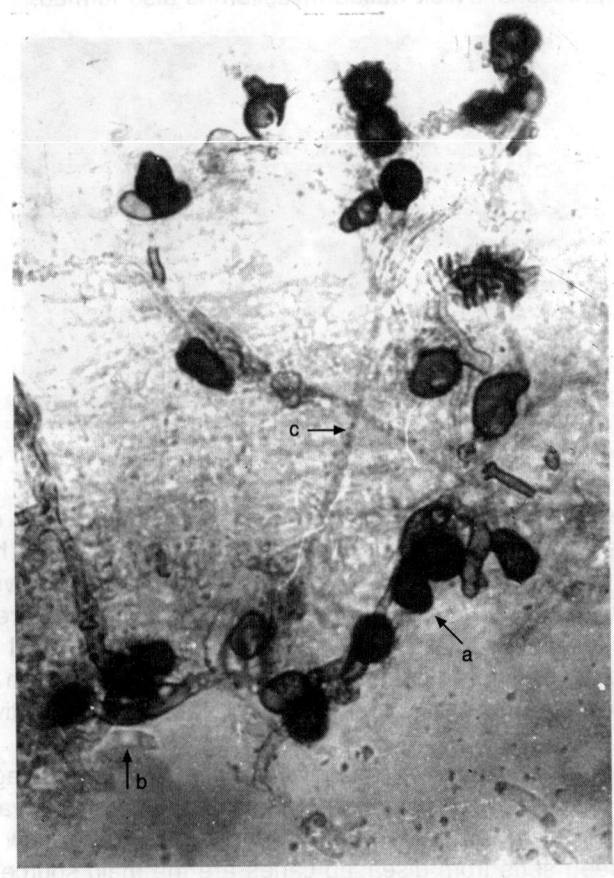

Fig. 119. Appressoria of *C. falcatum* in bud Tissues. (a) appressoria, (b) conidium, (c) mycelium.

of August and September when moderately resistant varieties may shift their reaction towards susceptibility. Under water-logged conditions, even resistant varieties succumb to red rot under. It is due to pre-disposition of the host as well as changes in the behaviour of the pathogen. Floating mycelial aggregates form acervuli and produce shorter than normal conidia under water-logged conditions. Hyaline, thin-walled conidia become brown, thick-walled conidia. Fusion of conidia occurs and the fused conidia germinate and produce secondary conidia. The floating mycelial aggregates and the conidia have better chances of getting attached to vulnerable sites of the host and also being transported to longer distances.

Management of the disease

Epidemics of red rot occur mostly due to accumulation of inoculum year after year if the above described causes are present. Field sanitation is one of the important measures for checking this inoculum build up. As soon as isolated cases of the disease are noticed in the field the entire stool should be dug out and burnt. Fallen leaves after harvest of a diseased crop should be burnt in the field. For a profitable crop low lying areas should be avoided. Repeated deep ploughings after crop harvest and before new plantings reduces inoculum. Once the disease has appeared in a field, a 2–3 year rotation is recommended. Rotation with wet rice culture reduces inoculum of the pathogen. Inclusion of a green manure in the rotation also suppresses the inoculum. Maintenance of soil moisture by irrigation delays onset of the disease. It is not advisable to cultivate same variety for several years in a locality.

Earlier, treatment of seed setts with organo mercurials such as Aretan, Agallol or Emisan (0.25% suspension for 5–10 min dip) was recommended but it only helped in eradication of superficial inoculum. Deep seated mycelium and other survival structures in the buds are not affected by this treatment. A 1-hour treatment of setts in 0.5% Bavistin suspension could reduce the incidence of red rot from infected setts. Beneficial effects of sett treatment with Bavistin and Thiram (0.1% each) were later demonstrated by others but treatment with Bavistin alone encourages *Sclerotium rolfsii* that causes bud rot in the field soil. Such chemical treatments for sugarcane setts on a field scale are difficult to apply but can be practical for seed nurseries.

The best approach to management of red rot is the judicious use of resistant varieties. However, even resistant varieties may fail to give complete security if the cultural practices given above are not followed. A large number of varieties have been evolved and found resistant to red rot but sooner or later, at one place or the other, they have becomes susceptible to the disease. *Saccharum spontaneum* has been the most important source of resistance since early years of the last century. Interspecific crosses between *S. officinarum* and *S. spontaneum* and intervarietal crosses have yielded many economically important cultivars which have high resistance or tolerance to red rot. Following sugarcane varieties have been recommended as resistant to red rot from time to time: Co 244, 285, 301, 349, 356, 370, 385, 393, 449, 475, 508, 513, 561, 562, 568, 846, 951, 975, 1007, 1148, 1261, 1336, 62101, 62399, 86249, 89029, 89003, 97015, 97017, 98016, CoK 30, 32, CoS 109, 443, 510, 8432, 8436, 92255, 96258, 96268, Co Pant 94211, CoH 99, Bo 3,7, 10, 11, 22, 24, 32, 91.

Varieties Co 1111, 1118, 1158, CoS 574 and 575 had moderate resistance to red rot. Stable resistance to red rot is reported in varieties Co 7314, CoS 767, CoLk 7702, 7710, CoS 8315, 422, 8432, 8436, 8016, 90269, 91269, and 92263, Bo 91, U.P. 5, 6 and 39.

Induction of systemic resistance by using chemical and biological inducers is reported. Acibenzolar-S-methyl (ASM) applied as soil drench or through a rooting mixture to seed setts induces high degree of resistance to penetration and colonization of the stalk by the red rot pathogen. The resistance persist for up to 30 days (Sundar *et al.,* 2002). The induction of resistance is accompanied by a significant increase in peroxidases and polyphenoloxidase activity. Viswanathan and Samiyappan (2002, 2007) induced systemic resistance by strains of *Pseudomonas fluorescens.* applied as sett treatment followed by two soil applications in the field. The treatment enhanced germination of setts and final cane yield. Induction of resistance was more pronounced in susceptible cultivars than in resistant cultivars. Pathogenesis related proteins such as chitinase, β-1,3-glucanase, thaumetin-like proteins and other enzymes such as peroxidase, polyphenol oxidase are involved in the *Pseudomonas*-mediated induction of resistance.

Treatment of seed setts with *Chaetomium* sp. and *Trichoderma harzianum* significantly improves germination of pathogen–inoculated setts. Folair sprays of these antagonists also suppress the disease. *Chaetomium* is more effective than *Trichoderma.* Viswanathan *et al.,* (2003) have reported that chitinolytic activities of *T. harzianum* isolated from sugarcane rhizospere and endosphere inhibit the growth of *C. falcatum* through lysis of pathogen mycelium. Integrated biological control of red rot as well as wilt by soil application of *Trichoderma* and *Aspergillus* + *Pseudomonas aeruginosa* is reported. Strains of *Pseudomonas putida* and *P. fluorescens*, native surgarcane rhizobacteria, were found to suppress red rot when applied as sett treatment or through soil.

Jayakumar *et al.,* (2007) have tested plant extracts and antagonistic bacteria singly or in combination for red rot control in pots and field plots. Plant extracts did not provide field control. Sett treatmernt with leaf extract of *Abrus precatorius* in combination with a spray of or soil application of *Pseudomonas fluorescens* strain Md 1 significiantl reduced red rot incidence in field trials.

RIPE FRUIT ROT AND DIE BACK OF CHILLIES

Anthracnose, ripe fruit rot and die-back is the most important fungal disease of chilli (pepper, *Capsicum annuum*) crop. Ripe fruit rot is more conspicuous as it causes severe damage to mature fruits in the field as well as during transit and storage under favourable conditions. In suitable weather it may cause 12–25% loss of fruits. The disease is present throughout India but is more common in Assam, North Bihar, Andhra Pradesh and parts of Uttar Pradesh. In Assam up to 32% of fruits have been found affected by the ripe rot. It is a serious disease in Sri Lanka also.

Symptoms

Fruit Rot: Only ripe fruits, turning red, are affected by the disease. A small black circular spot appears on the skin of the fruit and spreads in the direction of the long axis, thus becoming more or less elliptical. As the infection progresses, the spots get either diffused and black, greenish sharp black outline enclosing a lighter black or straw coloured area. Badly or dirty gray in color, or they are markedly delimited by a thick and affected fruits turn straw colored from the normal red. On this discoloured area numerous acervuli of the fungus may be found scattered. When a diseased fruit is cut open, the under surface of the skin is found covered with minute, elevated, spherical black stromatic masses or sclerotia of the fungus. In advanced stage the seeds are covered by a mat of fungus hyphae. Such seeds turn rusty in colour.

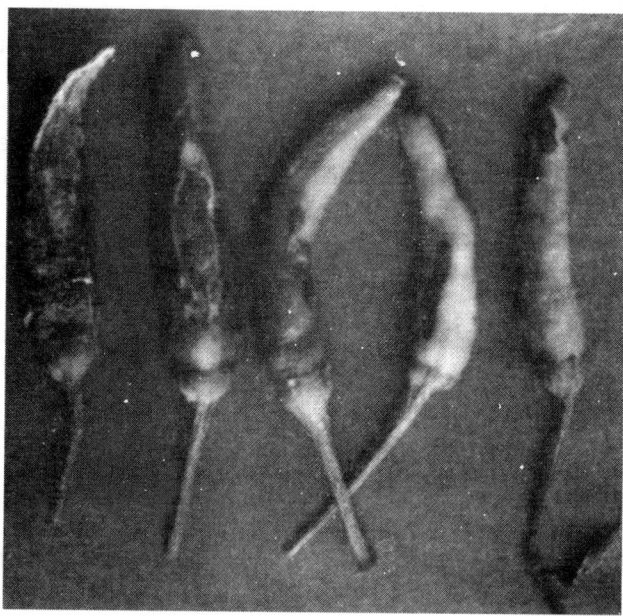

Fig. 120. Ripe rot of chilli fruits.

Die-back: As the name suggests, the disease causes necrosis of tender twigs from the tip backwards. The entire branch, or the entire top of the plant may wither away. The dead twigs are water-soaked to brown, becoming grayish white or straw colored in advanced stages of the disease. A large number of black dots (acervuli) are found scattered all over the necrotic surface of the affected twigs. Sometimes, the necrotic area is separated from the healthy area by a dark brown to black band. Only the top or few side branches may be finally killed or the entire plant may wither away. Partially affected plants bear few fruits of low quality.

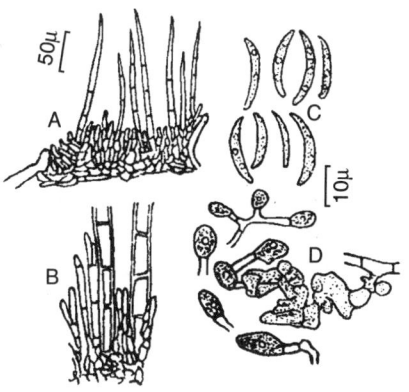

Fig. 121. *Colletotrichum capsici.*
A—Part of acervulus; B—Conidiophores;
C—Conidia; D—Appressorium formation
from hyphae.

The causal organism

The fruit rot and die-back are caused by *Colletotrichum capsici* (Syd.) Butler and Bisby. In addition, *Colletotrichum gloeosporioides* and *Colletotrichum acutatum* are also reported associated with the disease. *C. acutatum* especially causes anthracnose of unripe, green fruits.

On potato dextrose agar at 25°C colonies of *C. capsici* are at first white, rapidly becoming gray. Substrate colour is dark brown. Aerial mycelium forms light to dark gray felt. It is sometimes zonate with acervuli conspicuous due to dark setae in the lighter areas. The mycelium consists of septate, inter- and intracellular hyphae. Acervuli (conidiomata) formed on fruits, leaves and stems are rounded, elongated, approximately 350 μm in diameter, intra- and sub-epidermal, later disrupting the outer cell walls of the host epidermis. Setae are abundant, dark brown with light brown tip, 1–5-septate, rigid, hardly swollen at the base, slightly tapering towards the paler acute apex. They are up to 250 μm long and 5–8 μm wide. Conidiophores are aseptate and unbranched. Conidia in mass appear pinkish. They are borne singly at the tip of the conidiophores. Individually, they are hyaline, unicellular, uninucleate, curved with narrow ends (fusaroid) and measure 17–28 x 3–4 μm. Often, at the time of germination the conidia become 1-septate. They germinate in water within 4 hours. There is evidence that the conidia germinate better on the green or ripe fruit surface than in water. On coming in contact with a hard surface, the germ tubes soon forms appressoria which are medium brown, clavate to circular and measure 6–25 (10–14) x 4–10 (6.5–11.5) μm. They often become complex forming long closely branched chains.

The species has a wide host range and shows variability according to the substrate. In addition to chilli (*Capsicum annuum*) it can infect eggplant, tomato, turmeric, chickpea, cotton and jute. Minor differences in morphology and virulence occur between and within the strains but the species is generally regarded as unspecialized pathogen. Ability of isolates to adapt to new hosts has been demonstrated. The strains failing to do so are considered definite subspecies.

Disease cycle and environmental relations

The fungus survives in the field in plant debris. Seeds from diseased fruits also carry the primary inoculum. Secondary spread takes place through conidia formed in acervuli on stems, branches, and fruits and dispersed by rain-drop splashes and wind-borne. Infection of fruits can occur even when they are green. But it remains quiescent until the fruits start ripening. Esterase in unripe chilli fruits can inhibit appressorium formation of *C. gloeosporioides* giving protection to the fruits (Kim *et al.*, 2001). The density of waxes on the fruit skin also influences infection. The optimum temperature for germination of conidia of *C. capsici* is reported to be 30°C and no germination occurs at less than 100% relative humidity. The optimum temperature for growth of the fungus lies around 28°C and best growth occurs at 92% relative humidity. Maximum development of fruit rot takes place at 28°C and 95.7% relative humidity. Relative humidity is more important than temperature for disease development. The development of anthracnose after harvest on healthy looking fruits is from the incipient infections of the fruits in the field or from contamination by diseased fruits. Physiology and virulence of the fungus varies with isolates indicating existence of distinct strains within the species. The fungus produces a thermostable toxin that causes toxicity to host cell protoplasm without affecting the cell wall. The toxin is non-hostspecific and can suppress seedling growth in a number of crops including chilli, tomato and coriander (Saikia *et al.*, 2004).

The die-back phase of the disease is closely dependent on heavy and prolonged dew deposit after the rainy season. Since under shade this condition of humidity is not available, the shaded plants often escape injury from die-back. Intercropping reduces splash dispersal of conidia.

Management of the disease

Seeds should always be obtained from spotless fruits. Seed treatment with thiram, Brassicol and bisdithane has been found effective in elimination of seed-borne inoculum. Soak treatment with Benlate or carbendazim + maneb is highly effective. If only a few plants are attacked they may be destroyed. Good control of fruit rot had been obtained in Assam by spraying the crop with 0.2% Perenox (copper fungicide). In all 3–4 sprays were given, starting when fruits began ripening and were repeated at 15–21 days interval. Blitox-50 at 0.25% concentration is not effective in controlling infection of fruits but at 0.5 and 1.0% concentration it gives good control. Ziram is most effective against the disease. The same fungicides control the die-back also. Three sprays of 0.25% mancozeb at fortnightly intervals starting one month after transplanting provide good control of the disease. Best yield of ripe and dry fruits is also obtained in this treatment.

Sanitation is very important. Debris of a diseased crop should be collected and burnt. The pathogen has a large host range. Solanaceous weed hosts should not be allowed to grow in the vicinity of the field. On a small scale it is better to provide some sort of shade to the plants during night hours. Rice straw mulch or polyethylene sheet mulch improves, in general, the crop health. It helps in reducing anthracnose fruit rot of chilli and reduces the population of insect vectors of viruses affecting the crop.

In attempts to develop biological control of fruit rot infection in chillies, Ramamoorthy and Samiyappan (2001) reported reduction of fruit rot by soil and foliar application of a talc-based formulation of *Pseudomonas fluerescens* strain Pf 1. The reduction was attributed to some inhibition of mycelial growth of the pathogen but mainly to induction of defence responses. Seed dressing with *P. fluorescens* is more effective than seed dressing with *Trichoderma harzianum. P fluorescens* Pf1 and *Bacillus subtilis* improve seed germination and seedling vigor. The PGPR mixed formulation Pf1+ *B. subtilis* + neem extract + chitin was found to be best for reducing the fruit rot incidence and improving plant growth and yield parameters under field conditions. Chanchaichaivivat *et al.,* (2007) have isolated several yeasts from chilli fruits that give 66.4–93.3% protection against postharvest anthracnose. *Pichia guilliermondii* gives 93.6% and Candida musae 83.1% control.

Methanol extract of stems of *Catalpa avata* (Begnoniaceae) sprayed at 1.25 mg/ liter gives 95% control of fruit anthracnose. The active constituent of the extract is dehydro-*O*-lapachone. The extract is effective against rice blast, cereal powdery mildew and wheat leaf rust.

Resistance to anthracnose in pepper is reported in cultivated varieties. Cultivars K. Surkh, CH 107, Chamatkar, Saten Yellow, G-4, Bengal Green, H-1, H-4, H-6, S 20-1, Lorai, and Perennial are resistant to the disease.

MANGO ANTHRACNOSE

The anthracnose of mango fruits in association with leaf spots, blossom blight, and wither tip, all caused by *Colletotrichum gloeosporioides* Penz., is a widespread fungus disease of mango all over the world, particularly in humid areas. It is known to occur in Brazil, France, the Philippines, Indonesia, Trinidad, Peru, Mexico, the USA, Portugal and Hawaii. Blossom blight results in poor fruit set. Early infection of fruits causes pre-mature fruit drop. Ripe fruits are blemished hence fetch low price in the market and may soon rot.

Symptoms

The leaf anthracnose appears as irregularly-shaped black necrotic spots on both sides of the leaf. Under humid conditions these spots increase in size rapidly and coalesce to form irregular necrotic patches, frequently along the leaf margin. Young leaves are most susceptible to this infection. Symptoms of wither tip or die- back appear at the tip of very young branches. Black necrotic areas are formed on the affected twigs which dry from the tip downwards, accompanied by defoliation of the branch. In panicle anthracnose or blossom blight, inflorescence stalk and individual flowers both are affected. In the stalk, elongated dark gray to black lesions appear. Necrotic black dots appear on the flower which dry and turn brown to black. Fruits, smaller than pea-size, can be infected and aborted. Larger fruits that are aborted due to self–thinning or other physiological causes, are usually mummified. Mummies are saprophytically by *Colloetotriuchum gloeosporioides* and the fungus sporulates abundanrly on them.

Anthracnose of ripening fruits is characterized by the development of black spots of various forms which may be slightly sunken or may show surface cracks. These spots may coalesce to form larger spots and ultimately the whole fruit may be involved. The spots are often concentric at the stem end and sometimes in streaks (tear staiun) towards one side of the fruit. This suggests that the disease has spread through spores washed down by rain water from the stem end. The spots are usually restricted to the skin but in severe cases pulp may also be invaded. In advanced stages of the disease, the fungus produces acervuli and abundant orange to salmon pink masses of conidia.

The causal organism

Colletotrichum gloeosporioides Penz., anamorph of the Ascomycetous fungus *Glomerella cingulata* (Ston.) Spould and Shrenk, is a widely distributed fungus causing leaf spots and anthracnose on citrus, avocado, sugarcane, etc. In addition to *C. gloeosporioides*, many other fungi have been reported from time to time and at different places to cause tip blight and die-back. These include *Diplodia*, *Botryosphaeria ribis*, *B. theobrome*, *Physalospora rhodina*, *Ceratocystis fimbriata* and *Hendersonia toruloidea*. Many of these are only weak parasites. Wound infection of fruits by decay causing fungi, *viz. C. gloeosporioides*, *Lasiodiplodia theobromae*, cause release volatile metabolites which differ with the pathogens and have been used to distinguish them.

The mycelium of *C. gloeosporioides* is haploid and consists of rather narrow, sparsely septate hyphae which are at first hyaline but later take on slightly dark colour. Acervuli are formed abundantly on the affected host surface. These develop at first as tangled subepidermal masses of hyphae from which arise numerous closely packed conidiophores which partially raise the epidermis. One or more conidia are formed from the apex of each conidiophore. Setae are common on twigs but not on fruits. The conidia remain embedded in a viscid fluid. This mucilaginous matrix is composed of glycoproteins, polysaccharides, enzymes and other constiiuents. The mucilage prevents premature germination and desiccation of conidia and protects then against UV rays and toxic effects of host phenolics. In presence of moisture the matrix swells and ruptures the epidermis, exposing the conidial mass for dissemination by raindrop splashes and insects. The conidia are hyaline but in mass they look pinkish. They are broadly oval to oblong, with rounded ends, non-septate, and sometimes contain 1–2 globules. Their size is very variable, the average being around 12–16 x 4–6 μm. On germination conidial germ tubes form dark appressoria. Melanization of appressorium is essential for pathogenesis by the conidial germ tube. In infection of immature fruits, penetration is delayed and appressoria enter a period of quiescence. At molecular level, the contact of germ tube with a hard surface induces certain genes that contribute to the morphological differentiation of the conidia in to germ tubes and appressoria.

In the perithecial stage (*Glomerella cingulata*) the fungus produces perithecia on stromatic cushions. The perithecia are more or less compounded, subspherical, with prominent ostiolar hair. Asci are subclavate, often slightly pedicellate, fugacious, and 55–70 μm long. Ascospores are hyaline, 1-celled, allantoid and measure 12–22 x 3.5 μm. They are difficult to distinguish from conidia. The fungus is heterothallic.

Disease cycle and environmental relations

Diseased twigs and leaves, present on the tree or which fall on the ground are a prolific source of perennation and primary inoculum. *C. gloeosporioides* has been isolated from infected young or old leaves, separated from the tree, after 369 days when kept under ambient laboratory conditions and after 240 days when left in the field. The fungus has a long saprophytic survival ability on dead twigs. Fitzell and Peak (1984) had reported that conidia of *C. gloeosporioides* var. *minor* (*G. cingulata*), attacking mango, are produced in lesions on leaves, defoliated branch terminals mummified inflorescence and flower bracts. The pathogen spreads within tree canopies as water-borne conidia during the rains and causes symptoms on the young flush. There is abundant production of conidia during flush growth and flowering of the trees. Prolonged period of rains at this stage help severe outbreaks of the disease. Ascospores have no role in epidemics. Conidial production in acervuli is favored by temperatures of 10–30°C and relative humidity of 95–97%. The fungus does not grow at relative humidity below 95%. Thus, humid and misty conditions are considered most favourable for infection. Dispersal of conidia is through raindrop splashes.

On the fruits, under optimal conditions (95–100% RH, 25°C), germination and appressorium formation starts at 12 hour and 14 hours, respectively, after deposition of conidia on the peel. After 48 hours, 60% of fungal propagules present are appressoria. Most of the infections take place from the start of the blossoming period until fruits are more than half grown. The fungus enters through pores of the green fruits and enters quiescence prompted by antifungal compounds in immature fruits. Termination of quiescence occurs when fruits start ripening and the antifungal compounds are reduced. Ethylene production by the ripening fruit also helps in termination of quiescence.

An isolate of *C. gloeosporioides* from citrus is known to produce toxin(s) in liquid culture which may be responsible for die-back symptoms. The virulence of *Colletotrichum* spp. (*C. gloeosporioides*, *C. acutatum* on apple and *C. coccodes* on tomato) is associated with pectate lyase (PL) activity which, in turn, is related to external pH on the fruit surface. Fungi have pH regulation of gene expression. The pathogenicity genes are not expressed at pH 4. The fungus excretes and accumulates ammonia. The resulting alkalization increases pH and associated increase in PL making the fruit predisposed to infection (Prusky *et al.*, 2001; Drori *et al.*, 2003). Even avirulent strains can be induced to become virulent by addition of ammonia-releasing compounds.

Management of the disease

Good plant vigor is important for keeping the infection of twigs down. Proper fertilization and watering of the trees during summer must be done to maintain tree vigor. Tree sanitation is also essential. Diseased twigs should be pruned and burnt along with fallen leaves. Pruning should be followed by sprays of copper fungicides. Four to 5 sprays between January and July give satisfactory control of the disease. A combination of captan with zineb for spraying the trees has been found very effective against this disease. In some field trials mango anthracnose was effectively checked by giving the trees two sprays of zineb or a copper fungicide at the time of flowering. Subsequent sprays may be started just before the onset of monsoon and continued till harvest at 14-days interval.

Among systemic fungicides benomyl and thiabendazole have been found effective. In a study of the pathogen-fungicide interaction in South Africa, many isolates of *C. gloeosporioides* from mango were found resistant to benomyl used in pre-harvest sprays. No isolates were found resistant to thiabendazole or prochloraz (Sanders *et al.,* 2000). In the Philippines, spray of benomyl at 250 mg/L was found to significantly reduce the conidial density within the tree canopies. In a Brazilian study, Sales *et al.,* (2004) good control of mango anthracnose by sprays of azoxystrobin (75 mg a.i./L) + paraffin mineral oil (0.2%). However, they noted no significant different between chlorothalonil WP (1240 mg a.i/L), benomyl WP (500 mg a.i/L), copper oxychloride (2350 mg a.i./L) and azoxystrobin + paraffin mineral oil. The fungicides were sprayed 6 times at 15 days interval starting at panicle plain flux stage.

Post-harvest chemical treatments of fruits to prevent damage by fruit anthracnose are generally not very effective. Dodd *et al.,* (1991) had reported that a host benomyl dip (850 mg/L a,i)) at 52–55°C for 10 min. completely eradicated anthracnose on fruits treated on the day of harvest. Instantaneous dip of fruits in 1000 ppm benomyl or 2000 ppm thiabendazole before storage is reported to give good control of fruit rot. Prasanna Kumar *et al.,* (2002) have reported that post-harvest treatment of ripe mangoes with carbendazim (01%), tricyclazole (0.1%), benomyl (0.1%) caused significant reduction of rot caused by *C. gloeosporioides* and *Alternaria alternata*. Mango is a subtropical fruit and is tolerant to high temperatures. Hot water treatment of fruits is widely recommended. Combination of temperature and the time of exposure to high temperature is decided for individual varieties. Post-harvest fruit treatment in hot water at 50°C for 15 min. for the cultivars Langra and Dasehari was suggested in 1960s. Later, a hot water dip at 52°C for 30 min was also suggested. Hot water + 0.1% carbendazim gave 100% control of fruit rot. Hot water treatment and *Ocimum* leaf extract (5%) were next in efficacy. Exposure of fruits to vapour heat at 47°C for 15 min and/or hot water at 53°C for 5 min has been used in exotic mango cultivars. Hot water brushing (HWB) is more recent approach to control post-harvest decay and enhance shelf life of mangoes. It combines a 15–20 sec hot water spray and fruit brushing. New friendly compounds that control fungal infection have been developed which act by neutralizing changes in pH induced by several postharvest pathogens. Fruit treatment with imazalil, 500 mg/ L of water at 53°C gives better control than hot water alone. These treatments are effective against postharvest anthracnose rot but al to other fungal causes of fruit decay such as *Alternaria alternate, Lasiodiplodia* spp. and *Phomopsis* spp.

In biological control, *Bacillus cereus* and *Pseudomonas fluorescens* strains are known to provide good control of post-harvest fruit anthracnose. Vivekananthan *et al.,* (2004a) have reported biological control of mango anthracnose by spray of lytic enzyme (chitinase)-inducing biocontrol agents, *Pseudomonas fluorescensm, Bacillus subtilis* and *Saccharomyces cereviciae*. Preharvest spray was given at fortnightly and monthly intervals. Fortnightly spray of *P. fluorescens* with chitin gave the maximum induction of flowering, consequently better yield, and reduced latent infection. In field trials, *P. fluorescence* (FP7) + chitin treatment reduced anthracnose by 60% which was superior to standard fungicides, carbendazim (Vivekanandan *et al.,* 2004b). Prabakar *et al.,* (2007) have also reported control of mango anthracnose when inoculated fruits are dipped in spore/ cell suspemsion of *Triichoderma harzianum* and *Pseodomonas fluorescens*. In South Africa, Govender *et al.,* (2005, 2006) have reported control of postharvest anthracnose and stem end rot of mango by treating the fruits with a combination of quarter dose of prochloraz and *Bacillus licheniformis* (in hot water at 45°C). According to Silimela and Korsten (2008) moderate level of postharvest anthracnose, bacterial spot and soft rot of mango fruits is controlled by preharvest spray of *Bacillus licheniformis* alone or alternated with a copper fungicide at 3-week intervals starting from flowering until harvest. The yeast *Rhodotorula minuta* is a strong antagonist of *C. gloeosporioides* and suppresses mango fruit anthracose. Patido-Vera *et al.,* (2005) have reported that the yeast at 10^9 CFU/ml reduced anthracnose In attempt to develop a commercial preparation they formulated the yeast at 10^9 CFU/ml with glycerol (20% and xanthan (5g/l) which preserved the number of yeast cells for 6 months at 4°C. Yeast suspensions applied to mango trees reduced the fruit anthracnose severity at levels similar to better than chemical fungicides. Treatment of mango fruits with the plant activator benzothiadiazole (BTH) is reported to reduce postharvest fruit anthracnose. Peroxidase, polyphenol amonia lyase, chitinase and glucanase enzyme activities and total phenolic compound content are enhanced by the treatment (Zhu *et al.,* 2008).

ANTHRACNOSE OF BEAN

Anthracnose of bean (*Phaseolus vulgaris*) is a major disease of this crop throughout the world but causes greater losses in the temperate regions than in the tropics. In the hills of Uttaranchal in north India the disease appears in the second or third week of June and reaches the maximum damaging stage from the beginning of August to mid-September. The losses can approach 100% when badly contaminated seed is planted under conditions favourable for disease development. At relatively low soil temperature and when soil moisture is high there

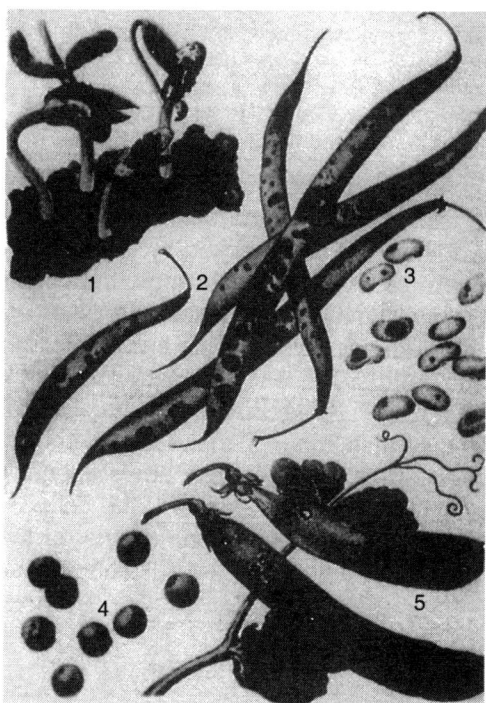

Fig. 122. Anthracnose of beans.

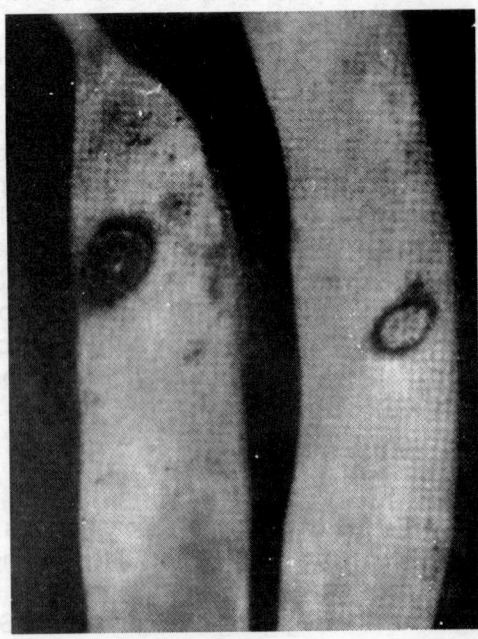

Fig. 123. Anthracose of bean. Pod lesions.

is very high seed rot and pre-emergence seedling rot. Plants of all stages of growth are susceptible but susceptibility of growing plants increases with increasing age. In addition to common bean the disease occurs on cowpea and mungbean also.

Symptoms

In soil the infected seeds generally undergo rot and show distinct fungal growth with pinkish spore masses on the seed coat. If germination occurs, the hypocotyl shows necrosis. After emergence, all aerial parts of the plant are attacked by the fungus but the most characteristic symptoms appear on immature pods. On the seedlings, the cotyledons show small to large, brown to black, sunken spots which may bear pink spore masses of the fungus in wet weather. Later, the young plants suffer from necrosis of veins and adjoining tissues. On growing plants the leaves show small, angular and brown lesions mostly adjacent to veins. On pods the spots appear as black sunken cankers with lighter colored or gray central area. These may reach a diameter of 5–8 mm. The central portion of these spots shows pinkish mass of spores, especially in wet weather. Later, the sides of these spots appear raised. Mature seeds from diseased pods may or may not be stained. If stained, the color is various shades of brown. Seedlings raised from such seeds show the typical lesions on the cotyledons. On stems of adult plants, the spots are eye shaped and longitudinal along the stem. They may be 5–7 mm long.

The causal organism

The anthracnose of bean is caused by the hemibiotrophic fungus *Colletotrichum lindemuthianum* (Sacc. & Magn.) Scribner. Its perfect stage was originally described as

Glomerella lindemuthianum (Sacc. & Magn.) Shear, Later, the name *Glomerella cingulata* f. sp. *phaseoli* has been proposed for the perithecial stage.

In cultures the mycelium is branched, septate, hyaline at first becoming dark colored with age. The conidia are formed singly at the ends of free hyphae or on stromatoid masses in acervuli. During infection of bean, *C. lindemuthianum* initially produces biotrophic primary hyphae that are of large diameter and entirely intracellular. They are followed by necrotrophic secondary hyphae that are narrower and inter- and intracellular. The cell walls of the secondary hyphae are 25–40 nm thick while those of primary hyphae 100–130 nm thick. The secondary hyphae are not surrounded by an extracellular matrix. Chitin is the only component shared by secondary hyphae and subsequent other fungal cell types (Perfect *et al.*, 2001).

On the host the conidial stage develops on stromata beneath the cuticle, later rupturing it and becoming erumpent. The conidia in acervuli are borne acrogenously on short, erect, hyaline, unbranched conidiophores which measure 40–60 µm long. These spores appear pink in mass. Setae are very few, mostly at the margins of the stroma on the host. They are brown, septate, pointed and stiff and measure 30–100 x 4–9 µm. Conidia measure 13–22 (11–20) x 2.5–5.5 µm. They are 1-celled, hyaline, oblong, cylindrical with rounded ends or with one end slightly pointed.

Conidia of Colletotrichum are surrounded by fibrillar spore coat, composed of several major glycoproteins. The spore coat is required for adhesion, appressorium formation and pathogenesis (Rawling *et al.*, 2007). The conidia of *C. lendemuthianum* germinate by 1–4 germ tubes. These germ tubes as well as hyphae form appressoria which are dark brown and serve as resting structures. Appressorium formation from germ tubes of *Colletotrichum* is influenced by the length of the host surface in direct contact with the germ tube and hydrophobicity of the surface. In non-pathogenic strains of the fungus, development of the appressorium can be blocked at three stages, *viz.*, stage of appressorium differentiation (no appressorium formed), appressorium maturation stage (low pigmentation and internal turgor pressure, and appressorium function stage when appressoria fail to penetrate the host tissue (Veneault-Fourrey *et al.*, 2005).

The perfect (sexual) stage has been found only in laboratory. Most isolates do not develop sexual reproductive structures. The perithecia are beaked and 120–210 µm in diameter. They contain filiform paraphyses and asci. The asci measure 48–68 x 8 µm. Each ascus contains 8 ascospores which measure 20 x 6.5 µm (if allantoid) or 10 x 4 µm (if ellipsoid). Although sexual fruit bodies have not been found in nature, large number of isolates can rapidly arise with different genetic and chromosomal compositions. Special kinds of hyphae occur in the fungus that create anastomoses directly between conidia. Conidial anastomoses can occur between two different *Colletotrichum* species Heterothallism is reported in the species in some countries.

Optimum growth of the fungus in culture occurs at 22.5°C. Conidial production is best at 14–18°C and is greatly inhibited at temperatures above 30°C. Sporulation is favored at pH 5.2–6.5 and is unaffected by aeration and light conditions. Isolates may lose pathogenicity when repeatedly transferred in culture; thus occasional re-isolation from inoculated hosts may be required.

A large number of races of the fungus are known, mostly from the temperate zones. In 1992, different races of *C. lindemuthianum* were designated alpha, beta, gamma, etc. Based on virulence to 12 differential cultivars of *Phaseolus vulgaris* 41 races were differentiated in collections from south, central and north America. Some were not common while some were widespread. Ansari *et al.*, (2004) studied pathogenic and genetic diversity of isolates of

C. lindemuthianum from 10 Central and South American, European and African countries and observed that there is some association between genetic diversity of the isolates and their country of their origin. They identified 30 different pathogenic races from 74 isolates of which 21 were restricted and exclusive to different countries. Race 9 was most widespread. In a study of pathogenic variability in *C. lindemuthianum* in Himachal Pradesh (India) 12 races were identified on international differentials. These included Alpha-Brazil, Beta, Gamma and Ind I to Ind IX. The nine races, Ind I–Ind IX, were new and distinct from identified races elsewhere including USA. The races of *C. lindemuthianum* differ in their level of pectin lyase activity.

In addition to *Phaseolus vulgaris* (common bean), other primary hosts of the pathogen are *Cajanus cajan* (pigeon pea) and *Vigna sinensis* subsp. *sesquipedalis*. Soybean, mungbean, urdbean, pea, cowpea and *Vicia faba* are secondary hosts.

Disease cycle and environmental relations

The fungus is soil- and seed-borne. In soil it survives on the diseased plant debris but cannot live long when separate from the debris. In cool climate areas decomposition of crop debris is slow and the fungus can survive for up to 2 years on the debris if it is not buried deep in soil. There is significant correlation between amount of debris left on soil surface and the final pod anthracnose in the field. In the seed the anthracnose fungus can survive as long as the seed is viable. The percentage of infected seed is an important determinant of the severity of bean anthracnose during the growing season since the secondary inoculum produced will be proportional to the amount of initial inoculum. Anthracnose appears randomly within the field early in the season indicating that inoculum was from bean debris left in the field or from infected seed. Later in the season, plant to plant spread results in a more clustered distribution of diseased plants. Where fields are ploughed during period between two bean crops, the role of debris is minimized.

When infected seeds are planted or from the soil-borne inoculum, first lesions develop and produce spore masses on the cotyledons. These initiate the disease in the crop. The spores are held in a gelatinous matrix and are not easily air-borne. Water is essential for their release from the matrix. Usually they are washed down by rain or dew to stems or they are splashed to new leaves by raindrops falling on the acervuli. The average distance of conidial spread is 3–4.6 meters per rainstorm of 10 mm and more. Insects, rabbits, man and other animals moving in the field, especially during wet weather, also help in dissemination of spores. During the early stages of invasion by *C. lindemuthianum* a biotrophic relationship is established in each successive cell colonized by the primary hyphae. During interaction with the host and formation of haustorium, the anthracnose fungus secretes and accumulates polygalacturonase in its cell walls. This results in extensive dissolution of host cell wall pectin during the development of secondary hyphae but not earlier around the infection peg (Herbert *et al.*, 2004).

The most important factors affecting the infection are temperature and moisture. Moderate temperatures between 13° and 26°C favour infection. No infection occurs at temperatures above 27°C and at 13°C also the disease is considerably reduced. A relative humidity of above 92% is necessary for infection, the optimum being close to 100%. A 10-hour wet period is reported to be necessary for conidial infection and new lesions usually appear in 3–7 days depending on prevailing temperature. Visible spots develop in 4–5 days after inoculation at 22–27°C, in 7 days at 15.5°C and in 9 days at 12°C. At 17°C there is abundant infection. Often, during relatively warm days the effects of high day temperatures are mitigated by the low night temperatures. Cool rainy days are conducive to development of epidemics.

Management of the disease

Control strategies include disease free seed, seed treatment, foliar spray of fungicides, crop rotation, cultivation of race specific resistant varieties and tillage methods. The seed must be disease free hence it should be collected only from healthy pods. Usually, seeds produced in dry areas are free from infection. The plant debris should be either removed to a place where beans are not likely to be grown in the near future or it should be deeply ploughed in. A 2–3-year rotation should be followed. The soil should be well-drained and good spacing and weed control should be employed to keep humidity low. If the disease assumes serious form the crop should be given fungicidal sprays. Captan, zineb or mancozeb at the rate 2 kg/ha are usually recommended. Benlate, Ziram, Vitavax, Ferbam and lime sulphur, in the order listed, had been recommended for foliar sprays. Bavistin, Vitavax and Agrosan GN were recommended for seed treatment. In field trials, seed treatment with thiram 75 WP, carbendazim 50 WP, carbendazim + thiram, mancozeb 75 WP, triadimenol 15 DS and metsulfovax 20 WP at 2.5 g/kg seed have given good control of seed-borne infection, the best being carbendazim and carbendazim + thiram. Seed treatment followed by 2–3 sprays of mancozeb (0.25%) after 45, 60 and 75 days of sowing effectively manage the disease resulting in manifold increase in yield of dry beans. On leaves treated with zosetric acid the spores fail to adhere to the surface, failing to produce appressoria and cause infection. (Stanley et al., 2000). The lesion development is delayed. Leaves treated with calcium or sodium silicate show significantly reduced anthracnose lesions without showing any accumulation of silicon barrier level (Moraes et al., 2006).

There are many reports of biological control of the anthracnose of bean through seed bacterization and through inoculation with avirulent strains of the pathogen (Sticher et al., 1997; Van Loon et al., 1998). Seed bacterization with a strain of Pseudomonas aeruginosa (rhizobacteria) and treatment with a derivative of benzothiadiazole also induce systemic resistance against the anthracnose fungus (Bigirimana and Hofte, 2002). The plant growth promoting rhizobacterium Pseudomonas putida, strain WCS 358, originally isolated from potato rhizosphere, induces systemic resistance in bean against anthracnose through its lippolysaccharides and fluorescent siderophores (pseudobactin). The latter also affects the pathogen by competing for iron nutrition. Pre-inoculation with a nonpathogenic or avirulent sytrain of the pathogen also imparts resistance through elicitation of defense responses. Appressorium maturation stage of the nonpathogenic forms is sufficient to induce most of defense responses. Inoculation of primary leaves of bean plant with the pathogen induces resistance in the trifoliate leaves when they are challenged with the pathogen. Penetration from appressoria and hyphal growth in epidermal cells are reduced. Wall appositions and encroachment of nuclei at attempted penetration sites occur more frequently. Early cell death (hypersensitive reaction) often follows in case of successful penetration.

Phaseolus polyanthus is highly resistant to the anthracnose fungus and its crosses with P. vulgaris provide valuable sources of resistance. These may aid in the development of stable resistance to anthracnose.

WILT OF PIGEONPEA

The wilt disease of pigeonpea (arhar, red gram, Cajanus cajan) is common throughout India, being very destructive in parts of Maharashtra, Madhya Pradesh, Uttar Pradesh and Bihar. This is, so far, the most destructive fungus disease of this most important pulse crop. Continuous cropping of pigeonpea in the same field causes as much as 50% plant mortality due to this disease. In Bihar and U.P., 5–10% loss of standing crop is common feature every year. In a survey conducted during 1975–1980 by ICRISAT the incidence of the disease was found to vary from 0.1% (Rajasthan) to 22.6% (Maharashtra). In Bihar and U.P. the average incidence was 18.3 and 8.2%, respectively. The production loss was estimated at about 32,000 tons in U.P., 44,000 tons in Maharashtra and 10,000 tons in Madhya Pradesh. The disease is common not only in India but also in several African countries such as Kenya (average 15.9% incidence), Malawi (36.3% incidence) and Tanzania (20.4% incidence). Generally, the incidence of the disease is maximum during flowering and pod formation stages of the crop. The damage by the wilt fungus is considerably enhanced if plants are infested by the cyst nematode (*Heterodera cajani*) and the root knot nematodes. The pigeonpea wilt fungus is host specific being pathogenic only on pigeonpea and its wild relative *Atylosia platycarpa* (Kannaiyan *et al.*, 1985). A strain of the species, *F. udum* f. sp. *crotalariae,* is reported to cause wilt of sunnhemp (cf. Sarkar *et al.*, 2000).

Symptoms

The pigeonpea plant is susceptible to attack of wilt throughout its development. However, the symptoms are more pronounced and the damage greater when the plants have grown up after the rainy season. The main symptom of the disease is wilting of seedlings and adult plants as if they have suffered from water shortage even though there may be plenty of moisture in the soil. The wilting is characterized by gradual, sometimes sudden, yellowing, withering and dying of leaves followed by drying of the entire plant or some of its branches. Patches of diseased plants are scattered throughout the field, indicating locations where colonies of the pathogen were present and infection started. These patches develop around the earliest attacked plants (foci) in a centrifugal manner.

Examination of the main roots and of the base of the stem shows that the tissues are blackened, either uniformly or, especially in early stages, in streaks. This blackening is sometimes visible through the bark but is best seen on removing the latter, the wood being the seat of most pronounced discoloration. In some cases, the black streaks may be traced up to a height of several feet on the stem. The earliest branches to wilt are those which arise from the discolored parts of the stem. In all cases, the streaks can be traced down to the roots and are found to arise from the main or lateral roots which show dry rot.

Partial wilting is also common. The stem on one side is blackened and few or all branches on that side wilt and wither away, the rest of the plant escaping injury. This type of symptom produced on the standing plants appears to be determined by the strain of the fungus and amount of inoculum present in the soil. As a rule most of the attacked plants succumb completely to the disease. Recovery of affected plants is rare. Aerial growth of the fungus may also occur on the base of the stem near the ground level as a white to pinkish cottony mass. This may cause secondary spread in the same season.

The causal organism

The disease is caused by *Fusarium udum* Butler. However, many types of fusaria are often obtained in cultures from infected roots. Some of them are variants of the same fungus while others are either saprophytes of different species or only weak parasites. A perithecial fungus isolated from wilted and dead pigeonpea plants was claimed as the teleomorph of *F. udum* in 1982 and named *Gibberella indica* Rai and Upadhyay. However, existence of perithecial stage is not yet universally accepted. Earlier *Gibberella udum* was described as the perithecial stage of the fungus. In asexual (anamorphic) phytopathogenic fungi such as *Fusarium oxysporum* and *Alternaria alternate* mating-type genes are reported which ptofucer the effects of sexual reproduction. These are designated as MAT1–1 and MAT1–2 idiomorphys (Arie *et al.*, 2000).

The fungus in the host, as a rule, is confined to vascular tissues and is both inter- and intra-cellular. The septate hyphae run across the cells, growing with great rapidity, along the inside of the walls of large vessels. These vessels appear plugged with the hyphae. However, there are instances where none or only few vessels show presence of hyphae although browning of the tissues is present.

The mycelium is hyaline and produces 3 types of spores within the host tissues as well as in cultures, depending on nutritional and other factors. Microconidia are small, elliptical or curved, unicellular or with 1–2 septa, and measure 5–15 x 2–4 μm. They are formed free on hyphal branches. In cultures a dozen or more of them may be held together in a ball or false head. Macroconidia are produced in small cushions of stromatic mycelium on the surface of the host near ground level. The stromatic bases (sporodochia) are tubercular in culture media. The macroconidia are long, curved (fusaroid), with prominent apical hook, and notched at the base, septate (3–4 septa), and measure 15–50 x 3–5 μm. The presence of the prominent apical hook distinguishes the species from *Fusarium oxysporum.*

Chlamydospores are also formed in the host as well as in old cultures. They develop from any cell of the hypha, often from cells of the macroconidium. The cells round off and become thick-walled to form chlamydospores. These spores are oval or spherical, single or in chains, terminal or intercalary and persist in the soil for long.

The perfect stage, *Gibberella indica* is usually found on exposed roots and collar region of the stem up to the height of 35 cm above the ground level. The mature perithecia are superficial, commonly aggregated, subglobose to globose, sessile, smooth walled, dark violet, and 350–550 μm in diameter. Asci are 8-spored, mostly subcylindrical, 60–80 x 6–10 μm, broader in the middle, with short stalk, a narrow apex, and a central apical pore. Ascospores are ellipsoidal to ovate, 10–17 x 5–7 μm, hyaline, commonly 2-celled, rarely 3–4-celled and constricted at the septa. In culture these spores germinate to produce short or long conidiophores bearing micro- and macroconidia which are pathogenic to pigeonpea. The fungus is heterothallic and single ascospore cultures do not produce perithecia. When cultures from different strains are grown together perithecia are formed after 25 days at 18°–22°C. The ascospores germinate to produce macro and microconidia.

F. udum shows wide variation in morphological, cultural and physiological characteristics, production of enzymes and in virulence on different cultivars of pigeonpea and physiologic races are known to occur. There is no relationship between cultrural characteristics and aggressiveness. Isolates may differ genetically (Kiprop, *et al.*, 2002a,b; Sivaramakrishnan *et al.*, 2002). The fungus is a mycoparasite of certain fungi such as *Rhizopus nigricans, Curvularia lunata, Cunninghamella echinulata, Rhizoctonia solani* and many others. The parasitism is by coiling around host hyphae, penetration, chlamydospore formation in host hyphae and lysis of hyphae.

Disease cycle and environmental relations

F. udum is soil-borne, basically a soil invader, which can survive in plant debris in soil after the host crop has been removed from the field. In a wilt sick soil, the fungus, though quantitatively reduced in volume, can survive for more than 10 years in absence of its host (pigeonpea). However, the saprophytic survival (growth as a saprophyte) in the soil is limited to the continued presence of the dead host roots and other parts pre-colonized by the fungus.

There is no definite evidence to show that the fungus can grow to any great distance in the soil due to antagonism of soil microflora. So long as the infected roots of the host are decaying in the soil the fungus exists as an active saprophyte but its vegetative growth continues to decline while producing resting structures such as chlamydospores. Since the roots of pigeonpea remain in soil for considerably long periods after harvest of the crop the fungus can persist and if the pigeonpea crop is again taken in the same field it can restart its parasitic life cycle. Maximum population of the pathogen is found in the top 15 cm of soil profile in the first crop and at 15–30 cm when a second crop follows the first crop (Chaudhary *et al.,* 2001). About 90% of the total fungal population in the rhizosphere of wilted plants has been found to comprise of *F. udum.* Apart from the saprophytic mycelium the conidia of the fungus also remain viable for long periods while chlamydospores persist in dormant state for even longer durations. Transfer of inoculum containing soil from plot to plot by farm implements, irrigation water, etc. spread the pathogen. Dissemination is also through seed-borne infection.

The above ground parts are never attacked. Secondary infection by conidia produced on aerial parts is rare. The infection occurs through the fine rootlets which are penetrated by germ tubes and rapidly blackened and shriveled. From the laterals the fungus passes into the larger roots which are not directly penetrated unless injured. After entering the roots the fungus mainly grows in and along the roots and rarely across the soil. The incidence of wilt is always high in fields where the disease has appeared in the past and no crop rotation has been followed. This helps in the build up of inoculum. Cultivation of the field helps in dispersal of the fungal propagules and diseased root pieces to increase the number of infection foci.

Resistance in pigeonpea to Fusarium wilt depends upon the extent of colonization of the plant by the fungus after infection. Pigeonpea plants wilt when colonization of the plant is more than 50%. In response to infection pigeonpea plants produce phytoalexins in the vascular system Four fungitoxic isoflavonoid phytoalexins-hydroxygenistein, genistein, cajanin and cajanol have been isolated from plants 15 days after inoculation. Cajanol, the major antifungal compound, accumulates rapidly in some resistant cultivars inoculated with conidia but the rapid accumulation of the phytoalexin is not positively correlated with resistance to wilt. Root-knot nematode infection retards accumulation of cajanol. The wilting of the plant due to infection of *F. udum* has been ascribed mainly to the choking of the water conducting vessels by the hyphae and their metabolic products and to the production of toxins within the host as well as in the soil around roots. The fungus produces 3 enzymes, pectin methyl esterase, polygalaturonase and cellulase, which are involved in pathogenesis. The breakdown of host tissues by action of pectic enzymes results in formation of gum plugs. In addition, fusaric acid has also been isolated from infected roots, stem and leaf and it plays important role in pathogenesis. Toxins have been detected and isolated in many fusaria (fusaric acid in *F. oxysporum* f. sp. *vasinfectum* and lycomarasmin in *F. oxysporum* f. sp. *lycopersici*). Culture filtrates of the pathogenic species of *Fusarium* have been found capable of producing wilt symptoms on the plant even when no fungus is directly associated with the plant. While the pathogen is host specific the toxin produced by it does not show so much selectivity in causing toxicity among different plant species.

Susceptibility of *F. udum* to soil antagonists including bacteria and fungi has been known since early 1950s. A higher incidence of wilt in sterilized than in unsterilized soil was the basis. The low incidence of wilt in unsterilized soil was attributed to the action of antagonistic organisms such as *Aspergillus niger, Rhizopus nigricans,* and *Bacillus subtilis.* In later studies, it was demonstrated that *Bacillus subtilis* produced an antibiotic, bulbiformin, that inhibited the growth of *F. udum.* The occurrence of severe wilt after October was attributed to the reduced population of this bacterium in the soil after rainy season. The bacterium develops well in presence of organic nitrogen with carbohydrates. Groundnut and cotton seed cake amendments considerably enhanced the production of bulbiformin in soil. The antibiotic activity remained stable in soil up to 35 days. The incidence of wilt was reduced in soil amended with groundnut cake, molasses, or sweet clover root material or inoculation with *B. subtilis*. The above reports that wilt incidence increases after October due to reduced population of *B. subtilis* in soil is at variance with a later report that even susceptible varieties sown in September-October show a much lower incidence of the disease. In a microbiological study of rhizosphere and nonhizosphere soil of pigeonpea, Pandey and Upadhyay (2000) found heavy colonization of rhizosphere soil of healthy plants with *Aspergillus niger* and *Penicillium* spp. They did not find antagonistic bacteria. Inhibition of *F. udum* in soil amended with organic matter such as oil-cakes by soil bacteria, lysis of the mycelium and induction of chlamydospore formation was demonstrated in studies reported in 1970–1980. Subsequently many other studies also reported the adverse effect of oil-cake amendment of soil on saprophytic survival of *F. udum*.

The competitive saprophytic ability of *F. udum* is highly suppressed when the pigeonpea substrate is pre-colonized by saprophytic fungi like *Penicillium citrinum*, *Aspergillus flavus*, *A. terreus*, *A. niger*, *Trichoderma viride*, *Streptomyces griseus* as well as *Bacillus subtilis* and *Myrothecium* spp., etc. When inocula of antagonists are mixed with the inoculum of the pathogen and added to the soil the competitive survival ability of *F. udum* is suppressed. Thus, resident fungal species in the soil of a pigeonpea field have a significant role in determining length of survival of *F. udum* in soil. *Trichoderma viride*, *T. harzianum* and *Gliocladium virens* are also strong antagonists of *F. udum* (Pandey and Upadhyay, 2000). Green manuring of the fields decreased wilt incidence. This could be due to increase in the antagonistic activity of soil microflora. The rhizosphere of resistant varieties of pigeonpea contains a greater proportion of *Actinomyces* sp. which were antagonistic to *F. udum* than the rhizosphere of susceptible varieties. The competitive saprophytic activity of the pathogen is adversely affected by high soil moisture. The optimum temperature for saprophytic activity of the wilt fungus is 20°–35°C.

The pigeonpea wilt can develop at a wide range of temperature. Temperatures between 17° and 29°C are reported conducive to disease development. High temperatures are unfavourable due to increased activity of antagonists. Neutral to slightly alkaline pH is favourable for the fungus. *F. udum* is capable of tolerating a pH of 4.6–9.0. Infection of pigeonpea roots by *F. udum is* independent of injury by root knot nimatodes. But if roots are pre-infected by root knot nematodes rate of root infection and progress of *fusarium* through the cortical parernchyma is facilitated. As a resut wilt symptoms appear early and are more severe.

Addition of solution of boron, manganese or zinc provide protection to seedlings during emergence. Spray or pre-soaking of seed in manganese solution induces resistance against infection. Colonization of stubbles in the soil by the fungus is also retarded when the soil is amended with these minor elements, especially zinc. Zinc at 20, 40 and 80 ppm enhances disappearance of the pathogen from soil. Application of zinc sulphate @ 25 kg/ha suppresses the disease in the field. Zinc abolishes the production of the phytotoxin fusaric acid, the *Fusarium* pathogenicity factor. It increases microconidial production but reduces the total biomass. It could also promote the efficiency of bacterial biocontrol agents like *Pseudomonas fluorescens* by suppressing *Fusarium* metabolites that are inhibitory to antibiotic production by the biocontrol agent.

Management of the disease

The control of soil-borne diseases such as the wilt caused by *Fusarium* species has always been a problem. Treatment of soil on a field scale is difficult. The only methods of wilt control are, therefore, through cultural practices including cultivation of resistant varieties.

Crop rotation is undoubtedly the best way of eliminating soil-borne inoculum. However, since the fungus survives on deep-seated roots of the host, below the depth of ordinary cultivation, the success of rotation will depend on field sanitation (removal of affected plants with their roots), hot weather cultivation (deep ploughing during summer), and similar practices. A 4–5 year rotation has been found to free the field completely of the wilt pathogen. The length of rotation can be reduced by ensuring removal of affected roots. In 1939, rotation of pigeonpea with tobacco was recommended as a possible means of control through adverse effect of root exudates of tobacco on the pathogen. Sorghum, pearl millet, cotton and resistant pigeonpea cultivars are recommended as rotation crops.

Mixed cropping with sorghum provides the most effective and practical solution of this problem. Many reasons were suggested for this reduction in incidence of wilt such as chances of contact between host and pathogen were reduced and root exudates of sorghum acted as barrier for the pathogen. Root exudates of sorghum are known to influence spore germination of *Fusarium* spp. in soil. Sorghum plant is well known for its cyanogenic compounds in roots. The exudates may thus contain HCN, a highlu toxic comound for *Fusariou*. Natarajan *et al.,* (1985) had reported results of a precise study of effect of cropping system on pigeonpea wilt. In continuous cropping of pigeonpea the incidence was as high as 64–90%. A rotation of sorghum and fallow reduced it to 16–31% and 2 cycles of sorghum followed by pigeonpea reduced the incidence to 16%. In their opinion also the root exudates of sorghum have a suppressive effect on the pathogen in soil thereby suppressing infection of pigeonpea. Although in soils with even high density of pathogen inoculum significant reduction in pigeonpea wilt incidence was reported in sorghum-intercropped fields. Madhukeshwara *et al.,* (2003) have reported that intercropping with sorghum, pearl millet or soybean and high or low levels of fertilizers fail to affect wilt incidence when soil level of the pathogen inoculum is high. Under conditions of low disease pressure, intercrops may reduce wilt incidence. In a similar wilt disease of sunnhemp caused by *F. udum* f. sp. *crotalariae* also the disease incidence was reduced by intercropping with sorghum. Line sowing of a sunnhemp-sorghum mixture (3:1) was best.

In the light of the role of green manuring, amendment of soil with oil-cakes and application of farmyard manure etc. and trace elements such as zinc and manganese, as stated earlier, introduction of a green manure crop in the rotations and increasing soil organic matter content by other means seems desirable. Infection of seedlings may be avoided by pre-soaking seeds in zinc or manganese solution. There are many reports of seed treatment with systemic fungicides for control of pigeonpea wilt. Treatments with carbendazim could suppress early infection of the plants. Kotasthane *et al.,* (1987) have reported that seed treatment with Benlate + Thiram (1:3) considerably reduced wilt incidence (20% wilting).

In studies on interactions between *F. udum* and pigeonpea rhizosphere saprophytic fungi, it was reported, in 1987, that when pure cultures of *F. udum* mixed with cultures of *Aspergillus niger. A. flavus* or *Micromonospora globosa* and added to soil pathogen population was significantly reduced. *Aspergillus nidulans* also suppressed the pathogen but only at relatively high temperature of 38°C. In a glasshouse study, Pandey and Upadhyaya (1999) observed that *Trichoderma harzianum* was the best biocontrol agent against the pigeonpea wilt pathogen. Although its combination with Bavistin in seed treatment was not beneficial, seed treatment with Thiram + *T. harzianum* significantly reduced wilt incidence. Direct soil application and seed coating with culture of *Bacillus subtilis*, neem oil with *Trichoderma viride* or *T. harzianum* significantly reduces wilt incidence. Prasad *et al.,* (2002) have claimed control of pigeonpea wilt by *T. harzianum* by using as seed treatment (10–20 g/kg seed) and soil application (10–20 g/ 9 sq meter. Soil treatment was more effective than seed treatment. The biocontrol agent multiplied more rapidly in soil than on seed. Inoculum of *T. harzianum* can be raised even on ordinary food wastes from large kitchens or on farm waste (FYM, sawdust, coir pith, etc). A strain of *Bacillus cereus* and of *Pseudomonas aeriginosa* are reported to induce systemic resistance in pigeonpea against F. udum (Dutta *et al.*, 2008). The induction of resistance is associated with increased levels of defense related enzyme.

Vidyasekaran *et al.*, (1997) isolated *Pseudomonas fluorescens* strains that inhibited mycelial growth of *F. udum*. Powder formulations were developed for seed treatment against pigeonpea wilt. Talc based formulations of the biocontrol agent remained effective after six months of storage. Shelf life of vermiculite, lignite or kaolin based formulations was short. Unformulated bacteria could not survive even for 10 days but in plant rhizosphere they remained effective for the entire crop growth period. Strains of this bacterial species have been successfully used against many root diseases of different crops. Anjaiah *et al.*, (2003) have reported that *Pseudomonas aeruginosa* strain PNA1, isolated from chickpea rhizosphere, protects pigeonpea and chickpea plants against wilt caused by *F. udum* and *F. oxysporum* f. sp. *ciceris*, respectively. However, protection until crop maturity was found only in moderately tolerant genotypes. The suppression by the antagonist was attributed to production of two phenazine antibiotics and level of root colonization by the antagonist. Fluorescent Pseudomonads are known to produce extracellular chitinase which might be contributing to suppression of wilt-causing *Fusarium oxysporum*. Bapat and Shah (2000) have reported that seed treatment with *Bacillus brevis* suppresses pigeonpea wilt. The bacterium produced an extracellular antagonistic substance which induced swelling of the *Fusarium* hyphal tips, cells were swollen and bulbous with shrunken and granulated cytoplasm and also inhibited conidial germination and caused death of hyphae. Pre-inoculation with *Fusarium oxysporum*, non-pathogenic to pigeonpea but pathogenic on other plants, provides cross-protection against the disease.

The use of resistant varieties is the most important approach to control wilt diseases. However, some varieties show resistance at one place but become susceptible when grown at other places. Varieties C-11, C-28, C-36, F-18, NP (WR) 15, NP 41, T 17, and NP 38 had been considered resistant to wilt. Singh, D.V. and Misra (1976) screened about 530 lines of pigeonpea and found none which could show less than 5% incidence. In multilocational trials pigeonpea lines ICP 4769, 0063, 9168, 10958, 11299 and cultivars C 11 (ICP 7118) and BDN 1 (ICP 7182) were found resistant in all the 5 years of testing at most of the 15 locations in Maharashtra, Bihar, U.P., W. Bengal, Delhi and Karnataka. The line ICP 8863 which showed stable and broad based resistance to wilt and good yield potential was released as cultivar Maruthi in Karnataka and abroad. Except the line 9168 all other lines have their origin in India. They are medium to long duration type. On the basis of multilocational trials in India and abroad the germplasm lines ICP 9145, ICP 9174, ICP 12745, ICPL 333, ICPL 8363, ICPL 88046, BWR 37, DPPA 85–2, DPPA 85–3, DPPA 85–8, DPPA 85–13, DPPA 85–14, Banda, Palera have been released for commercial cultivation in Malawa and have become very popular. Vishwadhar *et al.*, (2000) identified 11 pigeonpea lines that possessed resistance against wilt and sterility mosaic. The line KPL 43 was found to possess resistance to wilt, sterility mosaic and tolerance to Phytophthora blight as well as high yield potential. These lines are proposed to be used as donors of multiple disease resistance. At ICRISAT, three genotypes, ICPL 96047, ICPL 96061 and ICPL 99046 were found resistant to Fusarium wilt, sterility mosaic, powdery mildew and phyllody while six ICPL lines 93001, 96047, 96061, 99046,99055 and 87119 were resistant to Fusarium wilt and sterility mosaic (Saifulla and Byre, 2002).

WILT OF COTTON

Cotton plants suffer from two destructive vascular wilt diseases, one caused by a *Fusarium* and the other by *Verticillium dahliae* (*V. albo-atrum*). The Verticillium wilt, although reported in

India in 1971, is not common. The Fusarium wilt is generally found throughout the world wherever cotton is grown. It is believed to have originated in Mexico or Central America and has spread to South America, USA, West Indies, many African countries, Italy, Greece, part of France, Yugoslavia, Russian States, Romania and India. In India the disease was first reported from Nagpur in 1908 and is restricted to black cotton soils which are heavy clay with pH ranging from 7.6 to 8.0. It is rare in light alkaline or loam soils. On the other hand Fusarium wilt of cotton in USA is common and destructive in light sandy soil having pH of 5.5–5.9.

Symptoms

The symptoms of the disease are very similar to those of pigeonpea wilt, including the discoloration of tissues and plugging of vessels by hyphae. In very early stage of plant growth, vein clearing on cotyledonary and first leaves is also visible. Often the diseased plants are short with smaller leaves and bolls.

The causal organism

Fusarium oxysporum is the main species of *Fusarium* causing vascular wilt of plants. The species is classified in more than 120 *formae speciales* (f. sp.) on the basis of host specificity of the strains. These are further divided in to vegetative compatibility grous (VCG). Some formae speciales have a single VCG while some as many as 79 VCGs (Fravel *et al.,* 2003). A number of non-pathogenic strains of *Fusarium oxysporum* are known as excellent inducers of resistance to many root pathogens including pathogemc *F. oxysporum*. The non-pathogenic strains have no VCG.

The cotton wilt is caused by *Fusarium oxysporum* f. sp. *vasinfectum* (Atk.) Snyder and Hansen. Although many other fusaria may enter the plants roots they fai;l to colonize the xylem and cause disease. The aerial mycelium of *F. oxysporum* is white to grayish white or bluish purple and often forms a mat on the base of the affected plant. The hyphae are both inter- and intra-cellular. The conidiophores are verticillately branched and develop in sporodochia, reduced pionnotes, or sometimes directly on the mycelium. The microconidia are 1-celled (6–10 x 2–3 μm) or rarely 1-septate (13–20 x 3–5 μm), elliptical and scattered all over the mycelium. The macroconidia are multicellular, light buff in mass, fusiform-falcate, and curved inwards at both ends. These hyaline conidia are usually 3-septate, occasionally 4–5 septate. The 3-septate macroconidia measure 27–40 x 2–4 μm while the 5-septate conidia measure 32–50 x 3–4.5 μm. Chlamydospores are mostly globular, terminal or intercalary and 7–13 μm in diameter. No perfect stage of the fungus has so far been reported. A strain of the fungus causes wilt of okra also. Okra cultivar Pusa Sawani is highly susceptible.

Isolates from wilted cotton plants vary in growth rate and in pathogenic potentiality to cotton although they may have similar morphology. On the basis of a genealogic study, origin of the 8 races of *F. oxysporum* f. sp. *vasinfectum* was found as follows: races 1 and 2 from USA, race 3 from Egypt, race 4 from India, race 5 from Sudan, race 6 from Brazil and races 7 and 8 from China. Virulence of an isolate depends on susceptibility of the cultivar from which it is isolated and on age of the infected plants. Isolates from susceptible cultivars attack these cultivars only. Isolates from resistant cultivars attack both resistant as well as susceptible cultivars. Isolates from plants infected at maturity are less virulent than those obtained from seedlings.

Disease cycle and environmental relations

The fungus is soil-borne, initially surviving in diseased crop debris and then as free chlamydospores. It is a weak saprophyte. It can use the humic substances present in soil as carbon source. The fungus has been found up to a depth of 1 meter or more below the soil surface but the infection is most severe in first 40 cm. There is also evidence that the fungus is seed-borne. The percentage of seeds carrying the fungus varies from 1.2 to 7.5. In Tanzania (Africa), spread of the pathogen through seed from cotton ginning stations along with ginning residuals used for cattle feed is reported.

The conidia and chlamydospores germinate in the rhizosphere of 1–3 weeks old plants by germ tubes. These form hyphal strands on the root surface and grow along it forming spores. The fungus enters the roots directly without the aid of any specialized structures such as appressoria. In a study of infection process in cotton (*Gossypium barbadens*) it was observed that conidia of the fungus start germination on the root surface 6 hours after inoculation and form a compact mycelium covering the root surface. About 18 hours later, penetration hyphae branch off and infect the roots. The number of penetration hyphae increases with the number of conidia used for inoculation. The optimum temperature for penetration is between 28 and 30°C. The highest number of penetration hyphae are found in the meristem zone, 40% less in the elongation and root hair zones and none in the lateral root zone After entry the fungus grows through the cortex to the stele and invades the xylem where it starts sporulating. It becomes systemic and proliferates in the xylem vessels. The conidia are transported upward in the transpiration stream. These produce more mycelium. Wilt symptoms usually appear when the plants are about 5–6 weeks old.

The plant starts reacting as soon as fungal growth starts developing on the root surface. In the meristematic zone, a thickening of the plant cell wall occurs due to an apposition of dark and lightly staining material below the hyphae. This wall apposition increases in size around the hypha invading the plant cell and leads to the formation of a prominent wall apposition with finger-like projections into the host cytoplasma Such developments are less pronounced in the zone of elongation of the root.

Lipoidal substances (functioning as phytoalexins) and other compounds develop in contact cells and adjacent vascular parenchyma of cotton accompanied by changes in cytoplasmic organization after vessel infection by *F. oxysporum* f. sp. *vasinfectum.* These are secreted into vessels through pits. The secreted products coat the vessel walls and accumulate in vessel lumen (Shi *et al.,* 1992). The wilting and senescence occur as a result of water stress induced by the combined effect of mycelial growth in the xylem, fungal toxins and vascular occlusion by the host in an attempt to prevent systemic spread of the fungus. Fusaric acid, the wilt toxin, is known to be produced by the fungus in the plant tissues, in the rhizosphere as well as in the soil. Pectinolytic enzymes produced by the fungus cause breakdown of the cell wall components. The degradation products add to vascular occlusion. The scoring of vascular browning gives a more reliable indication of disease severity than the scoring of foliar symptoms.

The disease is favored by high soil temperature and low to moderate soil moisture. Optimum temperature for development of cotton wilt is 20°–27°C., the maximum being 31°C. Temperatures above 35°C are inhibitory to wilt development. Most of the deaths in cotton occur when the soil temperature at 15 cm depth is between 22° and 25°C and at 37 cm depth 24° and 25°C. However, temperatures of 25° to 30°C favor germination, germ tube elongation, growth and sporulation of the fungus. Although different reports suggest soil moisture levels of 20–40% moisture holding capacity or 80–90% saturation favorable for the disease and a warm and wet weather favors activity of the pathogen, there is no significant correlation between cotton wilt incidence and soil moisture.

Fusarium wilt of cotton, as also many other vascular wilt diseases, are often interrelated closely with avenues of entry into the host roots produced by different nematodes. However, the wilt incidence is reported to be significantly increased only by the root knot nematode (*Meloidogyne incognita*). Prior infection of cotton roots by the root knot nematodes hastens symptoms of wilt. Fungal invasion and extent of root galling are well correlated. Even resistant varieties lose resistance if nematodes have attacked the plant before the fungus reaches the roots. The predisposing effect is due to wounding of the roots facilitating penetration and nutrient accumulation at the nematode feeding site. Pre-infection by the nematode is the criterion for effect of the association. While moving through the cortical parenchyma, the nematode leaves easy passage for rapid movement of the fungus to the vascular bundles. Althjough the nematode does not penetrate the xylem, it brings about physiological changes that block accumulation of phytoalexins in the vessels (gossypol in cotton and cajanol in pigeonpea). This aggravates wilt severity. The loss of vitality in the plant due to nematode infestation also makes the plants susceptible. Germination of spores of *F. oxysporum* f. sp. *vasinfectum* as well as *Verticillium dahliae* is enhanced in the vascular fluid from cotton plants infected with root knot nematode (Katastonis *et al.*, 2005). The association does not qualify the wilt disease to be called a disease complex or wily complex. Vascular wilt occurs even if there are no root knot nematodes.

Potassium deficiency in soil favors disease development. High potash applications alone and fairly high potash applications in combination with nitrogen and phosphorus containing salts give definite reduction in cotton wilt. High nitrate application, application of nitrogen and phosphorus only, and cotton seed meal alone are not effective against the disease. In sandy alluvial soil, containing sufficient nitrogen and phosphorus but lacking potassium, the application of potash fertilizers reduces wilt. High nitrogen, especially ammonia nitrogen increases disease incidence. However, high dose of urea to the soil reduces the density of the pathogen in the soil. Planting of less susceptible cotton or non-hosts reduces the population of the pathogen.

A good amount of work has been done in south India on the relation of soil conditions to cotton and other wilt diseases. Trace elements such as zinc, molybdenum, lithium, aluminum, nickel, boron, cobalt, and manganese were inhibitory to conidial germination of *F. oxysporum* f. sp. *vasinfectum*, in the order mentioned. Manganese was least inhibitory. Boron, lithium, and molybdenum were toxic to cotton plants. Zinc and manganese were beneficial for plant growth but amendment of soil with only zinc reduced wilt incidence while that with manganese aggravated it. The presence of zinc in soil and plant tissues was supposed to stimulate formation of growth promoting auxins thus resulting in greater plant vigor. The resistance of cotton to wilt in zinc amended soil was ascribed to the greater reserves of carbohydrates, ascorbic acid, and reducing sugars in the vigorously growing plants. It has been suggested that these energy giving substances might be utilized in the plants for formation of a labile toxic substance which inhibits the development of the pathogen in the vascular system. Zinc at as low concentrations as 10 μg/ml abolishes the production of fusaric acid, the pathogenicity factor, increases microconidial production but reduces the total biomass of *Fusarium oxysporum*. It also promotes the efficiency of bacterial antagonists of the pathogen. High concentrations of plant gossypol inhibit conidial germination of *F. oxysporum* f. sp. *vasinfectum* but, over time, the pathogen can overcome the inhibitory effect of gossypol.

Qinolinol has been found effective in giving protection to cut shoots of cotton against toxemia in dialyzed culture filtrates of the fungus. The chemotherapeutant was also effective in sand cultures. Cystine is present in resistant cotton plant parts but absent in susceptible ones. It has been suggested that this amino acid might be related to the metabolism of wilt resistance.

Management of the disease

While practices like field sanitation, crop rotation, and mixed cropping are useful for reducing the incidence of wilt disease of cotton, the most effective control measure has been the use of resistant varieties. All American cotton varieties, even those highly susceptible to wilt in that country, are resistant to wilt in India. Similarly, susceptible varieties from India are immune to the disease in USA. These differences are due to different races and their origin. In Maharashtra, the severity of the disease was brought down to a great extent by replacing old varieties with wilt resistant varieties. Selections against wilt have been made from *Gossypium herbaceum* var. *frutescens, G. arboreum* var. *neglectum* f. *bengalensis* and f. *indica*. Cultivars Jayadhar, Jarila, Vijay, Verum and BDS are wilt resistant. Some accessions of *Gossypium sturtianum* and *Ĝ. australe* have been found to possess resistance to the wilt fungus.

The role of fertilizers and micronutrients was discussed above. Within limits, it is possible to reduce wilt incidence with proper application of nitrogen and potash fertilizers and zinc to the soil. Solarization of the soil, especially in combination with fumigation with Vapam reduces the inoculum in soil (Ben-Yophet *et al.,* 1988).

Biological control with use of endophytic bacteria is reported. Among many endophytes isolated from cotton tissues, *Aureobasidium sapordae, Bacillus pumilus, Phyllobacterium rubiacearum, Pseudomonas putida,* and *Burkholderia (Pseudomonas) cepacia* could survive for 28 days in stems of seedlings bacterized with them but provide only limited protection against cotton wilt. Strain CS85 of *Pseudomonas fluorescens,* isolated from rhizosphere of cotton seedlings is reported to act as both a plant growth promoting bacterium and a biocontrol agent against *F. oxysporum vasinfectum* and many other cotton pathogens (*R. solani, Verticillium dahliae, Colletotrichum gossypii*). It colonized the root surface well, though not uniformly, and maintained high level of population density for 2 weeks before gradually declining (Wang *et al.,* 2004). Strains of *Pseudomonas putida, Bacillus pumilus, Aureobacterium saperdae* can establish as endophytes in cotton stems and provide some protection against wilt.

Inoculations with non-pathogenic strains of the fungus induce cross protection. Dried biomass of *Penicillium chrysogenum* and its water extract induce systemic resistance in cotton seedlings against *F. oxysporum* f. sp. *vasinfectum* (Dong and Cohen, 2002a) and also against Verticillium wilt with concurrent enhanced growth of cotton plants (Dong and Cohen, 2002b). The effect is more pronounced in Fusarium wilt than in Verticillium wilt. Later, Dong *et al.,* (2005) have demonstrated field control of these wilt diseases in cotton by basal application of dry mycelium of *P. chrysogenum*. The control was dependent on application rates, At a dose of 30 g mycelial dry weight/ m² a significant control was achieved. Increased dose of the antagonist further increased the level of control. Basal application plus side-dressing gave better results than either alone but side dressing alone was ineffective. The expression of induced resistance to Verticillium wilt was somewhat cultivar-dependent and seedling age-dependent. The species abundantly secretes the small, highly basic and cysteine-rich protein which has antifungal property against many filamentous fungi including human and plant pathogens (Kaiserer *et al.,* 2003). It reduces fungal spore germination, causes distortion of hyphae and plasma membrane leakage. Since the biomass of *P. chrysogenum* is known to suppress root knot nematode invasion, it can be expected that this treatment will not only directly suppress the wilt fungus but also the nematodes that aid the wilt fungus in penetration of host roots. Aqueous extract of dry mycelium of *Penicillium chrysogenum* is also know to induce resistance in grapevines against *Plasmopara viticola* (downy mildew) and *Uncinula necator* (powderyt mildew), in apple against scab (*Venturia inaequalis*) and in onion against downy mildew caused by *Peronospora destructor* (Theurig *et al.,* 2006).

In China, a new systemic fungicide (miejuncuzhangi) was used in combination with carbendazim for foliar spray. It was reported to reduce infection by 89%. Seed treatment with carbendazim alone provides protection against early infection from soil. Up to 50% reduction in losses due to Fusarium and Verticillium wilts were obtained by an integrated management approach. By using an organic solvent for carbendazim efficacy of seed treatment was considerably increased. Resistant varieties and seed treatment were integrated to give these results.

FUSARIUM WILT OF CHICKPEA

Wilt of chickpea was first reported in India around 1910 and later elsewhere in Asia. It is also known to occur in many other countries including USA. It is a destructive disease. The yield loss is around 10% in India and Spain and 40% in Tunisia. Chickpea cultivars resistant to Ascochyta blight are susceptible to wilt and vice versa.

Symptoms

The foliage of diseased plants develops a grayish-green chlorosis. Lower leaves show the change of color first. Chlorosis gradually extends up the plant. Leaves eventually become dull yellow in color, wilt and the plant collapses and dies. In some cases there may be leaf vein clearing before wilt begins. The xylem tissues turn dark brown to almost black. Wilting may initially affect one side of the plant.

The causal organism

In 1940, G.W. Padwick named the *Fusarium* associated with chickpea wilt as *Fusarium orthoceras* var. *ciceri. F. orthoceras* is now considered synonymous with *F. oxysporum.* The species was erroneously named *Fusarium lateritium* when it was discovered in California (USA). The present name, *Fusarium oxysporum* f. sp. *ciceris* was given in 1962.

The species is not morphologically distinct from non-pathogenic strains of *F. oxysporum.* On synthetic, low-nutrient media microconidia form abundantly from short (8–25 μm long) cylindrical, unbranched phialides. They are mostly sessile and scattered on the vegetative mycelium. The microconidia are 0-septate or sometimes 1-septate, ellipsoid to cylindrical or allantoid and measure 5–15 x 2–5 μm. They collect in small slimy droplets. Macroconidia are 1–7 (mostly 3–5) septate and measure 25–55 x 2.5–6 μm. They are fusiform, curved, with a tapering, pointed, sometimes hooked, apical cell and distinctly pedicellate basal cell. The conidia form in the same manner as the microconidia from phialides except that often they are in clusters, borne on irregularly branching cells. In some strains, macroconidia are sparse while in others they are abundantly produced in pinkish sporodochia. Chlamudospores are globose, 7–15 μm in diameter, usually forming abundantly in mature colonies, terminal or interacalary, single or or in small groups or chains.

Morphological, cultural and pathogenic variability in *F. oxysporum* f. sp. *ciceris* is reported. Eight races (0–7) have been defined within the pathogen. Four races were reported in 1982 in India. These were Race 1(Hyderabad), Race 2 (Kanpur), Race 3 (Gurdaspur) and Race 4 (Hissar). Race 7 was identified from Himachal Pradesh. Races 0 and 6 occur in USA and

Spain, races 1 and 6 in Morocco and race 0 in Tunisia. The races are conventionally identified by inoculation of a set of 10 differential varieties. Race 1 is further subdivided into 3 groups, 1A, 1B and 1C. Of these races, race 1A and races 2 to 6 cause typical fusarial wilt and necrosis whereas races 0, 1B and 1C induce yellowing syndrome in the absence of wilting. Races 1 and 4 are proposed as same race. The pathogen produces extracellular xylanases which quantitatively differ among the races. In a given area the pathogrm shows great variability and several races may be ptesent in the same area.

Disease cycle and environmental relations

F. oxysporum f. sp. *ciceris* was reported to be seed-borne and up to 27% seed in a lot were found to contain the fungus. The fungus reaches the seed through the vascular system. The fungus can also survive for many years in soil as chlamydospores or as a saprobe in plant debris. Conidia are dispersed by water flow, rain splash and by movement of infected soil or seed externally contaminated during threshing.

Annual variations occur in the incidence of the disease due to temperature variations and inoculum density. Kabuli type of chickpea is more susceptible than the desi type. Yield losses are severe in India and Pakistan when soil moisture is scarce and day temperatures are high. At 10°C even if inoculum density is high there is no disease while at 25° or 30°C there is severe wilt at all inoculum levels. Landa *et al.,* (2001) had reported that Fusarium wilt development was greater at 25°C than at 20° or 30°C. At 25°C the inoculum density of the pathogen does not influence disease severity. But at 20° and 30°C the disease development was greater when 250–1000 chlmaydospore/g soil were used than when only 25–100 chlamydospores/g soil were used. Expression of ressistance to Fusarium wilt is related to temperature at which plants are growing after infection. In greenhouse tests, Landa *et al.,* (2006) found that a cultivar was moderately resistant when grown after inoculation at day/night temperature regime of 24/21°C but was highly susceptible when grown at 27°/25°C.

Root exudates influence wilt incidence. Exudates of some resistant cultivars contain antifungal compounds that suppressed spore germination of the pathogen. Wilt incidence as well as rhizosphere population of the fungus increase with decreased soil matric potential. Root knot nematodes seem to be involved in the Fusarium wilt of chickpea although they are not essential. Invasion by Meloidogyne artiellia is reported to fail in breakdown of resistance to pathogen races O, 1A and 2. Tolerant chickpea cultivars Jyoti, BEG 482, ICCC 37 and ICCV 2 reacted moderately susceptible in presence of *Meloidogyne incognita*. Resistannt cultivars BDN 9–3 and ICCC 4 became susceptible in presence of the nematode. However, the nematode could break resistance in cultivars Radhey and Avrodhi.

Disease Management

Field sanitation by collecting and burning diseased crop debris helps in reducing chances of survival in the field. Seed treatment with bavistin + thiram is recommended to eradicate the seed-borne inoculum. Chowdhury (2000) has reported that seed treatment with IAA, cycloheximide or cycocel has protective effect in susceptible chickipea against the wilt fungus. Resistant varieties are the most effective preventive measure. Some cultivars were listed above.

Races 0 and 1 of *F. oxysporum* f. sp. *ciceris* are less virulent while race 5 is most virulent. Seed treatment with race 0 or 1 provides resistance to race 5. Treatment of Kabuli chickpea seed with nonpathogenic *Fusarium oxysporum* strain 90105 results in significant increase in the incubation period of *F. oxysporum* f. sp. *ciceris* race 3 thus reducing the final disease incidence and intensity. The increase in incubation period varies between 11 and 25 days depending upon the genotype of chickpea used. Nonpathogenic *F. oxysporum*, *Trichoderma harzianum* and strains of *Bacillus subtilis*, alone or in combination, effectively colonize chickpea roots and suppress wilt development. Final disease severity index is reduced by 14–33%. However, combination of *F. oxysporum* and *B. subtilis* is not effective. Cachenero *et al.*, (2002) reported that not only race O of *F. oxysporum* f. sp. *ciceris* but other nonhost strains of *Fusarium oxysporum* induce resistance in chickpea against virulent strains of the wilt pathogen. The latter is more effective than race O. The resistance inducers cause accumulation of defence-related enzymes chitinase, beta-1,3-glucanse and peroxidase. The nonhost strains cause more accumulation than the race O. The nonpathogenic strains pof *Fusarium oxysporum* (Fo 47, Fo 52) also induce systemic resistance in chickpea against *F. oxysporum* f. sp. *ciceris* (Kaur and Singh, 2007).

Twenty four isolates of *Bacillus* sp. and *Pseudomonas chlororaphis* from chickpea roots have shown strong antagonism. *Bacillus* isolates are less inhibitory to chickpea wilt fungus than to other pathogenic *Fusarium oxysporum*. Cell-free culture filtrates of some *Bacillus* isolates inhibited conidial germination and hyphal growth of pathogenic and nonpathogenic *Fusarium oxysporum*. Seed + soil treatment with some selected antagonistic *Bacillus* isolates suppressed chickpea wilt caused by highly virulent strains of *F. oxysporum* f. sp. *ciceris.*

Vidyasekaren *et al.*, (1995) used formulations of *Psudomonas fluorescens* isolated from chickpea rhizosphere for control of the wilt. The biocontrol agent remained viable for 8 months in the formulations. When chickpea seed were treated with the talc-based formulation the antagonist survived on the seed for atleast 130 days. The strain NBRI 1303 of *P. fluorescens* was shown to be a good colonizer of chickpea roots. It was effective against *F. oxysporum* f. sp. *ciceris* as well as other destructive pathogens of chickpea. Seed bacterization resulted in 25% more seedling germination and 45% reduction in the number of diseased plants. Seed and soil treatment with selected strains of *Pseudomonas fluorescens*, *Bacillus megaterium*, and *Paenibacillus macerans* can suppress the wilt disease (Landa *et al.*, 2004). Earlier, they had reported that the suppressive effect of rhizosphere was influenced by temperature and inoculum density of the pathogen. Wilt development was greater at 25°C than at 20° and 30°C. *Paenibacillus lentimorbus* is another species having antagonistic effect on chickpea wilt *Fusarium* through production of chitinase and glucanase. The above listed control practices are partially effective by themselves. Integration of several practices has been suggested by Landa *et al.*, (2004). They combined sowing date, use of partially resistant cultivar and seed and soil treatent with biocontrol agents *Bacillus megatereum* RGAF 51, *B. subtilis* GB03, nonpathogenic *F. oxysporum* Fo 90105 and *Pseudomonas fluorescence* RG 26. The integrated approach was most effective treatment at suppressing Fusarium wilt or delaying onset of the disease and increasing seed yield. *Pseudomonas aeruginosa* induces resistance to chickpea wilt by production of lytic enzymes chitinase and glucanase which lyse the cell walls of the pathogen. The bacteria of *Bacillus subtilis* group are sensitive to fusaric acid produced by *Fusarium oxysporum*. Thus, their efficacy when used alone is insignificant.

In field experiments, most isolates of *Rhizobium* have failed to reduce chickpea wilt in susceptible cultivars but in moderately resistant cultivars they enhance expression of resistance (Arfaoui *et al.*, 2006, 2007). Seed treatment of moderately resistant cultivars with some isolates of *Rhizobium* is reported to reduce the percentage of wilted plants by enhancing the expression of phenyl propanoid defence-related genes in response to infection by *Fusarium oxysporum* f. sp., *ciceris*. A similar role of flavonoids in flax wilt is reported. Combination treatment of seed with Bavistin + *Rhizobium* enhances chickpea plant growth and suppresses wilt. Application of *Rhizobium*, *Glomus faciculatus* and *Trichoderma harzianum* to soil infested with the wilt pathogen has shown the efficacy of combined treatment in promoting plant growth, plant transpiration and suppression of the disease. However, the antagonist alone does not improve plant growth (Siddiqui and Singh, 2004).

WILT DISEASE OF LINSEED

The wilt disease of flax or linseed plant (*Linum usitatissimum*) occurs in almost all the countries where linseed is cultivated and is often responsible for 'soil sickness' problem when the soil becomes heavily infested with the fungus due to continuous cultivation of the crop. In India, the disease was first reported in 1923 from Madhya Pradesh.

Symptoms

The plants are attacked at all stages of growth. If the seedlings are attacked, further growth of the plant Deases and leaves and stems die. Lower leaves become shriveled and wither away. In the case of very young seedlings the edges of cotyledons roll inward and ultimately the whole cotyledon falls down. In older plants, the disease appears as small, ill-defined, dark green or brownish spots on the leaves, drooping of the plants, and shriveling of the leaves followed by death of the plants. The dead plants usually do not fall down but remain standing as dry and defoliated stems. If the disease appears very late, after pod formation, the only symptom is premature ripening of the pods. Transverse sections of roots and lower part of the stem show the browning and presence of fungus mycelium in the vascular tissues.

The causal organism

The flax or linseed wilt is mainly caused by *Fusarium oxysporum* f. sp. *lini* (Bolley) Snyder and Hansen. The morphology of the fungus is similar to that of other wilt fusaria described on the preceding pages. The mycelium is septate, branched, and intracellular. Conidia are hyaline and mostly 1–2 celled. The microconidia measure 4.8–14.4 x 2.2–4.8 μm and the macroconidia 21.0–53.0 x 2.4–5.6 μm.

In Punjab *Rhizoctonia bataticola* (*Macrophomina phaseolina*) was also reported to be associated with linseed wilt in addition to the *Fusarium*. Different isolates of *Fusarium oxysporum* f. sp. *lini* were found to differ in their temperature requirements.

Disease cycle and environmental relations

The pathogen is known to persist in soil for many years even in absence of the host. It is tolerant to such antagonists as *Trichoderma lignorum*, *Bacillus subtilis*, and *Penicillium* spp. Although bacterial antagonists including actinomycetes had been isolated from soil none could protect the plants against the disease in all stages of growth. The fungus is also known to be seed-borne, the mycelium persisting on or in the seed. *M. lini* was reported to survive for 50 years on wild species of *Linum* around fields in which the flax crop was not grown for this duration.

Incidence and severity of the disease is influenced by soil texture. Soils with high sand content are conducive while those with high clay content are suppressive Soil temperature and soil type have profound effect on the incidence of linseed wilt. The reaction of different varieties to the disease also varies according to these factors. Optimum temperature for infection is 24°C with no infection at 12°C and 38°C. At soil temperatures unfavorable for the pathogen even susceptible varieties escape much damage. In India, the disease is favored by low moisture and light sandy soils. Farmyard manure decreases wilt incidence.

Potassium content of the plants is reported to be correlated with their resistance to wilt. Plants of wilt resistant varieties contain more potassium than those of susceptible varieties. There is a relationship between HCN content and resistance. Zinc is reported to reduce fusaric acid production by the pathogen. Culture solution containing 3 ppm zinc considerably checked symptom development but had no effect on the spread of the pathogen within the host tissues. The pathogen consists of several cultural and pathogenic races which differ from each other in their cultural characters, pathogenicity, and temperature relations.

Management of the disease

Crop rotation provides some protection from the disease but is not very effective since the fungus can survive in soil for very long durations or it may exist on its alternate hosts. Partial control of seed-borne infection can be achieved by seed dressing with chemicals. Some of the rust resistant varieties of linseed such as RR 9 (in U.P.) possess wilt resistance also. In Maharashtra cv. NP 12, 21, 124, RR 5B, RR 80, RR 82, 439, 440, 5, 10, 37, 38, 62, 197, 199, 200, RR 202(1), 272, 328 are reported as resistant to wilt. In Punjab the varieties K 1 and K 2 are reported to possess resistance to wilt and rust. Among older cultivars, B 5128, NP 1, NP 124, NP (RR) 5, 0,10, 80 are resistant to linseed wilt. Transgenic linseed (flax) carrying β-1,3-glucanase gene shows resistance to the wilt fungus. In addition the plants show lower content of lignin thereby improving the quality of fiber. Transgenic flax engineered to express genes for enhanced synthesis of propanoids is resistant to wilt. The changes in phenulpropanoid accumulation affect cell wall composition that resists the invasion by the wilt fungus Flavonoids are a group of secondary plant metabolites important for plant growth and development.

The pathogen is suppressed by *F. oxysporum* strain Fo 47. Reduction in wilt by this antagonist is related to reduction in the population density and metabolic activity of the pathogen on the root surface (Duijff *et al.,* 1999). Olivain *et al.,* (2003) studied the early physiological responses of flax cells to colonization by *F. oxysporum* f. sp. *lini* and *F. oxysporum* Fo 47. Both strains colonized the roots but plant defence reactions, i.e., the presence of wall appositions, osmophilic material and collapsed cells were less frequent in roots colonized by the pathogenic strain than in roots colonized by the antagonist. Within few minutes of inoculation both strains triggered H_2O_2 production, a sign of resistance, but in inoculation with the antagonist the process was repeated. The bacterial antagonist *Pseudomonas putida* can enhance the action of the nonpathogenic *Fusarium* if its population is relative to the pathogen. It is more effective in wet soil. Plants inoculated with arbuscular mycorrhizal (AM) fungi show resistance to the wilt fungus. AM activates resistance mechanisms and also induces tolerance to wilt. Serra-Wittling *et al.,* (1996) obtained significant control of flax wilt in a wilt conducive soil by amending the soil with municipal solid waste compost. The suppression of wilt was attributed to microflora in the soil and the compost. While compost application gives significant control of flax wilt, results are variable in field trials even if same type of compost is used. The method of storage of the compost is important for uniformity in results. The cool storage (4°C) gives least variability in the results.

FUSARIUM WILT OF TOMATO

Fusarium wilt of tomato is more common in temperate climate regions but is known to occur in many warmer areas also, including India.

Symptoms

Two of the earliest symptoms on young plants are clearing of veinlets and drooping of the petioles. Later, the entire plant wilts. In the field the disease appears any time if conditions are favorable but symptoms are predominant in mature plants after flowering and at the beginning of fruit set. Lower leaves show yellowing and die. The chlorotic symptoms begin to appear on one side of the leaf and then all leaflets become yellow on one half of the leaf. Symptoms continue to appear on successive younger leaves. These symptoms appear on the entire plant or on only few branches. After the disease has advanced for few weeks, browning of the vascular system can be seen in a cross section of the lower stem or by removing stem tissue near the base with a knife. Young plants wilt suddenly but older plants linger on for some time. The plant as a whole is stunted. Due to attack on roots, the laterals show black rot condition. This hastens death of plants. Symptoms of wilt are more commonly observed during the warmer part of the day. In wet weather the fungal growth can be seen on dead plants as a pinkish layer. Wilt infection reduces photosynthetic activity of leaves. There is decrease in light-saturated rate of carbon dioxide assimilation. The infection also reduces the leaf area.

The causal organism

Fusarium oxysporum f. sp. *lycopersici* has septate, hyaline myceliumwhich later becomes cream colored in cultures On the host also the mycelium is at first whitish but after formation of macroconidia it looks pinkish. The macroconidia are relatively less abundant than microconidia. They are fusiform,, hyaline, mostly 2–3 septate, and measure 25–33 x 3.5–5.5 μm in size. No sporodochia, pionnotes, or sclerotia are produced. Microconidia are 1-celled, hyaline, ovoid or ellipsoidal, and measure 6–15 x 2.5–4 μm. The older mycelium forms chlamydospores.

At least three races (1, 2, and 3) of the pathogen are known. They attack only species of *Lycopersicon*. Within the race 1, the isolates can be divided into separate virulence groups. Simultaneous inoculation with two races, on a cultivar possessing resistance to one race, provides cross protection against the other race.

Fusarium crown and root rot is caused by *F. oxysporum* f. sp. *radicis-lycopersici*, not by the vascular wilt pathogen. The root and crown rot pathogen has a very wide host range. Symptoms are also different. It does not move through vascular system of the plant and has optimum temperature for growth at 18°C while the wilt fungus is favored by temperatures around 27°C.

Disease cycle and environmental relation

The pathogen survives in soil as chlamydospores or, for some time, as saprophytically growing mycelium in infected crop debris. Once it is established in soil it is difficult to eradicate in areas where high soil temperatures are uncommon. Approximately 3.5% of seeds from fruit produced on heavily infected plants contain the pathogen. The seed transmission is important in race 1 of the pathogen. One of the chief methods of distribution of the pathogen is the use of seedlings raised in infested soil. Wind-borne soil, surface drainage water and agricultural implements also distribute the inoculum from field to field. The pinkish superficial growth with spores on the plants and their dissemination may help secondary spread of the pathogen in the field.

Development of pathogenic strains of *F. oxysporum* on tomato and also the nonpathogenic *Fusarium oxysporum* used as biocontrol agent is stimulated in the vicinity of the roots, irrespective of plant species, but there is no chemotactic response toward or away from the root. Growth stimulation is mainly related to organic nitrogen (amino acids) in the root exudates. The signaling and recognition mechanisms between the host plant and the pathogens or non-pathogens occur on or in the roots not external to the roots. Adhesion of Fusarium conidia to the root surface is site-specific. There exist single class of specific, high-affinity adherence site on the root surface to which conidia of pathogenic and non-pathogenic *F. oxysporum* adhere. When seedlings with broken roots are planted in infested soil the fungus grows quickly through the wounds and slowly progresses up to the stem through xylem vessels. Nematode invasion also helps the pathogen in the same manner. In addition, the root-knot nematode invasion before invasion of the wilt fungus affects physiology of the host plant which loses vitality and become susceptible to wilt. In soil infested with the wilt fungus and root knot nematodes wilt occurs at a significant level at all temperatures from 16 to 24°C. However, there are conflicting reports suggesting that root knot nematodes do not affect resistance to Fusarium wilt. In the vessels, hyphae are normally in close proximity to vessel walls and aligned approximately parallel to the long axis of vessels. Vessel to vessel colonization is uncommon but occasionally occurs by penetration of the pit membranes by constricted hyphae and localized membrane degradation.

After infection the pathogen begins to secrete pectolytic enzymes which work upon the pectic materials in the tracheal walls and diffuse into the xylem parenchyma walls also. As a result of production of colloidal masses by these activities the xylem vessels are plugged, checking movement of water and minerals. Various toxins (fusaric acid, lycomarasmin) are also produced by the fungus which help in development of wilt symptoms. Roots and stems of tomato contain the saponin, an antifungal glycol alkaloid, alfa-tomatine, which acts as chemical barrier to infection by pathogens. In response to this barrier many pathogens of tomato, including *F. oxysporum* f. sp. *lycopersici,* produce extracellular enzymes, tomatinases, which deglycosylate alpha-tomatine to yield less toxic derivatives, tomatidine and lycotetraose, thus overcoming the chemical barrier. Tomatinase is not essential for pathogenivity of *Fusarium* but is required for its full virulence and enhamces severity of syptoms compared to symptoms caused by mutants lacking tomatinase (Pareja-Jaime *et al.,* 2008). Even non-pathogenic *Fusarium oxysporum* producing tomatinase can cause infection of tomato but without producing symptoms (Ito *et al.,* 2005).

Rodriguez-Malina *et al.,* (2003) studied the colonization of stem tissues by the pathogen in resistant and susceptible cultivars. One day after inoculation the propagules spread discontinuously irrespective of cultivar genotype. Five days later, the pathogen is limited to stem bases. For next one week the pathogen seems to undergo incubation at the stem bases. This condition becomes permanent in resistant cultivars and no symptoms develop. In susceptible cultivars, a gradual upward colonization of the stem is seen and symptom development occurs.

Optimum temperature for wilt development is 28°C. The disease is inhibited at soil temperatures of 33°C and below 21°C. If the temperature of the soil remains at 37°C for more than a few days the fungus is killed. Wilt incidence is more at 17°–20°C and at 32°C than at 42°C. Incidence also declines with increased time of exposure to higher temperatures. Soil moisture conducive to best vegetative growth of the host is optimum for disease development. Disease is more severe at low pH (below 6.4) and above pH 7.0. Nitrate nitrogen suppresses disease by increasing the pH of the medium while ammonia nitrogen does the reverse. Other conditions which pre-dispose the plant to wilt are short day length and low light intensity and presence of root knot nematodes. Fusarium wilt of tomato, grown in nutrient solution, is decreased with increasing nitrogen (420, 630 and 1050 $\mu g/ml$). High nitrogen levels that decreased wilt increased the protein content of leaf tissue. Decreased susceptibility to infection is probably the consequence of the reduced sensitivity of the host plant membranes to toxic substances produced by the fungus. However, in another study, wilt was enhanced by high doses of nitrogen (150 kg N/ha). Application of potash alone (100 kg/ha) or in combination with nitrogen (N 100 + K 100) decreases disease incidence by more than 60%.

Management of the disease

The pathogen may be seed-borne. Seed treatment with carbendazim (2.5 g/kg seed) is recommended against the seed-borne inoculum. Some varieties of tomato are resistant to Fusarium wilt, Verticillium wilt and root knot nematodes. Due to presence of pathogen races, often resistant varieties fail. In absence of resistant varieties cultural methods for disease suppression are recommended. Tomato fields should be ploughed in summer and the soil should be allowed to bake in hot dry weather. The diseased crop debris, including root material, should be burnt in the field. Long crop rotations (5–7 years) also help in reducing the soil-borne inoculum. The pathogen is significantly suppressed in the rhizosphere of onion, garlic, tobacco, okra, eggplant, spinach and cauliflower. Such crops can be included in the rotation Cowpea and sesbania are symptomless carriers. Very often root knot nematodes help in infection by *Fusarium*. This warrants use of nematode resistant varieties and control of the root knot nematodes. Soil fumigation is useful. As a substitute of methyl bromide, 1,3-dichloropropane + chloropicrin can be used.

Chemicals like cupric chloride, ferric chloride, zinc chloride, manganese sulphate and mercuric sulphate at low concentrations enhance resistance in tomato plants to the Fusarium wilt. Soil fertility and pH can have significant effect on the severity of Fusarium wilt. In some soil types adjusting soil pH to 6.5–7.0 and using calcium nitrate as fertilizer, rather than ammonium nitrogen source, can reduce wilt severity. Khan and Singh (2001) had reported that broadcast or in-row application to soil of fly ash in place of inorganic nitrogen fertilizers, checked the suppressive effect of the wilt pathogen on plant growth and yield. The treatment increased the soil contents of P, K, B, Ca, Mg, Mn and Zn carbonates. The antibiotics Validomycin A and Validoxylamine A do not show antifungal activity against *F. oxysporum* f. sp. *lycopersici in vitro*. However, when sprayed on tomato leaves, they induce resistance to wilt. by inducing expression of PR genes. There is accumulation of salicylic acid in the tissues. The treatment also controls late blight and powdery mildew on tomato.

Biological control of Fusarium wilt of tomato by soil amendments with compost of banana leaves, paddy straw, sugarcane bagasse, and spent mushroom compost is reported. Beneficial effect of oil-cake amendment of soil against the disease is also reported. Addition of composts (including FYM), oil cakes, dry or fresh crop residue to the soil stimulates the proliferation of antagonistic and growth promoting rhizobacteria most of which produce siderophores and enhance plant resistance. The suppression of Fusarium crown and root rot caused by *F. oxysporum* f. sp. *radicis-lycopersici* by pulp and paper compost is associated with increased plant resistance to fungal colonization through formation of physical barriers at the site of fungal penetration. The callose-enriched wall appositions and osmophilic deposits around the site of penetration prevent the hyphae reach vascular bundles. Hyphae coated with osmophilic material show marked cellular disorganization. Compost of sewage sludge with *Trichoderma asperellum* is reported to control tomato wilt. The use of vermicompost is highly beneficial and gives significant control of Fusarium wilt of tomato. The suppressive effect of vermicompost is attributed to biotic nature of the material. Composts from organic wastes are known to produce elicitors of defense responses with the help of microflora involved in decomposition (including the biocontrol agents). Good wilt control and increased plant growth and yields of tomato when phosphate solubilizing *Pseudomonas fluorescens, Aspergillus awamori, A. niger* and *Penicillium digitatum* were added to soil as biocontrol agents in field trials. The treatments decreased soil populations of *F. oxysporum* f. sp. *lycopersici* by 23–49%.

The biocontrol microbial agents effective against the tomato wilt fungus include non-pathogenic strains of *Fusarium oxysporum., F. solani, Trichoderma* spp., *Gliocladium virens, Penicillium oxalicum, Pseudomonas fluorescens, Burkholderia cepacia, Bacillus subtilis* and many others. *Fusarium oxysporum* and *Fusarium solani* are the most effective biocontrol agents in wilt suppressive soil. Pre-inoculation of tomato plants with *F. oxysporum* f. sp. *dianthi* provides cross protection against *F. oxysporum* f. sp. *lycopersici*. This cross protection is successful only when the protecting agent is inoculated a few days before the challenge pathogen and the inoculum density of the protecting agent is equal to or higher than the density of the pathogen inoculum. Pre-inoculation of tomato seedlings with a non-pathogenic strain of *Fusarium oxysporum* (Fo 47) is reported to induce resistance in tomato to the wilt pathogen. It also suppresses the pathogen through competition for nutrients on root surface. Non-pathogenic *F. oxysporum* strains are also known to produce fusaric acid and very low concentrations of fusaric acid elicit defense responses fro the plant. The suppression depends on the level of colonization of the root by the antagonist. Root portions not colonized by the antagonist may available for infection by the pathogen.

Elicitation of defense responses in the host is a major mechanism of action of biocontrol agents. Their metabolites act as elicitors. Culture filtrates and mycelial extracts act as elicitors. The responses include increased activity of defense-related enzymes (peroxidase, polyphenol oxidase, phenylalanine ammonia lyase), accumulation of phenols, and deposition of lignin on root cell walls. The cell wall is reinforced by the lignin deposition.

In a study of role of root exudates in the suppression of *F. oxysporum* f. sp. lycopersici by non-pathogenic *F. oxysporum*, Steinkellner *et al.,* (2008) observed that exudates of plants inoculated with non-pathogenic *F. oxysporum* or a strain of *F. oxysporum* f. sp., radicis-lycopersici were suppressive for microconidial germination of the pathogen. They concluded that pathogenic and non-pathogenic *F. oxysporum* trains alter the root exudation of tomato plants differently and consequently the fungal propagation of both fungi in the rhizosphere is affected differently.

Olivain and Alabouvitte (1999) compared the process of colonization of tomato roots by *F. oxysporum* f. sp. *lycopersici* and a non-pathogenic strain of *F. oxysporum*. The pathogenic strain rapidly colonized the root surface rforming a dense network of hyphae as early as 48 hours after inoculation. Defence responses of the host had started earlier 24 hours after inoculation. These responses were mainly in the hypodermis, but also in the cortex. Generally, the barriers failed to prevent the centrifugal growth of the pathogen towards the stele. The enzyme activity of the pathogen also moved with the growing hyphae and intense enzyme activity was seen in the stele seven days after inoculation. Although root tip surface was colonized by the dense hyphal mass, direct contact between hyphae and living cells was prevented by several layers of sloughed off cap cells. Rarely, when contact was established there was rapid destruction of the apical cells. Compared to this pattern of colonization by pathogenic strain, the nonpathogenic strain showed reduced growth in the cortex, lower frequency of apical death and failure to reach the stele. Olivain *et al.,* (2003) studied the early physiological responses of flax cells to colonization by *F. oxysporum* f. sp. *lini* and *F. oxysporum* Fo 47. Both strains colonized the roots but plant defence reactions, i.e., the presence of wall appositions, osmophilic material and collapsed cells were less frequent in roots colonized by the pathogenic strain than in roots colonized by the antagonist. Within few minutes of inoculation both strains triggered H_2O_2 production, a sign of resistance, but in inoculation with the antagonist the process was repeated.

According Olivain *et al.,* (2006) the suppression of the pathogenic strain due to competition for nutrients rather than competition for infection site.Bolwerk *et al.,* (2005) have studied the interactions between *Fusarium oxysporum* f. sp. *radicis-lycopersicae* (foot and root rot of tomato) and *Fusarium oxysporum* Fo 47 and have reported that for biocontrol at least 50-fold excess of Fo 47 over *Fo radicis-lycopersicae* was required to obtain control. The hyphae of the antagonist attach to0 tomato roots earlier than the pathogen if inoculated separately. The non-pathogenic *Fusarium* spp, used as biocontrol agents, also produce fusaric acid. Very low concentrations of fusaric acid are non-toxic and induce synthesis of phytoalexins that might prevent damage by pathogenic species Biocontrol efficiency of *F. oxysporum* Fo47 is enhanced when it is used with an antagonistic strain of *Pseudomonas fluorescens*. The control is more under controlled conditions than under uncontrolled conditions. Olivain *et al.,* (2004) have reported a method of producing both agents together on peat. While confirming the enhanced biocontrol efficacy of combined biocontrol agents, one report suggest that combination of non-pathogenic *Fusarium* and *T. harzianum* does not give better control of *F. oxysporum lycopersici* than their use singly. Some other nonpathogenic strains of *F. oxysporum* provide better control of the disease (Larkin and Fravel, 1999). Within 24 hours of inoculation with the non-pathogenic *F. oxysporum*, the root surface is colonized by a dense network of hyphae, with the exception of the apex, which is colonized only after 48 hours. A few hyphae penetrate into the epidermis, leading to the internal colonization of the root cortex. The colonization is always discontinuous since defense reactions of the plant limit the extension of the fungus. The barrier formed by thickening and coiling of the cell walls and hypertrophies cells is most frequent in the external cortex and sometimes deeper in the internal cortex, close to the vessels which are never colonized. Typical defense reactions, frequently observed, are cell wall appositions, intercellular plugging and intracellular osmophilic deposits. In a comparative study of strains of nonpathogenic *F. oxysporum* (C20 and CS-24) and *F. solani* (CS-1), it was observed that relative biocontrol efficacy could be influenced by environments, particularly temperature. Some strains are effective at all temperatures while others may not be effective at temperature range highly favorable for the pathogen.

Muslim *et al.,* (2003) have reported control of the tomato wilt with hypovirulent binucleate *Rhizoctonia* strains under glasshouse conditions. Singh *et al.,* (2004a, b) have reported significant control of tomato wilt by seed treatment with *Trichoderma viride, T. harzianum, Aspergillus nidulans* and *Gliocladium virens*. The antagonists could be grown on wheat meal, sorghum grain or pearl millet grains supplemented with 0.5% glucose. These substrates supported good growth and sporulation of the antagonists which could be stored for 3 months at room temperature. Effective control of wilt by seed treatment with *Trichoderma* species (superior to *Bacillus subtilus*) is reported. Efficacy is further improved if biocontrol agent is combined wth carboxin. The biocontrol agent is most effective when it colonizes vascular tissues extensively. When tried in an environment with uncontrolled microbial activity the efficiency decreases and minimum efficiency if found in open field trials (Shishido *et al.,* 2005). The non-pathogenic *Fusarium oxysporum* strain CS-20 is an effective biocontrol of *F. oxysporum* f. sp. *lycopersici* race 1. Tomato plants respond to the inoculation of this biocontrol agent within 24 hours by alterations in secondary metabolites, *viz.,* ferulic, caffeic and vanilic acids which are generally increased in leaves ard roots. The accumulation of these phenols contributes to resistance response (Panina *et al.,* 2007).

Minuto *et al.,* (2006) have report significant control of *F. oxysporum* f. sp. *lycopersici, F. oxysporum* f. sp. *radicis-lycopersicae* and *Verticillium dahliae* by application of

Streptomyces griseoviride to the soil. Mechanisms of disease suppression by strains of *Pseudomonas* include induced systemic resistance and antibiotic production such as 2,4 diacetylphloroglucinol. Some strains produce hydrogen cyanide (HCN) which highly toxic to *Fusarium*. In comparison to *phloroglucinol*, there is less enhancement of biocontrol activity of *P. fluorescence* by HCN. Chemotaxis of strains of *Pseudomonas fluorescens* toward fusaric acid is reported and this helps the antagonist to colonize hyphae of *Fusarium* (de Weert *et al.*, 2004). Suppression of Fusarium wilt by *Pseudomonas fluorescens* is ascribed to induced systemic resistance while suppression by nonpathogenic *F. oxysporum* is mainly ascribed to microbial antagonism and to a lesser extent to induced resistance. Ramamoorthy *et al.*, (2002) have reported that biocontrol strains of *Pseudomonas fluorescens* (Pf 1) suppress *F. oxysporum* f. sp. *lycopersici* through induction of defense enzymes and accumulation of phenolics and pathogenesis related proteins. Apart from the induction of systemic resistance and/or direct antagonism, the plant growth promoting rhizobacteria may suppress wilt symptoms through induction of physiolgical changes in the host. Some plant growth promoting bacteria (*viz., Achromobacter piechaudii*) are reported to induce resistance to water and salt stress in tomato and pepper (Mayak *et al.*, 2004). It is possible that PGPR that induce systemic resistance to vascular wilts might be additionally acting through this mechanism to attenuate the wilt symptoms. An antagonistic rhizobacterium strau\in of *Gluconobacter* sp. isolated from Eregenon (Asteraceae) causes growth inhibition and morphophysiological changes in *F. ofysporum* f. sp. *lycopersici*. There is extensive lateral branching of the pathogen hyphae and leakage of cytoplasm at the hyphal tips (*J. Pestic. Sci.* 33,2,138, 2008).

Often, the fluorescent *Pseudomonas* and non-pathogenic *Fusarium oxysporum* are individually not as effective as a combination of the two. Mixed inoculation of two biocontrol strains of different species of *Pseudomonas* improve biocontrol of Fusarium foot and root rot of tomato (*F. oxysporum* f. sp. *radicis–lycopersici*) and the gene encoding for colonization in an effective biocontrol strain can be transferred to other strains having less or no colonizing ability by genetic engineering. Pajand and Johnson (2000) reported that endophytic bacterial isolates recovered from internal tissues of roots and stems of oilseed rape not only significantly improved seed germination, seedling length and plant growth of oilseed rape and tomato but also, when used for seed treatment, significantly reduced wilt symptoms in tomato (*F. oxysporum* f. sp. *lycopersici*) and in oilseed rape (*Verticillium dahliae*).

Soil, seed and root dip application of *Trichoderma viride, T. harzianum* and an antagonistic *Pseudomonas* is also reported to give high degree of protection against the wilt pathogen. Commercial products containing. *Gliocladium virens* and *Trichoderma harzianum* give 62–68% control of wilt (Larkin and Fravel, 1998). Fusarium wilt of tomato was controlled by spraying a suspension of *Phytophthora cryptogea* zoospores on the green parts of the plant. There was no evidence of direct competition or antagonistic interaction. Systemic induced resistance was assumed to be responsible for the observed control.

Seed bacterization with selected strains of *Pseudomonas fluorescens* induces systemic acquired resistance. Density of fluorescent pseudomonads in tomato rhizosphere is significantly higher when the plants are treated with the resistance inducer acibenzolar-S-methy and there is bettrer control of fungal and bacterial diseases of tomato (Fakhouri and Buchenauer, 2002). Soil application of *Pseudomonas fluorescens* or *Pseudomonas chlororaphis* in combination with acibenzolar-S-methyl activates the resistance response in tomato against the wilt fungus (Fakhouri *et al.*, 2004). Inoculation with conidial suspension of *Penicillium oxalicum* before infection by *Fusarium* also induces systemic resistance against the wilt pathogen. The inducer leads to formation of additional secondary xylem vessels (De Cal *et al.*, 2000). Repeated application of *Penicillium oxalicum* to the growing substrate (peat and vermiculite), up to 4 times, prolongs the duration of control of Fusarium wilt, especially when disease incidence is high.

Larena *et al.,* (2003) reported effective biocontrol of Fusarium and Verticillium wilt of tomato by applying conidial suspension of *Penicillium oxalicum* before transplanting. The effective dose of the biocontrol agent was 1 to 10 million CFU/ g soil. Sabuquillo *et al.,* (2005) have used powder formulations of *P. oxalicum* for control of Fusarium wilt in greenhouse and Verticillium wilt in field trials. The powder was prep ared by maintaining conidial suspension with 60% sucrose or 1.5% sodium alginate for 10 min before drying. Formulations of *P. oxalicum* vary in their efficacy, giving a range of disease control (22–64%). The number of viable conidia in the formulation at the time of application and extent of development of the antagonist in the rhizosphere before seedlings are transplanted determine the efficacy (Sabuquillo *et al.,* 2006).

Akkopro and Demir (2005) have reported suppression of Fusarium wilt of tomato by inoculation with the mycorrhizal fungus *Glomus intraradices* together with some rhizobacteria (*Pseudomonas fluorescens, P. putida, Enteroibacter cloacae*) Colonization of roots with rhizobacteria increased, especially in triple inoculation (pathogen + *G. intrardices* + rhizobacteria). Nature of root exudates is altered by mycorrhization. Microconidial germination of *F. oxysporum* f. sp. *lycopersici* is favored by root exudates of host and nonhost plants but mycorrhization enhancers the effect. Spray of the plant activators acibenzolar-S-methyl and beta-aminobutyric acid on the foliage is reported to suppress wilt symptoms. Exposure if the fungus to vapors of essential oils of cinnamon, thyme or fennel for a week is fungicidal. Soil drenching with water extracts of the weed *Oxalis articulata* suppresses the wilt in tomato.

One accession of *L. cheesmanii* and two of *L. chilense* have shown highly level of resistance to two races. In general, polygenic resistance provides varying levels of tolerance to the wilt. The resistance vatries from plant to plant cv. Marglobe, grown from seed due to segregation and recombination of resistance genes. If plants are raised vegetatively the resistance is uniform in all plants of a population.

FUSARIUM WILT OR PANAMA DISEASE OF BANANA

Banana plantations suffer from many serious diseases such as Fusarium wilt, Bacterial wilt or Moko disease, Cercospora leaf spot or Sigatoka, and Bunchy top. The Fusarium wilt or Panama disease is a lethal disease of edible banana and is found throughout the tropical regions of the world. In India, the disease is an important and destructive fungal disease of this plant. Fusarium wilt is known to be a serious problem in the Central and South Americas, parts of Africa, Sri Lanka, Myanmar (Burma), Thailand, Malaysia, Indonesia, Hawaii, Fiji and the Philippines. It also occurs in Australia and New Zealand. Most authorities have believed that the disease coevolved with banana. Information on genetic diversity in the fungus and its ancestry confirm that it most likely originated in south and southeast Asia.

Symptoms

The symptoms are most pronounced on at least 5 months old plants although 2–3 months old plant are also killed under highly favorable conditions for disease development. The earliest signs of the disease are faint yellow streaks in the petioles of oldest lowermost leaf. Two types of symptoms follow this stage. In the yellowing type there is progressive yellowing of the old leaves and eventual collapse at the petiole. In the non-yellowing type, the leaf collapses at the petiole without leaf chlorosis. Often all the leaves but the youngest collapse, the heart alone remaining upright. Any new leaves that are produced are blotchy and yellow, often with wrinkling of the lamina. The pseudostem often shows a more or less conspicuous longitudinal splitting of the outer leaf sheaths that form its outer covering. But sometimes this symptom is not present. About 4–6 weeks after appearance of streaks on the petiole only the dead trunk of the pseudostem remains.

Fig. 124. Panama disease of banana. Collapse of leaves at base of the petiole.

Discolored vascular strands varying from light yellow to dark brown are the distinguishing internal symptoms. Usually the discoloration appears first in the outer or oldest leaf sheath and extends up to the pseudostem. It is pronounced in the rhizome but is not common in roots. However, roots of diseased rhizomes are frequently blackened and decayed. Longitudinal sections through diseased root bases show characteristic red strands passing into the rhizome stele.

The causal organism

Panama disease of banana is caused by *Fusarium oxysporum* f. sp. *cubense*. The mycelium is mainly intracellular, being typically found in the wood vessels, though intercellular hyphae may be observed in the cortex of roots and in parenchymatous tissues in proximity to the site of infection. In nature the conidia-bearing bodies (sporodochia) appear at a late stage on the surface of petioles and leaves of infected plants. These sporodochia emerge through the stomata on a globose mass of pseudoparenchymatous tissue measuring 26–30 μ in diameter. The conidiophores are verticillately. branched and measure, on an average, 70 μm in length. The side branches are usually 1-celled and measure up to 14 μm. Conidia are borne at the apical end of the main and lateral branches. Microconidia are 0- and 1-septate, ovate or somewhat elongated, and measure 5–7 x 2.5–3.0 μm. The macroconidia are pedicellate, sickle-shaped, mostly 3-septate, and measure 22–36 x 4–5 μm. Chlamydospores are oval or spherical and usually in pairs. Four races of the fungus are reported. Race 1 is most predominant.

Several clones differing in pathogenicity exist in nature and may be grouped into races. The disease had started on the cultivar Gros Michel which was the most popular export quality banana. This cultivar was almost completely wiped out. Fortunaterly another cultivar, Cavendish bananas, was available to sustain the banana export trtade. Races 1,2,3 were known to occur on banana cultivars and Cavendish bananas were thought to be resistant to these races. Later, a new tropical race (TR 4) appeared in the Southeast Asian banana cultivars and it could infect the Cavendish bananas. The race is supposed to be confined to Southeast Asia.

Disease cycle and environmental relations

The pathogen is a soil-borne fungus surviving in soil mainly as chlamydospores formed by the hyphal and conidial cells. As chlamydospores in plant debris the pathogen can survive for 30 years. On germination the chlamydospores develop mycelium on which conidia are formed in 8 hours and chlamydospores in 2–3 days. Infection is always through injured secondary or tertiary roots. There is no evidence of the fungus having ability to attack living cells of the main root of banana. Deep wounds to expose the xylem help in easy infection. Spore germination is stimulated in the vicinity of wounded root surface while it is inhibited by intact root surface. Nematodes help in exposing the roots to infection. After entry into the root, the fungus proceeds internally along the root to the rhizome where it develops extensively in vascular tissues before passing up systemically the vascular system into the pseudostem and the older leaf petioles. In tolerant or resistant cultivars, the entry of the pathogen triggers defense responses that include deposition of lignin and enzymic activities associated with it. In Panama disease, there is close relationship between zinc nutrition, plant, IAA level, tylose formation and wilt incidence. Zinc deficiency enhance wilt incidence (Fernandez-Falcon et al., 2004). In an Australian study, under field conditions, banana raised from mucropropagated plantlets were found to be more susceptible to Fusarium wilt than banana raised by conventional method.

Locally, the disease spreads mainly by contact of the root system of adjacent healthy plants with spores released by the diseased plants. Use of infected planting stock is another source of national and internation spread. Floods help in the local dispersal of the pathogen. The population of the fungus in soil increases considerably when wilted banana plants collapse and declines shortly after their removal.

Depth of soil penetration by the fungus, its survival and dispersal, and spread of the disease are influenced by soil conditions. Depth of soil penetration is more in sandy loam than in silt. Clay minerals retard the spread of the disease. Generally, light textured loam and sandy loam with an acid reaction favor more rapid disease spread than clay and silt loam with an alkaline reaction. Survival of the pathogen in field soil is best at 25% saturation. With increase in soil moisture survival ability is decreased. Appearance of banana wilt is severe under zinc deficient conditions. However, supply of zinc does not induce tylose formation which could result in less disease incidence.

Long distance spread of the pathogen is by movement of water and infected suckers. Spread within the field was by irrigation water, pruning practices, and planting of infected suckers. Mechanical wounding of roots helped infection. There was 10% disease at 17°C and 58% at 32°C. Rice rotation was very helpful in eliminating the pathogen which was mainly present in the top 20 cm soil. There is also report that the pathogen is very active in xylem at 34°C and is almost absent at 27°C.

Under the conditions in south India, fungus has been found to survive in soil amended with farmyard manure for more than 4 months. In limed soil its survival is reduced to two months. In soils where soil moisture is maintained at 20 and 100% of water holding capacity the viability is retained beyond 4 months but in soils submerged under 2-inch depth of water the survival is reduced to only one month.

Flood fallowing and soil amendments have been extensively studied to find out their suitability for eradication of the pathogen from soil and reclaiming the land for fresh plantations. In flooded fields, the fungus is presumed to be eliminated due to effect of toxins such as acetic acid and similar substances and due to low partial oxygen pressure to which the fungus is very sensitive. The carbon dioxide in submerged soils causes continued germination of conidia but prevents chlamydospore formation. Lack of these structures causes elimination of the fungus. Amendment of soil with carbohydrate sources and with plant tissues reduces population of the fungus by stimulation of chlamydospore germination, lysis of germ tubes and conidia, and inhibition of further chlamydospore formation.

Management of the disease

The control of banana wilt has been attempted in different ways. The only effective means of control is by host resistance. Application of sodium nitrate and many mercuric salts to the soil had been found to check growth of the fungus. But the method is not practicable on a field scale, especially among standing plants. In large plantation areas, near big rivers, silting of the infested land by diverting silt-laden water from the river, has been found to be a successful method of reclaiming the land for new planting. Flood fallowing has also been found effective method of reclaiming infested soil. The land is inundated under 2–5 feet of water for 6 months. Application of this method also is limited by availability of water and size and topography of the land.

Attempts to control the disease by applying fungicides to the soil and by deep ploughing to 75 cm or by weed and brush fallowing for 3–5 years have not been successful. Corm injection of 2% carbendazim plus 0.1% Agallol or Aretan soil drench has been claimed as a successful method by Lakshmanan and Mohan (1989).

Biological control: Certain actinomycetes and soil bacteria capable of producing the antibiotics musarin and monamycin and some lytic substances are found in the rhizosphere of resistant banana varieties. Addition of highly antagonistic organisms to soil around banana roots gives protection to plants against wilt. Cao *et al.,* (2004) isolated 242 actinomycetes from interior of leaves and roots of healthy and diseased banana plants and found most of them antagonistic to *F. oxysporum* f. sp. *cubense.* Their population was higher in roots of diseased plants than in roots of healthy Plants. The population in leaves of healthy and diseased plants did not differ. Ting *et al.,* (2007) have reported occurrence of the bacterium *Serratia* and the fungus *Fusarium oxysporum* as endophytes in banana roots. These endophytes are potential growth promoters of banana plant growth and render it tolerant to the wilt disease. Inoculation with an antagonistic strain of *Pseudomonas fluorescens* is reported to control the disease (Thangavelu *et al.,* 2003) through activation of defense enzymes (phenylalanine ammonia lyase, peroxidase, chitinase and glucanses) and elevated content of phenolics (Saravanan *et al.,* 2004). Sukhada *et al.,* (2004) have also reported that precolonization of banana roots by *Pseudomonas fluorescens* reduces Fusarium wilt by about 72%. The bacterium could enter the root and establish as an endophyte. The suppression of wilt is correlated with structural changes in the cortical cells, mainly with densely stained amorphous material and polymorphic wall thickenings. Massive deposition of unusual structures at sites of fungal entry was also noticed. The observations have suggested that the bacterized root cells were signaled to mobilize a number of defense structures for preventing the spread of pathogen in the tissue.

Ayyadurai *et al.,* (2006) have isolated a strain of *Pseudomonas aeruginosa* (strain FP10) from banana rhizosphere which reduced vascular discolouration caused by *F. oxysporum* f. sp. *cibense* and increases plant height. Encapsulation of banana shoot tips with the bacterium induced higher frequency of plantlet development in micropropagation.

Nei *et al.,* (2006a, b) have isolated a number of nonpathogenic strains of *F. oxysporum* including *F. oxysporum* Fo 47 from rhizosphere of health banana plants and reported that two of the nonpathogenic *F. oxysporum* isolates reduced wilt incidence by 75–87.4%. Fo 47 had no significant effect Only slight reduction in disease incidence was caused by *Trichoderma* isolates. *Pseudomonas fluorescens* strain WCS 417, known for its ability to suppress other Fusarium wilt diseases reduced banana wilt incidence by 87.4%. Colonization of roots and rhizomes by nonpathogenic *Fusarium oxysporum* isolates was studied by Paparu *et al.,* (2006). The isolates had endophytic phase and were able to establish in the hypodermis. Rhizomes were better colonized than roots but hyphal density in the tissues was higher in roots. There was more colonization of the root bases than of midsections or tips. Conditions should be created by suitable organic amendments of soil to induce development of such organisms in the root vicinity. Thangavelu *et al.,* (2004) have reported control of banana wilt by soil application of *Trichoderma harzianum.* They used dried banana leaves as the medium for mass production of inoculum of the antagonist.

Suppressive soils against *F. oxysporum* f. sp. *cubense* are known in central America since 1930. Saravanan *et al.,* (2003) have reported best control of banana wilt by basal application of neem (margosa) cake at 0.5 kg/ plant + sucker dipping in cell suspension of an antagonistic *Pseudomonas fluorescens* (*Pfm*) for 15 min + soil application of the antagonist at 10 g/plant at 3,5 and 7 months after planting. A combination of basal application of neem cake and soil application the antagonist at 3,5 and 7 months after planting was equally effective.

Natural sources of resistance exist in wild species of banana but because of infertility constraints in edible bananas these have not been exploited in breeding programmes. Cavendish bananas are resistant and Gros Michel bananas are susceptible to race 1. However Cavendish banana was found susceptible to race 4. In Taiwan, resistant clones of this cultivar of banana were isolated by somaclonal variation techniques and a resistant cultiuvar, Formosana, was released. The cultivar is much more resistant than the parent plants. In south India varieties Poovan, Moongil, Peyladen, Rajabale and Vamankeli were found resistant to wilt. Resistant banana cultivars respond to infection by initiation of tylose formation in the vessels. In an ultrastructural study of tylose formation, inoculation of a resistant plant had shown that tyloses weren formaned as extension of paratracheal parenchyma cells. By 80 days after inoculation the tyloses had almost completely filled and occluded lumina of the vessel.

A water soluble addition compound of vitamin K, menadione sodium bisulphate (MSB), first studied as a plant growth regulator, has been recently shown to induce resistance against Panama disease (Borges *et al.,* 2004). Querino *et al.,* (2005) have evaluated resistance inducer compounds ASM (acibenzolar-S-methy) and BABA (beta-amino-n-butyric acid) for reducing wilt severity. While BABA at 2100 mg/ml reduced disease severity ASM had no effect although it was capable of inhibition of conidial germination. Foliar spray od indole aceric acid (IAA) reduces wilt severity. The effect is more pronounced in zinc-sufficient soils than in zinc deficient soil. Zinc is known to abolish fusaric acid biosynthesis. In banana, it is reported not to to induce tylose formation but other events of colonization of the host tissue such as retraction of the plant plasma membrane, penetration of the fungus through plant cell wall and formation of protective layers in form of callose-like deposits are seen.

Quarantine and exclusion procedures are effective in restricting the movement of corms, suckers and soil that could carry the pathogen from infested to clean areas. Micro-propagated planting material from tissue culture planted in soil where bananas were not cultivated in the past remain disease-free for a considerable period.

In general, the integrated approach to Panama wilt management involves sanitation by immediate removal of diseased plants with surrounding soil from the field, use of healthy planting stock, care during cultivation to avoid root injury, control of nematodes, and cultivation of resistant varieties in the affected areas.

WILT DISEASE OF SUGARCANE

The wilt of sugarcane is a destructive disease in parts of India. It is only second to red rot in causing economic loss to the crop. The disease was first reported from Bihar in 1906 although a detailed study was published only in 1913. Up to mid-1940s the disease appeared to remain restricted to Bihar but now it occurs in U.P., Bihar, Haryana, Punjab, Gujarat, Tamil Nadu and Andhra Pradesh. Epiphytotics of wilt have been reported from many districts in Tamil Nadu. Outside India it is known to occur in Argentina, Bangladesh, Barbados, Columbia, Pakistan, the Philippines, Mexico, South Africa, Trinidad, USA and Zimbabwe. Often the disease occurs in association with red rot and/or borer injury, compounding the damage to the crop.

The spread of the disease to different states of India is considered to be due to unrestricted movement of diseased cane for seed, such as of the once important commercial variety Co 527. Many commercially important varieties had to be withdrawn from cultivation in north India due to this disease. In south India also many important varieties had to be withdrawn. The disease incidence in these varieties was 5-80%. The wilt disease adversely affects germinability of buds if infected canes are used as seed. cane yield and juice quality parameters reduction in yield May be as high as 65%.

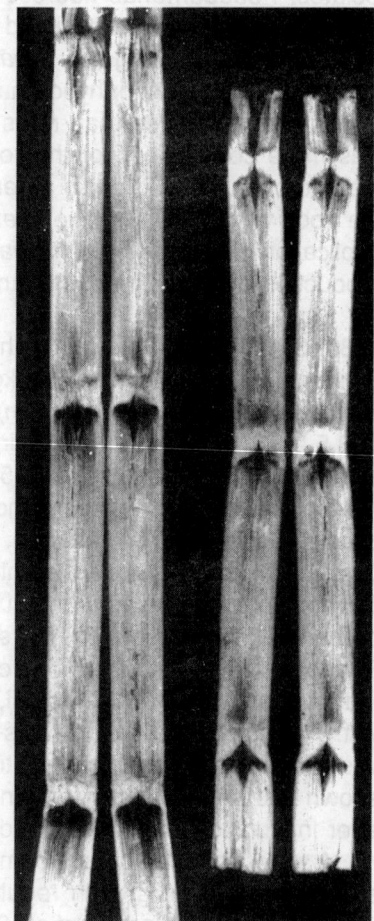

Fig. 125. Sugarcane wilt. Early internal discoouration.

Symptoms

The earliest symptom of the disease is stunting and unthrifty appearance of a few isolated plants or small groups of plants. This is usually noticed when the plants are about half grown. Root system, however, does not reveal damage or discoloration attributed to the disease. The most striking symptom is the yellowing and/or withering of the top when the crop is getting ready for harvest, followed by rapid drying of the canes. The canes become light and hollow.

When canes are split open in the early stage of disease development, the tissues, particularly of the lowest internode, have a brick red color with individual vascular strands a dark red. There are no white transverse bands as seen in the red rot disease. The reddening

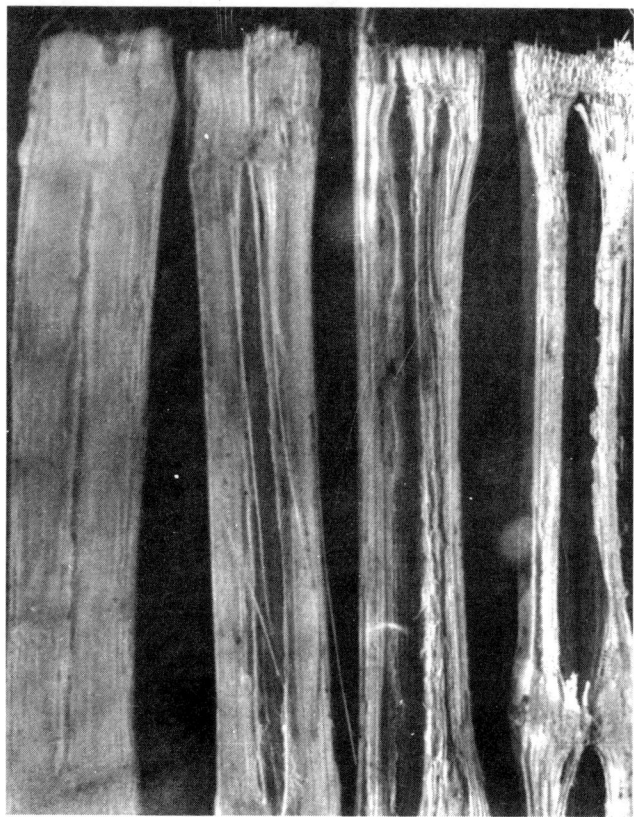

Fig. 126. Sugarcane wilt. Advanced stage of rotting of pity.

may be confined to a few joints or extended to the whole length of the cane. Brown patches on the rind are formed where the underlying tissues are dead. In the reddened pith fungus hyphae ramify throughout the cells in all directions. Gradually the pith becomes hollow and in the hollowed portion there is usually a copious grayish white, fluffy growth of hyphae, bearing large numbers of conidia at a later stage.

Sometimes, the canes may not show drying of the top and discoloration of the pith but reddening of the vascular strands may be present. Adventitious roots which sometimes develop from the above-ground nodes of affected canes as also the underground roots may redden and die when the stem tissues die. No discoloration of the roots is noticed in the earlier stages of the disease.

In the initial stages, the walls of the vessels are darkened and fungal hyphae are found inside. Later, gums appear in the lumen of the vessels, in the intercellular spaces of the parenchyma, and inside the cells in close proximity to the vessels.

The causal organism

The causal organism of sugarcane wilt was first described as *Cephalosporium sacchari* Butler. The basis of calling it a *Cephalosporium* was that it produced 1-celled conidia in false heads. But septation of conidia before their germination was reported in the first description also. From

Fig. 127. Sugarcane wilt. Foliage symptoms.

several countries association of *Fusarium moniliforme* var. *subglutinans* with sugarcane wilt was also reported and it was concluded that the actual pathogen was *Fusarium moniliforme*. Using a medium selective for *F. moniliforme* the fungus was found present in internal tissues near the nodes as soon as the earliest symptoms of wilt start appearing. In 1975, W. Gams coined a new combination as *Fusarium sacchari* (Butler) W. Gams to which both *C. sacchari* and *F. moniliforme* var. *subglutinans* were made synonymous. Subsequently, the fungus was divided into two varieties, *F. sacchari* var. *sacchari* and *F. sacchari* var. *subglutinans*. The variety *sacchari* produces mostly aseptate conidia in the aerial mycelium, no sporodochia, and bright red color on boiled rice grains while the variety *subglutinans* produces 1–3 septate conidia, macroconidia more commonly formed in sporodochia.

Isolations from wilted canes of Co. 527 had yielded 50–60% *F. moniliforme* and 5–20% *Acremonium terricola* (*A. implicatum*) and *A. furcatum* each. Pathogenicity was proved by artificial inoculations. Thus, sugarcane wilt appears to be caused by a number of fungi including two varieties of *F. sacchari* and two species of *Acremonium*. *Acremonium furcatum* was later renamed as recifei (Leao and Lobo) W. Gams.

The mycelium of the *Fusarium* associated with sugarcane wilt consists of effuse, white, sparsely septate creeping hyphae. The conidiophores are aseptate, tapering towards the apex, simple or verticillately branched. They measure 6–30 x 3–4 μm. Conidia are numerous, arising in succession at the apex of the branches and collect in false heads but separate easily. These conidia are hyaline, ovoid or oblong, without septation, and measure 4–12 x 1–3 μm. On the basis of growth habit, mycelial color, and degree of sporulation the fungus has been divided into several groups.

Disease cycle and environmental relations

The sugarcane wilt pathogens are primarily soil-inhabitants. *C. sacchari was* reported to survive in soil for 27–31 months. Later, there were reports that the wilt pathogen could colonize dead or living sugarcane tissues in soil but did not have strong competitive saprophytic ability. The nature of the crop permits uninterrupted survival of the pathogen in sugarcane fields on crop debris and also through seed setts. The infected seed setts permit growth of the pathogen in the shoots. Infection of healthy seed setts after planting mostly occurs through the cut ends when the pathogen grows in to the cane parenchyma. There have been reports that the fungi could enter the setts via root primordia and leaf scars. Wounds at underground parts of the plant such as those caused during Intercultural operations and those caused by insects and nematodes facilitate entry of the pathogen. After infecting the plants the fungus proliferates in the tissues particularly the vascular bundles where it can live for long periods without causing death of the plant or producing visible symptoms. Generally, the symptoms are expressed when the plant is put to some stress such as drought. While infected seed setts are the chief means of transmission of the pathogen, the soil-borne inoculum can also be disseminated by irrigation water, wind, and insects.

The incidence of wilt is influenced by soil conditions such as moisture, pH, composition, etc. Negligible wilt incidence is reported when the soil moisture is 8.1–9.1% and high incidence when moisture level drops to 3.5%. High C : N ratio in soil favors wilt incidence. In Bihar the wilt incidence is high in soils having neutral or slightly alkaline reaction while in Andhra Pradesh it is reported to be favoured by acid soils fertilized with heavy doses of nitrogen.

Management of the disease

No satisfactory method is known to destroy the soil- or seed-borne inoculum. Precautions such as destruction of crop debris, use of healthy seed pieces, proper maintenance of soil moisture and fertility can reduce incidence of the disease. Seed cane should be taken from apparently healthy crop without borer infestation. The seed pieces should be examined at both cut ends for any discoloration. Diseased crop should not be ratooned. In case of wilt sick soils a 3-year rotation has been recommended. Treatment of setts with 0.1% Agallol or other fungicides reduces superficial inoculum and protects the cut ends from infection by soil-borne inoculum. Sett treatment with 0.5% Agallol (MEMC) for reducing wilt incidence was recommended in 1992. Experiments have shown that application of boron or manganese to wilt sick soils reduces wilt incidence. Similarly, sett treatment with 40 ppm boron also helps in reducing disease incidence.

Many species of *Streptomyces* and *Bacillus* are natural antagonists of the wilt pathogens. Soil application of *Trichoderma* and *Aspergillus* with *Pseudomonas aeruginosa* is reported to suppress sugarcane wilt in the field. Resistant cultivars are reported to be CoS 767, Co 86027, Co 87010, Co 89014, Co 91010, Co 92003. Co 92009. Co 92023 and CoS 8436. Cultivars Co 62198, 8128, 85003, 85010, 86250, 87016, 87022, 87023 and 88029 show moderate resistance. Co canes 356, 370, 393, 395, 859, and 1158 had shown good tolerance to wilt.

CHARCOAL ROT OF SOYBEAN AND OTHER CROPS

Charcoal rot or summer wilt appears in hot, dry weather or when plant growth is checked by unfavorable environmental conditions. The causal fungus is widely distributed in soil and the disease is of worldwide occurrence. In tropical countries, where the disease causes a blight of seedlings also, plant losses up to 77% have been reported. The disease occurs in a wide range of crops including sorghum, maize, groundnut, cotton, potato, soybean, green gram, eggplant (brinjal), sugarbeet, jute, cowpea and watermelon.

Symptoms

Infected seedlings can show reddish brown discoloration at the emerging portion of the hypocotyl. If infection occurs through the root, the discoloration is evident at the soil line and above. The discolored area turns dark brown to black and infected seedlings may die in hot and dry weather. If wet and cool weather persists, infected seedlings survive but carry the latent infection and symptoms may reappear later when hot and dry weather returns or when the seedling has been exposed to some stress.

In older plants, the 'charcoal rot' phase appears after mid-season. The charcoal rot is characterized by light brown discoloration of the subepidermal tissues in the taproot and lower part of the stem. Infected plants at first do not show above ground symptoms. In advanced stages, leaves turn yellow and wilt, but remain attached to the plant. Superficial stem lesions may extend from the soil line upward. When the epidermis is removed, black sclerotia of the fungus may be so numerous as to give a grayish-black color to tissues. The sclerotia resemble a sprinkling of finely powdered charcoal. When split open the taproot and base of the stem show black streaks in the woody tissues and frequently sclerotia formed in the pith of the stem.

The causal organism

The disease is caused by the pycnidial fungus *Macrophomina phaseolina* (Tassi) Goid. Its synonyms are *M. phaseoli* (Maubl.) Ashby, *Rhizoctonia bataticola* (Taub.) Butler, *Sclerotium bataticola* Taub. and *Botryodiplodia phaseoli* (Maubl.) Thir. The fungus belongs to class Coelomycetes, order Sphaeropsidales, and family Sphaeropsidaceae of the Deuteromycotina. In nature, mostly the sclerortial form is common.

M. phaseolina is highly variable, differing in size of sclerotia and the presence or absence of pycnidia. The mycelium is superficial or immersed, hyaline to brown, branched, septate, often tree-like in form (dendroid). The pycnidial stage is uncommon on soybean but is widespread on jute and garden beans. Some strains produce pycnidia in culture media and others can be induced to do so with special techniques.

In cultures, sclerotia are most common. On the host these structures are formed within the roots, stems, leaves and fruits. They are jet black, smooth, hard, round to oblong or irregular in shape. They measure 100 μm to 1 mm in diameter (in culture 50 to 300 μm). However, size is highly variable within an isolate. Pycnidia are dark brown, solitary or gregarious on leaves and stem, immersed but becoming erumpent. They are more or less globose, membranous or sub-carbonaceous and become black with age. They measure 100–200 μm in diameter and open by a small truncate ostiole. Conidiophores (phialides) are hyaline, short obpyriform to cylindrical and develop from the inner wall of the pycnidium. They measure 5–13 x 4–6 μm. The conidia (pycnidiospores), cut off at the tip of conidiophores, are hyaline, ellipsoid to obovoid, and 14–30 x 5–10 μm. They are single celled. Host specialization in the fungus was reported by Su *et al.*, (2001). Isolates from maize colonize maize roots better than roots of soybean, sorghum or cotton. However, no such specialization is found in isolates from these latter crops.

Disease cycle and environmental relations

Sclerotia of the fungus can survive free or embedded in host tissues in soil. At least 75% of the sclerotia of *M. phaseolina* were found to survive for 1 year in moist natural soil with

50–55% of the moisture holding capacity when kept at 26°C. Sclerotia production is stimulated by low soil moisture (10% WHC). While no sclerotia are produced at high soil moisture, survival at 70% WHC in colonized crop residue is reported. Moisture and temperature effects are interrelated. Soil pH has little or no effect on sclerotial survival. Survival is longer in dead host tissue than in free state in the soil. In maize and sorghum stalk residues, the sclerotia are reported to survive for 18 and 16 months, respectively. In bean, the fungus is seed–borne. Seeds from plants with no disease symptoms contain 2,5% infection while seeds from severely infected plants have up to 13.5% infection.

Dry soils prolong survival. In very wet soil sclerotia cannot survive for more than 7–8 weeks while the mycelium cannot survive for more than 7 days. In wet soil viability of sclerotia is reduced to zero by a constant temperature of 55°C and above for 1 day or 50°C for 3 days. Exposure of sclerotia to 60°C cycle for 2 hours for 7 days also completely destroys them. High C : N ratio amendments, low bulk density of soil and oxygen concentration above 16% also favor survival. The fungus is a poor competitor in soil but readily colonizes plant debris.

Sclerotial populations of the fungus in soil increase when hosts are grown continuously in the same field and the disease becomes more severe in the succeeding crops. Higher counts of sclerotia in soil are found in the intercropping systems of sorghum and cowpea with pigeonpea than in single cropping systems. The highest counts of sclerotia were recorded in plots where sorghum was intercropped continuously with pigeonpea in both rainy and post-rainy seasons. An increase was also recorded when rainy season sorghum was followed by either safflower or chickpea. The fungus is highly susceptible to competition by *Aspergillus* and *Penicillium*, especially *A. fumigatus*.

Large number of seeds may carry the pathogen in the seed coat, particularly in the tropics. Infected seeds either fail to germinate or produce seedlings which may die soon after emergence. The seed-borne inoculum becomes important when there is intra-embryonic infection.

The disease is not evident at low temperature although growth of the pathogen may commence. Symptoms appear at temperatures of 28°–35°C. Seedling blight is seen in tropical countries only where soil temperatures are 30°C or above at planting time. Low soil moisture further enhances disease severity. Drip irrigation contributes to higher disease incidence than furrow irrigation (Nischwitz *et al.,* 2004).

Sclerotia germinate on the surface of roots and produce numerous germ tubes. Penetration of roots generally occurs from appressoria formed over the epidermal cells or natural openings. The fungus hyphae grow inter- and intracellularly through the xylem where they form sclerotia and plug the vessels. Sclerotia are not formed on young plants. Pathogenesis of the pathogen probably involves mechanical plugging of vessels by sclerotia, toxin production, and enzymatic action.

Management of the disease

In summer crops, irrigation lowers soil temperature and increases soil moisture. These conditions are unfavorable for the disease. Most efforts on control of *M. phaseolina* involve management of populations of microsclerotia. Crowding of seedlings should be avoided because this decreases vigor of the seedlings thus exposing them to infection. Proper nutrition for the crop must be provided to maintain plant vigor. Where soil becomes dry and temperature rises the field should be irrigated. The field may be flooded for 3–4 weeks before planting. Application of organic matter for decomposition in the soil, several weeks before planting, reduces soil inoculum. Long rotations between soybean crops reduce incidence and severity of the disease. Maize, sorghum and cotton are host of the fungus but they support lower populations of sclerotia in soil than soybean. Cultivation of these crops after or before soybean reduces disease incidence in the latter. Application of neem (margosa) seed cake at the rate of 150 kg/ha and farmyard manure at the rate of 10 t/ha is reported to significantly reduce charcoal rot in cowpea.

The antagonistic fungi *Trichoderma viride* and *T. harzianum* cause reduction in disease incidence by 50%. Pelleting of cowpea seed with *T. viride*, either alone or in combination with carbendazim, inhibits growth of the fungus, *in vitro*, improves seed germination and reduces post–emergence mortality. In field trials, *T. harzianum* is reported to reduce charcoal root rot of melons and corn by 22% and 28%, respectively. In addition, *T. harzianum* seemed to enhance plant growth and earlier development of fruit in melon. Talc-based formulation of *T. viride* or *T. harzianum* used as seed treatment (4 g/kg seed) reduce incidence of charcoal rot of sunflower and population of *M. phseolina* in the rhizosphere with concurrent increase in the population of the antagonist (Suriachandraselavn *et al.*, 2004). A bioformulation of *Pseudomonas fluorescens* Pf1 with chitin was reported by Sarvanakumar *et al.*, (2007) that reduced mungbean root rot caused by *M. phaseolina*. The treatment increased accumulation of defense enzymes peroxidase, phenylalanine lyase, polyphenol oxidase and chitinase. Farm practices which increase residue destruction immediately after harvest or those that enhance population of *Trichoderma* spp. may directly or indirectly lower the relative longevity of *M. phaseolina* and other soil-borne pathogens (Baird *et al.*, 2003). *Trichoderma* spp. help in disintegration of the crop residue. Fluorescent pseudomonads, *Paecilomyces lilacinus* and *Pseudomonas aeruginosa* are also reported as biocontrol agents against *M. phaseolina*. Chitinolytic strains of *Pseudomonas, Pantoea* and *Enterobacter* are reported to degrade and utilize the mycelia of *Macrophomina phaseolina*. *Pseudomonas fluorescens* is reported to control charcoal rot of chickpea. Inoculation of soybean seeds or roots with *Rhizobium japonicum* reduces disease severity. In peanut, siderophore-producing isolates of *Rhizobium meliloti* are antagonistic to *M. phaseolina* and can suppress charcoal rot. Rhizobacteria such as *Pseudomonas putida* and *Paenibacillus polymyxa* promote colonization of roots by AM fungi and give protection against *M. phaseolina*. *Bradyrhizobium* is also reported to be antagonistic to *M. phaseolina* in peanut. The rhizobitoxine obtained from culture filtrate of the *Rhizobium* has been shown to be toxic for *M. phaseolina*. In *Macrophomina* root rot of chickpea application of phosphate solubilizing microorganisms including arbuscular mycorrhizal fungus *Glomus intraradices, Pseudomonas fluorescens, Aspergillus* provides biocontrol of the disease (Siddiqui and Akhtar, 2007).

Lodha *et al.*, (2005) had obtained control of *Macrophomina phaseolina* by solarization or natural heating of irrigated soil amended with crucifer residue. Earlier, it was reported that although solarization markedly increased soil temperature it did not reduce soil inoculum at any depth. Lodha and Aggarwal (2002) found highest suppression of *M. phaseolina* inoculum by composting the crop residue amended with 4% urea-N and placed at 60 cm depth. The maximum suppression of inoculum was caused by pearl millet residue compost and highest yield promotion of cluster bean (*Cyamopsis tetragonoloba*) was given by cauliflower leaf compost. Soil application of benomyl or captan at 20 μg a.i./g soil appreciably reduce populations of sclerotia. Susceptibility of crops is related to the amount of carbohydrates released by seed. Susceptible plants release more carbohydrates in soil from seeds and this leads to pre-emergence damping off.

RHIZOCTONIA STEM CANKER AND BLACK SCURF OF POTATO

The disease is of worldwide occurrence. In India it was first reported in 1912-1915 and now affects potato tubers wherever the crop is grown in the country. The major damage to the crop is through the stem canker phase which reduces yield of tubers. The black scurf causes qualitative damage as it decreases the marketability of tubers both for table purposes and for seed. In seed tubers with 50% surface area affected, germination is reduced by 11%, number of stolons by 70%, yield by 68% while post emergence wilting is increased to 27% and number of necrotic stems to 99%.

Symptoms

There are two distinct phases of the disease-the stem canker and blight phase and the black scurf of tubers phase. In stem canker. when the seed tubers sprout the growing tip is particularly susceptible to the fungus and may be killed before emergence. The dormant bud near the base then sprouts to grow but may be attacked and killed. If favorable conditions continue, this may continue until there is no emergence from such tubers. But if conditions become unfavorable the later buds may produce sprouts that emerge and grow. However, the delayed emergence affects yield of tubers. Such stunted, yellowish and withering hills can be seen in the field. Normally, as the plant grows and tissue hardness sets in there is less infection. But even on growing plants, during the early part of the season, sunken or shallow brown cankers may develop. When stem cankers are serious, the plant may show some indirect effects of the attack. Due to damage to the stem normal downward flow of carbohydrates is disturbed. As a result carbohydrates accumulate in the tops. This

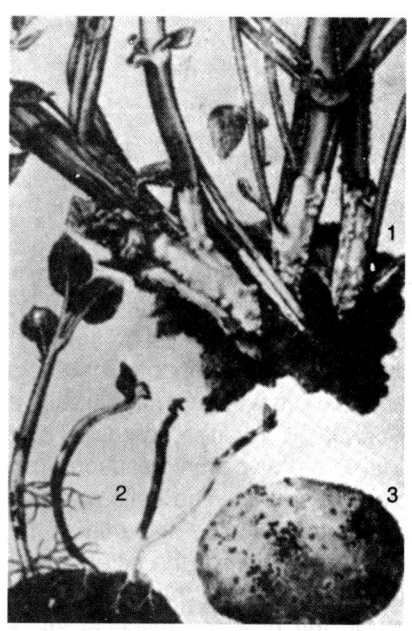

Fig. 128. Rhizoctonia stem canker of potato.

causes stunting, rosetting. enhancement of purpling due to excessive anthocyanin formation, and development of aerial tubers in the axes of branches and petioles.

The black scurf of tubers appears as black crust on the skin due to formation of sclerotia of the fungus. These sclerotia are normally skin deep and do not cause any damage to the tuber inside. However, if there is insect injury to the tuber the fungus may grow into the injured tissues and may induce rot. Black scurf is more common than stem canker in India.

The causal organism

Mycelial and sclerotia stage (anamorph): *Rhizoctonia solani* Kuhn.
Basidial stage (teleomorph): *Thanatephorus cucumeris* (Frank) Donk.

The mycelial stage producing sclerotia is the commonly encountered stage in India. In artificial culture media, the fungus produces a fast growing stout mycelium with septate hyphae. Young colonies may look white but older mycelium is invariably some shade of brown (no other color,

Fig. 129. (Fig 138 Page 497 of 8E)

a distinctive feature). Sclerotia may or may not be formed. In some isolates the sclerotia are light brown while in some they are so dark as to look black. The diameter of hyphae ranges between 5 and 14 μm. The septa are typically of dolipore type. The lateral branches from the main hypha are invariably constricted at the point of origin and a septum occurs in the branch near the junction with the main axis. In young advancing hyphae the branching almost invariably occurs near the distal septum but it may occur at any point in older hyphae. In many isolates the mycelium produces monillioid or barrel-shaped cells which have been termed as chlamydospores. The cells are as a rule multinucleate (in other species of *Rhizoctonia* they are binucleate). The sclerotia of the fungus are distinct from many other fungi in that there is no differentiation of the sclerotial tissue into a rind and internal medulla although outer cells may be darker and thicker walled.

On the host the mycelial stage of the fungus produces superficial, irregular, scab-like, black sclerotia growing on the surface of the usually subterranean parts of the plant. This is preceded or accompanied by a superficial, dark colored, short-celled, abundantly branched, stout mycelium. This mycelium is entirely different in appearance from the mycelium inside the host tissues which is slender, hyaline, and longer celled. No other spores are produced in the asexual stage of the fungus.

The basidial stage is in fact inconspicuous, mostly saprophytic stage of the pathogen. It appears as a flaky pellicle on the surface of the substrate or leaves near the ground level or dead parts of the stem under conditions of high humidity. The basidia are borne on small imperfectly symmetrical cymes. They are barrel-shaped, obpyriform or clavate and measure 9–25 x 5–12 (mostly 14 x 9) μm. They bear 4 sterigmata which arise as blunt knobs and later become horn shaped. On an average these sterigmata measure 13 μm in length and 3 μm broad at the base. Basidiospores are ellipsoid or oblong–ellipsoid, flattened on one side, hyaline, truncate, and measure 5–14.5 (9) x 4--8 (average 9 x 5.5) μm. These thin-walled spores are capable of producing secondary basidiospores of the same shape by repetition on germination.

In India at least 8 strains of the fungus on potato have been identified on the basis of cultural and sclerotial characters. The species is considered an assemblage of many

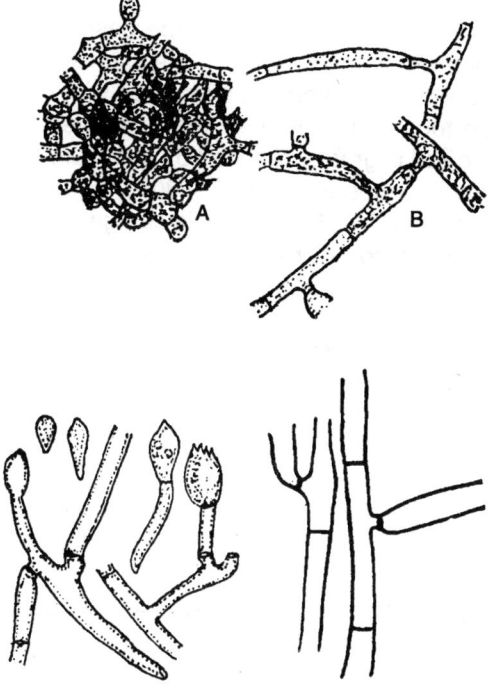

Fig. 130. *Thanatephorus cucumeris (Rhizoctonia Salani).*

morphologically and pathologically distinct strains or groups based on anastomosis behaviour. Hyphae of isolates representing the same anastomosis group (AG) are attracted to and connect, or fuse, with one another, whereas isolates from different anastomosis groups do not exhibit this behaviour. Although occasionally heterokaryon hyphae may be produced from one of the anastomosing cells, mostly 5–6 cells on either side of the fusion cells shrink and die indicating somatic incompatibility between isolates of the same anastomosis group. This limits outbreeding to a few compatible pairings. The anastomosis groups show some host specificity. Isolates of many of the 13 anastomosis groups are currently known to be associated with potato. However, AG-3 is is the principal cause of stem canker and black scurf of potato. Ag 2–1 and Ag 5 also infect potato but pruce only ligh lesions and little or no black scurf.

Most isolates from the host are AG-3 while lesser number of isolates from soil are AG-3. The AG 5 may also be associated with potato. Anastomosis group 1 (AG-1) causes seed and hypocotyl rot, web blight and sheath blight. AG-2 causes a canker of root crops, mostly crucifers. AG-3 affects mostly potato causing stem canker, stolon lesions and black scurf. AG-4 causes seed and hypocotyl rot on almost all angiosperms and stem lesions near the soil line in most legumes, cotton and sugarbeet. These groups have same significance as physiologic races in the breeding for resistance to *R. solani*.

Disease cycle and environmental relations

The fungus survives in soil or on seed tubers as sclerotia. It can survive on underground parts of weed plants growing in potato fields without any symptoms. This may increase the quantity of soil-borne inoculum. Some investigators regard generally low level of virulence from tuber-borne sclerotia while some claim that isolates from stem lesions are less virulent than isolates from stolons, tuber-borne sclerotia, and hymenia in the basidial stage. Black scurf incidence and disease symptoms on underground parts of the plants are significantly higher when inoculum is present in both seed tubers and soil compared with either of them separately. The level of disease intensity on tubers is correlated with the level of inoculum in soil and seed tubers. Viability of sclerotia in soil is greatly reduced by exposure to hot summer temperatures. The fungus requires an optimum temperature of 25°– 30°C for growth on culture media with the minimum at 8°C and maximum at 31°– 35°C. Sclerotia can germinate at 8°– 30°C with optimum at 23°C. The optimum for germination of basidiospores is 21°–25°C. However, the optimum temperature for infection and lesion development on potato is 18°C. The damage to the host is less at 24°C.

The fungus first attacks the clean and healthy sprouts. Initially the hyphae grow predominantyly in the longitudinal direction of the sprouts (runner hyphae). They tend to follow the junctions between the epidermal cells. They form side branches mainly half-way of the subterranean parts of the sprouts. The branches having short swollen cells form infection cushions Lesions develop only underneath the infection cushions. A few days after lesion formation, the depth of the lesions could reach the vascular elements. A few days after the appearance of lesions the pathogen forms brown, non-infective mycelium on and in the circumference of these lesions.

Colonization efficiency of *R. solani* as saprophyte and parasite in soil is influenced by soil structure. In high bulk density of clay and reduction in aggregate size reduces the number of large pores. The fungus colonies move slowly through small pores. Soil moisture is regarded as secondary to soil temperature in the development of stem canker. The black scurf on tubers develops more in sandy soil than in clay. However, this depends on the isolate of the pathogen. Some isolates survive equally well in sandy and clay soils and some in clay but poorly in sandy soils. Infection of roots by the cyst nematode *Globodera rostochiensis* enhances infection of stolons by *R. solani*.

Management of the disease

Tuber-borne inoculum is more important than soil-borne inoculum. The seed tubers should be scurf-free. In the past, organomercurials were effectively used for tuber treatment. Since mercurial compounds have been discontinued for use in plant protection attempts have been made to replace this treatment with safer chemicals. Dipping of tubers for 20 min in 1.75% sulphuric acid or 3% boric acid has been found very effective in suppressing sclerotia on the tuber surface. Boric acid can be applied as spray with a hand spray on the seed tubers. Pre-plant seed tuber treatment with a combination of sodium hypochlorite and thiophanate methyl is highly effective in controlling seed-borne black scurf. Errampalli and Johnston (2001) used sodium hypochlorite solution (500 ppm) for 8 min followed by thiophanate methyl (50 g a.i/100 kg seed) treatment and found significant reduction of *R. solani* as well as common scab. Similar results were obtained by tuber treatment with chlorine dioxide followed by thiophanate methyl (Errampalli *et al.*, 2006). Thind *et al.*, (2002) have compared a number of fungicides for seed tuber treatment against black scurf. Pencycuron, a phenylurea-based fungicide, causes total suppression of mycelial growth of *R. solani*. Its formulations Monceren 250 SC and Monceren 25 WP provide almost complete control of black scurf when used as seed tuber dip treatment at 0.5% and 0.75% concentration. The control is better than Emisan (an organo mercurial). Soil application of the fungicides is also effective. Leaf extract of *Salvia indica*

(Labiateae) is capable of suppressing growth of *Rhizoctonia solani* by more than 50%. It has potential for a seed treatment alternative.

Larkin and Griffin (2007) have evaluated different crucifers for use as green manure to control soil-borne pathogens of potato that include *R. solani, Spongospora subterranean, Streptomyces scabies.* Green manure of rape and canola proved to be most effective in suppressing *R. solani* while. Indian mustard was most effective against powdery scab and common scab. The suppression is attributed to sulphur compounds in the Brassicas including glucosinolates that release isothiocyanates on hydrolysis. Use of Indian mustard as a cover crop or addition of leaves andf stem of mustard ensure health of the soil and potato roots, suppressing soil borne pathogens. The proportion of healthy roots, indicated by white roots, is high in potato grown in such soil (Snapp *et al.,* 2007).

Yulianti *et al.,* (2007) have studied the effect of decomposition of green leaves of *Brassica nigra* and *Diplotaxin tenuifolia* (sources of glucosinolate) on survival of *Rhizoctonia solani* AG2-1 that causes damping off of canola. In nonamended soil the pathogen survived for up to 6 months at soil temperatures of 10 and 20°C. and soil holding capacity of 10%, 40% or 70%. At 30°C its survival was reduced to nondetectable level in the first week. Amendment of soil with green manure at 1% caused increased activity of *Rhizoctonia* and temporary increase in soil microbial activity. Amendments at 5% or 10% concentration suppressed *R. solani* and significant increase in soil microbial activity. The efficiency of hydrolysis of glucosinolates in the 5% treatment in the first week ranged from 1.6% for 2-propanyl isothiocyanate (*Brassica nigra* treatment) to 3.4% for 3-butenyl isothiocyanate (D. tenuifolia treatment). The propenyl isothiocyanate was not detected after 7 days of amendment.

Application of 20–30 kg/ha Brassicol (PCNB) is very effective as chemical soil treatment but is costly. As far as possible, a soil treatment should be combined with a seed treatment to get best results in badly infested fields.

The lack of resistant varieties and high cost of chemical soil treatments has prompted greater attention to control by cultural practices including indirect and direct biological control. *Rhizoctonia solani* is susceptible to antagonists in soils having good amount of decomposable organic matter. *Azotobacter chrococcum*, applied to soil, effectively prevents infection of potato seed tuber sprouts from infection of *R. solani* at 15°–25°C but not at 10°C. The fungal hyperparasite *Verticillium biguttatum* strrongly suppresses *R. solani* sclerotia in soil at 15° and 20°C but not at 10°C. This hyperparasite develops biotrophically in the host hyphae, drawing nutrienyts imncluding vitamins for growth. This causes a nutrient sink in the pathogen colony and suppression of sclerotia formation. Seed inoculation with the hyperparasite is also effective. A combination of *V. biguttatum* and *Azotobacter chrococcum* provides protection to sprouts, stems and stolons. Field control of *R. solani* on potato is reported by inoculation of seed tubers with *V. biguttatum, Gliocladium roseum* and *Azotobacter chrococcum* singly or in mixture. Combination of the fungicide pencycuron and the ycoparasite *V. guttatum* enables the reduction in quantity of pencycuron to 20% of the recommended dose and increases the biocontrol efficiency of the antagonist. *Microsphaeropsis* sp. (strain P130A) suppresses the germination and production of sclerotia of tuber-borne *R. solani.* Antibiosis and mycoparasitism both are involved. The antagonist induces rupture of the pathogen plasma membrane and grows in it. Seed tuber treatment with *T. harzianum* or *Gliocladium virens* reduce disease incidence in the field by 50% and 55%, respectively.

Increasing the organic matter content of the soil through green manuring or application of oil-cakes or sawdust has given good control of the disease. In 1972, control of the disease through organic amendment was reported. Amendment with neem (margosa) oil cake (25 Q/ha) or sawdust (25 Q/ha) followed by application of 120 kg N through urea in the latter case gave excellent control of black scurf with high increase in yields of clean tubers. The amendments were applied 3 weeks before planting. This was first report of suppression of *R. solani* on potato by neem cake or sawdust. Similar observations were subsequently reported by many other workers in India and other countries. The decomposing wood sawdust supports heavy growth of such antagonistic fungi as *Trichoderma*. Such fungi and other microorganisms including bacteria and mycophagous nematodes which are encouraged by decomposing oil-cakes and sawdust bring about decline of the pathogen in soil. Microorganisms in composted hardwood bark are reported to suppress *Rhizoctonia* in potting medium. Plant growth promoting rhizobacteria applied to seed tubers extend to roots and new tubers and protect the tubers from *Rhizoctonia* as well as soft rot bacteria.

Pythium oligandrum is a mycoparasite of such soil-borne plant pathogens as *Pythium ultimum, P. aphanidermatum, Fusarium oxysporum, Verticillium albo-atrum, Phytophthoraq megaseprma* and *Rhizoctonia solani* (Benhamou *et al.*, 1999). *Verticillium biguttatum* is a *Rhizoctonia*-specific biocontrol agent. It is compatible with selected chemical control systems. It may improve control efficiency in combination with *Rhizoctonia*-specific fungicides such as pencycuron and flutalonil. Its efficiency is not affected by other biocontrol agents like *Trichoderma, Gliocladium* and *Pseudomonas* (van den Boogert and Luttikholt, 2004). *Verticillium biguttatum* produce the enzymes chitinase and β-1,3-glucanase. These enzymes play a role in dissolving and penetrating cell walls of *R. solani* (McQuilken and Gemmell, 2004). The biocontrol agent can destroy sclerotia in about 6–8 weeks provided that (i) a direct contact between sclerotia of *R. solani* and conidia of *V. biguttatum* is obtained, (ii) the temperature during storage period is at least 15°C but preferably closer to 20°C during the first week and (iii) relative humidity of the air between the tubers is at least 99%. Brewer and Larkn (2005) have compared the efficacy of a large number of biocontrol agents including *Bacillus subtilis, Paenibacillus polymyxa, Pseudomonas fluorescens, Laetisaria arvalis, Verticillium biguttatum, Stillbella aciculosa, Penicillium* sp., *Trichoderma* sp. and *Rhizoctonia zeae* for their efficiency against stem canker and black scurf. *B. subtilis, R. zeae* and *S. aciculosa* and a chemical control (azoxystrobin) were effective in reducing stem canker severity by 40–49%. *L. arvalis, R. zeae* and azoxystrobin reduced black scurf by 54–60%. Combination of *B. subtilis* and *Trichoderma virens* gave somewhat better control of stem canker than each organism alone. Transgenic potato plants having the endochitinase encoding gene from *Trichoderma harzianum* show resistance to *Rhizoctonia solani* and *Alternaria solani*. Presence of *T. harzianum* in soil reduces black scurf on tubers per plant and also the proportion of small tubers (upto 20 g) but the number of tubers per plant is reduced (Wilson *et al.*, 2008).

Grosch *et al.*, (2005) have emphasized the high potential of endophytic bacteria as biocontrol agents against *R. solani*. They evaluated effectiveness three potato-associated ecto- and endophytically living bacterial strains, *Pseudomonas flueorescence* (B1 and B2) and *Serratia plymuthica* (B4). All the three reduced disease severity on potato sprouts when used for seed tuber bacterization. Best was the isolate B1, followed by B2 and then B4. Up to 78% and 85% control of Rhizoctonia canker of potato in glasshouse and fields with infested soil by binucleate *Rhizoctonia* isolates is reported.

Arbuscular mycorrhizal fungi are effective in reducing mortality due to *R. solani*. In micropropagation of potato, when plantlets are transplanted in mycorrhized soil and challenged with *R. solani* there can be 26 to 77% reduction in mortality depending on potato cultivar and species of the arbuscvular mycorrhizal fungus. The mycorrhizal fungi enhance accumulation of the phytoalexins rishitin and solavetivone which normally results in stressed situations for the potato plants. The mycorrhizae enhanced accumulation only when the plant is challenged by *R. solani* (Yao *et al.*, 2002, 2003).

Another approach to biological control of *R. solani* is the introduction of hypovirulent individuals of the pathogen. The fungus suffers from a decline problem caused by some infectious double stranded RNAs which reduce the survival and pathogenic abilities of the pathogen. When such individuals are introduced into a virulent population the RNA infects the latter during anastomosis. Isolates of *R. solani* AG3 from different scurf lesions on the same tuber show similar dsRNA profile with compatible hyphae but some isolates from same field on different tubers show somatic incompatibility (Robinson and Deacon, 2002). Charlton and Cubeta (2007) have reported horizontal transmission of the dsRNA between incompatible isolates of AG3 but the virus is not stable in the recipient mycelium and is lost in subcultures. Bharathan *et al.,* (2005) have studied the occurrence, distribution, and genetic relatedness of dsRNA components in nine anastomosis groups AG-2 to AG-13 and has reported considerable sequence heterogeneity within each isolate or isolates from same AG. According to Herr (1995) in the biocontrol of *R. solani* by hypovirulent strains or species of binucleate *Rhizoctonia* neither mycoparasitism nor antibiosis is involved. Postulated mechanisms include induction of systemic host resistance and/or competition for recognition and invasion sites or nutrients.

In many crops, not potato, mulching of the soil with a 2–3 cm thick layer of rice husk or with polethylene sheets has proved effective. Plastic mulch increases the density of rhizosphere, rhizoplane and endoroot microflora which provide resistance against the pathogen. Rotation has limited value. A 2–3 year rotation has been recommended. In the USA, Larkin and Honeycult (2005) studied the effect of 8 different 3–year rotations on soil microbial communities and health of potato crop. Crops included in the rotations were soybean, barley, sweet corn, green bean, canola and clover. Each crop in a sequence preceding potato had distinctive effect. Most rotation reduced stem and stolon cankers and black scurf. Potato following canola, barley or sweet corn provided best control of the Rhizoctonia disease. Shallow planting helps in reducing the damage to sprouts. Fertilizers containing sulfur (aonium sulphate, ammonium thiosulphate) applied 11 days after emergence reduce black scurf on potato tubers (Pavlista, 2005).

SHEATH BLIGHT OF RICE

The disease has been known since 1910 in the East and South East Asian countries but was considered a minor leaf disease (Ramakrishnan, 1971). It was known as oriental sheath and leaf blight. However, now it has increased in importance worldwide causing serious losses in both temperate and tropical rice producing countries including Brazil, Venezuela, Surinam, Madagascar, USA, Fiji, Papua New Guinea, Nigeria and Iran. The disease is considered second in importance after blast in Japan, Taiwan and USA. Major factors associated with the increased occurrence of sheath blight include the widespread cultivation of high yielding but highly susceptible varieties, the high tillering capacity of these mostly semi-dwarf varieties which creates a favourable microclimate for disease development and the need for application of high dosages of nitrogen fertilizers for these cultivars which results in increased susceptibility.

Seedlings and adult plants both are equally affected but loss is much more when the disease appears in seedlings. Older plants are attacked in flooded conditions and swampy grounds. When the disease develops before the flag leaf stage there may be up to 20% loss in grain yield. Depending on disease severity, various reports have give the losses at 5 to 16 per cent. In a study in Punjab, Chahal *et al.,* (2003) reported that sheath blight reduced yield by 32.3% in rice when diseased area of top 3 leaves is 54.3%. Plants having all leaves affected yield less than 50% of the healthy plants. The loss is closely correlated with the number of hills affected in a field. Losses may be 20–50% when all the sheaths in a stool are affected. There is also strong relationship between symptom severity and yield reduction which varies among cultivars. Although the incidence of sheath blight is high in Asian countries, the disease is reported from North and South America also and in the southern parts of the United States it is considered a major problem in long grained semi-dwarf rice varieties.

Symptoms

Lesions are formed on the leaf sheaths and culm at the water level. The spots on the leaf sheath are first ellipsoid or ovoid, about 10 mm long, and greenish gray. They enlarge and may reach 2–3 cm in length and become irregular in outline. The centre of the spots becomes white with brown or purplish margins depending on the host variety. Many such spots become confluent giving a characteristic banded appearance. Outer leaves may fall, plants look yellow and may ultimately wilt. In favourable weather infection may spread up the culm, killing the entire leaves. On the surface of the lesions and sometimes on the inner surface of the sheath and on the culm brownish silky wefts of mycelium are present. These wefts produce brown, and/ or dark sclerotia which eventually fall on the ground. It is the banded appearance and presence of these sclerotia which gives the name 'banded sclerotial disease' to such diseases caused by *Rhizoctonia*. Mycelial and sclerotial growth depends on the environment. High humidity is essential.

Fig. 131. Sheath blight of rice.

The causal organism

General description of the causal fungus, *Rhizoctonia solani*, is given in the preceding section under stem canker and black scurf of potato. However, the strain causing sheath blight is different. In India, sheath blight of rice is caused by anastomosis group 1 (AG-I-1) of the fungus having 3–16 nuclei per cell. Sheath blight or banded sclerotial disease is reported on many crops such as maize, sorghum and even wheat. In a study of *Rhizoctonia* isolates from sheath blight, Taheri *et al.,* (2007) found that out of 110 isolates, 99 were *R. solani* and 11 were *R. oryzae-sativae*. Of the *R. solani* isolates 96 were AG1-1A. one was AG1-1B and 2 were AG1-1C. In India, the rice sheath blight isolate does not seriously attack *Sorghum vulgare, S. halepensis* and *S. sudanensis* but maize, finger millet (*Eleusine coracana*), pearl millet, and *Echinochloa colona* var. *frumentosa* are highly susceptible to it. In USA, rice isolates causes aerial blight of soybean and inclusion of soybean in rotation with rice is considered one of the causes of serious incidence of sheath blight in that country. *Cynodon dactylon, Setaria glauca, Paspalum flavidum, P. scrobiculatum, Echinochloa colona, E. crus-galli, Eriochloa procera, Ergrostis pilosa, Panicum repens, Leptochloa chinensis, Paspalum distinctum, Cyperus rotundus, C. radiatus, C. iria, Chloris* sp., and *Digitaria* spp. are listed as hosts of rice sheath blight fungus. These weeds in and around the rice fields, in water channels, and in irrigation ponds may serve as a source of primary inoculum of the fungus. No wild rice variety is resistant to the disease.

The rice pathogen produces a host-specific toxin, RS-toxin. It is a carbohydrate containing glucose, mannose, N-acetylgalactosamine and N-acetylglucosamine. The toxin is also detected in infected leaves. Sensitivioty to the phytotoxins is correlated with susceptibility of the host cultivar. Highly viru lent isolates produced more toxin than less virulent isolates. Shanmugam *et al.,* (2001) have reported that the toxin is inactivated by a putative β-glucosidase from coconut leaves and also by isolates of *Trichoderma viride* (Sriram *et al.,* 2000).

Disease cycle and environmental relations

R. solani on rice is seed-borne. In addition to the weed hosts listed above and infected seed, the fungus mainly survives as sclerotia and/or mycelia in diseased plant debris left in the field. Positive correlation between the number of sclerotia present in soil at the time of planting rice and incidence of sheath rot has been demonstrated, The pathogen remains viable for at least 120 days in infected rice straw pieces at 10°C. At 28°C viability gradually declines. Viability in straw pieces is highest when left on soil surface (Basu and Sen Gupta, 2004). The viability of sclerotia in soil is influenced by environmental conditions. In undisturbed land, sclerotia are present only up to 2 cm depth. Population of sclerotia in ploughed land is higher at 6–12 cm depth than in the upper layers. The sclerotia buried deep in soil have better buoyancy and viability.

In tap water at 35°C the sclerotia lose viability within a month but at 20°C they remain viable for 2 years. However, in some studies prolonged survival of sclerotia in soil has been doubted. In a field with more than 50% crop infection the number of sclerotia may be very small and many of these sclerotia are colonized by bacteria and *Trichoderma*, thereby losing viability. After 3 months in soil very few sclerotia retain their viability. Survival of sclerotia is better in upland (drier) fields than in flooded fields. Under conditions of submergence (7 cm water) sclerotial germination drops down rapidly within 60 days. Continuous submergence (20 cm water) after appearance of the disease significantly checks its development.

There are reports that mycelium of the fungus in rice straw can remain viable for as long as the sclerotia. Sclerotia and mycelia which survive in plant debris are brought up on the soil surface during puddling, levelling, and other operations. They come in contact with new planted seedlings, germinate and cause infection. The fungus forms infection cushions and / or lobulate appressoria on the plant surface. Infection pegs develop from the infection cushions or appressoria which penetrate through the cuticle or stomata. After infection, mycelium moves up the plant by surface hyphae and develops new infections structures over the entire plant.

When *R. solani* is inoculated on inner surface of leaf sheath, it first colonizes the surface before producing the infection structures like lobate appressoria, bulbous appressoria and infection cushions. The most frequent penetration is by hyphal tips followed by lobate appressoria (Singh *et al.*, 2003). Colonization of epidermal and mesophyll cells is both intra- and intercellular. Intarcellular hyphae are thick and deformed. Surface hyphae from primary lesions penetrate the healthy tissue both by hyphal tips and branched lobate appressoria. Infection is early by mycelium as inoculum than by sclerotia. The pathogen becomes aggressive when suitable temperature and moisture conditions are available. Exposure to sunlight inhibits infection, mycelial development, and sclerotia formation.

The rice plants are most susceptible during the active tillering stage. With further maturity after tillering stage the plants become less susceptible. However, some workers have reported that the maximum susceptibility is at the heading stage. The infection is usually first initiated on the lower leaf sheaths but during the susceptible stage more number of tillers having infection of sheaths at upper portions are observed. Presence of injuries on the sheath increases severity of sheath blight. Long grain cultivars are highly susceptible while medium and short grain cultivars are moderately susceptible to moderately resistant. Sheath blight is basically a disease of wet conditions. It is especially destructive under conditions of high humidity and high temperature. Temperature near the leaf sheath is related to air temperature but humidity within the crop stand varies according to tillering nature of the variety, density of stand, and amount of fertilizers added. Therefore, very close stand and heavy doses of nitrogen, especially when the crop has passed tillering stage, predispose the plants to sheath blight. Silicon deficiency is implicated in susceptibility of rice to sheath blight and heavy nitrogen doses reduce silicon availability to plants (Rodrtiguez *et al.*, 2001). Rodriguez *et al.*, (2003) studied the effect of silicon and rice growth stages on tissue susceptibility. Silicon concentration in straw increased as its amount was increased in]soil. Incubation period was shorter at booting and panicle emergence stages. The number and size of lesions was reduced. Infestation of the brown leafhopper and the nematode, *Hirschmanniella oryzae*, aggravates sheath blight. In rice the disease is severe under shade.

Maximum infection occurs at 100% relative humidity and minimum at 85–88% humidity. Optimum temperature for growth of the fungus and sclerotia formation is about 30°C, the minimum being 10°C and maximum 40–42°C. A sudden rise and fall in temperature accelerates sclerotia formation. Infection can occur in the temperature range of 23°–35°C but optimum conditions are temperature of 30°–32°C and relative humidity 96–97%.

Management of the disease

Considering the factors responsible for survival of the pathogen and disease development it must be ensured that weed hosts are kept at the minimum within and around the rice fields and proper sanitation is maintained by removal of stubbles of a badly affected crop. Burning of stubbles may not totally destroy sclerotia in plant debris left in the field. Semi-dwarf cultivars suffer more than tall cultivars. Rotation with non-cereal crops may help in reduction of sclerotial density in soil. Close planting should be avoided. A spacing of 30 x 30 or 40 x 40 cm gives the best reduction in disease incidence. Although this spacing may not be feasible it emphasizes the need for proper spacing in high tillering varieties to reduce atmospheric humidity in the crop. Nitrogen management as stated above should include no high doses of N, especially after tillering.

Silicon fertilization is a promising method for sheath blight control in areas where soil is deficient in silicon and acceptable resistant varieties are not available. In silicon deficient soils, application of silicon (calcium silicate slag) increases the Si content of the plant tissue by 80%. This reduces disease severity. In different cultivars the percentage of infected tillers is reduced by 82% (Rodriguez et al., 2001). There is a report that sheath blight as well as sheath rot diseases are considerably reduced if the rice fields are amended with oil cakes of margosa, coconut, etc. at the rate of one ton per hectare. Wood sawdust and rice husk also give good control. In pot experiments, neem (margosa) cake applied at calculated field dose of 150 kg/ha reduced sheath blight incidence from 78.87% to 27.07% and disease severity from 18.84 to 2.54%. Farmyard manure was next best (Meena and Muthusamy, 1999). Neem-based pesticides Nemazol (0.3%), Wanis (0.3%), Achook (0.5%) and Neem Gold (0.3%), used for rice seed treatment increase seed germination, root and shoot growth and seedling vigor (Zope and Thrimurty, 2004). Application of Kava (*Piper methysticum*) roots to flooded paddy soil at the rate of 1 tonne/ha not only suppresses common weeds (*Echinochloa, Paspalum, Monochoria*) but also *R. solani* and *P. grisea*. Kava is a perennial plant in oceanic regions. It also suppresses such pathogens as *Taphrina deformans, Fusarium solani* and *Rhizopus stolonifer* (Xuan et al., 2003; 2005). Leaf extracts of *Datura metel* significantly reduce *in vitro* growth of *R. solani* (RS7, AG 1) and *Xanthomonas oryzae* pv. *oryzae*. Foliar application of the leaf extract effectively reduce the incidence of sheath blight and bacterial leaf blight. Pre-inoculation applications are better than post-inoculation applications (Kahale et al., 2004). The treatment induces systemic resistance also.

Application of the resistance activator compound acibenzolar-S-methyl for control of sheath blight was studied by Rohilla et al., (2002). The compund has limited fungitoxicity against *R. solani*. Mycelial growth, hyphal browing and sclerotia formation are reduced. Parasitic fitness of mycelia and sclerotia is also reduced. When applied as soil drench or foliar spray, the compound inhibits both disease development on inoculated sheath and spread to younger leaves. The effects are more pronounced with increased duration between BTH application and inoculation. The compound becomes effective within an hour when applied through roots but in foliar applications a gap of 24 hours between application and inoculation is required. The effect of the compound is through a combination of its host defense inducing activity and its adverse effect on the pathogen. BTH reduces frequency of penetration by *R. solani*, colonization of host tissue and spread of the hyphae from primary lesions to form secondary lesions. It induce swelling of hyphal tips on the sheath surface, formation of papillae, browning of penetrated epidermal cells and degeneration of inter-cellular hyphae colonizing epidermal and mesophyll.

Among synthetic fungicides, PCNB and carbendazim (Bavistin) have been very often used with success. Soil application of PCNB at the rate of 20–30 kg/ha reduces primary inoculum present in the field. Bavistin at the rate of 1 g per litre water sprayed 2–3 times gives good control. Benlate (0.18%) is equally effective. The same fungicides can be used for seed treatment at the rate of 2 g/kg seed. During spraying care should be taken to ensure that the liquid reaches the lower leaf sheaths. In the variety Jaya, soil treatment with either Kitazin (17%) or Thiram (each at 3 g/sq.m.) alone or in combination with foliar sprays of either carbendazim or benomyl or Kitazin (0.1% each) at 45, 55, and 65 days after transplanting has given good results In a study best disease control was obtained with Topsin M (thiophanate methyl) seed treatment followed by foliar sprays of the same fungicides (0.1%). In this study all fungicides tested had adverse effect on the disease. Chlorothalonil (Bravo) and iprodione (Rovral) are other fungicides which have been used as spray material. Sprays of propiconazole (Tilt 3.6 EC) at the rate of 0.18–0.48 kg a.i./ha at boot and heading stages of rice plants gives significant control of disease severity and result in significant increase in grain yield. Chahal et al., (2003) found propiconazole (0.1%), ediphenphos (0.1%), iprodione (0.3%) and carbendazim (0.1%) effective against sheath blight. Application of the strobilurin fungicide azoxystrobin at 0.17–0.22 kg a.i./ha, especially in early season gives good control in all cultivars and results in significant increase in yield (Slaton *et al.*, 2003). Spray after heading does not give significant control. Groth and Bond (2006) have reported that application of azoxystrobin at 0.17 kg a.i/ha 7 days after panicle differentiation and before 50% heading reduces disease incidence and severity and increases yield of good quality grains. Later, Groth (2008) reported that a single application of azoxystrobin at this rate applied at mid–boot stage in a moderately resistant cultivar was sufficient to increase yield of good quality grains. Four antibiotics, two developed in Japan (Validamycin and Polyoxin) and two in China (Jinganmycin and Chingfengmeisu) have been found effective against sheath blight. Jingamycin and Validamycin are commonly used antibiotics in China and Vietnam, respectively.

Seed and foliar treatments with antagonistic bacteria provide biological control of sheath blight. The bacterial biocontrol agents are strains of *Pseudomonas fluorescens*, *Bacillus* spp. and *Enterobacter*. The disease control by these agents is more pronounced in direct seed than in transplanted crop (Gnanamanickam *et al.*, 1992). Foliar spray of *Bacillus subtilis* suppresses sheath blight through increased activities of phenylalanine ammonia lyase (PAL), peroxidase, and accumulation of pathogenesis proteins (Jayaraj *et al.*, 2004). The antibiotic(s) produced by strain Pf7–14 of *Pseudomonas fluorescens* suppress sheath blight as effectively as fungicides. In seed treatment, root treatment + soil application and foliar spray of a peat-based formulation of *Pseudomonas fluorescens*, all the individual treatments were effective but a combination of all the four treatments resulted in the best sheath blight control under glasshouse conditions. Under field conditions, the formulation gave as good result as carbendazim. In evaluation of a talc-based powder formulation of *P. fluorescens* Pf1, seed treatment was found better that soil or root treatment. The bacteria applied through seed established well in the rhizosphere. Talc-based formulation of *Pseudomonas fluorescens* PF1 and FP7, applied separately or as mixture through seed, root, soil and foliar spray significantly reduce sheath blight as well as leaf folder insect. The mixture performs better than individual formulation (Radjacommare *et al.*, 2002). Induction of increased chitinase activity in plants treated with PF1 and FP7 has definite role in the disease suppression (Rajdacommare *et al.*, 2004). A strain of *Pseudomonas putida*, antagonistic to the sheath blight pathogen, persisted in IR 50 rice roots for 60 days in sterile and for 50 days in non-sterile soil after seeds were coated with the bacterium. It migrated to rice seedlings only for 10 days after seedling emergence. The bacterium survived in rice foliage for 40 days after it was sprayed on leaves and provided 61–62% sheath blight suppression in cv. IR 50. Methyl cellulose: talc formulation was found to maintain the bacteria in a viable state for up to 10 months affording 60% suppression of sheath blight in the field which was comparable with effective fungicides. The treatment had no adverse effect on seed germination. The seeds were treated for 16 hours. Peat and talc based formulations of *Pseudomonas fluorescens*, isolated from the crop rhizosphere had controlled the banded leaf sheath blight (*R. solani*) in maize. The formulations maintained the populations of the antagonist at high level for 40 days. The disease was effectively controlled by seed treatment with the peat based formulation @ 16 g/kg or as soil treatment @ 2.5 kg/ha or by applying a liquid formulation twice @ 5 g/l water. Floating pellet formulation and soluble granule formulation of *Bacillus megaterium* give as good control of sheath blight as standard fungicides. Pellet formulation of *B. megaterium* and *B. pumilus*, individually, suppress sheath blight of rice as well as freshly prepared bacterial suspension. The action is through antibiotic production which in some cases may get diluted in aquatic environment of the rice fields (Pengnoo *et al.*, 2000). As an answer to the problem of preparation of formulation and its shelf life, Mew *et al.*, (2003) have suggested, for farmers in developing countries, a programme. The biocontrol agent strain that is indigenous at a locality is mass produced at the research institution based on the total area required for application, as communicated by the farmers through the extension workers. The demand decides the amount to be produced and delivered to the farmers for seed bacterization at the time of planting. This process alleviates the biocontrol agent storage and shelf life problems.

While confirming an earlier report of reduction of sheath blight by application of *Trichoderma viride* and *Bacillus subtilis* as seed treatment and *Trichoderma* being more effective than the bacterium, Das and Hazarika (2000) reported that seed treatment with *T. harzianum* and *T. viride* alone or in association with 2% (w/w) methyl cellulose and/or 0.1M magsulph reduced the disease incidence in pot tests. The á-glucosidase secreted by *T. viride* degrades a phytotoxin associated with sheath blight (Shanmugam *et al.,* 2001). Mathivanan *et al.,* (2005) obtained suppression of sheath blight and increased plant growth and yield of rice by talc formulation of *Pseudomonas fluorescens* and *Trichoderma viride* when applied individually or in combination. The control achieved was at par with carbendazim. In *in vitro* studies maximum inhibition of mycelial growth and sclerotia production of *R. solani* is reported by *Gliocladium virens* compared to *Trichoderma viride* and *T. harzianum* Sclerotia were colonized by the antagonist. The antagonist hyphae coiled around the pathogen hyphae and caused lysis. All oilcakes tried by him completely inhibited the growth of *R. solani.* Laha and Venkataraman (2001) obtained significant control of sheath blight with strains of fluorescent *Pseudomonas* sp. and *Bacillus* sp. which were compatible with carbendazim.

A strains of *Bacillus* sp., isolated from wheat rhizosphere was found to be antagonistic to all AGs of *R. solani* in addition to *Pythium* and *Gaeumannomyces.* Apart from the antagonistic effect against the pathogens through antibiotic production, many strains of the above mentioned organisms induce systemic resistance also and are effective while confined to the plant rhizosphere. Nandkumar *et al.,* (2001a) reported that certain strains of *Pseudomonas fluorescens,* when applied through seed, root, soil or foliage induced systemic resistance against sheath blight. The suppression was accompanied by increased chitinase and peroxidase activity. A single method of application was not as effective as all the four methods. They (Nandakumar *et al.,* 2002b) used mixtures of PGPR applied through seed, root, folair and soil application and observed that while mixtures reduced disease by 45.1%, single strains gave only 29.2% control. The mixtures promoted plant growth in terms of increased plant height, number of tillers and grain yield. Better results were achieved by seed treatment followed by application to root or foliage. The antagonistic bacterium *Serratia marcescens* strain B2 produces a red-pigmented antibiotic, prodigiosin, which suppresses rice sheath blight if plants are treated with a bacterial suspension before inoculation with the pathogen (Someya *et al.,* 2003b, 2005). The phylloplane bacteria work as antagonists of this antagonst and inhibit the antibiotic synthesis (Someya *et al.,* 2003b). However, rice epiphytic bacterium *Erwinia ananas* transformed with chitinase gene of *S. marcescens* retains biocontrol activity even in presence of other bacteria Rice leaves harbor many endophytic bacteria which are antagonists of *R. solani* and suppress leaf blight. Culture filtrate and extracts of the filtrate of *Helminthosporium graminearum* have antagonistic effect on the sheath blight pathogen. The toxin in the filtrate has been identified as ophiobolin A. In field trials, treatment with the filtrates sheath blight was effectively checked. There was adverse effect on the plant and the toxin was not detected in the grains (Duan *et al.,* 2007).

Intensity of sheath blight and sheath rot both is reduced by foliar application of the herbicide benthiocarb at the rate of 2 kg a.i./ha. No commercial varieties in India are resistant to the disease. Tetep, a primitive cultivar from Vietnam, is most resistant to sheath blight. The resistance is expressed as a reduced number of infection cushions produced by the fungus and as the production of oxidized phenolic compounds which slow down the spread of the pathogen within the plant. This is manifested as a dark zone around the lesions. The result is fewer and smaller lesions. Genetic transformation has been attempted for management of the disease. A chitinase gene from *R. solani* was introduced in a *indica* rice cultivar. The transformants synthesized different levels of chitinase enzymes constitutively when challenged by the pathogen and showed different levels of enhanced resistance to sheath blight. Shreshtha *et al.*, (2008) have reported a negative correlation between level of chitinase activity of rice cultivar and number and length of leasions caused by *R. solani*.

BOTRYTIS DISEASES

Botrytis diseases are probably the most common and most widely distributed diseases of vegetables, fruits, ornamentals and even some field crops. The fungus has a necrotrophic lifestyle and attacks over 200 crop hosts worldwide. These diseases appear as blossom blights, fruit rot, damping off, stem cankers or rot, leaf spots and tuber and corm rot. Some of the major hosts affected by Botrytis diseases among vegetables and fruits are strawberry, grapes, apples, onion, bean, pea, cabbage, tomato and lettuce. In field crops chickpea gray mold has become a nationally important diseases in India. It was first reported from Pantnagar in 1969 (Joshi and Singh, 1969). They had also reported its occurrence in a nearby field on pea. Botrytis grey mold is now as important chickpea disease as Ascochyta blight and Fusarium wilt. The gray mold blight of chickpea had almost completely destroyed an entire field

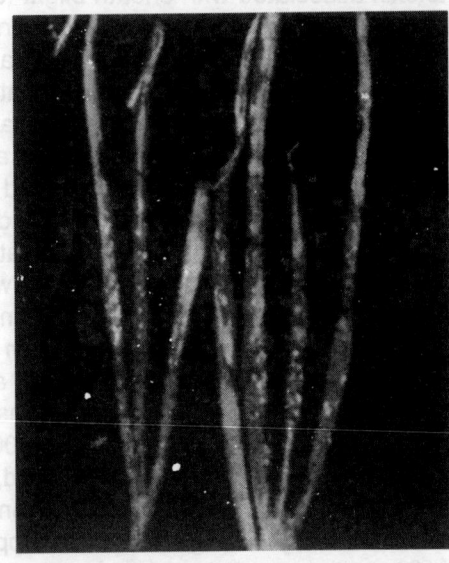

Fig. 132. Blast of onion caused by *Botrytis spp.*

when it was first reported. Losses varying from 70 to 100% are still reported when weather is favorable for the disease. In pea, 14.7–19.4% loss of flowers and 4.4–8.0 quintals per ha loss in grain ield is reported. Germinability of seed from infected pods is reduced by 85.1–88.1%. In wet and cool weather bunch rot may cause up to 30% destruction of bunches of grapes. The gray mold rot of apples is a serious problem in temperate climate. In Himachal Pradesh up to 10% incidence is reported. In lettuce, *Botrytis* causes damping off of seedlings, leaf rot and rot of lettuce heads during transit and in the market.

Symptoms

Symptoms of gray mold rot vary with hosts and plant organs affected. In chickpea, the disease causes leaf blight and rotting including rot of the stem. A gray aerial growth covers the affected parts. In early infection the plants are dead before flowering. Those that survive produce flowers which are soon attacked by the mold and there is blossom blight. Sclerotia of the fungus develop as black crust on affected stems. In the lettuce damping off phase, seedlings collapse and topple over. Soon the dead seedlings are covered with the gray mold which is made up of conidiophores and conidia of the fungus. On older plants in the field, spots appear on older leaves as large water-soaked areas which turn yellow and get covered with the gray fungal growth. Under wet conditions the entire leaf may be soon covered by this growth but in dry weather brown to black, dry, firm decay develops. The harvested lettuce heads carry the fungus with them. If wet and cool conditions exist, the rot continues during transit and storage or in the market.

In bunch rot of grapes, infection can occur on young as well as relatively older leaves which show irregularly shaped necrotic spots. In humid weather, these spots rapidly enlarge and coalesce. Infected flowers do not show any visible symptom but necrosis of stamens and growth of the fungus on the style and stigma can be seen under microscope. Most prominent symptoms are seen on the grape berries. Infected berries become dark colored and show the typical grayish, hairy mycelial growth all over the fruit surface. In severe attack, all the berries in a bunch are involved and totally lost. Infected berries are prone to attack of such secondary rot causing fungi as *Aspergillus, Penicillium* and *Trichothecium roseum*. These fungi give a bitter taste to the fruit. Losses are heavier if stalks of berries are infected.

The gray mold rot of strawberry is similar to bunch rot of grapes. It is also known as Botrytis rot, brown rot of strawberry and dry rot and is common in cool climate. The symptoms start from that part of the fruit which is in contact with the soil, dead fallen leaves and rotting fruits in the packing containers during transit. The decaying fruit tissue turns light brown and soft but there is no leakage of fluid from the fruits. The rot rapidly spread all over the fruit which dries and becomes hard. Greenish or gray fungal growth develops on such fruits. Infection can occur any time from growth to ripening stage of the fruits. Leaves, calices, pedicels and flowers are also attacked.

In onion, the disease is commonly known as blast and, in addition to *B. cinerea*, many other species (*B. allii, B. byssoidea, B. squamosa, B. cepae*) are also involved. In onion blast, hundreds of white specks are seen on the foliage. When a shower of conidia of Botrytis lands on the leaves, they germinate in presence of moisture and in doing so produce enough toxic metabolites which kill a few cells beneath the germ tubes. When a new shower of spores arrives, these injured points provide avenues for entry of the fungus into the tissues and then the blast phase occurs. If other diseases like downy mildew or thrips have also attacked the leaves they can help *Botrytis* establish as a pathogen. The disease then spreads very rapidly and tops of the entire crop may be killed.

The causal organism

The teleomorph of *Botrytis cinerea* Fr. or *Botrytis* of the *cinerea* group is known as *Botryotinia fuckeliana*. The fungus is classified under phylum Ascomycota, subphylum Pezizomycotina, class Leoiomycetes, order Helotiales and family Sclerotiniaceae, genus *Botryotinia*. *Botrytis allii*, the cause of neck rot of onion has its teleomorph in *Botryotinia allii*.

In culture, the mycelium is at first white but becomes dark with age. The hyphae are septate and 8–16 μm wide. The tall, erect, conidiophores arising from the mycelium are hyaline towards the tip but dark below. Just below the tip of these conidiophores several branches arise and re-branch once or more. Their tips round up and give rise to sporigenous ampullae. On each ampulla numerous conidia arise simultaneously on short denticles. The whole structure looks like a bunch of grapes. The conidia are aseptate, ovate, and measure 11–15 x 8–11μm (4–24 x 4–8 μm in grapes). The surface of dry conidia is rough with very minute protuverances which disaapear when conidia are hydrated and then dried.

The teleomorph is not common in nature. It develops in the black sclerotial crust. Where mating types and suitable condition are present for the sexual cycle, *B. cinerea* produces microconidia in abundance which function as spermatia to initiate the sexual cycle.

Generally the sclerotia of the fungus appear like a crust on the woody parts of the host but in some species fairly large, black sclerotia are produced. In addition to structures of survival, these sclerotia may produce spermatia that lead to the ascigerous stage. Presence of double stranded RNA molecules is hyphal cells of *B. cinerea* is reported and is related to avirulence of the strains.

Disease cycle and environmental relations

Botrytis can grow as a saprophyte on dead crop debris. It survives as sclerotia and on weed hosts. Sclerotia survive better in dry soil at 20°–25°C. Same temperature range is optimum for germination of sclerotia and infection of grapes. Survival through seed is reported in chickpea but it is mostly in the form of contaminant. Sclerotia–bearing debris is often carried with seed. Germination of sclerotia is myceliogenic producing conidia which are disseminated by wind. When living or non-living conidia land on a wet surface they adhere to it. They germinate quickly in presence of water (dew). Secretion of a gelatinous matric by the germlings ensures their adhesion to the host surfacve. The germ tubes enter the host through wound or directly. The fungus is a necrotroph. If the host shows hypersensitive cell death in response to *B. cinerea* infection or due to some other stress, *B. cinerea* gets benifitted as it easily colonizes the dead tissue. An elicitor from *B. cinerea* induces hypersensitive response in the host. (Govrin *et al.,* 2006). The dead host cells promote the growth of the fungus through factors present in them including glucose and phosphate. If the conidia fall on dead plant parts such as flowers, leaves etc. there also they grow and produce enormous numbers of conidia which are disseminated by wind to healthy leaves and blossom.

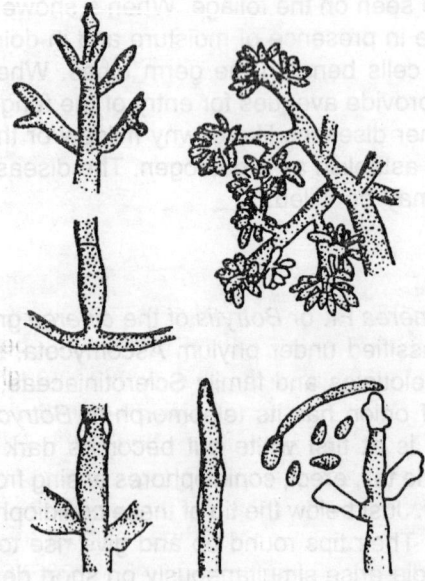

Fig. 133. Conidia and conidiophores of *Botrytis.*

Under field conditions, cool and foggy weather is ideal for development of gray mold. The blight of chickpea occurs when such weather persists during late December and January provided temperature is not too low. The optimum temperature for reproduction and infection by most species of *Botrytis* is from 10°–24°C. On lettuce, penetration of the host does not occur until the temperature reaches 12°C or above. Incubation period is 6 days at 6–8°C and only 2 days at 22°C. Optimum for growth of *B. cinerea* lies between 22° and 24°C. Conidia are produced only in light. On onion, lesion production by *B. squamosa* is optimal at 20°C, lower at 15°C and greatly reduced at 25°C. In storage of fruits, Botrytis gray mold can cause considerable damage even if the storage temperature is between 0 and 10°C.

Germinating spores adhere to the host surface with the help of a secreted ensheathing film that remains attached to the substrate even when germ tube is physically removed. They seldom penetrate actively growing tissue directly, but they can penetrate tissues through wound or after they have grown for a while and produced mycelium on old flower petals, dying foliage, dead bulb scales of onion. In grapes, the fungal growth outside the cuticle is enhanced by different nutrient sources such as pollen, debris of floral parts, or exudates from berries. High nitrogen treatment, especially application of ammonium nitrate, predisposes grapevines to infection of *Botrytis* and increases disease severity. In blossom blight of strawberry, anthers are the main route of entry into the developing receptacle of the fruit. Cultivars lacking anthers show less disease. Strains of *B. cinerea* isolated from different hosts show differences in adhesion pattern and in the percentage of germination on host cutin. Strains isolated from a particular host are more virulent on that host compared to other hosts.

Botrytis cinerea produces a toxin and the level of toxin production is related to the level of virulence of the fungus (Reino *et al.*, 2004). The enzyme endo-β-1,4-glucanase is expressed by the fungus during pathogenesis but is not required for the latter while endo-beta-1,4-xylanase is required for virulence.

Disease management

In crops where damping off is a problem, solarization of the nursery bed soil under plastic sheets is an economical and highly effective measure. Soil treatment with PCNB had also been recommended. Field sanitation practices such as removal of crop debris including dead or rotting fruits reduces primary inoculum. In the field, reduction in humidity is one of the most important step. Spacing in crops like chickpea and strawberry and canopy management in grapevines reduces humidity and allows quick evaporation of moisture on plant parts through aeration and sunlight. In chickpea gray mold, wide row spacing (60 cm), intercropping with dwarf or semi-dwarf wheat cultivars and uniform nipping of tender branches 45–60 days after planting not only reduce disease severity but also significantly increase the grain yield.

In grapevines a system of green pruning or leaf removal has been found very effective. A partial defoliation of basal portion of shoots near the clusters, performed shortly after bloom or done 2–3 times during the season improves wind speed through the vines. This can cause as much as 69% disease reduction. Row spacing and mulching of the soil with straw between rows of strawberry and lettuce also reduces humidity in the foliage and prevents contact between fruits or leaves and the contaminated soil. Effective control of grey mould of grape berries by spray of water extract of fermented compost is reported. The compost material used was cattle manure, chicken–cattle manure and grape residual after juice extraction.

B. cinerea quickly develops resistance to fungicides. Therefore, no highly effective and durable fungicide spray schedule has been found. Application of fungicides in crops like strawberry and lettuce is restricted because of limited time available for treatment. The chemicals cannot be applied when the crop is ready for harvest. In gray mold of grapes, during the late 1950s and until 1968 captan, captafol, folpet, chlorothalonil and thiram were commonly used in vineyards but they gave only 20–50% control. After 1968, benzimidazole fungicides (benomyl, thiophanate methyl and carbendazim) were found to give consistently better control.

By 1970, resistance to MBC in *B. cinerea* became common and benomyl and carbendazim are no more recommended. Thiophanate methyl continues to be in use. Since 1976, the carboximide fungicides iprodione (Rovral) and vinclozolin (Ornalin, Ronilan or Vorlon) have exhibited good efficacy against Botrytis diseases. Although the fungus developed resistance to these fungicides also, their efficacy has not broken down as rapidly as benzimidazole fungicides. Vinclozolin gives prolonged protection but its economics is not favorable. The post-harvest gray mold rot of grapes can be checked by periodical fumigation of bunches with sulphur dioxide. The use of neem (margosa) oil, horticultural oils, silicates, phosphates and carbonates is under trial for control of bunch rot. Extracts of *Allium, Capsicum,* cinnamon leaves and clove buds are highly antifungal against *Botrytis*. In *in vitro* studies sodium bicarbonate at concentrations as low as 20 mm is reported to inhibit colony growth of *Botrytis cinerea*.

In gray mold of strawberry also the above fungicides have been used effectively. Spray of captan and thiram has been commonly recommended. Early sprays are not effective. Spray of captan (0.2%) first when the flowers are opening and then two more sprays given until fruits change colour has been found beneficial. In Himachal Pradesh Bavistin (0.05%) and captan (0.3%) have provided best control of strawberry gray mold under field conditions. In view of resistance development against benzimidazoles, carboximides and even captan the recommendation is to use different fungicides in different sprays. The post-harvest gray mold rot of strawberry can be checked by hot air treatment of fruits. Exposure of fruits to hot air at 43°C under 93% relative humidity for 30 min controls the gray mold rot as well as Rhizopus rot of strawberry.

In onion, spray of captan, ferbam or ziram at 7–10 days interval protects the foliage from initial development of blast. Excellent onion blast control has been obtained by mixing iprodione (0.56 kg/ha) and chlorothaloni (0.87 kg/ha) with mancozeb (1.79 kg/ha). The mixture reduces the chances of resistance development in the fungus. Some isolates of *Botrytis squamosa* have shown resistance to iprodione (Roviral) and vinclozolin (Ronilan) but not to mancozeb and chorothalonil.

In gray mold of chickpea, fungicidal control involves the dicarboximide vinclozolin (Ronilan) and the benzimidazole (carbendazim). The fungus shows rapid development of resistance to benzimidazoles. Superiority of vinclozolin over other fungicides in suppression of gray mold is reported in some studies. However, when weather conditions are favorable for the disease chemicals alone fail.

Spray of bicarbonates for control of *B. cineria* is a cheaper and eco-friendly strategy. Salts of ammonia, sodium and potassium, at 25 mM concentration have been tries against *B. cineria* in agar plates. The best suppression of colony diameter was given by ammonium bicarbonate and ammonium carbonate. Sodium salts were not as effective.

Limitations associated with chemical control of gray mold of fruits and field crops have prompted greater effort on biological control using antagonistic strains of bacteria and fungi including yeasts. The bacterial species are *Bacillus mycoides, B cereus, B amyloliquefaciens, Pseudomonas syringae, Pseudomonas aeruginosa, Burkholderia (Pseudomonas) cepacia. Serratia marcescens* and *Paenibacilluys polymyxa.* Certain strains of *P. syringae* can provide control of *Botrytis* by suppressing liberation of conidia. Rhizobacteria-mediated resistance to *Botrytis* is also reported. *P. aeruginosa* produces salicylic acid that induces systemic resistance in bean leaves against *B. cinerea.* Even nanogram amounts of salicylic acid trigger the systemic acquired resistance pathway against *B. cinerea* in bean. *P. aeruginosa* 7NSK2 produces secondary metabolites such as pyrochelin (Pch), its precursor salicylic acid (SA) and the phenazine compound pyrocyanin. Pch and pyrocyanin are the determinants for induced resistance by the bacterium. Salicylic acid is probably very efficiently converted to Pch. Someya *et al.,* (2001) have reported strong antifungal activity of *Serratia marcescens* B12 against *B. cinerea* through chitinolytic anezymes, its red pigment prodigiosin and through induction of resistance. Chitinolytci activity of *Bacillus cereus* as a biocontrol of gray mold of chickpea is enhanced when the treatment is supplemented with chitin (Kishore and Pande, 2007). Inoculum density of bacterial and yeast biocontrol agents must be several times higher than density of Botrytid conidia and hyphae to achieve 80–90% control.

Up to 78% reduction of grey mold on grapes is reported by application of *Trichoderma harzianum* (0.5–1.0 g/L), Vinclozolin or iprodionee (90.5 g/L) or dithiofencarb (0.25 g/L). Since combination of the biocontrol agent and fungicides had no additional benefits, alternate application of fungicides and biocontrol agent was recommended. *Trichoderma harzianum* and the yeasts *Aureob0asidum pullulans* and *Cryptococcus albidus* were evaluated as microbial antagonists of *B. cinerea* (on cucumber and tomato). The disease incidence was reduced by 50% to 100% by all antagonists. The biocontrol efficiency was not affected by temperatures of 18° to 30°C although at 24° the efficacy showed decline when the relative humidity was 90% or 80% compared to 100%. Strains of the yeast *Candida guilliermondi* significantly reduce *Botrytis* infection on tomato plants through competition for nutrients. Exogenous application of nutrients reduces biocontrol efficacy (Saligkarias *et al.,* 2002). In onion, the yeast *Ulocladium atrum* suppresses sporulation of *Botrytis cinerea* and *B. aclada* on dead leaves. *Botrytis* spp. colonize the necrotic leaf tissues of onion. *U. atrum* is a strong competitor on such tissues and suppresses sporulation of *Botrytis* thus reducing spread of the disease. *Gliocladium roseum* is also a biocontrol agent of *Botrytis* on onion. *Microsphaeropsis ochrea* gives high degree of control of onion leaf blight caused by *B. squamosa*. It colonizes only the senescent and necrotic leaves and prevents formation of conidia by the pathogen (Carisse *et al.,* 2006).

Presence of quiescent fungi such as *Alternaria, Aureobasidium, Ulocladium* and *Cladosporium* in wound sites in grape berries provides natural protection against *B. cinerea* (Dugan *et al.,* 2002). *Botrytis cinerea* is a necrotrophic pathogen. Plant defence mechanisms against such pathogens are complex. Immediate release and$_2$ accumulation of hydrogen peroxide in infection sites is one defence pathway. Presence of other fungi in the wounds may cause accumulation of hydrogen peroxide (H_2O_2). *Clonostachys rosea (Gliocladium roseum)* is a very strong antagonist of *Botrytis*. It suppresses the pathogen by more than 93% on leaves, petals and stamens, better than even captan. In field plots weekly sprays are required. The control is through competition for nutrients. Control of gray mold of apple by spray of *Trichoderma harzianum* on flowers is also reported. Suppression of Botrytis disease by *T. harzianum* is attributed also to antagonistic activity through enzymes as well as through induction of systemic resistance. However, there is a report that even 9 sprays of Trichodex, a *T. harzianum* based commercial product, failed to control *B. cineria* in greenhouses. The yeast *Candida saitoana* and the bacteria *Burkholderia cepacia* and *Pseudomonas syringae* are also biocontrol agents against apple gray mold. Gray mold of grapes is also controlled by these biooagents. Spays of *Ulocladium atrum,* started at transplanting, result in better control of strawberry gray mold than sprays starting at beginning of flowering. Colonization and cell-wall degrading enzymes are implicated in the control of *Botrytis cinerea* by strains of *Ulocladium*. *Aureobasidium pullulans*, as a biocontrol agent, induces resistance in strawberry against Botrytis gray mold rot.

For better control of chickpea gray mold, combination of bencmyl and benomyl-tolerant strain of *Trichoderma viride*. Subsequently many studies found best control of chickpea gray mold with a combination of 0.2% Ronilan (vinclozolin) and *Trichoderma viride*. Spray of *T. viride* spore suspension at 20 days interval and Ronilan (0.2%) spray at 20 days interval were not as effective as a schedule of one spray of Ronilan (0.1%) + *T. viride* followed by one application of *T. viride* and then one application of Ronilan (0.1%) with *T. viride*. The two are compatible. Chickpea seed treatment with *Trichoderma* or *Gliocladium roseum* for biological control of chickpea gray mold is also reported. *G. roseum* (*Clonostachys rosea*) is versatile antagonist of *B. cinerea*. It suppresses *Botrytis* sporulation on crop debris. Its repeated application to diseased crop debris together with sanitation may be a potential management strategy (Morandi *et al.,* 2003) In field trials, this antagonist applied to the strawberry plant is reported to reduce infection of stamens and fruits by 79 t0 93% and 48 to 76%, respectively. The control given by *Clonostachys* was better than that give by other major biocontrol agents and standard fungicides.

Iinstead of using a single biocontrol agent, which may give erratic results, a combination of several antagonists with different mechanisms of action gives much better results. Supplementation of the treatments with trace elements such as molybdenum and zinc enhances biocontrol efficiency.

In some plant-*Botrytis* systems, spray of fresh water extract of composted manure-straw mixture is reported to inhibit the growth of the pathogen on the host surface. Commercial use of biocontrol agents for disease control has yet to reach the common farmer in spite of several commercial formulations made available in the market. A more recent approach to use of biocontrol agents is to combine them with fungicides thereby reducing the dosage of the fungicides and getting better disease control. This has been successfully demonstrated in field crops.

REFERENCES

Akkopro, A. and S. Demir. 2005. Biological control of Fusarium wilt in tomato caused by *Fusarium oxysporum* f. sp. *lycopersici* by AMF *Glomus intraradices* and some rhizobacteria. *J. Phytopathol.* 153(9): 544

Amadioha, A.C. 2000. Controlling rice blast *in vitro* and *in vivo* with extracts of *Azadirachta indica*. *Crop Protection* 19(5): 287

Amadioha, A.C. 2002. Fungitoxic effects of extracts of *Azadirachta indica* against *Cochliobolus miyabeanus* causing brown spot disease of rice. *Arch. Phytopath. Plant Prot.* 35(1): 37

Anjaiah, V., P. Corneiis and N. Koedam. 2003. Effect of genotype and root colonization in biological control of Fusarium wilts in pigeonpea and chickpea by *Pseudomonas aeruginosa* PNA 1. *Can. J. Microbiol.* 49(2): 85

Ansari, K.L., N. Palacios, C. Araya *et al.,* 2004. Pathogenic and genetic variability among *Colletotrichum lindemuthianum* isolates of different geographic origins. *Plant Pathology* 53(5):635

Arfaoui, A., B. Sifi, A.Boudabous *et al.,* 2006. Identification of *Rhizobium* isolates possessing antagonistic activity against *Fusarium oxysporum* f. sp. *ciceris*, the causal agent of Fusarium wilt of chickpea. *J. Plant Pathol.* 88(1): 67–75

Arfaoui, A., A. El-Hadrami *et al.,* 2007. Treatment of chickpea with *Rhizobium* isolates enhances the expression of phenylpropanoid defence-related genes in response to infection by *Fusarium oxysporum* f. sp., *ciceris*. *Plant Physiol. Biochem.* 45(6): 470

Arie, T., I. Kaneko,T. Yoshida *et al.,* 2000. Mating-type genes from asexual pytophatogenic ascomycetes *Fusarium oxysporum* and *Alternaria alternate*. *Mol. Plant Microbe Interact.* 13(12): 1330

Ayyadurai, N., P.R. Naik, M.S. Rao, R.S. Kumar, S.K. Samant, M. Manohar and N. Sakthivel. 2006. Isolation and characterization of a novel banana rhizosphere bacterium as fungal antagonist and microbial adjuvant in micropropagation of banana. *J. Appl. Microbiol.* 100(5): 925

Baird, R.E., C.E. Watson and M. Scruggs. 2003. Relative longevity of *Macrophomina phaseolina* and associate mycobiota on residual soybean roots in soil. *Plant Dis.* 87(5): 563

Bao, J.R. and G. Lazarovits. 2001. Differential colonization of tomato roots by nonpathogenic and pathogenic *Fusarium oxysporum* strains may influence Fusarium wilt control. *Phytopathology* 91: 449

Bapat, S. and A.K. Shah. 2000. Biological control of Fusarium wilt of pigeonpea by *Bacillus brevis*. *Can. J. Microbiol.* 46(2): 125

Basu, A. and P.K. Sen Gupta. 2004. Saprophytic perennation of rice sheath blight pathogen, *Rhizoctonia solani*, in infected plant parts. *J. Mycol. Pl. Pathol.* 34(2): 318

Bharathan, N., H. Saso, L. Gudipati, S. Bharathan, K. Whited and K. Anthony. 2005. Double-stranded RNA: distribution and analysis among isolates of *Rhizoctonia solani* AG-2 to -13. *Plant Pathology* 54(2): 196

Bharathi, R., R. Vivekanandan, S. Harish, A. Ramanathan and R. Samiyappan. 2004. Rhizobacteria-based bio-formulations for the management of fruit rot infection in chillies. *Crop Protection* 23(4): 835

Bigirimana, J. and M. Hofte. 2002. Induction of systemic resistance to *Coiletotrichum lindemuthian* in bean by a benzothiadiazole derivative and rhizobacteria. *Phytoparasitica* 30(2)

Boff, P., J. Kohl. M. Jensen, P.J.F.M. Horsten *et al.*, 2002. Biological control of gray mold with *Ulocladium atrum* in annual strawberry crop. *Plant Dis.* 86(3): 220

Bokshi, A.L., S.C. Morris and B.J. Deverall. 2003. Effects of benzothiadiazole and acetylsalicylic acid on â-1,3-glucanase activity and disease resistancein potato. *Plant Pathology* 52(1): 22

Bolwerk, A., A.L. Lagopodi, B.J.J. Lugtenberg and G.V. Bloemberg. 2005. Viasualization of interactions between a pathogenic and a beneficial *Fusarium* strain during biocontrol of tomato foot and root rot. *Mol. Plant Microbe Interact.* 18(7): 710

Borges, A.A., A. Borges-Perez and M. Fernandez-Falcon. 2004. Induced resistance to Fusarium wilt of banana by menadione sodium bisulphite treatment. *Crop Protection* 23(12): 1245

Brewer, M.T. and R.P. Larkin. 2005. Efficacy of several potential biocontrol organisms against *Rhizoctonia solani* on potato. *Crop Prot.* 24(11): 939

Cachenero, J.M., A. Hervas, R.M. Jimenez-Dias and M. Tena. 2002. Plant defence responses against Fusarium wilt in chickpea induced by incompatible race O of *Fusarium oxysporum* f. sp. *ciceris* and nonhost isolates of *Fusarium oxysporum*. *Plant Pathology* 51(6): 765

Cao, L., Z. Qiu, X. Dai *et al.*, 2004. Isolation of endophytic actinomycetes from roots and leaves of banana (*Musa acuminata*) plants and their activities against *Fusarium oxysporum* f. sp. *cubense*. *World J. Microbiol Biotechnol.* 20(5): 501

Carisse, O., D. Rolland and D.M. Tremblay. 2006. Effect of *Microsphaeropsis ochrea* on production of sclerotia-borne and air-borne conidia of *Botrytis squamosa*. *BioControl* 51(1): 107

Chahal, K.S., S.S. Sokhi and G.S. Rattan. 2003. Investigations on sheath blight of rice in Punjab. *Indian Phytopath.* 56(1): 22

Chanchaichichovat, A., P. Ruenwongsa and B. Panijpan. 2007. Screening and identification of yeast strains from fruits and vegetables: Potential for biological control of postharvest chilli anthracnose (*Colletotrichum capsici*). *Biol. Control* 42(3): 326

Charlton, N.D. and M.A. Cubeta. 2007. Transmission of the M2 double stranded RNA in *Rhizoctonia solani* anastomosis group 3 (AG-3). *Mycologia* 99(6): 859

Chaudhary, R.G., K. Kumar and Vishwa Dhar. 2001. Influence of dates of sowing and moisture regimes in pigeonpea on population dynamics of *Fusarium udum* at different soil strata. *Indian Phytopath.* 54: 44

Chen, L.Y., T.V. Price and Z. Park-Ng. 2003. Conidial dispersal by *Alternaria brassicicola* on Chinese cabbage (*Brassica pekinensis*) in the fild and under simulated conditions. *Plant Pathology* 52(5): 536

Cho, J.Y., H.Y. Kim *et al.*, 2006. Dehydro-O-lapachone isolated from Catalpa avata: activity against plant pathogenic fungi. *Pest Manag. Sci.* 62(5): 414

Chowdhury, A.K. 2000. Induction of resistance in chickpea plants against Fusarium wilt infection by seed treatment with non-conventional chemicals. *J. Mycol. Pl. Pathol.* 30(1): 53

Cools, H.J. and H. Ishii. 2002. Pre-treatment of cucumber plants with acibenzolar-S-methyl systemically primes a phenyl ammonia lyase gene (PAL 1) for enhanced expression upon attack with a pathogenic fungus. *Physiol. Mol. Plant Pathol.* 61(5): 227

Cota, I.L., R. Troncoso-Rohas *et al.*, 2007. Chitinase and B-1,3-glucanase enzymatic activity in response to infection by *Alternaria alternata* evaluated in two stages of development in different tomato varieties. *Sci. Hortic.* 112(1): 42

Culbreath, A.K., K.L. Stevenson and T.B. Brenneman. 2002. Management of late leaf spot of peanut with benomyl and chlorothalonil: A study in preserving fungicide utility. *Plant Dis.* 86(4): 349

Dang, J.K., M.S. Sangwan, N. Mehta and C.D. Kaushik. 2000. Multiple disease resistance against four fungal foliar diseases of rapeseed-mustard. *Indian Phytopath.* 53(4): 455

Das, B.C. and D.K. Hazarika. 2000. Biological management of sheath blight of rice. *Indian Phytopath.* 53: 433

Dayakar, V., N.N. Narayanan and S.S. Gnanamanickam. 2000. Cross compatibility and distribution of mating type alleles of the rice blast fungus *Pyricularia grisea* in India. *Plant Dis.* 84: 700

De Cal, A. and P. Melgarejo. 2001. Repeated application of *Penicillium oxalicum* prolongs biocontrol of Fusarium wilt of tomato plants. *Eur. J. Plant Pathol.* 107(8): 805

De Cal, A., R. Garcia-Lepe and P. Melgarejo. 2000. Induced resistance by *Penicillium oxalicum* against *Fusasrium oxysporum* f. sp. *lycopersici*: Histological studies of infected and induced tomato stems. *Phytopathology* 90: 260

Deepak, S.A., H. Ishii and P. Park. 2005. Acibenzolart-S-methyl primes cell wall strengthening genes and reactive oxygen species forming/scavenging enzymes in cucumber after fungal pathogen attack. *Physiol. Mol. Plant Pathol.* 69(1–3): 52

De Weert, S., I. Kuiper, E.L. Lagendijk et al., 2004. Role of chemotaxis toward fusaric acid in coloniziation of hyphae of *Fusarium oxysporum* f. sp. *radicis-lycopersici* by *Pseudomonas fluorescens* WCS365. *Mol. Plant-Microbe Interract.* 17(11): 1185

Di Pietro, A., M.P. Madrid, Z. Caracuelk et al., 2003. *Fusarium oxysporum*: exploring the molecular arsenal of a vascular wilt fumhis. *Mol. Plant Pathol.* 4(5): 315–325

Dita, M.A. et al., 2007. Histopathological studies of the *Alternaria solani* infection process in potato cultivars with different levels of early blight resistance. *J. Phytopathol.* 155(7–8): 462

Dong, H. and Y. Cohen. 2002a. Induced resistance in cotton seedlings against Fusarium wilt by dried biomass of *Penicillium chrysogenum* and its water extract. *Phytoparasitica* 30(1): 77

Dong, H. and Y. Cohen. 2002b. Dry mycelium of *Penicillium chrysogenum* induces resistance against Verticillium wilt and enhances growth of cotton plants. *Phytoparasitica* 30(2): 147

Dong, H., X. Zxhabg, Y. Coheb et al., 2005. Dry mycelium of *Penicillium chrysogenum* protects cotton plants against wilt diseases and increases yield under field conditions. *Crop Protection* 25(4): 324

Drori, N., H. Kramer-Haimovich, J. Rollins, et al., 2003. External pH and nitrogen source affect secretion of pectate lyase *Colletotrichum gloeosporioides*. *Appl. Environ. Microbiol.* 69(6): 3258

Duan, G., Z. Zhang, J. Zhang, Y. Zhou et al., 2007. Evaluation of crude toxin and metabolite produced by *Helminthosporium graminearum* Rebenh. For the control of rice sheath blight in paddy fields. *Crop Prot.* 26(7): 1036

Dugan, F.M., S.L. Lupien and G.G. Grove. 2002. Incidence, aggressiveness and *in planta* interactions of *Botrytis cinerea* and other filamentous fungi quiescent in grape berries and dormant buds in Central Washington state. *J. Phytopath.* 150: 375

Duijff, B.J., G. Recorbet, P.A.H.M. Bakker, J.E. Lopee and P. Lemansceau. 1999. Microbial antagonism at the root level is involved in the suppression of Fusarium wilt by the combination of nonpathogenic *Fusarium oxysporum* Fo47 and *Pseudomonas putida* ECS358. *Phytopathology* 89: 1073

Dutta, S., A.K. Mishra and B.S. Dileep Kumar. 2008. Induction of systemic resistance against Fusarium wilt in pigeonpea through interaction of plant growth promoting rhizobacteria and rhizobia. *Soil Biol. Biochem* 40(2): 452–461

Errampalli, D. and H.W. Johnston. 2001. Control of tuber-borne black scurf (*Rhizoctonia solani*) and common scab (*Streptomyces scabies*) of potatoes with a combination of sodium hypochlorite and thiophanate methyl preplanting seed tuber treatment. *Can. J. Plant Pathol.* 23(1): 68

Errampalli, D., R.D. Peters, K. MacIsaac et al., 2006. Effect of a combination of chlorine dioxide and thiophanate-methyl pre-plant seed tuber treatment on juuthe control of black scurf of potatoes. *Crop Protection* 25(12): 1231–1237

Fakhouri, W.D. and H. Buchenauer. 2002. Enhancement of population densities of fluorescent pseudomonads in the rhizosphere of tomato plants by addition of acibenzolar-S-methyl. *Can. J. Microbiol.* 48(12): 1069

Fakhouri, W.D., M. Neemann, F. Walker and H. Buchenauer, 2004. Application of fluorescent pseudomonadfs in combinaytion with acibenzolar-S-methyl induces disease resistance in tomato and tobacco. *J. Plant Dis. Prot.* 111(5):494

Fernandez-Falcon, M., A.A. Borges and A. Borgez-Perez. 2004. Response of Dwarf Cavendish banana plantlets to inoculation with races 1 and 4 of *Fusarium oxysporum* f. sp. *cubense* at different levels of Zn nutrition. *Fruits* 59: 319

Fravel, D., C. Olivain and C. Alabouvette. 2003. *Fusarium oxysporum* and its biocontrol. *New Phytologist* 157(3): 493

Fritz, M., I. Jakobsen, M.F. Lyngkjaer et al., 2006. Arbuscular mycorrhiza reduce susceptibility of tomato to *Alternaria solani*. *Mycorrhiza* 16(6): 413

Gnanamanickam, S.S. and T.W. Mew. 1992. Biological control of blast of rice (*Oryza sativa*) with antagonistic bacteria and its mediation by a Pseudomonas antibiotic. *Ann. Phytopath. Soc. Japan* 58: 380

Gnanamanickam, S.S., B.L. Candole and T.W. Mew. 1994. Influence of soil factors and cultural practices on biological control of sheath blight of rice with antagonistic bacteria. *Plant Soil* 144: 67 ABS

Gnanamanickam, S.S., N.N. Narayanan, P. Vasudevan, B.V. Priyadarshini, S. Kavitha et al., 2002. Role of biocontrol agents, resistance genes and transgenes for the management of rice diseases in India. *Indian Phytopath.* 55(3): 358(abstr)

Gomez-Rodriguez, O, E. Zavaleta-Meija, V.A. Gonzalez-Hernandez, et al., 2003. Allelopathy and microclimatic modification of intercropping with marigold on tomato early blight disease development. *Field Crop Res.* 83(1): 27

Govender, V., L. Korsten and D. Shivakumar. 2005. Semi-commercial evaluation of *Bacillus licheniformis* to control mango postharvest diseases in South Africa. *Postharvest Biol. Technol.* 38(1): 57

Govrin, E.M., S. Racchmiilevitch, B.S. Toiwari et al., 2006. An elicitor ftom *Botrytis cinerea* induces the hypersensitive response in *Arabidopsis thaliana* and other plants and promotes the gray mold disease. *Phytopathology* 96(3): 299

Grosch, R., F. Faltin, J. Lottmann, A. Kofoet and G. Berg. 2005. Effectiveness of 3 antagonistic bacterial isolates to control *Rhizoctonia solani* Kuhn on lettuce and potato. *Can. J. Microbiol.* 51(4): 345

Groth, D.E. 2008. Effects of cultivar resistance and single fungicide application on sheath blight, yield, and quality. *Crop Protection* 27(7): 1125

Groth, D.E. and J.A. Bond. 2006. Initiation of rice sheath blightepidemics and effect of application time of azoxystrobin on disease incidence, severity, yield and milling quality. *Plant Dis.* 90(8): 1073

Gupta, K., G.S, Saharan, N. Mehta and M.S. Sangwan. 2004. Identification of pothotypes of *Alternaria brassicae* from Indian mustard (*Brassica juncea L.*). *J. Mycol. Plant Pathol.* 34(1): 15–19

Gupta, R., R.P. Awasthi and S.J. Kolte. 2003. Influence of sowing dates and weather factors on development of Alternaria blight on rape seed-mustard. *Indian Phytopath.* 56(4): 398

Harish, S., D. Saravankumar, R.R adjacoommare *et al.*, 2008. Use of plant extracts and biocontrol agents for the management of brown spot disease in rice. *BioControl* 53(3): 555

Hayasaka, T., H. Fujii and T. Namai. 2005. Silicon content in rice seedlings to protect rice blast fungus at the nursery stage. *J. Gen. Plant Pathol.* 71(3): 169

Hazarika, D.K., L.N. Dubey and K.K. Das. 2000. Effect of sowing dates and weather factors on development of leaf spots and rust of groundnut. *J. Mycol. Pl. Pathol.* 30(1): 27

Herbert, C., R. O'Gaulin, V. Salesses *et al.*, 2004. Production of a cell wall-associated endopolygalacturonase by *Colletotrichum lindemuthianum* and pectin degradation during bean infection. *Fungal Genet. Biol.* 41(2): 140

Hwang, S-C. 2004. Cavendish Banana Cultivars Resistant to Fusarium Wilt Acquired through somaclonal variation in Taiwan. *Plant Dis.* 88(6): 580

Ishikawa, R., K. Shirosuzu, H. Nakashita *et al.*, 2005. Foliar spray of validomycin A and Validoxylamie A controls tomato Fusarium wilt. *Phytopathology* 95(10): 1209

Jadeja, K.B., D.M. Nandolia, I.U. Dhruj and R.R. Khandar. 1999. Efficacy of four triazole fungicides in the control of leaf spots and rust of groundnut. *Indian Phytopath.* 52: 421

Jayakumar V., R. Bhaskaran and S. Tsushima. 2007. Potential of plant extracts in combination with bacterial antagonist treatment as biocontrol agent of red rot of sugarcane. *Can. J. Microbiol.* 53(2): 196

Jayaraj, J., H. Yi, G.H. Liang, S. Muthukrishnan and R. Velazhahan, 2004. Foliar application of *Bacillus subtilis* AUBS1 reduces sheath blight and triggers defense mechanisms in rice. *J. Plant Dis. Prot.* 111(2): 115

Jeun, Y.C., K.S. Park, C.H. Kim, W.D.Fowler and J.W. Kloepper. 2004. Cytological observations of cucumber plants during induced resistance elicited by rhizobacteria. *Biol. Control* 29(1): 34

Joshi, M.M. and R.S.Singh. 1969. A Botrytis gray mold of gram. *Indian Phytopath.* 22: 125

Kaiserer, L., C. Oberparleiter, R. Weiler-Gorz *et al.*, 2003. Characterization of the *Penicillium chrysogenum* antifungal protein PAF. *Arch. Microbiol.* 180(3):204

Kalra, A., R.K.Grover, N. Rishi, S.M.P. Khurana and B.P. Singh. 2000. Effect of extracts, leachates and washings from potato virus X infected potato leaves on aggressiveness of *Alternaria solani. Indian Phytopath.* 53: 10

Kamble, A. and S. Bhargava. 2007. β-aminobutyrioc acid-induced resistance in *Brassica juncea* against the necrotrophic patjogen *Alternaria brassicae. J. Phytopathol.* 155(3): 152

Kapoor, A.S. and G.K. Sood. 2000. Effect of time of application and splitting of nitrogen on rice blast. *Indian Phytopath.* 53: 283

Kapoor, A.S., R. Prasad and G.K. Sood. 2004. Prediction of rice blast in district Kangra of Himachal Pradesh. *Indian Phytopath.* 57(4): 446

Karthikeyan, V. and S. Gnanamanickam. 2008. Biological control of Setaria blast (*Magnaporthe grisea*) with bacterial strains. *Crop Prot* 27(2): 263

Katastonis, D., R.J. Hillock and S. Gowen. 2005. Enhancement of germination of spores of *Verticillium dahliae* and *Fusarium oxysporum* f. sp. *vasinfectum* in vascular fluid from cotton plants infected with the root knot nematode. *Phytoparasitica* 33(3): 215

Kaur, R. and R.S. Singh. 2007. Study of induced systemic resistance in *Cicer arietinum* L. due to nonpathogenic *Fusarium oxysporum* using a modified split root technique. *J. Phytopathol.* 155(11–12): 694

Khan, M.R. and W. Nupendra Singh. 2001. Effects of soil application of fly ash on the fusarial wilt on tomato cultivars. *International J. Pest Management* 47(4): 293

Kim, S.G., K.W. Kim, E.W. Park and D. Choi. 2002. Silicon induced cell wall fortification of rice leaves: A possible cellular mechanism of enhanced host resistance to blast. *Phytopathology* 92(10): 1095

Kim, Y.A., H.H. Lee, M.K.Ko, C.E. Song, C-Y. Bae, Y.H. Lee and B.J. Oh. 2001. Inhibition of fungal appressorium formation by pepper (*Capsicum annuum*) esterase. *Mol. Plant-Microbe Interact.* 14(1): 80

Kiprop, E.K., J.P. Baudoin *et al.*, 2002a. Characteristics of Kenyan isolates of *Fusarium udfum* from pigeonpea (*Cajanus cajan*) by cultural characteristics, aggressiveness and AFLP analysis. *J. Phytopath.* 150(10)

Kiprop, E.K., A.W. Mwang'ombe, J.P. Baudoin, P.M. Kimanl and G. Mergeai. 2002b. Cultural characteristics, pathogenicity and vegetative compatibility of *Fusarium udum* isolates from pigeonpea (*Cajanus cajan* (L.) Millsp.) in Kenya. *Eur. J. Plant Pathol.* 108(2): 147.

500

Kishore, G.K. and S. Pande. 2005a. Integrated application of aqueous leaf extracts of *Datura metel* and chlorothalonil improved control of late leaf spot and rust of groundnut. *Aust. Plant Pathol.* 34(2): 261

Kishore, G.K. and S. Pande. 2005b. Integrated management of late leaf spot and rust diseases of groundnut (*Arachis hypogea*) with with *Prosopis juliflora* leaf extract and chlorothalonil. *Intern, J. Pest Manag.* 51(4): 325

Kishore, G.K. and S. Pande. 2007. Chitin-supplemented foliar application of chitinolytic *Bacillus cereus* reduces severity of Botrytis gray mold disease in chickpea under controlled conditions. *Lett. Appl. Microbiol.* 44(1): 98

Kishore, G.K., S. Pande and A.R. Podile. 2005. Chitin supplemented foliar application of *Serratia marcescens* GPS5 improves control of late leaf spot of groundnut by activating defence related enzymes. *J. Phytopath.*153(3): 169

Kishore, G.K., S. Pande and A.R. Podile. 2005a. Management of late leaf spot of groundnut (*Arachis hypogea*) with chlorothalonil-toilerant isolates of *Pseudomonas aeruginosa*. *Plant Pathology* 54(3): 401

Kishore, G.K., S. Pande and A.R. Podile. 2005b. Biological control of late leaf spot of peanut (*Arachis hypogea*) with chitinolytic bacteria. *Phytopathology* 95(10): 1157

Kishore, G.K., S. Pande and S. Harish. 2007. Evaluation of essential oils and their components for broad spectrum antifungal activity and control of late leaf spot and crown rot diseases in peanut. *Plant Dis.* 91(4): 375

Koga, H. and O. Nakayachi. 2004. Morphological studies on attachment of spores of *Magnaporthe grisea* to the leaf surface of rice. *J. Gen. Plant Pathol.* 70(1): 11

Koga, H., K. Dohi and M. Mori. 2004. Abscisic acid and low temperatures suppress the whole plant-specific resistance reaction of rice plants to the infection of *Magnaporthe grisea*. *Physiol. Mol. Plant Pathol.* 65(1): 3

Kotasthane, A.S. and U.S. Singh. 2000a. Perithecial morphogenesis in *Magnaporthe grisea*. *J. Mycol. Pl. Pathol.* 30: 172

Kotasthane, A.S. and U.S. Singh. 2000b. Mating behaviour of a hermaphrodite *Magnaporthe grisea* isolate from Foxtail millett. *J. Mycol. Pl. Pathol.* 30: 331

Krishnamurthy, K. 1998. Induction of systemic resistance and salicylic acid accumulation in *Oryza sativa* L. in the biological suppression of rice blast caused by treatment with *Pseudomonas* spp. *World J. Microbiol. Biotechnol.* 14(6): 935

Krishnamurthy, K. and S.S. Gnanamanickam. 1998. Biocontrol of rice sheath blight with formulated *Pseudomonas putida*. *Indian Phytopath.* 51: 233

Kumar, B. and S.J. Kolte. 2001. Progression of Alternaria blight of mustard in relation to components of resistance. *Indian Phytopath.* 54(3): 329

Kumar, S., N. Mehta, M.S. Sangwan and R. Kumar. 2004. Relative sensitivity of various isolates of *Alternaria brassicae* to fungicides. *J. Mycol. Plant Pathol.* 34(1): 28

Kurahashi, Y., S. Sakawa, H. Sakuma *et al.,* 1999. Effect of carpropamid on secondary infection by the rice blast fungus. *Pesticide Sci.* 55(1): 31

Laha, G.S. and S. Venkataraman. 2001. Sheath blight management in rice with biocontrol agents. *Indian Phytopath.* 54(4): 461.

Landa, B., B., J.A. Navas-Cortes, A. Hervas and R.M. Jimenez-Diaz. 2001. Influence of temperature and inoculum density of *Fusarium oxysporum* f. sp. *ciceris* on suppression of Fusarium wilt of chickpea by rhizosphere bacteria. *Phytopathology* 91: 807

Landa, B.B., J.A. Navas-Cortes and R.M. Jimenez-Diaz. 2004a. Influence of temperature on plant-rhizobacteria interactions related to biocontrol potential for suppression of Fusarium wilt of chickpea. *Plant Pathology* 53(3): 341

Landa, B.B., J.A. Navas-Cortes and R.M. Jimenez-Diaz. 2004b. Integrated management of Fusarium wilt of chickpea with sowing date, host resistance and biological control. *Phytopathology* 94(9): 946

Landa. B.B., J.A. Navas-Cortes *et al.,* 2006. Temperature response of chickpea cultivars to races of *Fusarium oxysporum* f. sp. *ciceris*, causal agent of Fusarium wilt. *Plant Disease* 90(3): 365

Larena, I., P. Sabuquillo, P. Melgarejo and A. De Cal. 2003. Biocontrol of Fusarium and Verticillium wilt of tomato by *Penicillium oxalicum* under greenhouse and field conditions. *J. Phytopath.* 151(9): 507

Larkin, R.P. 2008. Relative effects of biological amendments and crop rotations on soil microbial communities and soilborne diseases of potao *Soil Biol. Biochem.* **IN PRESS**

Larkin, R.P. and D.R. Fravel. 1999. Mechanism of action and dose-response relationship governing biological control of Fusarium wilt of tomato by nonpathogenic *Fusarium* spp. *Phytopathology* 89: 1152

Larkin,, R.P. and C.W. Honeycult. 2006. Effect of different 3-year cropping systems on soil microbial communities and Rhioctonia disease of potato. *Phytopathology* 96(1): 68

Larkin, R.P. and T.S. Griffin. 2007. Control of soilborne potato diseases using *Brassica* green manure. *Crop Prot.* 26(7): 1067

Latterell, F.M. and A.E. Rossi. 1986. Longevity and pathogenic stability of *Pyricularia oryzae*. *Phytopathology* 76 : 231

Lee, J,T, S.S. Moon and B.K. Whang, 2003a. Isolation and *in vitro* and *in vivo* activity against *Phytophthora capsici* and *Colletotrichum orbiculare* of phenazine-1-carboxylic acid from *Pseudomonas aeruginosa* strain GC-B26. *Pest Manage. Sci.* 59(8): 872

Lee, J.Y., S.S. Moon and B.K. Whang. 2003b. Isolation and antifungal and antioomycete activities of aerugine produced by *Pseudomonas fluorescens* strain MM-B16. *Appl. Environ. Microbiol.* 69(4): 2023

Lodha, S. and R.K. Aggarwal. 2002. Inactivation of Macrophomina phaseolina propagules during composting and effect of composts on dry root rot severity and on seed yield yield of clusterbean. *Eur. J. Plant Pathol.* 108(3): 253

Long, D.H., F.N. Lee and D.O. Te Beest. 2000. Effect of nitrogen fertilization on disease progress of rice blast on susceptible and resistant cultivars. *Plant Dis.* 84: 403

Madhivanan, N, V.R. Prabhavathy and V.R. Vijayanandraj. 2005. Application of tal formulations of *Pseudomonas fluorescens* and *Trichoderma viride* decrease the sheath blight disease and increase plant growth and yield of rice. *J. Phytopathol.* 153(11–12): 697

Madhukeshwara, S.S., B.C. Shankaralingappa, S.G. Mantur and T.B. Anilkumar. 2003. Effect of intercrops and fertility levels on wilt incidence of pigeonpea. *Indian Phytopath.* 56(1): 88

Malathi, P., R. Viswanathan, P. Padmanabhan, D. Mohanraj and A. Ramesh Sundar. 2002. Microbial detoxification of *Colletotrichum falcatum* toxin. *Current Science* 83(6): 749

Manandhar, H.K., H.J.L. Jergensen, S.B. Mathur and V. S-Petersen. 1998. Resistance to rice blast induced by ferric chloride, dipotassium hydrogen phosphate and salicylic acid. *Crop Protection* 17(4): 323

Mayak, S., T. Tirosh and B.R. Glick. 2004. Plant growth-promoting bacteria that confer resistance to water stress in tomatoes and peppers. *Plant Sci.* 166(2): 525

Mayee, C.D. and A.P. Suryawanshi. 1995. Structural defence mechanisms in groundnut to late leaf spot pathogen. *Indian Phytopath.* 48: 160

McQuilken, M.P. and J. Gemmell. 2004. Enzyme production by the mycoparasite *Verticillium biguttatum* against *Rhizoctonia solani*. *Mycopathologia* 157(2): 201

Meena, B., V. Ramamoorthy, T. Marimuthu and R. Velazhahan. 2000. *Pseudomonas fluorescens* mediated systemic resistance against late leaf spot of groundnut. *J. Mycol. Pl. Pathol.* 30(2): 151

Meena, B., R. Radhajeyalakshmi, T. Marimuthu, P. Vidyasekaran and R. Velazhahan. 2002. Biological control of groundnut late leaf spot and rust by seed and foliar application of a powder formulation of *Pseudomonas fluorescens*. *Biocontrol Sci. Technol.* 12(2): 195

Mehta, N., M.S. Sangwan, M.P. Srivastava and R. Kumar. 2002. Survival of *Alternaria brassicae* causing Alternaria blight of rapeseed-mustard. *J. Mycol. Pl. Pathol.* 32(1): 64

Mew, T.W., B. Cottyu, R. Pamphona *et al.,* 2004. Applying rice seed associated antagonistic bacteria to manage rice sheath blight in developing countries. *Plant Disease* 88(5): 557

Meziane, H., I. Van der Sluic, L.C. van Loon, M. Hofte and P.A.H.M. Bakker. 2005. Determinants of *Pseudomonas putida* WCS358 involved in inducing systemic resistance in plants. *Mol. Plant Pathol.* 6(2): 177

Minuto, A., D. Spadoro, A. Garibaldi *et al.,* 2006. Control of soilborne pathogens of tomato using a commercial formulation of *Streptomyces griseoviride* and solarization. *Crop Prot* 25(5):468–473 **ABS**

Mohanraj, D., M. Karunakaran and P. Padmanabhan. 2003. Pathogen toxin-induced electrolyte leakage and phytoalexin accumulation as ondices of red rot (*Colletotrichum falcatum* Went.) resistance in sugarcane. *Phytopath. Mediterr.* 42(2): 129

Monfort. W.S., A.K. Culbreath, K.L. Stevenson *et al.,* 2004.Effect of reduced tillage, resistant cultivars, and reduced fungicide inputs on progress of early leaf spot of peanut. *Plant Dis* 88(8); 858–864

Moraes, S.R.G., E,A, Poszza, E. Alves *et al.,* 2006. Effect of silicon sources on the incidence and severity of the common bean anthracnose. *Fitopat. bras.* 31(1): 69

Morandi, M.A.H., L.A. Maffia, E.S.G. Mizubuti *et al.,* 2003. Suppression of *Botrytis cinerea* sporulation by *Clonostachys rosea* on rose debris: a valuable component in Botrytis blight management in commercial greenhouse. *Biological Control* 26(3): 311

Muralidharan, K., C.S. Reddy, D. Krishnaveni and G.S. Laha. 2003. Evaluation of plant-derived commercial products for blast and sheath blight control in rice. *Indian Phytopath.* 56(2): 151

Muralidharan, K., C.S. Reddy, D. Krishnaven and G.S. Laha. 2004. Field application of fluorescent *Pseuidomonas* products to control blast and sheath blight disease in rice. *J. Mycol. Pl. Pathol.* 34(2): 411

Muslin, A., H. Horimouchi and M. Hyakumuchi. 2003. Biological control of Fusarium wilt of tomato with hypovirulent binucleate *Rhizoctonia* in greenhouse conditions. *Mycoscience* 33(2): 77

Mutto, M., V. Mulabaga, H.-C. Huang *et al.,* 2006. Toxicity of black nightshade (*Solanum nigrum*) extrats on *Alternaria brassicicola*, causal agent of black spot of Chinese cabbage (*Brassica pekinensis*). *J. Phytopathol.* 154(1): 45

Nandakumar, R., S. Babu, R. Viswanathan, T. Raguchander and R. Samiyappan. 2001a. Induction of systemic resistance in rice against sheath blight disease by *Pseudomonas fluorescens*. *Soil Biol. Biochem.* 33(4–5): 603

Nandakumar, R., S. Babu, R. Viswanathan, J. Sheela, T. Raghuchander and R. Samiyappan. 2001b. A new bio-formulation containing plant growth promoting rhizobacterial mixture for the management of sheath blight and enhanced grain yield in rice. *BioControl* 46(4): 493

Ndiaye, M., A.J. Termorshuien and A.H.C. van Bruggen. 2007. Combined effects of solarization and organic amendment on charcoal rot caused by *Macrophomina phaseolina* in the Sahel. *Phytoparasitica* 35(4): 392

Nel, B., C. Steinberg, N. Labuschagne and A. Viljoen. 2006a. Isolation and characterization of nonpathogenic *Fusarium oxysporum* isolates from the rhizosphere of healthy banana plants. *Plant Pathol.* 55(2): 207

Nel, B., C. Steinberg, N. Labuschagne and A. Viljoen. 2006b. The potential of nonpathogenic *Fusarium oxysporum* and other biological control orgasnisms for suppressing Fusarium wilt of banana. *Plant Pathol.* 55(2): 217

Nischwitz, C., M. Olsen, and S. Rasmussen. 2004. Effect of irrigation on inoculum density of *Macrophomina phaseolina* in melon fields in Arizona. *J. Phytopathol.* 152(3): 133

Ohtaka, N., H. Kawamata and K. Narisawa. 2008. Suppression of rice blast using freeze-killed mycelium of biocontrol fungus *candate* MKP5111B. *J. Gen. Plant Pathol.* 74(2): 101

Olivain, C. and C. Alaboutte. 1999. Process of tomato root colonization by a pathopgenic strain of *Fusarium oxysporum* f. sp. *lycopersici* in comparison with a non-pathogenic strain. *New Phytologist* 141(3): 497

Olivain, C., S. Trouvelot, M.N. Binet, C. Cordier, A. Pugin and C.O. Alabouvette. 2003. Colonization of flax roots and early physiological responses of flax cells inoculated with pathogenic and nonpathgogenic strains of *Fusarium oxysporum*. *Appl. Environ. Microbiol.* 69(9): 5453

Olivain, C., C. Alabouvette and C. Steinberg. 2004. Production of a mixed inoculum of *Fusarium oxysporum* Fo47 and *Pseudomonas fluorescens* C7 to control Fusarium diseases. *Biocontrol Sci. Technol.* 14(3): 227

Olivain, C., C. Humbert, J. Nahalkowa *et al.*, 2006. Colonization of tomato root by pathogenic and nonpathogenic *Fusarium oxysporum* strains inoculated together or separately into the soil. *Appl. Environ. Microbiol.* 72(2):1523

Otani, H., A. Kohnobe, M. Kodama and K. Kohmoto. 1998. Production of a host-specific toxin by germinating spores of *Alternaria brassicocola*. *Physiol. Mol. Plant Pathol.* 52: 285

Pajand, N. and P.J. Johonson. 2000. Endophytic bacteria induce growth promotion and wilt disease suppression in oilseed rape and tomato. *Biological Control* 18(2): 208

Pande, S. and J.N. Rao. 2001. Resistance of wild *Arachis* species to late leaf spot and rust in greenhouse trials. *Plant Dis.* 85:(8): 851

Pandey, K.K. and J.P. Upadhyay. 1999. Comparative study of chemical, biological and integrated approach for management of Fusarium wilt of pigeonpea. *J. Mycol. Pl. Pathol.* 29: 214

Pandey, K.K. and J.P. Upadhyay. 2000. Microbial populations from rhizosphere and non-rhizosphere soil of pigeonpea: Screening for resident antagonists and mode of mycoparasitism. *J. Mycol. Pl. Pathol.* 30(1):

Panina, T., D.R. Fravel, C.J. Baker and L.A. Scherbakova. 2007. Biocontrol and plant pathogenic *Fusarium oxysporum*-induced changes in phenolic compounds in tomato leaves and roots. *J. Phytopathol.* 155(7–8): 475

Paparu, P., T. Debois, C.S. Gold, B. Niere *et al.*, 2006. Colonization pattern of nonpathogenic *Fusarium oxysporum*, a potential; biological control agent, in roots and rhizomes of tissue cultured *Musa* plamtlets. *Ann. Appl. Biol.* 149(1): 1.

Parada, R.Y., K. Oka, D. Yamagishi *et al.*, 2008. Destruxin B produced by Alternaria brassicae does not induce accessibility of host plants to fungal invasion. *Physiol. Mol. Plant Pathol.* **IN Press**.

Pareja-Jaime, Y., M.I.G. Romcero and M.C. Ruiz-Roldan. 2008. Tomatinase from *Fusarium oxysporum* f. sp. *lycopersici* is required for full virulence on tomato plants, *Mol. Plant Microbe Interact.* 21(6): 728

Parveen, S. and V.R. Kumar, 2001. Studies on the seed-borne nature of leaf blight of wheat. *J. Mycol. Pl. Pathol.* 31(3): 292

Parveen, S. and V.R. Kumar. 2004. Antagonism of *Trichoderma viride* against leaf blight pathogen of wheat. *J. Mycol. Pl. Pathol.* 34(2): 220

Pasche, J.S., C.M. Wharam and N.C. Gudmestad. 2004. Shift in sensitivity of *Alternaria solani* in response to Q(o) fungicides. *Plant Dis.* 88(2): 181

Patido-Vera, M., B. Jimenez, K. Balderase *et al.*, 2005. Field-scale production and liquid formulation of *Rhodotorula minuta*, a potential biocontrol agent against mango anthracnose. *J. Appl. Microbiol.* 99(1): 540

Patel, M., I.L. Kothari and J.S.S. Mohan. 2004. Plant defense induced in *in vitro* propagated banana (*Musa paradisiaca*) plantlets by *Fusarium* derived elicitors. *Indian J. Exp. Biol.* 42(7): 728–731

Pavlista, A.D. 2005. Early season application of sulfur fertilizers increase potato yield and reduce tuber defects. *Agron. J.* 97: 599

Pengnoo, A., C. Kusongwiriyawong, L. Nilaratna and M. Kanjanamaneesathian. 2000. Greenhouse and field trials of the bacterial antagonist in pellet formulation to suppress sheath blight of rice caused by *Rhizocotnia solani*. *BioControl* 45(2): 245

Prabakar, K., T. Raguchander, D.S. Saravankumar *et al.*, 2008. Management of postharvest disease of mango: anthracnose incited by *Colletotrichum gloeosporioides*. *Arch. Phytopath. Plant Prot.* **IN PRESS**

Prasad, R.D., R. Rangeswaran, S.V. Hegde and C.P. Anuroop. 2002. Effect of soil and seed application of *Trichoderma harzianum* on pigeonpea wilt caused by *Fusarium udum* under field conditions. *Crop Protection* 21(4): 293

Prasanna Kumar, M.K., V.B. Nargund and A.N.A. Khan. 2002. Impact of post-harvest treayments on fruit diseases and physico-chemical properties of mango (*Mangifera indica* L.). *J. Mycol. Pl. Pathol.* 32(3): 372 (abstract)

Priyadarshini, V.B., G. Viji and S.S. Gananamanickam. 1999. Mating types distribution, fertility, and pathogenicity of rice isolates of *Magnaporthe grisea* in four rice growing regions of India. *Indian Phytopath.* 52: 28

Prusky, D., J.J. McEvoy, B. Leverentz and W.S. Conway. 2001.Local modulation of host pH by *Colletotrichum* species as a mechanism to increase virulence. *Mol. Plant-Microbe Interact.* 14(9): 1105

Querino, C.M.B., D. Laranjeira *et al.,* 2005. Effect of two resistance inducers on Panama disease severity. *Fitopat. Bras.* 30(3): 239

Radjacommare, R., R. Nandakumar, A. Kandan *et al.,* 2002. *Pseudomonas fluorescens* based bio-formulation for the management of sheath blight disease and leaffolder insect in rice. *Crop Protection* 21(8): 671

Radjacommare, R., A. Ramanathan, A. Kandan *et al.,* 2005. PGPR mediates induction of pathogenesis-related (PR) proteins against the infection of blast pathogen in resistant and susceptible ragi (*Eleusine coracana*) cultivars. *Plant Soil* 266(1–2): 165

Ramamoorthy, V. and R. Samiyappan. 2001. Induction of defense-related genes in *Pseudomonas fluprescens*-treated chilli plants in response to infection by *Colletotrichum capsici. J. Mycol. Plant Pathol.* 31(2): 146

Ramamoorthy, V., T. Raguvhander and R. Samiyappan. 2002. Induction of defense-related proteins in tomato roots treated with *Pseudomonas fluorescens* Pf 1 and *Fusarium oxysporum* f. sp. *lycopersici. Plant Soil* 239(1): 55

Rathour, R., B.M.Singh and P. Plaha. 2002. Host species-specific protoplasr damaging activity of spore germination fluids of blast pathogen *Magnaporthe grisea. J. Phytopath.* 150(10): 576

Reino, J.L., R. Hernandez-Galan *et al.,* 2004. Virulence-toxin production relationship in isolates of the plant pathogenic fungus *Botrytis cinerea. J. Phytopath.* 152(10): 5630

Robinson, M.L. and J.W. Deacon. 2002. Double-stranded RNA elements in *Rhizoctonia solani* AG 3. *Mycol. Res.* 106(1):12

Rodrigues, F.A. and L.E. Lawrence. 2005. Silicon and rice disease management. *Fitpat. Brasi.* 30(5): 457

Rodriguez, F.A., L.E. Datnoff, G.H. Komdorfer, K.W. Seebold and M.C. Rush. 2001. Effect of silicon and host resistance on sheath blight developement in rice. *Plant Dis.* 85: 827

Rodriguez, F.A., F.X.R. Vale, L.E. Datnoff, A.S. Prabhu and G.H. Korndorfer. 2003a. Effect of rice growth stages amd silicon on sheath blight development. *Phytopathology* 93(3): 256

Rodriguez, F.A., N. Benhamou, L.E. Datnoff, J.B. Jones and R.R. Belanger. 2003b. Ultrastructural and cytological aspects of silicon-mediated rice blast resistance. *Phytopathology* 93(5): 535

Rodriguez, F.A., D.J. McNally, L.E. Datnoff, J.B. Jones, C. Labbe, N. Benhamou, J.G. Menzies and R.R. Belanger. 2004. Silicon enhances the accumulation of diterpenoid phytoalexins in rice: A potential mechanism for blast resistance *Phytopathology* 94(2): 177

Rodriguez, F.A., W.M. Jurick, L.E. Darnoff, J.B. Joneas and J.A. Rollins. 2005. Silicon influences cytological and molecular events in compatible and in compatible rice- *Magnaporthe grisea* interactions. *Physiol. Mol. Plant Pathol.* 66(4): 144

Rodriguez-Molina, M.C., J. Medina, L.M. Torres-Vila and J. Cuartere. 2003. Vascular colonization patterns in susceptible and resistant tomato cultivars inoculated with *Fusarium oxysporum* f. sp. *lycopersici* races 0 and 1. *Plant Pathology* 52(2): 199

Rohilla, R. and U.S. Singh. 1999. Mode of action of carpropamid against rice blast caused by *Magnaporthe grisea. J. Mycol. Pl. Pathol.* 29: 159

Rohilla, R., U.S. Singh and R.L. Singh. 2002. Mode of action of acibenzolar-S-methyl against sheath blight of rice caused by *Rhizoctonia solani* Kuhn. *Pest Manage. Sci.* 58(1): 63-69

Sabuquillo, P., A. De Cal and P. Melgarejo. 2005. Dispersal improvement of a powder formulation of *Penicillium oxalicum*, a biocontrol agent of tomato wilt. *Plant Dis.* 89(12): 1317

Sabuquillo, P., A. De Cal and P. Melgarejo. 2006. Biocontrol of tomato wilt by *Penicillium oxalicum* formulations in different crop conditions/ *Biological Control* 37(3): 256

Saharan, G.S. 2000. Multiple disease resistance in rapeseed-mustard. *Indian Phytopath.* 53: 342

Saifulla, M. and M. Byre. 2002. Screening of pigeonpea genotypes against multiple disease resistance. *Indian Phytopath.* 55(3): 367 (abstr)

Saikia, R., P. Azad and D.K. Arora. 2004, Purification and partial characterization of toxin produced by *Colletotrichum capsici* that causes ripe rot of chilli. *J. Mycol. Pl. Pathol.* 34(2): 421

Sales Jr.,R., F.M. da Costa, E.M. Romaine *et al.,* 2004. The use of azoxystrobin in the control of mango anthracnose. *Fitopat. Bras.* 29(3): 193

Sanders, G.M., L. Korsten and F.C. Wehner. 2000. Survey of fungicide sensitivity in *Colletotrichum gloeosporioides* from different avocado and mango production areas in South Africa. *Eur. J. Plant Pathol.* 106: 745

Saravanan, T., R. Bhaskaran and M. Muthusamy. 2004. *Pseudomonas fluorescens*-induced enzymological changes in banana roots (cv. Rasthali) against Fusarium wilt disease. *Plant Pathol. J.* 3(2): 72–80

Sarkar, S.K. 2000. Effect of intercropping jowar on the incidence of sunnhemp wilt (*Fusarium udum* f. sp. *crotolariae*). *J. Mycol. Pl. Pathol.* 30(1): 41

504

Sarvankumar, D., S. Harish, et al., 2007. Rhizobacteria bioformulation for the effective management of Macrophomina phaseolina root rot in mungbean. Arch. Phytopath. Plant Prot. 40(5): 323

Seebold, K.W., T.A. Kaucharck, L.E. Datnoff et al., 2001. The influence of silicon on components of resistance to blast in susceptible, partially resistant and resistant cultivars of rice. Phytopathology 91: 63

Shanmugam, V., T. Raguchander, P. Balasubramanian and R. Samiyappan. 2001. Inactivation of Rhizoctonia solani toxin by a putative á-gucosidase from coconut leaves for control of sheath blight disease of rice. World J. Microbiol. Biotechnol. 17(6): 545

Shanmugam, V., S. Sriram, S. Babu, et al., 2001. Purification and characterization of an extracellular á-glucosidase protein from Trichoderma viride which degardes a phytotoxin associated with sheath blight disease in rice. J. Appl. Microbiol. 90(3): 320

Sharma, Roopali, S.Varshney, D. Joshi, S. Singhal and U.S.Singh, 2002. Farmyard manure: a promising substrate for mass multiplication and developing commercial formulation of Trichoderma harzianum. Indian Phytopath. 55(3): 382 (abstr)

Shi, J., W.C. Mueller and C.H. Beckman. 1992. Vessel occlusion and secretory activities of vessel contact cells in reisistant or susceptible cotton plants infected with Fusarium oxysporum f. sp. vasinfectum. Physiol. Mol. Plant Pathol. 40(2): 133

Shirakaishi, S., I. Watanabe, K. Kuno, H. Ishii and Y. Fujii. 2003. Soil drenching with water extracts of Oxalis articulata suppress Fusarium wilt of tomato. Weed Biol. Management 3(3): 184

Shrestha, C.L., I. Oha, S. Muthukrishnan and T.W. Mew. 2008. Chitinase levels in rice cultivars correlate with resistance to the sheath blight pathogen Rhizoctonia solani. Eur. J. Plant Pathol. 120(1): 69

Shtienberg, D. and W.E. Fry. 1990. Influence of host resistance and crop rotation on initial appearance of potato early blight. Plant Dis. 74: 849

Siddiqui, Z.A. and L.P. Singh. 2004. Effects of soil inoculants on the growth, transpiration and wilt disease of chickpea. J. Plant Dis, Prot. 111(2): 151

Siddiqui, Z.A. and M.S. Akhtar. 2007. Biocontrol of a chickpea root rot disease complex with phosphate solubilizing microorganisms. J. Plant Pathol. 89(1): 67

Silimela, M. and L. Korsten. 2007. Evaluation of pre-harvest Bacillus licheniformis sprays to control mango fruit diseases. Crop Prot. 26(10): 1474

Singh, Amita, R. Rohilla, S. Savary, L. Willocquet and U.S. Singh. 2003. Infection process in sheath blight of rice caused by Rhizoctonia solani. Indian Phytopath. 56(4): 434

Singh, F., I. Hooda and G.S. Sindhan. 2004a. Biological control of tomato wilt caused by Fusarium oxysporum f. sp. lycopersici. J. Mycol. Pl. Pathol. 34(2): 568

Singh, F., I. Hooda and G.S. Sindhan. 2004b. Substrate media for growth and sporulation of antagonists of Fusarium oxysporum f. sp. lycopersici. J. Mycol. Pl. Pathol. 34(2): 673

Slaton. N.A., R.D. Cartwright, J. Meng, E.E. Gbur ,Jr., and R.J. Normal. 2003. Sheath blight severity and rice yield as affected by nitrogen rate, application method and fungicide. Agron. J. 95: 1489

Snapp, S.S., K.U. Date, W. Kirk et al., 2007. Root, shoot tissues of Brassica juncea and Cereale secale promote potato health. Plant Soil 292(1–2): 55

Someya, N., M. Nakajima, K. Watanabe, T. Hibi and K. Akutsu. 2003b. Influence of bacteria isolated from rice plants and rhizospheres on antibiotic production by the antagonistic bacterium Serratia marcesxens strain B2. J. Gen. Plant Pathol. 69(5): 342

Someya, N., M. Nakajima, K. Watanabe, T. Hibi and K. Akutsu. 2005. Potential of Serratia marcescens strain B2 for biological control of rice sheath blight. Biocontr. Sci. Technol. 15(1):105

Sriram. S., T. Raguchander, S. Babu, R. Nanadakumar et al., 2000. Inactivation of phytotoxin produced by the rice sheath blight pathogen Rhizoctonia solani. Can. J. Microbiol. 46(6): 520

Stanley, M.S., M.E. Callow, R. Perry et al., 2002. Inhibition of fungal spore adhesion by zosteric acid as the basis for a novel non-toxic crop protection technology. Phytopathology 92(4): 378

Sticher, L., B. Mauch-Mani and J.P. Metraux. 1997. Systemic acquired resistance. Annu. Rev. Phytopathol. 35: 235

Steinkellner, S., R. Mammerler and H. Vierheilig. 2008. Germination of Fusarium oxysporum in root exudates from tomato plants challenged with different Fusarium oixysporum strains. Eur. J. Plant Pathol. IN PRESS

Sue, G., S.-O. Suh, R.W. Schneider and J.S. Russin. 2001. Host specialization in the charcoal rot fungus Macrophomina phaseolina. Phytopathology 91(2): 120

Sukhada, M., M. Manamohan, R.D. Rawal, et al., 2004. Interaction of Fusarium oxysporum f. sp. cubense with Pseudomonas fluorescens pre colonized in banana roots. World J. Microbiol. Biotechnol. 20(6): 651–655

Sundar, A.R., R. Velazhahan, R. Viswanathan, P. Padmanabhan and P. Vidyasekaran. 2001. Induction of systemic resistance to Colletotrichum falcatum in sugarcane by a synthetic signal molecule, acibenzolar-S-methyl. Phytoparasitica 29(3): 231–242

Suriachandraselvan, M., F. Salalrajan, K.E.A. Aiyanathan and K. Seetharaman. 2004. Seed treatment with Trichoderma spp. for the control of charcoal rot in sunflower caused by Macrophomina phaseolina. J. Mycol. Pl. Pathol. 34(2): 366

Taheri, P.S. Gnanamanicjam and M. Hofte. 2007. Characterization, genetic structure, and pathogenicity of *Rhizocotnia* spp. associated with rice sheath sheath blight in India. *Phytopathology* 97(3): 373

Thangavelu, R., A. Palaniswami, S. Doraiswamy and R. Velazhahan. 2003. The effect of *Pseudomonas fluorescens* and *Fusarium oxysporum* f. sp. *cubense* on induction of defence enzymes and phenolics in banana. *Biologia Plantarum* 46(1): 107

Thangavelu, R., A. Palaniswami and R. Velazhahan. 2004. Mass production of *Trichoderma harzianum* for managing Fusarium wilt of banana. *Agric. Ecosyst. Environ.* 103(1): 259

Thuerig, B., A. Binder, T. Boller *et al.,* 2006. An aqueous extract of the dry mycelium of *Penicillium chrysoigenum* induces resistance in several crops under controlled and field conditions. *Eur. J. Plant Pathol.* 114(2): 185

Ting, A.S.Y., S. Meon, J. Kadir, S. Radu and G. Singh. 2008. Endophytic microorganisms as potential growth promoters of banana. *BioControl* **IN PRESS**

Tsror, L. and I. Peretz-Alon. 2005. The influence of the inoculum source of *Rhizoctonia solani* on development of black scurf on potato. *J. Phytopath.* 153(4): 240

Ueno, M., J. Kihara, Y. Honda and S. Arase. 2004. Indole-related compounds induce the resistance to rice blast fungus *Magnaporthe grisea* in barley. *J. Phytopath.* 152(11–12): 606

Van den Boogert, P.H.J.F. and A.J.G. Luttikholt. 2004. Compatible biological and chemical control systems for *Rhizoctonia solani* in potato. *Euro. J. Plant Pathol.* 110(2): 111

Van Loon, L.C., P.A.H.M. Bakker and C.M.J. Pieterse. 1998. Systemic reistance induced by rhizosphere bacteria. *Annu. Rev. Phytopathol.* 36: 453

Veneault-Fourrey, C., R. Lauge and T. Langin. 2005. Nonpathogenic strains of *Colletotrichum lindemuthianum* trigger progressive bean defense responses during appressorium-mediated penetration. *Appl. Environ. Microbiol.* 71(8): 1761

Vergnes, D.M., M.-E..Renard, E. Duveiller and H. Maraite. 2006. Identification of *Alternaria* spp. on wheat by pathogenicity assays and sequencing. *Plant Pathology* 55(3): 442

Vidyasekaran, P., E.S. Borromeao and T.W. Mew. 1992. *Helminthosporium oryzae* toxin suppresses phenol metabolism in rice plants and aids pathogen colonization. *Physiol. Mol. Plant Pathol.* 41(5): 307

Vidyasekaran, P. and M. Muthamilan. 1995. Development of formulations of *Pseudomonas fluorescens* for control of chickpea wilt. *Plant Dis.* 79: 782–786

Vidyasekaran, P. and M. Muthamilan. 1995. Development of formulations of *Pseudomonas fluorescens* for control of chickpea wilt. *Plant Dis.* 79: 782

Vidyasekaran, P., R. Rabindran, M. Muthamilan *et al.,* 1997. Development of a powder formulation of *Pseudomonas fluorescens* for control of rice blast. *Plant Pathology* 46: 291

Vidyasekaran, P., K.Sethuraman, K. Rajappan and K.Vasumathi. 1997. Powder formulation of *Psedumonas fluorescens* to control pigeonpea wilt. *Biological Control* 8: 166–171

Vishwadhar, R.G. Chaudhury, R.A. Singh, Naimuddin and D.P. Srivastava. 2000. Sources of resistance to major diseases in pigeonpea. *Indian Phytopath.* 53(3) 353 (abstr.)

Viswanathan, R. and R. Samiyappan. 2000. Efficacy of *Pseudomonas* spp. strains against soil-borne and sett-borne inoculum of *Colletotrichum falcatum* causing red rot disease of sugarcane. *Sugar Tech.* 2(3): 26

Viswanathan, R. and R. Samiyappan. 2002. Induced systemic resistance by fluorescent pseudomonads against red rot disease of sugarcane caused by *Colletotrichum falcatum*. *Crop Protection* 21(1): 1

Viswanathan, R. and R. Samiyappan. 2008. Bio-formulation of fluorescent *Pseudomonas* spp. induces systemic resistance against red rot disease and enhances commercial sugar yield in sugarcane. *Arch. Phytopathol. Plant Prot.* 41(5): 372

Viswanathan, R., A.R.Sundar and S.M. Premkumari. 2003. Mycolytic effect of extracellular enzymes of antagonistic microbes to *Colletotrichum falcatum*, red rot pathogen of sugarcane. *World J. Microbiol. Biotech.* 19(9): 953

Vivekananthan, R., M. Ravi, A. Ramanathan and R. Samiyappan. 2004a. Lytic enzymes induced by *Pseudomonas fluorescens* and other biocontrol organisms mediate defence against the anthracnose pathogen in mango. *World J. Microbiol. Biotechnol.* 20(3): 235

Vivekananthan, R., M. Ravi, D. Saravanakumar, N. Kumar, V. Prakasam and R. Samiyappan. 2004b. Microbially induced defence related proteins against postharvest anthracnose infection in mango. *Crop Protection* 23(11): 1061

Wang, C., D. Wang and Q. Zhou. 2004. Colonization and persistence of a plant growth-promoting bacterium *Pseudomonas fluorescens* strai CS85 on roots of cotton seedlings. *Can. J. Microbiol.* 50(7): 475

Wilson, P.S., E.O. Ketola, P.M. Ahvennieni, *et al.,* 2008. Dynamic of soilborne *Rhizoctonia solani* in the presence of *Trichoderma harzianum*: effects on stem canker, black scurf and progeny tubers of potato. *Plant Pathol.* 57(1); 152

Xuan, T.D., O. Yuichi, C. Junko, E. Tsuzuki *et al.,* 2003. Kava root (*Piper methysticum*) as a potential natural herbicide and fungicide. *Crop Protection* 22(6): 873

Xuan, T.D., T. Shinkichi, T.D. Khanh and L.M. Chung. 2005. Biological control of weeds and plant pathogens in paddy rice by exploiting plant allelopathy-an overview. *Crop Protection* 24(1): 197

506

Yao, M., H. Desilte, M.T. Charles, R. Bolanger and R.J. Tweddell. 2003. Effect of mycorhization on the accumulation of rishitin and solavetivone in potato plantlets challenged with *Rhizoctonia solani*. *Mycorrhiza* 13(6): 333

Yulianti, T., K.S. Sivasitamparam and D.W. Turner. 2007. Saprophytic and pathogenic behaviour of *Rhizoctonia solani* AG2-a (ZG-5) in a soil amended with *Diplotaxis tenuifolia* or *Brassica nigra* manures and incubated at different temperatures and soil water content. *Plant Soil* 294(1–2): 277

Zhu, X., J. Cao, Q. Wang and W. Jiang. 2008. Postharvest infiltration of BTH reduces infection of mango fruits (*Mangifera indica* L. cv. Tainong) by *Colletotrichum gloeosporioides* and enhances resistance inducing compounds. *J. Phytopathol.* 156(2): 68

Zope, A.V. and V.S. Thrimurty. 2004. Effect of botanical pesticides on seed, rhizosphere microflora and seedling vigour in rice. *J. Mycol. Pl. Pathol.* 34(2): 576

CHAPTER

12

Virus and Viroid Diseases of Plants

PLANT VIRUSES

Among the non-cellular or mesobiotic causes of infectious plant diseases viruses and viroids are the only agents. They are more dangerous than fungi and bacteria as it is much more difficult to control them because of their chemical nature and special type of parasitism. Early history of viruses was briefly given in the chapter on history of plant pathology.

What is virus? Definitions

The term 'virus' had long been used for any slimy liquid, poison, venom, or infectious matter. Since about 1890 this term was used for a wide variety of infectious agents including bacteria and unidentified microbes although M.W. Beijerinck had used the term virus for the Tobacco mosaic agent to distinguish it from bacteria. When these agents were found to pass through bacteria-retaining filters, they were termed as 'filterable viruses' thus excluding bacteria from the group. It was in the second decade of the twentieth century that the term virus in its present form came into use for denoting a distinct group of filterable, obligately parasitic, submicroscopic or invisible infectious agents. The original meaning of "virus" can still be detected in the commonly used adjective, 'virulent', meaning extremely poisonous, venomous, or malignant. This adjective is used for diseases irrespective of the cause.

With advances in knowledge of the nature of virus particles more precise concepts about viruses emerged. Attempts were made to distinguish viruses from other filterable infectious agents such as mycoplasmas and RLB. The following definitions of virus have been chosen from major textbooks on the subject.

Viruses are submicroscopic entities capable of being introduced into specific living cells and reproducing inside such cells only. Later, viruses were defined as elements of genetic material that can determine in the cell, where they reproduce, the biosynthesis of a specific apparatus for their own transfer into other cells.

Subsequently, Luria *et al.,* (1978) defined viruses as "entities whose genomes are elements of nucleic acid that replicate inside living cells using cellular synthetic machinery (host ribosomes) and causing synthesis of specialized elements (complete virions or virus particles) that can transfer the viral genome to other cells".

Matthews (1991) considered a virus as a set of one or more template molecules normally encased in a protective coat or coats of protein or lipoprotein, which is able to organize its own replication only within suitable host cells where its production is:

(i) dependent on host's protein synthesizing machinery (ribosomes),

(ii) organized from pools of required material rather than binary fission, and

(iii) located at sites which are not separated from the host cell content by a lipoprotein bilayer membrane.

The salient features of viruses are contained in the definition given by Bos (1983). He defined virus as an infectious agent often causing disease, invisible with the optical microscope (submicroscopic), small enough to pass through a bacterial filter, lacking a metabolism of its own, and depending on a living host cell for multiplication. Viruses are small packages of host-alien genetic information of one type (RNA or DNA), either in one strand or in a few segments encapsulated together or separately and enclosed in a coat of one or more types of protein, sometimes with an extra coat (envelop) and some other constituents.

The above definitions vary but the central theme remains the same. Whatever may be the precise language to define a virus, the following characteristics will always be included:

1. Viruses contain one or more pieces of a single type of nucleic acid, either RNA or DNA, never both.

2. The nucleic acid carries the genome of the virus which differs from one virus to another.

3. The genome in the nucleic acid strand directs the synthesis of specific proteins for the protein coat which must be present in all viruses throughout their active phase except at the time of replication when protein coat and nucleic acid are separated (lose identity).

4. Viruses rely on living host cells for most of the enzymes necessary for their replication.

Viral parasitism

The definition of a parasite is "an organism which derives the materials it needs for growth from living cells". In other words living at the expense of another living being is parasitism. Strict adherence to this definition of a parasite may exclude a virus. They are not organisms. All organisms are cellular while viruses are not cellular. However, during replication of their genome inside the host cell viruses do use the internal cellular environment. This internal environment is created by the immediate function of genes and consists of such subcellular organelles as nuclei, mitochondria, ribosomes and cytoplasmic components. This environment, termed "genosphere", is where a virus displays the characteristics of reproduction and growth and at this level it behaves like an organism. Thus, the viral parasitism is purely at the genetic level. The definition of parasitism can be modified for viruses as "remaining active and multiplying at the expense of host's subcellular environment including ribosomes and specific enzymes therein". In the genosphere, viruses interfere with host cell function resulting in pathogenesis at the cellular level (cytopathogenesis). Genes guiding the cell functions (morphology, chloroplast synthesis, etc.) are inhibited from expression through silencing of related RNAs.

Are viruses living organisms?

The characteristic features of life are ability to (i) assimilate (metabolism) with the release of energy, (ii) excrete waste products of metabolism, (iii) grow, (iv) reproduce, and (v) exhibit some sort of response to the environment. Viruses have no metabolism, do not excrete, do not grow but multiply, and respond to external stimuli such as temperature, chemicals, etc. When the tobacco mosaic agent was first identified as a virus there was no controversy regarding its living or nonliving nature. At that time all known agents of disease (fungi and bacteria) were known as living organisms. But when virus particles could be purified and concentrated from the extract of infected cells and show uniform size, shape, and chemical composition even after crystallization by chemical treatment it was proposed that the agent could be a molecule, but a molecule is not capable of self replication and is not a living organism. DNA molecule by itself is not a living organism but is a source of life. Thus, the general opinion is that virus is a unique entity which is neither living nor non-living but a macromolecule on the threshold of life.

Biochemistry and morphology of plant viruses

Viruses are macromolecules. The mature particle of a plant virus is generally called virion but the precise term for the whole infective particle is nucleocapsid. The particle consists of two components, the nucleic acid and the protein coat or capsid. The nucleocapsid is the end product of a process initiated by the replication of virus nucleic acid or genome (all genes together) present in one or more pieces (usually called species), synthesis of protein as directed by the genome, and assembly of these proteins with the genome into a complete nucleocapsid. Complete loss of identity or structure, that is, breakdown of particles into their components (nucleic acid and protein) is a characteristic of viruses that differentiates them from cellular organisms.

Nucleic acid of viruses: All cellular organisms contain two types of nucleic acid, the deoxyribonucleic acid (DNA) which is the chromosomal nucleic acid or the genome, and ribonucleic acid (RNA) which is present in the cytoplasm as a workbench for synthetic activities. Viruses are not cellular organisms and neither have nucleus nor the cytoplasm. But either DNA or RNA (never both) is always present in the infective particle and serves as the genetic apparatus of the virus. It is the infective component of the particle.

Before 1968, all plant viruses that had been studied were found to contain only RNA. In 1968, *Cauliflower mosaic virus* particles were found to contain DNA, not RNA, A number of other viruses assigned to this group (caulimoviruses) contain DNA. The nucleic acid may be in a single strand or in a double strand. Most plant viruses contain single strand of RNA (ssRNA). These strands may be free at both ends (linear RNA strand) or the ends may be joined together to form a circular nucleic acid molecule (circular RNA strand). In double stranded viruses, the two strands are coiled around each other helically. The guanine (G) of each nucleotide in one strand is linked to a cytosine (C) apposed to it in the other strand and each (adenine (A) is similarly linked to a uracil (U) in the other strand. This forms a ladder-like double helix. The nucleic acid may be present as a single continuous strand or it may be present as two or more pieces in the same or different particles. When the genome is present in different particles the viruses are known as multipartite viruses or viruses with divided genome. These different particles are not infectious on their own and all types of particles with different segments of the genome must be present for successful infection and production of typical symptoms.

The physical and chemical types of plant viruses can, thus, be listed as below:

1. Linear ssRNA in one piece such as Tobacco mosaic virus, Tomato bushy stunt virus, and Potato virus X.
2. Linear ssRNA in more than one piece contained in the same particle such as Turnip yellows virus and Potato leaf roll virus.
3. Linear ssRNA in more than one piece, each piece in a separate particle such as Cucumber mosaic virus, Cowpea mosaic virus.
4. Linear dsRNA in many pieces but in one particle such as in sugarcane Fiji disease virus.
5. Circular ssDNA as in geminiviruses (Mungbean yellow mosaic virus, Maize streak virus)
6. Circular dsDNA as in caulimoviruses (cauliflower mosaic).

Infectivity of nucleic acid:

The demonstration, during 1956–1960, that the RNA of TMV was the actual infective component and that the whole RNA strand was necessary for infection were landmarks that established the necessity of the integrity of viral nucleic acid for infection. In the multipartite viruses each piece is complimentary to others for infection and particle formation. Although the TMV particles are highly stable at room temperatures, outside the cell the free RNA loses infectivity after 48 hours. This is because naked RNA is susceptible to inactivating agencies. Also, infectivity of the free RNA is less than the protein coated RNA. In *Beet curly top virus*, the coat protein is essential for infectivity and spread of the virus.

The virus protein and capsid structure: Protein constitutes the bulk of the virus particle mass. The major portion of this protein is in the shell or coat. Like any other protein, the virus protein also is made up of different amino acid sequences in different viruses. These amino acids are linked by peptide bonds. Since the nature of these proteins is dictated by the genome (nucleic acid) and the size of genome in the viruses is small, most plant viruses have only limited types of proteins.

The sequence of amino acids constitutes the primary structure of the protein. When the amino acids produce a helical structure it is known as secondary structure of the protein. A further level of organization occurs when the protein folds to give tertiary structure. The amino acid sequence and number, decided by the genome, gives the virus protein its intrinsic property to assemble in only a particular fashion so that particles of only specific shape for the particular virus are formed.

The protein subunits of the capsid are known as structural units (Su). When, as in isometric or spherical viruses, these structural units form groups of 2, 4, or 6 to accommodate more protein they are called morphological units (Mu) or capsomers.

The shape and size of the particles is determined not only by the intrinsic properties of the protein units but also by the size and amount of the nucleic acid. The number of units produced is large while the size of nucleic acid strand around which they assemble, is fixed. This leads to accommodation of a large number of protein units around a small strand. According to Crick and Watson (1956), in this situation only two types of structures are possible: elongated rods and polyhedral (icosahedral or octahedral) or isometric particles. The virus capsid shape is based on these two types.

The elongated or rod shaped capsids can be compared with a cylinder made up of a series of rings placed one above the other, leaving a central canal. These rods can be rigid tubes or long flexuous tubes. Not many plant viruses have elongated tubular shapes. Tobacco mosaic and *Potato virus S* are examples of rigid, tubular viruses. *Potato virus Y, Potato virus X*, and *Beet yellow mosaic* are examples of flexuous tubular viruses. The tubular viruses measure 128–250 nm or more in length and 12–28 nm in diameter. They have a low percentage of nucleic acid, 4–6% of the particle weight.

The spherical looking viruses are actually polyhedral, mostly based on icosahedral and sometimes octahedral symmetry. An icosahedran is a regular polyhedron with 20 equilateral triangular faces. In most polyhedral viruses the triangles are further subdivided in equilateral triangles. This enables the particles to have more protein subunits, both morphological and structural. Examples of polyhedral viruses are Cowpea mosaic, Cucumber mosaic, Tobacco ringspot and Turnip yellow mosaic.

Some viruses are usually accompanied by smaller spherical particles of another serologically unrelated virus called satellite virus, which is a defective virus that can multiply only in the presence of the main virus which is helper virus in such cases.

In addition to the above two major shapes, there are many viruses with complex structure. In *Cauliflower mosaic virus*, two types of coat proteins are present for two layers of the coat. The outer layer has more protein units while the inner layer has lesser number of units. Many ssDNA viruses, such as Beet curly top virus and Mungbean yellow mosaic virus show particles in pairs. These are called geminiviruses. The pair is made of two incomplete icosahedra and looks elongate.

Bacilliform or bullet-shaped viruses are the most complex of all plant viruses. They are rounded at one or both ends and contain ssRNA, carbohydrates and some lipid. Four types of protein are present.

Dilution end point

Dilution of the juice containing the virus may or may not affect its infectivity. TMV is reported to remain infective even in a dilution of 1:1000,000. Cucumber mosaic virus retains virulence at 1:1000 but few infections occur when the dilution is 1:10,000. Tomato mosaic virus remains infectious at 1:10000 dilution while Potato rugose mosaic virus causes poor infection when diluted to 1:10 or 1:100. However, dilution is a very variable character and depends upon many factors such as the medium used for dilution, temperature, etc. As such it cannot be considered a specific characteristic of plant viruses. Similarly, retention of infectivity in storage is another character that will vary with conditions of storage. If the juice from mosaic infected tobacco plant is kept from decomposition the virus can remain viable for a few hours to several months. At refrigerator temperatures or when frozen solid, viruses remain viable for much longer periods than at room temperatures. Most viruses cannot stand desiccation but Tobacco mosaic virus in dried leaves or in juice dried on filter paper remains infective for many years.

Purification and precipitation

The purification of viruses is necessary before the exact nature of the infectious principle could be determined. Since any protein can be precipitated and brought in pure form the viruses are no exception. By treating the infected juice with chemicals like acetone, ethanol, alcohol, calcium chloride, etc. the viral protein and RNA can be precipitated and made in crystalline form. In most cases where pure crystals of the virus have been obtained the infectivity of the RNA has been found to remain unaltered.

Effect of temperature

The fact that bacteria were killed by certain temperatures which varied with different species and with environments, led the early workers to test the effect of temperature on plant viruses. Since then a good amount of literature has accumulated on this aspect of virus studies showing that a number of viruses can be inactivated by thermotherapy. Tomato mosaic virus is destroyed by treating the virus containing juice at 85–90°C for 10 min. Lower temperatures seem to have no effect on the viability of plant viruses.

Mutability and strains

The presence of genetic material in the form of RNA in plant viruses ensures that new strains of the virus may develop probably by mutation of the RNA. In Tobacco mosaic virus alone there are more than 50 strains. Although mutation is supposed to be the most common method of origin of strains in viruses some workers are of the opinion that hybridization between two different kinds of RNA is also possible. However, the development of mutants or hybrids depends on the capacity of the mutant or different kinds of RNA to compete and coexist in the host cell.

Serological reactions

If a virus containing juice is injected into the body of a rabbit, the rabbit's blood forms antibodies that will react with viral protein to give precipitation. These reactions, known as serological reactions or precipitin type serological reactions, are specific, i.e., the antibodies obtained by inoculation of strain A of a virus will precipitate the juice containing the same specific type of virus.

Classification of viruses

Earlier, the viruses were broadly classified on the basis of the shape of their capsid and the type of nucleic acid they contained. Thus, the main groups were the elongated viruses with single stranded RNA, isometric viruses with single or double stranded RNA or DNA, and the bacilliform viruses. Brandes and Bercks (1969) had proposed a system of virus classification which had proved useful in grouping the elongate viruses. It was based on particle morphology and size and on serological affinities. The recognized groups were: Tobacco mosaic virus group, Tobacco rattle virus group, Potato virus X group, Potato virus Y group, and Potato virus S group.

In the early 1960s efforts had started to classify viruses on the same lines as for other biological entities such as bacteria, fungi, etc. in family, genus and species. This has been achieved only recently. In absence of this system, the viruses had been classified in 'virus groups' consisting of a coherent cluster of individual viruses sharing major characterizing properties. They bear vernacular names, most of which are in English. Majority of plant virologist still favour this classification of plant viruses in groups but there had been gradual shift in the approach towards a system of classification into genus and families. The International Committee on Taxonomy of Viruses (ICTV), responsible for providing guidance in classification and nomenclature of viruses, had published 5 reports on the subject till 1991. The first 16 groups of viruses were described and approved by ICTV in 1970. In the fourth report of the committee 26 groups had been recognized, most of which have been retained in the later report. In the fifth report published in 1991 (Francki et al., 1991) three of these groups were transferred to 3 families, Rhabdoviridae (negative sense RNA in one pieces), Bunyaviridae (negative sense ssRNA in 3 pieces) and Reoviridae (dsRNA in 10–12 pieces) and 9 groups were added. Thus, there are 32 groups in addition to 3 families consisting of subgroups and genera. The classification, as recognized until 1993, is given in the following table.

Table 11.1 Virus classification approved by ICTV (Francki *et al.,* 1991)

Classification	Genome properties*	Number of viruses**
GROUPS		
a) Rod-shaped rigid particles		
1. Tobamovirus	ssRNA (1)	14
2. Tobravirus	ssRNA(2)	3
3. Furovirus	ssRNA (2–4)	11
4. Hordeovirus	ssRNA (3)	4
b) Filamentous particles		
1. Capillovirus	ssRNA (1)	4
2. Carlavirus	ssRNA (1)	56
3. Closterovirus	ssRNA (1)	22
4. Potexvirus	ssRNA (1)	39
5. Potyvirus	ssRNA (1 or 2)	153
6. Tenuivirus	ssRNA (4)	7
c) Isometric particles		
1. Carmovirus	ssRNA (1)	17
2. Luteovirus	ssRNA (1)	21
3. Sobemovirus	ssRNA (1)	16
4. Tombusvirus	ssRNA (1)	12
5. Tymovirus	ssRNA (1)	19
6. Maize chlorotic dwarf virus	ssRNA (1)	3
7. Marafivirus	ssRNA (1)	3
8. Necrovirus	ssRNA (1)	4
9. Parsnip yellow fleck virus	ssRNA (1)	3
10. Comovirus	ssRNA (2)	13
11. Dianthovirus	ssRNA (2)	3
12. Nepovirus	ssRNA (2)	36
13. Fabavirus	ssRNA (2)	3
14. Pea enation mosaic virus	ssRNA (2)	1
15. Bromovirus	ssRNA (3)	6
16. Cucumovirus	ssRNA (3)	4
17. Caulimovirus	dsDNA (1)	17
18. Cryptovirus	dsDNA (2)	31
d) Quasi-isometric to Bacilliform particles		
1. Ilarvirus	ssRNA (3)	20
2. Alfalfa mosaic virus	ssRNA (3)	1
e) Geminate particles		
1. Geminivirus	ssDNA (1 or 2)	
Subgroup I		10
Subgroup II		5
Subgroup III		33
f) Bacilliform particles		
1. Badnavirus	dsDNA (1)	14
(Commelina yellow mottle virus)		
FAMILIES		
Rhabdoviridae	(–) ssRNA (1)	

Subgroup-A	9
Subgroup-B	4
Subgroup-C	
(non-enveloped particles)	4
Unclassifiedspecies	68
Bunyaviridae (−) ssRNA (3)	
Tospovirus genus	
(Tomato spotted wilt virus)	1
Reoviridae dsRNA (10–12)	
Phytoreovirus genus	3
Fijivirus genus	3
Unnamed genus	2

* Number of functional species of nucleic acid given in parentheses.
** Definite and possible numbers

The geminivirus group was later raised to status of family-Geminiviridae-with four genera on the basis of vector and plant host. These are:

Genus *Curtovirus: Beet Curly Top Virus.* Monopartite), hosts are from monocots, vectors are leafhoppers.

Genus *Mastrevirus: Maize Streak Virus.* Monopartite, hosts are from monocots, rarely from dicots, vectors are leafhoppers.

Genus *Begomovirus* containing mono- or bipartite geminate particles of leaf curl viruses (species *Tomato leaf curl virus, Tomato yellow leaf curl virus* etc.): Mono- or bipartite, hosts are.

Genus *Topocovirus: Tomato Pseudocurly top virus.* Bipartite, dicot hosts and tree hoppers are vectors.

The family-genus-species concept was finally achieved after 1993. All the plant viruses have now been classified in 13 families comprising of 41 genera and 17 ungrouped "floating" genera (Martelli, 1997).

The geminivirus group was later raised to status of family-Geminiviridae-with four genera on the basis of vector and plant host. The subgroup III constitutes the genus.

Genus *Mastrevirus: Maize Streat Virus.* Monopartite, hosts are from monocots, rarely from dicots, vecyors are leafhoppers.

Genus *Begomovirus: Bean Golden Mosaic virus.* Mono- or bipartite, hosts are dicots, vectors are white flies.

Genus *Topocovirus: Tomato Pseudocurly top virus.* Bipartite, dicot hosts and tree hoppers are vectors.

The family-genus-species concept was finally achieved after 1993. All the plant viruses have now been classified in 13 families comprising of 41 genera and 17 ungrouped "floating" genera (Martelli, 1997).

Standardization of names attracted the attention of botanists and microbiologists in the early 1930s. Before this, not much was known about the nature of viruses. Originally, virus names were common names or vernacular names. Because the viruses were first recognized by the diseases they caused the common names utilized the name of the host and the symptoms, such as Tobacco mosaic virus. In 1927, James Johnson proposed that the viruses should be named only after their host, not symptoms, since the latter are variable on different hosts infected with the same virus. In order to avoid the multiplicity of viruses on the same host a distinguishing number should be given to each virus. Thus, the first virus (mosaic) on tobacco was named Tobacco virus 1, the second Tobacco virus 2, and so on. In 1937, K.M. Smith proposed Latin names like.

Nicotiana virus 1 for Tobacco virus 1. With the increase in number of viruses on the same crop, sometimes 11 or 12, it became clear that this numerical code was difficult to understand by those who were not working on the same crop. Letter codes were also proposed such as Potato virus X, Y, S, etc. In 1939, F.O. Holmes had proposed a Linnaean-style latinized binomial system for virus nomenclature. In this system viruses were treated as having genera and species and Latin names were given accordingly. Thus, Tobacco mosaic virus was called *Marmor tabaci* in family Marmoraceae. In 1970, H.P. Hansen proposed another system based on the fundamental characters of the virus. In this system Tobacco mosaic virus was named *Minchordia nicotianae*. M for mechanically transmitted, chorda for rod shape and nicotianae for the host. However, none of these schemes for nomenclature of viruses was widely accepted and the internationally recognized system today is to call the viruses by their common or vernacular name and also mention the group or family to which they belong, for example, Potato virus X (potexvirus) or Tobacco mosaic virus (tobamovirus).

Symptoms of Virus Diseases of Plants

Symptoms, if properly studied in detail, may reveal the presence of a virus or similar agents in the plant and also help in differentiating strains of virus. Viruses produce visible or otherwise detectable abnormalities in plants which are recognized as symptoms. When infection is there but symptoms do not appear, the infection is said to be *latent* and symptoms are *masked*. The same isolate of a virus may produce different symptoms on different hosts (plant species or cultivars). Thus, symptoms are more a reflection of a host response to virus infection than property of the virus.

Symptom expression is governed by many factors such as:

   (i) type and strain of the virus,
  (ii) type and variety of the host plant,
 (iii) age and stage of plant growth,
 (iv) physiology of the host,
  (v) duration of infection,
 (vi) presence of other viruses and pathogens, and
(vii) environment and climatic conditions.

Mostly viruses cause *systemic infection*, that is, the virus spreads through the host. There may be *local lesion* hosts in which the virus is restricted to the lesion area. Same virus may cause systemic infection in one host and local lesions in another. Meristematic regions of root and shoot usually remain virus free. In some cases, initial rapidly produced *primary symptoms* may differ from those produced later. Usually, viruses express their symptoms in young parts which are more susceptible than older parts. Where symptoms result in prompt death of the invaded cells, thus, preventing further spread of infection (local lesions) the plants are said to show *hypersensitive reaction* and to be hypersensitive.

The most obvious symptoms of virus diseases, like any other disease, are external and appear as foliar colour changes and growth abnormalities. Internal symptoms are also produced and are more specific but these may be recognized only after optical and electron microscope observations.

External symptoms

Colour changes: The term variegation is used to accommodate all types of colour changes on leaves. **Mosaic** symptoms are characterized by mottled green (mottling means spots or blotches of colour) or patterns of green and yellow areas. When the yellow colour is uniform and unbroken it is known as chlorosis. In graminaceous plants, the mosaic pattern may be in the form of streaks and stripes. **Scorching** of entire leaf or its margins is burnt appearance. In **vein clearing** the tissues close to the veins turn yellow and the remaining lamina surface stays green. In **vein banding** the tissues close to veins remain green and the rest of the lamina surface turns yellow. Colour changes may be accompanied by some modification of the structural appearance of the organ. Pouch-like development of green parts of leaves is known as **puckering**. Drying of cells in a particular fashion leads to ringspots, black spots, etc.

Virus infected plants normally do not recover. However, **dark green island** symptom of mosaic diseases is often seen. These are clusters of green leaf cells that are free of the virus but surrounded by yellow virus-infected tissue. The dark green islands and recovery are considered to be related phenomena (Moore *et al.,* 2001).

Changes in growth pattern: **Leaf curl** is a symptom in which the leaves curl from the margins backward bringing the centre of the lamina upward. In **leaf roll** the margins roll inward forming a trough-like shape with midrib in the centre of the trough. The leaves may become thick and leathery due to accumulation of starch. Sometimes, outgrowths may occur on the backside of leaves (**enations**). **Galls** may also form on leaves and other organs. In some mosaic diseases, leaves may be abnormally lobed with a fern-like appearance, often forming filiform or shoe-string structures.

Reduction in size of leaves and shoots is a common symptom in many virus diseases. When leaves are small they give a bushy appearance to the plant. Shortening of internodes causes dwarfing or stunting of the plant. Overgrowth due to hypertrophy and hyperplasia results in symptoms of bunchy top or witches broom. Sometimes the stem may show a spike-like growth. Spindle-tuber and spindle shoots (spindle-like appearance) are other growth abnormalities.

External necrosis includes drying of buds and tip of shoots. Viruses do not cause rotting which may occur in drying tissues due to secondary invasion of bacteria and fungi.

Changes due to water deficiency (dysfunction of xylem) includes wilting, withering and etching.

Internal symptoms

Internal changes in virus affected plants can be observed by light and electron microscopy and include (a) anatomical and histological deviations, (b) cytological and ultra-structural changes, and (c) formation of inclusion bodies.

Anatomical and histological changes: Because the virus activity starts in and is confined to host cells, anatomical changes can be expected and do occur in one or other form in the virus affected plants. These changes are often responsible for particular external symptoms.

Cells may decrease in size (**hypotrophy**) or in number (**hypoplasia**) or may increase in size (**hypertrophy**) or in number (**hyperplasia**). Excessive hyperplasia leads to prolifation of tissues causing witches broom, bunchy top, etc. The cells may die (necrosis) or deviate in content such as degeneration of chloroplasts causing chlorosis, and nuclear swelling.

In leaves showing mosaic symptoms or other types of yellowing the palisade cells may be smaller with fewer chloroplasts. The lamina in chlorotic areas is thinner because of thinner and fewer mesophyll cells due to hypoplasia. Dark green areas grow normally while yellow areas have restricted growth. This leads to buckling, puckering, and blistering. Extreme hypertrophy of the lamina gives rise to 'shoe-string' leaves. Enlargement of cells adjacent to veins obliterates intercellular spaces and few chloroplasts are produced making the tissue translucent.

Anatomical changes occur also in the tissues other than in leaves. The phloem is often destroyed (phloem necrosis) as in potato leaf roll disease. In curly top of sugarbeet, there is hyperplasia resulting in death of phloem parenchyma. Abnormal meristematic activity of phloem parenchyma cells leads to formation of tumors.

Stem pitting or wood pitting symptoms are caused by development of localized areas of disorganized parenchyma instead of xylem and phloem.

Cytological and ultra-structural changes: The main cell organelles influenced by viruses appear to be the nucleus, chloroplasts, and mitochondria. These organelles may be disintegrated. In some cases of virus infection nuclei stain less densely and nucleoli may be swollen and distorted due to multiplication of the virus in the nucleus and nucleolus. The changes in leaf colour are due to changes in the chloroplasts or their pigments. Degeneration of chloroplasts also occurs depending on the type of virus-host combination. Fibril containing vesicles are formed in chloroplasts in the infection of tymoviruses, tobraviruses and closteroviruses. Normal or degenerated mitochondria may aggregate in viral infection.

Inclusion bodies: These have been defined as "intra-cellular structures produced *de novo*" as a result of virus infection. These structures are of special interest because they often consist of virus particles and may be visible with light microscopes. Inclusion bodies vary in form from amorphous structures to well defined crystalline structures. These can be in the nucleus or in the cytoplasm.

The nuclear inclusions may be in the nucleoplasm, in the nucleolus or between the membranes of nuclear envelop (perinuclear space). Inclusions in nucleoplasm are generally crystalline structures made up of protein crystals or arrays of virus particles which occur in the cytoplasm also having originated from the nucleus. Inclusions in nucleolus are also amorphous or crystalline. The amorphous inclusions appear as intensely electron dense masses at the periphery of the nucleolus. The protein crystals in nucleoli distort their shape.

The cytoplasmic inclusions consist of either scattered particles of the virus or as organized masses such as amorphous bodies, crystals, pinwheels, and laminated bodies and as viroplasm (the electron dense materials in which viral products are formed or in which virus assembly takes place).

Factors Affecting Virus Symptoms

The symptoms of majority of the plant virus diseases are most conspicuous on plants making rapid growth. Plants that are almost mature at the time of infection usually do not develop symptoms on any part except new growth. In some diseases, such as in the ringspot of tobacco, the symptoms appear in the mature plant parts at the point of inoculation.

Some infected plants with very slight or no symptoms of the disease may be carriers of the virus which may be transmitted from them to susceptible hosts in which symptoms will appear. They are called **symptomless carriers.** There are masked carriers in which the symptoms are simply masked due to unfavourable environmental conditions and will appear when the environments become favourable. Many viruses are known as latent virus because they may not produce symptoms even if present in the host. These hidden symptoms create great difficulties in the proper elimination of the disease from the affected fields.

Light: Our knowledge of the influence of light on virus symptoms is very indefinite. There are few data on the effects of different rays of the light spectrum. It has been demonstrated by some workers that different kinds of light affect the viruses and it is possible that they may affect the symptoms also. In many virus diseases it has been noticed that symptoms are masked in dark or shade and appear when the plant is removed to intense light.

Moisture: In sugarcane mosaic and streak and many other tropical plant viruses the symptoms of the disease are more evident when there is good rainfall to support good vegetative growth of the host. Dry conditions induce masking of the symptoms.

Temperature: Temperature has been studied more thoroughly than any other factor and appears to have a great influence on the symptoms of mosaic diseases. In many cases, the plants do no develop mosaic symptoms at certain temperatures. This phenomenon is known as **masking**. The optimum temperature for symptoms appears to be lower than the optimum for growth of the host plant. The same temperature may mask the effects of one virus and intensify those of another. It should, however, be borne in mind that disappearance of symptoms does not indicate a cure. The temperature probably affects viruses through effect on multiplication (replication) and translocation. In infection of barley by Brome mosaic virus, the infection is predominantly restricted to cells in and associated with veins at 24/20°C (day/night) but when the plants are exposed to 34°C for 2 hours this restriction is overcome.

The effect of temperature on symptoms of some virus diseases of plants as reported by different workers is listed below:

Virus disease	Optimum temperature for development of symptoms (C°)	Temperature at which symptoms are masked
Tobacco mosaic	13–18; 28–30	36–37
Potato mosaic	14–18	20–24
Bean mosaic	15–30	12–18
Cucumber mosaic	14–28	16

Host nutrition: It is now experimentally proved that in most of the virus diseases of plants the macro- and micronutrition of the plant influences the rate of virus multiplication and intensity of symptoms. In general, nutrition which promotes rapid vegetative growth of the host plant favours the virus multiplication also. Susceptibility is increased with increase in nitrogen supply. The same is true for phosphorus. However, only small doses of potash increase susceptibility which is decreased if high doses of the same fertilizers are used.

Survival and Transmission of Plant Viruses

The inability of newly synthesized particles to multiply in the same cell helps in their survival in the infected plants. Other important sources of survival of plant viruses are seeds including the embryo, weed hosts, vectors especially those in which the virus is persistent, and in some cases contaminated crop products (*viz.*, tobacco) and externally contaminated seed. Survival outside the active host cells is common only in viruses with highly stable particles.

Most of the severe plant viruses spread naturally and often have a wide host range. Means employed by viruses for achieving short and long distance dispersal are highly variable but the same virus may have more than one method of dispersal. The dispersal of viruses is mostly passive. Due to their dependence on living cells, they are usually carried by cell sap, either directly or indirectly through other means or agencies.

Movement within the plant: Systemic movement through the phloem of infected host plant is a key process in disease cycle of plant viruses. Cell to cell transport within the infected plant occurs with the cytoplasmic stream through plasmodesmata. Long distance passive movement within the plant is mostly within the phloem where particles move along with carbohydrates to energy requiring organs such as roots and developing shoots. It is known that time required for translocation of a virus from inoculated leaf to the other parts of the plant varies from 6 minutes to 10–21 days for different viruses in different host plants depending on the virus, age of leaf or plant at the time of inoculation, method of inoculation, environmental conditions, etc. Cell to cell movement of viruses is affected by sequences of the capsid protein. It has been demonstrated that different proteins including the coat protein (CP) of Potato virus X are required for cell-to-cell movement of the virus. In *Tobacco mosaic virus*, the movement of the virus particles between cells through plasmodesmata is mediated by a virus-encoded specialized movement protein, MP (Rhee *et al.*, 2000). The MP gene plays a crucial role in determining the virus specific CP required for cell-to-cell movement. *Brome mosaic virus* (BMV) requires the coat protein for cell-to-cell movement whereas Cowpea chlorotic mottle virus (CCMV) from the same genus (*Bromovirus*) does not. Interference in CP expression impairs the movement. Presence or induced synthesis of salicylic acid in the tissues inhibits replication of viruses and interferes with long distance virus movement within the plant. Gemini viruses reproduce only in intact host nuclei. Upon entrance in to a host cell, the virus enlists host machinery to facilitate its movement through the cytoplasm in to the nucleus. The entry is through nuclear pore complex. After replication and transcription of its genome, the infective form of the virus exploits host transport machinery together with virus encoded movement proteins to get out of the nucleus and move through the cytoplasm to and through the virus-modified plasmodesmata (Gafni and Epel, 2002).

Transmission from Plant to Plant:

Mechanical transmission: Mechanical transmission is accomplished by transfer of sap through contact between plants when diseased and healthy leaves rub together or indirectly through environmental contamination such as in surface water, soil on clothes, hands, and equipments or surface contamination of large beetles where, because of their stability *in vitro*, the viruses may persist for long. In sap or mechanical transmission some sort of tissue surface injury is essential. Thus, when contagious viruses like TMV spread by contact the rubbing causes minute injury on leaf hairs and this permits passage of infected sap into healthy sap. Experimental mechanical transmission involves use of abrasives (such as carborundum powder or diatomaceous earth) on the surface to be inoculated. The wounds must be so small that cells survive otherwise the virus will not come in contact with living protoplast.

Transmission by grafting and vegetative propagation materials: Plant viruses are mostly systemic in their true hosts. This ensures that the virus is present in all such parts as are likely to be used for vegetative propagation including grafts, bulbs, corms, tubers, roots, stem pieces (setts), etc. New plants developing from such materials are infected from the very beginning. In strict sense this is not transmission since the parts are the clones of the infected parent. However, long distance dispersal and also the survival is ensured by this method of transmission. Virus diseases of sugarcane and potato are the best examples of transmission by vegetative propagation material.

Transmission by dodder: Dodder (*Cuscuta* spp.) is a total parasite of stem. It produces haustoria that penetrate the vascular bundles. In plant species where vascular union by grafting has failed dodder has been successfully used to act as a bridge between diseased and healthy plant vascular bundles. Sometimes, the virus multiplies in dodder. Citrus tristeza virus, Potato leaf roll virus, and Cucumber green mottle mosaic virus have been transmitted by dodder.

Transmission through seed and pollen: Tomato mosaic virus had been suspected to be seed-borne since 1910 and seed transmission of common bean mosaic and wild cucumber mosaic was proved in 1918–1919. At present more than 100 viruses are known to be seed-borne. However, many of them are not true seed-borne. True seed-borne viruses are those that infect the embryo. Those that are present in seed tissues outside embryo are generally not infectious unless mechanically transmitted. Seed transmission of virus is property of the host rather than the virus. The same virus may be seed transmitted in one host but not in others. Most nematode transmitted viruses (nepoviruses) show low to high percentage of seed transmission but not pollen transmission.

Since there is no phloem connection between embryo and mother plant, the phloem-restricted viruses are not seed-borne. Even if they persist in the seed coat that has vascular connection with the mother plant they do not infect the seedlings because they are not mechanically transmitted. Embryo infection requires infection of the mother plant before production of gametes or at least before cytoplasmic separation of embryonic tissue from the mother plant.

Pollen and male gametes also become infected and lead to embryo infection even on healthy plants. About 39 viruses including *Ilarvirus, Nepovirus* and *Potyvirus* and 5 viroids are pollen transmitted. Infection of a healthy plant after flowering does not lead to seed or pollen transmission of a virus. Some viruses are transmitted in or on seed tissue outside the embryo. In such cases the virus must be highly resistant to desiccation. Examples of seed transmitted viruses are *Cowpea aphid-borne mosaic* (also pollen transmitted), *Cucumber mosaic, Tomato mosaic*, and *Common bean mosaic*.

Transmission by insects and mites: About 80% of plant viruses are transmitted by some vector Transmission of viruses by vectors including arthropods is an important part of virus study. No description of a virus is complete without information on the virus-vector relationship. This relationship is not merely a mechanical process but is highly specific for different groups of arthropods and may involve complex steps before actual transmission. Plant viruses are mostly transmitted by aphids (Class Insecta, Order Homoptera, Family Aphididae), leafhoppers (Family Cicadellidae) and whiteflies (Family Aleyrodidae). Beetles (Order Coleoptera, Family Coccinellidae and Chrysomelidae), thrips (Thysanoptera-Thripidae) as insect vectors also transmit some viruses. Thrips transmit plant viruses in the Tospovirus, Ilarvirus, Carmovirus, amd Sobemovirus genera. Mites (Class Arachnida, Order Acarina, Family Eriophidae) are non-insect vectors of some viruses.

Different taxonomic groups of arthropods show high specificity in virus transmission. Viruses transmitted by leafhoppers (phloem feeding) are not transmitted by aphids. Most insect-transmitted viruses are not transmitted by nematodes. Some viruses are transmitted only by white flies not by other insects. Strains of same virus may have different species of insect vectors.

In studies of insect transmission of plant viruses the following terms are often employed. The *acquisition feed* or *acquisition feeding period* is the time for which a virus-free vector actually feeds on a virus infected plant to acquire the virus. *Acquisition access period* is the time for which a vector is allowed to feed on a source of virus. During the acquisition feed, the vector acquires the virus in sufficient quantity after a certain length of time determined by access period. It may become capable of transmitting the virus (*viruliferous*) immediately or, more commonly, there is a waiting period or latent period before the virus can be transmitted. *Inoculation access period* is the time for which the vector, after acquiring the virus, is allowed to feed on a healthy plant and transmit the virus. The actual period of feeding is called *inoculation feeding period*. The minimum period of time that a virus needs for acquisition and subsequent transfer to a virus-free plant is the *transfer time* or the *transmission threshold period*.

On the basis of the length of time for which an insect remains viruliferous, the viruses can be categorized in 3 groups. In **non-persistent viruses** or non-circulative viruses, the acquisition and inoculation feed period are usually short (few seconds) and there is no incubation or latent period. Upon acquisition by the vector the virus is retained in the mouthparts or in the foregut. From there it is released in a new host plant through salivation or regurgitation. Persistence or retention of the virus in the vector is brief. Such viruses are mechanically transmissible and are lost by the vector during molting. About 75% of the aphid-transmitt6ed viruses are non-persistent, In **persistent viruses** the relationship is highly specific and there is an intimate biological relationship between the virus and the vector. After the virus is acquired by the vector, it passes through the alimentary canal, gut wall, and circulates in the body fluid (haemolymph) before reaching the salivary glands when it can be transmitted. Thus, such viruses have latent period in vector body and persist for long (days, weeks or for whole life) in the vector. Molting has no effect on persistence. Such viruses are often deep seated in the host such as in phloem and the adjoining tissues and acquisition and inoculation feed period are usually long. Sometimes the vector acts as a host of the virus and virus multiplies in the haemolymph. In such cases the virus can called insect virus also. Such persisten viruses are called **circulative-propagative viruses.** Some such viruses are carried in eggs of insect vectors. This transfer of virus to the progeny of the vector is known as **transovarial transfer**. The viruses that share the characteristics of both non-persistent and persistent viruses are known as **semi-persistent** viruses. The uptake of virus by the vector is from deep seated phloem with long feeding time but the vector immediately becomes infective, without latency. The virus persists in the vector for a few days. Infectivity is lost at molting. The virus particles are not found in the insect body fluid.

Specificity of vector transmission: The reason for specificity of vectors for picking up specific virus was not clearly understood and many possible reasons were advanced. In some cases it may be due to differences in the location of the virus in the tissues and the depth up to which the vector can probe and feed. Another possible reason may be the presence of some transmission factor (TF) in the virus capsid which enables the vector to pick up and transmit a specific virus from a mixture. This latter possibility has been explored extensively. Flasinsky and Cassidy (1998) reported that aphid transmission of potyviruses depends on the presence of specific sequence domains in two virus-encoded proteins, the coat protein and helper component-proteinase (HC-Pro). Froissart *et al.,* (2002) had reviewed the virus-vector relationship. Two molecular strategies were described. In the capsid strategy, the virus coat interacts directly with binding sites in the vector mouthparts. In the helper strategy, additional non-structural protein (helper component protein (HC-Pro), is required. The HC is characteristic of plant viruses. Potyviruses and caulimoviruses have been studied in detail. The potyvirus HC-protein performs many functions including aphid transmission, long-distance movement and suppression of post-transcription gene silencing. The HC of caulimovirus plays important role only in aphid transmission (Syller, 2006) The HC-Pro or the transmission factor on the capsid interact only with the vactor of the virus. The distal end of the stylet of the vector contains a very minute area where receptor proteins for the virus are presenmt. (Uzest *et al.,* 2007). These receptors interact with the HC=Pro, pick up the virus, which passes down the stylet food canal into the insect gut and is recycled back to the stylet for transfer to a new host. If due to some mutation, a single amino acid in the HC-Pro is nisplaced or lost, transmissibility is also lost Multiple pumctures by the vector with the receptor protein potentiate HC-Pro in the host tissues. It is possible that an HC acquired first by the vector assists the transmission of virus particles located in the same cell, in other cells or even in other host plants probed by the vector. The helper component help virus transmission by mites also The role of viral determinants and vector determinants in transmission specificity is reviewed by Andret-Link and Fuchs (2005).

The type of arthropods that transmit viruses can be divided into two categories:

(i) those that have biting and chewing mouth parts such as beetles and caterpillars, and

(ii) those that have piercing and sucking mouth parts such as whiteflies, leafhoppers and aphids.

Thrips have rasping and sucking mouth parts while mites have puncturing and sucking parts. The aphids are the largest and most important and successful insect vectors of viruses. They possess an outer and an inner pair of stylet, the former modified from the mandibles and the latter from the maxillae that are present in chewing insects. The bundle of stylets is contained in the anterior groove of the proboscis. The food canal and the salivary duct are in the maxillary stylets. In the resting position the proboscis is held along the underside of the thorax. During feeding the proboscis is brought forward and its tip is placed against the plant surface. Alternate protraction of mandibular (outer) and then the maxillary (inner) stylets causes penetration of the tissue. Penetration is assisted by secretions in the salivary duct.

Leaf hoppers are second only to aphids in importance as vectors of plant viruses. They transmit mycoplasma-like organisms also. One hundred and fourteen virus species are transmitted by whiteflies. *Bemisia tabaci* transmits 111 of these species while *Trialeurodes vaporariorum* and *T. abutilonia* transmit three species each. Most viruses transmitted by whiteflies (90%) are of Begomovirus group Hoppers have sucking mouth parts as the aphids have but being more robust they penetrate the tissues more rapidly, deeper, and cause more damage to cells. Viruses are transmitted by leaf hoppers in the semi-persistent and persistent manner. Usually acquisition and inoculation feeding periods are long. The persistent viruses are often circulative-propagative type. In whiteflies also the method of uptake of virus is similar. The viruses are circulative but not the propagative type.

Transmission of viruses by soil vectors: The soil-borne viruses are of two types-those losing infectivity when soil is allowed to dry at 20º C for a week or more, and those where soil remains infective after drying for several weeks or years. The first group contains nematode-transmitted viruses and the second group fungus-transmitted viruses.

Four genera of nematodes, all belonging to same group, Dorylaimoidea, are known to be vectors of many plant viruses which are either polyhedral (nepoviruses, or nematode transmitted polyhedral viruses) or tubular (netuviruses or tobraviruses, or nematode transmitted tubular viruses). Transmission of virus by nematodes resembles transmission by insects. Viruses are picked and injected into the host by stylet. There is acquisition feed and inoculation feed time and the viruses are retained in the nematode vector for different durations at specific locations in the body. The genera that transmit plant viruses are *Xiphinema*, *Longidorus*, *Trichodorus* and *Paratrichodorus*. The grapevine fanleaf virus transmitted by Xiphinema americanum can surive in the body of the vector, in absence of the host, for about 4 years. At molecular level, specificity of virus transmission by nematodes is related to virus proteins that are required for successful transmission.

There are 32 soil-borne viruses or virus-like agents that are transmitted by fungal vectors. The fungi that transmit plant viruses belong to class Plasmodiophoromycetes (family Plasmodiophoraceae) and class Chytridiomycetes (family Olpidiaceae and Synchytriaceae).

Plasmodiophorids are not true fungi. The species are *Spongospora subterranea* f. sp. *subterranea*, *Spongospora subterranea* f. sp. *nasturtii*, *Polymyxa graminis*, *Polymyxa betae*, *Olpidium brassicae* and *Olpidium bornovanus*. These fungi infect their host roots by zoospores. The two *Olpidium* spp. transmit 10 polyhedral viruses externally on the zoospores. Eighteen rod-shaped viruses are transmitted by *O. brassicae*, *P. graminis*, *P. betae* and *S. subterranea* *in vivo* manner in which the viruses survive within the resting spores. Persistence of the virus depends on longevity of the resting spores of the fungus. Recent progress in elucidating the interaction between a virus and uts zoospore vector suggests that specific sites on the virus capsid as well as on the zoospore are involved in transmission. Earlier *Synchytrium endobioticum* was also considered a virus vector in potato. Virus-like agents are suspected to be transmitted by *Pythium*.

Management of Plant Viruses

The general principles of virus disease control are almost the same as for other diseases caused by fungi and bacteria, i..e. clean seed, clean field and protection of the standing crop. Most of the methods described below are used for general plant disease control and can be used, in some cases, for controlling more than one disease at the same time. The applicability of these methods depends upon the value of the crop, severity of the disease, the cost of labor and supplies, and the care and efficiency with which they are used. The basic difference between steps adopted for the control of diseases caused by fungi, nematodes, bacteria, etc. and the diseases caused by viruses is that while the former is heavily dependent upon chemicals for direct control the chemical control of viruses has been almost totally ineffective perhaps because virus replication is so dependent upon the normal metabolism of the host cell. The viruses are nucleic acids functioning in the genosphere of the plant. Any chemical that will destroy the nucleic acid is likely to destroy the host nucleic acid. Large numbers of substances have been screened for direct control of plant viruses but so far without practicable results. Nucleic acid base analogues have shown some promise. The uracil analogues ribavirin and dioxohexahydrotriazine (DTH) have shown suppression of potato viruses in field trials. Plant virus control, therefore, has been and is still mainly based on prevention and sanitary practices.

1. *Selection of seed:* Although majority of the plant viruses are not true seed-borne, it is better to select seeds from disease free localities, field and plants. If disease free seed are not available in the market the cultivator can grow his own seed with special care for individual healthy plants.

2. *Cuttings, bulbs and tubers:* Majority of viruses of vegetatively propagated crops are carried by the propagating material. Sugarcane mosaic and viruses diseases of potato are good examples. In such cases, care should be taken to select the planting material from plants that are not infected. However, this method will not be successful when symptoms are not visible on the plant and when secondary infection through insect vectors is common. In such situations, attempt should be made to grow a crop for seed purpose only in a protected area with all possible means of insect control. The degeneration of potatoes caused by virus diseases has been considerably reduced where systematic roguing of diseased plants and planting of healthy, large sized tubers has been practiced.

3. *Tuber indexing:* This method is used to select virus-free potato tubers for multiplication on a large scale. It consists of selecting tubers from vigorous plants looking healthy. These tubers are numbered and one eye or bud from each is planted under suitable conditions. If the resulting plants are weak or unhealthy the parent tuber is rejected. If healthy, the parent tuber is kept for planting at the proper time in a carefully protected seed plot and all plants that

appear weak or diseased are removed during the growing season. When sufficient seed tubers have been produced in this way they are distributed to growers. The process takes several cropping seasons to produce seed stock sufficient for distribution to growers. Now attempts are being made to employ biotechnology for producing microtubers for field planting. The method takes less time and gives better elimination of viruses from seed stock.

New detection procedures like ELISA (enzyme linked immunosorbent assay), HADAS (heterologous antibody double antibody sandwich techniques), and IEM (immunosorbent electron microscopy techniques) are now available for quick and accurate detection of viruses which save the time and ensure accuracy not possible in indexing. For elimination of viruses from nucleus potato seed stock meristem culture, thermotherapy, thermotherapy followed by meristem culture and cryotherapy are used. In sugarcane also the similar techniques are being applied. Wang *et al.,* (2006) working on potato leaf roll virus and potato virus Y Wang *et al.,* (2006) have reported that both PLRV and PVY could be efficiently eliminate by cryogenic treatments with, respectively, 83–86% and 91–96% of frequencies of virus-free plantlets obtained. These figures are higher that those obtained by meristem culture (50% for PLRV and 62% for PVY and similar to those obtained by thermotherapy followed by meristem culture. In thermotherapy of planting material of vegetatively propagated crops like sugarcane the treatments recommended at present reduce germination due to adverse effect on buds. The virus is also not eliminated completely. However, the virus is not present in the distal parts of the shoots and these can be used for tissue culture for production of virus free seed stock on a large scale.

4. *Protection against insect vectors:* This is one of the most important steps in the prevention of plant virus diseases during growth of the crop. It can effectively check virus spread. Aphids and whiteflies (*Bemisia tabaci*) are very difficult to control. The approaches to control insect vectors include (i) insecticidal control, (ii) biological control with fungi, (iii) biological control with predators and parasitoids and (iv) modifications in cultural practices. The conventional insecticides have failed to provide convincing check of aphids and whiteflies. They are slow in action and do not cause instant kill of the insect. This gives it enough time to transmit the virus. More recent trend has been to emphasize the use of nicotinoids and insect growth regulators. Nicotinoids (*viz.,* imidacloprid) are systemic neurotxins that quickly incapacitate the insects. The insect growth regulators include buprofezin (chitin synthesis inhibitor) and pyriproxyfen (a juvenile hormone analog). The use of these insecticides has been successful in keeping the populations of *Bemisia tabaci* under check (Palumbo *et al.,* 2001). Mineral oil, rapeseed oil and soya oil have been used against aphids with success.

Lecanicillium (Verticillium) lecanii, Paecilomyces fumosoroseus and *Beaurveria bassiana* are fungi that parasitize aphids and whiteflies. Their potential for biological control of *Bemisia tabaci* has been demonstrated (Faria and Wraight, 2001; Cuthbertson *et al.,* 2005). *Lecanicillium muscarium* is highly effective against immature stages of *Bemisia tabaci*. *V. lecanii* (*Lecanicillium lecanii*) is reported as an effective parasite of *Macrosiphum euphorbiae* (Askary *et al.,* 1998) and *Myzus persicae*. According to Gindin *et al.,* (2000) pathogenicity of *Lecanicillium lecanii* to pupae of *Bemissia argentifolii* (B strain of *B. tabaci*) varies between 59 and 72% six days after inoculation. The maximum adult mortality is between 34 and 52%. Strans of *Lecanicillium lecanii* produce toxins which act as contact insecticides. Crude extracts of the toxin are ovicidal. Nymphs are most susceptible. Crude extract of the toxin applied to seedlings at 400 mg/L significantly reduces the hatching of *Bemissia* eggs and the subsequent survival; rate of the nymphs and emergence and fecundity of the progeny (Wang *et al.,* 2007). *Lecanicillium muscorium, L. longisporum., Isaria fumasorosea* and *I. javanica* are also reported emtomopathogens of *B. tabaci*. Some species of *Cladosporium* are also reported to cause mortality in whiteflies. Activity of the fungi varies with host plants on which they are feeding. Use of *Beaurveria bassiana* in combination with imidacloprid (the systemic neurotoxin) improves control of *Myzus persicae* by reducing resistance of the insect to the chemical (Ye *et al.,* 2005). Cotton plants produce a fungal inhibitor that protects the whiteflies from mycosis (Poprawski and Jones, 2001). *Pandora delphacis* (Entomophthorales-Entomophtghoraceae) has the potential as a useful fungal agent for control of aphids. Conidial showers of the fungus on aphids (apteras) greatly reduce fecundity (rate of reproduction) at 20°–30°C. More than 80% mortality occurs at 20–30°C and 95% or more RH (Xu *et al.,* 2002). *Alternaria alternata* is another fungal biocontrol agent of aphids. It can infect 26 aphid species. Infected aphids die in 2–4 days (Christias *et al.,* 2001). *Neozygites fresenii* (Entomophthorales), when introduced in *Aphis gossypii* populations, spreads in the population and provides biological control (Steinkraus *et al.,* 2002). However, the use of such biocontrol agents on commercial scale is limited although commercial products based on *V. lecanii* are available. They have the same disadvantage as the conventional insecticides. Their action is slow and the insect can shake them off easily. Alavo *et al.,* (2002) have reported that frequent spray of blastospores of the fungus *Lecanicillium lecanni* gives partial control of the aphid. The hindrance to effectiveness of the biocontrol agent is the inability of the fungus spores to adhere to the cuticle of the vector and cause infection. Due to rapid moulting in the vectors the fungus spores are soon displaced from the cuticle. Furthermore, mycoinsecticides are costlier than conventional insecticides and have low shelf life.

Predators of virus vectors are also known. Coccinelid predators of *Aphis craccivora* are often common on cultivars resistant to aphids (Ofuyam 1995). *Nephaspis oculatus* (Coleoptera: Coccinellidae), a beetle, is a predator of *Bemisia tabaci* and *B. argentifolii* (Ren *et al.,* 2002). Cultural practices such as crop-free period, alteration in date of planting crop rotation, intercropping, seed and crop residue disposal perform well against insect populations if adopted on a regional scale. Use of reflective mulches are advocated for repelling the insects. Effectiveness of reflective polyethylene and biodegradable, synthetic latex spray mulches for management of aphids and aphid-borne viruses in cucurbits was reported by Stapletone and Summers (2002). Thrips (Thysanoptera: Thripidae) are vectors of many tospoviruses. *Orius* is predator of Thrips. *Orius laevigatus* (Heteroptera: Anthocoridae) is commercially used in Europe for suppression of infestation of *Thrips tabaci* on vegetables.

5. *The host-free period:* This method is a voluntary effort made by all the growers in an area in which the growers agree to eliminate all host plants of the disease for a period. At the end of this period fresh planting may be started with disease-free seed or plants.

6. *Roguing:* The removal of all diseased plants, as soon as they are spotted, has proved advantageous in crops like potato and sugarcane. The treatment has proved successful in controlling mosaic disease of sugarcane provided (i) the infection is less than 5%, (ii) the roguing is started at the right time and done frequently, (iii) healthy seed cuttings are used, and (iv) the field and its surroundings are kept free from grass hosts of the virus. A precaution before roguing is to spray the crop with a suitable insecticide so that vectors are killed or immobilized and do not spread in the field during handling of the plants.

7. *Destruction of weed hosts:* It is well known that some viruses are carried in weeds. Every possible precaution should be taken to keep the field and its surroundings free from these weeds. Perennial plants should be dug out and destroyed. Practices like ratooning in sugarcane should be discouraged when the crop is diseased.

8. *Temperature treatment:* These treatments have been successfully used with some virus diseases such as those transmitted through cuttings, bulbs, and grafts. Many virus diseases of sugarcane are adversely affected by heat treatment of the seed cuttings. Sereh

disease of sugarcane has been controlled by immersing the cuttings in water at a temperature of 52°C for 30 min. Sugarcane mosaic has also been controlled by treating cuttings at 53°–54°C. Immersion of setts in water at 52°C for 20 min kills the virus of chlorotic streak of sugarcane. Similar results have been obtained with many virus diseases of fruit trees. Heat treatment combined with meristem tip culture eliminates all viruses (except capilloviruses) and viroids from graft propagated plants.

9. *Possible chemical control*: Chemical control of virus diseases in plants and animals has not been possible in spite of claims. Virazole (ribavirin, a ribofuranosy; thiazole carboxaide compound) has been tried on plant viruses. Callus of gladiolus infected with CMV and Bean yellow mosaic virus, grown in tissue culture, was freed of the viruses when treated with virazole (Singh *et al.*, 2007). Virus inhibitory substances are present in many plants and animal products like skimmed milk. Plant extracts have been found to reduce severity of many plant viruses. Oka *et al.*, (2008) have reported suppression of *Pepper mild mosaic virus* by leaf treatment with commercial cellulose from *Trtichoderma reesei* and *T. viride*. The suppressive effect is better than that of skimmed milk and culture filtrate of the mushroom *Lentinula edodes*. According to Shen *et al.*, (2008) the antiphytoviral compound, Bruceine-D, isolated from seeds of *Brucea javanica* (Simarcoubaceae) is reported to significantly inhibit infection and replication of *tobacco mosaic virus* (TMV). It also shows a strong inhibitory effect on the infectivity of *potato virus Y* and *cucumber mosaic virus* (PVY and CMV).

10. *Resistant varieties:* Durable resistance in plants is more common in viruses than in cellular plant pathogens and appearance of resistance breaking strains is low (Garcia-Arenal and McDonald, 2003). The development of resistant varieties is the most satisfactory method of plant disease control but comparatively few virus diseases have been controlled in this manner. The only satisfactory control that has been achieved through resistant varieties evolved by traditional breeding is that of sugarcane mosaic. Wild species of plants sometimes possess good tolerance to viruses and many other diseases. These sources can be used to develop resistance in cultivated plants. *Solanum brevidens*, a diploid non-tuber-bearing potato line is highly resistant to Potato Leaf Roll virus but difficult to cross sexually with cultivated potato. Protoplast fusion between the two has yielded somatic hybrids that carry the resistance and also bear fertile pollen and eggs. These can be used in traditional breeding programmes for developing potato resistant to potato leaf roll. The ideal resistant variety is the one which is resistant to the virus as well as the vector. Cytological and histological responses of potato to PVY inoculation are essentially the same in susceptible and resistant cultivars but the resistance response is delayed in susceptible cultivars allowing the virus to spread beyond the point of inoculation. The wild species *Solanum demissum* is extremely resistant to PVY and PVX.

With new developments in the techniques for detection and identification of viruses and possible use of genetic engineering for developing virus resistant plant lines the above listed conventional approaches to virus disease control have undergone a change to remove the limitations of time consuming steps such as tuber indexing and uncertainties in control through heat therapy which causes damage to host tissues. The perspectives in virus control list steps based on tissues culture and genetic transformation of the host to produce virus free planting material and protection of the healthy plant from re-infection. This may involve genetic engineering techniques to introduce into plants specific genes or genetic codes not only from plants, as is done now in plant breeding, but also from the viral genome itself.

Many higher plants contain antiviral factors in their sap. Some of these are *Pelargonium hortorum*, *Chenopodium album*, *C. amaranticolor*, *Azadurschta indica* (neem), *Vitis vinifera* (grape), *Rosa banktia* and *Mirabilis jalaapa*. The active compounds belong to the groups of furocoumarins, alkaloids, terpenoids, lignans and ribosome inactivating specific proteins (RIP) such as in *Mirabilis jalapa* (MAP). In most cases the effects of crude leaf extracts have been tried.

Some plant growth promoting rhizobacteria are reported to delay appearance of virus symptoms (Murphy *et al.,* 2000) There are reports suggesting that treatment of leaves with rhizobacteria and strains of *Streptomyces* spp. induces resistance and reduces intensity of symptoms caused by certain viruses. The suppression of symptoms appears to be due to the host gaining ability to repair the cellular damage caused by virus infection. Ryu *et al.,* (2007) have reported growth promotion and induction of systemic resistance in tobacco against *Cucumber mosaic virus* by *Pseudomonas chlororaphis*. Tobacco cultivars differed in their response. Elbadry *et al.,* (2006) have reported induction of systemic resistance in faba bean (*Vicia faba*) against *Bean yellow mosaic virus* (BYMV), a potyvirus, by seed bacterization with rhizobacteria *Pseudomonas fluorescens* strain FB11 and *Rhizobium leguminosarum* bv. *vicrae* There was significant increase in the level of salicylic acid and peroxidase activity in the host tissues.

Induction of synthesis of pathogenesis-related proteins (PR) in the plant through cross-protection with avirulent strains of the virus provides resistance to the virulent strain. Cross protection, though successful in many plant species, has many limitations such as use of live virus which may mutate and cause more serious disease, search and maintenance of mild strains and its use which requires expertise and labour. Except for *Citrus Tristeza Virus* cross protection has not been commercially accepted. Engineered protection, which is a form of parasite-derived resistance, using genetic engineering is now considered a better approach. Antiviral activity induced by chitosan is also reported.

Biotechnology is being employed to develop transgenic plants in which coat protein gene of a virus is incorporated in the genetic system of the plant and this provides resistance to the plant. Papaya ringspot, potato leaf roll, potato leafroll, tomato mosaic are some of the virus diseases in which this approach of control has been demonstrated on a field scale. Induction of resistance to Potato leaf roll virus in potatoes through biotechnology has resulted in reduced field spread of PLRV in potatoes transformed with the PLRV coat protein gene (Thomas *et al.,* 1997). In papaya, resistance to ringspot by transformation with viral coat protein genes has proved more effective than cross protection by inoculation with mild mutants of the virus (Gonsalves, 1998). Systemic acquired resistance (SAR) against plant viruses is effective against only localized infections where it limits cell-to-cell spread of the virus (Pennazio and Roggero (1998). Chemical induction of resistance to systemic virus infection by use of chitosan was reported by Chirkov *et al.,* (2001). Potato plants were sprayed with chitosan (1mg/ml) and leaves were inoculated with *Potato Virus X* after 1, 2, 3 or 4 days. The plants showed reduced infection by the virus. The induction of resistance was through callose formation and induction of ribonuclease. However this resistance was transient and disappeared within 2–3 days. Antiviral effect of chitosan can be also through host cell death.

11. *Immunization:* Although cross protection has been proved in many plant diseases, especially the virus diseases, field scale application of the method has been demonstrated. only in very few cases.

Although biological control of viruses is mainly indirect, through biocontrol of vectors, there has been a rapidly growing interest in the use of viral satellites as agents of biological control of plant viruses. They are natural inhibitors of crop-damaging viruses. The satellite of Tobacco necrosis virus (sTNV) greatly decreases accumulation of TNV in infected plants. Satellite of CMV (CARNA 5) could ameliorate the pathogenicity of its helper virus. Kundan *et al.,* (2005) have reported biological control of tomato spotted wilt virus (TSWV) by using strains of

Pseudomonas fluorescence as foliar, seed and soil treatment which increased polyphenol oxidase, β-1,3-glucanase and chitinase activities.

There are also reports of induction of systemic acquired resistance by using bacterial and other agents. There is no general agreement on this approach. Pennazio and Roggero (1998) had suggested that although SAR plays important role in management of other pathogens, it should be viewed with less enthusiasm in case of plant viruses. SAR operates against localized virus infection by limitimh cell-to-cell spread of virus through hypersensitive necrosis of host cells without inhibiting replication of the virus. It does not operate against systemic virus infection; it neither limits virus replication nor alleviates symptoms. Srinivas *et al.,* (2007) have reported control of *Sunflower Necrosis Virus* by application of some strains of *Bacillus, Pseudomonas* and *Streptomyces.* There are many such reports published earlier.

Role of salicylic acid in alleviation of effects of virus infection and induction of resistance in leaves has recently attracted much attention. In pumpkin or zuechini (*Cucurbita pepo*) infection of yellow mosaic virus causes serious metabolic changes in leaves. Oxidative stress leads to the development of symptoms and damage to the host. When leaves bearing symptoms are sprayed with salicylic acid, there is recovery from the undesirable effects of the infection. Leaves treated with 100 μM salicylic acid before inoculation have the appearance of healthy leaves (Radwan *et al.,* 2006). Salicylic acid also induces significant metabolic changes in the leaves. The resistance inducing capacity of exogenous application of SA was further confirmed by Radwan *et al.,* (2007).

VIRUS DISEASES OF POTATO

Viruses in potato have been known since long throughout the world as main cause of potato degeneration (successive increase in yield loss over years). The distribution and spread of the viruses varies from place to place depending on the climate, varieties under cultivation, vector population, seed source, etc. Occurrence of potato viruses in India had been noticed more than 70 years ago in Punjab. Almost 30 viruses are now reported on potato in India but only about 9 are responsible for significant losses. These include potato leaf roll virus (PLRV) and potato mosaic viruses such as Potato virus X (PVX), PVY, PVA, PVM, PVS and certain complexes of viruses such as rugose (PVX + PVY), crinkle (PVX + PVA) and aucuba (PVF + PVG).

A lower incidence of viruses (5–10%) does not much reduce the potato tuber yield. But a higher incidence coupled with early and severe infection causes severe depressions in the tuber yield. The severe strains of Potato virus Y and leaf roll virus may cause 60–75% loss in tuber yield while the mild ones like Potato viruses X, S, A and M may depress the yields by 10–30%. On the basis of a rough estimate, an average 40–45% infection causing up to 45% yield reduction may result in annual loss of around 5–6 thousand million rupees in India. Continuous use of old seed stocks for several years, in areas with high aphid (vector) population, and lack of adopting practices for management of viruses results in 100% infection of the seed stocks in 3–4 years bringing down their yielding ability to almost 50%.

Leaf Roll of Potato

Leaf roll has been the most common cause of potato degeneration. In U.K. the potato leaf roll was present before the late blight was introduced in the area in 1840s. In India, loss in yield due to leaf roll has been as high as 20–50%. Seven to 16% losses occur in autumn crop and 39–60% in spring crop. According to Hamm and Hane (1999) in some potato cultivars the use of infected seed tubers may result in 60% less total yield and 88% less marketable yield of tubers (> 85 g). The disease is common throughout India.

The name "leaf roll" is partly descriptive of foliar symptoms of the disease. In some varieties there is yellowing of margins or tips of leaflets. The rolling of leaves is characterized by curling of leaflet margins inward, thus forming a trough in which the midrib is at the bottom. In secondary infection (plants raised from infected tubers) this rolling of leaves starts in the lower leaves and progresses upward throughout the plant. In infections occurring during the season (primary infection), the rolling of leaves starts in the upper part of the plant and progresses downward. In some varieties individual leaflets have a tendency to be more erect, giving the diseased plants an upright appearance.

The rolled leaflets are stiff and rigid. They are thick and leathery, sometimes with pink margins, and have a characteristic rattle when brushed with hand. In plants developing from diseased seed tubers the internodes are short resulting in dwarfing or stunting of the entire plant. Often the tuber bearing stolons are shorter than normal and tubers appear attached directly to the underground stem. The number of tubers produced per plant and their size in a diseased crop are greatly reduced. This is the main cause of damage by this disease.

Fig. 134. Leaf roll of potato.

Internal necrosis of tubers is a symptom usually associated with primary infection (infection occurring in the current season) but can develop in tubers of plants with secondary infection (plants raised from infected tubers). Two types of necrosis may occur stem end browning and net necrosis. In the latter the necrosis appears as a brown ring in the vascular system. Phloem necrosis was known since 1913. Due to necrosis of the phloem translocation of food to the underground tubers is checked. Therefore, the tubers remain small in size and aerial tubers are formed above ground. The period of maximum net necrosis coincides with the period of most rapid tuber growth.

Leaf roll is caused by *Potato leaf roll virus* (PLRV), a luteovirus. It is the type species of the genus *Polerovirus* in the family *Luteoviridae*. Infected potato foliage contains about 1–2 mg PLRV particles per g of leaf tissue. Unlike most plant viruses, which infect majority of plant tissues, the luteoviruses are restricted to phloem. They are not seen in other tissues. This feature of the virus is focus of further investigations. The concentration of virus particle is maximum in roots and minimum in stems The particles, first purified in 1960s, are isometric and measure 24–26 nm in diameter (1 nm is 1/1000 μm). They contain a single strand of RNA, in one piece, which constitutes 30% of particle weight. In extracts of aphid vectors and in the sap of *Physalis floridana* the thermal inactivation point of the virus is 70°C (10 min.), dilution end point is 1:10,000, *in vitro* longevity 4 days at 2°C and 12–24 hours at 25°C. A number of strains of the virus have been differentiated on the basis of severity of symptoms induced in the host species and by their ease of transmission by aphids.

PLRV is not sap transmissible. In nature, transmission occurs through infected tubers and aphid vectors. *Myzus persicae* is the main aphid vector. Transmission by this vectors was disovered in 1920s. Other aphids are *Myzus circumflexus, M. convolvuli, M. solonafolii* and *Aphis gossypii.* All stages of *M. persicae* can acquire the virus but nymphs transmit more efficiently than adults. The vector requires a minimum feeding period of 2 hours on a diseased plant to acquire the virus but high level of infectivity is achieved by feeding for a day. It may become viruliferous within 2 min but a latency of more than half a day has been reported. Nymphs of Myzus persicae bruce honeydew on leaves of Physalis flotidana. This excretion is correlated with transmission efficiency. Nymphs provided with long acquisition period and producing more honey dew can cause about 100% transmission while nymphs not producing homey dew do not transmit the virus. Honeydew appears to make more virus available to the vector.

The vector retains the virus for its whole life and probably multiplies also in the insect body. Younger plants and younger leaves of old plants are better sources of virus inoculum. In India, the disease is more prevalent in the spring than in autumn crop due to high aphid population and motility. Spread of the virus is more within the row than across the rows. Although common on potato and rarely on tomato, the virus can infect *Datura stramonium, Datura tatula, Physalis angulata,* and *Pysalis floriata* (*Physalis pubescence*) which are its differential hosts. The non-solanaceous hosts include *Celosia argentea, Gomphrena globosa, Nolana lanceolata, Capsella bursa-pectoris* and *Montia perfoliata.* Epidemiology of the virus is largely associated with the ecology of its aphid vector, *Myzus persicae.*

Potato Mosaic

The mosaic diseases of potato have been categorized, according to symptoms, into latent or taint mosaic (PVX, PVS, PVT), mild mosaic PVM, PVA), severe green mosaic (PVY) and severe yellow mosaic (potato mop top virus- PMTV, potato yellow dwarf virus, tobacco rattle virus, and calico caused by alfalfa mosaic virus). Crinkle, rugose, and rosette are natural combinations of potato viruses. The mosaic symptoms caused by different viruses often overlap depending on host cultivar, virus strain, temperature, light, etc.

Mild or Latent Mosaic: This mosaic is caused by *Potato virus X* (potex virus group) also known as Potato mild mosaic virus, Potato latent virus and Potato mottle virus. At least 4 strains of the virus are known. A resistant breaking strain, PVX(hb), occurs in the Andean region of South America, Europe and India. It can infect PVX immune cultivars.

Most of the strains are latent in many cultivars where the plants carry the virus without showing any visible symptoms. Major symptom is interveinal to barely perceptible mosaic. In addition to mild mosaic and sometimes little dwarfing in potato, the virus causes mosaic and slight stunting in tomato and mottle and necrotic ringspot in tobacco. In potato cultivar Craig's Defiance top necrosis occurs. Symptoms are masked in vigorously growing plants at temperatures above 21°C. No symptom development and virus multiplication occurs at temperatures above 30°C. Although very widely present in most of the seed stocks the disease does not cause as much damage as the leaf roll and other viruses affecting potato in India.

The virus particles are flexuous filaments measuring 450–540 (or up to 575) x 13 nm. The ssRNA is in one piece and constitutes 8% of the particle weight. During early stages of infection the virus is mainly found in the palisade cells, less frequently in the epidermis. The particles are diffusely distributed in dense X-bodies which may fill a greater part of the cell. Particles and inclusion bodies are seen in plastids, dictyosomes and endoplasmic reticulum. After inoculation of the virus into the leaf, the virus requires a period of multiplication at the site of inoculation before movement from the area of inoculation begins. The translocation period varies with varieties and may be from 6 to 10 days.

Fig. 135. Mosiac of Potato.

In tobacco sap, thermal inactivation point for the virus lies between 68° and 76°C, depending on the strain. The dilution end point is 1:10,000 to 1:100,000. Infectivity is retained for several weeks at 20°C and for more than a year in the presence of glycerol.

Transmission of the virus easily occurs by contact. No insect vectors are known except that grasshoppers may mechanically transmit the virus. *Synchytrium endobioticum* was reported as a vector of PVX but it is not confirmed. Some strains of the virus can persist on surfaces of equipments used in the field. The host range includes potato, tomato, tobacco, eggplant (brinjal), pepper, *Solanum nigrum*, *Datura stramonium*, *Petunia* sp., *Gomphrena globosa* and *Chenopodium amaranticolor*. *Gomphrena globosa* and *Capsicum pendulum* are the local lesion hosts of PVX.

Vein-banding Severe Mosaic of Potato: This disease of potato is caused by *Potato virus Y* (PVY, potyvirus group). It was also known as Potato acropetal necrosis virus, Potato severe mosaic virus, and Tobacco vein banding virus Unlike PVX, this virus produces severe symptoms depending on the host variety and virus strain. The symptoms vary from mild to severe mottle and streak or "leaf drop streak" with necrosis along the veins on underside of the leaflets. The leaves become completely necrotic but remain hanging. The top most leaves are generally not affected by necrosis but are mottled and slightly crinkled. Plants from infected tubers are stunted with brittle and crinkled leaves.

Fig. 136. Potato leaf affected by *Potato virus Y.*

The particles of the virus are elongated, flexuous, helically constructed rods, measuring 730–740 x 11–15 nm. The ssRNA in the particle accounts for 5.4 to 6.4% of the particle weight. Cylindrical inclusion bodies are seen in the epidermal cells. In tobacco sap the thermal inactivation point lies between 50° and 62°C, the dilution end point between 100 and 1000000, and *in vitro* longevity at 18–22°C is 7–50 days. The Indian isolates of the virus have TIP of 52°–55°C and DEP of 1:100 to 1:1000. The virus mainly infects plants in the family Solanaceae but members of Leguminosae, Chenopodiaceae, Compositae and Amaranthaceae are also susceptible. Apart from potato, the solanaceous hosts are *Solanum demissum, Nicotiana tabacum, N. glutinosa* and *Datura stramonium.* There are several strains of the virus designated as Yo, Yn, Yc, etc.

The virus is mechanically transmitted by sap inoculation and by grafting. In nature it is transmitted by infected tubers and atleast 25 species of aphids in non-persistent manner. *Myzus persicae* is the most efficient vector. Other species are *Aphis fabae, Macrosiphum euphorbiae, Myzus certus, Phorodon humuli,* and *Rhopalosiphum insertum.* Pre-acquisition fasting periods of 15 min to one hour, sometimes 3–4 hours, increase transmission ability of the vectors. Optimum acquisition feed time is 15–60 sec. periods lasting more than 2 minutes are less favourable. Most aphids cease to transmit the virus within one hour after the end of acquisition feed but starved aphids may retain the virus up to 24 hours. Infectivity is not retained after molting. The maximum inoculation feed time is 30–60 sec.

In a study of PV.Yn strain that causes veinal necrosis in Samsun tobacco, it was observed that a single individual of *M. persicae* per plant is able to transmit the virus (20–76.6% transmission). The highest transmission is by 5–10 aphids per pant. Aphids pre-starved for one hour before acquisition feed cause maximum transmission. Prolonged pre-acquisition starvation beyond one hour reduces acquisition ability. Aphids acquire the virus within 15 sec of feeding but the percentage of transmission increases when 5 min of acquisition access period is allowed. Further increase in acquisition access period decreases transmission. The aphids poorly transmit the virus (33%) when inoculation access period is up to one minute which increases to 66% in 15 min inoculation access feed. The aphids do not infect more than 4 plants in a series of feeding. In susceptible cultivars of potato PVY^{NTN} is detected within 1–5 days after inoculation. From the inoculation point the virus spreads rapidly, first in to the stem and then more or less simultaneously to the upper leaves and the roots. Increasing nitrogen supply to the crop can influence total yield but does noit significantly mitigate the yield reduction due to PVY infections.

Rugose Mosaic of Potato: This disease causes very severe damage to individual plants. The foliage is not only mottled but also severely wrinkled, puckered, and markedly reduced in size. The leaflet margins are rolled downward and the entire plant is severely dwarfed. The lower leaves generally have black necrotic veins. While the symptoms of mottle may be masked by high temperature, the rugosity or roughness and abnormal hairiness of leaves and dwarfing of the entire plant persist. In severe cases, plants may even die before producing any tuber.

Fig. 137. Rugose mosaic of potato.

The germination of infected tubers is reduced by about 20%. The peak period of infection in Bihar and Bengal is middle of January to end of February. Although the disease causes severe injury to the plant, the total loss in yield is not as much as in leaf roll affected crop because fewer plants are attacked by the rugose mosaic.

The disease is caused by a combination of two viruses: *Potato virus X* (mild or latent virus) and *Potato virus Y*. Under field conditions the transmission of the virus is through small sized tubers formed on diseased plants and used as seed. *Myzus persicae* is capable of efficiently transmitting PVY, one of the components of the complex.

Crinkle of Potato: This disease resembles the rugose mosaic very closely but the yellowish patches on the foliage are bigger and more prominent. As death approaches, the colour becomes more pronounced and is accompanied by rusty brown spots beginning near the tip of the leaves. The foliage is brittle and easily injured.

Crinkle is caused by a combination of *Potato virus X* and *Potato virus A*. PVA is also a member of potyvirus group and causes super mild mosaic of potato. This virus complex also is tuber-borne. Sap inoculation transmits only one component, PVX, while *Myzus persicae* can transmit the other constituent, PVA.

Management of Potato Virus Diseases

The potato viruses are perpetuated and initiated in the field from diseased seed stock retained from preceding disease potato crop. Therefore, the use of disease-free seed tubers and elimination of chances of infection through insect vectors are the main principles of control of these diseases.

The chances of disinfection of tubers with chemicals are very remote. However, in the case of PVY spray of thiouracil or trichothrecin has been found partly effective. Heat therapy has been shown to inactivate many plant viruses and has been successfully used to free tubers of yellows diseases (MLOs) but not mosaic diseases. Further, heat therapy is not a practical solution where fleshy plant organs are involved.

In the absence of a direct method of control, it is better to apply indirect methods such as production of disease-free tubers for seed and control of vectors that spread the virus. Disease free crop for seed tubers can be raised in the following manner:

1. Obtain certified seed from a reliable source and plant early in the season.
2. Rogue out sick looking plants as soon as located.
3. Detop (remove foliage) the plants in the third or fourth week of December when aphid populations start building up. and then leave the tubers in soil to mature. The yield is less but it will have mainly the seed size tubers free from virus. The recommendations for Punjab had suggested that planting should be done between the first and tenth week of October and detoping done in the first week of January. Permissible level of virus infection is 1%. Critical level of aphid population is 20 aphids per 100 compound leaves.

Use of large sized tubers (about 5 cm dia.) and systematic roguing of virus affected plants from the field were practiced in the past with success in minimizing losses from virus diseases of potato.

Use of microtubers in potato for production of virus-free planting material is a promising future technique against potato viruses (Naik and Sarkar, 2000). Microtubers are miniature tubers developed under tuber-inducing conditions *in vitro* using the tissue culture technique. These small, dormant tubers can be easily handled and stored at low temperature for months. The variety Kufri Megha is resistant to PVX and PVS.

Sources of resistance to potato viruses have been fpound in wild tuber-bearimg species of *Solamum*. *S, stoloniferu,* shows resistance, through expression of HR, to all potyvituses. Diploid genotypes of potato are highly resistant to PLRV and some other potato viruses. In 1994, a clone (DW.84–1457) was describe which was highly resistant to PLRV, PVX and PVM. Another clone, G8107(1), reported by Solomon-Blackbirn *et al.,* (2008) has high found to show high resistance in graft and vector-transmission tests. The very high resistancde is attributed to resistance to vurus movement within leaves and from leaves to petioles. Hybrids highly resistant to PLRV have been developed by somatioc fusion between diploid *Solanum verrucosum* and dihaploid *S. tuberosum.*

Pardee *et al.,* (2004) have reported antivral activity of extracts of a number of marine algae belonging to the phylum Heterokontophyta. More than 80% reduction in infectivity of PVX was observed when tested on a local lesion host. The extracts caused aggregation of virus particles.

For aphid control, use of systemic insecticides such as carbofuran, aldicarb, and phorate is best. Application of these chemicals in granular form in the furrows at the time of planting keeps the plants in a position to ward off aphids at least for the first 50–60 days after planting. A further spray of Rogor can prolong this protection. Furrow application of aldicarb at the rate of 4–8 g a.i/100 m is reported to control aphids and give significant protection against leaf roll. Mowry (2005) have reported that transmission of PLRV is reduced by the aphicides imidacloprid (100%), thiamethoxam (99%) and pymetrozine (87%). The transmission of PLRV is reduced by 55–90% by neem (*Azadirachta indica*) metabolites (seed kernel extract and azadirachtin). The metabolites cause 100% mortality of *M. persicae* nymphs when applied at 2560 ppm. The biological control of aphids through fungal parasites was mentioned earlier. Straw mulching between rows reduces aphid infestation and PVY incidence (Saucke and Doring, 2004). Spray of 1 or 2% neem (margosa) oil imhibits aphod transmission of PVY. It interferes worth acquisition and inoculation capacity of the vectors.

Horticultural mineral; oil, refined rape seed oil and soya oil sprayed on plants cause over 80% mortality among aphids. Repeated sprays of oil emulsion (10 ml oil/liter water) on plants raised from imidacloprid treated seed tubers are reported to reduce number of PVY infected plants by 40% (rapeseed oils) and 60% (mineral oil).

VIRUS DISEASES OF TOMATO

The first record of virus diseases of tomato in India seems to be in 1930. The leaf curl of tomato was described in 1948.

Leaf curl of tomato

The disease has been reported from India, Sri Lanka, Nyasaland (Malawi) and South Africa. Leaf curl disease incidence is reported to be 83% in the winter crop planted in October and 14% in the summer crop planted in February (Tripathi and Varma, 2002). When the plants are infected within 20 days of transplanting the loss may be up to 92% while infections of 35 and 50 days old crops result in 74 and 29% loss, respectively.

Leaf curl is characterized by severe stunting of the plants due to shortening of internodes and downward rolling and crinkling of the leaves. The newly formed leaves show chlorosis. The older curled leaves become leathery and brittle. The whole plant looks pale and produces more lateral branches resulting in a bushy appearance. There is partial or complete sterility of the plant. There is synergistic effect of association between the root knot nematode (*Meloidogyne incognita*) and the tomato leaf curl virus (Goswami, 1993).

Fig. 138. Tomato leaf curl.

Leaf curl of tomato is caused by the *Tobacco leaf curl virus* (geminivirus group or family Geminiviridae, genus *Begomovirus*). The genome of geminiviruses is composed of circular, single stranded DNA, encapsulated by multiple subunits of a single capsid protein. Most are bipartite having two almost equal sized genomic components designated A and B, separately encapsulated in geminate particles. The Indian isolates from New Delhi and Karnataka have been separately named such as *Tomato leaf curl New Delhi virus* (ToLCNdV), a bipartite strain, and *Tomato leaf curl Bangalore virus* (ToLCBV), a monopartile strain. ToLCNdV is reported on cucumber, bottle gourd and muskmelon also in Thailand. Chakraborty *et al.,* (2003) have described.

Tomato leaf curl Gujarat virus (ToLCGV) from Varanasi in India which causes a severe leaf curl disease of tomato. In Bangladesh, several variants have been reported which have similarity to ToLC New Delhi and ToLCV Gujarat (Varanasi) viruses. Complete nucleotide sequence of this strain has been determined. Kon *et al.,* (2002, 2007) have similarly reported monopartite strains of ToLCV from the Philippines (ToLCV-Philippines) and Indonesia (ToLCJaV). Recombination of tomato-infecting begomoviruses is reported from many countries. *ToLC New Delhi Virus* is reported on bitter gourd in Pakistan. A begomovirus, AYVV-(Java), causes yellow vein disease of *Ageratum conyzoides* in Indonesia and is most closely related to ToLCJaV. Kon *et al.,* (2007b) have considered it a strain of ToLCJaV and named it *Tomato leaf curl Java-Ageratum virus*. Recombination of a severe species of ToLC virus with another distinct species of the same group may induce supervirulence.

The geminate particles of the leaf curl virus measure 25–30 x 15–20 nm. The ToLCV is neither seed nor sap transmissible but external contamination of seed may occur. Dodder transmission has been reported in India. The main agency of its transmission in nature is the whitefly, *Bemisia tabaci*. The first demonstration of transmission of a virus by this insect was made in tobacco leaf curl in early 1930s. The vector can transmit the virus after a 20-min acquisition access period and 60 min of inoculation access period with minimum latency period of 4–9 hours. In India, the virus is reported to be transmitted by the vector after 30 minutes of acquisition access and 30 min of inoculation access periods with a latency period of 6 hours Even a single viruliferous whitefly is able to transmit the virus. The virus is retained by the vector for more than 12 days or for the whole life but there is no evidence that it multiplies in the vector or is transferred to the progeny of the vector. The time required for symptom expression varies in different cultivars of tomato. Whitefly populations per plant also vary with host cultivar. The number of whiteflies per plant has no relationship with the time of appearance or severity of symptoms. At low temperatures symptoms are delayed. In tomato yellow leaf curl virus (TYLCV) the virus is first detected in the head of B. tabaci 10 min after acquisition access period (AAP) of 8 hours. It is present in the midgut after 40 min and in the haemolymph after 90 min. The virus is detected in the salivary glands 5.5 hours after it is first detected in the haemolymph. Although females are more efficient vectors than males, the virus is detected in the salivary glands of both after approximately the same acquisition access period.

Resistance to the virus in commonly cultivated tomato varieties is lacking. Wild tomato (*Lycopersicon peruvianum*) has high degree of resistance to the leaf curl virus. *L. glandulosum* also does not show symptoms of leaf curl (Hayati and Verma, 1984). Some accessions of the wild species *Lycopersicon hirsutum* have been identified as the best source of resistance to tomato leaf curl virus as well as tomato yellow leaf curl virus (Maruthi *et al.*, 2003a). The resistance is mediated through resistance to the vector. Accessions of *L. peruvianum* and *L. chilense* are resistant to ToLCBV but highly susceptible to whiteflies, whereas an accession of *L. hirsutum* is resistant to both. *Lycopersicon esculentum* genotypes H-24, FL 744–6–9 and FL 699 are tolerant to TpLCBV (Maruthi *et al.*, 2003b). Muigai *et al.*, (2003) have reported that certain accessions of *L. hirsutum f. typicum* are highly resistant to *Bemisia argentifolli* (a biotype of *B. tabaci*) as defined by density of all life stages of the whitefly on leaflets. Trichome density has some role to play. High density of trichomes is correlated with low density of the vector on the leaves. In their study *L. pennellii* was most resistant to *Bemisia*. In Punjab, the genotypes TLB 111.119, 122, 128, 129, 130 and 134 have been found resistant to ToLCV. Growing a cucurbit (squash) as a trap crop for *B. tabaci* could be a useful cultural manipulation in managing the vector and the leaf curl diseases transmitted by it. The method is more useful for small scale farm operations (Schuster, 2004).

Protection of seedlings from infection brought by the vectors is important. It can be achieved by using insecticides or insect repellents. Tripathi and Varma (2002) have demonstrated that seedlings protected by perforated polythene bags and transplanted in polysheet mulched fields have low incidence of the disease. Some workers have claimed good control of the disease through reducing vector population. Use of systemic insecticides such as methyl parathion (0.02%) and dimethoate (0.5%) as spray or carbofuran (1.5 kg a.i./ha) as soil application has been found effective in controlling build up of whitefly population and thus reducing the incidence of leaf curl. Spraying should be started soon after transplanting. Five to 6 sprays may be required. Single soil application of carbofuran or other granular systemic insecticides also gives significant relief for about 40 days. Carbofuran (Furadon 3G) at the rate of 3.3 g/sq.m. not only reduces disease incidence but also increases fruit yield. Nursery treatment, seedling soak, and plot treatments reduce populations of *Bemisia tabaci* and disease incidence. Increased plant population from 30–90 per m^2, three applications of insecticides (parathion 0.02%, demetonmethyl 0.02% or dimethoate 0.05%) as spray at 15-days interval with one soil application of phorate or carbofuran has been suggested for control of the disease. Pymetrozine is a feeding inhibitor compound. It interferes with activities of the whitefly on the leaf surface and reduces transmission of *Tomato Yellow Leaf Curl Virus*

(Polston and Sherwood, 2003). Spray of skimmed milk (1:10 dilution of milk powder) followed by application of kitchen ash sufficient to cover the foliage surface is reported to reduce vector activity and leaf curl incidence (Ali *et al.,* 2001).

Foliar application of the antibiotics Validomycin-A (0.02%), Avomycin-A (0.01%) and of guanidine, azoguanine, etc. suppresses the disease. The root dip of tomato seedlings in gibberellic acid and 2-thiouracil at 50 ppm reduces leaf curl incidence. Spraying of 2,4-D is highly effective in not only suppressing tomato leaf curl incidence but also in many-fold increase of fruit yield. However fruit size is reduced.

Although management of leaf curl has relied heavily on insecticides, the overlapping crop cycles, resistance to insecticides in the vectors and problem of environmental pollution has prompted greater emphasis on resistance or tolerance and biological control of the vectors through use of parasitoids (Stansly *et al.,* 2004).

Mosaic of Tomato

Tomato mosaic can be caused by many different viruses such as TMV (*Tobacco mosaic virus*), *Cucumber mosaic virus* (CMV), *Potato virus X*, and *Potato virus Y*. The common tomato mosaic was earlier thought to be the same as TMV but now it is considered to be a distinct virus (ToMV) of the same tobamovirus group. Viruses of this group cause significant yield losses in tomato. Their impact is more significant for tomato production under greenhouse conditions than under field conditions because these are highly infectious, extremely persistent and easily transmitted during handling of plants.

The symptoms of tomato mosaic are generally influenced by temperature, day length, light intensity, plant age, virus strain, and tomato variety. In the tropics or in warm weather with long day length and high light intensity leaves show light and dark green mosaic mottle, sometimes with distortion of young leaves. Green areas are sunken giving the leaf a rough appearance. In winter, with low light intensity, short day length, and temperatures not above 20ºC, plants often are severely stunted and leaves distorted to "fern-leaf" or tendril-shape but mottling may be slight. Seedling infection may kill the plants. Fruits are fewer, undersized, and often deformed. In some cases there is necrosis of stem, petioles, leaves and fruits.

The dark and light green pattern in mosaic is attributed to differences in the density of virions in the leaf tissues. The light green areas contain high amount of virions while the dark green areas contain low amount of virions Involvem of RNA silencing is suggested for the difference. The antiviral RNA silencing in a host causes more volume of the virion and little or no green islands.

In aucuba mosaic the symptoms appear as downward curling of the whole leaf with slight turning down at the margins. Surface of the leaf is rough, wrinkled or corrugated. Chlorosis appears as small points of yellow areas and gradually spreads. In extreme cases almost the entire lamina of old and new leaves becomes pale yellow to white, with scattered small islands of green which stand up as blisters. In less extreme cases the green areas are larger but not as large as the chlorotic areas. The plant is not killed but growth is retarded. The virus induces pollen sterility which results in low fruiting and low yield.

The virus causing the common mosaic of tomato is known as *Tomato mosaic virus* (ToMV, tobamovirus group). There are many strains of the virus producing different symptoms and often these have been described as different diseases. These strains are tomato aucuba mosaic, tomato enation mosaic, yellow ringspot strain and tomato rosetted strain. ToMV infects more severly the tomato plant than bell pepper while TMV (*Tobacco Mosaic Virus*) can infect tomato but is more severe on bell pepper.

Particles of the virus are straight tubules with helical construction and measure 300 x 18 nm. The single stranded RNA constitutes about 5% of the particle weight. The particles occur in all the tissues including the pollen and seed but not in the embryo. The inclusion bodies appear as crystalline structures, amorphous masses, fine needles, fibrous spikes, spindle bodies, and amoeboid X-bodies.

In tomato sap the thermal inactivation point is 85–90°C, dilution end point is 1:100000 to 1:1000000. In air-dried tomato leaves the virus has been found to remain infective even after 24 years at laboratory temperatures. The sap retains infectivity for 77 days or more at laboratory temperatures and for several years at 0°–2°C. The virus, like the TMV, is easily sap transmissible and is mainly transmitted by man through contact during cultivation. No natural vectors are known. Seed transmission also occurs mainly as external seed contamination. According to Chitra et al., (1999) TMV and ToMV become established in the seeds of tomato and bell pepper, irrespective of plant growth stage, at the time of inoculation. However, concentration of the virus is high in seeds from plants inoculated early. Dodder can transmit the virus. Diseased crop debris is also a source of primary inoculum. Infective particles of ToMV have been detected and identified in irrigation water in Europe (Boben et al., 2007).

Use of virus-free seedlings is the most important step for control of tomato mosaic. To produce healthy seedlings the seedbeds should be those in which no solanaceous crop susceptible to TMV had been grown for the last 4–6 months. Soil sterilization by heat is also recommended. Seeds should be treated in hot water at 50°C for 25 min or with 20% trisodium phosphate solution. There should be an interval of at least 5 months between susceptible crops in the same field. Field workers should avoid use of tobacco products while working in the field. If they remove a diseased plant they should wash their hands in soap solution.

The host gene Tm-22 is widely used in breeding programmes and has proved very effective against ToMV and TMV worldwide. However, breeding for resistance against tobamoviruses has been laborious for some cultivars due to undesirable traits linked to the single partially dominant Tm-22 gene. Engineered protection (transgenic plants) has been used to develop resistance to both these viruses in tomato and has yielded highly resistant varieties. The transgenes contain the coat protein gene of TMV or ToMV. Transgenes for control of CMV on tomato have also been developed.

YELLOW VEIN MOSAIC OF OKRA

This disease of okra (Abelmoschus esculentus) is the most serious handicap in the cultivation of this vegetable crop in India. It was first reported in 1924. Estimation of damage done by the disease has shown that if the plants are infected within 35 days of germination their growth is retarded, few leaves and fruits are formed, and the loss may be about 94%. The extent of damage declines with delay in infection of the plants in a field. Plants infected 50 and 65 days after germination suffer a loss of 84% and 49%, respectively.

Okra suffers from two mosaic diseases symptoms of which are somewhat similar but which are caused by two distinct viruses. One is okra mosaic reported from only West Africa and the other is yellow vein mosaic reported from India and Sri Lanka. The okra mosaic virus belongs to tymovirus group with ssRNA and is transmitted by sap inoculation while no vector is known. In this disease, the symptoms appear on the youngest leaves as light green and regular veinal necrosis. Symptoms appear on 2–3 subsequent leaves but leaves formed later do not show any symptom. Height of diseased plants is reduced by 19.5%, number of fruits by 64.7% and petiole length by 32.1% but stem girth is increased by 27%. On the other hand,

Fig. 139. Yellow Vein Mosaic of Okra.

in yellow vein mosaic leaves continue to show symptoms as they are formed throughout the crop growth. This virus is transmitted by whitefly and is not sap transmissible.

The main symptoms of yellow vein mosaic are vein clearing and veinal chlorosis of leaves. The yellow network of veins is very conspicuous and the veins and veinlets are thickened. In severe cases the chlorosis may extend to the interveinal areas and may result in complete yellowing of the leaf. Fruits are dwarfed, malformed, and yellowish green in color.

Not much is known about the nature and properties of the Yellow vein mosaic virus of okra. The virus is not sap transmissible but under artificial conditions it could be transmitted by grafting. They reported that in nature the virus is transmitted by the whitefly, *Bemisia tabaci*. Temperatures between 25°C and 36°C and relative humidity of 40% is congenial for multiplication *B. tabaci* and spread of the yellow vein mosaic virus. Temperatures of 38°C in May–June and low temperatures in January–February cause low vector population. In Bihar whitefly is not common in the crop during July to September while okra leafhopper, *Empoasca devastans*, is abundantly present in the diseased crop. The disease intensity is closely related to whitefly population which is mainly dependent on weather prevailing in the crop season. Thus, in crops sown in February–March incidence is low while in crops sown in April–June or later the incidence is high. Maximum rate of increase in disease development occurs in 35–50 days old crop (Bhagat *et al.*, 2001). There are several weed hosts of the virus such as *Croton sparsiflora, Malvastrum tricuspidatum* and *Ageratum* sp. which are commonly found on the roadside and harbor the virus.

Whiteflies are difficult to control with insecticides and are often resistant to pesticides. Barriers such as row covers and repellent mulches have some promise in delaying or reducing disease incidence but not effective when levels of vector and virus inoculum are high. Protection of the crop from whiteflies by spraying with Follidol (0.3%) or other suitable insecticides reduces disease incidence. Spray in the initial stages of crop growth, just after germination, is crucial. If the crop is not sprayed within 20 days after germination, disease incidence remains high. Four to six sprays of systemic insecticides such as Ekatox, Metasystox, Rogor, Dimecron and 1–2 applications of Thimet or Disyston granules to soil were recommended. Soil application of methyl phosphorodithioate (Furatox-10G) at the rate of 15 kg/ha followed by 4 foliar sprays of methyl dimeton (Metasystox 25 EC) at the rate of 0.03% at an interval of 15 days from the date of sowing significantly reduces whitefly population and disease incidence. Carbofuran 3G is better than phorate 10G and should be applied not only at the time of sowing but also within 20 days after sowing. Two applications of these insecticides are helpful in achieving a significant reduction in disease incidence and consequent improvement in yield. Significant reduction in vector population, disease incidence and increase in yield have been obtained by soil application of carbofuran 3G at the rate of 1 kg a.i./ha once at the time of sowing and then 20 days after sowing followed by one spray of oxydematon methyl at the rate of 250 ml a.i./ha at 45 days after sowing. Phorate 10G at the rate of 1 kg a.i./ha once at sowing time and then 20 days after sowing followed by oxydematon spray at 45 days after sowing was also effective.

Use of trap crops and crop-free periods can reduce vector population. Destruction of weed hosts, wherever possible, should also be given importance. No cultivated variety of okra is immune to the virus. Immunity has been reported in several wild species of *Abelmoschus* and *Hibiscus* and these were used in attempts to breed resistant varieties but the progenies were mostly infertile. *Abelmoschus* (*Hibiscus*) *manihot* and *A. manihot* subsp. *manihot* are resistant to Yellow vein mosaic virus. When okra cv. Pusa Sawani is crossed with these species the hybrids are resistant and partially fertile. Resistance is controlled by a single dominant gene.

In screening of 14 cultivars, cv. Parbhani Kranti has been found most promising. Some other varieties/lines reported to be resistant to the virus are: Arka Anamika (Selection 10), Arka Abhay (Selection 4) and Punjab-7 which remained free from the disease under field conditions. EM-9–8, GOH-4, BO-2 and Hybrid 6 are rated as resistant. Fugro and Rajput (1999) have claimed selection of three resistant lines (15–3, 19/1 and 27–1) on the basis of disease resistance, pod quality and yield potential.

Antiviral activity of some plant extracts has been tested against the Okra yellow Vein mosaic virus. Pun *et al.,* (1999) have reported that pre-inoculation spray of aqueous leaf extract of *Prosopsis chilensis* and *Bougainvillea spectabilis* suppressed infection and delayed symptom appeaance. They (Pun *et al.,* 2000) screened a number of virus inhibitory chemicals and neem products against the okra yellow vein mosaic virus and found barium chloride (1000 ppm) most effective in reducing the disease. Next were acetyl salicylic acid (200 ppm) and salicylic acid 500 ppm) which reduced virus infection by 73–75%. Neem oil and neem seed kernel extract reduced disease incidence by 88.3 and 86.7%, respectively. The incubation period was also raised from 10 days to around 19 days. The effectiveness of neem derivatives was explained in terms of their direct interference with vector behavior. Neem oil and other products are known as insect repellent and do not allow the whitefly to land on the leaf surface. Treatments with chemicals like polyacrylic acid, benzoic acid, salicylic acid, acetyl salicylic acid, mercuric chloride are known to induce systemic acquired resistance (SAR) against several viruses. This induction of SAR is found to be concomitant with synthesis of pathogenesis-related (PR) proteins. Salicylic acid treatment is known to cause endogenous accumulation of the acid and in turn acts as signal molecule to induce expression of a set of genes associated with the synthesis of PR proteins.

VIRUS DISEASES OF PAPAYA

Many mosaic diseases, a leaf curl disease, and a bunchy top disease have been described on papaya from different parts of the world. Bunchy top is now known to be caused by a leafhopper transmitted mycoplasma-like organism or a Rickettsia-like bacterium. Leaf curl is caused by the Tobacco leaf curl virus as described for tomato leaf curl. Typical papaya mosaic described only from USA and Venezuela is caused by *Papaya mosaic virus* (potexvirus group) and is sap transmissible, no vector is known. The mosaic of papaya that is more precisely known as ring spot and occurs in India is aphid transmitted. These two viruses of papaya can be distinguished readily by particle morphology, the type intracellular inclusion bodies and by serological tests.

Papaya ringspot

The papaya mosaic (papaya distortion mosaic, papaya distortion ringspot, papaya leaf distortion or papaya ringspot) in India was first reported in 1948 from Maharashtra. Prior to 1940s, it was referred to as papaya mosaic virus. Plants contact the infection at all stages of their growth but they are seriously affected when about a year old. The initial symptoms of the disease vary considerably according to age of the host, time of inoculation, season, and finally, under natural and artificial conditions of infection. In a survey, briefly reported in 1997, the highest severity of this disease was recorded in Bihar (60.6%), followed by Uttar Pradesh (57.4%) and Haryana (48.8%). The loss is 100% if the plants are infected in early stage of their growth. There are areas in the world where cultivation of papaya had to be stopped or shifted because of this disease.

Papaya ringspot appears as profuse mottling and puckering of leaves, especially the young ones. Within about 30–40 days of appearance of the first symptoms the infected plants show degeneration and a marked reduction in growth. The first infection of the plant occurs on the top leaves, the lower mature leaves showing no abnormality. However, further leaves produced after primary infection invariably show the symptoms. In plants allowed to remain standing after infection, symptoms appear every year with greater vigor after the summer is over. Symptoms, more conspicuous than mosaic, are reduced size of leaves, chlorotic and malformed appearance, and defoliation of older leaves leaving only a tuft of small leaves at the top. Leaves are often modified into tendril-like structures (shoe-strings). The number of lobes per leaf is increased and all these lobes are not normal but thin and distorted structures. In some leaves only half of the lamina is found to undergo these modifications while the other half remains normal.

Often conspicuous dark green spots and elongated streaks, appearing like water-soaked areas, are formed on the petioles and the stem. This symptom may not occur in many cases. Fruits generally remain much smaller and get deformed. On some fruits large mosaic patches are also formed. In severe infections, the plants have only a few, small, chlorotic tendril-like leaves and they fail to flower.

The *Papaya ringspot virus* (PRSV) is a species of the genus *Potyvirus* in the family Potyviridae. The virus is serologically related to Watermelon mosaic virus although the latter does not infect papaya. In addition to family Caricaceae (papaya) the virus can infect many cucurbit hosts such as *Cucurbita pepo, C. maxima, C. vulgaris, Cucumis sativus, Luffa acutangula,* and *Trichosanthes anguina.* The host range is related to the pathotype of the virus. Since 1974, two pathotypes have been distinguished. The pathotype P (PRSV-P) is exclusively or predominantly confined to Caricaceae and pathotype W (PRSV-W) is exclusively or predominantly confined to Cucurbitaceae (Roy *et al.,* (1999). The latter was also named as Watermelon mosaic virus-1 (WMV-1). Some authors treat the two strains as separate viruses which were derived from different evolutionary pathways in different geographic areas. Regional variations occur in the host range of the isolates. When a cucurbit plant is infected by PRSV-P it serves as a reservoir of inoculum for infecting papaya.

The virions are filamentous, non-enveloped and flexuous measuring 760–800 x 12 nm. The particles contain 94.5% protein and 5.5% nucleic acid. The PRSV genome consists of a unipartite, linear, single-stranded, positive sense RNA. In papaya sap, the *Papaya ringspot virus* (PRSV) has thermal inactivation point at 54°–56°C and dilution end point at 1:1000. *In vitro* longevity at room temperature is 8 hours. The Indian isolates are reported to have thermal inactivation point at 55°C and dilution end point of 1:1000 to 1:10000. *In vitro* longevity at room temperature is 26 hours. The particles are flexuous filaments measuring 800 x 12 nm (763 nm for Indian isolates). General properties of the virus are similar to other potyviruses. The virus is present in all aerial parts of the affected plants including flowers and pollen but is not seed-borne. Concentration of the virus in floral parts is low. *Chenopodium quinoa* is local lesion host of the virus.

The virus can be mechabically transmitted. No seed or dodder transmission is reported. In nature, more than 20 species of aphids transmit the virus in a non-persistent manner. Both the coat protein (CP) and helper component protein (HC-Pro) are required for vector transmission Most important is *Myzus persicae* while others are *Aphis malvae, A. gossypii, A. medicaginis, A. rumicis, A. coreopsidis, A. craccivora, A. fabae, Macrosiphum solanifolii, Rhopalosiphum maidis, Micromyzus formosanus* and *Toxoptera citricidus*. In a comparative study of transmission efficiency of *A. gossypii, A. craccivora* and *M. persicae*, Kalleshwarswamy and Krishna Kumar (2008) found *M. persicae* most efficient. All the three were 100% efficient when 5 onoculated aphids were used. With single aphid inoculation *M. persicae* was 56% efficient, which it maintained for 4 subsequent inoculation, *A. gossypii* 953% and *A. craccivora* 38%.

In *Rhopalosiphum maidis* pre-acquisition starvation increases the transmission efficiency. For acquisition and transmission of the virus 15 min of feeding, each, is optimum. Infection increases with the number of aphids per plant, maximum being 5 apterous aphids. Adults are more efficient vectors than nymphs. Vectors lose the virus after feeding on the first plant after acquisition.

Destruction of affected plants should be undertaken as far as possible to prevent PRSV-P. Insecticide sprays to keep vectors away can reduce disease incidence. In India, October-December planting shows less disease incidence and there is better fruit yield. The best fertilizer dose for reducing disease incidence consists of 200 g N, 200 g K and 2000 g P along with FYM and castor cake per plant in the field. Papaya seedlings raised in isolated areas delay onset of the virus disease. In Taiwan, raising the seedlings in screenhouses has proved a successful method of preventing introduction of the disease.

Resistance has not been observed in any good variety of papaya cultivated in India. In addition to papaya (*Carica papaya*), *C. microcarpa, C. cundinamarcensis* and *C. guoodotana* are also susceptible. *C. cundinamarcensis* is reported to be resistant to papaya mosaic in Venezuela. *C. cauliflora*, another species from Venezuela has been found immune to papaya mosaic in India. Magdalita *et al.,* (1997) had tested hybrids of *Carica papaya* x *Carica cauliflora* and reported them to be resistant to Australian PRSV-P isolates.

In Hawaii and many other places control of the virus by cross protection with a mild strain had been partially successful. The technique involves the inoculation of papaya seedlings with a mild strain of the virus which protects plants against damage by a severe strain in the field. Although the method improved papaya yields in Hawaii, feasibility of application of this method by farmers is doubted. The drawbacks include the adverse effect of the mild strain on papaya in cool, rainy conditions, additional cost of inoculating and indexing the seedlings, difficulties in propagation and preservation of the inoculum, possibility of occurrence of severe revertants, breakdown under severe disease pressure, and strain-specific protection.

Parasite-derived resistance (PDR) is a relatively new approach to control the disease (Gonsalves, 1998) which proved more reliable and effective than cross protection (pre-inoculation with attenuated or avirulent strain). This type of resistance is a phenomenon whereby transgenic plants containing genes or sequence of a parasite (in this case, the coat protein gene of a virus) are protected against detrimental effects of the same or related pathogens. This approach has been successfully implemented in Hawaii where it is commercially applied (Ferriera et al., 2002). Two transgenic papaya cultivars Raibow and SunUp are in cultivation there. Its potential for broad-spectrum resistance to isolates of the virus from different geographic areas has been recognized in a report from Taiwan (Bau et al., 2003; 2004). However, the coat-protein gene-derived protection is highly nucleotide sequence-specific. Indian isolates from different geographic regions are highly divergent within coat protein genes. Thus, transgenic resistance conferred by CP genes may not be durable in Indian conditions (Roy and Jain, 2002). Fernandez-Rodriguez et al., (2008) have expressed similar possibilities after studying 26 isolates of PRSV in Venezuela. The isolates were within a clade composed of isolates from the Americas and Australia. However, ability for overcoming the transgenic resistance is not solely correlated with higher degree of sequence divergence from the transgene (Tripathi et al., 2004).

Leaf Curl of Papaya

This disease was first reported in Tamil Nadu in 1939. A juice-transmissible leaf curl of papaya was described from Bihar in 1946. The leaf curl of papaya is characterized by severe curling, crinkling, and distortion of the leaves accompanied by vein-clearing and reduction in the size of leaves. Leaves become leathery and brittle and the interveinal areas are raised on the upper surface due to hypertrophy which produces rugosity. The most prominent symptom is the rolling of leaves downward and inward in the form of an inverted cup and thickening of veins. Sometimes all the leaves at the top of the plant get affected by these symptoms. The petioles are twisted in a zigzag manner. The affected plants fail to flower or bear any fruit. In advanced stages of the disease, defoliation takes place and growth of the plant is arrested.

The Gemini virus causing leaf curl of papaya is not mechanically transmitted. It is readily transmitted by grafting and juice inoculation. In nature the most important agent of its transmission is the whitefly, Bemisia tabaci. These insects are the most frequent visitors to papaya plants. Tests have shown that the leaf curl virus of papaya can be transmitted to tomato and tobacco producing symptoms of leaf curl and the Tobacco leaf curl virus can produce the disease in papaya when inoculated on the latter. On this basis it has been suggested that leaf curl of papaya is caused by the Tobacco leaf curl virus. Host range of the virus includes papaya, tomato, tobacco, zinnia, holyhock, Althea and many other plants.

Once the plants are affected they never recover and if allowed to stand in the orchard they simply help in the spread of the disease to healthy plants. Therefore, as soon as the disease is noticed the plant must be uprooted and destroyed. No other control measure for the disease can be recommended except that infestation of the insect vector should be kept at the minimum by spraying of insecticides. Foliar application of fly ash, waste product of thermal power plants, at 2 kg per plant significantly reduces the spread of the disease by controlling the insect vector.

SUGARCANE MOSAIC

Mosaic is most widely distributed and best known of the virus diseases of sugarcane. It was initially reported from Java in 1892 but now it is known to be of worldwide occurrence being common in India, North and South Americas, and many Pacific and Atlantic islands. The occurrence of sugarcane mosaic in India was first reported in 1921 at Pusa (Bihar). Although a source of potential danger to sugar industry in countries like the USA, the disease has not been regarded a menace to sugarcane in this country. Even a 100% mosaic affected crop shows a reduction of about 10–12% in yield and the juice quality remains unaffected. Perhaps, it is due to prevalence of some mild strain(s) of the virus occurring in most of the sugarcane varieties in India. There are indications that more virulent strains are also present in the country on certain varieties like Co. 313 in Punjab and Co. 527 in Tamil Nadu. The disease may cause as much as 21% loss in yield.

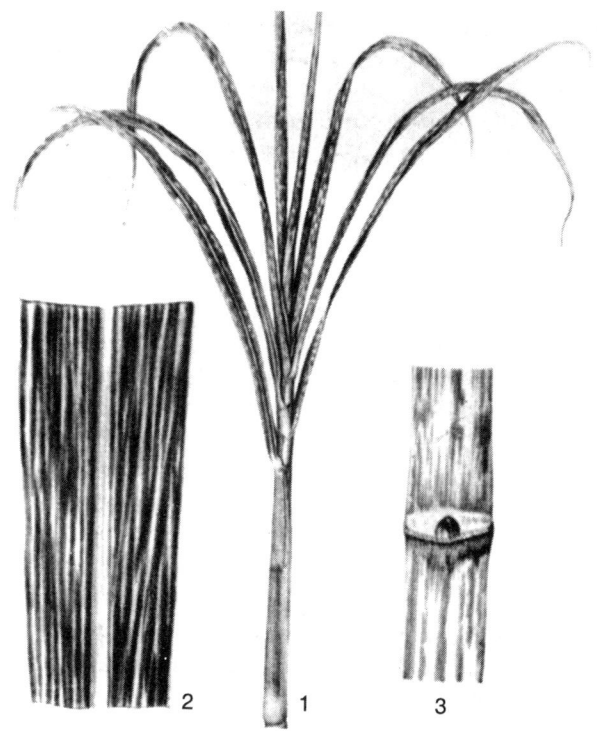

Fig. 140. Mosaic of sugarcane. Symptoms on leaves and internodes.

The first symptoms of sugarcane mosaic appear about 6 weeks after planting and continue to develop throughout the monsoon season after which they are obliterated on maturity of the plants. The primary and critical symptom of the disease is the appearance of pale patches or blotches in the green surface of the leaf. Small areas of the leaf are of a paler green colour than the rest. These patches are not uniform in size and shape. They may be large in some varieties and small in others. Usually they are oval or elongated, the longer axis lying parallel to the midrib. In other parts of the world, the patches on sugarcane leaf are not confined between veins but in India there is a clear demarcation of these patches by the leaf veins. The youngest unfolded leaves show the mottling very clearly while the symptoms are not very clear on older leaves. Sometimes leaves of young tillers are stiff, erect and crinkled. Mottling of the stem also occurs in some varieties and may lead to death of cells resulting in formation of cankers.

The cells of affected leaves always show, in one part of the cytoplasm, an area of proteolysis. This can be seen under microscope as a more heavily stained area than the rest and consisting of a vacuolated mass representing the X-body.

The disease is caused by *Sugarcane mosaic virus* (SCMV) which belongs to potyviruses group. Its synonym is Grass mosaic virus. The particles of the virus are flexuous filaments measuring 620–750 x 13–15 nm. Rishi and Rishi (1985) had reported the size of particles of strain A and F as 670–770 nm long. Maximum concentration of virus particles is found in young leaves and minimum in roots of old infected plants. Particle composition is not fully known. In addition to sugarcane, the virus can infect other graminaceous plants such as maize, sorghum, millets, wheat, barley, rye, and some grasses. Natural occurrence of the virus on maize, sorghum, pearl millet and elephant grass has been reported from India. *Brachiaria* is also a natural host of some strains of SMV. One strain that attacks *Musa textilis* has some hosts outside the gramineae.

The following serologically related potyviruses were considered strain of the Sugarcane mosaic virus: Sugarcane mosaic strains A, B, C, D, E, F, G, H, J, K (these strains rarely infect sorghum), Maize dwarf mosaic strain A and B, Sorghum red stripe strain, European maize mosaic strain and Abaca mosaic strain on *Musa textilis*. There was a suggestion that strains designated as H, I, and M be redesignated as strains of *Sorghum mosaic virus* on the basis of serological and chemical properties.

The virus is transmitted from sugarcane to sugarcane by at least 7 species of aphids such as *Dactynotus ambrosiae*, *Hysteroneura setariae*, *Rhopalosiphum maidis*, *Toxoptera graminum*. *R. maidis* and *Shizaphis graminum* are reported as vectors of Sugarcane mosaic virus in India. *R. maidis* and *Aphis gossypii* transmit the virus from maize to maize and *M. persicae* from sorghum to sorghum. *R. maidis*, *Melanophis sacchari*, *M. idiosacchari* and *Hysteroneura setariae* as vectors in India. Transmission is in the non-persistent manner. Seed transmission is reported only for Maize dwarf mosaic strain. Vegetative propagation is the main source of primary infection in sugarcane.

The thermal inactivation point of SCMV is 53–55°C and dilution end point 1:1000. *In vitro* longevity is 2–24 hours. In sorghum sap the thermal inactivation point is 56°C and dilution end point 1:100 to 1:1000. Using soybean as a new local lesion host, it is claimed that the virus is stable up to 1:100000 dilution and can withstand a temperature of 69°C.

For experimental purposes the virus can easily be transmitted by introducing it into the actively growing tissues. Young leaves are most suitable for testing. Most common method is to use a needle to puncture the tissue over which infected juice has been spread or infected leaf has been wrapped. SCMV moves from the point of inoculation to young leaves, roots and tillers and eventually to leaves that had emerged prior to inoculation. The pattern of SCMV distribution in moderately resistant and susceptible cultivars is not much different. However, the virus moves more slowly in the moderately resistant than in the susceptible cultivar (Putra et al.,, 2003).

Due to continuous evolution of strains of the virus, its presence on grass or cultivated collateral hosts, and long growing season of the sugarcane crop, resistant varieties are not permanent solution of the mosaic problem. Following practices have been recommended to minimize its incidence:

1. Use of selected healthy setts for seed.
2. Heat therapy (hot water or hot air) is effective against certain strains and can be used for raising disease-free nurseries. Although heat therapy reduces the virus title in the cane, it is no completely eliminated and about 8 weeks after planting treated setts,

virus titre starts rising (Balamuralikrishnan *et al.,*, 2003). Meristem cultures obtained soon after emergence thus can provide virus-free stock.

3. Systematic roguing of the infected canes if the incidence is not very high.
4. Elimination of grass hosts.
5. Use of resistant or tolerant varieties.

YELLOW MOSAIC OF URD, MUNG AND SOYBEAN

This virus disease is the most destructive disease of kharif legumes (urdbean and mungbean) in India. It was first reported in 1960 and is now known to occur throughout the country. Pigeonpea is another important legume that can be attacked by it. The loss in yield depends upon the stage at which the crop is infected. If the infection is early in the season there may be total loss of grain yield. Mungbean shows heavier losses than other legumes. Up to $ 300 million annual loss is estimated in these crops.

Symptoms

The diseased plants start appearing in the field when the crop is about a month old. Two types of symptoms, depending on host response, are seen. The yellow mosaic in the form of yellow mottle is a more common and aggressive symptom on mungbean and susceptible cultivars of urdbean. Necrotic mottle symptoms of the disease depict a resistant reaction of the host and are observed on tolerant urdbean cultivars, rarely on mungbean varieties.

Fig. 141. Yellow mosaic of urd bean.

The general pattern of development of both symptoms is the same. The first visible sign of the disease is the appearance of yellow spots scattered on the lamina surface. They are

Fig. 142. Necrotic mosaic of urd bean.

mostly round in shape. In yellow mottle, the spots are diffuse and expand rapidly. The leaves show yellow patches alternating with green areas that also turn yellow. Such completely yellow leaves gradually change to a whitish shade and ultimately become necrotic. These color changes of affected plants are so conspicuous that the disease can be spotted in the field from a distance. In case of necrotic mottle, the centre of yellow spots develops necrosis which is demarcated by finer veins.

The virus becomes systemic in the plant and all newly formed leaves show signs of mottle from the very beginning. In case of attack of yellow mottle alone or with necrotic mottle there may be reduction in the size of leaves. Number and size of pods per plant and seeds per pod are greatly reduced. The pods are deformed and contain shriveled, undersized seeds. The percent yield loss is higher in plants inoculated early in the season than in those inoculated in mid- or late season. Plants showing only necrotic mottle do not show reduction in size of leaves and pods.

The causal agent

Four viruses causing yellow mosaic disease of legumes across the South Asia have been identified as bipartite begomoviruses (genus *Begomovirus*, family *Geminiviridae*). The species are *Mungbean yellow mosaic virus*, *Mungbean yellow mosaic India virus*, *Horsegram yellow mosaic virus* and *Dolichos yellow mosaic virus* (Qazi *et al.*, 2007).

In India the disease is caused by *Mungbean Yellow Mosaic India Virus* (MYMIV). It has a large host range which includes, besides mungbean and urdbean, soybean, frenchbean, lima bean, pigeonpea, common bean, *Brachiaria ramosa*, *Eclypta alba*, *Xanthium strumarium*, and *Cosmos bipinatus*. The soybean strain of MYMV occurring in north India is distinct from the strain occurring in southern and western India (Usharani *et al.*, 2004). A strain of MYMIV, designated as MYMIV-Cp causes golden mosaic of cowpea. It has restricted host range and transmission by *Bemisia tabaci*. These viruses have evolved independently of the begomoviruses in plant species of other families.

The paired particles of the virus measure 30 x 18 nm. The particles contain two circular ssDNA molecules which account for 20% of the particle weight. The coat protein conatins one polypeptide with MW of 28.5 kDa. The isolate studied in Thailand has thermal inactivation at 40°–50°C, dilution end point of 1: 100 and in vitro longevity of 1–2 days at 20°C. The isolate was infective by mechanical inoculation. In India, MYMV is neither sap nor seed or soil transmitted. *Bemisia tabaci* (whitefly) is the only known vector. In mungbean leaf cells the virus particles often form loose aggregates which may fill the nuclei of the phloem cells. These aggregates together with hypertrophied nuclei and fibrillar bodies appear in the nuclei of the phloem cells as early as 2 days before visible symptoms appear on the leaves.

Female adults of the vector, *B. tabaci*, are more efficient vectors than males. Minimum acquisition feed time is 15 min and the same time is required for inoculation. Increasing feeding period up to 4 hours increases transmission ability. Latency (incubation period) in the vector is at least 3 hours, optimum being 5–6 hours. Pre-acquisition starvation of the vector increases the efficiency to acquire the virus. In urdbean, the vector is reported to acquire the virus 1–3 days before symptoms appear. A single viruliferous whitefly can transmit the virus but maximum infection is obtained with 10–20 whiteflies per plant. Neither female nor male adults can retain the virus throughout their life span. Normally the females adults retain infectivity for 10 days and male adults for 3 days.

Management of the disease

Recommended control measures, listed below, do not ensure complete protection but can check the spread and reduce losses.

1. *Local varieties in India are highly susceptible.* These should be replaced with improved varieties. Urdbean cultivars T9, UPU 1, Pant U19, 26, and 35, UG 298, and Vamban (Tamil Nadu) are mosaic resistant. In some trials Pant U30 has been found completely free from mosaic. According to Govindraj and Subramanian (1991) susceptibility in urdbean is dominant over resistance and is controlled by oligogenes. In mungbean, Pant 1, 2 and 3, T 1, T 44 are resistant. Cultivar PK 21–22 of soybean is tolerant to the disease. In a screening of 38 genotypes of urdbean and found KU 96 and TU 98–13 immune to the disease. Genotypes KU 96–1, UG 774, UG 757, UPU 97–10, KU 315, NP 6, PLU 96–8, PLU 44, PLU 131, PLU 463 and KU 96–8 were resistant. Several regenerants of mungbean in tissue culture have shown stable resistance to the virus. These may provide useful sources of resistance in the breeding programmes.

2. *Control of the disease through prevention of population build up of the vector* has also been recommended. Spray of 0.1% metasystox, starting when the crop is about a month old or as soon as a single diseased plant is seen in the field, can give relief from severe incidence of the disease. Anthio is effective at 0.2% when used as spray 3 times. Application of aldicarb alone or with endosulfan and captan reduces whitefly population and disease incidence. Recommendations whitefly control in okra yellow vein mosaic can be useful in this case also. Diseased plants should be rogued out after each spray of insecticides. Weed hosts should also be eradicated from the vicinity of the field.

Control of plant viruses through control of vectors is often not very effective due to the fact that common insecticides do not cause instant death of all individuals in the vector population and even a very few surviving population is capable of spreading the disease rapidly. Oil sprays can be more effective because they kill the insects within 15 min but they can be phytotoxic. Soil application of granular systemic insecticides, as recommended for yellow vein mosaic of okra seems to be the best chemical method for reducing vector population and delaying the appearance of the disease. Biological control of the vectors is being actively studied. There are many parasites and predators of aphids and whiteflies.

In a later study, Rashid *et al.,* (2004) have reported that priming seed of mungbean with water for 8 hours before sowing resulted in reduction in number of plants with severe and lethal symptoms from 70% to 14%. Priming improved germination and stand but subsequent mortality reduced the difference between treatment and check to nonsignificant level. Single or combines application of boron (2 kg/ha) and molybdenum (1.5 kg/ha) is reported to significantly reduce yellow mosaic severity. Plant spacing (25 cm x 5 cm) also reduce incidence and severity of the disease.

BEAN COMMON MOSAIC VIRUS

The common mosaic of bean is widely occurring virus disease of this group of crops. It is probably co-extensive with the host in India. The disease is of significant economic importance in areas where it commonly occurs.

Symptoms vary widely according to the host cultivar, time of infection, and environmental conditions. In mild infections the symptoms are not clear. The leaf symptoms can be of two types. Early symptoms or those produced on leaves expanding at the time of infection appear as crinkle, chlorosis and stiffness of the lamina. The leaves droop and petioles are much shortened. These symptoms may not always appear. There is no definite downward rolling of the leaflet and also there is no characteristic mosaic mottling of the lamina. On simple leaves

Fig. 143. Mosaic of common bean. (*Phaseolus vulgaris*).

showing secondary symptoms there may be a general chlorosis of the leaf blade or a definite pattern of light green and dark green areas. The light green areas are usually along the leaf margins. On the trifoliate leaves the mosaic pattern of light and dark green areas is present and the leaves look rough. The infected leaves are narrower and longer than healthy leaves. Abnormal growth of the tissue causes the veins to bend downward and the leaves may look cupped. A characteristic symptom of mosaic on the first trifoliate leaves and on the succeeding ones is the appearance of dark green blistered areas on the lamina. When infection occurs in very young plants they remain stunted and pale. Symptoms are not found on stems and seeds. Diseased plants produce fewer pods which are small in size. At high temperatures the symptoms are more pronounced.

The *Bean common mosaic virus* (BCMV) is a potyvirus. Many strains of the virus have been described on the basis of symptoms on different species of *Phaeolus* and the range of cultivars they attack in common bean. The virus particles are flexuous rods measuring 720–770 x 12–15 nm. The nucleic acid (5%) is single stranded RNA. The coat protein constitutes 95% of particle weight. The thermal inactivation point is variously reported to lie between 50° and 65°C (usually 56°–58°C). *In vitro* longevity at room temperature is 1–4 days. Cytoplasmic inclusion bodies are pinwheel associated scroll type. The virus particles are present in sepals, petals, stamens, pistil and pollen. Infection is highest in pods and embryo of immature dried seeds. Early infections result in more seed infection.

Natural hosts of BCMY are mostly restricted to *Phaseolus* species. However, there are many reports of its occurrence on other legumes including *Vigna* and *Crotalaria*. Non-legume hosts include species of *Chenopodium album, C. quinoa, C. amaranticolor* which develop only local lesions and *Nicotiana clevelandii* which develops systemic infection.

Seed transmission is common and most important source of initial infection in the crop. From 30–83% seeds from an infected crop may carry the virus which is present in the embryo and cotyledons. The virus disappears from the seed coat of mature dried seed. Plants infected after flowering produce healthy seeds. Not all seeds in the same pod are infected. The virus remains infective in bean seed for up to 30 years. Seed transmission levels of 3–4% were found in seed lots stored for more than 6 years at 2°–4°C.

Transmission in the field occurs through the agency of aphid vectors of which notable are *Myzus persicae, Aphis fabae* and *Acyrthosiphum pisum*. Other aphid vectors are *A. gossypii, A. medicaginis, Brevicoryne brassicae* and *Rhopalosiphum pseudobrassicae*. Acquisition and inoculation feed periods are less than one min and there is no incubation period in the vector body. Aphids transmit the virus more from the chlorotic areas than from the green areas of the leaf.

It is very difficult to get virus-free seed since the disease and its vectors are present everywhere. Roguing of diseased plants is not practical. Thus, only resistant varieties can provide relief from the disease. In frenchbean several varieties such as Kentucky Wonder, Contender and Pusa Parvati are highly resistant. Use of certified virus-free seed and cultivars with dominant resistance keep the disease under check until a new race of the virus appears. Control of insect vectors through the use of inse cticides does not help much. Sprays of leaf extracts of *Boerhavia diffusa* and *Bougainvillea spectabilis* are reported to reduce the disease caused by *blackeye cowpea mosaic* strain of BCMV in cowpea. (Prasad *et al.,* 2007).

MOSAIC DISEASES OF CUCURBITS

A variety of mosaic symptoms occur on different cucurbit plants. In India these mosaic diseases are common on almost all cucurbit vegetables and often cause heavy losses. The common mosaic is caused by the Cucumber mosaic virus (CMV). In addition, Cucumber green mottle mosaic virus (CGMMV), Watermelon mosaic virus (WMV) and Pumpkin yellow vein mosaic virus (PYVMV) are also common on certain cucurbits and cause significant losses. Some of these viruses attack, in addition to cucurbits, a large number of other host plants including tomato, chilli (pepper), eggplant, potato, spinach, cowpea and carrot.

Symptoms

The nature of symptoms depends on the host and the virus associated with the disease. If infection is early, the symptoms are aggressive and losses high. When the plant is attacked soon after emergence, cotyledons are yellow and seedlings show symptoms of wilt. In older plants, the symptoms appear first on young upper leaves. There are alternate green and yellow patches on the lamina. These spots are of irregular shape and enlarge rapidly, ultimately covering the entire leaf. The diseased leaves are mottled, deformed, small, and sometimes curled downward. Veins and veinlets also turn yellow. Sometimes, there are shallow depressions on the leaf. Internodes of stem are shortened, thus dwarfing the plant. Young fruits are rough, mottled, and deformed. Often these fruits are white and much smaller in size than normal fruits. Systemically infected vines yield few fruits. These symptoms are produced in the infection of CMV and CGMMV.

In the attack of Watermelon mosaic virus (which may be a strain of Papaya ringspot virus) the above symptoms appear but leaves are much deformed. Often they are filiform. Except the tissues adjacent to the veins and veinlets, the remaining lamina surface may be completely destroyed. The veins and veinlets often protrude beyond the leaf margin and leaves become spindle-shaped. The veins in the leaf originate from base of the leaf stalk. Poor flowering and defoliation are other effects. This virus is most dangerous among the cucurbit viruses. In India, the virus is common on bottlegourd and pumpkin.

Fig. 144. Effect of *Cucumber Mosaic Virus* (CMV) on leaves of *Curcurbita pepo*.

The Pumpkin yellow vein mosaic takes a heavy toll infecting the plant at all stages of its growth. It has been reported from Delhi, Uttar Pradesh, West Bengal and Kerala. There is vein clearing of the fine veins and chlorotic blotches are scattered all over the lamina surface. Young leaves show increasing vein clearing. Leaves of older plants display a mixed pattern of vein yellowing in smaller areas and chlorotic patches over larger areas of the lamina surface.

The causal viruses

Cucumber mosaic virus (Cucumovirus group) is a small icosahedral virus composed of single stranded RNA of positive sense polarity. It encapsulates a small satellite RNA together with its own tripartite genomic RNA and a fourth subgenomic RNA. The satellite is designated as CARNA 5 (CMV-associated RNA 5). Protection of tomato against CMV by pre-inoculation with CARNA 5 is reported. The satellite has been incorporated in transgenic plants.

The CMV particles are 28–30 nm in diameter. The nucleic acid constitutes about 19% of the particle weight. The virus particles are seen in the host cytoplasm, nuclei, and vacuoles but not in mitochondria or chloroplasts. Sometimes these particles aggregate in crystals, usually in the vacuoles, and can be seen by light microscopy. The virus is relatively unstable in expressed host sap being unable to withstand temperature above 70°C for 10 min. Some reports suggest thermal inactivation point between 60° and 70°C. Infectivity is lost within 1–5 days at room temperature while the longevity *in vitro* at 34°–39°C is 16–18 hours. Many strains of the virus are known to exist.

The virus is transmitted by more than 60 species of aphids, by seed, and by dodder. Aphids transmit the virus in non-persistent manner. Common aphid vectors are *Aphis gossypii, A. evonymi, A. craccivora* and *Myzus persicae*. The transmissibility by *M. persicae* may be lost altogether. All vectors can acquire the virus within 5–10 seconds of feeding but ability to transmit the virus declines after 2 min and is usually lost within 2 hours. The virus is seed-borne in 19 plant species. These include weeds which may play important role in perennation and spread of the virus. In muskmelon and pumpkin the seed-borne inoculum is important. The CMV in irrigation water is adsorbed on montmorillonite clay in water making the virus available for infection.

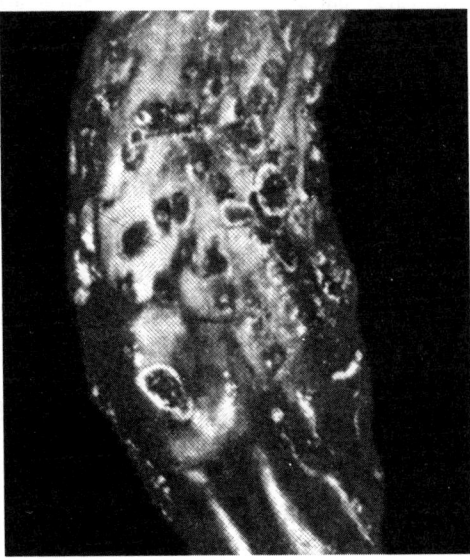

Fig. 145. (Fig 154 Page 568 of 8E).

Cucumber green mottle mosaic virus (Tobamovirus group): Particles of this virus are present in all the tissues of the infected plant including the pollen and occasionally seed embryo. The virus systemically moves through the phloem. Coat protein and RNA of the virus have been detected in phloem exudates of the infected plant but the RNA is not free suggesting that the virus moves as a whole particle (Simon-Buela and Garcia-Arenal, 1999). Cytoplasmic inclusions are also found. The particles are straight, rigid rods measuring 300 x 18 (280 x 17) nm. The nucleic acid is single stranded RNA constituting about 6% of particle weight. Infective particles of CGMMV are morphologically indistinguishable from those of other tobamoviruses and the virus is serologically related to several of them. However, tobacco mosaic virus does not provide cross protection in cucumber against CGMMV. Host range of the virus includes cucumber, watermelon, muskmelon, and bottlegourd but not vegetable marrow.

Fig. 146. Enlarged rigid particles of *Cucumber Green Mottle Mosaic Virus*.

CGMMV is sap transmissible through foliage contact and handling of plants during cultivation. Soil contamination occurs through diseased crop debris in which the virus remains infective. Fungus transmission is also suspected. No insect vectors are known. In cucumber up to 8% seeds may carry the virus one month after harvest but this falls down to 1% after 5 months of seed storage. In watermelon, 5% seeds are contaminated. Seed transmission also occurs in bottlegourd. Contamination of seed is mostly external. Some species of dodder including *Cuscuta campestris* also can transmit the virus.

All strains of CGMMV are extremely stable in expressed sap of infected plants. In cucumber, the type strain of CGMMV has thermal inactivation point at 90°C, the watermelon strain at 90°–100°C and some other isolates at 95°–96°C. Infectivity is retained for several months at room temperature and for several years at 0°C.

Watermelon mosaic virus (Potyvirus group). Particles of WMV are flexuous filamentous tubes measuring 725–765 (or up to 840) nm in length and 15 nm in diameter. The nucleic acid genome is singe stranded RNA. The host range includes only some members of the cucurbit family such as watermelon, vegetable marrow, pumpkin, cucumber, bottlegourd and spongegourd. However, in artificial inoculations the virus has been transmitted to plant species in 17 families.

WMV is transmitted by *Myzus persicae, Aphis gossypii, A. fabae* and several other aphids in a non-persistent manner. It is easily sap transmissible. Seed transmission is not confirmed. *A. gossypii* can acquire the virus after one minute of feeding. No incubation (latency) period is required and the vector becomes viruliferous immediately. Ten aphids per plant can cause 100% infection. In expressed sap of infected *Cucurbita pepo* the virus loses infectivity after 10 min at 59°–65°C and after 20–50 days at 20°C. The thermal inactivation point is 55°–60°C.

Pumpkin yellow vein mosaic virus (PYVMV) is neither seed nor sap trasmitted. In nature it is transmiited by whitefly, *Bemisia tabaci*, but not by aphids. Even a single viruliferous whitefly can transmit the virus to the extent of 21.6%. For 100% transmission the number of vectors is 20– 50. According to Jayashree *et al.,* (1999) this number could be only 15 per plant. The vector can acquire the virus by feeding on the infected host for only 5 min but acquisition feeding period of 6 hours or more results in 100% transmission. With increase in acquisition feeding period beyond 6 hours more vectors become viruliferous and the time required for expression of symptoms declines (Jayashree, et al., 1999). After feeding, an incubation period of 2 hours is required. Transmission feeding time of 30 min results in 50% transmission. An inoculation feed period of 3 hours or a pre-acquisition starvation period of 3 hours gives 100% transmission. The virus is semipersistent in the vector.

In absence of resistant varieties following measures can reduce severity of mosaic diseases in cucurbits:

1. Destruction of diseased hosts including weeds.
2. Use of seed from virus-free plants.
3. Use of insecticides such as Thiodan (0.1%) or Metasystox (0.1%) to suppress insect vectors.

The use of reflective (aluminium) polythene mulches for reducing watermelon mosaic virus in cucurbits was demonstrated in Australia in 1982. Black polythene mulch also reduced the incidence of virus. Induced resistance by rhizobacteria against Cucumber mosaic virus (CMV) on cucumber is reported to be suppressed by foliage treatment with selected strains of Streptomyces. The mechanism of suppression is not clear. Melon plants pre-inoculated with a benign viral satellite RNA in combination with a mild strain of CMV develop resistance against severe CMV strains and remain virus free when challenged by severe strains after vaccination (Montasser et al., 1998). CMV resistant cucurbits had been developed by incorporating coat proteins genes of the virus into the host but sincxe conventional breeding is effective in giving resistant varieties the transgenes are not favored.

Raupach et al., (1996) had reported biological control of CMV on cucumber and tomato by seed treatment with plant growth-promoting rhizobacteria Pseudomonas fluorescens and Cerratia marcescens). The treated plant remained symptomless throughout the period of the experiment. Spray of salicylic acid alleviates the harmful effects of yellow mosaic of some cucurbits.

RICE TUNGRO VIRUSES

Tungro, meaning degenerated growth, is the most important virus disease of the 14 viruses of rice in south and south-east Asia and also in the southern part of China. It is called cancer of rice because of the severe damage it causes and difficulties in its control. It was first dignosed in1963 but has a much older history. It had appeared as a devastating disease in Indonesia in 1859. Since mid-1960s, the changtes in cropping system of rice that included short-canopy rice cultivars and continuous planting of rice gave impetus to spread of the disease in many countries. The disease is reported from Philippines, Malayasia, Indonesia, Thailand, India and Pakistan. In India the disease is present in the south east to the north west parts of the country. In the 1984085 season Tungro affected area in India (Tamil Nadi and Andhra Pradesh) was 90,000 ha and during 90–91 season it was 1,80,000 ha in Andhra Pradesh and Orissa. Indonesia is perpetually affected by the disease, the area affected varying from 16000 to 25000 ha annually. The annual loss due to tungro is estimated at US $ 1.5 billion globally.

In rice the symptoms depend on host cultivar. In the cultivar TN 1, seedlings show stunting with mottling and yellowing of leaves. In growing plants there is reduced tillering. Yellowing and orange yellow coloration of leaves is a common symptom. Small and sterile panicles develop in mature plants. In early infection the young plants may die prematurely.

Tungro is a composite disease caused by two viruses: the rice tungro spherical virus (RTSV) and the rice tungro bacilliform virus (RTBV). The RSTV is a ssRNA virus which causes only very mild stunting of the host plants without leaf symptoms but it intensifies the symptoms of tungro caused by RTBV. The RTBV is a dsDNA bacilliform virus which causes mild stunting and yellowing of leaves. RTBV causes the main symptoms of tungro. It is dependent on RSTV for transmission by the insect vectors. In Philippines and South and South East Asia the RTSV occurs and spreads as an independent virus but generally as a latent disease.

RTSV is associated with maize chlorotic dwarf virus group. The particles of RTSV are isometric with a diameter of 30–33 nm. The ssRNA is in single piece (monopartite) and constitutes 12% of the particle weight. The RTBV belongs to Budnavirus group. The particles are bacilliform, 100–400 nm long and 30–35 nm in diameter. Both viruses are transmitted by same leafhopper vectors. The thermal inactivation point of RTSV is 60°C. The longevity *in vitro* at room temperature is 24 hours, one week at 4°C and one month in frozen samples. In the host cells dense granules are seen in the cytoplasm. Chloroplasts and other cell organelles in the infected tissues are degenerated.

Oryza species are the main host but at least 63 species of grasses could be infected. Natural occurrence on grasses in and around rice fields is reported. However, all except *Oryza* species are poor hosts of the vectors.

The rice tungro viruses are not transmitted by sap inoculation or by seed or dodder. In nature the main agency of tungro transmission is the green leafhopper, *Nephotettix cincticeps* (*N. virescens*). Other vectors are *N. nigropictus, N. malayanus, N. parvus, N. apicalis* and *Recilia* (*Anazuma*) *dorsalis. N. cincticeps* is the most effective vector and disperses the virus in the rice fields. All the five larval stages and adults of the leafhopper vector transmit the viruses in non-persistent manner. There is strong biological relationship of the vector with rice plants on which it has high adult longevity, nymph survival and population growth. Minimum acquisition feed period is 5–7 min but transmission increases with acquisition access feeding up to 4 days. There is no latency. The vector becomes viruliferous immediately after feeding. Nymphs cease to transmit the virus when they molt. All stages lose ability to transmit the virus within 5 days after the end of the acquisition feed. Weed hosts that may play a role in survival and dissemination of the rice tungro viruses include *Leersia hexandra, Eleusine indica, Echinochloa crusgalli, Echinochloa glabrescens, E. colona* and *Leptochloa chinensis*.

N. cincticeps is monophagous to rice. Its density can reach high levels depending on the environment. After rice is harvested, the density of this vector falls rapidly to a low level or to nil in the rice fields. Seasonal pattern of immigration and population dynamics of the major vectors (*N. cincticeps, N. nigropictus* and *Recilia dorsalis*) determines the incidence of tungro. In the Philippines, peaks in population density occur 50–65 days after transplanting in wet seasons but varied widely in dry seasons. *N. cincticeps* predominates because of its earlier colonization of rice plantings, more rapid population development and higher transmission ability (Chancellor *et al.,* 1996). Under Assam conditions, Nath and Bhagabati (2002) observed that leafhopper populations are first seen during June–July in the seedbeds. The population reaches the peak in the main field in October–November and then disappears from December to May. *N. virescens* populations were lower than *N. nigropictus* but higher than *R. dorsalis.* Tungro infected plants including rice stubbles or volunteer rice serve as source of RTBV and RTSV. The vector that feeds on source plants, moves to newly transplanted fields in surrounding areas and disperses the viruses. Probably the flight range of *N. cincticeps* is several kilometers. Initially few plants are infected and form the source for further spread. Generally, the plants with secondary infection from such sources form patches a few to several meters in diameter. Such patches later fuse with each other. In tungro-endemic areas major infection of rice plants with the viruses occurs after transplanting. Infection rate in the seedbeds is low. In transplanted fields, infection with RTSV alone precedes infection with RTBV. The disease incidence is generally low in fields planted in the early crop season when the vector population is low, but it is high in fields planted later when the vector population has build up. Infestation of mite *Oligonychus oryzae* does not adversely affect RTV acquisition by the leafhopper vector but RTV inoculation of mite infested plants results in only 50% infection (Niazi and Singh, 2001).

In many countries long fallow periods reduce the disease incidence. Plant extracts have been used to reduce the population of the vector and disease severity. Metabolites of the rice false smut fungus, *Claviceps oryzae sativae*, are reported to provide tolerance in rice to the tungro disease. Neem seed cake applied at 5 kg/0.032 ha of nursery followed by foliar spray of 5% neem kernel extract in the main field reduces vector population and incidence of tungro (Rajappan *et al.,* 2000). Two fungi, *Beauveria bassiana* and *Paecilomyces amoeneroseus*, are reported as parasites of rice green leaf hopper, *Nephotettix virescens*. Spray of conidial suspension of the fungi on rice plants kills the vectors (Niazi *et al.,* 2002).

Use of rice cultivars resistant to tungro has been a major approach for control of the disease. The resistance to the viruses is correlated with resistance to the vectors. The resistant cultivars become susceptible in few years because populations of the vector develop that can feed on these cultivars. A transgenic *japonica* rice plant is reported that contains RTSV replicase gene. Plants expressing full length Rep gene in the (+)-sense orientation show 100% resistance to RTSV even if challenged with a high level of inoculum. The rice cultivar Basmati 370 is reported to be resistant (Radhakrishnan and Niazi, 2000). Genes for resistance to tungro have been identified in *Oryza rufipogon* and transferred to rice. In the cultivar Matatag 9, an interspecies cross between IR 64 and *O. rufipogon*, tolerance to Tungro viruses and resistance to the vector green leafhopper contribute to the apparent resistance (Shibata *et al.,* 2007).

BUNCHY TOP OF BANANA

India grows bananas over an area of around 1,60,000 ha of which about 25% is in the state of Kerala alone. It is in this area that the bunchy top disease causes huge losses in the banana plantations. Reports suggest an annual loss of Rs. 60 million (US $ 150,000) in the state of Kerala.

Symptoms

The symptoms of bunchy top may become apparent at any stage of plant growth, from very young plants to the fruit-bearing adult plants. In badly infected plants, the leaves are typically bunched together at the apex, forming a dense rosette, hence the name bunchy top. The young infected plants are usually stunted and show more erect leaves than normal.

The symptoms vary with the type of infection. The primary infection refers to the infection which comes with the diseased suckers used for planting while the secondary infection is that which is caused during growth of the plant through the agency of insect vectors. The symptoms under these two conditions mainly differ in intensity. In acute primary infection growth of the plant is slow, the plant does not reach a height of more than 2 feet, and never produces fruits. Young plants receiving secondary infection may show the same symptoms as the plants infected through suckers but with lesser intensity.

The first evidence of bunchy top is seen in the leaves. Green streaks appear on the secondary veins on the underside of the lamina and on the midrib and petioles. These streaks are about 0.75 mm wide and 1–1.25 cm long. At first, one to several streaks may be present and are generally followed by others in the same region. The abnormal markings vary from a series of dark green dots to a continuous dark green line. In some cases, symptoms may be evident in the newly emerged or emerging tightly rolled leaf as pale whitish streaks along the secondary veins. These markings are accompanied by a slight transverse wrinkling along the length of the compactly rolled lamina. On opening these leaves, dark green streaks along the secondary veins may be seen. Older leaves show the same symptoms and also dwarfing, marginal necrosis and curling. The leaves gradually become brittle and petioles are incompletely elongated.

In plants with secondary infection, the fruit bunches become choked and may split the stem. They are reduced in size and the fruits are unsaleable. The virus is systemic and is probably confined to the phloem. In transverse sections the vascular strands show suppression of fibrous sheath. In advanced stages of the disease, the root system presents a decayed appearance which is a secondary effect caused by invasion of the weakened roots by bacteria, fungi and nematodes.

Hooks *et al.*, (2008) have studied the morphological and developmental difference between healthy and infected banana plants. Infected plants show significant reduction in lemgth of petioles, plant canopy and height, leaf area, pseudostem diameter and chlorophyll content. Growth differences do not appear until 40–50 days after inoculation with viruliferous vector. No significant effect on petiole width and leaf production is seen.

The causal agent

The causal agent of banana bunchy top is a nanovirus (genus *Babuvirus,* family *Nanoviridae)*, known as *banana bunchy top virus* (BBTV). It was considered a luteovirus in which the particles are isometric with ssRNA. However, in later studies it was reported as non-geminated ssDNA virus. The particles are icosahedral and 18–20 nm in diameter. The virus possesses a multicomponent genome consisting of at least 6 circular ssDNA molecules (DNA-1,2,3,4,5,6). These are present in all isolates from different regions. A virus of the same group causes bunchy to of abaca (*Musa textilis*). It has similar multicomponent genome but is distinct from BBTV in detailed genomic analysis. It is name *Abaca bunchy top virus*, ABTV (Sharmanz *et al.,* 2008).

A number of BBTV strains differing biologically and molecularly occur in nature (Su *et al.,* 2003). Most isolates of BBTV produce typical severe disease symptoms. However, mild or symptomless strains are also reported. Two broad groups of isolates have been identified on the basis of nucleotide sequence in the six recognized genome components. The "South Pacific Group" comprises isolates from Australia, Fiji, Tonga, Burundi, Egypt and India while the "Asian Group" comprises isolates from Philippines, Taiwan and Vietnam.

BBTV is systemic and restricted to phloem tissue. Following aphid inoculation, symptoms generally do not appear until two or more further leaves are formed. The virus replicates for a short period at the site of inoculation, then moves down the pseudostem. Then the basal mersitem and finally to the corm, roots and newly formed leaves. In the phloem, the virus induces hypertrophy and hyperplasia of the host tissue and a reduction in the development of the fibrous sclerenchyma sheath surrounding the vascular bundles. The cell surrounding the phloem contain abnormally large numbers of chloroplasts which give rise to the macroscopic dark green streak symptoms.

The virus is transmitted by the banana aphid, *Pentalonia nigronervosa*, in a circulative semi-persistent manner. It is not sap transmissible. All suckers produced by a diseased plant carry the virus and they constitute the most important source of spread of the disease from plantation to plantation. The aphids cause the secondary spread during the growth of the plants. The rate of multiplication of aphids is maximum when the minimum temperature is about 18°–20°C with relative humidity ranging between 41 and 84 %. All stages of the aphid acquire and transmit the virus but the nymphs are most efficient. To become infective, the aphid requires a minimum feeding period of 4–17 hours and transmits the virus to a susceptible plant by feeding on it for 30 min to two hours. They retain the infective capacity for a period of 13 days after removal from the infected plants. There is no transovarial transfer of the virus to the progeny. These aphids occur around the base of pseudostem at soil level and for some inches below the soil surface. They are also found between the outer leaf sheaths and the pseudostems. In heavy infestations the aphids are also present in groups at the apex of the plants around the central leaf and on the base of petioles of older leaves. The incubation period (from the time of inoculation with viruliferous vector to appearance of first symptoms) is 25 to 85 days.

In commercial plantations the average distance of secondary spread of the disease is only 15.5–17.2 meters. Nearly 70% of the new infections are within a distance of 20 meters from the source and 99% are within 86 meters from the source. Chances of spread to new plantations decline as the distance between an infected plantation and the new plantation increases. The chances are less than 5% if the distance is 100 meters. These conditions apply only when new plantations are raised from certified virus-free suckers.

In Australia, Geering and Thomas (1997) had aphid inoculated a number of cultivated and wild plant species with BBTV but did not find any evidence of their role as a reservoir of BBTV inoculum. The plant species tested included *Alocasia*, *Colocasia* and *Heliconia*. However, in India, *Colocasia esculenta* was claimed in 1985 as a reservoir of BBTV.

Management of the disease

Phytosanitation is of paramount importance in the management of bunchy top of banana. The movement of infected planting material from place to place should be prohibited. No planting material should be obtained from areas where the disease is reported to occur. Since the secondary infection of healthy plants around a plantation where the disease is present is possible through aphids it is advisable that planting material should not be obtained even from the neighbourhood of the affected area.

As soon as the diseased plants are located in a plantation they should be dug out with roots and burnt. This step is effective in regions where the disease has restricted distribution. Spray of power kerosene or parathion is recommended to control the banana aphid. It had also been recommended that a wide banana-free zone should be created around an infested area. This may help in preventing the spread of the disease through insect vectors.

In some countries chemical control of the vector with aphicides has shown decreased incidence of the disease. Biological control of the aphid vector through use of a parasitic wasp (*Aphidius coleman*) was tried but failed to give good results.

Kavino *et al.,* (2007a) subjected virus indexed micropropagated banana plantlets to root colonization followed by foliar sprays with *Pseudomonads fluorescens* strains Pf1 and CHAO and *Bacillus subtilis* strain EPB22. On transfer to the main field and allowed to grow for 3–7 months, the plants showed improved vegetative growth, PRT-proteins and phenolic contents beside reduction in bunchy top incidencve. In a later study, they (Kavino *et al.,* 2007b) reported that treatment of plants with *P. fluorescens* CHAO + chitin induced systemic resistance to bunchy top. It induced accumulation of defence enzymes (peroxidase, polyphenol oxudase, phenylalanine ammonia lyase, PR proteins, chitinase and glucanase. In a similar report, Harish *et al.,* (2008) have stated that biopriming or biohardening of tissue culture, for obtaining plantlets, with the above rhizosphere and endophytic bacterial strains induces systemic resistance against BBTV in the plantlets.

Immune or highly resistant varieties of *Musa* are not available. All the varieties yielding seedless fruits for eating are susceptible. In some plants of the cultivar Veimama (Fiji and Australia) show a partial recovery from bunchy top symptoms and produce fruits. In Taiwan the local cultivar Lan-ya-chiao is reported to be resistant.

STERILITY MOSAIC OF PIGEONPEA

Sterility mosaic disease (SMD), first described in 1931 from Pusa (Bihar, India), is a major constraint on pigeonpea (*Cajanus cajan*) production. The disease is confined to the Indian subcontinent. It is endemic in India, Bangladesh, Nepal and Myanmar. The sterility of the plant results in yield losses of up to 90%. Annual loss is about 2 million tons of grain in India in 1984. In 1993, the loss in India and Nepal reached US $.280 million from 76 million in 1984. Plants infected when less than 45 days old show 96–100% loss in grain yield. Infection older plants results in losses varying from 27 to 97% loss depending on the number branches infected per plant. Percent incidence of the disease in India is 21.4 in Bihar, 15.4 in U.P., 12.8 in Tamil Nadu, 12.2 in Gujarat and 9.8 in Karnataka.

Symptoms of the disease are of typical mosaic. Leaves show faint yellow mosaic pattern. They are short in size. Dwarfing of branches and small size of leaves gives a bushy appearance to top of the plant. Most characteristic symptoms are reduced or no flowering (sterility). Flowers if formed do not form pods.

Sterility mosaic of pigeonpea had been known in India since 1931 but the causal agent had remained elusive for decades despite intensive efforts. Now the causal agent has been identified as a virus with highly flexuous filamentous particles of 3–10 nm width and of undefined length. The RNA genome is divided in 5–7 RNA species. The virus has been named as *Pigeonpea Sterility Mosaic Virus* (PPSMV). It appears to be a new genus of plant viruses although some of its properties are similar to virus species in the genera *Tospovirus* and *Tenuivirus* (Kumar *et al.*, 2003). Kumar *et al.*, (2001) had identified an isolate of *Pothos Latent Virus* (PoLV) in family *Aureusivirus* which was associated with 10.7% of the sterility mosaic-diseased plants and with 8.1% of healthy plants. This virus has isometric particles of 30 nm dia. with three RNA species, only the largest of the three is infectious. It was not considered as the cause of typical sterility mosaic.

PPSMV is transmitted by the eriophyid mite *Aceria cajani* (Acari: Arhtropoda). The mite and virus are highly host-specific with a narrow host-range confined to pigeonpea and its wild relatives, *Cajanus scarabaeoides* and *C. cajanifolius*. Once established on a susceptible host, the mites can multiply to high densities within a few weeks. These vectors are very small, measuring 200–250 μm and have a very short life of about 2 weeks. They inhabit the lower surface of the leaves. Their feeding does not cause any direct damage to the pigeonpea plant. The transmission efficiency of a single eriophyid mite is 53% but is 100% when >5 mites per plant are used. The mite acquires the virus (PPSMV) after a minimum acquisition access period of 15 min and inoculates the virus after a minimum inoculation access period of 90 min. No latent period is observed. Starvation of the vector prior to or following acquisition reduce the minimum acquisition and inoculation periods to 10 min and 60 min, respectively. The vector can retain the virus for up to 13 hours. The transmission is semi-persistent. There is no transovarial transmission. (Kulkarni *et al.*, 2002). The sterility mosaic virus is highly unstable in expressed plant sap. Mechanical transmission has been tried but only few attempts have succeeded. The expression of symptoms takes a long time, more than 40 days (Kumar *et al.*, 2002). Some strains of the virus are mild while others are severe in their interaction with the host.

Several genotypes possess resistance to mild strains and a few possess resistance to both mild and severe strains of PPSMV. Singh *et al.,* (1999) screened 81 germplasm lines of which 10 were resistant, 7 showed mild resistance, 12 were susceptible and remaining 53 were highly susceptible. The resistant lines were ICPL 88046, Bahar, DA-35, DA-33, DA-32, DA-11, K-32–1, Pusa 19, Pusa 14, and GAUT–9005. At the Kanpur Pulses Research Center, 11 pigeonpea lines were identified that possessed resistance against sterility mosaic in addition to Fusarium wilt. The line KPL 43 was found to possess resistance to wilt, sterility mosaic and tolerance to Phytophthora blight as well as has high yield potential (Viswadhar *et al.,* 2000). At ICRISAT, three genotypes, ICPL 96047, ICPL 96061 and ICPL 99046 were found resistant to Fusarium wilt, sterility mosaic, powdery mildew and phyllody while six ICPL lines 93001, 96047, 96061, 99046,99055 and 87119 were resistant to Fusarium wilt and sterility mosaic (Saifulla and Byre, 2002). These sources of multiple disease resistance are being used in breeding programme.

PLANT DISEASES CAUSED BY VIROIDS

Viroids are small, low molecular weight ribonucleic acids (RNA) that can infect plant cells, replicate themselves and cause disease. Viroids differ from viruses in at least two main characteristics:

(i) the size of RNA which has a molecular weight of 110,000–130, 000 in viroids compared to 1,000,000–10,000,000 for self replicating viruses,

(ii) virus RNA is enclosed in a protein coat while the viroids lack a protein coat and exist as naked or free RNA molecules.

The small size of the RNA indicates that viroids contain about 250–400 nucleotides and, therefore, lack sufficient information to code for even one enzyme (replicase) that may be required to replicate the viroid. The methods of study of viroids (extraction, isolation, and purification) are different from those used in the study of viruses. Viroids are extremely difficult to be seen under electron microscopes. They appear to be associated with the cell nuclei, particularly the chromatin, and possibly with endomembrane system of the cell.

The circular, single stranded RNA molecule of viroids has extensive base pairing in pairs of RNA strand. The base pairing results in some sort of hairpin structure with single stranded and double stranded regions of the same viroid. Although viroids have many of the properties of single stranded RNAs, when seen under the electron microscope they appear about 50 nm in length and have the thickness of double stranded RNA. Replication of viroids is not clearly understood. Their small size is barely sufficient to code for a very small protein. Such a protein would be considerably smaller than known RNA polymerase (replicase) subunits and would, therefore, be unable to carry out the replication of the viroid. Besides, viroids have been shown to be inactive as a messenger RNA in several *in vitro* protein synthesizing systems and no new proteins could be detected in viroid infected plants. Recently, it has been proposed that viroids replicate by direct RNA copying, in which all components required for viroid replication, including the RNA polymerase, are provided by the host. During viroid replication, the circular (+) strand of the viroid is replicated while it acts as a rolling drum producing multimeric linear strands of (–) RNA. The linear (–) strand then serves as a template for replication of multimeric strands of (+) RNA. The (+) RNA is subsequently processed (cleaved) by enzymes that release linear unit-length viroid (+ + RNA) and these circularize and produce many copies of the original viroid RNA.

Viroids are officially grouped in two families, *Pospiviroidae*, the major family, and *Avsunviroidae*, a small family with only three genera and 4 viroids (species). *Pospiviroidae* has four genera and the largest number of species. The family names are based on the name of the type genus. *Potato spindle tuber viroid* (PSTVd) is the type genus of *Pospiviroidae*. This family has the important genera like *Pospiviroid* with species *Potato spindle tuber viroid*, *Citrus exocosrtis viroid* (CEVd) and *Cocadivirus* with species *Coconut cadang cadang viroid* (CCCVd). Variants are reported in almost all the species. *Potato spindle tuber viroid* has 109 variants with number of nucleotides varying from 341 to 364. *Citrus exocortis viroid* has 86 variants with 366–475 nucleotides. *Hop stunt viroid* (HSVd) has 144 variants with 294–303 nucleotides.

How viroids cause disease was also not known. Viroid diseases show a variety of symptoms that resemble those caused by virus infection. The amount of viroids formed in cells seems to be extremely small and therefore is unlikely to cause a shortage of RNA nucleotides in cells. Besides, as in virus diseases, many infected hosts show no obvious damage, although viroids seem to be replicated in them as much as in the sensitive hosts. So, viroids apparently interfere with the host metabolism in some ways resembling those of viruses. Strains of a viroid with subtle differences in nucleotide sequences cause dramatically different symptoms in infected plants (Itaya *et al.,* 2002) by differently altering the host genes encoding for various functions. Earlier, they (Itaya *et al.,* 2001) had detected small RNAs with sequence specificity to PSTVd in infected tomato plants which indicated presence of RNA silencing. However, this was not responsible for the differing symptoms induced by severe and mild strains of PSTVd. Detection of a viroid in the host plant is by itself not easy. Having no protein coat viroids are not immunogenic and routine serological techniques fail to detect and identify them.

Within the plant, the viroids move through the phloem and from cell to cell through plasmpdesmata. The long distance systemic movement in the phloem is facilitated by a particular protein (lectin) found in abundance in phloem exudates. Viroids are spread from diseased to healthy plants primarily by mechanical means (through sap on hands and tools, vegetative propagation). Some viroids such as potato spindle tuber, chrysanthemum stunt and chrysanthemum chlorotic mottle viroids are transmitted through sap quite readily while others such as citrus exocortis are transmitted through sap with some difficulty. Some viroids, such as potato spindle tuber, are transmitted through the pollen and seed in rates ranging from 0 to 100%. No specific insect or other vectors are known although viroids seem to be transmitted on the mouth parts or feet of some insects. Insects can transmit viroids indirectly if the viroid is encapsidated by some RNA virus and the insect is its vector.

Viroids survive in nature outside the host or in dead plant matter for a period of time varying from a few minutes to a few months. Generally, they seem to overwinter or oversummer in perennial hosts which include the main hosts of almost all known viroids. Usually viroids are quite resistant to high temperatures and cannot be inactivated in infected plants by heat treatment.

Control of viroid diseases is based on the use of viroid-free propagating stock, removal and destruction of infected plants, and washing of hand or sterilization of tools after handling infected plants before moving on to healthy plants. Nearly 10 plant diseases are known to be caused by viroids. These include potato spindle tuber, citrus exocortis, chrysanthemum stunt, chrysanthemum chlorotic mottle, hop stunt, tomato bunchy top, avocado sunblotch, coconut cadang cadang, and cucumber pale fruit.

POTATO SPINDLE TUBER DISEASE

The disease occurs in United States, Canada, Russia and South Africa. It has been reported from India also. It causes 16–64% loss in tuber yield and in some areas is one of the most destructive diseases of potato. It attacks all varieties, spreads very rapidly, and often occurs in combination with virus diseases. It also attacks tomato but in this crop it is not of economic importance. In addition to commercial cultivation of potato the spindle tuber disease poses a potentially serious threat to the production of seed potatoes and maintenance of potato germplasm collections. Crop losses are negligible in temperate climate if the incidence is kept low (approx.1%). A strain of potato spindled tuber viroid causing extensive veinal necrosis in four wild *Solanum* species was reported in 1992 from Shimla Hills in India. The strain could infect tomato producing symptoms caused by mild strain of the viroid. In addition to Potato spindle tuber viroid, tomato is naturall infected by many other viroids including Citrus exoxortis viroid. The potato spindle tuber viroid has been detected also in avocado (*Persea americana*). In Europe PSTVd has been detected in an ornamental Solanaceous plant (*Solanum hasminioides*) in which it does not produce any symptom but is mechanically transferable to tomato.

Infected potato plants appear erect, spindly, and dwarfed. Leaves are small and erect and the leaflets are darker green and sometimes show rolling and twisting. The tubers are elongated, with a cylindrical middle and tapering ends. They are more smooth than normal and have more tender skin and flesh. Eyes are more numerous, more conspicuous and shallower. Yields are reduced considerably, often by 25% or more. Susceptible tomato plants are stunted and have smaller rugose leaves with necrosis of petioles and veins. Diseased plants have a bunchy top appearance.

Potato spindle tuber viroid (PSTVd) is the first recognized viroid. It is classified under the genus *Pospiviroid* of family Paspiviroidae. While potato is the natural primary host of PSTVd, it affects tomato also. In addition, *Ipomoea batatas* (sweet potato) and *Solanum melongena* (eggplant) are its secondary hosts. Experimentally a large number of solanaceous plant have been found to be susceptible.

Molecular weight of the RNA is approximately 100,000 daltons. The RNA is a single stranded molecule of 359 nucleotides with extensive regions of base pairing. Under electron microscope purified but apparently denatured PSTVd appears as short strands, about 50 nm long. Sap from infected plants is still infective after dilution of 1:1,000 to 1:10,000 and after heating for 10 min at 75–80°C. The viroid is quickly inactivated in expressed sap of infected plants. The infectivity can be preserved by treatment with phenol. Phenol inhibits the activity of enzyme ribonuclease that breaks down the viroid RNA. About 109 variants of PSTVd are reported.

PSTVd is mechanically transmitted and spreads primarily by knives used to cut healthy and infected seed tubers and during handling and planting of the crop. In mechanized potato cultivation, in advanced stage of plant growth, tractors moving in between the ridges get contaminated from diseased plants and transfer the viroid to healthy plant. About 50% transmission is through contact of bruised sprouts and 80–100% infection by rubbing of healthy and diseased leaves. In potato and tomato the viroid is transmitted by true seed. It is also transmitted by pollen. Singh *et al.,* (1992) have reported that pollen on a PSTVd infected potato plant contained the viroid and when used for pollination of flowers of a healthy plant the viroid could infect leaves at base of the inflorescence, apical leaves and tubers. Several insects including some aphids, beetles, grasshoppers, bugs etc. also transmit PSTVd. The role of insects is incidental and the viroid is transported on mouth parts and legs of these insects. There are several reports of transmission of the virus by aphids but in all these cases a RNA virus infection of the host was associated. When the potato leaf roll virus, transmitted by aphids, is also present it encapsidates the viroid particles and the vectors carry both to a healthy host causing transmission by insects. If the plant is resistant to PLRV, only the viroid is expressed in the plant (Syller and Marczewski, 2001). Earlier, Singh *et al.,* (1992) had reported that PSTVd is not encapsidated *in vivo* by Potato virus Y particles.

After inoculation of a tuber with PSTV by means of a contaminated knife or of a growing plant with sap from infected plant, the viroid replicates and spreads systemically throughout the plant. There was no information on the spread of the viroid within the plant nor on the mechanism(s) by which the symptoms are developed. Itaya *et al.*, (2002) have stated that PSTVd in tomato alters the genes encoding for cell wall structures, chloroplast functions, protein synthesis, etc. Tobacco mosaic virus also alters many gene functions. Thus, there appears to be a similarity between mode of action of viruses and viroids in the host at the molecular level.

The disease can be effectively controlled by planting only viroid-free tubers in fields in which tubers from the last crop are not left. Precautions during processing of seed tubers at the time of planting (use of cutting knives) is also important.

CITRUS EXOCORTIS

Exocortis is worldwide in distribution and affects trifoliate orange, citranges, Rangpur and other mandarin and sweet limes, some lemons, and citrons. Orange, lemon, grapefruit, and other citrus trees grafted on exocortis sensitive rootstocks show slight to great reduction in growth and yields are reduced by as much as 40%. Mandarin and sweet orange trees in central India have been found to be viroid infected showing symptoms of bark scaling and leaf yellowing (Ghosh *et al.*, 2002).

Infected susceptible plants show vertical splits in the bark and narrow, vertical, thin strips of partially loosened outer bark that gives the bark a cracked and scaly appearance. Since many of the exocortis-susceptible plants, such as trifoliate orange, are used primarily as rootstocks for other citrus trees, and because the scions make poor growth on such rootstocks, the enlarged, scaly rootstocks have given the disease the name "scale butt". Infected exocortis-susceptible plants may also show yellow blotches on young infected stems, and some citrons show leaf and stem epinasty and cracking and darkening of leaf veins and petioles. All infected plants usually appear stunted to a smaller or greater extent and have lower yields.

Species of *Citrus* are natural hosts of five viroid species, namely, *Citrus exocortis viroid* (CEVd), *Citrus bent leaf viroid* (CBLVd), *Hop stunt viroid* (HSVd), *Citrus viroid III* (CVD-III) and *Citrus viroid IV* (CV-IV). CEVd and its variants cause the exocortis disease while HSVd causes the cachexia disease (Verniere *et al.*, 2004). Natural infection of CEVd on tomato is also reported. Variants of CEVd have been found in eggplant, carrot and turnip.

The *Citrus exocortis viroid* (CEVd) is similar to, but not identical with, potato spindle tuber viroid. It consists of 371 nucleotides arranged in a circular or linear form. It is readily transmitted from diseased to healthy trees by budding knives, pruning shears, or other cutting tools, by hand, and possibly by scratching and gnawing of animals. CEVd and other viroids of citrus can be mechanically transmited by a single slash of knife blade. It is also transmitted by dodder and by sap to *Gynura, Petunia* and other herbaceous plants. On contaminated knife blades CEVd retains infectivity for at least 6 days and, when partially purified, it remains infective at room temperature for several months. The thermal inactivation point of extracted sap is about 80°C for 10 min, but partially purified CEVd remains infective even after boiling for 20 min. The viroid also survives brief heating of contaminated blades in the flame of propane torch (blade temperature 260°C) and flaming blades dipped in alcohol. The viroid also survives on contaminated blades treated with almost all common chemical sterilants except sodium hypochlorite solution.

The viroid apparently enters the phloem elements and spread in them throughout the plant. It is associated with the host nuclei and internal membranes of cells and results in aberrations of the plasma membrane. Although the viroid apparently lacks the ability to serve as messenger molecule or as an amino acid acceptor. It brings about several metabolic changes in infected plants. These changes include an increase in oxygen uptake and respiration, and also in sugars and certain enzymes. Marked changes also occur in several amino acids.

Exocortis can be controlled only by propagating healthy nursery trees from certified foundation stock and use of sanitary budding, nursery, and field practices. Tools should be disinfected between cuts into different plants by dipping in 10–20% solution of household bleach (sodium hypochlorite).

REFERENCES

Alavo, T.B.C., H. Serman and H. Bochow. 2002. Biocontrol of aphids using *Verticillium lecanni* in greenhouse: factors reducing the effectiveness of the entomopathogenic fungus. *Arch. Phytopath. Plant Prot.* 34(6):407

Ali, M.L., M. Asadudddoulllah, B.K. Paramanic and M. Ashrafuzzaman. 2001. Management of leaf curl disease of tomato. *Pak. J. Biol. Sci.* 4(1): 1512

Andret-Link, P. and M. Fuchs. 2005. Transmission specificity of plant viruses by vectors. *J. Plant Pathol.* 87(3): 153

Anhalt, M.D. and P.P. Almeida. 2008. Effect of temperature, vector life stage and plant accession period on transmission of *Banana Bunchy Top Virus* to banana. *Phytopathology* 98(6): 743

Azzam, O. and T.C.B. Chancellor. 2002. The biology, epidemiology and management of rice Tungro disease in Asia. *Plant Dis.* 86(2): 88

Balamuralikrishnan, M., S. Doraiswamy, T. Ganapathy and R. Viswanathan. 2003. Impact of serial thermotherapy on sugarcane mosaic virus titre and regeneration in sugarcane. *Arch. Phytopath. Plant Prot.* 36(3–4): 173

Barbosa, C.J., J.A. Pina *et al.*, 2005. Mechanical transmission of citrus viroids. *Plant Dis.* 89(6): 749

Bau, H-J., Y-H Cheng, T-A Yu, H-S Yang and S-D Yeh. 2003. Broad-spectrum resistance to different geographic strains of *Papaya ringspot virus* in coat protein gene transgenic papaya. *Phytopathology* 93(1): 112

Bau, H-J., Y-H. Cheng, T-A. Yu *et al.*, 2004. Field evaluation of transgenic papaya lines carrying the coat protein gene of *Papaya ringspot virus* in Taiwan. *Plant Dis.* 88(6): 594

Beachy, R.N., S.L. Fries and N.E. Tumer. 1990. Coat protein-mediated resistance against virus infection. *Annu. Rev. Phytopathol.* 28: 451

Bhagat, A.P., B.P. Yadav and Y. Prasad. 2001. Rate of dissemination of okra yellow vein mosaic virus disease in three cultivars of okra. *Indian Phytopath.* 54(4): 488

Boben, J., P. Kramberger, N. Petrovic *et al.*, 2007. Detection and quantification of *Tomato mosaic virus* in irrigation waters. *Eur. J. Plant Pathol.* 118(1): 59

Bol, J.F., H.J.M. Linthorst and B.J.C. Cornelissen. 1990. Plant pathogenesis related proteins induced by virus infection. *Annu. Rev. Phytopathol.* 28: 113

Bos L. 1983. *Introduction to Plant Virology.* PUDOC, Wageningen.

Brown, D.J.F., W.M. Robertson and D.L. Tridgill. 1995. Transmission of viruses by plant parasitic nematodes. *Annu. Rev. Phytopathol.* 33: 223–249

Campbell, R.N. 1996. Fungal transmission of plant viruses. *Annu. Rev. Phytopathol.* 34: 87

Chakraborty, S., P.K. Pandey, M.K. Banerjee, G. Kalloo and C.M. Fauquiet. 2003. *Tomato leaf curl Gujarat virus,* a new *Begomovirus* species causing a severe leaf curl disease of tomato in Varanasi, India. *Phytopathology* 93(12): 1485

Chancellor, T.C.B., A.G. Cook and L. Heong. 1996. The within field dynamics of rice tungro disease in relation to the abundance of its major leafhopper vectors. *Crop protection* 15(5): 439

Chowda Reddy, R.V., J. Colvin, V. Muniappa and S. Seal. 2005. Diversity and distribution of begomoviruses infecting tomato in India. *Arch. Virol.* 150(5): 845

Christias, Ch., P. Hatzipapas, A. Dara, A. Kaliafas and G. Chrysanthis. 2001. *Alternaria alternata,* a new pathotype pathogenic to aphids. *BioContrtol* 46(1): 105

Cuthbertson, A.G.S., K.F.A. Walters and P. Northing. 2005. The susceptibility of immature stages of *Bemisia tabaci* to the entomopathogenic fungus *Lecanicillium muscarium* on tomato and verbena foliage. *Mycopathologia* 159(1): 23

Davis, M.J. and Z. Ying. 2004. Development of papaya breeding lines with transgenic resistance to *Papaya Ringspot Virus. Plant Dis.* 88(4): 352

Diener, T. O. 1984. Subviral pathogens. Viroids and Prions. *Plant Dis.* 68: 4.

Elbadry, M., R.M. Taha, K.A. Eldougdoug *et al.*, 2006. Induction of systemic resistance in faba bean (*Vicia faba*) to bean yellow mosaic potyvirus (BYMV) via seed bacterization with plant growth promoting rhizobacterria. *J. Plant Dis. Prot.* 113 (6)

Faccioli, G. and M.C. Colalongo. 2002. Eradication of potato virus Y and potato leaf roll virus by chemotherapy of infected potato stem cuttings. *Phytopath. Medit.* 41(1): 76

Faria, M. and S.P. Wraight. 2001. Biological control of *Bemissia tabaci* with fungi. *Crop Protection* 20(9): 767

Fauquet, C.M., D.M. Bisaro, R.W. Briddon, J. Brown, B.D. Harrison, E.P. Rybicki, D.C. Stenger and J. Stanley. 2003. Revision of taxonomic criteria for species allocation in the Geminiviridae family and an updated list of begomovirus species. *Arch. Virol.* 148: 405

Fernandez-Rodriguez, T., L. Rubio, O. Carba; O and E. Marys. 2008. Genetic variation of *papaya ringspot virus* in Venezuela. *Arch. Virol.* 153(2): 343

Ferriera, S.A., K.Y. Pitz, R. Manshardt, F. Zee, M. Fitch and D. Gonsalves. 2002. Virus coat protein transgenic papaya provides practical control of Papaya Ringspot Virus in Hawaii. *Plant Dis.* 86(2): 101

Flores, R.C. Hernandez, A.E.M. Alba *et al.,* 2005. Viroids and viroid-host interactions. *Annu. Rev. Phytopathol.* 43:117

Francki, R.I.B., C.M. Fauquet, D.L. Knudscn and F. Brown. 1991. Classification and nomenclature of viruses. 5th Report. Int. Comm. Taxon.Viruses. *Arch. Virol. Suppl.* 2. 450 pp.

Froissart, R., Y. Michalakis and S. Blance. 2002. Helper component transcomplementation in the vector transmission of plant viruses. *Phytopathology* 92(6): 576

Fugro, P.A. and J.C. Rajput,1999. Breeding okra for yellow vein mosaic virus resistance. *J. Mycol. Pl. Pathol.* 29: 25

Gafni, Y. and B.L. Epel. 2002. The role of host and viral proteins in intra- and intercellular trafficking of geminiviruses. *Physiol. Mol. Plant Path.* 60(5): 231

Gallitelli, D. and G.P. Accotta. 2001. Virus resistant transgenic plants: potential impact on the fitness of plant viruses. *J. Plant Pathol.* 83(2)

Garcia-Arenal, F. and B.A. McDonald. 2003. An analysis of the durability of resistance to plant viruses. *Phytopathology* 93(8): 941

Geering, A.D.W. and J.E. Thomas. 1997. Search for alternative hosts of banana bunchy top virus in Australia. *Aust. Plant Path.* 26(4): 250

Ghosh, D.K., S. Mathur, K.N. Gupta, Y.S. Ahlawat and P. Ramachandran. 2002. Viroid infection of citrus in central India. *Indian Phytopath.* 55(3): 290

Gindin. G., N.U. Geschtov, B. Raccah and I. Barash. 2000. Pathogenicity of *Verticillium lecanii* to different developmental stages of the silver whitefly *Bemissia argentifolii. Phytoparasitica* 28(3): 1

Gonsalves, D. 1994. Papaya ringspot virus, pp. 67–68. In *Compendium of Tropical Fruit Diseases.* R.C. Ploetz, *et al.,* (eds). APS Press, St. Paul, Minn. USA

Gonsalves, D. 1998. Control of papaya ringspot virus in papaya: A case study. *Annu.. Rev. Phytopathol.* 36: 415

Gonsalves. D. and S.M. Garnsey. 1989. Cross protection techniques for control of plant virus diseases in the tropics. *Plant Dis.* 73(7): 592

Goswami, B.K. 1993. Interrelationship of plant viruses and nematodes, pp. 254–268. In: *Handbook of Economic Nematology.* K. Sitaramaiah and R.S. Singh.(eds). Cosmo Publications, New Delhi.

Gray, S.M. and N. Banerjee. 1999. Mechanism of arthropod transmission of plant and animal viruses. *Microbiol. Mol. Biol. Rev.* 63(1): 128

Harish, S., M. Kavino, N, Kumar, D. Sarvankumar, K. Soorianathasundaram and R. Samiyappan. 2008. Biohardening with plant growth promoting rhizosphere and endophytic bacteria induces systemic resistance against *Banana bunchy top virus. Appl. Soil Ecol.* 39(2): 187

Hibino, H. 1996. Biology and epidemiology of rice viruses. *Annu. Rev. Phytopathol.* 34: 249

Hinrichs-Berger, J., M. Harfold, S. Berger and H. Buchemauer. 1999. Cytological responses of susceptible and extremely resistant potato plants to inoculation with potato virus Y. *Physiol. Mol. Plant Path.* 55(3): 143

Hohn, T. 2007. Plant virus transmission from the insect point of view. *Proc. Nayt. Acad. Sci* USA. 104(46): 17905

Itaya, A., A.Folimonov, Y. Matsuda, R.S. Nelson and B. Ding. 2001. *Potato spindle tuber viroid* as inducere of RNA silencing in infected tomato. *Mol. Pland-Microb Interactions* 14(11): 1332

Itaya, A., Y. Matsuda, R.A. Gonzales, R.S. Nelson and B. Ding. 2002. *Potato spindle tuber viroid* strains of different pathogenicity induces and suppresses expression of common and unique genes in infected tomato. *Mol. Plant-Microbe Interactions* 15(10): 990

Hooks, C.R.R., M.G. Wright, D.S. Kabasawa *et al.,* (2008). Effect of banana buncy top virus infection on morphology and growth characteristics of banana. *Ann. Appl. Biol.* **IN PRESS**

Jain, R.K., J Sharma, A.S. Shivakumar, *et al.,* 2004. Variability in the coat protein gene of *Papaya ringspot virus* isolates from multiple locations in India. *Arch. Virol.* 149(12): 2435

Jayashree, K., K.B. Pun and S. Doraiswamy. 1999a. Effect of plant extracts and derivatives, buttermilk and virus inhibitory chemicals on pumpkin yellow mosaic virus transmission. *Indian Phytopath.* 52: 357

Jones, A.T., P. L. Kumar, K.B. Saxena, N.K. Kulkarni, V. Muniyappa and F. Waliyar. 2004. Sterility mosaic disease. "Green Plague" of pigeonpea: Advances in understanding the etiology, transmission and control of a major virus disease. *Plant Dis.* 88(5): 436

Jones, D.R. 2003. Plant viruses transmitted by whiteflies. *Eur. J. Plant Path.* 109(3): 195

Jones, D.R. 2005. Plant viruses transmitted by thrips. *Eur. J. Plant Pathol.* 113(2): 119

Kalleswaraswamy, C.M. and N.K. Krishna Kumar. 2008. Transmission efficiency of *Papaya Ringspot Virus* by three aphid species of *Phytopathology*

Kalra, A., R.K. Grover, N. Rishi and S.M.P. Khurana. 1989a. Interaction between *Phytophthora infestans* and potato virus X and Y in potato. *J. Agric. Sci. Camb.* 12: 33

Kalra, A., R.K. Grover, N. Rishi and S.M.P. Khurana. 1989b. Influence of different factors on potato virus Y-induced resistance to *Phytophthora infestans. Zeits Pflanzenkrank und Pflanzenschutz* 96: 470

Kavino, M., S. Harish, N. Kimar, D. Saravankumar *et al.,* 2007a. Rhizosphere and endophytic bacteria for induction of systemic resistance of banana plantlets against bunchy top virus. *Soil Biol. Bioichem.* 39(5): 1087

Kavino, M., S. Harish, N. Kumar, D. Sarvankumar and R. Samiyappan. 2007b. Induction of systemic resistance in banana (*Musa* spp.) against *Basnana bunchy top virus* (BBTV) by combining chitin with root-colonizing *Pseudomonas fgluorescens* strain CHAO. *Eur. J. Plant Pathol.* **IN PRESS**

Kirthi, N., N.P. Maiya, M.R.N. Murthy, H.S. Savitri. 2002. Evidence for recombination among the tomato leaf curl virus strains/ species from Bangalore, India. *Arch. Virol.* 147(2): 255

Kon, T., L.M. Dolores, A. Murayama *et al.,* 2002. Genome organization of an infectious clone of *Tomato leaf curl virus* (Philippines), a new monopartite Begomovirus. *J. Phytopath.* 150(11–12): 587

Kon, T., P. Sharma and M. Ikegani. 2007a. Suppressor of RNA silencing encoded by the monopartite tomato leaf curl Java begomovirus. *Arch. Virol.* 152(7): 1273

Kon, T., K. Kuwabara, S.H. Hidayat and M. Ikegani. 2007b. A begomovirus associated with ageratum yellow vein disease in Indonesia: Evidence for natural recombination between tomato leaf curl Java virus and Agerratum yellow vein virus-Java. *Arch Virol.* 152(6): 1147

Kulkarni, N.K., P. Lav Kumar, V. Muniyappa, A.T. Jones and D.V.R. Reddy. 2002. Transmission of pigeonpea sterility mosaic virus by the eriophyid mite, *Aceria cajani* (Acari: Arthropoda). *Plant Dis.* 86(12): 1297

Kumar, P.L., A.T. Jones, P. Sreenivasulu and D.V.R. Reddy. 2000. Breakthrough in the identification of the causal virus of pigeonpea sterility mosaic disease. *J. Mycol. Pl. Pathol.* 30(2): 249

Kumar, P.L., A.T. Jones, P. Sreenivasulu, B. Fenton and D.V.R. Reddy. 2001. Characterization of a virus from pigeonpea with affinities to species in the genus *Aureusvirus,* family *Tombusviridae. Plant Dis.* 85: 208

Kumar, P.L., A.T. Jones and D.V.R. Reddy. 2002. Mechanical transmission of Pigeonpea Sterility Mosaic Virus. *J. Mycol. Pl. Pathol.* 32(1): 88

Kumar, P. L., G.H. Duncan, I.M. Roberts, A.T. Jones and D.V.S. Reddy. 2002, Cytopathology of *Pigeonpea sterility mosaic virus* in pigeonpea and *Nicotiana benthamiana*: similarities with those of eriophyid mite-borne agents of undefined aetiology. *Ann. Appl. Biol.* 140(1): 87

Kumar, P.L., A. Teifion Jones and D.V.R. Reddy. 2003. A novel mite-transmitted virus with a divided RNA genome closely associated with pigeonpea sterility mosaic disease. *Phytopathology* 93(1): 71

Kumar, P.L., T.K.S. Latha, N.K. Kulkarni, N. Raghavaendra, K.B. Saxena, F. Waliar, K.T. Rangaswamy *et al.,* 2005. Broad-based resistance to pigeonpea sterility mosaic disease in wild relatives of pigeonpea (*Cajanus*: Phaseoleae). *Ann. Appl. Biol.* 146(3): 371

Kundan, A., M. Ramesh, V.J. Vasanthi, R. Radjacommare, B. Nandakumar, A. Ramanathan and K. Samiyappan. 2005. Use of *Pseudomonas fluorescens*-based formulations for management of tomato spotted wilt virus (TSWV) and enhanced tomato yield. *Biocontrol Sci. Technol.* 15(6): 553–569

Luria, S.E, J.E. Darnell Jr., D. Baltimore and A. Campbell. 1978. *General Virology.* 3rd Ed. John Wiley

Magdalita, P.M., D.M. Persely, I.D. Godwin *et al.,* 1997. Screening *Carica papaya* x *Carica cauliflora* hybrids for resistance to papaya ringspot virus-type P. *Plant Pathology* 46: 837

Martelli, G.P. 1997. Plant virus taxa: Properties and epidemiological characteristics. *J. Pl. Pathol.* 79(3)

Martelli, G.P. 2006. Current status of plant virus taxonomy. *J. Plant Pathol.* 88(3) S13

Maruthi, M.N., H. Czosnek, F. Vidavski, S.Y. Tarba, *et al.,*. 2003a. Comparison of resistance to *Tomato Leaf Curl Virus* (India) and *Tomato Yellow Leaf Curl Virus* (Israel) among *Lycopersicon* wild species, breeding lines and hybrids. *Eur. J. Pl. Pathol.* 109(3): 1

Maruthi, M.N., V. Muniappa, S.K. Green, J. Colvin and P. Hanson. 2003b. Resistance of tomato and sweet pepper genotypes to Tomato leaf curl Bangalore virus and its vector *Bemisia tabaci. International J. Pest Management* 49(4): 297

Matthews, R.E.F. 1991. *Plant Virology,* 3rd Ed. Academic Press. 835 pp

Mayo, M.A. and A.A. Brunt. 2001. The current state of plant virus taxonomy. *Mol. Plant Pathol.* 2(2): 97

Montasser, M.S., M.E. Tousignant and J.M. Kaper. 1998. Viral satellite RNA for the prevention of Cucumber mosaic virus (CMV) disease in field grown pepper and melon plants. *Plant Dis.* 82: 1298

Moore, C.J., P.W. Sutherland, R.L.S. Forster, R.C. Gardner and R.M. Macdiarmid. 2001. Dark green islands in plant virus infection are the result of post-transcriptional gene silencing. *Mol. Plant-Microbe Interact.* 14(8): 939

Mowry, T.M. 2005. Insecticidal reduction of *Potato leafroll virus* transmission by *Myzus persicae. Ann. Appl. Biol.* 146(1): 81

Muigai, S.G., M.J. Bassett, D.J. Schuster and J.W. Scott. 2003. Greenhouse and field screening of wild *Lycopersicon* germplasm for resistance to the whitrfly *Bemisia argentifolii. Phytoparasitica* 31(1): 27–38

Muniappa, V., H.M. Venkatesh, H.K. Ramappa, R.S. Kulkarni, M. Zedan *et al.,* 2000. Tomato leaf curl virus from Bangalore (ToLCV-Ban4): sequence comparison with India ToLCV isolates, detection in plants and insects, and vector relationships. *Arch. Virol.* 145(8): 1583

568

Murphy, J.F., G.W. Zehnder, D.J. Schuster *et al.*, 2000. Plant growth promoting rhizobacterial mediated protection in tomato against *Tomato mottle virus*. *Plant Dis.* 84: 779

Nath, P. and K.N. Bhagabati. 2002. Population dynamics of keafhopper vector of Rice tungro virus in Assam. *Indian Phytopath.* 55(1): 92

Niazi, F.R. and J. Singh. 2001. Effect of mite *Oligonychus oryzae* infestation on Rice Tungro Virus acquisition by *Nephotettix virescens*. *Indian Phytopath.* 54(3): 380

Niazi, F.R., P.N. Chowdhry and J. Singh. 2002. Bioefficiency of *Beauveria bassiana* and *Paecilomyces amoeneroseus* against rice green leaf hopper. *Indian Phytopath.* 55(4): 522

Oka, M., T. Ohki, Y. Honda, K. Nagaoka and M. Takenaka. 2008. Inhibition of Pepper mild mosaic virus with commercial cellulases. *J. Phytopath.* 156(2): 65

Pardee, K.I., P. Ellis, M. Bouthillier, G.H.N. Towers and C.J. French. 2004. Plant virus inhibition from marine algae. *Can. J. Bot.* 82(3): 304

Pennazio, S. and P. Roggero. 1998. Systemic acquired resistance against plant virus infections : A reality? *J. Plant Pathol.* 80(3): 179

Prasad, H.P., U.A. Shankar, B.H. Kumar, S.H. Shetty and H.S. Prakash. 2007. Management of *Bean Common Mosaic Virus* strain *blackeye cowpea mosaic* (BCMV-BICM) in cowpea using plant extracts, *Arch.Phytopath. Plant Prot.* 40(2): 139–147

Pun, K.B., S. Doraiswamy and R. Jeyrajan. 1999. Screening of plant species for the presence of antiviral principles against okra yellow vein mosaic virus. *Indian Phytopath.* 52: 221

Pun, K.B., S. Doraiswamy and R. Jeyrajan. 2000. Screening of virus inhibitory chemicals and neem products against okra yellow vein mosaic virus. *Indian Phytopath.* 53(1): 95

Putra, L.K., H.J. Ogle, A.P. James and P.J.L. Whittle. 2003. Distribution of *Sugarcane mosaic virus* in sugarcane plants. *Aust. Plant Path.* 32(2): 305

Qazi, Z., M. Ilyas, S. Mansoor, R.W. Briddon. 2007. Legume yellow mosaic viruses: genetically isolated begomoviruses. *Mol. Plant Pathol.* 8(4): 343

Radwan, D.E.M., K.A. Fayez, S.Y. Mahmud *et al.*, 2006. Salicylic acid alleviates growth inhibition and oxidative stress caused by zuechnini yellow mosaic virus infection in *Cucurbita pepo*. *Physiol. Mol. Plant Pathol.* 69(4–6): 172

Radwan, D.E.M., K.A. Fayez, S.Y. Mahmoud *et al.*, 2007. Physiological and metabolic changes of *Cucurbita pepo* leaves in response to zuechnini yellow mosaic virus (ZYMV) infection and salicylic acid treatment. *Plant Physiol. Biochem.* 45(6): 480

Raj, S.K., R. Singh, S.K. Pandey and B.P. Singh. 2005. *Agrobacterium*-mediated tomato transformation and regeneration of transgenic lines expressing *Tomato leaf curl virus* coat protein gene for resistance against TLCV infection. *Current Science* 88(19): 1674

Rajappan, K., C. Ushamalini, N. Subramanian, V. Narasimhan and A.A. Kareem. 2000. Effect of botanical on the population dynamics of *Nephotettix virescens*, rice tungro disease incidence and yield of rice. *Phytoparasitica* 28(2):

Rashid, A., D. Harris, P. Hollington and S. Ali. 2004. On-farm seed priming reduces yield losses of mungbean (*Vigna radiata*) associated with mungbean yellow mosaic virus in the North West Frontier Province of Pakistan. *Crop Protection* 23(11): 1119

Raupach, G.S., L. Lin, J.P. Murphy, S. Tuzum and J.W. Kloepper. 1996. Induced systemic resistance in cucumber and tomato against Cucumber Mosaic Cucumovirus using plant growth promoting tjizobacteria/ *Plant Dis.* 80: 891

Ren, S-X., P.A. Stansly and T-X. Liu. 2002. Life history of the whitefly predator *Nephaspis oculatus* (Coleoptera: Coccinellidae) at six constant temperatures. *Biol. Control* 23(3): 262

Rhee, Y., T. Tzfira, M.H. Chen, E. Waigmann and V. Citovsky. 2000. Cell-to cell-movement of tobacco mosaic virus: enigmas and explanations. *Mol. Plant Pathol.* 1(1): 33

Roy, G. and R.K. Jain. 2002. Comparison of coat protein genes of two *Papaya ringspot virus* isolates from different geographic locations. *Indian Phytopath.* 55(3): 353

Ryu, C.-M., B.R. Kang, S.H. Han *et al.*, 2007. Tobacco cultivars vary in induction of systemic resistance against *Cucumber mosaic virus* and growth promotion by *Pseudomonas chlororaphis* O6 and its gacS mutant. *Eur. J. Plant Pathol.* 119(4): 383

Saucke, H. and T.F. Doring. 2004. *Potato virus Y* reduction by straw mulch in organic potatoes. *Ann. App. Biol.* 144(3): 347

Schuster, D.J. 2004. Squash as a trap crop to protect tomato from whitefly-vectored tomato yellow leaf curl. *Int. J. Pest Manage.* 50(4): 281

Sharma, O.P., P.K. Sharma, and P.N. Sharma. 1999. Seed transmission of BCMV as influenced by stage of inoculation, infection and location of seed in pods of *Phaseolus vulgaris* L. *Indian J. Virol.* 15(2): 107

Sharmanz, M., J.E. Thomas, S. Skabo and T.A. Holton. 2008. Abaca bunchy top virus, a new member of the genus *Babuvirus* (family *Nanoviridae*). *Arch. Virol.* 153(1): 135

Shen, J.-O., K. Zhang, Z-J. Wu *et al.*, 2008. Antipjutoviral activity of bruceine-D from Brucea javanica seed. *Pest Manag. Sci.* 64(2): 191

Shibata, Y., R.C. Cabunaga, P.O. Kabuata and L-R. Chor. 2007. Characterization of *Oryza rufipogon*–derived resistance to Tungrto disease in rice. *Plant Dis.* 91(11): 138

Shivpuri, A., K.K. Bhargava, H.P. Chhipa and R.P. Ghasiolta. 2004. Management of *Yellow Vein Mosaic Virus* of okra. *J. Mycol. Pl. Pathol.* 34(2): 353

Singh, B.R., V.K. Dube and Aminuddin. 2007. Inhibition of mosaic disease of gladiolus caused by Bean Yellow Mosaic and Cucumber Mosaic viruses by virazole. *Scientia Horticulturae* 114(1): 54

Singh, R.P., A. Boucher and T.H. Somerville. 1992. Detection of potato spindle tuber viroid in the pollen and various parts of potato plant pollinated with viroid infected pollen. *Plant Dis.* 76: 951

Singh, R.P., A. Boucher, R.G. Wang and T.H. Summerville. 1992. Potato spindle tuber viroid is not encapsidated *in vivo* by potato virus Y particles. *Can. J. Plant Path.* 14(1): 18

Solomon-Blackburn, R.M., J. Nikam and H. Barker. 2008. Mechanism of strong resistance to Potato leafroll virus infectionin a clone of potato (*Solanum tuberosum*). *Ann. Appl. Biol.* 152(3): 339

Sriniwas, K., M. Krishnaraj and N. Mathivanan. 2008. Plant growth promotion and control of sunflower necrosis virus disease by application of biocontrol agents in sunflower. *Arch. Phytopath., Plant Prot.* **IN PRESS**

Stansly, P.A., P.A. Sanches, J.M. Rodriguez *et al.*, 2004. Prospects for biological control of *Bemisia tabaci* (Homoptera: Aleyrodidae) in greenhouse tomatoes of southern Spain. *Crop Protection* 23(8): 701–712

Stapleton, J.J. and C.G. Summers. 2002. Reflective mulches for management of aphids and aphid-borne virus diseases in late-season cantaloupe (*Cucumis melo* var. *cantalupensis*). *Crop Protection* 21(10): 891

Steinkraus, D.C., G.O. Boys and J.K. Rosenheim. 2002. Classical biological control of *Aphis gossypii* (Homoptera: Aphididae) with *Neozygites fresenii* (Entomophthorales: Neozygitaceae) in California cotton. *Biological Control* 25(3): 297

Su, H.J., L.Y. Tsao, M.L. Wu and T.H. Hung. 2003. Biological and molecular characterization of strains of Banana bunchy top virus. *J. Phytopath.* 151: 290

Syller, J. 2006. The role and mechanisms of helper component proteins encoded by poty viruses and caulimo viruses. *Physiol. Mol. Plant Pathol.* **IN PRESS**

Syller, J. and Marczewski. 2001. Potato leaf roll virus-assisted aphid transmission of potato spindle tuber viroid to potato leafroll virus-resistant potato. *J. Phytopath.* 149(3–4): 195

Tabler. M. and M. Tsagris. 2004. Viroids: petite RNA pathogens with distinguished talents. *Trend Plant Sci.* 9(7): 339–348

Tripathi, S. and A. Varma. 2002. Eco-friendly management of leaf curl disease of tomato. *Indian Phytopath.* 55(4): 473

Tripathi, S., H-J. Bau, L-F. Chen and S-D. Yeh. 2004. The ability of *Papaya ringspot virus* strains overcoming the transgenic resistance of papaya conferred by the coat protein gene is not correlated with higher degree of sequence divergence from the transgene. *Eur. J. Plant Pathol.* 110(9): 871

Tripathi, S., J.Y. Suzuki, S.A. Ferrier and D. Gonsalves. 2008. *Papaya ringspot virus*-P: characteristics, pathogenicity, sequence variability and control. *Mol. Plant Pathol.* **IN PRESS**

Usharani, K.S., B. Surendranath, Q.M.R. Haq and V.G. Malathi. 2004. Yellow mosaic virus infecting soybean in northern India is distinct from the species infecting soybean in southern and western India. *Current Science* 86(6): 845

Uzest, M., D. Gargani, M. Drucker and E. Hebrard. 2007. A protein key to plant virus transmission at the tip of the insect vector stylet. *Proc. Nat. Acad. Sci. USA.* 104(46): 17959

Verchot-Lubicz, J. 2003. Soil-borne viruses: advances in virus movement, virus induced gene silencing and engineered resistance. *Physiol. Mol. Plant Path.* 62(2): 55

Vishwadhar, R.G. Chaudhury, R.A. Singh, Naimuddin and D.P. Srivastava. 2000. Sources of resistance to major diseases in pigeonpea. *Indian Phytopath.* 53(3) 353 (abstr.)

Wang, L., J. Huang, M. You, X. Guan and B. Liu. 2007. Toxicity and feeding deterrence of crude toxin extracts of *Lecanicillium* (*Verticillium*) *lecanii* (Hyphomycetes) against sweet potato whitefly, *Bemisia tabaci* (Homoptera: Aeyrodidae). *Pest Manag. Sci.* 63(4): 381

Wang, Q., Y. Liu, Y. Xie and M. You. 2006. Cryotherapy of potato shoot tips for effective elimination of Potato leaf roll virus (PLRV) and Potato virus Y (PVY). *Potato Research* 49(1): 119–129

Xu, J.H. and M.-G. Feng. 2002. *Pandora delphacis* (Entomphthorales-Entomophthoraceae) infection affects the fecundity and population dynamics of *Myzus persicae* (Homoptera: Aphididae) at varying regimes of temperature and relative humidity. *Biological Control* 25(1): 85

Ye, S-D., Y-H. Dun and M-G. Feng. 2005. Time and concentration dependent interactions of *Beauveria bassiana* with sublethal rates of imidacloprid against the aphid pests *Macrosiphoniella sanborni* and *Myzus persicae*. *Ann. Appl. Biol.* 146(4): 459

You, B-J., C-H. Chang, L-F. Chen *et al.*, 2005. Engineered mild strains of *Papaya ringspot virus* for broader cross protection in cucurbits. *Phytopathology* 95(5): 533

Diseases Caused by Plant Parasitic Nematodes

The nematodes (kingdom Metazoa, phylum Nematoda or Nemata, class Chromadorea) are organisms resembling roundworms found in human intestine. They are natural fauna of soil and water. Free-living (non-parasitic) nematodes are found almost everywhere. They occur in oceans, deserts, hot springs, lakes, and even polar seas, but water is essential for their survival and activity. Majority of them are found in seas. Fresh water environments are rather not very favorable for them and the land or terrestrial environments are the least favorable because of alternating wet and dry periods and fluctuations in temperatures. However, the nematodes constitute the largest and most ubiquitous groups of the animal kingdom in soil, comprising 80–90% of all the multicellular animals. The number of nematodes in a soil may be 1.8 to 120 million per square meter. Only a fraction of these terrestrial nematodes are destructive parasites and pathogens of plant.

The food of nematodes is invariably some source of protoplasm such as plant cells, fungal hyphae, algae, bacteria and actinomycetes, protozoa, and other animals including smaller nematodes. The dead organic matter, which usually shows very high population of nematodes, actually plays the role of a rich substrate for supporting the populations of microorganisms on which the nematodes feed. The organic matter or dead remains of plants and animals by themselves do not form part of the nematode diet. On the basis of feeding habits the nematodes can be categorized as:

    (a) Plant feeders or phytophagous nematodes which feed on living plants (the plant parasitic nematodes). Majority of phytophagous nematodes are root parasites, Few attack the shoot tissues.

(b) Microbial feeders or microbivorous nematodes feeding on bacteria and small algal cells. These nematodes are beneficial for plant growth because they feed on bacteria in the rhizosphere and promote root growth.

(c) Miscellaneous feeders.

(d) Predators feeding on protozoa, nematodes, etc.

Although nematodes belonging to the first category are responsible for plant diseases, other forms of nematodes also become significant in plant pathology because some of them sometimes adversely affect development of parasitic fungi and nematodes. Some nematodes are recognized as entomophilic feeding on insects including insect vectors of plant pathogens. The predatory nematodes are important and common biocontrol agents of parasitic nematodes in agricultural soils and in the rhizosphere.

The plant parasitic nematodes may be **ectoparasitic** feeding from the surface of the plant, or **endoparasitic** feeding from within the tissues. In the latter case, the entire or major portion of the nematode body is in the tissues while in the former case only the stylet is inserted into the tissues. These nematodes may be migratory, i.e., moving from point to point while feeding, or sedentary, i.e., remaining stationary at one site of feeding.

Economic Importance

The nematode parasites of man were known to Egyptians as early as 1553 B.C. and the first plant parasitic nematode (the causal organism of ear cockle of wheat) was reported as early as 1743. However, the importance of these tiny organisms in the scheme of human welfare is of much later realization. There had been stray cases when these organisms had been found to cause significant damage in the nineteenth century (such as the sugar beet cyst nematode in Europe) but it was only after the second world war (1945), when some easy to apply and relatively cheap nematicides were discovered, that their significance as widely occurring destructive plant parasites was demonstrated. On a global basis, out of the approximately 34% crop losses annually caused by crop enemies like fungi, bacteria, nematodes, viruses, insects, and weeds, the nematodes alone are responsible for losses of about 11%. The annual loss of crops due to nematodes is about 12.3%, more in the developing countries of Asia and Africa than in the developed countries of Europe and America. Losses in potato, tomato, eggplant (brinjal), okra (bhindi), and pepper (chilli) are 12.2, 20.6, 16.9, 20.4, and 12.2 %, respectively. Loss of yield in beans is estimated at 60–90%. Even in the USA about 7.2 % of the annual crop value was lost to nematode attacks during the mid-sixties. In England, the potato cyst nematode caused annual loss of about 20 million dollars. The root knot nematodes annually destroy 29–90% of vegetable crops. In India, reduction in yield of tomatoes due to root-knot is reported to range from 26.5 to 73.3%. In legume crops, in India, a loss of 13.7% in chickpea due to root knot and 13.2% in pigeonpea due to root-knot, cyst nematode (pearly root) and *Rotylenchulus reniformis* (dry root) is estimated. Citrus decline, pepper yellows, molya disease of wheat and barley (cereal cyst nematode), and ear cockle of wheat are some of the historical examples of the damage caused by plant parasitic nematodes. There is hardly any horticultural crop that is not attacked by nematodes. Even some grain crops suffer heavily from their attack.

General Characters of Plant Parasitic Nematodes

Nematodes are triploblastic (having three layers), bilaterally symmetrical, non-segmented animals with single cavity (non-coelomic). The body is more or less cylindrical, sometimes fusiform, pear shaped or otherwise modified, particularly in adult females of some genera. The anterior portion of the body is slightly tapering or blunt while the posterior portion or the tail end is sharply tapering. The mouth opening at the anterior end is usually surrounded by lips bearing sensory organs. The mouth is followed by a mouth cavity or stoma. In plant parasitic nematodes, the stoma contains a spear or stylet with which the nematode pierces the cell walls of its host and ingests the food. Below the stoma is the esophagus (pharynx) followed by intestine, and a rectum terminating into a ventral terminal or sub-terminal anus in females, or a cloacal opening in males.

The body is covered with cuticle beneath which there is a hypodermis and a single layer of muscles. The cavity between the body wall and the internal organs is filled with fluid and some large vacuolated cells. There is no segmentation of the body. The sexes are usually

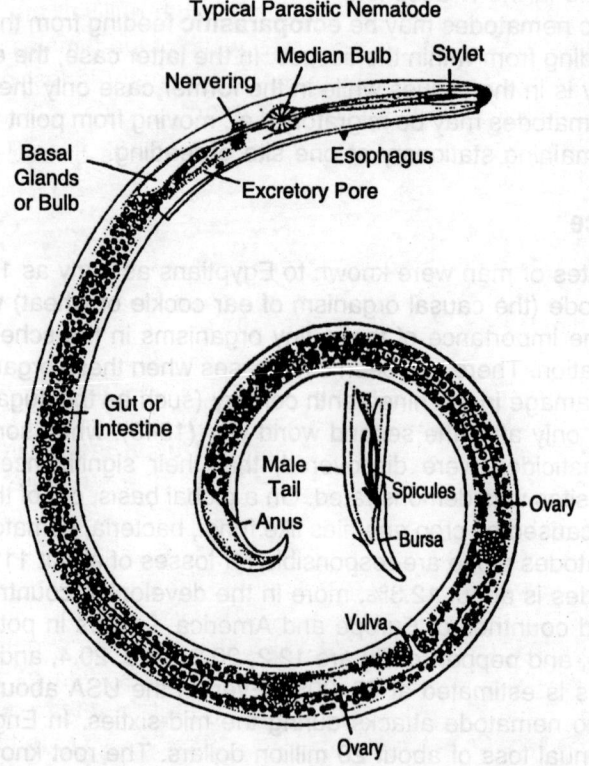

Fig. 147. Principal structures of a plant parasitic nematode.

separate. Males are always smaller than females. The male reproductive system consists of testis and vas deferens and is a single tube, sometimes two tubes, opening directly into the rectum forming the cloaca. The female reproductive system consists of either one or two ovaries which connect with the ventrally situated vulva by tubular uteri. Spermatheca are sometimes present in the uteri. In species with two ovaries, the branches may be placed in opposite direction or in one direction. The nematodes have excretory as well as nervous system but lack organs for circulation and respiration.

The cuticle is a semi-permeable, tough layer which is both skeletal and protective. It is either smooth or ornamented with annulations, punctations, etc. The presence of this layer enforces molting, four molts and five stages during the life cycle being typical of all nematodes.

The cuticle invades all the natural openings of the body such as mouth, esophagus, anus, excretory pores, etc. and the cuticular lining of these structures is shed with each molt.

Beneath the cuticle there is a thin cell layer known as hypodermis. It has numerous longitudinal thickenings which protrude in to the fluid filled cavity (pseudocoel). A layer of body wall musculature is located beneath the hypodermis. This somatic musculature is composed of a single layer of more or less spindle shaped cells attached to the hypodermis along the body length. The spindle shaped cells of the musculature are arranged longitudinally. This arrangement permits contraction of muscles only in the anterior-posterior (longitudinal plane). Thus, the nematode can move only in the forward direction. The musculature is responsible for the undulating swinging movement for locomotion in most nematodes. Due to the longitudinal arrangement of cells, the musculature can function only when the body retains its slender elongated shape. A strong increase in body width nullified the effect of musculature. Thus, in many nematodes, due to increase in body width of the females as they reach maturity and become filled with eggs, not only the bilateral symmetry is lost but the function of the musculature also stops and females become sedentary.

The digestive tract (alimentary canal) of the nematodes is divided into five parts:

(i)   stoma or buccal cavity,
(ii) esophagus or pharynx and esophago-intestinal valve,
(iii) intestine,
(iv) intestino-rectal valve and rectum in females and cloaca in males, and
(v)   anus.

In general appearance the alimentary canal is a simple internal tube extending from the oral opening to the anus. The stoma, esophagus, and the anus are lined by the cuticle and are, thus, well differentiated from the intestine. Depending upon the environments in which the nematode lives and the type of food source it utilizes, the structure of mouthpart is very diverse. In free-living, microbivorous nematodes the oral aperture is large and consists of well developed lips and papillae as well as short or large teeth but no spear or stylet. In plant parasitic nematodes, which feed on plant cells, the papillae are greatly reduced to facilitate movement in confined spaces, and the teeth are absent. Instead, all plant parasitic nematodes possess a piercing structure (stylet or spear). The stylet is like a minute hypodermic needle. Through it the nematode injects its saliva in to the plant cell and draws the cell contents back into its digestive tract. The passage (lumen) of the stylet is less than one micron in diameter and allows passage of only diluted fluid. The three parts of a stylet are the anterior conical tip, the central spear shaft, and the three basal knobs. The protractor muscles for operation of the stylet are located posterior to the basal knobs.

The esophagus is the muscular tube that leads from the stoma or base of the stylet to the esophago-intestinal valve which opens in the intestine. Internally, this organ is lined by cuticle and externally it is covered by a membrane that separates it from the pseudocoel (body cavity). The lumen of the esophagus consists of one or more muscular or glandular swellings called bulbs or esophageal glands which are anterior, median, or posterior according to location. At molecular level the gene products that determine parasitism of the nematode through secretions are located in the esophageal glands. The esophagus terminates in a plug which forms the esophago-intestinal valve. It is a muscular structure at the base of the esophagus which opens in the intestine.

The intestine of nematodes is a simple tube without musculature. It is composed of a single layer of epithelial cells. It opens in the rectum or the hind gut through the intestino-rectal valve or sphincter. The rectum is a narrow tube, flattened in dorso-ventral direction and is lined by the cuticle. In some nematodes such as in *Meloidogyne* a number of glands are present in the rectum which secrete a gelatinous matrix through the anus in which eggs are eventually embedded. In male nematodes the digestive and reproductive systems are joined together posteriorly to form the cloaca in place of rectum.

The nervous system is not well defined in plant parasitic nematodes. However, the nerve ring encircling the esophagus is easily identifiable. It is largely composed of nerve fibers and concentrations of nerve cell bodies (ganglia) forming a nerve center. The excretory system of nematodes consists of a glandular or tubular organ opening by a common excretory pore situated in the region of the nerve ring. Although no specific organs for respiration and circulation are seen in the nematodes they respire and intake of oxygen has been demonstrated. It is presumed that movement of the fluid in the body cavity controls these activities.

Reproduction of Nematodes

The nematodes, as a rule, are bisexual existing as separate males and females. Numerically, the males and females of plant parasitic nematodes exist mostly in equal numbers. The males are easily distinguished externally from the females by the presence of copulatory structures in the posterior part of the body near the anal region. Other characters of the males are their small size and curvature of the tail. Sexual dimorphism is very pronounced in some nematodes such as *Heterodera, Globodera* and *Meloidogyne*. The males remain slender and worm-like while the females swell to become pyriform, lemon-shaped, or saccate. In some genera such as in *Pratylenchus* the males fail to attain the full development of their stylet and esophagus. The females of all species of nematodes can be distinguished by the presence of a vagina and vulva located in the middle or posterior part of the body. In many genera of plant parasitic nematodes, though males may be present, reproduction is parthenogenetic in which fertilization of eggs is not necessary. Hermaphroditism is also found in which the gonad first produces sperms that are stored in spermatheca and fertilize the eggs developed by the same gonad later. Intersexes are known in some nematode genera such as *Meloidogyne* and *Ditylenchus*. An intersex is an individual which exhibits a blending of male and female characters. In most cases the intersexes are females which show secondary male characters. They may copulate with males and lay viable eggs.

The mature eggs are usually oval, elliptical or rounded. The morphology varies according to life habit of the nematode. The eggs are covered by three membranes, the outermost protein layer is secreted by the uterus, the middle chitinous layer or the egg shell is secreted by the egg, and the innermost lipoid membrane is present before fertilization of the egg. Majority of plant parasitic nematodes do not have the protein coat on the egg. The eggs are smooth in most phytonematodes. In species of *Meloidogyne* the eggs are laid in a gelatinous matrix secreted by the rectal matrix glands. This matrix protects the eggs from adverse environments. In *Heterodera* and other cyst forming nematodes the body wall of the female at maturity thickens, become resistant to decay and turns brown after death of the female. This forms the cyst which acts as protective shell for the eggs. The number of eggs produced by each female varies from 20–30 eggs (in *Aphelenchoides ritzemabosi*) to more than 2800 eggs in *Meloidogyne*. The rate of egg laying per day may be 24 egg (in *Meloidogyne* at 22°C) to only 2 eggs in *Aphelenchoides ritzemabosi*.

Classification of Plant Parasitic Nematodes

All nematodes belong to the phylum Nemata (**Nematoda**) of the Animal Kingdom. They are classified into class, order, suborder, superfamily, family and sometimes subfamily, genus and species. The phylum is divided into two classes: Secernentea and Adenophorea. Except the Dorylaimida group of plant parasitic nematodes all phytonematodes belong to the class Secernentea, especially the order Tylenchida. The plant parasitic nematodes of common occurrence in this order are subdivided as follows:

Order: Tylenchida
    Suborder: Tylenchina
        Superfamily: Tylenchoidea
            Family: Tylenchidae
                Genera: *Anguina, Ditylenchus*

            Family: Hololaimidae
                Subfamily: Hoplolaiminae
                    Genus: *Hoplolaimus*
                Subfamily: Rotylenchinae
                    Genera: *Rotylenchus, Helicotylenchus*
                Subfamiliy: Rotylenchulinae
                    Genus: *Rotylenchulus*

            Family: Tylenchorhynchidae
                Subfamily: Tylenchorhynchinae
                    Genus: *Tylenchorhynchus*

            Family: Pratylenchidae
                Subfamily: Pratylenchinae
                    Genus: *Pratylenchus*

        Superfamily: Heteroderoidea
            Family: Heteroderidae
                Subfamily: Heteroderinae
                    Genera: *Heterodera, Globodera*

            Family: Meloidogynidae
                Genus: *Meloidogyne*

        Superfamily: Criconemoidea
            Family: Criconematidae
                Subfamily: Paratylenchinae
                    Genuus: *Paratylenchus*:

            Family: Tylenchulidae
                Subfamily: Tylenchulinae
                    Genus: *Tylenchulus*

    Suborder: Aphelenchina
        Family: Aphelenchidae
            Genus: *Aphelenchus*
        Family: Aphelenchididae
            Genus: *Aphelenchoides*

Order: Dorylaimida
   Suborder: Dorylaimina
      Superfamily: Dorylaimoidea
         Family: Longidoridae
            Subfamily: Longidorinae
               Genera: *Longidorus, Paralongidorus*
            Subfamily: Xiphineminae
               Genus: *Xiphinema*

   Suborder: Diphtherophorina
      Superfamily: Trichodoroidea
         Family: Trichodoridae
            Genera: *Trichodorus, Paratrichodorus*

Movement, Orientation and Feeding

Nematode larvae by themselves move very slowly, hardly 1–2 meters in a year. Thus, their movement is related to their localized activities such as in finding or reaching the host surface and while on or in the host in migration from place to place in search of fresh cells for feeding. Undulatory locomotion or serpentine movement is the most common and is found in all nematodes except a few in which the larvae show a movement like earthworms. A film of water in the immediate surroundings is essential for nematode movement. They cease to move when the body surface completely dries. In presence of too much water also the larvae sink to the bottom. In soil where the larvae move through the soil pores the partially filled pores are the best for their movement provided the pore size is larger than the diameter of nematode body.

Soil moisture, texture, temperature and water currents are known to influence the direction of movement of the plant parasitic nematodes in soil. Many nematodes such as *Globodera rostochiensis*, *Ditylenchus dipsaci* and *Tylenchorhynchus* move towards the wet end of a moisture gradient. During this movement the soil pore size plays an important role not only because it holds water but also because it provides the space for movement of the larvae. Similarly, in temperature gradients also the larvae move to a specific temperature.

Chemical factors in soil are also important in directing the movement of the nematode larvae. These chemical substances emanate from the plant itself. There is general agreement that the chemical substances exuded by plant roots in gaseous or liquid form (for example carbon dioxide and root exudates) form a complex which acts as a source of attraction for the nematode larvae. These substances diffuse through the soil pores forming a gradient with the highest concentration at the root surface and weakest concentration at the other end. Larvae follow this gradient to reach the root urface whereas their movement in absence of these forces of attraction is at random.

After reaching the host root surface the larvae are ready to attempt at deriving food from the root. Penetration is the first step in this attempt. A large number of larvae normally aggregate at the same place. The nematodes first probe the cells in the vicinity of their head without actually puncturing them and move the head from side to side. The main object of this movement is to locate a weak spot where exudation is taking place. The sensory chemoreceptors located in the lip region help them in this act. The puncture of the host surface is caused at such locations. The stylet pierces the cell wall, perforation being achieved by rapid thrusts of the stylet. Thereafter, just the tip of the stylet is repeatedly inserted through the perforation hole to inject saliva and ingest cell contents. The head may move from side to side to give a rasping action to the stylet, the extent of thrust increasing as the stylet moves deeper. In many nematodes there is a period of immobility after penetration. During this period they inject the saliva in the cell. The saliva, which contains many enzymes, is secreted in the secretory glands of the esophagus and from the esophageal bulbs passes on to the stylet tip and from there to the host cell cytoplasm. The injected saliva has two functions: it performs extra corporeal digestion of cell contents with the help of enzymes, and it dilutes the cell contents, both effects helping in easy ingestion of food by the nematode. While in almost all the plant parasitic nematodes partially digested food is ingested, in the poorly advanced parasites like Trichodorids the cytoplasm along with cell organelles are swallowed. This is because the feeding tube of these nematodes is wide enough to permit passage of cell organelles.

Host-parasite Interactions

The saliva injected by the nematode contains cellulase, protease and amylase in addition to other enzymes depending on the nematode species and its food habit. The enzymes not only dissolve cell walls and help in ingestion and digestion of food, they induce metabolic changes also in the host. The enzyme activities in nematodes are influenced by parasitic habits. Thus, in migratory endoparasites the cellulase activity is several times more than in a sedentary endoparasite. The parasitism genes are located in the esophageal gland cell and encode secretions that control the complex process of parasitism. The secreted products of parasitism genes facilitate migration of endophytic migratory nematodes. They mediate the modification of root cells in to elaborate nurse cell systems (Gao *et al.,* 2003).

The root parasitic nematodes, whether primitive (such as *Trichodorus, Xiphinema,* etc.) or highly evolved (such as *Meloidogyne, Globodera* and *Heterodera*), establish a modified feeding site by disrupting or stimulating the normal root cell function to satisfy their nutritional demands for growth and reproduction. They create large volumes of cytoplasm near the feeding site and modify it for efficient withdrawal of nutrients. The establishment of specific feeding sites is closely linked with the feeding strategies. In sedentary endoparasites such as *Meloidogyne, Heterodera* and *Globodera* the strategy is highly advanced and involves induction of special types of nurse cell systems which serve as source of continuous supply of nutrients to the nematode.

In *Meloidogyne,* the second stage juveniles enter the roots and move quickly through the region of cell elongation and establish themselves to a permanent feeding site, usually with the head adjacent to or protruding into the vascular cylinder. The body of the nematode lies entirely within the cortex, usually parallel to the root axis and close to the stele with the anterior end pointed away from the root tip. After sometime they become sedentary and feeding is limited to a few cells that constitute the nurse cell system. A similar pattern is seen in *Heterodera* and *Globodera* but these nematodes do not induce gall or knot formation. In *Tylenchulus semipenetrans* (citrus nematode) the second stage juveniles penetrate epidermal and hypodermal cells and feed on the hypodermis and other layers of cortical parenchyma. They also establish a nurse cell system in the cortex and the developing young females remain attached to this site.

The sedentary plant parasitic nematodes can induce the re-differentiation of either differentiated or undifferentiated plant cells into specialized feeding cells through re-activation of the cell cycle. Differential activation of cell cycle gene promoters is supposed to be involved. The **nurse cell systems** of the sedentary endoparasites have, as a common feature, a high degree of metabolic activity. There is pronounced increase in cytoplasmic density and in ribosomes, polyribosomes and cell organelles. Nuclei are modified to increase the capacity of nuclear-cytoplasmic exchanges. Once introduced the nurse cell systems continue to grow and are probably maintained by a condition of nutrient sink created by the continuous nutrient withdrawal by the developing and reproducing nematode. Metabolic activity is most pronounced when the females start laying eggs. At this stage the nurse cells are packed with cytoplasm. When egg laying is complete, there are signs of disintegration of the system.

In *Meloidogyne* the nurse cell system consists of a variable number of discrete multinucleate **giant cells** which develop within the differentiating vascular cylinder by the expansion of individual cells accompanied by synchronous mitosis or repeated karyokinesis without cytokinesis. Such nurse cells are known to be induced and maintained by only *Meloidogyne* species which also induce copious gall formation around the infection site hence called root knot. In *Globodera* and *Heterodera* the nurse cell system is formed by the expansion of several cells at the feeding site. The wall expansion is accompanied by partial wall dissolution, mainly at the pit fields where the wall is thinnest. This causes mixing of contents of several cells and a multinucleate **syncytium** is established. Mitosis is not stimulated in the syncytia. The nematode feeds from this cell which draws nutrients from neighbouring cells. The nurse cells not only profoundly affect morphology and physiology of the root they also increase the susceptibility of the affected plant parts to other soil borne fungal and bacterial pathogens. In giant cells as well as syncytia there is extensive cell wall architectural modification including thickening and formation of numerous outgrowths that increase the plasmalemma surface area for solute uptake. These cell wall modifications arise from cell wall-modifying enzymes of the host plant rather than nematode origin.

Symptoms of Nematode Infection

Symptoms of nematode infection are mostly non-specific. Poor growth of the plant, stunting, patchiness of the crop stand, and discoloured foliage are such abnormalities that can be attributed to many other causes such as attack of fungi, bacteria and viruses as well as soil factors including nutritional disorders and unsuitable soil physical conditions. While some nematodes are primary pathogens, others, by feeding on the host plant, may allow entry of other pathogens or secondary invaders which cause the damage and visible symptoms. In many diseases, there is clear association between nematodes and other pathogens.

Above ground symptoms: The above ground symptoms include reduced growth, discoloration of foliage, and distortion of plant parts. Stunting and slow growth are common symptoms. Due to non-random distribution of nematodes in the field such plants appear in patches and generally pronounced stunting occurs when the nematode population is high. Leaf chlorosis often accompanies stunted growth of plants. The yellowing indicates nutritional deficiency caused by root destruction and immobility of nutrient from the root to the shoot. The discoloration may range from light yellow to deep red, purple, or even black. Distortion of aerial parts is generally caused by stem and leaf feeding nematodes. These distortions include swellings on leaves and stems, twisting of leaves, and replacement of grains with cockles.

Below ground symptoms: The root parasites induce conspicuous below ground symptoms which include reduced root system, root proliferation, root galls and cysts on roots. Depletion of the root system is the most common below ground symptom of nematode attack. In some cases root growth inhibition is the only symptom. The roots may show lesions in addition to growth inhibition. Some species of nematodes do not cause general decay of roots. The injury to roots induces the plant to grow more roots, in clusters, especially behind the damaged portion. Galling of roots is the most characteristic symptom of attack of root knot nematodes. However, many other nematodes such as *Xiphinema* and *Globodera* also cause swelling of underground parts of their host. In the attack of cyst nematodes, presence of white to brown cysts projecting on the root surface is characteristic symptom.

Relationship of Nematodes with Other Plant Pathogens

Nematodes are often involved in disease complexes in which the nematode is the primary pathogen and some fungus or bacterium is the secondary pathogen. Both can cause injury and disease independently but when in association there is synergistic effect and damage is much more along with alteration in symptoms and host-parasite relationship. Nematodes also provide avenues for entry of those pathogens which normally can not penetrate the intact host. In some cases there is obligatory relationship also.

Root knot nematodes are the most extensively studied nematode parasites that influence fungal root diseases. They are known to pre-dispose the host plants to attack of *Fusarium, Pythium, Phytophthora, Rhizoctonia, Sclerotium* and *Verticillium*. These fungi are independent root pathogens and can infect the host without the aid of the nematodes but when the nematodes are present the effects are aggravated. In vascular wilt diseases caused by *Fusarium oxysporum*, pre-infection of root knot nematodes is reported to alter chemical composition of vascular fluid and prevent accumulation of phytoalexins that could check fungal invasion.

The most common example of nematode-fungus diseases complex is wilt disease of various crops, especially in case of varieties resistant to the fungus. The incidence of Fusarium wilt of cotton, tomato, pea, sweet potato, and many other crops is increased when *Meloidogyne* spp., especially *M. incognita*, have invaded the roots.

The increase in wilt incidence is related to the number of nematodes infesting the root before invasion of the fungus, race of the nematode, and resistance of the host to the fungus, the effect being more pronounced in resistant varieties. Breakdown of resistance or inconsistency in the pattern of inheritance of resistance to wilt are most important effects of root knot nematode invasion. Changes in host physiology due to activities of the nematode are given as reason for resistance breakdown. The pre-colonization of tobacco roots by *M. incognita* greatly enhances the incidence of black shank disease caused by *Phytophthora nicotianae* var. *nicotianae*. The fungus extensively colonizes the hypertrophied and hyperplastic regions of the root galls. The interaction between *Meloidogyne* and *Rhizoctonia solani* is reported on cotton, soybean, pea, tobacco, tomato and okra.

Examples of nematode-bacterium associations in disease complexes are fewer. Among root diseases the interaction between root knot nematodes and bacterial wilt (*Ralstonia solanacearum*) is common. Bacterial wilt of potato, tomato, eggplant, and banana is more severe in root knot infested soil than in nematode free soil. The interaction is mainly the effect of wounding injury caused by the nematode that permits easier entry of bacteria. One of the best studied interactions is the Tundu or yellow slime or ear rot disease of wheat in which the nematode *Anguina tritici* and the bacterium *Rathayibacter tritici* are involved.

Nematodes as vectors of plant viruses: The first positive demonstration that a nematode could transmit a plant virus was in the fanleaf virus of grapevine which was transmitted by the nematode *Xiphinema index*. Since this demonstration in 1958 more than 20 plant viruses are now known to be transmitted by nematodes which belong to four genera in the nematode group Dorylaimida. These genera are *Xiphinema* and *Longidorus* which transmit isometric (polyhedral) viruses such as grapevine fanleaf, tobacco ringspot, tomato ringspot, tomato black ring, etc. and *Trichodorus* and *Paratrichodorus* which transmit tubular viruses such as tobacco rattle and pea early browning. These viruses are soil-borne and the soil becomes non-infective when dried at 20°C for a week or more because the nematode vectors are killed. Transmission of viruses by nematodes resembles the transmission by insect vectors. Viruses are picked up from the infected host and injected in the healthy host by the stylet during feeding process. There is acquisition feed time and inoculation feed time and the viruses are retained in specific areas of the nematode body for different durations.

Principles and Methods of Nematode Control

Efficient management of plant parasitic nematodes requires the carefully integrated combination of several methods. Although each individual method of management has a limited use, together they help in reducing the nematode populations over a long period. In general the measures for management of plant parasitic nematodes are based on the following principles of quarantine and other regulatory measures, cultural and land management practices, physical methods, biological control, resistant varieties, and direct control with chemicals.

Quarantine and sanitation: Regulatory methods that can prevent movement of soil and planting materials infested with nematodes can be one of the least expensive and the most efficient method of management of nematode diseases. Once the nematodes get into the field it becomes very difficult to remove them. Therefore, prevention of their entry in the field is very important step. This can be achieved by prevention of entry of the nematodes or their cysts with planting material and sanitary precautions such as cleaning of implements used in an infested field, avoiding movement of irrigation water from infested field to clean fields, use of seedlings free from nematode infestation, and such other practices.

Cultural and land management practices: The aim of such practices is to reduce the populations of nematodes in soil to a low level before a susceptible crop can be grown profitably. These practices cost very little because they form part of the normal cropping system.

Fertilizers and organic amendments that release ammonia have nematicidal effect. Crop rotation is an ancient agronomic practice used by farmers throughout the world to avoid crop failures due to soil problems. It is one of the most important component of any cropping system to combat the problem of soil-borne plant pathogens including nematodes. Crop rotation was recommended more than 100 years ago in Germany to control the sugarbeet cyst nematode, *Heterodera schachtii*. Control of phytonematodes by crop rotation is based on the fact that these nematodes are obligate parasites and can complete their life cycle only in presence of their host plant. Inclusion of a non-host crop or a highly resistant or immune variety of the host in the rotation causes significant decrease in the number of generations of the nematode and gradually their population falls down to an innocuous level. However, two important conditions are attached to success of crop rotation as a measure for nematode control. One, the nematode must have a limited host range, and second the farmer has a wide choice of crops for growing on his land. If the nematode has a very wide host range, chances are that there will be little choice of crops and there will not be effective check on multiplication of the nematode in the field.

Fallowing is another approach. This consists of ploughing the soil to keep it free from vegetation. It kills the nematodes by starvation and by desiccation or by exposure to high temperature during summer. If in a clean fallow the soil is ploughed and exposed to sun during summer root knot nematodes in 2.5 cm depth of soil are killed within 2 days. Even cysts forming nematodes such as *Heterodera avenae* (cereal cyst nematode) are adversely affected by desiccation in a fallow. However, there are some objections to the practice of fallows in a rotation. From an economic viewpoint, in countries where due to pressure on land maximum crop production per unit area is the aim keeping the land free from crops may not be acceptable. However, during May–June when water scarcity makes crop cultivation expensive fallowing can be practiced to advantage.

Flooding, a form of wet fallow or as a method of crop culture such as in rice, is also a method of reducing nematode populations in soil. Submergence of land under water for a specific period reduces populations of many nematodes, probably by production of nematotoxic substance from anaerobic decomposition of organic matter. Taking nematode susceptible crops after wet rice culture has been found effective in controlling root knot in vegetable crops.

Cover crops, trap crops, antagonistic or enemy crops or suppressive crops can also be used to reduce nematode populations, especially in home gardens and small vegetable plots. Cover crops usually have not much economic value but are highly resistant to nematode attack. *Crotalaria* spp. and sudan grass have been used for green manure and reduce root knot nematode damage. Sudan grass has been demonstrated to suppress infection and damage to vegetable crops susceptible to *Meloidogyne hapla* when incorporated as a green manure. Some cultivars of this grass contain a cyanogenic glucoside, hurrin, that is degraded through an intermediate step to hydroxyl-benzaldehyde and HCN. The latter fraction delays maturity of eggs and thereby reduces the number of infective second stage larvae (Widmer and Abawi, 2000). Decomposition of crucifer plants containing glucosinolates releases volatile isothiocyanates which is reported to suppress nematodes in soil including *Globodera rostochiensis, Meloidogyne* spp. and *Tylenculus semnipenetrans. Tylenchulus semipenetrans* is more sensitive than other nematodes (Zasada and Ferris, 2003; Zasada *et al.,* 2003) *Argemone mexicana,* a tropical annual weed, is another plant which when incorporated in soil, suppresses *M. javanica.* The nematicidal plants may harbour nematode-antagonistic bacteria in their roots. Insunza *et al.,* (2002) have reported such bacteria as *Pseudomonas* sp., *Bacillus mycoides* and *Stenotrophomonas maltophila* (*Pseudomonas maltophila*) which were isolated from roots of many plant species and were able to reduce densities of trichodorid nematodes in potato field.

Trap crops are plant species highly susceptible to a nematode of the major crop and are planted so as to get infested by the nematode at a particular stage of growth and both the crop and the nematodes are destroyed in a single operation before the nematodes can complete their life cycle. Enemy or antagonistic plants produce some toxic compound that destroys the nematodes in soil. Some grasses, varieties of mustard, marigold (*Tagetes*), species of *Crotalaria,* and asparagus have been listed as enemy plants. Use of forage pearl millet (*Pennisetum glaucum*) and marigold (*Tagetes erecta*) used as rotation crops suppress root lesion nematodes (*Pratylenchus penetras)* in supportive crops (Ball-Coehlo *et al.,* 2003). Nematode suppression by marigolds is thought to be due to thiophenes, heterocyclic sulphur-containing molecules (terthienyls) abundant in these plants. However, only prolonged cultivation of marigold in the field can suppress nematode populations. There is no noteworthy accumulation of biocidal agents in soil cropped to marigold and there is no general depression of soil microorganisms. The nematode control by marigold may not be due to only the release of a biocidal agent into the soil. Sturz and Kimpinski (2004) have reported that the antagonistic plants, African marigold (*Tagetes erecta*) and French marigold (*Tagetes patula*), harbor certain endophytic bacteria in their roots which possess activity against nematodes. The suppression of the effect of parasitic nematodes may be due to attenuation of nematode proliferation by these bacteria and also due to their transmission in soil to the host crop. Inclusion of forage or grain pearl millet in a short rotation controls *Pratylenchus penetrans* in tobacco (Belai *et al.,* 2004). El-Hamawi *et al.,* (2004) have also reported suppression of developmental stages of *M. incognita* and incidence of root knot in soybean by intercropping with marigold and *Ambrosia maritime.* Brassicaceous plants are, in general, resistant to root nematodes. It can be attributed to release of glucosinolates in the root exudates and their hydrolysis into isothiocyanates. Inclusion of crucifers in rotation and amendment of soil with crucifer residue is therefore beneficial in nematode control.

In small vegetable plots and in glasshouse grown vegetables mulching of the soil with straw, weeds, grasses, old news paper sheets, dry pine needles, tree bark, etc. have been found effective in significantly reducing root knot infestation. Often the populations of free-living nematophagous nematodes is highly increased in mulched soil and this provides biological control of the parasitic nematodes.

Organic amendments and fertilizers are reported as control measures against many nematodes especially the root knot nematodes (Singh, 1965; Singh and Sitaramaiah, 1971,1973; Stirling, 1989). Addition of farmyard manure and compost reduces the infestation level of *Globodera rostochiensis* in potato roots. These manures hamper the development of the nematode in plant roots. Deep ploughing during hot summer months followed by application of farmyard manure at planting spots increases the yield of tomato in root knot infested soil. Reduction in citrus nematode (*Tylenchulus semipenetrans*) populations by application of steer manure or chicken manure has also been reported. Chicken litter soil amendments suppress *Meloidogyne incognita* and increase cotton growth. The bacterial genera identified in litter-amended soil include *Arthrobacter, Bacillus, Cellulomonas, Micrococcus, Pseudomonas* and *Rhodococcus*. Organic amendments increase populations of the fungal antagonist *Trichoderma* in soil. *Trichoderma harzianum* is reported to parasitize larvae and eggs of *Meloidogyne javanica* and control root knot of tomato (Sharon *et al.,* 2001). Organic nitrogen in sewage sludge has been found more effective than ammonium nitrate for reducing populations of *Belonolaimus longicaudatus* in turf grass.

Application of decomposable organic wastes (including crop residue) to nematode infested soil is now well recognized method of nematode control and improving soil fertility. In addition to encouraging beneficial soil microflora, it mainly acts through a system of biofumigation. The volatile released during decomposition are toxic to nematodes. Chopped pineapple leaves and leaves of *Pongamia glabra* (karanj). *Azadirachta indica* (neem) and green manuring with sunnhemp (*Crotalaria*) reduce damage by root-knot. Soil application of powdered neem seed or neem cakeat 100 g/plant is reported to suppress *Radopholus similes, Pratylenchus* and *Meloidogyne* in banana. *Tagetes* spp. and green manuring with sunnhemp (*Crotalaria*) reduce root knot damage. Application of chopped cabbage leaves to soil reduces infestation of the cereal cyst nematode (*Heterodera avenae*). Green manure of rapeseed is reported to suppress *Meloidogyne chitwoodi* on potato and other nematodes. The toxic effect is due to glucosinolate content in plants of brassica family. On its enzymatic degradation isothiocyanates are released which have antinemic and antifungal properties. Glucosi nolate profiles differ among plant species and the isothiocyanate derivatives differ in the their toxicity to nematodes. The strategy is to select plants in the family Brassicaceae based on their glucosinolate profiles and the sensitivity of the target nematode species to the associated isothiocyanates. Benzyl or 2-phenyiethyl isothiocyanates and to some extent ethyl isothiocyanates are the most promising candidates for nematode management (Zasada and Ferris, 2003). Since 1965 and during the 1970s the role of organic amendment of soil with oil cakes of Karanj (*Pongamia glabra*) neem (margosa, *Azadirachta indica*) and sawdust were reported highly effective in reducing root knot (*Meloidogyne javanica*) in field trials (Singh, 1965, Sing and Sitaramaih, 1971, 1973). They had long-term residual effect. The amendment of soil with blood, fish meal, meat wastes is also reported to suppress plant parasitic nematodes in the soil. More recently, Abbasi *et al.,* (2005) have confirmed the efficacy of neem cake amendment of soil in controlling *Meloidogyne hapla* (in tomato) and *Pratylenchus penetrans* (in corn). The amendments suppress only the parasitic nematode. Free-living, nonpathogenic nematodes are not affected or their population may be enhanced Soil solarization of amended soil improves the effect of amendments by holding the toxic vapors in the soil. Chitin and its formulations have been used as amendments to inactivate nematodes and improve plant growth. Chitin has direct effect on nematode and also enhances microflora antagonistic to nematodes. Soil application of *Ricinus communis* (castor) fruit meal or ground fruits of wild cucumber (*Cucurbita myriocarpus*) is reported to suppress *Meloidogyne incognita* on tomato (Mashela, 2002; Mashela and Nithangeni, 2002; Mashela et al., 2008).

Physical methods: Soil steaming is a common practice in glasshouse raised crops to remove the nematode infestations of the soil. However, this method is not feasible on a field scale. Soil solarization (solar heating of soil) is a good substitute of soil steaming. It is cost effective, compatible with other pest management tactics, readily integrated into standard production systems and a valid alternative to pre-plant fumigation with costly and polluting chemicals. The soil is wetted and then covered with a thin, transparent polythene sheets for some weeks. This raises the soil temperature which is lethal for soil-borne pathogens. The system has been applied in temperate, tropical and subtropical regions, wherever there is sunshine, against root knot and other nematodes. Destruction of nematodes by desiccation or exposure to high temperatures during summer ploughing can be considered a physical method. For many fruit trees and for planting materials hot water treatment has been recommended. The process involves the correct time and temperature for killing the nematode but not the host.

Microbial control: Exploitation which consists of parasitism and predation is the main basis of biological control of plant parasitic nematodes in soil. There are many predaceous species of nematodes occurring in soil that continuously eat away the parasitic forms. Similarly, there are numerous nematophagous fungi, both parasites or toxin dependent and predators, using specialized structures to capture the nematodes. *Arthrobotrys oligospora* is a nematode trapping fungus found in soil. It is chemotropically attracted to roots of tomato and barley. It colonizes the roots and can be found in epidermal cells 3 months after inoculation. The trap-forming nematophagous fungi have specificity for the prey. The nematophagous fungi attract nematodes towards them and this attraction and predacity of the fungi are higly correlated. Trap formation in the nematophagous fungus *Arthrobotrys oligospora* increases the ability to attract nematodes. *Arthrobotrys dactylioides* can trap and kill larvae of *Melopidogyne incognita* and *Tylenchorhynchus braviae* but no *Hoplolaimus indicus* (Kumar and Singh, 2006). The nematophagous fungi *Pochonia robescens* (syn. *Verticillium suchlasoprium*), *Pochonia chlamydosporia* (syn. *Verticillium chlamydosporium*) and *Lecanicillium lecanii* (syn. *Verticillium lecanii*) parasitize nematode eggs and destroy the contents. *P. chlmydosporia* destroys eggs of *Meloidogyne* spp. and *Rotylenchulus reniformis* and cysts of nematodes. These fungal parasites adhere to the hydrophobic surface of the egg shell, produce appressoria and penetrate through action of protease enzymes (Lopez-Llorca et al., 2002). The egg-parasite *Pochonia chlamydosporia*, like *A. oligospora*, colonizes the root epidermis where it can be found up to 5–7 months after inoculation. In the case of *P. chlamydosporia*, knowledge of the plant's susceptibility to nematode attack and its ability to support the growth of *P. chlamydosporia* in its rhizosphere is essential. In general, increase in the density of *Meloidogyne* spp. in soil results in more nematodes in roots which, in turn, cause greater abundance of the nematophagous fungus in the rhizosphere. Addition of at least 500 second stage juveniles to soil is required to stimulate growth of the fungus on tomato roots (Bourne and Kerry, 1998). *P. chlamydosporia* has been isolated from nematode eggs up to nine months after application to soil It can survive in the rhizosphere for the entire growing season at one site but in low densities (Verdejo-Lucas *et al.,* 2003). *Clonostachys rosea* (*Gliocladium roseum*), the biocontrol agent of the fungal pathogen *Botrytis cinerea*, is reported to kill nematodes. Its conidia attach to nematode cuticle, germinate and penetrate into nematode body degrading and killing it (Zhang *et al.,* 2008).

There are certain bacterial parasites of nematodes also. By artificial introduction they have been shown to bring about significant reduction in the numbers of parasitic nematodes. However, field control of plant parasitic nematodes by introduced biocontrol agents has neither been found economical nor has it been demonstrated. Mass multiplication of antagonists on cheap substrates such as mixture of wheat or pulse bran, sawdust and water in 3 : 1 : 4 ratio and spot application in the field, nurseries or glasshouse pots is a feasible proposition. Saju *et al.,* (2002) haver suggested addition of *Trichoderma harzianum* to organic media such as FYM, margosa cake, coir pith and decomposed coffee pulp for better survival and increased biomass of the antagonist. Competition with other microorganisms in the biologically active substrates was not a limitation. The population of the nematophagous fungus *Catenaria anguillulae* is increased several fold in oil cake-amended soil (Singh *et al.,* 2002). This facultative endoparasite of nematodes has a wide distribution in soils from all over India and s present throughout the year (Vaish and Singh, 2000).

Metabolites of many fungi have antagonistic action against nematodes. *Fusarium equiseti* produces trichothecene which inhibits egg hatch and immobilizes the second stage larvae of *Meloidogyne incognita*. *Fusarium oxysporum* isolated from a cabbage field is reported to suppress cyst hatch of *Heterodera cruciferae* and even penetrate the cysts (Mennan *et al.,* 2005). Chitinases from the nematophagous fungi *Verticillium chlamydosporium* and *V. suchlasporium* damage the egg surface of *Globodera pallida*. The latter has better activity (Tikhonov *et al.,* 2002). Application of *Paecilomyces lilacinus* and *Pseudomonas aeruginosa* gives good control of root knot as well as root rot. The entomopathogenic fungi *Paecilomyces lilacinous, Verticillium* spp and *Beauveria* are phylogenitically closely related. There are methods of modification of soil environments which help in natural build up of populations of such predacious and parasitic biocontrol agents. These include mulching, flooding and application of undecomposed organic matter to the soil. The slow growing isolates of the nematode-trapping fungus *Arthrobotrys dactylioides* are more efficient than fast growing isolates in forming trapping structures and trapping the nematodes. The populations of nematode trapping fungi *Arthrobotrys dactyloides* and *Nematoctonus leisporus* are more frequent in plots given organic amendments. The same is not true for *Arthrobotrys thaumasia*. This nematode trapping fungus is suppressed by enchytraeids in the soil. *In vitro*, the antagonistic bacterium *Bacillus firmus* causes paralysis and mortality of *Radopholuys similes* (burrowing nematode of banana), *Meloidogune incognita* (root knot nematode) and *Dityelnchus dipsaci* (stem nematode). In sand infested with the nematodes addition of culture filtrtae of the bacterium, without cells, produced the same effects on the nematodes suggesting that the bacterium produced some toxic secondary metabolites that caused paralysis and mortality (Mendezaz *et al.,* 2008).

In addition to fungal biocontrol agents against nematodes, bacteria also play important role in suppression of nematodes. Nematode activity is mostly in the soil and in underground parts of the plant. They encounter numerous microorganisms in this environment. Bacteria are most abundant microflora in the soil. Many bacteria act as natural enemies of nematodes through diverse modes of action including parasitism, production of toxins, antibiotics, or enzymes, competing for nutrients, inducing host resistance, and pronotion of plant growth. As a group of enemies of nematodes, bacteria are divided into parasites, opportunistic parasites, rhizobacteria, crystal (Cry) forming bacteria, endophytes and symbionys of entomopathogenic nematodes (Tian *et al.,* 2007).

Endoparasitic bacteria (*Pasteuria* spp.) occupy the most important position among microbial biocontrol agents of nematodes, especially the root-knit nematodes. Members of the genus *Pasteuria* are obligate, mycelial, endospore-forming bacteria considered ancestral to *Bacillus*. *Pasteuria* was first described as a protozoan and later classified in the genus *Bacillus* and finally as a separate genus *Pasteuria*. So far, four species have been proposed. *Pasteuria penetrans* parasitizes primarily root-knot nematodes (*Meloidogyne* spp.), *Pasteuria thornei* primarily attacks root lesion nematodes (*Pratylenchus* spp), *Pasteuria nishizowae* occurs on cyst nematodes (*Heterodera* and *Globodera*) and *Pasteuria* occurring on *Belonolaimus longicaudatus* is proposed as *Pasteuria usgae*. The genetic nature of spore and spore coat of *P. penetrans* differs from those of other obligately parasitic species of *Pasteuria*.

Endospores of *Pas. penetrans* present in soil attach to the cuticle of second stage juveniles of *Meloidogyne*, germinate after the juveniles have entered the plant root and started feeding. The germ tubes can penetrate the nematode cuticle. Vegetative bacterial microcolonies then form and proliferate throughout the female body and degenerate the reproductive system. Finally, mature endospores are released into the soil. There is high degree of heterogenecity both within and among populations of the bacterium. Specificity of spore attachment to nematode cuticle is responsible for this.

Some soil bacteria are opportunistic parasites of nematodes. They are present in soil and on nematode body. They produce proteases that dissolve a hole in the cuticle. The bacteria feed on the dissolved tissue and may enter the nematpode body to proliferate there and kill the nematode.

Rhizobacterial population can be suppressive, neutral or stimulatory to the nematodes. Numerous *Bacillus* strains can suppress pests and pathogens of plants and promote plant growth. Some species are pathogens of nematodes. Bacillus subtilis is most thoroughly investigated. The proteases produced by Bacillus spp. Are involved ion the suppression of the nematodes (Lian *et al.,* 2007). Other bacteria genera in rhizosphere that suppress plant parasitic nematodes through diverse mechanisms are *Paenibacillus, Pseudomonas, Actinomycetes, Arthrobacter, Alcaligenes, Aureo-bacterium, Burkholderia, Clavicbacter, Clostridium, Enterobacte, Rhizobium* and many others. Species of *Pseudomonas, Bacillus, Paenibacillus* as well as the root nodule bacteria (*Rhizobium* spp.) are reported to suppress egg hatch and larval motility. Strains of *Bacillus cereus, B. megaterium, B. pumilus, B. thuringiensis, Enterobacter cloacae* cause high mortality in second stage juveniles of Meloidogyne exigua (coffee root knot nematode).

Most rhizobacteria act against plant parasitic nematodes by means of metabolic byproducts, enzymes, and toxins. The effects of these products include the suppression of nematode reproduction, egg hatch and juvenile survival, as well as direct killing of nematodes. Certain strains of *Streptomyces* which have suppressive effect against *S. scabies* (common scab of potato) reduce population densities of the root lesion nematode (*Pratylenchus penetrans*) alfalfa. The bacterium *Bacillus cereus* produces extracellular collagenolytic/ proteolytic enzymes which damage the nematode cuticle. *Bacillus* sp. strain B16 has strong nematotoxic activity through secretion of a serine protease which is its pathogenicity factor (Qiulong *et al.,* 2006). In addition to direct effect on nematodes, the rhizobacteria are reported to induce systemic resistance in plants by fortifying the physical and mechanical strength of cell wall through cell wall thickening, deposition of callose and accumulation of phenolic compounds.

Soil bacteria outside the rhizosphere also play important role in reducing poulations of nematodes in soil through liberation of nematicidal volatiles (Gu *et al.,* 2007). These volatiles are based on alcohols, ketones, alkenes and ethers. Maximum (.80%) nematicidal activity is shown by benzaldehyde, benzeneacetaldehyde cyclo-hexane and dimethyl disulfide, terpinol, benzenethanol propanone and phenylethanone.

Vesicular arbuscular mycorrhizae (VAM) are known to restrict nematode invasion of roots and reduce nematode populations in the soil. They can induce systemic resistance in the roots through surface carbohydrates. Combinations of antagonistic bacteria and VAM have also been found effective. Apple root stock inoculated with VAM (*Glomus* spp.) and planted in nematode-infested soil reduced population of the lesion nematode (*Pratylenchus penetrans*) in the soil. In banana also root inoculation with VAM prevents root galls (*Meloidogyne incognita*) and improves plant growth. Complete suppression of tomato root infection by *Meloidogyne javanica* by combined application of *Paecilomyces lilacinus* grown on chiocken litter and *Glomus mosseae* is reported. Species of the endoparasite *Hirsutella* spp. (fungus) are known to control *Meloidogyne, Heteroder* and *Globodera*.

Resistant varieties: In principle this is the best method of nematode control. Many nematode resistant (reduced nematode reproduction and good plant growth) or tolerant varieties of vegetable crops have been developed. The development of resistant varieties is a difficult task because of the soil borne nature of the pathogen, presence of biotypes of the pathogen, and large host range of many nematodes. The nematode-resistant varieties have the additional advantage of providing resistance to fungal or bacterial wilts in which nematode is the primary invader.

Chemical control: For medium and high income crops this is the best method available to-day. However, good nematode control based on use of synthetic nematicides can be achieved only for one cropping season and the treatment needs repetition in the next cropping season. There are a large number of nematotoxic chemicals available in the market and are being used to kill nematodes in soil and in the plant. These chemical are either soil fumigants or non-fumigant contact nematicides. Among fumigants carbon dissulphide, chloropicrin, DD, DBCP (Nemagon), Vapam, methyl bromide, etc. had been in common use in the past. Due to their polluting effect most of these fumigant nematicides are now not commonly recommended. There is more stress on systemic non-fumigant nematicides such as aldicarb (Temik), carbofuran (Furadon), oxamyl (Vydate), phenamiphos (Nemacur), phorate (Thimet), etc. In addition there are a number of highly effective contact nematicides/ insecticides also such as diazinon (Basudin), fensulphothion (Dasanit, Terrracur P), thionazin (Nemaphos, Zinophos, Cynem), ethoprophos and ethoprop (Mocap), VC-13, etc. These systemic and contact nematicides can be applied to the soil at planting time and also around established plants without any phytotoxic effect. Application can be made by spot treatment at the time of planting young seedlings in the main field or by row treatment. The disadvantages associated with chemical control through use of nematicides include high cost, reinfestation of soil after harvest, contamination of ground water and residues in fruits and vegetables. The trend is to reduce the amount of nematicides used and give more emphasis to cultural practices and biological control. Biocontrol is an integral part of the IPM approach to disease management. Certain essential oils, widely used as fragrance and flavors in the perfume and food industries, have long been known to repel insects. They have high potential for providing effective nematicidal compounds.

EAR COCKLES OF WHEAT

The ear cockle disease of wheat was first noticed in England in the year 1743. It is now reported from all the important wheat growing regions of the world. It occurs throughout Europe, Israel, Egypt, Syria. United States, Brazil, Canada, Pakistan, India, China, Australia and New

Zealand. This nematode disease is a problem in India, Eastern Europe, and in the Middle East. The losses from the ear cockle disease had been as high as 30–70% in the United States. In India the disease was first reported from Punjab in 1919 and is now known to be occasionally widespread in all the wheat growing states of the country. According to Vasudeva and Hingorani (1952) in India the disease accounts for a loss of 1–3% every year which may be as high as 50% in individual fields. More than often, the nematode is present in association with the yellow ear rot disease caused by the bacterium *Rathayibacter tritici*. Apart from the loss in yield, the disease. is known to cause toxicity in man and cattle when large quantities of galls are consumed with the flour or seed. Death of livestock feeding on seedheads of *Agrostic avenacea* and *Polypogon monspeliensis* infected with *Clavibacter toxicus* and the nematode *Anguina* sp. is reported (McKay *et al.,* 1993). Riley (1992) had reported that *Anguina tritici* is a potential vector of *Clavibacter toxicus*. In addition to *A. tritici, A. australis* is also reported to be a vector of *C. toxicus* (Schmitz *et al.,* 2001)

Symptoms

The effects of the disease are visible on the stems, leaves and floral organs. Affected plants are dwarfed and their leaves are twisted and wrinkled thus preventing the normal emergence of the young leaves from within, causing them to be buckled. The infected ears are shorter and broader and keep green longer than normal healthy ears. Grains in the affected ears are partially or wholly replaced by cockles (galls) that are hard, dark brown or black. These stony structures vary in size from region to region and are generally 3–5 x 2–3 mm. They are filled with nematode larvae which can be seen by soaking the galls in water and then macerating them. Each gall may contain 3600 to 32400 or more larvae.

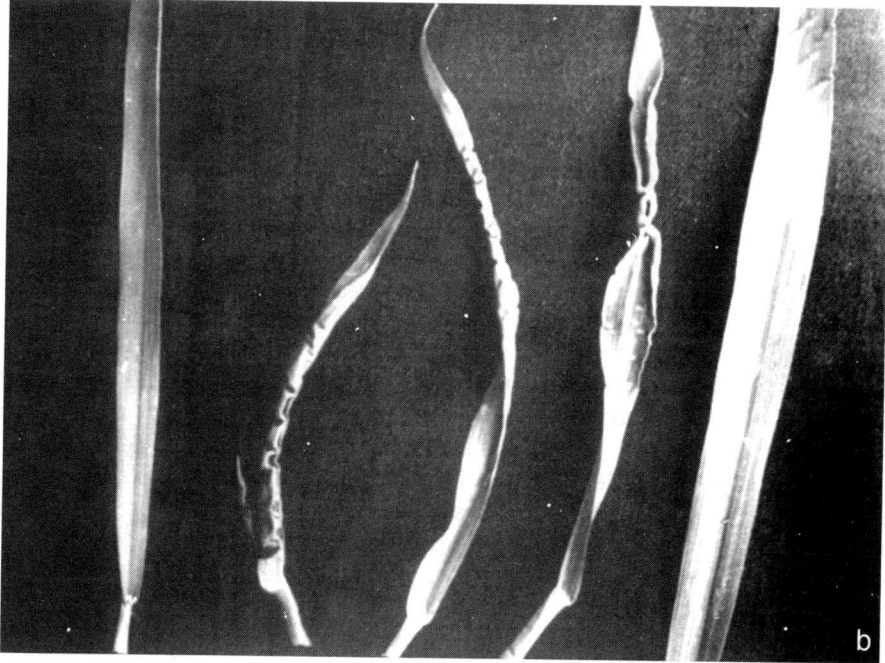

Fig. 148. Leaf twisting in wheat affected with ear cockle disease.

Fig. 149. Ear cockle of wheat. Healthy ear (*left*) and diseased ears (right).

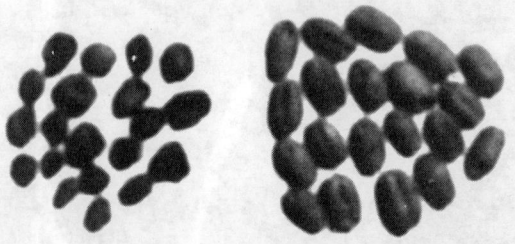

Fig. 150. Ear cockle of wheat. *left*—Cockles; *Right*—Healthy grains.

Wrinkling and twisting of leaves is caused by externally feeding larvae of the nematode at the leaf base and not due to penetration by them. These effects on leaves are probably due to irritant action of the larvae while they are surrounding the growing tip of the stem. Severely affected plants may even die. Seedling symptoms may be absent yet plants may bear galls in the ears. The plants show a spreading nature and tend towards more tillering. Early ear formation has also been noticed. Basal enlargement of the stem and formation of galls on awns, glumes, and staminate tissues have also been reported.

The causal organism

The cockles in wheat ears are produced by the attack of the nematode *Anguina tritici* (Steinbuch) Chitwood. The second stage larvae which emerge from eggs are 0.8–0.95 mm long.

The adult males are 2–2.5 mm long. The adult female is 2.6–4.4 mm long. The anterior and posterior portion is slender while the middle is swollen. The stylet is 9–11 μm long. The mature females are obese and immobile. The tail is very narrow compared to the obese body. Size of larvae varies according to geographic regions from where samples are collected. The eggs are, on an average, 87 x 44 μm in size. The seed galls contain 3–37 females and 1–41 males before copulation and egg production in the galls. One female can lay 2000 egs. The eggs hatch in the galls to produce thousands of larvae.

Disease cycle and environmental relations

The principal means of dispersal of A. tritici is by wheat seed containing the galls, which are the main source of survival also, and use of such seed in the field. Other means of spread include straw from an infected crop, rainfall, flooding, and animal manure. Natural migration of the nematode larvae is 20–30 cm horizontally and 7–9 cm vertically. The pathogen can be spread also by animal feet, farm implements and machinery. The larvae in wheat galls have extreme longevity. They can survive up to 35 years under dry conditions and up to 8–9 years or even 14 years under moist conditions. Free larvae, outside the galls, do not survive for more than a year. They can survive in faeces of rodents, sheep, frogs and fishes. In soil, wetting of the galls softens the walls and larvae are released. The moisture required for softening of the galls and germination of wheat seed is identical. For artificial inoculation a minimum of 10,000 juveniles/ kg soil is essential. The liberated larvae (juveniles) make their way to the growing point of the wheat plant while it is still near the soil line. It takes a few days for the larvae to reach embryonic growing point. A thin film of water on the seedling surface is required otherwise the larvae cannot move to the growing tip. The larvae crowd around the growing tip and feed ectoparasitically for about 45 days after inoculation. They are carried upward with the lengthening of the culms until the formation of embryonic flower tissues (about 70 days after inoculation). There, they undergo rapid metamorphosis, undergoing three successive molts within 3–5 days after invasion of the flower primordia and become adult males and females. Copulation follows and within 6–12 days after becoming adults several thousand eggs are laid by the females. After egg laying the adult females soon die. Eggs soon hatch to release second stage larvae which remain under forced dormancy (quiescence) imposed by anhydrobiosis. Under this condition, the water content of the nematode body is reduced from the normal level of 45–50% to 5%, the lipid droplets become wrinkled and the body becomes tightly coiled. These larvae can only revive when they come in contact with water. Occasionally, the larvae can enter the leaf tissues and form galls there. The total life cycle from cycle from second stage to second stage takes about 103–113 days in temperate climate countries. In India it is reported to be 164 days. The cockle are formed by modification of the seed into nematode-infected galls in the floral tissue.

In Australia, in the leaf gall nematode Anguina australis, Riley et al., (2001) have reported over-summer anhydrobiotic survival of the nematode in leaf galls. On rehydration and removal from galls, the larvae could undergo one or more cycles of dessication and rehydration. With moist incubation of galls reproductive activity was greatest at 20°C, limited at 10–15°C and 28°C and none at 5°C. At 20°C reproductive activity commenced after 3 weeks and continued for 7 days or more.

In addition to the necessity of moisture for release of larvae from galls, temperature and depth at which the seed galls are present in soil are also important. The optimum conditions for emergence of larvae from the galls are 15°C temperature and 2 cm depth at which galls are present and the soil moisture of 30% with 51% pore space. The nematode is susceptible to desiccation and to high temperature under moist conditions. The temperatures prevailing in the plains of India after the wheat crop is harvested may destroy the larvae if there is sufficient moisture (through rains or irrigation) in the soil to soften the galls.

The etiological relationship of this nematode with the bacterial ear rot of wheat has been mentioned. Infection of the plant with the nematode can occur only in the initial stage of plant growth, i.e., before the seedlings have emerged out of the soil. The nematode is essential for development of symptoms like basal enlargement of the stem, twisting and wrinkling of the leaves, formation of galls, and development of bacterial yellow rot. Larvae free from bacteria produce only the cockle disease. The bacterium is unable to produce the yellow rot independently. The development of ear cockle and yellow rot diseases together is dependent on inoculum concentration of the nematode and the bacterium, temperature, humidity, age of seedlings, and depth of placement of galls in the soil. There is significant increase in yellow ear rot in 25 days old inoculated plants which had been kept at 25°C and exposed to high humidity for 72 hours. Further, the symptoms of yellow ear rot appear only when the bacterium and the nematode have been inoculated simultaneously. Once inside the host, the bacterium multiplies rapidly and is independent of the nematode. Often, when its density is high the nematodes are killed. This is a case of elimination of one disease by another.

Other plant diseases in which association of *A. tritici* with the fungal pathogen is reported are *Tilletia indica, Tilletia laevis, Ustilago segetum tritici* and *Delophosphora alopecuri*. In the last named association, when the galls are filled with fungus spores, *A. tritici* is eliminated.

Management of the disease

Sowing of clean and healthy seeds: Since the parasite is introduced into new fields mostly through galls (cockles) mixed with seeds, use of clean seed is most important step in the management of the disease. During ear formation the disease is easily recognized by the nature of the spikes and the yellow ear rot often present with it. Such ear heads can be picked out and destroyed. Selection of healthy grains from a mixed lot can be done by using a sieve that can retain normal plump grains and allow the passage of small round galls. These galls can also be removed by floatation in brine solution. The difference in specific gravity of galls and healthy grains makes this possible. Seed lot is placed in 20% common salt solution in a suitable container and vigorously stirred for some time. The galls being lighter float on the surface and can be skimmed off and destroyed. The seeds are then washed with clean water thoroughly to remove traces of salt so that their germination is not adversely affected. This method had proved very effective in Europe where the disease has been brought under almost complete check. The hot water treatment recommended for loose smut of wheat is also good for management of the disease. The seed lot is first soaked in water for 4–6 hours at room temperature and then treated for 10 min in hot water at 54°C. Pre-soaking activates the larvae which come out of the galls. Exposure to 54°C for only 10 min destroys them. Drying of the seed after the treatment further helps in their destruction. In China mechanical separators are used which remove 90% of the galls from the seed lot.

Crop rotation: In fields where the disease has once appeared cultivation of wheat should be stopped for 2–3 years. Barley and oats are poor hosts and can be used in the rotation. During rotation seasonal wetting of galls ensures release of larvae which are killed in absence of the host.

Cultural practices and resistant varieties: The infection of wheat plants by the nematode is influenced by temperature. Lower the temperature at sowing time more is incidence of ear cockle. Hence, early sown crops usually escape infection. No wheat variety under cultivation in India is immune. Varieties Sonora 63, Lerma Rojo, NP 908 and S 227 had shown a certain degree of tolerance to the disease.

Chemical soil treatments: When the soil infestation is heavy granular insecticides such as Nemaphos (Zinophos) aldicarb or thionazin can be used for soil treatment. Nemaphos at the rate of 10 kg a.i./ha destroys active larvae. Seed treatment and soil application of aldicarb sulfone at the rate of 2 kg a.i./ha gives good control of ear cockle and yellow ear rot diseases. However, the cost of chemical treatment is high.

CEREAL CYST NEMATODE (MOLYA DISEASE)

The cereal cyst nematode or cereal root eelworm is reported from most of the wheat, barley, and oats growing areas of the world. It is one of the important plant diseases in Europe, Canada, Australia, and India. In India, this disease, of wheat and barley, known as molya, was first reported from Punjab in 1957. Its widespread occurrence in adjoining Rajasthan was reported in 1959–1961. The affected areas in Rajasthan were found to suffer upto 50% damage. The annual losses to the extent of Rs. 40 million in wheat and Rs. 30 million in barley are considered common in Rajasthan. The disease has now spread to neighboring areas of Haryana and Uttar Pradesh also.

Symptoms

In newly infested fields the molya disease occurs in small patches of 2–3 feet dia. These patches gradually increase in dimension every year with continuous cultivation of cereals in the same field until the whole field gets infested. A diseased crop looks stunted showing symptoms similar to those of nutritional deficiency. The plants are dwarfed and pale. The leaves are discoloured to yellow and often become \reddish from the tip. Tillering is also markedly reduced and badly affected plants fail to produce ear heads.

The growing point of roots is inhibited and often killed. The plant produces new roots near this point causing a characteristic rosetting effect. In severe cases the whole root system is dwarfed and matted. There are no long roots penetrating into deeper soil layers so that the affected plants are very susceptible to drought conditions. As the infection advances, slight swelling near the root tips can be seen. This condition is found generally within 4–6 weeks after sowing, i.e., by the end of November or early December in Rajasthan area in India. By the middle of February glistening white females of the nematode can be seen adhering to roots. A couple of weeks before harvest, these bodies (cysts) turn brown and may either remain attached to roots or fall off in the soil after the roots decay.

The causal organism

The cereal cyst nematode that causes the molya disease is known as *Heterodera avenae.* Many other species, *viz., H. filipjevi, H. hordecalis, H. australis,* are reported as cyst nematodes of different cereals and grasses. The body of females is swollen, pearly white and lemon-shaped and usually varies between 0.55 and 0. 75 mm in length and about two thirds as wide as long. Neck and vulva are protruding. They have paired ovaries. The cuticle bears zig-zag pattern. The females measure 418–740 x 285–534 μm. Cysts are typically lemon-shaped. New cysts have a white crystalline layer which is soon lost and only dark brown to black cysts are seen, Mature cysts of Indian populations measure 470–1010 x 370–730 μm. Elsewhere their size is reported to be 614–823– 382–627 μm. Each cyst contains 225–250 eggs. The oblong eggs measure 107–151 x 42.7 μm. On attaining its full development the embryo in the egg undergoes the first molt giving rise to the second stage larva. These larvae complete their full development within the egg inside the cyst. Under favorable conditions the larvae escape via the vulva and other apertures in the cyst wall. The free-living second stage larvae thus released migrate through the soil in search of suitable host plants. In general, 60% eggs hatch in the first year and hatching may extend to 3–4 years.

The second stage larvae are cylindrical with well developed stout stylet and basal knobs. Penetration of the host by these larvae occurs usually just behind the growing root tip causing a small elbow to develop which is the first sign of the presence of the nematode in the plant.

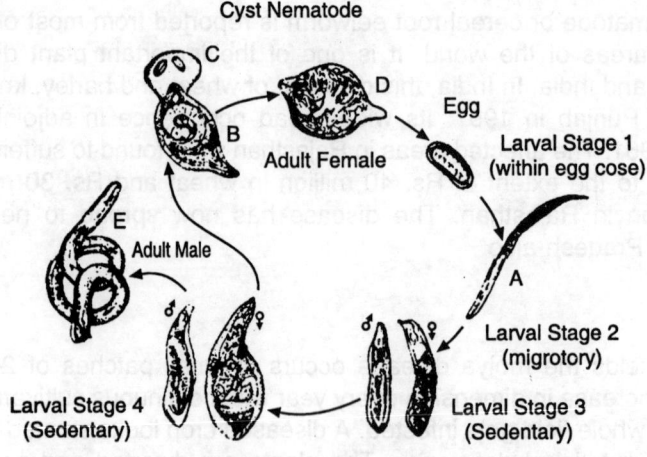

Fig. 151. Life cycle of a cyst nematode.

After penetration the larvae move towards the stele where they induce development of syncytium. The larvae grow rapidly and undergo 3 molts within the infected root. Sexes can be differentiated in the third stage (after second molt). Vermiform males develop within 3 weeks after invasion. Within a short time a fourth molt occurs. The fully developed fifth stage adult male, measuring 1070–1650 x 28–43 μm, leaves the root, wanders in soil for some time and then dies.

The third stage female larva develops 6–9 weeks after invasion. It is somewhat stouter than the second stage. Elongation of ovaries occurs in this stage. In the fourth stage (after third molt) flask-shaped female larvae, the ovaries almost completely fill the body cavity obliterating other structures. The fifth stage adult female larva grows into a typical lemon-shaped structure. The stylet of the female measures 26–32 μm long. A gelatinous matrix, usually covered with soil particles, surrounds the posterior end of the female body but eggs are not passed in to this matrix. Soon the fully formed eggs completely fill the body cavity. A conspicuously subcrystalline layer, secreted over the surface of the adult female, makes it look white but with maturity of the crop this layer is lost and the cyst looks chestnut brown. The cysts get separated from the roots and fall into the soil while still containing the eggs within. Although fully developed, these eggs do not hatch immediately. The intervening period between two crop seasons serves as a maturation period for the eggs and hatching of larvae. Thus, there is only one generation of the nematode in a year. The fallen cysts are the main source of primary inoculum for the host crop in the next season.

When temperatures start falling at the time of sowing of the next wheat or barley crop the infective second stage juveniles start emerging from the cysts. The peak emergence usually takes place by sixth week after sowing of the new crop. Best emergence of larvae occurs as a result of rise in temperature (day time) after a period of low temperature (night time). These larvae infect the roots to start the new life cycle.

Host range and pathotypes

The host range of *H. avenae* is confined to the plant family Gramineae except one single legume, *Senebiera pinnatifolia*, that grows in wheat and barley fields in Rajasthan. The hosts in Gramineae include wheat, barley, oats, rye, pearl millet, maize, and a large number of grasses. However, the degree of susceptibility (damage) and host efficiency (level of infection) differs. In India, wheat is more efficient than barley for increasing the cyst content of the soil. In other countries, oats and barley are more efficient hosts, carrying the highest infection level. The nematode can complete its life cycle in maize but many of the larvae that penetrate the roots die soon after entry as a result of root necrosis and starvation. Several pathotypes of the nematode on the basis of host preference are known to occur. Two races were identified in Denmark, Sweden and Ireland, three in England, four in Australia and four in Netherland and Germany. Five biotypes were identified in Rajasthan which were different from those reported in above countries. Biotype 3, which infects barley variety Herta and wheat line RS 31–1, is the most common biotype in Rajasthan.

Environmental relations

Active dispersal of *H. avenae* is very limited. The nematode spreads through soil very slowly unless its spread is promoted by external factors such as rainfall, runoff, wind and dust storms. Turbulent dust storms after wheat harvest in India are responsible for distribution of the nematode. Transportation of contaminated tools and infected culled plants are also responsible for passive dispersal of the nematode. The pathogen might have been distributed from its native Europe to other parts of the world in soil clods mixed with seed material.

Viability of cysts stored at 30°C is significantly reduced and after 6 months no viable content in the cysts is found. A highly significant correlation exists between the cyst numbers and soil pH. The cysts are susceptible to desiccation and are not long lived in dry soils. Very few larvae emerge from cysts which have been allowed to dry in the laboratory. Cyst contents do not survive if the soil has been air dried for 6–15 months but they do survive in plenty in undried soil or in soil that has been air dried only for 1–3 months. European populations of the nematode are not as resistant to dry conditions as Indian and Australian populations.

Hatching of larvae occurs as a result of rise in temperature after a period of low temperature. Under Indian conditions maximum emergence of larvae from cysts is reported to be at a constant temperature of 20°–22°C. At temperatures of 5° and 30°C practically there is no larval emergence. Although it is generally believed that root exudates do not influence larval hatch, root exudates of one week old barley seedlings stimulate hatch. Well aerated cysts hatch better. Under Indian conditions, maximum emergence of larvae occurs during November to January. Migration of larvae in the soil is facilitated by light textured soils having good moisture, aeration and drainage. Consequently, more damage to the crop occurs in well aerated loose soils than in poorly aerated compact soils. In adequately irrigated fields more number of larvae are required to cause damage than in poorly irrigated fields. Continuous cultivation of cereals in the same field promotes build up of cyst content year after year especially when the soil is not exposed to drying for more than 4 months. The population of the nematode in the field increase 5-fold by growing barley but can be decreased by certain grasses. The decrease in population after fallow or in the presence of non-host crops is about 60% in a year. However, the population remaining in the field after 2 years of fallow is capable of producing a large number of new cysts when the susceptible host crop is grown.

Management of the disease

Fallows and crop rotation: Since the cereal cyst nematode is very susceptible to desiccation, there can be considerable reduction in its population in soil if the infested field is kept fallow during the summer months and ploughed 2–3 times. This exposes the cysts and larvae to sun and hot summer winds thus reducing the nematode population. The advantage of summer heat can be obtained by soil solarization where practical. *H. avenae* is completely controlled by solarization at 90 cm or more inward from the edges of the plastic mulched plots but there is decreasing effect toward the edges.

Since the nematode is highly host specific (confined to wheat, barley, oats and rye), long rotations (at least 4 years) avoiding cereals, maize and millets definitely minimize or completely starve out the nematodes from the soil. In the presence of a susceptible crop the nematodes multiply 5–6 times and the population increases by 250–335% over the initial population. When a non-host crop is planted there is about 60% decrease in nematode population. If a non-host crop like chickpea or mustard is taken for 2–3 successive years before cereals are planted a profitable cereal crop can be raised in the field. The rotations can be more effective if summer fallow, light watering, and ploughing as stated above are also adopted. In Australia, the rotation fallow/wheat/ legume pasture is widely practiced.

As early as 1960, it had been demonstrated that green manuring of infested soil with chopped cabbage leaves results in heavy reduction of larvae on roots and causes significant increase in the height of plants. Green manuring increases the activity of nematode trapping fungi such as *Arthrobotrys oligospora* and *Dactylella thaumasia* in the soil if they are initially present. In addition, the decomposition products of the glucosinolates (sinigrin) in crucifer leaves (isothicyanates) are also detrimental to nematode activity. Inter- and mixed cropping with mustard may, therefore, be useful in managing the cereal cyst nematode Chopped leaves of margosa (neem) also reduce the nematode population in soil. Subject to economy of the farm, all possible steps should be taken to increase soil fertility, especially its humus content.

Resistant varieties: None of the wheat varieties have been found resistant to cyst nematode in India. However, in barley resistance is reported from many countries. Barley lines CI 3902 and Martin 403–2 possess high degree of resistance. BP 263, C 164, and BP 264 also are highly resistant. In these varieties the ratio of females to males significantly increases (1 : 106, 1 : 119, and 1 : 86, respectively). This results in lower reproduction and reduction in population. The nematodes penetrate the resistant and susceptible plants equally well. Later, differences appear in the development of the nurse cell system, In susceptible plants the nurse cell system (synchytium) develops in 4 days but in resistant plants it takes 14 days. In the latter case, the synchytia show vacuolation and degenration within a few days. In resistant cultivars, penetration by nematodes is followed by increased activity of peroxidase, esterase and superoxide dismutase. Thus, only few cysts develop in resistant wheat Festuga and resistant barley BP 263 and C 164. No cysts develop in barley variety BP 264. Field trials have shown that growing of a resistant barley variety in one season reduces the cyst population to a level where a normal wheat crop can be raised in the next season.

The resistance inducer compound DL-β-amino-*n*- butyric acid (BABA) is reported to reduce number of cysts on wheat and barley roots (Oka and Cohen, 2001). Foliar sprays of 80 mg/

L, BABA reduced the cyct numbers by 90%. Multiple sprays at 10-day intervals were effective at 2000 mg/L concentration. A soil drench of wheat with 125 mg/L BABA reduced *H. avenae* cysts by 43%. More adults were produced in the treated plants and high concentrations of the chemical inhibited development of adult males and females.

Chemical control: Soil fumigation with volatile chemicals or treatment with non-fumigant nematicides achieve only partial control of the nematodes in soil because eggs and larvae are protected inside the cysts. When volatile fumigants had been in use D-D fumigation could destroy 90% of the larvae but the remaining population could multiply rapidly in presence of a susceptible crop. In absence of soil fumigants, systemic non-fumigant nematicides have been tried. Bhatti *et al.,* (1981) had reported that early November sowing and application of aldicarb (2 kg a.i/ha) results in significant yield increase. Soil drenching with phosphonic acid containing 0.63 mg of phosphite per planmt reduces the number of *H. avenae* females. Potassium phosphate is not effecyive (Oka *et al.,* 2007).

Biological control: Some fungi such as *Verticillium chlamydosporum, Nematophthora gynophyla, Paecilomyces lilacinus, Catenaria auxiliaries, C. Vermicola, Hirsutella rhossiliensis* and *H. minnesotensis* parasitize eggs, larvae and cysts of *H. avenae*. However, parasitization actually begins with the females. Some female nematodes, even the young ones, containing no eggs, are colonized by fungi which colonize both roots and nematodes In early stages of infection, *Verticillium chlamydosporium* develops within the female nematodes between the organs, presumably using the female's body as a source of food. In immature females the uterus is destroyed. In old femalesm appressoria of *Verticillium* penetrate the eggs they contain and progressively destroy their contents until the egg shell is filled with hyphae and spopres. Infection of second stage juveniles within the egg, if formed, is rare. Newly formed cysts also are not commonly infected. Their infrection comes from the females. In some crops, soil suppressiveness develops after monocropping for several years and enhanced populations of *Hirsutella rhossiliensis* are associated with suppressiveness. The bacterium *Bacillus* (*Pasteuria*) *penetrans* is the most promising biocontrol agent. However, the use of such biocontrol agents on a field scale is yet to be demonstrated. As mentioned above, the populations of nematode trapping fungi, if present in the soil, can be increased by green manuring with suitable crops. The presence of such fungal enemies of nemarodes induce suppressive in soil against the nematodes. In beet cyst nematode, *Heterodera schachtii*, introduction of *Dactylella oviparasitica* and *Fusarium oxysporum* is reported to enhance suppressiveness.

POTATO CYST NEMATODES (GOLDEN NEMATODES OF POTATO)

At least 135 species of nematodes belonging to 45 genera are reported on potato from all over the world. In India, about 88 species are found on or around potato plants of which 22 are of common occurrence. Although the root knot nematode (*Meloidogyne*) was the first nematode parasite of potato reported in 1951, it was the golden nematode or the cyst nematode (*Globodera rostochiensis*), introduced in to the Nilgiri hills in 1961 from the U.K, that attracted attention of Indian scientists towards the seriousness of these plant pathogens. Of the 22 commonly occurring nematodes in potato fields, only the root knot nematodes and the cyst nematodes are considered most destructive. The dangerous nature of the cyst nematode of potato lies in the fact that once introduced in a field it is very difficult to eradicate it.

The golden nematode was first discovered in Germany in 1881 and has gradually spread to most of the cool climate countries along with potato tubers. In India, it has been kept confined to the Nilgiri and surrounding areas since its discovery in 1961 by strict quarantine measures. The same nematode can attack tomato, eggplant, and many species of *Solanum* and *Lycopersicon* but with lesser aggressiveness than in potato.

Symptoms

Visible and definite signs of cyst nematode injury become detectable in a field only when the cyst content of the soil has gone very high. In the first few crop seasons after entry of the nematodes in a field the disease goes undetected while the population of the cysts continues to rise. When the population of nematodes in soil is sufficiently high small patches of poorly growing plants may appear in the field. The plants appear as if suffering from poor nutrition. Temporary wilting of plants occurs during hotter part of the day. Typical symptoms of heavy infestation are stunted growth with unhealthy foliage, premature yellowing, poor development of the root system, and reduction in size and number of tubers. In early years when soil infestation is low there may not be very distinct symptoms on plants but the yield goes on decreasing. The development of lateral roots is often excessive. When several crops of potato are taken in the same field in the same year (as in the Nilgiri area) the patches of sick plants increase in dimension. The nematode feeds on the roots and heavily infested plants die early.

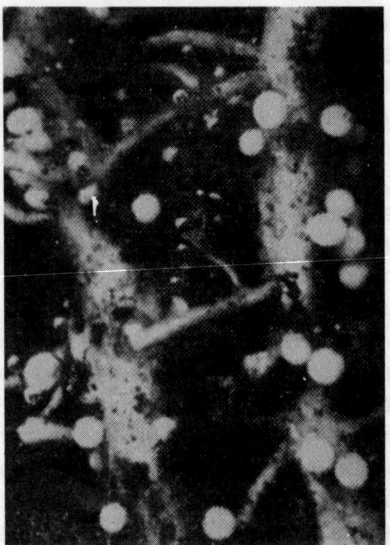

Fig. 152. Cysts of *Globodera rostochiensis* on potato roots.

The presence of golden nematode can be verified only by careful examination of soil and roots in the laboratory. If the roots of plants reaching blossoming age are examined under a magnifying glass, they show presence of white, yellow or golden females of the nematode adhering to the roots. As the season advances these females, which are full of eggs, turn reddish brown. Upon death, the body of the females becomes a protective cyst with the eggs inside and also the larvae if the eggs have hatched. These cysts persist in soil for several years and can be detected in absence of the host by proper screening of the soil.

The causal organisms

Two species of *Globodera*, *G. rostochiensis* (Wollen.,1923) Behrens,1975 and *G. pallida* (Stone,1973) Behrens, 1975 attack potato. Hybridization between the two occurs in nature. Earlier, these species were known as *Heterodera rostochiensis* Wollen. and *H. pallida* Stone. Prior to 1972, *H. pallida* was considered a pathotype of *H. rostochiens*. Both species are present in the infested areas of South India. *G. rostochiensis* predominates in cooler climate (61–91%) while *G. pallida* populations are abundant in warmer climate (89–100%). In the Nilgiri area of South India *G. pallida* is more important than *G. rostochiensis*. The pathotype Pa2 is most prevalent.

The cysts attached to roots or present in soil are very small, about the size of pinhead. Each cyst contains as many as 500 eggs in both species. Some of the eggs hatch within the cyst. At temperatures above 15°C and in presence of host root exudates (that contain the hatching factors), hatching of eggs occurs and the second stage larvae are released from the cysts. The level of hatching factors is more in root exudates of plants grown from tubers with their natural microflora than plants raised from tubers without natural microflora. In addition to the role of root exudates in stimulating hatch, in *G. rostochiensis,* the components of the cysts also provide some stimulation. Compared to sterile distilled water, hatching of custs is increased 5-fold by homogenates of cyst wall or rinsates of either cyst wall or of eggs. Cyst or egg homogenates increase larval hatch by 7.7–9.0-fold which is equivalent to 24.8–29.0 % of hatch in potato root exudates (Pridannikov *et al.,* 2007).

All the eggs do not hatch in the first year. Some are retained in the cyst and larval emergence occurs in the second, third, or fourth year, or even later. This explains the persistence of the nematode in soil once introduced. Hatching of cysts in *G. pallida* in absence of the host roots is very much slower than in *G. rostochiensis*. In this species, there may be spontaneous hatching and death of larvae within eggs when the rotation does no include potato.

Under the influence of root exudates the larvae move actively and migrate towards roots causing infection. There may be some role of ethylene produced by the host plant roots in attraction of larvae. In *Heterodera schachtii* (sugarbeet cyst nematode) it has been demonstrated that cultivars that overproduce ethylene are hypersusceptible to larvae of the nematode that are attracted by root exudates. Entry of the larvae in to the host is through roots

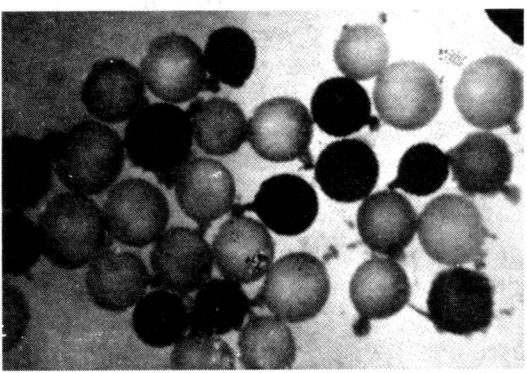

Fig. 153. Change in colour of cysts of *Globodera*.

(just behind the growing point or the laterals), stolons and tubers. From this point the juveniles move up and down. Chemical signals are exchanged with the host and the nematode juvenile (J2) selects a single cell for a initiating the development of the feeding site (syncytium). The larvae lie parallel to the axis of vascular bundles. The second stage juveniles that penetrate the pericycle cells of the plant are more likely to become males whereas the larvae that penetrate procambial cells tend to become females. Within a few hours of settling the larvae probe the cells. They insert the stylet into the cell while remaining motionless for several hours. The stylet is withdrawn and re-inserted into the same cell. A secretory product from the oesophageal glands is injected. This triggers the initial syncytial cell to develop into a full syncytium. In resistant plants, the juveniles may try to form a syncytial site but walls of the cells involved are thickened and cells may die. This starves the larvae. The development of syncytium involves lysis of inner cell walls and cell wall ingrowth formation next to the vessels. The nematodes draw their nutrition from the these syncytia. Induction and morphogenesis of these nurse cells are mediated by auxins (Groverse *et al.,* 2000) and gene expression changes in plant roots (Hermsmeier *et al.,* 2000).

In the early stages the males and females are cylindrical but after molting the females swell while the males remain vermiform. These males do not remain inside the roots but come out and are found in soil around the infected roots. They do not feed. Their only function is to fertilize as many females as possible before death. The females swell to become lemon-shaped. On full maturity the females break through the host surface as pale, transparent bodies which soon become opaque and pearly white. At this stage they are fertilized by the males. The color changes to golden or reddish brown. The female dies and its body becomes a cyst which contains a large number of fertilized eggs. The second stage larvae take about 37–39 days during summer (April–June) and 40–42 days during autumn (September–December) to become cysts. The multiplication rate is 7–13 times and 6–11 times during these two seasons, respectively, when the temperatures are between 15° and 21°C.

Disease cycle and environmental relations

Main source of survival of these nematodes is the cyst. The cysts are usually spread along with soil adhering to farm implements, harvested tubers, gunny bags used for transporting tubers, or anything that is contaminated with infested soil. Surface drainage water, compost, shoes and feet of workers, and wind-borne dry soil are important means of spread. When a susceptible crop is planted some hatching factor from the roots induces release of larvae. Hatching and emigration of larvae takes place actively at about 20°C. In-soil hatch of *G. pallida* is delayed compared to that of *G. rostochiensis*. Mycorrhizal inoculation of potato plants stimulates production of *G. pallida*-selective hatching factors or hatching factor stimulants (Ryan *et al.,* 2000). Subsequently, Ryan and Jones (2004) noted that potato root leachates from mycorrhizal plants had significantly more hatching factor stimulanmts active towards both species of *Globodera* than from non-mycorrhizal plants. Mycorrhization had no effect on hatching inhibitors. Host penetration and infection also are optimum at this temperature while at 13°C and 24°C these activities are inhibited. Hatching stimulants in root exudates vary according to host genotype. Some stimulate hatching of *G. rostochiensis* and others of *G. pallida*. This difference has implications in distribution of the two species. At 30 days after planting potato in a soil containing second and third stage juveniles on roots, there is sharp reduction in photosynthesis and transpiration of the plant. These effects become less pronounced with passage of time. There is a close relationship between the number of juveniles in the below-ground stem and the intensity of stolon infection by the stem canker and black scurf pathogen *Rhizoctonia solani* (Back *et al.,* 2006). The response of the potato plant to cyst nematode infection is expressed in the leaves as increased activity of defence related enzymes (beta-1,3-glucanases and beta-D- glucosidases).

Management of the disease

Regulatory measures: In countries where the potato cyst nematodes are established in the fields regulatory measures like quarantine and seed certification have little value. In countries where the nematode is not prevalent very strict regulations are followed to prevent its entry.

In India, where these nematodes are endemic to one particular region quarantine is of utmost importance. There should not be any movement of seed tubers and soil adhering to tubers and containers from this area to outside. Although environmental conditions may not favor establishment of the nematode in the crop in many areas, this precaution is essential. Not only this, some type of restriction is necessary locally wherein vegetable seedlings raised in infested soil should not be allowed to go to fields which are free from the cyst nematodes.

Cultural methods: The fact that viability of eggs in the cysts is maintained for several years and all eggs do not hatch in the first or second year requires application of cultural practices that permit the cysts to exhaust their contents before a solanaceous crop is taken in the field. While sanitation is a useful practice since it can reduce, to some extent, the cyst load by removal of diseased plant debris, suitable crop rotation is the most important cultural practice. Depending on the cyst content of the soil the length of rotation can be 2–8 years. No solanaceous vegetable should be grown during this period and the field should not contain any solanaceous weed. Non-solanaceous vegetables such as pea, cabbage, cauliflower, carrot, French bean, etc. help in reducing the cyst content of the soil. By following a 4-year rotation of potato and legumes like bean and pea the larval populations declines by 50–70% in the first year, 81–90% in the second year and 91–95% in the third year. Garlic followed by fallow, radish, cabbage, pea, French bean, and carrot also gives high degree of reduction in the inoculum level. There is significant reduction in the cyst population when 4–5 consecutive crops of barley are grown in the infested fields.

Trap cropping has been used successfully for the reduction of cyst nematode population in soil. Potatoes are grown in order to hatch the larvae. These are given time to infect roots and develop into young adults. The crop is destroyed 6–7 weeks after planting before new cysts are formed. About 80% reduction in cyst nematodes is reported. If crop destruction is delayed, the nematode density will increase. Scholte (2000a, b) considered use of potato as a trap crop not advisable. Instead, he recommended use of non-tuber bearing *Solanum sisymbriifolium* and *Solanum nigrum* as trap crop. However, these two are natural alternate hosts of *Phytophthora infestans* and may help in epidemic of late blight. Reduction in populations of the potato cyst nematodes by these trap crops at different soil depths depends on the length of growth duration and root length density. Soil solarization in very hot climates is also a good method. Sixty two days of solarization reduces the nematode population by 95%.

In the Nilgiri hills following measures had been suggested for preventing spread of the cyst nematode:

(i) Potato grown in infested field should not be used for seed.
(ii) Potato, tomato and eggplant should not be grown in the infested fields for 3–4 year.
(iii) Volunteer potato plants growing in the fields should be removed.
(iv) Vegetable nurseries and other transplantable plants should not be raised in infested fields.
(v) Movement of water and soil from infested fields should be avoided.
(vi) Crop rotation should be followed even in uninfested fields so that solanaceous crops are grown only once in 3–4 years.

Chemical control: Soil fumigants such as D-D, DBCP, Nemaphos etc are effective in reducing nematode populations in the soil. D-D applied at the rate of 500–1000 L/ha reduces nematode population by about 60% with concurrent increase in yield by 110–200% in the Nilgiri hills. In a massive programme of eradication of the cyst nematode the entire infested area was treated with Dasanit (phensulfothion), a granular nematicide which remains effective in soil for about 3 months at 20°C. The rate of application was 300 kg/ha in the first year and 150 kg/ha in subsequent years. Although the results were encouraging, lack of follow up action such as use of crop rotation, allowed the surviving nematode populations to multiply and again reach the destructive level. Other systemic nematicides/ insectcides which have been used are aldicarb, carbofuran, ethoprop, phorate, etc. Aldicarb (Temik) applied as spot treatment at the rate of 2 kg a.i./ha is most economical and effective. Minnis *et al.,* (2004) have suggested that best returns are obtained by combined use of a fumigant (1, 3-dichloropropene) in autmn and granular nematicide application in sprin.

Resistant varieties: High degree of resistance to cyst nematode exists in many wild species of *Solanum* and these have provided source of resistance genes in attempts to develop resistant cultivated varieties. Castelli *et al.,* (2003) tested 198 accessions from the Commonwealth Potato Collection which represented 63 different species. Of the accessions tested 56% contained *G. pallida*-resistant clones and 53% contained *G. rostochiensis*-resistant clones. While resistance to *G. rostochiensis* was common in accessions from South America, resistance to *G. pallida* was present across the geographic range of tuber-bearing *Solanum*. The wild species *Solanum sparsiphilum* possesses genes that express hypersensitive response to the invading larvae and halt their development into adult femalers. In India, many of the clones obtained from crosses of cultivated potato with wild species showed resistance but when exposed to field populations of the nematodes' they turned out to be susceptible because of the existence of two species and several pathotypes in each. A cross between a resistant clone obtained from Netherlands and Kufri Jyoti had given a resistant hybrid with desirable yield characters. This hybrid was later released as cv. Kufri Swarna in the year 1985. Transgenic resistance to the cyst nematode (*G. pallida*) can be effective under field conditions and plants in which a cysteine protease inhibitor is expressed are reported. However, food and environmental safety must be ensured before any practical application of the technique. Some clones of potato genetically transformed for resistance to late blight have shown moderate resistance to *H. rostochiensis.* The control of *G. pallida* is more difficult than that of *G. rostochiensis* because of its slow response to hatching stimulants and resistance to nematicides Resistance to both species of *Globodera* occurs in *Solanum vernei.* Cultivated potato possessing ther genes from this species show polygenic resistance to the cyst nematode.

Biological control: Attempts to develop biocontrol methods for potato cyst nematodes have been continuing since 1930 but so far no specific practical method has been developed. Many fungal species are natural enemies of the cyst nematode. These include *Acremonium, Cladosporium herbarum, Cylindrocarpon destrucans, Cylindrocarpon graciles, Fusarium oxysporum, Gliocladium roseum, Nectria radicicola, Paecilomyces lilacinus, Ulocladium atrum, Arthrobotrys* sp., *Colletotrichum coccodes* and *Verticillium chlamydosporum.* Most studies on biological control have involved some of these fungi, especially *C. destructans,* and *Pasteuria* (bacterium). *C. destructans* can reduce cysts by upto 76%. Jacobs *et al.,* (2003) tested the nematophagous fungi *Paecilomyces lilacinus Plectosphaerella cucumerina* and *Pochonia chlamydosporia* for their competitive ability against soil fungi including *R. solani* and *F. oxysporum,* tolerance to chemical pesticides and efficacy to control the potato cyst nematodes. *P. lilacinus* showed the greatest potential for use in combination with selected fungicides and nematicides for the control of the cyst nematodes. It also reduced symptoms of Rhizoctonia disease on potato stems when used for seed tuber treatnment. The chitinase-producing bacteria *Stenotrophomonas maltophilia* and *Chromobacterium* sp. suppress egg hatch of *G. rostochiensis* and reduce infection.

Pochonia chlamydosporia, the biocontrol agemtr of root-knot nematodes (*Meloidogyne* spp.) colonizes potato roots and, alone or in combination with a nematicides, reduces the multiplication rate of both potato cyst nematodes ((Tobin *et al.,* 2008).The endoparasitic nematophagous fungus *Hirsutella rhossiliensis* at 10^5 spore/g soil and *H. minesotensis* are reported to cause up to 24% reduction in the number of juveniles that penetrate the roots. These nematophagous fungi produce chitinolytic enzymes The endoparasites parasitize cysts of *Heterodera* and *Globodera* more than the eggs of *Meloidogyne* spp. *Hirsutella thompsonii* var. *thomsoni* is known to produce the insecticidal toxin hirsutellin A. In a study of population biology and biocontrol efficiency of *Hirsutella rhossiliensis* it was reported in 1992 that:

(i) parasitism of nematodes by the endoparasite is dependent on nematode density,

(ii) local populations of the fungus become extinct unless supplied with some minimum number of nematodes (the host threshold density), and

(iii) natural epidemics of the fungus in populations of nematodes develop slowly and only after long periods of high nematode density.

Mycorrhization of potato roots alters the nature of root exudates and hatching factor therein. This results in some adverse effect on the development of the populayion of *G. rostochiensis* and *G. pallida*.

In vitro or in-soil exposure of cysts of G. rostochiensis to Pseudomonas fluorescens that produces 2, 4-diacetylfluroglucinol is reported to double the ability of eggs to hatch but the percentage of mobile juveniles was reduced three-fold. In a study of effect of temperature on the biocontrol efficacy of *Pseudomonas oryzihabitans* against the potato cyst nematode, Andreoglou et al., (2003) observed that the bacterium significantly reduced invasion of potato roots by second stage juveniles of *G. rostochiensis* at 25° and 21°C. The biocontrol agent moved faster in wet than in dry soils. Organic amendment of soil with cowdung manure or farmyard manure induces tolerance to the cyst nematode. There is restricted development of the nematode in the tissues. At least 16 bacterial strains have been isolated from roots and cysts that cause more than 25% reduction in penetration potato roots by *G. pallida*. Seed tuber inoculation with some isolates in 0.2% methyl cellulose increased adhesion of the antagonists to tubers and promoted antagonistic activity and disease control. Living and heat-killed cells of the rhizobacterium *Rhizobium etli* (strain G12) induce systemic resistance in potato roots against invasion of *G. pallida*. The bacterial surface carbohydrates appear to be responsible. When exopolysaccharides (EPS) and lipopolysaccharides (LPS) extracted from *Rhizobium etli* G12 were applied to potato roots, the LPs induced resistance to infection of *G. pallida* while EPS was not effective.

A novel biological nematicide, DiTera (Valent Biosciences Corp, USA), has been successfully used against many plant parasitic nematodes. In potato cyst nematodes, it was specifically linked to the permeability of the eggshell membrane. Solutions of 1–10% inhibit the egg hatch. When cysts are exposed to it for 5 weeks, hatching is prevented and is irreversible. It interferes with the specific hatching mechanism in both species (Twomey et al., 2000). The effect on second stage juveniles appears as inhibition of activity and movement. Exposure to 1% of the formulation causes significant reduction of stylet thrust.(Twomey et al., (2002).

ROOT KNOT OF VEGETABLE CROPS

Root knot nematodes attack a large number of vegetable crops in India and elsewhere and cause considerable damage independently or in association with fungi and bacteria causing wilt and root rot, provided the fungal or bacterial pathogens contact the host roots after nematodes have invaded them. Simultaneous inoculation of nematodes and fungal pathogens or inoculation of fungus before nematodes does not enhance wilt due to nematode infection. Instead, in such situations the nematode populations may be decreased (Akinsanmi and Adelkunle, 2003). The hosts of the major species of root knot nematodes include at least 1700 plant species. These hosts include such common vegetables as cucurbits, potato, tomato, brinjal (eggplant, aubergine), chilli (pepper), okra (bhindi), carrot, radish, beans and colocasia. In addition, an extensive list of weed hosts covering a number of plant families has been reported from time to time in this country. Estimated crop losses range from 5 to 43% depending on the crop, nematode species prevalent in the area, and environmental factors. The reported yield losses in the tropics are 34–46% (tomato), 17–29% (eggplant), 42% (okra), 11–27% (beans), 22% (pumpkin), 33% (cucumber), 10–43% (cowpea), 38% (carrot), 26% (cabbage), and 10–20% (spinach). Losses to the extent of 31–62% in okra, 26–73% in tomato, 75% in eggplant and almost complete loss in cucurbits have been reported in India also.

Symptoms

The characteristic symptom of root knot is the formation of galls on the roots. Since these galls are underground the symptoms that appear on the aerial parts are mostly those that can be induced by any other cause including nutritional deficiency. In potato, the plants may be stunted, sick looking, and may show signs of premature and sudden drying. Often the attack of nematodes is associated with bacterial wilt and the premature sudden drying is due to the bacterial infection. The root knot nematode by itself does not usually kill the plants. The nematode infection of the tubers appears as tiny tubercles but heavy and localized infection stimulates cell division leading to gall formation. Females of the species may be seen as

Fig. 154. Root galls caused by *M. javanica*.

glistening white bodies embedded in the potato tuber tissues. The gall bearing tubers may undergo rotting due to invasion of other saprophytic organisms through the wounds caused by the nematodes.

In tomato, the symptoms are somewhat more pronounced. The plants show poor development and stunted growth if the infection is early and severe. The leaves are yellowish green to yellow. They tend to drooping followed by sudden wilting if wilt causing fungi and bacteria are associated. Sometimes, there is scorching of the leaves from the margins inward. The main root and the laterals bear spherical to elongated galls which are very small if *M. hapla* is associated otherwise all other species cause development of large galls of various size and shape. The presence of these galls is the most characteristic symptom on tomato, okra, eggplant, cucurbits, etc. In advanced stages the entire root system, whatever is left of it, is crowded with small and large galls. The volume of root system is considerably reduced if infection has occurred in early stages of plant growth and the plants have to obtain water and nutrients from only a limited volume of soil. The vascular elements are also broken causing disturbance in the translocation of whatever water and nutrients have been absorbed by roots. This shows in the aerial parts as symptoms of water and nutrient deficiency.

The causal organisms-Morphology and life cycle

The common species causing root knot in India are *Meloidogyne incognita, M. javanica,* and *M. arenaria.* Together with *M. hapla* they constitute 95% of the total root knot nematode species. These nematodes are sedentary endo-parasites of underground parts of the host plant. Although eggs and second stage juveniles may be present in the soil the active phase of the life cycle is always associated with the host.

The life cycle starts from the eggs deposited by the females within the host tissues or outside in a gelatinous matrix. These eggs are ellipsoid and measure 67–128 x 30–52 μm. The mature female laying these eggs is completely or partially embedded in the root tissues. The gelatinous matrix holds the eggs together in an egg sac the size of which can vary according to number of eggs contained. About 400–500 eggs are present in the sac but the number may go up to 1000 or even more. Then the size of the sac may be larger than the body of the

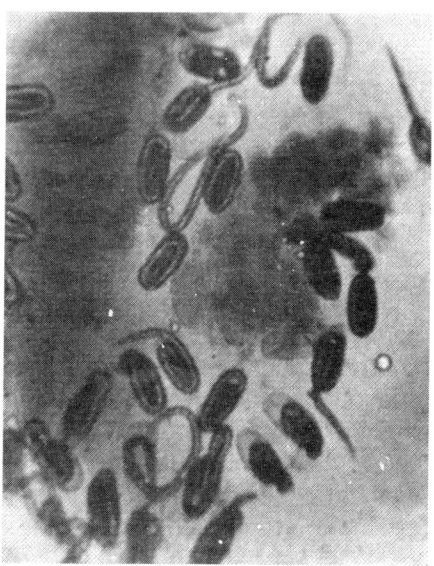

Fig. 155. Eggs of *M. javanica.*

female. After embryonic development a fully formed larva is formed in the egg. This first stage larva is highly resistant to dry conditions and can move within the egg. After the first molt, which occurs within the egg, second stage larva (J2) with well developed stylet is formed. Under suitable environmental conditions of temperature, moisture and osmotic pressure the eggs hatch and release the second stage larvae (J2) which are slender, small (0.3 to 0.45 mm long), with a hyaline conoid tail. These infective larvae move very slowly, without any particular direction and without feeding, in the soil. In absence of the host plant, migration of these larvae is influenced by soil pore size. Migration is greater in tilled than in non-tilled soil. Tilled soil has greater volume of pores with diameter > 100 microns. However, the speed of migration decreases as the pore size increases from 40 to 160 microns. (Eco *et al.,* 2007). When within few centimeters of a host root they are attracted by it and reach the root surface to start the parasitic life. These larvae can be found up to a depth of 2.5 meters in sandy soils. In root knot nematodes, the parasitism genes are expressed in the esophageal gland cells and encode proteins (enzymes) that are secreted into host root cells.

The second stage juveniles (J2) penetrate the roots and move mostly between the undifferentiated root cells. The genes and gene products related to parasitism are located in almost all organs of the nematode. After reaching the endodermis they come to rest with the head in the stele and the body in the cortex. Soon they become sedentary and feeding is confined to a few cells around the head. Parasitic J2 of *Meloidogyne* induces the formation of 3–6 multinucleate giant cells from individual root xylem parenchyma cells surrounding the nematode head. This serves as the feeding site. During feeding the nematode injects its secretions into the host cells and pericycle and cortical root cells immediately surrounding the giant cells are stimulated to divide (hyperplasia), giving rise to the galls. Oesophageal secretions play key role during parasitism from post-penetration migratory stages of juveniles and during formation of feeding site by the adults The genes for parasitism and expression of their products have been found in dorsal and subventral oesophageal glants, intestine, amphidial and rectal glands and in the hypodermis. Cell wall digesting enzymes in the secretions help the migratory juveniles in movement. In the nematode feeding site also the enzymes have been shown to be present at the tip of the stylet.

Initiation of this nurse cell system involves both host genes (susceptibility genes) and nematode genes. The root knot nematode seems to have adapted a part of the root nodule bacterial symbiosis pathway (the Nod factor or NF). It enhances its parasitic ability by producing a molecule with functional equivalence to NF (Weerasinghe *et al.,* 2005). This results in formation of giant cells which serve as the food tank for the nematode. Usually 4–6 giant cells are formed at each feeding site (per adult female). The number and area of giant cells do not vary among species of *Meloidogyne*. Cytoplasm in giant cells is aggregated alongside the thickened cell walls. Stellar tissues within galls appear disintegrated.

While the giant cells are forming and the nematode is drawing nutrients from the host, its body width increases and structural changes within the body also occur. After the second and third molts in the host tissue the fourth stage larva is formed. The sexes can be differentiated at this stage. The fifth stage adult males formed after the fourth molt measure 1.2–1.5 mm in length and 30–35 μm in diameter. They remain attached to the giant cells for some time and then emerge out of the root, wander in soil and soon die. The fourth stage female continues to grow in size and undergoes the fourth molt to become the adult females. These mature females are pyriform or saccate and measure 0.4–1.3 x 0.25–0.75 mm. The reproductive system consists of two ovaries. Fertilization of eggs is not essential and eggs develop partheno-genetically. The egg laying is completed in a period of about a week. The eggs hatch freely in water. When the eggs are laid within the host tissues the infective second stage larvae released by them move within the root tissues to establish new infections and form secondary galls. The larvae hatched in soil move in search of roots or fresh locations on the same root if the crop season is not over.

The length of life cycle and number of generations completed by these nematodes in one crop season depends to a great extent on temperature. Optimum temperature range for the species common in India is about 25°–30°C. There is very little activity by any species of *Meloidogyne* above 40°C and below 5°C. High temperatures of 40°– 50°C kill the larvae. The life cycle of *M. javanica* is completed in 21–25 days at 26°– 27°C and in 50–60 days or even 80 days at 14°–16°C. Thus, during winter in north India the incidence of root knot and the damage caused by it is low because the nematodes cannot produce many generations. Roots can be infected at any temperature from 12° to 35°C but little disease develops below 16°C. In general temperatures of 25°–28°C and light textured soil are best for rapid multiplication, larval movement, infection and increased size of galls.

Several races exist within some of these species. *M. incognita* has four races (1, 2, 3 and 4). All the four races are prevalent in U.P. and Haryana. In Assam and Karnataka races 1, 2 and 3 are present. In Andhra Pradesh, Delhi and Chandigarh only race 1 is present while in Bihar races 1 and 3 and in Himachal races 2 and 4 are prevalent Variability in nematodes is mostly a mutational event and erodes the efficacy of host resistance genes.

Management of the disease

Management of root knot nematodes was mainly based on chemical control with fumigants and granular nematicides. But now emphasis is use of resistant cultivars, soil management with organic amendments, use of enemy plants, and crop rotation with resistant or non-host cultivars.

Cultural and biological control methods: It is essential that all sources of existence of the nematodes, such as weed hosts, should be eliminated from the vicinity of the vegetable fields. The soil should be given deep ploughing during summer and allowed to dry. Infested seedlings from nursery beds are a major source of introduction of the nematode into hitherto nematode free main field. Therefore, care should be taken to raise seedlings in root knot nematode free nursery beds. The beds should either be chemically treated or the upper layer of soil sterilized by burning a stack of trash before preparation of seedbeds. If necessary seedling roots should be dipped in a nematicide suspension before transplanting. A 15-min. dip in 500 ppm thionazin has been recommended.

Flooding of the soil has been found effective in eliminating root knot nematodes from the soil. Lowland rice culture preceding the vegetable crops reduces the incidence of root knot. The natural control in flooded soils is attributed to metabolic fatty acids produced by bacteria under anaerobic conditions, to hydrogen sulphide and to reduced availability of oxygen. Although, the low molecular weight organic acids (formic, acetic, propionic, butyric and valeric acids) produced by decomposition of organic matter in flooded soils cause reduction in numbers of nematodes, their concentration declines by 54–97% within 10 hours hence other factors may also be involved in suppression of the nematodes (McBride *et al.,* 2002).

Ammonia is toxic to the root knot nematodes.. Ammonia releasing fertilizers are good for suppression of nematode activity. Their efficacy is improved if some alkaline amendment is added to the soil. It causes conversion of NH_4 to NH_3 which is the nematicidal form of ammonia. Oka *et al.,* (2006) have reported that application of slaked lime in combination with chicken manure significantly reduces the root-galling index of pepper plants infected with *M. incoignita* Many amendments, such as margosa seed cake, are inhibitors of nitrification and cause accumulation of ammonia in the amended soil.

Soil solarization is very effective in control of root knot but it does not completely eliminate the nematode inoculum. Solarization by covering the soil in a field with clear transparent polythene sheet for 6 weeks during the hot summer months caused 58% reduction in population density of *Meloidogyne incognita*. Solarization showed an increase in soil temperature by 8°C and conservation of moisture (5%). Combination of nursery soil treatment, summer ploughing of the land and solarization are most effective in the control of root knot, The requirement of high temperature and long duration of treatment can be reduced by amending the sopil with decomposable organic matter before subjecting to solarization. In many studies solarization has failed to provide effective reduction in nematode populations and combination with the nematicide oxamyl has been recommended.

In economical and effective approach to the management of root knot disease, especially in the tropical and sub-tropical regions is organic amendment of soil. Decomposition of organic matter in soil was known to reduce the activity of these nematodes since 1938. In potato, green manuring with rapeseed is reported to suppress *Meloidogyne chitwoodi*. Isothiocyanates are released from decomposing crucifer residue. The lethal concentration of phenylethyl isothiocyanate for *M. javanica* is 0.35 μmol/ml (Zasada and Ferris, 2003). Green manuring with sudan grass or sorghum-sudan grass hybrids also suppresses phytonematodes in soil. Sorghum and related plants contain cyanogenic compounds which are hydrolyzed by microbial activity to release HCN, toxic to nematodes and many fungal pathogens. Incorporation of *Argemone mexicana*, an annual weed, and *Lantena camera,* a forest and roadside bush, in the soil suppresses the effects of *Meloidognne javanica* in mungbean. *M. javanica* populations in the rhizosphere and in roots, and gall formation were significantly suppressed when 10, 30 or 50 g/kg *A. mexicana* was allowed to decompose in the soil. However, plant growth was significantly suppressed by 30g/kg dose of the amendment (Shaukat *et al.,* 2002). Ajaz and Tiyagi (2003) have further studied the beneficial role of *A. mexicana*. Significant reduction was observed in the population of *M. incognita* and other phytonematodes infesting eggplant and cauliflower when given root dip treatment in leaf extract of *A. mexicana* and *Solanum xanthocarpum*. The latter was more effective. In an evaluation of amendment of soil with harvested crop residue of marigold, mustard and sunflower in the control of *Meloidogyne incognita,* the amendments proved highly effective in suppressing root knot and enhancing plant growth. Lopez-Perez *et al.,* (2005) used crop residue of broccoli, melon and tomato and chicken litter as sources of biofumigation to control *M.incohnita* in tomato. At high soil temperature of 30°C each one drastically reduce root knot. The treatment was not as effective under cool conditions.

Studies carried out by Singh (1965a, b) and Singh and Sitaramaiah (1971a, b) had shown that use of oil cakes as manure at the rate of 2500 kg/ha provided a highly effective control of root knot, improves soil fertility and resulted in many-fold increase in yield of tomato and okra. They had also reported that application of wood sawdust, at the same rate as above, followed by application of 120 kg N/ha through urea also controlled root knot. The effect of the treatment lasted in the subsequent crop also. The reduced incidence of root knot with such treatments was attributed to:

   (i) stimulation of root growth,
   (ii) stimulation of predaceous fungi, and
   (iii) production of toxic metabolites during the decomposition of the organic matter.

The mechanisms of nematode suppression by organic amendments remain complex but the basic activity is biofumigation by volatiles released during the decomposition process. Oil cakes enhance the activity of predaceous fungi that feed on nematodes (Singh, K.P. *et al.,* 2002). The population of *Catenaria anguillulae* was increased several fold when the soil was amended with oilcakes of mustard, linseed, margosa or sesamum. Soil amendment with rapeseed meal is reported to reduce number of galls on tomato roots caused by *Meloidogyne arenaria*. Quantity of edible oil cakes effective against root knot nematodes can be reduced by combining with other biofertilizers, *viz.*, azolla. Abbasi *et al.*, (2005) have confirmed the efficacy of soil amendment with neem cake in control of root knot of tomato caused by *Meloidogyne hapla*. Rahman and Somers (2005) used green manure with *Brassica juncea* (Indian nustard cv. Nemfix) and its seed meal in vineyards and obtained good control of *M. javanica* when the meal was applied in the rows not between rows. Broccoli plant residues are highly effective amendments against root knot nematodes. The role of mustard can be explained by the release of isothiocyanates during decomposition of glucosinolates. The cost of oil-cakes has increased manifold in India and the treatment becomes as costly as the use of nematicides. Oil cakes of mustard, sesamum, linseed are cattle feed. However, cheaper materials in the form of farm and household decomposable waste are available and these should be allowed to decompose in the vegetable fields. Mashela and Nthangeni (2002) have used ground castor (*Ricinus communis*) fruits instead of castor oil cake for control of *M. incognita* in tomato. In microplot experiments, Mashela (2002) has demonstrated the root knot suppressing effect of soil application of ground fruits of the wild cucumber (*Cucumis myriocarpus*). Suppression of nematode activities by amendment of soil with green or dry crop residue and ne,matode suppression by extracts of same plants are two different approaches. In crop residue amendment biofumigation resulting from residue decomposition plays the main role. Suppression of nematode in extracts of leaves, stems, etc can be due to a phytochemical plus fluid density effect. In a study reported by Javed *et al.*, (2008) when second stage juveniles of *M. javanica* were exposed to neem leaf or cake extract (phytochemical effect), the 10% concentration of leaf or cake extract caused 83% and 85% immobilization and 35% and 28% mortality of the juveniles. Exposure of egg masses to the extracts completely stopped egg hatch but return of the egg masses to water revived hatching. A coomercial neem product Aza had no such effects. Soil amendment with neem leaf or cake extract or with Aza significantly reduced root knot infection. But amendmented were not effective if nematodes had entered the roots.

The efficacy of sawdust in controlling root knot reported in early 1970s by Singh and Sitaramaiah was later confirmed by Stirling (1989), Vawdrey and Stirling (1997), and Stirling and Nikulin (1998). They obtained good control of root knot of tomato and ginger in field trials. Sawdust amended soil was almost free of galls and had the lowest populations of root knot nematodes. However, chemical treatment gave better control. Powdered pine bark has also been used for control of *Meloidogyne arenaria* on soybean. Level of control increased with increasing amount of the amendment. The treatment was also effective against the soybean cyst nematode (*Heterodera glycines*). Gall and cyst formation was completely eliminated when the material was used at 5% rate. Fungal populations were increased by powdered pine bark. *Penicillium chrysogenum* and *Paecilomyces variotii* were the predominant fungal species. Nico *et al.*, (2004) have reported effective control of *M. incognita* and *M. javanica* by amendment of potting mixes with composted agro-industrial wastes such as dry cork, dry grapes residue after extraction of juice, dry rice husk etc.

Khan *et al.*, (1997) had reported management of root knot by addition of fly ash to the soil. Fly ash treatments adversely affected root invasion by juveniles, disease intensity and reproduction of the nematodes. A gradual increase in the ash content of soil caused a corresponding decrease in the number of invading juveniles, gall and egg masses per root system. Fly ash increased the availability of carbonates, bicarbonates, sulphur, chlorides, B, P, K. Ca, Mg, Mn, Cu and Zn in the soil.

Leaf extracts of *Inula viscose*, which have been found effective against many fungal foliar pathogens, have strong nematicidal activity, particularly against the root knot nematodes.

Addition of leaf powder to sand (0.1% w/w) greatly reduces the number of second stage juveniles of *M. javanicva*. Aqueous extract of the powder are less effective than the organic solvent extracts (Oka *et al.*, 2001). Extraction of dry leaves with a mixture of acetone and n-hexane or n-hexane alone yields an oily paste. After evaporation of the solvent, the emulsifiable product can be diluted in water. A concentration of 0.01% (paste w/w) killed juveniles of *M. javanica* and reduced galling on roots (Oka *et al.*, 2006). In a greenhouse study, application of horse manure or sesamum oil cake, or *Bacillus thuringiensis* singly or in combination with oxamyl *Meloidogyne incognita* development and improved plant growth.

Mycofumigation, the antimicrobial volatiles released by some fungi (*viz.*, *Muscodor albus*) and acting as fumigants, had been found useful in prevention of postharvest decay of fruits. Recently, Riga *et al.*, (2008) have reported suppression of plant parasitic nematodes by volatiles from *M. albus*. They exposed *Meloidogyne chitwoodii*, *M. hapla* and *Pratylenchus* spp. to cultures of *M. albus* grown on oat grains Exposure for 72 hours resulted in 95% mortality of juveniles of *M. chitwoodi* and about 82% in *Pratylenchus*. Mortality in *M. hapla* was only 21% but about 69% larvae displayed reduced mobility. Soil application of grain inoculum at 0.5 or 1.0% (w/w) gave similar high suppression of the nematodes (91–100% in *Pratylenchus*, 85–95% in *M. chitwoodii* and 100% in *M. hapla* in pepper roots and soil).

Crop rotation could be a very effective method if suitable resistant crops can be found for the cropping system in vegetable culture. Since most of the commercial vegetable crops are susceptible to root knot nematodes, rotation with vegetable crops often is not feasible. However, where vegetables can be excluded from the cropping system, cultivation of cereals for some years in the field considerably reduces root knot nematode populations. Cereals like rye (*Secale cereale*) produce hydroxaminic acid as secondary metabolites. The acid has been shown, *in vitro*, toxic to juveniles of *Meloidogyne*. Eggs are less affected. Even if vegetables are to be grown the rotation should include resistant cultivars. Onion, garlic and watermelon are resistant to *Meloidogyne incognita*. Cultivation of sesamum as an oilseed crop also suppresses root knot nematodes. The nematicidal effect of marigold (*Tagetes* spp.) is well documented. Its interculture with potato and tomato is reported to reduce root knot incidence. Tomato yield are reported to be 50% higher when grown after marigold than after fallow. In a comparison of co-culture of tomato and *Tagetes minuta, Datura metel* or *D. stromonium* with the nematicide aldicarb for suppression of *Meloidogyne javanica* it was found that although aldicarb gave most effective reduction of galling and highest shoot growth and fruit yield, the nematicidal plants also greatly enhanced shoot growth and fruit yield. *D. metel* had significantly greater suppressive effect on root galling than *T. minuta* and *D. stromonium*. Natarajan *et al.*, (2006) have reported that cold aqueous extracts (20% w/v, 100 ml aliquots) of pre- and post flowering whole plants, roots, and stems of *Tagetes erecta* suppressed *M. incognita* in tomato. There is varietal difference in efficacy of marigold. Rotation with suitable crops may induce soil suppressiveness to plant parasitic nematodes. Studies conducted by Vargas-Ayala *et al.*, (2000) indicate that the use of velvet bean (*Mucuna deeringiana*) in a cropping system alters the microbial communities of the rhizosphere and soil and this induces soil suppressiveness and control of root knot nematodes..

Kumar and Singh (2006) have studied the control of root knot (*Meloidogyne incognita*) by 5 isolates of *Arthrobotrys dactylioides*. The nematode trapping fungus trapped *M. incognita* and *Tylenchorhynchus brassicae* but not *Hoplolaimus indicus* suggesting choice of prey. The application of mass culture of the the fungus to soil infested with *M. incognita* reduced number of gallsd by 5.6–45.6%, of females by 44 to 72%, of egg masses by 44–51% and of juveniles by 38–82%. Stirling *et al.*, (1998a, b) and Stirling and Smith (1998) developed granule formulations of the nematophagous *Verticillium chlamydosporium* (*Pochonia chlamydosporia*) and nematode-trapping fungus *Arthrobotrys dactyloides* and used the granules in glasshouse and field control of the root knot nematode. Granules containing *V. chlamydosporium* when applied to soil at 10 g/liter soil resulted in 10^4 colony forming units/g soil in 7–14 weeks which colonized 37–82% of the first generation egg masses of *M. javanica*. Although *Pochonia chlamydosporia* is not highly effective when nematode density is very high, it slows down the build up of nematode inoculum and the effect lasts for 5–7 months. On application of granules containing *A. dactyloides,* the fungus grew from the granules in to the soil and immediately started forming traps. It reduced the number of *M. javanica* juveniles by more than 90%. In field tests, Stirling and Smith (1998) had used 55–880 kg granules/ha, which may not be practical. Further, they stated that formulations with greater biological activity will be needed if the level of nematode control obtained with chemical nematicides is to be achieved with *A. dactyloides.*

Oka *et al.,* (1997) had reported a hyphomycete of the genus *Scytalidium* which parasitized eggs of *Meloidogyne javanica*. Hyphae of the fungus proliferated in the gelatinous matrix of the egg mass and penetrated the egg shell via a penetration peg. It greatly reduced the rate of egg hatch. Application of the fungus did not reduce penetration by nematode juveniles but due to parasitization of eggs there was reduction in root knot severity. The biocontrol agents *Trichoderma harzianum* and *T. lignorum* have the potential to suppress root knot (*M. javanica*) by colonizing the second stage larvae and even by penetrating the eggs (Spiegen and Chet, 1998; Sharon *et al.,* 2001). In a short term experiment they have shown improved growth of nematode-infected plants and decreased root galling index and number of eggs per g root weight when the infested soil is pre-exposed to *Trichoderma harzianum*. In long term experiments, the antagonists improve plant growth and yield but there is no change in root gall index. Organic amendments increase populations of *Trichoderma* in soil. Conidia of strains of *Trichoderma asperellum, T. atroviride* and *T. harzianum* all can attach to eggs and second stage juveniles of *M. javanica* but the level of biocontrol varies with strains and life stage of the nematode. (Sharon *et al.,* 2007). *T. harzianum* is poor colonizer of the eggs in the gelatinous matrix while other species attach to the matrix, colonize the eggs in the matrix and released from the matrix. The gelatinous matrix stimulates immobilization of nematode juveniles by these fungi. Extraceuar chitinase activity of Trichoderma is iinvolved in its parasitism of nematodes. Shahebani and Hadavi (2008) have reported that with the help of chitimase *Trichoderma harzianum* can penetrate eggs and reduce the population of *M. javanica*.

Suppressive effect of culture filtrates of *Fusarium oxysporum* f.sp.*ciceri* and *Fusarium solani* on egg hatch and larval mobility of *M. incognita* was reported in 1984. Siddiqui and Shaukat (2003) have also reported the nematicidal effect of culture filtrates of *Fusarium solani*. Inoculation of the soil with conidia of the fungus significantly reduced nematode populations and severity of root knot. Non-pathogenic *F. solani* parasitizes eggs and females of *M. javanica* and, thereby reduces root knot severity. It enhances the tomato plant growth also. The rate of fungal infection of eggs and females increases with increase in nematode densities and fungal inoculum level. Culture filtrates of *Fusarium oxysporum* and *Sclerotium rolfsii* also cause inhibition of egg hatch and death of juveniles of *M. incognita*. Metabolites of a nonpathogenic *Fusarium oxysporum* are reported to reduce mobility of M. incognita within 10 min of exposure, After 60 min 98% of the juveniles are inactivated. Juveniles were initially inactivated within a few minutes but exposure of 5 hours killed 50% juveniles and 25-hour exposure killed 100% of the larvae. Filtrate of *S. rolfsii* caused 100% juvenile mortality on the fifth day after treatment (Adekunle and Akinsanmi, 2005). Filtrates of *Aspergillus sydowii, A. japonicus, A. niger* and *A. nidulans* are also reported to suppress egg hatch of *M. incognita* and cause mortality of the juveniles. Reduction to the extent of 65.9% in root galling and 97.5% reduction in nematode population when roots of seedlings, before transplantation, were dipped in culture filtrate of the blue-green alga *Microcoleus vaginatus.*(Khan *et al.,* 2005). Mittal and Goswami (2004a) isolated *Fusarium solani, Catenaria vermicola, C. anguillulae* and *Chaetomium abuense* from un-hatched black egg masses of *M. incognita* on cowpea roots. These inhibited hatching of the eggs.

Yeasts (fungi) are effective biocontrol agents of fungal pathogens associated witj fruit decay. Hashem *et al.,* (2008) evaluated 22 yeast strains for their effect on *Meloidogyne incognita* on grapevine. Four strains, Pichia guilliermondii, Candida albicans, Pachytrichospora transvaalensis, and Geotichum terrestre were found to significantly reduce the number of juveniles in the soil and improve plant growth and fruiting.

Strain 251 of *Paecilomyces lilacinus* is the most extensively studied fungal biocontrol agent of nematodes. This fungus colonizes the surface of eggs, even those containing un-hatched larva, forms appressoria and penetrates through the shell. The egg along with its contents is destroyed. Hyphae later grow out of the egg to continue growing or form conidiophores. Third and fourth stage juveniles as well as mature females are readily infected, with hyphae and conidiophores penetrating the body wall. Optimum temperatures for its growth are between 24° and 30°C and for its spore germination 28°–30°C. No growth is found at 36°C. At this temperature germ tube growth of the hyperparasite is significantly reduced. It is reported to reduce root gallin (*M. incognita*) by 66%, number of eggs by 74% and female nematode population within roots by 71%. However, in a single pre-plant application a concentraion of 1×10^6 CFU./g soil was required for substantial biocontrol (Kiewrnick and Sikora, 2006) Earlier, Bhatt *et al.,* (2002) had reported that the nematophagous fungus *Paecilomyces lilacinus* penetrates the eggs of *Meloidogyne incognita* and empties the contents leaving only empty shells. The combined effect of protease and chitinases of *P. lilacinus* causes changes in the egg shell structure. The lipid layer is destroyed, the chitin layer is hydrolysed and the vitellin layer loses integrity. The enzyme action reduces hatching of juveniles (Khan *et al.,* 2005). The fungus delays the life-cycle of *M. javanica* by 10 days. *Paecilomyces lilacinus* behaves like a parasitic fungus when growing on nematode eggs but not on plant roots when growing on root surface. It does not penetrate the roots (Holland *et al.,* 2003).It is a biocontrol agent for greenhouse insects and mites also. This nematophagous fungus produces two types of conidia, both of which exhibit similar nematophagous activity. The aerial conidia have rodlet layer on the surface which protects them from UV irradiation and provides better viability after drying and storage. The conidia produced in submerged cultures lack these properties (Holland *et al.,* 2002). Incorporation into soil of dry mycelium of *Penicillium chrysogenum,* a waste product of pharmaceutical industry, enhances plant growth andreduces root galling caused by *M. javanica* in cucumber (Gotlieb *et al.,* 2003). In *in vitro* studies water extract of dry mycelium was found to immobilize the juveniles and reduce egg hatching rate. These effects are reversible after washing the eggs and juveniles. *Verticillium (Lecanicillium) lecanii* reduces the number of eggs on roots when applied to the infested soil. However, it does not affect eggs embedded in tissues. Soil inoculated with the fungus retains suppressive effect even after autoclaving suggesting that the fungus produces some heat-stable substance (Meyer, 2006). For full efficacy of the nematophagous fungi (*viz, Verticillium chlamydosporium*) the required conditions are:

    (i) ability of the fungus to colonize the rhizosphere and
    (ii) ability of the host plant to support growth of the fungus.

Mycorrhizal fungi check root penetration by *Meloidogyne.* Plants growing with mycorrhizal roots show less penetration, reduced development of the nematodes in tissuies and reduced egg numbers. Inoculation of plants with *Glomus musseae* and *P. lilacinus* together or separately, in presence of chicken manure, completely suppresses infection of tomato by

M. javanica. Hirsutella minnesotensis is an endoparaite of *Meloidogyne hqpla* and suppresseses its population in soil., The level of suppression varies with populations of the nematode (Menan *et al.,* 2006). Diedhiou *et al.,* (2003) had reported that although arbuscular mycorrhizal fungus and non-pathogenic strains of *Fusarium oxysporum* both enhance plant growth and reduce *M. incognita* infection of tomato, their combined use does not give any additional benefit because the two have different mechanisms of action. *Fusarium* enhanced mycorrhizal growth on roots. A higher number of nematodes per gall was found by them in mycorrhizal than in nonmycorrhizal roots In papaya root knot (*M. incognita*) mycorrhization of seedlings with *Glomus mosseae* in the nursery has caused suppression of root knot after transplanting. Addition of plant growth promoting has not improved the beneficial effect of mycorrhization Tomato roots colonized by a endophytic, nonpathogenic strain of *Fusarium oxysporum* (Fig. 162) are reported to be resistant to infection by juveniles of *M. incognita*. The root exudates were found to be less attractive to the nematodes (Dababat and Sikora, 2007). Effect of the combination of *Glomus intraradices* and *Trichoderma viride* on root knot caused by *M. hapla* on different cultivars of tomato was studied by Masadeh *et al.,* (2004). The combination had no synergistic effect. Biocontrol effect was better visible in susceptible cultivar. In a cultivar that was unsuitable host (showing retarded development of giant cells) *Trichoderma viride* alone or in combination with arbuscular mycorrizal fungus did not produce any visible control. *Rhizobium etli*, a rhizxobacterium, is known to induce resistance to root knot nematodes. When coinoculated with the mycorrhizal fungus *Glomus intraradices* it promotes the establishment of the mycorrhiza thus enhancing biocontrol.

Greenhouse experiments have shown that selected strains of rhizobacteria (*Pseudomonas aeruginosa* and *P. fluorescens*) induce systemic resistance in tomato against *M. javanica* when used for bare root dip treatment or for soil drench (Siddiqui and Shaukat, 2002, 2005). The rhizobacterial strains of *Pseudomonas aeruginosa* and *Bacillus subtilis* used as seed dressing or as soil drench significantly suppress root rot-root knot complex (*Macrophomina phaseolina, Fusarium solani, Rhizoctonia solani* and *Meloidogyne javanica* in mungbean. They cause lysis of fungal mycelium. A strain of *Pseudomonas fluorescens* reduces nematode galling in tomato, eggplant, mungbean and soybean through antibiotic production. The antibiotic production is enhanced by addition of zinc @ 0.9–1.6 mg/kg soil (Siddiqui and Shaukat, 2002). Genetically modified derivatives of the biocontrol agent deficient in antibiotic production do not suppress root galls while mutants with capacity to overproduce the antibiotic are highly effective (Siddiqui and Shaukat, 2003). The antagonistic effect of the bacteria varies with plant species, host age and host genotypes. The bacteria also induce systemic resistance in tomato plants against *M. javanica* but there is no involvement of salicylic acid production in this induced resistance (Siddiqui and Shaukat, 2004; 2005). *In vitro* studies have shown that virulent strains of *Rhizoctonia solani* that produce phenylacetic acid suppress the biosynthesis of nematicidal compounds by which *Pseudomonas fluorescens* CHAO controls *M. incognita* (Siddiqui and Shaukat, 2005). Role of cyanide production by this strain in suppression of *M. javanica* in tomato is also reported (Siddiqui *et al.,* 2006). *Pseudomonas fluorescens* strain GRP3 with organic manure was found to be a good combination for management of root knot of tomato (Siddiqui, Z.A. *et al.,* 2001) The culture filtrates of certain endophytic bacteria are reported to immobilize second stage juveniles of *Meloidogyne javanica* and also inhibit egg hatch. The immobilized larvae could not recover even in water (Navesm *et al.,* 2004). Soaking of faba bean seed in preparation of biocontrol agents (*Streptomyces avermitilis, Bacillus thuringensis, Serratia marcescens*) were found to control root knot (El-Nagdi and Youssef, 2004). *Burkholderia cepacia* and most chytinolytic bacteria strongly suppress nematoes in soil. Addition of chitin or chitinous substrates enhances their activity. Siddiqui, Z.A. *et al.,* 2007) have evaluated a *Rhizobium* sp. and rhizobacteria *Pseudomonas putida, P. alcaligenes, Paenibacillus polymyxa* and *Bacillus pumilus* for suppression of *M., javanica* on lentil. *P. putida* caused the geatest inhibition of egg hatch and penetration, followed by *P. aureogenes, P. polymyxa* and *B. pumilus*. Combination of *Rhobium* with any of these rhizobacteria enhanced the suppression.

The spore forming bacterium *Pasteuria penetrans* infects larvae of *M. incognita* and multiplies in the nematode body. Infection of juveniles (J2) can occur at any temperature between 8° and 36°C. The percentage of infected females in the host decreases as the time for which the infected juveniles spent in the soil after infection. Inoculum of the bacterium can be multiplied by co-inoculation of a host plant with nematode larvae. When the infected larvae have grown, the plant roots can be harvested, dried and, powdered and used as a pesticide. Application of *Pasteuria* to field plots infested with *Meloidogyne* increases crop yield. *Pasteuria penetrans* was evaluated alone and with oxamyl, plant resistance and soil solarization in greenhouse for control of *Meloidogyne* spp. While all treatments suppressed root knot, combination of the bacterium with oxamyl had additive effect. In a similar study, Javed (2008) studied the efficacy of *P. penetrans* alone or in combination with dried neem leaves or oilcake against *M. javanica* in pot trials. The rates of infestation of root knot nematode populations by *P. penetrans* are influenced by variability in the bacterium, its concentration in the inoculum and soil conditions especially texture and chemical characteristics. In clay or silty clay soil the rate is high while in sandy soil it is low (Mateille *et al.,* 2002). The attachment of bacterial spores to nematode cuticle increases as the soil pH rises from neutral to alkaline and temperature rises above 15°C up to 30°C. There is no attachment at 5°C and 35°C (Zareen *et al.,* 2002; Javed *et al.,* 2002). The bacterium is unevenly distributed in the field soil. Irrigation, soil water holding capacity and texture affect its efficiency. Irrigation directly affects the distribution of the bacterium in soil pores. Excessive irrigation leaches the spores through soil profile and thus less nematode larvae are infected (Dabire *et al.,* 2005). Endospores of the bacteria can move to deeper soil layers through percolation of irrigation water. Attachment of the endospores to nematode body varies according to nematode species, developmental stage of the nematode and isolate of *P. penetrans* Comnbination of *P. penetrans* with *Enterobacter cloacae* or *Pseudomonas mendocina* enhances biocontrol efficacy of *P. penetrans.*

Giannakou *et al.,* (2004) have reported a bio-nematicide based on *Bacillus firmus* for suppression of root knot. Its effect was compared with the known biocontrol agent *Pasteuria penetrans* and some fumigant nematicides. The bionematicide was applied into top 0–20 cm of *Meloidogyne*–infested soil. Although in field studies the synthetic nematicides were generally superior to bionematicide, the latter maintained its effect till end of the crop season when numbers of nematode juveniles were kept low.

The nodule bacterium *Bradyrhizobium japonicum*, the arbuscular mycorrhizal fungus *Glomus mosseae* and the antagonist *Trichoderma pseudokoningii* used alone or in double or triple combinations suppress root knot in soybean. Siddiqui, Z.A. and Mahmood (2001) had evaluated the rhizobacteria *Pseudiomonas fluorescens, Azotobacter chroococcum* and *Azospirillum brasiliense* alone and in combination with root symbionts, *Rhizobium* sp. and *Glomus :mosseae* for control of *M. javanica* on chickpea. *G. mosseae* used alone was better than other tested organisms in increasing plant growth and reducing galling and nematode reproduction.

A predatory nematode, *Mononchoides fortidens*, has been used for the control of root galls caused by *Meloidogyne arenaria*. Application of the predator to soil infested with *M. arenaria* before transplanting of tomato significantly reduced root galls. Application after transplanting had no effect (Khan and Kim, 2005).

Resistant varieties: Resistance to root knot in plants is determined by the level of reproduction of the invading nematode. There is no difference between response of susceptible and resistant cultivars so far as attraction of second stage larvae, their entry in the host tissue and movement to the vascular cylinder of the host are concerned. However, in resistant plants, there is no development of the feeding site (nurse cell system). Instead, a localized region of necrotic cells (hypersensitive response, HR) develops near the head of the invading nematode at or near the site where the feeding cell would normally have been initiated. The earliest visible indication of HR occurs about 12 hours after inoculation of the roots. The nematode either dies or leaves the root.

Resistance to various species of the root knot nematodes does occur in many vegetable crops. In modern tomato cultivars the gene(s) conferring resistance to the major species of *Meloidogyne* is *Mi* derived from *Lycopersicon peruvianum*. The gene *Mi-1* is the most commonly used and the only resistance gene commercially available Generally, the resistance mediated by *Mi-1* is lost at high temperatures (above 28°C.). But other *Mi* genes, not as effective as Mi-1, are stable at higher temperature also. The gene *Mi-9,* obtained from a different species of *Lycopersicon* is a homolog of *Mi-1-1* and effective at high temperature. Resistance imparted by *Mi* genes is not uniform in all cultivars. Some show only tolerance. Plants carrying the *Mi* genes show resistance to bacterial wilt also in nematode infested soils. The resistance conferred by *Mi* genes can be overcome by biotypes of the nematodes (with modified genetic setup). Resistance may also vary due to incomplete transfer of the resistance genes. The tomato gene *Mi-1.2* confers resistance not only to root knot but also to aphids and whiteflies (*Macrosiphum euphorbiae, Bemissia tabaci*) thus providing some protection against viruses (Nombela *et al.,* 2003). The gene for resistance to aphids (*Meu 1*) is tightly linked to the Mi-1 gene and it is likely that it was introduced into tomato along with the *Mi* gene. It is also possible that *Mi* and Meu-1 are same (Rossi *et al.,* 1998). These genes can be transferred to other host species also. However, the transformed plants do not show resistance to insects. When eggplant (*Solanum melongena*) was transformed with Mi 1.2 it showed resistance to root knot but not to aphids and whitefly. This suggests that additive genes present in tomato and absent in eggplant decide resistance to insects (Goggin *et al.,* 2006). Although *Mi* resistance genes can be an effective and economic alternative to soil fumigation for root knot control, they should be used in an integrated management context to preserve their durability and prevent development of nematode strains that can overcome the resistance (Sorribas *et al.,* 2005). The non-cultivated sea beet (*Beta vulgaris* subsp. *maritime)* has resistance to *M. incognita* races 1, 2 and 4, *M. javanica, M. arenaria* races 1 and 2 and *M. hapla* and probably to other species. This could be a source of resistance for developing transgenic plants of crops other than sugarbeet also.

From time to time following varieties of some crops have been recommended as resistant:

TOMATO
Meloidogyne incognita: Pusa Ruby, S-120, NTDR-1, VFN-8, Nematex

M. incognita	Race 1	Karnataka Hybrid
	Race 2	Pusa 120, A-1-1-2
	Race 3	Punjab 6NR-7, VFN-Bush
	Race 4	Pusa 120, Punjab 6NR-7, VFN-8, VFN-Bush
M. javanica		S-120, SL-12, Resistant Bangalore, NTDR-1, Pusa 120, Punjab 6NR-7, Nematex, VFN -360, VFN-8, VFN-Bush

EGGPLANT (Brinjal)

M. incognita	Black Beauty, Giant of Banaras, Vijaya, Mysore Green, Pusa Purple Long

| *M. javanica* | Bhanta, Muktakeshi, Round Red, Coolie, Mathis B, Mysore Green, American Big Round, Arkasheel, R-34, Sonepat, BR-112 |

OKRA

| *M. javanica* | Long Green Smooth, IC-9273, IC-18960 |

CHILLI

M. incognita	Race 1	Pusa Jwala, Jwala
	Race 2	Jwala
	Race 3	Pusa Jwala, Jwala
	Race 4	Jwala
M. javanica:	Pusa Jwala, Suryamukhi Black, Jawala, Bull Nose, 579, CAP 63, Chilli NP-46-A, Chilli G-3.

CUCUMBER

| *M. javanica:* | Improved Long Green, S-445 (muskmelon) |
| *M. incognita:* | GY- 5937–587 |

Incorporation of these varieties in rotations with other vegetable crops brings down the nematode population in soil.

Induction of systemic acquired resistance (SAR) by elicitors like BTH does not prevent infection by infective juveniles but the number of galls per plant is significantly reduced. It has been suggested that probably the nematode suppresses the expression of genes responsible for accumulation of pathogenesis related proteins. Seedling root dip in salicylic acid solution or soil; drench around seedl;ing roots with the solution induces resistance resulting in reduction of number of eggs and root galls (Molinari, 2008).

Chemical control: Although direct control of root knot with chemicals had been considered an outstanding method available it is now realized all over the world that this approach can be recommended only for medium and high income crops. Earlier, soil fumigants like D-D mixture, Nemagon, methyl bromide, etc. had been found very effective and recommended on large scale. These were gradually withdrawn because of their adverse effect on environment. The non-fumigant, (Furadon), phorate (Thimet) and fensulfothion (Dasanit) have now replaced the fumigants although they are also highly toxic. Application of aldicarb to soil at the rate of 4–8 kg a.i./ ha is reported to completely check infection in tomato. In okra even a lower dose of 2–2.5 kg a.i./ha of aldicarb gives good control of root knot. Aldicarb has been used in combination with biocontrol agents. Phorate at the rate of 10 kg/ha also gives almost complete control when applied immediately or 7 days before transplanting. For nursery bed treatment, 7 g carbofuran per sq meter of bed is recommended. Vadhera *et al.,* (2000) have reported that carbofuran @ 0.6 g a.i. per sq. meter in nursery bed treatment improved the germination and seedling vigour of tomato and reduced intensity of root galls. There was 30% more yield. Seedling bare root dip for 6 hours in 1% carbofuran or triazophos increased the yield by 43% and 42%, respectively and reduced root galls. Similar observations on root dip treatment are reported for eggplant also. Dressing and soaking of okra seed with chemicals is reported to reduce infection of *Meloidogyne incognita* (Vadhera *et al.,* 1999). Tiwari *et al.,* (2002) have suggested nursery bed treatment, solarization and amendment with margosa cake for the control of *M. incognita*. In addition to their direct interaction with the nematodes, nematicides may induce suppressiveness in the soil to root knot species by altering the nature of microbial communities in the rhizosphere where antagonists of the nematode are promoted (Fernandez *et al.,* 2001).

Phytochemical-based strategies: Higher plants have yielded a broad spectrum of active compounds, including polythienyls (marigold), glucosinolates, yielding isothiocyanates (mustard), cyanogenic glycosides (sudan grass), polyacetylenes, alkaloids, lipids, terpenoids, sesquiterpenoids, diterpenoids, triterpenoids, quassinoids, steroids, simple and compound phenolics (cf. Chitwood, 2002). Such natural products have formed part of village medicines since ancient times against parasitic infections of man and animals. Examples are margosa and karanj (*Pongamia glabra*) oils. Some of these have now been exploited for control of fungal and nematodes infections of plants and repel insect vectors of viruses. Calvet *et al.,* (2001) have reported highly significant control of *Meloidogyne javanica* in saturated atmosphere of benzaldehyde, salicylaldehyde, borneol, cannamaldehyde and many other natural compounds. These compounds can be introduced in the soil through green manuring with leaves, amendmends with plant residues including oil cakes or seed treatment with plant extracts. Oka *et al.,* (2000) and Pandey *et al.,* (2000) have reported the importance of essential oils as source of potent nematicides. Essential oils have been shown to inhibit hatching and larval motility. Pandey (2005) evaluated distillation waste of many essential oil plants against *Meloidogyne incognita*. Least nematode infections were recorded in plants treated with distillation waste of *Murraya koengii* (curry leaves), *Cymbopogon martini* (palmarosa), *C. flexuosa* (lemongrass) and vermicompost of *Tagetes minuta* (marigold). Meyer *et al.,* (2008) have reported anti-*Meloidogyne incognita* activity of essential oil of clove (*Syzygium aromatica*). EC_{50} value for egg hatch was 0.097% of clove oil and for larval viability 0.14%. Volatiles of 5% clove oil also reduced egg hatch in water by 30% and viability of larvae by 100%. Chemical resistance activators have also been used as soil drench before transplanting seedlings of an aromatic plant for control of *M. incognita*. Maximum reduction in root knot severity and nematode population occurred with compounds of salicylic acid (Pandey and Kalra, 2005) Thymol oil applied to soil at the rate of 9.3 liters/ha reduces populations of most nematodes. Field conrol of root knot and bacterial wit of tomato by application of thymol as soil fumigant (73 kg.ha) and acibenzolar-S-methyl (ASM) at 25 mg/L as foliar spray is reported by Ji *et al.,* (2007). Sultana *et al.,* (2008) have recently reported that solvent extracted fractions of the seaweeds *Codium uyengerii, Jania capillacea, Stokeyia indica* and *Solieria robusta* caused more than 50% mortality of *M. javanica* juveniles within 24 hours at 10 mg/L concentration. At 1.0 mg/L concentration some extracts caused more than 50% mortality in 48 hours. Soil application of *S. indica* or *S. robusta* alone or in combination with *Pseudomonas aeruginosa* (growth promoting rhizobacteria) significantly suppressed infection of chilli roots.

Amendment of soil with residue of plants containing anti-nemic compound is not necessarily the effect of the compound. Application of marigold tops as green manure is not as effective against root knot nematodes as the ciltivation of marigold for about 8 weeks. The antinemic polytheinyls are constituents of root exudates of marigold and are effective after reaching a particular concentration. The reduction of galling due to green manuring with marigold tops is due to decomposition of vegetable matter and is only as good as that obtained by addition of any crop residue.

REFERENCES

Abbasi, P.A., E. Riga, K.L. Conn and G. Lazarovits. 2005. Effect of neem cake soil amendment on reduction of damping off severity and population densities of plant parasitic nematodes and soil borne plant pathogens.*Can. J. Plant Pathol.* 27(1): 38–45.

Ajaz, S. and S.A. Tiyagi. 2003. Utilization of noxious weeds for the management of plant-parasitic nematodes infesting some vegetable crops. *Arch. Phytopath. Plant Prot.* 36(3–4): 211

Andreoglou, F.I., I.K. Vagelas, M. Wood, H.Y. Samaliev and S.R. Gowen. 2003. Influence of temperature on the motility of *Pseudomonas oryzihabitans* and control of *Globodera rostochiensis. Soil Biol. Biochem.* 35(8): 1095

616

Back, M.A., P.P.J. Haydock and P. Jenkinson. 2006. Interaction between the potato cyst nematode *Globodera rostochiensis* and diseases caused by *Rhizoctonia soani* AG 3 in potato under field conditions. *Eur. J. Plant Pathol.* 114(2): 215

Bhatt, J., R.K. Chaurasia and S.K. Sengupta. 2002. Management of *Meloidogyne incognita* by *Paecilomyces lilacinus* and influence of different inoculum levels of *Rotylenchulus renformis* on betelvine. *Indian Phytopath.* 55(3): 348

Calvet, C., J. Pinochet, A. Camprubi, V. Estaun and R. Rodriguez-Kabana. 2001. Evaluation of natural chemical compounds against root-lesion and root-knot nematodes and side effects on the infectivity of arbuscular mycorrhizal fungi. *Eur. J. Plant Pathol.* 107(6): 601

Chitwood, D.J. 2002. Phytochemical based strategies for nematode control. *Annu. Rev. Phytopathol.* 40: 221

Dababat, A. El-Fattah and R.A. Sikora 2007. Influence of the mutual;ostic endophyte *Fusarium oxysporum* Fo 162 on *Meloidogyne incognita* attraction and invasion. *Nematology* 9(6): 771

Dabire, K.R., S. Ndiaye, J. Chotte, S. Fould, M.T. Diop and H. Mateille. 2005. Influence of irrigation on the distribution and control of the nematode *Melodiogyne javanica* by the biocontrol bacterium *Pasteuria penetrans* in the field. *Biol. Fert. Soils* 41(3): 205–211

Diedhiou, P.M., J. Hallmann *et al.,* 2003. Effect of arbuscular mycorrhizal fungus and a non-pathogenic *Fusarium oxysporum* on *Meloidogyne incognita* infection of tomato. *Mycorrhiza* 13(4): 199

Eco, J., N. Tomomi, *et al.,* 2007. Role of pore size on the migration of *Meloidogyne incognita* juveniles under different tillage systems. *Nematology* 9(6): 751

Giannakou, I.O. and R.G. Simon. 2004. Factors affecting biological control effectiveness of *Pasteuria penetrans* in *Meloidogyne javanica* and the bacterial development in the nematode body. *Nematropica* 34(2): 153

Giannakou, I.O., D.G. Karpouzas and D.P. Athanasiadou. 2004. A novel non-chemical nematicide for the control of root knot nematodes. *Appl. Soil Ecol.* 26(1): 69

Gotlieb, D., Y. Oka, B-H. Ben-Daniel and Y. Cohen. 2003. Dry mycelium of *Penicillium chrysogenum* protects cucumber and tomato plants against the root knot nematode *Meloidogyne javanica*. *Phytoparasitica* 31(3): 217

Gu, Y-Q., M-H.Mo, J-P. Zhou *et al.,* 2007. Evaluation and identification of potential organic nematicidal volatiles from soil bacteria. *Soil Biol. Biochem.* 39(10): 2567

Hashem, M., Y.A.M.M. Omran and N. M.A. Sallam. 2008. Efficacy of yeasts in the management of root-knot nematode, *Meloidogyne incognita*, in Flame Seedless grape vines and the consequent effect on the productivity of the vines. *Biocontrol Sci. Technol.* 18(3): 353.

Holland, R.J., T.S. Gunasekera, K.L. Williams and K.M.H. Nevalainen. 2002. Ultrastructure and properties of *Paecilomyces lilacinus* spores. *Can. J. Microbiol.* 48(10): 879

Holland, R.J., K.L. Williams and K.M.H. Novalainen. 2003. *Paecilomyces lilacinus* strain Bioact 251 is not a plant endophyte. *Aust. Plant Pathol.* 32(4): 473

Insunza, V., S. Alstrom and K.B. Eriksson. 2002. Root bacteria from nematicidal plants and their biocontrol potential against trichodorid nematodes in potato. *Plant Soil* 241(2): 271

Jablonska, B., J.S.S. Ammirajur, K.K. Bhattarai *et al.,* 2007. The *Mi-9* gene from *Solanum arcanum* conferring heat-stable resistance to root-knot nematodes is a homolog of *Mi-1*. *Plant Physiol.* 143: 1044

Javed, N., H.U. Khan, Z. Hussain and M. Ashfaq. 2002. Effect of temperature, soil pH, agitation intervals and soil type on the spore attachment of *Pasteuria penetrans* to root knot nematode *Meloidogyne javanica*. *Plant Pathol J.*. 1(2–4): 66

Javed, N, S.R. Gowen, S.A. El-Hassan *et al.,* 2008. Efficacy or neem (*Azadirachta indica*) formulation on biology of root knot nematodes (*Meloidogyne javaniica*). *Crop Protection* 27(1): 36

Javed, N,, S. El-Hassan, S. Gowen *et al.,* 2008. The potential of combining Pasteuria penetrans and neem (*Azadirachta indica*) formulations as a management system for root knot nematodes on tomato. *Eur. J. Plant Pathol.* 120(1): 53:

Ji, Oengsheng, T. Mamol, J.R. Rich *et al.,* 2007. Development of an integrated approach for managing bacterial wilt and root knot on tomato under field conditions. *Plant Dis.* 91(10): 1321

Khan, A., K.L. Williams and H.K.M. Nevalainen. 2005. Effects of *Paecilomyces lilacinous* protease and chitinase on the eggshell structure and hatching of *Meloidogyne javanica*. *Biological Control* 31(3): 346

Khan, A., K.L. Williams and H.K.M. Nevalaine. 2006b. Infection of plant parasitic nematodes by *Paecilomyces lilacinus* and *Monacrsporium lysipagum*. *Biocontrol* 51(5): 659–678

Khan, F.U., B.D. Nelson and T.C. Helms. 2005. Greenhouse evaluation of binucleate *Rhizoctonia* for control of *R. solani* in soybean. *Plant Dis.* 89(4): 373–379

Khan, M.R., M.W. Khan and K. Singh. 1997. Management of root-knot disease of tomato by the application of fly ash in soil. *Plant Pathology* 46(1): 33

Khan, Z. and Y.H. Kim. 2006. The predatory nematode. *Mononchoides fortidens*, (Nematoda: Diplogasterida), suppresses the root knot nematode, *Meloidogyne arenaria*, in potted field soil. *Biol.Control* 35(1): 78

Khan, Z., S.D. Park, S.Y. Shin, S.G. Bae, I.K. Yeon and Y.J. Seo. 2005. Management of *Meloidogyne incognita* on tomato by root–dip treatment in culture filtrate of the blue-green alga, *Microcoleus vagionatus*. *Bioresource Technol.* 96(12): 1338

Kiewnick. S. 2006. Effect of temperature on growth, germination, germ tube extension and survival of *Paecilomyces lilacinuas* strain 251. *Biocontrol Sci. Technol.* 16(5): 535

Kiewrnick, S. and R.A. Sikora. 2006. Biological control of the root knot nematode *Meloidogyne incognita* by Paecilopmyces lilacinus strain 251. *Biol. Control* 39(2): 179–182

Kumar, D. and K.P. Singh. 2006. Assessment of predacity and efficiency of *Arthronotrys dactylioides* for biocontrol of root knot disease of tomato. *J. Phytopathol.* 154(1): 1–5

Lian, L.H., B.Y. Tian, R. Xiong *et al.*, 2007. Proteases from *Bacillus*: a new insight into the mechanism of action for rhizobacterial suppression of nematode populations. *Lett. Appl. Microbiol.* 45(3): 262–269

Lopez-Llorca, L.V., C. Olivares-Bernabeu *et al.*, 2002. Pre-penetration events in fungal parasitism of nematode eggs. *Mycol. Res.* 106(4): 499

Lopez-Perez, J-A, R. Tatiana and P. Antoon. 2005. Effect of three plant residue and chicken manure used as biofumigants at three temperature on *Meloidogyne incognita* infection of tomato in greenhouse experiments. *J.Nematol.*37:489.

Masadeh, B., H. von Alten, G. Grundewaldt-Stoecher and R.A. Sikora. 2004. Biocontrol of root knot nematodes using the arbuscular mycorrhizal fungus *Glomus intraradices* and the antagonist *Trichoderma viride* in two tomato cultivars differing in their suitability as hosts of the nematodes. *J. Plant Dis. Prot.* 111(4): 322

Mashela, P.W. 2002. Ground wild cucumber fruits suppress ny\umbers of Meloidogyne incognita on tomato. in microplots. *Nematropica* 32(1): 13

Mashela, P.W. and M.E. Nthangeni. 2002. Efficacy of *Ricinus communis* fruit meal with and without *Bacillus* species on suppression of *Meloidogyne incognita* and growth of tomato. *J. Phytopath.* 150: 399

Mashela, P.W., H.A. Shimelis and F.N. Mudau. 2008. Efficacy of ground wild cucumber fruits, aldicarb and fenamiphos on suppression of *Meloidogyne incognita* in tomato. *J. Phytopathol.* 156(5): 264–267

McBride, R.G., R.L. Mikkelsen and K.R. Barker. 2000. The role of low molecular weight organic acids from decomposing rye in inhibiting root knot nematode populations in soil. *Appl. Soil Ecol.* 15(3): 243

Mendozaaz, A.R. S. Kiewnick and R.A. Sikora. 2008. *In vitro* activity of *Bacillus firmus* against the birrowing nematode *Radopholus similes*, the root-knot nematode *Meloidogyne incognita* and the stem nematode *Ditylenchus dipsaci*. *Biocontrol Sci. Technol.* 18(4): 377–389 **ABS**

Mennan, S., H.M. Aksoy and O. Ecevit. 2005. Antagonistic effect of *Fusarium oxysporum* on *Heterodera cruciferae*. *J. Phytopath.* 153(4): 221

Mennan, S., S. Chen and H. Melakebarhan. 2006. Suppression of *Meloidogyne hapla* populations by *Hirsutella minnesotensis*. *Biocontr. Sci. Technol.* 16(1): 181

Meyer,S.L.F. 2006. Efficacy of the fungus *Verticillium lecanii* for suppressing root knot nematode eggs on roots. *Hort Technol.* 9: 443

Meyer, S.L.F., D.K. Lakshman, I.A. Zasada *et al.*, Dose-response effects of clove oil from *Szyzgium aroimatica* on the root knot nematode *Meloidogyne incognita*, *Pest Manag. Sci.* 64(3): 223

Mittal, A. and B.K. Goswami. 2004a. Hatching inhibition in black egg masses of *Meloidogyne incognita* caused by hyphomycetous fungi. *J. Mycol. Pl. Pathol.* 34(2): 283

Mittal, A. and B.K. Goswami. 2004b. Effect of culture filtrates of *Aspergillus* species on mortality and hatching in *Meloidogyne incognita*. *J. Mycol. Pl. Pathol.* 34(2): 281

Molinari, S. 2008. Salicylic acid as an elicitor of resistance to root knot nematodes in tomato. *USHS Acta Hortic.* 789

Natrajan, N., N. Cork, N., Boomathi, R. Pandi, S. Velavan and G. Dhakshmnamoorthy. 2006. Cold aqueous extracts of African marigold, *Tagetes erecta*, for control of tomato root knot nematode, *Meloidogyne incognita*. *Crop Protection* 25(11):1210–1213

Nico, A.I., R.M. Jimenez-Diaz and P. Castillo. 2004. Control of root knot nematodes by composted agro-industrial wastes in potting mixes. *Crop Protection* 23(7): 581

Nombela, G., V.M. Williamson and M. Muniz. 2003. The root-knot nematode resistance gene *Mi-1, 2* of tomato is responsible for resistance against the whitefly *Bemisia tabaci*. *Mol. Plant Microbe Interact.* 16: 645

Oka, Y and Y. Cohen. 2001. Induced resistance to cyst and root knot nematodes in cereals by DL-b-amino-n-butyric acid. *Eur. J. Plant Pathol.* 107(2): 219

Oka, Y., S. Nacar *et al.*, 2000. Nematicidal activity of essential oils and their compounds against the root knot nematode. *Phytopathology* 90: 710

Oka, Y., B. Ben-Daniel and Y. Cohen. 2006 Control of *Meloidogyne javanica* by formulations of *Inula viscose* leaf extract. *J. Nematol.* 38(1): 46

Oka, Y., N. Tkachi, S. Shimshon, *et al.*, 2006. Field studies on the enhancement of nematicidal activity of ammonia-releasing fertilizers by alkaline amendments. *Nematology* 8(6): 881–893

Oka, Y., N. Tkach and M. Mor. 2007. Phosphite inhibits development of the nematodes *Heterodera avenae* and *Meloidogyne marylandii* in cereals. *Phytopathology* 97(4): 396

Oka, Y., N. Shapira and P. Fine. 2007. Control of root knot nematodes in organic farming system by organic amendments and soil solarization. *Crop Prot.* 26(10): 1556

618

Pandey, R. and A. Kalra. 2005. Chemical activators: A novel and sustainable management approach for *Meloidogyne incognita* (Kofoid and White) Chitwood in *Chamomilla recutita* L. *Arch. Phytopathol. Plant Prot.* 38(2): 107

Pandey, R., A. Kalra, I. Tandon, N.Mehrotra, H.N. Singh and S. Kumar. 2000. Essential oils as potent source of nematicidal compounds. *J. Phytopath.* 148(7-8): 501

Pradanniko, M. V., G.G. Petelina *et al.,* 2007. Influence of components of Globodera rostochiensis cysts pn the *in vitro* hatch of second stage juveniles. *Nematology* 9(6): 837

Preston, J.F., D.W. Diclson, J.E. Maruniak *et al.,* 2003. *Pasteuria* spp.: Systematics and phylogeny of these bacterial parasites of phytopathogenic nematodes. *J.Nematol.* 35(2): 198.

Qiulong, N., H. Xiaowei, T. Baoyi *et al.,* 2006. *Bacillus* sp. B16 kills nematodes with a serine protease which is identified as a pathogenicity factor. *Appl. Microbiol. Biotechnol.* 69(6): 722

Rahman, L. and T. Somers. 2005. Suppression of root knot nematode (*Meloidogyne javanica*) after incorporation of Indian mustard cv. Nemfix as green manure and seed meal in vineyards. *Aust. Plant Pathol.* 34(1): 77

Riga, E., L.A. Lacey and N. Guerra. 2008. Mascodor albus, a potential biocontrol agent against plant parasitic nematodes of economically important vegetabl crops in Washington State, USA. *Biol. Control* 45(3): 380

Riley, I.T. 1992. *Anguina tritici* is a potential vector of *Clavibacter toxicus*. *Aust. Plant Pathol.* 21(4): 147

Riley, I.T., D. Shedley and K. Sivasithamparam. 2001. Anhydrobiosis and reproduction in *Anguina australis*. *Aust. Plant Patrhol.* 30(4): 361

Ryan, N.A., E.M. Duffy, A.C. Cassells and P.W. Jones. 2000. The effect of mycorrhizal fungi on the hatch of potato cyst nematodes. *Appl. Soil Ecol.* 15(2): 233

Saju, K.A., M. Anandaraj and Y.R. Sarma. 2002. On farm production of *Trichoderma harzianum* using organic matter. *Indian Phytopath.* 55(3): 277

Schmitz, A., I.T. Riley and P. de Silva. 2001. *Anguina australis*, a vector of *Rathayibacter toxicus* in *Ehrharta longifolia*. *Aust. Plant Path.* 30(2): 171

Shahebanu, N. and N. Hadavi. 2008. Biological control of roopt-knot nrmatode *Meloidoghyne javanica* by *Trichoderma harziamu, Soil Biol. Biochem.* 40(8): 2016

Sharon, E., M. Bar-Eyal, I. Chet, A. Herrera-Estrella, O. Kleifeld and Y. Spiegel. 2001. Biological control of the root knot nematode *Meloidogyne javanica* by *Trichoderma harzianum*. *Phytopathology* 91: 687

Sharon, E., I. Chet, A.Viterbo, N. Bar-Eyal *et al.,* 2007. Parasitism of *Trichoderma* on *Meloidogyne javanica* and role of the gelatinous matrix. *Eur. J.. Plant Pathol.* 118 (3): 247

Shaukat, S.S., I.A. Siddiqui, G.H. Khan and M.J. Zaki. 2002. Nematicidal and allelopathic potential of *Argemone mexicana*, a tropical weed. *Plant Soil* 245(2): 239

Siddiqui, I.A. and S.S. Shaukat. 2002. Zinc and glycerol enhance the production of nematicidal compounds *in vitro* and improve the biocontrol of *Meloidogyne javanica* in tomato by fluorescent pseudomonads. *Letters Appl. Microbiol.* 35(3): 212

Siddiqui, I.A. and S.S. Shaukat. 2002. Rhizobacteria-mediated induction of systemic resistance (IRS) in tomato against *Meloidogyne javanica*. *J. Phytopath.* 150(8–9): 469

Siddiqui, I.A. and S.S. Shaukat. 2003a. Plant species, host age and host genotype effects on *Meloidogyne incognita* biocontrol by *Pseudomonas fluorescens* strain CHAO and its genetically modified derivatives. *J. Phytopath.* 151(4): 231

Siddiqui, I.A. and S.S. Shaukat. 2003b. Factors influencing the effectiveness of non-pathogenic *Fusarium solani* strain Fs5 in the suppression of root-knot nematode in tomato. *Phytopath, Mediterr.* 42(1): 17

Siddiqui, I.A. and S.S. Shaukat. 2004. Systemic resistance in tomato induced by biocontrol bacteria against the root knot nematode, *Meloidogyne javanica*, is independent of salicylic acid production. *J. Phytopath.* 152(1): 48

Siddiqui, I.A. and S.S. Shaukat. 2005. *Pseudomonas aeruginosa*-mediated induction of systemic resistance in tomato against root-knot nematode. *Plant Path Jour.* 4(1): 21

Siddiqui, I.A. and S.S. Shaukat. 2005. Phenylacetic acid-producing *Rhizoctonia solani* represses the biosynthesis of nematicidal compounds *in vitro* and influences biocontrol of *Meloidogyne incognita* in tomato by *Pseudomonas fluorescens* CHAO and its GM derivatives. *J. Appl. Microbiol.* 98(1): 43

Siddiqui, I.A., S. S. Shaukat, I.H. Sheikh and A. Khan. 2006. Role of cyanide production by *Pseudomonasw fluorescens* CHAO in the suppression of root-knot nematode, *Meloidogyne javanica*, in tomato. *World J.Microbiol. Biotechnol. IN PRESS*

Siddiqui, Z.A. and I. Mahmood. 2001. Effects of rhizobacteria and symbionts on the reproduction of *Meloidogyne javanica* and growth of chickpea. *Bioresource Technol.* 79(1): 41

Siddiqui, Z.A., G. Baghel and M.S. Akhtar. 2007. Biocontrol of *Meloidogyne javanica* by *Rhizobium* and plant-growth promoting rhizobacteria on lentil. *World J. Microbiol. Biotechnol.* 23(3): 435

Singh, K.P., P. Bandopadhyay, S.S. Vaish, T. Makeshkumar and R.C. Gupta. 2002. Growth and population dynamics of *Catenaria anguillulae* in relation to oil cakes. *Indian Phytopath.* 55(3):286

Singh, R.S. 1965. Control of root knot nematodes with organic soil amendments. *FAO Plant Prot. Bull.* 13: 35

Singh, R.S. and K. Sitaramaiah. 1966. Incidence of root knot of okra and tomato in oil-cake amended soil. *Plant Dis. Rep.* 50: 668

Singh, R.S. and K. Sitaramaiah. 1967. Effect of decomposition of green leaves, sawdust and of urea on the incidence of root knot of okra and tomato. *Indian Phytopath.*20: 349

Singh, R.S. and K. Sitaramaiah. 1970. Control of plant parasitic nematodes with organic amendment of soil. *PANS* 16: 287

Singh, R.S. and K. Sitaramaiah. 1971a. Control of root knot through organic and inorganic amendments of soil: Effect of sawdust and inorganic fertilizers. *Indian J. Nematol.* 1: 80

Singh, R.S. and K. Sitaramaiah. 1971b. Control of root knot through organic and inorganic amendment of soil : Effect of oil-cakes and sawdust. *Indian J. Mycol. Pl. Pathol.* 1: 20

Singh, R.S. and K. Sitaramaiah. 1973. Effect of organic amendments, green manuring and inorganic fertilizers on root knot of vegetable crops. *U.P. Agric. Univ. Exp. Sta. Bull.* 6:289 pp

Sorribas, F.J., C. Ornat *et al.,* 2005. Effectiveness and profitability of the *Mi*-resistant tomatoes to control root knot nematodes. *Eur. Jour. Plant Pathol.* 111(1): 29

Stirling, G.R. 1989. Organic amendments for control of root knot (*Meloidogyne incognita*) on ginger. *Aust. Plant Path.* 18(2): 39

Stirling, G.R. and A. Nikulin. 1998. Crop rotation, organic amendments and nematicides for control of root knot nematodes (*Meloidogyne incognita*) on ginger. *Aust. Plant Path.* 27(4): 234

Stirling, G.R. and L. J. Smith. 1998. Field tests of formulated products containing either *Verticillium chlamydosporium* or *Arthrobotrys dactyloides* for biological control of root knot nematodes. *Biological Control* 11(3): 231

Stirling, G.R. and A.M. Stirling. 2003. The potential of *Brassica* green manure crops for controlling root knot nematode (*Meloidogyne javanica*) on horticultural crops in a subtropical environment. *Aust. J. Exp. Agr.* 41(6): 623

Stirling, G.R., K.A. Licastro, L.M. West and L.J. Smith. 1998a. Development of commercially acceptable formulations of the nematophagous fungus *Verticillium chlamydosporium*. *Biological Control* 11(3): 217

Stirling, G.R., L.J. Smith, K.A. Licastro and L. M. Eden. 1998b. Control of root knot knot nematode with formulations of the nematode trapping fungus *Arthrobotrys dactyloides*. *Biological Control* 11(3): 224

Sultana, V., J. Ara and S. Ehteshamul-Haque. 2008. Suppression of root rotting fungi and rtoot knot nematode by seaweed and *Pseudomonas aeruginosa*. *J. Phytopathol.*156(7): 390–

Tian, B., J. Yang, and K-Q. Zhang. 2007. Bacteria used in the biological control of plant parasitic nematodes: populations, mechanisms of action, and future prospects. *FEMS Microbiology Ecology* 61(2): 197

Tobin, J.D., P. Hadock, M.C. Hare *et al.,* 2008. Effecty of the fungus *Pochonia chlamydosporia* and fosthiazate on the multiplication rate of potato cyst nematodes (*Globodera pallida* and *G. rostochiensis*) in potato crops grown under U.K. field conditions. *Biol. Control* 46((2): 194–202

Tiwari, S.P., I. Vadhera and B.N.Shukla. 2002. Management of *Meloidogyne incognita* in tomato through nursery bed treatment, solarization and neem cake. *Indian Phytopath.* 55(2): 244

Twomey, U., P. Warrior, B.R. Kerry and R.MN. Perry. 2000. Effects pf the biological nematicide, DiTera, on hatching of *Gobodera roistochiensis* and *G. pallida*. *Nematology* 2: 355

Twomey, U., R.N. Rolfe, P. Warrior, R.N. Perryt. 2002. Effects of the biological nematicide, DiTera, on movement and and sensory responses of second stage juveniles of *Globodera rostiochiensis* and styket activity of *Globodera rostochiensis* and fourth stage juveniles of *Ditylenchus dipsaci*. *Nematology* 4: 909.

Vaish, S.S, and K.P. Singh. 2000. Distribution of *Catenaria anguillulae,* a facultative endoparasite of nematodes in soils from different locations of India. *World Jour. Microbiol. Biotechnol.* 18(1): 65–67

Vargas-Ayala, R., R. Rodriguez–Kabana, G. Morgan-Jones, J.A. McInroy and J.W. Kloepper. 2000. Shift in soil microflora induced by velvet bean (*Mucuna deeringiana*) in cropping system to control root knot nematodes. *Biological Control* 17(1): 11

Vawdrey. L.L. and G.R. Stirling. 1997. Control of root knot nematodes (*Meloidogyne javanica*) on tomato with molasses and other organic amendments. *Aust. Plant Pathol.* 26(3): 179

Verdejo-Lucas,S., F.J. Sorribas, C. Ornat and M. Galeano. 2003. Evaluating *Pochonia chlamydosporia* in a double-cropping system of lettuce and tomato in plastic houses infested with *Meloidogyne javanica*. *Plant Pathology* 52(4): 521

Wang, K., R.D. Riggs and D. Crippen. 2005. Isolation, selection and efficacy of *Pochonia chlamydosporia* for control of *Rotylenchulus reniformis* on cotton. *Phytopathology* 95(8): 890

Weerasinghe, R.R., D. McK Bird and N.S. Allen. 2005. Root knot nematodes and bacterial Nod factors elicit common signal transduction events in *Lotus japonicus*. *Proc. Nat. Acad. Sci. USA* 102(8): 3147

Widmer, T.L. and G.S. Abawi. 2000. Mechanism of suppression of *Meloidogyne hapla* and its damage by a green manure of Sudan grass. *Plant Dis.* 84: 562

Zasada, I.A. and H. Ferris. 2003. Sensitivity of *Meloidogyne javanica* and *Tylenchulus semipenetrans* to isothiocyanates in laboratory assays. *Phytopathology* 93(6): 747

Zasada, I.A., H. Ferris, C.L. Elmore, J.A. Roncoroni, J.D/ MacDonals, L.R.0 Bolkan and L.E. Yakabe. 2003. Field application of Brassicaceous amendments for control of soilborne pests and pathogens. *Plant Health Progress*, November, 2003. APS

Zhang, L., J. Yang, Q. Niu, X/ Zhao *et al.,* 2008. Investigation on infection mechanism of the fungus *Clonostachys rosea* against nematodes using the green fluorescent protein. *Appl. Microbiol. Biotechnol.* 78(6): 983

CHAPTER

14

Phanerogamic Plant Parasites

Although most of the plant diseases are caused by fungi, bacteria, viruses, and nematodes, over 3000 species of flowering plants obtain nutrition as parasites. In many cases the damage caused by these parasites is only slight or the attacked host plants are of little economic importance. But there are many examples where these flowering plants attack valuable crops and trees causing considerable damage. Some of these parasites attack roots of the host while others parasitize the stem. Some are devoid of chlorophyll and entirely depend on their host for nutrition while others have chlorophyll but no true roots and obtain water and mineral constituents of food from the host. Such parasites also have no separate existence away from the host. There is much similarity between pathogenesis of fungal pathogens and parasitic weeds.

The common parasitic flowering plants can be grouped as follows:

1. Stem parasites
   (a) Holoparasites (entirely dependent)–*Cuscuta* spp.
   (b) Semi-parasites (partially dependent)–*Loranthus, Dendrophthoe*
2. Root parasites
   (a) Holoparasites–*Orobanche*
   (b) Semi-parasites–*Striga* spp.

There are many other parasitic weeds that harm cultivated crops but they have only local importance. The above root parasites are obligate parasite. There are some facultative root hemiparasites, *viz., Rhinanthus*, which can live actively in absence of the host plant. While in the parasitic association with the host their stomata function day and night to draw the maximum nutrients from the host xylem, on being detached from host roots they continue to live like any plant with their stomata remaining open only during the day time to carry on photosynthetic actiovity.

DODDER (Love-vine, Amarbel, Cuscuta spp.)

Dodder is a non-chlorophyll bearing, leafless, twining, parasitic seed plant which attaches its yellow, orange or pink, thread-like stems to the stems or other parts of cultivated or wild plants. The twining stem may be single but is mostly conspicuous as tangles mass of intertwining stems. The parasite not only draws its nutrition from the host plant, it also affects the photosynthesis, reducing nutrition of the host.

Cuscuta belongs to Convolvulaceae in the phanerogams. *Cuscuta campestris* (*C. arvensis*) and *C. reflexa* are commonly encountered species. *C. campestris* has a very wide host range. Almost all the hosts are dicots. The general characters of different species of *Cuscuta* are very similar.

The first appearance of the parasite in the field is noticed as small masses of branched, thread-like, leaf-less stems which are devoid of the green pigment and which twine around the stem or leaves of the host. Although *Cuscuta* species contain functional chlorophyll, the number of chloroplasts is extremely low and photosynthesis is only 1 or 2% of a normal green plant. The common color is cream yellow or orange but in some species the stem is frequently tinged with red or pink or it may be quite white. The leaves are represented by minute functionless scales which are evident on close examination. The tiny, white, pink or yellowish flowers occur in clusters. These flowers are gamosepalous and pentamerous. The stamens are inserted in the throat of the gamopetalous corolla, alternating with the lobes. Ovary is two loculate, each cell with two ovules. The styles may be distinct or united. The fruit is a capsule which may reach 2–3 mm across on maturity. The capsules do not readily dehisce and seeds remain on the plant long after maturity, Seeds are irregular in shape, rough surfaced and about 1 mm across. A single plant may produce as many as 3000 seeds.

Seeds of *Cuscuta* do not require any stimulus from the host plant and can germinate independently. Seeds buried in soil undergo a cycled induction of dormancy and breaking of secondary dormancy. Many seeds have a hard coat that requires gradual degradation or scarification. This ensures that all the seeds do not germinate at the same time. The germination occurs regardless of light or darkness. Germination is negligible, in *C. campetsris*, at 10°C and optimum at 30°C. The reserves of nutrient in the seed can support the shoot to 5–10 cm height, supported by a short stubby root, 1–2 cm long. A young seedling of the parasite is a slender, yellowish, unbranched thread. The top end is raised in more or less vertical position and the growing tip moves in anti-clockwise circle in search of suitable host. When a support is reached the young stem begins to twine around it in anti-clockwise fashion. If the support is a dead piece of wood or anything like that, the parasitic stem soon dies. However, on green host parts there may be tropism for light reflected by the green parts. On a susceptible host the stem of dodder soon produces haustoria to establish itself. The lower part then dies and stem becomes free from soil. Light and a relatively high temperature of 25°C are needed for the coiling and haustorium formation. Coiling is reduced under deep shade.

When the parasitic stem comes in contact with a host, minute root-like organs (haustoria) penetrate the host cortical parenchyma. The host parenchyma cells and the parasite are connected by plasmodesmata. The parasite ultimately reaches into the fibrovascular bundles entering the host phloem via plasmodesmata.. They serve as an anchor as well as organs of food absorption. The development of haustoria depends upon a supply of cytokinins which is partly available from the host. Although host-specificity is not known for *Cuscuta* spp., some sort of host-preference is reported. In mixed plant population a particular species of *Cuscuta* grows better on one plant species than on others.

In *C. campestris* the yellow to pale orange stems, about 0.3 mm in dia, generally do not twine and attach to the host stem directly but produce tendrils of similar appearance, arising opposite the scale leaves, which form coils and haustoria.

During the development of haustoria, first a sucker-like organ arises from the epidermis of the mother stem and adheres firmly to the host surface. The true haustorium develops later. It originates endogenously mainly from the cortical tissues just outside the pericycle like an adventitious root. The cells of the sucker or pre-haustorium dissolve their way into the host tissues, partly by pressure and partly by excreted enzymes, and into the space thus made the main haustorium grows.

The dodder perpetuates through seeds which ultimately fall on the ground and remain dormant until a favourable season returns. Portions of stems are also resistant to adverse weather. Many species are perennial. The parasite may be introduced into a field and spread by any of the following means:

1. As impurity in the crop seed: Seeds of dodder often go undetected in seeds of crops like berseem.
2. As seeds and stem pieces moved by irrigation water.
3. As stem pieces present on the dry straw from infested fields.
4. As seeds in the manure.
5. As stem pieces transported by cattle, birds, strong winds and farm implements.

Avoidance of above mentioned conditions should be ensured to check entry of the parasite and its spread in a field. The crop seed should be free from dodder seed, dodder infested cattle fodder should not be used, grazing animals should not be allowed to move through infested area, the badly infested crop should be burnt before the parasite produces seed, and the field should be left fallow after selected eradication measures have been completed and then a five year rotation beginning with a non host crop should be followed. Guar bean (*Cyamopsis tetragonoloba*) is not immune, but causes gross deformity and reduces the vigour of the parasite to the extent that it helps to protect a susceptible crop growin in mixture with guar. Deep shade suppresses the coiling and attachment of *Cuscuta*. Encouraging a dense crop canopy is, therefore, useful in reducing the parasite's activity. In soil solarization for 6 weeks, most dodder seeds are killed at soil surface in the first 10 days. Chicken manure reduces the solarization period from 6 weeks to 2 weeks.

If dodder is already present in the field, scattered patches may be sprayed early in the season with contact herbicides such as diesel oil fortified with DNBP (4,6-dinitro-o-butylphenol), PCP (pentachloro-phenol), or 2,4-D. This treatment as well as cutting and burning of patches, kill both the dodder and the host plant but prevent dodder from spreading and from producing seed. When dodder infestation in a field is widespread frequent tillage, burning, and use of soil herbicides such as chloropropham (one of the first to be used), DCPA, dichlobenil, dinoseb, pronamide, trifluralin, propyzamide, etc. have been recommended. These chemicals kill the dodder plant upon germination from seed but before it contacts the host plant. The amino acid inhibiting herbicide glyphosate is effective even after dodder has established connection with the host. However, there are reports that *Cuscuta campestris* has natural resistance to glyphosate.

Attempts at biological control of dodder have mainly involved the agromyzid fly *Melanogromyxa cuscutae* and gall forming weevils *Smicronyx* spp. Introduction of these parasites has been partially successful in controlling the weed at some places. Among fungi pathogenic on *Cuscuta*, *Alternaria cuscutacidae* is reported to have been used successfully on *C. campestris* in the former USSR, and a strain of *Colletotrichum gloeosporioides* has been used for many years in China for control of *C. chinensis* and *C. australis* on soybean.

GIANT OR TRUE LEAFY MISTLETOES (Banda; *Dendrophthoae*)

Dendrophthoe is a common parasite of fruit and wasteland and roadside trees. Its Sanskrit name, in ancient Indian literature, is "Vrikshabhaksha" meaning eater of trees describes the damage done by it and suggests that it was present in India in ancient time. In this country mango trees are the worst sufferers from this parasite. In northern India 60–90% of the old, desi type mango trees and a large number of other trees are heavily or moderately infested by this parasite. Unless taken care of at the right time, the entire tree becomes uneconomical in course of time.

Dendrophthoe spp. (Loranthaceae) are semi-parasites of tree trunks and branches. Their leathery and evergreen leaves possess chlorophyll and synthesize carbohydrate constituent of their food requirement. Since the parasite attacks the aerial parts of the host trees, situated far above the soil level, and since it is devoid of a true root system of its own, it is dependent on the host for water and mineral nutrients. Other manufactured food from the host also pass into the parasite. It obtains these by developing haustoria which grow into host tissues and become intimately associated with vascular elements. The continuous drain on nutrients by the parasite deprives the host of what is required for its own growth. Thus, in due course the attacked branch withers.

Dendrophthoe falcata, the common species reported in India, is a strongly branched and glabrous shrub. The stem is thick, erect or flattened at the nodes and appears to arise in clusters at the point of attack. This cluster forms a dense and bushy growth which can easily be spotted on the trees. The place at which the host is attacked and where the haustorium penetrates, often swells to form tumors which vary in size according to age of the parasite. Sometimes, the parasite, instead of confining its attack to one point, produces a creeping branch which grows closely along the host stem and forms haustoria at intervals. The flowers are borne in clusters. They are long and tubular in shape and usually greenish-white or red in color according to species. The fruit is fleshy and contains a solitary seed. It is sweet and eaten by birds, cattle and other animals.

The parasite is spread by dispersal of its seed mostly through birds and to some extent by other animals. The birds are attracted by the brilliant color of the fruit. The pulp is sticky and thus the seeds are easily carried by birds. When the seeds get deposited on other trees, at the junction of branches with the trunk, they germinate and give rise to haustoria, establishing the parasite. Droppings of birds containing seeds also help in dissemination of the parasite.

In early stages of the attack the damage to the tree may not be appreciable but later the parasite increases in vigor and the effects become apparent. Beyond the point of attack fresh growth of the host shoot is stunted. The damage done by the parasite is most marked in the production of new growth by the host. The quality and yield of fruits is considerably lowered. Leaves may be reduced in size and may show unhealthy green color. This is usually well marked in mango. The effect of the attack also depends upon the vigor of the host tree. A large tree, if mildly attacked, will not show any effect.

The commonly known method of control of the parasite is to top off the infected branches. It is important that branches should be cut sufficiently low, so that all vestiges of the haustorial system of the parasite are eradicated. In early stages of the growth of the parasite it can be easily detached from the host without damaging the latter. If the tumor is on one side of the branch then the wood just below the tumor may be sawed off.

Injection of copper sulphate and 2, 4-D into the affected branches has been found effective on many hosts. A spray of diesel oil emulsion in soap water is also effective in eradicating the parasite from mango trees.

BROOMRAPE (Orobanche spp.)

Orobanche is a member of plant family Orobanchaceae in Scrophulariales. It is a total root parasite affecting tobacco, eggplant, tomato, cabbage, cauliflower, turnip and many other Solanaceous and Crucifer plants. The weed is especially destructive to tobacco. In some areas of the world the parasite causes 10–70% loss. Plants attacked by broomrapes usually occur in small patches in the field and look stunted.

The parasite consists of a stout, fleshy stem, 15–50 cm tall. This stem is pale yellow or brownish red in color and is covered by small, thin, and brown scaly leaves, about 1 cm long. Although usually unbranched above ground, multiple stems sometimes arise from a single underground base. The base of the stem below ground is swollen and tuberous. The inflorescence occupying up to half the length of the stem carries many acropetally developing flowers. Flowers appear in axil of the scales and are white and tubular. The seeds are produced in ovoid pods, about 8–10 mm long, and are very small (0.2–0.4 mm) and black. A single plant carries 10 to 100 flowers (depending on species) and, hence, may produce 100000 or more (500000) seeds.

The parasite perennates through seed. The seed may remain viable for more than 10 years in absence of the host. Viability of *Orobanche* seed is retained after passage through digestive system of sheep. .Seed longevity varies with species of the parasite and is influenced by temperature and humidity. Lower these parameters, the greater is the longevity. Viability decline with passage of time. Seeds of *Orobanche cernua* buried in soil lost viability by 66% after 33 months and by 95% after 108 months. These seeds germinate only when roots of certain plants (host or non-host) grow near them. The growth regulators contained in the root exudates (indole-3-acetic acid, gibberellic acid and kinetin) stimulate germination, speed of germination and the structure of calli. Within a host species, root exudates of different genotypes differently affect *Orobanche* seed germination. Wild relatives of the cultivated host generally show resistance to invasion by *Orobanche*. In tomato, wild species *Lycipersicon hirsutum* and *L. chiliens* are highly resistant. In addition to the chemical stimulus, a moist environment for several days and suitable temperature is essential before the seed becomes responsive to the chemical stimulus. This preparatory period is known as conditioning or preconditioning. Conditioned seeds remain responsive to germination stimulants for a limited period beyond which secondary dormancy may be induced, especially at temperatures lower than optimum which is 15°–20°C for *O. cernua*. The ability to respond to germination stimuli also fades gradually when the seeds remain dry, and they then remain dormant until reconditioned. Within 24 hours of exposure to a germination stimulant *Orobanche* seeds release indole acetic acid (IAA) even before formation of a germ tube. Production of IAA is probably a part of a mechanism triggering germination of the seed.

On germination, the seed produces only a radicle which grows only a few mm long. towards the roots of host plants, becomes attached to it, and produces a shallow disc or cup-like appressorium which surrounds the host root, penetrates it with a mass of undifferentiated, polymorphic cells that extend to and, occasionally, into the xylem of the host root and absorb nutrients and water from it. Some of the cells differentiate into the xylem vessels of the parasite and establish contact with the xylem of the host. Other polymorphic cells becomes attached to the phloem cells of the host and obtain nutrition from them, transporting it back to the parasite. The penetration process through the cortex is assisted by the exudation of pectolytic enzymes. *Orobanche* plants have a long underground phase so that when they emerge most of the damage to the host has already occurred. A large number of parasitic stems may be seen breaking through the soil around the host plants. When the host is carefully uprooted, the parasite roots are seen intertwined with the host root system. The original root of the parasite continues to produce secondary roots which attach to other roots of the host. At the point of contact new stems of the parasite develop. As a result of drain on the food supply the growth of the host plant is checked, it remains stunted or may even die.

Response of the host to the parasite is affected by temperature. In sunflower crop infection occurs at 15° to 23°C but a delay in emergence of broomrape (*Orobanche cumana*) occurs at 15°C. At 27°C the level of infection is restricted. In *Orobanche minor* (small broomrape) on red clover, parasitism and temperature are strongly related. Low temperature delays infection and tubercle development. Resistance and susceptibility of sunflower to *O. cumana* and *O. aegyptiaca* is related to species of the weed and to temperature. A resistant variety becomes susceptible at low temperature (17/9°C, day/night) while susceptible variety is susceptible at all temperatures from 9° to 29°C. The soil below 22.5 cm depth is free from broomrape seeds. Very few seeds are present in 15–22 cm depth. The population of seeds of the parasite increases with decrease in soil depth and the maximum number of seeds are present in the top 5 cm of soil. Pathovars are reported in some species (*O. ramose*). A high proportion of *Orobanche* seedlings fail to penetrate roots in resistant plants. Resistance of a plant to invasion by *Orobanche* depends on lignification of host endodermis and pericycle cells at the penetration site. Callose deposition in cortical cells hampers the parasite growth and lignification of endodermal cells prevents entry into the vascular tissue at early infection stages. In addition, secretion of substances at infection site cause xylem occlusion which causes necrosis of established tubercles (Perez de Luque *et al.*, 2005; Moreno *et al.*, 2007). In *Vicia sativa*, resistance to development of *O. crenata* is attributed to formation of mucilage composed of carbohydrates and its accumulation in the host vessels, causing occlusion and prevention of the parasite development (Peroz de Lique *et al.*, 2006).

Although Orobanchaceae weeds are root parasites establishing connection with host vascular elements through their haustoria, Joel (2007) has reported a case where *Orobanche aegyptiaca* was directly attached to potato tubers (swollen part of underground stem). The adventitious roots of the parasite grew over the tubers and produced numerous haustoria that penetrated tuber skin, grew in the tuber tissues and produced xylem strands.

Sustainable control of the parasite can only be achieved by reducing the soil seed bank to levels of 1000–2000 seeds m^2 and maintaining it at that level during the subsequent years. Seeds of *Orobanche ramosa* succumb to drying. Six months of drying of soil in pots completely eliminates the parasite. Drying the soil for 3 months at 15–30° C considerably reduces the viability of seeds. Soil solarazation by mulching wetted soil with transparent polyethylene sheets for 50 days is reported to give complete control of *Orobanche creneta* in faba bean crop (Slaton *et al.*, 2001). Animal faeces (goat, sheep, cattle, chicken) applied to soil suppress *Orobanche ramosa* in potato ((Haider *et al.*, 2003). Effect of goat manure lasts throughout the growing season of the crop. Nitrogen in ammonium form is more inhibitory than nitrate to *Orobanche*. Ammonia toxicity does not affect seed germination but suppresses seedling radicle elongation.

The herbicides chlorsulfuron, pronamide and pendimenthalin effectively check *O. ramose* in fields for tomato. At 2.44 g a.i./ha, chlorsulfuron thoroughly mixed with soil prior to transplanting tomato completely prevented parasite infestation. However, there was some phytotoxicity. Lower dose of 0.61 g a.i/ha was not phytotoxic but gave significant but poorer control of the parasite. In sunflower, application of glyphosate, two applications of 40 g/ha or three applications of 20 g/ha at 12–14 days interval were found to effectively control *Orobanche cernua*. Application of the herbicides glyphosate at 80 g a.i/ha or imazaquin at 10 g a.i./ha completely suppressed *O. cranata* in faba bean (*Vicia faba*) but the yield was not increased. This was attributed to some adverse effect of the dosage of the herbicides on the host plants which counteracted the benefits of weed control.

Induced resistance is an innovative approach to manage *Orobanche ramose* (Gonstor *et al.,* 2004). Some plant activators such as benzothiadiazole, can activate resistance in the host against the weed. In sunflower application of benzothiadiazole (BTH) induces resistance against *Orobanche cumana*. The treatment does not interfere with broomrape seed germination or its chemotaxis but restricts its attachment to host roots and retards tubercle formation and development (Buschmann *et al.,* 2005). In the same host plant, soil application of the plant gtowth-retardant prohexadione-calcium induces resistance to *Orobanche cumana*. Strains of rhizosphere *Pseudomonas* spp., salicyclic acid derivatives and extracts of the alga *Ascophyllum nodosum* can decrease the broomrape infection to 80% by working as elicitors for resistance and activating immune responses. Exogenous application of salicylic acid to roots also reduces infection by Orobanche. Vurro *et al.,* (2006) observed suppression of seed germination of *O. ramosa* by exogenous application of some amino acids and suggested that:

    (i) appropriate amino acids applied exogenously to a root zone might result in the control of parasitic plants such as *Orobanche* spp.,

    (ii) amino acid excreting microbes introduced into the crop rhizosphere might control root parasites and

    (iii) amino acid producing biocontrol pathogens might be selected in order to enhance their virulence.

Fusarium toxins (fusaric acid and dehydrofusaric acid) inhibit seed germination or cause mortality of germinating seed of *Orobanche ramose, O. cernua, O.aegyptiaca,* and *O. cumana* (Dor *et al.,* 2007). In culture media the fungus starts producing the toxins within a week and maximum producti9on reaches in 4 weeks in non-shaking culture under illumination. These toxins are known to cause necrosis and wilting of plants attacked by vascular wilt fusaria. The hyperparasitic fusaria have gene encoding for indole-3-acetamide (AIM). Genetically transformed *F. oxysporum* with additional gene for encoding indole-3-acetic acid (IAA) has hypervirulence causing more suppression of number and size of *Orobanche* shoots. The auxin imbalance in the host plant works in a manner similar to low doses of auxin herbicides such as 2,4-dichlorophenoxy acetic acid. Thomas *et al.,* (1999) had emphasized the potential of *Fusarium oxysporum* f. sp. *orthoceras* as a practical biocontrol agent of *Orobanche* on sunflower. Low levels of conidial inoculum applied to the parasite post-emergence infected the parasitic shoots before they reached their flowering stage and produce seeds. Pre-plant soil inoculation with the conidial suspension reduced rate of emergence of *O. cumana* shoots by infecting the underground developmental stages including seed germ tubes. *O. cernua* on tobacco was also susceptible. Muller-Stover *et al.,* (2002) used chlamydospore inoculum of *F. oxysporum* f. sp. *orthoceras* for control of *O. cumana*. The inoculum was made granular in wheat flour-kaolin base and could be air-dried and stored. These workers (Shabana *et al.,* 2003) have proposed a formulation based on wheat-gluten matrix (pesta) as a suitable technique for the development of the fungus. Amending the pesta with yeast extract reduced the loss of viability of microconidia (40%) and chlamydospores (100%) during production of the formulation. The longest shelf-life (inoculum viability) was obtained when granules were stored at 3°C. Survival of *Fusarium* propagules in wheat-kaolin granules by addition of commercial iron fertilizer containg the iron chelator of EDDHA (Muller-Strover and Sauerborn, 2007). The amendment did not reduce loss of propagules during preparation of the formulatio/ Muller-Stover *et al.,* (2004) have compared pesta granules and sodium alginate pellets of *F. oxysporum* f. sp. *orthoceras* and noted that the pesta granules remained effective for longer time than the alginate pellets In a study of interaction between *O. aegyptiaca* and *Fusarium oxysporum* and *F. orthrosporioides* the fungal penetration of the weed was preceded by rapid loss of starch, with approx. 10% remaining in the weed at 9 hours. Penetration into the *Orobanche* tubercles began by 12 hours after inoculation. Hyphae penetrated the outer six cell layers by 24 hours reaching the center of the tubercle by 48 hours and infecting nearly all cells by 72 hours. Most of infected tubercle were dead by 96 hours. Breakdown of cell walls and the disintegration of cytoplasm in and around the infected cells occurred between 48 and 96 hours. However, these workers were of the opinion that these mycoherbicides lacked sufficient virulence to be effective in the field. Muller-Stover *et al.,* (2005) have reported that biocontrol efficiency of *F. oxysporum* f. sp. *orthroceras* can be increased by combining it with application of acibenzolar-S-methyl (the resistance inducer compound).

Abouzeid *et al.,* (2004) have studied 11 species of *Fusarium* for toxic metabolites against *O. ramose*. These included the known antagonists such as *F. oxysporum*, *F. proliferatum*, *F. solani*, *F. nygamai* and *F. verticilloides*. They found nine strains highly virulent. Four important criteria for *Fusarium oxysporum* to be effective as biocontrol agent in formulations are low loss of viability during the formulation process, satisfactory shelf life at room temperature, abundant chlamydospore formation and ability to colonize the rhizosphere. In addition to suppression of *Orobanche* by direct interaction, fusaric acid also plays some role. At nontoxic concentrations it could activate signal transduction components necessary for plant defense responses that could contribute biocontrol activity of *Fusarium* spp.(Bouizgarne *et al.,* 2006). In Egypt, Abodel-Kader and El-Mougy (2007) have claimed effective control of *Orobanche ramose* in tomato by using strains of *Trichoderma harzianum* and *T. viride* as mycoherbicide applied to soil followed by one spray of the herbicide glyphosate one day after transplanting.

Cells of the bacterium *Azospirillum brasilense* contain an alcohol soluble peptide that inhibits the germination and radicle growth of *Orobanche aegyptiaca* when applied to seed along with the germination stimulant (Dadon *et al.,* 2004). Muller-Stover and Kroschel (2005) have reported isolation of *Ulocladium botrytis* from disease *Orobanche crenata* which has potential to suppress this species but not others. El-Kassas *et al.,* (2005) isolated *Myrothecium verrucaria* from rhizosphere of *Vicia faba* grown in an Egyptian soil heavily infested with *Orobanche crenata*. The fungus was highly suppressive to seed germination of the weed. The inhibitor component was identified as verrucarin A. It suppressed *Orobanche* seed germination even at high dilution. Spores of the fungus could be added to soil for control of *Orobanche*. Zermane *et al.,* (2007) have isolated a strain of *Pseudomonas fluorescens* from faba bean rhizosphere which shows high biocontrol activity against underground pre-emergence structure of *O. crenata* and *O. foetida*. *Pseudomonas narginalis* strain also exhibited a tendency to reduce *O. crenata* Extracts of some higher plants stimulate germination of orobanche seeds. Some strains of the root nodule bacteria *Rhizobium leguminosarum* not only enhance growth of pea bust also protect roots from infection of *O. crenata* through necrosis of the parasite. The nodule bacteria increase polyphenol oxidase and hydrogen peroxide activity. There is accumulation of phenols and the phytoalexin pisatin in the roots (Mabrouk *et al.,* 2007).

Phytomyza orobanchia (Diptera, Agromyzidae) is a parasite of *Orobanche* only. Of the 140 *Orobanche* species described, the parasite is reported on 21 species (Klein and Kroschel, 2002). The larvae of *P. orobanchia* burrow in the shoot and capsules of *Orobanche* resulting in natural reduction of *Orobanche* seed production by 30–80%. Its efficacy is limited by low temperature and its own natural enemies.

The best way to eliminate broomrape is to destroy it before seed formation. If seeds have formed and shed on the soil it is difficult to eradicate the parasite for several years. Long day period, especially during summer, may help in reducing the soil population of seeds. Long crop rotations and use of catch crop may also help. Spraying the soil with 25% copper sulphate solution is reported to destroy the parasite. The herbicide glyphosate also effectively controls the parasite. In potato *O. aegyptiaca* and *O. ramosa* were controlled by split foliar application of the amino acid inhibiting herbicides imazapyr and rimsulfuron, starting 2 weeks after crop emergence and re-applied at 2 weeks interval. The latter is safe for potato tubers and can be used at 12–25 g/ha. The herbicide imazethapyr is considered safer and more effective than imazapyr. Rimsulfuron had been found to control *Orobanche ramose* but was phytotoxic to potato. Haidar *et al.,* (2005) have reported that application of rimsulfuron at 12.5 g ai/ha followed by sequential foliar application (3 times) of glyphosate at 100 g ai/ha controls the weed without phytoxicity to potato.

Intercropping with cereals reduces intensity of *Orobanche crenata* in legumes. Role of allelochemicals released by cereal roots is suggested as mechanism of inhibition of *Orobanche* seed germination (Fènandez-Aparicio *et al.,* 2007). Seeds of *Orobanche cernua* var. *desertorum,* are stimulated to germinate in presence of roots of several crops which do not support its growth. Their root exudates contain strigol, the germination stimulator. Chilli, mothbean, sorghum or cowpea can be used as trap crops to stimulate seed germination and destruction of the parasite. Zehhar *et al.,* (2003) tested 15 oilseed rape varieties and found them all susceptible to *O. ramosa.* Of the 4 carrot cultivars two were found resistant. In resistant carrot, the parasite germinated, became attached to the host root but became necrotic before emergence. In resistant plant species and cultivars, such as in legumes, the mechanisms involved in resistance have been identified as cell wall deposition, vessel occlusion and broomrape cellular disorganization. The disrupted cellular material of the host and secretions of the parasite inside the host vessels activate the response that leads to occlusion of host xylem vessels which causes starvation and necrosis of parasite tubercles. In a non-host plant (maize) thickening of xylem vessels, cell divisions in the central cylinder and formation of an encapsulation layer were associated with restricted development of haustoria. Parasitization by *Orobanche* initiates a rapid and sustained induction of a defence-related gene in the plant. Perez-de-Luque *et al.,* (2004) have reported systemic acquired resistance in pea against *Orobanche.* The resistance inducer compound benzothiodiazole (BTH) limits the success of attachment of the parasite with host roots and retards the development of established infections. Resistance in sunflower to *O. cumana* and *O. aegyptiaca* is temperature dependent. Resistance response of a resistant variety was expressed as the degeneration of the parasite tissues after its establishment on the plant roots. This stage was determined by temperature. As temperature rises, more tubercles degenerate and die. Degeneration rate of the parasite tubercles was about 5 times as high as that in the sensitive variety in the higher temperature range of 29°/21°C day/night.

WITCHWEED (Striga spp.)

Witchweed is a well known semi-root parasite of sugarcane, cereals, maize and millets in India as well as in Africa. There are about 30–35 species of *Striga* most of which are found in Africa where the weed causes an annual loss of about $ 7 billion. *Striga densiflora, S. euphrasioides* (*S. angustifolia*), *S. asiatica* and *S. lutea* are reported in India. *S. lutea* is now considered synonymous with *S. asiatica*. *S. densiflora* resembles *S. asiatica* and *S. hermonthica* in most aspects of biology and ecology. *S. hermonthica* (purple witchweed) is predominantly present throughout Africa where other species are also found. *S. angustifolia* is a facultative, not obligate, parasite. It can germinate and establish without the involvement of a host plant.

The attacked plants remain stunted and chlorotic. Heavily infected plants wilt and die. Apart from sapping the host of water and nutrients, *S. asiatica* has a profound physiological effect on the host from a very early stage, apparently as a result of some toxic or growth regulatory influence passing from the parasite to the host. Rank *et al.,* (2004) have reported that decimation of cereal growth and yields by *Striga* cannot be accounted for entirely by the removal of host plant nutrients. The production of cytotoxic compounds by the parasite has been suggested. In *S. hermonthica*, it has been shown that photosynthesis is substantially reduced, at least partly due to the closure of host stomata, associated with raised levels of inhibitors such as abscisic acid. These influence, occurring in *S. asiatica* also, cause stuntung of the plant and increased host root growth. In hosts affected by *S. asiatica* there is also a tendency for infected plants to show wilting symptoms even under most moist conditions. In sorghum, *Striga hermonthica* is reported to reduce stem height by 22% and weight by 25% at 38 days after planting. After 64 days the reduction is 34% and 36%, There is no effect on leaf weight. Root growth is stimulated by the parasite 38 days after planting but not after 64 days.

Most *Striga* species are favoured by relatively dry, infertile soils and prefer low humidity and high temperature. The parasite is a small plant, 15–30 cm tall, with bright green, slightly hairy stem and leaves. Usually these plants develop in clusters around the stem of the host. The stem appears branched. Leaves are narrow, long and in opposite pairs. The *Striga* species are abnormal in having almost permanently open stomata, ensuring a high transpiration rate and enhancing the flow of water and food materials from host to parasite. The flowers are small and usually brick red or scarlet, although some may be yellowish red, yellowish or almost white, always having yellow centers. Flowers appear just above the leaf attachment to the stem and are produced throughout the season. Seeds are formed in pods or capsules, about 5 mm long. Each capsule contains thousands of tiny brown seeds, about 0.3 mm long. A single plant of witchweed may produce 50,000 to 500,000 seeds. These seeds are extremely light each weighing about 7 micrograms. Viability of seeds is reported to be 12–40 years. All the seeds produced by the parasite do not germinate in the same season even in presence of the host. Short distance dissemination of the weed takes place by the seeds through rain and irrigation water while floods and strong winds are the chief agents of long distance dispersal of seeds. Only seeds of witchweed are responsible for long distance dispersal of the weed. Wind and cattle have no significant role.

The parasite survives through seeds which require a minimum dormancy of 15–18 months. However, seed viability and germination declines under moist conditions. Most *Striga* species have insufficient resources to establish seedlings independently and thus depend on attachment to a host root within a few days of germination. To improvethe chances of such attachment the seeds generally germinate only after inbibition in water and stimulation by a substance exuded from roots of a potential host or nonhost plant. The natural stimulants in root exudates have been identified as a group of closely related lactone compounds, including strigol and sorgolactone which appear to act on the seed through generation of ethylene. Level of such compounds varies with genotype of a trap plant species. Certain synthetic compounds also stimulate *Striga* seed germination. Germination in some species is better in light while in others seeds germinate better in darkness. However, in most species seed germination can occur at a depth of 40 cm in the soil. The endosperm nutrients in the weed seed can sustain the seedling only for 3–7 days in absence of the host. If there is no attachment with a host root within this period, the parasite dies. If a host root is within 2–3 mm of the parasite seedling, chemical signals are exchanged that direct the *Striga* radicle to the host, initiate haustorium formation and then successful attachment and xylem to xylem connection between the parasite and the host. One of the chemical signals is 2,6–dimethoxy-*p*-benzoquinone which is probably a product of enzymatic degradation of the host root.. As soon as the witchweed rootlet comes in contact with the host root, its tip swells into a conical or bulb-shaped haustorium which presses against the host root. The haustorium dissolves host cells by enzymic secretions and penetrates the host roots within 8-24 hours. The haustorium advances into the roots through dissolution of host cell walls. Finally, its leading cells, usually the tracheids, reach the vessels of the host roots. The walls of the host vessels are dissolved and the tracheids of the parasite enter the host vessels from which they absorb water and nutrients. In *Striga asiatica,* penetration of the epidermis of the host (sorghum) root and advance into the cortex occur within 24–48 hours of inoculation. Penetration of the endodermis is delayed for 72–96 hours after initial contact. After penetration, vascular continuity is established between the host and the parasite. In interactions with a non-host there is active resistance by the plant. The growth of the parasite is arrested in the cortex and its cells become necrotic 72 hours after inoculation. HR is involved in resistance of the nonhosts. It is characterized by expression of necrotic lesions at the haustorial attachment sites. In susceptible hosts there is no such reaction (Mahamed *et al.,* 2003).

Striga is obligate root parasite wholly dependent on the host from germination to flowering and reproduction. Being achlorophyllous and obligate parasite during its underground development, upon emergence from the soil it becomes green leafy plant. However, despite the presence of chlorophyll, the plants exhibit only low rates of photosynthesis.

After establishment of vascular connection with the host the parasite grows underground for 4–8 weeks prior to emergence aboveground. The underground stem contains buds in the axils of leaves. The parasite produces more roots from the initial rootlet which move parallel to host roots and send more haustoria into them. Several hundred separate witchweed plants may parasitize a single host plant but only few reach the soil surface. Numerous parasitic attachments occur on the same plant. Most of the damage occurs during this period. After emergence although the parasite develops leaves and photosynthesizes, it still cannot survive in the absence of the host attachment.

S. asiatica begins to flower 4 weeks after emergence. Seeds are mature 4 weeks after flowering. The parasite can be found in light as well as in heavy soils in "rabi" and "kharif" seasons. The parasite is more common in light sand, sandy loam and loam soils than in heavy clay. The growth and flowering of the parasite in different soil types are influenced by temperature. Sorghum roots releases resistance-conferring substances to the infection points after sensing infection. But in fully infected sorghum with widely distributed infection points there is less release of such substances making the plant more vulnerable. The species of *Striga* serve as virulence gene reservoirs and if they hybridize the hybrids are virulent and fertile. However, there is no 100% compatibility between the species.

Seed germination of *Striga* is associated with ethylene production. Biosynthesis of ethylene is stimulated by auxins and other compounds from the plant root Conditioning of seeds under warm and moist conditions prior to exposure to the stimulants is necessary for the stimulation of ethylene synthesis. The complex seed germination requirements and specialized physiology of *Striga asiatica*, together with certain ecological preferences can be exploited to some extent in cultural control of witchweed. Preference for relatively dry conditions can be countered by irrigation where applicable. Preference for infertile soil conditions can be countered by improving soil fertility through the use of nitrogenous manures, green manuring or rotation with leguminous crops. Long-term improvement in fertility by increasing organic content of the soil is preferable over inorganic nitrogen sources. Sorghum roots containing more than 2% nitrogen cause least stimulation of *Striga* seed germination. Preference for low humidity and high transpiration rates can be countered by dense crop planting, intercropping with leafy species or planting when rains have started. *Striga* seeds are present in the upper soil layer. If roots can avoid this layer, germination of the weed seed and infection of the crop host can be avoided. In a study in Kenya, deep planting, reducing the root length in the upper soil layer was found to reduce *Striga* infection in sorghum and maize. The effect was better in no-till cultivation. There is less *Striga* damage in transplanted maize under rain fed conditions

Striga hermonthica maintains its leaf temperature very low compared to the atmospheric temperature. It is because of high rate of transpiration and conseuqnt lowering of temperature through vaporization. When a film-forming compound is applied to the leaves, the process of transpiration and vsaporizastion is checked. As a result leaf temperature rises to a level where the leaf shrinks and dries. This could be physical method of control of *Striga*.

Weeding and interculture can be successful only when practiced in early stages of the parasitic growth. Within two months 4–6 weedings may be needed. This makes the cost of this method prohibitive. Use of trap crops such as legumes selected for their ability to cause suicidal germination of *Striga* seed in rotation with susceptible crops is effective but effective reduction of the parasitic seeds in field soils normally requires several seasons of the legume crop. For a maize-based farming system, intercropping with legumes combined with hand weeding of mature can control the weed. *Striga* plants for control of the weed. Interculture of cowpea with sorghum reduces striga infestation. In Kenya, the fodder legumes *Desmodium uncinatum* and *D. intortum* grown with maize to control insect damage was found to dramatically reduce the infestation of maize by *Striga* including *S. hermonthica*. It was better than other legumes used as intercrop. The allelopathic effect was due to water-soluble chemical components of the root exudates which stimulated germination of *S. hermonthica* seeds and also inhibited haustorial development (Khan *et al.*, 2002). The intercropping also checks damage by maize stem borer. Certain varieties of cowpea have proved to be trap crops for *Striga*. Cultivating cowpea in infested soil before susceptible maize is grown gives significant reduction in *Striga*. Cotton is not a host of *S. hermontheca* but its root exudates stimulate its seed germination and death of the parasitic weed. Genotypes of cotton differ in their effect on *Striga* (Botanga *et al.*, 2003). Catch crops consisting of host plants may be planted to force the germination of witchweed seed, and the parasite then can be destroyed by ploughing or by use of weedicides such as 2, 4-D. Trap crops, consisting mostly non-host plants, may be used to stimulate seed germination. The parasite cannot infect these plants and, therefore, is starved to death. However, the field is to be kept free from weed host plants. Crop rotation with non-host cultivars, selected to stimulate parasite seed germination, is a successful control measure against *Striga*. Such host or nonhost plants are selected on the basis of parasite-plant interaction in the locality. *Vigna unguiculata* stimulates germination of *Striga* seeds. *Celosia argentia* (Amaranthaceae), a non-host, stimulates suicidal seed germination of *Striga*. Such host or nonhost plants must be producing germination stiomulants in the root exudates. In a study in Uganda (Africa) inter-planting of *Celosia* with sorghum reduced *Striga* emergence by 55% and increased sorghum yield by 35%. In screen house tests, inter-planting of *Celosia* with sorghum (2:1 ratio) suppressed the weed by 48% and increased yield by 100% (Olupot *et al.*, 2003). Inclusion of soybean, with application of single super phosphate fertilizer, preceding maize in the rotation, was found to reduce *S. hermontica* infestation in maize in Nigeria. Sesame, an oilseed plant, known for its soil sanitizing effect since ancient times and reported to suppress root knot nematodes, has been shown in West Africa to suppress *S. hermonthica* in pearl millet. Interaculture of sesame with pearl millet or its use as a rotation crop reduce the parasitic weed in the pearl millet crop. The emerged *Striga* numbers and *Striga* fruiting were highly reduced (Hess and Dodo, 2004).

Berner *et al.,* (1999) used ethylene producing bacteria (*Pseudomonas syringae* pv *glycinea*) to effectively promote seed germination in *Striga* spp. Strains of the bacterium were as effective as a synthetic germination stimulant. The germination of seed is suicidal since the parasite cannot survive after germination in absence of nourishment from a host. Ahonsi *et al.,* (2003) have evaluated ethylene producing strains of *Pseudomonas* in combination with nitrogen-fixing *Bradyrhizobium japonicum* for *Striga hermonthica* control. Co-inoculation of one or two preceding crops of cowpea or soybean with ethylene producing *Pseudomonas syringae* pv.*glycinea* and the rhizobium has reduced *S. hermonthica* parasitism on subsequent maize crop. Ahonsi *et al.,* (2002b) studied the effects of soil pasteurization and soil N status on severity of *Striga hermonthica* (Del.) Benth. in maize. In natural soil, application of nitrogen to maize reduced the number of emerging *S. hermonthica* plants by 53% and increased maize dry matter by 154%. In pasteuriozed soil, nitrogen application increased *S. hermonthica* severity by 26% and also significantly increased maize dry weight. Differences in weed infestation in natural and pasteurized soil were attributed to soil biotic factors that reduce *S. hermonthica* in natural soil. In soil inoculated with the arbuscular mycorrhizal fungi (*Glomus clorum* and *Gigaspora margarita*), the number of *S. herminthica* shoots in sorghum and maize is reduced by 50% and 30%, respectively (Lenzemo *et al.,* 2005; Gworgwor and Weber, 2003).

Miche *et al.,* (2000) had reported that two strains of the rhizobacterium *Azospirillum brasilense* from sorghum rhizosphere prevented seed germination of *S. hermonthica* although they were stimulated to germinate in presence of sorghum roots. Culture filtrates of the bacterium also inhibited *Striga* seed germination. Ahonsi *et al.,* (2002) identified 15 isolates of *Pseudomonas fluorescens* and *P. putida* that significantly inhibited seed germination of *Striga hermonthica*. Bacterization of maize seed with any of these isolates reduced the number of *Striga* plants and increased maize plant growth. However, the bacteria could not retain their efficacy during the dry period after the crop harvest and needed periodic application.

Fungal toxins are known to inhibit witchweed seed germination. Some toxins cause 100% inhibition of germination at very low concentration. Many *Fusarium* spp. cause wilting of *Striga* plants. *Fusarium nygamai* is a hyperparasite of *Striga hermonthica* which penetrates the seed coat and destroys the endosperm. It also attacks the seed germ tubes. Marley *et al.,* (1999) isolated a strain of *Fusarium oxysporum* from wilted *Striga* plants and claimed that it was more effective than other species in controlling the weed. The isolate grown on sorghum grain and incorporated into soil completely inhibited the emergence of *S. hermonthica*. Ciotola *et al.,* (2000) used sorghum straw powder for obtaining maximum chlamydospores of *F.oxysporum* strain M12–4A, a biocontrol agent against *S. hermonthica*. Germination of chlamydospores was governed by the presence of exogenous carbon, nitrogen and sorghum root exudates. In field trials, application of chlamydospores reduced *S. hermonthica* emergence by 92%. Complete inhibition of *S. hermonthica* emergence occurred when the chlamydospore powder was added to the soil at sowing time and when sorghum seeds coated with chlamydospores were sown. Pesta granule (0.5–2 mm particle size) based chlamydospore formulation of a *Fusarium oxysporum* (Foxy 2) are reported to maintain 85–100% viability of chlamydospores for at least 1 year when stored at 4 C (Elzein *et al.,* 2004b,c). Marley and Shebayan (2005) obtained effective control of *Striga hermonthica* by spot application of 5–10 g of the mycoherbicide. *F.oxysporum* (Foxy 2) has potential to suppress both *S. hermonthica* and *S. asiatica* causing almost complete prevention of emergence of the parasite (Elzein and Kroschel, 2004a). Microconidia and fresh or dried chlamydospores of Microconidia and fresh or dried chlamydospores of *Foxy* 2 in gum Arabic as carrier applied to sorghum seeds remained viable for 8 months and colonized 27% *Striga* plants (Elzein *et al.,* 2006). An isolate of *F. oxysporum* was found to be host specific being more effective against species of Striga than on some other parasitic weeds ((Marley *et al.,* 2005) Isolates of *Fusarium solani* are also reported to inhibit *Striga* seed germination especially in presence of seed germination stimulants. The fungus produces trichothecens. Some of these metabolites can inhibit seed germination even at 24 μM (Sugimoto *et al.,* 2002). The amendment of soil with termite (*Cubitermis*) mound-powder is reported to provide indirect biological control of *Striga* in sorghum. The powder is not directly effective against the weed .but in amended soil it suppresses emergence of the weed through altered microbial community structure. Boari (2003) had reported an unidentified *Fusarium* sp. isolated from *Striga hermonthica* which controlled the weed. In liquid and solid media the fungus produced toxic compounds (fusaric acid and dehydrofusaric acid that caused total suppression of *Striga hermonthica* seed germination

Plant products have also been used to control *Striga*. Seed and leaf powder of neem (*Azadirachta indica*) and fruit and peel powder of parkia tree (*Parkia biglobosa*) are reported to suppress emergence of the weed (Marley *et al.,* 2004) and increase sorghum grain yield. Crude leaf extract of *Sesbania sesban* us reported to suppress *Striga* seed germination by 49%. Mulch of such leaves may be useful in the control of *Striga*.

Complete resistance to *Striga*, in most cultivated host plants· is lacking. However, resistance is present in wild relatives of the cultivated plants. In Africa, where Even though host-plant genotype plays a significant role in *Striga* reproduction, calculations have shown that only at very low infestation levels the uswe of the most resistant genotype is able to lower the *Striga* seedbank. *Striga hermonthica* is a serious weed in maize, *Tripsacum dactyloides,* a wild relative of maize, has been found to possess high level of resistance. Development of *Striga* is arrested after attachment to its roots. Vascular continuity is established between parasite and host but there is poor primary haustorial tissue differentiation. In sorghum, sources of resistance to damage by *Striga* are found in its wild relatives *Sorghum arundinaceum* and other species, subspecies and races of *Sorghum.* Wild sorghum species resist *Striga* through:

  (i) low germination stimulant production,
  (ii) germination inhibition, and
  (iii) low haustorial initiation activity.

Chemicals have been used for outright killing of *Striga* or for stimulation of its seed germination. Soaking the soil up to a depth of 10–15 cm with 2–3% copper sulphate solution was once recommended for the control of witchweed. Spraying with Fernoxone (80% sodium salt of 2, 4-D) at the rate of 400 g of the chemical in 500 lit of water is very effective against *Striga*. Application of tetrachlorodimethyl phenoxyacetic acid also kills the parasite outright. Sorghum seed treatment with 1.5M brine (sodium chloride) solution suppresses emergence of *Striga* (Gworgwor *et al.,* 2002). Sodium salts of 2, 4-D are toxic to many crops used in rotation with hosts of *Striga*. Such 2, 4-D sensitive crops are cotton, tobacco and soybean. Amine salt formulations of 2, 4-D are preferable. Application of the herbicides imazapyr (30–40 g/ha) or pyrithiobac (11–21 g/ha) suppresses witchweed emergence in herbicide-resistant maize. Seed dressing with the herbicides is also effective provided the host crop is resistant to the herbicide (Kanampiu *et al.,* 2003). Soil injection of methyl bromide under polyethylene cover has been successfully used to eliminate *Striga*. Injection of 2-chloroethyl phosphonic acid (ethylene) at the rate of 1 kg/ha into light sandy fallow soils at about 25°C which have been wetted stimulate seed germination of witchweed. Strigol (strigolacone), a natural stimulant, also works in the same manner as ethylene. Strigolactne is a sesquiterpene lactone and is found in roct exudates of many non-host plants which are known to reduce incidence of *Striga* by stimulating untimely germinatrion of seed. The compound is detected by *Striga* and *Orobanchre* as qell symbiont root fungi. A acetolactate synthase inhibitor, imazoquin, is effective as seed soak to protect the host from *Striga* infection.

Keeping the field under weed-free fallow increases emergence of *Striga*. Keeping the field flooded for sometime and then draining out the water also helps in the control of the parasite. In some areas, sowing delayed for 30 days from the normal time of sowing has shown reduced incidence of *Striga hermonthica* in maize and sorghum but yields are better in normal date of sowing with infestation of the weed.

REFERENCES

Abodel-Kader, M.M. and N.S. El-Mougy. 2007. Applicable control measure against *Orobanche ramose* in tomato plants. *Aust. Plant Pathol.* 36(2): 160

Abouzeid, M.A., A. Boari, M.C. Zonno, M. Vurro and A. Evidente. 2004. Toxicity profiles of potential biocontrol agents of *Orobanche ramose*. *Weed Science* 52(3): 326.

Ahonsi, M.O., D.K. Berner, A.M. Echehe and T.S. Lagoke. 2002a. Selection of rhizobacterial strains for suppression of germination of *Striga hermonthica* seeds. *Biological Control* 24(2): 143

Ahonsi, M.O., D.K. Berner, A.M. Emechebe and S.T. Lagoke. 2002b. Effect of soil pasteurization and soil N status on severity of *Striga hermonthica* in maize. *Soil Biol. Biochem.* 34(11): 1675

Ahonsi, M.O., D.K. Berner, A.M. Emechebe, S.T. Lagoke and N. Sanginga. 2003. Potential of ethylene-producing pseudomonads in combination with effective N_2-fixing bradyrhizobial strains as supplement to legume rotation for *Striga hermonthica* control. *Biological Control* 28(1): 1

Berner, D.K., N.W. Schaad and B. Volksch. 1999. Use of ethylene-producing bacteria for stimulation of *Striga* spp. seed germination. *Biological Control* 15: 274

Boari, A., A. Evidente, A E. Idris *et al.*, 2003. Identification of phytotoxic metabolites of a new *Fusarium* sp. inhibiting germination of *Striga hermonthica* seeds. *Phytopath. Mediterr.* 42(1): 65.

Botanga, C.J., S.O. Alabi, C.A. Echekwu and S.T.O. Lagoke. 2003. Genetics of suicidal germination of *Striga hermonthica* (Del.) Benth by cotton. *Crop Science* 43: 483

Bouizgarne, B., H.H. El-Maarouf-Bouteau, K. Madion *et al.*, 2006. A putative role for fusaric acid in biocontrol of the parasitic angiosperm *Oroibanche ramoda*. *Mol. Plant Microbe Interact.* 19(5): 550

Buschmann, H., Z.-W. Fan and J. Sauerborn. 2005. Effect of resistance-inducing agents on sunflower (*Helianthus annuum* L.) and its infestation with the parasitic weed *Orobanche cumana*. *J. Plant Dis. Prot.* 112(4): 386

Ciotola, M., A. DiTommaso and A.K. Watson. 2000. Chlamydospore production, inoculation methods and pathogenicity of *Fusarium oxysporum* M12-4A, a biocontrol for *Striga hermonthica*. *Biocontrol Sci. Technol.* 10(2): 129

Dadon, T., N.B. Nun and A.M. Mayer. 2004. A factor from *Azospirillum brasilense* inhibits germination and radicle growth of *Orobanche aegyptiaca*. *Israel J. Plant Science* 52(2): 83–86

Dor, E., A. Evidente, C. Amalfitano *et al.*, 2007. The influence of growth conditions on biomass, toxins and pathogenicity of *Fusarium oxysporum* f. sp. *orthoceras*, a potential agent for broomrape biocontrol. *Weed Research* 47(4): 345

El-Kassas, R., Z.K. El-Din, M.H. Beale, J.L. Ward and R.N. Strange. 2005. Bioassay-led isolation of *Myrothecium verrucaria* and verrucarin A as germination inhibitors of *Orobanche crenata*. *Weed Research* 45(3): 212

Elzein, A., J. Kroschel and D. Muller-Stover. 2004a Effects of inoculum type and propagule concentration on shelf life of Pesta formulations containing *Fusarium oxysporum* Foxy 2, a potent mycoherbicide agent for *Striga* spp. *Biol. Control* 30(2): 203

Elzein, A., J. Kroschel and D. Muller-Stover. 2004b Optimization of storage conditions for adequate shelf-life of Pesta formulation of *Fusarium oxysporum* Foxy 2, a potential mycoherbicide for *Striga*. Effects of temperature, granule size and water activity. *Biocontrol Sci. Technol.* 14(6): 545

Elzein, A. and J. Kroschel. 2004c. *Fusarium oxysporum* Foxy 2 shows potential to control both *Striga hermontheca* and *S. asiatica*. *Weed Research* 44(6): 433

Elzein, A., J. Kroschel and V. Leth. 2006. Seed treatment technology: an attractive delivery system for controlling root parasitic weed *Striga* with a mycogerbicide. Review. *Biocontrol Sci. Technol.* 16(1): 3

Fenandez-Aparicio, M., J.C. Sillero and D.R. Rubiales. 2007. Intercropping with cereals reduces infection by *Orobanche crenata* in legumes. *Crop Prot.* 26(6): 1166–1172

Gonstor, G., H. Buschmann, G. Szinicz *et al.*, 2004. Induced resistance- an innovative approach to manage branched broomrape (*Orobanche ramosa*) in hemp and tobacco. *Weed Science* 52(6): 1050

Gworgwor, N.A., A.I. Hudu and S.D. Joshua. 2002. Seed treatment of sorghum varieties with brine (NaCl) solution for control of *Striga hermonthica* in sorghum. *Crop Protection* 21(10): 1015

Gworgwor, N. and H. Weber. 2003. Arbuscular mycorrhizal fungi-parasite-host interaction for the control of *Striga hermonthica* (Del.) Benth.in sorghum. {*Sorghum bicolor* (L.) Moench}. *Mycorrixa* :13(5): 277–283

Haider, M.A., M.M. Sidahmed, R. Darwish and A. Lafta. 2005. Selective control of *Orobanche ramosa* in potato with rimsulfuron and sublethal doses of glyphosate. *Crop Protection* 24(8): 743

Hess, D.E. and H. Dodo. 2004. Potential for sesame to contribute to integrated control of *Striga hermonthica* in the West African Sahel. *Crop Protection* 23(6): 515

Joel, D.M. 2007. Direct infection of potato tubers by the root parasite *Orobanche aegyptiaca*. *Weed Research* 47(4): 276

Kanampiu, F.K., V. Kabambe, C. Masawe, L. Jasi, D. Frieser and J.E. Ransom. 2003. Multi-site, multi-season tests demonstrate that herbicide seed coating herbicide-resistant maize controls *Striga* sp. and increases yields in several African countries. *Crop Protection*. 22(5): 697

Khan, Z.R., A. Hassanali, W. Overholt, T.M. Khamis *et al.*, 2002. Control of witchweed *Striga hermonthica* by intercropping with *Desmodium* spp. and the mechanism defined as allelopathic. *Jour. Chem. Ecol.* 28(9): 1871

Klein, G. and J. Kroschel. 2002. Biological control of *Orobanche* spp. with *Phytomyza orobanchia*, a review. *BioControl* 47(3): 245

Lenzemo, V.W., Th. W. Kuyper *et al.*, 2005. Field inoculation with arbuscular mycorrhizal fungi reduces *Striga hermonthica* performance on cereal crops and has the potential to contribute to integrated *Striga* management. *Field Crop Res.* 91(1): 51

Mabrouk, Y., P. Simeir *et al.*, 2007. Molecular and biochemical mechanisms of defence induced in pea by *Rhibium leguminosarum* against *Orobanche crenata*. *Weed Research* 47(5): 452

Marley, P.S., S.M. Ahmed, J.A.Y. Shebayan and S.T.O. Lagoke. 1999. Isolation of *Fusarium oxysporum* with potential for biocontrol of witchweed (*Striga hermonthica*) in the Nigerian Savanna. *Biocontrol Sci. Technol.* 9(2): 159

Marley, P.S., J. Shebayan, D.A. Aba and N. Idem. 2004. Possibilities for control of *Striga hermonthica* in sorghum (*Sorghum bicolor*) using neem (*Azadirachta indica*) and parkia (*Parkia biglobosa*) based products. *Int. J. Pest Manage.* 50(4): 291

Marley, P.S. and J.A.Y. Shebayan. 2005. Field assessment of *Fusartium oxysporum* based mycoherbicide for control of *Striga hermonthica* in Nigeria. *BioContgrol* 50(2): 389

Marley, P.S., J. Kroschel and A. Elzun. 2005. Host specificity of *Fusarium oxysporum* Schlect (isolate PSM 197), a potential mycoherbicide for controlling Striga spp. *Weed research* 45: 407

Matusova, R.K., K. Ram, *et al.*, 2005. The strogolaccyon germination stimulatisnof the plant-parasite *Striha* and *Orobanche* spp. are derived from the carotenoid pathway. *Plant Physiol.* 139: 920.

Mayer, A.M. 2006. Pathogenesis by fungi and by parasitic plants: Similarotoes and differences. *Phytoparasitica* 34(1): 13

Miche, L., M.-L. Bouillant, R. Rohr, G. Salle and Bally. 2000. Physiological and cytological studies on the inhibition of *Striga* seed germination by the plant growth promoting bacterium *Azospirillum brasilense*. *Eur. J. Plant Pathol.* 106(4): 347

Moreno, P., S. Testillano and D. Rubiales. 2007. Resistance to broomrape (*Orobanche crenata*) in faba bean (*Vicia faba*): cell wall changes associated with prehaustorial defenseive mechanisms. *Ann. Appl. Biol.* 151(1): 89

Muller-Stover, D. and J. Saueborn. 2007. A commercial iron fertilizer increases the survival of *Fusarium oxysporum* f. sp. *orthocerasa* propagules in a wheat-kaolin formulation. *Bioxontrol. Sci. Technol.* 17(6): 597

Muller-Stover, D., J. Kroschel, H. Thomas and J. Sauerborn. 2002. Chlamydospores of *Fusarium oxysporum* Schlecht f. sp. *orthoceras* (Appel & Wollemw) Bilai as inoculum for wheat-flour kaolin granules to be used for the biological control of *Orobanche cumana* Wallr. *Eur. J. Plant Pathol.* 108(3): 221

Müller-Stöver, D., H. Thomas, J. Sauerborn and J. Kroschel. 2004. Two granular formulations of *Fusarium oxysporum* f. sp. *orthoceras* to mitigate sunflower broomrape *Orobanche cumana*. *BioControl* 49(5): 595

Muller-Stover, D., H. Buschmann and J. Saureborn. 2005. Increasing control reliability of *Orobanche cumana* through integration of a biocontrol agent with a resistance-inducing chemical. *Eur. J. Plant Pathol.* 111(3): 195

Muller-Stover, D. and J. Kroschel. 2005. The potential of *Ulocladium botrytis* for biological control of *Orobanche* spp. *Biological Control* 33(3): 301

Nadler-Hassar, T. and B. Robin. 2003. Natural tolerance of *Cuscuta campestris* to herbicides inhibiting amino acid biosynthesis. *Weed Research* 43(5): 341

Oluput, J.R., D.S.O. Osiru, J. Oryokot and B. Gebrekidan. 2003. The effectiveness of *Celosia argentia* (Striga-chaser) to control *Striga* on Sorghum in Uganda. *Crop Prot.* 22(3): 463

Perez-de-Luque, A., J.V. Jorrin and D. Rubiales. 2004. Crenate broomrape control in pea by foliar application of benzothiadiazole (BTH). *Phytoparasitica* 32(1): 21

636

Perez-de-Luque, A., D. Rubiales, J.L. Cubero, M.C. Press *et al.*, 2005. Interaction between *Orobanche crenata* and its host legumes: Unsuccessful haustorial penetration and necrosis of the developing parasite. *Ann. Bot.* 95(6): 935

Perez-de-Luque, A., M.D. Lozano *et al.*, 2006. Mucilage production during the incompatible interaction between *Orobanche crenata* and *Vucia sativa. J. Exp. Bot.* 57(4): 931

Rank, C., L.S. Rasmussen, S.R. Jensen *et al.*, 2004. Cytotoxic constituents of *Alectrea* and *Striga* species. *Weed Research* 44(4): 265

Shabana, Y.M., D. Muller-Stover and J. Sauerborn. 2003. Granular pesta formulations of *Fusarium oxysporum* f. sp. *orthoceras* for biological control of sunflower broomrape: efficacy and shelf life. *Biological Control* 26(2): 189

Slaton, N.A., R.D. Cartwright, J. Meng, E.E. Gbur Jr and R.J. Norman. 2003. Soil solarization, a non-chemical technique for controlling *Orobanche crenata* and improving yield of faba bean. *Agronomie* 21: 757

Sugimoto, Y., N.E. Ahmed, N. Yasuda and S. Inanaga. 2002. Trichothecene inhibitors of *Striga hermonthica* germination produced by *Fusarium solani. Weed Science* 50(5): 658

Thomas. H., A. Heller, J. Sauerborn and D. Muller-Stover. 1999. *Fusarium oxysporum* f. sp. *orthoceras*, a potential mycoherbicide, parasitizes seeds of *Orobanche cumana* (sunflower broomrape): a cytological study. *Annals of Botany* 83: 453

Vurro, M., A. Boari, A.L. Pilgram and D.C. Sands. 2006. Exogenous amino acids inhibit seed germination and tubercle formation by *Orobanche ramose)*. Potential appliocation for management of parasitic weeds. *Biological Control* 36(2): 235

Zehhar, N., P. Labrousse *et al.*, 2003. Study of resistance to *Orobanche ramose* in host (oilseed rape and carrot) and non-host (maize) plants. *Eur. J. Plant Pathol* 109 (1): 76

Zermane, N., T. Souissi, J. Kroschel and R. Sikora. 2007. Biocontrol of broomrape (*Orobanche crenata* Forsk and *Orobanche foetida* Poir.) by *Pseudomonas fluorescens* isolate Bf7–9 from the faba bean rhizosphere. *Biocontrol Sci Technol.* 17(5): 483

CHAPTER 15

Non-Infectious and Non-Parasitic Diseases

Plant diseases in which no foreign organism or parasite is associated with the cause are known as non-parasitic disease. They differ from virus diseases in being non-infectious. Such diseases are induced by certain disorders in the physiology of plants due to unfavorable environments including the soil conditions as listed below:

1. Unfavorable temperature

Plants usually grow at a temperature range of 1° to 40°C, the normal temperature range for growth of most plants being 15°–30°C. Whenever there is sudden and sharp variation from these limits a plant shows injury to its organs or the entire body.

(a) *Low temperature:* Damage to plants is much more by low temperature than by high temperature. Warm weather plants usually are more susceptible to low temperature. Freezing injury in potatoes and frost injury to winter crops. are common examples. In potato subfreezing temperatures cause ringlike necrosis of vascular elements which may expand to the flesh if there is prolonged exposure to freezing. The frost injury to winter field crops is characterized by death of meristematic tissues, the apical portions and leaves showing the maximum damage. Low temperature usually kills the tissues by ice formation in and between the cells which rupture.

(b) *High temperature:* Too high temperatures rarely occur in nature except for short periods during summer in the tropics. Generally the adverse high temperature effects are seen in conjunction with other abnormal environmental conditions such as excess of light, low oxygen supply, drought, high winds, etc. Sun scald of fruits and vegetables.are common example. The side of the fruits exposed to the sun shows necrosis of the skin which may go deeper in the flesh. High temperature usually injures the tissues by inactivating certain enzyme systems while accelerating others. This leads proteins, disruption of cytoplasmic membranes and release of toxic products into the cell.

2. Unfavourable light

Excess light is a rare phenomenon in nature and rarely injures plants except when it is combined with high temperature as during summer months. Low light conditions retard chlorophyll synthesis and promote slender growth. The leaves are pale, almost colourless if light is completely excluded. Generally dense plant stand or dense canopy of leaves excludes light from around the lower parts of the plants.

3. Unfavourable soil moisture

Stagnation of water or excessive moisture in the field may cause rotting or wilting of the plants, primarily due to lack of oxygen and accumulation of toxic materials around roots and base of the stem, and also due to non-availability of nutrients under conditions of poor aeration. On the other hand, absence of moisture causes the plants to die or wilt as a direct effect or due to rise in temperature.

4. Unfavourable oxygen relations

When there is excessive respiration in closed atmosphere the entire oxygen supply may be exhausted resulting in disintegration of cells due to enzymic action.

5. Atmospheric impurities

With increased industrialization and mechanization of different human activities including transport this aspect of plant pathology has assumed special significance. Presence of injurious gases in the atmosphere causes definite injuries to plants or plant parts. Sulphur dioxide and ozone injury are common examples. Other chemicals are hydrogen fluoride, nitrogen dioxide, etc. These pollutants are usually transported from a particular source of their origin to the vegetation by air currents. Being heavier than pure air the pollutants settle near the ground level. Some are released as such by brick kilns, combustion of fuel (automobiles, tractor engines, diesel operated pumping sets, and different processing plants in the industry) while others such as ozone are products of chemical reaction in the atmosphere. Hydrocarbons and nitrogen dioxide released by automobile exhausts reacts with oxygen in the atmosphere in presence of ultraviolet rays in the sunlight. This results in formation of ozone and nitric oxide. These two may again react together to form the original compounds but the presence of hydrocarbon radicals in the polluted atmosphere prevents it and ozone concentration builds up in the atmosphere.

Sulphur dioxide is toxic to plants at concentrations of 0.3–0.5 ppm. Low concentrations cause general chlorosis and at high concentrations it causes bleaching of interveinal tissues. Sulphur dioxide can combine with atmpospheric moisture to form toxic acids (popularly known as acid rain).

Nitrogen dioxide is toxic at 2–3 ppm. At low concentrations it suppresses plant growth while at high concentrations it causes bleaching and bronzing of plants. Actively growing wet leaves are highly susceptible to hydrogen fluoride present in the smoke released by ore and oil processing plants. Leaf margins of dicots and leaf tips of monocots turn brown to dark brown, die and may separate from the leaves.

Ozone is toxic to expanding leaves of almost all types of plants. It causes stippling, mottling and chlorosis of leaves, usually on the upper surface. Spots are small or large, bleached white to tan, brown or black. Premature defoliation and stunting of plants occurs. Burning of plastic, glass factories and oil refineries release chlorine and hydrogen chloride. It is toxic at 0.1 ppm. Plants near the source show bleached, necrotic areas between leaf veins. Leaf margins often appear scorched. There may be premature defoliation. Ethylene is a plant hormone with numerous functions in the plant system. However, its accumulation and continued presence in the atmosphere is toxic to plants at 0. 05 ppm. Apart from automobile exhausts and burning of fuel it is also released by ripening fruits. The toxicity of ethylene results in stunted growth of plants, abnormal leaves, premature senescence of leaves, low flowering and fruiting.

6. Toxic effects of decomposing organic matter

Crop residue decomposition, in the first 1–2 weeks, produces toxic substances such as fatty acids in the soil. The available nutrients especially nitrogen are held up by the microbial activity. The effects are transient but if the crop is grown during this period it may suffer from symptoms of damping off, root rot, wilt and nutritional deficiency.

7. Nutritional disorders

Deficiency of minerals, *viz.,* nitrogen, phosphorus, potash, manganese, magnesium, boron, zinc, copper, iron, etc. results in disorders in the plant metabolism and causes hunger signs in the crop. Excess of a particular mineral disturbs nutritional balance needed for good metabolism in the plant thus hindering the effects of essential elements. The deficiencies or excess of these minerals also reduce the resistance of plants to fungal, bacterial and other diseases.

BLACK HEART OF POTATO

Black heart is an important storage, transit and market disease of potatoes as a result of poor oxygen relations. The disease usually occurs in tubers stored in poorly ventilated rooms in closely packed conditions. It also occurs in the field when temperature of the soil goes above 32°C during growth and maturation of tubers. The disease occurs in transit when the temperature inside the carrying vans rises, for sometime, above 32°C. Thus, a set of three distinct environmental conditions can cause this disease, *viz.*

 (i) poor ventilation in the store,
 (ii) high temperature during transit, and
 (iii) high temperature of soil during growth and maturation of tubers in the field.

 Dark grey to purplish or inky black discolouration occurs in the central tissues of the tuber. In advanced stage, the affected tissues may dry out and separate thus forming cavities. The discolouration may extend to the surface also. Large tubers are more susceptible than small one to black heart.

 In poorly ventilated rooms even low respiration by tubers uses up the available supply of oxygen. This results in discolouration and disintegration of cells due to adverse enzymic action which continues after the supply of oxygen has diminished. High temperature brings about some sub-oxidation by stimulating respiration. The cells of the tuber disintegrate when the interior of the potato heap in the store cannot ensure a good supply of oxygen.

To avoid this disease the tubers should not be stored or transported at temperatures above 33°C (90°F). Storage of tubers should be done in well ventilated rooms and the bags of tubers should not be piled very high upon each other. In the field, if the soil temperature seems to go very high the field should be irrigated.

BLACK TIP OR MANGO NECROSIS

Black tip, tip rot, or mango necrosis, locally known as "koeli", is peculiar to India and does not occur in any other country. It is the most common disease of mango fruits in U.P. and Bihar and is also known to occur in West Bengal, Haryana and Punjab. South India appears to be free from this disease of mangoes. Some of the best mango varieties are highly susceptible. As many as 90% of the fruits on trees in close proximity to brick-kilns may bear the necrotic lesions and become useless for sale and consumption.

The disease is characterized by necrosis of tissues at the distal end of the fruit. The first symptom is the development of a small etiolated area at the distal end which gradually spreads, turns nearly black and covers the tip completely. The tip is flattened with the outer skin turning hard and sunken. The inner portion is soft and yields a dark brown liquid due to rotting induced by saprophytic bacteria. The disease commonly occurs when the fruits are about 6–8 weeks old or when they are reaching ripening stage.

During 1940s, the disease was most common in orchards in the vicinity of brick kilns. They suggested that the smoke of these kilns polluted the air with toxic gases like sulphur dioxide and that these gases caused necrosis of tissues of the fruit. In laboratory experiments the disease could be produced by exposing mango fruits to smoke and sulphur dioxide.

Later, in 1960, it was reported that boron was deficient in the mango fruits exposed to brick kiln fumes. Sprays of borax at the rate of 3–4 kg/500 lit water at an early stage of fruiting substantially reduced the disease incidence. One to two sprays were recommended by them.

DISEASES DUE TO LACK OF MINERALS

Deficiency of essential minerals in the soil or their non-availability to the plant even if not deficient in the soil develop hunger signs in the crop. This is especially true for certain crops like maize, soybean, rice, etc.

Nitrogen (N) is present in most substances of the plant cell. In deficiency of nitrogen plants grow poorly and leaves are of light green colour. Lower leaves turn yellow or light brown and the stems are short and slender. Overall effect is reduced yield.

Phosphorus (P) is related with nucleic acids, cell membranes and various oxidative processes. Deficiency of P leads to poor growth, bluish green leaves with purple tint. Lower leaves turn light bronze with purple or brown spots. Shoots are short, thin, upright and spindly.

Potash (K) is a catalyst of many reactions at cellular level. Potash deficiency results in thin shoots which may show die-back symptoms. Older leaves show chlorosis and browning of the tips, scorching of the margins and brown spots near the margins.

Copper (Cu) is part of many oxidative enzymes in the plant cells. Most of the diseases caused by copper deficiency develop symptoms of chlorosis. The leaves may fail to unroll and tend to appear wilted. Leaf margins turn yellow and leaf tips of cereals wither. Heading is reduced and sometimes, as in cereals, the ears remain sterile. They may be dwarfed and distorted. Citrus trees show die-back symptoms during summer.

Zinc (Zn) is part of enzymes involved in auxin synthesis and in oxidation of sugars. Due to zinc deficiency diseases like rosette of apple, mottle leaf of citrus, and little leaf of grapevine are caused. Leaves of zinc-deficient plants develop prominent veins, become pale yellow, and sometimes bear dead patches. There may be bronzing of the lamina surface also. Crinkling and reduction in size of leaves is also common. The internodes are shortened and give a branched or rosette appearance to the foliage of the trees. Zinc deficiencies are frequently encountered not only in soils lacking this element but also in soils in which, because of alkalinity or other causes, the zinc present in soil is not available to the plant. Spraying with zinc sulphate is very effective in giving quick relief.

The "khaira" disease of rice in the Tarai area of Uttaranchal (India) is an excellent example of the effects of non-availability of zinc to the plants in an otherwise normal soil. The disease usually appears 10–15 days after transplanting, a period which coincides with the peak period of decomposition of last year's stubbles in the flooded field. Leaves of diseased plants show chlorosis at the base. Large numbers of small brown or bronze spots appear on the lamina surface. These coalesce to form bigger spots and ultimately the entire leaf turns bronze colored and dries. The growth of diseased plants is stunted. Root growth is also restricted and usually the main roots turn brown. The finer roots are destroyed. In severe cases plants fail to grow further and produce ears but sometimes there is natural recovery of the plants and some of them may produce ears with few grains. The disease is completely controlled by two sprays of a mixture of 2 kg zinc sulphate and 1 kg of slaked lime in 400 litres of water per acre. The first spray is given as soon as early symptoms of the disease are noticed. The second spray is given 10 days later.

Iron (Fe) is mainly concerned with the process of photosynthesis. It is a catalyst of chlorophyll synthesis and part of many enzymes. Although plants need a very small quantity of iron they develop acute chlorosis when this element is deficient in the soil or it is not being absorbed by the plant. The chlorosis of leaves without any mottling is the most prominent effect. In calcareous soils available ferrous salts are usually converted into unavailable ferric salts due to alkalinity. In acid soils iron is usually available but if an excess of soluble phosphate is present, it may become unavailable. Soil or foliar application of ferrous sulphate often corrects this deficiency. Chelated iron compounds applied to the soil are much more effective than simple inorganic iron compounds in acid soils.

Magnesium (Mg) is present in the chlorophyll of plants and is important for phosphate metabolism. Plants lacking this element show a progressive loss of green color of leaves over the entire lamina surface. Usually this yellowing appears on the older leaves first because the magnesium present in these parts is drawn by the young growing parts if the element is not being supplied by the soil. Application of dolomite limestone or magnesium sulphate corrects this deficiency.

Manganese (Mn) is part of many enzymes involved in respiration, photosynthesis and nitrogen utilization. Chlorosis of spinach and bean, spots on potatoes and grey speck of oats are common diseases resulting from manganese deficiency. The deficiency is common in calcareous soils where manganese is not soluble. The symptoms on oats appear first as light green spots on the leaves, the areas enlarging and changing to buff or light brown colour. In many soils manganese is oxidized by soil bacteria thus rendered unavailable to the plants. Manganese sulphate may be used as foliar spray to correct this deficiency.

Boron (B) Due to lack of boron there is hypertrophy, degeneration, and disintegration of cambium cells in the meristematic tissues. Thus, necrosis of tissues is the most important symptom of boron deficiency. In heart rot of beet the disease is characterized by curling of central leaves which turn a blackish brown and die. Later, a dry rot starts from the crown and penetrates into the centre of the flesh. In apple fruits in orchards of Kumaon Hills in Uttaranchal brown necrotic areas in the flesh are caused by boron deficiency. Such deficiencies can be corrected by application of 20 kg/ha borax before sowing. However, in acid soils where calcium is deficient boron may be toxic to the plants.

Calcium (Ca) Apart from affecting many enzyme activities Ca regulates the permeability of cell membranes and forms compounds with pectin which may provide resistance to pathogen growth. In calcium deficiency the young leaves become distorted with their tips hooked back and the margins are curled. Leaves may be irregular in shape with scorching and spotting. Terminal buds die. Calcium deficiency causes blossom end rot of many fruit trees. Fruits and tubers deficient in calcium are prone to soft rot caused by bacteria. Apples deficient in calcium are prone to postharvest decay caused by molds. Usually calcium containing fertilizers such as calcium ammonium nitrate help in reducing the adverse effects.

Sulphur (S) is present in some amino acids and coenzymes. Sulphur deficiency symptoms resemble those of nitrogen deficiency. Sulphur deficiency is not very common. Use of inorganic fertilizers and sulphur based fungicides takes care of this deficiency.

Molybdenum (Mo) is an essential component of the nitrate reductase enzyme. In deficiency of Mo plants exhibit severe yellowing and stunting and fail to set fruits.

Nitrogen, phosphorus and potash are major elements in plant nutrition. Their excess is normally not toxic to plants. However, imbalance in ratio of NPK causes some adverse effect on productivity of the plant. Other elements listed above are required by the plants in traces but are essential. Their excess is toxic such as boron or copper injury. These are direct effects. Outside the plants excess of one or more of these trace elements may interfere with absorption or availability of other elements. Excessive sodium induces a deficiency of calcium in the plant. Excess of copper, manganese or zinc induces iron deficiency.

Silicon (Si) Silicon is the second most abundant elements in soils. The soil water or the soil solute contains silicon, mainly as silicic acid Silicon is readily absorbed so that the plants contain it in appreciably higher concventration such as in crops like rice, wheat and barley. Non-availability of silicon results in structural weakness of tissues and enhances proneness to fungal diseases. The role of silicon in suppression of cereal powdery mildew, rice blast and rice sheath blight is well documented. It protects plants against root diseases also. In addition, silicon has direct effect on pathogens and regulates gene expression.

REFERENCES

Darley, E.F. and J.T. Middleton. 1966. Problems of air pollution in plant pathology. *Annu. Rev. Phytopathol.* 4: 103

Epstein, E. 1994. The anomaly of silicon in plant biology. Proc. Nat. Acad. Sci. USA 91: 11

Evans, L.S. 1984. Acidic precipitation effects on terrestrial vegetation. *Annu. Rev. Phytopathol.* 22: 397

Krupa, S.V., G.C. Pratt and P.S. Teng. 1982. Air pollution: An important issue in plant health. *Plant Dis.* 66: 429

Laurence, J.A. and L.H. Weinstein. 1981. Effects of air pollution on plant productivity. *Annu. Rev. Phytopathol.* 19: 257

Nadler–Hassar, T. and B. Robin. 2003. Natural tolerance of *Cuscuta campestris* to herbicides inhibiting amino acid biosynthesis. *Weed Research* 43(5): 341

Pell, E.J. 1979. How air pollutants induce disease. In: *Plant Disease*, Vol. 4, pp. 273–292. Eds. J.G. Horsfall and E.B. Cowling. Academic Press, New York.

CHAPTER

16

Plant Diseases Management

Plant disease control, more appropriately called plant disease management, is based not on the principle of only eradication of the pathogen but mainly on the principle of maintaining the damage or loss below an economic injury level or at least minimizing occurrence of a disease above that level. Management suggests need for continuous adjustment in the cropping system while "control" suggests a final solution.

Disease management practices are desirable only when the cost, in terms of money and efforts, is materially less than the loss expected from the disease so that the grower is ensured a margin of profit from the plant protection measures. Since disease management is meant to increase productivity of a crop, the control or management measures should be planned in such a way as to fit in the programme of operations for raising the crop. As far as possible, the schedule of plant protection operations should include management of several diseases in one operation (such as use of healthy seed and clean field). Success is often not achieved by management measures simply because the operations are carried out in isolated fields while the pathogen is left unchallenged in the neighboring fields where it continues to produce inoculum unabated and reaches the treated fields at the first opportunity after operations are over. Therefore, for success of plant protection operations it is essential that they cover large contiguous areas and are adopted by all the farmers in the area.

The conventional approach to plant disease management involves the "immunization-prophylaxis system". It is based on the fact that cure of a diseased plant is not possible because the disease becomes visible only after injury to the plant has taken place. Therefore, preventive measure are most important. These preventive measures involve induction of resistance in the host (immunization) and protection of the plant by prophylactic measures. These groups of measures can be classified under manipulation of the "disease triangle". In the opening chapter of this book, under causes of plant diseases. it was emphasized that a plant disease is the function of interaction between three forces. the host, the pathogen, and the environment. None of them alone can operate to cause a disease. This constitutes the disease triangle. Since the three are intimately related in the causation of a disease the principles of management should tackle all the three. On the same basis, an appropriate approach for disease management will be the integration of methods directed against the pathogen, in favor of the host, and for modification of the environment. This is known as integrated pest or disease management (IPM).

Management of the host involves the practices directed to improve plant vigor and host-resistance through nutrition, induction of genetic resistance through conventional plant breeding and through genetic engineering (transgenes), induction of local o9r systemic acquired resistance (SAR) by biological or physical means, and providing protection against attack by chemical and/or biological means. Management of the environment involves water management, soil management (soil health), and crop management. These methods are described on the following pages. The chemical methods of plant disease management cover both the aspects, protection of the host and management of the pathogen (reduction and eradication of inoculum). The chemical method had been primarily based on toxic, synthetic compound (fungicides, bactericides, nematicides) but now non-synthetic, non-toxic, environment friendly chemicals are coming up. Similarly, cultural practices not only cover the aspects of pathogen management but also management of the host as well as environment.

MANAGEMENT OF THE PATHOGEN

The methods of control based on the principles of directly attacking the pathogen are:

I. Avoiding the contact between the pathogen and the host

The main objective in this case is to prevent entry of a pathogen into an area, field, or plant population. Most of the methods for **exclusion** of the pathogens are regulatory measures and are expected to be enforced by the government of a country or state:

1. *Quarantine regulations:* These are steps taken by a country or state through legislation with the object of preventing the entry of disease or pathogen carrying planting materials into the country, state, or area. The regulations anticipate that the disease or pathogen is not present in the country.

2. *Certification:* The object of this method is to ensure that as far as possible only seeds or other planting materials certified as being disease or pathogen free are allowed to enter the country or state or are offered to growers. While certified seed material arriving from abroad may still be put to quarantine inspection, the certified seed being distributed to the growers within the country aims to provide a healthy crop for the purchaser and is meant to prevent spread, multiplication, and inoculum build up of a existing pathogen.

3. *Notification of plant diseases:* The grower has to notify the authority about the occurrence of a disease of potential danger the pathogen of which might perennate and spread in the area in the near future. The authority will notify other growers to take necessary precautions against the disease. This requires education of the farmers.

4. *Prevention of sale of diseased plants:* The authority may ban by legislation the sale of any diseased plant or planting material to prevent spread or multiplication of a pathogen in an area hitherto free from the disease.

Theoretically, exclusion is the most effective approach to keep pathogens away from the area concerned provided they are already not present, In practice, it has some social and economic limitations such as ignorance of the farmers, regional trade interest, malpractices in trade, forced import of food grains to meet deficiency and misuse of the regulations.

II. Plant and field sanitation for reduction of inoculum

The object of these measures is to reduce the source of infection presuming that the disease or pathogen has entered the area and has established itself in the fields.

Eradication

Eradication of seed-borne inoculum: Many plant pathogens are internally or externally seed-borne. If the inoculum is allowed to survive it produces the disease in the next crop season. The externally seed-borne inoculum may be present as external contaminant in the seed lot (infected residues of the host or fungal and nematode resting bodies) or as spores, mycelium, etc. on the body of the seed. Physical methods such as sieving, steeping in water or brine, etc., to separate diseased seed or pathogen structures from healthy seed take care of the contaminants. Chemical seed treatments eradicate the inoculum present on the body of the seed. Internally seed-borne infection can be deep seated in the endosperm, embryo, etc. In such cases, hot water or solar energy treatments and use of selected antibiotics and systemic fungicides destroy the dormant inoculum. For externally seed-borne pathogens seed coating with antagonistic fungi or bacteria is now suggested as alternative to toxic chemicals.

    Eradication of diseased plants or parts of the plant: Plant hygiene envisages the removal of diseased parts or the entire plants from the vicinity of the healthy plants so that the latter may not catch infection and further build up of inoculum may be prevented.

    Eradication of alternative and collateral hosts: In the absence of the main host many diseases and / or their incitants perpetuate on weed hosts or wild plants. Eradication of such hosts from the field or locality is done to prevent perpetuation of the pathogen and spread of the disease.

Field sanitation

The objectives of field sanitation are to completely or partially destroy the source of infection present in the soil. Some of the measures which can help in achieving this objective covered by measures listed above or will be discussed later. For example, the eradication of diseased plants from the field also amount to field sanitation. However, we are mainly concerned with the perennation of the pathogens in the soil or in diseased plant debris left in the field after harvest. The pathogens may survive on these sources in the form of resting spores or active saprophytic mycelium. Various measure which can be adopted to destroy these sources are:

    (a) Removal of diseased plant debris and their burning.
    (b) Ploughing to bury the fallen diseased leaves, twigs, etc. deep in the soil.
    (c) Use of chemicals to disinfect the fallen plant debris.
    (d) Hastening decomposition of crop residue to promote biofumigation
    (e) Hot weather deep ploughing.
    (f) Rotation, flooding, organic amendments, etc.
    (g) Promotion or introduction of biocontrol agents on the plants and in the field

III. Protective Application of Synthetic Toxic Chemicals to Prevent Infection by the Parasite or Its Destruction if Superficially Present

These operations are undertaken with the assumption that the entry of the pathogen and its contact with the host cannot be avoided or is not feasible and that the disease is likely to appear through wind-borne primary or secondary inoculum from a nearby or remote source. The control measures under this category involve the application of toxic chemical compounds to the surface of the host plant through spray or dusting. The object is to cover the entire susceptible surface of the host exposed to infection with a thin covering of a suitable concentration of the toxicant before the pathogen has come in contact with the host and when it lands on the host it is destroyed. Some fungitoxicants are now available which can eradicate the pathogen even after it has penetrated the host tissues.

A good fungitoxicant (fungicide) should be toxic to the parasite or inhibit the germination of its spores without causing phytotoxicity. It should be reasonably easy to prepare for application and not too expensive. It should be capable of even distribution from the spraying or dusting machines on to the surface to be covered, should remain on the surface without running off (initial retention) and should stick to the surface after drying (tenacity). The spray fungicide should have a good wetting property so that they do not form large drops or air pocket. It should also be biodegradable in soil so that it does not persist for long to cause pollution of ground water and streams.

Formulations of Anti-pathogen Synthetic Chemicals

The fungicides, insecticides and nematicides are available in the market in any one or more of the following forms:

Wettable powder: A large number of fungicides are formulated as wettable powders which are most commonly used for spray mixtures. These contain rather a high percentage of the active ingredient (60–80%). Modern wettable powders are easily wetted and disperse well in water. A wetting agent is usually present in most wettable powders but the addition of a spreader-sticker is sometimes desirable, especially when used on plants with glossy or waxy leaves. Some commercial adjuvants not only improve coverage, they also increase absorption and efficacy of the fungicides. Agitation is generally necessary in the spray ranks to keep a uniform suspension.

Dusts: These formulations usually contain 4–10% of the active ingredient (a.i.) and are usually applied dry as dusts.

Emulsifiable concentrates: These are liquids in which the active ingredient is dissolved in a solvent. The fungicides and the solvents will often not mix with water, so an emulsifying material is included. When these emulsifiable concentrates are added to water, a milky mixture is formed which is a suspension of active ingredient and emulsifying solvent in water. Fungicides are not commonly formulated as emulsifiable concentrates.

Granules: These are formulations of toxicants with inert materials formed into particles about the size of coarse sugar. The percentage of active ingredients is usually low as in the case of dusts. Granules have the advantage that they can be measured in dry form more easily and accurately than dusts or wettable powders. Therefore, they are more conveniently used in furrow treatments of soil. Some degree of volatility or ability to diffuse is necessary for granule formulations to be effective in soil. Very few fungicides are sold in this form.

Solutions: True solutions are formulations in which the active ingredient or a combination of active ingredient and solvent is dissolved in water. Solutions have the advantage of requiring no agitation after the formulation is added to water. However, practically all fungicide chemicals are relatively insoluble in water. When these chemicals have a high degree of solubility they may be very effective in controlling the plant pathogens but too toxic to the plant also.

Suspensions or Slurries: These are formulations in which a dry form of the active ingredient is mixed with a liquid. Such formulations usually have a high percentage of active ingredient similar to wettable powders. They are mixed with water for final use and require agitation.

Types of Materials used for Chemical Control and Methods of Application

According to their chemical nature, formulation, and method of application the plant protection chemicals may be divided under following categories:

Spray materials: Fungicides as wettable powders or materials available in solution or emulsified forms are used as spray to provide a protective cover over the plant surface. Most of the fungicides are applied in this form. Sprays are applied by means of hand or machine operated sprayers which have mechanism for agitation of the suspension.

Dust materials: Insoluble powders are used to cover the plant surface as an alternative to spraying. Hand or machine operated dusters are available.

Seed treatment materials: These are used as dry powder, dips or slurries. The seeds are given a complete cover of the material and dried. Using fungicides on seed is an efficient and economical method of chemical disease control. On the basis of their tenacity these chemicals are divided into two groups. *Seed disinfestants* disinfest the seed but do not remain active for long after the seed has been planted. *Seed protectants* disinfest the seed surface and stick to the surface for sometime after the seed has been planted thus giving temporary protection to the seedlings during emergence. The same material may be disinfestant and protectant. Established infection in the seed is also eradicated by seed treatment with chemicals which have systemic action, i.e., they penetrate the seed and destroy the deep seated pathogen. The protectant seed dressing chemicals can be inorganic or organic compounds of copper, sulphur, mercury, etc.

Soil treatment chemicals: Fungicides are applied to soil to control soil-borne pathogens in certain situations. For nematode control, most of the nematicides are applied to the infested soil. Complete control of a plant disease is often not obtained or is expected by this method but the results may be very satisfactory. In the control of fungal disease the chemicals may not kill the fungus but their activity is considerably suppressed. The nematicides may destroy most of the larvae of the nematodes but usually the eggs escape harm. In both cases, when the pathogens are hidden in plant tissues the chemicals are less effective. Fungicides and nematicides applied to soil have some indirect role in disease management. They alter the microbial community structure of the soil including that of the rhizosphere. This may result in predominance of antagonists that provide biological control and may lead to development of soil suppressiveness against pathogens.

Cost of soil treatment with chemicals is very high on a per acre basis hence such treatments can be feasible only for high value crops. The potential value of the crop and the cost of the treatment should be calculated before full scale treatments are made. Following points should be considered at the time of deciding about use of soil treatment chemicals:

(a) Toxicity of the chemical to the crop.

(b) Effectiveness of the chemical against a particular pathogen.

(c) Possible stimulation of non-target parasites due to reduced intensity of competition among soil microflora.

(d) Optimum soil moisture and soil temperature needed for best results in the soil.

Most of the chemicals applied to soil as fungicides are diffused through soil particles to be highly effective. They are not very persistent in soil being degraded by microbial activity or by chemical reactions with soil particles. However, a few fungicides such as PCNB (quintozone) and some systemic fungicides are very persistent in soil. Use of such highly persistent chemicals is being discouraged. Soil fungicides are applied as soil drench or in furrows to reduce the cost of the material.

Soil fumigants: Soil fumigants are volatile chemicals that are mixed with the soil or injected into it to kill active fungal structures and nematode larvae. On a small scale the chemical is often applied to the soil which is then covered with polyethylene sheets to confine the gases for a specified time. On a large scale the soil surface may be packed following the treatment or a water seal may be applied. These chemicals are usually applied several days or weeks before planting to permit them to be effective and allow them to disappear before planting. Chemical soil fumigants have been on the way out and more emphasis is now placed on biofumigation or mycofumigation based on decomposion of crop residue or use of specific fungal agents.

Fungicidal pastes: In orchards, dressing of pruned surfaces of the trees is done with pastes prepared with suitable fungicides (copper sulphate plus lime, copper carbonate plus red lead, or antibiotics) in a suitable adhesive carrier such as linseed oil or lanoline. They have a prolonged residual effect and provide protection to the cut surface until natural healing of the wound.

Antibiotics: Many antibiotics such as streptomycin, actidione, griseofulvin, etc. have been used as plant protection chemicals. These can be antifungal or antibacterial. Antibiotic are used mostly as spray materials at very low concentration and have systemic action being absorbed by the leaves and translocated in the plant system. However, use of antibiotics requires certain precautions since very often they are phytotoxic and may kill the plant tissues if the concentration goes above the recommended limit.

Systemic fungicides: These are chemicals which when applied get into the plant system and persist in the tissues to ward off infection by a parasite or destroy the already established infection (curative effect). Potentially they are the most effective and economical substances for artificial and temporary immunization of the plant against disease. There has been greater emphasis on development and use of such compounds. Since these compounds affect the pathogens on one or few sites of life activities, the pathogens quickly develop resistance by mutation. There are some systemic fungicides which *per se* are not fungitoxic but bring about metabolic changes in the host to make it resistant to a particular pathogen or trigger systemic acquired resistance.

THE SYNTHETIC FUNGICIDES

Chemicals have been used as fungicides for centuries. In the beginning it were inorganic compounds like sulphur, copper sulphate, copper carbonate, etc which were recommended against some disease. Later, specific preparations like Bordeaux mixture and Burgundy mixture were introduced in later half of 18th century. Bordeaux mixture was used on large scale against a large number of foliar diseases for decades before 1940s. In late 1940s and during the 1950s a large number of synthetic compounds were introduced in the market as fungicides. These included chlorinated hydrocarbons, organophosphates, carbamates, phenoxys and acetamines. The number of chemicals available for plant disease management now runs into hundreds although all are not equally safe, effective, and popular. Salts of toxic metals and organic acids, organic compounds of mercury and sulphur, quinones and heterocyclic nitrogen compounds have been the major fungicides in use. Copper, mercury, zinc and to some extent chromium, nickel, cobalt and tin are the metals as base for inorganic and organic fungicides. Among non-metals substances used as fungicides are sulphur, chlorine, and phosphorus

SULPHUR FUNGICIDES

Use of sulphur in plant disease control is probably the oldest and continues to be in maximum use even today. Inorganic sulphur is used in the form of elemental sulphur or as lime-sulphur mixture. Elemental sulphur can be in dust form or as wettable powder, the latter being more commonly used. However, the most popular fungicides in sulphur group are the organic compounds known as dithiocarbamates.

Inorganic Sulphur

Elemental sulphur is man's oldest fungicide. Elemental sulphur dust is prepared by fine grinding of the mineral. The tenacity and fungicidal efficiency depends on particle size which should be within the range of 47–75 microns (200–300 mesh). Wettable sulphur can be suspended in water and used as spray material. It is used for control of powdery mildews and rusts in field and garden crops. In dry hot weather elemental sulphur may be phytotoxic. Commercial forms of wettable sulphur are Elosal, Cosan, Sulfex, etc.

Elemental sulphur is also formed by certain specialized prokaryotes. Recently it has been claimed that certain plant species from diverse families produce elemental sulphur as a localized component of active defence to vascular pathogens and elemental sulphur is constitutive in some cruicifers (Williams and Cooper, 2004).

Lime-sulphur mixture is prepared by boiling the two together. Ten kg of lime and 7 kg of sulphur are boiled in about 250 litres of water on open heat for about an hour. It gives a golden coloured liquid. The mixture can also be prepared by self boiling of sulphur in unslaked lime with water. Chemically the mixture is calcium polysulphide. In the home made lime-sulphur mixture the calcium polysulphide content is low. Standardized commercial formulations of lime-sulphur are commonly available in market. Lime sulphur is mainly used in orchards to control such diseases as peach leaf curl, powdery mildew of apple, anthracnose and brown rot. The dosage is 10–15 litres of the mixture in 500 litres of water. High concentration is used on dormant trees but on trees with opening buds and leaves the concentration is progressively lowered. Solbar (Bayer) is barium polysulphide and had been used for the same diseases as above. Sometimes, pure sulphuric acid is also used such as for delinting of cotton seed to eradicate seed-borne infections and also for groundnut pod disinfection.

Organic Sulphur Compounds

The organic compounds of sulphur are highly effective fungicides and are the most popular spray fungicides in use today. Thiram, ziram, ferbam, nabam, zineb and maneb are well known examples of organic sulphur fungicides. All these fungicides are derivatives of dithiocarbamic acid. The dithiocarbamic acid is unstable in free state. But when it is reacted with another molecule of the acid or with some metal, stable and highly fungitoxic compounds are obtained. These are the dithiocarbamates.

Thiram, introduced in 1942, consists of two molecules of dithiocarbamic acid joined together, with the amino hydrogens substituted by methyl groups. Chemically thiram is tetramethyl thiuram disulphide. It is sold under the trade names of Thiram, Thiride, Arasan, Arasan SF (slurry form), Thylate, Tersan 75, Spottrete, Nomersan, Panoram 75, Fermide,etc. It is a broad spectrum fungicide most intensively used for seed treatment at the rate of 250 g/quintal seed. In some diseases soil application at the rate of 15–25 kg/ha had been recommended. It has some antibacterial activity also.

Metallic dithiocarbamates such as ferbam and ziram are obtained by reaction of methylated dithiocarbamic acid molecules with a metal rather than with themselves. Ferbam consists of iron. Three molecules of dimethyl dithiocarbamic acid react with one atom of iron to form ferbam. Chemically, ferbam is ferric dimethyl dithiocarbamate and is sold under the trade names of Ferbam, Fermate, Fermocide, Ferberk, Coromet and Karbam black. Normally, ferbam is used as a spray material against downy mildew of chillies, anthracnose of citrus, blast of rice, early blight of tomato and diseases caused by *Botrytis* and *Rhizoctonia*.

Reaction of dimethyl dithiocarbamic acid with zinc yields ziram (zinc dimethyl dithiocarbamate) sold as Ziram, Zerlate, Karbam white, Corozate, Cuman, Methasan, Vancide 51Z, etc. It is effective against anthracnose of cucurbits, ripe fruit rot of chillies, early blight of potato and tomato, Alternaria blight of wheat, and Cercospora leaf spot of beet.

Another group of derivatives of dithiocarbamic acid is ethylene bisdithiocarbamate which includes nabam and zineb (introduced in 1943), maneb (introduced in 1955), mancozeb (introduced in 1961), etc. Nabam (disodium ethylene bisdithiocarbamate) is the first fungicide of this group. It contains sodium and is sold as Dithane A-40, Dithane D-14, Parzate liquid, etc. Nabam has been found effective against late and early blight of potato and tomato and seedling blight caused by *Pythium, Fusarium* and *Rhizoctonia*.

Zineb is more fungitoxic than nabam and is formed by substitution of sodium in nabam with zinc. Chemically, zineb is zinc ethylene bisdithiocarbamate. It is sold under the trade names of Zineb, Dithane Z-78, Lonacol, Parzate C., etc. Zineb had been one of the most popular spray fungicide intensively used against late and early blight of potato and tomato, blast of rice, helminthosporiose of rice, anthracnose and ripe fruit rot of chillies, leaf blight and downy mildews of maize and certain rusts.

Maneb contains manganese in place of zinc. Chemically it is manganese ethylene bisdithiocarbamate and is sold as Dithane M-22 or Manzate. When zinc is added to maneb it forms mancozeb (Dithane M-45). In many diseases maneb and mancozeb have been found more effective and economical than zineb. With nickel added to maneb Dithane S-31 is formed. It contains 2 parts maneb and one part nickel sulphate hexahydrate. Maneb and mancozeb are especially used as spray materials in vegetable crops. Anthracnose caused by *Colletotrichum* spp. in bean, cucurbits, jute, spinach, tomato and tobacco, downy mildews of beans, cucurbits, lettuce, peas, spinach and beet, rusts of pea and wheat, leaf spots and blight caused by *Alternaria, Phytophthora* and *Cercospora* on crops like potato, tomato, cauliflower, cabbage, carrot, cucurbits, beet, etc. and leaf blight of maize have been successfully managed by maneb and mancozeb. The dosage of the dithiocarbamates varies from 2 to 3 kg per hectare.

Another member of the dithiocarbamate group is Vapam (sodium methyl dithiocarbamate or SMDC) which is available in liquid form. It is soluble in water and is used for soil treatment. Vapam or Methane-sodium is effective against many fungi, nematodes and certain weeds. Diseases like cotton wilt (*Fusarium oxysporum* f.sp. *vasinfectum*), damping off of papaya (*Pythium aphanidermatum*), damping off of pea (*P. ultimum*), root rot of beet (*Sclerotium rolfsii*), citrus nematode (*Tylenchulus semipenetrans*), cyst nematode of potato (*Globodera rostochiensis*) have been reduced by soil injection of vapam at the rate of 1000 litres per hectare.

COPPER FUNGICIDES

The fungitoxic activity of copper was mentioned as early as 1807 but its large scale use as a fungicide started in 1885 after the discovery of Bordeaux mixture. Together with sulphur, the copper fungicides had been most extensively used in plant disease control before the advent of organic fungicides. All copper fungicides are inorganic materials. They are prepared from copper sulphate, copper carbonate, cuprous oxide and copper oxychloride. Copper oxide (Cuprocide, Copper Sandoz, Perenox, Fungimar or Fungicopper) and copper carbonate (Cobredon, 45% Cu) have been in use as seed dressing fungicides for long.

The oldest fungicide among copper compounds is Bordeaux mixture which is prepared by reaction of copper sulphate and calcium hydroxide in water. The list of diseases suppressed by Bordeaux mixture is very large. It was first discovered for the control of downy mildew of grapevines (*Plasmopara viticola*) in 1885 at Bordeaux in France by Millardet. Its most extensive use had been in the control of late blight of potato. Bordeaux paste, Burgundy mixture and Cheshunt compound are other copper based home made fungicides. Burgundy mixture was discovered in 1887 as a substitute for Bordeaux mixture for areas where fresh lime was not available. It contains sodium carbonate in place of lime. It is less effective than Bordeaux mixture. Cheshunt compound contains 2 parts of copper sulphate and 11 parts of ammonium carbonate. It is suspended in water and can be used as spray or for soil treatment. Chaubattia paste is used in apple orchards for the control of stem black, stem brown and pink diseases. It contains 800 g of copper carbonate, 800 g red lead and 1 litre of raw linseed oil or lanolin.

The "fixed" or "soluble" copper compcunds have the copper ions fixed to the molecule, chemically more securely, so that it is only slightly soluble. Therefore, these compounds are less phytotoxic and addition of lime is not required. However, due to some reason, their fungicidal efficacy is also low. They are easily available in the market and are easy to use. Hence they are more prevalent than Bordeaux mixture. Copper oxychloride is the main compound in this group. The formulations available in the market contain 4–50% metallic copper. The formulations with 4–12% metallic copper (Blimix 4 per cent) are used as dust. The preparations having 50% metallic copper such as Blitox-50, Blue copper or Fytolan are used as spray materials. These fungicides are effective against all those diseases that are controlled by Bordeaux mixture. Damping off and leaf blights are very effectively checked by copper oxychloride formulations.

MERCURY FUNGICIDES

Inorganic and organic compounds of mercury are highly fungitoxic and had been extensively used in plant protection, especially for seed treatment against seed-borne and some soil-borne pathogens. The inorganic compounds possess bactericidal properties also. Due to their persistence in soil and on plant surface (including grains) and their toxicity to animal life the use of mercury fungicides was discouraged and in many countries they are banned. The possibility of phytotoxicity had restricted their use as foliar sprays although they were used against helminthosporiose and blast diseases of rice.

Two inorganic compounds of mercury, mercuric chloride (Cal-Chlor) and mercurous chloride (Calomel, Calocure, Calogreen, Velsicol) were used at 1:1000 dilution for seed treatment. Sometimes the solution was used for soil drenching also for eradication of soil-borne plant pathogens. In fruit orchards wounds on stem and branches are disinfected with mercuric chloride solution. The main function of these treatments was eradication. Inorganic mercury has hardly any protective action.

The organic compounds of mercury had been more common in use, especially for seed treatment. They have both eradicant and protectant action. In India, the organo-mercurials are mostly based on ethyl, phenyl, or methoxyethyl mercury as cations and chloride or acetate as anions. Some of the main fungicides in this group are listed below: None of these are being recommended now.

Ethyl mercury chloride, EMC (Ceresan, Granosan)

Ethyl mercury phosphate (NI Ceresan)

Methoxy ethyl mercury chloride, MEMC (Ceresan wet, Aretan, Agallol) Phenyl mercury acetate, PMA (Ceresan dry seed dressing with 1% PMA)

Phenyl mercury acetate and ethyl mercury chloride mixture (Agrosan GN) N-(ethylmercuri)-p-toluene sulphonanilide (Ceresan M)

Phenyl mercury urea (Agrox)
Hydroxymercuri chlorophenol (Semesan Bel)
Methyl mercuric dicyandiamide (Panogen)
Ethyl-mercuri-thiosalicylic acid, sodium salt (Elicide)
Chloromethoxypropyl mercuric acetate (Quicksan)

The organo-mercurials used for dry seed treatment usually contain 1% Hg. They were used at a rate of 250 g/quintal seed. In wet seed treatment the compounds contain 3 or 6% mercury and 0.25–0.5% was used for seed, tuber and sugarcane sett treatment.

QUINONE FUNGICIDES

Quinones are naturally present in plants and animals and are the source of colouration. They are also produced by oxidation of phenolics. Quinones often exhibit antimicrobial activity and may be responsible for resistance in plants to pathogens.

Chloranil (Spergon); a benzoquinone, and Dichlone (Phygon), a naphthoquinone), are two well known fungicides in this group. Ceredan (4-benzoquinone n-butyl hydrazome oxime) is closely related to quinones. Chloranil is used only for seed treatment. It has been used against externally seed-borne smut of barley and sorghum, bunt of wheat, damping off and seed rot of beans, cabbage, cotton and pea. Dichlone is used for seed treatment and also for foliar spray. It has been used effectively against apple scab, peach leaf curl, damping off and seedling blight of pea, spinach and beet and grain smut of sorghum. Ceredan is recommended only for seed treatment against soil-borne pathogens. The dosages of chloranil and dichlone are 200–400 g and 50–100 g per quintal of seed, respectively.

AROMATIC COMPOUNDS (Benzene fungicides)

Diazoben (dimethyl amino benzene-diazo-sodium sulphonate) sold as Dexon is used as seed and soil treatment fungicide for the control of damping off and root rot caused by species of *Pythium* and *Phytophthora*.

Chlorothalonil (tetrachloro-isophthalonitrile), was imntroduced in 1964 and sold as spray material under the name Daconil, Bravo and Termil The fungicide is popular fungicide effective against many foliage diseases.

Daconil and Bravo have been used against leaf spots., late and early blights, downy mildews, rusts, anthracnoses and scab. In many areas Daconil is preferred over mancozeb for control of late blight of potato. Termil is a tablet formulation of chlorothalonil which is used in greenhouses for control of *Botrytis* on ornamentals and for several molds and blights of tomato.

Dichloran (dichloro nitroaniline), was introduced in 1960 and sold as Botran or DCNA, is used against fruit and vegetable diseases as a spray and soil treatment material. It is effective against sclerotia producing fungi and is also used as post-harvest dip or spray for fruits, vegetables and flowers.

Dinocap (methyl heptyl dinitrophenyl crotonate) is sold as Karathane, Mildex, Arathane and Crotothane. This fungicide was developed as an acaricide for suppression of mites but now it is better known as a fungicide. It is one of the most popular, widely used, and effective fungicide for control of powdery mildews of fruit trees and field crops.

Pentachloronitrobenzene (PCNB) is sold under the name Brassicol, Quintozone or Terrachlor and has been a very successful soil fungicide against clubroot (*Plasmodiophora brassicae*), and rots caused by *Sclerotinia*, *Sclerotium* and *Rhizoctonia*. It has no direct effect against *Pythium*, *Phytophthora* and *Fusarium*. Development of strains of *Rhizoctonia solani* resistant to PCNB are also reported. Hexachlorobenzene (HCB) is used for seed treatment against seed and soil-borne bunt of wheat and other grains. Because of their persistence and possible polluting effect the benzenes compounds are not favoured.

HETEROCYCLIC COMPOUNDS

This group of heterogenous but some of the best fungicides, also known as dicarboximide fungicides, includes captan, folpet, captafol, iprodione, vinclozolin, glyodin and dyrene. Captan (1952), has been the most widely used fungicide of this group. and has been sold under the trade names of Captan, Orthocide, Vancide, etc. For some time it was sold under the name Esso Fungicide 406 in India. Primarily, captan is a seed dressing fungicide against *Pythium* but is used also as soil drench and foliar spray. Folpet (1952) is closely related to captan. and is sold as Phaltan, Orthophaltan, etc. In addition to diseases controlled by captan, this fungicide is effective against powdery mildews also. Captafol (1962) sold as Difolatan, Difosan, Sanspor, etc. was originally meant for foliar spray but is often recommended for seed treatment also. This fungicide has high resistance to weathering and persists on the host surface for much longer period than other fungicides. Together with its low phytotoxicity this property enables use of three times higher dose as a single application treatment for control of apple scab. It is recommended against late blight of potato and tomato and several other foliar diseases. Due to its prolonged persistence use of Difolatan is being discouraged in some countries.

Captan, folpet, and captafol belong to old class of dicarboximides and are now known as phthalamide fungicides. The new members of dicarboximide group are iprodione, procymidone, vinclozolin, etc.

Iprodione, is sold as Rovral or Chipco-26019 and is a broad spectrum contact fungicide. It inhibits spore germination and mycelial growth but is only preventive in action. It can be curative in early stages of infection. Rovral is very effective against diseases caused by *Botrytis, Monilinia, Sclerotinia, Alternaria, Helminthosporium* and *Rhizoctonia*. It is generally used as a spray fungicide but can also be used for seed treatment and post-harvest dip of fruits. Vinclozolin is sold as Ornalin, Ronilan or Vorlan and is used as foliar spray against sclerotia forming fungi belonging to Ascomycotina such as *Botrytis, Monilinia* and *Sclerotinia.*

ORGANO-TIN COMPOUNDS

Du-Ter, Brestan and Brestanol are three fungicides based on organic tin compounds. Du-Ter is chemically triphenyltin hydroxide with the common name fertin hydroxide. It is very effective against *Cercospora, Helminthosporium, Alternaria, Pythium, Phytophthora* and *Rhizoctonia*. It also has suppressive and antifeeding properties on many insects.

Brestan is chemically triphenyltin acetate (TPTA) and is effective against *Cercospora, Alternaria, Septoria* and many other fungi. Brestanol is triphenyltin chloride with activity similar to Du-Ter and Brestan. These tin compounds have antibacterial activity also. Brestanol has some systemic activity. Due to phytotoxicity to the host these fungicides have restricted use.

MISCELLANEOUS PROTECTANT FUNGICIDES

Methyl arsenic sulphide (MAS) sold as Rhizoctol, Urbasulph, Rhizoctol-combi (Rhizoctol + Ceredan, a benzoquinone) is a seed treatment fungicide. It is highly effective against *Rhizoctonia solani* and controls black scurf disease of potato.

Polyram is a foliar and seed protectant fungicide used against rusts, downy mildews, leaf spots, and blights.

Dodine(a guanidine) introduced in 1957, is sold as Cyprex. It has been an excellent fungicide for the control of apple scab. It gives long lasting protection and is also a good eradicant. It appears to have local systemic action on leaves.

SYSTEMIC FUNGICIDES

The introduction of systemic fungicides in 1966 is a major landmark in the history of fungicidal management of plant diseases. The systemic activity of many insecticides, herbicides and antibiotics was known earlier. However, the successful use of synthetic systemic compounds of non-microbial origin was first demonstrated by von Schmeling and Kulka in 1966. This discovery of oxathiin fungicides was soon followed by confirmation of systemic activity of pyrimidines (Elias, *et al.,* 1968) and benzimidazoles (Delp and Klopping, 1968; Weinke, *et al.,* 1969). The effectiveness, ease in application, economy and popularity of these fungicides prompted discovery of many more systemic compounds such as metalaxyl, many triazoles, and organic phosphates.

Many protectant fungicides probably become systemic to a slight extent but not sufficiently to give worthwhile protection. These include salts of mercury, zinc and iron, sulfonamides, dithiocarbamates and quinones. However, they are unstable in other respects. They may be phytotoxic at fungicidal doses, their translocation within the plant may be unduly slow or limited, or they may not persist sufficiently long to give adequate protection.

A systemic fungicide is one which is taken up and translocated within the plant as a result of which the latter becomes fungitoxic. In these respect it resembles the antibiotics. But the latter are of microbial origin and differ in their translocability from the systemic fungicides. An ideal systemic fungicide should have the following characteristics:

1. The substance may either be toxic to the pathogen concerned or be converted in the host plant to become such a fungitoxicant.
2. Alternatively, the substance (or a derivative formed in the plant) may alter the metabolism of the host so that biochemical or physiological resistance to pathogen may be induced or enhanced.
3. Even though the substance may act as described under (2), it must not adversely affect the host plant to such an extent that the quantity and quality of the crop is reduced.
4. In systemicity the substance must be absorbed sufficiently and translocated from the point of application to the site of the pathogen and should have a considerable degree of stability within the plant.
5. If it is applied to an edible portion of the plant the mammalian toxicity must be low enough to avoid residue problems at the consumer stage.

There is hardly any systemic fungicide that meets all the requirements stated above. However, the fungicides that are in common use these days are considered safe enough. Majority of the products come under the category (1) stated. above. There are very few compounds that act indirectly as stated under category (2) although the development of systemic compounds which act by increasing host resistance is as important as the development of systemic fungicides which act directly. The uptake of hitherto known fungicides is passive. This is a disadvantage because the active ingredient can move only with the sap stream. Thus, when a systemic fungicide is applied to the roots the active ingredient passes intercellularly into the xylem vessels in which it is swept along by the sap stream towards the foliage. When applied to the foliage the fungicide gets into the xylem and spreads to the distal parts of the leaf but not in the opposite direction towards the roots. Coming up from the roots, such a compound will move into all organs that transpire, i.e., have \functional stomata. It will not pass into petals of a flower or into fruits as in tomato, and will accumulate in mature rather than in newly formed leaves. Claims have been made for downward translocation of a number of systemic fungicides, including thiabendazoles, but the amount of such movement, if any, is insignificant.

Oxathiins: The two products developed in 1966 for complete control of internally seed-borne infection of loose smut (*Ustilago nuda*) in barley were anilides, subsequently called oxathiins (carboxin and oxycarboxin). These were 5,6-dihydro-2-methyl-1,4-oxathiin-3-carboxinilide (DCMO, carboxin, tradename Vitavax) and 5,6-dihydro-2- methyl-1, 4-oxathiin-3-carboxinilido-4,4-dioxide (DCMOD, oxycarboxin, trade name Plantvax). Both chemicals are translocated from roots in sufficient quantity to control rust on primary leaves of bean inoculated with *Uromyces phaseoli*. However, when trifoliate leaves were inoculated only oxycarboxin could control the rust.

At 20°C carboxin has a water solubility of 170 ppm while oxycarboxin has a water solubility of 1000 ppm. These compounds are derivatives of salicylanilide but with quite opposite fungitoxic spectrum. The sulphide, carboxin, is liable to oxidation in water, soil and plants but the fully oxidized form, the sulphone (oxycarboxin), has better stability. The two compounds differ in their specificity in the fungitoxic spectrum.

Carboxin (Vitavax) has become the most popular fungicide for seed treatment to control loose smut in wheat and barley (250 g/100 kg seed). It has replaced the earlier recommendation of hot water or solar treatment against this disease. It also gives satisfactory control of bunt (*Tilletia*) and flag smut (*Urocystis*). Other Basidiomycetes are also affected by Vitavax and Plantvax. In combination with thiram or copper oxyquinolinate the efficacy of Vitavax is increased and the cost of treatment is decreased. When used with copper oxyquinolinate the synergistic effect controls *Septoria nodorum*, *Helminthosporiu gramineum* and *Fusarium nivale* which are not affected by Vitavax alone. There are numerous reports of toxicity of Vitavax to fungi other than Basidiomycetes and these have been mentioned in chapters on different diseases. Oxycarboxin (Plantvax) is reported toxic to (in addition to Basidiomycetes) *Helminthosporium sativum*, *Curvularia*, *Aspergillus*, *Cladosporium*, *Botrytis*, *Monilinia*, and *Cunninghamella*. Although oxathiins do not show in vitro toxicity to Pythiaceous fungi, some reports show that soil application controls damping off caused by *Pythium*. The exact mechanism is not understood but the effect seems to be indirect through modification of biotic environment of soil or through host metabolism.

Benzimidazoles: This group of systemic fungicides includes some important broad spectrum compounds such as benomyl, carbendazim, thiabendazole and thiophanate. They are effective against numerous types of diseases caused by a wide variety of fungi. Most benzimidazoles are converted on the plant surface to methyl benzimidazole carbamate (MBC, carbendazim). This compound interferes with nuclear division of sensitive fungi.

Benomyl (1968) [methyl-1-(butylcarbamamoyl)-2-benzimidazole carba- mate] is the best known benzimidazole fungicide. It is marketed as Benlate, Tersan 1991, and many other names. Benomyl is as effective as carboxin against smuts but has the additional advantage of being one of the most wide spectrum systemic fungicides. It has been found effective against powdery mildews (*Erysiphe*, *Sphaerotheca*, *Podosphaera*, *Uncinula*), apple scab (*Venturia inaequalis*), *Cercospora*, *Fusarium*, *Verticillium*, *Cephalosporium*, *Colletotrichum*, *Gloeosporium*, *Botrytis*, *Monilinia*, etc. It is highly effective against and suppresses infection by *Rhizoctonia, Thielaviopsis* and *Ceratocystis.* Benomyl has no effect against oomycetes and some dark coloured fungi such as *Alternaria* and *Helminthosporium,* some Basidiomycetes and bacteria. Bark placement of benomyl induces a remission of symptoms of the spike disease of sandal caused by a *Phytoplasma.* After the treatment the MLO bodies were not seen in the tissues. The fungicide can be applied as a seed treatment, foliar spray, trunk injection, root dip or row treatment and as a fruit dip. It has a tendency to induce mutations and this causes rapid development of races of the pathogen resistant to the fungicide.

Carbendazim or MBC is breakdown product of benomyl. In contact with moisture benomyl readily breaks down to two compounds – methyl-3-benzimidazole carbamate (MBC) and butyl carbamate. The latter is converted into a volatile toxic compound, butyl isothiocyanate. MBC has been in the market under the name of Bavistin which seems to have same fungitoxic properties as benomyl but is more stable.

Thiabendazole (2,4-thiazolyl benzimidazole or TBZ) is sold under the name Mertect. This is also a broad spectrum fungicides, same as benomyl, effective against many Fungi-imperfecti causing leaf spot diseases. However, quantitatively it is less effective than benomyl. It is taken up and translocated intact without hydrolysis to leaves and stems.

Benomyl and thiabendazole have been applied as seed treatment, soil drench, soil mix, and foliar sprays. Systemic activity has been demonstrated for all the methods of application. As foliar spray, benomyl and Bavistin control powdery mildews of apple (*Podosphaera leucotricha*), cereals (*Erysiphe* graminis), and cucurbits (*Sphaerotheca fuliginea*), apple scab, brown rot of stone fruits (*Sclerotinia fructigena*), eye spot of wheat (*Cercosporella herpotrichoides*) and gray mold (*Botrytis cinerea*). Dipping fruits and roots has controlled banana fruit rot, root rot of sweet potato, corm rot of gladiolus, etc. The most common use of Benlate and Bavistin has been as seed treatment material. In addition to control of smuts, seed treatment is reported to eradicate *Ascochyta pisi* from pea seeds and the nematode *Aphelenchoides besseyi* from rice seeds. Seed treatment of cereals protects the crop for more than 30 days against powdery mildew and takel-all diseases of wheat.

Many soil-borne pathogens are suppressed by soil application of Benlate, Bavistin, or TBZ. Bavistin is supposed to be fairly stable in soil and ensures continuous supply of the fungicide to the above ground parts of the plant. Through soil application benomyl is reported to control many vascular wilt diseases such as Fusarium wilt of tomato and melons, Verticillium wilt of potato, cotton and tobacco, and cereal rusts.

These fungicides have been used for selective isolation of Pythiaceous fungi from natural soil. A concentration of 15 ppm in the plating medium has been found to suppress growth of most soil fungi permitting growth of *Pythium* and *Phytophthora.* A secondary effect of Benlate in crops treated through seed or tuber is better vigor of the plants during early stages of growth. For the same amount of nitrogen added to the soil. the Benlate protected plants of potato were greener and more vigorous than unprotected plants. The mechanism is not clear. It could be through changes in the host metabolism or through some change in availability of nutrients from soil. Decreased nitrification without affecting respiration value of soil treated with benomyl has been reported.

Thiophanates are often classified under benzimidazoles although they are different in that they are based on thiourea. Chemically, thiophanate is 1,2-bis (3-ethoxycarbonyl-2-thio ureido) benzene. Thiophanate is sold under the names Topsin and Cercobin. They are effective against several root and foliar pathogens including powdery mildews.

Thiophanate methoxy (methoxy carbonyl thioureido benzene) is sold as Topsin M or Cercobin M. It is more effective than the ethoxy compound and is more commonly used now.

Acylalanines: The group includes metalaxyl, furalaxyl, and banalaxyl. Metalaxyl, the first systemic fungicides developed against Oomycetes in the early 1970s and introduced for commercial use in 1977 is highly effective against *Pythium, Phytophthora,* and many downy mildew fungi. It is sold under the name Ridomil for foliar application and as Apron for seed treatment. Another formulation Subdue is used for ornamentals and turf grass. Although originally meant for foliar application metalaxyl is widely used as a soil or seed treatment material for control of seed rot and damping off caused by *Pythium* and *Phytophthora*, as soil treatment for control of Phytophthora diseases of citrus and other perennial and annual crops, and downy mildews of maize, pearl millet and sorghum. The fungicides is also used as post harvest fruit dip to prevent brown rot of citrus. It has curative effect and can eradicate infection from treated plants. Metalaxyl is quite water soluble and readily translocated from roots to aerial parts but its lateral translocation is slight. It inhibits zoospore germination in *Plasmopara viticola*, has high inhibitory effect on sporangium formation in *Phytophthora* and also inhibits oospore and chlamydospore production and chlamydospore germination. It does not kill the target fungus but is fungistatic. In soil with no previous history of application, metalaxyl persists for several months. As secondary effects, metalaxyl induces a mild shift in host resistance expression, increases potato tuber resistance to *Alternaria solani*, enhances lytic capabilities of antagonistic microorganisms in rhizosphere and promotes growth of endo- and ectomycorrhizae.

Resistance to this compound in fungi is very common. Maximum cases of resistance have been found in *Phytophthora infestans*, many in downy mildews and one in *Pythium*. This has necessitated the use of combination products in which some broad spectrum protectant fungicide is combined with metalaxyl. Metalaxyl with dithiocarbamates is generally recommended against late blight of potato.

Cyprofuram (Vinicur) is also a acylanilide in acylaminobutyrolactone group. It is effective against the Oomycetes and is applied as foliar spray, as soil surface and seed treatment material.

STEROL BIOSYNTHESIS INHIBITORS

According to mode of action these sterol inhibiting (SI) fungicides are divided into two groups:

I. *Inhibitors of sterol C-14 demethylation* (**DMIs**) which affect cytochrome P-450 enzymes. This group includes pyrimidines (fenarimol, nuarimol, etc.), imidazoles (prochloraz), triazoles (bitertanol, triadimefon, propiconazole, penconazole, flutriazole, flusilazole, cyproconazole, difenoconazole) and piparazines (triforine).

II. *The second group includes* fungicides which interfere with the C-14 reductase such as morpholines (fenpropimorph and tridemorph).

The SI fungicides combine protective and curative properties and thus interfere with target fungi throughout their infection cycle after penetration. They are very good for use in a disease warning systems. Combination of elemental sulphur improves efficacy of DMI fungicides in the control of pathogen strains that have developed resistance.

Pyrimidines: This group includes diamethirimol (Milcurb), ethirimol (Milstem), bupirimate (Nimrod), fenarimol (Rubigan), nuarimol (Trimidal) and triarimol. Diamethirimol was first reported by Elias *et al.,* (1968) to be specifically systemic in cucurbits and fungitoxic to the powdery mildew fungus *Sphaerotheca fuliginea*.

Later, Babbington *et al.,* (1969) reported that ethirimol effectively controlled powdery mildew of barley. In India these compounds have been found to control powdery mildew of wheat. Diamethirimol. ethirimol and bupirimate are all effective against powdery mildews of various crops. These fungicides are translocated from seed (seed treatment) or roots (granular application in soil) and reach the leaves to control powdery mildews. Protected plants are more vigorous with greater root weight than the unprotected plants. Ethirimol is considered more effective as soil or seed treatment fungicide than as foliar spray material. Fenarimol and nuarimol are effective against powdery mildews and also several leaf spot fungi as well as rusts and smuts.

Triarimol (2–4-dichlorophenyl phenyl-5-pyrimidine methanol) has systemic and curative effect on apples scab disease. It has been found effective against *Ustilago striiformis* and *Urocystis agropyri* also.

Triazoles are the largest class of fungicides. Some of the best SI fungicides belong to this group. These include triadimefon sold as Bayleton, triadimenol sold as Bayton, bitertenol sold as Bacor, boutrizol sold as Indar or RH-124, propiconazole sold as Tilt, etaconazole (Vangard) and many others such as flusilazole, myclobutanil (trade names Eagle, Nova, Rally, Prohane and Systhane), defenoconazole (Ciba Geigy), penconazole and prothiaconazole (Byer). They show long term protective and curative activity (thus facilitating lesser number of applications and longer intervals between sprays) against many powdery mildews, rusts, smuts, leaf spots and blights. They are applied as foliar spray and as seed and soil treatment. Bayleton has been used against cereal rusts and powdery mildew in India. Tilt is very effective against cereal rusts even with one spray. Boutrizol (4-butyl-4H-1,2,4-triazole) which was introduced in India as RH-124 or Indar was found to give complete control of leaf rust of wheat (*Puccinia recondita*) and was considered the first fungicide to be ever applied for strategic control of disease. In many countries it has extensively been used against yellow or stripe rust (*Puccinia striiformis*)

Morpholines: This group includes the fungicides dodemorph (Meltatox) and tridemorph (Calixin). They are preventive and eradicant foliar fungicides effective against powdery mildews and leaf spots on cereals, ornamental and tropical plants. Tridemorph (N-tridecyl-2,5-dimethyl morpholine) is available in India as emulsifiable concentrate (750 g tridemorph per litre in Calixin) and has been used against powdery mildew of wheat, pea and cucurbits. The fungicides is also reported to be effective against the stripe or yellow rust of wheat and barley, Sigatoka or leaf spot of banana (*Mycosphaerella musicola*), pink disease of rubber trees (*Corticium salmonicolor*) and many other diseases of tea, coffee, cereals, etc. The uptake and translocation is acropetal. Dimethomorph, though a morpjoline, is quite distict because of its activity against oomycetes.

The **imidazole** SI fungicide prochloraz is effective against Ascomycetes and Fungi-imperfecti causing powdery mildews, leaf spots, blights and fruit rots. Triforine, a **piparazine** SI fungicide is a foliar spray fungicide sold as Cella, Funginex or Saprol. It is effective against powdery mildews, leaf and fruit spots, fruit rots, anthracnose, and some rusts.

Organic Phosphates: The systemic fungicides based on organic phosphate include fosetyl-Al, sold as Aliette, Kitazin (IBP), edifenphos (Hinosan) and pyrazophos (Afugan). The alkyl phosphonate fungicide, aluminium-O- ethyl phosphonate or phosethyl-Al, more commonly known as fosetyl-Al (Aliette), was developed as a fungicide against oomycetes after metalaxyl. It has a narrower spectrum of biological activity than metalaxyl but is more effective against some fungi. It has low in vitro and in vivo activity against late blight of potato, blue mold of tobacco, and Phytophthora root rot of soybean but is highly effective against Phytophthora diseases of pineapple (a single pre-plant application protects the crop for 18 months), avocado, citrus and walnut. The fungicide breaks down in soil or in plant tissues into carbon dioxide and phosphonic acid, the latter being the fungitoxic principle in the fungicide.. It persists in soil for several months. The fungicide is highly phloem mobile being translocated from leaves to roots. The breakdown product of fosetyl-Al is highly inhibitory to sporangia formation of *Phytophthora* *citrophthora* and *Phytophthora parasitica*. Zoospore release from sporangia is also sensitive. Chlamydospore production of *P. parasitic* is inhibited by the fungicide. Low *in vitro* activity against many species of *Pythium* and *Phytophthora* led to the hypothesis that the fungicide has indirect effect by triggering a resistance response in the host. However, the ability to develop strains resistance to phosphonic acid in laboratory and the fact that their resistance is expressed as a lack of control by phosphonates in the host, provides evidence that the fungicide acts directs on the target fungus. The indirect effects on host metabolism are of secondary importance.

Fosetyl-Al is very effective against foliar, root, and stem diseases caused by Oomycetes such as *Pythium* and *Phytophthora* and the downy mildews in a variety of crops. It can be applied as foliar spray, soil drench, root and post-harvest dip, as well as soil mix. Treatments remain effective for 2–6 months depending on the crop. Fosetyl-Al has also been used against some bacterial disease of ornamentals (Chase, 1993).

Kitazin and Hinosan have been found highly effective against the blast disease of rice (*Pyricularia grisea*). The latter has been extensively used in India. Afugan is effective against powdery mildews and diseases caused by *Helminthosporium*.

MISCELLANEOUS SYSTEMIC FUNGICIDES

Pyracarbolid is a compound closely related to oxathiins with the trade name Sicarol. It controls rusts, smuts and *Rhizoctonia solani*. It is reported to be more effective than oxathiins.

Chloroneb is sold under the name Demosan and used as seed and soil fungicide. It does not leach from the soil and is effective against seedling blight of cotton, bean, beet, etc. Often it is recommended as seed overcoat to seeds treated with standard fungicides.

Ethazol (Turban, Terrazole, Koban) is a seed and soil fungicide effective against damping off and root and stem rots caused by *Pythium* and *Phytophthora*. It is often sold combined with PCNB or with Topsin M (as Banrot) for broader spectrum effect against *Fusarium* and *Rhizoctonia*.

Imazalil (Fungaflor) has been found effective against powdery mildew, leaf spots, fruit rots and vascular wilts caused by Ascomycetes and Fungi Imperfecti. It has excellent curative and preventive properties and is used as foliar spray, seed treatment and post-harvest treatment.

Prothiocarb is a thiocarbamic acid ester developed in 1966 by Schering Ag and first released in 1974 for control of soil-borne Oomycetes. In vitro as well as in soil it has fungistatic activity against *Pythium* and *Phytophthora*. It is taken up by plant root slowly and moves acropetally into the foliage. It possesses systemic activity against foliar diseases such as tomato late blight.

Propamocarb (Banol, Previcur) is oxygen analog of prothiocarb and was first released for commercial use in 1978. It is effective against Oomycetes, especially some species of *Phytophthora*, and some rusts. It also exhibits good systemic activity when used as soil drench against downy mildews of cucurbits, crucifers,lettuce, onion, and hops. Like prothiocarb, it also is acropetally translocated in plants. The fungicide causes cell leakage in the target fungus.

Cyprodinil belongs to anilopyrimidine group which includes **pyrimethanil** and **mepannipyrin**. At 10 mg a.i./lit cyprodinil gives 90–100% control of apple scab. It significantly reduces fungal infection stages that are formed after penetration of leaf. It effective against powdery mildew of apple and *Botrytis* also. A formulation (Vision) containing 50 g/L fluquinconazole and 200 g/L pyrimethani is highly effective against scab and powdery mildew of apple.

Carpropamid is a new generation carboximide fungicide for specific control of rice leaf blast diseases. It has little in vitro effect on the pathogen but inhibits melanization of appressoria and checks the pre-penetration stages of the pathogen on the leaf.

Strobilurins are relatively a new class of fungicides introduced in 1996. The strobilurin fungicides are based on natural product (metabolites) of a fungus (*Strobilurus*) found growing on pine cones. They are also known as Qo inhibitors or QoIs. They act through inhibition of respiration by binding to the Qo center of the cytochrome *b*. They include azoxystrobin, kresoxim methyl, trifloxystrobin etc. These fungicides have very broad and balanced spectrum of activity on the foliage and have very favourable toxicological profile, rapidly dissipating from soil and surface water and are unlikely to cause undue hazard to non-target organisms. They have protective and curative effect against powdery mildews, Cercospora leaf spots, apple scab, downy mildew of grapevines snd cucurbits, and many cereal diseases. A strobilurin fungicide, Pyraclostrobin, enhances the resistance of tobacco against infection by either TMV or the wildfire bacteria *Pseudomonas syringae* pv. *tabaci*. It primes the plants for potential activation of subsequently pathogen-induced cellular defense responses (Herms *et al.*, 2002) In addition to their antifungal activity strobilurins are reported to act indirectly by promoting plant growth through auxin-like effect. Although, initially it was thought that resistance to this group of fungicides is uncommon, subsequently it was found that repeated use (which became common because of its efficiency) leads to deverlpment of resistance to many important fungal plant pathogens. This has prompted less number of its application and use in alternation with other systemic fungicides having different modes of action.The resistance to strobilurins in fungi is mainly through development of alternative mechaniksms of respiratin.

Phenylamides include mefenoxam which acts by inhibiting polymerization of r-RNA.

Carboxylic acid amide (CAA) include some highly effective anti-oomycete fungicides such as dimethomorph and mandiprpomid. Mode of their action is not clear although they are site specific.

Cyano-acetamide oxime includes cymoxanil used against downy mildews Its mechanism of action is noty known although it is site specific.

Benthiavalicarb-isopropyl, introduced in 2003, is a carboxy acid amide (CAA) It is a novel member of the amino acid amide carbamate group of fungicides. It is effective against all oomycetes except*Pythium*. The fungicide effectively controls late blight of potato and tomato causedby metalaxyl-sensitive and metalaxyl-resistant strains of *Phytophthora infestans*. It does not stop the discharge of zoospores from zoosporangia or their motility but strongly inhibits mycelial growth, sporulation and germination of sporangia and cystspores. It has both preventive and curative effrect. The application dose is 2.5–7.5 g. a.i/L

Surfactants: Synthetic surfactants are inert components of synthetic fungicides to reduce surface and interfacvial tension.. They have been found to have antifungal activity against zoospores eich are lysed. Synthetic surfactants have been used on spinach for control of white rust and in hydroponic cultures for control of *Pythium aphanidermatum.* Many biocontrol bacteria and fungi produce biosurfactants (glycolipis, phospholipids, lipopeptides) which add to their efficiency. Strains of *Pseudomonas aeruginosa* producing phenazines but not surfactant (rhamnolipid) are ineffective biocontrol agents against *P. myrioitylum* but those that produce the surfactarit also are effective. Strains of *Bsacillus subtilis* produces the surfactants surfactin, iturin and fengycin, all lepidopeptides); *Bacillus licheniformis* produces lichenysin, *Pseudomonas fluorescens* produces viscosin, and *Serratia* produces serrawettin,

ANTIBIOTICS

The use of antibiotics as systemic plant disease control agents precedes systemic fungicides. These compounds have been used since the 1950s to control certain bacterial diseases of high value fruit, vegetable and ornamental plants. The antibiotics are defined as substances produced by microorganisms which, in very dilute concentrations, have the capacity to inhibit the growth of, or even destroy, other microorganisms. Most antibiotics known to-date are products of Actinomycetes. Some are produced by fungi (penicillin, gliotoxin). The antibiotics can be antibacterial (viz., penicillin, streptomycin, tetracycline, etc.) or antifungal (viz., actidione, aureofungin, griseofulvin, pimaricin, blasticidin, kasugamycin, etc.). The chemical nature and formula of antibiotics are quite complex and not fully known in most cases. They differ from each other.

Antibiotics have systemic action in living organisms. In plants they are absorbed by roots or foliage and translocated, to varying extents, to other parts. This capacity depends on the antibiotic and the plant. On being absorbed the antibiotics either directly kill the pathogen in the tissues or check its development through altered physiology of the host. Their effect remains stable for some time, thus they act not only as eradicants but also as protectants and provide resistance against invasion of pathogens. The properties of systemic action and inhibition or eradication of the pathogen after invasion puts these agents under the category of chemical therapeutants (chemotherapeutants).

Phytotoxicity is one major disadvantage with antibiotics. They are used in very small quantities. If the dosage increases beyond permissible limits plants are damaged. This is why most antifungal antibiotics such as griseofulvin and actidione could not become popular plant disease control chemicals in spite of their strong fungitoxic action. Another disadvantage with antibiotics is their narrow spectrum of antipathogenic action. The antibacterial antibiotics are mostly ineffective against fungi while the antifungal antibiotics have no antibacterial activity (with few exceptions) and are effective only against specific fungal pathogens. These limitations have restricted the use of antibiotics in plant disease control. In the USA, antibiotics applied to plants account for less than 0.5% of total antibiotic use (McManus et al., 2002).

The antibiotics used in plant protection mainly belong to groups known as streptomycin, tetracyclines, polyenes, cycloheximide and griseofulvin. Streptomycin is a metabolite of Streptomyces griseus. As a plant protection chemical it is available under the trade names of Agrimycin, Phytomycin, Orthostreptomycin, Streptocycline, etc. Agrimycin and Streptocycline are mixed preparations of streptomycin and tetracycline. The antibiotic binds to the bacterial ribosomes and prevents protein synthesis. Maximum use of streptomycin, as streptomycin sulphate, has been made in the control of bacterial diseases of fruit trees such as fire blight of apple and pear (Erwinia amylovora) and citrus canker (Xanthomonas axonopodis pv. citri). Mostly it has been used as a spray material. Soil drenching is also recommended in some cases. The antibiotic is recommended for eradication of seed-borne bacterial infection in potato, cotton, cereals and cauliflower. Although streptomycin is an antibacterial antibiotic certain fungal diseases (caused by Mastigomycotina) are also suppressed by it. Glycerine, copper oxychloride, etc. are often added to streptomycin to improve its efficacy.

Tetracyclines were discovered in 1940s. They are broadspectrum antibiotics active against Gram-negative and Gram-positive bacteria, chlamydiae, mycoplasmas, rickettsiae and protozoan parasites. Antibiotics in the tetracycline group are produced by many species of Streptomyces. The group includes the antibiotics known as Terramycin (oxytetracycline), Aureomycin (chloretetracycline) and Achromycin (tetracycline) which have been used against plant diseases. Oxytetracycline, obtained from Streptomyces rimosus is most common and is often used with streptomycin. Resistance in bacteria to tetracyclines was first reported in 1953. The mode of action is same as of streptomyces. The antibiotic binds to bacterial ribosomes and inhibits protein synthesis.

Cycloheximide was isolated from *Streptomyces griseus* and is obtained as a by-product in the production of streptomycin. It is commercially available under the name of Actidione, Actispray, Actidione RZ (5% cycloheximide and 75% PCNB) This antibiotic is antifungal being strongly toxic to many phytopathogenic fungi at very low concentration. As spray material even 2 ppm Actidione can control some diseases. It has been used against powdery mildew of bean (5 ppm), covered smut of oats (10 ppm), powdery mildew of rose (20–0100 ppm), bunt and leaf and stem rust of wheat. However, its use has been restricted due to its toxicity to many plants.

Griseofulvin was isolated from the fungus *Penicillium griseofulvum* nearly 60 years ago and was found effective against several plant pathogens, notably *Botrytis*. It is no more a promising fungicide. Under field conditions it gives only a limited degree of control. It had been found effective against powdery mildew of rose (*Sphaerotheca pannosa*), downy mildew of cucurbits (*Pseudoperonospora cubensis*), powdery mildew of cucurbits (*Erysiphe cichoracearum*), powdery mildew of beans (*Erysiphe polygoni*), early blight of tomato (*Alternaria solani*), brown rot of apple (*Sclerotinia fructigena*), Ascochyta blight of pea (*Ascochyta pisi*) and Botrytis rot of lettuce. The dosage of foliar spray varies from 100 to 1000 ppm.

Aureofungin is a heptaene antibiotic produced in the submerged cultures of *Streptomyces cinnamomens* var. *terricola*. It was developed by Hindustan Antibiotics Ltd, Pimpri. This antibiotic had been claimed to control many diseases, even those that are caused by Pythiaceous fungi which are generally insensitive to polyene antibiotics. It is recommended for seed treatment of rice against helminthosporiose and blast. Pimaricin and nystatin are other antifungal polyene antibiotics obtained from species of *Streptomyces*. Antibiotics developed in Japan for the control of rice blast have been mentioned in the chapter on blast of rice.

NATURE OF FUNGICIDAL ACTION

The basis of fungicidal action is selectivity of chemicals to various forms of life. The metabolic processes of plants and fungi, although different in some respects, are basically similar, so that fungicides are likely to damage both the fungus and the host plant (phytotoxic fungicides). However, there is some degree of selective toxicity although many fungicides are rather nonspecific in their toxic action. The complete mechanisms by which the various chemicals applied to plants control the disease are as yet unknown for a large number of chemicals.

Majority of fungicides kill or inhibit the fungi responsive to them through direct effect on fungal cells or spores after entering them. Solubilization of fungicides on the host surface is facilitated by free water, carbon dioxide and ammonia in rain water or dew, guttation fluids and other exudates from the plants, spore exudates, and the ability of spores to accumulate fungicides from very dilute solutions. The killing of fungal cells by fungicides may brought about by

(i) injurious effects on cell walls and on cell division,
(ii) effect on the permeability of cell membrane,
(iii) effect on enzyme system of the fungal cells,
(iv) chelation and precipitation of chemicals, and
(v) by antimetabolism.

The protectant fungicides produce these effects at the point of entry of the pathogen. Sulphur interferes with electron transport along the cytochromes of fungi and is then reduced to hydrogen sulphide which is toxic to most cellular proteins. Copper ion is toxic to all cells because it reacts with sulfhydryl group of certain amino acids and causes denaturation of proteins and enzymes. Many organic fungicides also inactivate fungal proteins and enzymes in the same manner. Dithiocarbamates and ethazol, when taken up by fungal cells, release thiocarbonyl which inactivates sulphydryl (-SH) groups, The chlorinated aromatic and heterocyclic compounds such as PCNB, chlorothalonil, chloroneb, captan, and vinclozolin inactivate enzymes in the fungal cells.

The systemic fungicides and antibiotics are absorbed by the host plant, are translocated internally through the plants and are effective against the pathogen at the site of infection both before and after infection has become established. Once in contact with the pathogen these chemotherapeutants appear to affect pathogens in the same manner as the protectant fungicides, but the systemic fungicides are much more specific and affect only one or two functions in the pathogen rather than a variety of them as in case of protectant wide-spectrum fungicides. Oxathiins inhibit only the enzyme succinic dehydrogenase which is essential for mitochondrial respiration. Benzimidazoles interfere with nuclear division by binding to protein subunits of the spindle microtubules. The organic phosphorus fungicides Kitazin and Hinosan primarily act by inhibiting chitin synthesis in the pathogen. Fungi that have no chitinous walls are not responsive to these fungicides. Metalaxyl and related compounds affect ribosomal RNA of the fungi and interfere with protein synthesis. Triazole fungicides have become well known as sterol inhibiting fungicides because they inhibit ergosterol biosynthesis. Ergosterol is a cellular compound that plays a crucial role in the structure and function of the membranes of many fungi. These fungicides penetrate the leaf cuticle and, therefore, have curative post-infection applications.

Not much is known about the possible role of fungicides in stimulating resistance responses of the host. However, in the case of fosetyl-Al this has been implicated. In late blight of potato (*Phytophthora infestans*), this fungicide does not directly affect the pathogen but the disease is controlled. The mechanism(s) of action is not fully known. Phosphites are known to affect mitosis and meiosis in plants The disease control s supposed to be by increasing the resistance of the host to the pathogen, This is probably done by altering the constitution of the host cell walls, by limiting the availability of essential coenzymes in the host, or by altering the rate or direction of metabolism in the host, which may thus be in a better position to defend itself. The secondary effects of metalaxyl mentioned in earlier section imply that induction of mild shift in resistance expression, promotion of mycorrhizal growth, and enhancement lytic capabilities of the antagonistic microflora in the rhizosphere may substantially contribute to host resistance in specific diseases. In many crops application of systemic fungicides and nematicides has resulted in a perceptible change in the appearance of the host and its vigor even if it was disease or pathogen free. The indirect effect of fungicides applied to soil through changes in biotic environments or in the pathogen that facilitates antagonism by soil microbiota is possible. In the case of root rot caused by *Armillaria mellea* it has been demonstrated that fumigation of soil weakens the pathogen and it fails to produce antibiotics to fight the antagonist (*Trichoderma*). Changes in uptake of nutrient by the plant may also occur. These observations suggest that indirect effects of fungicides also play some role in disease suppression. However, as Cohen and Coffey (1986) had suggested these secondary effects may be only of secondary importance. The major role of fungicides is direct effect on the pathogens.

DEVELOPMENT OF RESISTANCE TO FUNGITOXICANTS

The development of resistance in fungi to fungicides is based on the same principle on which plants develop resistance to a pathogen by modification in their genetic make up and insects, mites and bacteria develop resistance to chemicals used intensively to destroy them. For many years when only protectant fungicides such as thiram, maneb, or captan were used, no resistant strains of fungal pathogens were reported. In the 1960s, resistance to some fungicides, all containing a benzene ring, started appearing. *Penicillium* strains resistant to diphenyl, *Tilletia* strains resistant to hexachlorobenzene, and *Rhizoctonia* strains resistant to PCNB were found to occur in nature. Later, a strain of the apple scab fungus, *Venturia inaequalis*, was found resistant to dodine, an excellent fungicide at that time for control of the disease. Resistance of plant pathogenic bacteria to oxytetracycline is rare, but the emergence of streptomycin-resistant strains of *Erwinia amylovora*, *Pseudomonas* spp. And *Xanthomonas campestris* is not uncommon.

One of the major problems associated with frequent use of site specific systemic fungicides is the development of resistance to them in the fungal populations With widespread use of systemic fungicides, especially benomyl, large scale appearance of resistant strains became common. As newer systemic fungicides such as metalaxyl and triazoles came in extensive use the resistance development problem became more acute. The present position is that resistance in fungi to most systemic fungicides including benzimidazoles, almost all the sterol synthesis inhibitor fungicides and fosetyl-Al has become a common phenomenon. The resistance has developed within one season of frequent application in the field or within 2–4 years of large scale use. These resistant strains may be resistant to other members of the same group with similar mode of action and they spread to fields with no history of application of the fungicide. Some of the fungi listed as having developed resistance to such fungicides are *Phytophthora*, *Pseudoperonospora cubensis*, *Plasmopara*, *Sphaerotheca*, *Venturia inaequalis*, *Fusarium*, *Cercospora*, *Colletotrichum*, *Verticillium* and *Ustilago*.

The development of resistance to systemic fungicides is apparently because these fungicides interfere with only one or sometimes two functions in the body of the fungus which it easily overcomes by either a single mutation or by selection of resistant individuals in a population. On the other hand, the non-systemic protectant fungicides affect too many functions in the fungus body and to develop resistance the fungus will have to make too many gene changes.

The mechanisms by which a pathogen can develop resistance to a toxic chemical are:

   (i) decreased permeability of pathogen cell membranes to the chemical,
   (ii) detoxification of the chemical through modification of its structure or through binding it to a cell constituent,
   (iii) decreased conversion of the chemical to real toxic compound,
   (iv) decreased affinity at the reaction site in the cell,
   (v) bypassing of a blocked reaction or function through a shift in metabolism, and
   (vi) compensation for the effect of inhibition by producing more of the inhibited product.

There are ways and means to preserve the efficacy of systemic fungicides and prevent appearance of resistant strains of the pathogens. Most important is the modification in the method of their use. Alternate spraying with systemic and protectant fungicides is one approach which has been recommended in many diseases such as control of apple scab. attempt is made not to repeat the systemic fungicide application during the season. Where 5–6 sprays are required a number of protectant fungicides should also be included in the schedule. The pesticide industry is now coming up with products which have combination of systemic and nonsystemic, broad spectrum, protectant fungicides. Examples are mixture of benomyl and either captan or dichloran or iprodione for control of *Sclerotinia* and related fungi; mixture of metalaxyl (Ridomil) and maneb or zineb for control of downy mildews of different crops and late blight of potato. In this system, while the systemic fungicide by its specific action destroys most of the infection, the protectant fungicides maintain the level of control and destroy any resistant strain that may develop. The combination of fungicides with phytochemicals also reduces chances of resistance development in the pathogen. Combined use of fungicides and antibiotic with plant extracts or with biocontrol agents is also an approach to reduce resistance development in pathogens.

THE NEMATICIDES

Nematicides in commercial use basically belong to two groups: fumigants (volatile soil fumigants) and non-fumigants (contact and systemic nematicides). The fumigants consist of compounds belonging to hydrogenated hydrocarbon and isothiocyanate groups while the non-fumigants are mostly organo-phosphates and carbamates. The former directly kill the nematode larvae while the latter do not cause direct kill.

The Soil Fumigants: Fumigant nematicides such as methyl bromide, thylene dibromide, D-D mixture, methyl isothiocyanate, etc. possess high vapor pressure, get dissolved in soil moisture in concentrations high enough to kill nematode larvae and disperse through the soil pores. They are not very specific in their toxic action and can destroy many types of nematodes (parasitic and non-parasitic), many insects, and even fungi and bacteria. The halogenated hydrocarbon group includes methyl bromide, ethylene dibromide, Telone, D-D mixture, DBCP (Nemagon) and chloropicrin. The isothiocyanate group includes fumigants that release methyl isothiocyanate such as Vapam and Dazomet, Mylone or Basamid. Vapam is a recommended fungicide also and belongs to dithiocarbamates.

D-D mixture: D-D is a mixture of 1,3-dichloropropene and 1,2-dichloropropane in 1:1 ration with some other hydrocarbon impurities. It was discovered in 1943 and had been a very widely used nematicide. It kills nematodes, soil insects, wireworms, and also some fungi at high dosages. The nematicide is sold as undiluted or as an emulsifiable concentrate in drums. D-D is the trade name of the product of Shell Chemical Co. Other trade names of the nematicide are Viden-D, Nemafume and Nematox. The dosage rate is 400–500 lit. per ha as a pre-plant soil fumigant. The best results are obtained when soil is treated at 10–27°C. Since the fumigant causes temporary reduction in the number of nitrifying bacteria and inhibition of nitrification results in high levels of ammonia, it is desirable that nitrogen in the form of nitrate is added to the soil especially if it contains high levels of organic matter. Crops like sugarcane and pineapple can tolerate ammonia hence in such cases post treatment application of nitrates can be avoided.

Telone: This nematicide was introduced in 1956 by the Dow Chemical Company. Telone is a mixture of cis- and trans-isomers of 1,3-dichloropropene. Dorolone is a mixture of Telone and ethylene dibromide. The dosage rate is 200 lit./ha.

EDB or ethyl dibromide (1,2-dibromoethane): This chemical has excellent nematicidal properties. It contains bromine and may cause phytotoxicity in bromine sensitive plants. The biological activity is similar to that of D-D mixture.. Rate of application is 80–120 lit./ha depending on the soil type and the crop. This nematicide also reduces nitrification.

Nemagon (DBCP): Nemagon is 1,2-dibromo 3-chloropropane and had been very useful as pre-plant and post-plant treatment of established plantation crops. Due to its low phytotoxicity it was also used through irrigation water. However, because of its residue problem its use has been discontinued.

Methyl bromide: The chemical is compressed in to liquid and solid in metal containers. Since it is toxic to humans, and is odorless, a warning agent such as chloropicrin is usually added to it. For seed and nursery beds the treated surface is immediately covered with polythene sheets. Unless the treated surface is covered, MB diffuses in the atmosphere and depletes the ozone layer. There is an international agreement to reduce its use and phase it out by the year 2005. This chemical also destroys nitrifying bacteria, The trade names are Dowfume MC2, Metho-O-Gas and Terabol. There have been concerted efforts to find substitutes for methyl bromide. Metham sodium plus chloropicrin or 1,3-dichloropropene plus chloropicrin often give better nematode control than methyl bromide alone. Combination of suitable cultural practices with calcium cyanamide has been shown to serve as alternative to methyl bromide. However, these compounds are also hazardous. Biofumigation is a hazard-free alternative. Organic amendments with crucifers result in the release of toxic methyl isothiocyanates that biofumigates the soil. Thus has been suggested as an alternative of methyl bromide.

Chloropicrin (trichloronitromethan): The chemical was first developed as a soil fungicide for control of fungal wilts of plants but was found extremely useful for controlling root knot nematodes. It is also known as tear gas or Picfume.

Dazomet: On application to soil this isothiocyanate compound breaks into methyl aminomethyl dithicarbamate which yields isothyocyanate effective against nematodes and fungi. Rate of applicaton is 20–30 g/sq m as a pre-plant soil fumigant. Other names of the nematicide are Mylone and Basamid.

Vapam or Metham sodium (Sodium M-methyldithio carbamate) is another soil fumigant of the isothiocyanate group. The chemical is used against fungi, nematodes, weeds and some soil insects. Rate of application is 700–1000 liters.

The fumigants are applied to soil through injection with an injector gun or other suitable equipments. The field is prepared well in advance of the treatment into a fine seedbed condition. Marker lines are drawn at 30–35 cm distance in either direction. The nematicide is injected into the soil at the intersecting points to a depth of 22.5 to 30 cm. These fumigants are usually injurious to seeds or plants if they are planted immediately after soil fumigation. Therefore, generally 2–3 weeks waiting period is required before planting is done.

Mode of action of fumigants: The distinguishing feature of the action of fumigant nematicides from the non-fumigant nematicides is that when there is sufficient exposure of the nematodes to the fumigants they are killed whereas non-fumigant nematicides at field dosages usually incapacitate the nematodes in one or more functions and the effects are reversible. The nematicides may enter the nematode body directly through the cuticle or through such openings as mouth, anus, and vulva. Eggs are less susceptible than juveniles and adults to the effect of nematicides. the first reaction of the nematode to the entry of chemicals in the body is hyperactivity followed by gradual decrease in activity, leading to eventual paralysis. Within the body the chemical may act in many ways such as by precipitating proteins, blocking nerve endings, destroying nerve sheaths, cellular membranes and amphids.

THE NON-FUMIGANT NEMATICIDES

The introduction of granular contact and systemic insecticide-nematicides in the 1970s has provided very effective chemicals for nematode control. These nematicides are now more commonly used than the fumigants. Some of the organo-phosphate and carbamate contact and systemic nematicides are listed below.

Phenamiphos: Sold under the name Nemacur, this is a highly effective systemic nematicide against all the economically important phytonematodes. The chemcal also affords good protection against sucking insects for 6–8 weeks. The chemical is readily absorbed by roots as well as by leaves and is translocated both up and down in the plant system.

Phorate: Commonly sold as Thimet, phorate is a systemic organic phosphate soil insecticide and nematicide. Absorbed by roots it is translocated in the plant system and persists for severa weeks protecting it not only against nematodes but also sucking insects on the foliage.

Fensulphothion: Sold as Terracur P and Dasanit, this non-systemic organo-phosphate can be used as nematicide as well as insecticide. The chemical remains effective in soil for about 3 months at a temperature of 20°C.

Ethoprop, Ethoprofos sold as Mocap is a non-fumigant broad spectrum phosphatic compound effective against nematodes and soil insects through contact action. It can be applied immediately before or at the time of planting or used for established plants. It is appled with irrigation water or mixed with soil.

Thionazin (Nemafos, Cynem or Zinophos) and **Diazinon** (Basudin, Neocidol) are other organic phasphate non-fumgan nematocodes with contact action. The latter is often used as an insecticide.

Aldicarb: This is usually sold under the name Temik and was introduced in 1965 by Union Carbide Corporation and is used extensively. It is a strong systemic carbamate insecticide-nematicide and acaricide. When applied to soil at the time of planting it protects plants against nematodes and aphids for 40–45 days in potato, tomato, and okra.

Carbofuran: This carbamate systemic nematicide, sold under the name of Furadon or Curatter, Oxamyl (Thiooxamyl, Vydate, Du Pont 1410) is an oxime carbamate having systemic action. In addition to soil application, it has been used for foliar spray also. **Carbosulfan** is another carbamate nematicidal compound.

Some of the organophosphate and carbamate nematicides are absorbed only by roots and translocated up to the foliage. Few such as oxamyl, phenamiphos and carbofuran have the capacity to move from roots to the foliage and from foliage to the roots. These non-fumigant nematicides can be applied to soil by spot or row treatment. The rate of application varies with the method of application, nematode species, soil conditions, environments and cultural practices. Broadcasting the granules or spraying the emulsifiable concentrates on the soil surface followed by thorough mixing with the soil is the normal method of application for achieving maximum protection of the plant. The soil application places the nematicide in the top few inches of the soil and then the chemical moves downwards and reaches greater depths

Mode of action of contact and systemic nematicides: It is generally accepted that at filed dosages the organic phosphate and carbamate nematicides do not kill the nematodes directly. At the recommended field dose a five day exposure of *Meloidogyne incognita* to Thionazin has no effect. They act by impairing the nervous system of the nematodes, reducing the rate of hatching, motility, orientation to host roots, feeding, and development. Soil application of thionazin reduces fecundity of *Meloidogyne incognita* females. In the nematode nervous system these nematicides inhibit the enzyme acetyl cholinesterase at the cholinergic synapses. The same effect occurs in insects and mammals.

BIOLOGICAL CONTROL OF PATHOGENS

Synthetic toxicants used as fungicides or nematicides have dominated the plant protection strategies for over 110 years. They have the advantage of being sure and quick acting.

However, with increasing awareness about environmental pollution and their rising cost there is now greater attention to hazard-free alternatives, althpugh the industry continues to develop newer fungicides. Instead of synthetic chemical plant pathologists have turned to natural products and natural enemies of the pathogens. Biological control is an approach towards this. The approach is not new if we look at the ancient agricultural systems and literature. Every biological entity has its adversary in nature which keeps its population under check. Biological control is basically natural control, a system by which nature maintains the biological equilibrium and during the process checks the populations of plant pathogenic organisms. In this system, nature employs interactions between microorganisms, interaction between microorganisms and plants and interactions between organisms and environment. This may be the basis of survival of plant species for thousands of years irrespective of challenges of pathogens and, probably a contributory factor in origin of resistant genotypes. The present day approaches to the biological control of plant diseases appear to be only a part of the whole natural system.

Biological control was defined, in 1974, as the reduction of inoculum density or disease-producing activities of a pathogen or parasite in its active or dormant state, by one or more organisms, accomplished naturally or through manipulation of the environment, host, or antagonists, or by mass introduction of one or more antagonists. Subsequently, Cook and Baker (1983) shortened the definition to 'Biological control is the reduction of the amount of inoculum or disease producing activity of a pathogen accomplished by one or more organisms other than man." This definition does not specifically say that only antagonistic microorganism are involved in biological control. The host is also a biological entity and if it prevents infection it is also biological control. Thus, development of resistant cultivars and elicitation of defense responses leading to acquired resistance during the lifetime of the plant are also biological control. In an immunized plant, the plant plays the role of antagonist. However, literature on biological control is confined mostly to microbial control involving interaction between the pathogens and other microbes as biocontrol agents. This is microbial control, a part of biological control. It operates through various components of antagonism and/or through the host by inducing resistance.

MICROBIAL CONTROL OF PLANT PATHOGENS

The microbial control of plant pathogens has been more amenable to its natural occurrence than to introduction or manipulation by man (including mass introduction of antagonists) unless the antagonist belongs to the same niche as the pathogen. The entities involved in biological control are:

1. avirulent or hypovirulent individuals or populations within the pathogenic species itself (cross protection or immunity),
2. the host plant manipulated genetically, by cultural practices or with microorganisms toward greater or more efficient resistance to the pathogen, and
3. antagonists (competitors, parasites and predators), defined as microorganisms that interfere with survival or disease-producing activities of the pathogen.

Fungi, bacteria, fungivorus and entomopathogenic nematodes, protozoa, and even viruses have been reported to protect plants against pathogens. Most such agents of biological control are saprophytic in nature. Some bacterial agents, *viz.,* Pasteuria, are obligate parasites of nematodes. The saprophytic fungi and bacteria can be cultivated on artificial substrates and mass multiplied. Such substrates may be farmyard manure, crop residues, oil cakes, coire, etc. The obligate parasites are multiplied on the living plants inoculated with the pathogen infested with the biocontrol agent.

Mechanisms of pathogen (disease) suppression by microbial agents include:
(a) Competition with the pathogen for available resources (space and nutrition)
(b) Antagonism that includes (i) antibiosis (role of microbial metabolites, antibiotics, (ii) parasitism and lytic enzymes or sometimes toxins, (iii) predations.
(c) Plant growth promotion (probiotic action)
(d) Elicitation of defense responses and induction of systemic acqjkred resistance
(e) Suppression of toxin production by the pathogen.

Biocontrol microbial agents may employ one, few or most of the above mechanisms. Organisms employing larger number of mechanisms are more efficient biocontrol agents. *Trichoderma* is not a predator but otherwise it has the faculty of employing other mechanisms. Furthermore, *Trichoderma* spp. are compatible with many fungicides and are tolerant to biofumigation with crucifer residue or seed meal amendment. *Pythium oligandrum* is a mycopartasite but also produces the toxin oligandrin. Major biocontrol microbes are *Pythium oligandrum, Trichoderma* spp (including *T. harzianum, T. hamatum* and *T. viride*). *Clonostachys rosea* (*Gliocladiu, roseum*), many epiphytic yeasts, *Penicillium oxalicum. Bacillus* spp.and *Pseudomonas* spp especially *Pseudomonas fluorescens* strains. Use of mixture of biocontrol agents is required to ensure larger number of mechanisms. Successful biocontrol depends on having the biocontrol agent delivered to the right place, at the right time, in the appropriate phsiologica state and relative in density to the density of the pathogen. In addition to these considerations, application must be compatible with the crop production system being followed (Fravel *et al.,* 2003).

There are three approaches to microbial control of plant pathogens: biological control of inoculum, biological protection of plant surfaces, and cross protection/induced resistance.

BIOLOGICAL CONTROL OF INOCULUM

Biological control of inoculum includes:
(i) destruction of inoculum by parasites and predators,
(ii) prevention of formation of inoculum,
(iii) weakening or displacement of the pathogen from the food base (infected residue), and
(iv) reduction of vigor or virulence of the pathogen by such agents as mycoviruses (dsRNA).

(i) Destruction of dormant propagules

Natural destruction of fungal propagules in soil is common. Oospores of many *Pythium* spp. are parsitized by *Pythium oligandrum*. Sclerotia are destroyed by parasitism of *Sporodesmium sclerotivorum, Trichoderma harzianum,* and *Coniothyrium minitans, Pythium oligandrum* and other fungi. Oospores of *Phytophthora, Pythium* and *Aphanomyces* are parasitized by many chytridiales, hyphomycetes, actinomycetes and pseudomonads. Conidia of *Helmintho- sporium sativum* and chlamydospores of *Thielaviopsis basicola* are killed by predatory vampyrellid amoebae. Cells of *Xanthomonas campestris* pv. *campestris* are ingested by protozoa. Certain species of Collembolans are known to feed on hyphae of *Rhizoctonia solani*. They also feed on eggs of the root knot nematode, *Meloidogyne javanica*. Nematode trapping fungi abound in soil and are known to feed on plant parasitic nematodes including cysts by parasitization and predation. Bacteria, such as *Pasteuria penetrans*, parasitize root knot nematodes and eventually kill them. The fungi *Verticillium chlamydosporium, Fusarium oxysporum* and *Cylindrocarpon destrustans* are known to parasitize cysts of *Heterodera schachtii* (soybean cyst nematode) which is also destroyed by strains of *Pseudomonas fluorescens*.

Such naturally occurring enemies of other microbes including plant pathogens play the role of maintaining biological equilibrium in nature. The objective of microbial control of plant pathogens is to hasten the death of pathogenic or parasitic propagules with the help of such organisms and for this several methods have been suggested to strengthen their numbers. The incorporation of decomposable organic matter such as farm yard manure, green manures, oil-cakes, sawdust etc. is one such approach. Among many effects it produces in soil, one is to increase the microbial biomass. Under the influence of nutrients from the decomposition of organic matter dormant propagules of many pathogens, *viz.,* sclerotia of *Sclerotium*, are induced to germinate and the germlings (germ tubes, hyphae) are then killed by lysis through action of soil microbiota. The phenomenon of germination-lysis is probably a common one in soils to which regular addition of organic matter is made. The reduction in the activity of cereal cyst nematodes in soils green manured with chopped cabbage leaves, reduction in the incidence of root knot in soils amended with oil-cakes and sawdust, reduction in the numbers of phytonematodes and *Sclerotinia sclerotiorum* or *Fusarium oxysporum* by flooding or wet rice culture in the rotation are examples of biological destruction of inoculum.

The problem of mass production of biocontrol agents sufficient for soil application to reduce pathogen inoculum is being solved in recent studies. There are reports of using such organic substrates as FYM, wheat, maize or pulse bran, sawdust, coir pith, spent coffee grounds or coffee pulp, kitchen waste, etc. for mass production of antagonists. It may be possible to add antagonists like *T. harzianum* to manure pits and use the manure as base for development of antagonists in the field. Hyseik *et al.,* (2002) had reported that a *T. harzianum*-based commercial product (Supresivit) applied at 0.5 g per kg of mineral fertilizers could suppress foliage diseases in wheat, barley, maize, oil rape and potato and increase yields.

Bacteriophages are common in soil envireonment and are easy to mass produce at low cost. The use of phages for disease control is a fast expanding area of plant protection with potential of replacing chemicals for control of soil-borne bacterial plant pathogens. However, the efficacy of phages as biocontrol agents depends on prevailing environmental factors and on susceptibility of the targest bacteria

Indirect destruction of propagules or active growth of pathogens in soil can be achieved by **biofumigation**. Decomposition of crucifer residue in soil releases volatile isothiocyanates and other sulphur compounds which are antifungal, anti-nematode and also antibacterial. In organic amendments or decomposition of crop residues in soil biofumigation is the major mechanism and in biofumigation accumullation of organic acids such as acetic and butyric acidplays important role. specific antagonist is involved. Soil bacteria attack the sinigrin in such residue and cause release of sulphur compounds through myrosinase enzyme activity. The deompsoition of crucifer residue also generates alcohols, aldehydes, and sulfides which add to the suppressive effect. Solarization of the amended soil enhances release of these products. Similar to biofumigation is the role of **mycofumigation** in which specific fungi produce volatiles that kill or inhibit the pathogens without direct contact with the [athogen.

(ii) Prevention of inoculum formation

This approach to biological control can be more efficient than mass action of biocontrol agents on biomass of the pathogens. The logic behind this approach is to incapacitate the inoculum producing organs, such as females and cysts of nematodes, to prevent a pathogenic fungus from colonizing plant residue in soil where it could multiply inoculum, encouraging development of antagonists on aerial parts of the plant where they could destroy the inoculum, or using decoy or trap crops that could be destroyed before the inoculum is formed and released in soil.

The nematophagous fungus *Nematophthora gynophila* parasitizes females and cysts of *Heterodera avenae* (cereal cyst nematode), the nematode trapping fungus *Dactylella oviparasiticus* parasitizes females and eggs of *Meloidogyne* species and the bacterium *Bacillus penetrans* parasitizes root knot nematodes preventing production of larvae as inoculum. *Verticillium chlamydosporum* parasitizes eggs, larvae and cysts of the cereal cyst nematode (*Heterodera avenae*) thus reducing nematode population and preventing inoculum formation for the next season. This fungus has been cited as a good biocontrol agent against the root knot nematodes also. In tomato it is reported to reduce the disease by 90% at 25°C and by 60–70% at either 20°C or 30°C. Many fungal pathogens such as *Pythium*, *Phytophthora* and *Armillaria* are unlikely to colonize host plant residues in soil if the latter are pre-colonized by saprophytes such as *Fusarium roseum* 'Culmorum'. Parasitism of the honey dew stage and sclerotia of the ergot fungi (*Claviceps* species) by species of *Fusarium* makes the sclerotia nonviable and no spores are produced as inoculum. A strain of *Pseudomonas fluorescens* is reported to inhibit teliospore germination of *Tilletia laevis* and suppress the common bunt of wheat.

One of the methods recommended for management of phytonematode populations is to plant susceptible trap crops which are infested by the nematodes present in the soil and then destroying the crop before the nematodes complete their life cycle. This stops multiplication of inoculum and the subsequent economic crop is safe from nematode attack.

(iii) Weakening or displacement of the pathogen in crop residue

Many root pathogens (species of *Helminthosporium*, *Gaeumannomyces graminis*, *Fusarium* that cause vascular wilt, *Armillaria mellea* and many phytopathogenic bacteria) use crop debris for short or long duration perpetuation. They are primary colonizers (pioneers) of the host residue and are difficult to displace by secondary invaders or saprophytes. The displacement is possible only if the pioneer has finished utilizing the substrate, it becomes dormant because of adverse environments, or if the secondary colonizer is capable of killing the primary colonizer by antagonism (mostly antibiosis, parasitism and predation) or some other means. Such secondary colonizers or antagonists are few and there are not many examples where the pathogen, protected by host tissues, has been displaced by soil microbiota. However, conditions can be created by the soil microbiota for rapid exhaustion of the substrate thereby shortening the length of survival time of the pathogen. Starvation of *Gaeumannomyces graminis* var. *tritici* by depriving it of external source of nitrogen is reported.

(iv) Reduction of vigor or virulence of the pathogen

This approach involves the reduction of vigor, aggressiveness, fitness, pathogenicity, virulence or other attributes of the pathogen essential to its saprophytic or parasitic activities accomplished through factors inherent (or carried) in the pathogen itself. Thus, altered sex ratio in nematodes determined by availability of food, hypovirulence in some fungi and similar phenomena are candidates for means of this type of biological control. However, not much work has been done to exploit these phenomena. While explaining the reduction in root knot by amendment of soil with oil cakes and sawdust, Singh and Sitaramaiah (1973) had stated that in extracts of amended soil larvae of *Meloidogyne javanica* became very active with jerking motion. Since fitness of the larvae to cause penetration of host roots depends on their strength derived from lipid reserves in their body (they do not feed in soil outside the host), they considered that the chemical environments created by microbial activity in amended soil caused extraordinary agility in the larvae which exhaust their lipid reserves in this movement and lose fitness or vigor to cause penetration when they came in contact with host roots.

In some fungi hypovirulent strains have been detected which are supposed to result from infection of one or more dsRNA determinants. If hyphae of a hypovirulent strain can be anastomosed with hyphae of a virulent strain the latter loses vigor for infection and producing disease. A dsRNA mycovirus-like agent has been found in some strains of *Rhizoctonia solani*. A mixture of normal and hypovirulent strains significantly reduces damping off caused by the fungus. However, the hypovirulent strains have less longevity. Similarly, in *Gaeumannomyces graminis tritici* (take-all of wheat) mycovirus infection is common and has been found associated with decline of the fungus in soil. Presence of dsRNA associated with hypovirulence is reported in *Sclerotinia sclerotiorum*, *S. homoeocarpa* and *S. minor* also. The number of pathogenic fungi in which such RNA molecules have been detected is increasing. They have been found in even Karnal bunt fungus. But their role in Karnal bunt fungus is yet to be determined.

BIOLOGICAL PROTECTION AGAINST INFECTION

The approach involves establishment of an antagonist in or around the site of infection so as to provide protection of the area against attack of a pathogen. The host is not involved in the interaction between the pathogen and the antagonist. The resident antagonists on the host surface providing control of a disease, effective biological control achieved by organic treatments and the phenomenon of suppressive soils characterized by lack of propagule germination for penetration and growth in the rhizosphere fall in this category.

(i) *Protection of planting material:* The control of crown gall of stone fruit trees (*Agrobacterium radiobacter* pv. *tumefaciens*) by avirulent, bacteriocin producing *Agrobacterium radiobacter* pv. *radiobacter* strain K84 is known as one practical method of biological control on a field scale. Some facts learnt from this successful method are:

(a) only the strains that produce the bacteriocin (Agrocin 84) are effective; the strain must be an active colonizer of the host surface; simply ability to produce the antibiotic is not enough;

(b) necessary attributes of an effective antagonist are ability to produce the desired antibiotic and aggressiveness as a root colonizer and these attributes can be introduced into a microorganism by genetic manipulation, and

(c) such biocontrol agents may not be effective against all strains of the pathogen.

There are numerous examples of biological control achieved by protective covering of seed, rhizomes, tubers, etc. with propagules of an antagonist. *Bacillus subtilis*, some species of *Pseudomonas*, *Penicillium*, *Chaetomium* and *Trichoderma* are often as effective as seed protectant chemicals such as thiram and captan. In pre emergence seed rot of pea caused by *Pythium ultimum*, the pathogen derives nutrients for colonization of seed and subsequent invasion from seed exudates released during swelling of the seed in soil. If the seed is coated with spores of *Penicillium oxalicum* the antagonist exhausts this source of nutrients before the pathogen could utilize it. Species of *Trichoderma* have also been used similarly to provide protection to seeds during germination against seed rot fungi. *Trichoderma hamatum* and *T. harzianum* are effective seed protectants against *Pythium* spp. and *Rhizoctonia solani*. Seedling roots, corms, bulbs, tubers, etc. can also be treated with spore- or cell -suspension of such antagonists. *Bacillus subtilis* has been used against *Fusarium* species that cause rot of cuttings and bulbs. This bacterium has been used for many years in attempts to control plant pathogens and increase plant growth. Seed treatments with this bacterium have been shown to control various diseases caused by *R. solani, Helminthosporium* in rice and tomato damping off. Production and harvest of *B. subtilis* in quantities necessary for field scale use is easy. It forms endospores hence can be formulated in dusts, wettable powders, etc without losing efficacy. Commercial biocontrol formulations based on *B. subtilis* strain A13, isolated from lysed mycelium of *Sclerotium rolfsii,* other strains of *B. subtilis, B. cereus,* etc., *Trichoderma* spp, *Gliuocladium* sp. and many others are already sold in the market in some countries. *B. subtilius* stimulates germination, causes better emergence, enhances plant nutrition, reduces cankers caused by *R. solani* AG-4 in peanut, increases root growth, and finally yield. A similar control of wilt of chickpea caused by *Fusarium oxysporum* f.sp. *ciceris* by *Pseudomons fluorescence* is reported. In a talc based formulation the bacteria survived for 8 months of storage. On seeds treated with formulation, the bacteria survived for 6 months. Control was further improved when along with seed treatment, the bacterial formulation was applied to the rhizosphere also.

The use of bacteria for biocontrol of soil-borne pathogens is receiving much attention these days not only because they provide protection against major pathogens through competition for nutrients and space, antibiotic production and induced resistance but also because many of them stimulate plant growth which results in better yields. The fluorescent species of *Pseudomonas* have received special attention. Strains of *Pseudomonas fluorescens* and *P. putida* applied to potao seed pieces and to sugarbeet and radish are reported to increase the yield of potato by 5–33%, of sugarbeet by 4–8 tons per hectare, and root weight of radish by 6–144%. These bacterial strains have been given the name plant growth-promoting rhizobacteria (PGPR). The term rhizobacteria was coined for bacteria with the ability to colonize roots aggressively. The PGPR colonize the root system and prevent infection by major pathogens as well as deleterious rhizobacteria and fungi which damage the root tips and root hair without producing disease symptoms but have adverse effect on root functioning. Some suppress cyanide producing organisms in the root zone of potato. Siderophores, low molecular weight, high afffinity iron (III) chelators that transport iron into bacterial cells, are also involved in biocontrol by bacteria and consequent iimproved growth of the plant. They deprive the pathogens of iron supply.

An ideal antagonist for use as a seed treatment material should combine the attributes of aggressiveness as a competitor and seed surface colonist with ability as a hyperparasite, producer of antibiotics, or both. *Trichoderma hamatum* which suppresses *Rhizoctonia* has these abilities. It is an aggressive colonist of the seed coat and it produces enzymes that destroy hyphal walls of both *R. solani* and *Pythium* spp. Against *Rhizoctonia solani* the hyphae of such antagonists as *Trichoderma* spp. coil around the host hyphae, cause perforations by enzymic action, penetrate the lumen of host hypha, dissolve the cell walls but do not cause antagonism by antibiotic production. In general, the greater the number and diversity of methods used by an antagonist to inhibit a pathogen, the more successful it is in biological control. Such effective antagonists are usually found where the target pathogen is present and can cause the disease but is unable to do so because of the antagonists such as in suppressive soils. Biological seed treatments often provide protection to roots also against root parasites.

Roots of majority of plant species have myorrhizal associations. The mycorrhiza help in better nutrient uptake by the plant. Vesicular arbuscular mycorrhizae (VAM) and ectomycorrhizae are initiated as a parasitic interaction between plant roots and certain fungi. These roots are often resistant to attack of certain pathogens including nematodes. There are many reports suggesting that mycorrhizae act as a barrier to fungal infections of roots. They also increase the efficiency of growth promoting mucoid rhizobacteria and antagonists naturally present or introduced in the soil by providing more surface for adherence to roots. They are also likely to play the role of resistance inducers. In *Rhizobium*-plant symbiosis, the NOD-factor from the rhizoibium is reported to induce systemic resistance to disease. .

(ii) *Protection of foliage and flowers:* Existence of epiphytic microflora on plant surfaces including leaves and flowers is a natural phenomenon. These organisms do not harm the plant. There are many studies where their presence has been cited to explain reduction of disease incidence (Blakeman and Fokema, 1982) Brown leaf spot of rice (*Helminthosporium oryzae*), leaf spot of rye (*Helminthosporium sativum*), fire blight of apple and pear (*Erwinia amylovora*), Alternaria spot of tobacco, and many other foliage diseases are less severe when the normal epiphytic microflora is allowed undisturbed than when this microflora has been eliminated or reduced by some treatment such as spray of broad spectrum fungicides. The epiphytic microflora can be introduced on the host surface by spraying their propagules provided they have been selected or developed from the same site or similar plant surface. Bacteria are better colonizers than fungi in this respect. Recently, a novel mycofumigation technique was reported from Japan by Koitabashi (2005). He isolated a volatile antifungal compound-producing fungus and placed its potato dextrose agar culture at 30 cm intervals in middle of the rows of parsley. The volatiles release by the test fungus significantly reduced powdery mildew (*Oidium* sp.).

(iii) *Prevention of post-harvest decay of fruits:* Attempts to check various types of fruit rct after harvest had been mostly through heat and chemical treatment. In recent years there have been successful demonstrations of biological control of post harvest fruit rots by using bacteria and fungi including yeasts. Application of *Burkholderia cepacia* to lemon fruits after harvest gives 80% control of green mold (*Penicillium*) without any visible injury to the fruits. *Bacillus subtilis* gives control of peach brown rot (*Monilinia fructicola*), *Enterobacter cloacae* reduces peach Rhizopus rot (*Rhizopus stolonifer*) and *Pseudomonas cepacia*, the fungus *Acremonium brevae*, and some species of yeasts control decay of apple fruits and prevent such diseases as Botrytis rot in strawberry and other fruits.

Mycofumigation is exploiting the ability of some fungi to produce antifungal volatiles, for fumigation of the environmemnt of the plants or its product It is different fro, **biofumigatiom** in which the fumigating volatiles are produced decomposition of organic matter by complex of microbes (mostly bacteria). Mycofumigatiom has been used in the control of decay of pome and citrus fruits. *Muscodor albus* (Xylariaceae) produces volatiles that inhibit the growth and kill *Botrytis, Colletotrichum, Geotrichum, Monilinia, Penicillium* and *Rhizopus* Mercier and Jimenez, 2004). Fumigation of apples for 7 days with culture of *M. albus* grown on autoclaves grain gave complete control of blue mold (*P. expansum*) and gray mold (*B. cinerea*). There was no direct contact between the pathogens and the *M. albus* culture. Shorter fumigation durations of 24 or 72 hours also controlled these molds. In a recent study of the volatiles of *M. albus* on plant parasitic nematodes, Riga *et al.,* observed that Exposure of juveniles to the volatiles for 72 hours caused up to 95 mortality in *Meloidogyne chitwoodii*. Similar results were obtained when the grain culture of *M, albus* was added to soil at 0.5 or 1.0% (w/w). The application of grain culture of *M. albus* to soil controls root rot of *Brassica oleracea* caused by *Pythium ultimum* and improves seed;img emergence (Worapong and Strobel, 2008).

Teo volatiles are isobutyric acid (IBA) and 2-metrthjyl-1-butanol. A third major component is ethyl butyrate IBA suppresses or kills nB. Cinerea at 25 micro liter/liter and merthyl butanol does so at 100 microliter/liter. Strobel *et al.,* (2001) had reported that *M. albus*, an endophytic xylariaceaous fungus, produces a mixture of volatile compounds which include alcohols, esters, ketones, acids and lipids. Although each has inhibitory effect, collectively they act synergistically to kill a broad range of plant- and human-pathogenic fungi and bacteria. Individually, the esters (1-butanol,3-methyl acetate) are most effective. Placement of inoculum of *M. albus* grown on grains is placed among the fruits for control of fruit decay. The mixture of the inhibitory compounds is deposited on surface of the hyphae of *Muscador*.

Nitrous oxide (N_2O) is capable of inhibiting growth of *Monilinia fructicola*, *Colletotrichum acutatum*, *Botrytis cinerea*, *Penicillium expansum*, *P. italicum*, *Rhizopus stolonifer* and *Phytophthora citrophthora* which are highly sensitive to it. It can be explored for postharvest control of decay fungi. The isothiocyanates derived from glucosinolates in *Brassica* spp. have been found to suppress *M. laxa*, *B. cinerea* and *Mucor piriformis* in apples and pears. Peracetic acid and chlorine dioxide have sanitizing effect on stone fruits. Peracetic acid at 500 μg/ml kills conidia of *M. laxa* after 5 min of contact while chlorine dioxide at 50 μg/ml kills the conidia after 1 min of contact The antibiotic volatile producing fungi have mostly been isolated from tropical rainforest trees. Recently an isolate of *M.albus* has been isolated from an unidentified vine in central Sumatra (Indonesia) which produces compounds not present in other isolates of the fungus. Thwe plant is used by local inhabitants of the area for snakebite. It also inhibits growth of plant pathogenic fungi (Atmosukarto *et al.*, 2005).

(iv) *Inoculation of pruning wounds with antagonists*: This method has been successfully demonstrated in case of certain wood and stump rot causing fungi.

CROSS PROTECTION, TRANSGENES AND INDUCED RESISTANCE

The host plays the major role in this type of biological control. Cross immunity or cross protection is well known in virus diseases. When an avirulent or mild strain of a virus is introduced into the host it may induce resistance in the host against a virulent strain of the same virus. Strains of the same virus from a different host may also provide protection against the virus in a particular host. The resistance inducer strain itself is not pathogenic to the particular host. In plant virus infections, systemic acquired resistance operates only against localized infection by limiting cell-to-cell spread of the virus through hypersensitive reaction without preventing virus replication There are many examples of cross protection in fungal diseases. *Phialophora graminicola* colonizes wheat root cortical tissues and provides protection against the take-all fungus *Gaeumannomyces graminis* var. *tritici*. In wilt disease of mint caused by *Verticillium dahliae* inoculation of roots at least two days earlier with *Verticillium nigrescens* reduces the incidence of wilt. Inoculation with mild or avirulent strain of *Fusarium oxysporum* f. sp. *vasinfectum* provides protection to cotton against Fusarium wilt. In *Fusarium oxysporum* f. sp. *ciceris* (wilt of chickpea) seven races are known of which races O and 1 are less virulent while the race 5 is most virulent. Seed inoculation with race O or 1 provides potection to the plants against race 5.

With developments in the application of biotechnology in plant pathology, it has been possible to develop transgenic plants by using genes of the pathogen. In potato leaf roll, cucumber mosaic virus in tomato, papaya ringspot transgenic resistance has been achieved. However, transgenic resistance is highly linked to the nucleotides equence of the genes in the pathogen used. The sequence may vary in different isolates and then the resistance will not be stable.

Systemic acquired resistance (SAR) is the resistance acquired by the plant during its life time. It is also known as induced systemic or local resistance. The inducer can be an inorganic chemical (carbonates, phosphates, riboflavin, and sometimes fungicides), biochemical (salicylic acid or plant hormones jasmonic acid, ethylene, thiamine (vitamin B_1)) or microorganisms (antagonistic and/or plant growth promoting rhizobacteria). They elicit defense responses from the host plant and trigger some sleeping resistance genes in the plant to become active. More than often, the systemic acquired resistance works simultaneously against many pathogens, sometimes against insect pests also. Induced systemic resistance works like the multigenic resistance in the plants. In multigenic resistance plants constitutively express gene products that activate defense mechanisms (lignification, hydrogen peroxide accumulation etc). Induced systemic resistance inducts constitutive accumulation of these and other gene products prior to challenge by the pathogen(s). In ISR there is no specificity in accumulation of gene products and multiple disease resistance is found. Plants can acquire systemic tolerance (SAT) in the same manner as SAR (Block *et al.,* 2005).

MICROBIAL GENES IN PLANTS FOR DISEASE CONTROL

A large number of studies have been carried out to introduce the genes of biocontrol that confer the property of biocontrol into the plants so that application of biocontrol agents could be avoided. This has provided a much larger gene pool for developing resistance in plants without the need for genes from related plant genotypes. The plant accepts such genes which are expressed when challenged by a pathogen.(Lorito and Scala,1999).

The endochitinase gene of *Trichoderma harzianum* has been transferred to apple, potato, tobacco, grapevine 0for the control of foliar and fruit diseases. *Aspergillus niger* (glucose oxidase), *Rhizopus oligoporus* (chitinase), *Ustilago maydis* (the killer toxin) are some fungal donors of genes for enhancement of defense responses in the receiver plants. Genes from many bacteria have been used for activayion of defence responses or protection from pathogen toxins.

THE ROLE OF ENDOPHYTES

In the preceding pages the role of introduced or pre-existing antagonistic microorganisms in the soil or the rhizosphere or on the phylloplane in biological control has been emphasized. There is another,equally common, group of organisms (bacteria and fungi) which are pre-existing inside the plant and have been shown to provide biological control of plant diseases, nematodes and imsects. These are endophytes. Endophuytes are Endophytic fungi and bacteria are foumd in virtually all plant species. These are nonpathogenic or weak patjhogens and reside in cells of their host plant in a variety of relationships ranginmg fro, symbiotic to slight;u pathogenic They promote plant growth and increase yield, suppress pathogens insduce resistance, solubilize phosphates and contribute to assimilation of nutrients.. Biosynthesis of many drugs in the plants has been attributed to endopjhytes. In addition, it has been shown that bacterial endophytes have the potential to remove soil contaminants by enhancing phytoremediation and may play role in soil fertility through phosphate solubilization and nitrogen fixation. In recent years greater attention is being paid to study of endophytes as the source of novel compounds for use in medicine and agrochemicals (fungicides/bactericides). The role of these organisms in the observed protection against pathogens by plant extracts is not yet clear.

As biocontrol agents the endophytes have become emerging new tools for exploration.. Species of *Musacor* are endophytes. Some emdophytic fumgi reduce the densityu of pustules of *Puccinia recondite* on wheat leaves.The role of endophytes in suppressing pathogens of *Theobroma cacao* has been extensively studied The endophytes in this host include *Colletotrichum gloeosporioides, Clonostachys roseaa* (a known biocontrol agent), and *Botryosphaeria ribis* which provide field control of *Phytophthora* on *Theobroma*. Species and strtains of *Trichoderma* are also enmdophytic in manmy plants

PROBLEMS AND PROSPECTS OF BIOLOGICAL CONTRO

During the 80 years or so since G.B.Sanford in 1926 first indicated the scientific possibility that potato scab (*Streptomyces scabies*) was suppressed under field conditions by green manuring through activities of antagonists, enough data have been produced to show that biological control is a feasible proposition for disease management and could be an answer to environmental pollution caused by fungicides and nematicides and high cost of the chemical control. However, during the same period the practical application of the methods for field control of diseases has been demonstrated in relatively few instances. There are, of course, many examples of field control of root diseases by alteration of soil conditions by appropriate crop husbandry measures so as to enhance natural biological control by the resident soil microflora and fauna. Results from inoculation of natural soil with selected antagonists have not been very encouraging although under laboratory or glasshouse conditions the same antagonists had given good control of the disease. Singh and Reddy (1979) had isolated a number of *Streptomyces* spp. from a variety of soil types and sites and screened them for antagonistic activity against selected plant pathogenic fungi. *Streptomyces diastaticus* was found to give one hundred per cent control of damping off of tomato (*Pythium* spp.) and root and collar rot of chickpea (*Sclerotium rolfsii*) when the antagonist was added to autoclaved soil in pots inoculated with the pathogens. However, when the antagonist was applied to natural soil having infestation of these pathogens there was absolutely no reduction in disease incidence. Such experimental results are too many. Soil is biologically buffered by the resident microbial population and artificially introduced populations soon dwindle to negligible size so that the status quo is restored. The biocontrol agent mass is not effective against all races or strains of a pathogen. *S. rolfsi* strains with large sclerotia are less susceptible than small sclerotia strains to the antagonist *Gliocladium virens*. Soil fungistasis in which germination of fungal spores is suppressed by soil is known to affect germination of spores of many nematophagous fungi. These incapabilities of even very strong antagonists to be effective in natural soil, from which they were not isolated, is a major constraint in the direct use of antagonists for biological control. To be effective the antagonist must be selected from the same ecological niche where it is intended to work against a given pathogen. This is possible only when observations have revealed situations where there is natural suppression of a disease (suppressive soils, plots of monoculture of the susceptible crop, etc.).

Once this has been achieved, lack of proper system of delivery of the antagonist to the field is another problem. Preparation of large volumes of culture of the antagonist sufficient for application on field scale is not only tedious but uneconomical also. Thus, instead of direct application of the antagonists, the cultural practices to enhance the microbial biomass in soil is a more feasible proposition. This can very well fit in the normal schedules of crop production and may provide wide variety of antagonistic microflora and fauna. The use of compost and manure for mass production of antagonists was mentioned earlier.

Inoculation of host surfaces with a protective coating of a competitive or antagonistic microorganism is a much more sound procedure and in some cases it is being widely practiced. Inoculation of seed surfaces with cells of antagonistic bacteria or spores of antagonistic fungi has met with success in biological control and is becoming more widely extended. In such cases the problem of introducing an alien in a new environment does not arise. The antagonist gets established on the seed surface before it has to face soil microorganisms as well as the pathogen.

Biological control systems will take some time before they become an alternative to chemicals but, together with cultural practices they can definitely supplement the chemical control in integrated pest management strategies. They can reduce cost of disease management and also mitigate the harms done by excessive and repeated use of pesticides.

However, it may be noted that introduction of antagonistic bacteria in the ecosystem requires caution since many of them are associated with diseases of man.

STRENGTHENING OF THE HOST

The importance of strengthening of the host to resist attack of pathogens was recognized in the early 1900s. However, the farmers had been practicing similar measures much earlier in the so-called traditional agriculture. They tried to store seeds from vigorous and healthy looking plants from the crop not knowing that by doing so they was probably selecting plants that were resistant to one or more diseases in the plant population. Resistance to diseases and pests is a naturally occurring phenomenon in plants. That is how they have been facing the attack of these enemies probably since their evolution. This is evident from the fact that plant species have survived the attack of parasites and have not been completely wiped out from the face of the earth. May be that all these resistant types are not of economic use to common farmer but they are definitely a sure tool in the hands of scientists who need variability in crop plants to develop resistant varieties. For example, the species of *Solanum*, the ancestors of potato, which originated somewhere in the Andes mountains of South America and which have been under pressure from the late blight disease, are a source of resistance genes against the late blight pathogen. Breeding resistant varieties with the help of such gene sources, which is but one part of the broader plant breeding programme, is, therefore, more popular and intensive now than it was in the past.

Occurrence of Variability in Plants

Variability in plants, expressed by their response to external factors including resistance to pathogens, is an essential requirement in any crop improvement programme because it provides the sources of genes for the desirable characters. This variability is naturally occurring or is created during the crop improvement programmes by the breeders.

The evolution of present day crop plants from their wild ancestors has occurred slowly over millions of years in one or more geographic areas and has produced countless genetically diverse forms of these plants (genetic modification). Most of these forms look useless but, as stated above, they are important sources of genes for desirable resistance characters since they have survived the pressure of diseases and, therefore, carry numerous genes for resistance.

In those days when agriculture was not supported by scientifically developed resistant varieties, man selected and cultivated some of the wild plants for their desirable characters. In this way he produced numerous cultivated lines and varieties. The most productive of these varieties got favourable treatment and was perpetuated from year to year. Rice is an example of this selection and cultivation of varieties in different localities.

Nature and pathogens have been eliminating the weak and susceptible ones from the population while the farmer selected the best yielders from the survivors. These surviving populations carried different sets of major or minor genes for resistance. By their genetic diversity these varieties or lines have survived sudden and serious outbreaks of diseases and are a source of genes for resistance although lacking the ability to meet the high yield targets of to-day.

With the introduction of scientific breeding of crop plants the activities during breeding also contribute to the stock of genetic variability. In the search for new useful genes plant breeders cross existing, local cultivated varieties with each other and with those of other localities and with wild relatives of the crop plant from wherever they can be obtained. Induced mutation with the help of chemical or physical mutagens is used to obtain material for selection and breeding. During the process of breeding many crosses are discarded which become discarded breeders' stock and can be used as a source of variability at any time.

As the breeding work advances and several of the most useful genes are located subsequent steps in breeding tend to eliminate variability by combining the best genes in a few cultivated varieties and discarding or leaving behind varieties or lines that have no usefulness at that time. The process ultimately leads to narrowing down of the choice of cultivated varieties to a few which become accepted over large areas. The genetic base of these varieties is narrow. Hence they are vulnerable and when a new race of a pathogen appears they suffer more than the lines that had been left behind or discarded.

Classical Plant Breeding for Disease Resistance

As stated above sources of genes for disease resistance are:

- (i) native or foreign commercial varieties,
- (ii) older varieties abandoned earlier,
- (iii) discarded breeders' stock,
- (iv) wild plant relatives, especially from areas of origin of the plant species or cultivated varieties from such areas, and
- (v) induced mutations.

The traditional or classical techniques for crop improvement include:

1. *selection* of individuals with desired characters from a population,
2. induction of *mutation* by using mutagens to select plants of desired characters, and
3. most importantly *hybridization* between plants, from sources as given above, with desired characters. Selection and mutation are, thus, complimentary to the hybridization programme.

Selection: In fields where natural infection of a pathogen occurs regularly or in artificially created sick plots (having populations of a pathogen with a large number of its races), mass selection of seeds can be made from the most highly resistant individuals that survive. The method is simple but improves plants only slowly and does not work well in cross pollinated plants where there is no control over the source of pollen.

Pure line or pedigree selection involves separate propagation of highly resistant individuals and their progenies and repeated inoculations to test for resistance. This method is easy and most efficient with self pollinated crops but most difficult with cross pollinated crops.

Mutation: This involves use of natural or artificially induced mutants that show increased resistance. Change in the number of chromosomes in a plant and production of euploids (4n, 6n) or aneuploids also can be brought about by using chemicals or radiation.

Hybridization (Recurrent selection or Back crossing): The basis of hybridization is to combine the desired genes from the two parents used in crosses to obtain plants possessing good characters of both. The same methods used to breed for any heritable characteristic are used for breeding for disease resistance also and depend upon the mating system of the plant (self or cross pollinated). However, breeding for disease resistance is more complicated because (i) it can be assayed only by exposing the plant to the disease agents, thereby, employing another living and variable organism that must interact with the plants, and (ii) the existence of physiologic races of the pathogen which may not allow stability of the disease resistance characters in the progenies. Stable or durable resistance is the final aim in resistance breeding.

When a desirable but susceptible variety of a crop is crossed with another cultivated or wild relative that carries genes for resistance to one or more pathogens the resulting progenies are tested for resistance. The resistant individuals are back crossed to the desirable variety. This is repeated several times until the resistance is stabilized in the genetic background of the desirable variety. The method of evolving disease resistant varieties by hybridization is spread over several crop seasons. The time gap sometimes results in ineffectiveness of the efforts since during the lapsing time the pathogen may develop races to which the progenies have not been exposed. Hybridization is easier in cross pollinated than in self pollinated crops.

Other traditional methods of breeding resistant varieties include the use of F1 hybrids of two different but homozygous lines carrying different genes for resistance. This involves the phenomenon of heterosis and widens the base of resistance in the plant.

Breeding for Resistance using Biotechnology

Conventional plant breeding has been of utmost importance for nearly 100 years for the production of better and more useful plants. It will continue to play the role in future also irrespective of the new technologies for gene manipulation. However, in conventional plant breeding there are two major constraints:

1. Only closely related species can be crossed sexually because of incompatibility problems leading to limited gene pool available for new combinations. This limits the pool of required genes.
2. Sexual crossing means a massive randomization of two compatible genomes. Therefore, extensive back crossing becomes a necessity. Considering the seasonal nature of most crop plants (one reprodctive stage in a year) the production of a new variety through conventional plant breeding is considerably slowed down. The time needed for breeding a new variety may be somewhere between 8 and 15 years depending upon the type of plant species. In fruit trees this may be much longer.

The pathogens breed faster producing new races that may be ready to overcome the resistance in the new variety. Therefore techniques that allow the shortening of the time needed to produce a new variety and / or to enlarge the gene pool for conventional plant breeding have received high interest during the last 10–15 years. Techniques in biotechnology have the potential to overcome both constraints listed above for plant breeding.

Biotechnology is defined as the manipulation, genetic modification, and multiplication of living organisms through novel technologies, such as tissue culture and genetic engineering, resulting in the production of improved or new organisms and products than can be used in a variety of ways.

The possibility of culturing plant somatic cells in a medium was first speculated in 1902. In 1934 first tissue cultures from root tip were grown. Tissue culture has been exploited to create genetic variability from which crop plants can be improved, to improve the state of health of the planted material and to increase the number of desirable germplasms available to the plant breeder. The technique has been used (in plant pathology) for study of disease mechanism, production of pathogen free clones such as in potato and sugarcane, and in attempts to generate novel disease resistant lines of crop plants.

In tissue culture, sections (**explants**) of shoots or root tips, shoot nodes, germinated seedlings, and in some plants sections of leaves are trimmed, surface sterilized, washed repeatedly in sterile distilled water, and then placed in test tubes containing suitable culture medium in either liquid or, more often, semi-solid form. Depending on the kinds and relative amounts of growth regulators added to the medium, the explants produce numerous shoots, roots, or both. These explants are usually further subcultured at frequent intervals by subdividing single mother culture into several daughter cultures. Complete plants are finally set out in greenhouse or in the field. Pollen can also be similarly regenerated into haploid plants.

During regeneration, the tissues multiply and form a mass of unorganized, undifferentiated mass of cells called **callus**. If the culture medium is kept in constant, slow motion, *single cells* can be obtained. When the callus or single cells are suitably treated with enzymes the cell wall can be removed and membrane bound protoplasts can be obtained. Protoplasts from different sources can be fused to bring about a new type of plant. Advances in plant tissue culture including meristem tip propagation, callus and single cell culture, haploid plant production, and protoplast isolation, culture, transformation, fusion and regeneration into whole plants have opened new possibilities and methodologies for crop improvement including resistance to disease.

The potentials of the above tissue culture methods have further increased with the use of recombinant DNA and other technologies (broadly called **genetic engineering**) which include detection, isolation, modification, transfer, and expression of single genes or groups of related genes from one organism to another. Genetiic enginneering is considered one of the major scientific revolutions of this century, the others being unlocking of the atom, escaping the earth's gravity to land on the moon, and the computer revolution. The recombinant DNA technologies are being used in many ways especially in areas dealing with industrial and pharmaceutical microbiology to develop new microorganisms or improve the existing one. In plant sciences also their use is being made in attempts to develop new varieties of plants. Development of genetic engineering technology and molecular characterization of plant defense responses have provided strategies for controlling plant diseases by methods additional to the chemicals and classical plant breeding. In general, these alternative strategies are based on over-production of one component of the plant's own defence response.The important advantage of this strategy for gene manipulation is the unlimited gene pool. Genes from not only unrelated plants but genes from the microbes (bacteria and fungi) have been used to develop transgenic plants. Transfer of microbial genes to plants has been successful in inducing local or systemic resistance through enabling the plant to trigger its defense responses including ability to neutralize pathogen toxins.

In clonally propagated crops such as sugarcane, potato, banana, etc., in which propagation by seed is not done, hundreds and thousands of plants of a disease resistant variety can be produced from prolific plantlets. Pathogen-free stocks of a susceptible but otherwise good commercial variety can also be obtained in the same way. Plants regenerated from callus, single cells, or protoplasts in the culture show considerable variation (**somaclonal variation**) and much of it is useless. However, plants possessing resistance to a disease may also emerge. These plants can be isolated and propagated for inclusion in the programme of traditional breeding or can be further propagated as a resistant line.

In tissue culture, the immature pollen cells (microspores and less often megaspores) can be induced to develop into haploid (1n) plants in which single copies (alleles) of each gene are present in numerous combinations. By vegetative propagation and proper screening for disease resistance, the most highly resistant haploids can be selected. These can be subsequently treated with colchicine to induce diploidization of the nuclei. Thus, dihaploid plants homozygous for all genes, including genes for resistance, are produced.

A very useful possibility is increasing disease resistance by protoplast fusion in crops in which interspecific hybridization is not easy. Under proper conditions protoplasts of plants of different species or even closely related genera can be induced to fuse. The fusion produces

hybrid cells containing nuclei (chromosomes) and the cytoplasm of both protoplasts or **cybrid cells** may be produced in which nuclei are from one protoplast and cytoplasm from the other protoplast. Fusion of protoplasts from unrelated cells generally does not succeed because the hybrids either abort or fail to develop into plants. Fusion is more successful between protoplasts of related cells in which many or most of the chromosomes of one cell are eliminated during cell division but one or few of the chromosomes survive and may be incorporated in the genome of the other cell. In this way plants with more chromosomes, thus, new characteristics, can be regenerated from the products of protoplast fusion. The technique has great potential when fusion is induced between protoplasts of different, highly resistant haploid lines of the same variety or species. The product is diploid and the plants combine the resistance genes of two highly resistant haploid lines.

However, it must be noted that tissue culture is neither a fast nor an easy way of developing resistant varieties. It may take longer than the conventional methods (Daub, 1986) except where haploids are used to supplement the conventional methods of breeding. A plant selected for resistance by tissue culture must usually still go through an extensive screening and breeding process before a new disease resistant variety with desired commercial qualities can be released.

Genetic transformation of plant cells for disease resistance is a great step towards producing plants with genes from outside. The transgenic plants can be resistant to one specific pathogen or to a group of pathogens. Genetic material (DNA) can be introduced into plant cells or protoplasts by several methods such as direct DNA uptake, microinjection of DNA, liposome (liquid vesicle-mediated delivery of DNA), delivery by means of centromere plasmids (minichromosomes) and use of vectors as transfer vesicles such as certain DNA viruses and bacteria (*Agrobacterium tumefaciens*). In these methods, small or large pieces of DNA are introduced into plant cells or protoplasts and the DNA may or may not integrate with plant chromosomal DNA. When the introduced DNA carries appropriate regulatory genes recognized by the plant cell, or is integrated near appropriate regulatory genes along the plant chromosomes, it is transcribed into mRNA (expressed) which is then translated into proteins.

The crown gall bacterium *Agrobacterium tumefaciens* induces a tumor inducing factor through its Ti plasmid (extrachromosomal DNA) which gets integrated with the host genome and leads to multiplication of cells and tumor formation. A specific gene from other sources (resistant plants, avirulence gene in the pathogen, gene encoding biochemicals in bacteria and fungi) can be isolated and spliced into an appropriate area of Ti plasmid and when the bacterium is allowed to infect a plant it can transmit the gene into the host cells and then the gene will be expressed by the plant as a new character.

Advantages and Disadvantages of Vertical and Horizontal Resistance

Genetic resistance is governed by two types of genes. Some genes (major genes) control important steps in disease development while others (minor genes) play a role in only peripheral events of lesser importance in disease development. One or few major role genes are sufficient to make a plant resistant to a pathogen. This is known as **monogenic** or **vertical** resistance. In this type of resistance the plant is completely resistant to a few races of the pathogen. But this complete resistance is likely to breakdown easily if new races of the pathogen attack the plant. The genetic base for resistance in varieties with vertical resistance being narrow, pathogenic races find it easy to bypass the resistance mechanism(s). However, vertical resistance is easy to manipulate in a breeding programme and was earlier generally preferred to horizontal resistance. This type of resistance is most effective when:

(i) it is incorporated in annual crops that are easy to breed such as grain crops,

(ii) it is directed against pathogens that do not reproduce and spread rapidly (*Fusarium*) or mutate very frequently (*Puccinia graminis*),

(iii) it consists of strong genes that confer complete resistance to the plant, and

(iv) the host population does not consist of a single genetically uniform variety grown over large acreages. If one or more of these conditions are not met vertical resistance becomes short lived.

When many minor-effect genes make a plant resistant to a disease the resistance is **polygenic** or **horizontal**. A plant with horizontal resistance is never completely resistant or completely susceptible. But the resistance is more stable because of a broad genetic base. Horizontal resistance involves physiological processes of the host that act as mechanism of resistance and that are beyond the capacity or probable limits of the variability in the pathogen. Plants with horizontal resistance are protected against all races of a pathogen including the most virulent types. More pathogenic the races present, the greater the selection for general resistance in the host. This general resistance is eroded if the pathogen is not present because there is no selection pressure for resistance. The horizontal resistance is universally present in wild and domesticated plants but is at its highest in wild plants and lowest in greatly improved varieties. The wide natural prevalence of this type of resistance is because of the selection pressure due to exposure to wide variety of races of the pathogen.

The drawbacks in vertical resistance have been overcome in some crops by using multilines. Multilines are either mixtures of individual varieties that are agronomically similar but differ in their resistance genes or varieties that are derived from crossing several to many varieties that contain different genes for resistance and then selecting from those that contain mixture of genes. This greatly enhances the genetic base for resistance. The usefulness of multiline cultivars and cultivar mixtures for disease control has been demonstrated for rusts and powdery mildews of small grain crops. There are now examples of mixtures providing disease control in a wide range of pathosystems including crops with large plants and pathogens that have low host specificity or are splash dispersed, soil-borne or insect transmitted (Mundt, 2002). However, agronomic and marketing aspects must be considered before mixtures are used for crop health.

Disadvantage of Growing Genetically Uniform Crops Over Large Areas

Genetic uniformity is disadvantageous so for as resistance to pathogens having potential for variability and adaptation is concerned. Varieties with vertical resistance are often attacked suddenly and rapidly by a new virulent race which results in severe epidemics. Whenever one or a few highly improved varieties have been grown over large acreages continuously for years breakdown of resistance has been common. One example is that of wheat cultivar S227. The breakdown of resistance is due to appearance of races to which the variety did not have resistance genes. This threat of resistance breakdown and incidence of epidemics is the most unfavorable point against genetic uniformity in a crop. Risk in the widespread use of genetically uniform material is as much in classical plant breeding as in biotechnologically developed material.

TRANSGNES WITH GENES FROM BIOCONTROL AGENTS

The development of plant transformation and gene cloning techniques has opened new avenues for augmenting disease and insect resistance in plants. Fungi and bacteria with biocontrol potentials have provided a gene pool that has been used to transform plants showing enhanced or activated defence responses to pathogens Some examples are listed below.

Genes for avirulence in the pathogen (viz, *Phytophthora infestans, Cladosporium* etc) are incorporated in the host genetic system to activate defense response or protect the plant from pathogen toxins. This works like cross protection. The avirulent gene from the pathogen, in a compatible or incompatible host-pathogen system, induces pathogenesis related (PR) proteins in the host which act like antibodies. Creation of transgenic plants by introducing coat protein genes of viruses has been demonstrated in many host-virus systems. Genes encoding for chitinase and endochiytinase in *Trichoderma harzianum* have been incorporated in tobacco, potato, apple etc. to enhance defense responses (Lorita and Scala, 1999). . In the control of apple scab chitinase encoding gene from *Trichoderma harzianum* was transferred to apple. The resulting plants showed good resistance to leaf scab but the presence of a foreign gene in the plant genetic system disturbed metabolic activities of the plant.

INDUCTION OF RESISTANCE BY MEANS OTHER THAN GENETIC MANIPULATION

Host nutrition often influences the development of a disease. In many diseases nitrogen enhances disease while potash suppresses it. In others nitrogen suppresses the disease and potash enhances it. Such examples have been cited along with different diseases described in this book. However, examples of control of a disease through host nutrition are very few. A resistant variety usually responds better to protection given by nutrition than a susceptible variety. At the same time a susceptible variety can show moderate resistance to some diseases under certain sets of environmental conditions (see red rot of sugarcane). Calcium and phosphorus have often been implicated in resistance response of a plant to a soil-borne disease. Application of calcium chloride or calcium nitrate reduces Phytophthora stem rot of soybean. The effective element was found to be calcium which decreases zoospore release.

In systemic acquired resistance (SAR) chemical and biotic agents induce expression of resistance genes which were not operative and this results in induction defence responses Inducers of SAR or LAR (local acquired resistance) do not create or induce a new gene. They activate the already existing dormant genes. B-3-aminobutyric acid (BABA), acibenzolar-S-methyl (ASM), ethylene, salicylic acid, jasmonic acid and many other chemicals (including some inorganic chemicals) are used as elicitors of defense responses in plants.

Induction of short time resistance by use of systemic fungicides, antibiotics and insecticides (nematicides) and herbicides has been cited. So long as these compounds or their breakdown toxic products remain in the plant system, it remains resistant to a particular disease.

THE MODIFICATION OF ENVIRONMENTS FOR DISEASE MANAGEMENT

Physical, chemical and biological environment is closely associated not only with the plants in the field but also with the pathogen, its survival, infection, and disease development. The measures under modification of environments are basically cultural practices aimed at changing the environment in such a way that growth of the plant is favorably affected but there is adverse effect on the life-cycle of the pathogen or its disease-cycle. Katan (2000) categorized cultural practices into:

(i) cultural practices for regular purposes which can also be used for disease management, i.e., irrigation,

(ii) cultural practices which are used solely or mainly for disease control, e.g., sanitation, and

(iii) cultural practices which can be used for both agricultural purposes and disease control, e.g., crop rotation.

Cultural practices must be environmentally, technologically and economically feasible. They must minimize the use of fungicides or other pesticides. Cultural practices include crop rotation, fallow, flooding, deep ploughing, flaming, soil solarization which involves a combination of physical and biological processes, adjusting planting date, irrigation, fertilization (both soil application and foliar fertilizers), compost and organic amendments, weed control, herbicide application, field sanitation, tillage and some others. Some of these measures such as field sanitation and artificial immunization have been mentioned on the preceding pages. Others are of a more general nature and affect the disease as a whole without indicating as to whether the pathogen or the host is most concerned. Some of the practices which directly or indirectly affect the pathogen and/or disease development are listed below:

Crop rotation: Crop rotation is an ancient cultural practice for raising healthy crops. The early principle was to rotate a shallow feeding crop (like cereals) with a deep rooted crop (legume). The idea was to make best use of nutrients in different depths of the soil. The major mode of action of crop rotation is through biological control of soil-borne pathogens. The interposition of 2 or 3 non-susceptible crops between the cultivation of susceptible crops in the same field is effective in reducing soil-sickness and populations of a large number of soil-borne pathogens such as *Fusarium* spp. causing vascular wilt of cotton, pigeonpea, chickpea, linseed and tomato, *Verticillium* spp. causing wilt of different crops, *Rhizoctonia solani* causing black scurf and stem canker of potato, cyst nematode causing Molya disease of wheat and barley, etc. Due to economic reasons the basic concept of totation in India has been ignored and cereal after cereal has become popular. In most other countries such rotation is avoided. The contribution of rice-wheat rotation to incidence and severity of Karnal bunt was mentioned in connection with the description of the disease. Studies are reported where it has been emphasized that there is no long term adverse effect of rice-wheat rotation on the soil. But such studies have taken into consideration the mineral status of the soil, ignoring the organic status and microbial diversity which play a major role in disease suppression.

For rotation to be effective it is essential that plant debris, on which the pathogens lead a saprophytic life, must be removed or destroyed. The weed hosts of the pathogens should also be eliminated. Rotations are most effective in case of soil-invaders which have a limited saprophytic survival in soil. The soil-inhabitants (native flora of the soil) are difficult to eradicate by crop rotation. Rotations are also not practicable in case of pathogens which have a very wide host range such as root-knot nematodes. However, in such cases resistant or immune cultivars of the host can be included in the rotation.

The suppression of the pathogens through crop rotation may be a result of starvation of the pathogen in absence of the host, toxic effects of root exudates of the intervening crops or due to adverse soil microflora and fauna enhanced by root exudates and debris of the intervening crops.

Mixed cropping: The system of mixed cropping originated due to needs of the farmers. They wanted to harvest several grain crops in the same season from same land. Mixed cropping is different from inter-cropping. In the latter two or more crops are grown simultaneously in the same field but in separate rows or blocks. This method of growing several crops together is not as effective as random mixing of crops such as sowing of a mixture of seeds of sorghum and pigeonpea for management of wilt disease of the latter. Growing a susceptible crop mixed with a non-susceptible crop has several advantages. In root rot of cotton caused by *Rhizoctonia solani* and *Macrophomina phaseolina* it had been found long ago in Punjab that the incidence of the disease was reduced when cotton was grown mixed with *Phaseolus aconitifolius* (mothbean), a bushy, low height plant. The latter reduces temperature and retains high soil moisture due to its shading effect. In wilt disease of pigeonpea (*Fusarium udum*), mixture with sorghum reduces the disease incidence probably by increasing the distance between host plants and between infected and healthy roots, by creating root barriers between the roots of diseased and healthy plants, and through toxic root exudates of sorghum. Mixed crops reduce the spore trapping efficiency of the susceptible crop due to reduced host surface area. If the susceptible crop is lost to the disease, the other crop partly makes up for the loss. In mixtures of tall and low height bushy plants, the latter trap spores moving from the soil to the aerial parts.

Plant spacing: Spacing between plants also helps in reducing disease incidence. Due to better ventilation and sunlight the atmospheric humidity in the crop is reduced and those organisms that flourish in high humidity, such as downy mildews, are discouraged. Spacing also decreases the chances of contact between diseased and healthy foliage or roots and dissemination of spores and bacterial cells through raindrop splashes.

Alteration of date of planting and methods of sowing: Some diseases are very destructive when susceptible age of the host and optimum soil and atmospheric conditions for aggressiveness of the pathogen coincide. Alteration in the date of sowing in such a way that the susceptible stage of the plant growth does not coincide with the favorable environment for the pathogen helps in reducing losses from such diseases. Infection by many soil- and seed-borne pathogens is favoured by deep placement of seeds at the time of sowing. Delayed emergence of seedlings above the soil surface gives the pathogen more time for causing infection. Regulation of depth of sowing in such cases helps young plants escape infection.

Irrigation and drainage: Management of irrigation can be used for management of many diseases. Excess soil moisture enhances incidence of several diseases such as damping off. In such diseases reduced watering helps in checking disease development. Charcoal rot of potato and many other crops (*Macrophomina phaseolina*) is favoured by high soil temperature coupled with dry condition of the soil. In such cases irrigation helps in lowering the temperature and avoiding the disease. In common scab of potato, irrigation to keep the soil continuously wet during tuber formation and development helps in reducing disease incidence.

Irrigation system is also important. Sprinkler irrigation spreads bacterial cells and fungal spores in the same manner as raindrop splashes. In diseases where there is heavy sporulation of the pathogen or exudation of bacterial cells on the foliage surfaces sprinkler irrigation should be avoided. The soil-borne pathogens like *Pythium* and *Phytophthora* spread from plant to plant through irrigation water. If the plants are on the main channel they can easily get infected. In such cases care should be taken to irrigate the plants via side channels.

Use of pond water for irrigation may promote some diseases. Ponds receive drained off water from the fields. Pathogens present in the field thus reach the pond and may survive and multiply on weeds growing there at some stage. Rhizoctonia solani causing sheath blight of rice is an example.

Accumulation of water in a field helps in accumulation of inoculum from other fields. If water accumulation occurs when there is a standing crop it helps in accumulation of toxic substances around the roots and base of the stem and also reduces oxygen supply to roots. These conditions weaken the roots and stems making them liable to easy infection. Sometimes resistance of a crop to a disease is lost under water-logged conditions. Drainage is, therefore, essential to keep the field free from pathogens and making the plants resistant.

Hot weather deep ploughing: In the tropics deep ploughing during summer is of special importance as it exposes the propagules of pathogens to hot temperatures and desiccation thus reducing their population. In ear cockle disease of wheat, wetting of the soil before summer ploughing causes more rapid destruction of larvae as wetting causes swelling of cockles present in the soil releasing the larvae which are then killed by desiccation.

Soil solarization: A new system of use of solar heat to destroy inoculum of pathogens (fungi, nematodes) and weed seed in the soil is trapping of solar heat under a plastic mulch spread over the soil. In soil solarization, the soil is first wetted and then covered with a thin, clear plastic sheet. The trapped heat raiuses the soil temperature to a level where most pathogen structrures including sclerotia, and weed seeds are killed. If the soil is amended with organic matter before solarization the effects are enhanced. The suppression orf pathogen propagules is not only due to raised temperature. Solarization of soil amended with cabbage residue enhances generation of many volatiles such as alcohols, aldehydes, sulfides and isothiocyanates which add to killing of pathogen proagules. Soil solarization significantly affects nutrient composition of the plant, especially nitrogen and this leads to better plant growth, disease or no disease.

Green manures and organic amendments of soil: Decomposition of organic matter in the field produces a number of toxic volatiles, especially fatty acids, that work as soil fumigants sanitizing the soil by causing death or inhibition of parasitic organisms. Green manures and amendments increase biodiversity in the soil thus increasing the chances of biological control. Combining organic amendment with soil solarization is a very effective non-chemical approach to management of soil-borne pathogens. Solarization of soil amended with suitable organic materials, actuates a chain of reaction of chemical and microbial degradation, which leads to the generation of toxic compounds in vapor and liquid phases in the soil. The generation of toxic compounds increases with temperature. These compounds accumulate under the plastic mulch, and enhance toxicity against soil flora and fauna, especially soil-borne plant pathogens. The plastic mulch traps the volatile compounds and creates an atmosphere in the soil that enhances degradation of organic matter. At the end of the process the soil contains contains less pathogens, and different microflora, which may suppress re-establishment of pathogens (Gamliel *et al.,* 2000). The period of solarization is less for organically amended soils than for non-amended soils. The change in soil texture and fertility resulting from increased humus content helps in better growth and quicker development of roots. This enables the plants to produce new roots to replace diseased and dead roots. The losses are consequently reduced.

Composted organic waste, including cowdung or farmyard manmure are supposed to be fully decomposed materials if matured. They lack the biofumigation propert which has occurred during the process of composting. But they have the diverse and rich microbial communities that contain antagonists and plant growth promoting forms. In addition, they carry the stable forms of of the volatiles like aldehydes, acetates amd butyrates.

The use of film-forming compounds against foliar pathogens: The leaf surface provides the first barrier that fungi must overcome in order to gain access to the leaf. The use of fungicides and biocontrol agents help the leaf by direct action against the pathogen. The leaf also provides chemical and physical cues that are necessary for the development and adherence of infection structures for many fungal pathogens. Exudation of chemicals as attractants or nutrients and topography of the surface are examples. The film-forming polymers can coat the leaf surface, acting not just as an extra barrier to infection, but also disguiseing the cues necessary for germ tube development. Thus, the use of such materials, a physical activity, disturbs the ability of rust or powdery mildew germ tubes to identify the topography and follow straight route to the infection cite. The coating covers the stomata and fails to give the chemical and physical advantages to the germ tubes. The use of film-forming or anti-transpirant materials has been successful in reducing infection of *Botrytis* on foliage of *Capsicum, Phaseolus*, tomato, and cucurbits, *Puccinia recondite* (leaf rust of wheat), rice blast (*Pyricularia oryzae*), powdery mildew of wheat, crown gall infection of cuttings, and witchweed (*Striga*). Film coating is not effective when there is latent infection in the leaves.

The role of plant nutrienyts: Nutrients influence plant growth and plant may get benefit from its vigor in tolerating the injury caused by diseases. However, the direct role, if any, in disease management is not very clear. The role of major nutrients (NPK) and minor nutrients (*viz.,* zinc) have been mentioned in description of diseases in the preceding chapters. Of particular interest is the role of silicon which is considered, as of now, neither a major nor a minor nutrient.

INTEGRATION OF PLANT DISEASE MANAGEMENT PRACTICES

The four basic requirements for management of plant diseases are: clean and healthy seeds, clean field or pathogen free soil, prevention of entry of and infection by a pathogen in a standing crop and precautions during harvesting and storage of the produce. An ideal schedule for controlling a disease is to integrate measures covering these four requirements. A combination of several measures (chemical, physical or biological) for dealing with one of the above four is not an effective integration. Usually an integrated programme is aimed against all diseases affecting a crop. Sometimes it is aimed against one or two most destructive diseases of the crop.

In an integrated disease management programme in an annual crop (potato, cereals, etc.) the first step is the choice of healthy and clean seed of a resistant variety. In potato, the seed tubers carry pathogens of late blight, bacterial wilt, and almost all the viruses. In wheat, seed is the only source of loose smut. In sugarcane, seed setts carry the red rot, wilt and smut and ratoon stunting disease pathogens and also the viruses. The healthy clean seed can be obtained as certified seed from a reliable agency who ensure proper raising of a seed crop and proper treatment of the seed. The next step is to treat the seed chemically if the same has not been done by the supplier. There may be need for cleaning the seed physically for removal of contaminants. The healthy, pathogen free seed is then planted in a relatively clean field in which the precautions listed in the preceding section have been taken. In potato, the field should be free from old crop debris. If root knot nematodes are a problem, furrow application of a systemic granular nematicide should be given. This will not only check nematode infection but prevent infestation of aphids for the first 1–2 months thus preventing virus infection from external sources. Similar precautions should be taken for sugarcane. Crop rotation is another precaution. During growth of the crop periodical rogueing of diseased plants, if any, should be done. The inoculum of a pathogen may arrive from external sources. This inoculum should be prevented from causing infection by sprays of suitable chemicals. Maintenance of the standing crop should be backed by cultural practices as listed earlier. At the time of harvest, care should be taken to discard diseased plants, do the harvesting in a clear weather, and store the seed/ produce in proper storage conditions.

REFERENCES

Abawi, G.S. and T.L. Widmer. 2000. Impact of soil health management practices on soil-borne pathogens, nematodes and root diseases of vegetable crops. *Appl. Soil Ecol.* 15(1): 37

Ahn, P., S. Kim and Y-H. Lee. 2005. Vitamin B₁ functions as an activator of plant disease resistance. *Plant Physiol.* 138: 1505

Al-Mughrabi, I.K.L., T.A. Aburjal, G.L. Anfoka and W. Shahroor. 2001. Antifungal activity of olive cake extract. *Phytopath. Medit.* 40(3): 240

Arora, D.K. and A.K. Pandey. 1989. Soil solarization for control of soil borne diseases: Theory and application, pp. 429 – 438. In: V.P.Agnihotri, *et al.,* (eds.) *Perspectives in Plant Pathology.* Today and Tomorrow's Printers and Publishers, New Delhi.

Atmosukarto, L., U. Castillo, W.M. Hess, J. Sears and G. Stobel. 2005. Isolation and characterization of *Muscodor albus* I41,3s, a volatile antibiotic producing fungus. *Plant Sci.* **IN PRESS ABS**

Bailey, B.A., H. Bae, M.D. Strem *et al.,* 2008. Antagonism, mycoparasitism and colonizatiom success for endophytic *00* isolates with biological control potential in T*heobroma cacao.* Biol. Control 46(1): 21

Baker, K.F. 1987. Evolving concepts of biological control of plant pathogens, *Annu. Rev. Phytopathol.* 25: 67

Bakker, P.A.H.M., L.X. Ran, C.M.J. Pieterse and L.V. van Loon. 2003. Understanding the involvement of rhizobacteria-mediated induction of systemic resistance in biocontrol of plant diseases. *Can. J. Plant Path.* 25(1): 5–9

Bartlett, D.W., J.M. Clough, J.R. Hall *et al.,* 2002. The strobilurin fungicides. *Pest Manage. Sci.* 58: 649

Block, A., E. Schnekz, P.J. O'Donnell, J.B. Jones and H.J. Klee. 2005. Systemic acquired tolerance to virulent bacterial pathogens in tomato. *Plant Physiol.* 138: 481

Borneman, J. and J.O. Becker. 2007. Identifying microorganisms involved in specific pathogen suppression in soil *Annu. Rev. Phytopathol.* 45: 153

Caruso, C., G. Closi, C. Caporale *et al.,* 2000. Pathogenesis related proteins and genes. *J. Plant Pathol.* 82(1)

Chaube, H.S. 1989. Suppressive soils and plant pathogens, pp. 409–428. In: V.P. Agnihotri *et al.,* (eds). *Perspectives in Piant Pathology.* Today and Tomorrow Printers and Publishers, New Delhi.

Chellemi, D.O. 2002. Non-chemical management of soil-borne pests in fresh market vegetable production systems. *Phytopathology* 92(12): 1367

Compant, S., B. Duffy, J. Nowak *et al.,* 2005. Use of plant gtowth-promoting bacteria for biocontrol of plant diseases; Principles, Mechanisms of action, and future prospects. *Appl. Environ. Microbiol.* 71(9): 4951

Cook, R.J. 2000. Advances in Plant Disease Management in the Twentieth Century. *Annu. Rev. Phytopathol.* 38: 95

Cook, R.J. and K.F. Baker. 1983. *The nature and practice of biological control of plant pathogens.* Am. Phytopath. Soc., St.Paul, Minn., U.S.A.

Cooper, R.M. and J.S. Williams. 2004. Elemental sulphur as an induced antifungal substance in plants. *J. Exp. Bot.* 55(404): 1947

Daub, M.E. 1986. Tissue culture and the selection of resistance to pathogens. *Annu. Rev. Phytopathol.* 24: 159

De Boer, S.H. 2003. Perspectives on genetic engineering of agricultural crops for resistance to disease. *Can. J. Plant Path.* 25(1): 10–20

Doran, J.W. and M.R. Zeiss. 2000. Soil health and sustainability: managing the biotic component of soil quality. *Appl. Soil Ecol.* 15(1): 3

Duffy, B., A. Schouten, J. M. Raaijmakers. (2003). PATHOGEN SELF-DEFENSE: Mechanisms to Counteract Microbial Antagonism. *Annual Review of Phytopathology* 41, 501

El-Hasan, A., F. Walker and H. Buchenauer. 2008. *Trichoderma harzianum* and its metabolite 6-pentyl-alpha pyrone suppress fusaric acid production by *Fusarium moniliforme J. Phytopathol.* 156(2): 79

Fravel, D., C. Olivain and C. Alabouvette. (2003). *Fusarium oxysporum* and its biocontrol *New Phytologist* 157(3): 493

Gamliel, A., M. Austerweil and G. Kritzman. 2000. Non-chemical approach to soil-borne pest management-organic amendments. *Crop Protection* 19: 847

Garg, G.K., U.S. Singh, R.K. Khetrapal and J. Kumar. 1988. Application of tissue culture in plant pathology, pp. 83–49. In: *Experimental and Conceptual Plant Pathology.* Eds. R.S. Singh, U.S. Singh, W.M. Hess and D.J. Webber. Oxford and IBH Publishing Co. Pvt. Ltd., New Delhi.

Gent, D.H., H.F. Schwartz and S.J. Nissen. 2003. Effect of commercial adjuvants on vegetabie crop fungicide coverage, absorption and efficacy. *Plant Dis.* 87(5): 591

Gonslaves, D. and S.M. Garnsey. 1989. Cross protection techniques for control of virus diseases in the tropics. *Plant Dis.* 73: 592

Harman, G.E. 2000. Myths and dogmas of biological control. Changes in perceptions derived from research on *Trichoderma harzianum* T-22. *Plant Dis.* 84(4): 377

Herma, S., K. Seehaus, H. Koehle and U. Conrath. 2002. A strobilurin fungicide enhances the resistance of tobacco against Tobacco Mosaic Virus and *Pseudomonas syringae* pv. *tabaci. Plant Physiol.* 130: 120

Hoitink, H.A.J. 1980. Composted bark, a light weight growth medium with fungicidal properties. *Plant Dis.* 64: 142

Hoitnik, H.A.J. and P.C. Fahy. 1986. Basis for the control of soil-borne plant pathogens with compost. *Annu. Rev. Phytopathol.* 24: 93

Honee, G. 1999. Engineered resistance against fungal plant pathogens. *Eur. J. Plant Path.* 105(4): 319

Hornby, D. 1983. Suppressive soils. *Annu. Rev. Phytopathol.* 21: 65–85.

Howard, R.L. 1996. Cultural control of plant diseases: a historical perspectiuve. *Can. J. Plant Pathol.* 18: 145

690

Howell, C.R. 2003. Mechanisms employed by *Trichoderma* species in the biological control of plant diseases: The history and evaluation of current concepts. *Plant Dis.* 87(1): 4

Hutchinson, S.W. 1998. Current concepta of active defence in plants. *Annu. Rev. Phytopathol.* 36: 59

Hyseik, J., M. Vach *et al.*, 2002. The influence of the application of mineral fertilizers with the biopreparation Supresivit (*Trichoderma harzianum*) on the hyealth and yield of different crops. *Arch. Phytopath.Plant Prot.* 35(2): 115

Jacobsen, B.J. and P.A. Backman. 1993. Biological and cultural plant disease controls: Alternatives and supplements to chemicals in IPM strategy. *Plant Dis.* 77: 311

Janisiewicz, W.J. and L. Korsten. 2002. Biological control of postharvest diseases of fruits. *Annu. Rev. Phytopathol.* 40: 411

Jeyrajan, R. 1989. Biological control of plant diseases, pp. 375–390. In: V.P. Agnihotri, *et al.*, (eds). *Perspectives in Plant Pathology*. Today and Tomorrow Printers and Publishers, New Delhi.

Jones, J.B., L.E. Jackson, D. Balogh *et al.*, 2007. Bacteriophages for plant disease control. *Annu. Rev. Phytopathol.* 45: 245

Katan, J. 1987. Soil solarization, pp. 77–105. In: *Innovative Approaches to Plant Disease Control*. I. Chet (Ed.). Wiley-Interscience, New York.

Katan, J. 2000. Physical and cultural methods for the management of soil-borne pathogens. *Crop Protection* 19: 735

Kim, B.S. and B.K. Hwang. 2008. Microbial fungicides in the control of plant diseases. *J. Phytopathol.* **IN PRESS**

Kirkegaard, J.A. and M. Sarwsar. 1998. Biofumigation potential of brassicas. I. Variation in glucosinolate profiles of diverse field grown brassicas. I 201(1): 71

Koitabashi, M. 2005. New biocontrol method for parsley powdery mildew by the antifungal volatile-producing fungus Kyu-W63. J. Ger. Plant Pathol. 71(4): 280

Koller, W., C. Avila-Adome, G. Olaya and D. Zheng. 2001. Resistance to strobilurin fungicides. pp. 215–229. In: *Agrochemical Resistyance: Extent, Mechanisms and Detection.* J.M. Clark and I. Yamaguchi eds. Am. Chem. Soc. Washington

Kuc, J. 1987. Plant immunization and its applicability for disease control, pp. 255–274. In: *Innovative Approaches to Plant Disease Control*. I. Chet (Ed.). Wiley-Intersceince. New York.

Kuc, J. 1995. Induced systemic resistance-an overview, pp. 169–175. In: *Induced Resistance to Disease in Plants*. R. Hammerschmidt and J. Kuc (eds). Dordrecht, Kulwer.

Lazarovits, G. 2001. Organic amendments as a disease control strategy for soil-borne diseases of high value agricultural crops. *Aust. Plant Pathol.* 30(2): 111

Lorito, M. and F. Scala. 1999. Microbial genes expressed in transgenic plants to improve disease resistance. *J. Plant Pathol.* 81(2): 73–88

McManus, P.S., V.O. Stockwell, G.W. Sundin and A.L. Jones. 2002. Antibiotic use in plant agriculture. *Annu. Rev. Phytopathol.* 40: 443

Mehjia, L.,C., E.I. Rojas, Z. Maynand *et al.*, 2008., Endophytic fungi as biocontrol agents of *Theobroma cacao* pathowegens. *Biological Contro* 46(1): 4I

Metrauxs, J. 2001. Systemic acquired resistance and salicylic acid: Current state of knowledge. *Eur. J. Plant Pathol.* 107(1): 13

Minuto, A., D. Spadoro *et al.*, 2007. Control of soilborne pathogens of tomato using commercial formulation of *Streptomyces griseobiride* and solarization. *Crop Protection* **IN PRESS**

Miyake, Y., J. Sakai, M. Shibata *et al.*, 2005. Fungitoxic activity of benthiavalkicarb isopropyl against *Phytophthora infestans*. *J. Pestic. Sci.* 30(4): 390

Mundt, C.C. 2002. Use of multiline cultivars and cultivar mixtures for disease management. *Annu. Rev. Phytopathol.* 40: 381

Osboun, A.E. 2001. Plant mechanisms that give defence against soil-borne diseases. *Aust. Plant Pathol.* 30(2): 99

Palti, J. 1981. *Cultural Practices and Infectious Crop Diseases*. Springer-Verlag,Berlin.

Papavizas, G.C. and R.D. Lumsden. 1980. Biological control of soil-borne fungal propagules. *Annu. Rev. Phytopathol.* 18: 389–412.

Pennazio, S. and P. Roggero. 1998. Systemic acquired resistance against plant virus infections: A reality. *J. Plant Pathol.* 80(3)

Pieterse, C.M.J., J.A. Van Pelt, S. C.M. Van Wees *et al.*, 2001. Rhizobacteria-mediated induced systemic resistance: Triggering, signaling and expression. *Eur. J. Plant Path.* 107(1): 51

Raajmakers, J.M., T.C. Paulitz, C. Steinberg., C. Albouvitte and Y. Moenne-Locoz. 2008. The rhizosphere: a playground and battlefield for soilborne pathogens and beneficial microorganisms. *Plant Soil* **IN PRESS**

Reuvini, R. and M. Reuvini. 1998. Foliar fertilizer therapy- a concept in integrated pest management. *Crop Prot.* 17: 111

Riga, E., L.A. Lacey and N. Guerra. 2008. Muscodor albus, a potential biocontrol agent against plant parasitic nematodes of economically important vehetable crops in Washington State, USA. *Biol. Control* **IN PRESS**

Ron, E.Z. and E. Rosenberg. (2001). Natural roles of biosurfactanta. Min RTeview. Environmental Microbiology 3(4): 229

Russell, P.E. 2005. A century of fungicide evolution. *J. Agric. Sci.* (Cambridge) 143(1): 11

Ryan, R.P., K. Germaine, A. Frank *et al.,* 2008. Bacterial endophytes: recent developments and applications. *FEMS Microbiology Letters* 278(1): 1

Sanford, G.B. 1926. Some factors affecting the pathogenicity of *Actinomyces scabies. Phytopathology* 16: 525–547.

Schneider, R.W. (ed.). 1982. *Suppressive Soils and Plant Disease.* Am. Phytopath. Soc., St. Paul, Minn., U.S.A.

Sequeira, L. 1999. Mechanism of induced resistance in plants. *Annu Rev. Phytopathol.* 37: 51

Sikora, R.A. 1992. Management of the antagonistic potential in agricultural ecosystems for the biological control of plant parasitic nematodes. *Annu. Rev. Phytopathol.* 30: 245–270.

Sikora, R.A.,L. Pocasangreb, A. Zum *et al.,* 2008. Mutualistic endophytic fungi and in-planta suuppressiveness to plant parasitic nematodes. Biol. Control 46(1): 15

Singh, R.S. 2002. *Introduction to Principles of Plant Pathology.* 4th Ed. Oxford and IBH Publishing Co. (P.) Ltd., New Delhi.

Singh, R.S. 2002. *Plant Disease Management.* Oxford and IBH, New Delhi

Staub,T. and D. Sozzi. 1984. Fungicide resistance: A continuing challenge. *Plant Disease* 68: 1026

Sticher, L., B. Mauch-Mani and J.P. Metraux. 1997. Systemic acquired resistance. *Annu. Rev. Phytopathol.* 35: 235

Thurston, H.D. 2004. Ten thousand years of experience with sustainable plant disease control. *Plant Dis.* 88(5): 550

Tuzun, S. 2001. The relationship between pathogen-induced systemic resistance (ISR) and multigenic (horizontal) resistance in plants. *Eur. J. Plant Path.* 107(1): 85

Van Loon, K.C., P.A.H.M. Bakker and C.M.J. Pieterse. 1998. Systemic resistance induced by rhizosphere bacteria. *Annu. Rev. Phytopathol.* 36: 453–483

Walters, D.R. and L.J. Bingham. 2007. Influence of nutrition on disease development caused by fungal pathogens: implications for plant disease control. *Ann. Appl. Biol.* 151(3): 307

Weller, D.M. 1988. Biological control of soil-borne pahogens in the rhizosphere with bacteria. *Annu. Rev. Phytopathol.* 26: 379

Weller, D.M., J.M. Raaijmakers, B.M. Mcspadden Gardener and L.S. Thomashaw. 2002. Microbial population responsible for specific soil suppressiveness to plant pathogens. *Annu. Rev. Phytopathol.* 40: 309

Williams, J.S. and R.M. Cooper. 2004. The oldest fungicide and newest phytoalexin-a reappraisal of fungitoxicity of elemental sulphur. *Plant Pathology* 53(3): 263

Willmitzer, L. 1987. Genetic engineering for plant disease resistance, pp. 353–364. In: *Innovative approches to plant disease control.* I Chet. (ed). Wiley-Interscience, New York

Worapong, J. and G. Strobel. 2008. Biocontrol of a root rot of kale by *Muscodor albus* strain MFC 2 *Biocontrol* **IN PRESS**

Zitter, T.A. and J. N. Simons. 1980. Management of viruses by alteration of vector efficiency and by cultural practices. *Annu. Rev. Phytopathol.* 18: 289–310

Glossary

Acquired immunity: Immunity developed during the life time.

Agglutination: A serological test in which viruses or bacteria suspended in a liquid collect into clumps whenever the suspension is treated with antiserum containing antibodies specific against these viruses or bacteria.

Allantoid: Slightly curved with rounded ends; sausage-like:

Allele: One of two or more alternate forms of a gene occupying the same locus on a chromosome.

Allelochemical: Chemical released by the plant that moves, without change, to the target and harmas it.

Alternate (alternative) host: One of the two kinds of plants on which a parasitic fungus (mostly rusts) must develop to complete the life cycle. A host of a pathogen growing during off-season for the main host crop.

Anastomosis: The union of a hypha or a vessel with another resulting in intercommunication of their contents.

Antibiotic: Chemical compound (metabolite) of an organism (generally microorganisms) that inhibits or kills other microorganisms.

Antibody: A protein produced in warm blooded animals in reaction to an injected foreign antigen and capably of reacting specifically with that antigen.

Antigen: Foreign proteins, and occasionally complex lipids, carbohydrates, and some nucleic acids, which upon injection into a warm blooded animal induce the production of antibodies.

Antiserum: The blood serum of a warm blooded animal that contains antibodies.

Avirulent: Lacking virulence.

Axenic: That type of culture of an organism in which no other organisms are present.

Bacterial mycophagy: Phenomenon defined as a set of behaviour that enable bacteria to obtain nutrients from living fungi through necrotrophy, extracellular biotrophy and endocellular biotrophy.

Bacteriocin: Bactericidal proteinaceous metabolites produced by certain strains of bacteria and active against one or more strains of the same or closely related species. Their production is genetically controlled by extrachromosomal DNA (plasmids).

Bacteriophage: A virus that infects bacteria and usually destroys them.

Bioassay: The use of a test organism to measure the relative infectivity of a pathogen or toxicity of a substances.

Biofilm: Assembly of bacterial cells on a surface. It includes all extracellular materials of the bacteria and any other material entrapped in it.

Biosurfactants: Surface active emulsifying compounds produce by bacteria. These may be essential for the action of their antibiotics in biological control.

Biotechnology: The use of genetically modified organisms and/or modern techniques and processes with biological systems for industrial production, crop improvement, disease control, etc.

Biotype: A subgroup within a species usually characterized by the possession of a single or a few characteristics in common.

Blastospore: A spore which has been budded off.

Callus: A mass of thin-walled undifferentiated cells developed as a result of wounding or culture on nutrient media.

Capsomer: The aggregated protein subunits in virus protein coat.

Chemotherapy: Control of a plant disease with chemicals (chemotherapeutants) that are absorbed and translocated internally.

Chronic disease: A disease which persists in the host for a long period of time.

Chronic symptoms: Symptoms that appear over a long period of time.

Cistron: The sequence of nucleotides within a certain area of RNA or DNA that codes for a particular protein.

Clone: The group of genetically identical individuals produced asexually from one individual.

Coding: Process by which the sequence of nucleotides within a certain area of RNA determines the sequence of amino acids in the synthesis of the particular protein.

Codon: The coding unit consisting of three adjacent nucleotides which codes for a pecific amino acids.

Compatible interaction: Interaction between susceptible host and virulent pathogen

Conjugation: A process of sexual reproduction involving fusion of morphologically identical gametes.

Cyst: An encysted zoospore in fungi; in nematodes the carcass of the dead females in the genera *Heterodera* and *Globodera* which may contain viable eggs.

Cytokinins: A group of plant growth regulating substances that regulate cell division.

Dikaryotic: Mycelium or spores containing two sexually compatible nuclei per cell.

Disease assessment: Quantitative measurement of disease (=phytopathometry) or the process of measuring disease quantitatively.

Disease cycle: The chain of events involved in disease development including the stages of development of the pathogen and the effects of the disease on the host.

Disease incidence: Number of plants units sampled that are diseased expressed as a percentage or proportion of the total number of units assessed. Proportion or percentage of plants diseased in a population.

Disease intensity: General term for amount of disease present in a population.

Disease prevalence: Incidence of fields with diseased plants in a defined geographic area (country, state, etc.).

Elicitors: Molecules produced by the pathogen (or thehost) cell wall that induce

ELISA (enzyme-linked immunosorbent assay): A serological test in which one antibody carries with it an enzyme that releases a coloured compound.

Endoparasite: A parasite that enters the host and feeds from within.

Epiphyte: A plant which grows on another plant without being a parasite.

Etiology (aetiology): Study of the cause of the plant disease and nature of the causal agent

Exanthema: Eruption or discharge of gum or other substances from diseased tissues.

Falcate: Curved like the blade of a scythe (sickle).

Fasciation: A plant disease, due to cell injury in the bud, resulting in flattened and sometimes spirally curved shoots.

Fission: Transverse splitting of bacterial cell into two daughter cells; asexual reproduction.

Forma specialis (f.sp.): A group of biotypes of a pathogen species that can infect only plants within certain species of genus.

Fructification: Production of spores by fungi. Fungus fruiting body or spore bearing structure.

Fungistasis: Phenomenon wherein fungus spores fail to germinate or hyphae fail to grow even if temperature and moisture conditions are favourable for these activities.

GABA: γ-aminobutyric acid, plant metabolite involved in induction of resistance to stress

Gene: A linear portion of the chromosome which determines or conditions one or more hereditary characters. The smallest functional unit of the genetic

Gene cloning: The isolation and multiplication of an individual gene sequence by its insertion into a bacterium where it can multiply.

Genetic engineering: The alteration or manipulation of the genetic composition of cell by various procedures such as transformation, protoplast fusion, etc. in tissue culture.

Genome: All genes together.

Genomics: Study of genes and genome

Genotype: The genetic constitution of an organism.

Gram-negative: Bacteria not stained by Gram's stain.

Gram-positive: Bacteria retaining the Gram's stain.

Growth regulator: A natural substance that regulates the enlargement, division, or activation of plant cells.

Guttation: Exudation of water from plants, particularly along the leaf margins.

Haploid: A cell of an organism whose nuclei have a single complete set of chromosomes (1n).

Haustorium: Special branch of fungus hypha, especially intercellular hypha, within the living cell to absorb nutrients. Also root-like absorbing organ connecting parasitic seed plant to the vascular system of the host.

Hermaphrodite: An individual bearing both functional male and female reproductive organs.

Heteroecious: Requiring two different kinds of plants (hosts) to complete its life cycle (particularly in rust fungi).

Heterokaryosis: The condition in which a mycelium contains two genetically different nuclei per cell.

Heterothallic fungi: Fungi producing compatible male and female gametes on physiologically distinct mycelia.

Homothallism: The phenomenon in which a fungus produces compatible male and female gametes on the same mycelium.

Hybrid: The offspring of two individuals differing in one or more heritable characteristics

Hybridization: The crossing of two individuals differing in one or more heritable characteristics.

Hybridoma: A hybrid animal cell produced by the fusion of a spleen cell and a cancer cell and able to multiply and to produce monoclonal antibodies.

Hydathodes: Structures with one or more openings that discharge water from the interior of the leaf to its surface.

Hyperparasite: An organism that is parasitic on another parasite.

Hypersensitivity: Violent reaction of the host tissues to attack of a pathogen in which there is prompt death of the invaded tissue preventing further spread of the pathogen; associated with resistance.

Hypoplastic: Term applied to a disease in which there is subnormal cell division.

Hypovirulence: Reduced virulence of a pathogen strain as a result of the presence of transmissible double stranded RNA.

Immunity: Freedom from a given disease due to lack of qualities permitting infection or possession of qualities that do not permit establishment of infection.

Imperfect fungus (Anamorphic fungus): A fungus that is not known to produce sexual spores.

Incompatible interaction: Interaction between resistant host and avirulent pathogen

Incubation period: The period of time between penetration of a host by a pathogen and the first appearance of symptoms on the host. In virus diseases, the time lapsing between the acquisition of the virus by the vector and ability of the vector to transmit the virus.

Indexing: A procedure to determine whether a given plant is infected by a virus. It involves the transfer of bud, scion, sap, etc. from one plant to one or more kinds of plant that are sensitive to virus (indicators).

Indicator plants: Plants that react to certain viruses or environmental conditions with production of specific symptoms and are used to detect and identify these factors.

Infection court: The location where infection of the host parts may take place.

Infestation: Presence of pathogen in a population of plants or of pathogen in a position or material (seed, soil, etc.) where it has the possibility of causing infection and producing disease.

Inoculum potential: Combined energy of propagules to cause infection. It is a function of the number of propagules (inoculum density) and their pathogenic capacity (inoculum capacity).

In vitro: In culture, outside the host.

In vivo In the host; under living conditions.

L-form bacteria: Bacteria that have temporarily or permanently, lost their ability to produce a cell-wall as a result of exposure to antibiotics or conditions that inhibit cell-wall synthesis.

Latent infection: The state in which a host is infected but does not show symptoms.

Latent period: The period required by a virus to multiply and accumulate within the vector before the latter becomes infective.

Latent virus: A virus that does not induce symptom development in the host.

Lectins: A group of plant proteins that bind to specific carbohydrates.

Lipids: Substances whose molecules consist of glycerine and fatty acids and sometimes certain additional types of compounds.

Local lesion: A localized spot produced on a leaf upon mechanical inoculation with a virus.

Malignant: Used for a cell or tissue that divides and enlarges autonomously, i.e., its growth can no longer be controlled by the organism on which it is growing.

Masked symptoms: Virus disease symptoms that are absent under certain environmental conditions but appear when the host is exposed to certain conditions of light and temperature.

Messenger RNA (mRNA): A chain of ribonucleotides that codes for a specific protein.

Monoclonal antibodies: Identical antibodies produced by a single clone of lymphocytes.

Monocyclic: Having one cycle per season.

Monoecious fungus: Usually for rusts that have all the spore stages on one host plant, except the basidiospores.

Mosaic: Symptom of certain virus diseases of plants characterized by intermingles patches of normal and light green or yellowish colour.

Mutant: An individual possessing a new, heritable character as a result of an accidental change in genes or chromosomes.

Mutation: An abrupt appearance of a new characteristic in an individual as the result of an accidental change in genes or chromosomes.

Mutualism: Mutually beneficial relationship between organisms.

Mycofumigation: Use of fungi producing volatile antifungal compounds that suppress pathogens from a distance.

Mycotoxicoses: Diseases of animals and humans caused by consumption of feed and foods invaded by toxin producing fungi.

Mycotoxin: Toxic substances produced by fungi that are harmful to animals and humans.

Nanometer (nm): A unit of length equal to 1/1000 microns.

Nucleic acid: An acidic substance containing pentose, phosphorus, pyrimidine and purine bases. Nucleic acids determine the genetic properties of organisms and viruses.

Nucleoside: The combination of a sugar and a base molecule in nucleic acid.

Nucleotide: The phospshate ester of a nucleoside. Building blocks of DNA and RNA.

Operon: A cluster of functionally related genes regulated and transmitted as a unit.

Oxidative burst: Rapid generation of active oxygen species like superoxide anion (O_2-). hydroxide radical (OH), H_2O_2. One of the very early defense mechanisms triggered by infection.

Oxidative phosphorylation: The conversion of energy released by oxidative reaction of respiration to high energy ATP bonds.

Parasexualism: A mechanism whereby recombination of hereditary properties occurs within fungal heterkaryons.

Pelleting: Coating seed with a plastic substance in which a fungicide can be mixed.

Pectinase: Enzyme that breaks down pectin.

Phage: A virus that attacks bacteria and usually destroys them. The bacteriophage.

Phenotype: The external visible appearance of an organism.

Phytoalexins: Substances which inhibit the development of a fungus on hypersensitive tissue, formed when host plant cells come in contact with the fungus.

Phytoanticipitins: Pre-forme or pre-existing biochemical defense compounds in plants

Plasmalemma: The cytoplasmic membrane found on the outside of the protoplast adjacent to the cell-wall.

Plasmid: A self replicating, extrachromosomal, hereditary circular DNA found in many bacteria, less frequently in fungi; generally not required for survival of the organism; occurring usuall in the form of supercoiled, covalently closed circular dsDNA.

Plasmodesma (pl. plasmodesmata): A fine protoplasmic thread connecting two protoplasts and passing through the wall which separates the two protoplasts.

Plasmolysis: The shrinking and separation of of the cytoplasm from cell-wall due to exosmosis of water from the protoplast.

Polycyclic: The pathogen which completes many life cycles and disease cycles in one season.

Polymerase: An enzyme that joins single small molecules into chains of such molecules as in RNA and DNA.

Polysome (Polyribosome): A cluster of ribosomes associated with a messenger RNA.

Precipitin: The reaction in which an antibody causes visible precipitation of antigens (proteins).

Quorum sensing is a regulatory mechanism that connects gene expression to cell density in bacteria. Numerous functions, including pathogenicity are regulated in this way.

Race: A genetically and often geographically distinct mating group within a species; also a group of pathogens that infect a given set of plant varieties.

Recognition factor: Specific molecules or structures on the host (or pathogen) that can be recognized by the pathogen (or the host).

Resistance gene clusters: Resistance genes are not spread throughout the genome but are clustered at specific loci

Restrictive enzyme: A group of enzymes from bacteria which brak internal bonds of DNA at highly specific points.

Rhizosphere: The soil around the plant roots under the influence of exudates from the roots.

Ribonuclease (Rnase): An enzyme that breaks down RNA.

Ribonucleic acid (RNA): A nucleic acid involved in protein synthesis in organisms and viruses; also the genetic material of most viruses that attack plants.

Ribosome: A subcellular particle involved in protein synthesis.

Serology: A method using specificity of the antigen-antibody reaction for the detection and identification of antigenic substances and the organisms that carry them.

Serum: The clear, watery portion of the blood remaining after coagulation.

Shock symptoms: The severe, often necrotic, symptoms produced on the first new growth following infection with some viruses; also acute symptoms.

Sieve plate: Perforated wall area between two phloem sieve cells through which they are connected

Sieve tube: A seies of phloem cells forming a long cellular tube through which food materials are transported in plants

Soil inhabitants: Microorganisms able to survive in the soil indefinitely as saprophytes.

Soil invaders: Plant parasitic fungi and bacteria that pass a part of their life cycle in soil.

Soil transients: Parasitic microorganisms that can live in soil for short period; soil invaders.

Somaclonal variation: Variability in clones generated from a single mother plant, leaf, etc. by tissue culture.

Somatic hybridization: Production of hybrid cells by fusion of two protoplasts with different genetic makeup

Sporodochium: A fruiting structure consisting of a cluster of conidiophores woven together on a mass of hyphae.

Staghead: Term applied to diseased trees in which there are leafless branchesamong and above leaf bearing ones; the inflorescence infection by white rust or mildew in brassicas.

Stem-pitting: A symptom of some virus diseases of trees characterized by depressions on the stem under the bark.

Sterilization: Complete elimination of pathogens and other living organisms from soil, containers, etc. by means of heat or chemicals.

Strain: The descendant of a single isolation of pure culture; an isolate. Also a group of similar isolates; a race. In plant viruses, a group of virus isolates having most of their antigens in common.

Suppressive soils: Soils in which certain diseases are suppressed because of the presence of antagonists of the pathogen in the soil or due to some abiotic soil factors..

Symbiosis: A mutually beneficial association of two or more organisms of different kinds.

Symptomless carrier: A plant which, although infected with a pathogen (usually a virus), produces no visible symptoms.

Syncytium: A multinucleate giant cell formed as a result of infection of some nematodes.

Syndrome: A group of signs and symptoms that occur together and characterize the disease.

Synergism: The added and enhanced effect of two organisms or factors together. The effects produced are greater than the effects produced by each pathogen or factor alone.

Systemic: Spreading throughout the plant body internally.

Therapeutic: Term applied to measures designed to cure a diseased plant.

Therapy: Treatment to cure diseased plant.

Tolerance: The ability of a plant to sustain the effects of a disease without dying or suffering serious injury or crop loss.

Toxin: A compound produced by a microorganism and being toxic to a plant or animal.

Transcription: Copying of a gene into RNA. Also copying of a viral RNA into a complimentary RNA.

Transduction: The transfer of genetic material from one bacterium to another by means of a bacteriophage.

Transfer RNA (tRNA): The RNA that moves amino acids to the ribosomes to be places in the order prescribed by the mRNA.

Transformation: The change of a cell through uptake and expression of additional genetic material.

Transgenic: Genetically modified by introducion of foreign gene

Translation: Copying of mRNA into proteins.

Translocation: Transfer of nutrients or virus particles through the plant.

Transposable element: A segment of chromosomal DNA that can move around (transpose) in the genome and integrate at different sites on the chromosome. **Tumor:** An uncontrolled overgrowth of tissue or tissues.

Tylosis (pl. tyloses): An overgrowth of the protoplast of a parenchyma cell into an adjacent xylem vessel or tracheid.

Variability: The property or ability of an organism to change its characteristics from one generation to the other.

Vector: An organism (mostly insects) that transmits a pathogen (mostly virus); a living agent used to transfer genetic material carrying a particular characters in genetic engineering.

Vertical resistance: Complete resistance to some races of the pathogen but not to

Virescent: A normally white or coloured tissue (plant organ) that develops chloroplasts and becomes green.

Virion: A complete virus particle.

Virulence: Relative ability to cause disease; the degree of pathogenicity of a pathogen.

Virulent: Capable of causing a severe disease; strongly pathogenic; highly aggressive.

Viruliferous: Virus carrying; said of a vector containing the virus and capable of transmitting it.

Witches' broom: Broom-like overgrowth or massed proliferation caused by dense clustering of branches of a plant.

Xylem: A plant tissue consisting of tracheids, vessels, parenchyma cells, and fibers; wood.